Electromagnetic Fields and Waves

Fundamentals of Engineering

Sedki M. Riad, Ph.D., P.E.
Professor Emeritus
Virginia Tech

Iman M. Salama, Ph.D.
Northeastern University

McGraw Hill

New York Chicago San Francisco
Athens London Madrid
Mexico City Milan New Delhi
Singapore Sydney Toronto

Library of Congress Control Number: 2019946268

Electromagnetic Fields and Waves: Fundamentals of Engineering

1 2 3 4 5 6 7 8 9 QVS 24 23 22 21 20 19

ISBN 978-1-260-45714-8
MHID 1-260-45714-1

This book is printed on acid-free paper.

Sponsoring Editor
Lara Zoble

Editorial Supervisor
Stephen M. Smith

Production Supervisor
Lynn M. Messina

Acquisitions Coordinator
Elizabeth M. Houde

Project Manager
Apoorva Mathur, Cenveo® Publisher Services

Copy Editor
Surendra N. Shivam, Cenveo Publisher Services

Proofreader
Manish Tiwari, Cenveo Publisher Services

Indexer
Robert Swanson

Art Director, Cover
Jeff Weeks

Composition
Cenveo Publisher Services

Dedication

To The Creator Whose Bounties are far beyond numeration and accounting! I choose to highlight some of His Special Bounties in my life:

- *The love to learn and educate*
- *The perseverance to sustain hardships and prevail*
- *The persistence to strive for perfection and pay attention to detail*
- *The life full of challenges and fulfillment—truly satisfying and truly rewarding*

Also,
To my family
To my special Heba
To my colleagues, students, and professors

Sedki Riad
November 2013–November 2019

About the Authors

Sedki M. Riad, Ph.D., P.E., is Professor Emeritus in The Bradley Department of Electrical and Computer Engineering at Virginia Tech.

His academic experience is in research and development work in the areas of communications and electronics engineering, with particular emphasis in the areas of radio-frequency (RF) and microwave measurements, instrumentation, signal processing, and physical modeling. Dr. Riad has taught electromagnetic (EM) fields–related courses for over 45 years during his tenure at several academic institutions, including Cairo University in Cairo, Egypt; the University of Central Florida in Orlando, Florida; King Saud University in Riyadh, Saudi Arabia; and, for over 35 years, Virginia Tech in Blacksburg, Virginia.

He has been a Fellow of the Institute of Electrical and Electronics Engineers (IEEE) since 1992, cited for "contributions to time-domain measurements through physical modeling of sampling devices, simulation, and deconvolution."

Dr. Riad has extensive academic and administrative experience through his many years of teaching, research, and project management. He led several specialized committees that were focused on curriculum development and assessment. He has held several lead positions at Virginia Tech related to international collaboration and globalization. He has received several awards and noted honors.

Iman M. Salama, Ph.D., is Associate Teaching Professor in the Electrical and Computer Engineering Department at Northeastern University. From 2008 to 2016, she held a similar position with Drexel University, and has also worked as an RF consultant with Nortec RF and Microwave, LLC. She received her Ph.D. degree in Electrical Engineering from Virginia Tech.

In the Time Domain and RF Measurement Laboratory at Virginia Tech, Dr. Salama performed research and development work in the general area of microwave measurements, time-domain reflectometry, modeling, and simulation. Applications included the analysis and modeling of RF and microwave electronic circuits and material characterization in both the time domain and frequency domain. In her Ph.D. work, she developed a new, computationally efficient frequency-domain numerical technique for the solution of Maxwell's equations. The technique was based on the transmission line matrix (TLM) method.

Dr. Salama's teaching interests include RF and microwave circuit analysis and design, applied electromagnetics, radio-wave propagation, electronic devices, analog and digital communication systems, network analysis and synthesis, signals and systems, probability, and random processes.

Contents

Preface

Why Another EM Fields Book?

True, there are a number of electromagnetic (EM) fields books on the market for undergraduate engineering students. I personally have used many of them in my classes as texts and as occasional references. I liked some for their simple presentation and readability and others for their thorough analysis and comprehensive coverage. However, whichever book I used, the students always had difficulty with the subject. One reason is that many students lack adequate background preparation for the material (math and physics). Available textbooks and most professors do not take that into account in their presentation. For this reason, this book is an attempt to address this issue and present the material with students' realistic preparation in mind.

In addition, this book pays attention to material that always falls in the cracks between the courses students take and that is vital to the comprehension of the overall picture. The book is written for both readability and adequate material coverage. It focuses on tackling three major issues that are typically encountered in teaching EM fields.

First Issue

Let us face it: EM fields undergraduate education leaves the majority of the students with more questions than answers. Their biggest question is "Why are we studying this material?" In a way, they are right. If we do not tell them what Coulomb's law, Faraday's experiment, or Maxwell's equations should mean to them, then this material is an unnecessary burden and a waste of their time. Typically, all they get is a bunch of vector analysis exercises and complicated integrations—another "math" course at best.

The fact of the matter is that EM fields is pivotal to all aspects of electrical engineering (EE). That should be our theme when teaching undergraduates the topic. It is not just a matter of showing some practical applications for electrostatics or wave propagation; we need to go far beyond. We should show them how Maxwell's equations provide the foundation for the topics of circuit theory, electronics, communications, power generation and transmission, microwaves, and antennas, just to name a few application areas. Borrowing from the computer community, EM fields is like the machine language and the way we deal with these EE topics is like writing the code in a higher-level language.

Second Issue

Teaching EM fields to undergraduates has the potential of arming them with "problem-solving" skills. It is a fringe benefit that comes naturally with the topic, far beyond what comes through from other courses in their curricula. This is manifested in the ability to take a physical problem through the steps of deriving a physically based model for which a mathematical model is then developed. Next, we do the math analysis and obtain results for which we find relevant explanations for physical phenomena.

Third Issue

The other point is that the majority of our students lack adequate background preparation for the material required for the study of EM fields (in both math and physics). The challenge is that the subject of EM fields requires good mastery of *both* math and physics skills, math being a coded language of expression of the physical phenomena and how things evolve. It is just like language and culture; they are both critical to communication skills. For successful communication, one needs a good appreciation of both the language of communication and the culture of the people communicated with. Neither language nor culture can work well by themselves. The same is the case for math and physics. Some of our students (and professors) are more skilled in one side and pay less attention to the other. We need to be continually reminded that math is just an abstract language of expression (a coded one) and we must constantly be mindful of the physical phenomena and their implications as we go through the mathematical derivations and analyses.

EM fields is one of the toughest topics for undergraduate EE students. The reason is twofold: (1) lack of student preparation and (2) inadequate textbooks and teaching of the material. The topic requires a holistic presentation of the math and physics to ensure both student comprehension and appreciation of its importance to their education. In this text, we provide detailed explanations and always relate the math to the physics.

We recognize the need to keep the physics involved in the entire process of developing a mathematical model of a physical problem, in deciding the correct math analysis approach, in successfully carrying out solutions, and finally in interpreting (decoding) the mathematical results and translating their implications to physical terms. Unfortunately, most of our EM fields education does not give adequate emphasis to maintaining the physics insight and ignores the last and most important step in the process of translating the obtained mathematical results into physical phenomena.

So, in this book we shall attempt to keep both the students and professors mindful of the physics aspects of all topics. Both will find many eye-opening conclusions as a result. There will be many challenges, as most of us are not accustomed to that. But, at the end of the day, the benefit of achieving lasting comprehension is worth much more than training to carry out procedures.

Book Plan

The plan is to allow maximum flexibility to enable a wide sector of the academic community to customize this book to their curriculum needs. The book chapters can be arranged in three blocks (see the book outline table below); each block can be taught independently of the others, and as a result, the textbook can be customized by rearranging the blocks. Each block will have an introductory section to transition the learners into the material presented in that block.

There are three typical approaches to EM fields coverage, which I believe this book can serve very well. Referring to the outline table shown below, the three approaches, and the corresponding proposed block coverage arrangement, are

- Early transmission line approach: Blocks A-B-C
- Classical static-dynamic approach: Blocks B-(C/A)
- Maxwell's equation approach: Blocks (C/A)-B

There are two other features of the book that add to the customization flexibility:

- "Chapter Learning Objectives" are provided at the beginning of each chapter to help the instructor and learner pick and choose relevant study material.
- Chapter material is presented with "potentially secondary" material moved to addenda at the end of the chapter. This allows the instructor and learner to decide which and how much of the material in these addenda to cover.

This book contains adequate material for a full-year (two-semester) EM fields program of study. However, with the flexibility features built in as discussed above, the book can be used for both one-semester and two-semester curricula. The following table demonstrates a possible scenario:

Block	Topic	Two-Semester Plan (All)	One-Semester Plan
	Chapter 1: Introduction		
A	Chapter 2: Transmission Lines—Wave Equations		✓
		A: Driving Point Impedance	✓
		B: Impedance Matching	✓
		C: The Frequency-Domain Bounce Diagram	
		D: The Time-Domain Bounce Diagram	✓
		E: The Smith Chart	
B	Chapter 3: Transition to Electrostatics		✓
		A: Coordinate Systems	
		B: Vector Calculus and Vector Identities	
		C: Spatial Distributions and Densities	
		D: Line, Surface, and Volume Integrations	
B	Chapter 4: Electrostatic Fields: Electric Flux and Gauss' Law		✓
		A: Application of the Integral Form of Gauss' Law to Symmetrical Charge Distributions	
B	Chapter 5: Electric Force, Field, Energy, and Potential		✓
		A: Electric Field due to Famous Examples of Charge Distributions	
		B: Electric Potential (and Field) due to Famous Examples of Charge Distributions	
B	Chapter 6: Materials: Conductors and Dielectrics		✓
		A: Resistance Evaluation	
		B: Capacitance Evaluation	
		C: Resistors and Capacitors as Circuit Elements	
B	Chapter 7: Poisson's and Laplace's Equations and Solution Methods		✓
		A: The Method of Images	
		B: Numerical Methods	
B	Chapter 8: Magnetic Fields and Flux		✓
		A: Analogies with Electrostatic Quantities	
		B: Magnetic Field Evaluation Using the "Incrementation" Scheme	
		C: Magnetic Field Evaluation Using Ampere's Circuital Law Scheme	
B	Chapter 9: Magnetic Materials, Magnetic Circuits, and Inductance		✓
		A: Evaluation of Self-Inductance	
		B: Evaluation of Mutual Inductance	
		C: Magnetic Forces in Air Gaps: Magnetic Pull	
C	Chapter 10: Time-Varying Fields—Faraday's Law—Maxwell's Equations		✓
		A: Inductance under Time-Varying Currents	
C	Chapter 11: Wave Propagation—Transmission Lines Revisited		✓
		A: Derivation of the Laplacian Form of the Wave Equation	
		B: Skin Effect and Shielding	✓
		C: Skin Effect in Coaxial Transmission Lines	
		D: Loss Tangent for Energy-Storage Media (Materials) and Devices	✓

(Continued)

Block	Topic	Two-Semester Plan (All)	One-Semester Plan
C	Chapter 12: Wave Polarization and Propagation in Multiple Layers		✓
		A: Derivation of Reflection and Transmission Coefficients for Normal Incidence on Multiple Layers	
		B: Derivation of Snell's Law	
		C: Total Reflection: Physical Applications	
		D: Derivation of Brewster Angle Expressions	
		E: The "Complex" Snell's Law and the Mystery of the "Complex" Angle of Refraction	
C	Chapter 13: Waveguides		✓
		A: Wave Equation Solution for Metallic Rectangular Waveguides	
		B: Phase and Group Velocities	✓
		C: Wave Equation Solution for Metallic Rectangular Waveguides	
		D: Field Maps for the TE_{10} and TM_{11} Modes	
		E: Active and Dominant Mode Identification	
		F: Physical Insight in Guided Wave Propagation	
		G: Metallic Rectangular Cavity Resonators	
	Appendix A: Symbols and Units		
	Appendix B: Constants and SI Units		
	Appendix C: Material Properties		
	Appendix D: Vector Identities		
	Appendix E: Summary of EM Relationships		
	Appendix F: Historical Review of EM Scientists		

Acknowledgments

I am deeply indebted to my dear friend, colleague, and office neighbor for more than 20 years, Dr. Ioannis Besieris. He supported this effort through many discussions and thorough review of the manuscript.

This project would not have been possible without Dr. Iman Salama, my co-author and former doctoral student. Her invaluable contributions were instrumental in bringing this project to life.

Finally, I would like to acknowledge the support of all my dear friends and former graduate students who helped me in many aspects of the production of this manuscript.

Sedki M. Riad

Note: A full suite of digital ancillaries to accompany this book, including all figures in color, is available for download at www.mhprofessional.com/riad.

CHAPTER 1

Introduction

Jean-Baptiste Joseph Fourier
(1768–1830)

Chapter Learning Objectives

- Analyze transmission line parameters, characteristic impedance, complex propagation coefficients, reflections and standing waves, impedance matching, and power transfer from generator to load.
- Use the Smith chart to track changes of both the impedance and complex reflection coefficient at different locations of a transmission line.
- Perform transmission line analysis using the Smith chart including the design of impedance matching stub tuners and quarter wave transformers.
- Use Gauss', Coulomb's, Ampere's, and Biot-Savart laws to derive expressions and evaluate electric and magnetic field intensities and flux densities in a variety of simple configurations.
- Describe and understand the physical insight of the terms "field intensity," "flux density," and "potential" and their relationships to each other in both electric and magnetic fields.
- Distinguish between materials, based on their electromagnetic (EM) properties.
- Evaluate the resistance, capacitance, and inductance of simple material configurations. Utilize the same tools to design such configurations.
- Use Poisson's equation to solve boundary value problems, using theoretical, numerical, and graphical techniques. Use the method of images to identify equivalent problems to facilitate the solutions when applicable.
- Describe and understand the physical insight of Maxwell's equations and be able to manipulate and apply them to EM problems.
- Use Maxwell's equations to describe planar EM wave propagation in dielectrics. Describe wave propagation characteristics, for example, wave impedance, phase velocity, and wavelength.
- Describe and understand the physical insight of the transmission and reflection of uniform plane waves for normal and oblique incidence for the cases of dielectric-dielectric and dielectric-conductor interfaces. Explain related mathematical expressions.
- Develop equations governing wave propagation in rectangular waveguide.
- Analyze rectangular waveguides propagating transverse electric (TE), transverse magnetic (TM), or transverse electromagnetic (TEM) modes. Describe propagation modes, cutoff frequency, impedance, and wave propagation in rectangular waveguide.
- Describe and understand the physical insight of modal dispersion in multimode propagation in waveguides.
- Describe and understand the physical insight of the characteristics of rectangular cavity resonators and their associated quality factor and bandwidth.

1.1 Why Study EM Fields?

Most certainly, the course of EM fields is perceived as one of the toughest for undergraduate engineering students. The typical reason is that the topic requires a holistic presentation of both the math and physics to ensure student comprehension and appreciation of its importance to their education.

The fact of the matter is that EM fields are pivotal to all aspects of electrical engineering (EE). Maxwell's equations provide the foundation for the topics of circuit theory, electronics, communications, power generation and transmission, microwaves and antennas, just to name a few application areas. Borrowing from the computer community, learning EM fields is like learning machine language and the way we deal with these EE topics is like writing the code in a higher-level language.

There are many classes of problems that require a decent appreciation of EM analysis tools. Examples include (but not limited to)

- Antenna analysis and design
- Wireless communication systems and devices
- Semiconductor devices and integrated circuits
- Computer chips (processors, memory, etc.)
- Circuit boards (planar, multilayer, flexible, etc.)
- Cables, connectors, and adaptors
- Discrete components (R, C, L, transformers, etc.)
- Power systems and machines (transformers, motors, generators, etc.)
- Magnetic devices (sensors, relays, actuators, etc.)
- Micro-electro-mechanical systems (MEMS) and microwave (MW) devices
- Radio detection and ranging (RADAR) and remote sensing
- Microwave heating and curing
- Electromagnetic interference (EMI)/electromagnetic compatibility (EMC) analysis
- Nanotechnology
- Photonics

In addition to developing insight into, and appreciation of, the roots of the different EE curriculum topics, we cite two additional benefits of studying EM fields as an undergraduate. In particular, we cite the added skills in both the areas of design and problem solving.

1.2 Competent Design Skills?

As an engineer, you are likely to get involved in the design, analysis, or testing of an electrical or electronic device (or circuit). These devices and circuits are typically made of "conductors," "dielectrics," and possibly "magnetic" media. These different materials are shaped and assembled (or integrated) together in specific manners to perform certain functionality. When powered up, the electrical charge distribution within the device structure is altered to produce the desired device function.

Studying EM fields is about gaining the skill and acquiring the tools that enable the engineer to relate to the physics of electrical charge distributions and currents and their interaction with different materials. In other words, this study is all about developing the appreciation to the properties and limitations of the elements and components we use in electrical and electronic devices.

Some would say that we learned how to design circuits by putting together passive and active components "wired" in certain fashions to produce the desired function. The point is that the models we *claim* for the components we use in circuit design are approximate (simplified) ones that ignore *secondary* features of these components. These models are typically developed through an EM field study of the components' physical structures with certain degrees of approximations.

In other words, the *basic* models of passive components, such as resistors, capacitors, inductors, and even interconnecting wires (transmission lines), all are developed based on EM field studies of material structures and their interaction with charge distributions and currents. Similarly, the models of active components (devices), such as diodes, transistors, and various integrated circuits components, require an EM field analysis of how their physical structure and material properties react to different EM field excitations.

Not only the models of these components are developed through an EM study of their physical structures but also the tools we use in circuit design are developed based on an EM study. Even primary tools like Ohm's law and/or Kirchhoff's current and voltage laws have their roots in EM field studies.

To demonstrate, we take an example of a simple resistor component. Typically, designers would use the simplified model of a *pure* resistance in the circuit schematic as well as in the circuit analysis or design equations. The fact of the matter is that the *simple resistor* contains other features that may be ignored in some cases but would be critical to the circuit performance in others. Those other features may include lead inductance, packaging capacitance, material nonlinearities, and more. In fact, the model for other features as such requires a field analysis as well, and the resulting component is again an approximation to the actual physical performance of the structure.

Another example is the interconnecting *wires* in a circuit. Typically, those are ignored as having no effect on the circuit performance other than connecting the circuit components to each other. This simplification could lead to serious errors in some applications. Connecting wires contribute to resistive, inductive, and capacitive circuit elements that could alter the circuit performance at high frequencies and high currents.

From the above discussion, we can conclude that EM field studies are key to the development of component models we can use in circuit/device design and analysis. We can also conclude that several degrees of modeling "fidelity/accuracy" are possible, and in some cases, approximations suffice more than in others. Consequently, we can extrapolate to say that there will be cases where circuit models will not provide adequate representation of the physical devices, in which case we must resort to the fundamental tool of EM field analysis.

In essence, a good designer needs to fully appreciate the EM properties of different material. Not only that, but a good designer needs to get into other interdisciplinary areas such as mechanical, thermal, and chemical properties, in addition to including environmental, economical, and human factors in the device design.

This book focuses on the basics of EM field studies. It does not provide all the tools required for the learner to become a "good designer" in this regard. However, it provides the foundations of EM field needed to build on with more advanced course work. It certainly attempts to provide the student with the insight of what needs to be learned to be a good designer.

1.3 Problem-Solving Skills

Studying EM fields has the potential of arming the learner with "problem-solving" skills. This fringe benefit comes naturally with the topic, far beyond what comes through from other courses in his/her curricula. This is manifested in the ability to take a physical problem through the steps of deriving a physically based model for which a mathematical model is then developed. Next, we do the math analysis and obtain results for which we find relevant physical explanations and applications; see Figure 1.1.

1.4 Why Start with Transmission Lines First?

Traditionally, EM fields are taught starting with electrostatics followed by magnetostatics, after which dynamics, time-varying fields, and EM propagation are introduced. This approach has been criticized by many as "dry," as most of the interesting EM applications are not introduced until the dynamics part is covered. As a result, it is believed that many students are turned away from the subject because of the abstract nature of the *static* part.

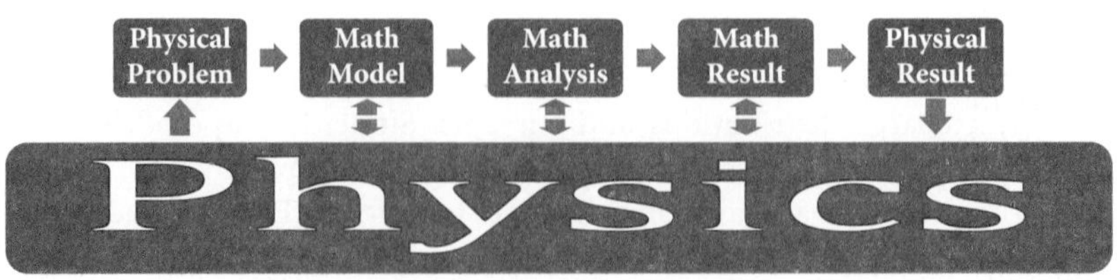

FIGURE 1.1 Math and physics holistic approach in problem solving.

The idea of reversing the order of coverage has been entertained by many educators in the past with varying degrees of success. The impeding factors are the difficulty in presenting the dynamics part without the prerequisite foundation established during the study of the statics part.

In this book, we start with Transmission Line (TL) analysis, as a topic close to the students' familiarization with many applications encountered in their daily lives. A logical question follows regarding the nature of the transmission line characteristic parameters: internal resistance, inductance, conductance, and capacitance (R, L, G, and C). What do these parameters physically imply, and what does it take to control their values, and hence control the performance of the transmission line? This question can be used to lead the students to the following electrostatic and dynamics chapters.

So, we find ourselves in need of studying EM fields topics that enable us to explain the physics of each of these *circuit* elements. It is the only way to understand how they work, but also how can we design them, control their parameters, and make them do what we desire for them to do. Some may wonder, is it the chicken and egg dilemma? We start with circuit theory and end up in fields. The fact of the matter is that "fields" is the chicken that laid for us the "circuits" egg (and a few other eggs) that we encounter in our physics and EE studies.

Finally, we close the loop by coming back to the TL analysis based on field theory, thus demonstrating to the students the limitations of the use of the Resistance-Inductance-Conductance-Capacitance (RLGC) circuit theory model.

1.5 Review of Transient and Harmonic Analysis Techniques and the Use of Complex Variables

As we prepare to study transmission lines and carry out relevant analyses using circuit theory techniques, we need two critical reviews first. The first review, which is the topic of Addendum 1A, is that of transient and harmonic analysis techniques. In this addendum, we will review circuit and system analysis methods using transient techniques as well as the transformation into the phasor domain to enable mathematical solutions of certain classes of differential equations that we encounter in such analyses. We will also discuss harmonic analysis and their phasor domain representations as well. We will confine our discussions in this addendum to linear systems which is a prerequisite for performing the Fourier/Laplace transformations between the time and phasor domains.

The second review, which is the topic of Addendum 1B, is that of complex variables and the *mystery of j*. In that addendum, we will try to bring the imaginary world of complex numbers closer to our reality. In this regard, we will present discussions with physical insight of the imaginary numbers and the square root of (−1), which we refer to as *j*.

ADDENDUM 1A
TRANSIENT AND HARMONIC ANALYSIS OF LINEAR SYSTEMS

1A.1 Introduction

We live in the time domain. All occurrences happen in the time domain. Events happened in the past, happen in the present, and other events will happen in the future. We witness physical phenomena in the real world and express their properties in real numbers. However, at times, we use imaginary tools and fictitious descriptions to facilitate the analysis, especially in difficult physical situations.

Much of this is true in our EE studies. Examples of imaginary tools are the use of complex variables to describe voltage phasors, impedances, and complex power. Examples of fictitious (nonphysical) descriptions include noncausal operations, integrations from zero to infinity, and "steady-state" harmonic analysis.

With this introduction, you may think that this book intends to discredit the foundation of all our EE education and more. The answer is "nope." What is intended here is to remind the reader of some *basic* foundations that get overlooked with time and raise relevant warning flags. It is important that we stay mindful of the foundations used in developing the tools that we constantly use and apply. In addition, it is most important to be aware of the physics of all procedures we carry out and tools we use.

Our subject in this chapter is transient and harmonic analyses. Transient analysis deals with the system (or circuit) response to a time-varying excitation during the transition between two steady states. One example is the famous battery-switch-RC circuit when we throw the switch on or off to charge or discharge the capacitor. Another example would be studying the transient as we turn on (or off) an AC (110/220 V 50/60 Hz) appliance.

The other type of analysis we refer to here is what we call harmonic analysis, and there is a good reason for choosing this wording over other familiar wordings! You see, the complementary state to transient is *steady*. However, steady may mean steady constant or steady variable, that is, periodic. Considering steady constant as a special case of periodic with a period that is too long (mathematically, we say infinite), we can identify the alternate state to transient as the *periodic* one, the simplest form of which is the harmonic time variation. Also, noting that for linear systems, the Fourier analysis can enable us to express other forms of periodic variations as a superposition of harmonics of different periodicities (different frequencies). Hence, we can claim that by covering transient and harmonic analyses, we are in fact covering tools that can encompass all forms of linear system analyses that we can think of.

1A.2 Time Domain and Frequency Domain

Now, we have two possible domains that we can work in: the time domain and the frequency domain. The time domain is where we deal with all analysis procedures as functions of time. In this type of analysis, we typically develop time-dependent differential equations to express the behavior of physically based system (or circuit) models. Solving these equations can get too complicated, and we may end up resorting to mathematical procedures and transforms to enable a solution. One typical transform that we often use in solving electrical circuit problems in the time domain is the Laplace transform. The Laplace transform converts the time derivatives into powers of "s" in a complex "s" domain, and we end up with algebraic equations that are typically solvable. Once we get a solution in the "s" domain, the next task is to use the inverse Laplace transform to yield the final results of the analysis as functions of time. Our challenge at that time is to find physical interpretations to the obtained expressions and relate them to the physical phenomena. Figure 1A.1 demonstrates the process of performing analysis in the time domain.

An alternate approach to working in the time domain is to convert the problem to the frequency domain early in the process and carry out the solution there and convert the solution back to the time domain at the end (see Figure 1A.2).

The use of the downward arrows on the time domain side is to indicate the difficulties that this approach faces. As we said before, we can use the Laplace transform once we reach the point where we have "complicated" differential equations; however, the frequency-domain approach avoids the need to develop the differential equations in the first place.

In our EE education, the frequency-domain approach dominates the EE curriculum to the point that most of us have lost our relationship with the time domain altogether.

What happened was that we started out with the time-domain approach and did a few examples and simple cases there. Later on, we introduced the frequency domain and learned how to do the whole process from the physical problem in the time domain concluding by the inverse transformation of the final results again in the time domain. Soon afterward, we learned how to jump from the physical model into the frequency-domain circuit model, run all our analyses, and reach the results in the frequency domain and end there. We simply lose touch with the physics of the problem and the physics of the analyses. It takes additional training and experience to be able to relate the frequency-domain results to the physics of the system without going back to the time domain (see Figure 1A.3).

In this regard, here are a couple of interesting analogies that you may find useful:

- Working in the time domain is like being awake in real life where we witness events as they occur and experience problems and issues as they happen. Suppose we encounter this complication that we cannot resolve. Next, we go to "sleep" and dream a solution for this problem during our sleep, and we use

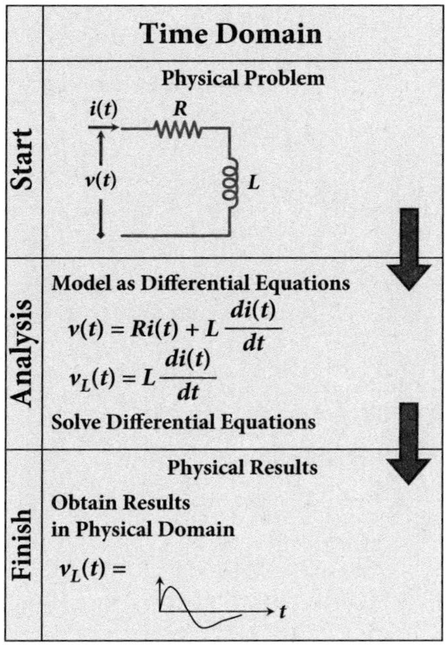

FIGURE 1A.1 Process of performing analysis in the time domain.

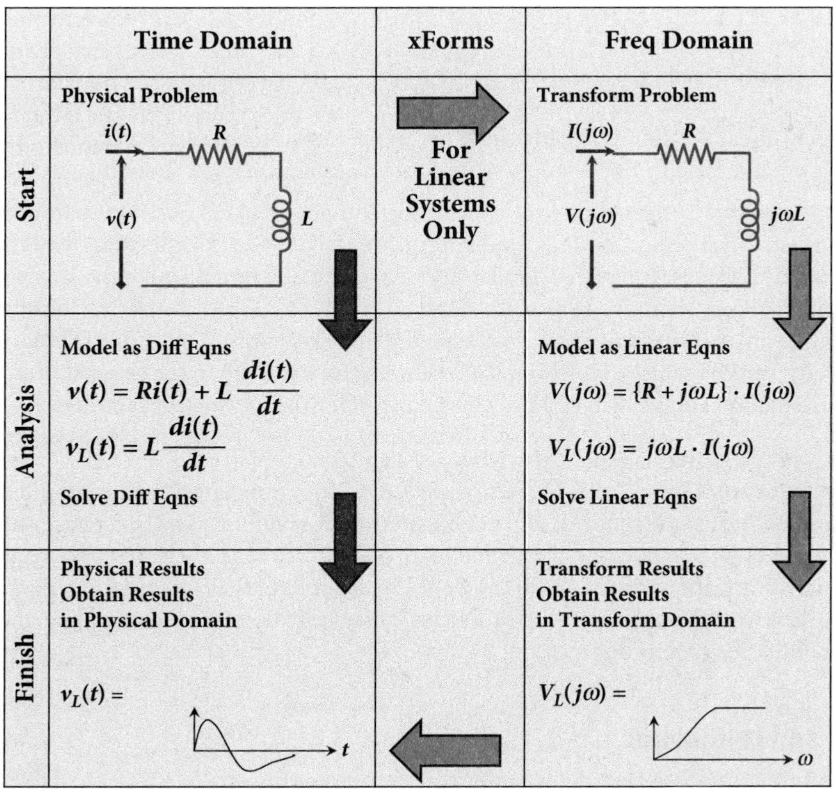

FIGURE 1A.2 An alternate approach is to perform the analysis in the frequency domain. The conversion from and to the time domain is demonstrated.

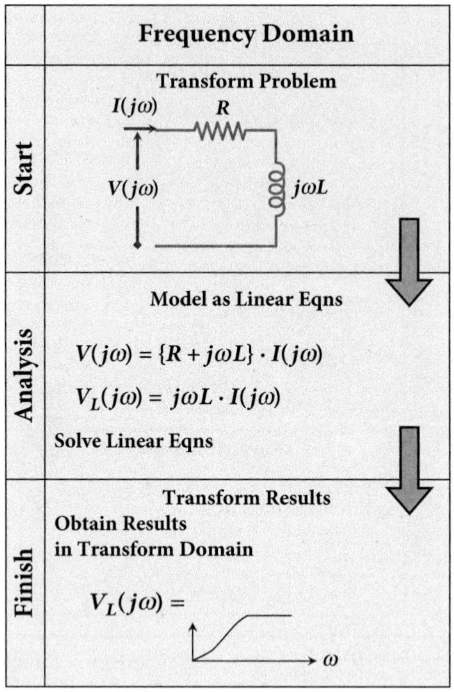

FIGURE 1A.3 Performing the analysis in the frequency domain. The conversion from and to the time domain is omitted.

magical tools and reach great results to the *complicated* issue in our "dream world." However, after we wake up, we need to be able to translate those dream world solutions to physical solutions that are meaningful in real life. Some would argue, the frequency domain is not a dream world, I argue back by asking: what do you think "$j = \sqrt{-1}$" is? j is an imaginary quantity in an imaginary domain.

- The second analogy is that we are in a "shore" city. Our goal is to get from one part of town to another. With congested traffic and complicated routes, we decide to take the waterway that is faster and straightforward. Hence, we move from land to sea, make our way through the waterway, and when it is the right location, we move back to the land again. If we choose a waterway close to the shore, we can maintain an eye on the city's landmarks and do not lose our orientation. That is, we stay in touch with the city and its realities. However, if we take a waterway deep in the sea, we stand a chance of losing our orientation and we will certainly lose our feel for the city and its landmarks.

It is true that with extra training and experience, we can relate the "dream" to reality or stay on the course without monitoring the city's landmarks. Therefore, we can always come up with physical interpretations to all the frequency-domain terms, procedures, and conclusions. For example, j is a 90° phase shift, resonance at certain frequency means energy bouncing back and forth between storage in electric and magnetic forms, and a product of transfer functions corresponds to the impulse response of their cascade, etc. However, it is always a plus to stay in touch with the physics of the problems. This yields deeper understanding and stronger comprehension and often leads to new discoveries.

1A.3 States and Languages

Now, which domain should we use, and how to decide? Here is another analogy (refer to Table 1A.1). It is typical for people living in England to communicate in English, while residents of France do so in French. However, it is possible in some cases, although untypical, that people in France would communicate in English, and those in England do so in French.

Language	State: England	State: France
English	Conventional	Unconventional but valid
French	Unconventional but valid	Conventional

Language	State: Transient	State: Steady State
Time Domain	Conventional	Unconventional but valid
Frequency Domain	Unconventional but valid	Conventional

TABLE 1A.1 A Demonstration of States and Languages

Likewise, transient state and steady state are states while time-domain and frequency-domain analyses are forms of communication or expression languages. Hence, it is typical to do transient analysis in the time domain and steady-state (harmonic) analysis in the frequency domain. However, for certain special cases, we do the opposite.

1A.4 Phasors and Frequency-Domain (Harmonic) Analysis

It is known that harmonic excitation of linear systems yields harmonic solutions of the same frequency as that of the excitation. In such case, all signals throughout the entire system (circuit) will be of the same frequency with differing magnitudes and phases.

The general form of a typical time harmonic signal in the time domain would be $A\cos(\omega t+\varphi)$ (see Figure 1A.4). In the frequency domain, we represent this time-varying function by a "phasor" of magnitude A and a phase angle φ rotating at an angular speed ω (see Figure 1A.5).

$$\bar{A} = A\angle(\omega t+\varphi) \tag{1A.1}$$

Using analogies to the mechanical harmonic motion and the conversion of rotational motion into a reciprocating one, the relationship between the time-domain representation (the cosine function) and the phasor

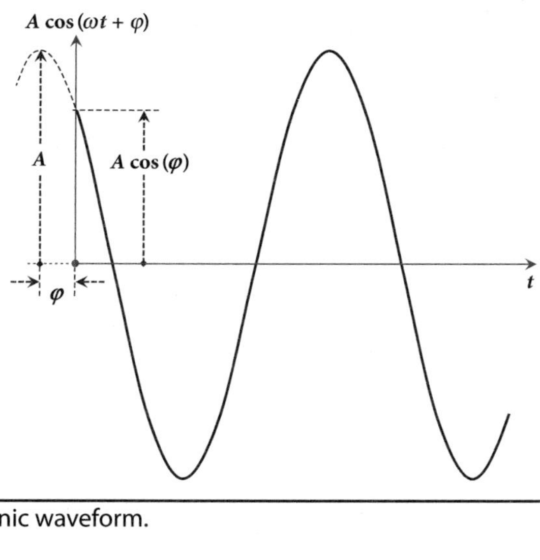

FIGURE 1A.4 A simple harmonic waveform.

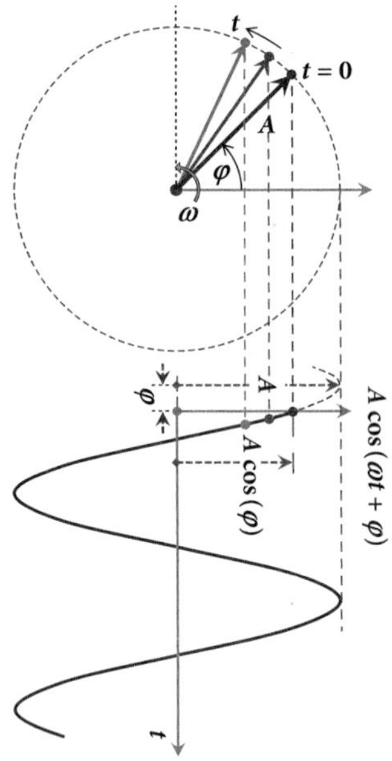

FIGURE 1A.5 Representing the simple cosine harmonic waveform by a rotating phasor.

representation is better demonstrated in Figure 1A.6. In this figure, the time-domain waveform is presented as the projection of the rotating phasor on the *horizontal* axis.

With this graphical phasor representation of the harmonic signal, an intriguing question arises. Why do we view only the *horizontal* projection of the phasor diagram? What can we make out of its *vertical* projection? The answer is that the vertical projection is in fact complementary to the horizontal projection and it coexists whether we pay attention to it or not. In other words, its inclusion in the picture does not add any new parameters to the physics of the problem.

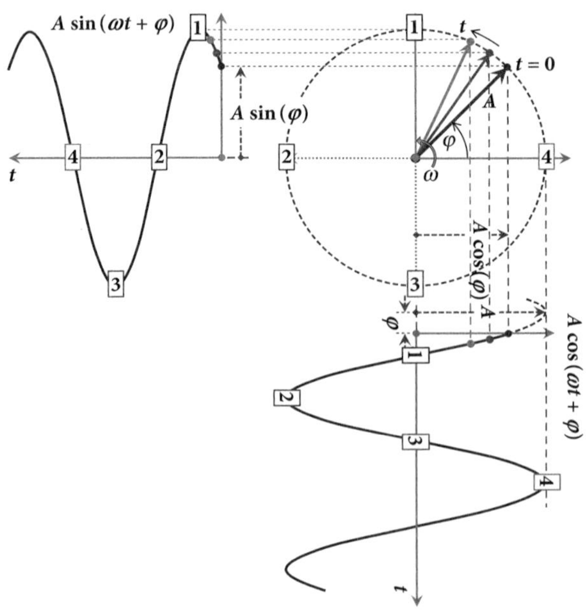

FIGURE 1A.6 A holistic picture of the rotating phasor with two projections: a cosine wave and a sine wave.

To examine the holistic picture, we demonstrate both projections in Figure 1A.6. The figure reveals to us the presence of "vertical" projection in the form of another "complementary" harmonic function $A\sin(\omega t+\varphi)$ that is in quadrature to the original one, for example, when the original function is at its zero value (positions 1 and 3) in the figure, the complementary function reveals the location of the phasor as it has $+A$ and $-A$ on the vertical axis. In addition, when the original function is at positions 2 and 4 ($-A$ and $+A$ in the horizontal projection), the vertical projection is zero.

The question now is, if we were to include this complementary term in our mathematical analysis without altering the nature of the original harmonic signal term or having the two terms mix or interfere, or affect the accuracy of the analysis or alter the final result, would that be unacceptable? The logical answer is no. As will be demonstrated later in this section, this inclusion (in the proper manner) greatly facilitates the mathematical analysis.

Referring to the addendum at the end of this chapter, we can pack the harmonic function and its complement, into a one-pack term to express the harmonic phasor in all relevant mathematical analyses. This term is known as the complex phasor and can be expressed as the sum of both real and imaginary components. Its expression in the complex domain is of the form

$$\bar{A}=[A\cos(\omega t+\varphi)]+j[A\sin(\omega t+\varphi)] \tag{1A.2}$$

This phasor can also be expressed in its magnitude and phase angle format while dropping the ωt part of the angle for efficient presentation:

$$\bar{A}=A\angle(\varphi) \tag{1A.3}$$

Furthermore, the use of the Euler formula provides a convenient mathematical format that greatly facilitates relevant mathematical operations and procedures:

$$\bar{A}[=A\angle(\varphi)]=A\cdot e^{j(\varphi)}=A\cos(\varphi)+j\cdot A\sin(\varphi) \tag{1A.4}$$

Consequently, to recover the time-domain expression of the harmonic signal, all needed is to extract the real part (Re) of the phasor *pack*, hence, using generic notations, we can write,

$$C\cos(\gamma)=\text{Re}\{C\cdot e^{j(\gamma)}\} \tag{1A.5}$$

Now, it is relevant to emphasize that

- As we use the pack phasor term expression in our analysis, our goal is to recover the physically acceptable result by simply eliminating the imaginary term.
- The use of Euler's formula offers the great convenience of dealing with exponentials in contrast with potentially complicated trigonometric identities.

To demonstrate the convenience of using Euler's exponential form for phasor expressions, let us review some of the fundamental mathematical operations in this format.

Addition and Subtraction

Time-domain representation:

$$A\cos(\omega t+\varphi_A)\pm B\cos(\omega t+\varphi_B)=C\cos(\omega t+\varphi_C) \tag{1A.6}$$

Phasor representation:

$$[A\cdot e^{j(\omega t+\varphi_A)}]\pm[B\cdot e^{j(\omega t+\varphi_B)}]=[C\cdot e^{j(\omega t+\varphi_C)}] \tag{1A.7}$$

Recovering the time-domain representation:

$$\begin{aligned}\mathrm{Re}[C\cdot e^{j(\omega t+\varphi_C)}] &= \mathrm{Re}\{[A\cdot e^{j(\omega t+\varphi_A)}]\pm[B\cdot e^{j(\omega t+\varphi_B)}]\} = \mathrm{Re}[A\cdot e^{j(\omega t+\varphi_A)}]\pm \mathrm{Re}[B\cdot e^{j(\omega t+\varphi_B)}]\\ &= A\cos(\omega t+\varphi_A)\pm B\cos(\omega t+\varphi_B)\end{aligned} \tag{1A.8}$$

Multiplication by a "Real" Constant

Time-domain representation:

$$a\cdot A\cos(\omega t+\varphi_A) = C\cos(\omega t+\varphi_C) \tag{1A.9}$$

Phasor representation:

$$\{a\cdot[A\cdot e^{j(\omega t+\varphi_A)}]\} = [\{a\cdot A\}\cdot e^{j(\omega t+\varphi_A)}] \tag{1A.10}$$

Recovering the time-domain representation:

$$\mathrm{Re}\{a\cdot[A\cdot e^{j(\omega t+\varphi_A)}]\} = a\cdot \mathrm{Re}[A\cdot e^{j(\omega t+\varphi_A)}] = a\cdot A\cos(\omega t+\varphi_A) \tag{1A.11}$$

Time Derivatives

Time-domain representation:

$$\frac{d[A\cos(\omega t+\varphi_A)]}{dt} = -\omega A\sin(\omega t+\varphi_A) = \omega A\cos\left(\omega t+\varphi_A+\frac{\pi}{2}\right) \tag{1A.12}$$

Phasor representation:

$$\frac{d}{dt}[A\cdot e^{j(\omega t+\varphi_A)}] = [j\omega A\cdot e^{j(\omega t+\varphi_A)}] = [\omega A\cdot e^{j(\omega t+\varphi_A+\pi/2)}] \tag{1A.13}$$

Recovering the time-domain representation:

$$\begin{aligned}\mathrm{Re}\left\{\frac{d}{dt}[A\cdot e^{j(\omega t+\varphi_A)}]\right\} &= \mathrm{Re}\left\{j\omega\cdot[A\cdot e^{j(\omega t+\varphi_A)}]\right\} = \omega A\cdot \mathrm{Re}[e^{j(\pi/2)}\cdot e^{j(\omega t+\varphi_A)}]\\ &= \omega A\cos(\omega t+\varphi_A+\pi/2) = -\omega A\sin(\omega t+\varphi_A) = \frac{d}{dt}[A\cos(\omega t+\varphi_A)]\end{aligned} \tag{1A.14}$$

Electric Power in Time Harmonic Circuits (Using Both Voltage and Current Phasors)

Time-domain representation:

Time-averaged power (averaged over one period) is given by

$$P_{av} = \mathrm{avg}\{V_{pk}\cos(\omega t+\varphi_V)\cdot I_{pk}\cos(\omega t+\varphi_I)\} = (1/2)V_{pk}I_{pk}\cos(\varphi_V-\varphi_I) \tag{1A.15}$$

Phasor representation:

$$P_{av} = \mathrm{Re}\left\{\left[\left(V_{pk}/\sqrt{2}\right)\cdot e^{j(\omega t+\varphi_V)}\right]\cdot\left[\left(I_{pk}/\sqrt{2}\right)\cdot e^{j(\omega t+\varphi_I)}\right]^*\right\} \tag{1A.16}$$

Recovering the time-domain representation:

$$\begin{aligned}P_{av} &= \mathrm{Re}\left\{\left[\left(V_{pk}/\sqrt{2}\right)\cdot e^{j(\omega t+\varphi_V)}\right]\cdot\left[\left(I_{pk}/\sqrt{2}\right)\cdot e^{j(\omega t+\varphi_I)}\right]^*\right\}\\ &= \mathrm{Re}\left\{(1/2)\cdot\left[V_{pk}\cdot I_{pk}\cdot e^{j(\varphi_V-\varphi_I)}\right]\right\} = (1/2)V_{pk}I_{pk}\cos(\varphi_V-\varphi_I)\end{aligned} \tag{1A.17}$$

1A.5 Use of Phasors in Circuit Analysis (in the Frequency Domain)

Performing circuit analysis using phasor packs is typically known as frequency-domain analysis. This terminology reflects the fact that the analysis is performed for one frequency at a time (single harmonic). The Euler expression of the phasor displays only the phasor's magnitude and phase and not its frequency; however, the phasor frequency must be defined and is an integral part of the analysis and its solution.

1A.6 Demonstration of Circuit Analysis in the Frequency Domain

1A.6.1 Starting with the Time-Domain Form

To demonstrate the above-discussed concepts, let us consider a simple *R-L* circuit excited by a harmonic voltage source (Figure 1A.7). The desired results are to evaluate the current in the loop, the voltages across the resistor and inductor, as well as the power dissipated in the resistor.

We assume that the voltage source is a harmonic source with an amplitude of V_s, radial frequency ω, and a phase angle of φ_{vs} at the instant $t = 0$ (the time reference for this analysis). Hence, we can express $v_s(t)$ in the time domain as

$$v_s(t) = V_s \cos(\omega t + \varphi_{vs}) \tag{1A.18}$$

Now, we write Kirchhoff's voltage law (KVL) around the only loop in the circuit, thus

$$v_s(t) = R \cdot i_{\text{Loop}}(t) + v_L(t); \qquad v_L(t) = L \cdot \frac{di_{\text{Loop}}(t)}{dt} \tag{1A.19}$$

Next, we write the time-domain expression for the loop current in the form:

$$i_{\text{Loop}}(t) = I_{\text{Loop}} \cos(\omega t + \varphi_i) \tag{1A.20}$$

where I_{Loop} is the peak amplitude and φ_i is the corresponding phase.

Now, the KVL expression (1A.19) in the time domain takes on the form:

$$\begin{aligned}\{v_s(t)\} &= V_s \cos(\omega t + \varphi_{vs}) = RI_{\text{Loop}} \cos(\omega t + \varphi_i) + L\frac{d[I_{\text{Loop}} \cos(\omega t + \varphi_i)]}{dt} \\ &= RI_{\text{Loop}} \cos(\omega t + \varphi_i) - \omega LI_{\text{Loop}} \sin(\omega t + \varphi_i) \\ &= RI_{\text{Loop}} \cos(\omega t + \varphi_i) + \omega LI_{\text{Loop}} \cos(\omega t + \varphi_i + \pi/2)\end{aligned} \tag{1A.21}$$

Working backward toward the phasor representation, we can replace the cos terms as the real part of the complex Euler (exponential) form:

$$\{v_s(t)\} = V_s \operatorname{Re}[e^{j(\omega t + \varphi_{vs})}] = RI_{\text{Loop}} \operatorname{Re}[e^{j(\omega t + \varphi_i)}] + \omega LI_{\text{Loop}} \operatorname{Re}[e^{j(\omega t + \varphi_i + \pi/2)}] \tag{1A.22}$$

$$\begin{aligned}\{v_s(t)\} &= \operatorname{Re}\left[V_s e^{j(\omega t + \varphi_{vs})}\right] = \operatorname{Re}\left[RI_{\text{Loop}} e^{j(\omega t + \varphi_i)}\right] + \operatorname{Re}\left[\omega LI_{\text{Loop}} e^{j(\omega t + \varphi_i)} e^{j(\pi/2)}\right] \\ &= \operatorname{Re}\left[RI_{\text{Loop}} e^{j(\omega t + \varphi_i)}\right] + \operatorname{Re}\left[j\omega LI_{\text{Loop}} e^{j(\omega t + \varphi_i)}\right]\end{aligned} \tag{1A.23}$$

$$\{v_s(t)\} = \operatorname{Re}\left[V_s e^{j(\omega t + \varphi_{vs})}\right] = \operatorname{Re}\left[(R + j\omega L) I_{\text{Loop}} e^{j(\omega t + \varphi_i)}\right] \tag{1A.24}$$

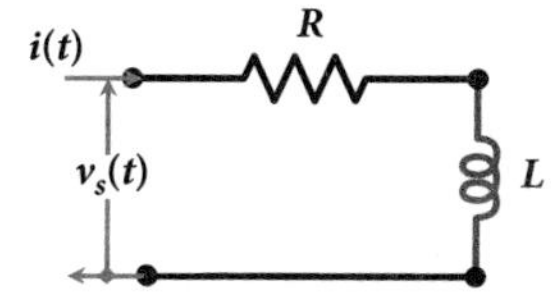

FIGURE 1A.7 Demonstration of circuit analysis in the time domain using *RL* circuit.

$$\{v_s(t)\} = \mathrm{Re}\left[V_s e^{j(\varphi_{vs})} e^{j(\omega t)}\right] = \mathrm{Re}\left[(R + j\omega L) I_{\mathrm{Loop}} e^{j(\varphi_i)} e^{j(\omega t)}\right] \tag{1A.25}$$

$$\{v_s(t)\} = \mathrm{Re}\left[\overline{V}_s e^{j(\omega t)}\right] = \mathrm{Re}\left[(R + j\omega L)\overline{I}_{\mathrm{Loop}} e^{j(\omega t)}\right] \tag{1A.26}$$

where the voltage and current phasors notations $\overline{V}_s$ and $\overline{I}_{\mathrm{Loop}}$ are used in place of $V_s e^{j(\varphi_{vs})}$ and $I_{\mathrm{Loop}} e^{j(\varphi_i)}$, respectively.

Examining Equation (1A.26), we recognize that it represents the real part of the complex form:

$$\left[\overline{V}_s e^{j(\omega t)}\right] = \left[(R + j\omega L)\overline{I}_{\mathrm{Loop}} e^{j(\omega t)}\right] \text{ or} \tag{1A.27}$$

$$\overline{V}_s = (R + j\omega L)\overline{I}_{\mathrm{Loop}} \tag{1A.28}$$

which is the phasor form of the KVL equation for the circuit represented in the frequency domain.

Now, using Equation (1A.27), we can solve for $\overline{I}_{\mathrm{Loop}}$ as

$$\left[\overline{I}_{\mathrm{Loop}}\right] = \frac{\left[\overline{V}_s\right]}{(R + j\omega L)} \tag{1A.29}$$

From which the time-domain expression for i_{Loop} can be obtained through multiplying the phasor expression by $e^{j(\omega t)}$ then extracting the real part of the product:

$$\{i_{\mathrm{Loop}}(t)\} = \mathrm{Re}\left[\overline{I}_{\mathrm{Loop}} e^{j(\omega t)}\right] \tag{1A.30}$$

Likewise, we can derive expressions for the time-domain form of the voltage drops across the resistor and the inductor as well as the power dissipated in the resistor.

Now, let us tackle this example from the frequency-domain end. We start by converting the circuit to its frequency-domain model; voltage and currents assume their phasor formats, while circuit elements take on their impedance/admittance expressions.

$$\left[\overline{V}_R\right] = (R)\left[\overline{I}_{\mathrm{Loop}}\right] \xrightarrow{\text{yields}} \{v_R(t)\} = \mathrm{Re}\left[\overline{V}_R e^{j(\omega t)}\right] \tag{1A.31}$$

$$\left[\overline{V}_L\right] = (j\omega L)\left[\overline{I}_{\mathrm{Loop}}\right] \xrightarrow{\text{yields}} \{v_L(t)\} = \mathrm{Re}\left[\overline{V}_L e^{j(\omega t)}\right] \tag{1A.32}$$

$$P_R = \mathrm{Re}\left\{(1/2)\cdot\left[\overline{V}_R\right]\left[\overline{I}^*_{\mathrm{Loop}}\right]\right\} \tag{1A.33}$$

1A.6.2 Starting with the Frequency-Domain Form

The short version for carrying out this analysis starts by converting the circuit into its frequency domain model (Figure 1A.8).

Next, we write KVL for the loop in phasor terms:

$$\left[\overline{V}_s\right] = (R + j\omega L)\left[\overline{I}_{\mathrm{Loop}}\right] \tag{1A.34}$$

which we typically write in the frequency-domain form

$$V_s(j\omega) = \{R + j\omega L\}\cdot I_{\mathrm{Loop}}(j\omega) \tag{1A.35}$$

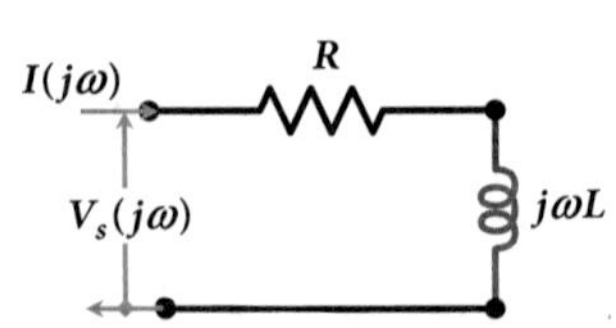

FIGURE 1A.8 Demonstration of circuit analysis in the frequency domain using *RL* circuit.

where

$$V_R(j\omega) = R \cdot I_{\text{Loop}}(j\omega) \text{ and } V_L(j\omega) = j\omega L \cdot I_{\text{Loop}}(j\omega) \tag{1A.36}$$

Hence, we continue the solution as before

$$I_{\text{Loop}}(j\omega) = \frac{V_s(j\omega)}{(R + j\omega L)} \xrightarrow{\text{yields}} \{i_{\text{Loop}}(t)\} = \text{Re}\left[I_{\text{Loop}}(j\omega)e^{j(\omega t)}\right] \tag{1A.37}$$

$$V_R(j\omega) = (R)I_{\text{Loop}}(j\omega) \xrightarrow{\text{yields}} \{v_R(t)\} = \text{Re}\left[V_R(j\omega)e^{j(\omega t)}\right] \tag{1A.38}$$

$$V_L(j\omega) = (j\omega L)I_{\text{Loop}}(j\omega) \xrightarrow{\text{yields}} \{v_L(t)\} = \text{Re}\left[V_L(j\omega)e^{j(\omega t)}\right] \tag{1A.39}$$

$$P_R = \text{Re}\{(1/2) \cdot [V_R(j\omega)][I^*_{\text{Loop}}(j\omega)]\} \tag{1A.40}$$

which yields identical results to those obtained earlier in Equations (1A.29) to (1A.33).

1A.7 The Frequency Domain and the Laplace Transform

In this section, we briefly remind the reader of the Laplace transform and how it relates to the frequency domain. As you recall from other signals and systems studies, the Laplace transform allows us to convert time-domain expressions (signals as well as system responses) into the Laplace or s-domain. The Laplace transform expression of a time-domain function $f(t)$ is given by

$$F(s) = \int_0^\infty f(t)e^{-st}\,dt \tag{1A.41}$$

The obtained transform $F(s)$ is expressed in the complex s-domain where $s = \sigma + j\omega$. This transform tool enables us to solve time-domain transient problems through the conversion of complicated differential equations to much more manageable algebraic equations. The transform enables us to solve steady-state harmonic analyses as well, since by letting the real part $\sigma = 0$, $F(s)$ converges to $F(j\omega)$, which is the frequency-domain phasor expression for $f(t)$ function.

If we were to solve the same example in Figure 1A.8 in the s-domain, we would replace the $j\omega$ in the KVL loop equation [Equation (1A.34)] by s. Hence,

$$V_s(s) = (R + sL)I_{\text{Loop}}(s) \tag{1A.42}$$

ADDENDUM 1B
THE MYSTERY OF *J* AND IMAGINARY NUMBERS

Referring to the earlier discussion in Addendum 1A, and to Figure 1A.6 in particular, we will explore the mystery of the imaginary term j. In this figure, the time-domain waveform is presented as the projection of the rotating phasor on the horizontal axis. However, as per our discussion in Addendum 1A, we realize the vertical projection as complementary to the horizontal projection. We also concluded that its inclusion in the picture does not add any new parameters to the physics of the problem.

Now, suppose we package both the horizontal and vertical projection terms of the rotating phasor, A, into a one pack term, for example, P_A. This pack term is to be used to express the phasor in all relevant mathematical analyses. Hence, the P_A term needs to take the simple form of a linear mix of the two projections in order to facilitate the separation process when needed. We will use the notations α, β, and γ for the angles of the phasors A, B, and C, respectively, for convenience.

Hence, we write

$$P_A = [A\cos(\alpha)] + q[A\sin(\alpha)] \tag{1B.1}$$

where q is the mixing coefficient.

In a similar way, we consider a second harmonic function as follows:

$$P_B = [B\cos(\beta)] + q[B\sin(\beta)] \tag{1B.2}$$

Now, we examine the results obtained by putting the packed terms P_A and P_B through different mathematical procedures that are typically required in the analysis of harmonic signals. The goal is to decide if the obtained results for the packed terms maintain the separability that allows us to retrieve the proper harmonic analysis results. We will demonstrate here some of these typical mathematical procedures, namely, scaling (multiply or divide by a scalar), addition/subtraction, and multiplication.

Scaling

Let us start by the scaling operation by considering the pack P_A scaled by the scalar "a," hence

$$a \cdot P_A = a \cdot \{[A\cos(\alpha)] + q[A\sin(\alpha)]\} = \{[a \cdot A\cos(\alpha)] + q[a \cdot A\sin(\alpha)]\} \tag{1B.3}$$

Thus, concluding that the scaling operation did not affect the separability of both the cosine and sine terms.

Addition/Subtraction

The sum $a \cdot P_A \pm b \cdot P_B$ yields

$$\begin{aligned} a \cdot P_A \pm b \cdot P_B &= \{[a \cdot A\cos(\alpha)] + q[a \cdot A\sin(\alpha)]\} \pm \{[b \cdot B\cos(\beta)] + q[b \cdot B\sin(\beta)]\} \\ &= \{[a \cdot A\cos(\alpha) \pm b \cdot B\cos(\beta)] + q \cdot [a \cdot A\sin(\alpha) \pm b \cdot B\sin(\beta)]\} \end{aligned} \tag{1B.4}$$

It is clear that the sum of the cosine terms is separable from that of the sine terms, which makes our packaging scheme plausible so far.

Multiplication

Recognizing that the product of two phasors is supposed to yield a product phasor whose magnitude is the product of the two individual magnitudes and its phase is the sum of the two individual phasor phases. Using the notation

$$\bar{A} = A\angle\alpha, \qquad \bar{B} = B\angle\beta, \qquad \text{and} \qquad \bar{C} = C\angle\gamma \tag{1B.5}$$

Then, the product $\bar{C} = \bar{A} \cdot \bar{B}$, is defined as $C = A \cdot B$, and $\gamma = \alpha + \beta$.

Hence

$$P_C = \{A \cdot B \cdot \cos(\alpha+\beta) + q \cdot A \cdot B \cdot \sin(\alpha+\beta)\} \tag{1B.6}$$

$$P_C = \{A \cdot B \cdot [\cos(\alpha)\cos(\beta) - \sin(\alpha)\sin(\beta)] + q \cdot A \cdot B \cdot \sin(\alpha+\beta)\} \tag{1B.7}$$

Now, multiplication of the two packs P_A and P_B is put through to yield P_C, hence

$$\begin{aligned} P_C = P_A \cdot P_B &= \{A\cos(\alpha) + q \cdot A\sin(\alpha)\} \cdot \{B\cos(\beta) + q \cdot B\sin(\beta)\} \\ &= \{A \cdot B \cdot \cos(\alpha)\cos(\beta) + q^2 \cdot A \cdot B \cdot \sin(\alpha)\sin(\beta) \\ &\quad + 2q \cdot A \cdot B \cdot [\sin(\alpha)\cos(\beta) + \cos(\alpha)\sin(\beta)]\} \end{aligned} \tag{1B.8}$$

$$P_C = \{A \cdot B \cdot [\cos(\alpha)\cos(\beta) + q^2 \cdot \sin(\alpha)\sin(\beta)] + q \cdot A \cdot B \cdot \sin(\alpha+\beta)\} \tag{1B.9}$$

For the two results [Equations (1B.7) and (1B.9)] to agree, the quantity q^2 must equal -1, and hence q should equal to the square root of "-1." Which is not a physical quantity; however, examining q and q^2 in the light of phasor expressions, we recognize

$$q^2 = 1\angle\pi \qquad \text{and} \qquad q = 1\angle(\pi/2) \tag{1B.10}$$

This q is what we customarily refer to as the imaginary term "j," (also termed as "i" in the literature).

$$j^2 = 1\angle\pi \qquad \text{and} \qquad j = 1\angle(\pi/2) \tag{1B.11}$$

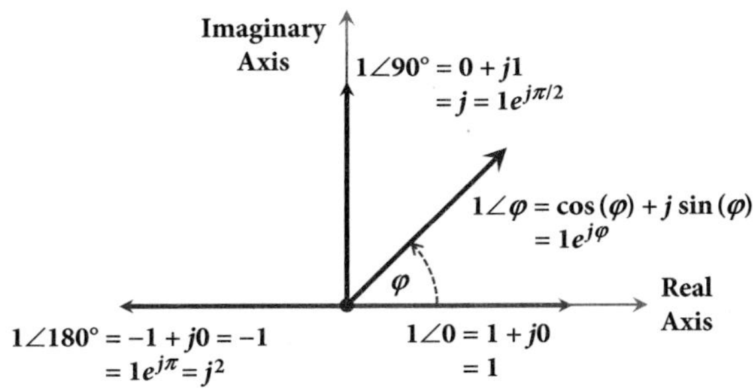

FIGURE 1B.1 A graphical demonstration of the phasor concepts through expressing a phasor of unit magnitude and different phase angles.

The naming of the "complementary" q (now j) term as "imaginary" is befitting since the square root of "–1" is not physical or "real," and hence "imaginary." We also recognize that in our harmonic analysis, the original harmonic term is the one that physically exists and that the complementary term does not. True, we packed this complementary imaginary term in our analysis for mathematical convenience, but at the end of our analysis, only the physical signal is what matters and what results.

Consequently, as we use pack term expressions, for example, $P_A = [A\cos(\alpha)] + j[A\sin(\alpha)]$, etc. in our analysis, our goal is to recover the physically acceptable term for the analysis result by simply eliminating the imaginary term, hence

$$C\cos(\gamma) = \text{Re}\{C\cos(\gamma) + jC\sin(\gamma)\} \tag{1B.12}$$

where Re is short for "real part of."

With the presence of both the real and imaginary components in the resulting phasor expressions, we find ourselves working in what we refer to as the complex domain.

At this point, it is relevant to state that the packing formula that we proposed earlier to combine the original "real" harmonic signal and the "imaginary" complementary term is in harmony with the Euler formula:

$$\bar{C}[= C\angle(\gamma)] = C\cdot e^{j(\gamma)} = C\cos(\gamma) + j\cdot C\sin(\gamma) \tag{1B.13}$$

and hence,

$$C\cos(\gamma) = \text{Re}\{C\cdot e^{j(\gamma)}\} \tag{1B.14}$$

The use of Euler's formula offers the great convenience of dealing with exponentials in contrast to potentially complicated trigonometric identities.

It is relevant to state that the imaginary quantity j, Equation (1B.11), represents the packing factor of the "quadrature" complementary term. "j" as a phasor represents a unity magnitude and a $\pi/2$ phase (90° CCW rotation):

$$j = 1\cdot e^{\frac{j\pi}{2}} \quad \text{and} \quad j^2 = 1\cdot e^{2\frac{j\pi}{2}} = 1\cdot e^{j\pi} = -1 \tag{1B.15}$$

and hence the famous identity

$$j = \sqrt{-1} \tag{1B.16}$$

Finally, Figure 1B.1 is a graphical demonstration of the concepts discussed above through expressing a phasor magnitude of "1" with different phase angles.

CHAPTER 2

Transmission Lines—Wave Equations

Phillip Hagar Smith
(1905–1987)

Chapter Learning Objectives

- Identify the conditions under which it becomes necessary to analyze a circuit using the transmission line (TL) model.
- Enumerate the characteristic parameters of a controlled geometry TL (Z_o, α, β).
- Represent a controlled geometry TL using a lumped circuit model and solve for the 2 wave (incident and reflected) solution that describes the voltage and current at any point across the TL.
- Calculate reflection coefficient, transmission coefficient, and standing wave ratio (SWR) for terminated TLs.
- Distinguish between the different modes of waves on a TL (propagating, standing, propagating and standing, decaying, etc.). In addition, what is the impact of the TL parameters and load in determining those modes?
- Calculate the different components of power in the circuit and differentiate between delivered power to load, dissipated power, and stored energy.
- Identify impedance matching conditions that result in max power delivery to the load.
- Explain how TL parameters affect the fidelity of a transmitted signal and what are the conditions that need to be satisfied by the TL parameters to obtain distortion-less signal delivery.
- Design impedance matching networks using both lumped and distributed components, including quarter wave transformers and stubs.
- Use Smith charts to find the different circuit quantities at any point along the TL, which include reflection coefficient, impedance, voltage, and current and conversely find locations of voltage and current maxima on the TL.
- Design impedance matching stubs using Smith chart.

2.1 Introduction

Transmission lines (TLs) is the name we use to refer to those wires and cables that connect electrical and electronic devices (parts or components) to each other. The wires used to connect the light bulb to the electrical mains is a TL, the wire that connects the landline phone to the wall outlet is a TL, and the cable that connects the TV to the satellite receiver, the wall outlet, or antenna is also a TL. The wires you used in the circuits and electronics laboratories and the wires you used in *breadboarding* an analog or a digital circuit are all TLs. The conducting strips on (and within) the motherboard of the computer as well as the *interconnects* inside the semiconductor chips (processor, memory, etc.) are also TLs. Figure 2.1 demonstrates a few samples of different kinds of TLs.

"OK. So, what is so important about TLs to make this topic the opening chapter of an EM book?" We used those wires in our circuits and electronics labs with no issues whatsoever. We also wired so many of these breadboards and things worked well. The answer is that it has to do with the length of the wire relative to the wavelength of the signals propagating, as well as the wire's internal resistance. Let us take these two items one at a time. First, let us talk about wire length and wavelength.

To remind you about wavelength, we visualize a signal/wave propagating in one dimension (Figure 2.2). Well, as the figure shows, the signal travels a distance as time elapses. The distance between two consecutive locations of similar phase on the wave (e.g., two consecutive peaks) is called the wavelength, λ. Simple physics tells us that distance $\lambda = c/f$, where c is the wave's phase velocity and f is its frequency.

Now, let us assume that this signal propagates along a conducting wire (cable/TL) of length ℓ. Figure 2.3 demonstrates that the signal phase varies along the TL length. The amount of phase variation along the TL length, ℓ, depends on the length of this wire relative to the wavelength, λ. The ratio ℓ/λ is the same as (Δ phase/360°).

Hence, the change in signal phase between that at the beginning of the TL segment ℓ and its end, Δ phase, equals $360 \times \ell/\lambda$. This change in phase corresponds to a change in signal amplitude (Figure 2.3), and hence the signal at the beginning of the TL segment ℓ and its end are not identical. Consequently, if you are using this

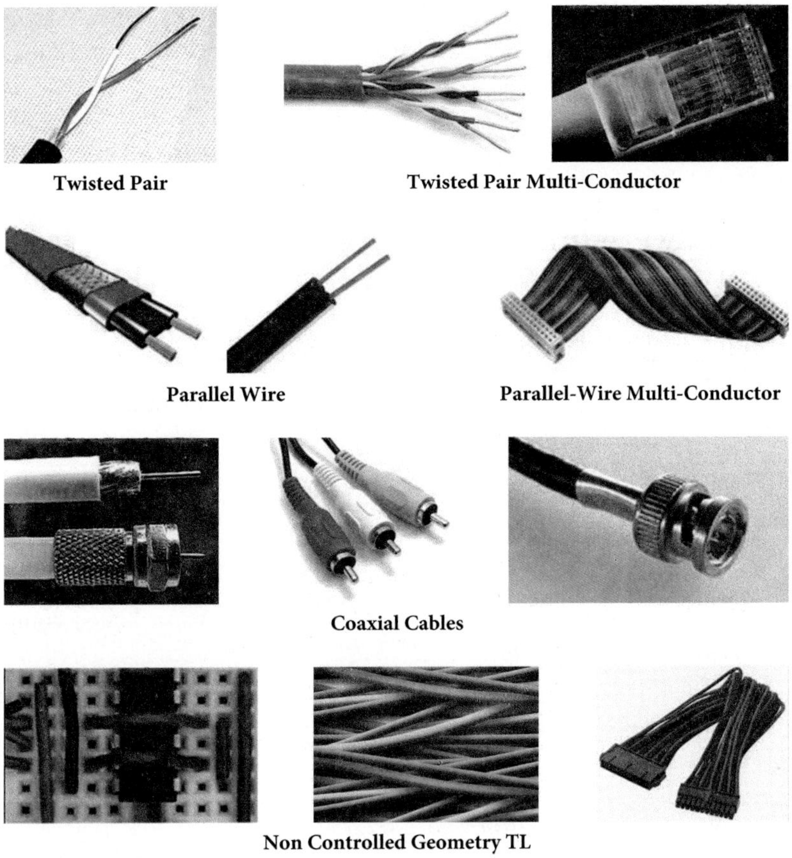

FIGURE 2.1 Samples of different kinds of transmission lines.

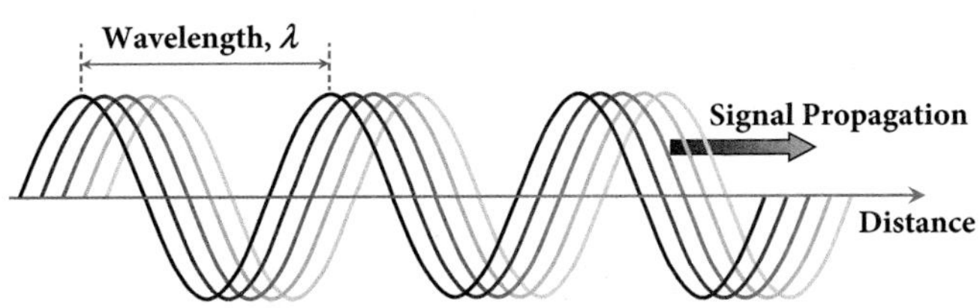

FIGURE 2.2 Waveform propagation and wavelength.

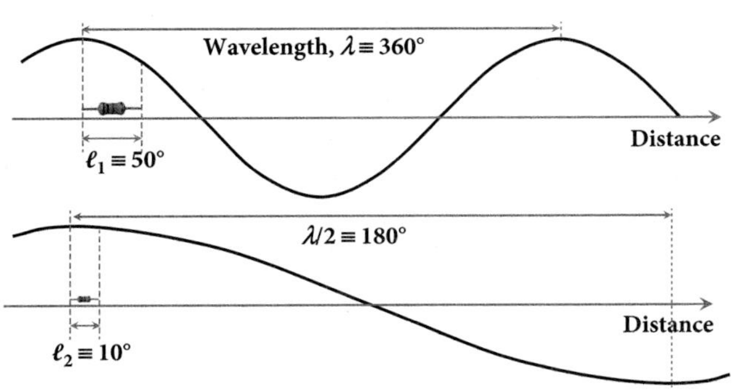

FIGURE 2.3 Demonstration of the component size relative to the wavelength.

line to connect the output side of a device or a component and the input of another, then, one cannot assume that the signal that came out the first device entered into the second. Rather one should account for the changes introduced by the interconnecting TL.

Working at low frequencies, the wavelength is long and most typical wiring arrangements are too short compared to it. In these cases, the Δ phase is infinitesimally small and so is the change in signal amplitude. [Notice that for DC signals (zero frequency), the wavelength is infinite.]

Now, we turn briefly to the second item, "the wire's internal resistance," which may not be ignored even if ℓ/λ is infinitesimally small. At low frequencies and at DC, we would only be concerned about the signal drop as a result of the wire's internal resistance.

Back to the cases where the ratio ℓ/λ matters. The question is how much ℓ/λ can we tolerate (ignore) in our analysis (or design)? The answer to this question lies in deciding how much signal level (and hence phase) change can be tolerated. Considering the wave changes, the fastest around its zero crossing where $\sin(x) \approx x$, the worst-case analysis holds. Table 2.1 demonstrates the orders of magnitudes of circuit dimensions that

% Change in signal amplitude [x]		10.0%	1.0%	0.1%	
Corresponding Δ phase [$\sin^{-1}(x)$]		0.100	0.010	0.001	Radians
Corresponding Δ phase (degrees)		5.739	0.573	0.057	Degrees
Corresponding ℓ/λ [Δ phase/2π]		0.016	0.0016	0.00016	
(a) Metric Units					
Frequency	**Wavelength in Air**	**Tolerated Length (Air)**	**Tolerated Length (Air)**	**Tolerated Length (Air)**	**Metric Unit**
1 Hz	300.0	4.78264	0.47747	0.04775	Mm
60 Hz	5.0	0.07971	0.00796	0.00080	Mm
1 kHz	300.0	4.78264	0.47747	0.04775	km
1 MHz	300.0	4.78264	0.47747	0.04775	m
10 MHz	30.0	0.47826	0.04775	0.00477	m
100 MHz	3.0	0.04783	0.00477	0.00048	m
1 GHz	300.0	4.78264	0.47747	0.04775	mm
10 GHz	30.0	0.47826	0.04775	0.00477	mm
100 GHz	3.0	0.04783	0.00477	0.00048	mm
1 THz	300.0	4.78264	0.47747	0.04775	µm
(b) English Units					
Frequency	**Wavelength in Air**	**Tolerated Length (Air)**	**Tolerated Length (Air)**	**Tolerated Length (Air)**	**English Unit**
1 Hz	186,411.4	2,971.80	296.688	29.6683	Mile
60 Hz	3,106.9	49.5299	4.94480	0.49447	Mile
1 kHz	186.4	2.97180	0.29669	0.02967	Mile
1 MHz	984.3	15.6911	1.56651	0.15665	Feet
10 MHz	98.43	1.56911	0.15665	0.01566	Feet
100 MHz	118.1	1.88293	0.18798	0.01880	Inch
1 GHz	11.81	0.18829	0.01880	0.00188	Inch
10 GHz	1.181	0.01883	0.00188	0.00019	Inch
100 GHz	118.1	1.88293	0.18798	0.01880	Mil
1 THz	11.81	0.18829	0.01880	0.00188	Mil

TABLE 2.1 Tolerance Levels for Ignoring Component Dimensions in Circuit Analysis

can be tolerated at different operating frequencies. Three *tolerance* levels are given corresponding to different levels of phase errors.

So, as you can see, if you are building a circuit (or device) to operate at 1 GHz, your wiring interconnects should be shorter than 0.04775 cm (0.0188 in), otherwise you must analyze the performance of such interconnect using TL theory to avoid errors more than 1%.

2.2 Transmission Line Analysis (Theory)

The accurate and comprehensive way to analyze the performance of TLs is to do the analysis as an electromagnetic (EM) fields problem. This requires the use of Maxwell's equations to develop a full EM fields model and study the field waves as they propagate along the TL. This process will be presented in Chapter 11 of this book. This means that there is a lot for us to learn before we can get into such analysis. However, we can use a circuit theory approach to develop a circuit model and get a simplified analysis for the TL for the time being.

The interesting point here is that the circuit theory tools that we will be using to develop TL analysis are not but the derivatives of the field theory and Maxwell's equations. We learned these tools in earlier circuit courses and we are familiar with them; specifically, Ohm's law and Kirchhoff's voltage and current laws.

Representing a TL by an equivalent circuit, containing circuit elements that physically model the TL, requires future knowledge that will be developed in later chapters as well. However, let us assume that we guess what these circuit elements are like (qualitatively) and have some physical insight of their nature. For example, we can guess that the line conductors do have some internal resistance R, the TL as a "loop" of two conductors has magnetic "inductive" behavior L, the two conductors separated by a dielectric represent a capacitance C, and the imperfect dielectric isolation between the two conductors would cause leakage conduction G. While we are familiar with these components in circuit analysis, the EM fields properties and physical insight of these components will be the subject of later chapters in this book (capacitance, resistance, and conductance in Chapter 6 and inductance in Chapter 9).

2.3 Circuit Theory Analysis of a Two-Conductor Controlled Geometry TL

In this book, we confine our study of TL theory to the simplest of cases; that of two conductors with uniform cross section (material and geometry do not vary along the TL length). This is what we call controlled geometry two conductor TL. The lines will be assumed straight with no curves or bends.

We will further limit ourselves to the case where the TL material, that is, both conductors and the insulating dielectric(s), are linear (properties do not change with field intensity) and frequency independent (properties do not change with frequency).

Examples of such TLs include the parallel wire TL, the coaxial TL, and the microstrip TL (Figure 2.4).

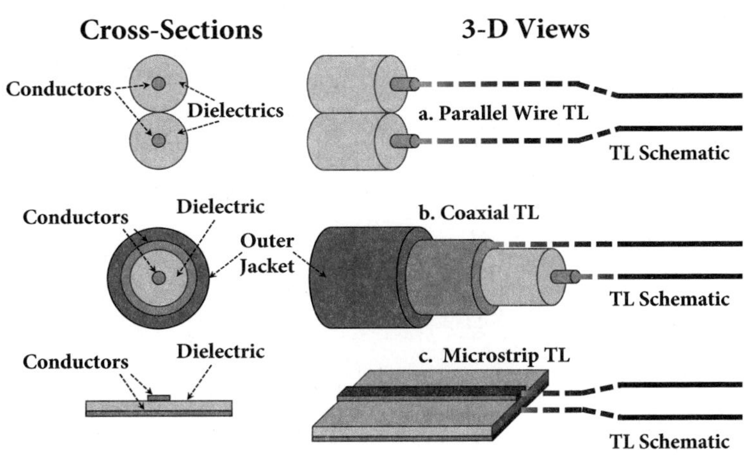

Figure 2.4 Examples of controlled geometry TLs.

2.4 RLGC Model

To develop a physically based model, we have to realize that it is a distributed one, that is, it is distributed over the length of the line and is not lumped at any specific location. To explain, we talk about the internal resistance of the two conductors forming the line. That resistance is determined by the conductor material, the cross-sectional area of the conductor, and the length of the wire. The resistance is distributed along the length of the wire and is not lumped at either end or anywhere else. So, we can model it as a distributed resistance and we can define a measure R_z as the resistance per unit length of the TL. This would be the sum of the total *circuit* resistance per unit length of both TL conductors. The units of R_z would be Ω/m.

Likewise, we can talk about the distributed inductance of the TL circuit. We will use L_z to denote the inductance per unit length, henries/meter. For the insulating dielectric(s), we model two different physical parameters: C_z (farads/meter) and G_z (siemens/meter). The capacitance per unit TL length, C_z, is the physical model of the dielectric(s) insulation between the two conductors and the conductance per unit TL length, G_z, is the physical model of the imperfect insulation of the dielectric causing leakage current through the dielectric. Figure 2.5 depicts the nature of these RLGC elements. This physically based circuit model of the TL is not perfect or accurate, but it will help us set our feet into this mysterious domain of TL theory. This circuit model is typically referred to as the RLGC model.

It is to be noted that most references name the distributed RLGC parameters as R, L, G, and C with no subscript. This practice may lead to some confusion as it implies a reference to *whole* resistance, inductance, conductance, and capacitance quantities and not the per unit length parameters. We choose to add the *z* subscript to distinguish these quantities as per unit *z* length parameters. Hence, the incremental *whole* resistance of a line increment of length Δz would be $\Delta R = [R_z][\Delta z]$. Likewize, $\Delta L = [L_z][\Delta z]$, $\Delta G = [G_z][\Delta z]$, and $\Delta C = [C_z][\Delta z]$ for the incremental *whole* inductance, conductance, and capacitance, respectively.

2.5 Transmission Line Circuit Analysis Using the Distributed RLGC Model

For this analysis, we will assume the line to extend along a linear axis, let us call it the *z* axis (Figure 2.6). At $z = 0$, we would have the input side (or the sending/transmitter end) to the TL (typically, a signal source with an internal impedance) and at the end of the line, say $z = \ell$, the line output side, a load would be connected (receiving/receiver end).

Now, as we replace the line by its distributed RLGC circuit equivalency, we expect the voltage and currents to vary with position along the line. We will use the notations $V(z)$ and $I(z)$, for the voltage and current phasors as functions of position, respectively [or $v(z,t)$ and $i(z,t)$ for instantaneous (time-domain) expressions]. We use the upper case notations for the phasors in the frequency-domain analysis, while the lower case notations represent the time-domain variables. At $z = 0$ and at $z = \ell$, the voltage and current terms would be $V(0)$, $I(0)$, $V(\ell)$, and $I(\ell)$, [or $v(0,t)$, $i(0,t)$, $v(\ell,t)$, and $i(\ell,t)$], respectively.

Next, we need to develop expressions for the voltages current relationships as functions of position and time. In order to develop a thorough and insightful analysis of the TL network, we intend to carry the analysis in both the time and frequency domains simultaneously. Moreover, we will also be monitoring the solutions for both the transient and the steady (harmonic analysis) states as we go. As we take the analysis back and forth

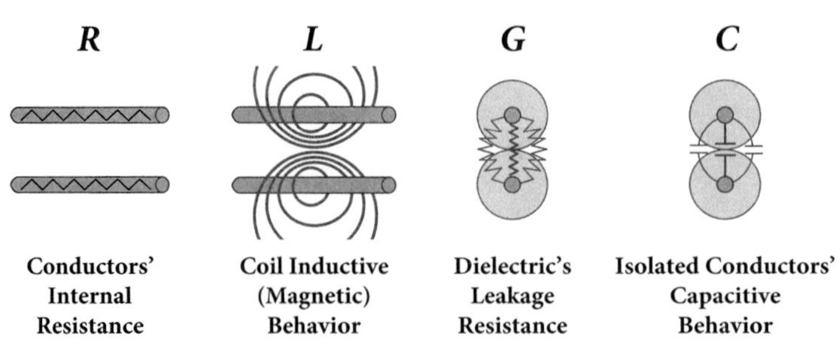

FIGURE 2.5 Graphical demonstration of the RLGC physical model.

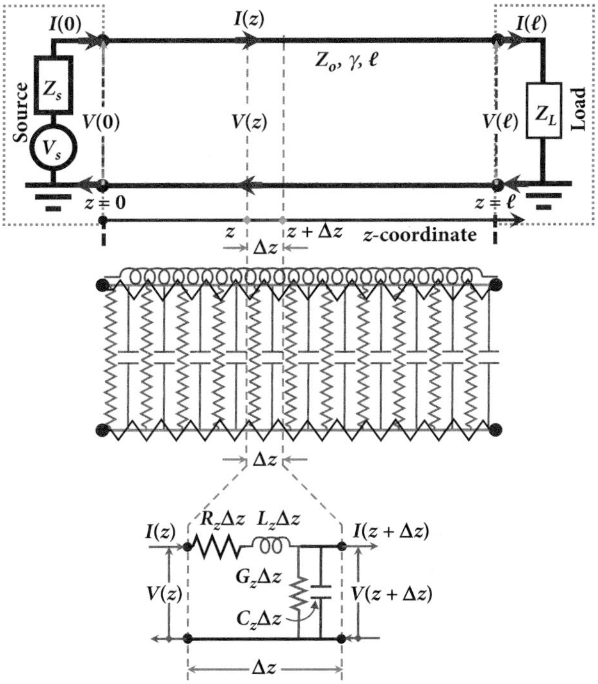

Figure 2.6 Using distributed RLGC to model a transmission line circuit.

between the two domains, some readers may find it a bit confusing. However, going over this section a couple of times, and with focused attention, the results will be all positive.

2.6 Steady-State Harmonic Analysis

In the following, we will start with the steady-state harmonic analysis in both time and frequency domains. We will assume an excitation in the form of a time harmonic voltage

$$v_s(t) = |V_s|\cos(\omega t + \varphi_s) \xleftrightarrow[\text{freq dom}]{} V_s(j\omega) = |V_s|e^{j\varphi s} \tag{2.1}$$

Since we are limiting our discussion to a linear system (without any nonlinear elements), all solution voltages and currents are expected to be of the same frequency as that of the source; for example, the voltage on the line at a general location z at time t would assume the following form:

$$v(z,t) = |V(z)|\cos(\omega t + \varphi(z)) \xleftrightarrow[\text{freq dom}]{} V(z, j\omega) = |V(z)|e^{j\varphi(z)} \tag{2.2}$$

In the following, we will develop TL equations and solutions in both frequency and time domains side by side. The left column would be for the time-domain development while the right one will be for the frequency domain. In doing so, we are attempting not to get "disoriented/lost in the ocean of the frequency domain and lose our ability to get back to shore." Instead, we will constantly relate to the physics of the development and the solution by "swimming close to shore" and monitoring the time-domain phenomena while doing the frequency-domain development.

To analyze the "electric circuit" shown, we need to build the distributed equation for $V(z, j\omega)$ and $I(z, j\omega)$ [or $v(z,t)$ and $i(z,t)$] along the line. We do that by considering an infinitesimally small increment of the line length Δz (to start at z and end at $z + \Delta z$). Writing the circuit (input-output) relationships between the voltages and currents for the Δz increment enables us to build the differential equation describing the behavior of this physically based circuit model. To do that, we use Kirchhoff's voltage and Kirchhoff's current laws (KVL and KCL) for the Δz "equivalent" circuit (Figure 2.6).

Time-Domain Analysis: $v(z,t)$ and $i(z,t)$	Frequency-Domain Analysis: $V(z,j\omega)$ and $I(z,j\omega)$	Eq#
Forming the differential equations: We start with the differential relationship, while skipping the writing of "t" and "$j\omega$" in all voltage and current expressions.		
$v(z+\Delta z,t)=v(z,t)+\Delta v(z,t)$	$V(z+\Delta z)=V(z)+\Delta V(z)$	(2.3)
$i(z+\Delta z,t)=i(z,t)+\Delta i(z,t)$	$I(z+\Delta z)=I(z)+\Delta I(z)$	(2.4)
Now, we apply KVL		
$v(z,t)=v_R+v_L+[v(z+\Delta z,t)]$	$V(z)=V_R+V_L+[V(z+\Delta z)]$	(2.5)
where v_R and v_L [V_R and V_L] are the voltage drops across $R_z\Delta z$ and $L_z\Delta z$, and using Equations (2.3)		
$v(z,t)=(R_z\Delta z)i(z,t)+(L_z\Delta z)\dfrac{di(z,t)}{dt}$ $+[v(z,t)+\Delta v(z,t)]$	$V(z)=(R_z\Delta z)I(z)+(j\omega L_z\Delta z)I(z)$ $+[V(z)+\Delta V(z)]$	(2.6)
$\dfrac{\Delta v(z,t)}{\Delta z}=-\left\{R_z i(z,t)+L_z\dfrac{di(z,t)}{dt}\right\}$	$\dfrac{\Delta V(z)}{\Delta z}=-\{R_z I(z)+j\omega L_z I(z)\}$	(2.7)
$\dfrac{dv(z,t)}{dz}=\lim\limits_{\Delta z\to 0}\dfrac{\Delta v(z,t)}{\Delta z}=-\left\{R_z i(z,t)+L_z\dfrac{di(z,t)}{dt}\right\}$	$\dfrac{dV(z)}{dz}=\lim\limits_{\Delta z\to 0}\dfrac{\Delta V(z)}{\Delta z}-\{R_z+j\omega L_z\}I(z)$	(2.8)
Next, we apply KCL		
$i(z,t)=i_G+i_C+[i(z+\Delta z,t)]$	$I(z)=I_G+I_C+[I(z+\Delta z)]$	(2.9)
where i_G and i_C [I_G and I_C] are the current drops through $G_z\Delta z$ and $C_z\Delta z$, and using Equations (2.4)		
$i(z,t)=(G_z\Delta z)v(z+\Delta z,t)+(C_z\Delta z)\dfrac{dv(z+\Delta z,t)}{dt}$ $+[i(z,t)+\Delta i(z,t)]$	$I(z)=(G_z\Delta z)V(z+\Delta z)+(j\omega C_z\Delta z)V(z)$ $+[I(z)+\Delta I(z)]$	(2.10)
$\dfrac{\Delta i(z,t)}{\Delta z}=-\left\{G_z v(z+\Delta z,t)+C_z\dfrac{dv(z+\Delta z,t)}{dt}\right\}$	$\dfrac{\Delta I(z)}{\Delta z}=-\{G_z V(z+\Delta z)+j\omega C_z V(z+\Delta z)\}$	(2.11)
$\dfrac{di(z,t)}{dz}=\lim\limits_{\Delta z\to 0}\dfrac{\Delta i(z,t)}{\Delta z}-\left\{G_z v(z,t)+C_z\dfrac{dv(z,t)}{dt}\right\}$	$\dfrac{dI(z)}{dz}=\lim\limits_{\Delta z\to 0}\dfrac{\Delta I(z)}{\Delta z}-\{G_z+j\omega C_z\}V(z)$	(2.12)
We have two simultaneous differential equations in the current and voltage. So, we need to combine them in a way to end up with a new equation in one variable only.		
Differentiate Equation (2.8) with respect to (z) and Equation (2.12) with respect to (t)	Differentiate Equation (2.8) with respect to z and using Equation (2.12)	
$\dfrac{d^2v(z,t)}{dz^2}=-\left\{R_z\dfrac{di(z,t)}{dz}+L_z\dfrac{d^2i(z,t)}{dt\,dz}\right\}$	$\dfrac{d^2V(z)}{dz^2}=-\{R_z+j\omega L_z\}\dfrac{dI(z)}{dz}$	(2.13)
$\dfrac{d^2i(z,t)}{dz\,dt}=-\left\{G_z\dfrac{dv(z,t)}{dt}+C_z\dfrac{d^2v(z,t)}{dt^2}\right\}$		(2.14)
Combining Equations (2.13) and (2.14)	Combining Equations (2.12) and (2.13)	
$\dfrac{d^2v(z,t)}{dz^2}=-\left\{R_z\left[-\left\{G_z v(z,t)+C_z\dfrac{dv(z,t)}{dt}\right\}\right]+L_z\left[-\left\{G_z\dfrac{dv(z,t)}{dt}+C_z\dfrac{d^2v(z,t)}{dt^2}\right\}\right]\right\}$		(2.15)
$\dfrac{d^2v(z,t)}{dz^2}=\left\{R_zG_zv(z,t)+[R_zC_z+L_zG_z]\dfrac{dv(z,t)}{dt}\right.$ $\left.+L_zC_z\dfrac{d^2v(z,t)}{dt^2}\right\}$	$\dfrac{d^2V(z)}{dz^2}=\{R_z+j\omega L_z\}\{G_z+j\omega C_z\}V(z)$	(2.16)

Table 2.2 Analysis of TL Circuit—Forming the Basic Equations

We now rewrite Equations (2.16) into forms that display the analogy between the time-domain and frequency-domain forms, which will become more apparent as we apply the Laplace transform to the time-domain form.

Time-Domain Analysis: $v(z,t)$ and $i(z,t)$	Frequency-Domain Analysis: $V(z,j\omega)$ and $I(z,j\omega)$	Eq#
$\frac{d^2v(z,t)}{dz^2}=\left\{R_zG_z+[R_zC_z+G_zL_z]\left(\frac{d}{dt}\right)+L_zC_z\left(\frac{d^2}{dt^2}\right)\right\}v(z,t)$	$\frac{d^2V(z)}{dz^2}=\{R_zG_z+(j\omega)[R_zC_z+G_zL_z]+(-\omega^2)L_zG_z\}V(z)$	(2.16)

Table 2.3 Analysis of TL Circuit—Developing the Wave Equation

2.6.1 Solution

Now, having developed the wave equations, Equations (2.16), we have one equation in one variable: the voltage. It is important to recognize that each of these equations is a second-order differential equation (DE), which resulted from combining two first-order DEs thus yielding the one second-order DE.

In doing so, we had to differentiate one of the two first-order DEs. Taking the derivative eliminates the constant term in that quantity. This elimination means that we lost one piece of information and that information is not represented as part of the obtained second-order DE. The only way to recover that missing information is by using the original first-order DEs before differentiation.

Therefore, what we need to do is solve the second-order DEs [Equations (2.16)] for the voltage and get back to the first-order DEs [Equations (2.8)] to get the current.

Time-Domain Analysis: $v(z,t)$ and $i(z,t)$	Frequency-Domain Analysis: $V(z,j\omega)$ and $I(z,j\omega)$	Eq#
To solve the DE in Equation (2.16), we use the Laplace transform		
$\frac{d^2V(z)}{dz^2}=\{R_zG_zV(z)+[R_zC_z+L_zG_z]sV(z)+L_zC_zs^2V(z)\}$		(2.17)
$\frac{d^2V(z,s)}{dz^2}=\{R_z+sL_z\}\{G_z+sC_z\}V(z)$	$\frac{d^2V(z,j\omega)}{dz^2}=\{R_z+j\omega L_z\}\{G_z+j\omega C_z\}V(z)$	(2.18)
Defining γ^2		
$\gamma^2(s)=R_zG_z+s[R_zC_z+L_zG_z]+s^2L_zC_z$	$\gamma^2(j\omega)=\{R_z+j\omega L_z\}\{G_z+j\omega C_z\}$	(2.19)
$\frac{d^2V(z)}{dz^2}=\gamma^2V(z)$	$\frac{d^2V(z)}{dz^2}=\gamma^2V(z)$	(2.20)
Solving the DEs in Equation (2.20)		
$V(z)=V^+e^{-\gamma z}+V^-e^{+\gamma z}$	$V(z)=V^+e^{-\gamma z}+V^-e^{+\gamma z}$	(2.21)
where V^+ and V^- are constants		
To get the current expression, as discussed earlier, we need to get back to the first-order DE, Equation (2.8). So, we apply the Laplace transform to Equation (2.8)	From Equation (2.8)	
$\frac{dV(z)}{dz}=-\{R_z+sL_z\}I(z)$		(2.22)
$I(z)=-\frac{1}{\{R_z+sL_z\}}\frac{dV(z)}{dz}$	$I(z)=-\frac{1}{\{R_z+j\omega L_z\}}\frac{dV(z)}{dz}$	(2.23)
$I(z)=\frac{\gamma}{\{R_z+sL_z\}}\{V^+e^{-\gamma z}-V^-e^{+\gamma z}\}$	$I(z)=\frac{\gamma}{\{R_z+j\omega L_z\}}\{V^+e^{-\gamma z}-V^-e^{+\gamma z}\}$	(2.24)
$\frac{\gamma}{\{R_z+sL_z\}}=\frac{1}{Z_o}$	$\frac{\gamma}{\{R_z+j\omega L_z\}}=\frac{1}{Z_o}$	(2.25)

Table 2.4 Analysis of TL Circuit—Solving the Wave Equation

2.6.2 Conclusion

Time-Domain Analysis: $v(z,t)$ and $i(z,t)$	Frequency-Domain Analysis: $V(z,j\omega)$ and $I(z,j\omega)$	Eq#
Discussion of possible solutions:	Summary:	
The inverse Laplace transform of these equations back to the time domain in the general case is too complicated. However, possible solutions can be achieved for either the case of harmonic signals ($s = j\omega$) and the case of lossless TL ($R_z = 0$ and $G_z = 0$)	$I(z)=\frac{1}{Z_o}\cdot\{V^+e^{-\gamma z}-V^-e^{+\gamma z}\}$	(2.26)
	$Z_o = R_o + jX_o = \sqrt{\{R_z + j\omega L_z\}/\{G_z + j\omega C_z\}}$	(2.27)
	$\gamma = \alpha + j\beta = \sqrt{\{R_z + j\omega L_z\}\cdot\{G_z + j\omega C_z\}}$	(2.28)
	$V(z)=V^+e^{-\gamma z}+V^-e^{+\gamma z}$	(2.29)
	$V(z)=V^+e^{-\alpha z}e^{-j\beta z}+V^-e^{+\alpha z}e^{+j\beta z}$	(2.30)
	Transfer back to time domain:	(2.31)
Case of steady-state harmonic signals ($s = j\omega$): See right column starting with Equation (2.26)	$v(z,t)=\lvert V^+\rvert e^{-\alpha z}\cos(\omega t-\beta z+\varphi_v^+)+\lvert V^-\rvert e^{+\alpha z}$ $\cos(\omega t+\beta z+\varphi_v^-)$	(2.32)
where φ_{Zo} is the phase angle associated with Z_o	$i(z,t)=\lvert V^+/Z_o\rvert e^{-\alpha z}\cos(\omega t-\beta z+\varphi_v^+-\varphi_{Zo})$ $-\lvert V^-/Z_o\rvert e^{+\alpha z}\cos(\omega t+\beta z+\varphi_v^--\varphi_{Zo})$	(2.33)

TABLE 2.5 Analysis of TL Circuit—Conclusion and Discussion

2.6.3 Case of Lossless TL ($R_z = 0$ and $G_z = 0$)

Time-Domain Analysis: $v(z,t)$ and $i(z,t)$	Frequency-Domain Analysis: $V(z,j\omega)$ and $I(z,j\omega)$	Eq#
Equations (2.19) and (2.25) reduce to	Equations (2.28) and (2.25) reduce to	
$\gamma^2(s)=s^2L_zC_z$ and $Z_o=\sqrt{L_z/C_z}$	$\gamma=j\beta=j\omega\sqrt{L_zC_z}$ and $Z_o\sqrt{L_z/C_z}$	(2.34)
which reduces Equations (2.21) and (2.24) to	which reduces Equations (2.30) and (2.26) to	
$V(z)=V^+e^{-s\sqrt{L_zC_z}z}+V^-e^{+s\sqrt{L_zC_z}z}$	$V(z)=V^+e^{-j\beta z}+V^-e^{+j\beta z}$	(2.35)
$I(z)=\frac{1}{\sqrt{L_z/C_z}}\{V^+e^{-s\sqrt{L_zC_z}z}-V^-e^{+s\sqrt{L_zC_z}z}\}$	$I(z)=\frac{1}{\sqrt{L_z/C_z}}\{V^+e^{-j\beta z}-V^-e^{+j\beta z}\}$	(2.36)
Transfer back to time domain:	Also, Equations (2.32) and (2.33) reduce to	
$v(z,t)=v^+(t-\sqrt{L_zC_z}z)+v^-(t+\sqrt{L_zC_z}z)$	$v(z,t)=\lvert V^+\rvert\cos(\omega t-\beta z+\varphi_v^+)+\lvert V^-\rvert\cos(\omega t+\beta z+\varphi_v^-)$	(2.37)
$i(z,t)=\frac{1}{\sqrt{L_z/C_z}}\left\{v^+\left(t-\sqrt{L_zC_z}z\right)-v^-\left(t+\sqrt{L_zC_z}z\right)\right\}$	$i(z,t)=\lvert V^+/Z_o\rvert\cos(\omega t-\beta z+\varphi_v^+)-\lvert V^-/Z_o\rvert\cos(\omega t+\beta z+\varphi_v^-)$	(2.38)

TABLE 2.6 Analysis of TL Circuit—Case of Lossless TL

So, we have arrived at a solution for the steady-state harmonic case using both the time-domain and the frequency-domain approaches. The obtained solutions are expressed by Equations (2.32) and (2.33) in the time domain and in Equations (2.26) to (2.30) in the frequency-domain phasor format. We did not carry out the solution in the time domain; instead, we made use of the Laplace transform to solve the second-order DE. In a way, that was similar to resorting to the frequency domain for help. The Laplace domain is but a generalization of the frequency domain as it uses $s = \sigma + j\omega$ and hence allows the study of transients. When $\sigma = 0$, the Laplace domain coincides with the frequency domain and the problem reduces to its steady-state harmonic solution.

In the special case of a lossless TL, we were able to carry out the solution further in the time domain to nonharmonic cases. These solutions are given by Equations (2.37) and (2.38). The corresponding harmonic solutions are given on the right-hand column of the same lines.

2.7 Physical Implications of Solutions: α, β, λ, and c_{ph}

At this point, it is important to take a pause to examine the obtained results and make sure that we relate to their physical implications.

We examine the harmonic solution obtained in the time domain:

$$v(z,t)=\left|V^+\right|e^{-\alpha z}\cos\left(\omega t-\beta z+\varphi_v^+\right)+\left|V^-\right|e^{+\alpha z}\cos\left(\omega t+\beta z+\varphi_v^-\right) \tag{2.39}$$

This solution has two terms. Let us take them one at a time: $\left|V^+\right|e^{-\alpha z}\cos\left(\omega t-\beta z+\varphi_v^+\right)$ describes a voltage wave that changes with both t and z, a graph of which is shown in Figure 2.7.

It is evident from the figure that this term represent a traveling wave that moves in the positive z direction (and hence the plus superscript in denoting V^+). The wave amplitude attenuates (decays) exponentially with distance with an attenuation coefficient of $e^{-\alpha z}$.

The wave experiences a full cycle in a distance λ, and hence the corresponding phase change $\beta\lambda$ should equal to 2π, that is, $\beta\lambda=2\pi$, or $\lambda=2\pi/\beta$.

The velocity of its phase propagation can be derived by monitoring the motion of a specific phase on the wave,

$$\omega t_1-\beta z_1+\varphi_v^+=\omega t_2-\beta z_2+\varphi_v^+ \text{ and hence } \omega\Delta t=\beta\Delta z, \text{ or } \Delta z/\Delta t=c_{\text{ph}}=\omega/\beta \tag{2.40}$$

$$\text{Phase velocity } c_{\text{ph}}=\frac{\omega}{\beta}=\frac{2\pi f}{\frac{2\pi}{\lambda}}=\lambda f \tag{2.41}$$

With this definition for the phase velocity, the terms of Equation (2.32) can be rewritten as

$$v(z,t)=\left|V^+\right|e^{-\alpha z}\cos\left[\omega\left(t-\frac{z}{c_{\text{ph}}}\right)+\varphi_v^+\right]+\left|V^-\right|e^{+\alpha z}\cos\left[\omega\left(t+\frac{z}{c_{\text{ph}}}\right)+\varphi_v^-\right] \tag{2.42}$$

which makes Equation (2.37) for the lossless line reduce to

$$v(z,t)=\left|V^+\right|\cos\left[\omega\left(t-\frac{z}{c_{\text{ph}}}\right)+\varphi_v^+\right]+\left|V^-\right|\cos\left[\omega\left(t+\frac{z}{c_{\text{ph}}}\right)+\varphi_v^-\right] \tag{2.43}$$

Now, comparing this form to that obtained in the time domain for the lossless line in Equation (2.37), we conclude that the time argument $\left(t-\sqrt{L_zC_z}\,z\right)$ must be equivalent to $\left(t-\frac{z}{c_{\text{ph}}}\right)$ obtained in Equation (2.43), which implies that

$$c_{\text{ph}}=\frac{1}{\sqrt{L_zC_z}} \tag{2.44}$$

Consequently, the time-domain solution in the lossless line case, Equation (2.37), becomes

$$v(z,t)=v^+\left(t-\frac{z}{c_{\text{ph}}}\right)+v^-\left(t+\frac{z}{c_{\text{ph}}}\right) \tag{2.45}$$

Now, back to the second term of the $v(z,t)$ solution $\left\{\left|V^-\right|e^{+\alpha z}\cos\left(\omega t+\beta z+\varphi_v^-\right)=\left|V^-\right|\cos\left[\omega\left(t+\frac{z}{c_{\text{ph}}}\right)+\varphi_v^-\right]\right\}$, Figure 2.8. This term represents a traveling wave that moves in the negative ($-z$) direction (and hence the minus superscript in denoting V^-). The wave amplitude attenuates (decays) exponentially with distance with an attenuation coefficient of $e^{(-\alpha)(-z)}=e^{+\alpha z}$. This term has the same exact α and β of the V^+ term.

The two terms combined represent the $v(z,t)$ solution obtained in Equation (2.32). The complete solution is shown in Figure 2.9.

By the way, and before we leave this point, in the lossless TL case, the waves on the TL would not suffer any attenuation. The corresponding wave shapes are shown in Figure 2.10.

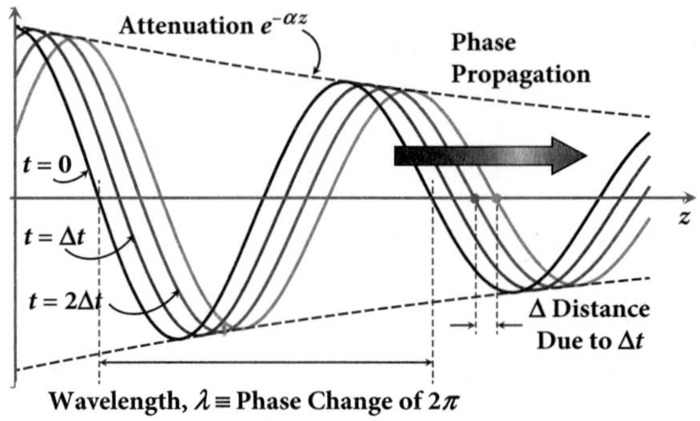

FIGURE 2.7 Wave propagation for a lossy TL.

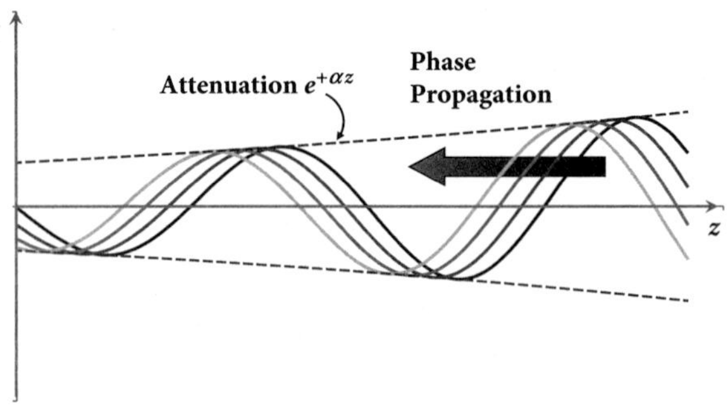

FIGURE 2.8 Reflected wave for a lossy TL.

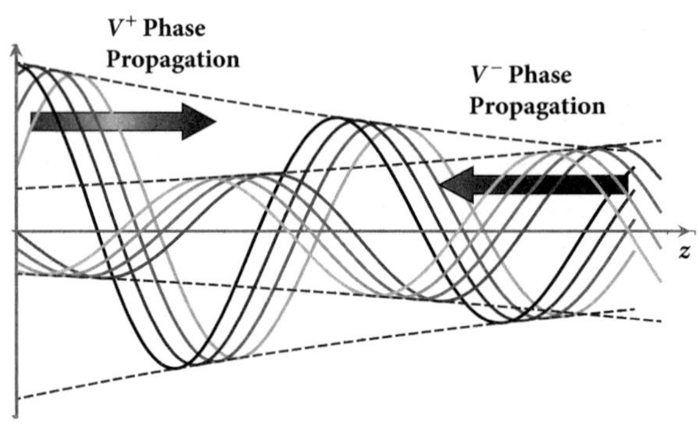

FIGURE 2.9 Forward and backward waves for a lossy TL.

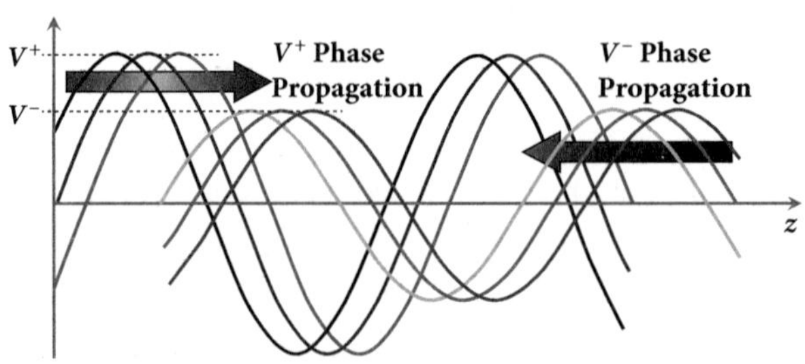

FIGURE 2.10 Forward and backward waves for a lossless TL.

Example 2.1 The per unit length parameters of a TL were measured as 0.1 Ω/m, 200 nH/m, 125 pF/m, and 50 μS/m. The operating frequency is 1 GHz. Determine

a. The characteristic impedance of the line.

$$R_z = 0.1\ \Omega/\text{m} \qquad L_z = 200\ \text{nH/m} \qquad C_z = 125\ \text{pF/m} \qquad G_z = 50\ \mu\text{S/m}$$

$$Z_o = \sqrt{\frac{(R_z + j\omega L_z)}{(G_z + j\omega C_z)}} = \sqrt{\frac{(0.1 + j2\pi \times 10^9 \times 200 \times 10^{-9})}{(50 \times 10^{-6} + j2\pi \times 10^9 \times 125 \times 10^{-12})}} = 40 - j0.0003\ \Omega$$

b. The complex propagation constant of the line

$$\gamma = \sqrt{(G_z + j\omega C_z)(R_z + j\omega L_z)} = 0.0022 + 31.42$$

$$\alpha = 0.0022\ \text{Np/m} \qquad \beta = 31.42\ \text{rad/m}$$

c. The phase velocity

$$c_{\text{ph}} = \frac{\omega}{\beta} = \frac{2\pi \times 10^9}{31.42} = 2 \times 10^8\ \text{m/s}$$

d. The wavelength

$$\lambda = \frac{2\pi}{\beta} = \frac{2\pi}{31.42} = 0.2\ \text{m}$$

e. The attenuation coefficient of the line in dBs/m

$$1\ \text{Np} = 8.686\ \text{dB} \rightarrow \alpha = 0.0022\ \text{Np/m} = 0.0195\ \text{dB/m}$$

2.8 Physical Implications of Solutions: γ, Z_o, V^+, and V^-

Now as we saw the physical meaning of α and β, let us move back to the frequency-domain solution in order to interpret the physical meaning of the parameters: γ, Z_o, V^+, and V^-.

Let us start with γ which is a complex quantity with a real part α and an imaginary part β.

$$\gamma = \alpha + j\beta = \sqrt{\{R_z + j\omega L_z\} \cdot \{G_z + j\omega C_z\}} \tag{2.46}$$

The real part, α, appears in the exponent of the voltage and current solutions as to affect the amplitude of the wave along the TL (coordinate z). This real part becomes zero for lossless lines $R_z = G_z = 0$ and, hence, it is related to losses and causes the signals to attenuate as they travel along the line. For this reason, α will be called the *attenuation coefficient*, and its units are nepers/meter. [1 Np (neper) = 20/ln(10) ≅ 8.686 dB (decibels).]

The imaginary part, β, also appears in the exponent of the voltage and current solutions as to affect the phase of the wave along the TL (coordinate z). Hence, β will be called the *phase constant*, and its units are radians/meter. For lossless lines, $\beta = \omega\sqrt{L_z C_z}$.

γ is called the complex propagation constant of the TL.

$$Z_o = R_o + jX_o = \sqrt{\{R_z + j\omega L_z\}/\{G_z + j\omega C_z\}} \tag{2.47}$$

Next, we discuss Z_o which is a complex quantity with real part R_o and an imaginary part X_o. For lossless lines, it reduces to a real value

$$Z_o = R_o = \sqrt{L_z/C_z} \tag{2.48}$$

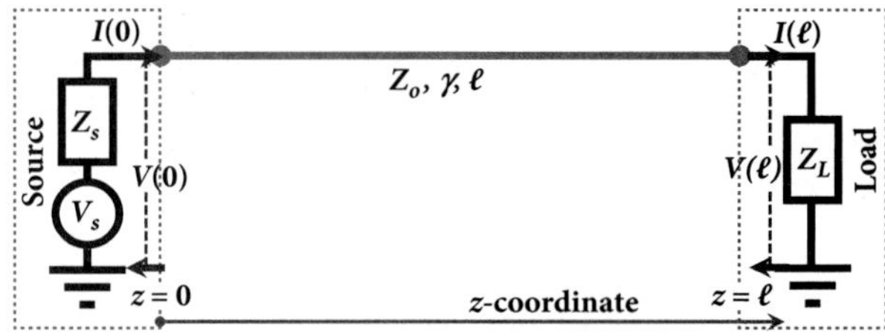

FIGURE 2.11 A demonstration of the boundary conditions of the TL circuit.

This may appear rather puzzling as we talk about a *lossless* TL with R_o equal to so many ohms, for example, a cable TV TL has a characteristic impedance of 75 Ω. The thing is that there are no 75 Ω resistor(s) in existence anywhere on this line and the 75 Ω does not represent an internal resistance. It simply means that the 75 Ω is nothing but a characteristic number representing the ratio between the traveling voltage and current waves.

Z_o is called the characteristic impedance of the TL.

It is worthy to note that both Z_o and γ fully characterize the TL model. Given Z_o and γ and using Equations (2.27) and (2.28), the TL's RLGC model can be fully determined.

While both Z_o and γ are characteristic parameters of the TL, V^+, and V^- are not. $V^+=\left|V^+\right|e^{j\varphi_v^+}$ and $V^-=\left|V^-\right|e^{j\varphi_v^-}$ are two arbitrary constants that resulted from solving the TL differential equation without applying the boundary conditions. These constants can be fully determined once the TL conditions at its boundaries ($z=0$ and at $z=\ell$) are implemented.

Referring to Figure 2.11 and using Equations (2.21) and (2.26), the boundary conditions at the source end are given by

$$V(0)=V^+ + V^- = V_s - Z_s\cdot I(0) = V_s - Z_s\cdot\frac{1}{Z_o}(V^+ - V^-) \tag{2.49}$$

and the boundary condition at the load end are

$$V(\ell)=V^+e^{-\gamma\ell}+V^-e^{+\gamma\ell}=Z_L\cdot I(\ell)=Z_L\cdot\left\{\frac{1}{Z_o}(V^+e^{-\gamma\ell}-V^-e^{+\gamma\ell})\right\} \tag{2.50}$$

These are two "complex-variable" equations that can be solved to yield both V^+ and V^- (magnitudes and phases). The resulting expressions can be written in the form:

$$V^+=\left\{\frac{V_sZ_o}{Z_s+Z_o}\right\}\left\{\frac{1}{1-\left(\frac{Z_s-Z_o}{Z_s+Z_o}\right)\left(\frac{Z_L-Z_o}{Z_L+Z_o}\right)e^{-2\gamma\ell}}\right\} \tag{2.51}$$

$$V^-=V^+\left(\frac{Z_L-Z_o}{Z_L+Z_o}\right)e^{-2\gamma\ell} \tag{2.52}$$

Physically, these are the levels of the waves traveling on the line. As will be discussed later, the V^+ is the level of the voltage wave traveling in the positive z direction, while the V^- is that of the negative z wave. Considering that we only have one source of power (at the source end), there is only one explanation for the presence of a negative z traveling wave; that it must have resulted as a result of waves reflecting (bouncing back) from the load end. A useful analogy here is watching the water waves at the seashore. The sea is the source of water that travels toward the shore. After the water waves hit the shore, we notice water waves traveling from the shore back to the sea. These are reflections of the original sea waves.

2.9 Physical Implications of Solutions: Γ

The V^- wave results from reflections of the V^+ wave as it bounces off the TL "discontinuity" at the load end. Hence, we define the ratio of the "reflected" ($-z$) wave to the "incident" ($+z$) wave as the reflection coefficient.

$V(z)=V^+e^{-\gamma z}+V^-e^{+\gamma z}$	(2.21)
$I(z)=\dfrac{1}{Z_o}\cdot\{V^+e^{-\gamma z}-V^-e^{+\gamma z}\}$	(2.26)
Defining $\Gamma(z)$	
$\Gamma(z)=\dfrac{V^-e^{+\gamma z}}{V^+e^{-\gamma z}}=\dfrac{V^-}{V^+}e^{+2\gamma z}$	(2.53)
$V(z)=V^+e^{-\gamma z}[1+\Gamma(z)]$	(2.54)
$I(z)=\dfrac{V^+e^{-\gamma z}}{Z_o}[1-\Gamma(z)]$	(2.55)
Hence, the "driving point impedance*" at any location on the line, $Z(z)$, is given by	
$Z(z)=\dfrac{V(z)}{I(z)}=Z_o\dfrac{[1+\Gamma(z)]}{[1-\Gamma(z)]}$ and hence, $\Gamma(z)=\dfrac{[Z(z)-Z_o]}{[Z(z)+Z_o]}$	(2.56)
At both TL ends, $Z(z)$ reduces to	
$Z(0)=\dfrac{V(0)}{I(0)}=Z_o\dfrac{[1+\Gamma(0)]}{[1-\Gamma(0)]}$	(2.57)
$Z(\ell)=\dfrac{V(\ell)}{I(\ell)}=\{Z_L\}=Z_o\dfrac{[1+\Gamma(\ell)]}{[1-\Gamma(\ell)]}$	(2.58)
This gives us an insight regarding $\Gamma(\ell)$	
$Z_L=Z_o\dfrac{[1+\Gamma(\ell)]}{[1-\Gamma(\ell)]}\Rightarrow\Gamma(\ell)=\dfrac{[Z_L-Z_o]}{[Z_L+Z_o]}$	(2.59)
It means that $\Gamma(\ell)$ exists as a result of the impedance discontinuity (mismatch) $Z_L\neq Z_o$	
Using Equation (2.53) twice at both z and ℓ	
$\Gamma(z)=\dfrac{V^-e^{+\gamma z}}{V^+e^{-\gamma z}}$ and $\Gamma(\ell)=\dfrac{V^-e^{+\gamma\ell}}{V^+e^{-\gamma\ell}}$	(2.60)
$\Gamma(z)=\Gamma(\ell)\cdot e^{+2\gamma(z-\ell)}=\Gamma(\ell)\cdot e^{-2\gamma(d)}$	(2.61)
Here, $d=\ell-z$ represents a distance "d" starting at the load end and extending opposite to the "z" coordinate (Figure 2.12)	
The use of the coordinate "d" instead of "z" is typical in solving impedance matching problems in TL analysis and design	

*More on "driving point impedance" is given in Addendum 2A.

Table 2.7 Analysis of TL Circuit—Driving Point Impedance and Reflection Coefficient

From Equation (2.59), we conclude that if it was not for the mismatch Z_L-Z_o between the load impedance Z_L and the TL characteristic impedance Z_o, and as a consequence the reflection coefficient $\Gamma(z)$, the reflection ($-z$ wave) $V^-e^{+\gamma z}$ would not exist. Therefore, we can say that waves traveling on a uniform TL with a matched end are fully absorbed by the load and will never experience any reflections. Likewise, if the TL is uniform and infinite in length, with no discontinuities, the source-end-launched waves would continue to travel indefinitely resulting in the absence of reflections ($-z$ waves). This is exactly what happens to a light beam projected in the open space. If it meets a reflective object (mismatched impedance), a reflection takes place. However, no reflections occur in one of two possible cases, the case of a fully absorbing object (matched impedance) and the case of open space with no intercepting objects.

Before we leave this section, let us finalize the expressions for the voltages and currents at any z location on the line as well as at both the sending and receiving ends.

We start by rewriting Equation (2.51) in terms of the Γ's (reflection coefficients) associated with the source and load impedances:

$$\Gamma_L = \Gamma(\ell) = \frac{[Z_L - Z_o]}{[Z_L + Z_o]} \tag{2.62}$$

$$\Gamma_s = \frac{[Z_s - Z_o]}{[Z_s + Z_o]} \tag{2.63}$$

Note that Γ_s is different from $\Gamma(0)$. Γ_s is defined in Equation (2.63) as the reflection coefficient seen by the $(-z)$ signal traveling back toward the source with the source impedance acting as the load to the Z_o line. On the other hand, $\Gamma(0)$ is the reflection coefficient seen at the input port of the line by the $(+z)$ wave, $\Gamma(z=0) = V^-/V^+$, Equation (2.60).

Now, back to Equation (2.51),

$$V^+ = \left\{\frac{V_s Z_o}{Z_s + Z_o}\right\}\left\{\frac{1}{1-\left(\dfrac{Z_s - Z_o}{Z_s + Z_o}\right)\left(\dfrac{Z_L - Z_o}{Z_L + Z_o}\right)e^{-2\gamma\ell}}\right\}$$

$$= \left\{\frac{V_s[1-\Gamma_s]}{2}\right\}\left\{\frac{1}{1-\Gamma_s\Gamma_L e^{-2\gamma\ell}}\right\} = \left\{\frac{V_s}{2}\right\}\left\{\frac{1-\Gamma_s}{1-\Gamma_s\Gamma_L e^{-2\gamma\ell}}\right\} \tag{2.64}$$

$$V^- = V^+\left(\frac{Z_L - Z_o}{Z_L + Z_o}\right)e^{-2\gamma\ell} = \left\{\frac{V_s}{2}\right\}\left\{\frac{1-\Gamma_s}{1-\Gamma_s\Gamma_L e^{-2\gamma\ell}}\right\}\left(\frac{Z_L - Z_o}{Z_L + Z_o}\right)e^{-2\gamma\ell} = \left\{\frac{V_s}{2}\right\}\left\{\frac{(1-\Gamma_s)\Gamma_L e^{-2\gamma\ell}}{1-\Gamma_s\Gamma_L e^{-2\gamma\ell}}\right\} \tag{2.65}$$

Using Equations (2.54), (2.61), and (2.64), we can write the voltage as a function of position as follows:

$$V(z) = V^+ e^{-\gamma z}[1+\Gamma(z)] = \left\{\frac{V_s}{2}\right\}\left\{\frac{1-\Gamma_s}{1-\Gamma_s\Gamma_L e^{-2\gamma\ell}}\right\}e^{-\gamma z}\left[1+\Gamma_L e^{+2\gamma(z-\ell)}\right] \tag{2.66}$$

$$V(z) = \left\{\frac{V_s}{2}\right\}\left\{\frac{(1-\Gamma_s)\left[1+\Gamma_L e^{+2\gamma(z-\ell)}\right]}{1-\Gamma_s\Gamma_L e^{-2\gamma\ell}}\right\}e^{-\gamma z} \tag{2.67}$$

Substituting $z = 0$ and $z = \ell$ in this general expression, we can write the voltages at the sending and receiving ends as

$$V(0) = \left\{\frac{V_s}{2}\right\}\left\{\frac{(1-\Gamma_s)\left(1+\Gamma_L e^{-2\gamma\ell}\right)}{1-\Gamma_s\Gamma_L e^{-2\gamma\ell}}\right\} \tag{2.68}$$

$$V_L = V(\ell) = \left\{\frac{V_s}{2}\right\}\left\{\frac{(1-\Gamma_s)(1+\Gamma_L)\,e^{-\gamma\ell}}{1-\Gamma_s\Gamma_L e^{-2\gamma\ell}}\right\} \tag{2.69}$$

Alternatively, in terms of impedances, we can write

$$V(0) = V_s\left\{\frac{Z_o\{(Z_L+Z_o)+(Z_L-Z_o)e^{-2\gamma\ell}\}}{(Z_s+Z_o)(Z_L+Z_o)-(Z_s-Z_o)(Z_L-Z_o)e^{-2\gamma\ell}}\right\} \tag{2.70}$$

$$V_L = V_s\left\{\frac{2Z_L Z_o}{(Z_s+Z_o)(Z_L+Z_o)-(Z_s-Z_o)(Z_L-Z_o)e^{-2\gamma\ell}}\right\}e^{-\gamma\ell} \tag{2.71}$$

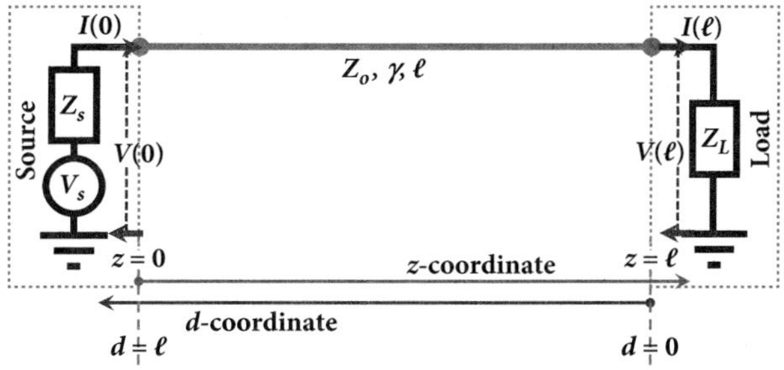

Figure 2.12 Use of coordinate "d" as the distance from the load end instead of "z."

Example 2.2 Referring to Figure 2.12, consider a lossless TL section of length ℓ and characteristic impedance $Z_o = 50\ \Omega$ is connected to a sinusoidal source, $v_s(t) = 5\cos\left(\omega t + \frac{\pi}{4}\right)$. The source has an internal impedance $Z_s = 30\ \Omega$. The transmission line section is connected from the other end to a 150-Ω load resistance. For the following cases, find the load reflection coefficient, the amplitude of the forward going wave at the source, V^+, the total voltage and current at the source end ($z = 0$) and the load end ($z = \ell$) and the driving point impedance at the input of the line ($z = 0$)

a. $\ell = 0.1\lambda$ b. $\ell = 0.25\lambda$ c. $\ell = 0.5\lambda$

$$\Gamma_L = \Gamma(\ell) = \frac{Z_L - Z_o}{Z_L + Z_o} = \frac{150 - 50}{150 + 50} = 0.5$$

$$\Gamma_s = \frac{Z_s - Z_o}{Z_s + Z_o} = \frac{30 - 50}{30 + 50} = -0.25$$

$$V(z) = V^+ e^{-\gamma z} + V^- e^{+\gamma z}$$

Since the line is lossless, $\gamma = j\beta$

$$V(z) = V^+ e^{-j\beta z} + V^- e^{+j\beta z} = V^+ e^{-j\beta z}[1 + \Gamma(z)]$$

$$I(z) = \frac{V^+ e^{-j\beta z}}{Z_o}[1 - \Gamma(z)]$$

$$\Gamma(z) = \Gamma(\ell)e^{j2\beta(z-\ell)}$$

$$V(z) = V^+ e^{-j\beta z}[1 + \Gamma(\ell)e^{j2\beta(z-\ell)}] = V^+ e^{-j\beta z}[1 + 0.5e^{j2\beta(z-\ell)}]$$

$$I(z) = \frac{V^+}{Z_o} e^{-j\beta z}[1 - \Gamma(\ell)e^{j2\beta(z-\ell)}] = \frac{V^+}{Z_o} e^{-j\beta z}[1 - 0.5e^{j2\beta(z-\ell)}]$$

$$V_s = 5\angle\frac{\pi}{4}$$

Using Equation (2.64)

$$V^+ = \left\{\frac{V_s}{2}\right\}\left\{\frac{1 - \Gamma_s}{1 - \Gamma_s \Gamma_L e^{-2\gamma\ell}}\right\} = \left\{\frac{V_s}{2}\right\}\left\{\frac{1 - \Gamma_s}{1 - \Gamma_s \Gamma_L e^{-2j\beta\ell}}\right\}$$

a. $\ell = 0.1\lambda$

$$\beta\ell = \frac{2\pi}{\lambda} \times 0.1\lambda = 0.2\pi,\ e^{-j2\beta\ell} = e^{-j0.4\pi} = \cos(-0.4\pi) + j\sin(-0.4\pi)$$

$$V^+ = \left\{\frac{V_s}{2}\right\}\left\{\frac{1 - \Gamma_s}{1 - \Gamma_s \Gamma_L e^{-2j\beta\ell}}\right\} = \left\{\frac{5e^{\frac{j\pi}{4}}}{2}\right\}\left\{\frac{1 - (-0.25)}{1 - (-0.25)(0.5)e^{-2j(0.2\pi)}}\right\} = 2.99e^{j0.9}$$

$$V(z) = V^+ e^{-j\beta z}[1 + \Gamma(z)] = V^+ e^{-j\beta z}[1 + 0.5e^{j2\beta(z-\ell)}] = 2.99e^{j0.9} e^{-j\beta z}[1 + 0.5e^{j2\beta(z-\ell)}]$$

At the source end $z=0$

$$V(0)=2.99e^{j0.9}\,[1+0.5e^{j2\beta(0-\ell)}]=2.99e^{j0.9}\,[1+0.5e^{-j2\times0.2\pi)}]=3.73\,\angle0.51$$

$$v(0,t)=3.73\cos(\omega t+0.51)\text{ V}$$

$$I(0)=\frac{2.99e^{j0.9}}{50}[1-0.5e^{j2\beta(0-\ell)}]=\frac{2.99e^{j0.9}}{50}[1-0.5e^{-j2\times0.2\pi}]=0.058\,\angle1.41$$

$$i(0,t)=0.058\cos(\omega t+1.41)\text{ A}$$

$$Z_{\text{in}}=\frac{V(0)}{I(0)}=\frac{3.73\angle0.51}{0.058\angle1.41}=64.36\angle-0.9\ \Omega=39.85-j50.54\ \Omega$$

At the load end $z=\ell$

$$V(\ell)=2.99e^{j0.9}e^{-j0.2\pi}[1+0.5e^{j2\beta(\ell-\ell)}]=2.99e^{j0.9}e^{-j0.2\pi}[1+0.5]=4.48\,\angle0.27\text{ V}$$

$$v(\ell,t)=4.48\cos(\omega t+0.27)\text{ V}$$

$$I(\ell)=\frac{2.99e^{j0.9}}{50}e^{-j0.2\pi}[1-0.5e^{j2\beta(\ell-\ell)}]=\frac{2.99e^{j0.9}}{50}e^{-j0.2\pi}[1-0.5]=0.03\,\angle0.27\text{ A}$$

$$i(0,t)=0.03\cos(\omega t+0.27)\text{ A}$$

b. $\ell=0.25\lambda$

$$\beta\ell=\frac{2\pi}{\lambda}\times0.25\lambda=0.5\pi\qquad e^{-j2\beta\ell}=e^{-j\pi}=-1\qquad\text{and}\qquad e^{-j\beta\ell}=e^{-j0.5\pi}=-j$$

$$V^{+}=\left\{\frac{V_s}{2}\right\}\left\{\frac{1-\Gamma_s}{1-\Gamma_s\Gamma_L e^{-2j\beta\ell}}\right\}=\left\{\frac{5e^{\frac{j\pi}{4}}}{2}\right\}\left\{\frac{1-(-0.25)}{1-(-0.25)(0.5)(-1)}\right\}=3.57e^{\frac{j\pi}{4}}$$

$$V(z)=V^{+}e^{-j\beta z}[1+\Gamma(z)]=V^{+}e^{-j\beta z}[1+0.5e^{j2\beta(z-\ell)}]=3.57e^{\frac{j\pi}{4}}e^{-j\beta z}[1+0.5e^{j2\beta(z-\ell)}]$$

At the source end $z=0$

$$V(0)=3.57e^{\frac{j\pi}{4}}\,[1+0.5e^{j2\beta(0-\ell)}]=3.57e^{\frac{j\pi}{4}}\,[1+0.5e^{-j2\times0.5\pi)}]$$

$$=3.57e^{\frac{j\pi}{4}}\,[1-0.5]=1.79\angle\frac{\pi}{4}\text{ V}\rightarrow v(0,t)=1.79\cos\left(\omega t+\frac{\pi}{4}\right)\text{V}$$

$$I(0)=\frac{3.57e^{\frac{j\pi}{4}}}{50}[1-0.5e^{j\beta(0-\ell)}]=\frac{3.57e^{j\frac{\pi}{4}}}{50}[1-0.5e^{-j2\times0.5\pi}]$$

$$I(0)=\frac{3.57e^{j\frac{\pi}{4}}}{50}[1+0.5]=0.11\angle\frac{\pi}{4}\text{ A}\rightarrow i(0,t)=0.11\cos\left(\omega t+\frac{\pi}{4}\right)\text{A}$$

$$Z_{\text{in}}=\frac{V(0)}{I(0)}=\frac{1.79\angle\frac{\pi}{4}}{0.11\angle\frac{\pi}{4}}=16.67\ \Omega$$

At the load end $z=\ell$

$$V(\ell)=3.57e^{\frac{j\pi}{4}}e^{-j0.5\pi}\,[1+0.5e^{j2\beta(\ell-\ell)}]=3.57e^{\frac{j\pi}{4}}e^{-j0.5\pi}\,[1+0.5]=5.36\angle-\frac{\pi}{4}\text{ V}$$

$$v(\ell,t)=5.36\cos\left(\omega t-\frac{\pi}{4}\right)\text{V}$$

$$I(\ell)=\frac{3.57e^{\frac{j\pi}{4}}}{50}e^{-j0.5\pi}[1-0.5e^{j2\beta(\ell-\ell)}]=\frac{3.57e^{\frac{j\pi}{4}}}{50}e^{-j0.5\pi}[1-0.5]=0.036\angle-\frac{\pi}{4}\text{ A}$$

$$i(\ell,t)=0.036\cos\left(\omega t-\frac{\pi}{4}\right)\text{A}$$

c. $\ell=0.5\lambda$

$$\beta\ell=\frac{2\pi}{\lambda}\times0.5\lambda=\pi \qquad e^{-j2\beta\ell}=e^{-j2\pi}=1 \quad \text{and} \quad e^{-j\beta\ell}=e^{-j\pi}=-1$$

$$V^{+}=\left\{\frac{V_s}{2}\right\}\left\{\frac{1-\Gamma_s}{1-\Gamma_s\Gamma_L e^{-2j\beta\ell}}\right\}=\left\{\frac{5e^{\frac{j\pi}{4}}}{2}\right\}\left\{\frac{1-(-0.25)}{1-(-0.25)(0.5)(1)}\right\}=2.78e^{\frac{j\pi}{4}}$$

$$V(z)=V^{+}e^{-j\beta z}[1+\Gamma(z)]=V^{+}e^{-j\beta z}[1+0.5e^{j2\beta(z-\ell)}]=2.78e^{\frac{j\pi}{4}}e^{-j\beta z}[1+0.5e^{j\beta(z-\ell)}]$$

At the source end $z=0$

$$V[0]=2.78e^{\frac{j\pi}{4}}[1+0.5e^{j2\beta(0-\ell)}]=2.78e^{\frac{j\pi}{4}}[1+0.5e^{-j2\times\pi)}]=4.17\angle\frac{\pi}{4}\text{ V}$$

$$I(0)=\frac{2.78e^{\frac{j\pi}{4}}}{50}[1-0.5e^{j\beta(0-\ell)}]=\frac{2.78e^{\frac{j\pi}{4}}}{50}[1-0.5e^{-j2\times\pi}]=0.028\angle\frac{\pi}{4}\text{ A}$$

$$Z_{\text{in}}=\frac{V(0)}{I(0)}=\frac{4.17\angle\frac{\pi}{4}}{0.028\angle\frac{\pi}{4}}=150\ \Omega$$

At the load end $z=\ell$

$$V(\ell)=2.78e^{\frac{j\pi}{4}}e^{-j\pi}[1+0.5e^{j2\beta(\ell-\ell)}]=2.78e^{\frac{j\pi}{4}}e^{-j\pi}[1+0.5]=4.17\angle-\frac{3\pi}{4}\text{ V}$$

$$I(\ell)=\frac{2.78e^{\frac{j\pi}{4}}}{50}e^{-j\pi}[1-0.5e^{j2\beta(\ell-\ell)}]=\frac{2.78e^{\frac{j\pi}{4}}}{50}e^{-j\pi}[1-0.5]=0.028\angle-\frac{3\pi}{4}\text{ A}$$

Note that in all cases we can confirm $\frac{V(\ell)}{I(\ell)}=Z_L$.

2.10 Two Special Cases: The Infinite Line and the Matched Load Line

Let us assume a TL with infinite length. As we just stated in the previous section, there will be no reflections from the load resulting in $\Gamma(\ell)=0$, and hence $\Gamma(z)=0$ and $V^{-}=0$ [Equations (2.61) and (2.60)].

Consequently, according to Equations (2.21) and (2.26),

$$V(z)=V^{+}e^{-\gamma z} \quad \text{and} \quad I(z)=\frac{1}{Z_o}\cdot\{V^{+}e^{-\gamma z}\} \tag{2.72}$$

which makes $Z(z)=Z_o$ at all z locations including $z=0$ (TL input impedance), $Z_{\text{in}}=Z(0)=Z_o$.

Solving the boundary condition at the source end,

$$V(0)=V^{+}=V_s-Z_s\cdot I(0)=V_s-Z_s\cdot\frac{1}{Z_o}(V^{+}) \tag{2.73}$$

which yields

$$V(0) = V^+ = V_s \frac{Z_o}{Z_s + Z_o} \quad \text{and} \quad I(0) = \frac{V^+}{Z_o} = V_s \frac{1}{Z_s + Z_o} \tag{2.74}$$

Therefore, physically speaking, both an infinite line and a match terminated line exhibit matching at line locations including the line input (at $z = 0$). Only waves traveling in $(+z)$ direction can exist on the line.

2.11 Standing Waves and Standing Wave Ratio

The combination of waves traveling in both directions on the same TL is likely to cause standing waves to form. To explain, let us start with the lossless line time-domain combined solution [Equation (2.37)]. By adding and subtracting the term $|V^-|\cos(\omega t - \beta z + \varphi_v^+)$ in the right-hand side of the equation, we obtain

$$\begin{aligned} v(z,t) &= |V^+|\cos(\omega t - \beta z + \varphi_v^+) + |V^-|\cos(\omega t + \beta z + \varphi_v^-) \\ &= \left\{\left[|V^+| - |V^-|\right]\cos(\omega t - \beta z + \varphi_v^+)\right\} + \left\{|V^-|\left[\cos(\omega t - \beta z + \varphi_v^+) + \cos(\omega t + \beta z + \varphi_v^-)\right]\right\} \\ &= \left\{\left[|V^+| - |V^-|\right]\cos(\omega t - \beta z + \varphi_v^+)\right\} + \left\{2|V^-|[\cos(\omega t + \varphi_1)\cdot\cos(\beta z - \varphi_2)]\right\} \end{aligned} \tag{2.75}$$

where φ_1 and φ_2 replace the quantities $\frac{\varphi_v^+ + \varphi_v^-}{2}$ and $\frac{\varphi_v^+ - \varphi_v^-}{2}$, respectively.

The first term of the resulting equation, $\left\{\left[|V^+| - |V^-|\right]\cos(\omega t - \beta z + \varphi_v^+)\right\}$, has the same traveling form as the $(+z)$ traveling wave but with a reduced amplitude of $\left[|V^+| - |V^-|\right]$. On the other hand, the second term, $\left\{2|V^-|[\cos(\omega t + \varphi_1)\cdot\cos(\beta z - \varphi_2)]\right\}$, has an amplitude of $2|V^-|$ and describes a "standing wave" that does not travel in either direction on the line and is characterized by stationary "peaks/maxima" and "valleys/minima." This can be seen mathematically by examining the expression and recognizing that the spatial dependence term $[\cos(\beta z - \varphi_2)]$ forces the term to zero at certain z locations irrespective of the time-dependent term. This is demonstrated in Figure 2.13.

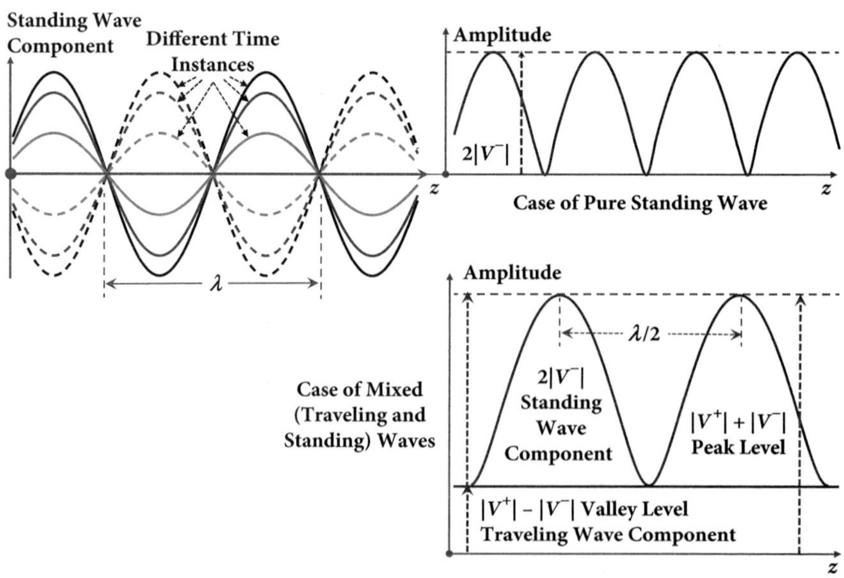

FIGURE 2.13 Demonstration of standing waves and standing wave ratio.

Hence, expression (2.75), as seen in the figure, is a mixture of a traveling wave component and a standing wave one. The amplitude of the standing wave component $2\left|V^-\right|$ is determined by the level of the reflected signal. In the absence of reflections, there will be no standing waves and we get pure traveling waves on the line. However, a 100% reflection where $\left|V^-\right|=\left|V^+\right|$ results in a pure 100% standing waves on the TL since the traveling wave term $\left[\left|V^+\right|-\left|V^-\right|\right]$ reduces to zero.

2.11.1 Standing Wave Ratio

An important indicator of the level of standing waves in a TL network is the standing wave ratio (SWR) the definition for which is the ratio of the peak (max) amplitude of the signal and its valley (min) value:

$$\text{SWR}=\frac{|V|_{\text{max}}}{|V|_{\text{min}}}=\frac{\left[\left|V^+\right|+\left|V^-\right|\right]}{\left[\left|V^+\right|-\left|V^-\right|\right]}=\frac{1+|\Gamma|}{1-|\Gamma|} \tag{2.76}$$

SWR = 1 for a pure traveling waves ($|\Gamma|=0$) and SWR = ∞ for a pure standing wave ($|\Gamma|=1$).

Figure 2.14 is a graphical demonstration of the relationship between the SWR and $|\Gamma|$.

For lossy lines, the line attenuation causes the signals to decay down the TL path, resulting in the signal forms displayed in Figure 2.15.

The figure shows the presence of a mixture of traveling and standing waves, just as in the lossless case. It is important to note here that since all the levels are changing, it is not possible to define a single value for the SWR. It is typical in such cases; to identify the location at which the SWR is given, that is, the SWR of the load would be that of the nearest minimum and maximum to the TL load.

Another way of looking at the combination of the two traveling waves and their "interference" is to consider the corresponding phasor diagrams. For convenience, we will consider the case of a lossless TL only. In this case, the two phasors representing the two traveling waves are

$$V^+e^{-j\beta z} \quad \text{and} \quad V^-e^{+j\beta z} \tag{2.77}$$

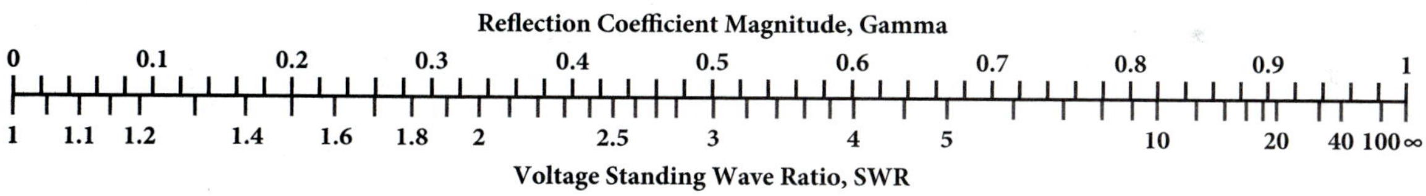

FIGURE 2.14 Graphical demonstration of the relationship between the SWR and $|\Gamma|$.

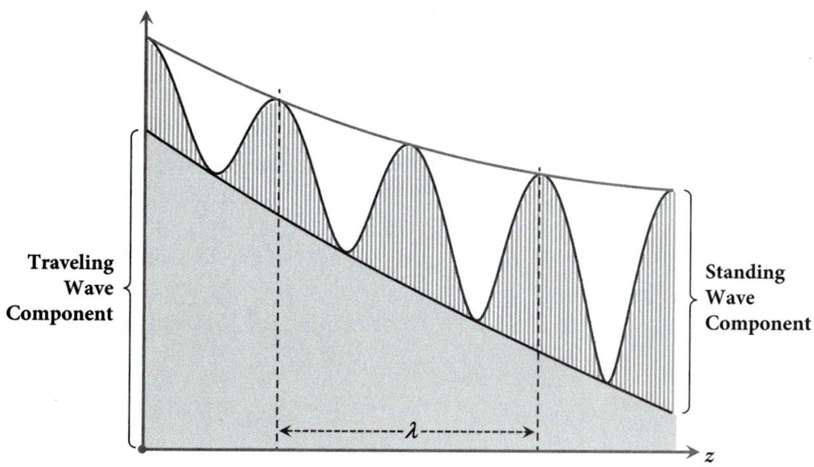

FIGURE 2.15 Demonstration of standing waves for lossy lines.

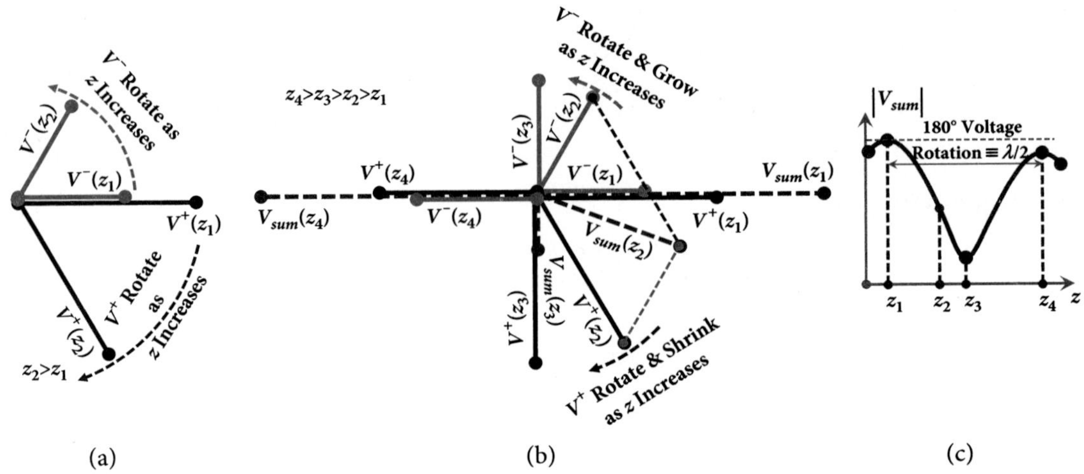

FIGURE 2.16 Phasor demonstration of standing waves as the sum of both forward and backward traveling waves.

Referring to Figure 2.16, we will start with a graph of the two phasors at some location z_1 where they line up in their phase. As we move from z_1 to z_2 (where $z_2 > z_1$), the phase of the V^+ phasor changes by $-\beta(z_2 - z_1)$ (decreases) while the phase of the V^- phasor changes by $+\beta(z_2 - z_1)$ (increases), as seen in part (a) of the figure. Similarly, we monitor the phasors' rotation as we move to z_3 and z_4. The sum of the two phasors is displayed for all four positions in part (b) of the figure, and its magnitude is plotted vs distance in part (c) of the figure. We notice that the sum exhibits the same shape as that demonstrated earlier in Figure 2.13. We also notice that the peaks (maxima) occur when the two phasors line up with equal phases. The spacing between two consecutive peaks corresponds to a 180 degree rotation which is equivalent to a distance of $\lambda/2$.

2.11.2 Standing Wave Maxima and Minima

Referring to Figure 2.16, let us derive expressions for the locations of the maxima and minima on the line. These expressions are helpful to designers to decide the location of special components or devices in a TL circuit. They are also useful in TL measurement techniques where measurements of wavelength, and impedance related parameters as at stake.

We start with combining Equations (2.51) and (2.61) to write the magnitude of voltage as a function of position in the form

$$|V(z)| = \left|V^+ e^{-\gamma z}[1+\Gamma(z)]\right| = \left|V^+ e^{-\gamma z}\right|\left|1+\Gamma(\ell)\cdot e^{-2\gamma d}\right| \tag{2.78}$$

Now, examining Equation (2.78), we realize that the voltage maxima and minima are determined by the magnitude $\left|1+\Gamma(\ell)\cdot e^{-2\gamma d}\right|$, which in turn depends on the polarity associated with the amplitude of the term $\Gamma(\ell)\cdot e^{-2\gamma d}$. When the term $\Gamma(\ell)\cdot e^{-2\gamma d}$ becomes a positive real, the magnitude $\left|1+\Gamma(\ell)\cdot e^{-2\gamma d}\right|$ will correspond to a maximum (peak) voltage, and when the term $\Gamma(\ell)\cdot e^{-2\gamma d}$ becomes a negative real, the magnitude $\left|1+\Gamma(\ell)\cdot e^{-2\gamma d}\right|$ will correspond to a minimum (valley) voltage. This is in line with the discussion of Figure 2.16 where the maximum correspond to the alignment of the phases of the forward and backward waves determines the maxima and minima locations.

Hence, we need to examine the term $\Gamma(\ell)\cdot e^{-2\gamma d}$ to determine the locations at which it becomes positive and negative reals

$$\Gamma(\ell)\cdot e^{-2\gamma d} = |\Gamma(\ell)|\, e^{j\varphi_L}\cdot e^{-2\alpha d} e^{-j2\beta d} = |\Gamma(\ell)|\, e^{-2\alpha d}\cdot e^{j[\varphi_L - 2\beta d]} \tag{2.79}$$

where φ_L is the phase of reflection coefficient at the load, $\Gamma(\ell) = \Gamma_L$.

Based on Equation (2.79), we conclude that

$$\Gamma(\ell)\cdot e^{-2\gamma d}=\begin{cases}\text{positive real if } [\varphi_L-2\beta d]=\pm \text{ even multiples of } \pi\\ \text{negative real if } [\varphi_L-2\beta d]=\pm \text{ odd multiples of } \pi\end{cases} \tag{2.80}$$

Consequently, we can finally write that maxima and minima locations as

$$\left.\begin{aligned} d_{\text{maxima}}&=\frac{\varphi_L \pm \text{even multiples of } \pi}{2\beta}=\frac{\varphi_L\pm 2n\pi}{2\beta}=\frac{\varphi_L\pm 2n\pi}{4\pi}\lambda=\frac{\varphi_L}{4\pi}\lambda\pm\frac{n}{2}\lambda\\ d_{\text{minima}}&=\frac{\varphi_L \pm \text{odd multiples of } \pi}{2\beta}=\frac{\varphi_L\pm(2n+1)\pi}{2\beta}=\frac{\varphi_L}{4\pi}\lambda\pm\frac{(2n+1)}{4}\lambda\end{aligned}\right\} \tag{2.81}$$

where n is an integer and both d_{maxima} and d_{minima} are positive distances between zero and ℓ. (Reminder: since $d=\ell-z$, and $0<z<\ell$ then $0<d<\ell$.)

Equation (2.81) confirms that the maxima are spaced $\lambda/2$ apart so are the minima. It also confirms that the separation between a maximum and its adjacent minimum is $\lambda/4$.

Example 2.3 A slotted line probe is an instrument used to measure the unknown impedance of a load. The instrument employs a slotted coaxial air TL which contains a narrow slit in the outer conductor. A small probe is inserted in the slit and used to measure the amplitude of the electric field, which is directly related to the magnitude of the voltage on the line. By moving the probe along the TL, it's possible to measure $|V_{\max}|$ and $|V_{\min}|$ and the distance from the load at which they occur.

In a 50-Ω slotted line, the distance between two successive voltage minima was found to be 20 cm. The voltage standing wave ratio (SWR) was measured to be 3. The first minimum was located at a distance 6 cm from the load. Determine the load impedance and the distance from the load to the first voltage maximum.

Distance between two successive minima $=\lambda/2\rightarrow\lambda/2=20\text{ cm}\rightarrow\lambda=40\text{ cm}$

$d_{\text{minima}}=\frac{\varphi_L}{4\pi}\lambda\pm\frac{(2n+1)}{4}\lambda$. For the first minimum $\rightarrow n=0$

$$6\text{ cm}=\frac{\varphi_L}{4\pi}\lambda\pm\frac{\lambda}{4}\rightarrow\quad \varphi_L=\left(\frac{4\pi}{\lambda}\right)\left(6\text{ cm}-\frac{\lambda}{4}\right)=\frac{4\pi}{40}\left(6-\frac{40}{4}\right)=-\frac{2}{5}\pi=-0.4\pi$$

In the above equation, taking the positive or negative sign in the term $\pm\frac{\lambda}{4}$ will yield the same angle φ_L

$$\text{SWR}=\frac{\left[|V^+|+|V^-|\right]}{\left[|V^+|-|V^-|\right]}=\frac{1+|\Gamma|}{1-|\Gamma|}\rightarrow\qquad |\Gamma|=\frac{\text{SWR}-1}{\text{SWR}+1}=\frac{3-1}{3+1}=0.5$$

$$\Gamma_L=|\Gamma|e^{j\varphi_L}=0.5e^{-j0.4\pi}=0.5(\cos(-0.4\pi)+j\sin(-0.4\pi))=0.1545-j0.4755$$

$$Z_L=Z_o\frac{[1+\Gamma_L]}{[1-\Gamma_L]}=50\frac{[1+0.1545-j0.4755]}{[1-(0.1545-j0.4755)]}=(39.85-j50.54)\,\Omega$$

$$d_{\text{maxima}}=d_{\text{minima}}\pm\frac{\lambda}{4}=6\text{ cm}\pm\frac{40}{4}=6\text{ cm}+\frac{40}{4}=16\text{ cm}$$

The first maximum from the load will be located 16 cm away from the load.

Example 2.4 A 50-Ω lossless TL is terminated in a load with impedance $Z_L = (100 + j50)\ \Omega$. The line length is 2 cm and the operating wavelength is 10 cm. Find the following:

a. The reflection coefficient at the load:

$$\Gamma_L = \frac{Z_L - Z_o}{Z_L + Z_o} = \frac{(100+j50)-50}{(100+j50)+50} = \frac{(50+j50)}{(150+j50)} = \frac{70.7e^{j\frac{\pi}{4}}}{158.1e^{j0.32}} = 0.4 + j0.2 = 0.45e^{j0.464}$$

$$|\Gamma_L| = 0.45 \quad \varphi_L = 0.464 \text{ rad}$$

b. The reflection coefficient at the input of the line:

$$\Gamma(z) = \Gamma_L e^{j2\beta(z-\ell)} \quad \Gamma(0) = \Gamma_L e^{-j2\beta\ell} \quad \beta\ell = \frac{2\pi}{\lambda}\ell = \frac{2\pi}{10} \times 2 = 0.4\pi \rightarrow$$

$$\Gamma(0) = \Gamma_L e^{-j2\beta\ell} = 0.45e^{j0.464}e^{-j2\times 0.4\pi} = 0.45e^{-j2.049} = -0.207 - j0.399$$

c. The input impedance of the line:

$$Z_{\text{in}} = Z_o \frac{Z_L + jZ_o \tan(\beta\ell)}{Z_o + jZ_L \tan(\beta\ell)} = Z_o \frac{[100 + j50 + j50 \tan(0.4\pi)]}{[50 + j(100 + j50)\tan(0.4\pi)]} = 50\frac{(100 + j203.88)}{(-103.88 + j307.76)} = (24.8 - j24.6)\ \Omega$$

The input impedance can also be found from the input reflection coefficient :

$$Z_{\text{in}} = Z_o \frac{1+\Gamma(0)}{1-\Gamma(0)} = 50 \times \frac{(1+[-0.207 - j0.399])}{(1-[-0.207 - j0.399])} = (24.8 - j24.6)\ \Omega$$

d. The standing wave ratio on the line:

$$\text{SWR} = \frac{1+|\Gamma|}{1-|\Gamma|} = \frac{1+0.45}{1-0.45} = 2.6$$

e. If the TL is connected to a 5 V voltage source with source impedance $Z_s = 50\ \Omega$, find the magnitude of each of the standing wave and the traveling wave.

$$\Gamma_s = \frac{Z_s - Z_o}{Z_s + Z_o} = \frac{50-50}{50+50} = 0$$

$$\beta\ell = 0.4\pi \quad V^+ = \left\{\frac{V_s}{2}\right\}\left\{\frac{1-\Gamma_s}{1-\Gamma_s\Gamma_L e^{-2j\beta\ell}}\right\} = \left\{\frac{5}{2}\right\}\left\{\frac{1-0}{1-2}\right\} = 2.5 \text{ V}$$

$$V^- = V^+\left(\frac{Z_L - Z_o}{Z_L + Z_o}\right)e^{-2j\beta\ell} = \Gamma(0)V^+ = 0.45e^{-j2.049}(2.5) = 1.125e^{-j2.049}$$

The magnitude of the traveling wave $= |V^+| - |V^-| = 2.5 - 1.125 = 1.375$ V

The magnitude of the standing wave $= 2|V^-| = 2(1.125) = 2.25$ V

f. The position of the voltage maximum nearest the load:

$$d_{\text{maxima}} = \frac{\varphi_L \lambda}{4\pi} + n\frac{\lambda}{2} \rightarrow n = 0 \rightarrow d_{\text{maxima}} = \frac{0.464 \times 10}{4\pi} = 0.37 \text{ cm}$$

g. The position of the current maximum nearest the load:
The position of current maximum is the same as the position of the voltage minimum

$$d_{\text{minima}} = d_{\text{maxima}} \mp \frac{\lambda}{4} = 0.37 \pm \frac{10}{4} = 0.37 + 2.5 = 2.87 \text{ cm}$$

Example 2.5

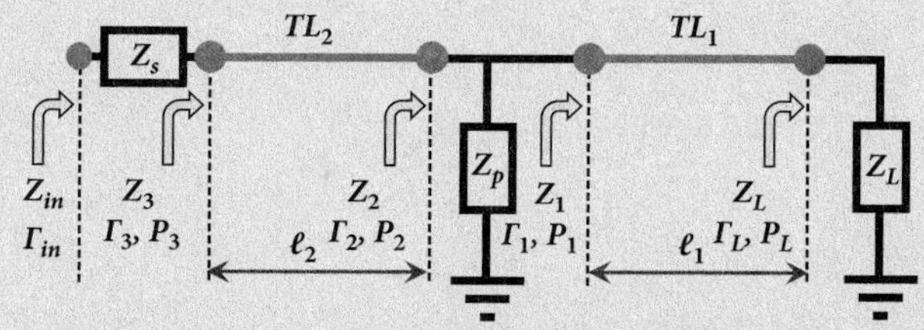

For the given TL circuit, evaluate P's (upper case Greek rho) and Γ's and the impedance seen at all marked positions in the TL given $\ell_1 = 7.5$ cm, $\ell_2 = 3.75$ cm, and $Z_L = 50 - j30\ \Omega$, $Z_p = 100 + j50\ \Omega$, and $Z_s = 20 - j20\ \Omega$. Both TL sections have a characteristic impedance of 50 Ω and a phase velocity of $c_{\text{ph}} = 0.75c$, where c is the speed of light in air. The operating frequency is 1.5 GHz.

Recall that P is defined as the local reflection coefficient at the immediate location of the "discontinuity" assuming no further reflections beyond the current location, where as Γ defines the "steady-state" reflection coefficient that accounts for all "past and future" discontinuities. Only at the load location where there are no further discontinuities P and Γ become the same.

The operating wavelength $\lambda = \dfrac{c_{\text{ph}}}{f} = \dfrac{0.75c}{f} = \dfrac{0.75 \times 3 \times 10^8}{1.5 \times 10^9} = 15$ cm

$$\ell_1 = 7.5 \text{ cm} = 0.5\lambda \text{ and } \ell_2 = 3.75 \text{ cm} = 0.25\lambda$$

At the load location

$$\Gamma_L = P_L = \frac{Z_L - Z_o}{Z_L + Z_o} = \frac{(50 - j30) - 50}{(50 - j30) + 50} = \frac{-j30}{(100 - j30)} = \frac{30e^{-\frac{j\pi}{2}}}{104.4e^{-j0.29}} = 0.287e^{-j1.28}$$

At the location of Z_1, Z_1 is the impedance seen by a TL section of characteristic impedance Z_o and length ℓ_1 terminated with a load impedance Z_L

$$\beta\ell_1 = 2\pi\frac{\ell_1}{\lambda} = 2\pi \times 0.5 = \pi \rightarrow \tan(\beta\ell_1) = 0$$

$$Z_1 = Z_o\frac{Z_L + jZ_o\tan(\beta\ell_1)}{Z_o + jZ_L\tan(\beta\ell_1)} = Z_L = 50 - j30\ \Omega$$

which agrees with the fact that the impedance seen at the input of an $n\dfrac{\lambda}{2}$ TL section terminated in Z_L is the same as $Z_L \therefore \Gamma_1 = \Gamma_L = 0.287e^{-j1.28}$.

P_1 only defines the local discontinuity given by the TL section assuming the TL section is infinitely extended and therefore has an input impedance equal to the characteristic impedance $P_1 = \dfrac{Z_o - Z_o}{Z_o + Z_o} = 0$.

$$\text{The impedance } Z_2 = Z_p \| Z_1 = \left(\frac{1}{Z_p} + \frac{1}{Z_1}\right)^{-1} = \left(\frac{1}{100 + j50} + \frac{1}{50 - j30}\right)^{-1} = 42.14 - j8.95\ \Omega$$

$$\Gamma_2 = \frac{Z_2 - Z_o}{Z_2 + Z_o} = \frac{42.14 - j8.95 - 50}{42.14 - j8.95 + 50} = -0.075 - j0.105 = 0.129e^{-j2.19}$$

To find P, the impedance defining the local discontinuity is given by

$$Z_{2,\text{local}} = Z_p \| Z_o = \left(\frac{1}{Z_p} + \frac{1}{Z_o}\right)^{-1} = \left(\frac{1}{100 + j50} + \frac{1}{50}\right)^{-1} = 35 + j5\ \Omega$$

$$P_2 = \frac{Z_{2,\text{local}} - Z_o}{Z_{2,\text{local}} + Z_o} = \frac{35 + j5 - 50}{35 + j5 + 50} = -0.172 + j0.069 = 0.186e^{j2.76}$$

At the location of Z_3, Z_3 is the impedance seen by a TL section of characteristic impedance Z_o and length ℓ_2 terminated with a load impedance Z_2

$$\beta\ell_2 = 2\pi\frac{\ell_2}{\lambda} = 2\pi \times 0.25 = \frac{\pi}{2} \rightarrow \tan(\beta\ell_2) = \infty$$

$$Z_3 = Z_o\frac{Z_2 + jZ_o\tan(\beta\ell_2)}{Z_o + jZ_2\tan(\beta\ell_2)} = Z_o\frac{\dfrac{Z_2}{\tan(\beta\ell_2)} + jZ_o}{\dfrac{Z_o}{\tan(\beta\ell_2)} + jZ_2} = \frac{Z_o^2}{Z_2} = \frac{50^2}{42.14 - j8.95}$$

$$Z_3 = 56.76 + j12.06\ \Omega$$

which agrees with the fact that a TL section of characteristic impedance Z_o and length equal to $(2n+1)\frac{\lambda}{4}$ acts as an impedance inverter for a load (Z_L) and has an input impedance given by Z_o^2/Z_L

$$\Gamma_3 = \frac{Z_3 - Z_o}{Z_3 + Z_o} = \frac{56.76 + j12.06 - 50}{56.76 + j12.06 + 50} = 0.075 + j0.105 = 0.129e^{j0.95}$$

Since there is no local discontinuity at the location of Z_3, $P_3 = 0$.

Finally $Z_{in} = Z_3 + Z_s = 56.76 + j12.06 + 20 - j20 = 76.76 - j7.94\ \Omega$

$$\Gamma_{in} = \frac{Z_{in} - Z_o}{Z_{in} + Z_o} = \frac{76.76 - j7.94 - 50}{76.76 - j7.94 + 50} = 0.214 - j0.049 = 0.22e^{-j0.23}$$

2.12 Standing Waves and the Bounce Diagram[1]

In this section, we further our exploration of the physics of the multiple reflections/standing waves in mismatched TL networks. In this regard, we will consider the general case of impedance mismatches at both the source and load ends; hence, we can see that multiple wave reflections will occur on the line with waves bouncing back and forth between the (generally) mismatched TL ends. Figure 2.17 depicts the TL network and the expected "bounce" diagram as a result.

Using this bounce (reflection-transmission) diagram, let us now examine the physical meaning of the wave reflection and the corresponding reflection coefficient as defined in the frequency domain in Equation (2.60), $\Gamma(z) = \frac{V^- e^{+\gamma z}}{V^+ e^{-\gamma z}}$. We claim that the $V^- e^{+\gamma z}$ wave resulted from the reflections of the wave $V^+ e^{-\gamma z}$. Meanwhile, we do not see any discontinuities at a general location z that can cause reflections at that location. So, where did the $V^- e^{+\gamma z}$ come from? The answer lies in examining the transients build up before the steady state (harmonic) is reached.

In this regard, and based on the time-domain solution obtained earlier in Equation (2.32), a $v^+(z,t)$ wave, was initially launched at the sending end of the line. This wave traveled along the TL experiencing both attenuation and phase shift but no reflections until it met the first discontinuity at the load end. At the load *discontinuity*, this wave suffered its first reflection triggering the backward traveling wave, $v^-(z,t)$. The reflected wave would then travel back to the source end experiencing only attenuation and phase change until it met the source end discontinuity. A source end reflection would then be in order and the bouncing back and forth continued. With each of these reflections, in general, some of the signal energy would be absorbed in the reflecting elements (the TL end termination impedances, Z_L and Z_s).

[1]More on the frequency-domain bounce diagram is available in Addendum 2C. In addition, the time-domain bounce diagram is covered in Addendum 2D.

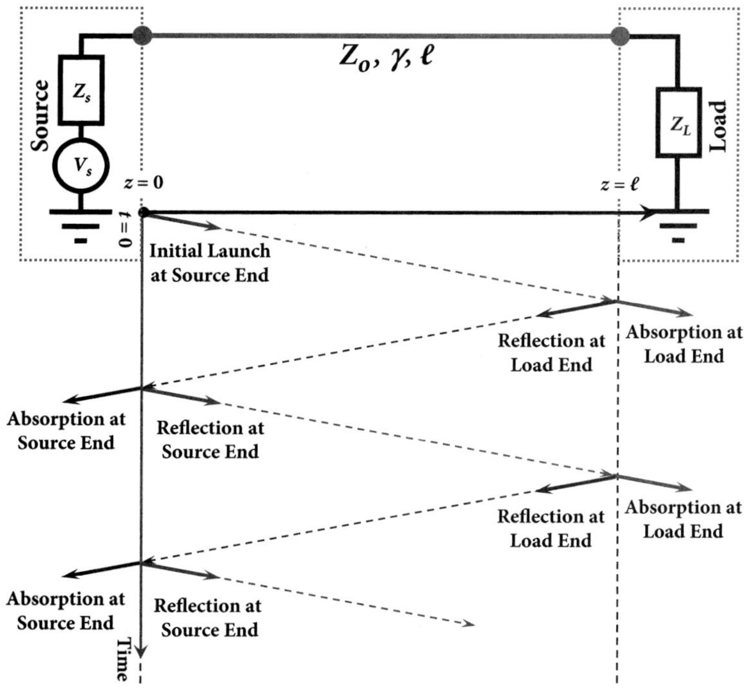

FIGURE 2.17 Conceptual demonstration of the bounce diagram for mismatched TL.

At steady state, there will continue to be a multiple reflections at both line ends resulting in a series of $(+z)$ traveling waves and another series of $(-z)$ traveling waves. The sum of terms of each of these two series correspond to the two terms of the steady-state harmonic solution, $V^- e^{+\gamma z}$ for the $(+z)$ series and $V^+ e^{-\gamma z}$ for the $(-z)$ series. (The proof of this statement will be carried out later in this chapter.) The ratio between these two terms is what we defined as the reflection coefficient $\Gamma(z)$. So, as you can see, $\Gamma(z)$ does not physically exist as a localized phenomenon at the z location of the line; rather, it is a result of the comprehensive performance of the whole setup as seen by the line at the z location.

It is interesting to observe that $\Gamma(z)$ has a localized physical meaning at $z = \ell$. At this location, and using Equation (2.59), $\Gamma_L = \Gamma(\ell) = \dfrac{[Z_L - Z_o]}{[Z_L + Z_o]}$ which is determined by the discontinuity mismatch at that location and is totally independent of the rest of system.

2.13 The Issues of Reflections and Standing Waves

In this section, we investigate the issues with reflections and the resulting standing waves on TLs. To do that, we need to recognize that the purpose of a TL is to deliver the signal/power of the source/transmitter to the load/receiver end. Having said that it is logical that a reflection from that end indicates the lack of a signal/power "full/adequate" delivery and that "some" of that signal/power was "denied/altered" and traveled back the "wrong" way. In terms of power, we can guess that more power could have been delivered to the load if none was reflected. Also, in terms of signal, the receiving end is not seeing the "one" signal that was intended but received multiple deliveries over time.

At this point, we need to demonstrate the concepts discussed above analytically, for both power and signal deliveries.

2.13.1 Power Delivery

First, we need to understand the concept of power available from a source. To do so, let us assume a simple source load arrangement as shown in Figure 2.18. We will carry the analysis in the frequency domain for

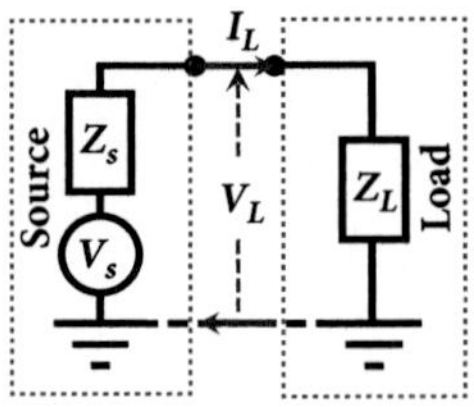

FIGURE 2.18 Basic circuit schematic for power and signal delivery analysis.

the convenience of avoiding differential equation formation and solutions. After all, we will be "swimming close to the shore" as we have done earlier and maintain close ties with the physics involved.

Now, let us evaluate the power delivered to the load in the simple source load circuit shown in the figure. In this analysis, we will use the root mean square (rms) values, for both the current and voltage.

For the basic circuit shown in the figure, the current in the load is $I_L = V_s/(Z_s + Z_L)$, and hence, the power delivered to the load is given by

$$P_L = |I_L|^2 \cdot \text{Re}[Z_L] \tag{2.82}$$

Using the notations $Z_s = R_s + jX_s$ and $Z_L = R_L + jX_L$, where R_s, X_s, R_L, and X_L are the resistive and reactive parts of the source and load impedances, respectively, we can rewrite the expression in Equation (2.82) as follows:

$$P_L = |V_s|^2 \frac{[R_L]}{\left[(R_s + R_L)^2 + (X_s + X_L)^2\right]} \tag{2.83}$$

This power is maximum when the following conditions are BOTH met:

$$X_s + X_L = 0 \text{ and } R_s = R_L, \text{which means } Z_L = Z_s^* \tag{2.84}$$

The reactance condition is simply telling us that the load reactance should null out (tune out) that of the source (or vice versa). So, we learn that in power transmission, since most of the loading (homes, factories, etc.) is inductive (motors, appliances, etc.), the load reactance is inductive and hence a capacitive source impedance is necessary to improve power delivery to these loads. This typically is referred to as capacitive power factor correction.

With tuned out reactances, the second condition signifies that the resistive components of the source and load must be matched to each other, $R_s = R_L$.

With both constraints met, the power delivered to the load is the maximum power possible to obtain from the source and hence the term power available, P_{av}

$$P_{\text{av}} = [P_L]_{\max} = \frac{|V_s|^2}{4R_s} \tag{2.85}$$

Now, how does this apply to our TL network? First, for the source to deliver its *maximum* available power, the impedance seen by the source $[Z_{\text{in}} = Z(0)]$ is the loading impedance to the source. Hence, the requirement becomes

$$Z_s^* = Z_{\text{in}} \xrightarrow{\text{yields}} \Gamma_s^* = \Gamma_{\text{in}} = \Gamma_L \cdot e^{-2\gamma\ell} \tag{2.86}$$

Likewise, for the line to deliver maximum power to the load, the condition is

$$\Gamma_L^* = \Gamma_s \cdot e^{-2\gamma\ell} \xrightarrow{\text{yields}} \Gamma_s^* = \Gamma_L \cdot [e^{2\gamma\ell}]^* \tag{2.87}$$

Comparing the two conditions, we find out that both conditions can be simultaneously satisfied if and only if $e^{-2\gamma\ell} = [e^{2\gamma\ell}]^*$, which can only be true for lossless lines only, $\alpha = 0$.

2.13.2 Signal Delivery

In this section, we discuss the effect of multiple reflections (and hence the standing waves) on the quality (fidelity) of signal delivery. This is of prime concern in communication applications of TLs. For the receiving end not to get the "full" delivery of the transmitted signal and having to "reflect" part of it could be thought of as a reduction in the level of the received signal. The fact of the matter is that it can go beyond affecting the level to actually causing distortion to the received signal.

If we examine the expression for the receiving end signal, Equation (2.69), we can see that the relationship between V_L and V_s includes, in general, complex coefficients. This implies that the signal arriving at the receiving end suffers both amplitude and phase changes. Typically, this is of no concern for single harmonic transmission applications. However, for communication applications, where a composite signal with components of multiple frequencies are propagated on the line, signal components of differing frequencies are expected to have different amplitude and phase modifiers.

The same can be said regarding digital communications where the information signal is coded in pulse format. These "time-domain" pulses are rich in harmonics and thus contain a wide spectrum of frequencies. If the frequency components are to experience different treatment as they travel from the source to the load, the arrival composite of altered frequency components would result in *distorted* pulses, thus compromising the quality and fidelity of the signal delivery. This form of pulse distortion is called *pulse* spreading or *dispersion*.

In addition to pulse spreading as discussed, multiple reflections result in multiple arrivals of time-domain signals at the receiving end (Figure 2.17). An example of this issue is demonstrated in Figure 2.19. In this example, a single-pulse transmission on a line with multiple reflections is shown. The receiving end signal does not arrive to the load all at once; instead, the load receives multiple arrivals at different delay times. Therefore, the arrival at the receiving end is spread over an extended period due to the multiple occurrences in the time domain.

One obvious consequence is the serious limitation this form of dispersion causes to digital communication applications. The rate of sending information pulses has to be slowed down to allow adequate decay of the multiple reflections of one pulse before sending the next one.

Ideally, for communication purposes where the fidelity reception of the transmitter signal is very important, multiple reflections are undesirable. To eliminate multiple reflections, either Γ_s or Γ_L (or both) should be reduced to zero. Hence, according to Equation (2.69),

$$\left.\begin{aligned} \Gamma_s = 0 \quad &\text{yields} \quad V_L = \left\{\frac{V_s}{2}\right\}(1+\Gamma_L)e^{-\gamma\ell} \\ \Gamma_L = 0 \quad &\text{yields} \quad V_L = \left\{\frac{V_s}{2}\right\}(1-\Gamma_s)e^{-\gamma\ell} \\ \Gamma_s = \Gamma_L = 0 \quad &\text{yields} \quad V_L = \left\{\frac{V_s}{2}\right\}e^{-\gamma\ell} \end{aligned}\right\} \tag{2.88}$$

Furthermore, for signal fidelity, all frequency components of V_L should have the same propagation treatment as they travel from the source to the load. In other words, they all travel as the same speed and experience the same level of attenuation. In view of Equation (2.88) we can say that V_L/V_s should have a frequency

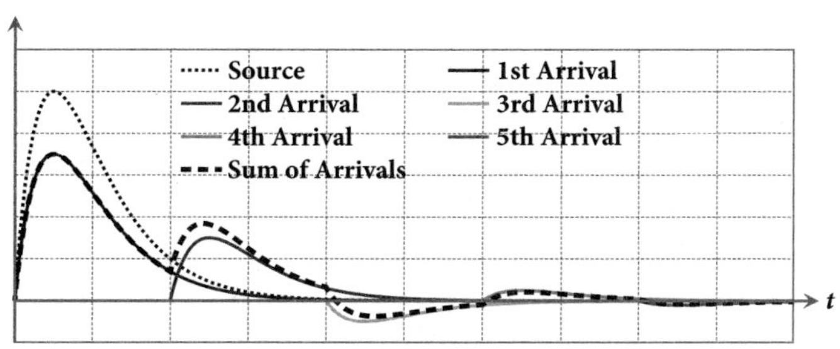

Figure 2.19 Demonstration of the effect of multiple reflections on the fidelity of signal delivery.

independent magnitude with no nonlinearity in its phase. Assuming $e^{-\gamma\ell}$ of the transmission line to produce the same attenuation and signal delay for all frequencies, we can rewrite Equation (2.88) in the form

$$\left.\begin{array}{ll} \Gamma_s=0 \text{ yields } V_L=\left\{\dfrac{V_s}{2}\right\}(1+\Gamma_L)e^{-\gamma\ell} & \text{– Fidelity constraint: } \Gamma_L \text{ should be real} \\ \Gamma_L=0 \text{ yields } V_L=\left\{\dfrac{V_s}{2}\right\}(1-\Gamma_s)e^{-\gamma\ell} & \text{– Fidelity constraint: } \Gamma_s \text{ should be real} \\ \Gamma_s=\Gamma_L=0 \quad \text{ yields } V_L=\left\{\dfrac{V_s}{2}\right\}e^{-\gamma\ell} & \text{– Fidelity constraint: none} \end{array}\right\} \tag{2.89}$$

The requirements for signal fidelity can be summarized as follows: either $\Gamma_s = 0$ or $\Gamma_L = 0$, while the nonzero one should be real.

2.14 Combined Power and Signal Delivery Constraints

According to the conditions in Equations (2.86) and (2.87), if either Γ_s or Γ_L is zero, the other should be zero as well. Hence, to combine the two sets of requirements for maximum available power delivered to the load while maintaining signal fidelity, we need to have both Γ_s and Γ_L equal to zero, meanwhile, we should be using a lossless (or near lossless) TL.

In practical cases, we can design the source to have a matching internal impedance to the TL used and hence maintain Γ_s as zero. However, in many applications, we may not have control over the load impedance and consequently Γ_L. In such cases, a matching network becomes necessary to be implemented in order to enable the source-TL matched network to see an *equivalent* matched load. This is discussed in detail in Addendum 2B.

ADDENDUM 2A DRIVING POINT IMPEDANCE

2A.1 TL Driving Point Impedance and Input Impedance

It is obvious from the discussion above that the impedance relationships on a TL plays a dominant role in assessing its performance. For this reason, in the following section we focus on developing some critical impedance relationships.

First, let us derive an expression for what we know as the "driving point impedance." This is the impedance seen by an observer at a given location on the line looking toward the load side (Figure 2A.1).

Using Equations (2.56) and (2.61),

$$Z(z)=\frac{V(z)}{I(z)}=Z_o\frac{[1+\Gamma(z)]}{[1-\Gamma(z)]}=Z_o\frac{\left[1+\Gamma(\ell)\cdot e^{-2\gamma d}\right]}{\left[1-\Gamma(\ell)\cdot e^{-2\gamma d}\right]}=Z_o\frac{\left[1+\dfrac{[Z_L-Z_o]}{[Z_L+Z_o]}\cdot e^{-2\gamma d}\right]}{\left[1-\dfrac{[Z_L-Z_o]}{[Z_L+Z_o]}\cdot e^{-2\gamma d}\right]}$$

$$Z(z)=Z_o\frac{[[Z_L+Z_o]+[Z_L-Z_o]\cdot e^{-2\gamma d}]}{[[Z_L+Z_o]-[Z_L-Z_o]\cdot e^{-2\gamma d}]}=Z_o\frac{[[Z_L+Z_o]\cdot e^{+\gamma d}+[Z_L-Z_o]\cdot e^{-\gamma d}]}{[[Z_L+Z_o]\cdot e^{+\gamma d}-[Z_L-Z_o]\cdot e^{-\gamma d}]} \tag{2A.1}$$

$$Z(d)=Z(z)=Z_o\frac{[2Z_L\cdot\cosh(\gamma d)+2Z_o\cdot\sinh(\gamma d)]}{[2Z_L\cdot\sinh(\gamma d)+2Z_o\cdot\cosh(\gamma d)]}=Z_o\frac{[Z_L+Z_o\cdot\tanh(\gamma d)]}{[Z_o+Z_L\cdot\tanh(\gamma d)]} \tag{2A.2}$$

Hence,

$$Z_{\text{in}}=Z(z=0)=Z_o\frac{[Z_L+Z_o\cdot\tanh(\gamma\ell)]}{[Z_o+Z_L\cdot\tanh(\gamma\ell)]} \tag{2A.3}$$

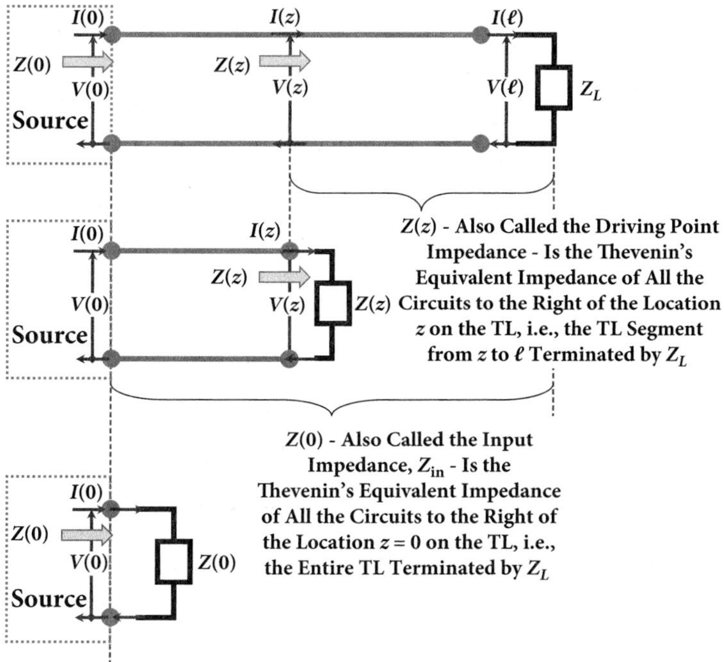

Figure 2A.1 Demonstration of the concept of driving point impedance.

To gain some insight into this expression, let take the lossless TL case where

$$\gamma = j\beta \text{ and } Z_o = R_o \tag{2A.4}$$

$$Z(d) = R_o \frac{[Z_L + R_o \cdot j\tan(\beta d)]}{[R_o + Z_L \cdot j\tan(\beta d)]} \tag{2A.5}$$

and since $\tan[\beta(d \pm n\lambda/2)] = \tan\left(\frac{2\pi d}{\lambda} \pm \pi\right) = \tan(2\pi d/\lambda) = \tan(\beta d)$, then

$$Z(d \pm n\lambda/2) = Z(d) \text{ or } Z(z \pm n\lambda/2) = Z(z) \tag{2A.6}$$

In other words, the impedance values repeat on a lossless line every half-wavelength while, as we recall, the voltage waves currents repeated every full wavelength.

Furthermore, since $\tan\,[\beta(d \pm n\lambda/4)] = \tan\left(\frac{2\pi d}{\lambda} \pm \pi/2\right) = -\{\tan(\beta d)\}^{-1}$, then

$$Z\left(d \pm n\lambda/4\right) = R_o \frac{[Z_L + R_o \cdot \{j\tan(\beta d)\}^{-1}]}{[R_o + Z_L \cdot \{j\tan(\beta d)\}^{-1}]} = R_o \frac{[R_o + Z_L \cdot j\tan(\beta d)]}{[Z_L + R_o \cdot j\tan(\beta d)]} = R_o^2/Z(d) \tag{2A.7}$$

Therefore, every quarter-wavelength the lossless TL driving point impedance inverses its nature.

2A.2 Some Special Cases

For a lossless line (almost true for low loss lines as well), we can summarize some features derived from Equation (2A.5) in Table 2A.1.

It is interesting to know that some of the features cited in Table 2A.1 have useful applications. One typical application is the use of a quarter-wavelength TL section to act as an impedance transformer (adapter).

$$Z_{in}\big|_{\lambda/4} = R_o^2/Z_L \tag{2A.8}$$

Z_L	$Z(d)$, all values	$Z(d =$ multiples of $\lambda/2)$	$Z(d =$ odd multiples of $\lambda/4)$
R_o (matched)	R_o (matched)	R_o (matched)	R_o (matched)
0 (short circuit)	Reactive values: $\infty \leq X(z) \leq +\infty$	Same as Z_L 0 (short circuit)	∞ (open circuit)
∞ (open circuit)	Reactive values: $\infty \leq X(z) \leq +\infty$	Same as Z_L ∞ (open circuit)	0 (short circuit)
$Z(d)$ has positive reactance (inductive)	Reactive values: $\infty \leq X(z) \leq +\infty$	Same as Z_L	Negative reactance (capacitive)
$Z(d)$ has negative reactance (capacitive)	Reactive values: $\infty \leq X(z) \leq +\infty$	Same as Z_L	Positive reactance (inductive)
$\lvert Z(d)\rvert < R_o$		Same as Z_L	$\lvert Z(d)\rvert > R_o$
$\lvert Z(d)\rvert > R_o$		Same as Z_L	$\lvert Z(d)\rvert < R_o$

TABLE 2A.1 TL Impedances for Different TL Terminations

Examples: A quarter-wavelength TL section terminated by a capacitor works as an inductor for use in high-frequency devices. This is a highly preferred approach since wound coils do not offer the desired performance at radio and microwave frequencies. We find a similar approach in using a short-circuited quarter-wavelength TL section to act as an open circuit at high frequencies.

Another typical application is to use a quarter-wavelength TL section with the right value of R_o to provide impedance matching between differing source and load resistive impedances.

$R_o^2 = R_{\text{in}}\big|_{\lambda/4} \cdot R_L = R_s \cdot R_L$, for example, a 150-Ω quarter-wavelength TL can match a 75-Ω source to a 300-Ω load.

ADDENDUM 2B
IMPEDANCE MATCHING

2B.1 How to Achieve Matching

Referring to Figures 2B.1 to 2B.4, matching the load impedance to the source impedance implies making the projected load impedance at the source end equal to that of the source. In other words, making the input impedance of the load *network* equal the source impedance. It is worthy to mention here that although the discussion in this section is focused on matching the load to the source, the opposite is possible and doable through the same approaches presented here. It is also important to cite that in typical practical cases, the source impedance is pure resistive, and hence that is what we confine our discussion in this section to.

To bring the "projected" load impedance as viewed at the source port to the source impedance (resistance) value, a matching network is needed to act as an impedance transformer between the two values.

One may argue that a simple series or shunt impedance matching network would achieve the desired result (Figure 2B.2). For example, if the load impedance (resistance) level is lower than that of the source, then add a series impedance in the amount of $Z_{\text{series}} = Z_s - Z_L$, thus making $Z_{\text{in}} = Z_s$. On the other hand, if the load impedance (resistance) level is higher than that of the source, then add the matching network impedance in shunt, and

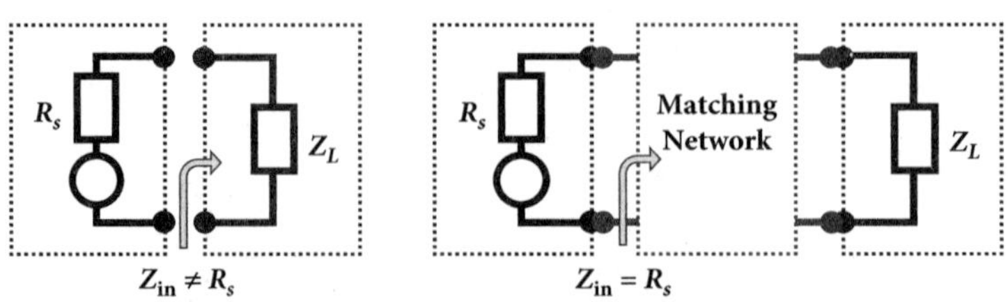

FIGURE 2B.1 Conceptual demonstration of the use matching networks.

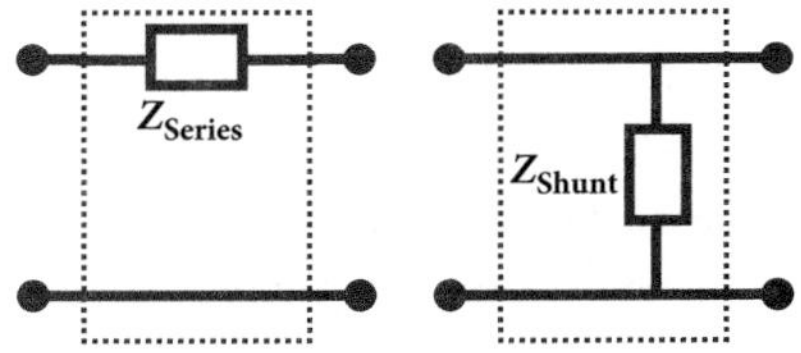

FIGURE 2B.2 Series and shunt impedances as matching networks.

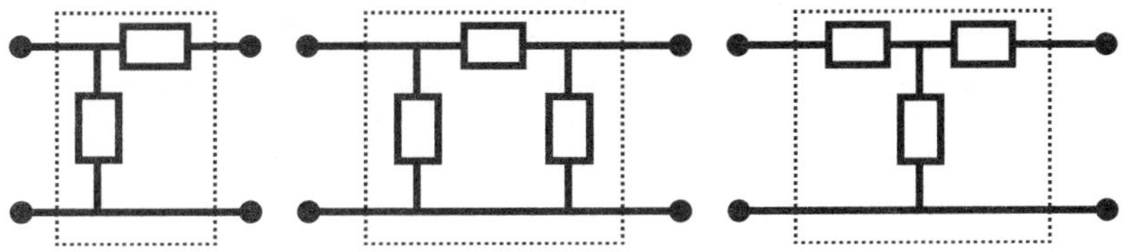

FIGURE 2B.3 The use of L, PI, and T networks for matching.

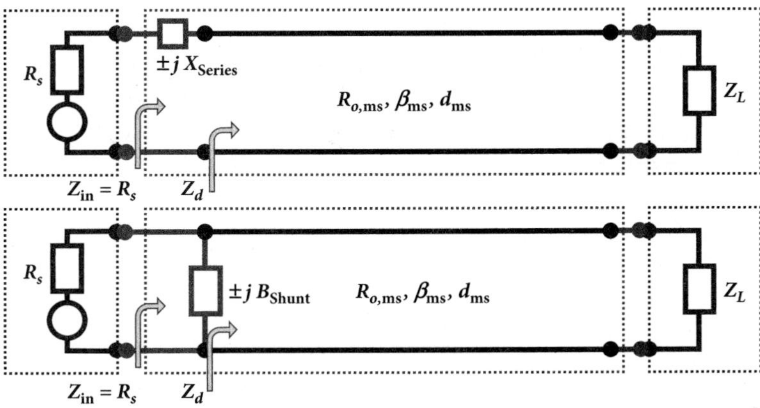

FIGURE 2B.4 Demonstration of the use of series and shunt impedances for matching (ms stands for TL matching section).

hence $Z_{in} = Z_L/Z_{shunt}$. The problem with these configurations is that either Z_{series} or Z_{shunt} are likely to contain resistive components and hence introducing loss elements to the circuit that would degrade its performance.

As the networks of Figure 2B.2 prove inadequate, some of the typically used matching network forms are shown in Figures 2B.3, 2B.4, and 2B.5.

2B.2 L-PI-T Matching Networks

Figure 2B.3 shows three matching network configurations: the "L-section," the "Π-network," and the "T-network."

The literature has many references regarding the design approaches of such networks as well as comparisons of their performances. Depending on the network configuration, the design may be optimized in terms of bandwidth or other possible performance parameters.

For optimum performance, all matching network components must not add any loss to the circuit (or at least be "low loss" components). In other words, the network elements need to be high quality "high-Q" reactances. Depending on the operating frequency of the circuit, discrete components cannot be used for lumped circuit elements sought in these networks as they may cause performance issues. If the size of the discrete components is not much smaller than the operating wavelength, as discussed earlier in Table 2.1, the component must be treated as a distributed TL device instead of a lumped element. Moreover, the physical devices for these elements will not be the pure reactive components sought in the design, as they will most certainly contain other parasitics.

2B.3 Stub Matching

At high frequencies, lossless TLs (or practically, low loss lines) are used instead of discrete components. Typical TL matching networks are shown in Figures 2B.4 and 2B.5. The design principle of the networks shown is to use a TL on the load side with the proper line parameters that would project a resistive value at its input port so that it would match the resistive values of the source impedance. Tuning reactances can then be added to achieve the desired full impedance match.

To further explain, assume we are working with the top configuration of Figure 2B.4. We will need to choose a "matching" section TL with parameters $R_{o,\text{ms}}$, β_{ms}, and d_{ms} such that the real part of ($Z_d = Z_{\text{in,ms}}$) is equal to R_s (where ms stands for TL matching section). Using Equation (2A.5), we can write

$$\text{Real part of}\left\{Z_d = R_{o,\text{ms}}\frac{[Z_L + R_{o,\text{ms}}\cdot j\tan(\beta_{\text{ms}}d_{\text{ms}})]}{[R_{o,\text{ms}} + Z_L\cdot j\tan(\beta_{\text{ms}}d_{\text{ms}})]}\right\} = R_s \tag{2B.1}$$

The added series reactance $\pm jX_{\text{series}}$ is to tune out the imaginary part (reactance) of Z_d, and hence

$$\pm jX_{\text{series}} + \text{Imaginary part of}\left\{Z_d = R_{o,\text{ms}}\frac{[Z_L + R_{o,\text{ms}}\cdot j\tan(\beta_{\text{ms}}d_{\text{ms}})]}{[R_{o,\text{ms}} + Z_L\cdot j\tan(\beta_{\text{ms}}d_{\text{ms}})]}\right\} = 0 \tag{2B.2}$$

Similarly, the equations for the bottom configuration with the shunt reactance can be written in terms of the conductance and susceptance terms instead of resistance and reactance terms; hence,

$$\text{Real part of}\left\{Y_d = \left[R_{o,\text{ms}}\frac{[Z_L + R_{o,\text{ms}}\cdot j\tan(\beta_{\text{ms}}d_{\text{ms}})]}{[R_{o,\text{ms}} + Z_L\cdot j\tan(\beta_{\text{ms}}d_{\text{ms}})]}\right]^{-1}\right\} = [R_s]^{-1} \tag{2B.3}$$

and an added shunt susceptance $\pm jB_{\text{shunt}}$ is to tune out the imaginary part (susceptance) of Y_d; therefore,

$$\pm jB_{\text{shunt}} + \text{Imaginary part of}\left\{Y_d = \left[R_{o,\text{ms}}\frac{[Z_L + R_{o,\text{ms}}\cdot j\tan(\beta_{\text{ms}}d_{\text{ms}})]}{[R_{o,\text{ms}} + Z_L\cdot j\tan(\beta_{\text{ms}}d_{\text{ms}})]}\right]^{-1}\right\} = 0 \tag{2B.4}$$

To realize the series reactance or shunt susceptance needed in the matching networks of Figure 2B.4, series and parallel TL stubs as shown in Figure 2B.5 are utilized. The idea is that an open- or short-circuited TL at one of its ends presents a pure reactive/susceptive input impedance at the other end; refer back to Table 2A.1. Typically, a short-circuited stub is used instead of an open-circuited one to avoid open-end radiation and associated losses at the open end of the line.

For a short-circuited stub, the corresponding parameters are to be chosen so that (using Equation (2A.5) with $Z_L = 0$)

$$\pm jX_{\text{series}} = jR_{o,\text{st}}\cdot\tan(\beta_{\text{st}}d_{\text{st}})\ \text{for the series stub, and} \tag{2B.5}$$

$$\pm jB_{\text{shunt}} = [jR_{o,\text{st}}\cdot\tan(\beta_{\text{st}}d_{\text{st}})]^{-1} \tag{2B.6}$$

If an open stub is preferred, the equations for the stub reactances can be obtained for Equation (2.67) by setting Z_L to ∞ (or an estimate to its physical value). The results would be

$$\pm jX_{\text{series}} = -jR_{o,\text{st}}\cdot\cot(\beta_{\text{st}}d_{\text{st}})\ \text{for the series stub, and} \tag{2B.7}$$

$$\pm jB_{\text{shunt}} = [-jR_{o,\text{st}}\cdot\cot(\beta_{\text{st}}d_{\text{st}})]^{-1} \tag{2B.8}$$

Later, in Addendum 2E of this chapter, we will introduce the famous TL graphical technique known as the Smith chart. This technique will be demonstrated to not only produce relatively simple computational

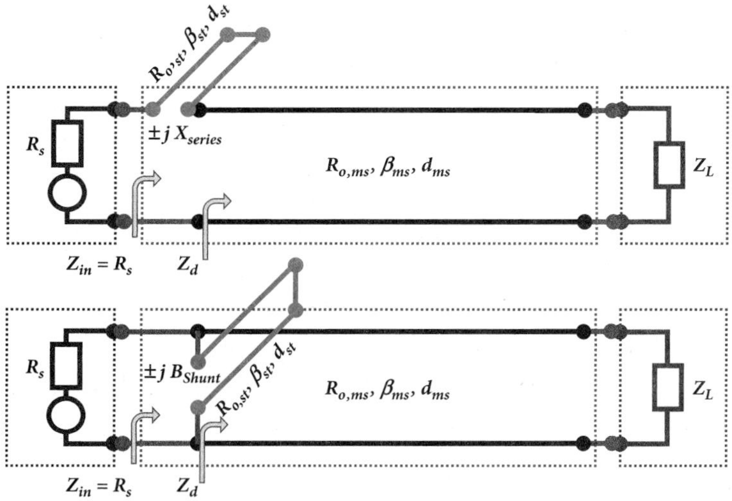

Figure 2B.5 Demonstration of the use of series and shunt stubs for impedance matching (st stands for stub TL section).

solutions to TL problems but also provide physical insight into the solution and the different parameters involved. As an application to the Smith chart, we will do the TL matching networks discussed above.

2B.4 The Quarter Wave Transformer as a Matching Network

In some special cases, a quarter wave TL section can be used to provide the desired matching between the source and the load.

Using Equation (2A.5) again and setting $d = \lambda/4$, we can write

$$Z_{\text{in}} = \frac{R_o^2}{Z_L} \tag{2B.9}$$

The resulting projected impedance Z_{in} is to be equated to R_s for matching with the source impedance (resistance); hence,

$$R_s = \frac{R_o^2}{Z_L} \tag{2B.10}$$

which implies that Z_L should be real. In other words, a lossless/low-loss quarter wave transformer is good for matching real (resistive) load impedances to resistive source impedances provided that we choose the characteristic impedance of the quarter wave section to be

$$R_o = \sqrt{R_s \cdot R_L} \tag{2B.11}$$

Figure 2B.6 demonstrates the use of a $\lambda/4$ section for matching purposes.

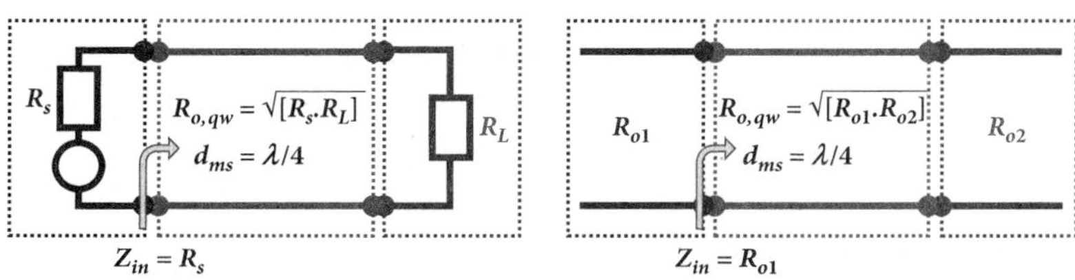

Figure 2B.6 Demonstration of the use of a quarter-wave TL section for impedance matching (qw stands for quarter-wave TL section).

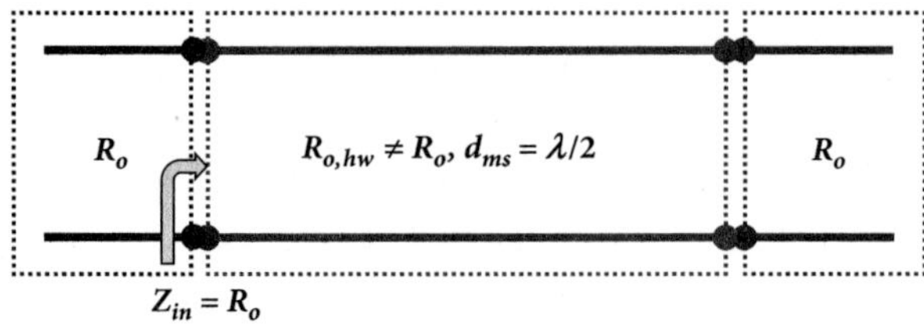

FIGURE 2B.7 Demonstration of the use of a half wave TL section for impedance matching (hw stands for half-wave TL section).

2B.5 The Half Wave Transformer as a Matching Network

Now, we wonder, what can a half wavelength section be used for? Can it be used for matching as well? The answer starts by using Equation (2A.5) while setting $d = \lambda/2$, which yields

$$Z_{\text{in}} = Z_L \tag{2B.12}$$

which is independent of R_o (the characteristic impedance of the half wave section). This means that the load must be already matched to the source, which may lead us to think that the $\lambda/2$ section has no practical use in matching problems.

Well, let us consider the case where a matched source and load pair must be connected, for some practical reason, by a TL whose characteristic impedance is different. In such case, a $\lambda/2$ section of this connecting line would be the ideal length, as it would maintain the existing matching between the source and the load. Such cases exist in real life where connecting TL sections must take a nonuniform shape such as a bent or a twist, etc. These sections need to be designed to have a half wavelengths in order to maintain matching (Figure 2B.7).

ADDENDUM 2C
THE FREQUENCY-DOMAIN BOUNCE DIAGRAM

The title of this addendum may sound provocative to some. We talk about the evolution of the signals on the TL, that is, transients, not in the time domain, but in the frequency domain. This is what we cited in Chapter 1 when we said you might find people communicating in French in the state of England. There must be a good reason for doing so, and the following will demonstrate the idea.

So, we start by recognizing that an initial launch was let by the source. The initial voltage wave corresponding to this launch is labeled "A" in Figure 2C.1. This phasor voltage appears across the input terminals of the TL. Referring to the figure, the launched signal at $t = 0$ sees only the input side of the TL and does not see the TL discontinuity of the load side till the signal arrives there after a delay of ℓ/c_{ph}. No reflections appear at this moment and hence only $V^+e^{-\gamma z} = V^+$ exists. The corresponding current would be $[V^+e^{-\gamma z}]/Z_o = V^+/Z_o$, which means that the input impedance of the TL at that instant ($t = 0$) is equal to Z_o. The initial launch voltage phasor, A, would then be given by a voltage division of the source voltage V_s between the source internal impedance Z_s and the TL input impedance Z_o.

$$A = V(0)|_{t=0} = V_s \cdot \frac{Z_o}{[Z_s + Z_o]} \tag{2C.1}$$

Referring to Figure 2C.1, the *Local* reflection coefficients [$P(j\omega) \equiv$ upper case rho] defined at the load and source *discontinuities* are, in general, different from $\Gamma(j\omega)$ defined earlier. This local $P(j\omega)$ is the reflection coefficient at the immediate location of the *discontinuity* assuming no further reflections beyond the current location. Said another way, $P(j\omega)$ is the *instantaneous* reflection coefficient of the harmonic wave at the time it arrives at a discontinuity not seeing any *future* discontinuities. This is different from $\Gamma(j\omega)$, which defines the "steady-state" reflection coefficient that accounts for all "past and future" discontinuities. For the example

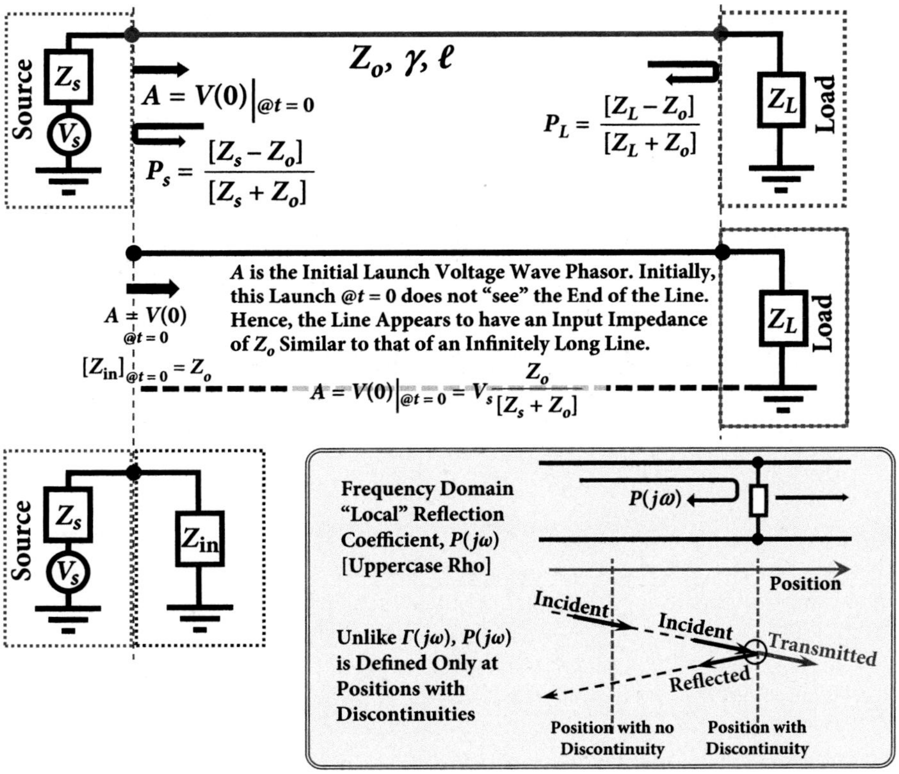

Figure 2C.1 Definition of the "local" reflection coefficient for use in developing the frequency-domain bounce diagram.

considered in Figures 2C.1 to 2C.3, $P_L = \Gamma_L[= \Gamma(\ell)$ in this case$] = \frac{[Z_L - Z_o]}{[Z_L + Z_o]}$ since there are no future discontinuities beyond the load end of the line.

In the following, we need to refer to both Figures 2C.2 and 2C.3. The difference between the two figures is that Figure 2C.2 shows the multiple reflections at both line ends while Figure 2C.3 adds the line signals at an interim location (z). As demonstrated in both figures, the initial launch (A) travels down the TL experiencing modifications in the form of attenuation, $e^{-\alpha z}$, and phase change, $e^{-j\beta z}$. So, at location z on the line the modified initial launch will take the form of the phasor $A \cdot e^{-\gamma z}$, which we will label at $A \cdot T_z$, and by the end of the line at $z = \ell$, the phasor will take the form $[V^+ e^{-\gamma\ell}]_{@t=\ell/c} = A \cdot e^{-\gamma\ell} = A \cdot T$. This positive traveling wave will bounce off the Z_L discontinuity causing a negative z-traveling wave at $z = \ell$, viz.,

$$[V^- e^{+\gamma\ell}]_{@t=\ell/c} = P_L \cdot [V^+ e^{-\gamma\ell}]_{@t=\ell/c} = P_L \cdot A \cdot T \tag{2C.2}$$

The total voltage phasor at that location and at that instant would be

$$V(\ell) = [V^+ e^{-\gamma\ell}]_{@t=\ell/c} + [V^- e^{+\gamma\ell}]_{@t=\ell/c} = A \cdot T + P_L \cdot A \cdot T = [1 + P_L] \cdot A \cdot T \tag{2C.3}$$

This is the voltage phasor across the load termination, and hence, we may call it the transmitted voltage phasor (at that location and at that instant). This component will be *absorbed* by the load.

Next, we examine the reflected phasor for the load side. Starting with the $[V^- e^{+\gamma\ell}]_{@t=\ell/c} = \{P_L \cdot A \cdot T\}$, this phasor changes through the multiplier $e^{-\gamma \text{ distance traveled}}$, so, at location z, the distance traveled would be $d = \ell - z$ and at the source end the distance traveled would be ℓ. Hence the ($-z$) traveling wave will have its phasor expressions as $e^{-\gamma \cdot d} \cdot \{P_L \cdot A \cdot T\} = T_d \cdot \{P_L \cdot A \cdot T\}$ and $e^{-\gamma \cdot \ell} \cdot \{P_L \cdot A \cdot T\} = T \cdot \{P_L \cdot A \cdot T\} = P_L \cdot A \cdot T^2$. This ($-z$) wave arrives at the source end at $t = 2\ell/c$. When it does, it will be faced with the discontinuity of the source

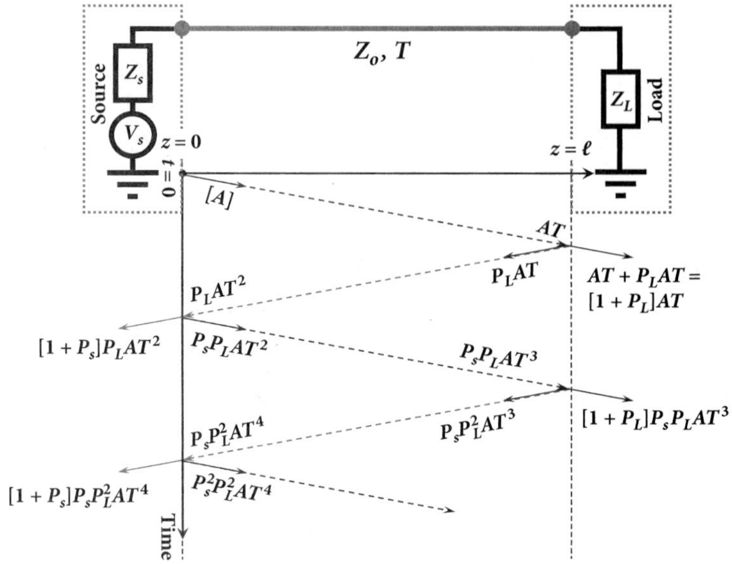

Figure 2C.2 An example of the frequency-domain bounce diagram for a simple TL circuit, $T = e^{-\gamma}$.

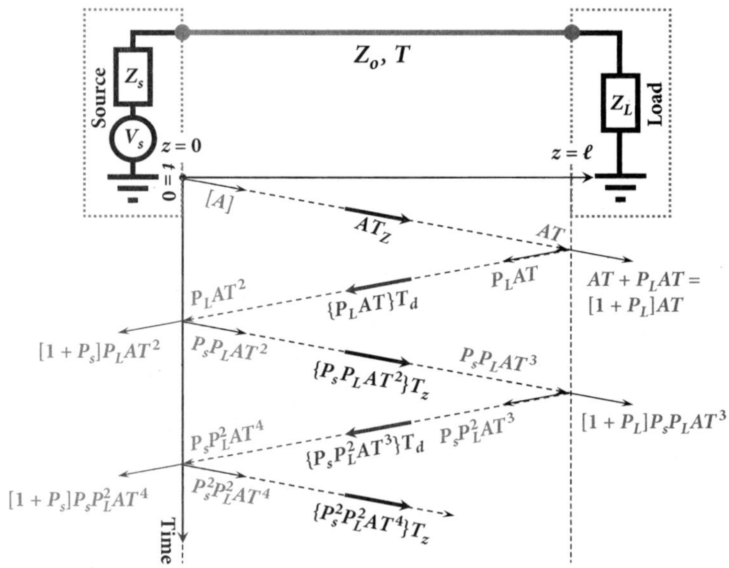

Figure 2C.3 The bounce diagram showing phasor expressions at a general location on the TL, $T = e^{-\gamma\ell}$, $T_z = e^{-\gamma z}$, $T_d = e^{-\gamma d}$, and $d = \ell - z$.

impedance, causing a reflection and a transmission (similar to what happened at the load end). The reflection coefficient facing the $(-z)$ propagation at the source end will be given by $P_s = \frac{[Z_s - Z_o]}{[Z_s + Z_o]}$. This makes the reflected and transmitted phasors $[P_s \cdot P_L \cdot A \cdot T^2]$ and $[1 + P_s] \cdot P_L \cdot A \cdot T^2$, respectively. Following the same approach, we can construct Table 2C.1 for the phasor bounce diagram:

$$A = V_s \cdot \frac{Z_o}{[Z_s + Z_o]} = V_s \cdot \frac{[1 - P_s]}{2} \tag{2C.4}$$

@ Source end			@ location z ($d = \ell - z$)				@ Load end
To Z_s		$+z$ Travel		$+z$ Travel		$+z$ Travel	To Z_L
	$-z$ Travel		$-z$ Travel		$-z$ Travel		
		A		AT_Z		AT	$[1 + P_L]AT$
$[1 + P_s]P_LAT^2$	P_LAT^2		$\{P_LAT\}T_d$		$\{P_LAT\}$		
		$[P_sP_LAT^2]$		$[P_sP_LAT^2]T_z$		$P_sP_LAT^3$	$[1 + P_L]P_sP_LAT^3$
$[1 + P_s]P_sP_L^2AT^4$	$P_sP_L^2AT^4$		$\{P_sP_L^2AT^3\}T_d$		$P_sP_L^2AT^3$		
		$[P_s^2P_L^2AT^4]$		$[P_s^2P_L^2AT^4]T_z$		$P_s^2P_L^2AT^5$	$[1 + P_L]P_s^2P_L^2AT^5$
$[1 + P_s]P_s^2P_L^3AT^6$	$[P_s^2P_L^3AT^6]$		$\{P_s^2P_L^3AT^5\}T_d$		$P_s^2P_L^3AT^5$		
		$[P_s^3P_L^3AT^6]$		$[P_s^3P_L^3AT^6]T_z$			

Table 2C.1 Phasor Expressions for the Bounce Diagram at Different Locations on the TL

		Phasor series	Expression sum
@ Source	To Z_s	$A+[1+P_s]P_LAT^2+[1+P_s]P_sP_L^2AT^4+[1+P_s]P_s^2P_L^3AT^6+\cdots$	$\frac{[1+P_LT^2]}{[1-P_sP_LT^2]}A$
	$-z$	$P_LAT^2+P_sP_L^2AT^4+P_s^2P_L^3AT^6+\cdots$	$\frac{P_LT^2}{[1-P_sP_LT^2]}A$
	$+z$	$A+P_sP_LAT^2+P_s^2P_L^2AT^4+[P_s^3P_L^3AT^6]+\cdots$	$\frac{A}{[1-P_sP_LT^2]}$
@ z	$-z$	$V^-(z)=\{P_LAT\}T_d+\{P_sP_L^2AT^3\}T_d+\{P_s^2P_L^3AT^5\}T_d+\cdots$	$\frac{P_LT\,T_d}{\left[1-P_sP_LT^2\right]}A$
	$+z$	$V^+(z)=[A]T_z+P_sP_LT^2]T_z+[P_s^2P_L^2T^4]T_z+\cdots$	$\frac{T_z}{\left[1-P_sP_LT^2\right]}A$
@ Load	$-z$	$\{P_LAT\}+\{P_sP_L^2AT^3\}+\{P_s^2P_L^3AT^5\}+\cdots$	$\frac{P_LT}{\left[1-P_sP_LT^2\right]}A$
	$+z$	$AT+P_sP_LAT^3+P_s^2P_L^2AT^5+\cdots$	$\frac{T}{\left[1-P_sP_LT^2\right]}A$
	To Z_L	$[1+P_L]AT+[1+P_L]P_sP_LAT^3+[1+P_L]P_s^2P_L^2AT^5+\cdots$	$\frac{[1+P_L]T}{\left[1-P_sP_LT^2\right]}A$

Table 2C.2 Total Voltage Phasors at Different Locations of the TL

The expressions in the table can be used to construct answers for the voltage phasors as functions of position, that is,

At the source end:

$$V^+ = \frac{A}{\left[1-P_sP_LT^2\right]} = \left\{V_s \cdot \frac{[1-P_s]}{2} \cdot \frac{1}{\left[1-P_sP_LT^2\right]}\right\} \tag{2C.5}$$

$$V^- = \frac{P_LT^2A}{\left[1-P_sP_LT^2\right]} = \left\{V_s \cdot \frac{[1-P_s]}{2} \cdot \frac{1}{\left[1-P_sP_LT^2\right]} \cdot P_LT^2\right\} = V^+P_LT^2 \tag{2C.6}$$

$$V(0) = \{V^+\} + \{V^-\} \tag{2C.7}$$

At location z:

$$V^+(z)=\frac{T_z A}{\left[1-P_s P_L T^2\right]}=\left\{V_s\cdot\frac{[1-P_s]\,T_z}{2\cdot\left[1-P_s P_L T^2\right]}\right\}=\{V^+\}\cdot T_z\ =\{V^+\}\cdot e^{-\gamma\cdot z} \tag{2C.8}$$

$$V^-(z)=\frac{P_L T T_d A}{\left[1-P_s P_L T^2\right]}=\left\{V_s\cdot\frac{[1-P_s]P_L\,T T_d}{2\cdot\left[1-P_s P_L T^2\right]}\right\}=\{V^-\}/T_z\ =\{V^-\}\cdot e^{+\gamma\cdot z} \tag{2C.9}$$

$$V(z)=\{V^+\}\cdot e^{-\gamma\cdot z}+\{V^-\}\cdot e^{+\gamma\cdot z} \tag{2C.10}$$

At the load:

$$V(\ell)=\{V^+\}\cdot e^{-\gamma\cdot \ell}+\{V^-\}\cdot e^{+\gamma\cdot \ell} \tag{2C.11}$$

It is important to mention that the above obtained expressions for $\{V^+\}$ and $\{V^-\}$ are in agreement with those obtained earlier in Equations (2.51) and (2.52). This can be shown upon the substitution of $P_s=\frac{[Z_s-Z_o]}{[Z_s+Z_o]}$, $P_L=\frac{[Z_L-Z_o]}{[Z_L+Z_o]}$, and $\frac{V_s Z_o}{Z_s+Z_o}=V_s\cdot\frac{[1-P_s]}{2}$.

ADDENDUM 2D
THE TIME-DOMAIN BOUNCE DIAGRAM

2D.1 Introduction

When we discussed the bounce diagram in the frequency domain, we were thinking of harmonic signals traveling back and forth on the line (in the steady state). Although we were discussing a steady state, we used a *transient* approach to construct the steady-state answer. To do the same exact thing in the time domain, that is, use the transient approach to construct the time-domain steady-state answer, we will limit the discussion to the lossless line case. This is because the time-domain analysis of lossy lines gets beyond our mathematical capabilities and our interest at this point.

Two main advantages to working the bounce diagram in the time domain are the following: First, is the ability to work out transients of nonharmonic signals in simple expressions (only for lossless lines). The second reason is related to practical application in time-domain reflectometry (including practical transmission media, which are not lossless).

In time-domain reflectometry, a pulse is launched in a transmission medium and its reflections are monitored. The delay of the reflected signals is used to determine the distance (and hence location) of the medium discontinuity that caused the reflection. Meanwhile, the nature of the reflection (waveform) can be used to reveal information about the nature of the discontinuity. Practical devices are built around these concepts in many application fields. Among those devices are radars that are used to locate, monitor, and identify airplanes and other flying objects. Sonars are other devices that have applications in medical applications as well as beneath the earth explorations. The most direct application in TLs is the time-domain reflectometer (TDR) used for cable testing and fault location. The optical version of that is called optical time-domain reflectometer (OTDR), which is used for fiber cables.

With this background in mind, we can see the importance of discussing time-domain bounce diagrams even though we will limit the analysis to lossless media. Furthermore, we will initially confine our discussion to pure resistive discontinuities including the source and load *impedances*.

2D.2 Time-Domain Bounce Diagram for Lossless Lines and Resistive Discontinuities

In this discussion, we will be working with the signals in their time-domain form, and hence, we will be using the lower-case notations, $v(z,t)$ and $i(z,t)$. Furthermore, we will be dealing with time-domain (local/instantaneous) reflection coefficients for which we will use the notation $\rho(t)$.

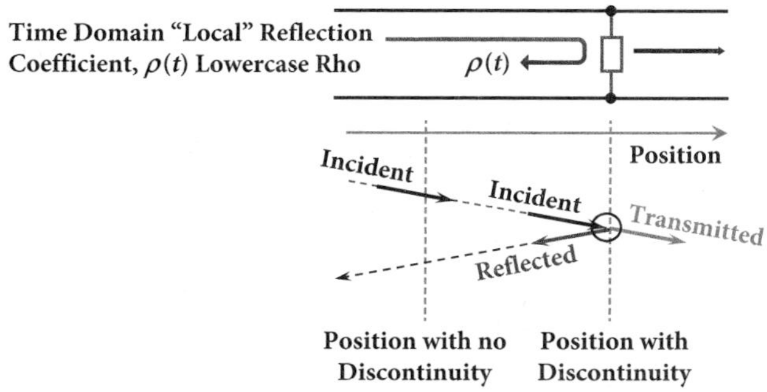

FIGURE 2D.1 The time-domain *local* reflection coefficient for use in developing the time-domain bounce diagram.

It is relevant to state that $\rho(t)$ is the time-domain form of the frequency-domain *local* reflection coefficient, $P(j\omega)$. Hence, $\rho(t)$, can only be found at the discontinuities at the instant the *incident* signal reaches that discontinuity. This concept is demonstrated in Figure 2D.1. At any instance of time, and in locations with no discontinuities, only one signal may exist. However, at discontinuity locations, reflections off the discontinuity result in two coexisting signals: the incident signal and its reflection.

Now, consider the setup shown in Figure 2D.2 with a source and load as shown. Both source and load impedances (R_s and R_L) are purely resistive. The TL connecting the two sides is assumed lossless, and hence has a pure resistive characteristic impedance R_o. Consequently, we can write the local reflection coefficients at the source and load ends of the line as

$$\rho_s = \frac{[R_s - R_o]}{[R_s + R_o]} \quad \text{and} \quad \rho_L = \frac{[R_L - R_o]}{[R_L + R_o]} \tag{2D.1}$$

Also, the initial launch $a(t) = v(z=0, t=0)$ can be written as the voltage divider of $v_s(t)$ between the line's characteristic resistance (impedance), R_o, and the source's resistance, R_s (Figure 2D.3):

$$a(t) = v(0)|_{t=0} = v_s \frac{R_o}{[R_s + R_o]} \tag{2D.2}$$

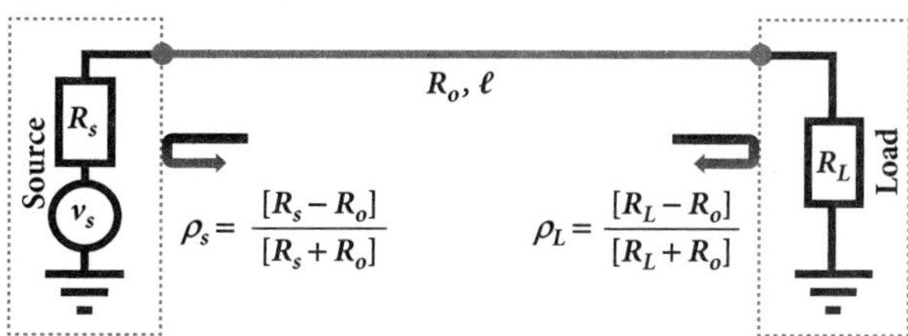

FIGURE 2D.2 The time-domain *local* reflection coefficients at the source and load ends of the TL circuit.

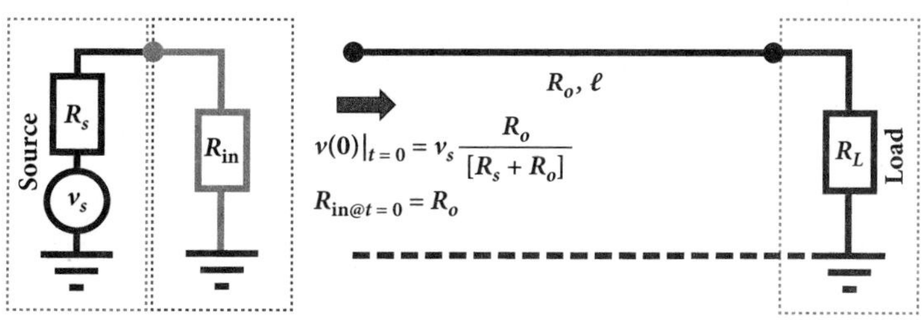

FIGURE 2D.3 The initial launch voltage for the time-domain bounce-diagram.

Now, we can construct the time-domain bounce diagram as we follow the signal travel (delay) and reflect (and transmit) at discontinuities (load and source resistances). As discussed earlier, Equation (2.45), the time delay corresponding to a distance traveled is given by $\tau_{\text{distance}} = \text{distance}/c_{\text{ph}}$.

Using the notation, $\tau = \ell/c_{\text{ph}}$, $\tau_z = z/c_{\text{ph}}$, and $\tau_d = (\ell - z)/c_{\text{ph}}$, the first positive z waveform is the initial launch $a(t)$, which becomes $a(t - \tau_z)$ after traveling a distance z and $a(t - \tau)$ at the load. The first reflection at the load yields $[\rho_L]\cdot[a(t - \tau)]$ as the first negative z traveling wave. This reflection travels back toward the source experiencing further delays to arrive at the source after a delay of $2\ell/c_{\text{ph}} = 2\tau$, and hence can be written as $[\rho_L]\cdot[a(t - 2\tau)]$. Reflecting from the source, a second positive z travel is initiated through a reflection coefficient of $[\rho_s]$. This signal is thus initiated as $[\rho_s]\cdot[\rho_L]\cdot[a(t - 2\tau)]$ and travels toward the load accumulating more delays. The multiple reflections go back and forth with further delays (time shift) of distance/c_{ph} and amplitudes changing by the reflection coefficients multipliers $[\rho_s]$ and $[\rho_L]$.

Now, we examine the transmission (through) waveforms that take place at the load and the source ends upon each reflection. Upon the arrival of the signal at the discontinuity, a reflection voltage develops at that discontinuity in the form of $[\rho_{s \text{ or } L}]\cdot$[incident voltage]. The total voltage at that location is then equal to [incident voltage] + $[\rho_{s \text{ or } L}]\cdot$[incident voltage] = $[1 + \rho_{s \text{ or } L}]\cdot$[incident voltage]. This voltage transmits to the other side of the discontinuity whether it is the source or the load. Consequently, the first arrival signal to the load side is of the form $[a(t - \tau)] + [\rho_L]\cdot[a(t - \tau)] = [1 + \rho_L]\cdot[a(t - \tau)]$ and the first arrival to the source is $[\rho_L]\cdot[a(t - 2\tau)] + [\rho_s]\cdot[[\rho_L]\cdot[a(t - 2\tau)] = [1 + \rho_s]\cdot[\rho_L]\cdot[a(t - 2\tau)]$.

The process we just discussed in the above paragraphs is demonstrated in Figure 2D.4 and the resulting waveforms are summarized in Table 2D.1.

From Table 2D.1, we can write expressions for the signal as a function of position and time as

$$v^+(z,t) = a(t - \tau_z) + \rho_s\rho_L a(t - 2\tau - \tau_z) + \rho_s^2\rho_L^2 a(t - 4\tau - \tau_z) + \rho_s^3\rho_L^3 a(t - 6\tau - \tau_z) + \cdots \tag{2D.3}$$

$$v^-(z,t) = \rho_L \cdot \left\{a(t - \tau - \tau_d) + \rho_s\rho_L a(t - 3\tau - \tau_d) + \rho_s^2\rho_L^2 a(t - 5\tau - \tau_d) + \cdots\right\} \tag{2D.4}$$

$$v(z,t) = v^+(z,t) + v^+(z,t) \tag{2D.5}$$

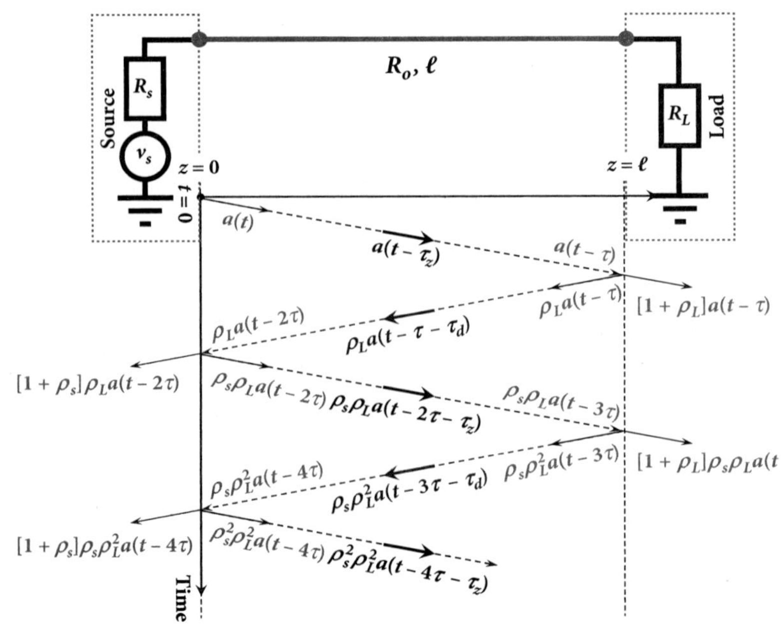

FIGURE 2D.4 The time-domain bounce diagram.

@ Source end			@ location z $(d=\ell-z)$				@ Load end
To Z_s		$+z$ Travel		$+z$ Travel		$+z$ Travel	To Z_L
	$-z$ Travel		$-z$ Travel		$-z$ Travel		
		$a(t)$		$a(t-\tau_z)$		$a(t-\tau)$	$[1+\rho_L]a(t-\tau)$
$[1+\rho_s]\rho_L a(t-2\tau)$	$\rho_L a(t-2\tau)$		$\rho_L a(t-\tau-\tau_d)$		$\rho_L a(t-\tau)$		
		$\rho_s\rho_L a(t-2\tau)$		$\rho_s\rho_L a(t-2\tau-\tau_z)$		$\rho_s\rho_L a(t-3\tau)$	$[1+\rho_L]\rho_s\rho_L a(t-3\tau)$
$[1+\rho_s]\rho_s\rho_L^2 a(t-4\tau)$	$\rho_s\rho_L^2 a(t-4\tau)$		$\rho_s\rho_L^2 a(t-3\tau-\tau_d)$		$\rho_s\rho_L^2 a(t-3\tau)$		
		$\rho_s^2\rho_L^2 a(t-4\tau)$		$\rho_s^2\rho_L^2 a(t-4\tau-\tau_z)$		$\rho_s^2\rho_L^2 a(t-5\tau)$	$[1+\rho_L]\rho_s^2\rho_L^2 a(t-5\tau)$
$[1+\rho_s]\rho_s^2\rho_L^3 a(t-6\tau)$	$\rho_s^2\rho_L^3 a(t-6\tau)$		$\rho_s^2\rho_L^3 a(t-5\tau-\tau_d)$		$\rho_s^2\rho_L^3 a(t-5\tau)$		
		$\rho_s^3\rho_L^3 a(t-6\tau)$					

TABLE 2D.1 Time-Domain Expressions for Bouncing Signal at Different Locations on the TL

		Time-domain series
@Source	To Z_s	$a(t)+[1+\rho_s]\rho_L\cdot\{a(t-2\tau)+\rho_s\rho_L a(t-4\tau)+\rho_s^2\rho_L^2 a(t-6\tau)+\cdots\}$
	$-z$	$\rho_L\cdot\{a(t-2\tau)+\rho_s\rho_L a(t-4\tau)+\rho_s^2\rho_L^2 a(t-6\tau)+\cdots\}$
	$+z$	$a(t)+\rho_s\rho_L a(t-2\tau)+\rho_s^2\rho_L^2 a(t-4\tau)+\rho_s^3\rho_L^3 a(t-6\tau)+\cdots$
@z	$-z$	$v^-(z,t)=\rho_L\cdot\left\{a(t-\tau-\tau_d)+\rho_s\rho_L a(t-3\tau-\tau_d)+\rho_s^2\rho_L^2 a(t-5\tau-\tau_d)+\cdots\right\}$
	$+z$	$v^+(z,t)=a(t-\tau_z)+\rho_s\rho_L a(t-2\tau-\tau_z)+\rho_s^2\rho_L^2 a(t-4\tau-\tau_z)+\rho_s^3\rho_L^3 a(t-6\tau-\tau_z)+\cdots$
@Load	$-z$	$\rho_L\cdot\left\{a(t-\tau)+\rho_s\rho_L a(t-3\tau)+\rho_s^2\rho_L^2 a(t-5\tau)+\cdots\right\}$
	$+z$	$a(t-\tau)+\rho_s\rho_L a(t-3\tau)+\rho_s^2\rho_L^2 a(t-5\tau)+\cdots$
	To Z_L	$[1+\rho_L]\cdot\left\{a(t-\tau)+\rho_s\rho_L a(t-3\tau)+\rho_s^2\rho_L^2 a(t-5\tau)+\cdots\right\}$

TABLE 2D.2 Time-Domain Total Voltage Expressions at Different Locations on the TL

2D.3 Time-Domain Reflectometry and the Bounce Diagram

As discussed earlier, time-domain reflectometry is a technique used for extracting information of a transmission medium by monitoring and analyzing reflections off the medium due to an incident pulse. Typical pulse shapes used in TDR devices are either a step-like waveform [a practical version of an ideal step function, $u(t)$], or an impulse [a practical version of a Dirac-delta function, $\delta(t)$]. In this section, we work with the ideal waveforms for mathematical convenience.

2D.4 Time-Domain Reflectometry for Ideal Step Waveform Excitations

In this case, we assume $v_s(t) = V_s u(t)$, where V_s is the amplitude of the voltage step. To demonstrate the concept, let us assume the simple network of Figure 2D.4 with the following parameters:

$$v_s = 1u(t) \text{ volts} \tag{2D.6}$$

$$R_o = 75 \text{ ohms} \tag{2D.7}$$

$$R_L = 125 \text{ ohms} \tag{2D.8}$$

$$R_s = 50 \text{ ohms} \tag{2D.9}$$

$$\text{Line length} = 15 \text{ cm} \tag{2D.10}$$

$$\text{Speed of light in TL material} = 1.5 \times 10^8 \text{ m/s} \tag{2D.11}$$

Using these parameters, we conclude that

$$a(t) = 1u(t)(75)/(75+50) = 0.6\,u(t) \text{ volts} \tag{2D.12}$$

$$\rho_s = (50-75)/(50+75) = -0.2 \tag{2D.13}$$

$$\rho_L = (125-75)/(125+75) = 0.25 \tag{2D.14}$$

$$\tau = 0.15/1.5 \times 10^8 = 1 \text{ ns} \tag{2D.15}$$

Next, we use these values to substitute in the bounce diagram of Figure 2D.4 and extract the essential information from Table 2D.1, namely, the TDR (time-domain reflectometry) signal (appearing at the source end, and TDT (time-domain transmission) signal appearing at the load end. The results are shown below:

	Time-domain series
@Source end (TDR)	$0.6u(t)+0.12u(t-2)-0.006u(t-4)+0.0003u(t-6)$
@Load end (TDT)	$0.75u(t-1)-0.0375u(t-3)+0.001875u(t-5)+\cdots$

Table 2D.3 Time-Domain Total Voltage Expressions at the Source and Load Ends for the Step Waveform Excitation of the Example of Section 2D.4

Consequently, the TDR and TDT waveforms are as shown in Figure 2D.5.

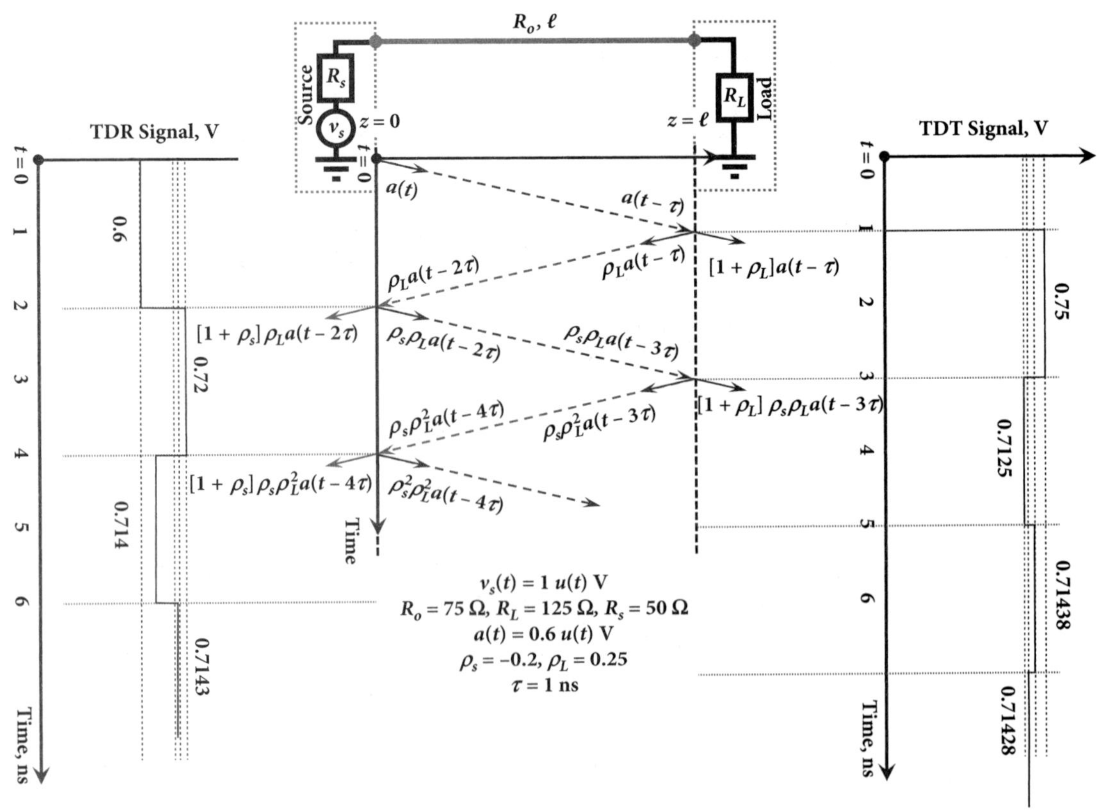

Figure 2D.5 The bounce diagram and voltage waveforms for the step waveform excitation of the example of Section 2D.4.

It is worth noting that both TDR and TDT signals converge to the same amplitude. As time approaches infinity, the "final value theorem" tells us that the solution is the same as the DC solution. For DC, the TL acts like a zero length wire, and hence both voltages at the sending and receiving ends are the same as the voltage divider between the load and source impedances:

$$v_{t=\infty} = V_s R_L/(R_s + R_L) = (1)\cdot(125)/(50+125) = 0.7142857\ \text{V} \tag{2D.16}$$

2D.5 Time-Domain Reflectometry for Ideal Dirac-Delta Impulse Waveform Excitations

To demonstrate, let us consider the same example we just did for the step waveform while considering $v_s(t) = 1\ \delta(t)$ volts, hence $a(t) = 0.6\ \delta(t)$ volts and the expressions for the TDR and TDT signals becomes:

	Time-domain series
@Source end (TDR)	$0.6\delta(t)+0.12\delta(t-2)-0.006\delta(t-4)+0.0003\delta(t-6)$
@Load end (TDT)	$0.75\delta(t-1)-0.0375\delta(t-3)+0.001875\delta(t-5)+\cdots$

TABLE 2D.4 Time-Domain Total Voltage Expressions at the Source and Load Ends for the Dirac-Delta Excitation of the Example of Section 2D.5

Consequently, the TDR and TDT waveforms are as shown in Figure 2D.6.

The final value in this case is a zero since $\delta(t)$ vanishes when t approaches infinity.

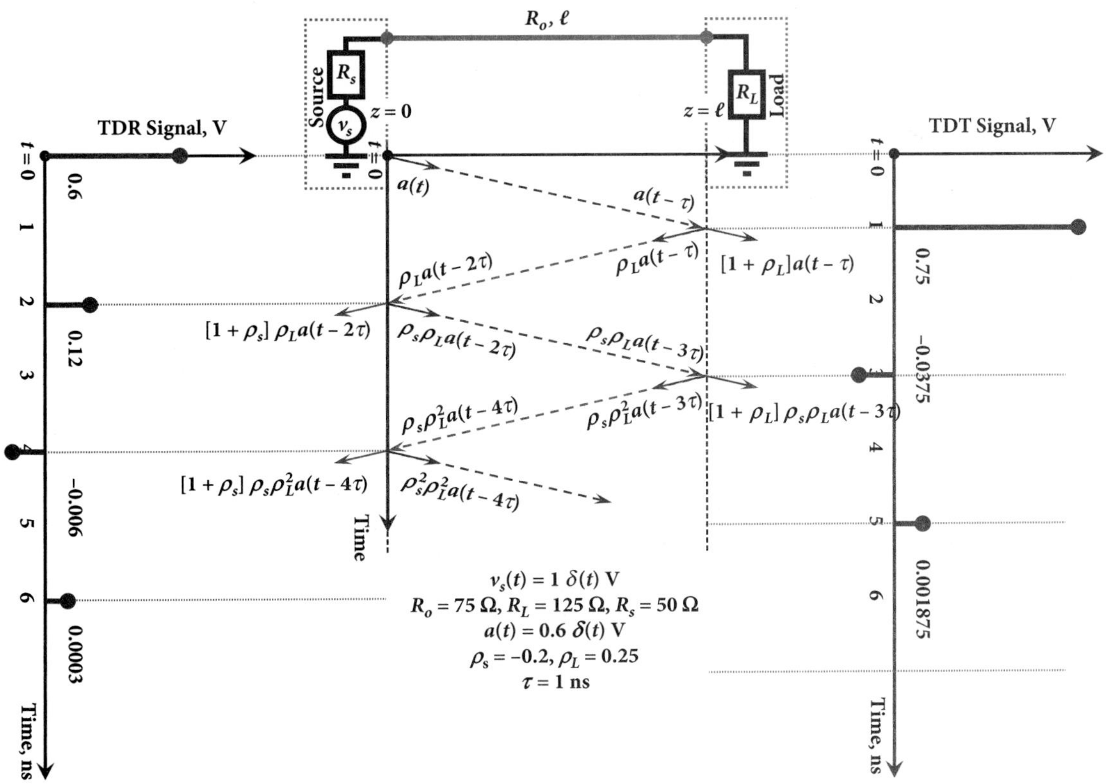

FIGURE 2D.6 The bounce diagram and voltage waveforms for the Dirac-delta excitation of the example of Section 2D.5.

ADDENDUM 2E
THE SMITH CHART

2E.1 Introduction

It was evident from the presentation of the previous sections that the impedance analysis is of prime importance in studying TL performance. However, the corresponding equations looked a bit complicated and are not easy to visualize. TL and RF circuit designers need a tool to enable visual assessment and prediction of problem areas and assessing impedance levels to work on impedance matching problems; such a tool exists and is known as the Smith chart.

The Smith chart is a graphical tool that maps the impedance values versus the corresponding reflection coefficient values. Since both are complex numbers, we don't expect a single trace of one versus the other, instead, the Smith chart is composed of multiple traces with parametric variables.

The Smith chart is simply the graph that describes the mathematical relationship between the normalized impedance $Z(z \text{ or } d)/Z_o$ and $\Gamma(z \text{ or } d)$. Using Equation (2.56), we can write

$$Z_n(d)=\frac{Z(d)}{Z_o}=\frac{[1+\Gamma(d)]}{[1-\Gamma(d)]} \quad \text{and} \quad \Gamma(d)=\frac{[Z_n(d)-1]}{[Z_n(d)+1]} \tag{2E.1}$$

Although this equation can be written with either (d) or (z) as the distance parameter, we chose to use the (d) as it is typical to carry most Smith chart procedures with the (d) parameter. Spelling out all complex quantities in terms of their real and imaginary components, Equation (2E.1) takes the form

$$Z_n(d)=\frac{R(d)+jX(d)}{Z_o}=r(d)+jx(d) \tag{2E.2}$$

$$\Gamma(d)=\Gamma_{\text{re}}(d)+j\Gamma_{\text{im}}(d) \tag{2E.3}$$

By dropping the (d) notation for convenience,

$$r+jx=\frac{[1+\Gamma_{\text{re}}+j\Gamma_{\text{im}}]}{[1-\Gamma_{\text{re}}-j\Gamma_{\text{im}}]} \tag{2E.4}$$

Equating both the real and imaginary parts of the above equation yields

$$r=\frac{[1-(\Gamma_{\text{re}})^2-(\Gamma_{\text{im}})^2]}{[(1-\Gamma_{\text{re}})^2+(\Gamma_{\text{im}})^2]} \tag{2E.5}$$

$$x=\frac{[2\Gamma_{\text{im}}]}{[(1-\Gamma_{\text{re}})^2+(\Gamma_{\text{im}})^2]} \tag{2E.6}$$

Manipulation of these two equations yields the following convenient forms:

$$\left[\Gamma_{\text{re}}-\frac{r}{1+r}\right]^2+[\Gamma_{\text{im}}]^2=\left[\frac{1}{1+r}\right]^2 \tag{2E.7}$$

$$[\Gamma_{\text{re}}-1]^2+\left[\Gamma_{\text{im}}-\frac{1}{x}\right]^2=\left[\frac{1}{x}\right]^2 \tag{2E.8}$$

Both equations describe families of circles on a plane formed by Γ_{re} and Γ_{im} as its horizontal and vertical axes.

The general circle equation centered at $(\Gamma_{\text{re},o}, \Gamma_{\text{im},o})$ with a radius Γ_{radius} in this plane takes the form:

$$[\Gamma_{\text{re}}-\Gamma_{\text{re},o}]^2+[\Gamma_{\text{im}}-\Gamma_{\text{im},o}]^2=[\Gamma_{\text{radius}}]^2 \tag{2E.9}$$

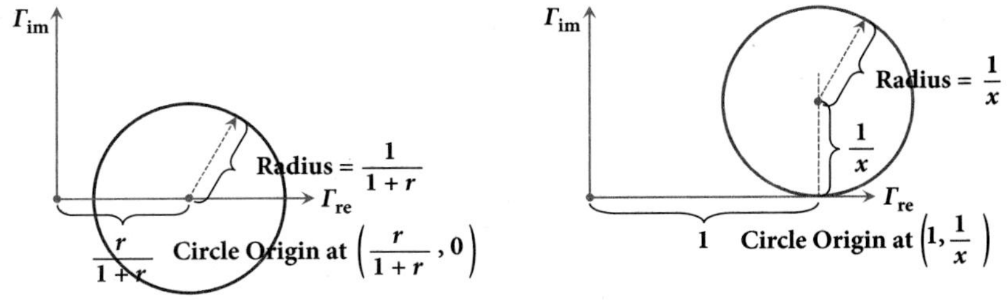

Figure 2E.1 The basic concept of the Smith chart.

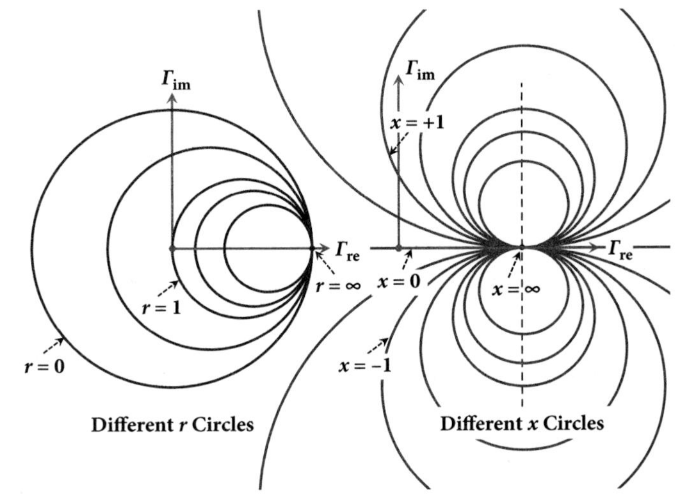

Figure 2E.2 The different r and x circles of the Smith chart.

Hence, for the first Equation (2E.7), the r circles will have their origins located at $\left(\frac{r}{1+r}, 0\right)$ and the corresponding radius is given by $\left(\frac{1}{1+r}\right)$, while for the x circles, Equation (2E.8), the origins are located at $\left(1, \frac{1}{x}\right)$ and the corresponding radius is given by $\left(\frac{1}{x}\right)$ (Figure 2E.1). Figure 2E.2 shows sets of these circles for various values of r and x.

In the left part of Figure 2E.3, both the r and x circles are assembled on the same chart. Finally, the chart on the right is the familiar Smith chart that appears in most prints. In that chart, the parts corresponding to $r < 0$ are eliminated.

2E.2 Scales on the Smith Chart

There are several scales typically included in a printed Smith chart that can be used to facilitate data entry and extraction. In this section, we will discuss the main scales, which are displayed in Figure 2E.4.

2E.2.1 The Magnitude of the Reflection Coefficient ($|\Gamma|$) Scale

The magnitude of Γ is the length of the phasor originating from the origin of the chart and hence is the radius of the circle centered at the origin and passes by the point representing the phasor. The scale for $|\Gamma|$ is typically positioned on the bottom left of the chart with its zero aligned vertically with the chart origin and has the same "unit length" as that of the "$r = 0$" circle. (Note that $r = 0$ implies $|\Gamma| = 1$.)

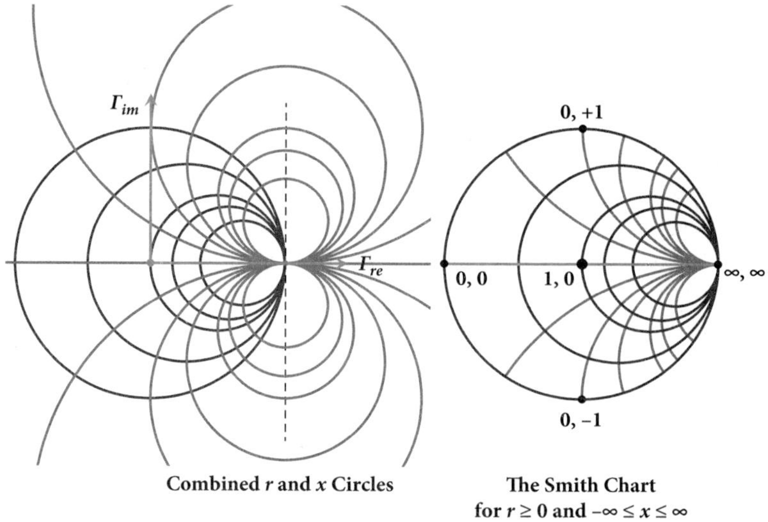

FIGURE 2E.3 The combined r and x circles of the Smith chart.

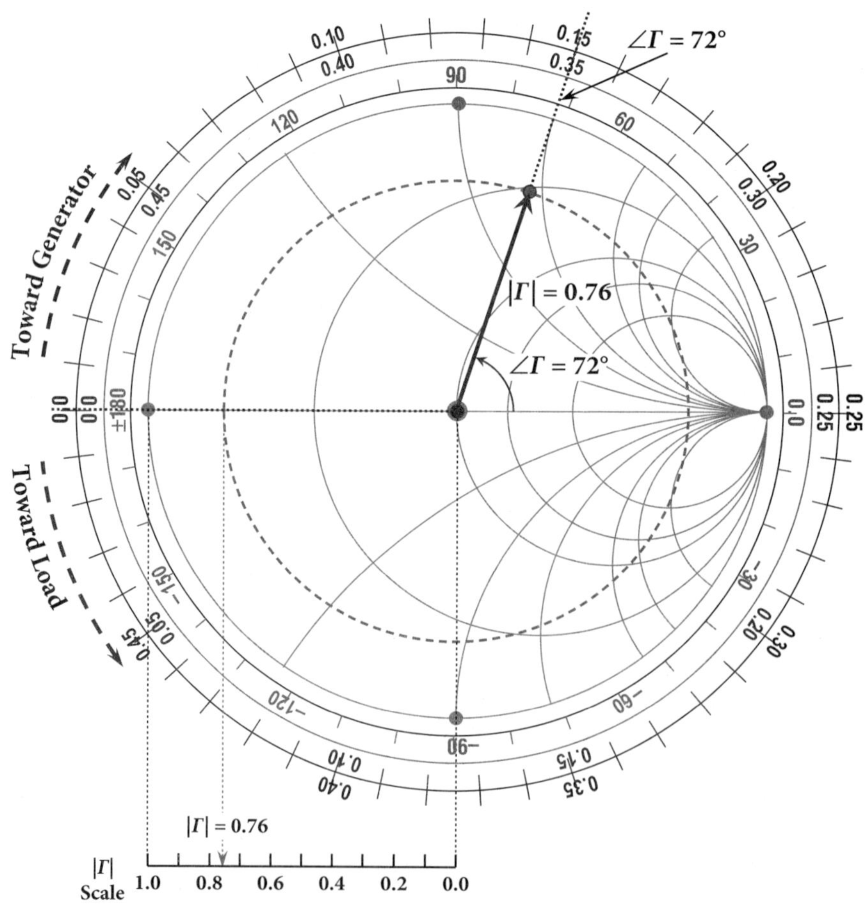

FIGURE 2E.4 Smith chart scales: Γ magnitude and phase and normalized distance.

To demonstrate the use of this scale, an example of $|\Gamma| = 0.76$ is shown in the figure. From the location of $|\Gamma| = 0.76$ on the scale a vertical line is drawn till it intercepts the horizontal axis of the chart. Next, a circle (centered at the origin) is drawn to pass by the point of interception. Consequently, the radius of this circle is 0.76, and all points on this circle will have $|\Gamma| = 0.76$.

2E.2.2 The Phase Angle of the Reflection Coefficient ($\angle\Gamma$) Scale

The phase angle of Γ is the angle the Γ phasor makes with the "horizontal" Γ_{re} axis. The scale for $\angle\Gamma$ is positioned on the outer perimeter of a circle outside the "$r = 0$" circle of the chart. The scale is calibrated from −180° to +180° with its zero aligned with the positive side of the horizontal axis of the chart.

To demonstrate the use of this scale, assume a phase angle of +72° for the $|\Gamma| = 0.76$ example discussed earlier. As shown in the figure, a radial line is placed on the chart for the origin to the +72° mark on the scale. The point of intercept represents the Γ phasor $0.76\angle{+72°}$.

2E.2.3 Normalized Distance Moved Scale

This scale has to do with the relative locations of the Γ phasors corresponding to different locations on the line. Referring to Figure 2E.4, $\Gamma(z_2)$ and $\Gamma(z_1)$ can be related to each other through the use of Equation (2.53):

$$\Gamma(z_1)=\frac{V^-e^{+\gamma z_1}}{V^+e^{-\gamma z_1}}=\Gamma(0)\cdot e^{+2\gamma z_1} \tag{2E.10}$$

$$\Gamma(z_2)=\Gamma(0)\cdot e^{+2\gamma z_2}=\Gamma(z_1)\cdot e^{+2\gamma(z_2-z_1)} \tag{2E.11}$$

$$\text{Hence,} |\Gamma(z_2)|=|\Gamma(z_1)|e^{+2\alpha\,(z_2-z_1)} \text{ and } \angle\Gamma(z_2)=\angle\Gamma(z_1)+2\beta(z_2-z_1)$$

$$|\Gamma(z_2)|=|\Gamma(z_1)|e^{+2\alpha(z_2-z_1)} \text{ and } \angle\Gamma(z_2)=\angle\Gamma(z_1)+4\pi\cdot\left[\frac{(z_2-z_1)}{\lambda}\right] \tag{2E.12}$$

Consequently, moving on the line from z_1 to z_2 implies a change of the angle of the phasor (rotation) by $4\pi\cdot[(z_2-z_1)/\lambda]$. If $z_2 > z_1$, the change is an increase in the angle and hence a counterclockwise rotation, and for $z_2 < z_1$, the rotation is clockwise. One can use the phase angle scale to enter these rotations on the chart, however, two normalized distance scales are given to facilitate this process without multiplying the 4π by the $[(z_2 - z_1)/\lambda]$. The normalized distance scales provide direct calibration of the $[(z_2 - z_1)/\lambda]$ quantity. One of these scales increases counterclockwise to enable tracking "increasing z" locations as we move *toward* the load, while the other scale increases clockwise for tracking "decreasing z" locations (increasing d) *toward* the source (generator). Both scales are provided along the perimeter of a circle located outside the phase angle scale with the "TOWARD GENERATOR (WTG)" scale marked on the outer side of the circle and the "TOWARD LOAD (WTL)" scale marked on the inner side of the circle.

Both scales carry tick marks from 0 to 0.5 in a full circle, 360°. This is because a $\lambda/2$ distance change corresponds to $4\pi\cdot[(z_2-z_1)/\lambda]=2\pi$ which is 360°.

2E.3 Transmission Line Trace on the Smith Chart

The Smith chart is a graphical representation that can enable us to monitor the TL performance parameters in terms of its $\Gamma(z)$ and $Z(z)$ at different location of the line. Referring to Figure 2E.5, we will demonstrate how the chart can be used to achieve this objective.

Since the chart is a plot of the Γ phasor, then tracking the line trace implies tracing its $\Gamma(z)$ on the chart. Using Equation (2E.11), we can write the expression for the $\Gamma(z)$ in terms of $\Gamma(0)$ and $\Gamma(\ell)$ as

$$\Gamma(z)=\Gamma(0)\cdot e^{+2\gamma z}=\Gamma(0)\cdot e^{+2\alpha z}e^{+2j\beta z}=\Gamma(0)\cdot e^{+2\alpha z}e^{+j4\pi\cdot(z/\lambda)} \tag{2E.13}$$

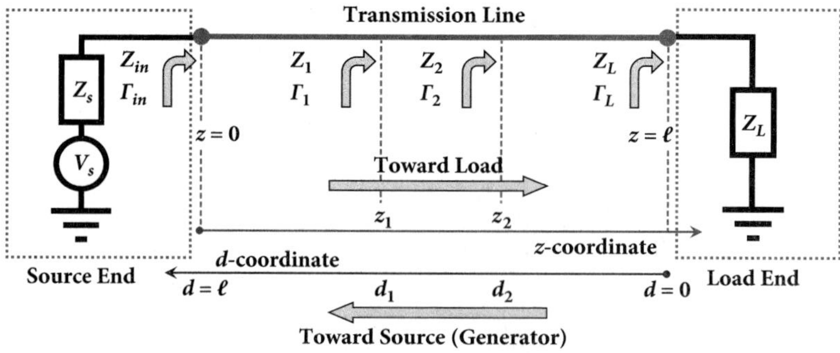

FIGURE 2E.5 Demonstration of TL circuit labels for tracking Γ and Z values on the line using the Smith chart.

and,

$$\Gamma(z)=\Gamma_L\cdot e^{-2\gamma(\ell-z)}=\Gamma_L\cdot e^{-2\alpha(\ell-z)}e^{-j4\pi\cdot[(\ell-z)/\lambda]} \tag{2E.14}$$

The latter can be rewritten as a function of d (the distance from the load end), using $d=\ell-z$:

$$\Gamma(d)=\Gamma_L\cdot e^{-2\gamma d}=\Gamma_L\cdot e^{-2\alpha d}e^{-2j\beta d}=\Gamma_L\cdot e^{-2\alpha d}e^{-j4\pi\cdot(d/\lambda)} \tag{2E.15}$$

Equations (2E.13) and (2E.15) can be rewritten in terms of their magnitude and phase components:

$$|\Gamma(z)|=|\Gamma(0)|e^{+2\alpha z} \quad \text{and} \quad \angle\Gamma(z)=\angle\Gamma(0)+4\pi\cdot(z/\lambda) \tag{2E.16}$$

and

$$|\Gamma(d)|=|\Gamma_L|e^{-2\alpha d} \quad \text{and} \quad \angle\Gamma(d)=\angle\Gamma_L-4\pi\cdot(d/\lambda) \tag{2E.17}$$

Equations (2E.16) and (2E.17) are telling us that the magnitude of the Γ phasor changes with position on the line depending on the attenuation coefficient, α. They also show us the continuous (linear) phase change of the phasor.

2E.3.1 Case of Lossless TL, $\alpha = 0$

In this case, and as can be shown from Equations (2E.16) and (2E.17), the magnitude of the reflection coefficient is the same (constant) at all line locations:

$$|\Gamma(z)|=|\Gamma(0)|=|\Gamma_L| \tag{2E.18}$$

and, hence, the trace of $|\Gamma(z)|$ is a circle centered at the origin and has a radius of $|\Gamma(0)| = |\Gamma_L|$. On the other hand, the phase angle increases as we move toward the load and decreases with motion toward the source. Consequently, the left chart of Figure 2E.6 displays the trace of $\Gamma(z)$ starting at $z = 0$, moving toward the load, passing by points z_1 and z_2 till the end of the line at $z = \ell$. The phase shifts from $z = 0$ to $z = z_1$, $z = z_2$, and finally $z = \ell$ are represented by angles of counterclockwise rotations starting at the phase at $z = 0$ in the amounts 4π (z_1/λ), 4π (z_2/λ), 4π (ℓ/λ), respectively. These rotation angles correspond to rotations of (z_1/λ), (z_2/λ), and (ℓ/λ), respectively, on the counterclockwise normalized distance WTL scale.

Likewise, the right chart in Figure 2E.6 demonstrates the same line but starting the trace at the load end on the TL. The trace is thus starting at $d = 0$ ($z = \ell$), and moves toward the source, passing by points d_2 (or z_2) and d_1 (or z_1) till the source end of the line at $d = \ell$ ($z = 0$). The phase shifts from $d = 0$ to $d = d_2$, $d = d_1$, and finally $d = \ell$ are represented by angles of clockwise rotations in the amounts 4π (d_2/λ), 4π (d_1/λ), 4π (ℓ/λ), respectively. Again, these rotation angles correspond to rotations of (d_2/λ), (d_1/λ), and (ℓ/λ), respectively, on the clockwise normalized distance WTG scale.

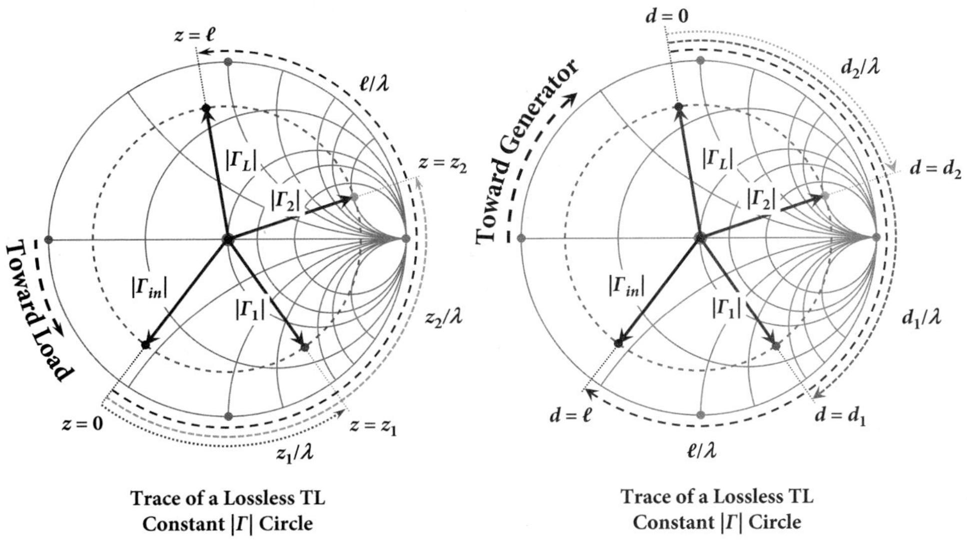

FIGURE 2E.6 Tracking Γ and Z values on a lossless TL using the Smith chart.

Example 2E.1 A load impedance of 40 + j70 terminates a 100-Ω TL that is 0.3λ long. Find the reflection coefficient at the load, the input reflection coefficient, the input impedance, the SWR on the line, the return loss. The distance to the first voltage minimum and maximum from the load.

$$Z'_L = \frac{Z_L}{Z_o} = \frac{40 + j70}{100} = 0.4 + j0.7$$

- Locate the normalized load impedance on the Smith chart, point A, draw the constant SWR circle, with center at the origin.
- Project the SWR circle on the Γ scale to find the magnitude of the reflection coefficient at the load. $\Gamma_L = 0.59$.
- Extent the line connecting the center of the chart and the normalized impedance point to the Γ angle scale to read the angle of reflection coefficient *angle* = $\angle 104°$.
- On the constant SWR circle, move a distance of 0.3λ toward generator, WTG, to point B, read normalize impedance and multiply by Z_o.
- $Z_{\text{in}} = (0.365 - j0.61) \times 100 = (36.5 - j61.1)\ \Omega \quad \Gamma_{\text{in}} = 0.59\angle -112°$.
- Input load admittance is point E, diagonally across from point B on the SWR circle.
- $Y'_{\text{in}} = 0.72 + j1.2 \rightarrow Y_{\text{in}} = \dfrac{Y'_{\text{in}}}{Z_o} = \dfrac{0.72 + j1.2}{100} = (0.0072 + j0.0012\,)\,\Omega^{-1}$.
- Distance to the first voltage maximum = from point A to point C, read on the WTG scale.
- $d_{\text{max}} = 0.25\lambda - 0.106\lambda = 0.144\lambda$.
- The value of the real impedance at the intersection of the constant SWR circle with the real axis is the VSWR = 3.87.
- Distance to the first voltage minimum = From point A to point D, read on the wavelength toward generator scale.
- $d_{\text{min}} = 0.5\lambda - 0.106\lambda = 0.394\lambda$.

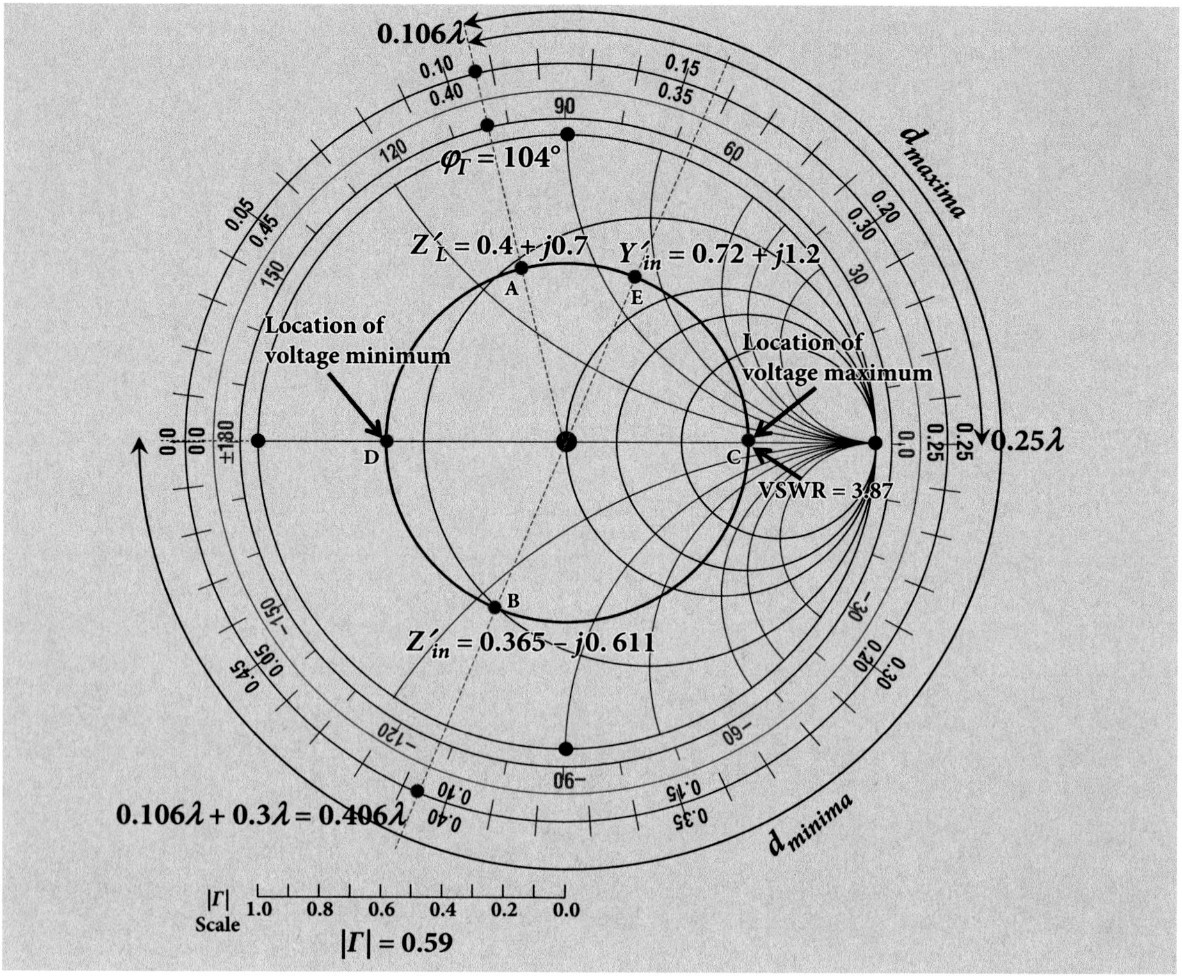

2E.3.2 Case of Lossy TL, $\alpha \neq 0$

In this case, and as can be shown from Equations (2E.16) and (2E.17), the magnitude of the reflection coefficient varies with position, and hence, the $\Gamma(z)$ is no longer a circle of a constant radius. Varying the magnitude while rotating the phase results in a spiral trace as demonstrated in Figure 2E.7. The prescribed spiral pitch depends on the value of the attenuation coefficient. For a low loss line, the pitch is small while it becomes large for larger attenuation coefficients. The spiral trace is inward as we move toward the source while appearing outward for motions toward the load.

Figure 2E.7 demonstrates the same TL traces discussed earlier in Figures 2E.5 and 2E.6 in the presence of line loss, $\alpha \neq 0$.

2E.4 How Does the Smith Chart Work?

2E.4.1 Starting with $Z(z \text{ or } d)$

In this case, we assume that the driving point impedance is known at some point on the line located at a distance z from its source end (distance d from its load end). The objective is to determine the frequency-domain reflection

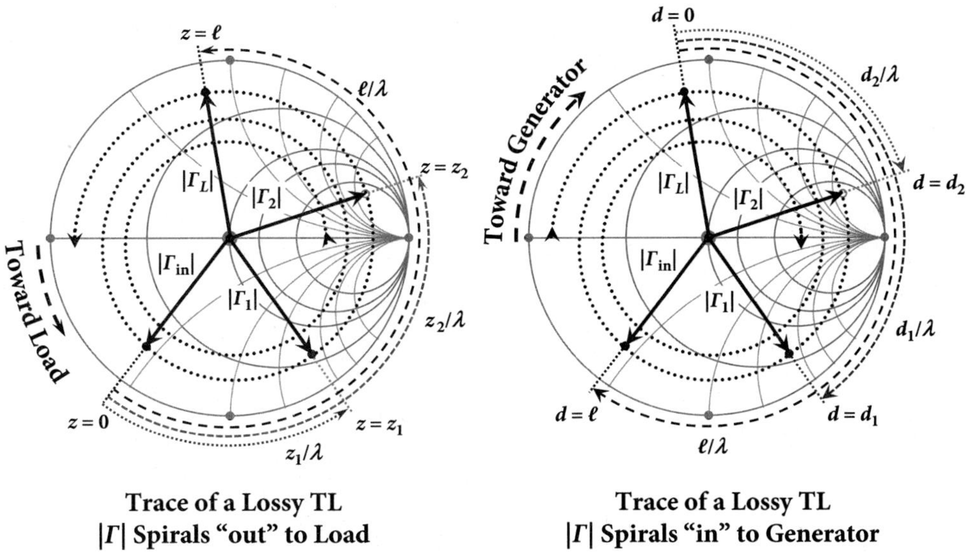

FIGURE 2E.7 Tracking Γ and Z values on a lossy TL using the Smith chart.

coefficient at that location. The mathematical expression to be used in this case if we were to do the calculations mathematically is that of Equation (2.56 or 2E.1):

$$\Gamma(z \text{ or } d) = \frac{[Z(z \text{ or } d) - Z_o]}{[Z(z \text{ or } d) + Z_o]} \tag{2E.19}$$

The Smith chart (graphical) procedure goes as follows, refer to Figure 2E.8:

1. Choose a normalization impedance, Z_o, for the problem at hand. Typically, the characteristic impedance of the *dominant* TL of the circuit is a preferred choice.
2. Compute the normalized impedance by dividing $\dfrac{Z(z \text{ or } d)}{Z_o} = Z_n(z \text{ or } d) = r + jx$.
3. Locate the r circle for the obtained r value.
4. Locate the x circle for the obtained x value.
5. The intersection of the two circles defines the $\Gamma(z \text{ or } d)$ point.
6. The line connecting the origin to the $\Gamma(z \text{ or } d)$ point is the phasor representing $\Gamma(z \text{ or } d)$.
7. The length of the $\Gamma(z \text{ or } d)$ phasor is the magnitude of $\Gamma(z \text{ or } d)$.
8. Draw a circle centered at the origin passing by the Γ point and project the radius of this circle on the $|\Gamma|$ scale to read the magnitude of Γ.
9. The angle the $\Gamma(z \text{ or } d)$ phasor makes with the horizontal (real axis) is the angle of the $\Gamma(z \text{ or } d)$ phasor. Read the phase angle of Γ on the provided angle scale.
10. In rectangular format, the projections of the $\Gamma(z \text{ or } d)$ phasor on the horizontal and vertical axes are the real and imaginary parts of the $\Gamma(z \text{ or } d)$ phasor, respectively.

2E.4.2 Starting with the $\Gamma(z \text{ or } d)$ Phasor

In this case, we assume that the frequency-domain reflection coefficient is known at some point on the line located at a distance z from its source end (distance d from its load end). The objective is to determine the

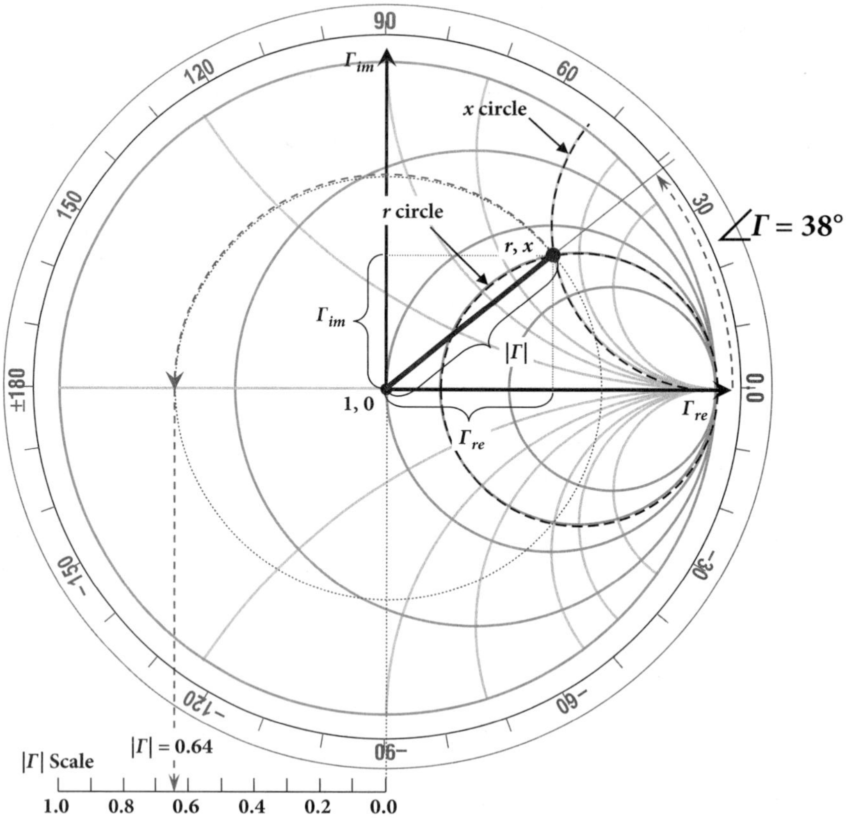

FIGURE 2E.8 The use of the Γ magnitude and phase scales on the Smith chart.

driving point impedance of the line at that location. The mathematical expression to be used in this case if we were to do the calculations mathematically is that of Equation (2.56 or 2E.1):

$$Z(z \text{ or } d) = Z_o \frac{[1 + \Gamma(z \text{ or } d)]}{[1 - \Gamma(z \text{ or } d)]} \tag{2E.20}$$

The Smith chart (graphical) procedure goes as follows, again, refer to Figure 2E.8:

1. The normalization impedance Z_o must be given to us along with the $\Gamma(z \text{ or } d)$.
2. Locate the $\Gamma(z \text{ or } d)$ phasor (magnitude and phase) or real and imaginary.
3. For $|\Gamma|$, Γ_{re}, and Γ_{im} use the $|\Gamma|$ scale provided to enter the proper line lengths to be graphed on the chart. For phase, use the phase angle scale to enter the $\angle\Gamma$.
4. Locate the r circle that passes by the $\Gamma(z \text{ or } d)$ phasor point—read the value of r.
5. Locate the x circle that passes by the $\Gamma(z \text{ or } d)$ phasor point—read the value of x.
6. Compose the obtained r and x values to obtain a numerical value for the complex normalized impedance $Z_n(z \text{ or } d) = Z/Z_o = r + jx$.
7. Denormalize $Z_n(z \text{ or } d)$ to compute the actual driving point impedance of the line $Z(z \text{ or } d)$ at the position (z or d). To denormalize, you multiply the normalized impedance by the given normalization impedance Z_o. Thus, $Z(z \text{ or } d) = Z_n(z \text{ or } d) \cdot Z_o$.

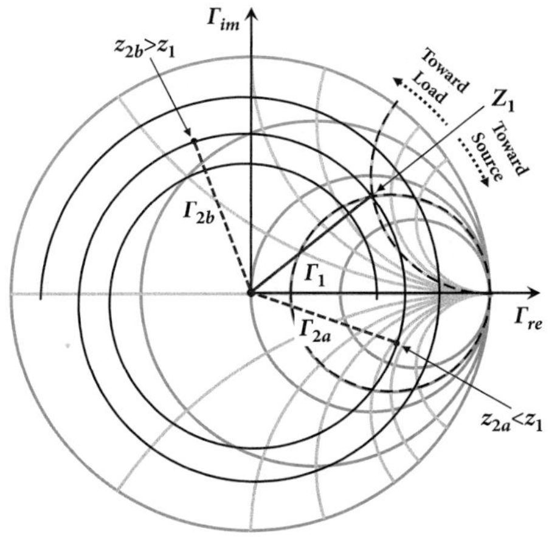

FIGURE 2E.9 Finding Γ_2 and Z_2 knowing Γ_1 and Z_1 for lossy TL using the Smith chart.

2E.4.3 Finding $\Gamma(z_2$ or $d_2)$ [and $Z(z_2$ or $d_2)$] Knowing $\Gamma(z_1$ or $d_1)$ or $Z(z_1$ or $d_1)$

In this case, we assume that line information (either Z or Γ) is known at some point on the line (e.g., z_1 or d_1). The objective is to determine the corresponding Z and Γ at another point on the line (e.g., z_2 or d_2). The mathematical expressions for this case was given earlier in Equation (2E.12), that is,

$$|\Gamma(z_2)| = |\Gamma(z_1)|e^{+2\alpha(z_2-z_1)} \text{ and } \angle\Gamma(z_2) = \angle\Gamma(z_1) + 4\pi\cdot[(z_2-z_1)/\lambda] \quad (2E.12)$$

Once $\Gamma(z_2)$ is determined, we would use Equation (2.56) to get $Z(z_2)$.

The Smith chart (graphical) procedure is illustrated in Figure 2E.9:

1. Locate the line position z_1 (or d_1) on the chart though the Γ phasor or impedance Z information as demonstrated earlier.
   - Moving toward the load, $z_2 > z_1$, then according to Equation (2E.12), $|\Gamma(z_2)| > |\Gamma(z_1)|$ and $\angle\Gamma(z_2) = \angle\Gamma(z_1)$.
     - The corresponding trace for line points on the chart is a counterclockwise growing spiral. The spiral radius growth is determined by the exponent $e^{+2\alpha(z_2-z_1)}$. This computation is to be performed manually since the spiral is not part of the chart.
   - However, if $z_2 < z_1$ (or $d_2 > d_1$), that is, we move toward the source (generator), the corresponding motion on the (same) spiral would be in the opposite direction.
2. Compute the normalized distance of rotation $[(z_2 - z_1)/\lambda]$ (or $[(d_2 - d_1)/\lambda]$ and enter this amount on the chart using the normalized distance scale starting at point z_1 to locate point z_2.
3. Read $\Gamma(z_2)$ (and $Z_{n2} = r_2 + jx_2$).

2E.4.4 Special Case (Lossless TL), $\alpha = 0$

This case is similar to the case above except for having a constant radius circle replacing the spiral trace. Figure 2E.10 demonstrates the graphical procedure on the chart in a similar manner.

2E.5 The Admittance Smith Chart

At high frequencies, it is often preferred to work with admittances instead of impedances. One reason is that most circuit components are mounted in parallel with one terminal grounded. Series components are floating devices and would lead to the circuit suffering from additional undesired parasitic. Working with shunt devices makes admittance expressions more attractive since combining admittances in parallel is achieved through simple addition of their values.

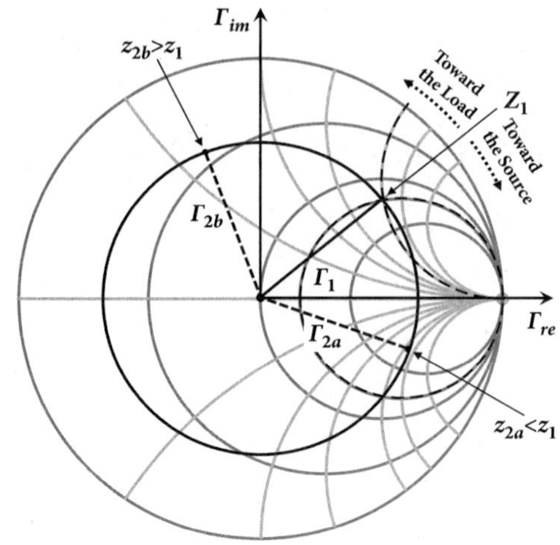

FIGURE 2E.10 Finding Γ_2 and Z_2 knowing Γ_1 and Z_1 for lossless TL using the Smith chart.

Examining Expression (2E.1), we can rewrite it in terms of admittance components as follows:

$$Z_n(z \text{ or } d) = \frac{Z(z \text{ or } d)}{Z_o} = \frac{[1+\Gamma(z \text{ or } d)]}{[1-\Gamma(z \text{ or } d)]} \text{ and } \Gamma(z \text{ or } d) = \frac{[Z_n(z \text{ or } d)-1]}{[Z_n(z \text{ or } d)+1]} \tag{2E.1}$$

$$Y_n(z \text{ or } d) = \frac{Y(z \text{ or } d)}{Y_o} = \frac{1}{Z_n(z \text{ or } d)} = \frac{[1-\Gamma(z \text{ or } d)]}{[1+\Gamma(z \text{ or } d)]} \text{ and } -\Gamma(z \text{ or } d) = \frac{[Y_n(z \text{ or } d)-1]}{[Y_n(z \text{ or } d)+1]} \tag{2E.21}$$

Comparing Equations (2E.1) and (2E.21), we notice the similarity except for replacing the Z_n by Y_n and at the same time replacing Γ by $-\Gamma$. Likewise, if we express Y_n in terms of its real and imaginary conductance g_n and susceptance b_n, $Y_n = g_n + jb_n$, we can show that the constant g and b circles are of the form:

$$\left[-\Gamma_{\text{re}} - \frac{g}{1+g}\right]^2 + [-\Gamma_{\text{im}}]^2 = \left[\frac{1}{1+g}\right]^2 \tag{2E.22}$$

$$[-\Gamma_{\text{re}} - 1]^2 + \left[-\Gamma_{\text{im}} \frac{1}{b}\right]^2 = \left[\frac{1}{b}\right]^2 \tag{2E.23}$$

which are, again, similar to Equations (2E.7) and (2E.8) by replacing r and x by g and b and at the same time replacing Γ_{re} and Γ_{im} by $-\Gamma_{\text{re}}$ and $-\Gamma_{\text{im}}$, respectively.

The conclusion is that for a given Γ phasor on the chart (corresponding to a given value of $Z_n = r + jx$), the $-\Gamma$ phasor on the same chart corresponds to the $Y_n = g + jb$ value at that location on the line.

This concept is demonstrated in the chart of Figure 2E.11. The Γ phasor is shown for the normalized impedance $Z_n = 01.5 + j2.0$. Thus, the corresponding $-\Gamma$ phasor reveals to us the corresponding Y_n, which is $Y_n = 0.24 - j0.32$. (It is interesting to observe that the chart can be used as a complex number inverter since $Y_n = 1/Z_n$. Hence, if you place a complex number as the Z_n, the corresponding Y_n point is the inverse of that complex number.)

Now that we identified the placement of admittances on the chart, we can actually work out series and parallel combinations in conjunction with segments of TLs in a relatively convenient way. To demonstrate, let us consider the example of Figure 2E.12.

The figure shows a two-TL segments TL_1 and TL_2, both having the same Z_o and γ but different lengths. The two lines are connected in cascade with a shunt impedance Z_p in between. TL_1 is terminated in Z_L while a series impedance Z_s is connected at the input side of TL_2. It is desired to find the combined circuit input impedance Z_{in} and reflection coefficient phasor Γ_{in}.

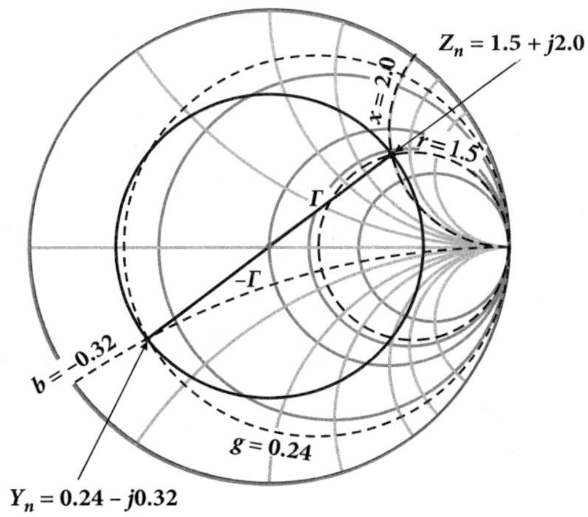

FIGURE 2E.11 Admittance representation on the Smith chart.

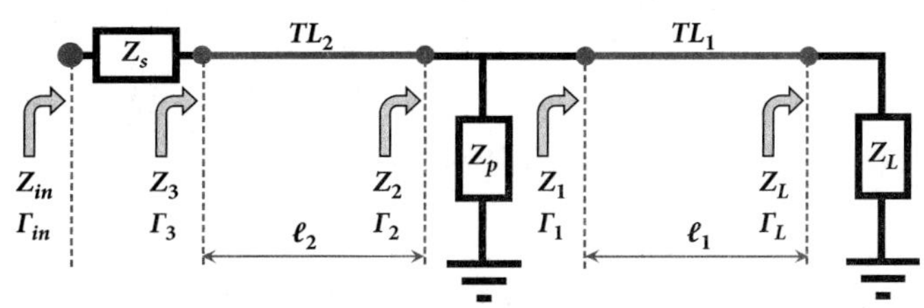

FIGURE 2E.12 TL example requiring impedance-admittance conversions for Smith chart analysis.

To solve this problem analytically, we would start with TL_1 terminated in Z_L. Using Equation (2A.3), we obtain the input impedance to that segment, Z_1. Next, we add Z_p in parallel to Z_1 to get Z_2. Again, we use Equation (2A.3) for TL_2 terminated in Z_2 as a load and we get its input impedance Z_3, and finally, add Z_s in series to get Z_{in}. To get Γ_{in}, we use Equation (2.56).

To do this problem using the Smith chart, we follow the procedure outlined below and demonstrated in Figure 2E.13.

1. Choose the normalization impedance to be Z_o of the TLs.
2. Compute the normalized $Z_{Ln} = Z_L/Z_o$.
3. Locate Z_{Ln} on the chart and identify the Γ_L phasor point [1].
4. Draw a circle centered at the origin passing by Γ_L.
5. Move on the circle of step 4 above "toward generator" a normalized distance of ℓ_1/λ. The new point represents Γ_1 [2]. You may read Z_{1n} and Γ_1 from this point.
6. Extend the Γ_1 phasor through the origin to reach the $-\Gamma_1$ point [3]. This is the Y_{1n} point. You may read Y_{1n} off this point.
7. Compute Y_{pn} as $1/Z_{pn}$ (or Z_o/Z_p).
8. Add Y_{pn} to Y_{1n} to obtain Y_{2n} Locate Y_{2n} on the chart [5]. You may read $-\Gamma_2$ off that point.
9. Draw a circle centered at the origin passing by $-\Gamma_2$.
10. Extend the $-\Gamma_2$ phasor through the origin to reach the $+\Gamma_2$ point [6]. This is the Z_{2n} point. You may read Z_{2n} and Γ_2 off this point.
11. Move on the circle of step 9 above "toward generator" a normalized distance of ℓ_2/λ. The new point represents Γ_3[7]. You may read Z_{3n} and Γ_3 from this point.

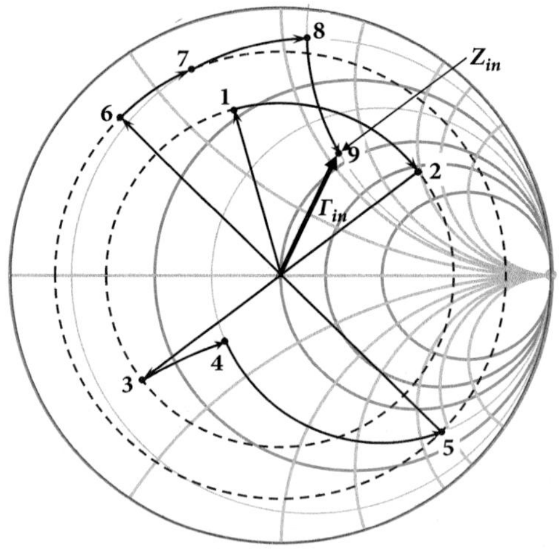

FIGURE 2E.13 Smith chart analysis of the example of Figure 2E.12 .

12. Compute Z_{sn} as Z_s/Z_o.
13. Add Z_{sn} to Z_{3n} to get Z_{in}. Locate Z_{in} on the chart and identify the Γ_{in} phasor point [9]. You may read Z_{inn} and Γ_{in} from this point.
14. Denormalize Z_{inn} to obtain $Z_{\text{in}} = Z_o * Z_{\text{inn}}$.

It is worth noting that the impedance and admittance additions in steps 8 and 13 above can be performed on the chart. This is simply done by the imaginary part first then the real part (or vice versa). To add the imaginary part, you move on a constant real part circle the proper amount that corresponds to the added imaginary part. Likewise, to add the real part, you move on a constant imaginary part, circle the proper amount that corresponds to the added real part. This is demonstrated in Figure 2E.13 in the steps 3–4, 4–5, 7–8, and 8–9.

Example 2E.2 A 100-Ω TL of length 1.3λ is terminated in an unknown load impedance. Measurements revealed that the SWR on the line is 3.5 and the first minima occurred at a distance of 0.38λ from the load. Find the load impedance, reflection coefficient at the load, the input reflection coefficient, and the input impedance.

- Locate the SWR value on the real *r* axis (point *C* on the chart).
- Draw the constant Γ circle centered at the origin and passes by the point *C*. This is the locus of Γ's and *Z*'s of the TL.
- Point *D* on this circle corresponds to a voltage minimum on the line. This point is 0.38λ away from the load. This point corresponds to 0λ on the distance scale of the chart.
- Move 0.38λ away from the point *D* toward the load (counterclockwise) to reach $0\lambda + 0.38\lambda = 0.38\lambda$ on the distance scale (the inner calibration on the distance scale circle). We arrive at point *A*. This point corresponds to the location of the load.
- Read the normalized load impedance on the Smith chart, point *A*, $Z_{L_n} = 0.5 + j0.8$.
- Project the SWR circle on the Γ scale to find the magnitude of the reflection coefficient at the load. $\Gamma_L = 0.55$.
- Extent the line connecting the center of the chart and the normalized impedance point to the Γ angle scale to read the angle of reflection coefficient angle = ∠94°.
- On the constant SWR circle, move a distance of 1.3λ (which is the same as 0.3λ) toward generator (clockwise) to arrive at point *B*. This point corresponds to the input impedance of the TL. Read the normalized impedance and multiply by Z_o.

- $Z_{\text{in}} = (0.365 - j0.5) \times 100 = (36.5 - j50)\ \Omega \quad \Gamma_{\text{in}} = 0.55\angle -122°$.
- The input load admittance is point *E*, diagonally across from point *B* on the SWR circle.
- $Y_{\text{in}_n} = 0.96 + j1.3 \rightarrow Y_{\text{in}} = \dfrac{Y_{\text{in}_n}}{Z_o} = \dfrac{0.96 + j1.3}{100} = (0.0096 + j0.0013)\ \Omega^{-1}$.

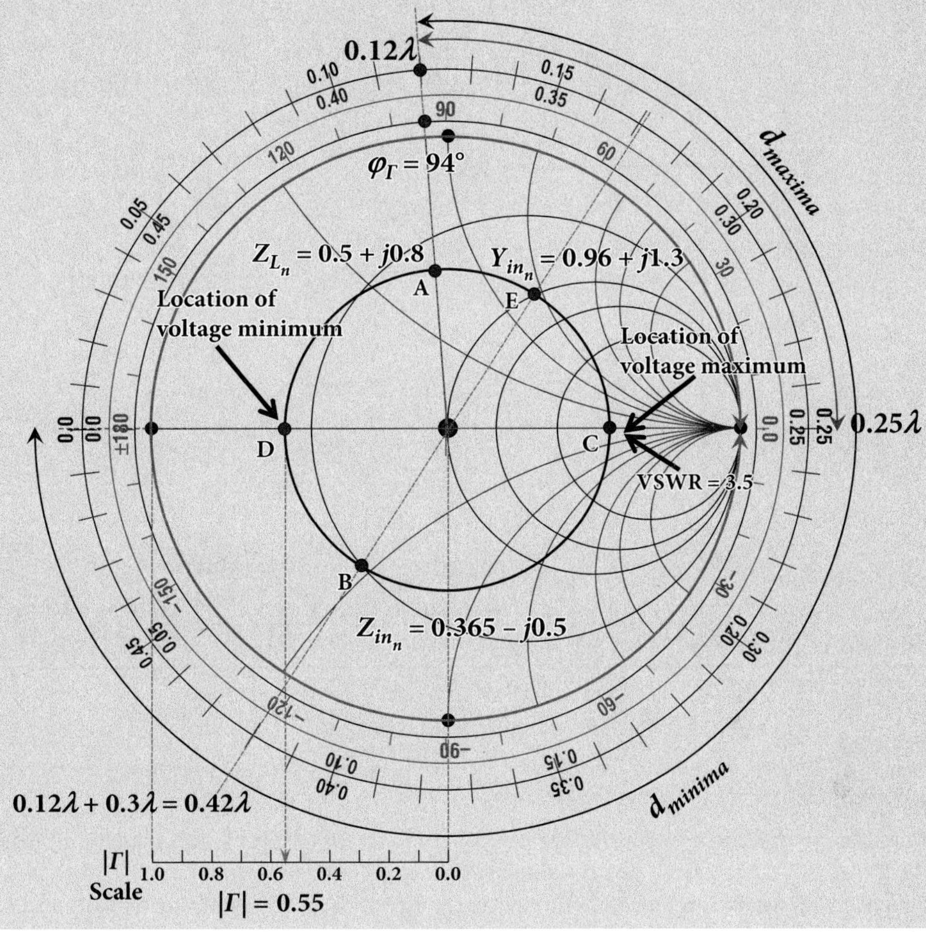

2E.6 Smith Chart Features and Short Cuts

1. The commonly known Smith chart prints display circles and portion of circles for the $r \geq 0$. Typically, $r < 0$ (negative resistance) occurs in TL circuits with active devices. For cases requiring $r < 0$ analysis, a computer produced Smith chart is typically used.
2. A few other scales, Figure 2E.14, are typically included in (commercially available) printed Smith charts; among them are
   a. Scale for the standing wave ratio. This scale goes from 1 to ∞ corresponding to the Γ range from 0 to 1, respectively.
   b. Scales for normalized voltages and currents. These scales correspond to $1 + \Gamma$ and $1 - \Gamma$. A magnitude scale for both is typically given the name "transmission coefficient," which ranges from 0 to 2. The corresponding phase angle scale, ranging from −90° to +90°, is given on the inner side of the Γ phase angle circle.
3. It is important to notice that a full rotation around the chart (change of $\angle\Gamma = 2\pi$ is equivalent to a distance of $\lambda/2$). Thus, half rotation of 180° corresponds to $\lambda/4$, etc.

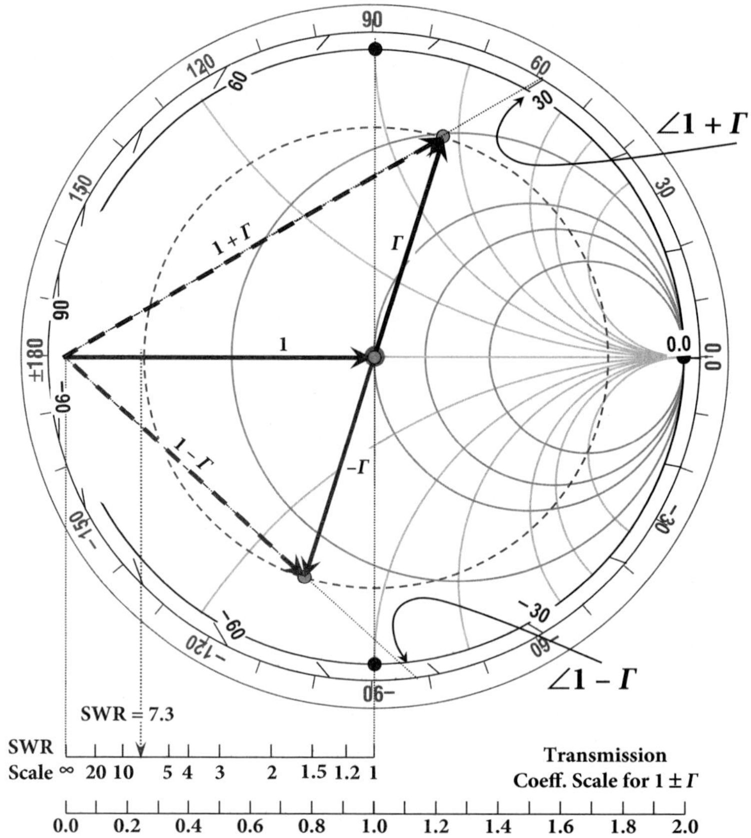

FIGURE 2E.14 Smith chart features and short cuts as discussed in Section 2E.6.

2E.7 Matching Using the Smith Chart

To demonstrate the concept, let us consider the case of short-circuited shunt stub matching presented in the lower part of Figure 2B.5. As you recall, the equations for this case were Equations (2B.3), (2B.4), and (2B.6). The Smith chart offers an alternative graphical method to working out these equations analytically and numerically.

The objective is to find answers for the location of the stub on the TL, d_{ms}, as well as the stub length, d_{st}. In the following, we will demonstrate the use of the chart for this matching case. Please refer to Figure 2E.15:

1. Choose the normalization impedance to be Z_o to be matched to (R_s in this case).
2. Compute the normalized $Z_{Ln} = Z_L/Z_o$.
3. Locate Z_{Ln} on the chart and identify the Γ_L phasor point [1].
4. Draw a circle centered at the origin passing by Γ_L.
5. Extend the Γ_L phasor through the origin to reach the $-\Gamma_L$ point [2]. This is the Y_{Ln} point.
6. Move on the circle of step 4 above "toward generator" until it intercepts the $g = 1$ ($r = 1$) circle. There will be two intercept points [3] and [4]. This step is the realization of Equation (2B.3), where the normalized real part of the line admittance is 1, and hence the real part of the admittance equals that of the desired matching value. Having the real part of the line admittance matched, what remains is to tune out (null) the imaginary part using the stub susceptance. Hence, the identified intercept points define the proper location for adding the shunt stub.
7. Determine the normalized distances moved from the load location to the two-intercept points: d_{ms1}/λ [5] and d_{ms2}/λ [6], respectively.

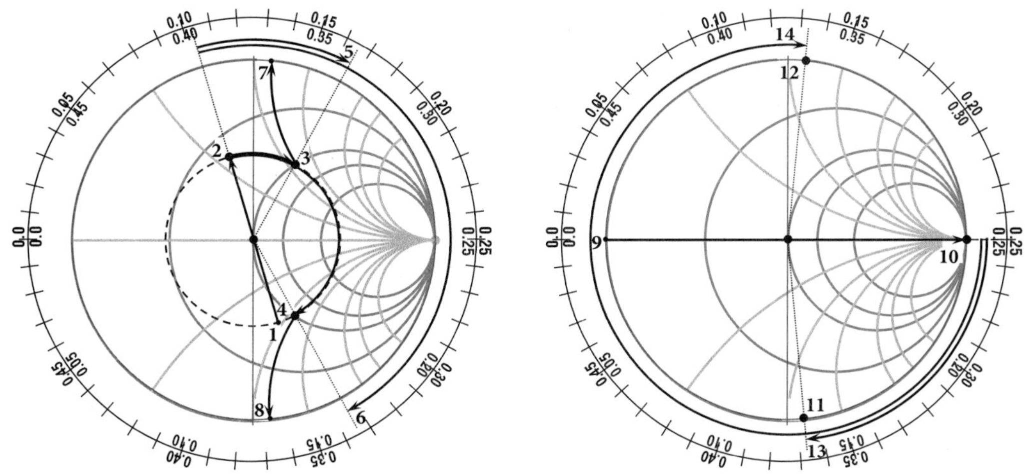

FIGURE 2E.15 Impedance matching using the Smith chart as discussed in Section 2E.7.

8. In principle, additional rotations of integer multiples of $\lambda/2$ will bring us to the same points, which means that we have many possible solutions in the form of $\{d_{ms1}/\lambda + 0.5n\}$ and $\{d_{ms2}/\lambda + 0.5n\}$, where n is an integer. Typically, the first two intercepts are presented as the proposed stub locations.
9. Determine the line normalized susceptances for both solutions. Notice that both values are symmetrically located and thus have the same magnitude but opposite polarities, for example, jb_{ms1} [7] and jb_{ms2} $(-jb_{ms1})$ [8] corresponding to the d_{ms1} and d_{ms2} solutions, respectively.
10. The chosen stub is needed to null the line susceptance and hence the stub susceptance should be equal but opposite in polarity to that of the line. Hence, the desired stub susceptances should be $jB_{st1} = -(jb_{ms1} * Y_o)$ and $jB_{st2} = -(jb_{ms2} * Y_o)$, respectively.
11. Compute the normalized stub susceptances with respect to its characteristic impedance: $jb_{st1} = jB_{st1}/Y_{o,st}$ and $jb_{st2} = jB_{st2}/Y_{o,st}$, respectively.
12. To determine the proper stub length(s), we need to do another chart for the stub as a TL with a zero-load impedance (infinite load admittance) and known input admittance (susceptance). This chart can be done on the same chart we used for the matching section line without interfering with the first solution. This is because the circle trace of the stub line is the circle that passes by zero load impedance (infinite admittance), which is the $r = 0$ circle (the biggest circle on the chart). If you prefer, you may use a separate chart for the stub section.
13. Locate the zero-impedance load for the stub [9]. In admittance terms, the short-circuited load end is on the opposite side of the chart (the point ∞,∞) [10].
14. Locate the two possible stub susceptances on the chart, jb_{st1} and jb_{st2}, [11] and [12].
15. Starting at the (∞,∞) point and moving toward the generator, determine the normalized distances to be moved to arrive at the desired stub susceptances, d_{st1} [13] and d_{st2} [14], respectively.
16. Again, additional rotations of integer multiples of $\lambda/2$ yield valid answers which means that again we have many possible solutions in the form of $\{d_{st1}/\lambda + 0.5m\}$ and $\{d_{st2}/\lambda + 0.5m\}$, where m is an integer. Typically, the first two solutions $(m = 0)$ are presented as the proposed stub lengths.

CHAPTER 2 PROBLEMS

- Problems 2.1 to 2.7: Problems dealing with TL parameters and characteristics
- Problems 2.8 to 2.26: Problems dealing with TL reflections and standing waves
- Problems 2.27 to 2.35: Problems dealing with TL bounce diagram
- Problems 2.36 and 2.37: Problems dealing with TL power and signal delivery
- Problems 2.38 to 2.45: Problems dealing with use of the Smith chart for TL circuit analysis
- Problems 2.46 to 2.55: Problems dealing with TL impedance matching

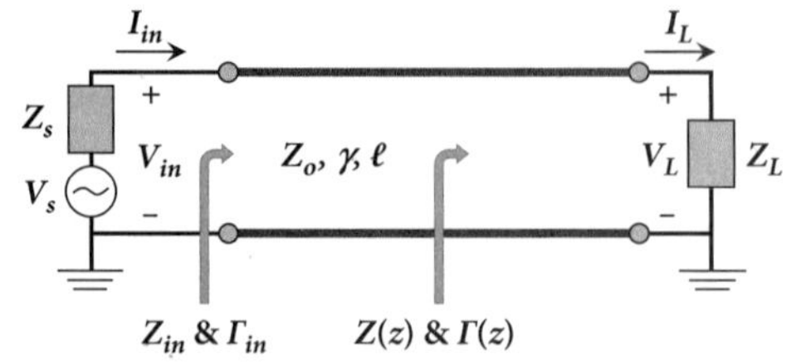

FIGURE 2P.1 Basic transmission line circuit.

Note: Figure 2P.1 should be used with all problems in this section unless otherwise stated.

2.1 Evaluate the frequency dependent expressions for Z_o and γ for the TL with the following per unit length model parameters: $R_z = 0.05\ \Omega$/m, $L_z = 0.2\ \mu$H/m, $C_z = 80$ pF/m, $G_z = 0.5$ mS/m. Plot both real and imaginary parts for both Z_o and γ versus frequency for the range 100 MHz to 10 GHz.

- Repeat while neglecting the loss elements, that is, $R_z = 0$ and $G_z = 0$.
- Evaluate the phase velocity and plot versus frequency for both cases above.

2.2 Evaluate the frequency dependent expressions for Z_o and γ for the TL with the following per unit length model parameters: $R_z = 0.01\ \Omega$/m, $L_z = 0.16\ \mu$H/m, $C_z = 66$ pF/m, $G_z = 0.1$ mS/m.

- Plot both real and imaginary parts for both Z_o and γ versus frequency for the range 100 MHz to 10 GHz.
- Plot the phase velocity versus frequency for the same range.

2.3 The per unit length parameters of a TL were measured as 0.1 Ω/m, 160 nH/m, 100 pF/m, and 50 μS/m. The TL connects a source with an open circuit voltage of 50 mV, source impedance of 50 Ω and operates at 1 GHz to a 75 Ω load. Determine

a. characteristic impedance of the line.

b. complex propagation constant of the line.

c. phase velocity.

d. wavelength.

e. attenuation coefficient of the line in dBs/m.

2.4 For a TL with $\alpha = 0.1$ dB/m and $\beta = 60$ rad/m, plot the normalized voltage (or current) time-domain terms $\exp(-\alpha z)\cdot\cos(\omega t - \beta z)$ and $\exp(\alpha z)\cos(\omega t + \beta z)$ versus $z(0 - 1\text{ m})$ for the frequency $f = 1$ GHz, at time intervals of 0.1 ns for the duration $t = 0 - 0.5$ ns.

2.5 Given a TL with the following per unit length RLGC model parameters: $R_z = 0.01\ \Omega$/m, $L_z = 0.5\ \mu$H/m, $C_z = 0.1$ nF/m, $G_z = 1$ mS/m. At an operating frequency of $f = 3$ GHz, find the attenuation coefficient, the phase constant and the characteristic impedance.

2.6 The following table summarizes the relationships for the RLGC parameters of a coaxial TL at both low and high frequencies. At low frequencies, the currents flow in the conductors is assumed to have uniform density over its entire cross section. However, at high frequencies, these currents are confined to a *thin* layer of the skin of the conductor and are modeled to exist only within what we refer to as the skin depth, δ.

	DC and Low Frequencies	High Frequencies $\delta \cong 1/\sqrt{\pi f \mu \sigma}$	Units
R_z	$\frac{1}{\sigma_i(\pi a^2)} + \frac{1}{\sigma_o(\pi[c^2 - b^2])}$	$\cong \frac{1}{\sigma_i(2\pi a\delta_i)} + \frac{1}{\sigma_o(2\pi b\delta_o)}$	Ω/m
L_z	$\frac{\mu}{2\pi}\ln(b/a)$	$\cong \frac{\mu}{2\pi}\ln(b/a)$	H/m
G_z	$\frac{2\pi\sigma_d}{\ln(b/a)}$	$\frac{2\pi\sigma_d}{\ln(b/a)}$	S/m
C_z	$\frac{2\pi\varepsilon}{\ln(b/a)}$	$\frac{2\pi\varepsilon}{\ln(b/a)}$	F/m

In this table, a, b, and c refer to the radii of outer surface on the inner conductor, and the inner and outer surfaces of the outer conductor, respectively. The conductivities σ_i, σ_o, and σ_d, are for the inner and outer conductors, and the dielectric medium in-between, respectively. Finally, ε and μ, are the dielectric material's permittivity and permeability, respectively.

a. Using these relationships, calculate the low frequency values for R, L, G, and C of a coaxial TL with copper conductors and Teflon dielectric, given that
   - Outer diameter of center conductor is 1 mm
   - Inner diameter of outer conductor is 3 mm
   - Outer diameter of outer conductor is 4 mm

b. Repeat part a for $f = 10$ GHz

2.7 Consider a TL with the following per unit length model parameters: $R_z = 0.05$ Ω/m, $L_z = 0.2$ μH/m, $C_z = 80$ pF/m, $G_z = 0.5$ mS/m terminated with an open circuit termination and operating at a frequency of 5 GHz. Plot the driving point impedance (magnitude and phase—real and imaginary) at different point over a section of length 30 mm.

2.8 Consider the circuit consisting of a TL with $Z_o = 78$ Ω and $\gamma = 0.01 + j6.6$/m, a voltage source $V_s = 2$ V, $Z_s = 50$ Ω, and $Z_L = 50$ Ω.

a. Find the value of both the forward and backward waves on the line at $z = 0$ and $z = 4$ cm.

b. Plot the magnitude of each of the individual voltage components (forward and backward) and the magnitude of their (phasor) sum over the length of the TL (0 to 4 cm).

c. Plot the magnitude of the ratio of the two voltages over the length of the TL.

2.9 For a TL having the following parameters: Z_o (characteristic impedance) = 50 Ω, $Z_s = 50$ Ω, $Z_L = 75$ Ω, length = 40 cm, and $f = 20$ GHz, calculate the reflection coefficient at the load, reflection coefficient at the source, driving point impedance at the source side and the SWR.

2.10 Using the TL [$Z_o = 50$ Ω, $Z_s = 50$ Ω, $Z_L = 75$ Ω, $\ell = 40$ cm, and $f = 20$ GHz], and using a voltage source of 2 V, find the magnitude of each of the traveling wave components and the standing wave component in the TL voltage waves?

2.11 For the TL [$Z_o = 50\ \Omega$, $Z_s = 50\ \Omega$, $Z_L = 0$ (short circuit), $\ell = 40$ cm, and $f = 20$ GHz] and using a voltage source of 2 V:

a. Find the magnitude of the traveling wave and the standing wave components.

b. Find the location of the first voltage peak away from the load.

2.12 For a TL with $Z_o = 50\ \Omega$ terminated in $Z_L =$ infinity (open circuit) operating at $f = 3$ GHz, what should the length of the TL be for the driving point impedance at the source end equal to $0\ \Omega$?

2.13 A TL with characteristic impedance $50\ \Omega$ and length $\ell = 0.25\lambda$ has a forward propagating wave $v^+(z,t) = 2\cos(\omega t - \beta z)$ and a reflected backward propagating wave $v^-(z,t) = 0.5\cos(\omega t + \beta z)$. Determine the value of the load impedance and the SWR on the line.

2.14 A lossless TL is terminated with a $100\text{-}\Omega$ load. If SWR = 2, find the possible values of the characteristic impedance.

2.15 For a TL with $Z_o = 50\ \Omega$, $\ell = 0.3\lambda$ terminated in $Z_L = 20 - j10\ \Omega$, find

a. The reflection coefficient at the load

b. The reflection coefficient at the source

c. The driving point impedance at the source end

2.16 Consider the TL circuit with the following parameters: $V_s = 2$ V, $Z_s = 10\ \Omega$, $Z_o = 50\ \Omega$, $\ell = 0.3$, and $Z_L = 75\ \Omega$. Find the magnitude of each of the forward and backward traveling waves and the total voltage at the load end.

2.17 If a load impedance of $75 + j40\ \Omega$ is connected to a $50\text{-}\Omega$ TL, find the nearest two locations from the load (in fractions of λ) to get a purely resistive driving point impedance.

2.18 For a $75\text{-}\Omega$ TL terminated with a short circuit, find the locations on the TL where

a. driving point impedance is a capacitive reactance of $+j75\ \Omega$.

b. driving point impedance is an inductive reactance of $-j75\ \Omega$.

c. magnitude of voltage is maximum.

2.19 A lossless TL is terminated with a $75\text{-}\Omega$ load. If the SWR on the line is 1.5, find the two possible values of the characteristic impedance of the line.

2.20 For the basic TL, Figure 2P.1, given that $Z_L = 100\ \Omega$, $Z_o = 50\ \Omega$, and $\ell = 0.35\lambda$, find the following quantities:

a. SWR on the line

b. Reflection coefficient at the load

c. Driving point impedance at the input of the line

d. The nearest voltage minimum to the load

2.21 Given a 2-ft TL with characteristic impedance $Z_o = 50\ \Omega$, a 5-V voltage source with source impedance $Z_s = 50\ \Omega$ and load impedance $Z_L = 50\ \Omega$. Find the voltage magnitude at the source and load ends of the line. For a propagation constant $\gamma = j0.2$ rad/ft, find the magnitude of the voltage at a distance 1 ft from the load.

2.22 Given a 10-m TL with $Z_o = 75\ \Omega$ and $\gamma = j0.6$ rad/m with end impedances of $Z_s = 50\ \Omega$ and $Z_L = 25\ \Omega$. The source voltage is 10 mV and operates at 200 MHz.

a. Find the reflection coefficients at the load and at the source.

b. Find the magnitude of each of the traveling wave and standing wave components.

c. Repeat for and $\gamma = 0.01 + j0.6$/m. Comment on the obtained result.

2.23 Measurements on a 50-Ω lossless line at 1.5 GHz showed an SWR of four and three consecutive voltage minima at 1.5, 9, and 16.5 cm away from its load end. Find

a. the operating wavelength on the TL.

b. the phase velocity of the wave on the line. Is it the same as the speed of light in free space? Explain.

c. $|\Gamma|$ on the line.

d. the termination load impedance.

Hint: Voltage minima occur at $\Gamma = |\Gamma|\exp(j\pi)$.

2.24 Derive expressions for ρ and Γ at all four marked positions on the TL circuit shown, given $\ell_1 = \frac{\lambda}{2}$ and $\ell_2 = \frac{\lambda}{4}$.

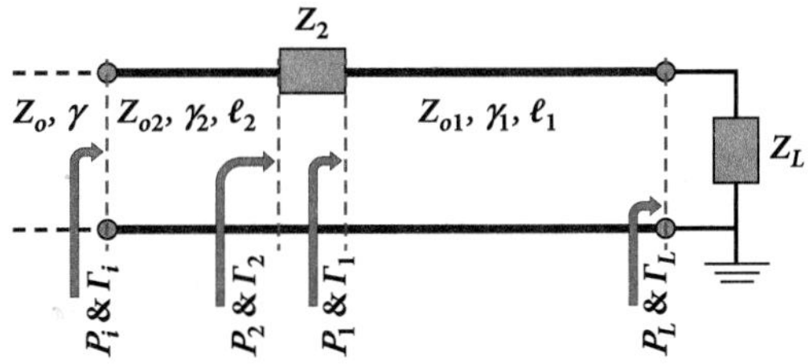

2.25 Evaluate P's (upper case rho) and Γ's at all marked positions in the TL circuit shown given

$\ell_1 = 0.9\lambda$	$\ell_2 = 0.4\lambda$	$Z_o = 50\ \Omega$	$Z_1 = 100 + j30\ \Omega$
$Z_{o1} = 50\ \Omega$	$Z_{o2} = 100\ \Omega$	$Z_L = 50 - j20\ \Omega$	$Z_2 = 10 - j10\ \Omega$

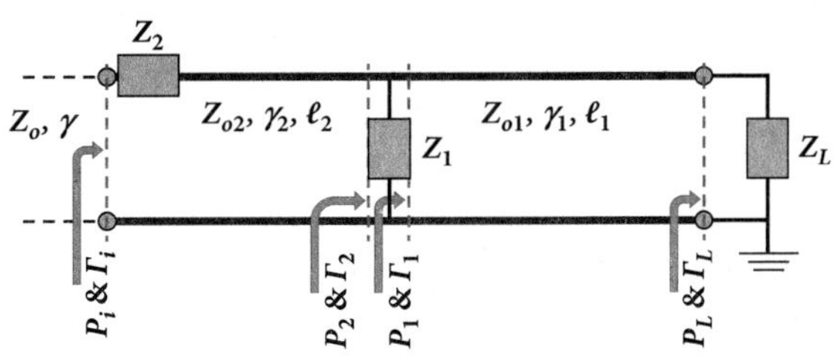

2.26 Derive expressions for the Thevenin's open circuit voltage and impedance equivalents of the circuit of Figure 2P.1 at the load side. Use the result to derive an expression for the voltage at the load. Use $V_s = 10$, $Z_s = 50\ \Omega$, $Z_o = 75\ \Omega$, $Z_L = 125\ \Omega$, and $\ell = 0.5\lambda$.

2.27 Given a lossy TL with $Z_o = 50\ \Omega$, $\alpha = 0.05$ dB/wavelength and length 2.3λ. The line has $V_s = 2$ V, $Z_s = 40\ \Omega$, and $Z_L = 75\ \Omega$. Using the bounce/multi-reflection analysis, find the magnitude of the first voltage component traveling the ($+z$) direction, the first voltage component traveling in the ($-z$) direction and the second voltage component traveling in the ($+z$) direction at a point 0.5λ away from the load. Find the SWR at the same point.

2.28 Given a lossy TL with $Z_o = 50\ \Omega$, $\alpha = 0.05$ dB/wavelength and length 2.3λ. The line has $V_s = 2$ V, $Z_s = 40\ \Omega$, and $Z_L = 75\ \Omega$.

a. Plot versus distance the magnitude of the first ($+z$) component, first ($-z$) component and second ($+z$) component along the TL.

b. Plot the magnitude of the total voltage ($V^+ + V^-$) along the TL.

2.29 For a TL circuit with parameters: $V_s = 10$ V, $Z_s = 10\ \Omega$, $Z_o = 50\ \Omega$ and $Z_L = 25\ \Omega$. Using the bounce/diagram analysis, find the magnitude of the first/earliest component traveling in the (+z) direction at the source end.

2.30 A TDR device is used to inject a pulse, $u(t)$, into a lossless (air dielectric) cable. If the forward propagating waveform is given by $u\left(t - \frac{z}{c}\right)$, where c is the speed of light in free space, 3×10^8 m/s, and the reflected waveform is given by $u\left(t + \frac{z}{c} - \tau\right)$, $\tau = 6$ ns. Assuming the cable has no faults, what is the load connected to the end of the cable and how long is the cable?

2.31 If a fault exists half way along a 2-m cable terminated in $Z_L = 2Z_o$, what will the voltage waveform observed at $z = 0$ look like if

a. the fault is a short circuit.

b. the fault is an open circuit.

Given that the source excitation is of the form $2u(t)$ and has a matched internal impedance.

2.32 A 50-Ω TDR device is connected to a 75-Ω cable of length 1 m and phase velocity of 0.75 c_o. The TDR operates by injecting a unit pulse $u(t)$ into the cable. If reflections plot the voltage waveform at $z = 0$ for a load impedance $Z_L = 10\ \Omega$.

2.33 Develop the bounce diagram for a 75-Ω airline of length 30 cm terminated in a 100-Ω load given the source:

- $6\delta(t)$ with 50 Ω internal impedance
- $6u(t)$ with 50 Ω internal impedance

For both cases, develop an expression and plot the voltage waveforms at both the sending and receiving ends.

2.34 A 45-cm section of 75-Ω air-filled TL connects between a 10cos(ωt)-V source with a 50-Ω source impedance and a 25-Ω load. Develop the bounce diagram and derive expressions for $V(0)$, $V(45)$, and $V(10)$ voltage phasors. Also, Find V^+ and V^-.

2.35 A source with $10\delta(t)$ V and $R_s = 125\ \Omega$ is connected to the sending end of a 15-cm airline. The airline has $R_o = 75\ \Omega$ and is terminated in $R_L = 50\ \Omega$.

a. Construct the bounce diagram (also label the units on the time axis).

b. Determine and plot the voltages at the sending end of the line.

c. What is the final value for both voltages?

d. Repeat for a source with 10 $u(t)$ V.

2.36 For a TL circuit with the following parameters: $V_s = 2$ V, $Z_s = 50\ \Omega$, $Z_o = 50\ \Omega$, and $Z_L = 10\ \Omega$. Find the power delivered to the load. What is the efficiency of the power delivery?

If the voltage source is replaced with a digital pulse signal source, will the signal be transmitted with perfect fidelity? Why or why not?

2.37 For the circuit shown, find the length of the section of the TL (in fractions of λ) and the series reactance (Z_2) that will result in maximum power delivery from a source having an internal resistance of 50 Ω to a load $Z_L = 200 - j100\ \Omega$. The characteristic impedance of the TL is 50 Ω.

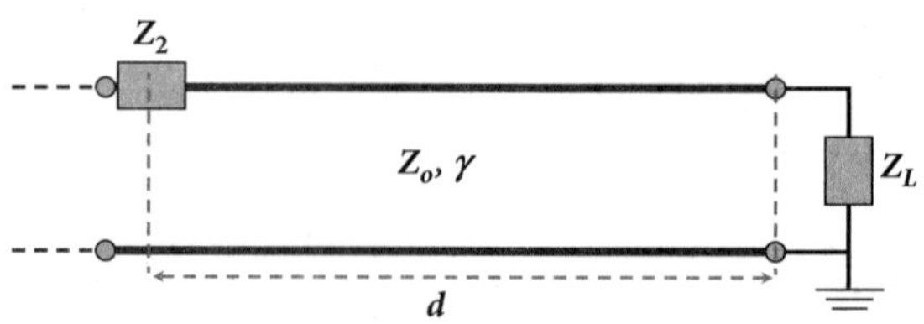

2.38 What do the points at coordinates (1,0) and (–1,0) on the Smith chart represent? What is the length of a TL section required to transform one to the other?

2.39 Show that the r scale in the region between $r = 1$ and infinity on the Smith chart is identical to that of the SWR scale.

2.40 Based on the Smith chart, how much TL length is needed so that the reactive part of the translated impedance of any arbitrary load is changed from capacitive to inductive and vice versa?

2.41 The input impedance of a 100-Ω TL terminated in an unknown load Z_L was measured as $0 + j25\ \Omega$. When the load impedance was replaced with a 50-Ω termination, the measured input impedance changed to $200 + j0\ \Omega$.

a. Use the Smith chart to determine the length of the TL and the unknown Z_L.

b. Redo part "a" above analytically (use equations not the Smith chart).

2.42 For a 50-Ω TL of length 0.4λ and load $Z_L = 40 + j30\ \Omega$, using the Smith chart find the reflection coefficient at the source.

2.43 For a 50-Ω TL of length 0.4λ and connected to a load $30 + j20\ \Omega$.
Using an impedance Smith chart, find the admittance at the load and at the source end locations.
Find the length of the TL required so that the normalized driving point admittance at the source end of the TL has a real component of unity.

2.44 For a TL circuit of Figure 2P.1, given $Z_o = 50\ \Omega$, $Y_L = .02 + j.02$ S, and $\ell = 0.6\lambda$. Use the Smith chart to evaluate:

a. Γ_L and Γ_{in}. Verify Γ_L using analytical expressions.

b. The input impedance and standing wave ratio.

2.45 For the TL circuit shown $Z_2 = 20\ \Omega$, $Z_1 = 10\ \Omega$, $Z_L = 30\ \Omega$, $Z_{o1} = Z_{o2} = Z_o = 50\ \Omega$, $\ell_1 = 0.3\lambda$, and $\ell_2 = 0.2\lambda$. Using the Smith chart, find the following:

a. Driving point impedance Z_b (just to the right of Z_1).

b. Driving point impedance Z_d (just to the right of Z_2).

c. Input impedance for the TL circuit, Z_e.

Assuming a source with a voltage of 1 V and an internal impedance of 50 ohms, find

d. voltage across the impedance Z_1.

e. voltage at the load impedance Z_L.

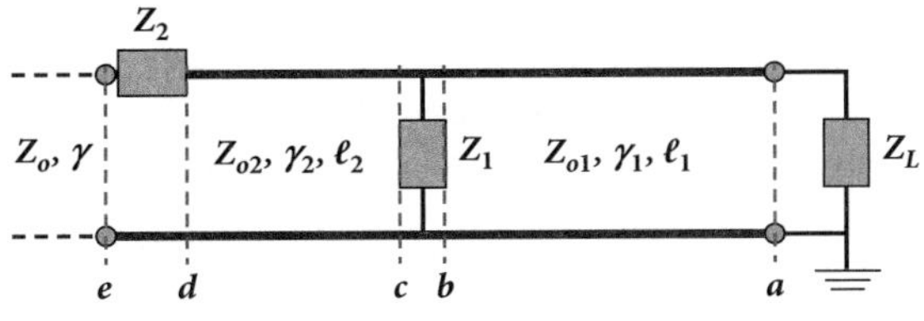

2.46 For the circuit of Figure 2P.1, with a 50-Ω TL, find the location and value of the shunt admittance that will result in a perfect match to an inductive load of $75 + j25$.

2.47 For the TL circuit of Figure 2P.1, use the Smith chart to design an inductive short-circuited shunt stub tuner for the given load, given $Z_o = 75\ \Omega$, $Z_L = 125 - j100\ \Omega$.

2.48 For the TL circuit of Figure 2P.1, given $Z_o = 50\ \Omega$, $Y_L = 0.02 + j0.02$ S, and $\ell = 0.6\lambda$. Use the Smith chart to find the location and length of a series inductive short circuit stub required to match the load to the line.

2.49 For the TL circuit of Figure 2P.1, given $Z_o = 75\ \Omega$, $Z_L = 100 + j50$ S, and $\ell = 0.45\lambda$. Use the Smith chart to find the location and length of a shunt capacitive short circuit stub required to match the load to the line.

2.50 Starting with the normalized load impedance of $0.95 + j0.75$, find the eight-stub tuner solutions:

- Inductive series shorted and opened, inductive shunt shorted and opened
- Capacitive series shorted and opened, capacitive shunt shorted and opened

2.51 Find the design parameters of the matching stub in each of the following circuits assuming a source and TL characteristic impedances of 50 Ω each. $Z_L = 10 - j20\ \Omega$.

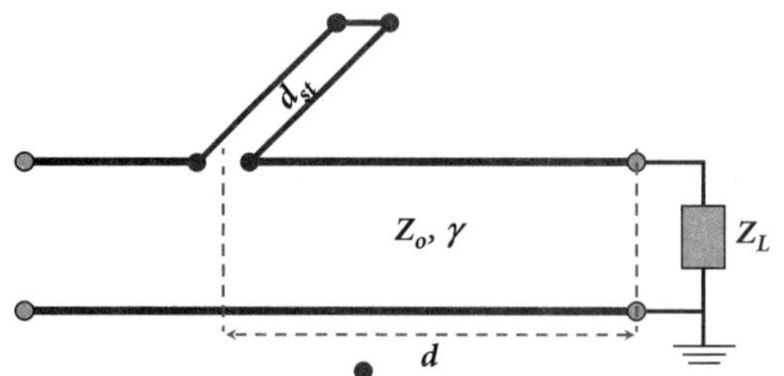

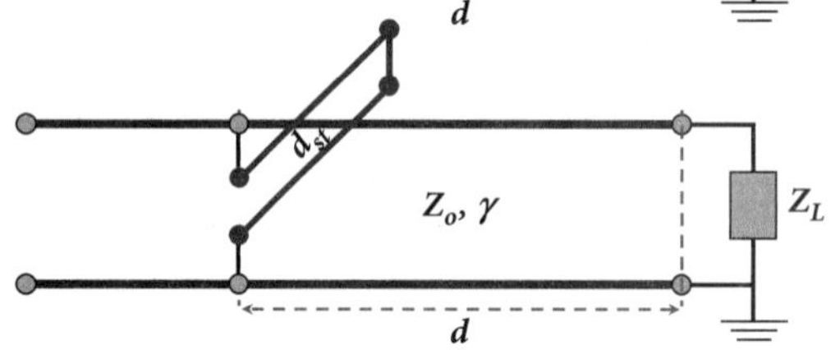

2.52 For the circuit shown, assuming a source impedance of 20 Ω, find the values of all the unknown parameters that will result in a perfect match.

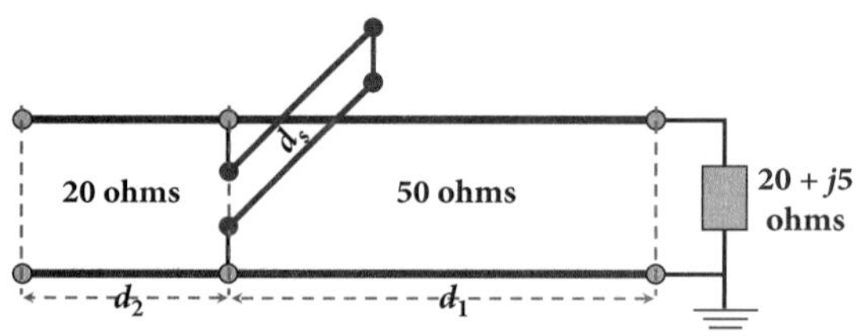

2.53 A single stub tuner can be used to match any termination load to the impedance of the line. However, if the load is changed to a different impedance, not only do we need to readjust the length of the stub but also we must readjust the stub location. This is a big disadvantage in using a single stub tuner with varying load applications. As it turned out, we can (partially) overcome this problem if we were to use two fixed location stubs; typically, $\lambda/8$ apart and $\lambda/8$ from the load. Show that the two stub tuners can't match all possible loads and identify the load impedances that are not *matchable*.

2.54 Find the range of impedances that can be tuned using a tunable stub located $\lambda/8$ away from the load. Use both the Smith chart as well as analytical methods.

2.55 Using the Smith chart, show that three stubs spaced $\lambda/8$ apart can be used to tune any load termination. Discuss the cases of $\lambda/4$.

CHAPTER 2 SUMMARY[2]

Transmission Lines Overview

Examples of TLs: controlled geometry lines

Circuit component size compared to operating wavelength, λ

TL analysis is needed if size is not too small compared to wavelength

Circuit Theory Analysis of Controlled Geometry Lines

RLGC physically based model (linear)—distributed parameters

Use KVL and KCL to develop voltage-current relationships

Analysis in both time and frequency domains

Second-order differential equation for voltage as function of position and time

Solution in Frequency Domain

$$V(z) = V^+ e^{-\gamma z} + V^- e^{+\gamma z}$$

$$I(z) = (1/Z_o)\cdot\{V^+ e^{-\gamma z} - V^- e^{+\gamma z}\}$$

$$Z_o = \sqrt{\{R_z + j\omega L_z\}/\{G_z + j\omega C_z\}}$$

$$\gamma = \alpha + j\beta = \sqrt{\{R_z + j\omega L_z\}\cdot\{G_z + j\omega C_z\}}$$

Harmonic Solution in Time Domain

Using Laplace transform

$$v(z,t) = \left|V^+\right| e^{-\alpha z}\cos(\omega t - \beta z + \varphi_v^+) + \left|V^-\right| e^{+\alpha z}\cos(\omega t + \beta z + \varphi_v^-)$$

$$i(z,t) = \left|V^+/Z_o\right| e^{-\alpha z}\cos(\omega t - \beta z + \varphi_i^+) - \left|V^-/Z_o\right| e^{+\alpha z}\cos(\omega t + \beta z + \varphi_i^-)$$

Transient Solution in Time Domain

Only for lossless line $R_z = G_z = 0$

$$v(z,t) = v^+\left(t - \sqrt{L_z C_z}\, z\right) + v^-\left(t + \sqrt{L_z C_z}\, z\right)$$

$$i(z,t) = \left(1/\sqrt{L_z/C_z}\right)\left\{v^+\left(t - \sqrt{L_z C_z}\, z\right) - v^-\left(t + \sqrt{L_z C_z}\, z\right)\right\}$$

Physical Insight

Two traveling waves: $(+z)$ wave and $(-z)$ wave both have same Z_o and γ

Z_o: Characteristic impedance, ohms — $\gamma = \alpha + j\beta$: Complex propagation constant, m^{-1}

α: Attenuation coefficient, nepers/m — β: Phase constant, rad/m

Phase velocity $c_{ph} = \Omega/\beta$, m/s — Wavelength $\lambda = c_{ph}/f = 2\pi/\beta$, m

V^+ and V^- are determined by boundary conditions at TL source and load ends

---

[2] ***Important Warning to Students:*** Please do not use this summary without thorough study of the relevant material in this chapter. Using this summary and its formulas without adequate appreciation of the physics involved and conditions of validity can lead to wrong implementations and invalid conclusions.

Reflection Coefficient

Define reflection coefficient $\Gamma(z) = (V^- e^{+\gamma z})/(V^+ e^{-\gamma z}) = ((-z)\text{ wave})/((+z)\text{ wave})$

$\Gamma(z) = \Gamma(\ell)e^{+2\gamma(z-\ell)}$ — $\Gamma(\ell) = \Gamma_L = [Z_L - Z_o]/[Z_L + Z_o]$

$\Gamma(z) = [Z(z) - Z_o]/[Z(z) + Z_o]$ — $Z(z) = Z_o([1+\Gamma(z)]/[1-\Gamma(z)])$

Reflections are caused by impedance discontinuities "mismatch—not equal to Z_o"

Traveling and Standing Waves

$Z_L = Z_o$, $\Gamma_L = 0 \rightarrow \Gamma(z) = 0 \rightarrow Z(z) = Z_o$: $V^- = 0$, only $(+z)$ wave, pure traveling wave, $|V(z)| = |V^+|e^{-\alpha z}$

Magnitude Constant Except for Decay due to Attenuation (Loss)

$Z_L = 0 + jX_L$, $|\Gamma_L| = 1 \rightarrow |\Gamma(z)| = 1 \rightarrow Z(z) = 0 + jX(z)$: $|V^+| = |V^-|$, pure standing waves, $|V(z)| = 2|V^+|$ $\cos(\beta z + \varphi)$

Wave Does Not Travel, Magnitude Has Peaks (=2$|V^+|$) and Valleys of Zeroes (Nodes)

$Z_L \neq Z_o$, $\Gamma_L \neq 0 \rightarrow \Gamma(z) \neq 0 \rightarrow Z(z) \neq Z_o$: combination of $(+z)$ and reflected $(-z)$ waves

Mix of traveling $\{|V^+|-|V^-|\}$, and standing $\{2|V^-|\}$ waves

Standing wave ratio, SWR $= V_{max}/V_{min} = [1 + |\Gamma|]/[1 + |\Gamma|]$—$|\Gamma| = 0 \rightarrow$ SWR $= 1$, $|\Gamma| = 1 \rightarrow$ SWR $= \infty$

Reflections: Power and Signal Delivery

Reflections (standing waves) reduce power delivered to load—also degrade signal quality

Mismatch Causes Dispersion to Communication Signals—Reflections Limit Bandwidth of Digital Communications

For optimum power & signal delivery, impedance matching at both source and load ends is necessary

Addendum A: Driving Point Impedance

$$Z(z) = Z_o \frac{[1+\Gamma(z)]}{[1-\Gamma(z)]} = Z_o \frac{\{Z_L + Z_o \tanh[\gamma(\ell - z)]\}}{\{Z_o + Z_L \tanh[\gamma(\ell - z)]\}}$$

For Lossless Lines

$\tanh[\gamma(\ell - z)] \rightarrow j\tan[\beta(\ell - z)]$ – *ALSO*: $Z(z \pm n_{int}\lambda/2) = Z(z)$ & $Z(z \pm n_{odd}\lambda/4) = R_o^2/Z(z)$

Z_L	Γ_L	Z_{in}	Z_L	Γ_L	Z_{in}	Z_L	Γ_L	Z_{in}
Z_o (matched)	0	Z_o	0 (short circuit)	–1	$jZ_o\tan(\beta\ell)$	∞ (open circuit)	+1	$-jZ_o\cot(\beta\ell)$

Addendum B: Impedance Matching

- Matching networks: L, T, and Π (pi) section, using lumped elements work for low RF frequencies
- TL matching network using series/shunt stubs needed at higher RF frequencies
  - Add TL section at load "far" end—length chosen to yield reactive input impedance at "near" end
  - Add TL stub (series or shunt) to tune out the obtained reactance at the "near" end
  - Preferred stub: short-circuited at its far end and connected in shunt at its near end

Addendum C: Frequency-Domain Bounce Diagram

- Unconventional approach to build steady state frequency domain solution through monitoring transients
- Initial launch by source: voltage divider between the "transient" input impedance of the line (Z_o) and the source impedance
- Initial launch is monitored as it travels down the line ($e^{-\gamma z}$), reflects off the load, $P_L(j\omega)$ and reflects back off the source side $P_s(j\omega)$. Multiple reflections cumulate to form a geometric series expression. (P is "local" reflection coefficient in frequency domain.)

Addendum D: Time-Domain Bounce Diagram

- Conventional approach to monitor multiple reflection transients in the time domain. Networks with lossless lines and resistive networks are presented in this chapter
- Initial launch by source: voltage divider between the "transient" input resistance of the line (R_o) and the source resistance
- Initial launch is monitored as it travels down the line (time delay $= z/c_{\text{ph}}$), reflects off the load, ρ_L and reflects back off the source side ρ_s. Multiple reflection cumulate to form a sequence of successful delays. The "local" reflection coefficient in time domain $\rho(t)$ is the time domain form of the "local" reflection coefficient $P(j\omega)$ in frequency domain
- Time-domain reflectometry, TDR, is a practical tool for locating and characterizing cable faults and discontinuities

Addendum E: The Smith Chart

- A graphical tool to solve TL impedance problems
- Two-way transform between the normalized complex impedance, Z_n/admittance, Y_n (real and imaginary) and the complex reflection coefficient (real and imaginary or magnitude and phase)
- $Z_n = Z/Z_o = r + jx$, $Y_n = Y/Y_o = g + jb$, $\Gamma = \Gamma_{re} + j\Gamma_{im} = |\Gamma|\angle\varphi_\Gamma$
- The chart maps Γ_{im} vs Γ_{re} with r and x (g and b) as parameters:
  - Constant r (or g) circles have their origins on $\Gamma_{\text{im}} = 0$ axis and all pass by the $\Gamma = 1 + j0$
  - Constant x (or b) circles have their origins on $\Gamma_{\text{re}} = 1$ axis and all pass by the $\Gamma = 1 + j0$
  - The phasor stretching from the origin to the (r, x) intercept represent the corresponding Γ phasor. The phasor to the (g, b) intercept is the $-\Gamma$ phasor
  - Scales are provided to read Γ magnitude (also re and im) as well as $\angle\varphi_\Gamma$
  - Scale is provided to read the SWR
- The trace of $\Gamma(z)$ is a circle centered at $\Gamma = 0 + j0$ for lossless lines and a spiral for lossy lines. Spiral pitch depends on the attenuation coefficient α
- Scales for normalized distance $\Delta z/\lambda$ are given to guide moving from one line location to another on this trace for both directions of motion: toward the load and toward the generator (source)
- A full rotation around the chart correspond to a distance of half wavelength
- Applications of the Smith chart are given:
  - Find Γ given Z or Y and vice versa
  - Find $Z(z_2)$, $Y(z_2)$, and $\Gamma(z_2)$ given $\{Z(z_1), Y(z_1), \text{or } \Gamma(z_1)\}$ and $\{z_2 - z_1\}$
  - Stub impedance matching problems

CHAPTER 3

Transition to Electrostatics

Muḥammad ibn Mūsā al-Khwārizmī (780–850)

Chapter Learning Objectives

- Recognize that the electromagnetics field's theory provides the foundation for the circuits' theory and circuit analysis tools.
- Identify the limitations of analyses using typical approximations and oversimplifications of models used in circuit analyses.
- Identify the need to study electromagnetics and its indispensable utilization in wide and diverse applications and fields.
- Recognize the role of scientists and inventors in the historical advancements of the field of electromagnetics and its applications.

3.1 Introduction

In the previous chapter, we studied the performance of transmission lines (TLs) as circuit elements using an RLGC model. In other words, we used the circuit elements model to study the wave propagation properties of the transmission media. This RLGC model is in fact an approximation and oversimplification of the physical TL configuration. The only way to appreciate this fact is for us to examine how such a model is developed. This will be the topic of the following chapters of this book where we will be studying relevant electromagnetic topics leading to the tools that allow us to determine these "circuit" model parameters. In the process, we will arrive at more powerful static and dynamic electromagnetic analysis tools. These tools will enable us to study phenomena that are more complex without resorting to the oversimplification of using circuit element modeling.

You see, the issue with the use of approximate models is that you typically cannot predict when the use of these models will fail in yielding reliable results. In the early discussion we had in Chapter 1, we showed the need to resort to TL analysis techniques when the size of the components is no longer negligible compared to the wavelength. One can predict as a result that *approximate* circuit model would fail as the frequency goes high. Considering the cases of mm waves and optical waves, it is not practical to represent such cases using circuit models at all. The only way to deal with these cases is through electromagnetic analysis.

There are other classes of problems that require a decent appreciation of electromagnetic analysis tools. Examples include (but not limited to) the following:

- Antenna analysis and design
- Wireless communication systems and devices
- Semiconductor devices and integrated circuits
- Computer chips (processors, memory, etc.)
- Circuit boards (planar, multilayer, flexible, etc.)
- Cables, connectors, and adaptors
- Discrete components (R, C, L, transformers, etc.)
- Power systems and machines (transformers, motors, generators, etc.)
- Magnetic devices (sensors, relays, actuators, etc.)
- MEMS and MW devices
- RADAR and remote sensing
- Microwave heating and curing
- EMI/EMC analysis
- Nanotechnology
- Photonics

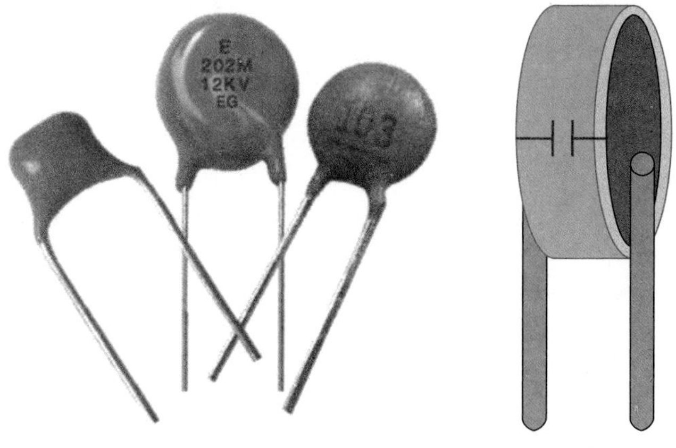

Figure 3.1 Examples of physical disc capacitors with a sketch of the internal physical structure and the corresponding simplified model.

To appreciate the argument we are making here, let us discuss an example of capacitance; one of the fundamental circuit elements that you have learned about before and will be discussed in more detail in later chapters.

A capacitor is simply formed of two conductors isolated from each other by a dielectric (insulator). Physical capacitors that we often see in circuit boards and electrical and electronic devices take many shapes; typical shapes are depicted in Figure 3.1. The figure shows the physical shapes of some "ceramic disk" capacitors and next to them is a simplified sketch of their internal physical structure.

The typical circuit analysis for such a device is a capacitive circuit element in the nominal capacitance value of the capacitor. Careful analysis shows that such simplified model is not accurate enough for certain applications, especially at high (RF) frequencies. Other "packaging" and "parasitic" elements are present that must be accounted for. Examples are dielectric leakage conductance (G), packaging capacitance (C_p), lead inductance (L_p, due to induced magnetic fields), and lead (wire) resistance (R_p). These are shown in Figure 3.2.

That being said, to be able to develop the details of such model, we need to understand and relate to the physical phenomenon of resistance, inductance, capacitance, and how the geometrical configuration and material properties among other factors influence their behavior. This is exactly what we are going to introduce to the learner in this text book. We use the term "introduce" since the topics involved are far wide and cannot be put all in one book, especially at the introductory level to the subject of EM.

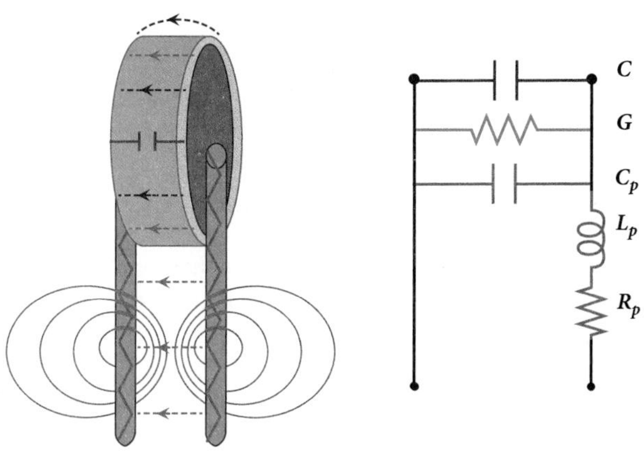

Figure 3.2 Packaging and parasitic elements for the disk capacitor that are critical for accurate high-frequency analysis.

3.2 More about the RLGC Parameters

As we recall from past studies of circuit theory, resistance and conductance, are both two sides to the same phenomenon. Both have to do with the ability of the charges to move and the "resistivity" or "conductivity" of the media they move or "flow" in. The conductance is simply the mathematical inverse of the resistance. In this regard, we learned that materials are divided into two main categories: conductors and insulators. A third category was added later in our studies as the semiconductors, which is by default an insulator that conducts in certain ways under certain conditions. Conductors are those materials that have amble of mobile "free" charges that can move if forced to. The motion of the free charges is known as the electric current. The amount of *impedance* this flow encounters is called the resistance, which depends on the properties of the material and the geometrical configuration and dimensions. The current flow against the resistance causes an electric potential drop and hence power dissipation.

As for the capacitance, it is associated with the presence of an insulating material (dielectric) that isolates two conductor surfaces. The capacitor can be charged (and discharged) by altering the charge amount on the conductor terminals of the capacitor. This, in turn, results in a potential difference across the capacitor terminals, which is an indication of energy storage in the capacitor. Energy can be stored (and retrieved) in the capacitor without loss as long as the insulating material is lossless and the two conductors are perfect. An ideal insulating material does not allow direct current flow in a capacitive circuit element. Yet in a dynamic situation, and as a result of charge fluctuations, a dynamic current is modeled in the circuit. We will learn later about this current as the *displacement* current in the dielectric insulator.

The inductance has to do with the magnetic behavior of the media as seen by an electric circuit. Typically, a conductor interaction (linkage) with a medium (magnetic or nonmagnetic) represents an inductance as modeled in the electrical circuit containing that conductor. A time varying current flowing in the inductor results in a potential difference across its terminals while a direct current does not produce any. Both direct and time varying currents result in static and dynamic energy storage in the inductance with no associated dissipation or loss. The inductance value depends on the magnetic (or nonmagnetic) properties of the medium, its geometrical configuration and dimensions, as well as the geometrical configuration and dimensions of the conductor (coil) and its shape of interaction (linkage) with the medium.

With all this background information about the capacitance, resistance, and inductance, it is clear that we need to study the electric and magnetic phenomena associated with static and dynamic charges and their interactions with dielectrics, conductors, and magnetic/nonmagnetic materials. We will do so in the next few chapters starting with static charges, charges moving with constant velocity generating direct currents, and finally the general case of time varying currents. The first two steps of this study will fall under the *statics* category, while the third one will be under *dynamics*.

Before we indulge ourselves in this static/dynamic study of electromagnetics, we will need to review some mathematical tools that we cannot do without. These tools are covered in the addenda of this chapter, while we start our statics study in the next chapter.

ADDENDUM 3A
COORDINATE SYSTEMS

3A.1 Introduction

Coordinate systems are used to define positions/locations in space relative to some arbitrary origin. Three orthonormal coordinates sufficiently define points and locations in three-dimensional space. There are three acknowledged coordinate systems in this regard which we will review here in this chapter; they are the Cartesian, the cylindrical, and the spherical coordinate systems.

3A.2 The Cartesian Coordinate System

The Cartesian coordinate system is the most commonly used as it utilizes three orthonormal constant-direction coordinates. Also, because most of the physical structures are based on rectangular shapes. The convenience of

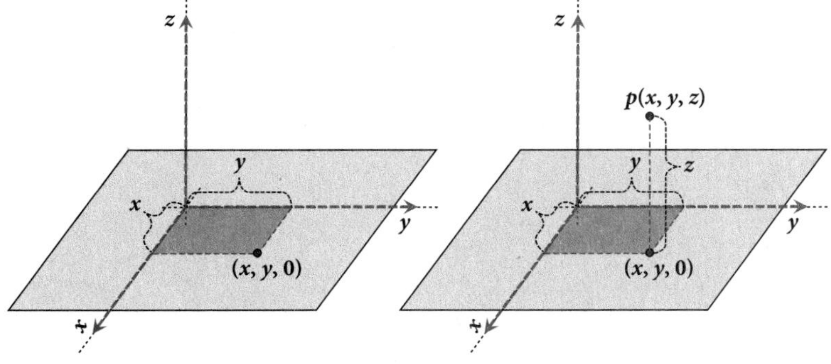

FIGURE 3A.1 The basics of the Cartesian coordinate system.

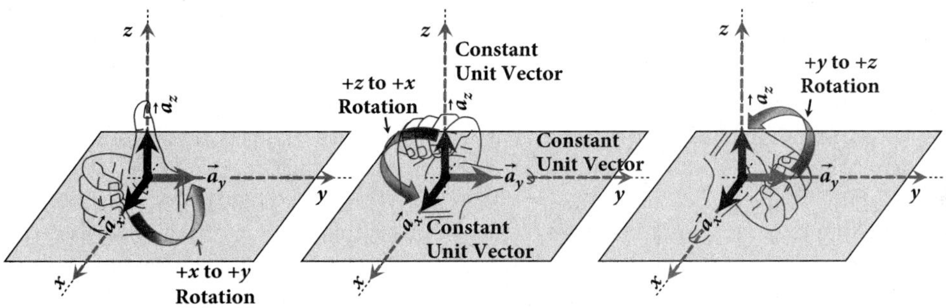

FIGURE 3A.2 The three-unit vectors are orthonormal and form right-hand sets.

having fixed direction for its three coordinated pays off in many analyses as we will see in our studies in this book and elsewhere.

In Cartesian coordinates, we establish three orthonormal lines/directions/coordinates that stem from an arbitrary origin. Typically, two of these coordinates are laid flat on a horizontal plane; these are typically chosen as the *x* and *y* coordinates (Figure 3A.1). A third coordinate is made to emerge orthonormal to the *x*-*y* plane and is called the *z* coordinate. The positive directions of the three coordinates are chosen to form a right-hand system in the *x*-*y*-*z*-*x* order (Figure 3A.2). The figure also shows the three-unit vectors $\vec{a}_x$, $\vec{a}_y$, and $\vec{a}_z$ for this coordinate system.

In the Cartesian coordinate system, a point *p* is specified by the three coordinates *x*, *y*, and *z*; these are the normal distances between the point *p* and the *y*-*z* plane, *z*-*x* plane, and the *x*-*y* plane, respectively (Figure 3A.3).

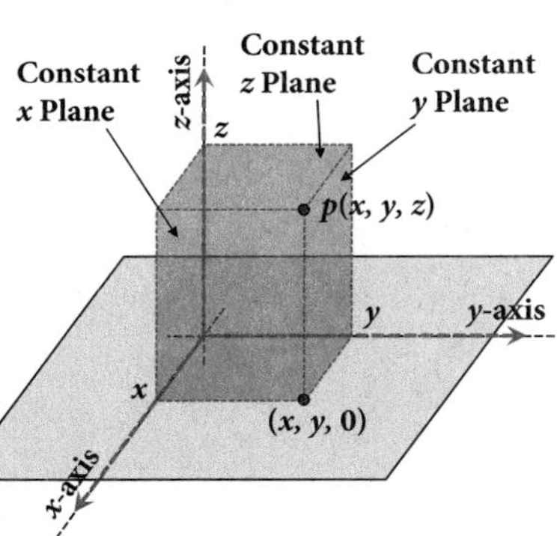

FIGURE 3A.3 Constant *x*, *y*, and *z* planes.

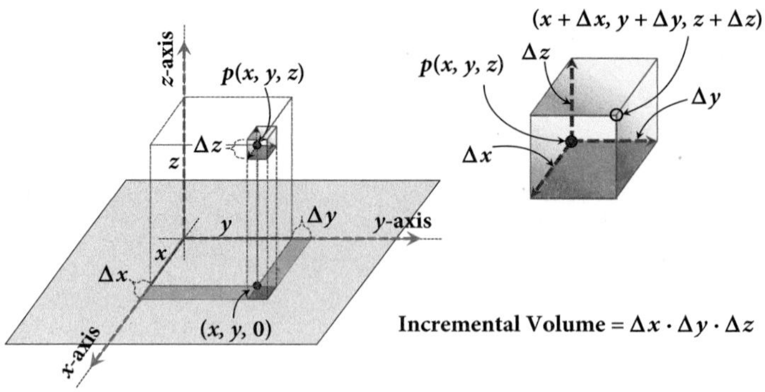

Figure 3A.4 The incremental volume in the Cartesian system.

For integration purposes, incremental elements at p correspond to increasing the coordinate values x, y, and z by Δx, Δy, and Δz, respectively (Figure 3A.4). The corresponding volume for the incremental rectangular prism is the product of all three sides $\Delta v = \Delta x \cdot \Delta y \cdot \Delta z$.

The vector representation of the three incremental lengths is given in Figure 3A.5 while the corresponding three incremental surface area vectors are shown in Figure 3A.6. More discussion of vectors will follow in the next addendum of this chapter.

We need to recognize that the unit vectors $\vec{a}_x$, $\vec{a}_y$, and $\vec{a}_z$ are constant for all points in space for this coordinate system. They are constant in magnitude (which is unity) and constant in directions as well. Length vectors have directions along the lengths themselves, while the direction of an area vector is in the outward normal of the face of that area.

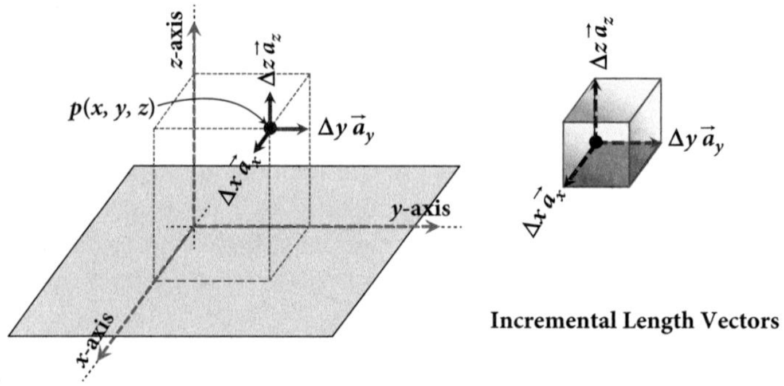

Figure 3A.5 The incremental length vectors in the Cartesian system.

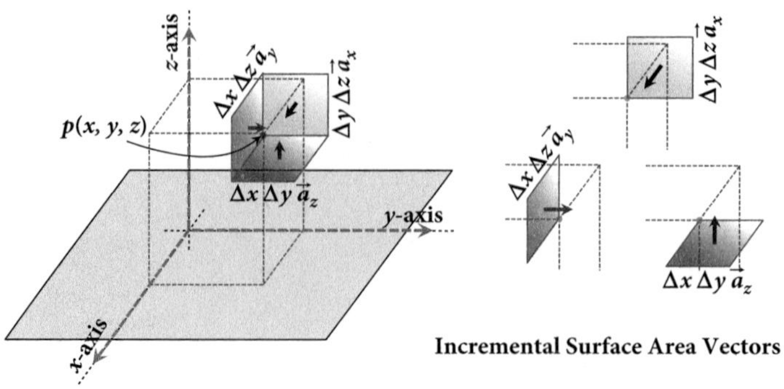

Figure 3A.6 The incremental surface area vectors in the Cartesian system.

3A.3 The Cylindrical Coordinate System

Cylindrical coordinates are useful to express positions and location when a spatial configuration is essentially cylindrical. In this case, Figure 3A.7 shows the three orthonormal cylindrical coordinates ρ, φ, and z as laid over the corresponding Cartesian system chosen with a common z axis. In this figure, ρ is the radius of a vertical cylinder passing by the position, φ is the angle between the vertical plane containing the z axis and the position and the z-x plane; finally, z is the height of the position above the x-y plane. The unit vectors along the three coordinates are demonstrated in Figure 3A.8. The positive directions of the three coordinates are chosen to form a right-handed system in the ρ-φ-z-ρ order (Figure 3A.9).

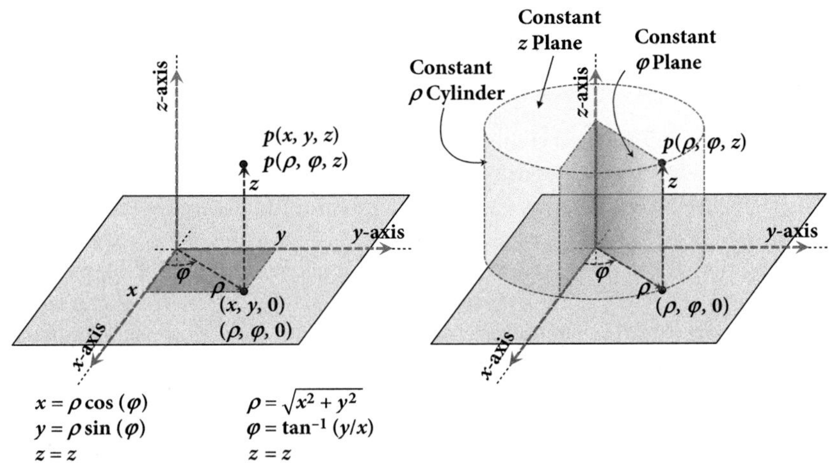

FIGURE 3A.7 The basics of the cylindrical coordinate system and the constant ρ, φ, and z surfaces.

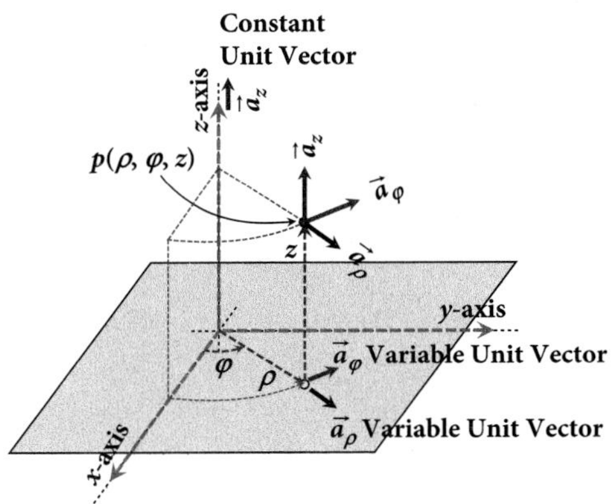

FIGURE 3A.8 The three-unit vectors in the cylindrical system.

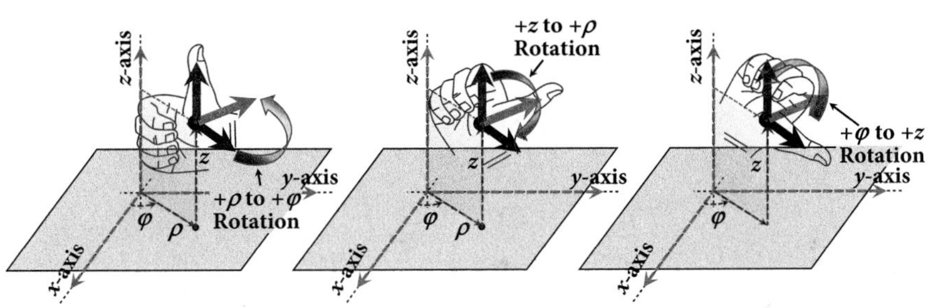

FIGURE 3A.9 The three coordinates are orthonormal and form right-hand sets.

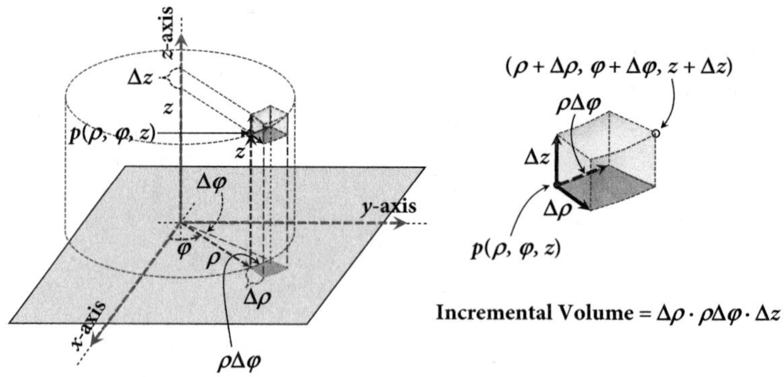

FIGURE 3A.10 The incremental volume in the cylindrical system.

For integration purposes, incremental elements at p correspond to increasing the coordinate values ρ, φ, and z by $\Delta\rho$, $\Delta\varphi$, and Δz, respectively (Figure 3A.10). The side lengths of the corresponding incremental *rectangular* prism are $\Delta\rho$, $\rho\Delta\varphi$, and Δz, respectively. The corresponding incremental volume is the product of all three sides $\Delta v = \Delta\rho \cdot \rho\Delta\varphi \cdot \Delta z$.

The vector representation of the three incremental lengths is given in Figure 3A.11, while the corresponding three incremental surface area vectors are shown in Figure 3A.12. Unlike the case with Cartesian coordinates, the unit vectors $\vec{a}_\rho$ and $\vec{a}_\varphi$ are variable vectors as they both vary in direction as we move around in space. However, the unit vector $\vec{a}_z$ is the same as that of Cartesian coordinates and hence is a constant vector for all points in space.

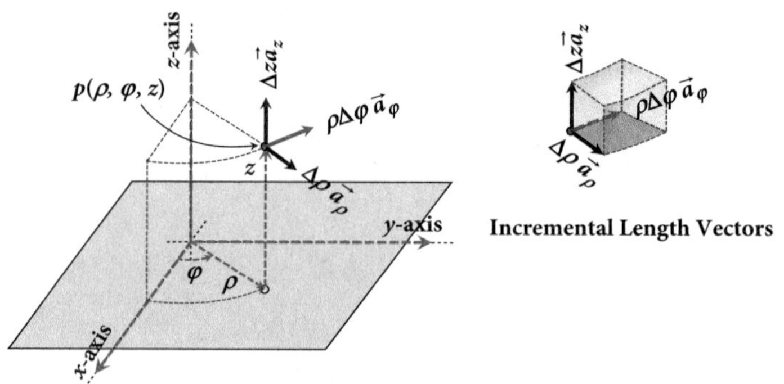

$$\vec{a}_\rho = \cos\varphi\,\vec{a}_x + \sin\varphi\,\vec{a}_y \qquad \vec{a}_x = \cos\varphi\,\vec{a}_\rho - \sin\varphi\,\vec{a}_\varphi$$
$$\vec{a}_\varphi = -\sin\varphi\,\vec{a}_x + \cos\varphi\,\vec{a}_y \qquad \vec{a}_y = \sin\varphi\,\vec{a}_\rho + \cos\varphi\,\vec{a}_\varphi$$
$$\vec{a}_z = \vec{a}_z \qquad \vec{a}_z = \vec{a}_z$$

FIGURE 3A.11 The incremental length vectors in the cylindrical system.

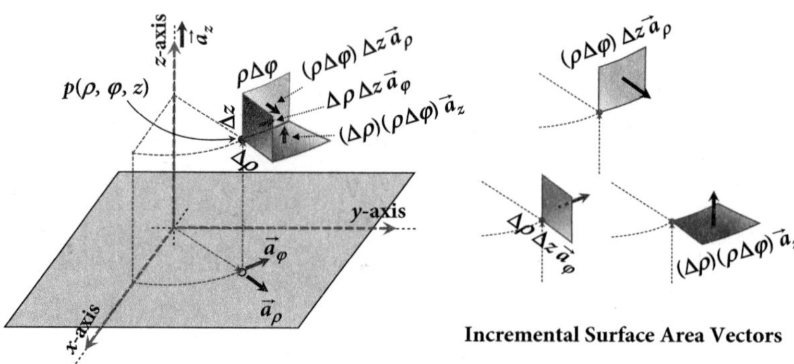

FIGURE 3A.12 The incremental surface area vectors in the cylindrical system.

3A.4 The Spherical Coordinate System

Spherical coordinates are typically used in cases where a spatial configuration is essentially spherical. The three orthonormal spherical coordinates are r, θ, and φ (Figure 3A.13). This figure shows all three coordinate systems simultaneously in order to demonstrate their relationship to each other.

As shown in the figure, r is the radius of a sphere passing by the position p while being concentric with the origin. The angle θ is the second coordinate parameter and is the angle between the radial line connecting the position to the origin and the z axis. The angle θ can also be defined as the angle of the cone whose head is the origin, coaxial with the z axis and passing by the position p. Finally, the angle φ is the same as defined for cylindrical coordinates, that is, the angle between the vertical plane containing the z axis and the position p and the z-x plane.

In Figure 3A.14, we demonstrate the three orthonormal unit vectors $\vec{a}_r$, $\vec{a}_\theta$, and $\vec{a}_\varphi$. Again, the positive directions of the three coordinates are chosen to form a right-handed system in the r-θ-φ-r order (Figure 3A.15).

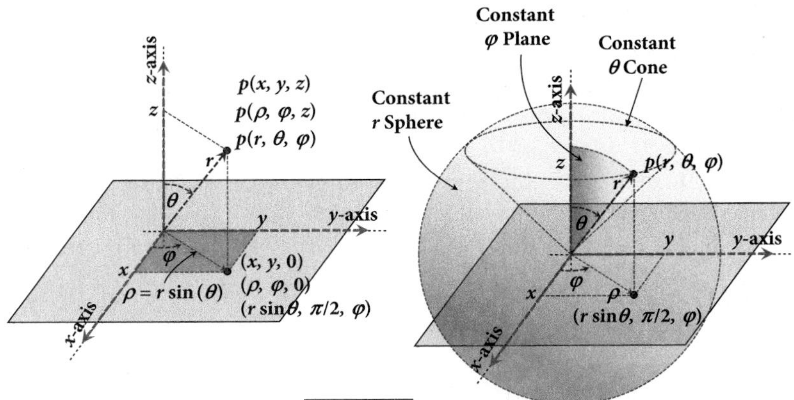

$$x = r\sin(\theta)\cos(\varphi) \qquad r = \sqrt{x^2+y^2+z^2}$$
$$y = r\sin(\theta)\sin(\varphi) \qquad \theta = \cos^{-1}(z/r)$$
$$z = r\cos(\theta) \qquad \varphi = \tan^{-1}(y/x)$$

FIGURE 3A.13 The basics of the spherical coordinate system and the constant r, θ, and φ surfaces.

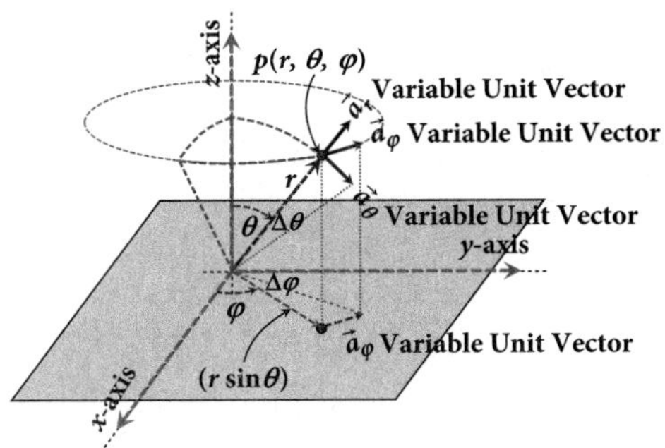

FIGURE 3A.14 The three-unit vectors in the spherical system.

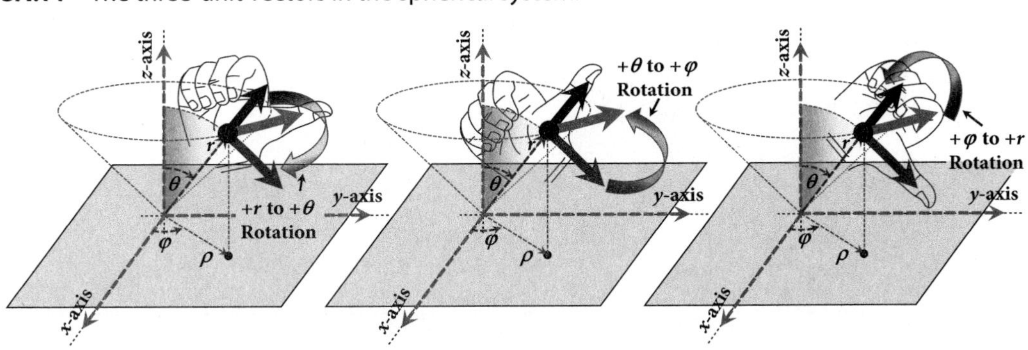

FIGURE 3A.15 The three-unit vectors are orthonormal and form right-hand sets.

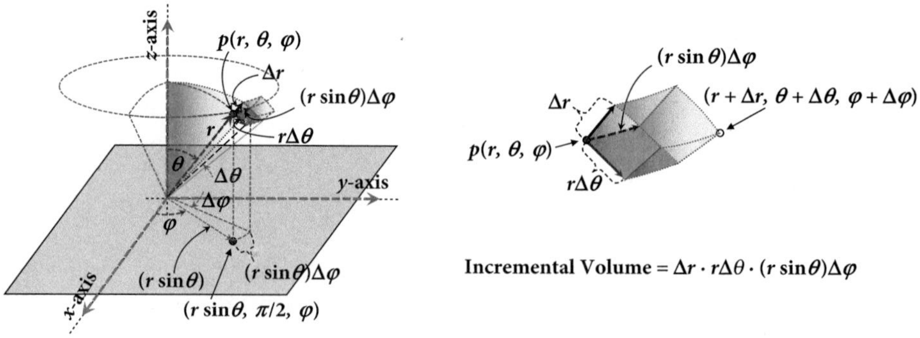

FIGURE 3A.16 The incremental volume in the spherical system.

For integration purposes, incremental elements at p correspond to increasing the coordinate values r, θ, and φ by Δr, $\Delta\theta$, and $\Delta\varphi$, respectively (Figure 3A.16). The side lengths of the corresponding incremental *rectangular* prism are Δr, $r\Delta\theta$, and $r\sin(\theta)\Delta\varphi$, respectively. The corresponding incremental volume is the product of all three sides $\Delta v = \Delta r \cdot r\Delta\theta \cdot r\sin(\theta)\Delta\varphi$.

The vector representation of the three incremental lengths is given in Figure 3A.17 while the corresponding three incremental surface area vectors are shown in Figure 3A.18. Needless to say, all three unit-vectors $\vec{a}_r$, $\vec{a}_\theta$, and $\vec{a}_\varphi$ are variable vectors as they all vary in direction as we move around in space.

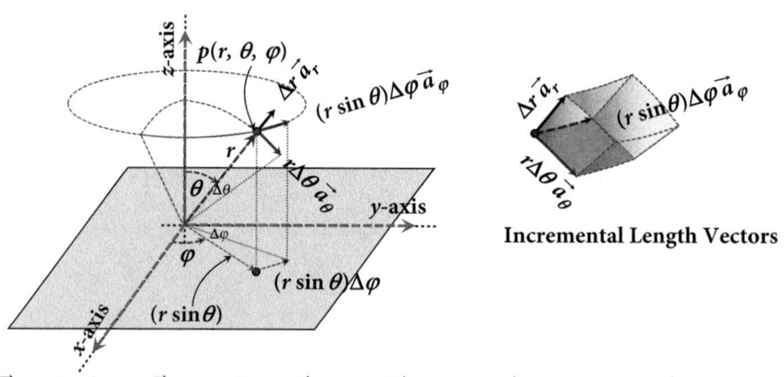

$$\vec{a}_r = \sin\theta\cos\varphi\,\vec{a}_x + \sin\theta\sin\varphi\,\vec{a}_y + \cos\theta\,\vec{a}_z$$
$$\vec{a}_\theta = \cos\theta\cos\varphi\,\vec{a}_x + \cos\theta\sin\varphi\,\vec{a}_y - \sin\theta\,\vec{a}_z$$
$$\vec{a}_\varphi = -\sin\varphi\,\vec{a}_x + \cos\varphi\,\vec{a}_y$$

$$\vec{a}_x = \sin\theta\cos\varphi\,\vec{a}_r + \cos\theta\cos\varphi\,\vec{a}_\theta - \sin\varphi\,\vec{a}_\varphi$$
$$\vec{a}_y = \sin\theta\sin\varphi\,\vec{a}_r + \cos\theta\sin\varphi\,\vec{a}_\theta + \cos\varphi\,\vec{a}_\varphi$$
$$\vec{a}_z = \cos\theta\,\vec{a}_r - \sin\theta\,\vec{a}_\theta$$

FIGURE 3A.17 The incremental length vectors in the spherical system.

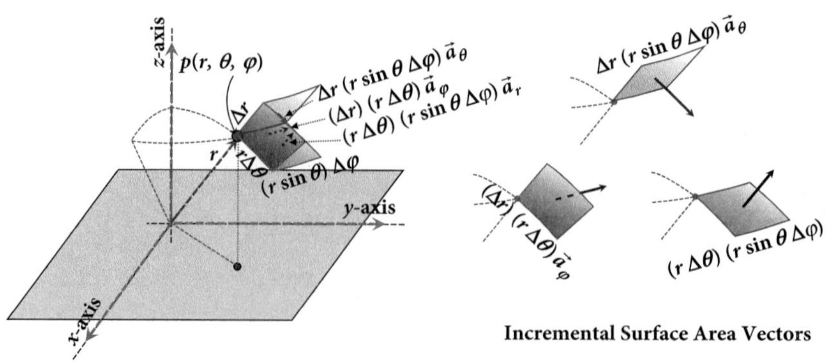

FIGURE 3A.18 The incremental surface area vectors in the spherical system.

3A.5 Relationships between Coordinate Systems

Frequently, in the course of analytical procedures, we would be faced with the need to convert from one coordinate system to another. Typically, the conversion would be needed between cylindrical and Cartesian and between spherical and Cartesian. In some cases, this is needed for convenience in carrying out a *difficult* integration or to identify with a specific coordinate-related phenomenon. However, in many cases it is done to avoid dealing with variable unit vectors. Converting to Cartesian coordinates replaces the variable unit vectors with constant ones that facilitates integration procedures involving vectors.

Examining Figures 3A.7 and 3A.8 for cylindrical coordinates and Figures 3A.12 and 3A.13 for spherical coordinates, we can write down the relationships for the coordinate parameters and the corresponding unit vectors as laid down in Tables 3A.1 and 3A.2.

3A.6 Vector Expressions

Also, for a vector $\vec{F}$ expressed in Cartesian, cylindrical, and spherical coordinates as (Table 3A.3)

$$\vec{F} = F_x\vec{a}_x + F_y\vec{a}_y + F_z\vec{a}_z$$

$$\vec{F} = F_\rho\vec{a}_\rho + F_\varphi\vec{a}_\varphi + F_z\vec{a}_z$$

$$\vec{F} = F_r\vec{a}_r + F_\theta\vec{a}_\theta + F_\varphi\vec{a}_\varphi$$

Cylindrical–Cartesian	Spherical–Cartesian	Cylindrical–Spherical
$\rho=\sqrt{x^2+y^2}$ $\varphi=\tan^{-1}(y/x)$ $z=z$	$r=\sqrt{x^2+y^2+z^2}$ $\theta=\cos^{-1}(z/r)$ $\varphi=\tan^{-1}(y/x)$	$\rho=r\sin\theta$ $\varphi=\varphi$ $z=r\cos\theta$
$x=\rho\cos\varphi$ $y=\rho\sin\varphi$ $z=z$	$x=r\sin\theta\cos\varphi$ $y=r\sin\theta\sin\varphi$ $z=r\cos\theta$	$r=\sqrt{\rho^2+z^2}$ $\theta=\cos^{-1}(z/r)$ $\varphi=\varphi$

TABLE 3A.1 Coordinate Relations

Cylindrical–Cartesian	Spherical–Cartesian	Cylindrical–Spherical
$\vec{a}_\rho=\cos\varphi\vec{a}_x+\sin\varphi\vec{a}_y$ $\vec{a}_\varphi=-\sin\varphi\vec{a}_x+\cos\varphi\vec{a}_y$ $\vec{a}_z=\vec{a}_z$	$\vec{a}_r=\sin\theta\cos\varphi\vec{a}_x+\sin\theta\sin\varphi\vec{a}_y+\cos\theta\vec{a}_z$ $\vec{a}_\theta=\cos\theta\cos\varphi\vec{a}_x+\cos\theta\sin\varphi\vec{a}_y-\sin\theta\vec{a}_z$ $\vec{a}_\varphi=-\sin\varphi\vec{a}_x+\cos\varphi\vec{a}_y$	$\vec{a}_r=\vec{a}_\rho\sin\theta+\vec{a}_z\cos\theta$ $\vec{a}_\theta=\vec{a}_\rho\cos\theta-\vec{a}_z\sin\theta$ $\vec{a}_\varphi=\vec{a}_\varphi$
$\vec{a}_x=\cos\varphi\vec{a}_\rho-\sin\varphi\vec{a}_\varphi$ $\vec{a}_y=\sin\varphi\vec{a}_\rho+\cos\varphi\vec{a}_\varphi$ $\vec{a}_z=\vec{a}_z$	$\vec{a}_x=\sin\theta\cos\varphi\vec{a}_r+\cos\theta\cos\varphi\vec{a}_\theta-\sin\varphi\vec{a}_\varphi$ $\vec{a}_y=\sin\theta\sin\varphi\vec{a}_r+\cos\theta\sin\varphi\vec{a}_\theta+\cos\varphi\vec{a}_\varphi$ $\vec{a}_z=\cos\theta\vec{a}_r-\sin\theta\vec{a}_\theta$	$\vec{a}_\rho=\vec{a}_r\sin\theta+\vec{a}_\theta\cos\theta$ $\vec{a}_z=\vec{a}_r\cos\theta-\vec{a}_\theta\sin\theta$ $\vec{a}_\varphi=\vec{a}_\varphi$

TABLE 3A.2 Unit Vectors Relationships

Cylindrical to Cartesian	Spherical to Cartesian
$F_\rho=F_x\cos\varphi+F_y\sin\varphi$ $F_\varphi=-F_x\sin\varphi+F_y\cos\varphi$ $F_z=F_z$	$F_r=F_x\sin\theta\cos\varphi+F_y\sin\theta\sin\varphi+F_z\cos\theta$ $F_\theta=F_x\cos\theta\cos\varphi+F_y\cos\theta\sin\varphi-F_z\sin\theta$ $F_\varphi=-F_x\sin\varphi+F_y\cos\varphi$
$F_x=F_\rho\cos\varphi-F_\varphi\sin\varphi$ $F_y=-F_\rho\sin\varphi+F_\varphi\cos\varphi$ $F_z=F_z$	$F_x=F_r\sin\theta\cos\varphi+F_\theta\cos\theta\cos\varphi-F_\varphi\sin\varphi$ $F_y=F_r\sin\theta\sin\varphi+F_\theta\cos\theta\sin\varphi+F_\varphi\cos\varphi$ $F_z=F_r\cos\theta-F_\theta\sin\theta$

TABLE 3A.3 Vector Expression Relationships

ADDENDUM 3B
VECTOR CALCULUS AND VECTOR IDENTITIES

3B.1 Vector Definition and Examples

A vector is a quantity characterized by having a specific magnitude and a specific direction at each point in space. In this book, we use the accented notation to denote vector, for example, a vector "F" will appear as $\vec{F}$. A vector can be expressed as a scalar multiplied by a unit vector that has the same direction as the vector itself:

$$\vec{F} = F\vec{a}_F \tag{3B.1}$$

Examples of vectors include distance, velocity, and force among many other physical quantities. Table 3B.1 demonstrates some of these vector quantities.

Vector	Examples
Length	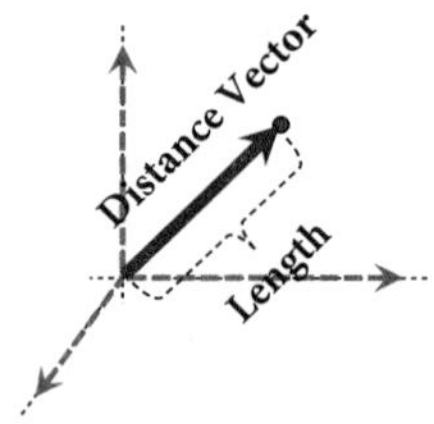
Velocity	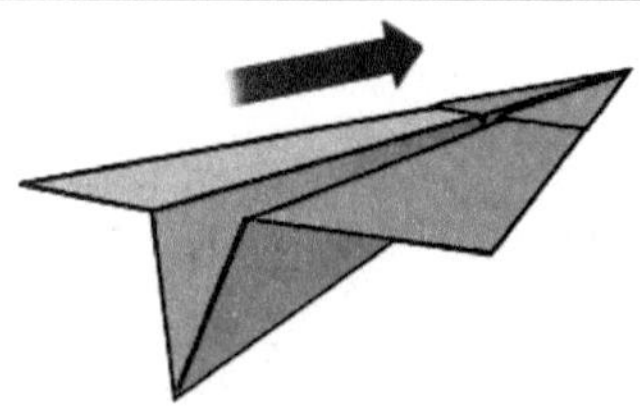
Weight	
Surface area	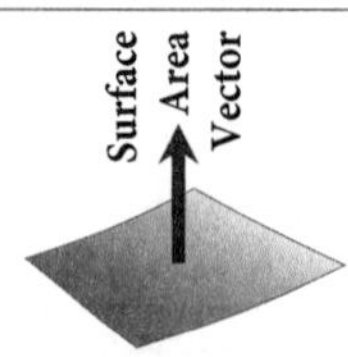
Force	
Torque	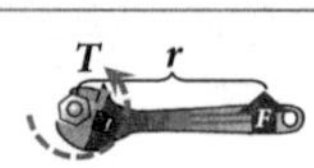

TABLE 3B.1 Examples of Vectors

3B.2 Vector Representations in Coordinate Systems

To express a vector in a coordinate system is to write the vector in terms of the three orthonormal components of that coordinate system. This is done by projecting the vector along the three directions of the coordinate system and finding the magnitude of each projection. If we denote the three projections of a vector $\vec{F}$ along the three coordinates of a "1-2-3" coordinate system as F_1, F_2, and F_3, respectively, we can write

$$\vec{F} = \vec{F}_1 + \vec{F}_2 + \vec{F}_3 = F_1\vec{a}_1 + F_2\vec{a}_2 + F_3\vec{a}_3 \tag{3B.2}$$

where $\vec{a}_1$, $\vec{a}_2$, and $\vec{a}_3$ are the corresponding unit vectors of the coordinate system (Figure 3B.1).

In the following, we overview vector representation in the three coordinate systems. Two types of vectors are demonstrated: distance vectors and others. The reason for this classification is that distance vectors are expressed in terms of the coordinate system dimensions and directions while the others will have different units but only expressed in terms of the coordinate system directions.

3B.2.1 Vector Representation in a Cartesian Coordinate System

Figure 3B.2 shows two types of distance vectors as represented in Cartesian coordinates; distance from the origin and distance between two points. The distance from the origin vector that we denote with the lower case r is, in fact, the same as the spherical coordinate vector $\vec{r}$, which can be written as

$$\vec{r} = x\vec{a}_x + y\vec{a}_y + z\vec{a}_z \tag{3B.3}$$

The distance between two points (1 and 2) will be denoted by the upper-case vector $\vec{R}$. This can be expressed as the difference between two $\vec{r}$ vectors as follows:

$$\vec{R}_{12} = \vec{r}_2 - \vec{r}_1 = (x_2\vec{a}_x + y_2\vec{a}_y + z_2\vec{a}_z) - (x_1\vec{a}_x + y_1\vec{a}_y + z_1\vec{a}_z) = (x_2 - x_1)\vec{a}_x + (y_2 - y_1)\vec{a}_y + (z_2 - z_1)\vec{a}_z \tag{3B.4}$$

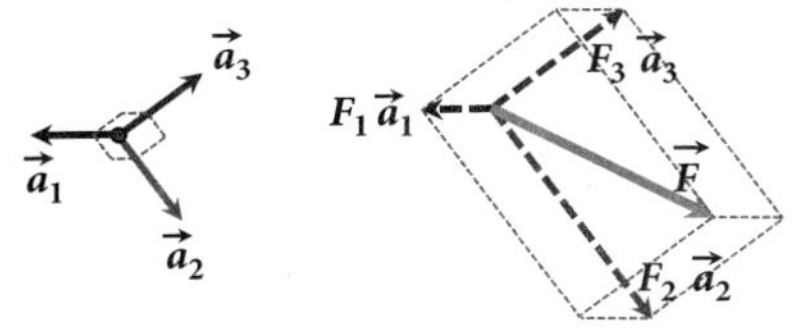

FIGURE 3B.1 A generic coordinate system with three orthonormal unit vectors.

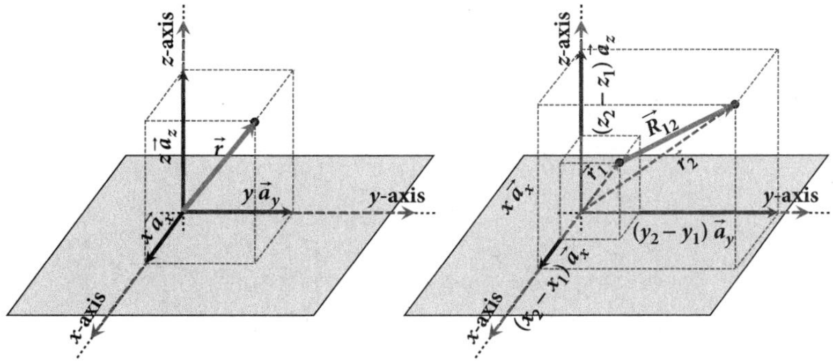

FIGURE 3B.2 Distance vector representation in the Cartesian coordinate system.

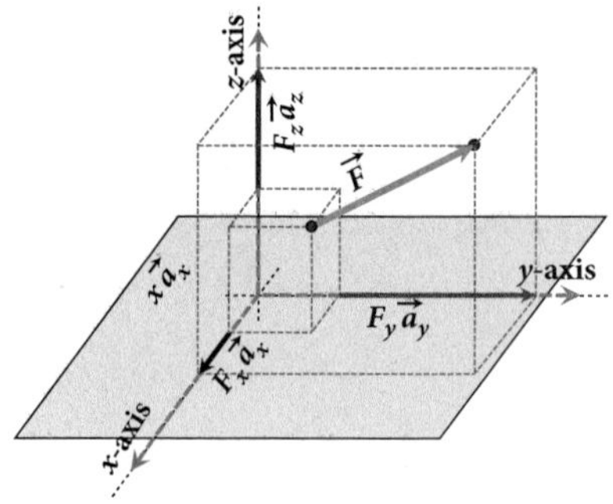

$$\vec{F}=F_x\vec{a}_x+F_y\vec{a}_y+F_z\vec{a}_z$$

Figure 3B.3 General vector representation in the Cartesian coordinate system.

3B.2.2 Vector Representation for a General Form Vector

Next, we view the vector representation for a general form vector $\vec{F}$ in Cartesian coordinates. This is demonstrated in Figure 3B.3 yielding the analytical expression:

$$\vec{F}=\vec{F}_x+\vec{F}_y+\vec{F}_z=F_x\vec{a}_x+F_y\vec{a}_y+F_z\vec{a}_z \tag{3B.5}$$

3B.2.3 Vector Representation in Cylindrical Coordinates

Figure 3B.4 shows the two types of distance vectors as represented in cylindrical coordinates. The distance from the origin vector $\vec{r}$ can be written as

$$\vec{r}=\rho\vec{a}_\rho+z\vec{a}_z \tag{3B.6}$$

It is to be noted here that the $\vec{a}_\rho$ unit vector is a variable one that may cause some analytical challenges in some cases especially when integration is involved. When these challenges dominate, it may be *wiser* to switch to the Cartesian coordinate representation where all the unit vectors are constants.

The distance vector $\vec{R}_{12}$ between two points is also demonstrated in the figure. When expressed as the difference between two $\vec{r}$ vectors we will be challenged by the vector subtraction of $\vec{a}_{\rho 2}$ and $\vec{a}_{\rho 1}$:

$$\begin{aligned}\vec{R}_{12}=\vec{r}_2-\vec{r}_1&=(\rho_2\vec{a}_{\rho 2}+z_2\vec{a}_z)-(\rho_1\vec{a}_{\rho 1}+z_1\vec{a}_z)=(\rho_2\vec{a}_{\rho 2}-\rho_1\vec{a}_{\rho 1})+(z_2-z_1)\vec{a}_z\\&=(x_2-x_1)\vec{a}_x+(y_2-y_1)\vec{a}_y+(z_2-z_1)\vec{a}_z\end{aligned} \tag{3B.7}$$

Again, resorting to Cartesian coordinates offers a simple way to deal with this challenge.

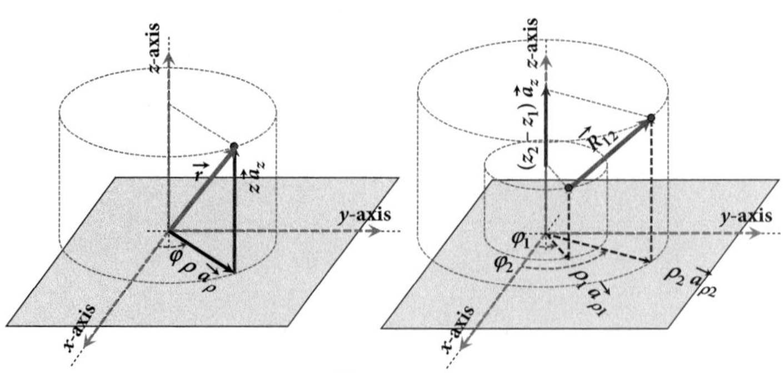

$$\vec{r}=\rho\,\vec{a}_\rho+z\,\vec{a}_z$$

$$\begin{aligned}\vec{R}_{12}&=\vec{r}_2-\vec{r}_1\\&=(\rho_2\,\vec{a}_{\rho 2}+z_2\,\vec{a}_z)-(\rho_1\,\vec{a}_{\rho 1}+z_1\,\vec{a}_z)\\&=(\rho_2\,\vec{a}_{\rho 2}-\rho_1\,\vec{a}_{\rho 1})+(z_2-z_1)\,\vec{a}_z\\&=(x_2-x_1)\,\vec{a}_x+(y_2-y_1)\,\vec{a}_y+(z_2-z_1)\,\vec{a}_z\end{aligned}$$

Figure 3B.4 Distance vector representation in the cylindrical coordinate system.

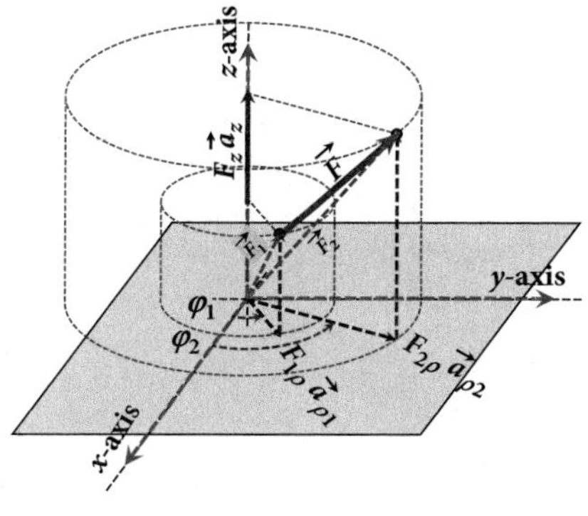

$$\begin{aligned}\vec{F} &= \vec{F}_2 - \vec{F}_1\\ &= (F_{2\rho}\,\vec{a}_{\rho 2} + F_{2z}\,\vec{a}_z) - (F_{1\rho}\,\vec{a}_{\rho 1} + F_{1z}\,\vec{a}_z)\\ &= (F_{2\rho}\,\vec{a}_{\rho 2} - F_{1\rho}\,\vec{a}_{\rho 1}) + (F_{2z} - F_{1z})\,\vec{a}_z\\ &= F_x\vec{a}_x + F_y\vec{a}_y + F_z\vec{a}_z\end{aligned}$$

FIGURE 3B.5 General vector representation in the cylindrical coordinate system.

Similarly, the general form vector $\vec{F}$ in cylindrical coordinates is demonstrated in Figure 3B.5. The corresponding analytical expression is

$$\begin{aligned}\vec{F} = \vec{F}_2 - \vec{F}_1 &= (F_{2\rho}\vec{a}_{\rho 2} + F_{2z}\vec{a}_z) - (F_{1\rho}\vec{a}_{\rho 1} + F_{1z}\vec{a}_z)\\ &= (F_{2\rho}\vec{a}_{\rho 2} - F_{1\rho}\vec{a}_{\rho 1}) + (F_{2z} - F_{1z})\vec{a}_z\\ &= F_x\vec{a}_x + F_y\vec{a}_y + F_z\vec{a}_z\end{aligned} \tag{3B.8}$$

3B.2.4 Vector Representation in Spherical Coordinates

The case of spherical coordinates (Figure 3B.6) has challenges similar to those discussed in cylindrical coordinate representation; all three-unit vectors are variables. Switching to Cartesian offers the same convenience as discussed above.

$$\vec{F} = \vec{F}_2 - \vec{F}_1 = F_x\vec{a}_x + F_y\vec{a}_y + F_z\vec{a}_z \tag{3B.9}$$

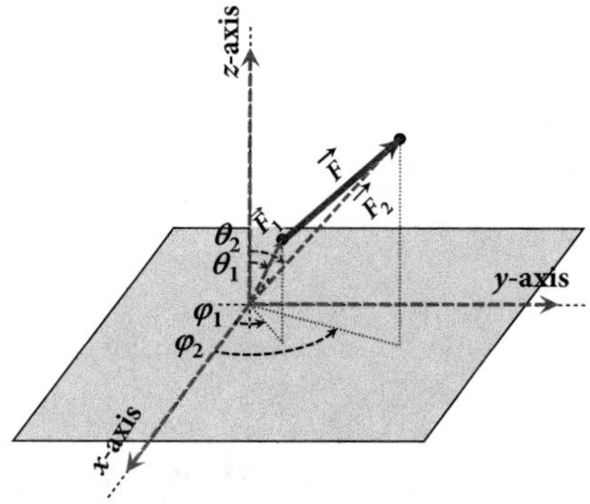

$$\begin{aligned}\vec{F} &= \vec{F}_2 - \vec{F}_1\\ &= F_x\vec{a}_x + F_y\vec{a}_y + F_z\vec{a}_z\end{aligned}$$

FIGURE 3B.6 General vector representation in the spherical coordinate system.

3B.3 Vector Operations

Tables 3B.2 to 3B.5 review various vector operations and give examples of their graphical and physical applications, as well as their representations in the three coordinate systems. The arrows in these tables indicate cases where variable vectors present mathematical challenges. In such cases, we may resort to the Cartesian representation where the unit vectors have constant directions at all times.

Addition and subtraction	 $\vec{A}\pm\vec{B}=\pm\vec{B}+\vec{A}$
Cartesian	$\vec{A}\pm\vec{B}=A_x\vec{a}_x+A_y\vec{a}_y+A_z\vec{a}_z\pm(B_x\vec{a}_x+B_y\vec{a}_y+B_z\vec{a}_z)$ $=(A_x\pm B_x)\vec{a}_x+(A_y\pm B_y)\vec{a}_y+(A_z\pm B_z)\vec{a}_z$
Cylindrical	$\vec{A}\pm\vec{B}=(A\vec{a}_A\pm B\vec{a}_B)\rightarrow=(A_x\pm B_x)\vec{a}_x+(A_y\pm B_y)\vec{a}_y+(A_z\pm B_z)\vec{a}_z$
Spherical	$\vec{A}\pm\vec{B}=(A\vec{a}_A\pm B\vec{a}_B)\rightarrow=(A_x\pm B_x)\vec{a}_x+(A_y\pm B_y)\vec{a}_y+(A_z\pm B_z)\vec{a}_z$

Table 3B.2 Vector Addition and Subtraction

Scaling	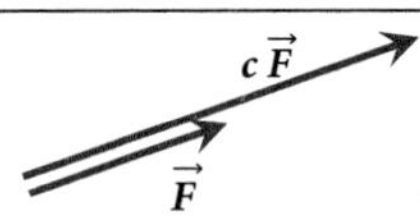
Cartesian	$c(A_x\vec{a}_x+A_y\vec{a}_y+A_z\vec{a}_z)=cA_x\vec{a}_x+cA_y\vec{a}_y+cA_z\vec{a}_z$
Cylindrical	$c(A\vec{a}_A)=cA\vec{a}_A$
Spherical	$c(A\vec{a}_A)=cA\vec{a}_A$

Table 3B.3 Vector Scaling

Scalar (dot) Product	$W=$ work done by force $\vec{F}$ along the path $\vec{\ell}$ 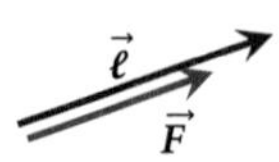 $W=\vec{F}\cdot\vec{\ell}=F\ell$ 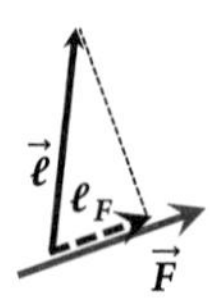 $W=\vec{F}\cdot\vec{\ell}=F\ell_F$
	$I=$ Current flow through the surface $\vec{S}$ due to area density $\vec{J}$ 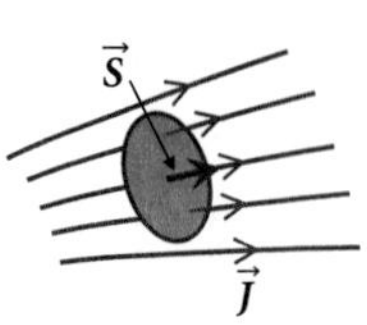 $I=\vec{J}\cdot\vec{S}=JS$ 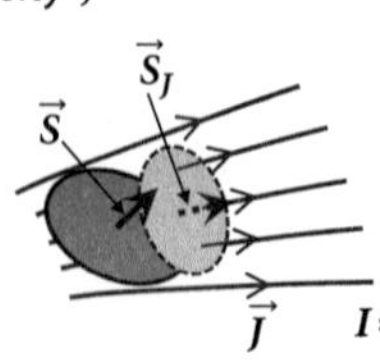 $I=\vec{J}\cdot\vec{S}=JS_J$

Table 3B.4 Vector Dot Product

	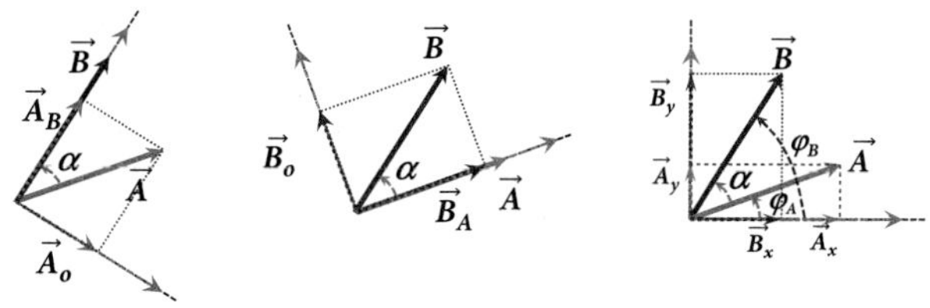
	$\vec{A}\cdot\vec{B}=[A_B]B=[A\cos(\alpha)]B=AB\cos(\alpha)$ $\vec{A}\cdot\vec{B}=A[B_A]=A[B\cos(\alpha)]=AB\cos(\alpha)$ $\vec{A}\cdot\vec{B}=A_xB_x+A_yA_y=[A\cos(\varphi_A)B\cos(\varphi_B)+A\sin(\varphi_A)B\sin(\varphi_B)]=AB[\cos(\varphi_A)\cos(\varphi_B)+\sin(\varphi_A)\sin(\varphi_B)]$ $=AB[\cos(\varphi_B-\varphi_A)]=AB\cos(\alpha)$
Cartesian	$\vec{A}\cdot\vec{B}=(A_x\vec{a}_x+A_y\vec{a}_y+A_z\vec{a}_z)\cdot(B_x\vec{a}_x+B_y\vec{a}_y+B_z\vec{a}_z)=A_xB_x+A_yB_y+A_zB_z$
Cylindrical	$\vec{A}\cdot\vec{B}=(A\vec{a}_A)\cdot(B\vec{a}_B)=AB(\vec{a}_A\cdot\vec{a}_B)$
Spherical	$\vec{A}\cdot\vec{B}=(A\vec{a}_A)\cdot(B\vec{a}_B)=AB(\vec{a}_A\cdot\vec{a}_B)$

Table 3B.4 Vector Dot Product (*Continued*)

Vector (Cross) Product	$\vec{T}$ = Torque vector due to a force $\vec{F}$ and an arm $\vec{d}$ 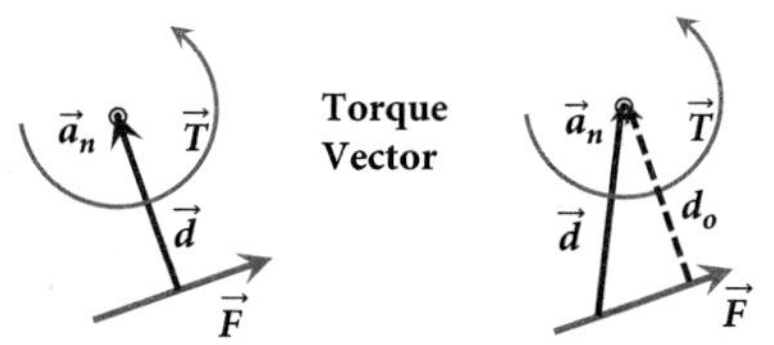 $\vec{T}=\vec{F}\times\vec{d}=F\,d\,\vec{a}_n$ $\quad$ $\vec{T}=\vec{F}\times\vec{d}=F\,d_o\,\vec{a}_n$
	$\vec{S}$ = Area vector of a parallelogram with sides $\vec{w}$ and $\vec{\ell}$ 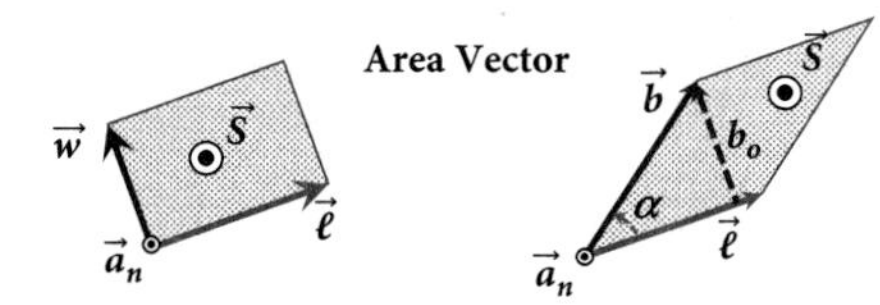 $\vec{S}=\vec{\ell}\times\vec{w}=\ell\,w\,\vec{a}_n$ $\quad$ $\vec{S}=\vec{\ell}\times\vec{b}=\ell\,b_o\vec{a}_n$
	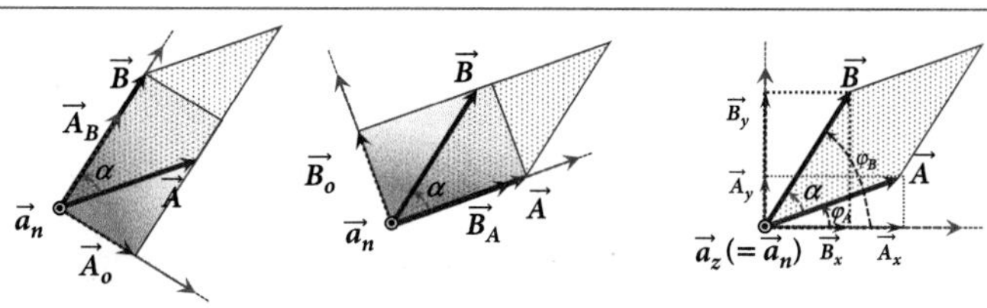 $\vec{A}\times\vec{B}=[A_o]B\vec{a}_n=[A\sin(\alpha)]B\vec{a}_n=AB\sin(\alpha)\vec{a}_n$ $\vec{A}\times\vec{B}=A[B_o]\vec{a}_n=A[B\sin(\alpha)]\vec{a}_n=AB\sin(\alpha)\vec{a}_n$ $\vec{A}\times\vec{B}=[A_xB_y\vec{a}_n+A_yB_x(-\vec{a}_n)]=[A\cos(\varphi_A)B\sin(\varphi_B)-A\sin(\varphi_A)B\cos(\varphi_B)]\vec{a}_n$ $=AB[\cos(\varphi_A)\sin(\varphi_B)-\sin(\varphi_A)\cos(\varphi_B)]\vec{a}_n=AB[\sin(\varphi_B-\varphi_A)]\vec{a}_n=AB\sin(\alpha)\vec{a}_n$
	$\vec{A}\times\vec{A}=0,\ \vec{A}\times\vec{B}=-\vec{B}\times\vec{A}$

Table 3B.5 Vector Cross Product

Cartesian	$\vec{a}_x \times \vec{a}_y = \vec{a}_z, \quad \vec{a}_y \times \vec{a}_z = \vec{a}_x, \quad \vec{a}_z \times \vec{a}_x = \vec{a}_y$ $\vec{A}\times\vec{B} = (A_x\vec{a}_x + A_y\vec{a}_y + A_z\vec{a}_z)\times(B_x\vec{a}_x + B_y\vec{a}_y + B_z\vec{a}_z) = \begin{vmatrix} \vec{a}_x & \vec{a}_y & \vec{a}_z \\ A_x & A_y & A_z \\ B_x & B_y & B_z \end{vmatrix}$
Cylindrical	$\vec{a}_\rho \times \vec{a}_\varphi = \vec{a}_z, \quad \vec{a}_\varphi \times \vec{a}_z = \vec{a}_\rho, \quad \vec{a}_z \times \vec{a}_\rho = \vec{a}_\varphi$ $\vec{A}\times\vec{B} = (A\vec{a}_A)\times(B\vec{a}_B) = AB(\vec{a}_A \times \vec{a}_B)$
Spherical	$\vec{a}_r \times \vec{a}_\theta = \vec{a}_\varphi, \quad \vec{a}_\theta \times \vec{a}_\varphi = \vec{a}_r, \quad \vec{a}_\varphi \times \vec{a}_r = \vec{a}_\theta$ $\vec{A}\times\vec{B} = (A\vec{a}_A)\times(B\vec{a}_B) = AB(\vec{a}_A \times \vec{a}_B)$

TABLE 3B.5 Vector Cross Product (*Continued*)

ADDENDUM 3C
SPATIAL DISTRIBUTIONS AND DENSITIES

In the following, we discuss spatial distributions and densities of both static and dynamic quantities.

3C.1 Static Distributions and Densities

Static quantities, such as mass, charge, or energy, may exist in several possible spatial distributions (Figure 3C.1):

1. A concentration in an infinitesimally small volume, which we typically represent as a *point.*
2. A distribution over a volume with infinitesimally small cross section, which we typically represent as a *line.*
3. A distribution over a volume with infinitesimally small thickness, which we typically represent as a *surface.*
4. A distribution over a volume with nonzero dimensions.

These distributions may or may not be uniform. To express their spatial features, we use appropriate density expressions such as volume, area, or linear density of the quantity distribution. Densities of static quantities are scalar in nature, that is, have only magnitudes and do not have directions. We will use the subscripted ρ_v, ρ_s, and ρ_ℓ to denote volume, area, and linear densities, respectively. Hence, for the static quantity Q, we can write

$$\rho_v = \frac{dQ}{dv}, \quad \rho_S = \frac{dQ}{dS}, \quad \text{and} \quad \rho_\ell = \frac{dQ}{d\ell} \tag{3C.1}$$

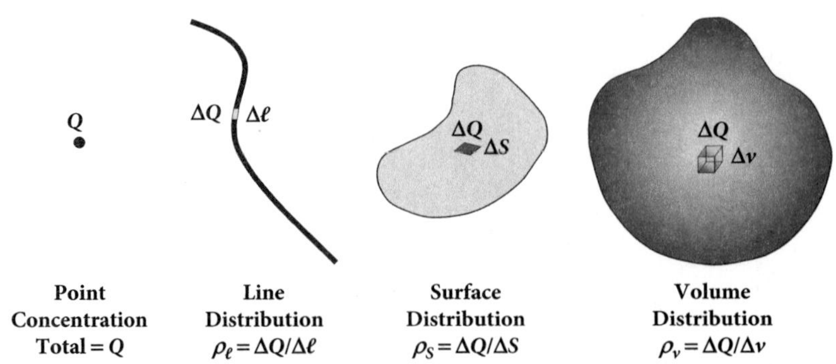

FIGURE 3C.1 Static distributions and densities.

Configuration	Volume Density = Quantity per Unit Volume	Surface Density = Quantity per Unit Area	Linear Density = Quantity per Unit Length
Point concentration	Zero volume $\rightarrow \rho_v = \infty$	Zero area $\rightarrow \rho_S = \infty$	Zero length $\rightarrow \rho_\ell = \infty$
Line distribution	Zero volume $\rightarrow \rho_v = \infty$	Zero area $\rightarrow \rho_S = \infty$	ρ_ℓ
Surface distribution	Zero volume $\rightarrow \rho_v = \infty$	ρ_S	(Atypical?)
Volume distribution	ρ_v	(Atypical?)	(Atypical?)

TABLE 3C.1 Static Distributions and Densities

For uniform distributions, the density is constant at all the distribution locations, otherwise, the density would be a function of position. Depending on the case, these densities may or may not have physical relevance and their definition could be meaningless. Examples are defining a volumetric density for a point concentration where the volume is zero, or defining the linear density for a spherical volume distribution. Table 3C.1 summarizes the corresponding densities for the four distribution forms.

3C.2 Conversions between Static Density Expressions

In some distribution configurations, more than one form of density can be simultaneously defined. An example is a cylindrical volume distribution for which a volumetric density can be defined while a linear density along the cylindrical axis can be defined as well. For such cases, it is useful to have conversion expressions between the different forms. In Table 3C.2, we provide examples of such conversion relationships.

	Nonuniform Distribution	Uniform Distribution
Sheet distribution $\rho_S = \dfrac{dQ}{dS}$ is defined	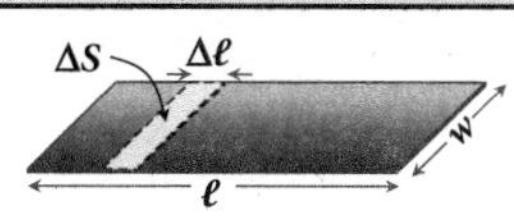	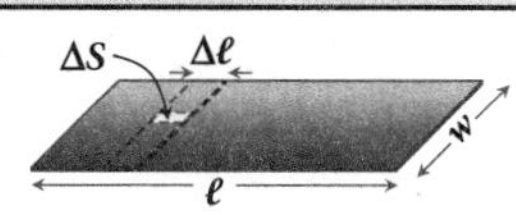
	$\rho_\ell = \dfrac{\int_w \rho_S\, dS}{\Delta\ell} = \int_w \rho_S\, dw$	$\rho_\ell = \rho_S w$
Rectangular prism (Slab) distribution $\rho_v = \dfrac{dQ}{dv}$ is defined	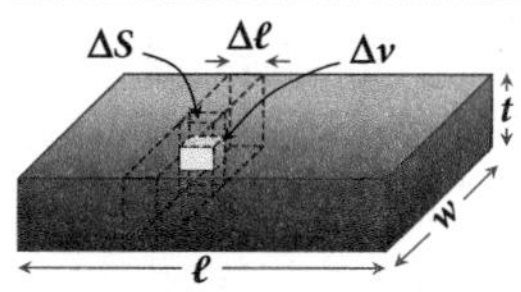	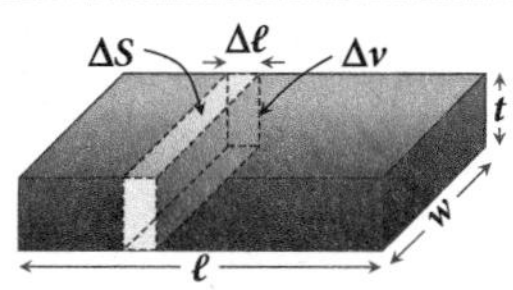
	$\rho_S = \dfrac{\int_t \rho_v\, dv}{\Delta S} = \int_t \rho_v\, dt$	$\rho_S = \rho_v t$
	$\rho_\ell = \dfrac{\int_w \int_t \rho_v\, dv}{\Delta\ell} = \int_w \int_t \rho_v\, dt\, dw$	$\rho_\ell = \rho_v t w$
Circular prism (Cylinder) distribution $\rho_v = \dfrac{dQ}{dv}$ is defined	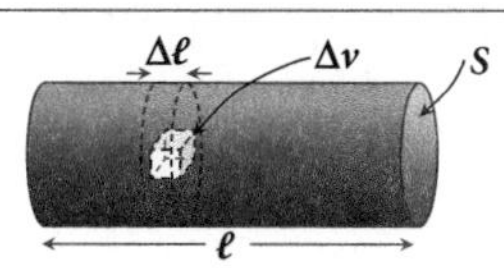	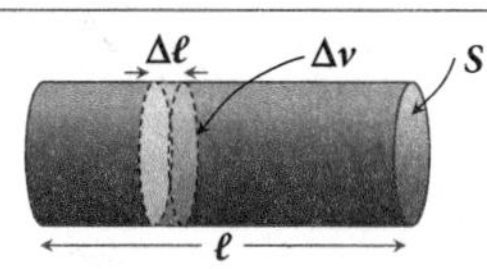
	$\rho_\ell = \dfrac{\int_S \rho_v\, dv}{\Delta\ell} = \int_S \rho_v\, dS$	$\rho_\ell = \rho_v S$

TABLE 3C.2 Conversions between Static Density Expressions

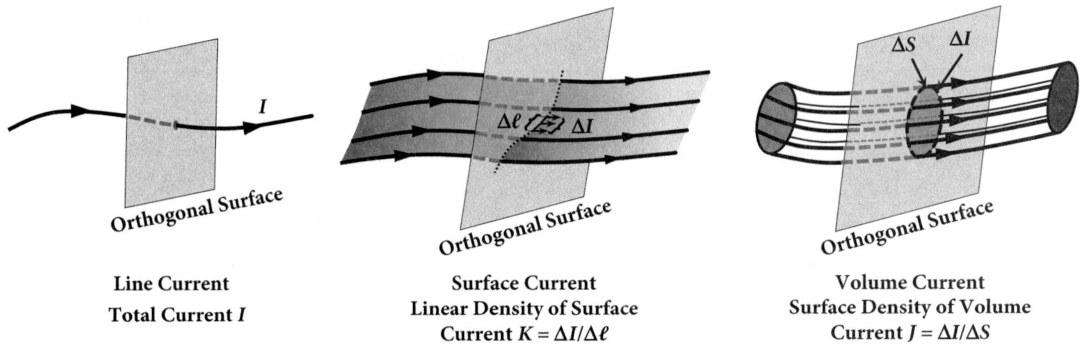

FIGURE 3C.2 Dynamic distributions and densities.

3C.3 Dynamic Distributions and Densities

Examples of dynamic quantities include electric current, air current, fluid flow, and energy flow. Spatial distributions for a dynamic flow can only exist in one of the following forms (Figure 3C.2):

1. A stream distribution with infinitesimally small cross section, which we typically represent as a "line current/flow."
2. A stream distribution with infinitesimally small thickness, which we typically represent as a "surface current/laminar flow."
3. A stream distribution with nonzero cross-sectional area, "volume current/flow."

These distributions may or may not be uniform. To express their spatial features, in both magnitude and direction, we use appropriate vector density expressions.

$$\vec{J} = \frac{dI}{dS_n}\vec{a}_I \quad \text{and} \quad \vec{K} = \frac{dI}{d\ell_n}\vec{a}_I \tag{3C.11}$$

where the n subscripts indicate that both $d\ell$ and dS must be orthogonal to the flow and $\vec{a}_I$ is the unit vector along the current/flow direction. For uniform distributions, the densities are constant at all the distribution locations; otherwise, the density would be a function of position. Again, depending on the case, these densities may or may not have physical relevance and their definition could be meaningless. Table 3C.3 summarizes the corresponding densities for dynamic distributions.

Configuration	Area Density = Quantity per Unit Area	Linear Density = Quantity per Unit Length
Line stream concentration	Zero area → $\vec{J} = \infty$	Zero length → $\vec{K} = \infty$
Surface stream distribution	Zero area → $\vec{J} = \infty$	$\vec{K}$
Volume stream distribution	$\vec{J}$	(Atypical?)

TABLE 3C.3 Dynamic Distributions and Densities

	Nonuniform Distribution	Uniform Distribution
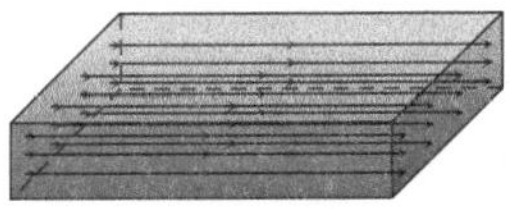	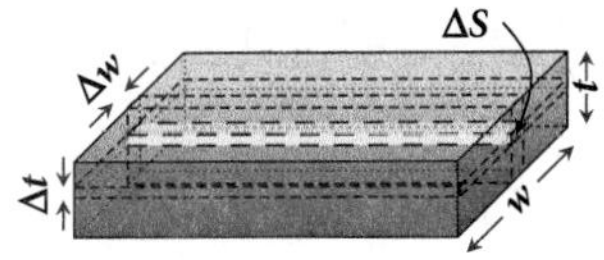	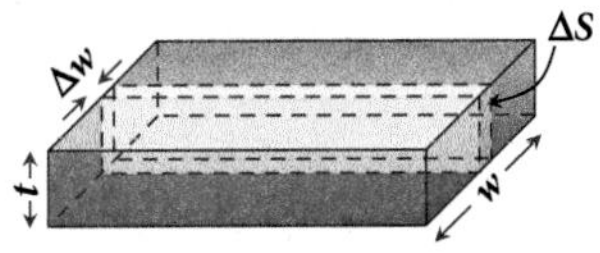
Slab stream flow/current $\vec{J} = \frac{dI}{dS_n}\vec{a}_I$ is defined	$\vec{K} = \frac{\int_t J\,dS}{\Delta w}\vec{a}_I = \int_t J\,dt\,\vec{a}_I$	$\vec{K} = Jt\vec{a}_I$

Table 3C.4 Conversions between Dynamic Density Expressions

3C.4 Conversions between Dynamic Density Expressions

In this section, we demonstrate the relationship between the two forms of dynamic distribution densities in cases where both can be simultaneously defined. In Table 3C.4, we show the example of a stream distribution in a slab configuration and the corresponding conversion relationships.

ADDENDUM 3D
LINE, SURFACE, AND VOLUME INTEGRATIONS

3D.1 Introduction

In the course of this book, as we deal with various physical phenomena, analyses involving integration of scalar and vector quantities is common. We often need to carry out contour (or line) integrations, integrations over an area of a surface, a closed surface as well as volume integrations. Our background in mathematics should enable us to carry most of these integrations once they are set up properly. We can also resort to integration tables, computer software packages, or even numerical tools for *difficult* integrations. Hence, in this addendum, we will focus on two aspects of the issue that are often the obstacle. One is how to set up the integration equation starting with the physical problem, and the second is how to deal with vector quantities within the integrand.

The first obstacle of setting up the integral is an *integral* part of setting up the proper mathematical model of the physical problem. This is an acquired skill that the learners in this field acquire with practice. The learner needs to get exposed to a variety of cases and a variety of analysis tools to appreciate what works and what does not and when to use a specific model and what are the constraints of that use. Gaining this skill requires proper appreciation to the physics of the subject and good command of relevant mathematical tools. This will be demonstrated and emphasized throughout the different chapters of this book.

We now turn to dealing with integrations containing vector quantities. This will be followed by an overview/survey of line, surface, and volume *scalar* integrations.

3D.2 Integrating Vector Quantities

To integrate vector quantities is simply performing vector summation of incremental vector elements. Since the sum of vectors is controlled by the directions of the vectors involved, the process of vector integration must take into account the variability of the direction of the vectors being integrated. Let us start with a *sarcastic* but true statement by saying that "The proper way of handling vector integration is not to do vector integration." What is meant here is that we should not start the integration process before we remove all vector quantities from within the integrand.

The process involves finding ways to get all vector terms outside the integration sign leaving only scalar quantities inside the integration. To extract a term outside the integral operator, this term must be a constant. Hence, what we must do is express all variable-direction vectors in terms of fixed-direction vectors that can then be extracted outside the integration. Let us cite some examples to demonstrate.

Example of a Cartesian Coordinate Integral

$$\int \vec{x}\,dx = \int x\vec{a}_x\,dx = \vec{a}_x \int x\,dx = \vec{a}_x\left(\frac{x^2}{2}+c\right) \tag{3D.1}$$

$$\int \vec{y}\,dx = \int y\vec{a}_y\,dx = \vec{a}_y \int y\,dx \tag{3D.2}$$

Example of a Cylindrical Coordinate Integral

$$\int \vec{\rho}\,d\rho = \int \rho\vec{a}_\rho\,d\rho = \vec{a}_\rho \int \rho\,d\rho = \vec{a}_\rho\left(\frac{\rho^2}{2}+c\right) \tag{3D.3}$$

For an integration along $d\rho$, the "$\vec{a}_\rho$" has a constant direction and hence it is a constant vector that can be taken outside the integration. However, if the same integration was along $d\varphi$, the "$\vec{a}_\rho$" will vary as we vary φ and hence "$\vec{a}_\rho$" needs to be expressed in terms of other constant-direction vectors to be able to carry out the integration. Logically, we use the Cartesian unit vectors "$\vec{a}_x$, $\vec{a}_y$, and $\vec{a}_z$" which are always constant vectors in this regard,

$$\begin{aligned}\int \vec{\rho}\,d\varphi &= \int \rho\vec{a}_\rho\,d\varphi = \int \rho(\cos\varphi\vec{a}_x + \sin\varphi\vec{a}_y)d\varphi = \int \rho(\cos\varphi\vec{a}_x)d\varphi + \int \rho(\sin\varphi\vec{a}_y)d\varphi \\ &= \vec{a}_x\int \rho(\cos\varphi)d\varphi + \vec{a}_y\int \rho(\sin\varphi)d\varphi = \vec{a}_x(\rho\sin\varphi + c_1) - \vec{a}_y\rho(\cos\varphi + c_2)\end{aligned} \tag{3D.4}$$

Adding the limits between 0 and 2π, this integration reduces to

$$\int_0^{2\pi} \vec{\rho}\,d\varphi = \vec{a}_x\rho\int_0^{2\pi}(\cos\varphi)d\varphi + \vec{a}_y\rho\int_0^{2\pi}(\sin\varphi)d\varphi = 0 \tag{3D.5}$$

This result makes physical sense since adding the variable direction vector $\vec{\rho}$ around a complete 2π (360°) rotation will produce zero net (Figure 3D.1).

Example of a Spherical Coordinate Integral

$$\int \vec{r}\,dr = \int r\vec{a}_r\,dr = \vec{a}_r\int r\,dr = \vec{a}_r(r^2/2) \tag{3D.6}$$

$$\begin{aligned}\int \vec{r}\,d\theta &= \int r\vec{a}_r\,d\theta = \int r(\sin\theta\cos\varphi\vec{a}_x + \sin\theta\sin\varphi\vec{a}_y + \cos\theta\vec{a}_z)d\theta \\ &= \vec{a}_x\int r\sin\theta\cos\varphi\,d\theta + \vec{a}_y\int r\sin\theta\sin\varphi\,d\theta + \vec{a}_z\int r\cos\theta\,d\theta\end{aligned} \tag{3D.7}$$

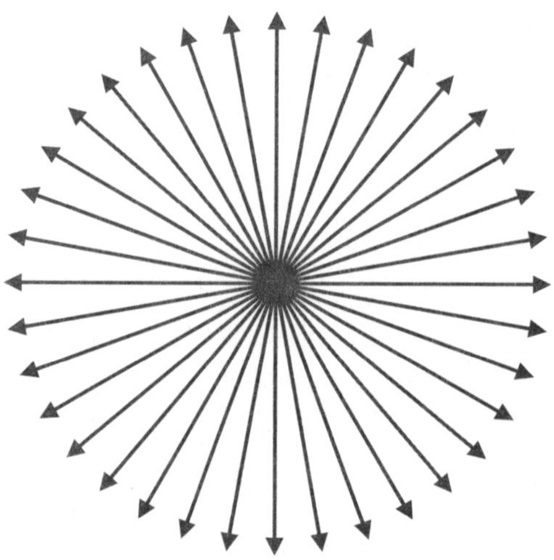

FIGURE 3D.1 Integrating the variable vector $\vec{\rho}$ around a complete φ rotation produces a zero net.

3D.3 Integrating Scalar Quantities

In this section, we will survey a few examples of line, surface, and volume integrations that we will find relevant in the chapters ahead.

Examples of Density Integrations

Two types of densities are reviewed here; static and dynamic. Examples of static densities include mass and charge distributions, while examples of dynamic densities include fluid flow and electric current.

Static Linear Density If ρ_ℓ (*position*) is the linear density of the distribution of a quantity Q, we can obtain the total Q by integrating the linear density over the length of the distribution.

$$Q = \int dQ = \int \frac{dQ}{d\ell} d\ell = \int \rho_\ell \, d\ell \tag{3D.8}$$

Static Surface Density If ρ_S (*position*) is the surface density of the distribution of a quantity Q, we can obtain the total Q by integrating the surface density over the area of the distribution.

$$Q = \int dQ = \iint \frac{dQ}{dS} dS = \iint \rho_S \, dS \tag{3D.9}$$

Static Volumetric Density If ρ_v (*position*) is the volumetric density of the distribution of a quantity Q, we can obtain the total Q by integrating the volumetric density over the volume of the distribution.

$$Q = \int dQ = \iiint \frac{dQ}{dv} dv = \iiint \rho_v \, dv \tag{3D.10}$$

Dynamic Linear Density If K_ℓ (*position*) is the linear density of the distribution of a current (flow) *I*, we can obtain the total *I* by integrating the linear density across the cross-sectional length of the distribution. The cross-sectional length is a length orthogonal to the flow distribution (Figure 3D.2).

$$I = \int dI = \int \frac{dI}{d\ell} d\ell = \int K_\ell \, d\ell \tag{3D.11}$$

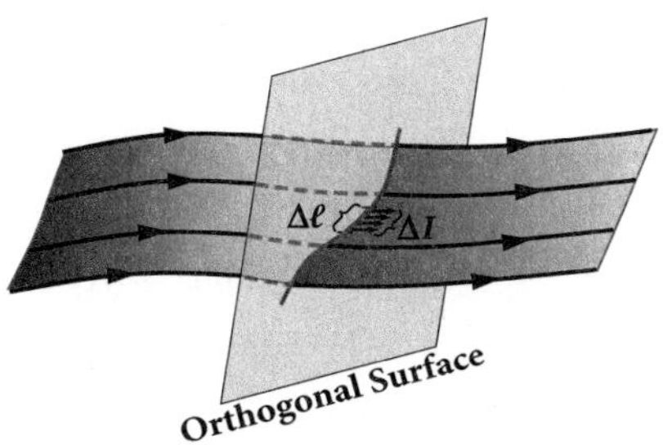

FIGURE 3D.2 Dynamic linear density.

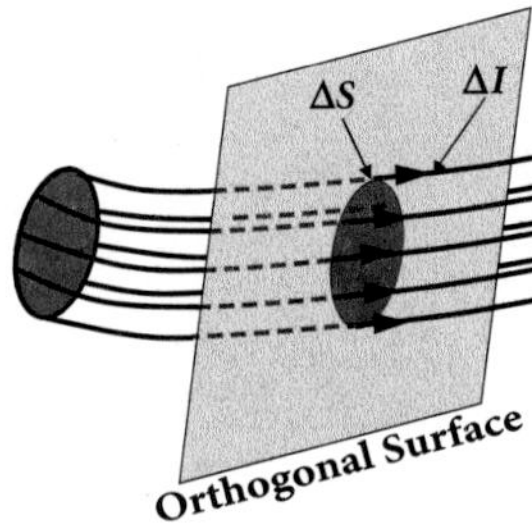

FIGURE 3D.3 Dynamic surface density.

Dynamic Surface Density The total flow (current), I, flowing through a specific surface is the integration of the flow surface density ($J = dI/dS$) over the cross-sectional surface area of interest. The cross-sectional area is that of a surface orthogonal to the flow distribution (Figure 3D.3).

$$I = \int dI = \iint \frac{dI}{dS} dS = \iint J\, dS \tag{3D.12}$$

3D.4 Examples of Work and Energy Integrations

The work done by a force acting on an object as it moves an unconstrained object a certain distance (along the action line of the force) (Figure 3D.4).

$$W = \int dW = \int F\, d\ell \tag{3D.13}$$

The total energy stored in a specific volume is the integration of the volumetric energy density ($w = dU/dv$) over the volume of interest.

$$U = \int dU = \iiint \frac{dU}{dv} dv = \iiint w\, dv \tag{3D.14}$$

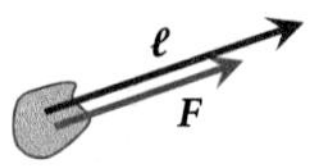

FIGURE 3D.4 The work done by a force to move an object is the product of the force and the distance moved along the force direction.

In addition, Table 3D.1 lists several examples of integrations in different coordinate systems.

	Line: $\int F\, d\ell$	Surface: $\iint J\, dS$	Volume: $\iiint w\, dv$
Cartesian	$\int F(x,y,z)dx$	$\iint J(x,y,z)dy\,dz$	$\iiint w(x,y,z)dx\,dy\,dz$
	$\int F(x,y,z)dy$	$\iint J(x,y,z)dz\,dx$	
	$\int F(x,y,z)dz$	$\iint J(x,y,z)dx\,dy$	
Cylindrical	$\int F(\rho,\varphi,z)d\rho$	$\iint J(\rho,\varphi,z)\rho\, d\varphi\, dz$	$\iiint w(\rho,\varphi,z)d\rho\; \rho\, d\varphi\, dz$
	$\int F(\rho,\varphi,z)\rho\, d\varphi$	$\iint J(\rho,\varphi,z)dz\, d\rho$	
	$\int F(\rho,\varphi,z)dz$	$\iint J(\rho,\varphi,z)d\rho\, \rho\, d\varphi$	
Spherical	$\int F(r,\theta,\varphi)dr$	$\iint J(r,\theta,\varphi)r\, d\theta\, r \sin\theta\, d\varphi$	$\iiint w(r,\theta,\varphi)dr\, r\, d\theta\, r \sin\theta\, d\varphi$
	$\int F(r,\theta,\varphi)r\, d\theta$	$\iint J(r,\theta,\varphi)dr\, r \sin\theta\, d\varphi$	
	$\int F(r,\theta,\varphi)r \sin\theta\, d\varphi$	$\iint J(r,\theta,\varphi)dr\, r\, d\theta$	

TABLE 3D.1 Examples of Integrations in Different Coordinate Systems

CHAPTER 4

Electrostatic Fields: Electric Flux and Gauss' Law

Carl Friedrich Gauss
(1777–1855)

Chapter Learning Objectives

- Visualize and mathematically describe different charge distribution functions.
- Use the mathematical tool of vector fields to describe flux density of different physical quantities such as electric flux density.
- Calculate the total amount of flux passing through an arbitrary surface for a given electric flux density.
- Use the integral form of Gauss' law to find the electric flux density generated from symmetric charge density distributions using the method of hypothetical Gaussian surface (HGS) enclosing surfaces.
- Use the point form of Gauss' law to find the electric flux density generated from arbitrary charge density distributions.
- Identify the different symmetric charge distributions and the appropriate coordinate system for each.
- Identify the conditions under which the symmetric charge density distribution idealizations can be applied to practical devices such as capacitors in their various configurations, for example, parallel plate, coaxial, etc.

4.1 The Electric Charge

The source of electromagnetic fields is the electrical charge. Stationary charges produce electrostatic phenomena, while dynamic charges produce electrical currents that cause magnetostatic and electromagnetic phenomena depending on the nature of the dynamic behavior. This chapter is dedicated to the study of electrostatic behavior of electrical charges, in other words, stationary charges.

The electric charge is a basic property of elementary particles of matter, which is behind all electric and magnetic phenomena and applications. We experience many of these phenomena and applications daily in our lives without thinking too much of their origin or how they are related to the electric charge property. To name a few, let us consider the following examples:

- When we plug a device into the wall outlet to power it up, we are counting on the charges to flow and currents to activate the device functionality.
- When you turn the ignition key in the car to start its engine, we are using the charges stored in the car battery to flow into the starter causing magnetic forces to rotate its rotor and thus initiating the car engine's rotation.
- Electrostatic printers rely on forming an ink image to be deposited on the printing paper using forces between electrically charged objects.
- All forms of "electronic" communications rely on electromagnetic waves that are generated as a result of controlling charges.
- Lightning is another phenomenon where electrically charged clouds "discharge" causing transients of extremely large arcs of current.
- A child with silky hair carries a charge as a result of rubbing against his nylon made tent in a dry weather, see the picture to the right.

The atoms of matter contain charged elementary particles, namely electrons and protons. The electron carries a negative elementary charge and the proton carries a positive charge. These two types of charge are equal and opposite. This elementary (or unit) charge has a magnitude of $1.6021765 \times 10^{-19}$ coulombs, where the coulomb, or C, is the unit for electric charge in the meter-kilogram-second (MKS) International System of Units. The electronic charge is typically denoted by "e"

$$e = -1.6021765 \times 10^{-19} \text{ coulombs (or C)} \tag{4.1}$$

Unbalancing the count of electrons and protons in an atom (or in a body of matter) causes the atom (or the body) to be electrically charged. The resulting charge can only occur in multiples of the *elementary charge.* A neutral atom (or body of matter) contains equal numbers of electrons and protons. The electric charge is typically denoted by q or Q.

4.2 Charge Distributions and Charge Densities

As presented under 3C.1 "Static Distributions and Densities," Addendum 3C, charges may exist in several possible forms:

1. An infinitesimally small volume, which we typically represent as a "point" charge (or charged point). An example would be one electron.
2. An infinitesimally small cross-sectional volume, which we typically represent as a "line" of charges (or charged line). An example would be a charged nylon thread (e.g., rubbed with wool cloth).
3. An infinitesimally small thickness volume, which we typically represent as a "surface" of charges (or charged surface). An example would be a charged nylon ball (e.g., rubbed with wool cloth).
4. A volume of charges (or charged volume). An example would be a charge plasma cloud.

These forms are demonstrated in Figure 3C.1.

The corresponding charge densities, as defined in Equation (3C.1), take the forms:

$$\rho_v = \frac{dQ}{dv}, \quad \rho_S = \frac{dQ}{dS}, \quad \text{and} \quad \rho_\ell = \frac{dQ}{d\ell} \tag{3C.1}$$

Charges may be distributed with uniform or nonuniform densities.

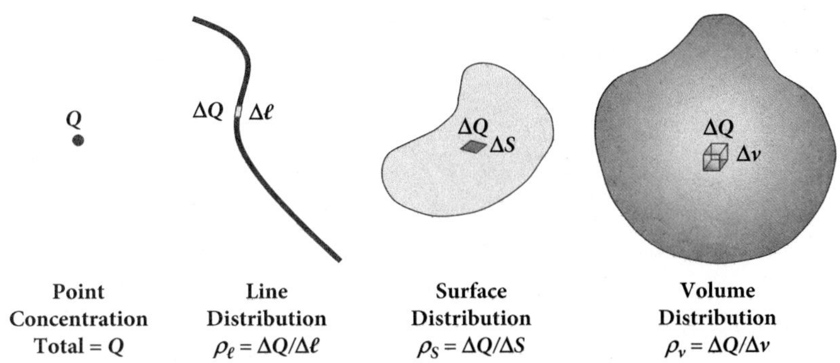

FIGURE 3C.1 Different forms of charge distributions.

Table 4.1 summarizes the corresponding charge densities for the four charge distribution forms.

Configuration	Notes	Volume Density ρ_v, C/m³	Surface Density ρ_S, C/m²	Linear Density ρ_ℓ, C/m
Point charge	Zero volume Zero area Zero lengths	$\frac{d(\text{charge})}{d(\text{volume})}=\infty$	$\frac{d(\text{charge})}{d(\text{area})}=\infty$	$\frac{d(\text{charge})}{d(\text{length})}=\infty$
Line charge	Zero volume Zero area	$\frac{d(\text{charge})}{d(\text{volume})}=\infty$	$\frac{d(\text{charge})}{d(\text{area})}=\infty$	$\frac{d(\text{charge})}{d(\text{length})}=\rho_\ell$
Surface charge	Zero volume	$\frac{d(\text{charge})}{d(\text{volume})}=\infty$	$\frac{d(\text{charge})}{d(\text{area})}=\rho_S$	
Volume charge		$\frac{d(\text{charge})}{d(\text{volume})}=\rho_v$		

TABLE 4.1 Charge Distributions and Their Corresponding Densities

Example 4.1

a. A square plate in the xy plane in free space is defined by -2 cm $\leq x \leq 2$ cm and -2 cm $\leq y \leq 2$ cm. If the surface charge density on the plate is given by $\rho_S = -4y^2$ (C/m²), find the total charge on the plate.
b. A spherical shell centered at the origin extends between $R_1 = 2$ cm and $R_2 = 4$ cm. If the volume charge density in the shell is given by $\rho_v = 2r\times10^{-4}$(C/m³), find the total charge contained in the shell.

Solution

Part (a)

$$Q=\int\rho_S\,dS=\int_{y=-0.02}^{y=0.02}\int_{x=-0.02}^{x=0.02}-4y^2\,dx\,dy=\int_{x=-0.02}^{x=0.02}dx\int_{y=-0.02}^{y=0.02}-4y^2\,dy$$

$$=x\Big|_{-0.02}^{0.02}\left[\frac{-4y^3}{3}\right]_{y=-0.02}^{y=0.02}=-0.04\times4\times2\times\frac{(0.02)^3}{3}=8.53\times10^{-7}\,\text{C}$$

Part (b)

$$Q=\int\rho_v\,dv=\int_v\rho_v r^2\sin\theta\,dr\,d\theta\,d\varphi=\int_0^{2\pi}\int_0^{\pi}\int_{0.02}^{0.04}2r\times10^{-4}r^2\sin\theta\,dr\,d\theta\,d\varphi$$

$$=2\times10^{-4}\int_0^{2\pi}d\varphi\int_0^{\pi}\sin\theta\,d\theta\int_{0.02}^{0.04}r^3\,dr=2\times10^{-4}(2\pi)[-\cos\theta]\Big|_0^{\pi}\left[\left(\frac{r^4}{4}\right)\right]_{0.02}^{0.04}$$

$$=2\times10^{-4}(2\pi)(2)\left(\frac{0.04^4-0.02^4}{4}\right)=1.51\times10^{-9}\,\text{C}$$

4.3 Electric Flux

The charge as a source of electrostatic phenomena is responsible for producing an electrical *influence* in the surrounding region. The nature of the charge *influence* is divergent similarly to light (optical) sources and heat sources. Divergent sources are the ones that will have their influence emerging out of (or into) them in a way similar to radiation. This influence can be represented by a hypothetical *flux* that *diverges* out of (or into) the source. Sources that receive the flux are negative sources and typically are referred to as sinks.

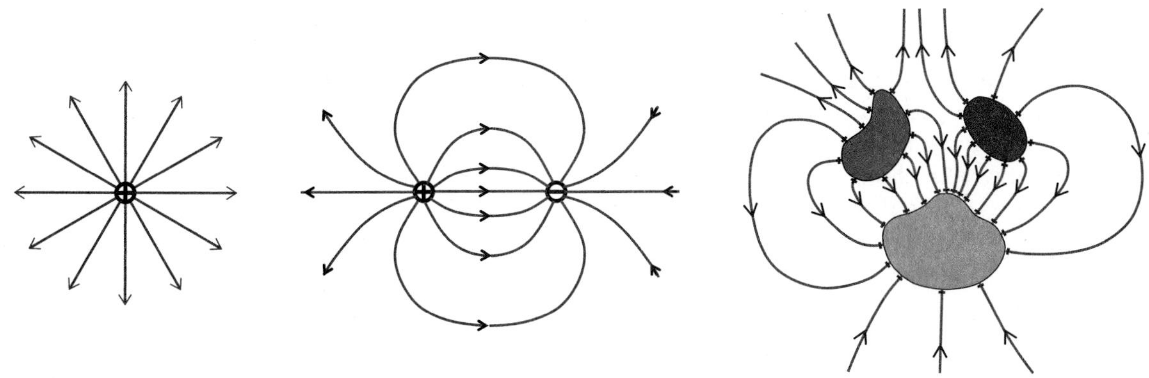

FIGURE 4.1 Graphical illustration of the hypothetical electric flux.

The term "flux" is used to represent the flow of a physical property in space. In a way, it is similar to the term "flow" and the term "current." It is often used when the physical property is not associated with physical motion or transport. So, typically we say electrical flux, heat flux, and optical flux, while we use current for air and water flow. Electrical current is associated with the flow of charges.

We hypothesize this flux emerging out of a positively charged source and landing onto a charge sink (negatively charged source). Figure 4.1 depicts electric flux maps for a few charge distributions. We also hypothesize that the total amount of flux (scalar) emerging from a source is equal to the net charge of that source and has the same units. Therefore, the flux put out by a (source) charge $+Q$ coulombs is equal to $+Q$ coulombs and is considered a positive flux. Likewise, the flux landing on a (sink) charge of $-Q$ coulombs is equal to $-Q$ coulombs and is considered a negative flux.

Since the universe is assumed neutral with zero charge sum, then a collection of charges that has a nonzero sum is simply representing an incomplete charge system. The balance of the charges must exist and should be included in our analysis. If these charges are too far distant from the primary charge collection, it is typical to assume them located at infinity. Corresponding flux lines would then extend from our charge collection to the balance of charges at infinity (Figure 4.2).

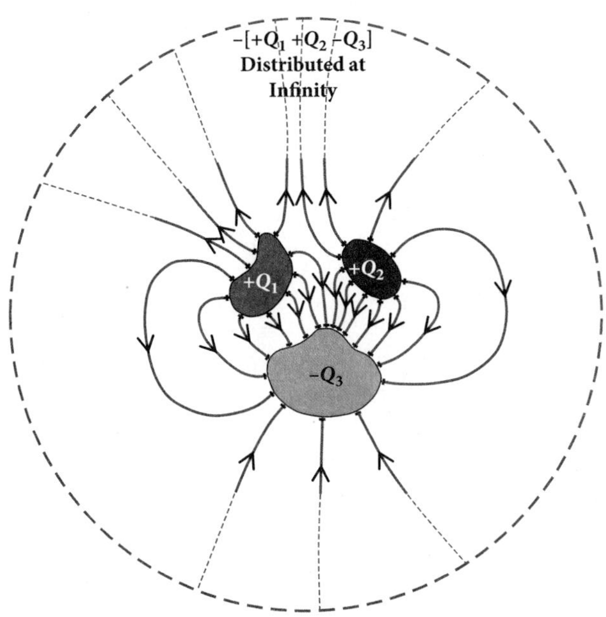

FIGURE 4.2 Flux lines extend from a system of charges to the counter of the system's net charge at infinity.

4.4 Faraday's Concentric Spheres Experiment

Around 1837, Michael Faraday's (director of the Royal Society in London) interest in electrostatics led him to conduct his famous concentric spheres experiment (Figure 4.3). In this experiment, he assembled two concentric metallic spheres separated by an insulating/dielectric material. He charged the inner sphere with a known positive charge and left the outer one neutral. After assembling the two isolated spheres, he discharged the outer sphere by connecting it momentarily to ground. He then disassembled the outer sphere and checked it out for induced charges. Mr. Faraday found the outer sphere to have a negative induced charge equal in magnitude to the original charge he had placed on the inner sphere.

From that experiment, Faraday concluded that there was some form of "charge displacement" on the outer sphere because of the charge on the inner sphere. This conclusion led the way to what we call now the displacement flux or electric flux. To explain, the flux emerging from the inner spherical charge landed on an equal but negative charge (sink) on the inner surface of the outer sphere. This, in turn, left a balance of the same amount of positive charge on the outer sphere, which was discharged upon connecting it (momentarily) to the ground.

So, in reality, the flux leaving the positive charge on the inner sphere is terminated to the negative charge on the inner surface of the outer sphere. Any closed surface in between will intercept this flux, and the flux emerging from (or diverging out of) is the same as the flux emerging from (or diverging out of) the charge on inner sphere. Hence, we can conclude that the flux emerging from (or diverging out of) a closed surface is equal to the flux emerging from (or diverging out of) the charges enclosed by that surface (which is equal to the total charge enclosed by that surface).

In mathematical terms:

$$\psi_{\text{closed surface}} = \sum Q_{\text{enclosed by the surface}} \tag{4.2}$$

This relationship is an early form of Gauss' law that we will develop in the next section, which in turn is one of the four famous Maxwell's equations.

It is important to recognize that the relationship is for closed surfaces only and cannot be extended to surfaces that are not completely closed. Any gap in the surface closure may pass though some of the enclosed charge flux and hence unbalance the equality.

For further clarification of this equality, we examine Figure 4.4. Two closed surfaces are shown for which their flux is shown to equal to the sum of their enclosed charges.

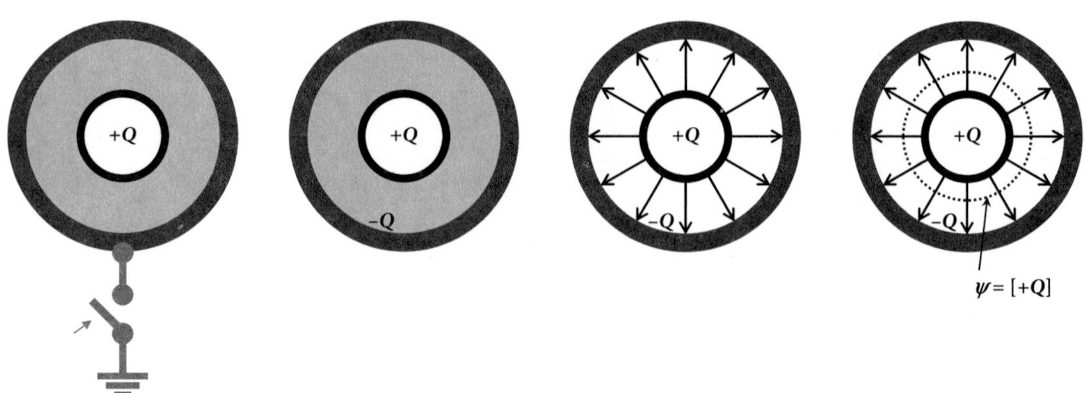

FIGURE 4.3 Faraday's concentric spheres experiment.

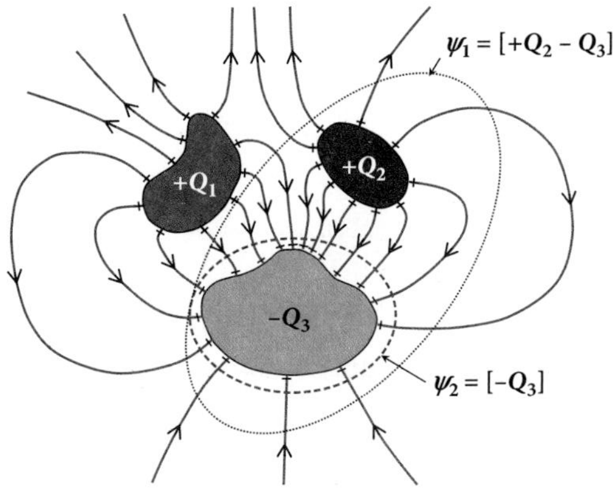

FIGURE 4.4 Gauss' law: the total flux diverging out of a closed surface is equal to the algebraic sum of charges enclosed by that surface.

4.5 Electric Flux Density

We notice from the examples of Figure 4.1 that the flux map varies in density depending on the configuration of the charge distribution. In the simple case of a *point* charge, we notice a symmetrical (spherical) distribution of the flux *uniform density*. Using the information presented earlier about densities and distributions, we can define a surface density for the flux emerging out of that surface. We will denote the surface density of the electric flux D; the units of which are C/m^2.

For a given flux distribution, several density amounts can be defined depending on the angle at which the flux intercepts the surface, and hence a unique definition can be obtained if we were to define the density as the flux per unit area of an orthogonal surface.

Examining Figure 4.5, we realize that the amount of flux flowing through a given surface area depends on the way the surface area intercepts the flux, for example, dS_1, dS_2, and dS_3 are equal areas that intercept differing amounts of flux $d\psi_1$, $d\psi_2$, and $d\psi_3$, respectively. It is clear that dS_3 being the area that intercepts the flux in a normal fashion yields the maximum and true value for the flux density, refer to Figure 4.6.

$$D = \frac{d\psi_3}{dS_3} = \left\{\frac{d\psi}{dS}\right\}_{\max} = \frac{d\psi}{dS_{\text{normal}}} = \frac{d\psi}{dS_p} \tag{4.3}$$

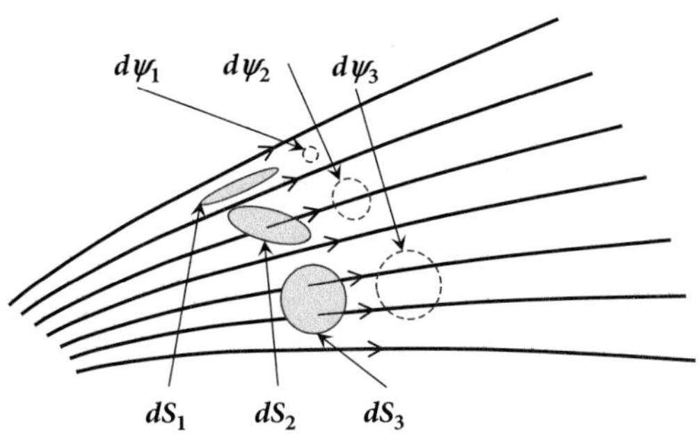

FIGURE 4.5 The flux flowing through a given surface area depends on the projection of the area that intercepts the flux.

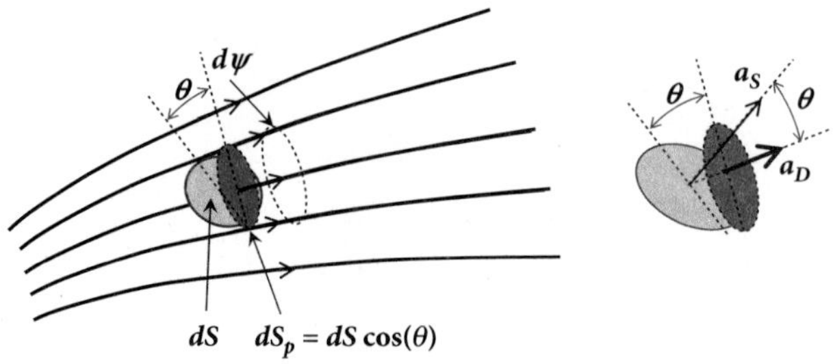

FIGURE 4.6 The flux density is defined as the amount of flux per unit normal area.

Hence, we can write

$$d\psi = [D][dS_p] \tag{4.4}$$

This relationship can be generalized for any area increment, normal or otherwise, as follows:

$$d\psi = [D][dS_p] = [D][dS]\cos\theta = [D][dS]\vec{a}_D \cdot \vec{a}_S = [D\vec{a}_D]\cdot[dS\vec{a}_S] \tag{4.5}$$

Hence,

$$d\psi = \vec{D}\cdot\overrightarrow{dS} \tag{4.6}$$

and the total flux on any surface S would be the result of integrating the incremental amounts $d\psi$ over the surface area:

$$\psi = \iint_S d\psi = \iint_S \vec{D}\cdot\overrightarrow{dS} \tag{4.7}$$

4.6 Gauss' Law: The Integral Form

Applying Equation (4.7) to a closed surface and using Equation (4.2), we arrive at an expression of Gauss' law in terms of the flux density $\vec{D}$ integration over a closed surface:

$$\psi_{\text{closed surface } S} = \iint_{\text{closed surface } S} d\psi = \oiint_S \vec{D}\cdot\overrightarrow{dS} = \sum Q_{\text{enclosed by } S} \tag{4.8}$$

Depending on the shape of the charge distribution inside the closed surface, we may have a sum of discrete charges or an integration of a charge distribution; in general, we may have some or all of the following terms:

$$\sum Q_{\text{enclosed by } S} = \left\{\sum Q_{\text{discrete}} + \sum\iiint_{\text{volume distribution}} \rho_v\, dv + \sum\iint_{\text{surface distribution}} \rho_S\, dS + \sum\int_{\text{line distribution}} \rho_\ell\, d\ell\right\} \tag{4.9}$$

This general form is often abbreviated to the volume integral form, however, when appropriate, the above full form should be used.

Consequently, the **"integral" form of Gauss' law** (one of the four Maxwell's equations) is

$$\{\psi_{\text{closed surface } S}\} = \oiint_S \vec{D}\cdot\overrightarrow{dS} = \sum Q_{\text{enclosed by } S} = \iiint_{\text{enclosed volume by } S} \rho_v\, dv \tag{4.10}$$

Example 4.2 Assuming that the electric flux density is given by $\vec{D}=\vec{a}_x 100\varepsilon_o \overline{x}$ (C/m^2), find the total electric charge contained inside in

a. A cubical volume, 100 mm on a side, centered symmetrically at the origin.
b. A cylindrical volume around the z axis having a radius of 100 mm and a height of 200 mm.

Solution

Solution Part (a)

$$\oint \vec{D}\bullet\overline{dS}=Q_{\text{enclosed}}$$

The only faces that will contribute to the integration of $\vec{D}$ over the closed surface of the cube are those where the flux falls normal. In this case, the electric flux density is in the x direction, so the only 2 faces where the flux falls normal are the front and back faces parallel to the yz plane.

$$\oint \vec{D}\bullet\overline{dS}=\int_{\text{Frontface } x=0.05\text{ m}}\vec{D}\bullet\overline{dS}+\int_{\text{Backface } x=-0.05\text{ m}}\vec{D}\bullet\overline{dS}$$

Frontface $x=0.05$ m $\quad \vec{D}=\vec{a}_x 100\varepsilon_o x=\vec{a}_x 100\varepsilon_o\times 0.05=\vec{a}_x 5\varepsilon_o \quad \overline{dS}=\vec{a}_x\, dy\, dz$

$$\Rightarrow \vec{D}\bullet\overline{dS}=\vec{a}_x 5\varepsilon_o\bullet\vec{a}_x\, dy\, dz=5\varepsilon_o\, dy\, dz$$

$$\int_{\text{Frontface}}\vec{D}\bullet\overline{dS}=\int_{-0.05}^{0.05}\int_{-0.05}^{0.05}5\varepsilon_o\, dy\, dz=5\varepsilon_o\times 0.1\times 0.1=0.05\varepsilon_o$$

Backface $x=-0.05$ m $\quad \vec{D}=\vec{a}_x 100\varepsilon_o x=\vec{a}_x 100\varepsilon_o\times -0.05=-\vec{a}_x 5\varepsilon_o \quad \overline{dS}=-\vec{a}_x\, dy\, dz$

$$\Rightarrow \vec{D}\bullet\overline{dS}=-\vec{a}_x 5\varepsilon_o\bullet -\vec{a}_x\, dy\, dz=5\varepsilon_o\, dy\, dz$$

$$\int_{\text{Backface}}\vec{D}\bullet\overline{dS}=\int_{-0.05}^{0.05}\int_{-0.05}^{0.05}5\varepsilon_o\, dy\, dz=5\varepsilon_o\times 0.1\times 0.1=0.05\varepsilon_o$$

$$\oint \vec{D}\bullet\overline{dS}=\int_{\text{Frontface } x=0.05\text{ m}}\vec{D}\bullet\overline{dS}+\int_{\text{Backface } x=-0.05\text{ m}}\vec{D}\bullet\overline{dS}=0.05\varepsilon_o+0.05\varepsilon_o=0.1\varepsilon_o$$

$$Q_{\text{enclosed}}=\oint \vec{D}\bullet\overline{dS}=0.1\varepsilon_o\times 1=0.1\varepsilon_o\text{ C}$$

Solution Part (b) To perform the integration over the surface of the cylinder, we need to convert the expression of $\vec{D}$ to cylindrical coordinate system

$$\begin{Bmatrix} D_\rho \\ D_\varphi \\ D_z \end{Bmatrix}=\begin{Bmatrix} \cos\varphi & \sin\varphi & 0 \\ -\sin\varphi & \cos\varphi & 0 \\ 0 & 0 & 1 \end{Bmatrix}\begin{Bmatrix} D_x \\ D_y \\ D_z \end{Bmatrix}=\begin{Bmatrix} \cos\varphi & \sin\varphi & 0 \\ -\sin\varphi & \cos\varphi & 0 \\ 0 & 0 & 1 \end{Bmatrix}\begin{Bmatrix} 100\varepsilon_o x \\ 0 \\ 0 \end{Bmatrix}$$

$$D_\rho=\cos\varphi\times 100\varepsilon_o x=\cos\varphi\times 100\times\rho\cos\varphi=100\varepsilon_o\rho\cos^2\varphi$$

$$D_\varphi=-\sin\varphi\times 100\varepsilon_o x=-\sin\varphi\times 100\varepsilon_o\times\rho\cos\varphi=-100\varepsilon_o\rho\sin\varphi\cos\varphi$$

$$\oint \vec{D}\bullet\overline{dS}=Q_{\text{enclosed}}$$

The only surface of the cylinder that has the electric flux density perpendicular is the side surface and therefore it will be the only surface contributing to the integration. On the side face ρ is equal to 0.1 m

$$\oint \vec{D} \cdot \vec{dS} = Q_{\text{enclosed}}$$

$$\vec{dS} = a_\rho \rho \, d\varphi \, dz \; \vec{D} \cdot \vec{dS} = (\vec{a}_\rho D_\rho + \vec{a}_\varphi D_\varphi) \cdot \vec{a}_\rho \rho \, d\varphi \, dz = D_\rho \rho \, d\varphi \, dz$$

$$\int_{\text{SideSurface}} \vec{D} \cdot \vec{dS} = \int_0^{0.2} \int_0^{2\pi} 100\varepsilon_o \rho \cos^2\varphi \, \rho \, d\varphi \, dz = 100\varepsilon_o \rho^2 \int_0^{0.2} \int_0^{2\pi} \cos^2\varphi \, \rho \, d\varphi \, dz$$

$$= 100\varepsilon_o \rho^2 \int_0^{2\pi} \frac{1}{2}(1 + \cos 2\varphi) \, d\varphi \int_0^{0.2} dz = 100\varepsilon_o \rho^2 \times \frac{2\pi}{2} \times 0.2$$

$$= 100\varepsilon_o (0.1)^2 \times \frac{2\pi}{2} \times 0.2 = 0.2\pi\varepsilon_o$$

$$\oint \vec{D} \cdot \vec{dS} = 0.2\pi\varepsilon_o = Q_{\text{enclosed}} \Rightarrow Q_{\text{enclosed}} = 0.2\pi\varepsilon_o \text{ C}$$

4.7 Application of Gauss' Law in the Integral Form: Electric Flux due to Symmetrical Charge Distributions

Examining Equation (4.10) from a mathematical perspective, we observe that $\vec{D}$ is an integrand term inside an integral operation. This implies that it has to be *known* in order to be able to process this equation. In which case, given D at all points on a closed surface can enable us to evaluate the total flux out of that surface as well as the total charge enclosed by it.

The logical thing at this point in our study is that we are interested in the opposite; that is, given the charge distribution, we need to derive knowledge about its corresponding flux density vector and its distribution. This implies that we need to "undo" the integration process to obtain the integrand instead of the integral.

This seems like posing the problem in the form of a "puzzle" rather than a mathematical procedure. So, the question we are after here is what is the flux density vector function that when integrated over a certain closed surface would result in the known charge distribution within that surface. It is obvious that the typical strategy to dealing with this type of question is the strategy of guessing and trial and error, which is not of interest to us here in this study.

Well, as it turns out, there are special cases in which we can solve for the integrand and not the integral. Those are the cases where the integrand is a *constant* that can be factored outside the integration sign and hence the integration can be carried out without any ambiguities.

In other words, and again examining Equation (4.10), if we choose our closed surface in such a way that $\vec{D}$ is factored out and is no longer an integrand, the equation now would relate $\vec{D}$ to the charge distribution and the geometrical parameters of the problem. In such cases, $\vec{D}$ can be evaluated through an analytical procedure with no ambiguities or trial and error guessing.

As it turned out, this concept can be implemented in a few symmetric cases as listed below and discussed in full detail later in Addendum 4A at the end of this chapter. These cases are limited to the following:

A. Charges distributed with spherical symmetries (see Addendum 4A for full treatment)
   1. Point charge
   2. Case of spherical surface charge distribution with constant ρ_S (no variations with θ or φ)
   3. Case of spherical shell charge distribution with ρ_v allowed to vary with r only (no variations with θ or φ)
   4. Case of spherical volume charge distribution with ρ_v allowed to vary with r only (no variations with θ or φ)

Example 4.3 Find the electric flux density everywhere for the following spherically symmetric charge distribution. There are no charges for r less than R_1 or greater than R_3. Q coulombs of charge is distributed uniformly over the volume between R_1 and R_2. Similarly, $(-Q)$ coulombs of charge is distributed uniformly over the volume between R_2 and R_3.

Find the charge densities in the regions between R_1 and R_2 and between R_2 and R_3.

Find the electric flux density everywhere.

Solution From symmetry, the electric flux density is along the r direction. The Gaussian surface is considered to be a sphere of radius r.

Between R_1 an R_2, the volume charge density is given by

$$\rho_{v1} = \frac{Q}{v_1} = \frac{Q}{\frac{4\pi}{3}\left(R_2^3 - R_1^3\right)}$$

Between R_2 an R_3, the volume charge density is given by

$$\rho_{v2} = \frac{-Q}{v_2} = \frac{-Q}{\frac{4\pi}{3}\left(R_3^3 - R_2^3\right)}$$

Region 1 $r < R_1$, $\quad Q_{\text{enclosed}} = 0$ and therefore $\vec{D} = 0$

Region 2 $R_1 < r < R_2$

$$\oint \vec{D} \cdot \overrightarrow{dS} = Q_{\text{enclosed}} \qquad Q_{\text{enclosed}} = \rho_{v1} v = \rho_{v1} \frac{4\pi}{3}\left(r^3 - R_1^3\right)$$

$$\vec{D} \cdot \overrightarrow{dS} = \vec{a}_r D_r \cdot \vec{a}_r dS_r = D_r dS_r$$

$$\oint \vec{D} \cdot \overrightarrow{dS} = D_r \int dS_r = D_r S = D_r 4\pi r^2$$

$$D_r 4\pi r^2 = Q_{\text{enclosed}} = \rho_{v1} \frac{4\pi}{3}(r^3 - R_1^3) \Rightarrow D_r = \rho_{v1} \frac{\left(r^3 - R_1^3\right)}{3r^2}$$

Region 3 $R_2 < r < R_3$

$$\oint \vec{D} \cdot \overrightarrow{dS} = Q_{\text{enclosed}} \qquad Q_{\text{enclosed}} = Q + \rho_{v2} v = Q + \rho_{v2} \frac{4\pi}{3}\left(r^3 - R_2^3\right)$$

$$\vec{D} \cdot \overrightarrow{dS} = \vec{a}_r D_r \cdot \vec{a}_r dS_r = D_r dS_r$$

$$\oint \vec{D} \cdot \overrightarrow{dS} = D_r \int dS_r = D_r S = D_r 4\pi r^2$$

$$D_r 4\pi r^2 = Q_{\text{enclosed}} = Q + \rho_{v2} \frac{4\pi}{3}\left(r^3 - R_2^3\right) \Rightarrow D_r = \frac{Q}{4\pi r^2} + \rho_{v2} \frac{\left(r^3 - R_2^3\right)}{3r^2}$$

Region 4 $r \geq R_3$

$$\oint \vec{D} \cdot \overrightarrow{dS} = Q_{\text{enclosed}} \qquad Q_{\text{enclosed}} = Q + (-Q) = 0$$

$$\oint \vec{D} \cdot \overrightarrow{dS} = D_r 4\pi r^2$$

$$D_r 4\pi r^2 = Q_{\text{enclosed}} = 0 \Rightarrow D_r = 0$$

B. Charges distributed with cylindrical symmetries (see Addendum 4A for full treatment)

1. Line charge with uniform linear charge density ρ_ℓ over infinite length.
2. Cylindrical surface charge distribution with constant ρ_S (no variations with φ or z), again over infinite length.
3. Cylindrical shell charge distribution with ρ_v allowed to vary with ρ only (no variations with φ or z), again over infinite length.
4. Cylindrical volume charge distribution with ρ_v allowed to vary with ρ only (no variations with φ or z), again over infinite length.

C. Charges distributed with planar symmetries (see Addendum 4A for full treatment)

5. Planar surface charge distribution with constant ρ_S (no variations with x or y), with infinite extension in both x and y coordinates.
6. Planar *slab* of charge distribution with ρ_v allowed to vary with z only constant (no variations with x or y), again with infinite x and y extensions.

Only in these cases, and combinations thereof, we can use Gauss' law in its integral form to derive expressions of $\vec{D}$ given the charge distribution. Although this approach is not valid in the general cases for general charge distributions, the exercise of its implementation in symmetrical charge distributions is useful for us, as early learners in this field to get started. Another great benefit of this exercise is that it gives us a great deal of insight into the relationship between charge and flux density distributions.

Our next step at this point is the development of the point form tool of Gauss' law. As we will demonstrate, it presents an amazing leap forward in the pursuit of flux density distributions due to general cases of charge distributions. As we will demonstrate, this tool will enable us to solve problems that are more complicated and deal with general charge distributions.

Example 4.4 Two infinitely long concentric coaxial cylindrical surfaces having $\rho = a$ and $\rho = b(b > a)$ carry surface charge densities ρ_{Sa} and ρ_{Sb}, respectively.

Determine the electric flux density, $\vec{D}$, in the regions of $a < \rho < b$, and $\rho > b$.

What must be the relation between a and b for $\vec{D}$ to vanish for $(\rho > b)$?

Solution From symmetry, the electric flux is along the ρ direction. Consider a cylindrical Gaussian surface of radius ρ and height h.

Solution Part (a)

Region 1, $\rho < a$ $\quad \oint \vec{D} \cdot \overrightarrow{dS} = Q_{\text{enclosed}} \qquad Q_{\text{enclosed}} = 0 \rightarrow \vec{D} = 0$

Region 2, $a \le \rho < b$ $\quad \oint \vec{D} \cdot \overrightarrow{dS} = Q_{\text{enclosed}} \qquad Q_{\text{enclosed}} = \rho_{Sa} S_a = \rho_{Sa} 2\pi a h$

$$\vec{D} \cdot \overrightarrow{dS} = \vec{a}_\rho D_\rho \cdot \vec{a}_\rho \, dS_\rho = D_\rho \, dS_\rho$$

$$\oint \vec{D} \cdot \overrightarrow{dS} = D_\rho \int dS_\rho = D_\rho S = D_\rho 2\pi \rho h$$

$$D_\rho 2\pi \rho h = Q_{\text{enclosed}} = \rho_{Sa} 2\pi a h \Rightarrow D_\rho = \frac{\rho_{Sa} a}{\rho}$$

Region 3, $\rho \geq b$ $\oint \vec{D} \cdot \vec{dS} = Q_{\text{enclosed}}$

$$Q_{\text{enclosed}} = \rho_{Sa} S_a + \rho_{Sb} S_b = \rho_{Sa} 2\pi ah + \rho_{Sb} 2\pi bh$$

$$\oint \vec{D} \cdot \vec{dS} = D_\rho 2\pi \rho h$$

$$D_\rho 2\pi \rho h = Q_{\text{enclosed}} = \rho_{Sa} 2\pi ah + \rho_{Sb} 2\pi bh \Rightarrow D_\rho = \frac{\rho_{Sa} a + \rho_{Sb} b}{\rho}$$

Solution Part (b) For the flux to be zero in the region $\rho \geq b$

$$D_\rho = \frac{\rho_{Sa} a + \rho_{Sb} b}{\rho} = 0 \Rightarrow \rho_{Sa} a + \rho_{Sb} b = 0 \Rightarrow \rho_{Sa} = -\rho_{Sb} \frac{b}{a}$$

Example 4.5 An infinitely long cylindrical shell extends between $\rho = 1$ m and $\rho = 3$ m in free space. The region between $\rho = 1$ m and $\rho = 3$ m contains a volume charge density, ρ_v, that increases linearly with the distance from the origin such that $\rho_v = 0$ at $\rho = 1$ and $\rho_v = 2\,(\text{C/m}^3)$ at $\rho = 2$ m. Find the electric flux density intensity in the following regions

a. Region I $0 < \rho < 1$ m
b. Region II $1 \text{ m} < \rho < 3$ m
c. Region III $\rho > 3$ m

Solution

$$\rho_v = k(\rho - 1) \qquad \rho_v = 0 \text{ at } \rho = 1 \qquad \text{and} \qquad \rho_v = 2 \text{ at } \rho = 2 \rightarrow k = 2$$

$$\rho_v = 2(\rho - 1)$$

From symmetry, the electric flux density is along the ρ direction. Consider a cylindrical Gaussian surface with radius ρ and height h. The only surface that will contribute to the integration of $\vec{D} \cdot \vec{dS}$ is the side surface of the cylinder. The side surface of the cylinder is the only surface of the three surfaces of the cylinder that has the electric flux density normal

Part (a) region 1, $\rho < 1$ $\oint \vec{D} \cdot \vec{dS} = Q_{\text{enclosed}}$ $Q_{\text{enclosed}} = 0 \rightarrow \vec{D} = 0$

Part (b) region 2, $1 \leq \rho < 3$ $\oint \vec{D} \cdot \vec{dS} = Q_{\text{enclosed}}$

$$Q_{\text{enclosed}} = \int_v \rho_v \, dv = \int_0^h \int_0^{2\pi} \int_1^\rho 2(\rho - 1)\rho \, d\rho \, d\varphi \, dz = 2\int_0^h dz \int_0^{2\pi} d\varphi \int_1^\rho \rho(\rho - 1) d\rho$$

$$= 2(h)(2\pi)\left(\frac{\rho^3}{3} - \frac{\rho^2}{2}\right)_1^\rho$$

$$= 2(h)(2\pi)\left(\frac{\rho^3}{3} - \frac{1}{3} - \frac{\rho^2}{2} + \frac{1}{2}\right) = 2\pi h\left(\frac{2\rho^3}{3} - \rho^2 + \frac{1}{3}\right) = \frac{2\pi h}{3}(2\rho^3 - 3\rho^2 + 1)$$

$$\vec{D} \cdot \vec{dS} = \vec{a}_\rho D_\rho \cdot \vec{a}_\rho dS_\rho = D_\rho dS_\rho$$

$$\oint \vec{D} \cdot \vec{dS} = D_\rho \int dS_\rho = D_\rho S = D_\rho 2\pi \rho h$$

$$D_\rho 2\pi \rho h = Q_{\text{enclosed}} = \frac{2\pi h}{3}(2\rho^3 - 3\rho^2 + 1) \Rightarrow D_\rho = \frac{(2\rho^3 - 3\rho^2 + 1)}{3\rho}$$

Part (c) Region 3, $\rho \geq 3$ Notice that in this case the enclosed charge can be found using the same integral used in region 2, except in this case the upper limit of ρ would be fixed and equal to 3. Therefore, the enclosed charge can be found by simply using the same expression in region 2 and substituting $\rho = 3$

$$\oint \vec{D} \bullet \vec{dS} = Q_{\text{enclosed}}$$

$$Q_{\text{enclosed}} = \int_v \rho_v dv = \int_0^h \int_0^{2\pi} \int_1^3 2(\rho - 1)\rho\, d\rho\, d\varphi\, dz = \int_0^h dz \int_0^{2\pi} d\varphi \int_1^3 \rho(\rho - 1) d\rho$$

$$= 2(h)(2\pi)\left(\frac{\rho^3}{3} - \frac{\rho^2}{2}\right)_1^{\rho}$$

$$= \frac{2\pi h}{3}(2\rho^3 - 3\rho^2 + 1)_{\rho=3} = \frac{2\pi h}{3}(2 \times 3^3 - 3 \times 3^2 + 1) = \frac{2\pi h \times 28}{3}$$

$$\vec{D} \bullet \vec{dS} = \vec{a}_\rho D_\rho \cdot \vec{a}_\rho dS_\rho = D_\rho dS_\rho$$

$$\oint \vec{D} \bullet \vec{dS} = D_\rho \int dS_\rho = D_\rho S = D_\rho 2\pi\rho h$$

$$D_\rho 2\pi\rho h = Q_{\text{enclosed}} = \frac{2\pi h \times 28}{3} \Rightarrow D_\rho = \frac{28}{3\rho}$$

4.8 Gauss' Law in the Point Form (Differential Form)

4.8.1 Point Form versus Integral Form

Before we develop the point form of Gauss' law, it is important to explain what the "point/differential" form is all about and how it is different from the integral form. To do so, let us consider two alternate forms for revealing information about a functional relationship (or distribution) $f(x)$ versus x. The goal is to enable the use of the *given* information to arrive at full knowledge of the functional relationship $f(x)$. The two considered forms are

1. The integral form, in which the integral expression $\int_a^b f(x)dx = K$ is given

2. The differential form, in which the derivative $\frac{df(x)}{dx} = g(x)$ is given

None of these two forms describes a unique solution for $f(x)$ and additional information must be available to enable a unique definition of $f(x)$. However, it is mathematically obvious that the differential form offers a straightforward approach to obtaining the $f(x)$ solution. This is doable through solving the given differential equation analytically (or numerically/graphically) and applying relevant boundary conditions. On the other hand, to use the integral form to obtain $f(x)$ requires the ability of extracting the integrand $f(x)$ from within the integration operation, which is not possible unless a functional form for $f(x)$ is known (or assumed). An example of that is the exercise of solving for the flux distribution in the specific cases of symmetrical charge distributions (Addendum 4A).

From a different perspective, the integral equation provides a gross (aggregate) relationship of a region while the differential/point form is more specific in providing localized relationships. Hence, it can be stated that the point form offers a step forward and a more useful mathematical form than the integral form.

An interesting observation is that, starting with the integral form, we will develop the differential form by applying the integral form to an infinitesimally small *differential* space element. In doing so, we are relating both D and ρ_v for a point rather than a collective integrated value for a region. In the following, we will demonstrate the concept in Cartesian coordinates. The development in other coordinate systems is similar and final expressions in all three coordinate systems are given in the "Vector Identities" section of Appendix E.

4.8.2 Cartesian Coordinates Differential Form of Gauss' Law

We start with Gauss' law in the integral form, and apply it to a small differential increment $\Delta x - \Delta y - \Delta z$ centered at the point (x, y, z) as our HGS (Figure 4.7). For that, we can write

$$\oiint_S \vec{D} \cdot \overrightarrow{dS} = \sum Q_{\text{enclosed by } S} = \iiint_{\text{volume enclosed by } S} \rho_v \, dv \tag{4.10}$$

The HGS has six sides, each one has its own expression for $\overrightarrow{dS}$, and hence, we need to break the integration into six separate parts:

$$\begin{aligned}
\oiint_S \vec{D} \cdot \overrightarrow{dS} &= \iint_{\text{front}} \vec{D}\left(x+\frac{dx}{2}, y, z\right) \cdot [(dy\,dz)(\vec{a}_x)] + \iint_{\text{back}} \vec{D}\left(x-\frac{dx}{2}, y, z\right) \cdot [(dy\,dz)(-\vec{a}_x)] \\
&+ \iint_{\text{right}} \vec{D}\left(x, y+\frac{dy}{2}, z\right) \cdot [(dz\,dx)(\vec{a}_y)] + \iint_{\text{left}} \vec{D}\left(x, y-\frac{dy}{2}, z\right) \cdot [(dz\,dx)(-\vec{a}_y)] \\
&+ \iint_{\text{top}} \vec{D}\left(x, y, z+\frac{dz}{2}\right) \cdot [(dx\,dy)(\vec{a}_z)] + \iint_{\text{bottom}} \vec{D}\left(x, y, z-\frac{dz}{2}\right) \cdot [(dx\,dy)(-\vec{a}_z)] = dQ_{\text{enclosed by } S}
\end{aligned} \tag{4.11}$$

Using the Cartesian coordinate representation of the flux density vector,

$$\vec{D} = D_x \vec{a}_x + D_y \vec{a}_y + D_z \vec{a}_z \tag{4.12}$$

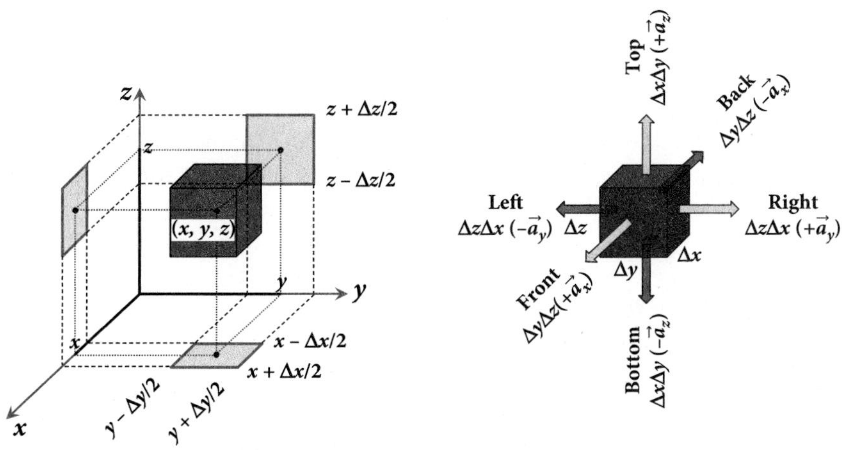

FIGURE 4.7 Closed surface of an incremental volume in Cartesian coordinates used to develop the differential form of Gauss' law.

and removing the zero dot product terms (e.g., $\vec{a}_x \cdot \vec{a}_y$, etc.), the expression (4.11) reduces to

$$\oiint_S \vec{D}\cdot\vec{dS} = \left\{ D_x\left(x+\frac{dx}{2},y,z\right) - D_x\left(x-\frac{dx}{2},y,z\right)\right\}(dy\,dz) + \left\{ D_y\left(x,y+\frac{dy}{2},z\right) - D_y\left(x,y-\frac{dy}{2},z\right)\right\}(dz\,dx) + \left\{ D_z\left(x,y,z+\frac{dz}{2}\right) - D_z\left(x,y,z-\frac{dz}{2}\right)\right\}(dx\,dy) = dQ_{\text{enclosed by } S} \tag{4.13}$$

Dividing both sides by "*dx dy dz*" and taking the limit as *dx*, *dy*, and *dz* reduce to zero,

$$\frac{\left\{ D_x\left(x+\frac{dx}{2},y,z\right) - D_x\left(x-\frac{dx}{2},y,z\right)\right\}}{dx} + \frac{\left\{ D_y\left(x,y+\frac{dy}{2},z\right) - D_y\left(x,y-\frac{dy}{2},z\right)\right\}}{dy} + \frac{\left\{ D_z\left(x,y,z+\frac{dz}{2}\right) - D_z\left(x,y,z-\frac{dz}{2}\right)\right\}}{dz} = \frac{dQ_{\text{enclosed by } S}}{dx\,dy\,dz} \tag{4.14}$$

$$\frac{\partial D_x}{\partial x} + \frac{\partial D_y}{\partial y} + \frac{\partial D_z}{\partial z} = \rho_v(x,y,z) \tag{4.15}$$

which can be written as

$$\left[(\vec{a}_z)\frac{\partial}{\partial x} + (\vec{a}_z)\frac{\partial}{\partial y} + (\vec{a}_z)\frac{\partial}{\partial z}\right]\cdot[D_x(\vec{a}_x) + D_y(\vec{a}_y) + D_z(\vec{a}_z)] = \rho_v(x,y,z) \tag{4.16}$$

or

$$\vec{\nabla}\cdot\vec{D} = \rho_v(x,\ y,\ z) \tag{4.17}$$

Here, the "del" or "nabla" inverted delta symbol, $\vec{\nabla}$, which we will refer to as the "gradient" vector operator, is used to express the vector differential operator as given in its Cartesian coordinates form by

$$\vec{\nabla}_{\text{Cartesian}} = \left[(\vec{a}_z)\frac{\partial}{\partial x} + (\vec{a}_z)\frac{\partial}{\partial y} + (\vec{a}_z)\frac{\partial}{\partial z}\right] \tag{4.18}$$

The expression $\vec{\nabla}\cdot\vec{F}$ is called the divergence of the vector $\vec{F}$ and yields a scalar quantity.

Before going any further, we need to recognize that a similar development can be made to yield results in both cylindrical and spherical coordinates. The conclusion is the same as Equation (4.17) above:

$$\vec{\nabla}\cdot\vec{D} = \rho_v(\text{position}) \tag{4.17}$$

where the $\vec{\nabla}\cdot\vec{D}$ vector operation takes on one of the following forms:

Cartesian:

$$\vec{\nabla}\cdot\vec{D} = \left[\frac{\partial}{\partial x}D_x + \frac{\partial}{\partial y}D_y + \frac{\partial}{\partial z}D_z\right] \tag{4.19}$$

Cylindrical:

$$\vec{\nabla}\cdot\vec{D}=\left[\frac{1}{\rho}\frac{\partial}{\partial\rho}(\rho D_\rho)+\frac{1}{\rho}\frac{\partial}{\partial\varphi}D_\varphi+\frac{\partial}{\partial z}D_z\right] \tag{4.20}$$

Spherical:

$$\vec{\nabla}\cdot\vec{D}=\left[\frac{1}{r^2}\frac{\partial}{\partial r}\left(r^2 D_r\right)+\frac{1}{r\sin\theta}\frac{\partial}{\partial\theta}(\sin\theta\, D_\theta)+\frac{1}{r\sin\theta}\frac{\partial}{\partial\varphi}D_\varphi\right] \tag{4.21}$$

We now resume our discussion of differential form of Gauss' law. As you recall, we started with the integral form of Gauss' law [Equation (4.10)]:

$$\oiint_S \vec{D}\cdot\vec{dS}=\sum Q_{\text{enclosed by } S}=\iiint_{\text{volume enclosed by } S}\rho_v\, dv \tag{4.10}$$

which we applied to an infinitesimal increment of space ($dx\,dy\,dz$) and the end result was the divergence differential "point" form of Gauss' law. This process can be viewed in different mathematical terms by dividing Equation (4.10) by the incremental volume (Δv) and taking the limit as dv becomes infinitesimally small (approaches zero):

$$\lim_{\Delta v\to 0}\left\{\frac{\oiint_S \vec{D}\cdot\vec{dS}}{\Delta v}\right\}=\lim_{\Delta v\to 0}\left\{\frac{\sum Q_{\text{enclosed by } S}}{\Delta v}\right\}=\lim_{\Delta v\to 0}\left\{\frac{\iiint_{\text{enclosed volume by } S}\rho_v\, dv}{\Delta v}\right\} \tag{4.22}$$

The right side of this form reduces to the volumetric charge density within that vanishing increment of space Δv, or in other words, the volumetric charge density at the point of space where dv is located, that is, ρ_v as a function of position.

$$\lim_{\Delta v\to 0}\left\{\frac{\oiint_S \vec{D}\cdot\vec{dS}}{\Delta v}\right\}=\rho_v(\text{position}) \tag{4.23}$$

Hence, by combining the two forms [Equations (4.17) and (4.23)], we arrive at what is known as the divergence theorem as applied to Gauss' law:

$$\vec{\nabla}\cdot\vec{D}=\lim_{\Delta v\to 0}\left\{\frac{\oiint_S \vec{D}\cdot\vec{dS}}{\Delta v}\right\}=\rho_v(\text{position}) \tag{4.24}$$

The physical meaning of this form is defining the divergence of electric flux density vector as per unit volume flux emerging from an incremental volume (point) due to the charge (source) distribution in space. The insight revealed to us through the name "divergence" is that we are referring here to a flux that diverges (or converges) as determined by the charge distribution. In the absence of ρ_v, and according to Equation (4.24), the divergence of the electric flux is zero, which indicates a no change uniform density flow that does not diverge (or converge). On the other hand, the presence of a charge distribution at the location ($\rho_v \neq 0$) will cause the flux to diverge (positive ρ_v) or converge (negative ρ_v causes negative divergence).

It is also important to notice that the divergence operation applies to a vector field and yields a scalar quantity. The operation is a form of a *dot* product operation combined with performing spatial differentiation.

4.9 The Divergence Theorem

Combining Equations (4.10) and (4.24), we arrive at the divergence theorem form as applied to the electric flux density vector:

$$\oiint_S \vec{D}\cdot\overrightarrow{dS} = \iiint_{\text{volume enclosed by } S} \rho_v \, dv = \iiint_{\text{volume enclosed by } S} \vec{\nabla}\cdot\vec{D}\, dv \tag{4.25}$$

or

$$\oiint_S \vec{D}\cdot\overrightarrow{dS} = \iiint_{\text{volume enclosed by } S} \vec{\nabla}\cdot\vec{D}\, dv \tag{4.26}$$

It is important to realize that the divergence theorem [Equation (4.26)] has many applications in many other fields and is not limited to the electric flux density vector presentation. The theory is mathematically and physically sound and its general form for the divergence of a generic vector $\vec{A}$ is of the form:

$$\oiint_S \vec{A}\cdot\overrightarrow{dS} = \iiint_{\text{volume enclosed by } S} \vec{\nabla}\cdot\vec{A}\, dv \tag{4.27}$$

4.10 Application of Gauss' Law in the Point Form

Now, we examine the usefulness of the point/differential form of Gauss' law as arrived at through the divergence theorem development. Simply put, the expression (4.17) provides us with a differential relationship that expresses $\vec{D}$ at any point in space in terms of the charge density at that point. In contrast, the integral form [Equation (4.10)] gives an aggregate description of the total flux for a region of space as related to the total charge in that region. It is obvious that the point form provides a tool that is more *general* in nature and can be used to express fine details of the physical distribution of the charge and the corresponding flux density.

Mathematically speaking, the point form has a differential equation format that we can solve either analytically or numerically/graphically to obtain a solution subject to the implementation of the problem/region's boundary conditions. To demonstrate, we will consider one of the cases that we discussed earlier with the spherical symmetry, Figure 4.8, specifically the case of a spherical volume charge distribution with ρ_v allowed to vary with r only (no variations with θ or φ).

Referring to the figure, the spherical charge distribution is confined to the region $0<r<b$. Beyond this region, $b<r<\infty$, there are no charge distributions, and hence we need to write two separate mathematical

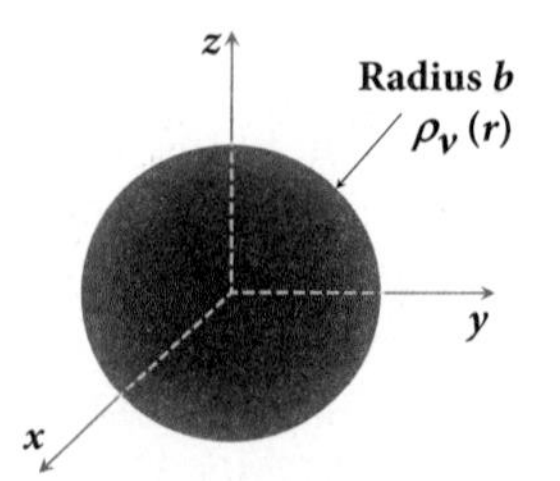

FIGURE 4.8 A spherical volume charge distribution with spherical symmetry.

expressions for each of the two regions. We will refer to $0<r<b$ as region I and $b<r<\infty$ as region II, and consequently the naming of the flux density vector $\vec{D}$ as $\vec{D}_I$ and $\vec{D}_{II}$ in the two regions, respectively.

Applying the point form of Gauss' law to the two regions, we write

$$\vec{\nabla}\cdot\vec{D}_I=\rho_v(r),\ 0<r<b \tag{4.28}$$

where $\rho_{v(r)}$ describes the volumetric charge density distribution within the $r=b$ sphere (region I), and

$$\vec{\nabla}\cdot\vec{D}_{II}=0,\ b<r<\infty \tag{4.29}$$

Next, we express the divergence of D in spherical coordinates using Equation (4.21)

$$\vec{\nabla}\cdot\vec{D}=\left[\frac{1}{r^2}\frac{\partial}{\partial r}\left(r^2D_r\right)+\frac{1}{r\sin\theta}\frac{\partial}{\partial\theta}(\sin\theta D_\theta)+\frac{1}{r\sin\theta}\frac{\partial}{\partial\varphi}D_\varphi\right] \tag{4.30}$$

Considering the spherical symmetry of the charge distribution, we can set both $\frac{\partial}{\partial\theta}$ and $\frac{\partial}{\partial\varphi}$ derivatives to zero, and hence:

$$\vec{\nabla}\cdot\vec{D}=\left[\frac{1}{r^2}\frac{\partial}{\partial r}\left(r^2D_r\right)\right] \tag{4.31}$$

Combining this form with the region I Gauss' formula, we write

$$\vec{\nabla}\cdot\vec{D}_I=\left[\frac{1}{r^2}\frac{\partial}{\partial r}\left(r^2D_{Ir}\right)\right]=\rho_v(r) \tag{4.32}$$

which can be rewritten as

$$\left[\frac{\partial}{\partial r}\left(r^2D_{Ir}\right)\right]=r^2\rho_v(r) \tag{4.33}$$

$$(r^2D_{Ir})=\int r^2\rho_v(r)dr+C_1 \tag{4.34}$$

where C_1 is the integration constant, and finally

$$D_{Ir}=\frac{1}{r^2}\int r^2\rho_v(r)dr+\frac{C_1}{r^2} \tag{4.35}$$

Similarly, for region II where $\rho_v=0$, the expression is of the form:

$$D_{IIr}=0+\frac{C_2}{r^2} \tag{4.36}$$

As we can see, the solution of the differential equation yielded some unknown constants which require additional information to determine and arrive at a unique solution for the particular problem at hand. This is typical of differential equations as they quantify rates of changes (slopes, grades, divergences, etc.) rather than the absolute quantities. In other words, solutions of differential equations yield families of functional relationships of all possible forms that satisfy these equations. Only when relevant boundary conditions are applied we are able to specify the particular solution that uniquely describes the particular problem solution.

The boundary conditions for the problem here are needed to evaluate the two solution constants C_1 and C_2. As we have two regions, each one has two boundaries with one of these boundaries is common to both regions. In other words, we have three boundaries to work with

BC#1: At $r=0$ (in region I)

$D_{Ir}=\frac{1}{r^2}\int r^2\rho_v(r)dr+\frac{C_1}{r^2}$ at $r=0$, the term $\frac{C_1}{r^2}$ blows up to infinity which is not physically possible in the absence of a singularity (e.g., $\rho_{v(0)}=\infty$). This implies that this term must be rejected as it does not represent a physically acceptable solution, and hence $C_1=0$.

BC#2: At $r=b$ (common to both regions I and II)

At $r=b$, we expect a continuity in the flux density vector between the two regions assuming no singularities in the charge distribution at this interface (e.g., surface or point charges that would cause $\rho_{v(b)}=\infty$).

Hence,

$$\left[D_{Ir}=\frac{1}{r^2}\int r^2\rho_v(r)dr\right]_{r=b}=\left[D_{IIr}=\frac{C_2}{r^2}\right]_{r=b} \tag{4.37}$$

This should enable us to evaluate C_2, hence

$$C_2=\left[\int r^2\rho_v(r)dr\right]_{r=b} \tag{4.38}$$

BC#3: At $r=\infty$ (in region II)

$D_{IIr}=\frac{C_2}{r^2}$ at $r=\infty$, the term $\frac{C_2}{r^2}$ diminishes which is consistent with our physical problem but does not reveal any new knowledge.

Hence, the final solution is

$$D_r=\begin{cases}\left[\frac{1}{r^2}\int r^2\rho_v(r)dr\right], & 0<r<b\\ \frac{1}{r^2}\left[\int r^2\rho_v(r)dr\right]_{r=b}, & b<r<\infty\end{cases} \tag{4.39}$$

This solution is identical to that obtained using the integral form of Gauss' law in Equation (4A.14).

Now, let us apply that to the simple case of $\rho_{v(r)}=\rho_o$ (a constant).

$$D_{Ir}=\frac{1}{r^2}\int r^2\rho_v(r)dr=\frac{\rho_o}{r^2}\left(\frac{r^3}{3}\right)=\left(\frac{\rho_o r}{3}\right) \tag{4.40}$$

and

$$D_{IIr}=\frac{\left[\int r^2\rho_v(r)dr\right]_{r=b}}{r^2}=\left(\frac{\rho_o b^3}{3r^2}\right) \tag{4.41}$$

Hence, the final solution in its vector form:

$$\vec{D}=\begin{cases}\left(\frac{\rho_o r}{3}\right)\vec{a}_r, & 0<r<b\\ \left(\frac{\rho_o b^3}{3r^2}\right)\vec{a}_r, & b<r<\infty\end{cases} \tag{4.42}$$

ADDENDUM 4A
APPLICATION OF THE INTEGRAL FORM OF GAUSS' LAW TO SYMMETRICAL CHARGE DISTRIBUTIONS

4A.1 Electric Flux Distributions for Charges of Spherical Symmetries (No Variations with θ or φ)

To remind the reader with the spherical coordinate system and how does spherical symmetry imply no variations with θ or φ, we need to closely examine Figure 4A.1. The figure presents the spherical coordinate system, where r is the distance from the origin, θ is the angle the radial coordinate r makes with the z-axis, and φ is the angle between the projection of r on the x-y plane and the x-axis. (Similarly, it is the angle the plane formed by the radial r and the z-axis on one side and the x-z plane on the other.)

In this coordinate system, a spherical distribution (shape/body) centered at the origin will be symmetrically viewed from all angles around it provided that this shape is uniform for all θ and φ angles. The only allowed variation in the distribution configuration is with respect to r. In such case, if we were to rotate this distribution around any major axis, we should not observe any change in the problem configuration.

Charges distributed with spherical symmetries may take one or more of the following forms (Figure 4A.2):

1. Point charge located at the origin
2. Case of spherical surface charge distribution with constant ρ_S (no variations with θ or φ)

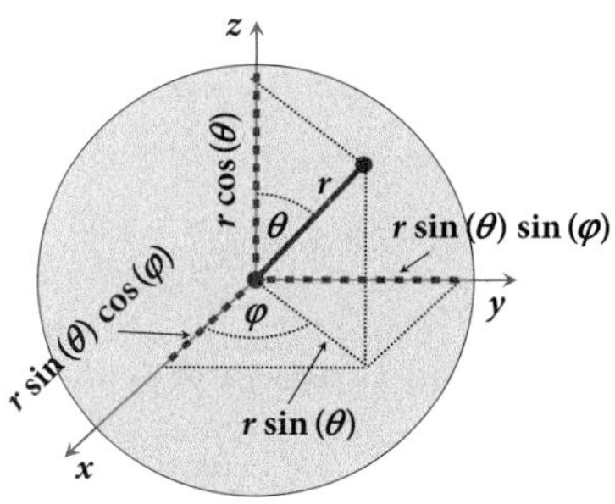

Figure 4A.1 Review of the spherical coordinate system.

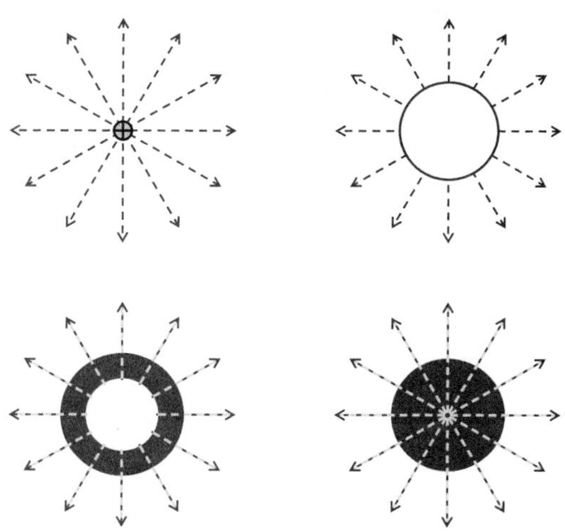

Figure 4A.2 Different forms of charge distributions with spherical symmetries.

3. Case of spherical shell charge distribution with ρ_v allowed to vary with r only (no variations with θ or φ)
4. Case of spherical volume charge distribution with ρ_v allowed to vary with r only (no variations with θ or φ)

For one or more concentric combinations of the above, we expect the electric flux distribution to be independent of θ and φ variations and hence its density would only vary with r. The symmetry also implies that $\vec{D}$ is radial as it emerges from the spherical symmetry of the charge distribution.

Hence,

$$\vec{D} = D(r)\vec{a}_r \tag{4A.1}$$

Now, let us apply Gauss' law to these spherical symmetries. To do so, we introduce an imaginary closed surface, S, in the form of a sphere concentric with the charge distribution (Figure 4A.3). We will often use the abbreviation of HGS for this surface as short for "hypothetical Gaussian surface."

$$\{\psi_{\text{closed surface } S}\} = \oiint_S \vec{D}\cdot\vec{dS} = \iiint_{\text{enclosed volume by } S} \rho_v \, dv \tag{4A.2}$$

Referring to Figure 4A.4, the left-hand side (LHS) of this equality reduces to

$$\{\psi_S\} = \text{LHS} = \oiint_S \vec{D}\cdot\vec{dS} = \oiint_S D(r)\vec{a}_r \cdot \vec{dS} \tag{4A.3}$$

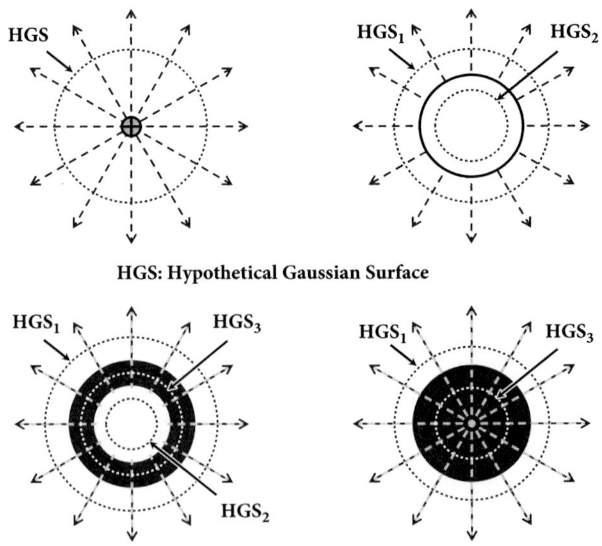

FIGURE 4A.3 Choosing closed Gaussian surfaces for cases of charge distributions with spherical symmetries.

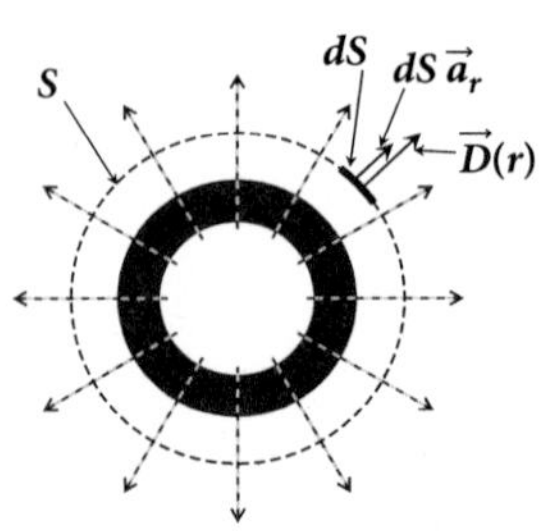

FIGURE 4A.4 Application of Gauss' law to the closed Gaussian surface for the spherical symmetry cases.

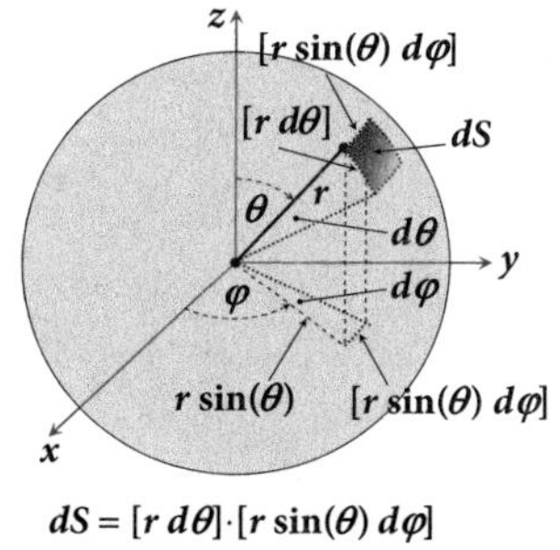

FIGURE 4A.5 The differential area element of the spherical surface.

In spherical coordinates, the differential area element of the spherical surface (Figure 4A.5) is given by

$$\vec{dS} = dS\vec{a}_r = [r\,d\theta][r\sin(\theta)d\varphi]\vec{a}_r \tag{4A.4}$$

So,

$$\text{LHS} = \oiint_S D(r)\vec{a}_r \cdot \vec{dS} = \oiint_S D(r)\vec{a}_r \cdot dS\vec{a}_r = \oiint_S D(r)dS \tag{4A.5}$$

where $\vec{a}_r \cdot \vec{a}_r = 1$. Using the expression for dS and integrating over the entire closed surface we obtain

$$\{\psi_{\text{closed surface } S}\} = \text{LHS} = \oiint_S D(r)dS = \int_{\varphi=0}^{2\pi}\int_{\theta=0}^{\pi} D(r)[r\,d\theta][r\sin(\theta)d\varphi]$$

$$= r^2 D(r)\int_{\varphi=0}^{2\pi}\int_{\theta=0}^{\pi}\sin(\theta)d\theta\,d\varphi = 4\pi r^2 D(r) \tag{4A.6}$$

We could have guessed this answer for the LHS based on the fact that $D(r)$ is uniform over the entire area of the (constant r) Gaussian (spherical) surface, and hence, the total flux is simply the product of the density $D(r)$ times the surface area of the sphere $4\pi r^2$.

Now, we tackle the right-hand side (RHS) of the equation. It is clear that the result of that depends on the charge distribution we deal with.

4A.1.1 Case of Point Charge q Located at the Origin (Notice, We Use q or ΔQ for Point Charge Notation)

It is clear that any Gaussian sphere of radius r encloses the entire point charge, and hence the RHS is equal to the point charge value, q. Equating the sides RHS = LHS, we arrive at

$$4\pi r^2 D(r) = q \tag{4A.7}$$

which then yields the form of distribution of the electric flux density $D(r)$ due to a single point charge q located at the origin as

$$D(r) = \frac{q}{4\pi r^2} \quad \text{and} \quad \vec{D}(r) = D(r)\vec{a}_r = \frac{q}{4\pi r^2}\vec{a}_r \tag{4A.8}$$

Examining this result, we can see that the charge q is letting out a total flux q that diverges radially and uniformly over all angles of space ($0 \le \theta \le \pi$ and $0 \le \varphi \le 2\pi$). In such case, the flux density varies only with the distance r from the charge and hence, points at equal (r) (concentric spheres) have equal flux density magnitudes. The constant amount of flux emerging from q spreads over increasing spherical surface areas as the distance from the charge (r) increases. The flux density as the ratio between the flux and the area it passes

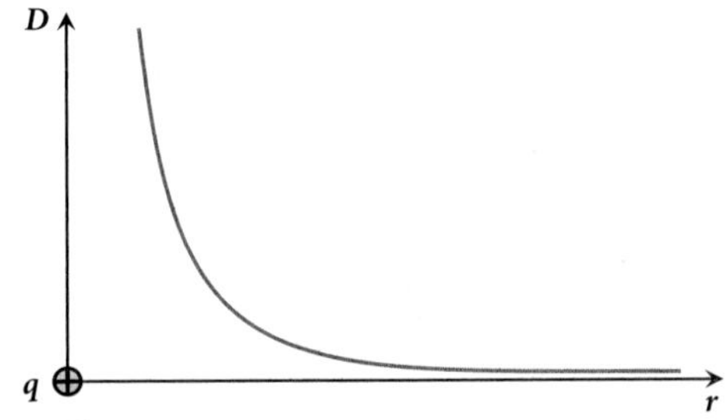

FIGURE 4A.6 Flux density versus distance for the case of a point charge at the origin.

through, becomes equal to the total flux q divided by the area of the sphere at any given radial distance (r) and is equal to $\frac{q}{4\pi r^2}$.

Figure 4A.6 displays the functional dependence of $D(r)$ with respect to r, which is of the form of an inverse square relationship, $1/r^2$. As we can see, D is infinite at $r=0$ due to the "singularity" of having a point charge with finite charge amount with a zero-charge surface. Hence, q coulombs of flux are diverging out of a zero surface-area resulting in an infinite density. At infinity (far away from the charge center), however, the same amount of flux is spread over an infinite surface yield an infinitesimally small (approaching zero) flux density.

4A.1.2 Case of Spherical Surface Charge Distribution with Uniform ρ_S

In this case, we assume that a total charge Q is uniformly distributed over the spherical surface of radius a centered at the origin. The surface charge density in this case is

$$\rho_S = \frac{Q}{4\pi a^2} \tag{4A.9}$$

This charge distribution divides the space into two spherical regions:

$r < a$ and $r > a$, and hence we would expect to work our analysis for each region separately as the flux density may have different functional dependence. Consequently, we will choose two different HGS, one for each region; HGS1 for $r > a$ and HGS2 for $r < a$. Applying Gauss' law to both cases, we find that the expression of the LHS is identical to the one developed earlier for the case of the point charge. The differing results will be in the RHS of the equation due to the different amounts of charge enclosed.

HGS1: $RHS = Q$ as the surface HGS1 encloses all of the charge Q

HGS2: $RHS = 0$ since there are no charges enclosed by HGS2

Therefore, we can write

$$\vec{D}(r) = \begin{cases} 0, & \text{for } r < a \\ \frac{Q}{4\pi r^2}\vec{a}_r = \rho_S \frac{a^2}{r^2}\vec{a}_r, & \text{for } r > a \end{cases} \tag{4A.10}$$

Figure 4A.7 displays the functional dependence of $D(r)$ with respect to r, which is zero for $r < a$, and of the form of an inverse square relationship, $1/r^2$, for $r > a$. The physical interpretation for the zero flux inside $r = a$ is simple, as the symmetry implies the flux to be radial out of the charge Q. Flux lines starting on the Q charge pointing inward need a sink of –Q to terminate on. In the absence of such charge sink, these flux lines cannot exist yielding a zero value for D inside $r = a$.

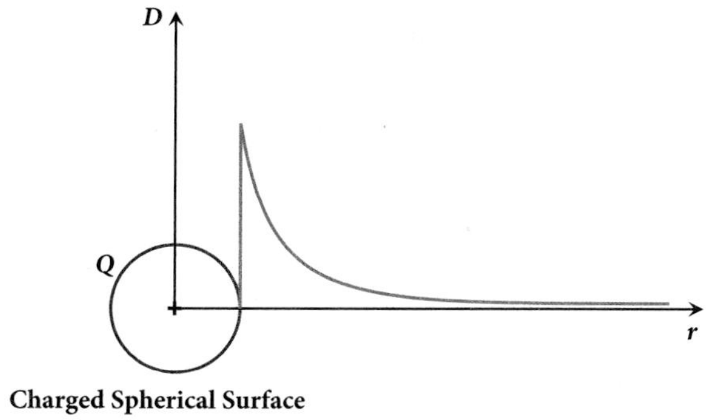

FIGURE 4A.7 Flux density versus distance for the case of a spherical surface charge distribution with uniform ρ_S.

4A.1.3 Case of Spherical Shell Charge Distribution with ρ_v Varying with r Only

In this case, we assume a total charge Q distributed over the spherical shell of inner radius a and an outer radius b centered at the origin. The volumetric charge density varies only with r.

$$Q=\iiint \rho_v(r)dv=\int_{\varphi=0}^{2\pi}\int_{\theta=0}^{\pi}\int_{r=a}^{b}\rho_v(r)[dr][r\,d\theta][r\sin(\theta)d\varphi]=\int_{r=a}^{b}\rho_v(r)4\pi r^2\,dr \tag{4A.11}$$

This charge distribution divides the space into three spherical regions: $r<a$, $a<r<b$, and $r>b$, and hence, we need to repeat the analysis for each region separately. In this case, we will use three different HGSs, one for each region; HGS1 for $r>b$, HGS2 for $r<a$, and HGS3 for $a<r<b$. Applying Gauss' law in all three cases yields the same LHS that we developed for the two earlier cases. The differing results will be in the RHS of the equation due to the different amounts of charge enclosed.

HGS1: $RHS=Q$ as the surface HGS1 encloses all of the charge Q

HGS2: $RHS=0$ since there are no charges enclosed by HGS2

HGS3: $RHS=$ the part of Q enclosed by the surface HGS3 between a and r

Therefore, we can write

$$\vec{D}(r)=\begin{cases} 0, & \text{for } r<a \\ \dfrac{\int_{r=a}^{r}\rho_v(r)4\pi r^2\,dr}{4\pi r^2}\vec{a}_r, & \text{for } a<r<b \\ \dfrac{\int_{r=a}^{b}\rho_v(r)4\pi r^2\,dr}{4\pi r^2}\vec{a}_r=\dfrac{Q}{4\pi r^2}\vec{a}_r, & \text{for } r>b \end{cases} \tag{4A.12}$$

Figure 4A.8 displays the functional dependence of $D(r)$ with respect to r, which is zero for $r<a$, and grows gradually (depending on the nature of the functional dependence of ρ_v on r). Starting $r=b$, the function is again an inverse square relationship, $1/r^2$. There are no discontinuities in the electric flux density D at $r=a$ or $r=b$ as there are no "singularities" in the charge density distribution.

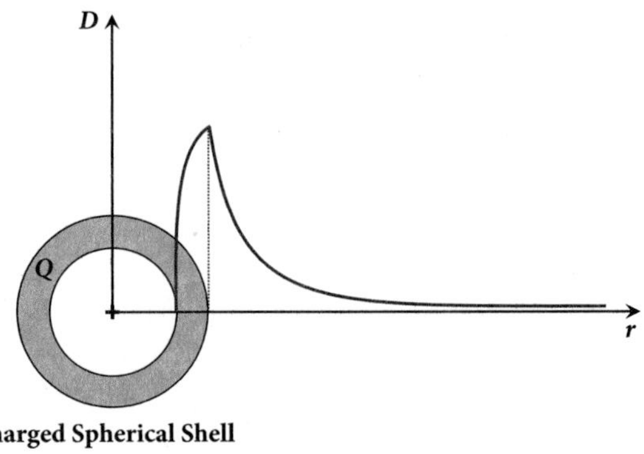

Figure 4A.8 Flux density versus distance for the case of a spherical shell charge distribution with ρ_v varying with r only and centered at the origin.

4A.1.4 Case of Spherical Volume Charge Distribution with ρ_v Varying with r Only

This case is similar to the spherical shell case with the inner sphere of radius a eliminated, that is, $a=0$. Adjusting the results obtained in the shell case accordingly, we can write

$$Q=\iiint \rho_v(r)dv=\int_{\varphi=0}^{2\pi}\int_{\theta=0}^{\pi}\int_{r=0}^{b}\rho_v(r)[dr][r\,d\theta][r\sin(\theta)d\varphi]=\int_{r=0}^{b}\rho_v(r)4\pi r^2\,dr \tag{4A.13}$$

$$\vec{D}(r)=\begin{cases}\dfrac{\int_{r=0}^{r}\rho_v(r)4\pi r^2\,dr}{4\pi r^2}\vec{a}_r=\dfrac{1}{r^2}\displaystyle\int_{r=0}^{r}\rho_v(r)r^2\,dr\,\vec{a}_r, & \text{for } r<b\\[2ex] \dfrac{\int_{r=0}^{b}\rho_v(r)4\pi r^2\,dr}{4\pi r^2}\vec{a}_r=\dfrac{1}{r^2}\displaystyle\int_{r=0}^{b}\rho_v(r)r^2\,dr\,\vec{a}_r=\dfrac{Q}{4\pi r^2}\vec{a}_r, & \text{for } r>b\end{cases} \tag{4A.14}$$

Figure 4A.9 displays the functional dependence of $D(r)$ with respect to r, which grows gradually starting at $r=0$ (depending on the nature of the functional dependence of ρ_v on r). While starting $r=b$, the function is again an inverse square relationship, $1/r^2$. Again, there are no discontinuities in the electric flux density D at $r=b$ as there are no "singularities" in the charge density distribution.

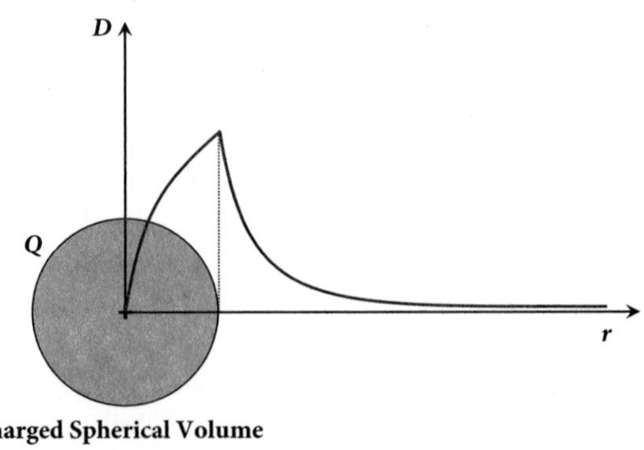

Figure 4A.9 Flux density versus distance for the case of a spherical volume charge distribution with ρ_v varying with r only and centered at the origin.

4A.2 Electric Flux Distributions for Charges of Cylindrical Symmetries (No Variations with φ or z)

Similar to our discussion of the spherical coordinate system, the cylindrical coordinate system is reviewed in Figure 4A.10. The figure presents the cylindrical coordinate system, where ρ is the radius of the cylinder passing by the point of interest and coaxial with z axis, φ is the angle between the projection of ρ on the x-y plane and the x axis. (Similarly, it is the angle the plane formed by the radial ρ and the z axis on one side and the x-z plane on the other). Finally, z is the height of the point above the x-y plane.

In this coordinate system, the cylindrical symmetry of interest here implies no variations with φ or z. Hence, a cylindrical distribution (shape/body) placed coaxially with the z axis and infinite in extension along that axis will be symmetrically viewed from all angles around it provided that this shape is uniform for all φ angles. Furthermore, it will be symmetrically viewed from all z positions provided that this shape is uniform for all its infinite extension. The only allowed variation in the distribution configuration is with respect to ρ. In such case, if we were to rotate this distribution around z axis or move it along that axis, we should not observe any change in the problem configuration.

Charges distributed with cylindrical symmetries may take one or more of the following forms (Figure 4A.11):

1. Line charge with uniform linear charge density ρ_ℓ over infinite length.
2. Cylindrical surface charge distribution with constant ρ_S (no variations with φ or z), again over infinite length.

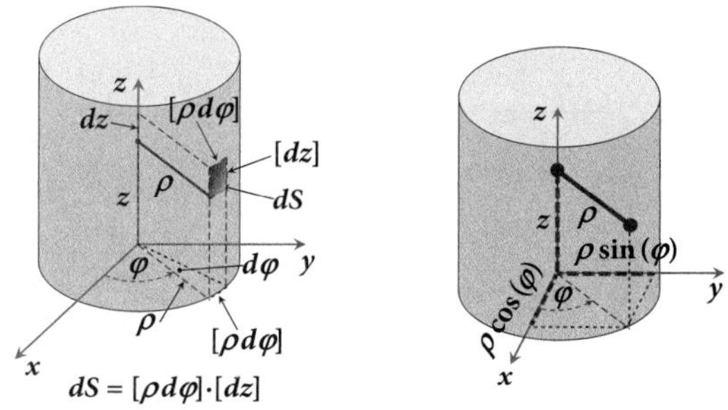

Figure 4A.10 Review of the cylindrical coordinate system and the differential area element of a cylindrical surface.

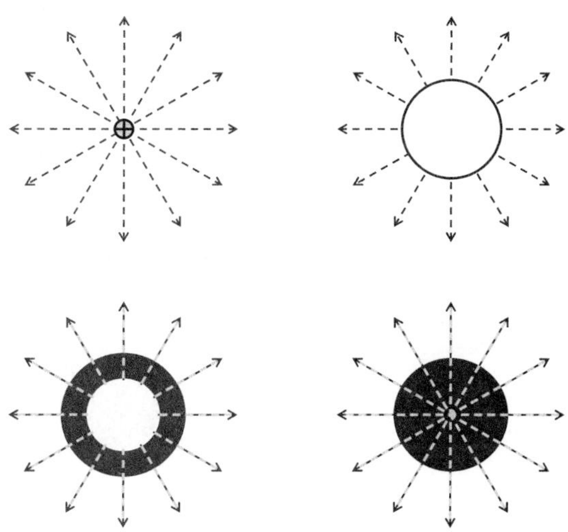

Figure 4A.11 Different forms of charge distributions with cylindrical symmetries.

3. Cylindrical shell charge distribution with ρ_v allowed to vary with ρ only (no variations with φ or z), again over infinite length.
4. Cylindrical volume charge distribution with ρ_v allowed to vary with ρ only (no variations with φ or z), again over infinite length.

For one or more concentric combinations of the above, we expect the electric flux distribution to be independent of φ and z variations and hence its density would only vary with ρ. The symmetry also implies that $\vec{D}$ is cylindrically radial as it emerges from the cylindrical symmetry of the charge distribution.

Hence,

$$\vec{D} = D(\rho)\vec{a}_\rho \tag{4A.15}$$

Now, let us apply Gauss' law to these cylindrical symmetries. To do so, we introduce an HGS in the form of a closed finite height cylinder concentric with the charge distribution, (Figures 4A.12 and 4A.13).

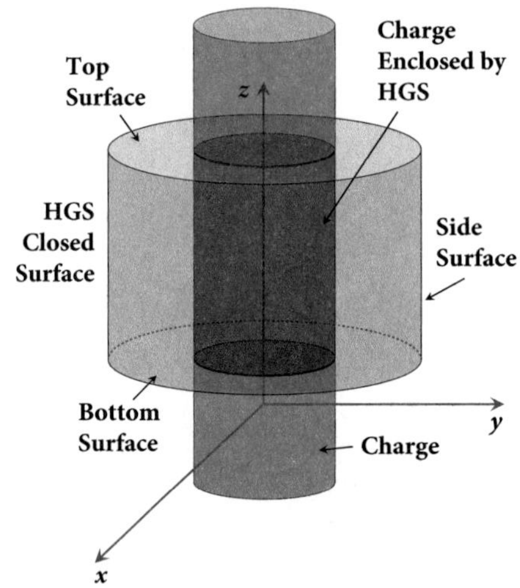

FIGURE 4A.12 Choosing a closed HGS for cylindrical symmetry cases.

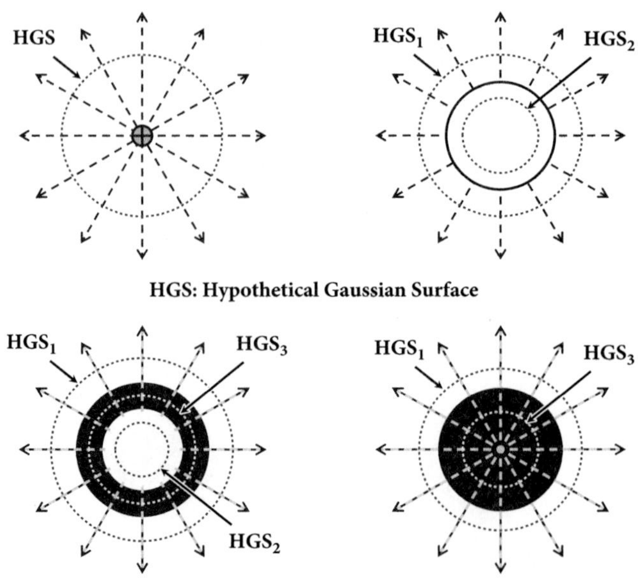

FIGURE 4A.13 Choosing closed Gaussian surfaces for cases of charge distributions with cylindrical symmetries.

This closed surface is composed of the circular cylindrical side (radius ρ and height h) and two flat circular discs radii ρ each to serve as the top and bottom closures.

$$\{\psi_{\text{closed surface } S}\} = \oiint_S \vec{D}\cdot\overrightarrow{dS} = \iiint_{\text{enclosed volume by } S} \rho_v \, dv \tag{4A.16}$$

Substituting $\vec{D} = D(\rho)\vec{a}_\rho$, the left-hand side (LHS) of this equality becomes

$$\{\psi_{\text{closed surface } S}\} = \text{LHS} = \oiint_S \vec{D}\cdot\overrightarrow{dS} = \oiint_S D(\rho)\vec{a}_\rho\cdot\overrightarrow{dS} \tag{4A.17}$$

To perform the closed surface integration on this HGS, the integration takes the form of the sum of three integrals; one on the curved side and two on the flat sides. In cylindrical coordinates, the differential area element of the curved cylindrical side is

$$\overrightarrow{dS} = dS\,\vec{a}_\rho = [\rho d\varphi][dz]\vec{a}_\rho \tag{4A.18}$$

While on the HGS top and bottom discs:

$$\overrightarrow{dS}_{\text{top}} = dS\vec{a}_z = [d\rho][\rho\, d\varphi]\vec{a}_z \qquad \text{and} \qquad \overrightarrow{dS}_{\text{bottom}} = dS(-\vec{a}_z) = -[d\rho][\rho d\varphi]\vec{a}_z, \text{ respectively.}$$

Consequently,

$$\begin{aligned}\text{LHS} = \oiint_S D(\rho)\vec{a}_\rho\cdot\overrightarrow{dS} &= \iint_{\text{side}} D(\rho)\vec{a}_\rho\cdot dS\vec{a}_\rho + \iint_{\text{top}} D(\rho)\vec{a}_\rho\cdot dS\vec{a}_z \\ &+ \iint_{\text{bottom}} D(\rho)\vec{a}_\rho\cdot[-dS\vec{a}_z]\end{aligned} \tag{4A.19}$$

$$\text{LHS} = \iint_{\text{side}} D(\rho)dS \tag{4A.20}$$

since $\vec{a}_\rho\cdot\vec{a}_\rho = 1$ and $\vec{a}_\rho\cdot\vec{a}_z = 0$. Using the expression for dS and integrating over the HGS side surface we obtain

$$\{\psi_{\text{closed surface } S}\} = \text{LHS} = \iint_{\text{side}} D(\rho)dS = \int_{z=z}^{z+h}\int_{\varphi=0}^{2\pi} D(\rho)[\rho\, d\varphi][dz] = 2\pi\rho h D(\rho) \tag{4A.21}$$

We could have guessed this answer for the LHS based on the fact that $D(\rho)$ is uniform over the entire area of the (constant ρ) cylindrical (side) surface, and hence, the total flux is simply the product of the density $D(\rho)$ times the area of the cylinder's side $2\pi\rho h$.

Now, we tackle the right-hand side (RHS) of the equation. It is clear that the result of that depends on the charge distribution we deal with.

4A.2.1 Case of an Infinite Line Charge Uniformly Stretched along the z Axis

It is clear that the chosen (concentric) HGS of radius ρ and height h encloses a portion on length h of the line charge, and hence the RHS is equal to the length of the enclosed charge multiplied by the constant density of the charge distribution, $\rho_\ell h$. Equating the sides RHS = LHS, we arrive at

$$2\pi\rho h D(\rho) = \rho_\ell h \tag{4A.22}$$

which then yields the form of distribution of the electric flux density $D(\rho)$ due to a line charge Q stretched along the z axis as

$$D(\rho) = \frac{\rho_\ell}{2\pi\rho} \qquad \text{and} \qquad \vec{D}(\rho) = D(\rho)\vec{a}_\rho = \frac{\rho_\ell}{2\pi\rho}\vec{a}_\rho \tag{4A.23}$$

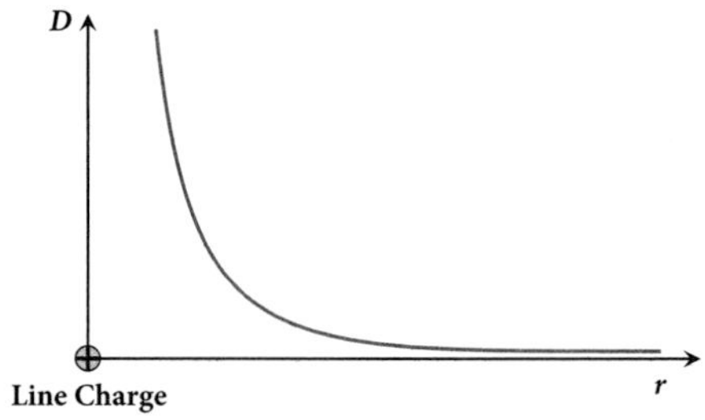

FIGURE 4A.14 Flux density versus distance for the case of an infinite line charge uniformly stretched along the z axis.

Examining this result, we find the flux to diverge radially with no φ or z dependence. The flux density is uniform over all φ angles, $(0 \le \varphi \le 2\pi)$, and all z heights, $(-\infty < z < +\infty)$. It only varies with the distance (ρ) from the line charge and hence, points of equal (ρ) (concentric cylinders) have equal flux density magnitudes. The constant amount of flux emerging from a segment of the line height spreads over increasing cylindrical surface areas of the same height. As the cylindrical surface area increases linearly with ρ, the flux density decreases in the same rate resulting in the $(1/\rho)$ expression.

Figure 4A.14 displays the functional dependence of $D(\rho)$ with respect to ρ. D is infinite at $\rho = 0$ due to the "singularity" of having a zero surface-area from which the flux emerges. While at infinity (far away from the charge line), the same amount of flux is spread over an infinite surface yielding an infinitesimally small (approaching zero) flux density.

4A.2.2 Case of an Infinite Height Cylindrical Surface Charge Distribution with Uniform ρ_S (or ρ_ℓ)

Similar to what we have done for the case of the spherical surface, we need to deal with two regions, inside the surface and outside it. Hence, we will implement two (concentric) HGS, one for each region, one at a time.

The HGS inside the surface with a radius $\rho < a$ and height h encloses no charges, resulting in a zero RHS and consequently zero $D(\rho)$.

For the outside HGS (radius $\rho > a$ and height h), a portion of height h of the charged surface is enclosed, and hence the RHS is equal to $\rho_S 2\pi a h$ (or $\rho_\ell h$) and hence

$$2\pi\rho h D(\rho) = \rho_S(2\pi a h) = \rho_\ell h, \quad \text{and} \quad D(\rho) = \frac{a\rho_S}{\rho} = \frac{\rho_\ell}{2\pi\rho} \tag{4A.24}$$

So, we can write the final result in this case as

$$\vec{D}(\rho) = \begin{cases} 0, & \text{for } \rho < a \\ \dfrac{a\rho_S}{\rho}\vec{a}_\rho = \dfrac{\rho_\ell}{2\pi\rho}\vec{a}_\rho, & \text{for } \rho > a \end{cases} \tag{4A.25}$$

which is graphed in Figure 4A.15.

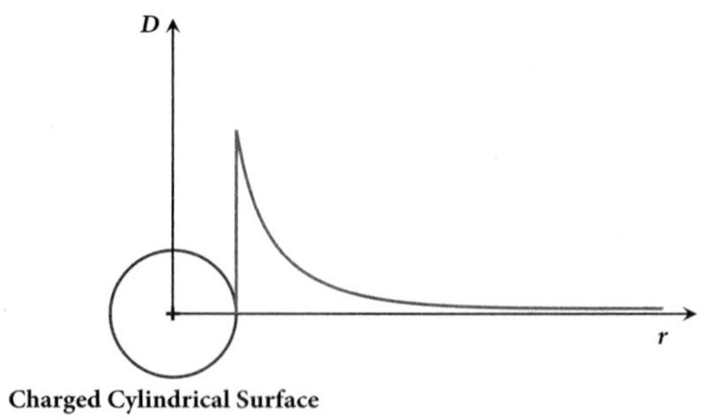

FIGURE 4A.15 Flux density versus distance for the case of an infinite height cylindrical surface charge distribution with uniform ρ_S (or ρ_ℓ) placed coaxial with the z axis.

4A.2.3 Case of a Cylindrical Shell of Charge Distribution with Infinite Height and ρ_v Varying with ρ Only

In this case, we need to deal with three regions, inside the inner surface, within the charge volume, and outside the outer surface. Hence, we will work with three (concentric) HGS, one for each region, one at a time.

The HGS inside the inner surface with a radius $\rho < a$ and height h encloses no charges, resulting in a zero RHS and consequently zero $D(\rho)$.

For the HGS within the charge volume, (radius $a<\rho<b$ and height h) a portion the volume charge is enclosed, and hence the RHS is equal to $\int_{z=z}^{z+h}\int_{\varphi=0}^{2\pi}\int_{\rho=a}^{\rho}\rho_v(\rho)[d\rho][\rho d\varphi][dz]$ and hence

$$2\pi\rho h D(\rho)=\int_{z=z}^{z+h}\int_{\varphi=0}^{2\pi}\int_{\rho=a}^{\rho}\rho_v(\rho)[d\rho][\rho d\varphi][dz]=2\pi h\int_{\rho=a}^{\rho}\rho_v(\rho)[\rho d\rho] \tag{4A.26}$$

Finally, for the outside HGS, (radius $\rho>b$ and height h) a portion on height h the charged is given by $\int_{z=z}^{z+h}\int_{\varphi=0}^{2\pi}\int_{\rho=a}^{b}\rho_v(\rho)[d\rho][\rho d\varphi][dz]$:

$$2\pi\rho h D(\rho)=2\pi h\int_{\rho=a}^{b}\rho_v(\rho)[\rho d\rho] \tag{4A.27}$$

So, we can write the final result in this case as

$$\vec{D}(\rho)=\begin{cases} 0, & \text{for } \rho<a \\ \dfrac{1}{\rho}\displaystyle\int_{\rho=a}^{\rho}\rho_v(\rho)[\rho d\rho]\vec{a}_\rho, & \text{for } a<\rho<b \\ \dfrac{1}{\rho}\displaystyle\int_{\rho=a}^{b}\rho_v(\rho)[\rho d\rho]\vec{a}_\rho, & \text{for } \rho>b \end{cases} \tag{4A.28}$$

and for the case of constant ρ_v,

$$\vec{D}(\rho)=\begin{cases} 0, & \text{for } \rho<a \\ \rho_v\dfrac{(\rho^2-a^2)}{2\rho}\vec{a}_\rho, & \text{for } a<\rho<b \\ \rho_v\dfrac{(b^2-a^2)}{2\rho}\vec{a}_\rho, & \text{for } \rho>b \end{cases} \tag{4A.29}$$

which is graphed in Figure 4A.16.

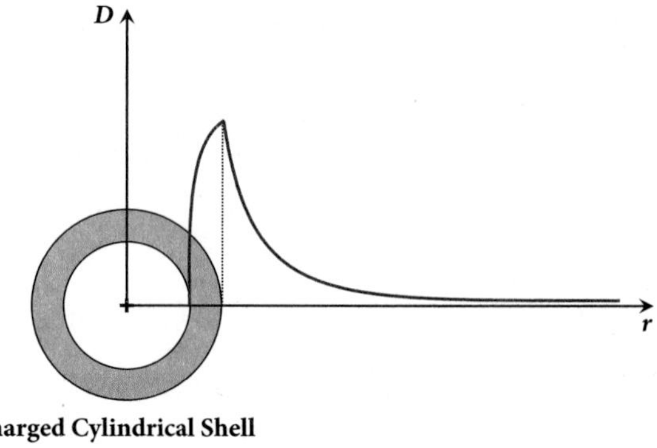

FIGURE 4A.16 Flux density versus distance for the case of an infinite height cylindrical shell charge distribution with ρ_v varying with ρ only placed coaxial with the z axis.

4A.2.4 Case of "Full" Cylindrical Charge Distribution with Infinite Height and ρ_v Varying with ρ Only

We can write the answers for this case by simply setting a in cylindrical shell case to zero, and hence, we have only two regions.

$$\vec{D}(\rho)=\begin{cases} \dfrac{1}{\rho}\displaystyle\int_{\rho=0}^{\rho}\rho_v(\rho)[\rho\, d\rho]\vec{a}_\rho, & \text{for } \rho<b \\ \dfrac{1}{\rho}\displaystyle\int_{\rho=0}^{b}\rho_v(\rho)[\rho\, d\rho]\vec{a}_\rho, & \text{for } \rho>b \end{cases} \tag{4A.30}$$

and for the case of constant ρ_v,

$$\vec{D}(\rho)=\begin{cases} \rho_v\dfrac{\rho}{2}\vec{a}_\rho, & \text{for } \rho<b \\ \rho_v\dfrac{b^2}{2\rho}\vec{a}_\rho, & \text{for } \rho>b \end{cases} \tag{4A.31}$$

which is graphed in Figure 4A.17.

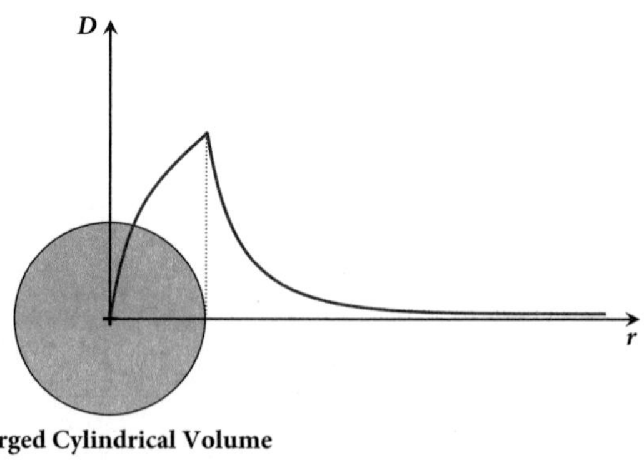

FIGURE 4A.17 Flux density versus distance for the case of an infinite height cylindrical volume charge distribution with ρ_v varying with ρ only placed coaxial with the z axis.

4A.3 Electric Flux Distributions for Charges of Planar Symmetries (No Variations with x or y)

Charges distributed with planar symmetries may take one or more of the following forms:

1. Planar surface charge distribution with constant ρ_S (no variations with x or y), with infinite extension in both x and y coordinates.
2. Planar "slab" of charge distribution with ρ_v allowed to vary with z only (no variations with x or y), again with infinite x and y extensions.

For one or more parallel combinations of the above, we expect the electric flux distribution to be independent of x and y variations and hence its density would only vary with z. The symmetry also implies that D is along the z axis [perpendicular to the charge plane(s)] as it emerges from the infinitely extended planar charge distribution.

Furthermore, for the cases demonstrated in Figure 4A.18, where the charge distribution is located with symmetry with respect to the $z=0$ origin, the symmetry implies symmetrical distribution of the flux density vector with respect to $z=0$ as well. In other words, the flux should flow out of (or into) the planar charge in the same way with the same density for both $z>0$ and $z<0$ regions. Moreover, for this case, $D(+z)=D(-z)$, hence

$$\vec{D}(-z)=D(-z)(-\vec{a}_z)=D(+z)(-\vec{a}_z)=-D(+z)(\vec{a}_z)=-\vec{D}(+z) \tag{4A.32}$$

Now, to utilize Gauss' law in order to derive flux density expressions for these cases, we need to choose one (or more) hypothetical Gaussian surface(s), HGS for the task. The chosen HGS needs to be shaped with the problem symmetry in mind; hence, we choose a vertical cylindrical shape for the HGS (Figure 4A.19). As will be apparent in the analysis, the cylindrical cross-section shape does not matter whether it is round, rectangular, or otherwise.

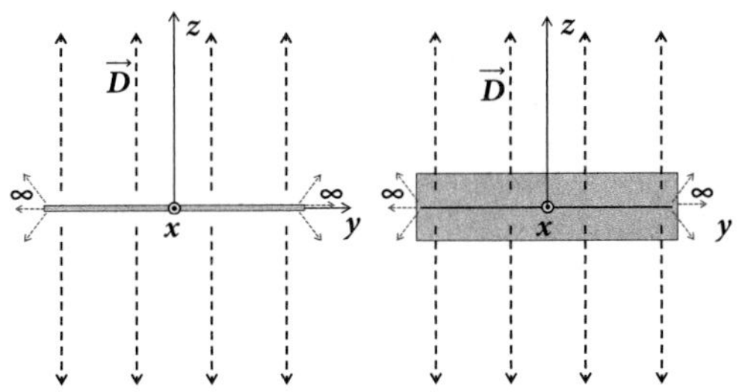

FIGURE 4A.18 Cases of charge distributions with planar symmetries (no variations with x or y).

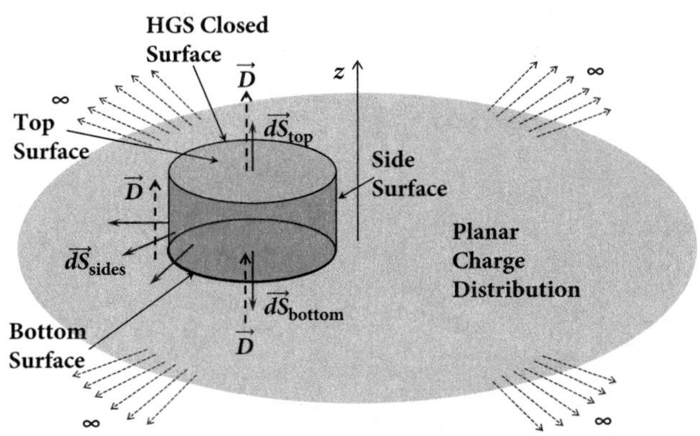

FIGURE 4A.19 Choosing a *closed* HGS for charge distributions with planar symmetries.

This HGS closed surface is composed of top and bottom surfaces that are parallel to the x-y plane, and hence perpendicular to the flux flow, and side surface(s) that are parallel to both the z axis and the flux flow. Consequently, there is no flux flow through the side surface(s), and the LHS of the Gaussian equation reduces to

$$\{\psi_{\text{closed HGS}}\} = \text{LHS} = \iint_{\text{top}} \vec{D}_{\text{top}}(z)\cdot\overrightarrow{dS}_{\text{top}} + \iint_{\text{bottom}} \vec{D}_{\text{top}}(z)\cdot\overrightarrow{dS}_{\text{top}} \tag{4A.33}$$

which is supposed to be equated to the RHS related to the charge enclosed by the HGS depending on the location of the chosen closed surface with respect to the charge.

It is important to recognize that $\overrightarrow{dS}$ for a closed surface is defined in the direction of the vectors emerging out of the surface and perpendicular to it, and hence

$$\overrightarrow{dS}_{\text{top}} = dS\vec{a}_z = dx\,dy\,\vec{a}_z \qquad \text{and} \qquad \overrightarrow{dS}_{\text{bottom}} = dS(-\vec{a}_z) = -dx\,dy\,\vec{a}_z \tag{4A.34}$$

Consequently, Equation (4A.33) reduces to

$$\begin{aligned}\text{LHS} &= \iint_{\text{top}} D_{\text{top}}(z)\vec{a}_z\cdot dx\,dy\,\vec{a}_z + \iint_{\text{bottom}} D_{\text{bottom}}(z)\vec{a}_z\cdot(-dx\,dy\,\vec{a}_z) \\ &= D_{\text{top}}(z)S_{\text{top}} - D_{\text{bottom}}(z)S_{\text{bottom}}\end{aligned} \tag{4A.35}$$

Since the upright cylindrical shape of the HGS has equal cross-sectional area throughout including its top and bottom, then

$$S_{\text{top}} = S_{\text{bottom}} = S_{\text{HGS cross section}} \tag{4A.36}$$

and finally, the LHS reduces to

$$\text{LHS} = [D_{\text{top}}(z) - D_{\text{bottom}}(z)]S_{\text{HGS cross section}} \tag{4A.37}$$

and the full Gauss' law relationship for the planar symmetry cases under study for the chosen HGS becomes

$$\text{LHS} = [D_{\text{top}}(z) - D_{\text{bottom}}(z)]S_{\text{HGS cross section}} = \text{RHS} = \text{Charge enclosed by the HGS} \tag{4A.38}$$

4A.3.1 Case of Planar Surface Charge Distribution with Constant ρ_S (No Variations with x or y), with Infinite Extension in Both x and y Coordinates

This charge distribution is demonstrated in Figure 4A.20. Let us choose our coordinate system with $z=0$ located at the charge plane, hence as discussed earlier, we can write the $z=0$ symmetry equation as

$$\vec{D}(-z) = -\vec{D}(+z) \tag{4A.39}$$

Now, let us start with an HGS positioned in a general location outside the charge region as referred to in Figures 4A.21 and 4A.22 as HGS-A. Applying Equation (4A.38) to HGS-A with no enclosed charge, we find

$$[D_{\text{top}}(z) - D_{\text{bottom}}(z)]S_{\text{HGS cross section}} = 0 \tag{4A.40}$$

which implies that $D_{\text{top}}(z) = D_{\text{bottom}}(z)$ in any region that does not enclose (or intersect) the planar charge distribution. This applies to both regions $\infty > z > 0$ and $-\infty < z < 0$, hence $D(z)$ is independent of z in both regions. When combined with Equation (4A.39), we can conclude that $\vec{D}(+z) = -\vec{D}(-z) = D\vec{a}_z$, where D is a constant at all z locations.

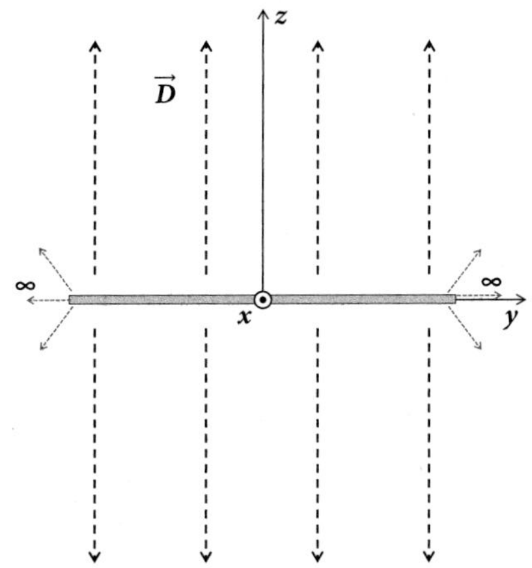

FIGURE 4A.20 Case of planar surface charge distribution with constant ρ_S (no variations with x or y) with infinite extension in both x and y coordinates.

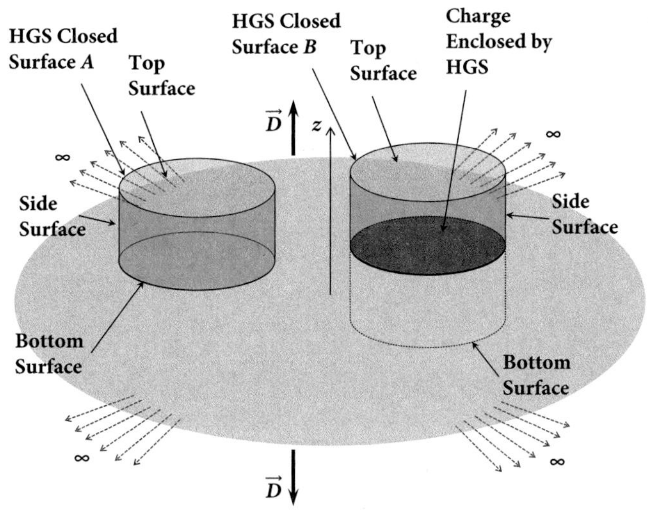

FIGURE 4A.21 Choosing two closed HGS for the planar surface charge distribution with infinite extension in both x and y coordinates.

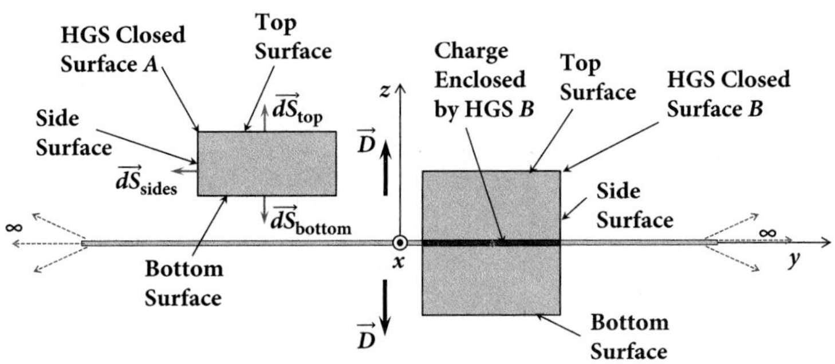

FIGURE 4A.22 Projection view of the HGS for the planar surface charge distribution with infinite extension in both x and y coordinates.

To find the constant D, we choose a second HGS-B similar to HGS-A except that this new HGS encloses a portion of the charge distribution. In other words, the top of the upright cylinder is above the charge (located in the $+z$ region) and the bottom surface is in the $-z$ region below the charge plane. For HGS-B, the Gauss' law equality yields

$$[D_{\text{top}}(z) - D_{\text{bottom}}(z)]S_{\text{HGS cross section}} = \text{Charge enclosed by the HGS} = \rho_S S_{\text{HGS cross section}} \tag{4A.41}$$

$$[D - (-D)] = \rho_S \qquad \text{or} \qquad D = \rho_S/2 \tag{4A.42}$$

and the final answer for the flux density distribution is given by

$$\vec{D}(+z) = -\vec{D}(-z) = [\rho_S/2]\vec{a}_z \qquad \text{or} \qquad \vec{D}(z) = [\rho_S/2]\left[\frac{z}{|z|}\right]\vec{a}_z \tag{4A.43}$$

which is graphed in Figure 4A.23.

4A.3.2 Case of Planar "Slab" of Charge Distribution with ρ_v Varying with z Only (No Variations with x or y), Again with Infinite x and y Extensions

The planar charge distribution is assumed to be of the form shown in Figure 4A.24. The charge extends infinitely along the x and y axes while is confined in height along the z axis. The volumetric charge density may vary with z and hence is of the form $\rho_v(z)$.

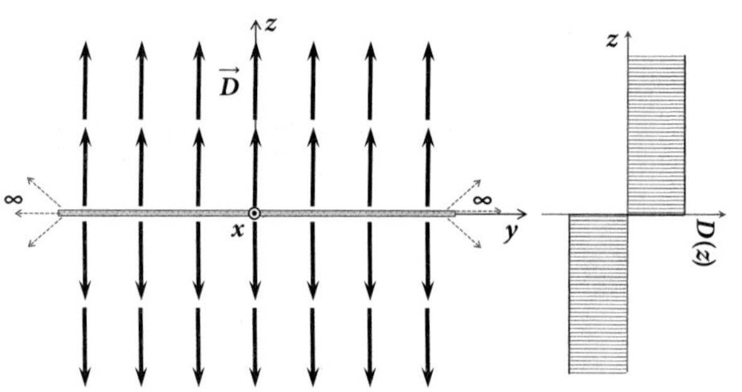

FIGURE 4A.23 Flux density versus distance for the case of planar surface charge distribution with constant ρ_S with infinite extension in both x and y coordinates.

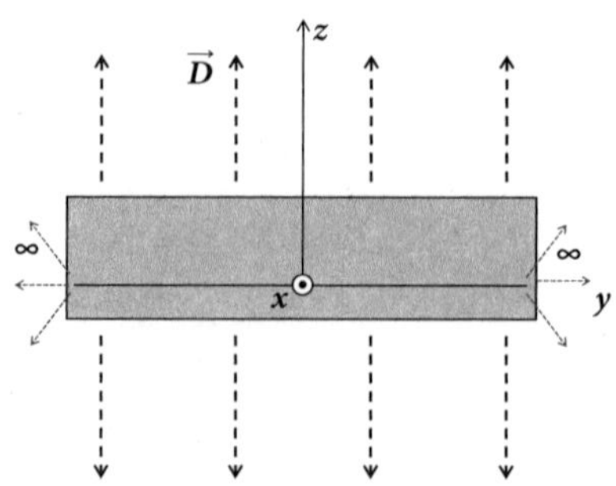

FIGURE 4A.24 Case of planar "slab" of charge distribution with ρ_v varying with z only (no variations with x or y) with infinite x and y extensions.

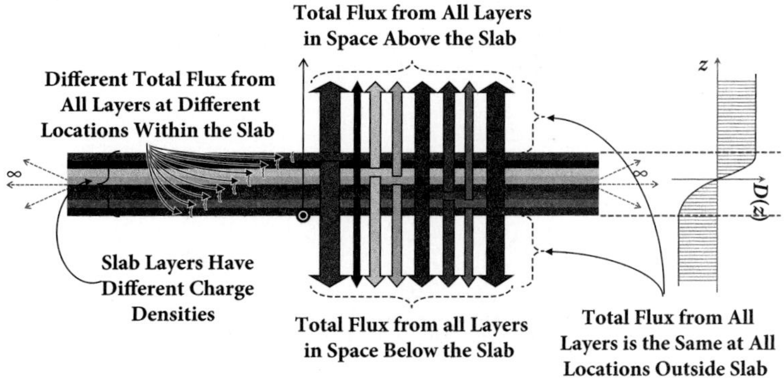

FIGURE 4A.25 Using a superposition scheme for the analysis of the case of the planar "slab" of charge distribution.

To analyze the flux density distribution, we take advantage of the results obtained above for the planar surface charge case. Therefore, let us divide the slab into many stacked thin layers. Each of these layers can be treated as a planar surface charge as depicted in Figure 4A.25. The equivalent surface charge density for each layer can be written in terms of the actual volumetric charge density of that layer. This is done by writing the expression of an increment of the layer charge of surface area ΔS using both densities formats:

$$\text{Increment of charge} = \rho_{S,\text{layer}} \cdot \Delta S = \rho_{v,\text{layer}} \cdot \text{layer thickness} \cdot \Delta S \tag{4A.44}$$

and hence

$$\rho_{S,\text{layer}} = \rho_{v,\text{layer}} \cdot \text{layer thickness} \tag{4A.45}$$

The flux flowing from each of these layers is divided into two equal halves to flow outward from the layer (one-half flows upward and the other half flows downward). Each layer's flux density is given by

$$D_{\text{layer}}(z) = \rho_{S,\text{layer}}/2 = [\rho_{v,\text{layer}} \cdot \text{layer thickness}]/2 \tag{4A.46}$$

Examining the figure, we can conclude

The total flux from all layers in space outside the slab is the same, and hence, the flux density at all points of space outside the slab can be written as

$$D_{\text{all layer}}(z \text{ outside slab}) = \sum_{\text{all layers}} [\rho_{v,\text{layer}} \cdot \text{layer thickness}]/2 = \left[\int_{\text{slab thickness}} \rho_{v,\text{layer}}\, dz\right]/2 \tag{4A.47}$$

The total flux from all layers in space inside the slab varies with location, and hence, the flux density within the slab is a function of position. The figure demonstrates that at any given location within the slab the flux due to the layers below is pointed upward and the flux due to the layers above points downward and the resultant flux would be the difference between the two fluxes.

To obtain an expression for the flux density vector at points within the slab, we resort to Gauss' law through the implementation of a closed surface positioned with either its top or bottom surfaces at that point. Figure 4A.26 demonstrates HGS-C positioned with its bottom surface at location z_p within the slab and its top surface anywhere outside the slab. For HGS-C, Equation (4A.38) takes the form:

$$[D_{\text{top}}(z) - D_{\text{bottom}}(z)]S_{\text{HGS cross section}} = \text{Charge enclosed by the HGS} \tag{4A.48}$$

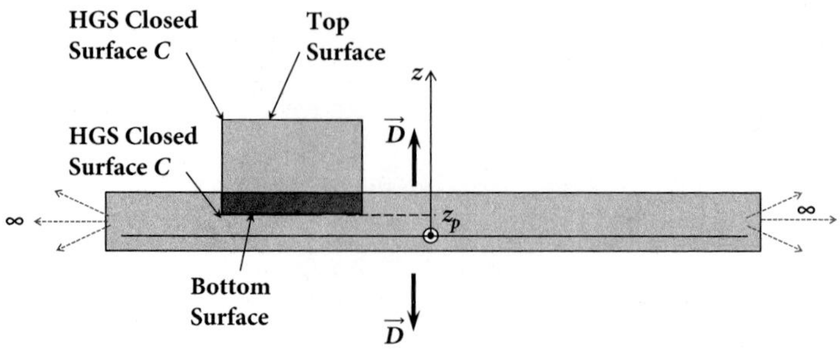

FIGURE 4A.26 Using the closed HGS scheme for the analysis of the case of the planar "slab" of charge distribution.

or

$$\left[(1/2)\int_{\text{slab thickness}} \rho_{v,\,\text{layer}}\, dz - D(z_p)\right] S_{\text{HGS cross section}} = \left[\int_{z_p}^{z_{\text{slab upper surface}}} \rho_{v,\,\text{layer}}\, dz\right] S_{\text{HGS cross section}} \tag{4A.49}$$

$$D(z_p) = \left[(1/2)\int_{z_{\text{slab lower surface}}}^{z_{\text{slab upper surface}}} \rho_{v,\,\text{layer}}\, dz - \int_{z_p}^{z_{\text{slab upper surface}}} \rho_{v,\,\text{layer}}\, dz\right] \tag{4A.50}$$

For the simple case of ρ_v = constant, extended from $z = -h/2$ to $+h/2$ (Figure 4A.27), the above expression reduces to

$$\vec{D}(+z \text{ outside slab}) = (\rho_v h/2)\vec{a}_z, \quad \text{for } |z| > h/2 \tag{4A.51}$$

$$\vec{D}(+z \text{ inside slab}) = (\rho_v z_p)\vec{a}_z, \quad \text{for } |z| < h/2 \tag{4A.52}$$

$$\vec{D}(-z) = \begin{cases} -(\rho_v z_p)\vec{a}_z, & \text{for } |z| < h/2 \\ -\left(\rho_v \dfrac{h}{2}\right)\vec{a}_z, & \text{for } |z| > h/2 \end{cases} \tag{4A.53}$$

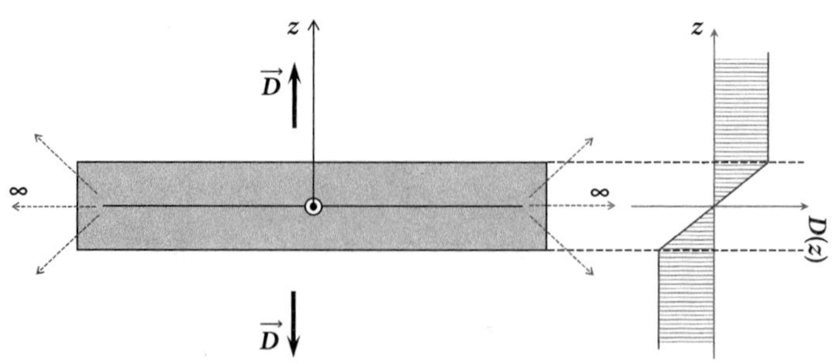

FIGURE 4A.27 Flux density versus distance for the case of the planar "slab" of charge distribution with constant ρ_v, extending from $z = -h/2$ to $+h/2$.

4A.4 Flux Density Distribution in Some Familiar Combinations of Symmetrical Charge Distributions

4A.4.1 Two Concentric Spherical Surfaces (Spherical Capacitor)

This case is demonstrated in Figure 4A.28. Two equal-but-opposite concentric surface charges, +Q and –Q are uniformly distributed over their respective spherical surfaces, of radii a and b, respectively. By applying Equation (4A.10) to each charge and superimposing the two solutions, we can write

$$\vec{D}_{+Q}(r)=\begin{cases} 0, & \text{for } r<a \\ \dfrac{+Q}{4\pi r^2}\vec{a}_r, & \text{for } r>a \end{cases} \tag{4A.54}$$

$$\vec{D}_{-Q}(r)=\begin{cases} 0, & \text{for } r<b \\ \dfrac{-Q}{4\pi r^2}\vec{a}_r, & \text{for } r>b \end{cases} \tag{4A.55}$$

$$\vec{D}(r)=\begin{cases} 0, & \text{for } r<a \\ \dfrac{Q}{4\pi r^2}\vec{a}_r, & \text{for } b>r>a \\ \left[\dfrac{+Q}{4\pi r^2}+\dfrac{-Q}{4\pi r^2}\right]\vec{a}_r=0, & \text{for } r>b \end{cases} \tag{4A.56}$$

This result is graphed versus position in Figure 4A.29.

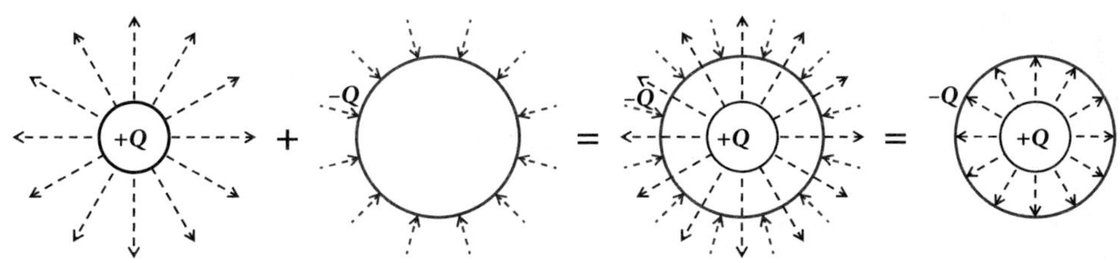

FIGURE 4A.28 Using superposition to analyze the case of two concentric spherical charge surfaces.

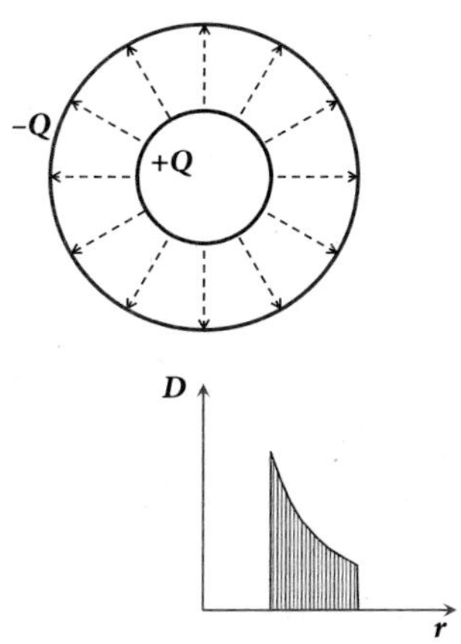

FIGURE 4A.29 Flux density versus distance for the case of two concentric spherical charge surfaces.

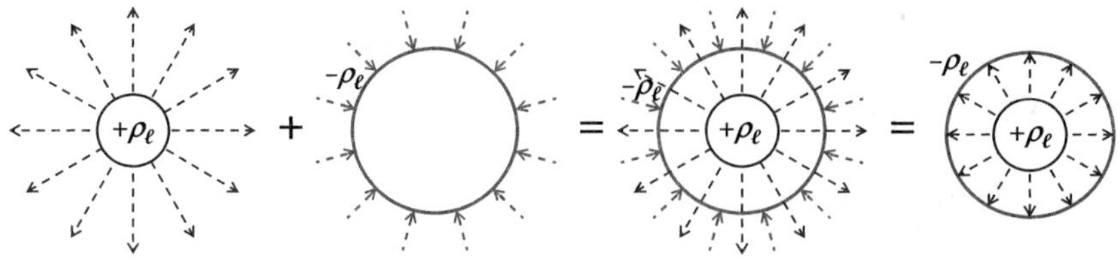

FIGURE 4A.30 Using superposition to analyze the case of two concentric cylindrical charge surfaces.

4A.4.2 Two Coaxial Cylindrical Surfaces (Cylindrical Capacitor/Coaxial Capacitor/Coaxial Transmission Line)

This case is demonstrated in Figure 4A.30. Two equal-but-opposite concentric surface charges, $+\rho_\ell$ and $-\rho_\ell$ are uniformly distributed over their respective cylindrical surfaces, radii a and b, respectively. By applying Equation (4A.25) to each charge and superimposing the two solutions, we can write

$$\vec{D}_{+\rho_\ell}(\rho)=\begin{cases} 0, & \text{for } \rho<a \\ \dfrac{+\rho_\ell}{2\pi\rho}\vec{a}_\rho, & \text{for } \rho>a \end{cases} \tag{4A.57}$$

$$\vec{D}_{-\rho_\ell}(\rho)=\begin{cases} 0, & \text{for } \rho<b \\ \dfrac{-\rho_\ell}{2\pi\rho}\vec{a}_\rho, & \text{for } \rho>b \end{cases} \tag{4A.58}$$

$$\vec{D}(\rho)=\begin{cases} 0, & \text{for } \rho<a \\ \dfrac{+\rho_\ell}{2\pi\rho}\vec{a}_\rho, & \text{for } b>\rho>a \\ \left[\dfrac{+\rho_\ell}{2\pi\rho}+\dfrac{-\rho_\ell}{2\pi\rho}\right]\vec{a}_\rho=0, & \text{for } \rho>b \end{cases} \tag{4A.59}$$

This result is graphed versus position in Figure 4A.31.

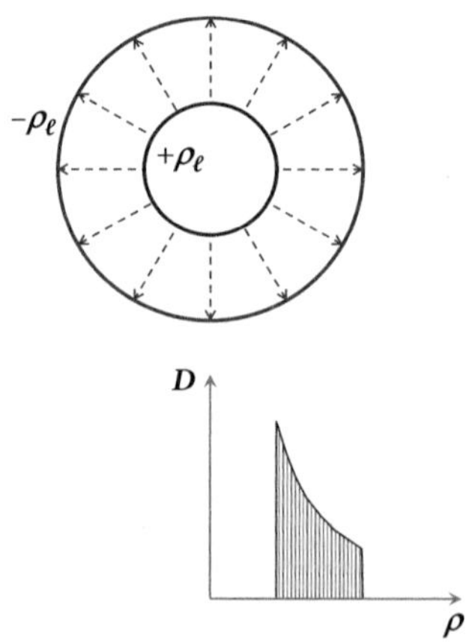

FIGURE 4A.31 Flux density versus distance for the case of two concentric cylindrical charge surfaces.

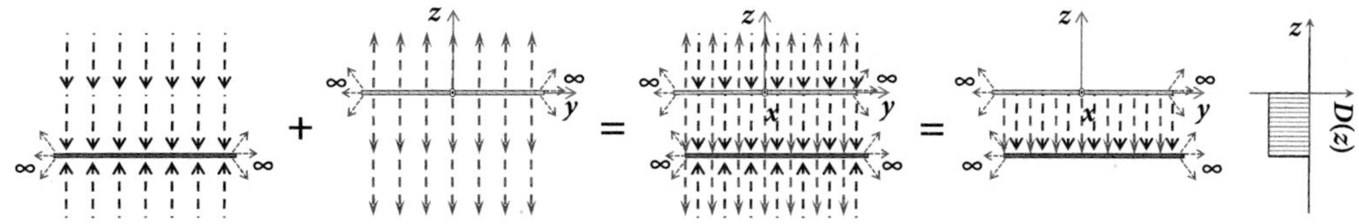

Figure 4A.32 Using superposition to analyze the case of two parallel planar charge surfaces.

4A.4.3 Two Parallel Planar Surfaces (Planar Capacitor/Parallel Plate Capacitor)

This case is demonstrated in Figure 4A.32. Two equal-but-opposite infinitely extended surface charges, $+\rho_S$ and $-\rho_S$ are uniformly distributed over their respective planar surfaces, placed at $z=0$ and h, respectively. By applying Equation (4A.43) to each charge and superimposing the two solutions, we can write

$$\vec{D}_{+\rho_s}(z)=\begin{cases}\dfrac{\rho_S}{2}\vec{a}_z, & \text{for } z>0\\ -\dfrac{\rho_S}{2}\vec{a}_z, & \text{for } z<0\end{cases} \tag{4A.60}$$

$$\vec{D}_{-\rho_s}(z)=\begin{cases}\dfrac{-\rho_S}{2}\vec{a}_z, & \text{for } z>h\\ -\dfrac{-\rho_S}{2}\vec{a}_z, & \text{for } z<h\end{cases} \tag{4A.61}$$

$$\vec{D}(z)=\begin{cases}\left[\dfrac{\rho_S}{2}+\dfrac{-\rho_S}{2}\right]\vec{a}_z=0, & \text{for } z>h\\ \left[\dfrac{\rho_S}{2}+\dfrac{-(-\rho_S)}{2}\right]\vec{a}_z=\rho_S\vec{a}_z, & \text{for } 0<z<h\\ \left[-\dfrac{\rho_S}{2}+\dfrac{-(-\rho_S)}{2}\right]\vec{a}_z=0, & \text{for } z<0\end{cases} \tag{4A.62}$$

This result is also graphed versus position in Figure 4A.32.

It is worthy to register a note here that the above cases of combined positive and negative charges will be visited again in future chapters when we discuss capacitance. As you may recall from your past studies of the "over-simplified" concept of a capacitor formed from two "conductors" that carry opposite charges. This topic has been dealt with in adequate detail in Chapter 6.

CHAPTER 4 PROBLEMS

- Problems 4.1 to 4.6: Problems dealing with finding the electric flux due to a given charge distribution
- Problems 4.7 to 4.9: Problems dealing with finding the electric flux given the flux density distribution
- Problems 4.10 to 4.13: Problems dealing with finding the charge distribution given the flux density distribution
- Problems 4.14 to 4.23: Problems dealing with finding the electric flux density due to a given charge distribution

4.1 A volume charge density is located in free space as $\rho_v = 3e^{-1100\,r}$ nC/m^3 for $0 < r < 1.1$ mm, and $\rho_v = 0$ elsewhere.

a. Find the total charge enclosed by the spherical surface $r = 1.1$ mm.

b. Calculate the electric flux through the spherical surface of $r = 0.5$ mm.

4.2 Three infinite line charges with linear charge densities $\rho_\ell = 5$ nC/m each are placed on the three coordinate axes x, y, and z. Find the total flux on the hypothetical spherical surface centered at the origin with radius 25 cm.

4.3 A cubical surface with sides $a = 20$ cm carry uniform surface charge density of 10 nC/m^2. What is the total electric flux flow through a hypothetical spherical surface with radius 10 cm. Repeat for radii of 12 and 15 cm.

4.4 Spherical surfaces at $r = 2$, 5, and 7 cm carry uniform surface charge densities of 15 nC/m^2, –5 nC/m^2, and ρ_S, respectively.

a. Calculate the electric flux through the spherical surfaces of radii $r = 1$, 3, and 6 cm.

b. Determine ρ_S such that the electric flux is zero through the spherical surfaces of radius $r = 8$ cm.

4.5 A very long uniform line of charge has charge per unit length 4.9 μC/m and lies along the x axis. A second long uniform-line of charge has charge per unit length –2.3 μC/m and is parallel to the x axis at $y = 0.4$ m. What is the net electric flux density at the following points on the y axis:

a. $y = 0.2$ m

b. $y = 0.6$ m?

4.6 A hollow, conducting spherical shell with an outer radius of (0.24 m) and an inner radius of (0.19 m) has a uniform surface charge density of (6.4×10^{-6}) C/m^2. A charge of (–0.45 μC) is now introduced into the cavity inside the sphere.

a. What is the new charge density on the outside of the sphere?

b. What is the electric flux through a spherical surface just inside the inner surface of the sphere?

4.7 A uniform electric flux density vector points vertically downward with $\vec{D} = -100\vec{a}_z$ C/m^2. A cylindrical box with radius 15 cm and height 20 cm has nonconducting top, nonconducting sides, and a perfect conducting bottom is placed coaxial with the z axis.

a. What is the total electric flux through the box? Explain your answer.

b. What is the total electric charge enclosed by the box? Where is this charge located?

c. Repeat (a) and (b) above for a hemisphere with nonconducting spherical surface and a conducting base placed on the x-y plane.

4.8 Given a uniform flux density vector $\vec{D} = \vec{D}_o\vec{a}_z$. Evaluate the total flux flowing through the hypothetical hemisphere of radius R laying on its flat side on the x-y plane.

4.9 An electric flux points vertically downward along the z axis with a density of $D = 0.9e^{-9}$ C/m^2 for $z > 0$ and $D = 0$ for $z < 0$.

a. Evaluate the total electric flux through a closed cylindrical surface of radius 25 cm and height extending between $-25 < z < +25$ cm?

b. What is the total electric charge enclosed by the surface in part (a)? Where is this charge located?

c. What are the answers to (a) and (b) for a spherical box of the same radius, placed within the same z range?

4.10 Calculate the divergence of $\vec{D} = \left[\dfrac{8xyz\,\vec{a}_x + 3x^2z\,\vec{a}_y + (z^3 - 4x^2y)\vec{a}_z}{z^2}\right]$ at the location $P(-1, 3, 4)$.

4.11 Let $\vec{D} = 6\rho\sin(\varphi)\vec{a}_\rho + 6\rho\cos(\varphi)\vec{a}_\varphi$ C/m^2.

a. Find div $\vec{D}$.

b. Find the volume charge density at $P(2.5, 35°, -6)$ in cylindrical coordinates.

c. How much charge is located inside the region defined by the cylindrical coordinates $0 < \rho < 1.7$, $21° < \varphi < 71°$, and $2.3 < z < 3$?

4.12 If the electric flux density in free space is given by $\vec{D}(x, y, z) = \vec{a}_y(2y^3 + 1)\varepsilon_o$ C/m^2, find the total electric charge enclosed inside the volume shown in the figure. The volume extends in the x direction from 0 to 2 m, in the y direction from -1 m to 1 m and in the z direction from 0 to 4 m.

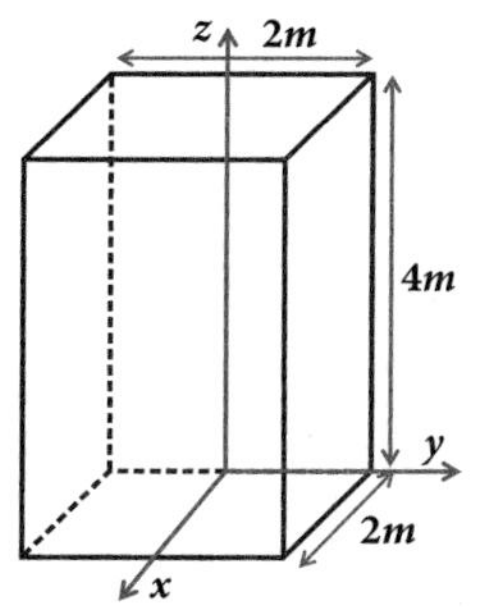

FIGURE 4P.12

4.13 Within the spherical shell, $2 < r < 3$ m, the electric flux density is given as $\vec{D} = 4(r-2)^2\vec{a}_r$ C/m^2.

a. What is the volume charge density at $r = 3$ m?

b. What is the electric flux density at $r = 3$ m?

c. How much flux leaves the sphere at $r = 3$ m?

d. How much charge is contained within the sphere $r = 3$ m?

4.14 A charge Q is uniformly distributed along a rod of length L and negligible cross section. Find the electric flux density vector at all points in the orthogonal plane passing by the mid-point of the rod.

4.15 Derive an expression for the flux density vector at all points in space due to the spherical charge distribution:

$$\rho_v(r) = \begin{cases} \rho_o(1 - r/a) & \text{for } r < a \\ 0 & \text{for } r > a \end{cases}$$

4.16 For Problem 4.6, calculate the magnitude of the electric flux density just outside the sphere.

4.17 Given a spherical insulator with an outer radius of 15 cm and concentric spherical cavity with a radius of 5 cm. The insulator body is uniformly charged with charge density of (7.4×10^4) C/m^3 and a point charge of $(-2\,\mu C)$ is located in the center of a spherical cavity. Calculate the electric flux density inside the insulator at a distance of (10 cm) from the center of the cavity.

4.18 Find the flux density vector at the origin due to two point-charges of –2 C each located at (2, 1, 0) m and (1, –2, 0) m.

4.19 Three infinite line charges with linear charge densities $\rho_\ell = 10\ \mu\text{C/m}$ each are placed on the three coordinate axes x, y, and z. Find the flux density vector at the point (1, 1, 1) m.

4.20 A long cylindrical plastic rod has radius a and carries an electric charge uniformly distributed within its volume with a linear density of ρ_ℓ C/m.

a. Sketch maps showing the rod and the electric flux pattern.

b. Derive expressions for the electric flux density for all radii inside and outside a.

c. Derive expressions for the total electric flux through a coaxial cylindrical-surfaces of radius a/3 and length 2 m and radius 2a and length 3 m.

4.21 For a sphere of radius a containing uniform charge density (charge per unit volume) of ρ_v, the magnitude of the electric flux density, D, at a distance r from the center is given by

$$\vec{D} = \frac{\rho_v r}{3}\vec{a}_r \text{ for } r < a \qquad \text{and} \qquad \vec{D} = \frac{\rho_v a^3}{3r^2}\vec{a}_r \text{ for } r > a$$

a. Sketch the variation of D with radial distance.

b. Derive an expression for the total flux on spherical surfaces with $r < a$ and $r > a$.

c. Discuss the nature of the charge distribution in the light of the findings of part (b) above.

4.22 For the case given in Problem 4.21 given that $D = 3.33\times10^{-6}$ C/m^2 at $r = 10$ mm, and $D = 1.66\times10^{-6}$ C/m^2 at $r = 40$ mm. Determine

a. The charge density, ρ_v

b. The radius, a

c. The total charge, Q

4.23 The electric charge of the proton is modelled as an exponentially varying charge density $\rho_v = \rho_o \exp(-ar)$. Use Gauss' law to develop an expression for the electric flux density as a function of r.

CHAPTER 4 SUMMARY[1]

Electric Charge, Q

Positive and negative—measured in coulombs (C)

Electron charge: $e = -1.6021765 \times 10^{-19}$ C

Universe is neutral—default state of a material body is neutral

Charge Distributions

Point charge (infinitesimally small volume)

Line distribution—linear density ρ_ℓ C/m

Surface distribution—surface density ρ_S C/m^2

Volume distribution—volume density ρ_v C/m^3

Electric Flux, ψ

Flux equals charge producing the flux (coulombs)

Emerges out of positive charge (source)

Lands on negative charge (sink)

$\psi_{\text{closed surface}} = \text{Total } Q_{\text{enclosed by the surface}}$

Flux Density Vector, $\vec{D}$

$D = \{d\psi/dS\}_{\max} = d\psi/(dS_{\text{normal}})$ C/m^2

Vector direction along flux lines

$d\psi_{dS} = \vec{D} \cdot \overrightarrow{dS}$ and $\psi_S = \iint_S d\psi = \iint_S \vec{D} \cdot \overrightarrow{dS}$

Gauss' Law (Integral Form)

$\{\psi_{\text{closed surface } S}\} = \oint\!\!\oint_S \vec{D} \cdot \overrightarrow{dS} = \iiint_{\text{enclosed volume}} \rho_v \, dv$

Gauss' Law (Point Form)

$\vec{\nabla} \cdot \vec{D} = \rho_v(\text{position})$

$$\vec{\nabla} \cdot \vec{D} = \left[\frac{\partial}{\partial x} D_x + \frac{\partial}{\partial y} D_y + \frac{\partial}{\partial z} D_z\right]$$

$$\vec{\nabla} \cdot \vec{D} = \left[\frac{1}{\rho}\frac{\partial}{\partial \rho}(\rho D_\rho) + \frac{1}{\rho}\frac{\partial}{\partial \varphi} D_\varphi + \frac{\partial}{\partial z} D_z\right]$$

$$\vec{\nabla} \cdot \vec{D} = \left[\frac{1}{r^2}\frac{\partial}{\partial r}\left(r^2 D_r\right) + \frac{1}{r\sin\theta}\frac{\partial}{\partial \theta}(\sin\theta D_\theta) + \frac{1}{r\sin\theta}\frac{\partial}{\partial \varphi} D_\varphi\right]$$

[1] ***Important Warning to Students:*** Please do not use this summary without thorough study of the relevant material in this chapter. Using this summary and its formulas without adequate appreciation of the physics involved and conditions of validity can lead to wrong implementations and invalid conclusions.

The Divergence Vector Operator

Cartesian: $\vec{\nabla}\cdot\vec{D}=\left[\frac{\partial}{\partial x}D_x+\frac{\partial}{\partial y}D_y+\frac{\partial}{\partial z}D_z\right]$

Cylindrical: $\vec{\nabla}\cdot\vec{D}=\left[\frac{1}{\rho}\frac{\partial}{\partial\rho}(\rho D_\rho)+\frac{1}{\rho}\frac{\partial}{\partial\varphi}D_\varphi+\frac{\partial}{\partial z}D_z\right]$

Spherical: $\vec{\nabla}\cdot\vec{D}=\left[\frac{1}{r^2}\frac{\partial}{\partial r}\left(r^2D_r\right)+\frac{1}{r\sin\theta}\frac{\partial}{\partial\theta}(\sin\theta D_\theta)+\frac{1}{r\sin\theta}\frac{\partial}{\partial\varphi}D_\varphi\right]$

The Divergence Theorem

$$\oiint_S \vec{D}\cdot\overrightarrow{dS}=\iiint_{\text{volume enclosed by } S}\vec{\nabla}\cdot\vec{D}dv$$

$$\oiint_S \vec{A}\cdot\overrightarrow{dS}=\iiint_{\text{volume enclosed by } S}\vec{\nabla}\cdot\vec{A}dv$$

Application of Gauss' Law

Evaluate Charge—Integral Form

Total enclosed in a volume, given flux density distribution

$$Q_{\text{vol}}=\oiint_{S\text{ enclosing vol}}\vec{D}\cdot\overrightarrow{dS}$$

Evaluate Charge—Point Form

Evaluate volumetric charge density distribution, given flux density distribution

$$\rho_v\,(\text{position})=\vec{\nabla}\cdot\vec{D}$$

Total charge in a volume, given flux density distribution

$$Q_{\text{vol}}=\iiint_{\text{vol}}\rho_v\,dv=\iiint_{\text{vol}}(\vec{\nabla}\cdot\vec{D})dv$$

Evaluate Flux Density Vector—Integral Form

- **ONLY** for certain symmetrical charge distributions and combinations thereof
  - Charges distributed with spherical symmetries (no variations with θ or φ)
    - Point charge—spherical surface charge—spherical shell charge—spherical volume charge
  - Charges distributed with cylindrical symmetries (no variations with φ or z)
    - Line charge—cylindrical surface charge—cylindrical shell charge—cylindrical volume charge
  - Charges distributed with planar symmetries (no variations with x or y)
    - Planar surface charge—planar "slab" of charge predict flux distribution (map) based on charge symmetry
- Choose appropriate closed hypothetical Gaussian surfaces (HGS) with surface(s) either normal or tangential to expected flux lines
- Perform the closed surface integration $\oiint_S \vec{D}\cdot\overrightarrow{dS}$ with the integrand D factored outside the integration
- Equate the flux density integral to the charge enclosed by HGS, $\iiint_{\text{HGS}}\rho_v\,dv$
- Evaluate $\vec{D}$

Evaluate Flux Density Vector—Point Form

- Given charge density distribution
- Divide space in regions as per the charge distribution
- Solve $\vec{\nabla}\cdot\vec{D}=\rho_v$ (position) differential equation for each region
- Boundary conditions determine integration constants (Continuity of flux density except at singularities)
- Example: Spherical volume distribution, $\rho_v(r < b)$

$$\vec{D}=\begin{cases}\vec{a}_r(1/r^2)\int r^2\rho_v(r)dr & 0<r<b\\ \vec{a}_r(1/r^2)\int^{r=b} r^2\rho_v(r)dr & b<r<\infty\end{cases}$$

Flux Density Vector, $\vec{D}(r)$, for Charge Distributions with Spherical Symmetries

Point

$$\frac{q}{4\pi r^2}\vec{a}_r$$

Spherical Surface

$$\begin{cases}0, & r<a\\ \dfrac{Q}{4\pi r^2}\vec{a}_r, & r>a\end{cases}$$

Spherical Shell

$$\begin{cases}0, & \text{for } r<a\\ \dfrac{\int_{r=a}^{r}\rho_v(r)4\pi r^2\,dr}{4\pi r^2}\vec{a}_r, & a<r<b\\ \dfrac{Q}{4\pi r^2}\vec{a}_r, & \text{for } r>b\end{cases}$$

Spherical Volume

$$\begin{cases}\dfrac{1}{r^2}\int_{r=0}^{r}\rho_v(r)r^2\,dr\,\vec{a}_r, & r<b\\ \dfrac{Q}{4\pi r^2}\vec{a}_r, & r>b\end{cases}$$

Flux Density Vector, $\vec{D}(r)$, for Charge Distributions with Cylindrical Symmetries

Infinite Line

$$\frac{\rho_\ell}{2\pi\rho}\vec{a}_\rho$$

Infinite Cylindrical Surface

$$\begin{cases} 0, & \rho < a \\ \dfrac{\rho_\ell}{2\pi\rho}\vec{a}_\rho, & \rho > a \end{cases}$$

Infinite Length Cylindrical Shell

$$\begin{cases} 0, & \rho < a \\ \rho_v \dfrac{(\rho^2 - a^2)}{2\rho}\vec{a}_\rho\vec{a}_\rho, & a < \rho < b \\ \rho_v \dfrac{(b^2 - a^2)}{2\rho}\vec{a}_\rho, & \rho > b \end{cases}$$

Infinite Length Cylindrical Volume

$$\begin{cases} \rho_v \dfrac{\rho}{2}\vec{a}_\rho, & \rho < b \\ \rho_v \dfrac{b^2}{2\rho}\vec{a}_\rho, & \rho > b \end{cases}$$

Flux Density Vector, $\vec{D}(r)$, for Charge Distributions with Planar Symmetries

Surface Charge with Density ρ_S

$$\vec{D}(+z) = -\vec{D}(-z) = [\rho_S/2]\vec{a}_z$$

"Slab" of Charge with Thickness h and ρ_v Varying with z Only

$$(\rho_v z_p)(z/|z|)\vec{a}_z, |z| < h/2$$

$$\left(\rho_v \frac{h}{2}\right)(z/|z|)\vec{a}_z, |z| > h/2$$

CHAPTER 5

Electric Force, Field, Energy, and Potential

Charles-Augustin de Coulomb (1736–1806)

Chapter Learning Objectives

- Calculate the mutual force between two point charges.
- Calculate the electric field distribution generated by arbitrary charge distributions using the electric field incrementation method.
- Derive analytical expressions for symmetric charge distributions with finite dimensions (which was not possible using the hypothetical Gaussian surface (HGS) method described in Chapter 4).
- Describe the physical process of the development of electric potential in a system of charges.
- Calculate the spatial distribution of electric potential resulting from a given system of charges using the potential field incrementation method.
- Derive the electric field for a given electric potential distribution and vice versa.
- Calculate the energy density in a given system of charges for a given electric potential distribution.

5.1 Introduction

We are still in the pursuit of capacitance; the device (circuit element) that accumulates charges and allows energy storage. In the previous chapter, we learned about the concept of electric flux due to the presence of charges. This chapter, we will continue that study by examining the forces between electrical charges, the energy in a system of charges, and the resulting electrical potential. In doing so, we are in fact studying all about capacitance without using the term. By the conclusion of this chapter, we will be ready to define the capacitance and discuss its characteristics. The actual definition of capacitance will be presented in the following chapter after studying the electromagnetic behavior of materials (conductors and dielectrics).

5.2 Coulomb's Forces

As a result of experimenting with two very small charged objects, the French physicist, Charles-Augustin de Coulomb, published his "famous" Coulomb's inverse square law in 1785. This law simply states that the force between two (very small/point) charges is proportional to the two charges while inversely proportion to the square of the separation distance. Referring to Figure 5.1, the mathematical expression for the Coulomb force takes the form:

$$F = k_e \frac{q_1 q_2}{R_{12}^2} \tag{5.1}$$

This force is along the line connecting the charges and is mutual between the two charges, that is, q_1 exerts on q_2 the same magnitude force that q_2 exerts on q_1. These forces are repulsion forces if the two charges are of similar polarities and reverse to attraction forces if charges are of opposite polarities.

Coulomb's law is limited to charges that are confined to an infinitesimally small volume. Mathematically, we refer to them as point charges (zero volume). In his experiments, Coulomb used very small particles and placed them a large distance apart in an effort to approximate the point charge case where the particle size is much too small compared to the separating distance thus approximating a point.

Using the International System of units (SI), the charges q_1 and q_2 are expressed in coulombs (C) and the distance R in meters, the coulomb constant k_e in vacuum (free space) is equal to $8.9875517873681764 \times 10^9$ and has Nm^2/C^2 for its units.

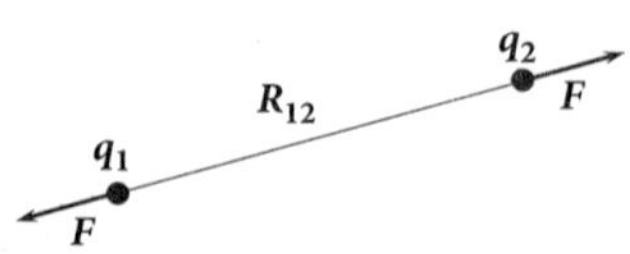

FIGURE 5.1 Coulomb's forces.

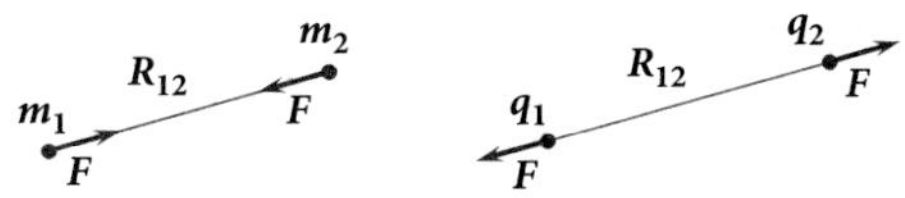

FIGURE 5.2 Analogy between Coulomb's forces and Newton's forces.

It is relevant to notice the analogy between the Coulomb's law for electrical forces and that of Newton's for universal gravitation. Newton's law describes the force between two (very small/point) masses m_1 and m_2 separated by a distance R_{12} in a similar inverse square law relationship

$$F = k_g \frac{m_1 m_2}{R_{12}^2} \tag{5.2}$$

where k_g is the gravitational constant. One main difference with the gravitational force is that masses have only positive polarity and the resulting force is always an attraction one (Figure 5.2). We will find this analogy useful later in this chapter when we discuss fields, energy, and potential.

We now get back to Coulomb's law and Equation (5.1). For reasons that will become apparent later, k_e is replaced by $1/(4\pi\varepsilon_o)$, and hence introducing the new constant ε_o for free space (this constant is different for other material as will be discussed in the following chapter about materials)

$$\varepsilon_o = 8.854187817 \times 10^{-12}\ \text{C}^2/\text{Nm}^2 \,(\text{or farads/m}) \simeq (1/36\pi) \times 10^{-9}\ \text{F/m} \tag{5.3}$$

Hence, the famous form of Coulomb's force in free space in its SI units' format takes the form:

$$F = \frac{q_1 q_2}{4\pi\varepsilon_o R_{12}^2}\ \text{N (newtons)} \tag{5.4}$$

and the vector force expressions (Figure 5.3) become

$$\vec{F}_2 = \frac{q_1 q_2}{4\pi\varepsilon_o R_{12}^2}\vec{a}_{12} \quad \text{and} \quad \vec{F}_1 = \frac{q_2 q_1}{4\pi\varepsilon_o R_{21}^2}\vec{a}_{21} \tag{5.5}$$

The unit vector $\vec{a}_{12}\,(=-\vec{a}_{21})$ is a vector of unity magnitude and direction pointing from q_1 to q_2 (same as the vector $\vec{R}_{12}$). Hence $\vec{a}_{12}$ can be expressed as

$$\vec{a}_{12} = \frac{\vec{R}_{12}}{R_{12}} = \frac{\vec{r}_2 - \vec{r}_1}{|\vec{r}_2 - \vec{r}_1|} = \frac{(x_2 - x_1)\vec{a}_x + (y_2 - y_1)\vec{a}_y + (z_2 - z_1)\vec{a}_z}{\sqrt{(x_2 - x_1)^2 + (y_2 - y_1)^2 + (z_2 - z_1)^2}} \tag{5.6}$$

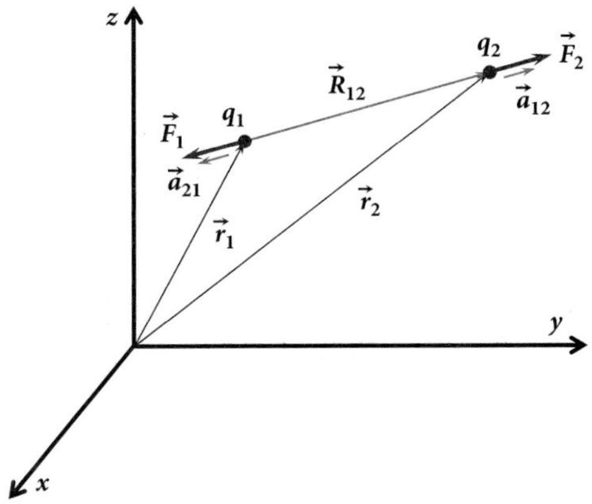

FIGURE 5.3 Vector representation of Coulomb's force.

Examining the force expression [Equation (5.5)], we observe the similarity in expressions between the force vector and the electric flux density vector, $\vec{D}$, derived in the previous chapter, Equation (4A.8). If we rewrite Equation (4A.8) for $\vec{D}$ at point 2 due to the charge q_1, we would get

$$\vec{D}_2 = \frac{q_1}{4\pi R_{12}^2}\vec{a}_{12} \tag{5.7}$$

Hence, the force on q_2 located at the point 2 (in the flux of q_1) can be expressed in terms of $\vec{D}$ at the same point (point 2) as follows:

$$\vec{F}_2 = \frac{q_1 q_2}{4\pi\varepsilon_o R_{12}^2}\vec{a}_{12} = q_2\frac{1}{\varepsilon_o}\frac{q_1}{4\pi R_{12}^2}\vec{a}_{12} = q_2\frac{1}{\varepsilon_o}\vec{D}_2 \tag{5.8}$$

Now, if we were to replace q_2 by a unit charge (1C), the resulting (normalized) force per unit charge at point 2 would be

$$\vec{F}_{\text{normalized (per unit charge)}} = \frac{q_1(1)}{4\pi\varepsilon_o R_{12}^2}\vec{a}_{12} = \frac{1}{\varepsilon_o}\frac{q_1}{4\pi R_{12}^2}\vec{a}_{12} = \frac{1}{\varepsilon_o}\vec{D}_2 \tag{5.9}$$

which gives rise to the definition of the electric field intensity, $\vec{E}$, vector as the (normalized) force per unit charge; that is, the force on a "test" charge of unity magnitude located in the "field" of the "main" charge (or charge distribution).

5.3 The Electric Field

The generalized form of the definition of $\vec{E}$ is demonstrated in Figure 5.4 where a system of charges produces a force $\vec{F}$ on a "test" charge (of 1C) located at a "position" in the field of that system, $\vec{E}$ at this position is given by

$$\vec{E}(\text{position}) = \vec{F}_{\text{normalized (per unit charge)}}(\text{position}) = \frac{1}{\varepsilon_o}\vec{D}(\text{position}) \tag{5.10}$$

While in the special case of a point charge as the cause of this field,

$$\vec{E} = \frac{q}{4\pi\varepsilon_o R_{q\text{-position}}^2}\vec{a}_{R_{q\text{-position}}}, \text{ where } \vec{R}_{q\text{-position}} \text{ is the distance vector from } q \text{ to the position} \tag{5.11}$$

So, we are defining a new physical quantity called the electric field intensity vector, $\vec{E}$, as an indicator to the "influence" of a charge (or a system of charges) in its surrounding "field" as detected by other charges in the form of a force. This normalized influence indicator is complementary to the electric flux density indicator, $\vec{D}$, which was also another influence indicator. As we will see later on, the two indicators differ in how they react to the presence of dielectric materials and their dielectric properties.

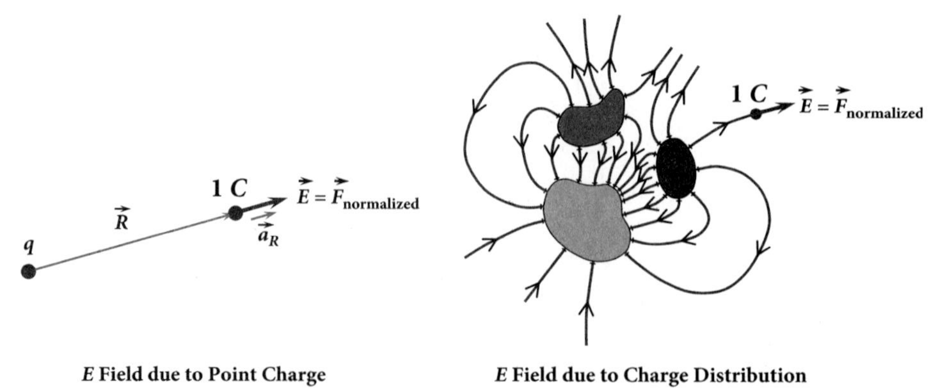

Figure 5.4 The electric field is a normalized force.

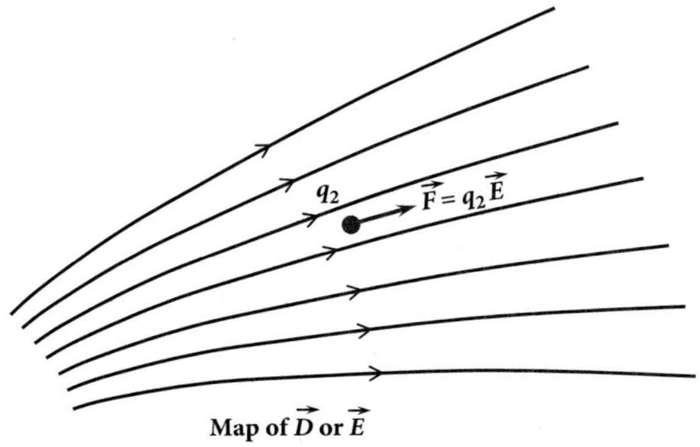

FIGURE 5.5 Electric force on a charge due to an applied electric field.

For now, we are dealing with charges in free space, and according to Equation (5.10), we can write

$$\vec{E}(r) = \frac{1}{\varepsilon_o}\vec{D}(r) \quad \text{and} \quad \vec{D}(r) = \varepsilon_o \vec{E}(r) \tag{5.12}$$

Also, we can write the force expression on a point charge q_2 located in the field (or flux) of a charge distribution (Figure 5.5) as

$$\vec{F}_{q_2} = q_2\vec{E}(\text{at } q_2) = q_2\left[\frac{1}{\varepsilon_o}\vec{D}(\text{at } q_2)\right] \tag{5.13}$$

Equations (5.12) and (5.13) enable us to evaluate both the $\vec{E}$ and $\vec{F}$ vectors knowing the $\vec{D}$ vector at any location due to a charge distribution. In Chapter 4, we have developed expressions for $\vec{D}$ for special charge distributions with certain symmetry features. For those cases, Equations (5.12) and (5.13) can be used to extend the derived $\vec{D}$ expressions to $\vec{E}$ and $\vec{F}$.

Example 5.1 Two point charges Q_1 and Q_2 are located at $r_1(1,2,0)$ and $r_2(2,0,0)$, respectively. Find the relation between Q_1 and Q_2 so that the total force on a test charge at point $P(-1,1,0)$ will have

a. No x component,
b. No y component.

Hint: For the force to be zero in a certain direction, the total electric field at point P due to Q_1 and Q_2 must be zero in the same direction $\vec{F} = q\vec{E}$.

Solution

$$\vec{E}_p = \vec{E}_1 + \vec{E}_2 = \frac{Q_1}{4\pi\varepsilon_o|\vec{r}_p - \vec{r}_1|^2}\vec{a}_{1p} + \frac{Q_2}{4\pi\varepsilon_o|\vec{r}_p - \vec{r}_2|^2}\vec{a}_{2p} = \frac{Q_1(\vec{r}_p - \vec{r}_1)}{4\pi\varepsilon_o|\vec{r}_p - \vec{r}_1|^3} + \frac{Q_2(\vec{r}_p - \vec{r}_2)}{4\pi\varepsilon_o|\vec{r}_p - \vec{r}_2|^3}$$

$$(\vec{r}_p - \vec{r}_1) = [\vec{a}_x(-1) + \vec{a}_y 1] - [\vec{a}_x 1 + \vec{a}_y 2] = \vec{a}_x(-2) + \vec{a}_y(-1)$$

$$|\vec{r}_p - \vec{r}_1| = \sqrt{2^2 + 1^2} = \sqrt{5}$$

$$(\vec{r}_p - \vec{r}_2) = [\vec{a}_x(-1) + \vec{a}_y 1] - [\vec{a}_x 2 + \vec{a}_y 0] = \vec{a}_x(-3) + \vec{a}_y 1$$

$$|\vec{r}_p - \vec{r}_2| = \sqrt{3^2 + 1^2} = \sqrt{10}$$

$$\vec{E}_1 + \vec{E}_2 = \frac{Q_1[\vec{a}_x(-2) + \vec{a}_y(-1)]}{4\pi\varepsilon_o(\sqrt{5})^3} + \frac{Q_2[\vec{a}_x(-3) + \vec{a}_y 1]}{4\pi\varepsilon_o(\sqrt{10})^3}$$

For the x component to be zero

$$\vec{E}_{1x} + \vec{E}_{2x} = 0 = \frac{Q_1[-2]}{4\pi\varepsilon_o(\sqrt{5})^3} + \frac{Q_2[-3]}{4\pi\varepsilon_o(\sqrt{10})^3} = 0 \Rightarrow Q_1 = \frac{Q_2[-3/2](\sqrt{5})^3}{(\sqrt{10})^3} \Rightarrow Q_1 = \frac{-3Q_2}{4\sqrt{2}} = -0.53Q_2$$

For the y component to be zero

$$\vec{E}_{1y} + \vec{E}_{2y} = 0 = \frac{Q_1[-1]}{4\pi\varepsilon_o(\sqrt{5})^3} + \frac{Q_2[1]}{4\pi\varepsilon_o(\sqrt{10})^3} = 0 \Rightarrow Q_1 = \frac{Q_2(\sqrt{5})^3}{(\sqrt{10})^3} \Rightarrow Q_1 = \frac{Q_2}{2\sqrt{2}} = 0.35Q_2$$

5.4 Electric Field Evaluation Using the "Incrementation" Scheme

With the availability of Coulomb's law, we have in fact a new tool that can enable us to derive flux density expressions for nonsymmetrical charge distribution cases that we were unable to do using Gauss' law. This can be done by dicing up the charge distribution into small increments and treating each increment as a point charge. Applying expression [Equation (5.11)] for a point charge to each charge increment and integrating over the entire charge distribution, we get an expression for the total field intensity vector from which the flux density vector is obtainable. We will refer to this as the electric field "incrementation" scheme.

It is important to state that this scheme is valid only for linear systems (media) where the concept of superposition applies. We will discuss material linearity in Chapter 6 when we study about electromagnetic material properties.

Denoting the charge increment as dQ, we can rewrite Equation (5.11) for the corresponding increment of the electric field vector, $\overrightarrow{dE}$, at a point in space located at the end of a vector $\vec{R}$ initiated at dQ (Figure 5.6) as

$$\overrightarrow{dE}_{\text{point}} = \frac{dQ}{4\pi\varepsilon_o R_{dQ-p}^2}\vec{a}_{R_{dQ-p}} \tag{5.14}$$

and hence

$$\vec{E}_{\text{point}} = \int_{\text{Charge distribution}} \frac{dQ}{4\pi\varepsilon_o R_{dQ-p}^2}\vec{a}_{R_{dQ-p}} \tag{5.15}$$

and in the most general form for a general system of charges composed of possibly point charges, linear charge densities, surfaces charge densities, and volumetric charge densities, the expression for $\vec{E}$ at a point p in space, $\vec{E}_p$, takes the form:

$$\begin{aligned}\vec{E}_p = &\sum\nolimits_{\text{Point Charges}} \frac{q}{4\pi\varepsilon_o R_{q-p}^2}\vec{a}_{R_{q-p}} + \sum\nolimits_{\text{Linear Densities}} \int \frac{\rho_\ell\, d\ell}{4\pi\varepsilon_o R_{dQ-p}^2}\vec{a}_{R_{dQ-p}} \\ &+ \sum\nolimits_{\text{Surface Densities}} \int \frac{\rho_S\, dS}{4\pi\varepsilon_o R_{dQ-p}^2}\vec{a}_{R_{dQ-p}} + \sum\nolimits_{\text{Volumetric Densities}} \int \frac{\rho_v\, dv}{4\pi\varepsilon_o R_{dQ-p}^2}\vec{a}_{R_{dQ-p}}\end{aligned} \tag{5.16}$$

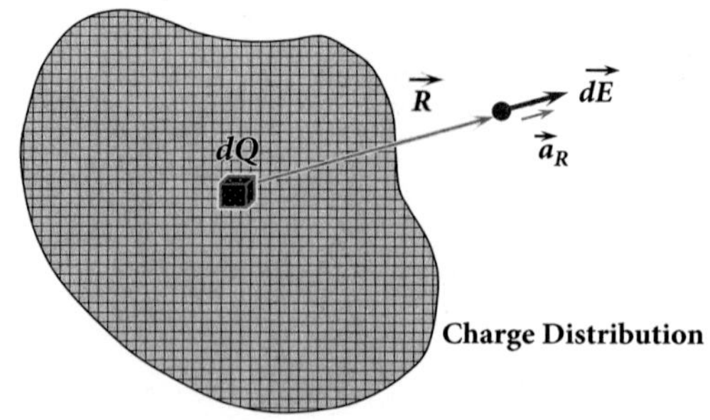

FIGURE 5.6 Electric field evaluation using the "Incrementation" scheme.

5.5 Electric Field due to Famous Examples of Charge Distributions

The expression we just concluded, Equation (5.16), gives us a powerful tool to evaluate the electric field due to "any" charge distribution. The process involves dicing the charge distribution into infinitesimal charge increments; write down the expressions for the corresponding electric field due to each charge increment; then perform the summing integration(s). In the following, we will take a simple case to demonstrate this process. However, Addendum 5A covers several other cases that typically demonstrate the use of this "electric field incrementation scheme" tool in deriving expressions for $\vec{E}$ and hence $\vec{D}$.

It is important to realize that writing an analytical form for the field expressions may not be that simple and the integrations may be too complicated in general cases of charge distributions. Obviously, we can always resort to numerical schemes for evaluating such integrations. All the cases we demonstrate in this chapter vary in their degree of complexity; however, they all have analytical results.

5.5.1 Case of Charges Distributed Uniformly in a Finite Length Straight Line

The line of charge distribution is demonstrated in Figure 5.7. Our goal is to solve for the electric field intensity vector, $\vec{E}$, at all points of space. We choose our coordinate system such that the charges line up on the z axis from ($z = z_1$ to $z = z_2$). The point at which $\vec{E}$ expression to be developed is named "p" and assumes a general location in space denoted in the spherical coordinates by $\vec{r}_p$. The corresponding Cartesian and cylindrical coordinates of the point p are denoted by (x_p, y_p, z_p) and (ρ_p, φ_p, z_p), respectively.

The total line charge is denoted by "Q" and is uniformly distributed over the line length $(z_2 - z_1)$, hence the linear density of the charge distribution is given by $\rho_\ell = Q/(z_2 - z_1)$. To derive an expression for the field due to this charge, we resort to the Coulomb's approach where we dice up the charge distribution into small increments, evaluate the electric field contribution of each increment, and perform needed summation/integration as given above in Equation (5.16).

It is obvious that the line-charge integration part of that equation is the one to be implemented here. In addition, since we have one charge distribution, only one integration will be performed, and hence, for this case, the equation reduces to

$$\vec{E}_p = \int_{\text{Line Charge}} d\vec{E}_p = \int_{\text{Line Charge}} \frac{\rho_\ell \, d\ell}{4\pi\varepsilon_o R_{dQ-p}^2} \vec{a}_{R_{dQ-p}} \tag{5.17}$$

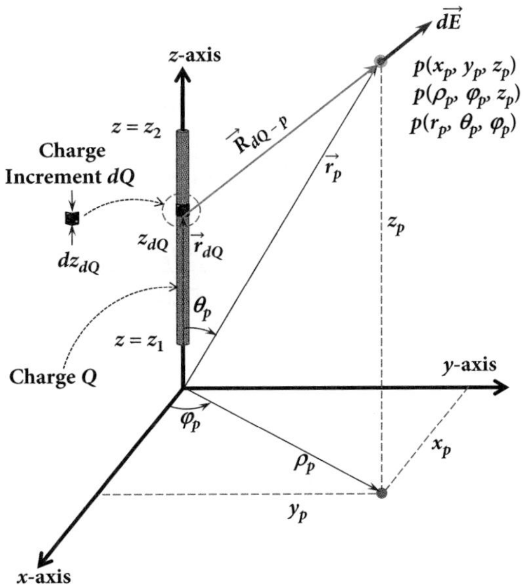

FIGURE 5.7 Electric field due to a finite length line of charges.

What we need to do now is develop expressions for the integration parameters and variables; $\rho_\ell, d\ell$, and $\vec{R}_{dQ-p}$. Both R_{dQ-p} and $\vec{a}_{R_{dQ-p}}$ are related to $\vec{R}_{dQ-p}$ as

$$R_{dQ-p} = \left|\vec{R}_{dQ-p}\right|, \quad \text{and} \quad \vec{a}_{R_{dQ-p}} = \frac{\vec{R}_{dQ-p}}{\left|\vec{R}_{dQ-p}\right|} \tag{5.18}$$

thus yielding

$$\frac{\vec{a}_{R_{dQ-p}}}{R_{dQ-p}^2} = \frac{\vec{R}_{dQ-p}}{\left|\vec{R}_{dQ-p}\right|^3} \tag{5.19}$$

hence

$$\vec{E}_p = \int_{\text{Line Charge}} \frac{\rho_\ell d\ell}{4\pi\varepsilon_o R_{dQ-p}^3} \vec{R}_{dQ-p} \tag{5.20}$$

Next, we examine the integrand expression to decide which of the three coordinate system, Cartesian, cylindrical, or spherical, to use. In this regard, we present the following guidelines to help with this decision:

1. Only scalar integrand expressions are allowed and the integrand should not include any vectors. If the integrand contains any constant vectors, they can be factored outside the integration sign leaving a scalar integrand. However, variable vectors within the integrand must be expressed in terms of constant vectors in the appropriate coordinate system.
2. After factoring out all constants in the integrand, the remaining scalar integrand expression is of a form that can be integrated. An unfamiliar expression in one coordinate system may correspond to a more familiar one in another coordinate system. It takes some experience to recognize the need to switch the coordinates to simplify the integration. Trial and error approach may be the tool to use when uncertain.
3. The integration limits may provide a clue as to which coordinate system to use.

In our case, the integrand of Equation (5.17) contains the vector $\vec{R}_{dQ-p}$, which can be expressed in spherical coordinates as $\vec{R}_{dQ-p} = \vec{r}_p - \vec{r}_{dQ}$. Although $\vec{r}_p$ is a constant vector, $\vec{r}_{dQ}$ is not since it varies in magnitude as we integrate the contribution of the various dQ's. (Please note that $\vec{r}_p$ is a constant vector since the point p does not vary throughout the integration process.) Meanwhile, the variable vector $\vec{r}_{dQ}$ is variable in magnitude only and maintains a constant direction along $\vec{a}_z$, on the z-axis. Hence, $\vec{r}_{dQ} = r_{dQ}\vec{a}_z$, and consequently, we now rewrite Equation (5.20) to break $\vec{R}_{dQ-p}$ down into its fixed and variable parts:

$$\vec{E}_p = \int_{\text{Line Charge}} \frac{\rho_\ell d\ell}{4\pi\varepsilon_o R_{dQ-p}^3}(\vec{r}_p - \vec{r}_{dQ}) = \int_{\text{Line Charge}} \frac{\rho_\ell d\ell}{4\pi\varepsilon_o R_{dQ-p}^3}\vec{r}_p - \int_{\text{Line Charge}} \frac{\rho_\ell d\ell}{4\pi\varepsilon_o R_{dQ-p}^3}\vec{r}_{dQ} \tag{5.21}$$

$$\begin{aligned}\vec{E}_p &= \vec{r}_p \int_{\text{Line Charge}} \frac{\rho_\ell d\ell}{4\pi\varepsilon_o R_{dQ-p}^3} - \int_{\text{Line Charge}} \frac{\rho_\ell d\ell}{4\pi\varepsilon_o R_{dQ-p}^3}\vec{r}_{dQ} \\ &= \vec{r}_p \int_{\text{Line Charge}} \frac{\rho_\ell d\ell}{4\pi\varepsilon_o R_{dQ-p}^3} - \int_{\text{Line Charge}} \frac{\rho_\ell d\ell}{4\pi\varepsilon_o R_{dQ-p}^3} r_{dQ}\vec{a}_z\end{aligned} \tag{5.22}$$

$$\vec{E}_p = \vec{r}_p \int_{\text{Line Charge}} \frac{\rho_\ell d\ell}{4\pi\varepsilon_o R_{dQ-p}^3} - \vec{a}_z \int_{\text{Line Charge}} \frac{\rho_\ell d\ell}{4\pi\varepsilon_o R_{dQ-p}^3} r_{dQ} \tag{5.23}$$

In Equation (5.23), we factored the constant vector $\vec{r}_p$ outside its integral and replaced the variable vector $\vec{r}_{dQ}$ by its variable magnitude along $\vec{a}_z$. We also factored out $\vec{a}_z$ outside its integration, and we now have an all-scalar integrand in Equation (5.23) and the first guideline is now satisfied. Now, we move to the second guideline for choosing the coordinate system and we start by factoring out all constant terms:

$$\vec{E}_p = \left(\frac{\rho_\ell}{4\pi\varepsilon_o}\right)\left[\vec{r}_p \int_{\text{Line Charge}} \frac{d\ell}{R_{dQ-p}^3} - \vec{a}_z \int_{\text{Line Charge}} \frac{r_{dQ}\, d\ell}{R_{dQ-p}^3}\right] \tag{5.24}$$

leaving the two remaining integrand expressions $\frac{d\ell}{R_{dQ-p}^3}$ and $\frac{r_{dQ}\, d\ell}{R_{dQ-p}^3}$. Both expressions have R_{dQ-p}^3 in their denominators. Expressing R_{dQ-p} in spherical coordinates yields the "complicated" expression $|\vec{r}_p - \vec{r}_{dQ}|$, which is not easy to manipulate thus eliminating the choice of that coordinate system.

Now, we consider the two remaining choices of cylindrical and Cartesian systems. Both of them use the z coordinate, which greatly simplifies the R_{dQ-p} expression:

$$R_{dQ-p} = |\vec{r}_p - \vec{r}_{dQ}| = |[\vec{\rho}_p + \vec{z}_p] - [0 + \vec{z}_{dQ}]| = |\rho_p \vec{a}_{\rho_p} + (z_p - z_{dQ})\vec{a}_z| \tag{5.25}$$

$$R_{dQ-p} = \sqrt{(\rho_p)^2 + (z_p - z_{dQ})^2} \tag{5.26}$$

This form for R_{dQ-p} has only one variable z_{dQ} while both ρ_p and z_p are constants. This finding does not yield any preference to the coordinate system choice. This is also the case if we consider the other remaining integrand terms $d\ell = dz_{dQ}$ and $r_{dQ} = z_{dQ}$.

Expressing $\vec{r}_p$ in Equation (5.24) in terms of its equivalent cylindrical coordinates,

$$\vec{r}_p = \vec{\rho}_p + \vec{z}_p = \vec{\rho}_p + z_p \vec{a}_z \tag{5.27}$$

Consequently, Equation (5.24) takes the form

$$\vec{E}_p = \left(\frac{\rho_\ell}{4\pi\varepsilon_o}\right)\left[\vec{\rho}_p \int_{z_1}^{z_2} \frac{dz_{dQ}}{\left[(\rho_p)^2 + (z_p - z_{dQ})^2\right]^{3/2}} + \vec{a}_z \int_{z_1}^{z_2} \frac{(z_p - z_{dQ})dz_{dQ}}{\left[(\rho_p)^2 + (z_p - z_{dQ})^2\right]^{3/2}}\right] \tag{5.28}$$

in which the integration limits for z_{dQ} are defined as z_1 and z_2.

The two integrations in Equation (5.28) are among the famous integrations that are typically available in integration tables. They simply require the introduction of two new variable $u = z_p - z_{dQ}$ and α as the angle between the vector $\vec{R}_{dQ-p}$ and the z axis.

Starting with Equation (5.28) and making the substitution $u = z_p - z_{dQ}$, we get

$$\vec{E}_p = \left(\frac{\rho_\ell}{4\pi\varepsilon_o}\right)\left[\vec{\rho}_p \int_{u=z_p-z_1}^{u=z_p-z_2} \frac{-du}{\left[\rho_p^2 + u^2\right]^{3/2}} - \vec{a}_z \int_{u=z_p-z_1}^{u=z_p-z_2} \frac{u\,du}{\left[\rho_p^2 + u^2\right]^{3/2}}\right] \tag{5.29}$$

It can be seen that the use of cylindrical coordinates simplified the expression that we can now perform the two integrations yielding the following result:

$$\vec{E}_p = \left(\frac{\rho_\ell}{4\pi\varepsilon_o}\right)\left[\vec{\rho}_p \left[\frac{-u}{\rho_p^2\sqrt{\rho_p^2 + u^2}}\right]_{u=z_p-z_1}^{u=z_p-z_2} - \vec{a}_z \left[\frac{-1}{\sqrt{\rho_p^2 + u^2}}\right]_{u=z_p-z_1}^{u=z_p-z_2}\right] \tag{5.30}$$

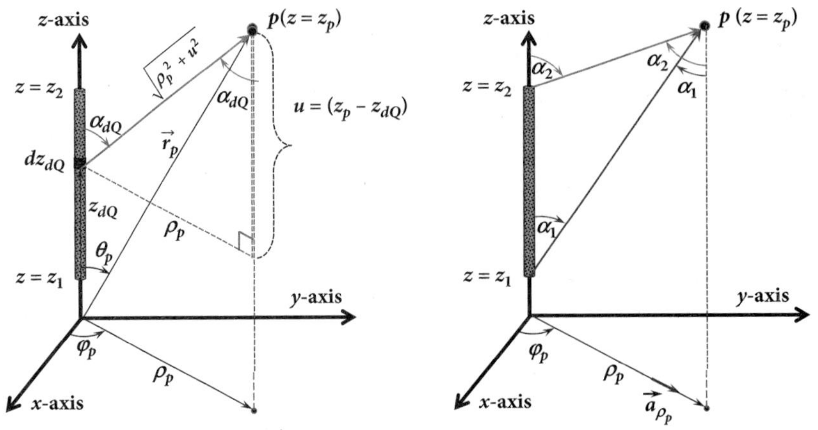

FIGURE 5.8 Electric field due to a finite length line of charges expressed in terms of the α angles.

$$\vec{E}_p = \left(\frac{\rho_\ell}{4\pi\varepsilon_o\rho_p}\right)\left[-\vec{a}_{\rho_p}\left[\frac{u}{\sqrt{\rho_p^2+u^2}}\right]_{u=z_p-z_1}^{u=z_p-z_2} + \vec{a}_z\left[\frac{\rho_p}{\sqrt{\rho_p^2+u^2}}\right]_{u=z_p-z_1}^{u=z_p-z_2}\right] \tag{5.31}$$

Now, introducing α as the angle between the vector $\vec{R}_{dQ-p}$ and the z axis, then with the aid of Figure 5.8, we can write

$$\cos\alpha = \frac{u}{\sqrt{\rho_p^2+u^2}} \quad \text{and} \quad \sin\alpha = \frac{\rho_p}{\sqrt{\rho_p^2+u^2}}. \tag{5.32}$$

Hence, Equation (5.31) becomes

$$\vec{E}_p = \left(\frac{\rho_\ell}{4\pi\varepsilon_o\rho_p}\right)\left[-\vec{a}_{\rho_p}[\cos\alpha]_{\alpha_1}^{\alpha_2} + \vec{a}_z[\sin\alpha]_{\alpha_1}^{\alpha_2}\right] \tag{5.33}$$

which finally yields the result for the electric field intensity vector $\vec{E}_p$ as

$$\vec{E}_p = \left(\frac{\rho_\ell}{4\pi\varepsilon_o\rho_p}\right)\{-\vec{a}_{\rho_p}[\cos\alpha_2 - \cos\alpha_1] + \vec{a}_z[\sin\alpha_2 - \sin\alpha_1]\} \tag{5.34}$$

This equation expresses the field vector in cylindrical coordinates. Now we ask ourselves, what if we worked in Cartesian coordinates instead. The answer is that the process would not be much different except for the term ρ_p would be decomposed to its x_p, and y_p components:

$$\rho_p = \sqrt{x_p^2 + y_p^2} \quad \text{and} \quad \vec{a}_{\rho_p} = \cos(\varphi_p)\vec{a}_x + \sin(\varphi_p)\vec{a}_y \tag{5.35}$$

5.5.2 Case of Charges Distributed Uniformly in an Infinite Length Straight Line

The line of charge in this case is similar to the case of finite length except for the infinite extension (Figure 5.9). Hence, we can use the same expression developed for the finite length Equation (5.28) and simply substitute $z_1 = -\infty$, and $z_2 = +\infty$. The corresponding angles would be $\alpha_1 = 0$ and $\alpha_2 = \pi$. Thus, we conclude that

$$\vec{E}_p = \left[\frac{\rho_\ell}{4\pi\varepsilon_o\rho_p}\right]\{[\cos(0) - \cos(\pi)]\,\vec{a}_{\rho_p} + [\sin(\pi) - \sin(0)]\,\vec{a}_z\} = \frac{\rho_\ell}{2\pi\varepsilon_o\rho_p}\vec{a}_{\rho_p} \tag{5.36}$$

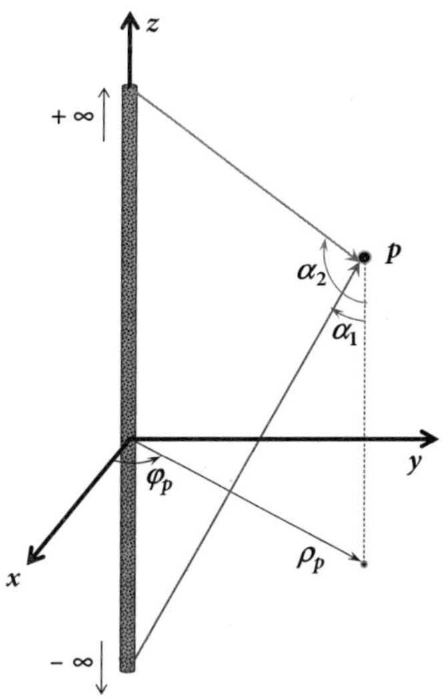

Figure 5.9 Electric field due to an infinite length line of charges.

It is important to note that this result is in full agreement with the flux density vector result obtained earlier in Equation (4A.23) since

$$\vec{D}_p = \varepsilon_o \vec{E}_p = \frac{\rho_\ell}{2\pi\rho_p}\vec{a}_{\rho_p} \tag{5.37}$$

5.6 Energy in a System of Charges

With the recognition of forces between charges, we are ready now to discuss the energy stored in a system of charges. We will use analogies to the gravitational system to help us in this discussion; Figure 5.10 depicts this analogy. In the figure, m_o exerts a force on m in a way similar to q_o exerting a force on q (except for polarity). The electric field intensity is defined as the force per unit q so was the gravitational

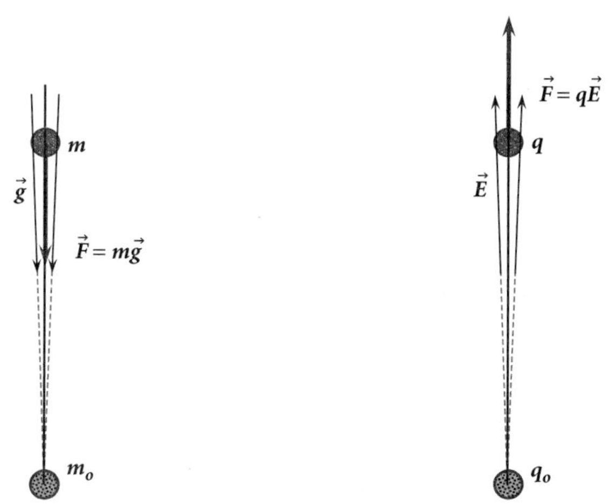

Figure 5.10 Analogy between electrical forces and gravitational forces.

acceleration (field intensity) is defined as the gravitational force per unit m. The forces and fields in both cases are proportional to the sources (charges/masses) and inversely proportional to the square of the separating distance.

Since we have an appreciation to the definitions of energy and potential in the gravitational system, we will use the analogy between the two systems to arrive at the definition of energy, and potential in the charge system.

The definition of the energy in a system is the work needed to construct the system from a "zero-energy/reference" state. An alternate definition is the amount of work (or energy) released as the system disintegrates and returns to a "zero-energy/reference" state. The "zero-energy" state has to be defined and hence the definition of energy is always a relative one.

In the gravitational system, the "zero-energy" state is often defined as masses resting at the sea level. However, in the electrical charge system, the "zero-energy" state is defined as case of a neutral space with charges, if any, located at infinity (far away from our space).

In both definitions, the energy is the amount of work done in the transition between states. In the *construction* definition, it is the work done by the building force while in the *disintegration* definition it is the work done by the system's internal forces. To clarify, we think of the energy in erecting a stack of building blocks is equal to the work done by the "builder" in lifting each block against gravitation forces from its "zero-energy" position to its final position. Theoretically, this would be exactly equal to the work done by gravity in returning all blocks to their initial "zero-energy" positions.

In a system of charges, infinity is assumed the zero-energy location for all charges, and hence, the total energy in a system of charges is defined as the work done by an external forces (against the internal "Coulomb" forces) to assemble the charge system as we bring its components from their "zero-energy" states at infinity. Alternatively, it is the same exact amount of work done by the internal Coulomb forces in returning all charge components back to their "zero-energy" states at infinity. In mathematical terms, we will use the notation W_e for the electrostatic energy, it is a scalar quantity, its MKS unit is joules, J, and can be expressed as

$$W_{e,\text{ system of charges}} = \sum_{\text{Charge Elements}} \int_{\infty}^{\text{Position}} dW_{e,\text{ by external force to move each charge element}} \tag{5.38}$$

$$W_{e,\text{ building system of charges}} = \sum_{\text{Charge Elements}} \int_{\infty}^{\text{Position}} \vec{F}_{\text{external}} \cdot \vec{d\ell} \tag{5.39}$$

$$W_{e,\text{ system of charges}} = \sum_{\text{Charge Elements}} \int_{\text{Position}}^{\infty} dW_{\text{by Coulomb force to move each charge element}} \tag{5.40}$$

$$W_{e,\text{ destructing system of charges}} = \sum_{\text{Charge Elements}} \int_{\text{Position}}^{\infty} \vec{F}_{\text{Coulomb}} \cdot \vec{d\ell} \tag{5.41}$$

It is clear that the two terms yield identical results since the external "building" force is to counter the internal force to allow the system construction, and hence $\vec{F}_{\text{external}} = -\vec{F}_{\text{coulomb}}$, while the limits on the integration are opposite to each other yielding another negative polarity to make the two equations identical. Consequently, using the electric field intensity vector, $\vec{E}$, instead of the force, we can write

$$W_{e,\text{ system of charges}} = \sum_{\text{all discrete } q\text{'s}} q \int_{\text{Position}}^{\infty} \vec{E} \cdot \vec{d\ell} \quad \text{or} \tag{5.42}$$

$$W_{e,\text{ system of charges}} = \int_{\text{all continuous } Q\text{'s}} dQ \int_{\text{Position}}^{\infty} \vec{E} \cdot \vec{d\ell} \tag{5.43}$$

5.7 Examples of Energy in a System of Charges

5.7.1 Energy in a System of Point Charges

We assume that building a system of point charges is to bring these charges to their position in the system from their zero-energy states. We assume the zero-energy state for these point charges is at infinity where they are infinite distance away from each other and hence no mutual or interactive forces. Now, we evaluate the energy associated with moving the point charges one by one from infinity to its final position in the system.

Moving the very first point charge involves zero work done simply because that motion takes place under zero forces since all other charges are still at infinity and thus our first point charge is infinite distance away from all other charges throughout its path to its final position (Figure 5.11).

Moving the next point charge is a different situation. Initially, there are no forces on this point charge being located at infinity. However, as it approaches its final position in the system, it will be moving under the force of the first charge which was brought earlier to its post, Figure 5.12. Therefore, the total energy of the

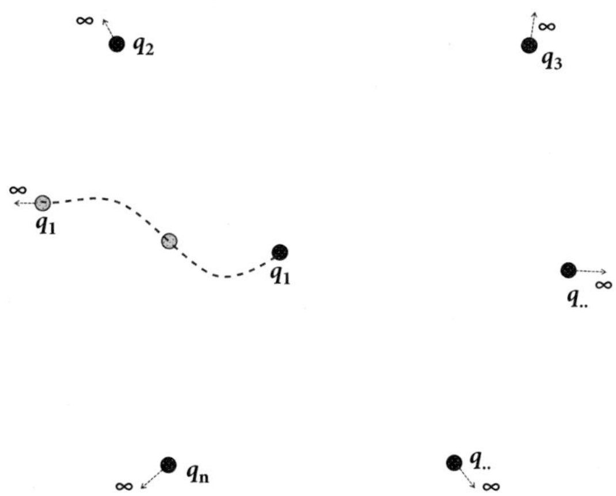

FIGURE 5.11 Energy in a system of charges is equal to the work done to build the system of charges.

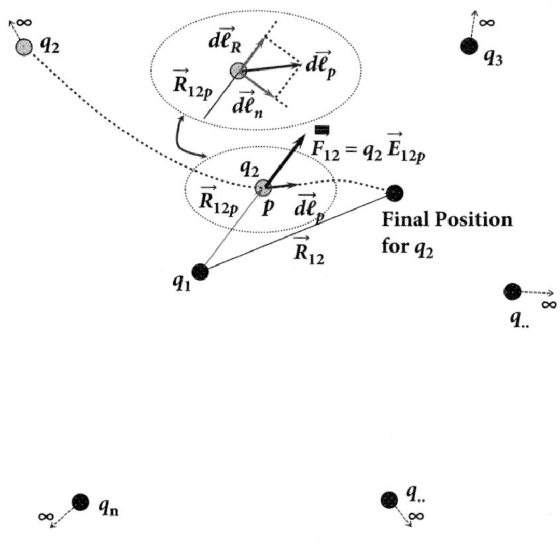

FIGURE 5.12 Evaluating the work done to move a charge from infinity (zero potential energy) to its final location in the system of charges.

system of the two charges will be the integration of the work done as the second charge moves under the influence of the force between the two point-charges.

$$W_{e,q_1-q_2} = \int_{\infty}^{\text{Position}} \vec{F}_{\text{external}} \cdot \vec{d\ell} = \int_{\text{Position}}^{\infty} q_2 \vec{E}_p \cdot \vec{d\ell}_p = \int_{\text{Position}}^{\infty} \frac{q_2 q_1}{4\pi\varepsilon_o R_{12p}^2} \vec{a}_{R_{12p}} \cdot \vec{d\ell}_p \tag{5.44}$$

Breaking $\vec{d\ell}_p$ into two orthogonal components, $\vec{d\ell}_R$ along $\vec{R}_{12p}$ and $\vec{d\ell}_n$ normal to it,

$$\vec{d\ell}_p = \vec{d\ell}_R + \vec{d\ell}_n = dR_{12p}\vec{a}_{R_{12p}} + \vec{d\ell}_n \tag{5.45}$$

and recognizing that $\vec{a}_{R_{12p}} . \vec{d\ell}_n = 0$ since they are orthogonal to each other, the energy of the two-charge system becomes

$$W_{e,q_1-q_2} = \frac{q_2 q_1}{4\pi\varepsilon_o} \int_{\text{Position}}^{\infty} \frac{\vec{a}_{R_{12p}} \cdot \left[dR_{12p}\vec{a}_{R_{12p}} + \vec{d\ell}_n \right]}{R_{12p}^2} = \frac{q_2 q_1}{4\pi\varepsilon_o} \int_{R_{12}}^{\infty} \frac{dR_{12p}}{R_{12p}^2} = \frac{q_2 q_1}{4\pi\varepsilon_o R_{12}} \tag{5.46}$$

Adding more charges to the system requires additional work and hence additional energy in the system. This addition is algebraic as the polarity of the system forces, and hence the work done could lead to either positive or negative addition to the system energy.

$$\begin{aligned} W_{e,q_1-q_2-q_3} &= \frac{q_2 q_1}{4\pi\varepsilon_o R_{12}} + \int_{\text{Position}}^{\infty} \left[\frac{q_3 q_1}{4\pi\varepsilon_o R_{13p}^2} \vec{a}_{R_{13p}} + \frac{q_3 q_2}{4\pi\varepsilon_o R_{23p}^2} \vec{a}_{R_{23p}} \right] \cdot \vec{d\ell}_p \\ &= \frac{q_2 q_1}{4\pi\varepsilon_o R_{12}} + \frac{q_3 q_1}{4\pi\varepsilon_o R_{13}} + \frac{q_3 q_2}{4\pi\varepsilon_o R_{23}} \end{aligned} \tag{5.47}$$

Consequently, we can write the total energy in a system of n point charges as

$$W_{q_1 \cdots q_n} = \sum_{j=i+1}^{n} \sum_{i=1}^{n} \frac{q_j q_i}{4\pi\varepsilon_o R_{ij}} \quad \text{or} \tag{5.48}$$

$$W_{e,q_1 \cdots q_n} = \frac{1}{2} \sum_{j=1}^{n, j\neq i} \sum_{i=1}^{n} \frac{q_j q_i}{4\pi\varepsilon_o R_{ij}} = \frac{1}{2} \sum_{j=1}^{n, j\neq i} q_j \sum_{i=1}^{n} \frac{q_i}{4\pi\varepsilon_o R_{ij}} \tag{5.49}$$

where the ½ is included to compensate for the repetition of *ij* and *ji* terms.

It is interesting to observe that the final answer for the energy in the system is independent of the paths taken by the charges to assemble in their final position. Only their relative positions appear in the energy of the system. This is true for other forms of charge distribution as well and not only for point charges (proof provided later in this chapter). Hence, this form of energy is a position energy, which we technically term as the potential energy of the system. Again, this is analogous to the potential energy in the gravitational system, which as we know is determined by the relative positions of the masses and not how they got there.

5.7.2 Energy in Other Forms of Charge Distributions

To derive an energy expression for a "continuous" charge distribution, let us start by dicing that distribution into infinitesimally small increments and treat each increment as a point charge. If we do that, then we can use Equation (5.49) while replacing each point charge by one of the charge increments, ΔQ.

$$W_{e,Q} = \frac{1}{2} \sum_{j=1}^{n, j\neq i} \Delta Q_j \sum_{i=1}^{n} \frac{\Delta Q_i}{4\pi\varepsilon_o R_{ij}} \tag{5.50}$$

Now, taking the limit as ΔQ approaches zero, the summations above converge to integrals, thus we can write

$$W_{e,Q} = \frac{1}{2}\int_{\text{all}\,Q}\left\{dQ_j\left[\int_{\text{all}\,Q}\left(\frac{dQ_i}{4\pi\varepsilon_o R_{ij}}\right)\right]\right\} \tag{5.51}$$

We will revisit both Equations (5.49) and (5.51) later in this chapter after we introduce the definition of the electric potential.

5.8 The Electric Potential

Again, in analogy with the gravitational system, we define the electric potential. The gravitational system potential is the static "potential" energy at a given location per unit mass located at that location. Similarly, we define the electric potential as the electric energy at a given location per unit charge located at that location, and hence, using Equation (5.43)

$$\Delta W_e = \Delta Q\int_{\text{Position}}^{\infty} \vec{E}\cdot\overrightarrow{d\ell} \quad \text{or} \tag{5.52}$$

From which

$$V_p = \frac{\Delta W_e}{\Delta Q} = \int_{\text{Position}}^{\infty} \vec{E}\cdot\overrightarrow{d\ell} \tag{5.53}$$

The quantity $V_p = \dfrac{\Delta W_e}{\Delta Q}$ defines what is known as the electric potential, V, of a given point on space relative to that at infinity, and its MKS unit is volt. Again, with infinity being the location of zero electrical energy, then it is also the reference of zero electric potential. Hence,

$$V_p = \int_p^{\infty} \vec{E}\cdot\overrightarrow{d\ell} \tag{5.54}$$

An alternate way to define the electric potential is through the work done (by an external force) of moving a unit charge from infinity "zero potential" to the location, or the work done by the Coulomb force to move the unit charge from its location to infinity.

We also define the electrical potential difference as work done in moving the unit charge between two locations. Suppose we move from locations A to location B, then we can write the potentials at A, B, and the potential difference as V_A, V_B, V_{BA}, respectively as follows:

$$V_A = \int_A^{\infty} \vec{E}\cdot\overrightarrow{d\ell} \quad \text{and} \quad V_B = \int_B^{\infty} \vec{E}\cdot\overrightarrow{d\ell} \tag{5.55}$$

$$V_{BA} = V_B - V_A = \int_B^{\infty} \vec{E}\cdot\overrightarrow{d\ell} - \int_A^{\infty} \vec{E}\cdot\overrightarrow{d\ell} = \int_B^{A} \vec{E}\cdot\overrightarrow{d\ell} = -\int_A^{B} \vec{E}\cdot\overrightarrow{d\ell} \tag{5.56}$$

Example 5.2 A spherical surface of radius R_1 has a uniformly distributed surface charge density ρ_S C/m^2. The region between $r = 0$ and $r = R_1$ is filled with air. The region between $r = R_1$ and $r = R_2$ is filled with a dielectric with a permittivity, ε, and $R_2 > R_1$. Determine the following:

a. The electric field intensity for $r < R_1$, for $R_1 \le r < R_2$ and for $r \ge R_2$.

b. The electrical potential at the center of the spherical surface.

Solution

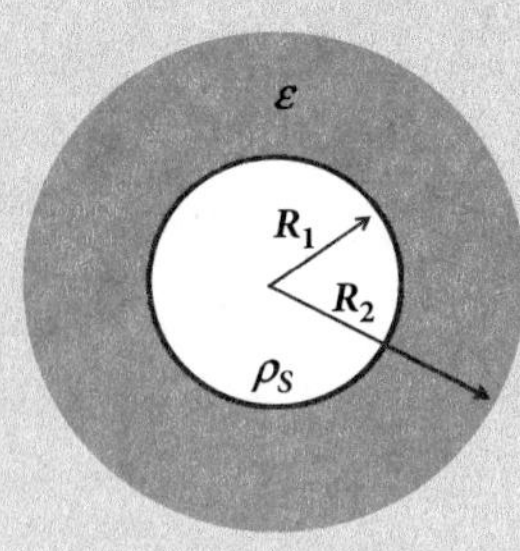

a. "Region" 1 $r < R_1$ $\oint \vec{E}\cdot\vec{dS} = \frac{Q_{\text{enclosed}}}{\varepsilon_o}$ $Q_{\text{enclosed}} = 0 \to E = 0$

"Region" 2 $R_1 \le r < R_2$

$$\oint \vec{D}\cdot\vec{dS} = Q_{\text{enclosed}} \qquad Q_{\text{enclosed}} = \rho_S 4\pi R_1^2$$

$$\vec{D}\cdot\vec{dS} = \vec{a}_r D_r \cdot \vec{a}_r dS_r = D_r dS_r$$

$$\oint \vec{D}\cdot\vec{dS} = D_r \int dS_r = D_r S = D_r 4\pi r^2$$

$$D_r 4\pi r^2 = \rho_S 4\pi R_1^2 \Rightarrow D_r = \frac{\rho_S 4\pi R_1^2}{4\pi r^2} = \frac{\rho_S R_1^2}{r^2} \Rightarrow E_r = \frac{D_r}{\varepsilon} = \frac{\rho_S R_1^2}{\varepsilon r^2}$$

"Region" 3 $r \ge R_2$ $Q_{\text{enclosed}} = \rho_S 4\pi R_1^2$

$$D_r 4\pi r^2 = \rho_S 4\pi R_1^2 \Rightarrow D_r = \frac{\rho_S 4\pi R_1^2}{4\pi r^2} = \frac{\rho_S R_1^2}{r^2}$$

$$D_r = \frac{\rho_S 4\pi R_1^2}{4\pi r^2} = \frac{\rho_S R_1^2}{r^2} \to E_r = \frac{D_r}{\varepsilon_o} = \frac{\rho_S R_1^2}{\varepsilon_o r^2}$$

b. The electrical potential at the center of the spherical surface.

$$V(r=0) = -\int_{\infty}^{0} E_r\, dr = -\left(\int_{\infty}^{R_2} E_{r_{\text{region3}}}\, dr + \int_{R_2}^{R_1} E_{r_{\text{region2}}}\, dr + \int_{R_1}^{0} E_{r_{\text{region1}}}\, dr\right)$$

$$= -\left(\int_{\infty}^{R_2} \frac{\rho_S R_1^2}{\varepsilon_o r^2} dr + \int_{R_2}^{R_1} \frac{\rho_S R_1^2}{\varepsilon r^2} dr + \int_{R_1}^{0} 0\, dr\right) = -\rho_S R_1^2 \left[\left(\frac{-1}{\varepsilon_o r}\right)_{\infty}^{R_2} + \left(\frac{-1}{\varepsilon r}\right)_{R_2}^{R_1}\right]$$

$$V(r=0) = \rho_S R_1^2 \left(\left(\frac{1}{\varepsilon_o} - \frac{1}{\varepsilon}\right)\frac{1}{R_2} + \frac{1}{\varepsilon}\frac{1}{R_1}\right)$$

5.8.1 The Electric Potential due to the Field of a Point Charge

In this section, we practice the application Equations (5.54) and (5.56) in the case of a point charge q to derive an expression for the corresponding electric potential as a function of position.

Using Equation (5.54) and referring to Figure 5.13, we can write

$$V_P = \int_P^{\infty} \vec{E}\cdot\vec{d\ell} = \int_P^{\infty} \frac{q}{4\pi\varepsilon_o R_{q-P}^2} \vec{a}_{R_{q-P}} \cdot \vec{d\ell}_P \tag{5.57}$$

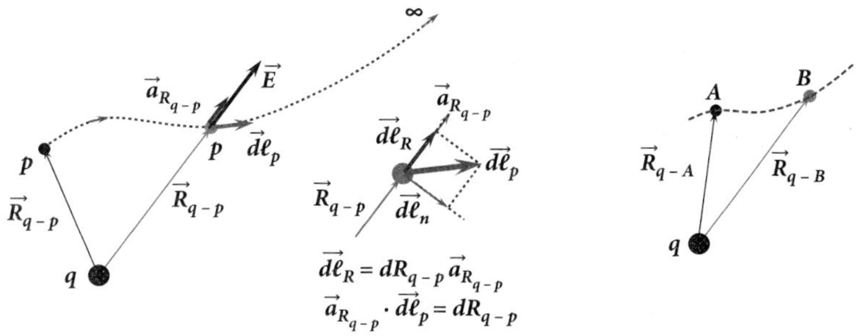

FIGURE 5.13 The electric potential due to the field of a point charge.

Expressing $\vec{d\ell}_p$ in terms of two orthogonal components, one along, $\vec{R}_{q-p}$ and the other is normal to it, only the component along will survive the dot product and the normal component will have zero contribution, hence,

$$V_p = \int_p^\infty \frac{q}{4\pi\varepsilon_o R_{q-p}^2} dR_{q-p} = \left[\frac{-q}{4\pi\varepsilon_o R_{q-p}}\right]_p^\infty = \frac{q}{4\pi\varepsilon_o R_{q-p}} \tag{5.58}$$

Also, the potential difference between two positions A and B as given by Equation (5.56) becomes

$$V_{BA} = V_B - V_A = -\int_A^B \vec{E}\cdot\vec{d\ell} = \frac{q}{4\pi\varepsilon_o R_{q-B}} - \frac{q}{4\pi\varepsilon_o R_{q-A}} = \frac{q}{4\pi\varepsilon_o}\left[\frac{1}{R_{q-B}} - \frac{1}{R_{q-A}}\right] \tag{5.59}$$

5.9 Potential Gradient

Applying this relationship, Equation (5.56), to an incremental distance $\Delta\ell$, and referring to Figure 5.14, we write

$$\Delta V = (V+\Delta V) - V = V_{(\ell+\Delta\ell)} - V_\ell = -\int_\ell^{\ell+\Delta\ell} \vec{E}\cdot\vec{d\ell} = -\vec{E}\cdot\vec{\Delta\ell} \qquad \text{or} \qquad dV = -\vec{E}\cdot\vec{d\ell} \tag{5.60}$$

If we express the vector $\vec{E}$ in terms of two main components, $\vec{E}_\ell$ along $\vec{d\ell}$ and the other, $\vec{E}_n$, normal to it then we can rewrite Equation (5.60) as $dV = -(\vec{E}_\ell + \vec{E}_n)\cdot\vec{d\ell} = -E_\ell\, d\ell$ and finally

$$E_\ell = -\frac{\partial V}{\partial \ell} \tag{5.61}$$

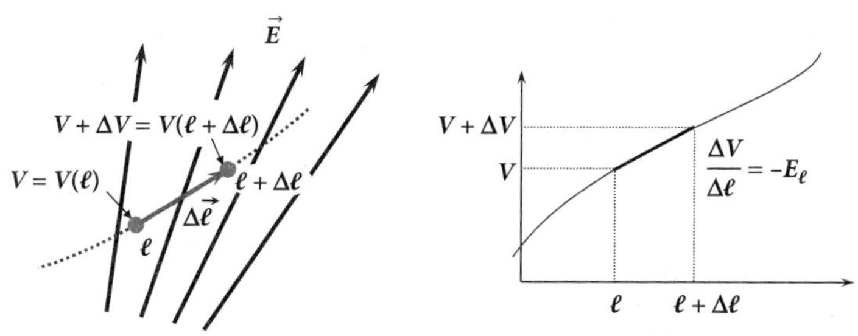

FIGURE 5.14 The electric field is the (negative) gradient of the potential function.

Equation (5.61) is another complementary tool for the evaluation of the electric field (and hence the flux density). This may not be an obvious use of Equation (5.61), hence, let us explain this further. If we somehow were able to derive an expression for the potential function as a function of position, we can then differentiate that function with respect to three orthogonal dimensions (e.g. x, y, and z) to get full knowledge of the three spatial components of the electric field vector. In Cartesian coordinates, we write

$$E_x = -\frac{\partial V}{\partial x}, \quad E_y = -\frac{\partial V}{\partial y}, \quad \text{and} \quad E_z = -\frac{\partial V}{\partial z} \tag{5.62}$$

Consequently,

$$\vec{E} = E_x\vec{a}_x + E_y\vec{a}_y + E_z\vec{a}_z = -\left[\frac{\partial V}{\partial x}\vec{a}_x + \frac{\partial V}{\partial y}\vec{a}_y + \frac{\partial V}{\partial z}\vec{a}_z\right] \tag{5.63}$$

This is the famous potential gradient expression, which we typically express in the form:

$$\vec{E} = -\vec{\nabla}V = -\text{grad } V \tag{5.64}$$

where the gradient in Cartesian coordinates is given by

$$\vec{\nabla} = \left[\vec{a}_x\frac{\partial}{\partial x} + \vec{a}_y\frac{\partial}{\partial y} + \vec{a}_z\frac{\partial}{\partial z}\right] \tag{5.65}$$

The cylindrical and spherical forms for the gradient are

$$\vec{\nabla} = \left[\vec{a}_\rho\frac{\partial}{\partial \rho} + \vec{a}_\varphi\frac{1}{\rho}\frac{\partial}{\partial \varphi} + \vec{a}_z\frac{\partial}{\partial z}\right] \tag{5.66}$$

$$\vec{\nabla} = \left[\vec{a}_r\frac{\partial}{\partial r} + \vec{a}_\theta\frac{1}{r}\frac{\partial}{\partial \theta} + \vec{a}_\varphi\frac{1}{r\sin\theta}\frac{\partial}{\partial \varphi}\right] \tag{5.67}$$

The question now is how are we going to able to derive an expression for V without knowing $\vec{E}$? [As suggested by Equation (5.54)] The answer lies in using a potential incrementation scheme similar to that used for the electric field vector in Equation (5.15).

Example 5.3 Three charges ($+q$, $-2q$, and $+q$) are arranged along the z axis at $z = d/2$, 0, and $-d/2$, respectively.

a. Determine V and $\vec{E}$ at a distant point, very far away from the origin.
b. Find the equation for the equipotential surfaces.

Hint: Use the same approximation used in Equations (5B.21) and (5B.22) to calculate the distances from the charges to the point where the field is calculated.

Solution

Part a $r >> d$

$$R_1 \approx r - \frac{d}{2}\cos\theta \qquad R_2 \approx r + \frac{d}{2}\cos\theta$$

$$V = \frac{q}{4\pi\varepsilon_o R_1} - \frac{2q}{4\pi\varepsilon_o r} + \frac{q}{4\pi\varepsilon_o R_2}$$

$$= \frac{q}{4\pi\varepsilon_o}\left[\frac{1}{r - \frac{d}{2}\cos\theta} - \frac{2}{r} + \frac{1}{r + \frac{d}{2}\cos\theta}\right]$$

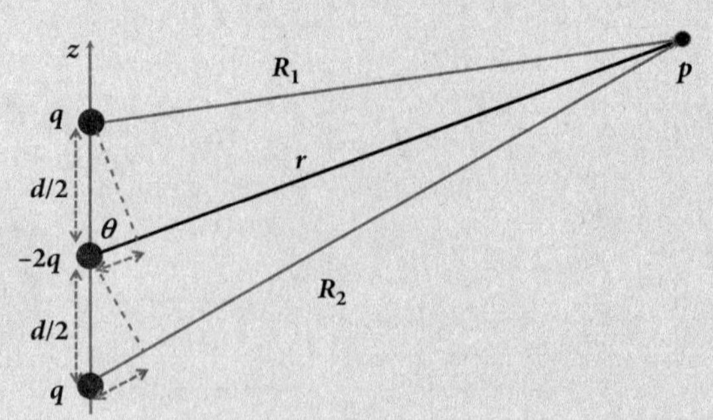

$$= \frac{q}{4\pi\varepsilon_o}\left[\frac{2r^2 - 2\left(r^2 - \left(\frac{d}{2}\right)^2\cos^2\theta\right)}{r\left(r^2 - \left(\frac{d}{2}\right)^2\cos^2\theta\right)}\right] \approx \frac{qd^2\cos^2\theta}{8\pi\varepsilon_o r^3}$$

Part b For the constant potential or equipotential surfaces:

$$V = \frac{qd^2\cos^2\theta}{8\pi\varepsilon_o r^3} = k \rightarrow r^3 = \left(\frac{qd^2}{8\pi\varepsilon_o k}\right)\cos^2\theta \rightarrow r^3 = C_v\cos^2\theta$$

where k is a constant representing a certain value of the potential and C_v is also a constant $C_v = \left(\frac{qd^2}{8\pi\varepsilon_o k}\right)$

$$\vec{E} = -\vec{\nabla}V = -\left(\vec{a}_r\frac{\partial V}{\partial r} + \vec{a}_\theta\frac{1}{r}\frac{\partial V}{\partial\theta} + \vec{a}_\varphi\frac{1}{r\sin\theta}\frac{\partial V}{\partial\varphi}\right)$$

$$= -\left(\vec{a}_r\frac{-3qd^2\cos^2\theta}{8\pi\varepsilon_o r^4} + \vec{a}_\theta\frac{1}{r}\frac{qd^2 2\cos\theta(-\sin\theta)}{8\pi\varepsilon_o r^3}\right)$$

$$\vec{E} = \frac{qd^2\cos\theta}{8\pi\varepsilon_o r^4}(\vec{a}_r 3\cos\theta + \vec{a}_\theta 2\sin\theta)$$

5.10 Electric Potential Evaluation Using the "Incrementation" Scheme

The way this scheme works is to start with a charge increment ΔQ out of the whole charge system and evaluate the corresponding ΔV_p at a general location (p) in space. In this regard, we use the point charge form, Equation (5.58), and replace q by ΔQ, hence

$$\Delta V_p = \frac{\Delta Q}{4\pi\varepsilon_o R_{\Delta Q-p}} \quad \text{or} \quad dV_p = \frac{dQ}{4\pi\varepsilon_o R_{dQ-p}} \tag{5.68}$$

Consequently, the potential due to a system of charges can be expressed as

$$V_{p,\text{system of charges}} = \sum\nolimits_{\text{all discrete } q\text{'s}} \frac{q}{4\pi\varepsilon_o R_{q-p}} \quad \text{or} \quad = \int_{\text{all continuous } Q\text{'s}} \frac{dQ}{4\pi\varepsilon_o R_{dQ-p}} \tag{5.69}$$

It is important to remind the reader that this incrementation scheme is valid only for linear systems (media) where the concept of superposition applies.

Once a potential function is determined at a general location in space, the gradient operation can be used to derive the electric field intensity vector using Equation (5.64). This scheme may provide an easier way to get V and $\vec{E}$ in contrast with the electric field "incrementation" scheme. The reason for that is obtaining V first involves a scalar integration followed by the differentiation of the gradient operation. On the other hand, starting with $\vec{E}$ first requires a vector integration followed by another integration.

Addendum 5B demonstrates the process of deriving expressions for V for a few "famous examples" through the potential Incrementation scheme. It also demonstrates the derivation of the electric field intensity vector from the obtained potential function using the gradient approach.

Example 5.4 An $L \times L$ square plate lying in the x-y plane from $x = 0$ to L and from $y = 0$ to L has a surface charge density of $\rho_S = \frac{xy}{L^2}$. Determine V and $\vec{E}$ at a point along the z axis.

Solution

$$\vec{r}_p = \vec{a}_z z \qquad \vec{r}_q = \vec{a}_x x + \vec{a}_y y$$

$$V = \frac{1}{4\pi\varepsilon_o}\int_{S'} \frac{\rho_S}{R_{q-p}} dS \qquad R_{q-p} = |\vec{r}_p - \vec{r}_q| = \sqrt{(0-x)^2 + (0-y)^2 + (z-0)^2}$$

$$V = \frac{1}{4\pi\varepsilon_o}\int_0^L\int_0^L \frac{xy\,dx\,dy}{L^2\sqrt{x^2+y^2+z^2}}$$

$$\int_0^L \frac{x\,dx}{\sqrt{x^2+y^2+z^2}} = \frac{1}{2}\frac{\left(\sqrt{x^2+y^2+z^2}\right)_0^L}{1/2} = \sqrt{L^2+y^2+z^2} - \sqrt{y^2+z^2}$$

$$V = \frac{1}{4\pi\varepsilon_o L^2}\int_0^L y\left[\sqrt{L^2+y^2+z^2} - \sqrt{y^2+z^2}\right]dy$$

$$= \frac{1}{4\pi\varepsilon_o L^2}\left[1/2\frac{(L^2+y^2+z^2)^{3/2}}{3/2} - 1/2\frac{(y^2+z^2)^{3/2}}{3/2}\right]_0^L$$

$$V = \frac{1}{12\pi\varepsilon_o L^2}[(2L^2+z^2)^{3/2} - (L^2+z^2)^{3/2} - (L^2+z^2)^{3/2} + |z|^3]$$

$$V = \frac{1}{12\pi\varepsilon_o L^2}[(2L^2+z^2)^{3/2} - 2(L^2+z^2)^{3/2} + z^3] \qquad z \ge 0$$

$$V = \frac{1}{12\pi\varepsilon_o L^2}[(2L^2+z^2)^{3/2} - 2(L^2+z^2)^{3/2} - z^3] \qquad z < 0$$

$$\vec{E} = -\vec{\nabla}V = -\vec{a}_z\frac{\partial V}{\partial z}$$

$$\vec{E} = \vec{\nabla}V = \frac{-1}{12\pi\varepsilon_o L^2}[3z(2L^2+z^2)^{1/2} - 6z(L^2+z^2)^{1/2} + 3z^2] \qquad z \ge 0$$

$$\vec{E} = -\vec{a}_z\frac{z}{4\pi\varepsilon_o L^2}[(2L^2+z^2)^{1/2} - 2(L^2+z^2)^{1/2} + z] \qquad z \ge 0$$

$$\vec{E} = -\vec{a}_z\frac{z}{4\pi\varepsilon_o L^2}[(2L^2+z^2)^{1/2} - 2(L^2+z^2)^{1/2} - z] \qquad z < 0$$

5.11 Conservative Nature of Electrostatic Potential

Examining Equation (5.58), we notice that the result of the line integration from the position (p) to infinity is independent of the path taken. The result depends on the distance between the point charge and the position (p). This is an important finding, which indicates that the static electric potential is conservative.

Examining Equation (5.59) can help give us more insight in this matter. According to Equation (5.59), a closed loop integration, for example, from A and back to A, results in a zero potential-difference. In other words, V_A is single valued and that value is unique and is solely determined by the location and not the path taken to get there.

$$\int_A^A \vec{E}\cdot\vec{d\ell} = \oint \vec{E}\cdot\vec{d\ell} = 0 \tag{5.70}$$

This is analogous to the gravitational potential, which is also a conservative one. As an example, if we lift an object from one position (level) to another in the gravitational system and then return it back to the same position, the net change in the gravitational potential is zero.

This finding is not limited to electrostatic fields and potential resulting from point charges, it is generally valid for all other charge distributions as well. By virtue of Equation (5.69), the potential function of a given charge distribution is the superposition of the incremental potential contributions of the incremental (point) charges constituting the charge distribution. Hence, adding increments that are only a function of position yields a sum that is of the same nature.

In other words, the summation of the potential contributions does not alter the conservative nature of the potential function. Simply because the sum of conservative potential-increments will certainly be conservative.

5.12 Energy Density in Electrostatic Fields

Earlier in this chapter, we discussed the energy of a system of charges. Equations (5.49) and (5.51) described the energy of a discrete system of point charges and that of a continuum of charge distribution, respectively.

$$W_{e,q_1\ldots q_n} = \frac{1}{2}\sum_{j=1}^{n,\,j\neq i} q_j \sum_{i=1}^{n} \frac{q_i}{4\pi\varepsilon_o R_{ij}} \tag{5.49}$$

$$W_{e,Q} = \frac{1}{2}\int_{\text{all Q}} \left\{ dQ_j \left[\int_{\text{all Q}} \left(\frac{dQ_i}{4\pi\varepsilon_o R_{ij}} \right) \right] \right\} \tag{5.51}$$

In this section, as promised, we will revisit these two expressions in an effort to re-express them in the light of the potential definition as well as develop expressions for the energy density per unit volume.

First, using Equation (5.69), we can rewrite Equations (5.49) and (5.51) in terms of the potential function as follows:

$$W_{e,q_1\ldots q_n} = \frac{1}{2}\sum_{j=1}^{n,\,j\neq i} q_j \left[\sum_{i=1}^{n} \frac{q_i}{4\pi\varepsilon_o R_{ij}} \right] = \frac{1}{2}\sum_{j=1}^{n,\,j\neq i} q_j [V_j] \tag{5.71}$$

$$W_{e,Q} = \frac{1}{2}\int_{\text{all Q}} \left\{ dQ_j \left[\int_{\text{all Q}} \left(\frac{dQ_i}{4\pi\varepsilon_o R_{ij}} \right) \right] \right\} = \frac{1}{2}\int_{\text{all Q}} \{dQ_j [V_j]\} \tag{5.72}$$

Replacing the charge increment dQ by the product of the volumetric charge density and the increment volume dv, and dropping the subscript Q from the W notation, we rewrite Equation (5.72) as

$$W_e = W_{e,Q} = \int_{\text{volume}} \frac{1}{2}\{\rho_v\, dv [V]\} = \int_{\text{volume}} \frac{1}{2}\{V[\rho_v]\} dv \tag{5.73}$$

Now using Equation (4.17), $\vec{\nabla}\cdot\vec{D}=\rho_v$

$$W_e=\int_{\text{volume}}\frac{1}{2}\{V[\vec{\nabla}\cdot\vec{D}]\}dv \tag{5.74}$$

Next, we use the identity:

$$\vec{\nabla}\cdot(V\vec{D})=V\vec{\nabla}\cdot\vec{D}+\vec{D}.\vec{\nabla}V \quad \text{or} \quad V\vec{\nabla}\cdot\vec{D}=\vec{\nabla}\cdot(V\vec{D})-\vec{D}.\vec{\nabla}V \tag{5.75}$$

$$W_e=\int_{\text{volume}}\frac{1}{2}\{\vec{\nabla}\cdot(V\vec{D})-\vec{D}.\vec{\nabla}V\}dv=\int_{\text{volume}}\frac{1}{2}\{\vec{\nabla}\cdot(V\vec{D})\}dv-\int_{\text{volume}}\frac{1}{2}\{\vec{D}.\vec{\nabla}V\}dv \tag{5.76}$$

The final step in this development is to prove that the first integral in this expression vanishes leaving the second integral as the answer. Using Equation (5.64), $\vec{E}=-\vec{\nabla}V$, on the second integral, the final answer is

$$W_e=\int_{\text{volume}}\frac{1}{2}\{\vec{D}.\vec{E}\}dv=\int_{\text{volume}}\frac{1}{2}\varepsilon_o E^2\,dv \tag{5.77}$$

To prove the first integral is zero, we resort to the divergence theorem (4.27) which converts the volume integral into a closed surface integral:

$$\int_{\text{volume}}\vec{\nabla}\cdot\vec{A}\,dv=\oiint_S \vec{A}\cdot\overrightarrow{dS} \tag{5.78}$$

Hence

$$\int_{\text{volume}}\frac{1}{2}\{\vec{\nabla}\cdot(V\vec{D})\}dv=\oiint_S V\vec{D}\cdot\overrightarrow{dS} \tag{5.79}$$

The choice of the closed surface integration is arbitrary as long as it encloses all the volume charges. Hence, taking the surface to infinity, the enclosed charges would occupy a volume that is much too small compared to the surface and hence can be viewed as a point charge for which V varies as $1/r$, D varies as $1/r^2$ while dS varies as r^2. Consequently, the product $V\vec{D}\cdot\overrightarrow{dS}$ varies as $(1/r)(1/r^2)(r^2)$ would reduce to $(1/r)$ which vanishes on the closed surface at $r=\infty$.

Now, to derive an expression for the energy density per unit volume, we apply Equations (5.73) and (5.77) to a small incremental volume of space, the corresponding dW_e is

$$dW_e=\frac{1}{2}\{\rho_v V\}dv=\frac{1}{2}\{\vec{D}.\vec{E}\}dv=\frac{1}{2}\varepsilon_o E^2 dv \tag{5.80}$$

Hence, the electric energy volumetric density, w_e, defined as the static electric energy stored per unit volume is given by

$$W_e=\frac{dW_e}{dv}=\frac{1}{2}\{\rho_v V\}=\frac{1}{2}\{\vec{D}.\vec{E}\}=\frac{1}{2}\varepsilon_o E^2 \tag{5.81}$$

Finally, we write the expression for the total energy stored in a given volume in terms of the energy density as

$$W_e=\int_{\text{volume}}w_e\,dv=\int_{\text{volume}}\frac{1}{2}\{\rho_v V\}dv=\int_{\text{volume}}\frac{1}{2}\{\vec{D}.\vec{E}\}dv=\int_{\text{volume}}\frac{1}{2}\varepsilon_o E^2\,dv \tag{5.82}$$

Example 5.5 Calculate the amount of electrostatic energy of a uniform sphere with radius b and volume charge density ρ_v stored in the following regions:

a. Inside the sphere.
b. Outside the sphere.

Solution

a. Region I: Inside the sphere.

$$\oint \vec{D} \cdot \vec{dS} = Q_{\text{enclosed}} = D_r \int dS_r = D_r S = D_r 4\pi r^2$$

$$D_r 4\pi r^2 = \rho_v \frac{4}{3}\pi r^3 \Rightarrow D_r = \frac{\rho_v r}{3} \Rightarrow E_r = \frac{D_r}{\varepsilon_o} = \frac{\rho_v r}{3\varepsilon_o}$$

$$W_{eI} = \frac{1}{2}\varepsilon_o \int_v (E_r)^2 \, dv = \frac{1}{2}\varepsilon_o \int_v \left(\frac{\rho_v r}{3\varepsilon_o}\right)^2 r^2 \sin\theta \, dr \, d\theta \, d\varphi$$

$$= \frac{\rho_v^2}{18\varepsilon_o} \int_0^b r^4 \, dr \int_0^\pi \sin\theta \, d\theta \int_0^{2\pi} d\varphi \frac{\rho_v^2}{18\varepsilon_o} \frac{b^5}{5} \times 2 \times 2\pi = \frac{2\pi\rho_v^2 b^5}{45\varepsilon_o}$$

b. Region II: Outside the sphere

$$\oint \vec{D} \cdot \vec{dS} = Q_{\text{enclosed}}$$

$$\oint \vec{D} \cdot \vec{dS} = D_r \int dS_r = D_r S = D_r 4\pi r^2$$

$$D_r 4\pi r^2 = \rho_v \frac{4}{3}\pi b^3 \Rightarrow D_r = \frac{\rho_v b^3}{3r^2} \Rightarrow E_r = \frac{D_r}{\varepsilon_o} = \frac{\rho_v b^3}{3\varepsilon_o r^2}$$

$$W_{eII} = \frac{1}{2}\varepsilon_o \int_v (E_r)^2 \, dv = \frac{1}{2}\varepsilon_o \int_v \left(\frac{\rho_v b^3}{3\varepsilon_o r^2}\right)^2 r^2 \sin\theta \, dr \, d\theta \, d\varphi = \frac{\rho_v^2 b^6}{18\varepsilon_o} \int_b^\infty \frac{1}{r^2} dr \int_0^\pi \sin\theta \, d\theta \int_0^{2\pi} d\varphi$$

$$= \frac{\rho_v^2 b^6}{18\varepsilon_o} \left(-\frac{1}{r}\right)_b^\infty \times 2 \times 2\pi = \frac{2\pi\rho_v^2 b^5}{9\varepsilon_o}$$

ADDENDUM 5A
ELECTRIC FIELD DUE TO FAMOUS EXAMPLES OF CHARGE DISTRIBUTIONS

5A.1 Charges Distributed Uniformly in a Circular Ring

The ring of charge distribution is demonstrated in Figure 5A.1. For this distribution, solving for $\vec{E}$ at all locations of space is too complicated and does not serve us at this point. However, we can demonstrate solving for $\vec{E}$ at locations on the ring axis (the z axis in the figure). Either the Cartesian or the cylindrical coordinate systems can be used to develop the sought electric field expressions. For convenience sake, we chose to locate the ring of charges in the x-y plane and centered at the origin.

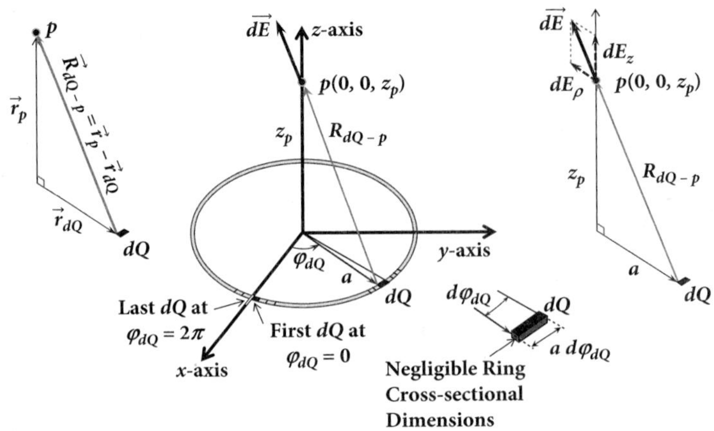

FIGURE 5A.1 The electric field due to a uniformly charged ring.

We assume a uniform distribution of a total charge Q coulombs on a circular ring (line of charges) with infinitesimal cross section (width and thickness). We further assume that ring has a radius of a meters, which means a linear charge density of $\rho_\ell = \frac{q}{2\pi a}$. To develop an analytical expression for $\vec{E}$, we start by dicing the ring of charges into infinitesimal increments, which in this case would be along the ring circumference. If we consider one of those increments to write the corresponding electric field, we need to choose an increment in first quadrant at a general angle φ to ensure that we do not write our expressions for a "special" case and thus potentially yielding the wrong answer. The increment chosen is highlighted in the figure and is located at the cylindrical coordinates ($\rho = a$, φ, $z = 0$). The length of the increment is the curved length ($a d\varphi$) corresponding to an incremental angle change $d\varphi$, and the corresponding charge increment dQ is given by

$$dQ = \rho_\ell d\ell = \frac{Q}{2\pi a} a d\varphi \tag{5A.1}$$

Next, we choose a general location point on the z axis for electric field vector evaluation; for example, the point p located on the positive side of the axis (again to avoid solving for a special case). The cylindrical coordinates for this point are ($\rho = 0$, φ is irrelevant, $z = z_p$).

Now, we examine the general integral form of Equation (5.16) to our charge distribution to recognize that the line charge expression is what applies to our case here:

$$\vec{E} = \int \frac{\rho_\ell d\ell}{4\pi\varepsilon_o R^2} \vec{a}_R \tag{5A.2}$$

where $\vec{R}$ in this case is $\vec{R}_{dq-p}$ the vector initiated at dQ and ends at z_p. To write an analytical expression for R and $\vec{a}_R$, we have three choices:

In spherical coordinates:

$$\vec{R}_{dq-p} = \vec{r}_p - \vec{r}_{dq} \tag{5A.3}$$

In cylindrical coordinates, since both points are in the same φ plane:

$$\vec{R}_{dq-p} = (\rho_p - \rho_{dq})\vec{a}_\rho + (z_p - z_{dq})\vec{a}_z \tag{5A.4}$$

In Cartesian coordinates:

$$\vec{R}_{dq-p} = (x_p - x_{dq})\vec{a}_x + (y_p - y_{dq})\vec{a}_y + (z_p - z_{dq})\vec{a}_z \qquad (5A.5)$$

It is obvious that the cylindrical coordinate format provides the simplest looking expression of the three above; however, it will face an obstacle when it comes to performing the integration. To explain, we recognize the presence of the vector $\vec{a}_R$ in the integrand. This is a variable vector, that is, it varies in direction as we integrate around the ring adding the contributions of the different charge increments. Integrating a variable scalar is a familiar process; however, the integration of a variable vector is a different story. The integration of a vector means summing vectors of differing directions, which is not feasible except through the addition of components along fixed coordinate directions. Using cylindrical coordinates $\vec{a}_R$ could be expressed in terms of both $\vec{a}_\rho$ and $\vec{a}_z$. $\vec{a}_z$ is a constant vector that does not vary with the location of dQ, however, $\vec{a}_\rho$ does. Consequently, we must resort to the Cartesian coordinate expression to allow for integration of components along the fixed direction coordinates x, y, and z, and avoid dealing with the variable vectors $\vec{a}_r$ in spherical and $\vec{a}_\rho$ in cylindrical.

$$\vec{R}_{dQ-p} = (0 - a\cos\varphi_{dQ})\vec{a}_x + (0 - a\sin\varphi_{dQ})\vec{a}_y + (z_p - 0)\vec{a}_z = -a\cos\varphi_{dQ}\vec{a}_x - a\sin\varphi_{dQ}\vec{a}_y + z_p\vec{a}_z \qquad (5A.6)$$

$$\left|R_{dQ-p}\right| = \sqrt{(-a\cos\varphi_{dQ})^2 + (-a\sin\varphi_{dQ})^2 + (z_p)^2} = \sqrt{a^2 + z_p^2} \qquad (5A.7)$$

and

$$\vec{a}_{R_{dQ-p}} = \frac{\vec{R}_{dQ-p}}{\left|R_{dQ-p}\right|} = \frac{-a\cos\varphi_{dQ}\vec{a}_x - a\sin\varphi_{dQ}\vec{a}_y + z_p\vec{a}_z}{\sqrt{a^2 + z_p^2}} \qquad (5A.8)$$

hence

$$\vec{E}_p = \int_{\varphi_{dQ}=0}^{\varphi_{dQ}=2\pi} \frac{\frac{Q}{2\pi a}\, a\, d\varphi_{dQ}}{4\pi\varepsilon_o\left(a^2 + z_p^2\right)} \left[\frac{-a\cos\varphi_{dQ}\vec{a}_x - a\sin\varphi_{dQ}\vec{a}_y + z_p\vec{a}_z}{\sqrt{a^2 + z_p^2}}\right] \qquad (5A.9)$$

$$\vec{E}_p = \frac{\frac{Q}{2\pi a}\, a\, 2\pi}{4\pi\varepsilon_o\left(a^2 + z_p^2\right)} \frac{z_p\vec{a}_z}{\sqrt{a^2 + z_p^2}} = \frac{Qz_p}{4\pi\varepsilon_o\left(a^2 + z_p^2\right)^{3/2}}\vec{a}_z \qquad (5A.10)$$

It is interesting to notice that the resultant field after integration is along $\vec{a}_z$ only. The $\vec{a}_\rho$ componenets ($\vec{a}_x$ and $\vec{a}_y$) all canceled out due to the integration from 0 to 2π. If we were to work with a portion of the ring (an arc), the integration in the a_x and a_y components may not vanish. The symmetry of the charge distribution from 0 to 2π is the reason for the cancelation of all x and y components leaving a resultant along the z axis only.

It is also important to notice the z axis symmetry. The ring of charges is placed in symmetry with respect to both the $+z$ and $-z$ directions. In other words, if the field in the upper $+z$ region is pointing up (away from the charge), then for the lower region in the $-z$ axis, the field should be pointing down (again, away from the charges). The obtained mathematical expression expresses that point properly as it includes the product $z_p\vec{a}_z$ which is upward for $+z_p$ and downward for $-z_p$.

5A.2 Charges Distributed Uniformly in a Circular Disc

Let us now attempt the case of a surface (disc) of charges of uniform charge density ρ_S as shown in Figure 5A.2. The disc has a radius a and is laid down in the x-y plane centered at the origin. Again, we will concern ourselves with deriving the expression for $\vec{E}$ at locations on the z axis.

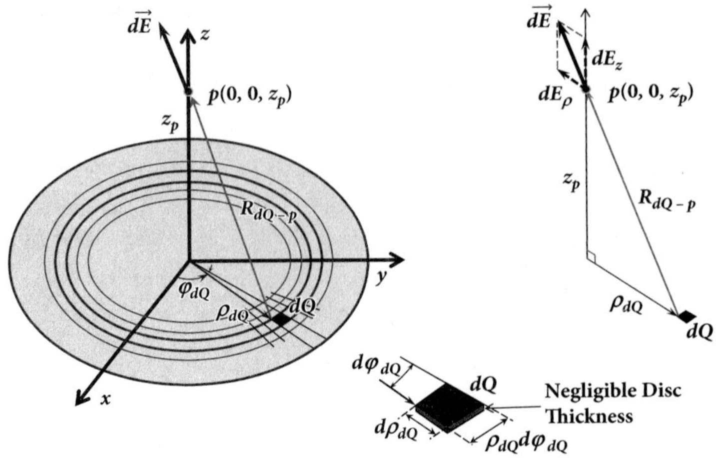

FIGURE 5A.2 The electric field due to a uniformly charged disc using the incrementation scheme.

We start by dicing up the charge into small increments. The round shape of the disc suggests using cylindrically radial lines (constant φ) and circles (constant ρ) to define the charge increments. An increment dQ located at $(\rho_{dq}, \varphi_{dQ}, 0)$ would have a radial length $d\rho_{dQ}$ and a width $\rho_{dQ}\, d\varphi_{dQ}$ along the perimeter of the radius ρ_{dQ} circle Figure 5A.3.

Assuming the total disc charge is Q, the surface density of the uniformly distributed charge is $\rho_S = \dfrac{Q}{\pi a^2}$. The charge increment dQ would be given by $dQ = \rho_S(d\rho_{dQ})(\rho_{dQ}\, d\varphi_{dQ})$.

Similar to what we did in the ring case, we use the surface charge portion of the expression in Equation (5.16), hence we write

$$\vec{E} = \iint \frac{\rho_S dS}{4\pi\varepsilon_o R^2}\vec{a}_R \tag{5A.11}$$

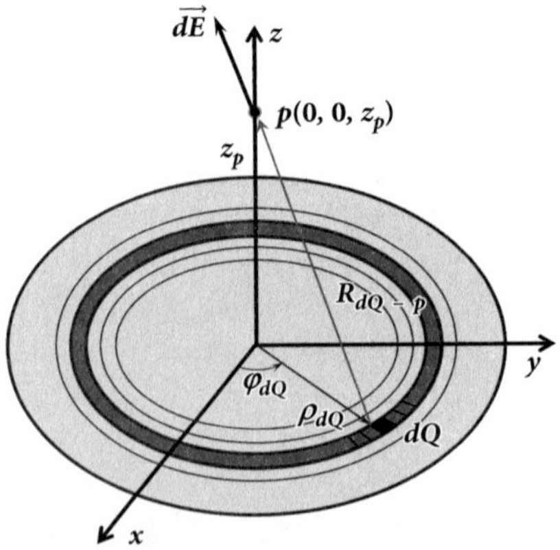

FIGURE 5A.3 The electric field due to a uniformly charged disc with the φ incrementation followed by the ρ incrementation.

Borrowing the R expression from the ring case with replacing a by ρ_{dq}.

$$\vec{R}_{dQ-p} = (0-\rho_{dQ}\cos\varphi_{dQ})\vec{a}_x + (0-\rho_{dQ}\sin\varphi_{dQ})\vec{a}_y + (z_p - 0)\vec{a}_z \tag{5A.12}$$

$$\left|R_{dQ-p}\right| = \sqrt{\rho_{dQ}^2 + z_p^2} \tag{5A.13}$$

and

$$\frac{\vec{a}_{R_{dQ-p}}}{(R_{dQ-p})^2} = \frac{\vec{R}_{dQ-p}}{\left|R_{dQ-p}\right|^3} = \frac{-\rho_{dQ}\cos\varphi_{dQ}\vec{a}_x - \rho_{dQ}\sin\varphi_{dQ}\vec{a}_y + z_p\vec{a}_z}{\left[\rho_{dQ}^2 + z_p^2\right]^{3/2}} \tag{5A.14}$$

We can now write φ_{dQ}

$$\vec{E}_p = \int_{\rho_{dQ}=0}^{\rho_{dQ}=a}\int_{\varphi_{dQ}=0}^{\varphi_{dQ}=2\pi} \frac{\rho_S(d\rho_{dQ})(\rho_{dQ}\,d\varphi_{dq})}{4\pi\varepsilon_o\left[\rho_{dQ}^2 + z_p^2\right]^{3/2}}[-\rho_{dQ}\cos\varphi_{dQ}\vec{a}_x - \rho_{dQ}\sin\varphi_{dQ}\vec{a}_y + z_p\vec{a}_z] \tag{5A.15}$$

$$\vec{E}_p = \int_{\rho_{dQ}=0}^{\rho_{dQ}=a} \frac{\rho_S(\rho_{dQ}\,d\rho_{dQ})(2\pi)z_p\vec{a}_z}{4\pi\varepsilon_o\left[\rho_{dQ}^2 + z_p^2\right]^{3/2}} \tag{5A.16}$$

$$\vec{E}_p = \frac{\rho_S z_p\vec{a}_z}{2\varepsilon_o}\left[\frac{-1}{\sqrt{\rho_{dQ}^2 + z_p^2}}\right]_{\rho_{dQ}=0}^{\rho_{dQ}=a} = \frac{\rho_S z_p}{2\varepsilon_o}\left[\frac{1}{\sqrt{z_p^2}} - \frac{1}{\sqrt{a^2 + z_p^2}}\right]\vec{a}_z \tag{5A.17}$$

Again, the cylindrical symmetry leaves only an $\vec{a}_z$ resultant.

You may have noticed that we did not reduce the Expression (5A.17) any further. The temptation is to reduce the expression to $\vec{E}_p = \frac{\rho_S}{2\varepsilon_o}\left[1 - z_p/\sqrt{a^2 + z_p^2}\right]\vec{a}_z$ in which the quotient $\frac{z_p}{\sqrt{z_p^2}}$ is reduced to 1 which is not correct. The correct reduction for this quotient is $sgn(z_p)$ or $\frac{z_p}{|z_p|}$ in which the polarity of z_p is maintained.

Hence,

$$\vec{E}_p = \frac{\rho_S z_p}{2\varepsilon_o}\left[\frac{1}{|z_p|} - \frac{1}{\sqrt{a^2 + z_p^2}}\right]\vec{a}_z \tag{5A.18}$$

5A.3 Alternative Integration Approaches to the Finite Disc Case

In the solution process above, you noticed that we completed the integration on φ_{dq} first, then the ρ_{dQ}. In fact, we could have done it the other way as well. In completing the φ_{dQ} integration first, we composed a partial answer to $\vec{E}_p$ due to a ring of charge. Continuing with the ρ_{dQ} integration yields the addition of the contributions of all the rings from $\rho_{dQ} = 0$ to $\rho_{dQ} = a$, Figure 5A.4.

If we were to do the ρ_{dQ} integration first, we would obtain a partial answer to $\vec{E}_p$ as the contribution of a wedge of charge (pizza/pie-like slice) which is to be followed by the φ_{dQ} integration to add up the contribution of all the other wedges from $\varphi_{dQ} = 0$ to $\varphi_{dQ} = 2\pi$.

One would argue that we did not need to start from scratch in this case as we already have an expression for $\vec{E}_p$ due to a ring of charges, hence we could have started with our disc cut into incremental rings

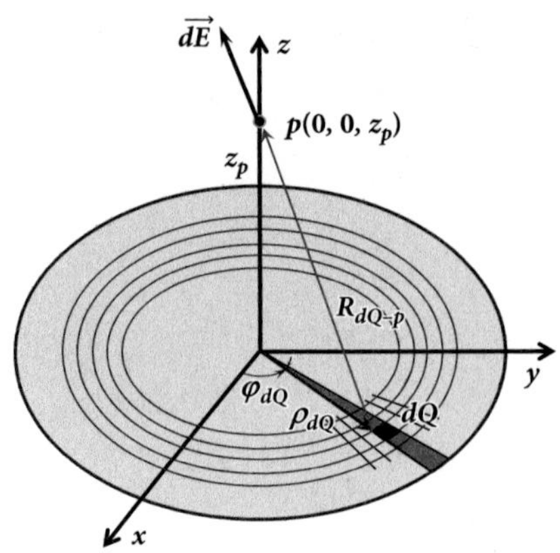

Figure 5A.4 The electric field due to a uniformly charged disc with the ρ incrementation followed by the φ incrementation.

and do one integration on the radius, Figure 5A.5. This scenario would start with the ring expression [Equation (5A.10)].

$$\vec{E}_{p-\text{ring}} = \frac{Q_{\text{ring}}\, z_p}{4\pi\varepsilon_o\left(a_{\text{ring}}^2 + z_p^2\right)^{3/2}}\vec{a}_z \tag{5A.19}$$

Apply that to the incremental ring off the disc:

$$dQ = Q_{\text{ring}} = \rho_S(\text{area of ring}) = \rho_S(2\pi\rho_{dQ}\, d\rho_{dQ}) \tag{5A.20}$$

Also,

$$a_{\text{ring}} = \rho_{dQ} \tag{5A.21}$$

hence

$$d\vec{E}_{p-dQ} = \frac{\rho_S(2\pi\rho_{dQ}\, d\rho_{dQ})z_p}{4\pi\varepsilon_o\left(\rho_{dQ}^2 + z_p^2\right)^{3/2}}\vec{a}_z \tag{5A.22}$$

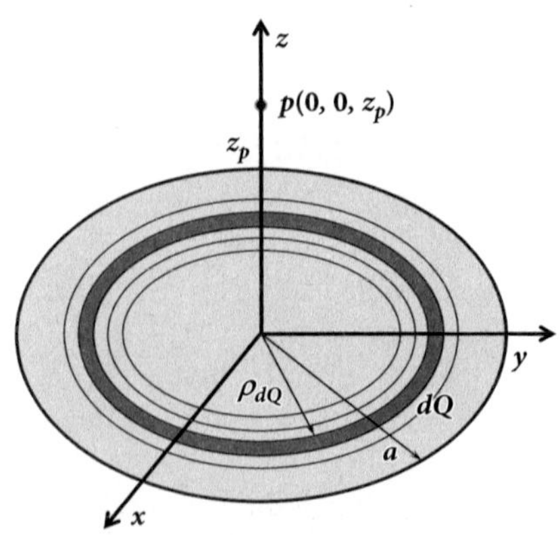

Figure 5A.5 The electric field due to a uniformly charged disc evaluated as the superposition of charged rings.

$$\vec{E}_p = \int_{\rho_{dQ}=0}^{\rho_{dQ}=a} \frac{\rho_S(\rho_{dQ}\, d\rho_{dQ}) z_p \vec{a}_z}{2\varepsilon_o\left(\rho_{dQ}^2 + z_p^2\right)^{3/2}} \tag{5A.23}$$

which is identical to Equation (5A.16).

5A.4 Charges Distributed Uniformly in an Infinitely Extended Sheet of Charges (Figure 5A.6)

It is interesting to observe that the case of an infinite sheet of charges that was analyzed in Chapter 4 is identical to that of the disc of charges of Figure 5A.4 while extending its radius (a) to infinity (Figure 5A.7).

Hence, using the disc result of Equation (5A.18) while setting the limit of (a) to approach infinity, we can write

$$[\vec{E}_p]_{\text{Infinite Sheet of Charge}} = \frac{\rho_S}{2\varepsilon_o}\left[\frac{z_p}{|z_p|}\right]\vec{a}_z \tag{5A.24}$$

which is in full agreement with Equation (4A.43).

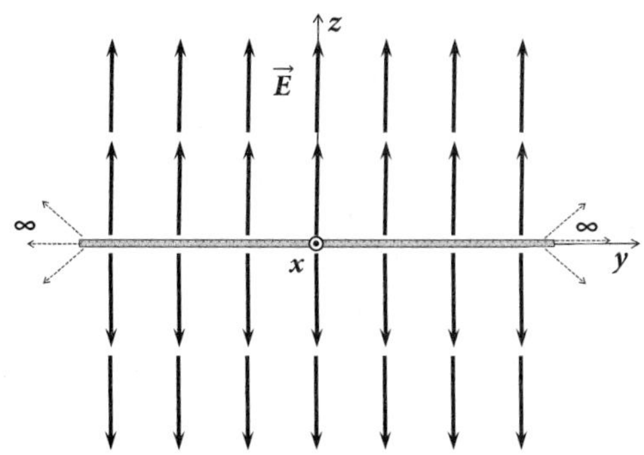

Figure 5A.6 The electric field due to a uniformly charged infinite plane.

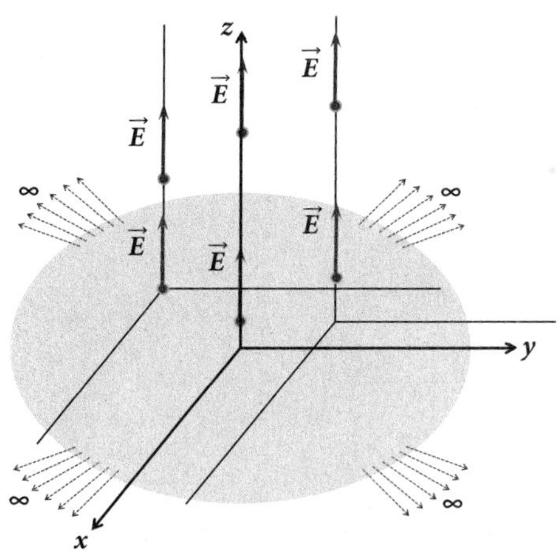

Figure 5A.7 The electric field due to a uniformly charged infinite plane using a disc of an infinite extension.

5A.5 Important Remark

The above cases are some of the relatively simple ones that can be solved by applying the Coulomb's law "incrementation" scheme. Things would get more complicated if we were to try other shapes, for example, a sphere of charges or a cylinder of charges. You may want to get a feel for such cases by trying to write out the integration on your own.

ADDENDUM 5B
ELECTRIC POTENTIAL (AND FIELD) DUE TO FAMOUS EXAMPLES OF CHARGE DISTRIBUTIONS

5B.1 Charges Distributed Uniformly in a Circular Ring

Similar to our handling of this charge distribution in Addendum 5A, solving for V at all locations of space is a bit complicated and does not serve us at this point. However, we can demonstrate solving for V at locations on the ring axis (the z axis in the figure). Either the Cartesian or the cylindrical coordinate systems can be used to develop the sought electric potential expressions. For convenience sake, we will use the same charge layout used in Addendum 5A, Figure 5B.1.

Similar to what we did for the $\vec{E}$ evaluation, we start by dicing the ring of charges into infinitesimal increments. Next, we write the potential expression one of the first quadrant's charge increments:

$$dQ = \rho_\ell \, d\ell = \frac{Q}{2\pi a} a \, d\varphi \tag{5B.1}$$

Next, we choose a general location point on the z axis for electric potential evaluation; for example, the point p located on the positive side of the axis. The cylindrical coordinates for this point are ($\rho = 0$, φ is irrelevant, $z = z_p$).

Now, we use the integral form of Equation (5.70) to write

$$V_p = \int_{\text{all}\,Q} \frac{dQ}{4\pi\varepsilon_o R_{dQ-p}} = \int_{\text{all}\,Q} \frac{\rho_\ell \, d\ell}{4\pi\varepsilon_o R_{dQ-p}} \tag{5B.2}$$

where

$$R_{dQ-p} = \sqrt{a^2 + z_p^2} \tag{5B.3}$$

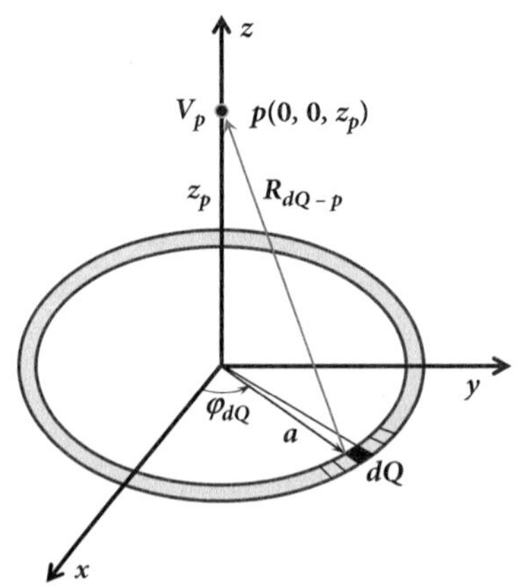

FIGURE 5B.1 The electric potential due to a uniformly charged ring.

It is obvious that the all the denominator terms in these integrations are constants and do not change from one charge increment to the next, hence

$$V_p = \frac{1}{4\pi\varepsilon_o R_{dQ-p}} \int_{\text{all}Q} \rho_\ell \, d\ell = \frac{Q}{4\pi\varepsilon_o R_{dq-p}} = \frac{Q}{4\pi\varepsilon_o \sqrt{a^2 + z_p^2}} \tag{5B.4}$$

Now that we have an expression for V_p, can we use that to derive its gradient to get the electric field vector E? The answer is yes and *NO*. As we recall, we derived the expression for V at a special location and not a general location point. This lack of generalization made the derived expression confined to variations with z only and does not contain information about x and y variations (only $x = y = 0$). Hence, the obtained V_p expression can be used to derive gradients with respect to z only, and consequently, using Equation (5.64), we can write

$$\vec{E} = -\vec{\nabla} V \tag{5B.5}$$

where the gradient in Cartesian coordinates is given by

$$E_x = -\frac{\partial V}{\partial x}, \qquad E_y = -\frac{\partial V}{\partial y} \tag{5B.6}$$

(Available V_p expression does not have x and y dependence information.) On the other hand, we have z dependence information:

$$E_z = -\frac{\partial V}{\partial z} = -\frac{\partial}{\partial z}\left[\frac{Q}{4\pi\varepsilon_o\sqrt{a^2+z^2}}\right] = \left[\frac{Qz}{4\pi\varepsilon_o(a^2+z^2)^{3/2}}\right] \tag{5B.7}$$

which is identical to the expression obtained earlier in Addendum 5A, Equation (5A.10).

A common error that some may do is to attempt the use of the full gradient approach in this case and arrive at zero answers for E_x and E_y [which are coincidentally zeros as we have seen in Equation (5A.10).] This pure coincidence does not justify the wrong approach.

Comparing the development effort between the evaluation of $\vec{E}$ first then obtaining V though integrating $\vec{E}.\vec{d\ell}$ and deriving V first then differentiating to obtain $\vec{E}$, it is obvious that the latter is much easier. Obvious constraints include the nature of the problem and whether or not we can get a generalized form for the potential function.

5B.2 Charges Distributed Uniformly in a Circular Disc (Figure 5B.2)

Similar to what we did in the ring case, we integrate the contributions of the incremental charges using Equation (5.69), hence we write

$$V_p = \iint \frac{\rho_S \, dS}{4\pi\varepsilon_o R_{dq-p}} \tag{5B.8}$$

$$R_{dq-p} = \sqrt{\rho_{dq}^2 + z_p^2} \tag{5B.9}$$

hence

$$V_p = \int_{\rho_{dQ}=0}^{\rho_{dQ}=a} \int_{\varphi_{dQ}=0}^{\varphi_{dQ}=2\pi} \frac{\rho_S (d\rho_{dQ})(\rho_{dQ} \, d\varphi_{dQ})}{4\pi\varepsilon_o\sqrt{\rho_{dQ}^2 + z_p^2}} \tag{5B.10}$$

$$V_p = \left[\frac{\rho_S(2\pi)\left(\sqrt{\rho_{dQ}^2 + z_p^2}\right)}{4\pi\varepsilon_o}\right]_{\rho_{dQ}=0}^{\rho_{dQ}=a} = \frac{\rho_S}{2\varepsilon_o}\left[\sqrt{a^2+z_p^2} - \sqrt{z_p^2}\right] \tag{5B.11}$$

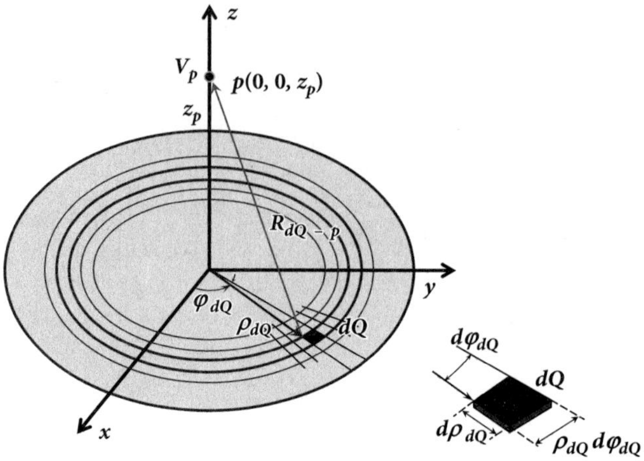

Figure 5B.2 The electric potential due to a uniformly charged disc.

Again, we can only get $\vec{E}_z$ from this result,

$$E_z = -\frac{\partial V}{\partial z} = -\frac{\partial}{\partial z}\left[\frac{\rho_S}{2\varepsilon_o}\left[\sqrt{a^2+z^2}-\sqrt{z^2}\right]\right] = \frac{\rho_S}{2\varepsilon_o}\left[1-\frac{z}{\sqrt{a^2+z^2}}\right] \tag{5B.12}$$

which is identical to Equation (5A.17).

5B.3 Electric Dipole (Field and Potential)

This is an important case not just to practice using the Coulomb and Gauss tools, but because it allows us to understand the properties of the electric dipole. As will be discussed next chapter, charge polarization of material results in the formation of electric dipoles, which in turn influence the electric field and flux distributions.

An electric dipole is a pair of equal yet opposite charges (+/− q) separated by a distance (d). The dipole is typically denoted by a dipole moment vector (Figure 5B.3).

$$\vec{p}_{\text{dipole}} = q\vec{d} \tag{5B.13}$$

Typically, this dipole moment is formed on the atomic or molecular level, and our interest is in its contribution to bulk behavior of the matter. In other words, we are interested in developing expressions for the dipole moment's contribution to the electric potential, field, and flux at distances much larger than its size, referring to Figure 5B.4, $r >> d$.

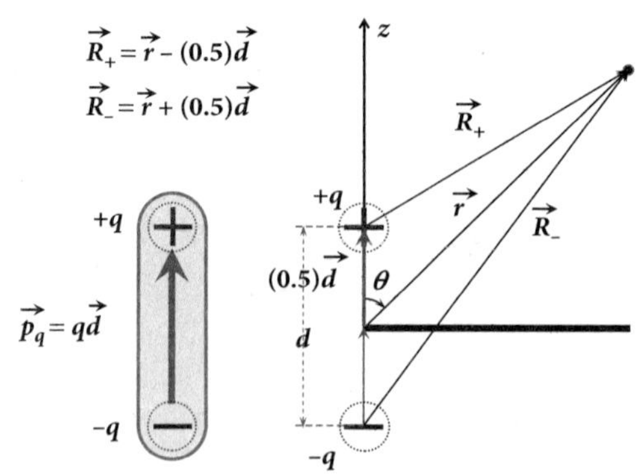

Figure 5B.3 The electric dipole of charges.

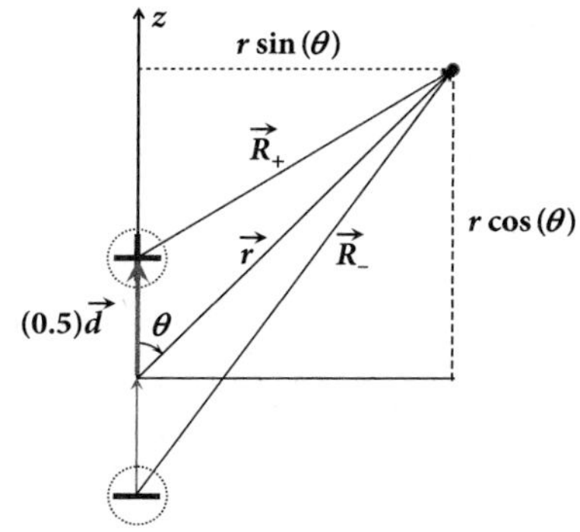

FIGURE 5B.4 The electric field and potential of a dipole of charges.

Deriving the electric potential first in this case would be much easier than starting with the electric field vector. With two point-charges in the dipole, the expression for V can be written as

$$V = \frac{+q}{4\pi\varepsilon_o R_+} + \frac{-q}{4\pi\varepsilon_o R_-} \tag{5B.14}$$

$$R_+ = \sqrt{[r\sin\theta]^2 + [r\cos\theta - d/2]^2} = \sqrt{[r]^2 + [d/2]^2 - rd\cos\theta} \tag{5B.15}$$

$$R_- = \sqrt{[r\sin\theta]^2 + [r\cos\theta + d/2]^2} = \sqrt{[r]^2 + [d/2]^2 + rd\cos\theta} \tag{5B.16}$$

Hence the electric potential of Equation (5B.14) takes the form

$$V = \frac{+q}{4\pi\varepsilon_o}\left[\frac{1}{\sqrt{[r]^2 + [d/2]^2 - rd\cos\theta}} - \frac{1}{\sqrt{[r]^2 + [d/2]^2 + rd\cos\theta}}\right] = \frac{+q}{4\pi\varepsilon_o}\left[\frac{1}{\sqrt{-}} - \frac{1}{\sqrt{+}}\right] \tag{5B.17}$$

where

$$\sqrt{-} = \sqrt{[r]^2 + \left[\frac{d}{2}\right]^2 - rd\cos\theta} \quad \text{and} \quad \sqrt{+} = \sqrt{[r]^2 + \left[\frac{d}{2}\right]^2 + rd\cos\theta} \tag{5B.18}$$

Next, we derive an expression for the electric field intensity vector by applying the gradient to Equation (5B.17),

$$\vec{E} = -\vec{\nabla}V = -\left[\frac{\partial}{\partial r}\vec{a}_r + \frac{1}{r}\frac{\partial}{\partial\theta}\vec{a}_\theta + \frac{1}{r\sin\theta}\frac{\partial}{\partial\varphi}\vec{a}_\varphi\right]V = -\frac{+q}{4\pi\varepsilon_o}\left[\frac{\partial}{\partial r}\left[\frac{1}{\sqrt{-}} - \frac{1}{\sqrt{+}}\right]\vec{a}_r + \frac{1}{r}\frac{\partial}{\partial\theta}\left[\frac{1}{\sqrt{-}} - \frac{1}{\sqrt{+}}\right]\vec{a}_\theta + 0\vec{a}_\varphi\right]$$

$$= -\frac{+q}{4\pi\varepsilon_o}\left\{\left[\frac{-0.5(2r - d\cos\theta)}{(\sqrt{-})^3} - \frac{-0.5(2r + d\cos\theta)}{(\sqrt{+})^3}\right]\vec{a}_r + \frac{1}{r}\left[\frac{rd\sin\theta}{(\sqrt{-})^3} - \frac{-rd\sin\theta}{(\sqrt{+})^3}\right]\vec{a}_\theta\right\} \tag{5B.19}$$

$$\vec{E} = -\frac{+q}{4\pi\varepsilon_o}\left\{-r\left[\frac{1}{(\sqrt{-})^3} - \frac{1}{(\sqrt{+})^3}\right]\vec{a}_r + [d\cos\theta\vec{a}_r + d\sin\theta\vec{a}_\theta]\left[\frac{1}{(\sqrt{-})^3} + \frac{1}{(\sqrt{+})^3}\right]\right\} \tag{5B.20}$$

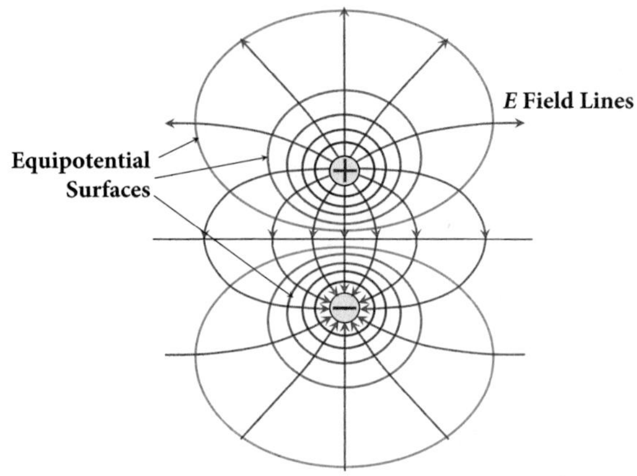

FIGURE 5B.5 Field lines and equipotential surfaces for a dipole of charges.

The following Figure 5B.5 is a plot of the equipotential lines (V = constant) and the electric field lines for the dipole.

Now, we turn our attention to the case of $r >> d$, Equations (5B.15) and (5B.16) approximates to

$$R_+ = \sqrt{[r]^2 + [d/2]^2 - rd\cos\theta} \cong r - 0.5d\cos\theta \tag{5B.21}$$

$$R_- = \sqrt{[r]^2 + [d/2]^2 + rd\cos\theta} \cong r + 0.5d\cos\theta \tag{5B.22}$$

These two approximations for R_+ and R_- could have been arrived at using trigonometric/graphical methods. Referring to Figure 5B.6, since $r >> d$, the r, R_+, and R_- are approximated as parallel lines. Projecting r on R_+ and R_-, we can see that r is larger than R_+ by $(d/2)\cos\theta$ while it is smaller than R_- by the same amount.

In this case, Equation (5B.17) reduces to

$$V \cong \frac{+q}{4\pi\varepsilon_o}\left\{\frac{1}{r - 0.5d\cos\theta} - \frac{1}{r + 0.5d\cos\theta}\right\} \cong \frac{+qd\cos\theta}{4\pi\varepsilon_o[r^2 - (0.5d\cos\theta)^2]} \tag{5B.23}$$

or

$$V \cong \frac{+qd\cos\theta}{4\pi\varepsilon_o r^2} \tag{5B.24}$$

and the expression for the electric field intensity vector can be obtained by applying the gradient to Equation (5B.24), thus

$$\vec{E} = -\vec{\nabla}V = -\left[\frac{\partial}{\partial r}\vec{a}_r + \frac{1}{r}\frac{\partial}{\partial\theta}\vec{a}_\theta + \frac{1}{r\sin\theta}\frac{\partial}{\partial\varphi}\vec{a}_\varphi\right]V = \left[\frac{2qd\cos\theta}{4\pi\varepsilon_o r^3}\vec{a}_r + \frac{qd\sin\theta}{4\pi\varepsilon_o r^3}\vec{a}_\theta + 0\vec{a}_\varphi\right] \tag{5B.25}$$

$$\vec{E} = \frac{qd}{4\pi\varepsilon_o r^3}[2\cos\theta\vec{a}_r + \sin\theta\vec{a}_\theta] \tag{5B.26}$$

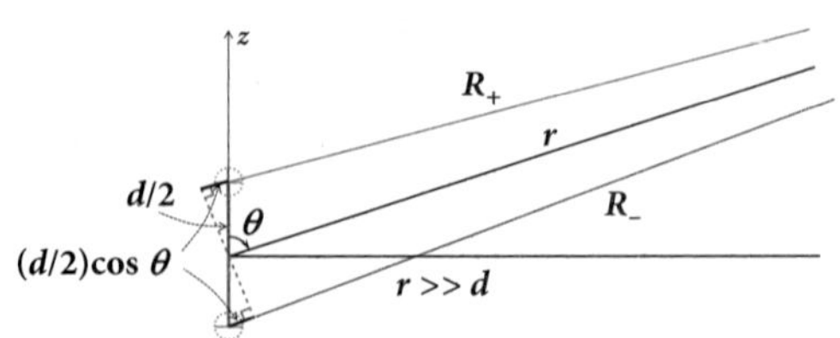

FIGURE 5B.6 Far field analysis of the electric field and electric potential for a dipole of charges.

CHAPTER 5 PROBLEMS

- Problems 5.1 to 5.18: Problems dealing with finding E and V for a given charge distribution
- Problems 5.19 to 5.24: Problems dealing with finding V given E
- Problems 5.25 to 5.28: Problems dealing with work and potential
- Problems 5.29 to 5.31: Problems dealing with finding the flux on a surface for a given potential distribution
- Problems 5.32 to 5.37: Problems dealing with finding E for a given potential distribution

5.1 A particle has a charge –3 nC. (a) Find the magnitude and direction of the electric field due to this particle at a point located a distance 0.25 m directly above it. (b) At what distance from this particle does its electric field have a magnitude of 12 N/C?

5.2 A point charge ($q_1 = -4$ nC) is at the point $A(0.6, 0.8)$ m, and a second point charge ($q_2 = +6$ nC) is at the point $B(0.6, 0)$ m. Calculate the magnitude and direction of the net electric field at the origin due to these two point charges.

5.3 Using incrementation, derive an expression for the electric field vector at point $P(\rho, \varphi, 0)$ due to a uniformly distributed line charge with linear charge density ρ_ℓ lying on the z axis between $z = -h$ and $z = +h$.

5.4 Using incrementation, derive an expression for the electric field vector at point $P(0, 0, z)$ due to a uniformly distributed line charge with linear charge density ρ_ℓ in the form of a ring of radius a lying in the x-y plane.

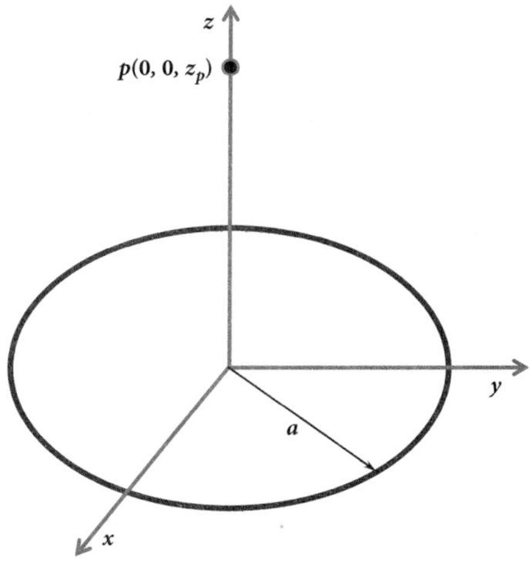

FIGURE 5P.4

5.5 Consider a circular ring in the x-y plane between r_1 and r_2 as shown. The ring extending from ρ_1 to ρ_2 has a surface charge density, ρ_S, uniformly distributed on the surface of the ring. Find the electrostatic potential, V, at any point, p, on the z axis above the x-y plane.

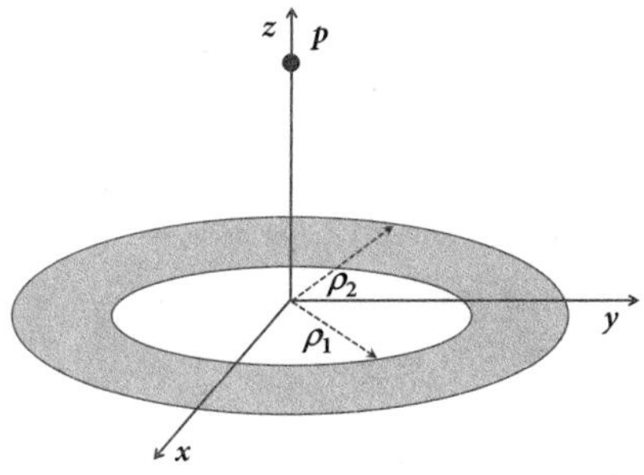

FIGURE 5P.5

5.6 From the two shown charged rings, each has a charge of $+Q$ and radius a. Find the net electric field at the point p, which is located at a distance a from the center of each ring.

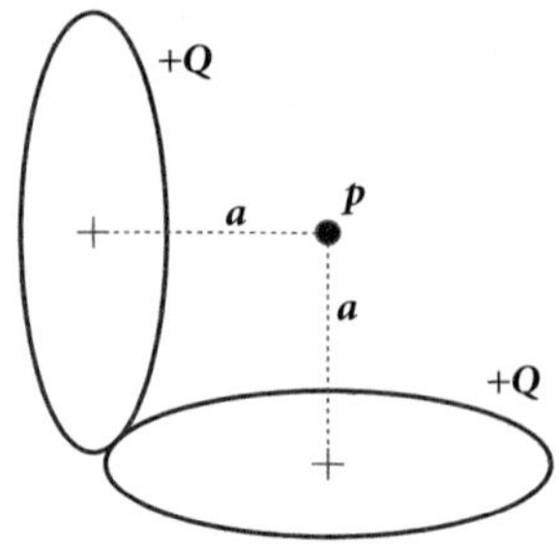

FIGURE 5P.6

5.7 Using incrementation, derive an expression for the electric field vector at point $P(0, 0, z)$ due to a uniformly distributed line charge with linear charge density ρ_ℓ in the form of a half ring of radius a lying in the x-y plane.

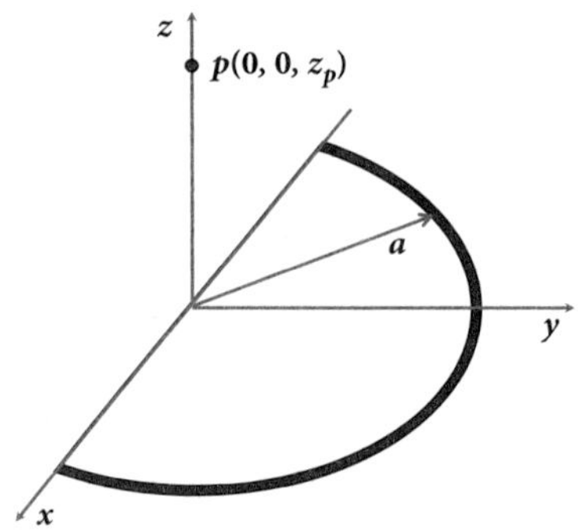

FIGURE 5P.7

5.8 Derive an expression for the electric potential function at points on the z axis due to a uniformly distributed line charge in the form of a quarter circle, radius a, lying in the x-y plane with its center located at the origin. The total charge is Q and is located between $\varphi = \frac{\pi}{2}$ and $\varphi = \pi$.

5.9 Derive an expression for the electric potential function at points on the z axis due to a uniformly distributed line charge in the form of a half a circle lying in the x-y plane and centered at the origin. The circle radius is a, and its linear charge density is ρ_ℓ. The circular charge starts at $\varphi = 0$ and ends at $\varphi = \pi$.

- Repeat if charge distribution above is expanded to cover half a disk of radius a between $\varphi = 0$ and $\varphi = \pi$ with uniform surface density.

5.10 Two cylindrical conductors with infinite extension along the z axis (normal to the page), the conductor radii are given by a, b, and c as shown. The inner conductor was charged with a linear charge density ρ_ℓ C/m along the z axis. As a result, $-\rho_\ell$ C/m is now uniformly distributed on the inner surface of the outer conductor and $+\rho_\ell$ C/m is uniformly distributed on its outer surface.

a. What is the electric field intensity vector in all regions: within inner conductor, within dielectric, within outer conductor, and outside outer conductor.

b. Find the potential differences V_{ab} and V_{12}, where $P_1\left(\rho_1 = \frac{a}{2}\right)$, and $P_2\left(\rho_2 = \frac{(b+c)}{2}\right)$.

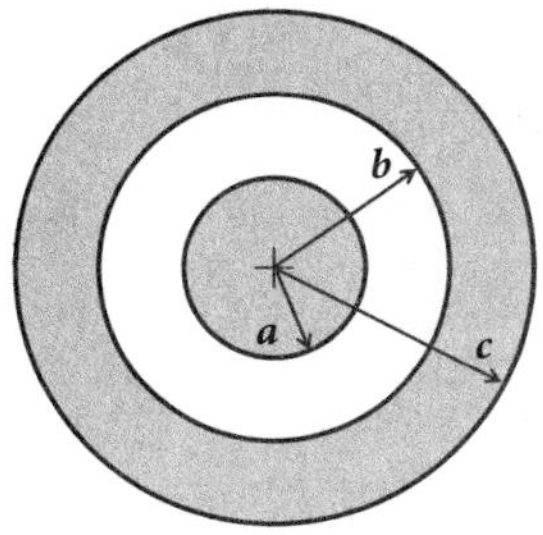

FIGURE 5P.10

5.11 Two concentric spherical surfaces of radii R_1 and R_2 have uniformly distributed surface charge densities ρ_{S1} C/m^2 and ρ_{S2} C/m^2, respectively, as shown in the figure. The region between $r = 0$ and $r = R_1$ is filled with air. The region between $r = R_1$ and $r = R_2$ is filled with a dielectric with permittivity, ε, and $R_2 > R_1$. Determine the following:

The electric field intensity for $r < R_1$, for $R_1 \leq r < R_2$, and for $r \geq R_2$.

a. The electrostatic potential at the center.

b. The relationship between ρ_{S1} and ρ_{S2} so that the electrostatic field outside $r = R_2$ vanishes, find the electrostatic potential of the surface with radius R_2 in this case.

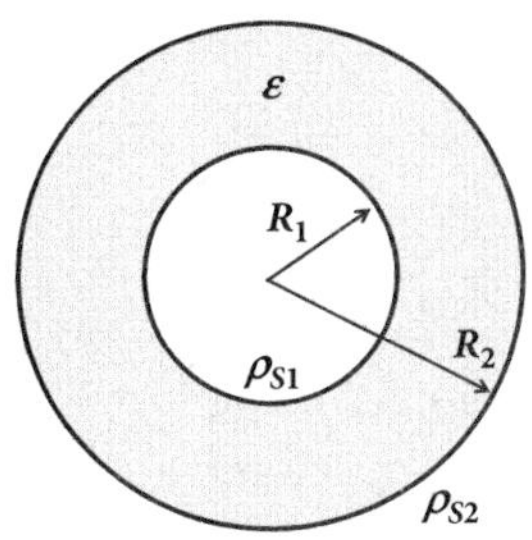

FIGURE 5P.11

5.12 Find an expression for the electric field generated from a metallic sphere of charge $-Q$ and of radius a at a point located at a distance r from the center of the sphere in different cases "inside $r < a$, outside $r > a$, and on the surface of the sphere $r = a$". Then, draw the distribution of electric field versus the distance r.

5.13 A charge Q is uniformly distributed along a rod of length L and negligible cross section. Find the potential and electric field vector at all points in the orthogonal plane passing by the mid-point of the rod.

5.14 Find the force on a 1 C charge located at the origin due to two point-charges of –2 C each located at (2,1,0) m and (1,–2,0) m.

5.15 Three infinite line charges with linear charge densities $\rho_\ell = 10\ \mu$C/m each are placed on the three coordinate axes x, y, and z. Find the electric field and potential at the point (1,1,1) m.

5.16 An infinite plane of surface charge density 5 μC m^{-2} is located in the x-y plane.

a. Find the potential difference V_{AB} given the coordinates of A and B in cm as (1,3,2) and (0,0,5), respectively. Also, find the work done to move a point charge of –3 μC from A to B.

b. Repeat for the coordinates (1,3,–2) and (4,–3,–2), respectively.

c. Repeat for the coordinates (1,3,–2) and (0,0,5), respectively.

5.17 A thin insulating shell of radius a carries a total charge Q distributed uniformly over its surface. Derive expressions for the potential function at all points inside and outside the surface.

5.18 Consider two parallel charged (infinitely thin) wires with charges $\rho_{\ell 1}$ and $\rho_{\ell 2}$ per unit length at a distance d from each other. Determine the condition on $\rho_{\ell 1}$ and $\rho_{\ell 2}$ and d that has to be satisfied in order to get cylinders among the equipotential surfaces of these charges. Find the location of these cylindrical equipotential surfaces.

5.19 For the electric field: $\vec{E}=5\vec{a}_x-3\vec{a}_y+2|z|\vec{a}_z$, find the potential difference between

a. Cartesian coordinate points $P_1(-1,0,1.5)$ and $P_2(1,2.5,-1.5)$

b. Cylindrical coordinate points $P_3\left(\rho=2.4,\ \varphi=\frac{\pi}{2}, z=3\right)$, and $P_3\left(\rho=0.4,\ \varphi=\frac{3\pi}{2}, z=-3\right)$

5.20 Find V_{ab} given $\vec{E}=3\vec{a}_\rho+5\vec{a}_z$. The points a and b are given in cylindrical coordinates as $a\left(1,\frac{\pi}{2},1\right)$ and $b\left(3,\frac{-\pi}{2},-2\right)$.

5.21 Given that $\vec{E}=\left(\frac{\alpha}{\rho}\right)\vec{a}_\rho$, $a<\rho<b$, and $\vec{E}=0$ otherwise, find the potential differences V_{ab} and V_{12}, where $P_1\left(\rho_1=\frac{a}{2}\right)$, and $P_2(\rho_2=2b)$.

5.22 Given that $\vec{E}=\left(\frac{\alpha\rho}{a}\right)\vec{a}_\rho$ for $\rho<a$, $\vec{E}=\left(\frac{\alpha a}{\rho}\right)\vec{a}_\rho$ for $a<\rho<b$, and $\vec{E}=0$ otherwise, find the potential differences V_{ab} and V_{12}, where $P_1\left(\rho_1=\frac{a}{2}\right)$, and $P_2(\rho_2=2b)$.

5.23 Given that $\vec{E}=0\vec{a}_\rho$, $\rho<1$, $\vec{E}=\left(\frac{5}{\rho}\right)\vec{a}_\rho$, $1<\rho<2$, and $\vec{E}=0$, $\rho>2$, find the potential differences V_{ab}, where $P_a\left(\rho=2.4,\ \varphi=\frac{\pi}{2}, z=3\right)$, and $P_b\left(\rho=0.4,\ \varphi=\frac{3\pi}{2}, z=-3\right)$.

5.24 Given that $\vec{E}=0\vec{a}_r$, $r<1$, $\vec{E}=\left(\frac{5}{r^3}\right)\vec{a}_r$, $1<r<2$, and $\vec{E}=0$, $r>2$, find the potential differences V_{ab}, where $P_a\left(r=2.4, \theta=\frac{\pi}{2},\ \varphi=\frac{\pi}{2}\right)$, and $P_b\left(r=0.4, \theta=\frac{-\pi}{2},\ \varphi=\frac{3\pi}{2}\right)$.

5.25 How much work is needed to assemble an atomic nucleus containing three protons (such as Lithium). We model it as an equilateral triangle of side 2×10^{-15} m with a proton at each corner [$q_p=1.6\times10^{-19}$ C]. What is the net electric potential at the center of the triangle?

5.26 Using the work done approach, derive an expression for the potential difference between points P_1 and P_2 due to the electric field intensity vector given by $E_r=0$ for $r<a$, $E_r=\frac{\alpha}{r}$ for $a<r<b$, and $E_r=0$ for $r>b$.

5.27 Using the work done approach, derive an expression for the potential difference between points P_1 and P_2 due to the electric field intensity vector given by $E_r = 0$ for $r < a$, $E_r = \frac{\alpha}{r}$ for $a < r < b$, and $E_r = 0$ for $r > b$.

a. $P_1\left[r = \frac{a}{2}, \varphi = 0, z = -a\right]$ and $P_2\left[r = 2b, \varphi = 0, z = \frac{(a+b)}{2}\right]$

b. $P_1\left[r = \frac{a}{2}, \varphi = 0, z = 0\right]$ and $P_2[r = 2b, \varphi = 0, z = 0]$

c. $P_1\left[r = \frac{a}{2}, \varphi = \pi, z = 0\right]$ and $P_2[r = 2b, \varphi = 0, z = 0]$

5.28 Using the work done approach, derive an expression for the potential difference between points P_1 and P_2 due to the electric field intensity vector given by $E_r = 0$ for $r < a$, $E_r = \frac{\alpha}{r^2}$ for $a < r < b$, and $E_r = 0$ for $r > b$.

a. $P_1\left[r = \frac{a}{2}, \varphi = \frac{-\pi}{3}, z = \frac{3\pi}{4}\right]$ and $P_2\left[r = 2b, \varphi = \frac{\pi}{2}, z = \frac{-\pi}{3}\right]$

b. $P_1\left[r = \frac{(a+b)}{2}, \varphi = \frac{\pi}{2}, z = \frac{-\pi}{3}\right]$ and $P_2\left[r = b, \varphi = \frac{\pi}{2}, z = \frac{-\pi}{3}\right]$

5.29 Find the electric flux on the outer surface of a hemisphere with radius 1 cm centered at the origin and based in the x-y plane given that the potential function covering this region is $V = 3x + y^2 + 2z$.

5.30 Find the electric flux on the closed surface of the sphere of unit radius centered at the origin given that the potential function covering this region is $V = x + 3y^2 + z^3$.

5.31 Consider infinite conducting plate separated into two halves at $y = 0$ by infinitely thin insulating line. Potential of one half is kept at $+V_o$, another at $-V_o$. Find the resulting distribution of charge on these conductors.

5.32 Find expressions for the electric field strength vector for the following potential functions:

a. $V = 5, r < 1, V = \frac{(16 - r^2)}{3}, 1 < V < 2$, and $V = \left(\frac{8}{r}\right), r > 2$,

b. $V = \frac{3\cos\theta}{r^2}$

5.33 In a certain region of space, the electric potential is given by $V = Ax^2y - By^3$ where A, and B are constants.

a. Find an expression for the x, and y components of electric field?

b. If the x component of electric field at point (x = 1 m, y = 1 m) is –4 V/m while the y component of electric field at the same point is zero, Find A, and B values?

5.34 Derive expressions for the electric field intensity vector given: $V = \frac{qd\cos\theta}{4\pi\varepsilon_o r^2}$, where r and θ are in spherical coordinates.

5.35 Starting with the potential expression for a dipole of charges, derive an expression for the electric field intensity vector. State any constraints related to these expressions.

5.36 Given the electric potential function as $V(x,y,z)=2x^2z+2y^2z-4z^3$ V with (x, y, and z are in meters)

a. Find the electric field vector $\vec{E}(x,y,z)$.

b. Find the work required to move a charge of −2 C from the origin to position (1, 2, 3).

c. Describe the equipotential surfaces in this field.

5.37 Consider the electrostatic potential function: $V(x,y,z)=\cos(x)\sin(y)\sin(z)$

a. Find V at $x=0$, $y=\pi/2$, and $z=\pi/2$, and at $x=y=z=\pi/4$.

b. Find the electric field intensity, $\vec{E}(x,y,z)$.

c. Find $\vec{E}$ at $x=\pi/2$, $y=\pi$, and $z=\pi/2$, also at $x=y=z=\pi/4$.

d. Calculate $\vec{\nabla}\times\vec{E}$, is this electric field conservative?

e. Calculate $\vec{\nabla}\cdot\vec{E}$. What physical quantity does $\vec{\nabla}\cdot\vec{E}$ measure?

CHAPTER 5 SUMMARY[1]

Coulomb's Force Law in Free Space

$$\vec{F}_2 = \frac{q_1 q_2}{4\pi\varepsilon_o R_{12}^2}\vec{a}_{12} \quad \text{and} \quad \vec{F}_1 = \frac{q_2 q_1}{4\pi\varepsilon_o R_{21}^2}\vec{a}_{21}, \quad \varepsilon_o = 8.854187817\times 10^{-12} \simeq (1/36\pi)\times 10^{-9}\ \text{F/m}$$

Electric Field Intensity Vector

$$\vec{E}_p = \vec{F}_{\text{normalized (per unit charge)}} = \frac{1}{\varepsilon_o}\vec{D}_p \quad \text{and} \quad \vec{D}_p = \varepsilon_o \vec{E}_p$$

Electric Field "Incrementation" Scheme

$$\vec{E}_p = \sum\nolimits_{\text{Point Charges}} \frac{q}{4\pi\varepsilon_o R_{q-p}^2}\vec{a}_{R_{q-p}} + \sum\nolimits_{\text{Linear Densities}} \int \frac{\rho_\ell\, d\ell}{4\pi\varepsilon_o R_{dQ-p}^2}\vec{a}_{R_{dQ-p}}$$

$$+ \sum\nolimits_{\text{Surface Densities}} \int \frac{\rho_S\, dS}{4\pi\varepsilon_o R_{dQ-p}^2}\vec{a}_{R_{dQ-p}} + \sum\nolimits_{\text{Volumetric Densities}} \int \frac{\rho_v\, dv}{4\pi\varepsilon_o R_{dQ-p}^2}\vec{a}_{R_{dQ-p}}$$

Energy of a System of Charges

$$W_{e,q,Q} = \sum\nolimits_{q,Q} \int_\infty^p \vec{F}_{\text{external}} \cdot \vec{d\ell} = \sum\nolimits_{q,Q} \int_p^\infty \vec{F}_{\text{Coulomb}} \cdot \vec{d\ell}$$

$$W_{e,q,Q} = \sum\nolimits_{q\text{'s}} q \int_p^\infty \vec{E}\cdot\vec{d\ell} + \int_{Q\text{'s}} dQ \int_p^\infty \vec{E}\cdot\vec{d\ell}$$

$$W_{e,q_1-q_2} = \frac{q_2 q_1}{4\pi\varepsilon_o R_{12}}$$

$$W_{e,q_1\cdots q_n} = \frac{1}{2}\sum\nolimits_{j=1}^{n,j\neq i} q_j \sum\nolimits_{i=1}^{n} \frac{q_i}{4\pi\varepsilon_o R_{ij}} = \frac{1}{2}\sum\nolimits_{j=1}^{n,j\neq i} q_j [V_j]$$

$$W_{e,Q} = \frac{1}{2}\int_{\text{all Q}} \left\{ dQ_j \left[\int_{\text{all Q}} \left(\frac{dQ_i}{4\pi\varepsilon_o R_{ij}} \right) \right] \right\} = \frac{1}{2}\int_{\text{vol}} \{\rho_v [V]\}\, dv$$

$$W_e = \int_{\text{vol}} \frac{1}{2}\{\vec{D}.\vec{E}\}\, dv = \int_{\text{vol}} \frac{1}{2}\varepsilon_o E^2\, dv$$

$$W_e = \frac{dW_e}{dv} = \frac{1}{2}\{\rho_v V\} = \frac{1}{2}\{\vec{D}.\vec{E}\} = \frac{1}{2}\varepsilon_o E^2$$

Electrostatic Potential

$$V_p = \frac{\Delta W_e}{\Delta Q} = \int_{\text{Position}}^\infty \vec{E}\cdot\vec{d\ell} \quad \text{and} \quad V_{BA} = V_B - V_A = \int_B^A \vec{E}\cdot\vec{d\ell} = -\int_A^B \vec{E}\cdot\vec{d\ell}$$

[1]***Important Warning to Students:*** Please do not use this summary without thorough study of the relevant material in this chapter. Using this summary and its formulas without adequate appreciation of the physics involved and conditions of validity can lead to wrong implementations and invalid conclusions.

Electric Potential "Incrementation" Scheme

$$V_{p,\text{system of charges}} = \sum\nolimits_{\text{all discrete } q\text{'s}} \frac{q}{4\pi\varepsilon_o R_{q-p}} \quad \text{or} \quad = \int_{\text{all continuous } Q\text{'s}} \frac{dQ}{4\pi\varepsilon_o R_{dQ-p}}$$

$$\vec{E} = -\vec{\nabla} V = -\text{grad}\, V$$

Cartesian, Cylindrical, and Spherical Forms for the Gradient Are

$$\vec{\nabla} = \left[\frac{\partial}{\partial x}\vec{a}_x + \frac{\partial}{\partial y}\vec{a}_y + \frac{\partial}{\partial z}\vec{a}_z\right]$$

$$\vec{\nabla} = \left[\frac{\partial}{\partial \rho}\vec{a}_\rho + \frac{1}{\rho}\frac{\partial}{\partial \varphi}\vec{a}_\varphi + \frac{\partial}{\partial z}\vec{a}_z\right]$$

$$\vec{\nabla} = \left[\frac{\partial}{\partial r}\vec{a}_r + \frac{1}{r}\frac{\partial}{\partial \theta}\vec{a}_\theta + \frac{1}{r\sin\theta}\frac{\partial}{\partial \varphi}\vec{a}_\varphi\right]$$

Conservative Electrostatic Field

The electrostatic potential is single valued and that value is unique and is solely determined by the location and not the path taken to get there

$$\oint \vec{E}\cdot\vec{d\ell} = 0$$

Electric Dipole: Far Field Electric Field and Potential

$$V \cong \frac{+qd\cos\theta}{4\pi\varepsilon_o r^2} \text{ and } \vec{E} = \frac{qd}{4\pi\varepsilon_o r^3}[2\cos\theta\,\vec{a}_r + \sin\theta\,\vec{a}_\theta].$$

CHAPTER 6

Materials: Conductors and Dielectrics

Georg Simon Ohm
(1789–1854)

Gustav Robert Kirchhoff
(1824–1887)

Chapter Learning Objectives

- Explain the physical behavior of the different categories of materials (conductors, semiconductors, insulators) when exposed to static and dynamic electric fields.
- Identify the different forms of current distributions and the corresponding current densities.
- State the continuity equation and its relationship to KCL.
- Evaluate the resistance and capacitance for different material configurations (coaxial line in particular).
- Evaluate the polarization vector and bound charge distribution for a dielectric immersed in an electric field.
- Define the physical insight in the dielectric constant as it relates to the polarizability of dielectric materials.
- Evaluate the energy stored in capacitances. Explain the physics involved.
- Identify the different boundary conditions and surface charge distributions at the different combinations of material interfaces. Emphasis on physical insight.

6.1 Introduction

In the previous chapters, we ignored the presence of material and its influence on the defined electrostatic quantities: flux, field, force, and energy. In this chapter, we will discuss the different categories of material based on their electrical properties and their influence of the mentioned electrostatic quantities. One warning here before we continue is that we will confine our discussion to a simplified categorization based on a simplified atomic model that serves our purpose in this textbook fairly well.

In this regard, we recognize in simple terms that a broad material classification is based on the nature of the atomic structure of the material and the "mobility" of its charge carriers (typically electrons). Material with mobile free charge carriers are classified as conductors while materials without mobile free charge carriers are classified as insulators or *dielectrics*. We will explain the use of the term "dielectric" later in this section.

To start, we state that the atomic structure of a material determines the electromagnetic properties of that material among its other properties. The material atoms are typically neutral in terms of its electrical charge. The atom in its simple illustration is composed of a positively charges nucleus (a cluster of protons) and a cloud of orbiting and spinning negatively charged electrons. The binding forces between the nucleus and the orbiting electrons vary from one material to another.

For solid conducting material, some of these electrons have weak bondage to the nucleus and are practically free to move within the material boundaries as *mobile* free charges. On the other hand, in insulators, all electrons have a strong bond to the nucleus and have no mobility at all. The most that can happen under the influence of an applied force is that these electrons exhibit limited displacement in accordance of the material properties and the intensity of the applied force. In this regard, the resulting displacement is similar to the mechanical strain that solid material undergoes under the influence of an applied stress.

In some cases, the displacement/strain is elastic that is completely recovered with the removal of the applied force. In other cases, some residual (plastic) displacement/strain stays after the removal of the applied force. While at certain levels of applied forces, breakdown (or yield) occurs resulting in freeing the electrons from their atomic bond in which case the insulator is considered to have failed (as an insulator). In their breakdown state, insulators typically assume new physical properties and cannot be returned to their original state.

Materials in the liquid and gas states are typically insulators; however, ionization results in charged ions that can be mobile and hence would be considered "conductors" in this case.

Another material category is semiconductors. Semiconductor materials are typically insulators. However, under certain conditions of exciting energy (heat/electric), charge carriers are mobilized resulting in constrained conduction (typically in a one-way fashion) and hence the name semiconductors. There are other forms of categorization such as superconductors, photoconductors, etc. that may be beyond the scope of this chapter.

6.2 Conductors

Examples of conductors include copper, aluminum, iron, silver, and gold to name a few. More conductor materials are given along with some of their properties in Appendix C at the end of the book. In conductors, a number of mobile free charge carriers are present with certain volumetric density in the material and are mobile to move under the influence of a driving force. The motion of these charges constitutes what is known as the conduction current.

That motion is not typically uniform or have a constant velocity like what could possibly happen with water or air currents flowing in friction-less channels. The charge carriers' motion is typically erratic as they encounter multiple collisions with the material's stationary heavy atomic structure. Hence, the mobile carriers would ascribe a motion with nonuniform velocities and nonuniform paths. From an engineering point of view, we typically consider an average *drift* velocity, v_d, for the moving charges in our analysis. This average drift velocity is determined by the applied electric field (causing the mechanical force) and a charge-carrier mobility factor, μ_c, as determined by the material atomic structure. Hence, we write

$$\text{Average drift velocity, } v_d = [sgn(q)]\mu_c E, \text{ where } sgn(q) = \left[\frac{q}{|q|}\right] \tag{6.1}$$

where q is the charge of the carrier, and $sgn(q)$ function adds the charge carrier's polarity to the direction of the drift velocity since negative charge carriers, for example, electrons, move opposite to the applied field.

6.2.1 Conductors under Static Conditions

Conductors in their steady state while being exposed to static electromagnetic fields are considered in a *static* condition. The reason is that the mobility of the free charge carriers will always level out the potential within the conductor material. With the entire conductor body assuming the same potential, no potential gradients, and hence no internal electric fields or forces would exist. Hence, the steady-state charge distribution would be *static* in the absence of internal forces. If the conductor is exposed to an external dynamic field, the resulting dynamic forces would keep the mobile charges in a dynamic motion in response to the *forced* dynamic state.

The analogy here is to zero-viscosity liquids, where the liquid *particles* have unlimited mobility. In their steady states under static forces (not varying with time), the liquid levels out. Only if the liquid is exposed to an external dynamic force that there would be liquid currents in response to the applied dynamic forces.

Consequently, conductors under static conditions are characterized by the following features:

1. The entire bulk of the conductor is equipotential (same potential) at all locations.
2. Electric fields are nulled within the conductor bulk and can only exist external to the conductor (normal to its surface as will be discussed under boundary conditions later in this chapter).
3. Charges can only exist on the surface and never within the bulk of the conductor. To elaborate, let us assume that a charge is placed within the body of a conductor under static conditions. This charge will induce an electric field within the bulk of the conductor thus disturbing its *static* zero internal field state. The induced internal field would produce internal forces on the conductor's mobile free charge carriers forcing their motion. The resulting motion will be bound by the conductor body boundaries and hence the added charge will be forced to the conductor surface. The steady-state *final* result will be reached when the conductor surface charges reach a distribution that produces zero internal fields in which case there will be no further motion of charges within the body of the conductor.

6.2.2 Conductors under Dynamic Conditions

Conductors exposed to time varying electromagnetic fields are considered under dynamic conditions whether the variations are continuous or only transients (or transitions) between steady states. Moreover, conductors exposed to static fields are considered under dynamic conditions as well if the conductor is connected as part of a closed electric circuit with an electro-motive-force, emf, source (sustained potential difference).

Conductors under dynamic conditions are characterized by the following features:

1. Charges flow in the body of the conductor causing an electric current.
2. An internal electric field sustains the motion of the charge carriers to overcome the conductor's internal resistance due to the carriers' multiple collisions (resistance will be discussed later in this chapter).
3. A potential difference appears across the terminals of the conductor as determined by the current flow/internal electric field.

Obviously, in the absence of an applied field, the average drift velocity of the charge carriers is zero, and hence the conductor is considered in a static mode. Consequently, conductors under static conditions do not sustain any electric fields throughout the body of their material.

6.3 Electric Current and Current Densities

The electric current is an expression of the flow (motion) of the electric charge. Hence, the nature and distribution of the electric current depends on the nature of the charge distribution and its motion. A point charge in motion will result in a line current (straight or otherwise), while a string of charges may cause a surface current, and in general, a surface or volume charge distribution in motion would result in a volume current. The electric current flowing across a given boundary (line or surface) is the rate of charge flow across that boundary. Table 6.1 and Figure 6.1 demonstrate these various cases.

At this point, we can take advantage of the similarity between the definition of the electric current density vector and the electric flux density vector presented earlier in Chapter 4, Equation (4.3). This similarity

	Line Current	Surface (Sheet) Current	Volume Current	Units
Current, I	$I = \frac{dQ}{dt}$	$I = \frac{dQ}{dt}$	$I = \frac{dQ}{dt}$	Ampere, A
Linear density vector, $\vec{K}$	∞	$\vec{K} = \frac{dI}{d\ell_n}\vec{a}_I$	NA	A/m
Surface density vector, $\vec{J}$	∞	∞	$\vec{J} = \frac{dI}{dS_n}\vec{a}_I$	A/m^2

TABLE 6.1 Different Forms of Current Distributions and Related Densities

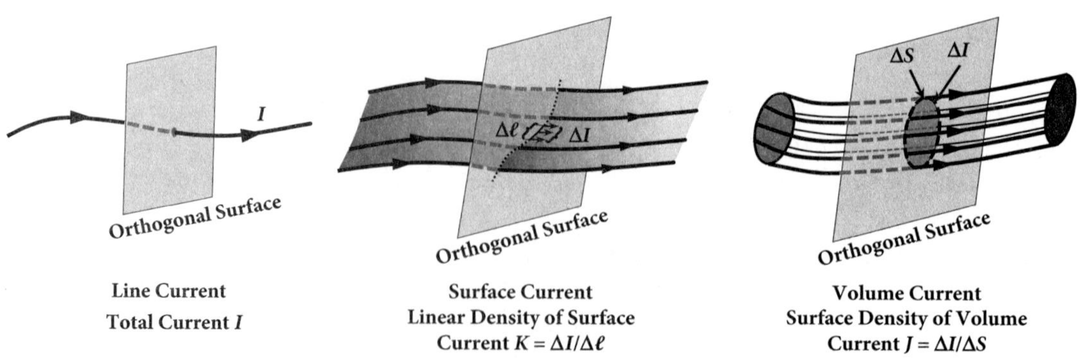

FIGURE 6.1 Different forms of current distributions and related densities.

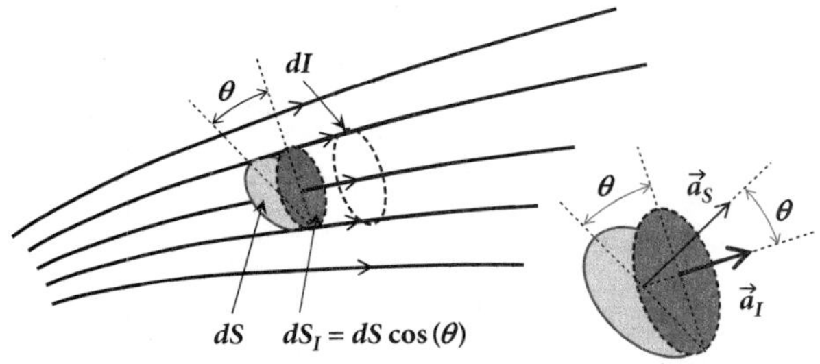

FIGURE 6.2 The current density is defined as the amount of current per unit normal area.

exists for the surface density of the volume current only as we do not have analogous definitions for either a line flux or a surface flux. Hence, referring to Figure 6.2, we can write

$$dI = [J][dS_I] = [J][dS]\cos\theta = [J\vec{a}_I]\cdot[dS\vec{a}_S] = \vec{J}\cdot\overrightarrow{dS} \tag{6.2}$$

and the total current on any surface S would be the result of integrating the incremental amounts dI over the surface area:

$$I = \iint_S dI = \iint_S \vec{J}\cdot\overrightarrow{dS} \tag{6.3}$$

6.4 The Continuity Equation

In the case of a closed surface, the current flowing out of the closed surface is the rate of charge flow out of the closed surface, which in turn, and according to charge conservation, is equal to the rate of decrease of the charge enclosed by the surface (Figure 6.3).

$$I_{\text{closed surface } S} = \oiint_S \vec{J}\cdot\overrightarrow{dS} = \frac{dQ_{\text{flowing out of } S}}{dt} = -\frac{dQ_{\text{enclosed by } S}}{dt} \tag{6.4}$$

Consequently, the "integral" form of the "continuity equation" is

$$\{I_{\text{closed surface } S}\} = \oiint_S \vec{J}\cdot\overrightarrow{dS} = -\frac{d\left[\iiint_{\text{enclosed volume by } S}\rho_v\,dv\right]}{dt} = \iiint_{\text{enclosed volume by } S}\left[-\frac{\partial\rho_v}{\partial t}\right]dv \tag{6.5}$$

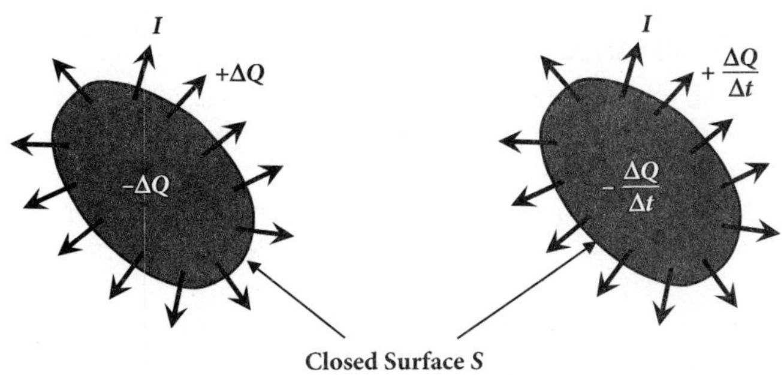

FIGURE 6.3 The continuity equation: current flowing out of a closed surface equals the rate of decrease of the charge enclosed within.

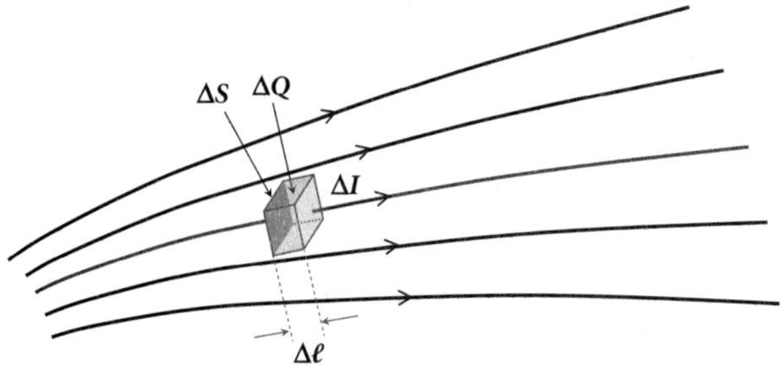

FIGURE 6.4 The differential form of the continuity equation.

and using the divergence theorem,

$$\lim_{\Delta v \to 0}\left[\frac{\oiint_S \vec{J}\cdot d\vec{S}}{\Delta v}\right] = \vec{\nabla}\cdot\vec{J} \tag{6.6}$$

and applying the integral form of the continuity equation to an incremental volume Δv, the point "differential" form of the continuity equation takes the shape:

$$\vec{\nabla}\cdot\vec{J} = \lim_{\Delta v \to 0}\left[\frac{\oiint_S \vec{J}\cdot d\vec{S}}{\Delta v}\right] = \lim_{\Delta v \to 0}\left[\frac{\iiint_{\text{enclosed volume by } S}\left[-\frac{\partial \rho_v}{\partial t}\right]dv}{\Delta v}\right] = -\frac{\partial \rho_v}{\partial t} \tag{6.7}$$

or

$$\vec{\nabla}\cdot\vec{J} = -\frac{\partial \rho_v}{\partial t} \tag{6.8}$$

Referring to Figure 6.4, let us denote the amount of charge flowing through an incremental area ΔS during a time increment Δt as ΔQ. During that Δt, ΔQ has moved a distance $\Delta\ell$ due to the charge carrier velocity v_ℓ, consequently, the resulting current ΔI is given by

$$\Delta I = \frac{\Delta Q}{\Delta t} = \frac{\rho_v \Delta S \Delta\ell}{\Delta t} = \rho_v \Delta S\,\frac{\Delta\ell}{\Delta t} = \rho_v \Delta S v_\ell \tag{6.9}$$

Hence

$$J = \frac{\Delta I}{\Delta S} = \rho_v v_\ell \quad \text{and in vector form,} \quad \vec{J} = \rho_v \vec{v}_\ell \tag{6.10}$$

6.5 Conductivity and Resistance

Considering the general case of both positive and negative mobile charge carriers (Figure 6.5) under the force of an applied electric field vector $\vec{E}$. Both charges will move at differing drift velocities (v_d^+ and v_d^-, respectively), due to their differing mobilities.

Hence, we define the charge carrier mobilities (μ^+ and μ^-, respectively) as the coefficient of proportionality between the carriers' average drift velocities and the applied electric field:

$$\overrightarrow{v_d^+} = \mu^+\vec{E} \quad \text{and} \quad \overrightarrow{v_d^-} = -\mu^-\vec{E} \tag{6.11}$$

Consequently, the total conduction current is given by the sum of the currents of both carriers:

$$\vec{J} = \rho_v^+\overrightarrow{v_d^+} + \rho_v^-\overrightarrow{v_d^-} = \rho_v^+\mu^+\vec{E} - \rho_v^-\mu^-\vec{E} = \left[\rho_v^+\mu^+ - \rho_v^-\mu^-\right]\vec{E} = \sigma\vec{E} \tag{6.12}$$

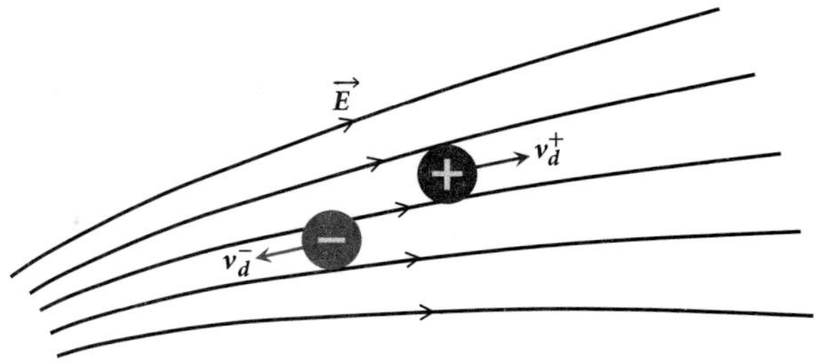

Figure 6.5 Drift velocities of mobile charges due to an applied electric field.

Defining the conductor "electric conductivity, σ" as the coefficient of proportionality between the conduction current density and the applied electric field:

$$\vec{J} = \sigma\vec{E} \quad \text{and} \quad \sigma = \left[\rho_v^+\mu^+ - \rho_v^-\mu^-\right] \tag{6.13}$$

The units for σ are those of J/E, or $(\text{A/m}^2)/(\text{V/m}) = (\text{A/V})/\text{m} = \text{S/m} = 1/(\Omega\ \text{m})$.

The inverse of the electric conductivity σ defines the electric resistivity, ρ, of the conducting material:

$$\rho = \frac{1}{\sigma} = \frac{1}{\left[\rho_v^+\mu^+ - \rho_v^-\mu^-\right]}, \quad \Omega\ \text{m} \tag{6.14}$$

Both σ and ρ define the bulk material property as determined by the volumetric density of the conductors' mobile charges and the mobilities of their carriers. For example, in a metallic conductor, electrons are the only mobile charges and their number and mobility vary from one metal to another. It is a known fact that temperature affects the mobility of electrons and hence affects the metal's conductivity. The conductivities of some materials are given in Appendix C.

6.5.1 Power Dissipated due to Conductivity/Resistivity

There are no energy storage mechanisms in a conduction/resistance body of material. However, we can evaluate the power consumed/dissipated in the conduction flow mechanism through evaluating the work needed to push the charges through the media's conductivity. Hence, with the help of Figure 6.4, we can write the work done to push ΔQ a distance $\overline{\Delta\ell}$ through the electric field $\vec{E}$ as

$$\Delta W = \vec{F}_{\Delta Q}\cdot\overline{\Delta\ell} = \Delta Q\vec{E}\cdot\overline{\Delta\ell} \tag{6.15}$$

and the corresponding power dissipation ΔP in the incremental volume $\Delta v = \Delta S\Delta\ell$ is given by

$$\Delta P = \frac{\Delta W}{\Delta t} = \frac{\Delta Q\,\vec{E}\cdot\overline{\Delta\ell}}{\Delta t} = \Delta I\left[\vec{E}\cdot\overline{\Delta\ell}\right] = \left[\vec{J}\cdot\overline{\Delta S}\right]\left[\vec{E}\cdot\overline{\Delta\ell}\right] = JE\Delta v = \sigma E^2\Delta v \tag{6.16}$$

and for a chunk of material with volume v we can write

$$P_{\text{total}} = \int_{\text{vol}} JE\,dv = \int_{\text{vol}} \sigma E^2\,dv \tag{6.17}$$

6.5.2 Resistance and Conductance

Having defined the conductivity and resistivity of bulk material, we are now ready to define the electrical resistance and conductance of a material body, such as the wires used in the transmission line cable or the insulator separating these wires.

The resistance of a given body of material is defined as the ratio between an applied potential difference between its terminals (two equipotential surfaces) and the resulting current flow (between the two terminals). In this regard, let us assume that an electric filed is applied to a given body of material by applying a potential difference between two surfaces (terminals), an electric current will flow between these two terminals as a

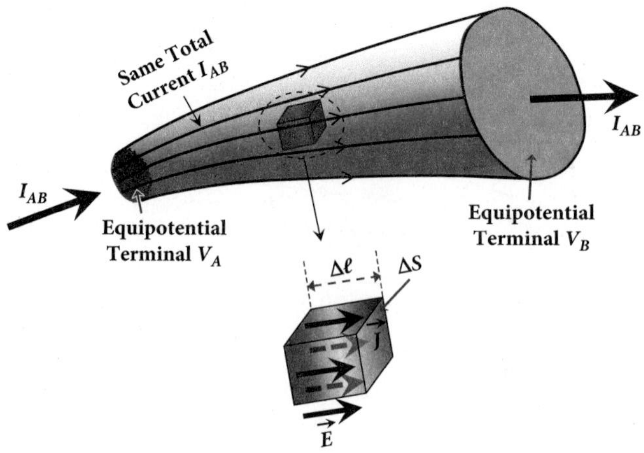

FIGURE 6.6 The flow on an electric current between two equipotentials as a result of the applied electric field.

result of the motion of its charge carriers under the influence of the forcing filed (Figure 6.6). The ratio between the terminal potential difference and the resulting current defines the resistance of the material body. If we use the notation V_{AB} and I_{AB} to denote the respective potential difference and current flow, we can express the resistance R_{AB} between the two terminals "A" and "B" of the material body as

$$R_{AB} = \frac{V_{AB}}{I_{AB}} \tag{6.18}$$

The conductance, G_{AB}, is defined as the inverse of the resistance, and hence

$$G_{AB} = \frac{1}{R_{AB}} = \frac{I_{AB}}{V_{AB}} \tag{6.19}$$

Equation (6.19) is what is known as Ohm's law in circuit theory.

Based on Equation (6.17) and using the integral forms for V_{AB} and I_{AB}

$$R_{AB} = \frac{V_{AB}}{I_{AB}} = \frac{\int_A^B \vec{E}\cdot\overrightarrow{d\ell}}{\iint_S \vec{J}\cdot\overrightarrow{dS}} = \frac{\int_A^B \vec{E}\cdot\overrightarrow{d\ell}}{\iint_S \sigma\vec{E}\cdot\overrightarrow{dS}} \tag{6.20}$$

$$G_{AB} = \frac{1}{R_{AB}} = \frac{I_{AB}}{V_{AB}} = \frac{\iint_S \vec{J}\cdot\overrightarrow{dS}}{\int_A^B \vec{E}\cdot\overrightarrow{d\ell}} = \frac{\iint_S \sigma\vec{E}\cdot\overrightarrow{dS}}{\int_A^B \vec{E}\cdot\overrightarrow{d\ell}} \tag{6.21}$$

where S is any cross section of the current flow in the material body.

6.5.3 The Resistance as a Circuit Element

Resistances are always present in electrical and electronic circuits. They are there in the form of added discrete or lumped components in the circuit, as the internal resistances of interconnecting wires/conductors or as the leakage conductances (resistances) of insulators. The acknowledged schematic we use to represent resistance in electrical circuits is shown in Figure 6.7.

Resistors may take any general geometrical configuration. Simply put, any bulk of material (of any shape) represents a resistance between its two terminals. Material conductivities range from zero for ideal insulators (infinite resistance) to infinite conductivity for ideal conductors (zero resistance).

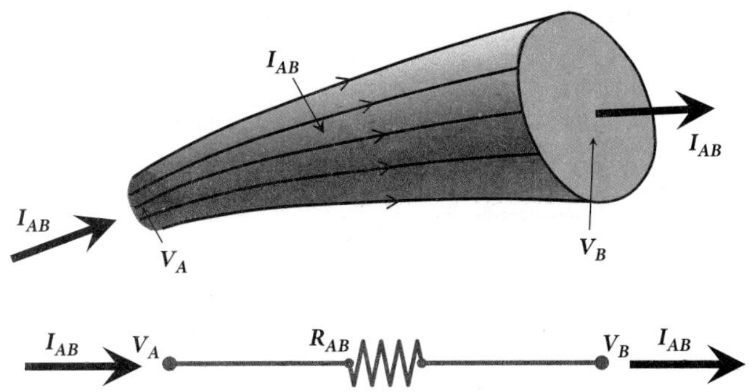

FIGURE 6.7 Schematic representation of the resistance of a body of material between two equipotentials.

In Addendum 6C, we discuss the resistance's power relationship in addition to developing the equivalencies of combined resistances in series and parallel.

6.6 Dielectrics (Insulators) and Polarization

As was discussed earlier in this chapter, (ideal) insulators are materials that lack the presence of mobile charge carriers, and hence no conduction current can take place. They are typically called dielectrics as an expression of their polarization under the influence of an applied electric field. The field forces the material's positive and negative charge carriers in opposite directions causing limited charge displacements according to their bondage forces in the atomic structure and the intensity of the applied field. This causes charge polarization, which in turn, results in an opposing field and thus a reduced resultant field within the body of the material.

Although there are different polarization mechanisms that occur in different dielectric material, our macroscopic analysis of the phenomenon will be the same. Typically, polarization is mostly *elastic* and vanishes with the removal of the applied field while in some dielectrics, a residual polarization remains after the field is removed. In addition, some dielectrics are microscopically polarized in their original state; however, their microscopic polarizations are random in orientations resulting in a zero macroscopic effect. With the application of an external field, these polarizations respond in a fashion that results in a macroscopic polarization behavior. Both types of polarizations are depicted in Figure 6.8.

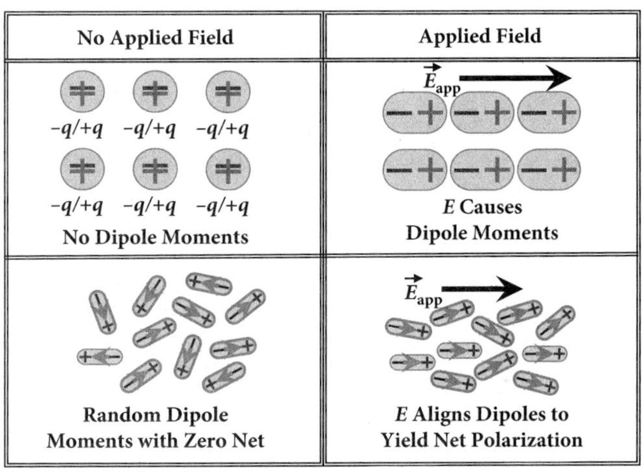

FIGURE 6.8 Dielectric polarization and dipole alignment under the influence of an applied electric field.

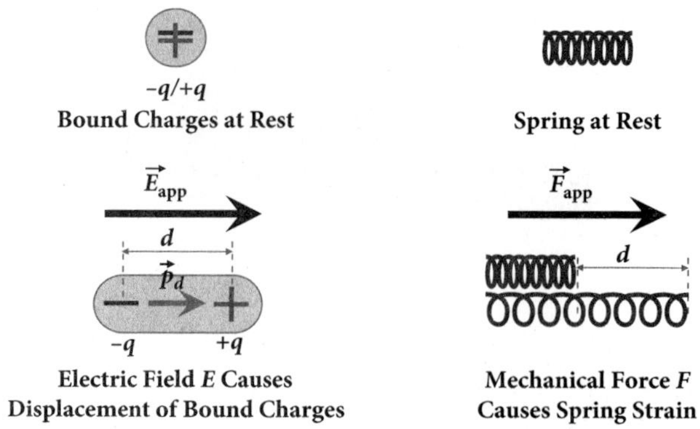

FIGURE 6.9 Analogy between the electric dipole moment and the mechanical strain of a spring.

Polarization is an important phenomenon that deserves to be carefully studied, as it is a form of energy storage and hence contributes to *capacitance*. The fact is that as we alter the arrangement of charges inside a body of material, we are changing its internal energy also. Thinking that an applied field causes polarization that can be released when the field is removed is in a way similar to stretching a mechanical spring or straining a rubber band (Figure 6.9).

The engineering interest in the polarization topic is macroscopic. We will briefly start with a single dipole (one polarized charge couple) and quickly move into the macroscopic polarization density vector. To start, we model the dielectric material as a distribution of dipoles in free space. Each one of these dipoles is a polarized pair of bound charges ($+q$ and $-q$) separated by distance d. In the absence of an external field, the separation distance d is zero and it increases with the increase of the applied field. Next, we define the dipole moment $\vec{p}_d$ as an indicator of a microscopic polarization of two bound charges, with a vector direction from $-q$ to $+q$ (Figures 5.31 and 6.9).

$$\vec{p}_d = q\vec{d} \text{ (C·m)} \tag{6.22}$$

6.6.1 The Polarization Vector

We also define the dielectric polarization vector $\vec{P}$ (flux density of bound charges) as the macroscopic resultant of the microscopic dipoles per unit volume (C/m^2), and hence

$$\vec{P} = \lim_{\Delta v \to 0} \frac{\sum_{\Delta v} \vec{p}_d}{\Delta v} \tag{6.23}$$

Referring to Figure 6.10, let us now consider an incremental element of a polarized dielectric material as a result of an applied electric field vector, $\vec{E}_{app}$, which is assumed uniform throughout the volume. The result is a number of "uniform (in average)" dipole moments $\vec{p}_d$ forming in the bulk of the material at certain density determined by the material properties and the intensity of the applied field. In a linear dielectric material, the polarization vector $\vec{P}$ as defined earlier is proportional to the applied field, hence, we write

$$\vec{P} \propto \vec{E}_{app}, \quad \text{or} \quad \vec{P} = \chi\varepsilon_o\vec{E}_{app} \tag{6.24}$$

where χ is positive and is called the electrical susceptibility of the dielectric. The choice of $\chi\varepsilon_o$ for the constant makes χ dimensionless and adds a convenience that will be apparent later.

Now, we turn to $\vec{P}$ and how it relates to the bound (polarized) charges. In this regard, let us examine the incremental volume Δv in Figure 6.10. The length of the volume increment is chosen as $\Delta\ell$ along the applied field and a cross section ΔS normal to the field. Examining the figure, we can see that due to the uniform presence of the polarized dipoles of charges, all internal bound charges are distributed to cancel each other within the volume of the increment leaving a surface distribution of bound charges only at the two ends of the

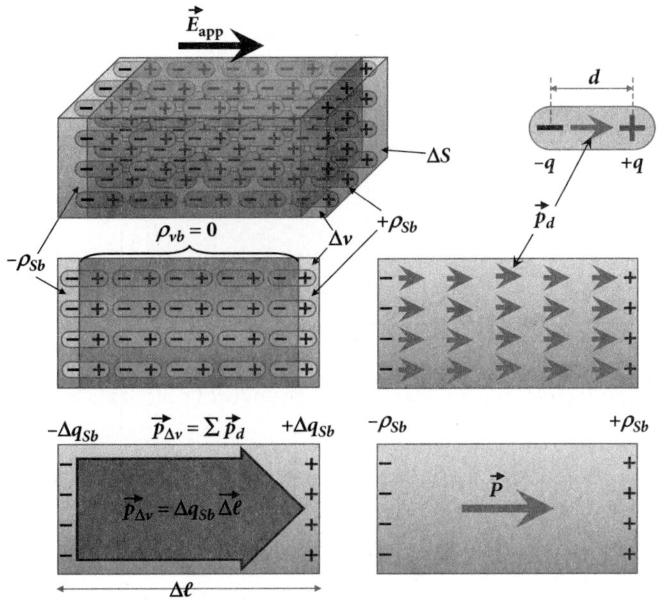

FIGURE 6.10 The dielectric polarization vector $\vec{P}$ is the macroscopic resultant of the microscopic dipoles per unit volume.

increment. Hence, the superposition of all the contained dipole moments, $\vec{p}_{\Delta v}$, is equivalent to the dipole formed of the end surface charges, $\pm\Delta q_{Sb}$, that is

$$\vec{p}_{\Delta v}\left\{=\Delta q_{Sb}\overline{\Delta \ell}\right\}=\sum_{\Delta v}\vec{p}_d\left\{=\sum_{\Delta v}q\vec{d}\right\} \tag{6.25}$$

This can be rewritten in terms of the surface charge density of the bound surface charges

$$\vec{p}_{\Delta v}=\left[\rho_{Sb}\,\Delta S\right]\overline{\Delta \ell}=\rho_{Sb}\,\Delta v\,\vec{a}_{\Delta \ell} \qquad \text{and hence} \qquad \vec{P}=\frac{\vec{p}_{\Delta v}}{\Delta v}=\rho_{Sb}\,\vec{a}_{\Delta \ell} \tag{6.26}$$

Please note that both $\vec{p}_{\Delta v}$ and $\vec{P}$ vectors are directed from the negative charge to the positive charge. Hence, referring to Figure 6.11, the surface density of the bound surface charge on the positive side, $+\rho_{Sb}$, has the polarity of $\vec{P}\cdot\vec{a}_{S+}$, while the negative side charges, $-\rho_{Sb}$, has the polarity of $\vec{P}\cdot\vec{a}_{S-}$.

Hence, we can rewrite Equation (6.26) above to yield the surface charge polarity as follows:

$$\vec{P}=\frac{\vec{p}_{\Delta v}}{\Delta v}=\rho_{Sb}\,\vec{a}_{\Delta \ell}\rightarrow\rho_{Sb}=\vec{P}\cdot\vec{a}_S \tag{6.27}$$

and the total surface bound charge on ΔS,

$$\Delta q_{Sb}=\vec{P}\cdot\overline{\Delta S} \tag{6.28}$$

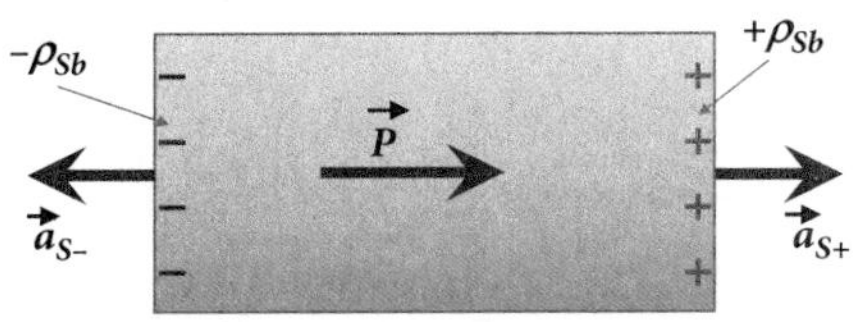

FIGURE 6.11 Surface densities of bound surface charges are related to polarization vector $\vec{P}$.

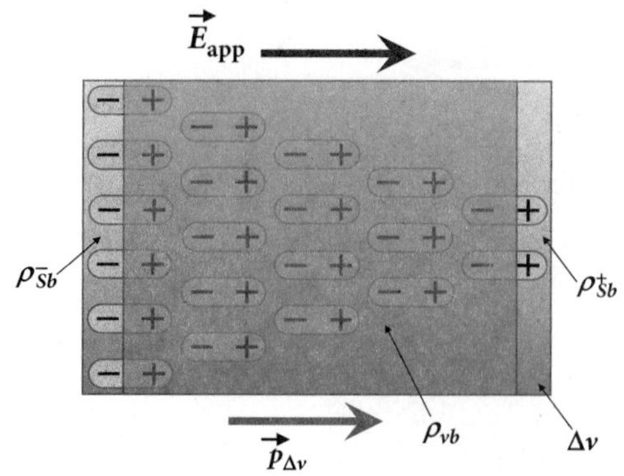

FIGURE 6.12 Nonuniform applied electric field results in nonuniform polarization within the dielectric material.

Now, let us consider the case of a dielectric volume where E is not uniform throughout and hence P is a function of position (Figure 6.12). The internal bound charges would not cancel each other and ρ_{Sb} on the surfaces enclosing the volume would be different.

Since the sum of all charges within the volume dipoles is zero, then

$$\oiint \rho_{Sb}\, dS + \iiint_{\text{volume}} \rho_{vb}\, dv = 0 \tag{6.29}$$

and hence,

$$\rho_{vb} = -\lim_{\Delta v \to 0} \frac{\oiint \rho_{Sb}\, \Delta S}{\Delta v} = -\lim_{\Delta v \to 0} \frac{\oiint \vec{P} \cdot \overline{\Delta S}}{\Delta v} = -\vec{\nabla} \cdot \vec{P} \tag{6.30}$$

$$\vec{\nabla} \cdot \vec{P} = -\rho_{vb} \tag{6.31}$$

Now, we have an expression of the polarization vector and its relationship to the volumetric density of the bound charges, the question is how does that affect the electric field inside the dielectric material? To discuss this relationship, let us assume that a region of free space is populated by both free and bound charges with volumetric densities of ρ_{vf} (representing the presence of free charges) and ρ_{vb} (representing the dielectric material's bound charges), respectively, the total volumetric charge density in this region would be given by

$$\rho_{vt} = \vec{\nabla} \cdot \varepsilon_o \vec{E} = \rho_{vf} + \rho_{vb} = \vec{\nabla} \cdot \vec{D} - \vec{\nabla} \cdot \vec{P} \tag{6.32}$$

and hence,

$$\vec{\nabla} \cdot \vec{D} = \vec{\nabla} \cdot \varepsilon_o \vec{E} + \vec{\nabla} \cdot \vec{P} \tag{6.33}$$

$$\vec{D} = \varepsilon_o \vec{E} + \vec{P} = \varepsilon_o \vec{E} + \chi \varepsilon_o \vec{E} = (1+\chi)\varepsilon_o \vec{E} = \varepsilon_r \varepsilon_o \vec{E} = \varepsilon \vec{E} \tag{6.34}$$

where

$$\varepsilon = \varepsilon_r \varepsilon_o \quad \text{and} \quad \varepsilon_r = (1+\chi) = \frac{\varepsilon}{\varepsilon_o} \tag{6.35}$$

ε_r is called the relative permittivity of the dielectric, and ε is the dielectric permittivity in F/m.

To appreciate the physical insight of what happens to the electric field inside the dielectric when it is exposed to an "external" applied field $\vec{E}_{app}$, let us consider the simple illustration shown in the Figure 6.13 for an incremental volume of dielectric material.

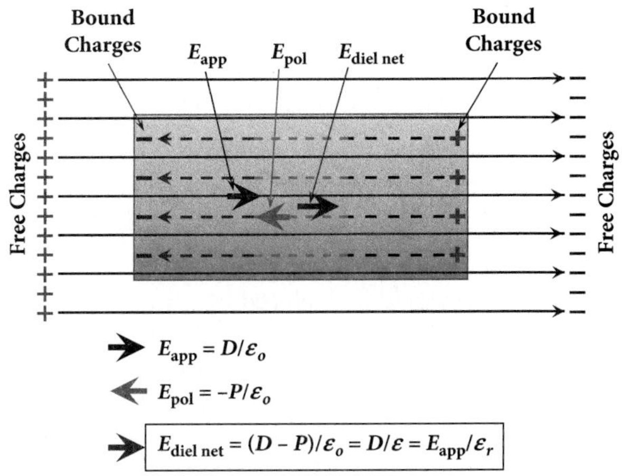

FIGURE 6.13 The relationship between E, P, and D.

In the figure, a dielectric increment (cylindrical with orthogonal end surfaces) is immersed in an applied E field. Because of the polarization of the dielectric's bound charges, an opposing polarization field results in a reduced net electric field inside the dielectric material.

$$E_{app} = D/\varepsilon_o \tag{6.36}$$

$$E_{pol} = -P/\varepsilon_o \tag{6.37}$$

$$E_{diel\ net} = (D - P)/\varepsilon_o = D/\varepsilon = E_{app}/\varepsilon_r \tag{6.38}$$

To summarize, we rewrite the relationships between D, E, and P in possible alternate forms as

$$\left.\begin{aligned}
&\vec{D} = \varepsilon_o\vec{E} + \vec{P}\\
&\vec{D} = \varepsilon_r\varepsilon_o\vec{E} = \varepsilon\vec{E}\\
&\varepsilon = \varepsilon_r\varepsilon_o\\
&\vec{D} = \varepsilon\vec{E} = \frac{\varepsilon_r}{(\varepsilon_r - 1)}\vec{P}\\
&\vec{P} = (\varepsilon_r - 1)\varepsilon_o\vec{E} = \frac{(\varepsilon_r - 1)}{\varepsilon_r}\vec{D}\\
&\vec{E} = \frac{1}{\varepsilon}\vec{D} = \frac{1}{\varepsilon_o(\varepsilon_r - 1)}\vec{P}
\end{aligned}\right\} \tag{6.39}$$

6.6.2 Energy in Dielectric Polarization

Starting with a dielectric material under zero electric field influence, we claim that it has zero energy. This is because the internal bound charges experience no forces in the absence of an external applied field as well as zero internal (macroscopic) field. Hence, the internal bound charges are in their lowest energy state with no dielectric polarization.

With the application of an external field, polarization of bound charges occur as bound dipole moments develop in alignment of the applied field. This bound charge polarization corresponds to an energy stored in dielectric's developed dipole moments. Figure 6.9 demonstrates this behavior in analogy with the mechanical spring. With no forces applied, the spring is at its lowest energy state (at rest). Upon the application of a stretching mechanical force, the spring is strained with energy stored in the process.

Recalling the expressions for electric energy density and total energy due to electrostatic fields [Equations (5.81) and (5.82)],

$$W_e = \frac{1}{2}\{\vec{D}\cdot\vec{E}\} = \frac{1}{2}\varepsilon E^2 \qquad \left[= \frac{1}{2}\varepsilon_o E^2 \quad \text{for free space}\right] \tag{6.40}$$

$$W_e = \int_{\text{vol}} \frac{1}{2}\{\vec{D}\cdot\vec{E}\}dv = \int_{\text{vol}} \frac{1}{2}\varepsilon E^2\,dv \qquad \left[= \int_{\text{vol}} \frac{1}{2}\varepsilon_o E^2\,dv \quad \text{for free space}\right] \tag{6.41}$$

In the presence of a dielectric material, ε_o would be replaced by $\varepsilon = \varepsilon_r\varepsilon_o$, and hence, the energy stored in the field would be larger in the presence of a dielectric as compared to the free space case since $\varepsilon_r > 1$ (for all known materials/non-metamaterials). This increased energy storage is owed to the developed dipole moments and their energy storage mechanisms. The analogy to mechanical springs of Figure 6.9 can help us develop some insight in this increased energy storage.

6.7 Capacitance

In principle, capacitance is defined for an "electric flux bundle" that "flows" in a dielectric medium (or free space) between two equipotential surfaces (or terminals). Referring to Figure 6.14, the flux bundle shown is initiated from the equipotential (terminal) V_A and lands on V_B. The total flux in the bundle (from "A" to "B") is labeled ψ_{AB}. The capacitance is defined as the ratio between the electric flux bundle ψ_{AB} and the corresponding potential difference between these terminals, $V_{AB} = V_A - V_B$. This definition is similar to that of the conductance except for using the flux *flow* instead of current flow:

$$C_{AB} = \frac{\psi_{AB}}{V_{AB}} \qquad \left\{\text{similar to } G_{AB} = \frac{I_{AB}}{V_{AB}}\right\} \tag{6.42}$$

Since

$$I_{AB} = \iint_S \vec{J}\cdot\vec{dS} = \iint_S \sigma\vec{E}\cdot\vec{dS}, \text{ while } \psi_{AB} = \iint_S \vec{D}\cdot\vec{dS} = \iint_S \varepsilon\vec{E}\cdot\vec{dS} \tag{6.43}$$

Again, similar to the case of conductance, each of the two-capacitance terminals is defined to be equipotential. In addition, the body of the capacitance confines the same total flux throughout from one terminal to the other.

Practically speaking, the two terminals are typically made of conducting surfaces to maintain their equipotential status. Also, the flux bundle occurs in the nonconducting (insulating dielectric or free space) region between the two terminals. Consequently, some references define the capacitance for a two-conductor system with a dielectric insulation in between.

In consideration of the similarities between the capacitance and the conductance, all math development that was done for the conductance earlier would hold for the capacitance if we replace σ by ε.

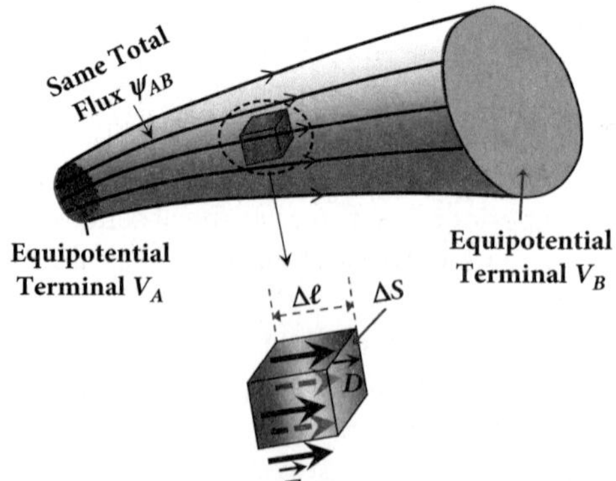

FIGURE 6.14 An electric flux flows in a dielectric medium between two equipotentials.

Hence, Equations (6.17) and (6.19) take the following forms for capacitance. Consequently, and using the integral forms for V_{AB} and ψ_{AB},

$$C_{AB} = \frac{\psi_{AB}}{V_{AB}} = \frac{\iint_S \vec{D}\cdot\overrightarrow{ds}}{\int_A^B \vec{E}\cdot\overrightarrow{d\ell}} = \frac{\iint_S \varepsilon\vec{E}\cdot\overrightarrow{ds}}{\int_A^B \vec{E}\cdot\overrightarrow{d\ell}} \tag{6.44}$$

In Addendum 6B, we demonstrate the implementation of Equation (6.44) for capacitance evaluation. We also apply the presented approach to derive expressions for the capacitance of a few typical geometrical configurations. Two geometrical configurations typically discussed in the literature because of analytical convenience are the parallel plate capacitor and the coaxial capacitor.

6.7.1 The Capacitance as a Circuit Element

Capacitances are typically present in majority of electrical and electronic circuits. They are typically included in circuit designs for many purposes including wave shaping, energy storage, and filtration. They also exist in circuit structures as packaging, stray, and parasitic elements. The flux of the capacitance flows in a dielectric body while the equipotential terminals are typically conductors.

As discussed earlier, the capacitance of a flux bundle is defined as the ratio between the bundle flux flowing between its terminals and the corresponding potential difference between these terminals (equipotential surfaces). In practice, the electromagnetic structure of a capacitance is configured with two conducting surfaces to act as the equipotential terminals and an insulating region in between to host the flux bundle. The insulating region may include one or more dielectric media (or free space).

It is important to remind ourselves that conductors can be equipotentials under static conditions only. If dynamic conditions exist, the conductor potential would vary from one location to another due to the potential drop caused by the electric current passing through a nonzero conductor resistance. Additionally, the size of the conductor relative to the wavelength of the operating frequency decides how uniform the conductor's potential is.

Now, recognizing that the bundle flux is equal to the charge responsible for producing it, we can rewrite the definition of the capacitance [Equation (6.44)] as

$$C_{AB} = \frac{\psi_{AB}}{V_{AB}} = \frac{Q_{AB}}{V_{AB}} \tag{6.45}$$

where Q_{AB} is the charge affiliated with the flux bundle ψ_{AB} (algebraically, $+Q_{AB}$ on terminal "A" and $-Q_{AB}$ on terminal "B"). These charges may or may not be the total charges on the bodies of their respective conductors. The example of Figure 6.16 of the three-conductor system demonstrates this concept.

The acknowledged schematic we use to represent capacitance in electrical circuits is shown in Figure 6.15.

Capacitors may take any general geometrical configuration. Simply put, any two conductors (of any shape) separated by an insulator (dielectric or free space) form a capacitor.

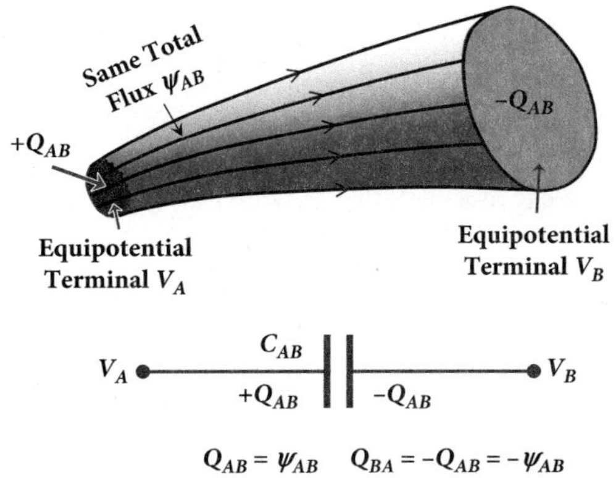

FIGURE 6.15 Schematic representation of the capacitance of a dielectric body of material between two equipotentials.

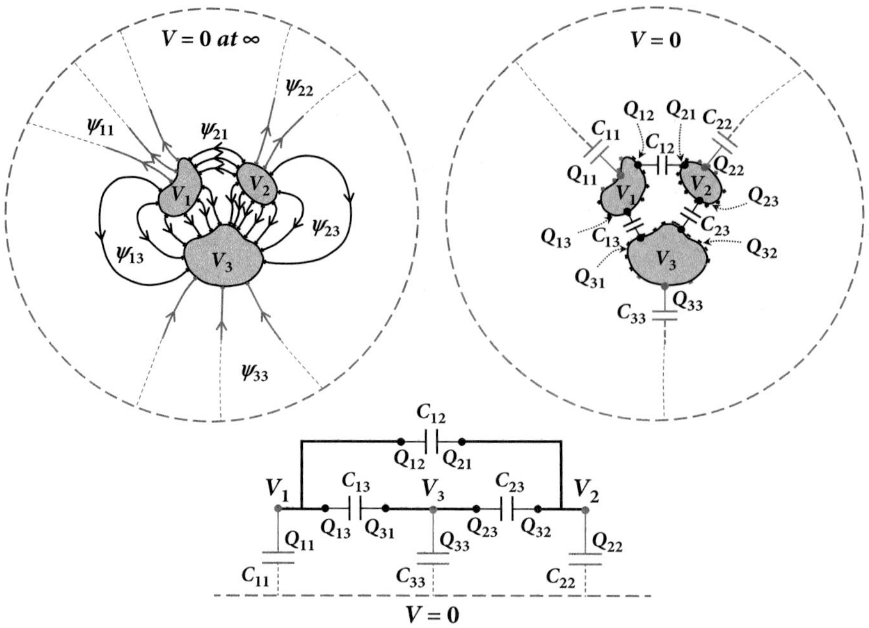

FIGURE 6.16 Schematic representation of capacitances between conductors separated by dielectrics and between conductors and infinity (zero potential or ground plane).

Figure 6.16 demonstrates this concept through a case of three conductors with mutual flux bundles as well as other (stray) flux bundle that tie each conductor to infinity where the balance of the three conductor charges exist. In practice, the "self" flux bundles would go to the ground (or earth) conductor ($V = 0$) that would be present as part of the configuration.

The charges and potentials of the three conductors are denoted by Q_1, V_1, Q_2, V_2, Q_3, and V_3, respectively.

In this figure, ψ_{21}, ψ_{23}, and ψ_{13} are mutual flux bundles, while ψ_{11}, ψ_{22}, and ψ_{33} are what we refer to as "stray" or "self" flux bundles. Capacitances are defined for each one of the six flux bundles in the example and hence, the schematic shown in the bottom of the figure. In this figure, each of the common flux bundles emerge (as a positive flux) from one of the bodies and land (as a negative flux) on another, and hence

$$Q_{21} = \psi_{21} = -\psi_{12} = -Q_{12}, \quad Q_{23} = \psi_{23} = -\psi_{32} = -Q_{32}, \quad Q_{13} = \psi_{13} = -\psi_{31} = -Q_{31} \tag{6.46}$$

In addition, the total charges (and hence total fluxes) on the individual conductors are given by

$$\left.\begin{aligned} Q_1 &= \psi_1 = \psi_{11} + \psi_{12} + \psi_{13} = Q_{11} + Q_{12} + Q_{13} \\ Q_2 &= \psi_2 = \psi_{21} + \psi_{22} + \psi_{23} = Q_{21} + Q_{22} + Q_{23} \\ Q_3 &= \psi_3 = \psi_{31} + \psi_{32} + \psi_{33} = Q_{31} + Q_{32} + Q_{33} \end{aligned}\right\} \tag{6.47}$$

Now, using Equation (6.44), we can express each flux bundle as a *capacitance* and hence the expressions:

$$\left.\begin{aligned} &C_{11} = \frac{\psi_{11}}{V_{1\infty}} = \frac{Q_{11}}{V_{1\infty}}, \quad C_{12} = \frac{\psi_{12}}{V_{12}}, \quad C_{13} = \frac{\psi_{13}}{V_{13}} \\ &C_{21} = \frac{\psi_{21}}{V_{21}}, \qquad C_{22} = \frac{\psi_{22}}{V_{2\infty}}, \quad C_{23} = \frac{\psi_{23}}{V_{23}} \\ &C_{31} = \frac{\psi_{31}}{V_{31}}, \qquad C_{32} = \frac{\psi_{32}}{V_{32}}, \quad C_{33} = \frac{\psi_{33}}{V_{3\infty}} \end{aligned}\right\} \tag{6.48}$$

Using i and j as integer counters (in this case from 1 to 3), we can write $V_{i\infty} = V_i$, is the electric potential of the conductor i, and $V_{ij} = V_i - V_j$ is the potential difference between the two conductors i and j.

Combining the equation sets [Equations (6.47) and (6.48)], we can write

$$\left.\begin{aligned} Q_1 &= C_{11}V_1 + C_{12}V_{12} + C_{13}V_{13} \\ Q_2 &= C_{21}V_{21} + C_{22}V_2 + C_{23}V_{23} \\ Q_3 &= C_{31}V_{31} + C_{32}V_{32} + C_{33}V_3 \end{aligned}\right\} \tag{6.49}$$

or in general, with multiple conductors, we can generalize

$$Q_i = \sum\nolimits_{j=1}^{n} C_{ij}V_{ij} \tag{6.50}$$

where $V_{ii} = V_i$, C_{ij} is the *mutual* capacitance between the two conductors i and j, while C_{ii} is the self-capacitance of the conductor i (stray capacitance to the $V = 0$ location which is typically the ground conductor).

In Addendum 6C, we develop the capacitance current-voltage relationship, as well as discuss its power and energy relationships. In the addendum, we also develop the equivalencies of combined capacitances in series and parallel.

Example 6.1 Find the breakdown voltage of a parallel plate capacitor assuming the conducting plates are 50 mm apart and the medium between them is air. Dielectric strength of air (breakdown electric field) = 3×10^6 V/m.

a. Find the breakdown voltage if the entire space between the conducting plates is filled with plexiglass, which has a relative dielectric constant, ε_r of 3 and dielectric strength of 20 kV/mm.
b. If a 10-mm-thick plexiglass layer is inserted between the plates, what is the maximum voltage that can be applied to the plates without a breakdown?

Hint: For part (b), you can assume the plexiglass layer is placed anywhere between the two plates.

The dielectric strength is the maximum field intensity the dielectric material can withstand before the material breaks down (ionization and conduction).

Solution

- The breakdown voltage when the space is filled with air = $V_{B\text{-air}}$ and $E_{\text{max-air}} = 3\times10^6$ V/m.
- $V_{B\text{-air}} = E_{\text{max-air}} \times d \rightarrow V_{B\text{-air}} = 3\times10^6 \times 50 \text{ mm} = 3\times10^6 \times 50/1000 = 150 \text{ kV}$

a. $V_{B\text{-plexiglass}} = E_{\text{max-plexiglass}} \times d \rightarrow V_{B\text{-plexiglass}} = 20\times10^6 \times 50 \text{ mm} = 20\times10^6 \times 50/1000 = 1000 \text{ kV}$
b. From the boundary condition (Section 6.8), the normal component of the electric displacement, D, is continuous across the boundary. Notice that in this case, there two different types of material in the spacing between the two conductors. The maximum electric field that can be used will be the maximum electric field that the weaker material can withstand. Therefore, the maximum electric field in this case will be the dielectric strength of air.

$$D_{n\text{-air}} = D_{n\text{-plexiglass}} \rightarrow \varepsilon_o E_{\text{air}} = \varepsilon_o \varepsilon_r E_{\text{plexiglass}} \rightarrow E_{\text{plexiglass}} = E_{\text{air}}/\varepsilon_r \qquad d_{\text{air}} = 40 \text{ mm} \qquad d_{\text{plexiglass}} = 10 \text{ mm}$$

$$V_B = E_{\text{air}} d_{\text{air}} + E_{\text{plexiglass}} d_{\text{plexiglass}} = E_{\text{air}}\left(d_{\text{air}} + \frac{1}{\varepsilon_r} d_{\text{plexiglass}}\right) = 3\times10^6\left(40\times10^{-3} + \frac{1}{3}\times10\times10^{-3}\right) = 130 \text{ kV}$$

Example 6.2 The parallel plate capacitor shown consists of two parallel dielectric layers. Determine the following:

a. The electric field intensity in each dielectric region in terms of the applied voltage V_o.
b. The surface charge density on the upper conductor surface in terms of V_o.
c. The capacitance of the parallel plate capacitor.
d. The electrostatic energy stored.

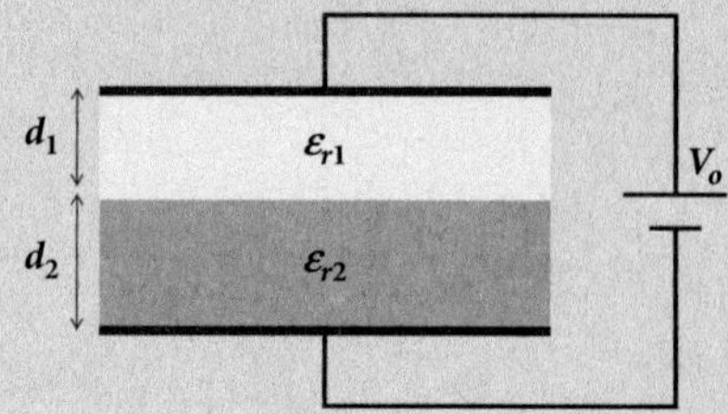

Solution

a. The electric field intensity in each dielectric region in terms of the applied voltage V_o.

$$V_o = E_1 d_1 + E_2 d_2 \quad D_{1n} = D_{2n} = D \rightarrow \varepsilon_o \varepsilon_{r1} E_1 = \varepsilon_o \varepsilon_{r2} E_2 \rightarrow E_1 = \frac{\varepsilon_{r2}}{\varepsilon_{r1}} E_2 \text{ (boundary conditions, Section 6.8)}$$

$$V_o = E_2 d_1 \frac{\varepsilon_{r2}}{\varepsilon_{r1}} + E_2 d_2 = E_2\left(d_1 \frac{\varepsilon_{r2}}{\varepsilon_{r1}} + d_2\right)$$

$$E_2 = \frac{V_o}{\left(\frac{\varepsilon_{r2}}{\varepsilon_{r1}} d_1 + d_2\right)} \qquad E_1 = \frac{\varepsilon_{r2}}{\varepsilon_{r1}} E_2 = \frac{V_o \frac{\varepsilon_{r2}}{\varepsilon_{r1}}}{\left(\frac{\varepsilon_{r2}}{\varepsilon_{r1}} d_1 + d_2\right)} \qquad E_1 = \frac{V_o}{\left(d_1 + \frac{\varepsilon_{r1}}{\varepsilon_{r2}} d_2\right)}$$

b. The surface charge density on the upper conductor surface in terms of V_o.

$$\rho_S = \varepsilon_o \varepsilon_{r1} E_1 = \frac{\varepsilon_o \varepsilon_{r1} V_o}{\left(d_1 + \frac{\varepsilon_{r1}}{\varepsilon_{r2}} d_2\right)}$$

c. The capacitance of the parallel plate capacitor.

The two dielectric regions represent two capacitances connected in series

$$\frac{1}{C} = \frac{1}{C_1} + \frac{1}{C_2} = \frac{d_1}{\varepsilon_o \varepsilon_{r1} S} + \frac{d_2}{\varepsilon_o \varepsilon_{r2} S} \Rightarrow \frac{1}{\varepsilon_o S}\left(\frac{\varepsilon_{r2} d_1 + \varepsilon_{r1} d_2}{\varepsilon_{r1} \varepsilon_{r2}}\right) \rightarrow C = \frac{\varepsilon_o \varepsilon_{r1} \varepsilon_{r2} S}{\varepsilon_{r2} d_1 + \varepsilon_{r1} d_2}$$

d. The electrostatic energy stored.

$$W_e = \frac{1}{2} C V_o^2 = \frac{1}{2}\left(\frac{\varepsilon_o \varepsilon_{r1} \varepsilon_{r2} S}{\varepsilon_{r2} d_1 + \varepsilon_{r1} d_2}\right) V_o^2$$

Example 6.3 Assuming the earth to be a large conducting sphere (radius = 6.37×10^3 km) surrounded by air.

a. Find the capacitance of the earth.
b. Find the maximum charge that can exist on the surface of the earth before the air around the earth breaks down.

Hint: Calculate the capacitance of two concentric conducting spheres, then take the limit as the radius of the outer sphere tends to infinity, this will give you the capacitance of an isolated conducting sphere. The dielectric strength (breakdown electric field) of air is 3×10^6 V/m.

Solution

a. The earth capacitance

$$\oint \vec{D} \bullet \overrightarrow{dS} = Q_{\text{enclosed}} \text{ and } \vec{D} \bullet \overrightarrow{dS} = \vec{a}_r D_r \bullet \vec{a}_r dS_r = D_r dS_r$$

$$\oint \vec{D} \bullet \overrightarrow{dS} = D_r \int dS_r = D_r S = D_r 4\pi r^2$$

$$D_r 4\pi r^2 = Q \Rightarrow D_r = \frac{Q}{4\pi r^2} \Rightarrow E_r = \frac{D_r}{\varepsilon_o} = \frac{Q}{4\pi \varepsilon_o r^2}$$

$$V_{12} = -\int_{R_2}^{R_1} E_r\, dr = -\int_{R_2}^{R_1} \frac{Q}{4\pi \varepsilon_o r^2} dr = \frac{Q}{4\pi \varepsilon_o}\left(\frac{1}{R_1} - \frac{1}{R_2}\right)$$

$$C = \frac{Q}{V_{12}} = \frac{4\pi \varepsilon_o}{(1/R_1 - 1/R_2)}$$

$$R_2 \to \infty \to C_{\text{earth}} = C_{R_2 \to \infty} = \frac{4\pi \varepsilon_o}{(1/R_1)} = 4\pi \varepsilon_o R_1 = 4\pi \times 8.854 \times 10^{-12} \times 6.37 \times 10^6 = 707\ \mu\text{F}$$

b. Find the maximum charge that can exist on the surface of the earth before the air around the earth breaks down.

The maximum charge should be calculated where the electric field is maximum. Looking at the expression of the electric field, the field decreases as r increases. Therefore, the maximum electric field due to charges on the earth surface will be on the surface of the earth at $r = R_{\text{min}} = R_{\text{earth}}$. We can then equate the maximum electric field to the dielectric strength of air.

$$E_{\text{max}} = \frac{Q_{\text{max}}}{4\pi \varepsilon_o R_{\text{earth}}^2} \Rightarrow 3 \times 10^6 = \frac{Q_{\text{max}}}{4\pi \varepsilon_o R_{\text{earth}}^2}$$

$$Q_{\text{max}} = 3 \times 10^6 \times 4\pi \varepsilon_o R_{\text{earth}}^2 = 3 \times 10^6 \times 4\pi \times \frac{10^{-9}}{36\pi} \times (6.37 \times 10^6)^2 = 13.53 \times 10^9\ \text{C}$$

Example 6.4 A cylindrical capacitor of length h consists of coaxial conducting surfaces. The radius of the inner surface is ρ_i and that of the outer surface is ρ_o. Two dielectric media of different dielectric constants ε_{r1} and ε_{r2} fill the space between the conducting surfaces as shown in the figure. Determine its capacitance.

Solution The electric field in the two regions is tangential to the boundary between the two dielectrics, therefore from the boundary conditions (Section 6.8).

$$E_{1t} = E_{2t} = E \Rightarrow \frac{D_{1t}}{\varepsilon_o \varepsilon_{r1}} = \frac{D_{2t}}{\varepsilon_o \varepsilon_{r2}}$$

$$D_{\rho_1} = D_{1t} = \varepsilon_o \varepsilon_{r1} E_\rho \qquad D_{\rho_2} = D_{2t} = \varepsilon_o \varepsilon_{r2} E_\rho$$

Now, consider a closed Gaussian surface as shown in the figure,

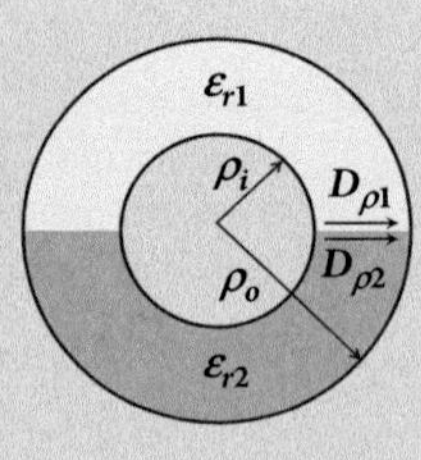

$$\oint \vec{D} \bullet \vec{dS} = Q_{\text{enclosed}}$$

$$\vec{D} \bullet \vec{dS} = \vec{a}_\rho D_\rho \bullet \vec{a}_\rho dS_\rho = D_\rho dS_\rho$$

$$\oint \vec{D} \bullet \vec{dS} = D_{\rho_1} \int_{\text{upper half}} dS_\rho + D_{\rho_2} \int_{\text{lower half}} dS_\rho = D_{\rho_1} \pi \rho h + D_{\rho_2} \pi \rho h$$

$$= \varepsilon_o \varepsilon_{r1} E_\rho \pi \rho h + \varepsilon_o \varepsilon_{r2} E_\rho \pi \rho h$$

$$\varepsilon_o \varepsilon_{r1} E_\rho \pi \rho h + \varepsilon_o \varepsilon_{r2} E_\rho \pi \rho h = Q \Rightarrow E_\rho = \frac{Q}{\pi h(\varepsilon_o \varepsilon_{r1} + \varepsilon_o \varepsilon_{r2})\rho}$$

$$V_{12} = -\int_{\rho_o}^{\rho_i} E_\rho \, d\rho = -\int_{\rho_o}^{\rho_i} E_\rho \, d\rho = -\int_{\rho_o}^{\rho_i} \frac{Q}{\pi h(\varepsilon_o \varepsilon_{r1} + \varepsilon_o \varepsilon_{r2})\rho} d\rho = \frac{Q}{\pi h(\varepsilon_o \varepsilon_{r1} + \varepsilon_o \varepsilon_{r2})} \ln\left(\frac{\rho_o}{\rho_i}\right)$$

$$C = \frac{Q}{V_{12}} = \frac{\pi h(\varepsilon_o \varepsilon_{r1} + \varepsilon_o \varepsilon_{r2})}{\ln\left(\frac{\rho_o}{\rho_i}\right)}$$

Another Solution Notice that the way the dielectrics are arranged realizes two capacitors connected in parallel each having one half the surface area of the entire cylinder

$$C_1 = \frac{2\pi h \varepsilon_o \varepsilon_{r1}}{\ln\left(\frac{\rho_o}{\rho_i}\right)} \times \frac{1}{2} = \frac{\pi h \varepsilon_o \varepsilon_{r1}}{\ln\left(\frac{\rho_o}{\rho_i}\right)} \qquad C_2 = \frac{\pi h \varepsilon_o \varepsilon_{r2}}{\ln\left(\frac{\rho_o}{\rho_i}\right)}$$

$$C = C_1 + C_2 = \frac{\pi h(\varepsilon_o \varepsilon_{r1} + \varepsilon_o \varepsilon_{r2})}{\ln\left(\frac{\rho_o}{\rho_i}\right)}$$

6.8 Boundary Conditions

So, what is this topic about and why do we need to learn about "boundary conditions"? The answer is that the topic of boundary conditions is about what happens at the interface between two different material and how does the different electromagnetic quantities behave across that interface. Having talked about two categories of materials so far; conductors and dielectric, we will have three possible cases of interfaces; namely dielectric/dielectric, conductor/conductor, and conductor/dielectric.

6.8.1 Dielectric-Dielectric Interface

Let us assume the case of two dielectric materials sharing a surface interface as shown in Figures 6.17 and 6.18. We further assume that the electric field intensity vectors in the two material regions 1 and 2 are denoted by E_1 and E_2, respectively. To develop the relationship between E_1 and E_2, we typically use two tools. The first tool is the closed loop integration of the electric field and the second is the closed surface integration of the flux density vector. Both the closed loop path and the closed surface volume share common regions in both materials as shown in the figures.

In the first figure in this process (Figure 6.17) we hypothesize a closed loop of rectangular shape of length $\Delta\ell$ and a height of Δh, half of which lies in material 1 and the other half in material 2. The closed loop integration around this rectangular loop must be written in terms of different mathematical expressions since there is no single mathematical expression that applies to the entire loop. For example, for the top section of the loop the electric field intensity vector is E_1 and $\overrightarrow{\Delta\ell}$ is tangential to the interface while for the upper halves of Δh (within region 1), the electric field intensity vector is still E_1 but the $\overrightarrow{\Delta\ell}$'s are normal to the interface, etc.

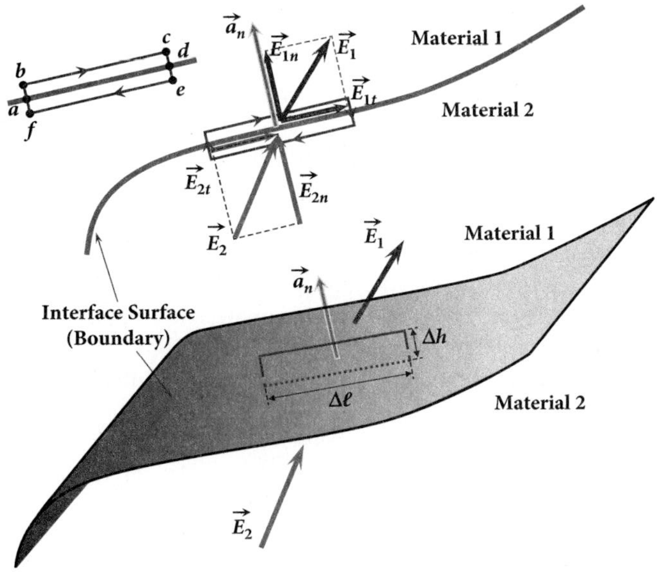

FIGURE 6.17 Deriving the electric field boundary conditions at a dielectric-dielectric interface.

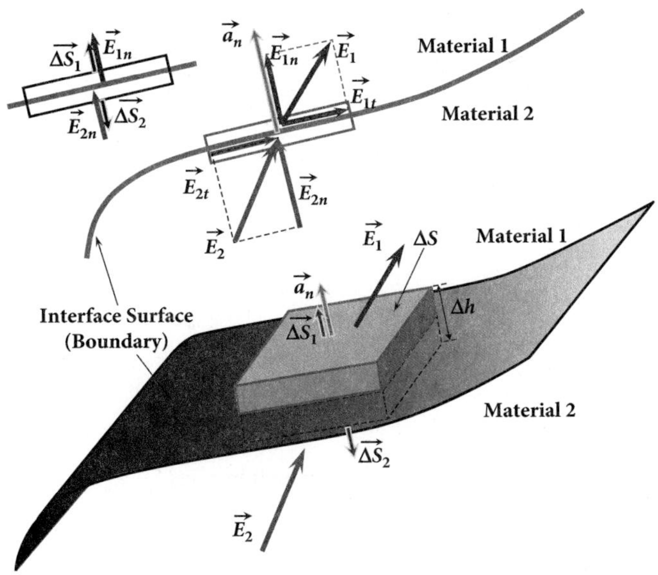

FIGURE 6.18 Deriving the electric flux density boundary conditions at a dielectric-dielectric interface.

So, according to Equation (5.70),

$$0=\oint \vec{E}\cdot\overrightarrow{d\ell}=\int_a^b \vec{E}_1\cdot\overrightarrow{d\ell}+\int_b^c \vec{E}_1\cdot\overrightarrow{d\ell}+\int_c^d \vec{E}_1\cdot\overrightarrow{d\ell}+\int_d^e \vec{E}_2\cdot\overrightarrow{d\ell}+\int_e^f \vec{E}_2\cdot\overrightarrow{d\ell}+\int_f^a \vec{E}_2\cdot\overrightarrow{d\ell} \tag{6.51}$$

However, in the limit as Δh goes to zero, the integrals *a-b*, *c-d*, *d-e*, and *f-a* will vanish leaving:

$$0=\int_b^c \vec{E}_1\cdot\overrightarrow{d\ell}+\int_e^f \vec{E}_2\cdot\overrightarrow{d\ell} \tag{6.52}$$

which reduces to (since E_{1t} is along $\overrightarrow{d\ell}$ and E_{2t} is opposite):

$$0=[E_{1t}-E_{2t}]d\ell, \qquad \text{or} \qquad E_{1t}=E_{2t} \tag{6.53}$$

Next, we hypothesize a closed Gaussian surface in the form shown in Figure 6.18. The closed surface integration on this surface is to be equated to the charge enclosed within. As we shrink the *height* of this enclosure, Δh, to zero, this Gaussian surface would only enclose the charges on the interface surface and none of the materials bulk (if any).

The shape is of the form of a cuboid (or any other upright cylindrical shape) of surface area ΔS and a height of Δh, half of which lies in material 1 and the other half in material 2. Again, the closed surface integration around this shape must be written in terms of three surface integrations: one on the top surface, another on the bottom surface, and a third one on the orthogonal sides. This later integration on the sides will vanish as we take the limit of Δh approaches zero.

$$\sum\nolimits_{\text{enclosed}} Q = \oiint_{\text{cuboid}} \vec{D}\cdot\overrightarrow{dS} = \iint_{\text{top surface}} \vec{D}_1\cdot\overrightarrow{dS} + \iint_{\text{bottom surface}} \vec{D}_2\cdot\overrightarrow{dS} + \iint_{\text{side surfaces}} \vec{D}\cdot\overrightarrow{dS} \tag{6.54}$$

$$\lim_{\Delta h\to 0}\left\{\sum\nolimits_{\text{enclosed}} Q\right\} = \Delta Q_{\text{interface surface}} = \rho_S\,\Delta S \tag{6.55}$$

$$\lim_{\Delta h\to 0}\left\{\oiint \vec{E}\cdot\overrightarrow{dS}\right\} = D_{1n}\,\Delta S - D_{2n}\,\Delta S \tag{6.56}$$

Now, equating the closed surface integration to the charge enclosed yields

$$D_{1n}\,\Delta S - D_{2n}\,\Delta S = \rho_S\,\Delta S, \qquad \text{or} \qquad D_{1n}-D_{2n}=\rho_S \tag{6.57}$$

And in the absence of any surface charges,

$$D_{1n}=D_{2n}, \qquad \text{or} \qquad \varepsilon_1 E_{1n}=\varepsilon_2 E_{2n} \tag{6.58}$$

Hence, to summarize, at the interface between two dielectrics:

$$E_{1t}=E_{2t} \qquad \text{and} \qquad D_{1n}-D_{2n}=\rho_S \tag{6.59}$$

And in the absence of free surface charges,

$$E_{1t}=E_{2t} \qquad \text{and} \qquad D_{1n}=D_{2n} \qquad \text{which implies} \qquad \varepsilon_1 E_{1n}=\varepsilon_2 E_{2n} \tag{6.60}$$

Hence, in the case of no surface charges at the interface,

$$\frac{E_{1t}}{\varepsilon_1 E_{1n}}=\frac{E_{2t}}{\varepsilon_2 E_{2n}} \qquad \text{or} \qquad \frac{\tan\theta_1}{\varepsilon_1}=\frac{\tan\theta_2}{\varepsilon_2} \tag{6.61}$$

So, the angle of the E field vector (and hence D) changes as the field crosses the interface boundary into a different medium; see Figure 6.19. The angle of "refraction" depends on the relationship between the media permittivities. In higher permittivity media, the normal field component is smaller because of the stronger polarization in the normal direction. With the tangential field component unchanged, the angle of refraction into a higher permittivity region is larger.

6.8.2 Surface Charges at the Interface: Free and Bound (Polarization)

In this section, we further examine the interface between the two dielectrics. In the discussion above, we have looked at the field and flux density continuities, and now, we look at the charge distribution at the boundary.

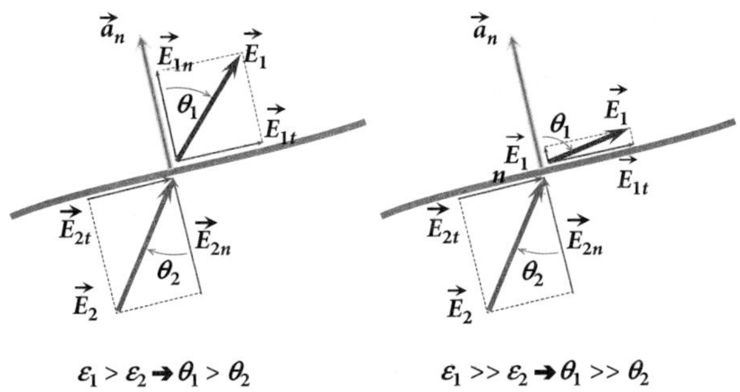

FIGURE 6.19 The boundary conditions at a dielectric-dielectric interface in the absence of surface charges.

For that, we resort to Equation (6.27) which relates the bound surface charge density to the polarization vector at the dielectric surface:

$$\rho_{Sb} = \vec{P} \cdot \vec{a}_S \tag{6.27}$$

Applying Equation (6.27) to the interface, and referring to Figure 6.20, and using Equation (6.39), we can write

$$\left.\begin{aligned} \rho_{Sb1} &= \vec{P}_1 \cdot [-\vec{a}_n] = -P_{1n} = -\frac{\varepsilon_{r1}-1}{\varepsilon_{r1}} D_{1n} \\ \rho_{Sb2} &= \vec{P}_2 \cdot [+\vec{a}_n] = +P_{2n} = \frac{\varepsilon_{r2}-1}{\varepsilon_{r2}} D_{2n} \end{aligned}\right\} \tag{6.62}$$

Hence, the total bound charge at the interface is given by

$$\rho_{Sb} = \rho_{Sb1} + \rho_{Sb2} = P_{2n} - P_{1n} = \frac{\varepsilon_{r2}-1}{\varepsilon_{r2}} D_{2n} - \frac{\varepsilon_{r1}-1}{\varepsilon_{r1}} D_{1n} \tag{6.63}$$

In the typical case of $\rho_S = 0$, and according to Equation (6.57), $D_{1n} = D_{2n} = D_n$, hence

$$\rho_{Sbt} = \left[\frac{\varepsilon_{r2}-1}{\varepsilon_{r2}} - \frac{\varepsilon_{r1}-1}{\varepsilon_{r1}}\right] D_n \tag{6.64}$$

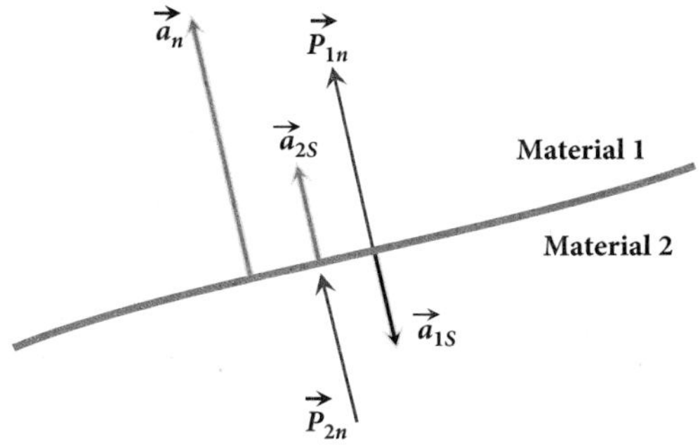

FIGURE 6.20 The boundary conditions at a dielectric-dielectric interface for the polarization vector and bound charges.

6.8.3 Conductor-Conductor Interface

As was discussed in the early part of this chapter regarding conductors under static conditions, the electric field within a conductor is zero and the entire conductor region is equipotential. However, if currents are allowed to flow when the conductor is placed as part of a closed electrical circuit or exposed to a time varying field, an electric field may exist within the conductor material. In this case, the conductor region will seize to become equipotential; instead, a potential gradient will exist. Hence, there are two possible scenarios for the conductor-conductor interface; the static case with zero currents, and the dynamic case where currents are allowed to flow (Figure 6.21).

Static Case

In this case, the conductors are both in static situation with no currents are allowed to flow and hence

$$J_1 = J_2 = 0 \text{ or } \sigma_1 E_1 = \sigma_2 E_2 = 0, \text{ and hence } E_1 = E_2 = 0 \tag{6.65}$$

Dynamic Case

In this case, electric currents are allowed to flow in the conductors. To study the field relationships across the interface, we resort to the same two tools used earlier for the dielectric-dielectric interface.

Reviewing the earlier development in this regard, we should be able to conclude that the results of Equation (6.53) will hold for the conductor-conductor interface in the same exact way they held for the dielectric-dielectric interface, and hence

$$E_{1t} = E_{2t} \tag{6.66}$$

However, for the normal components' relationship, we examine the current on a closed surface as opposed to examining the flux in the dielectric-dielectric case. This is because the dominant phenomenon in conductors is conduction. Hence, we repeat the same exact closed surface integration process on the cuboid enclosure but this time we apply the conduction current continuity principle instead of the electric flux analysis. In this regard,

$$\oiint_{\text{cuboid}} \vec{J} \cdot \vec{dS} = -\frac{d[\Delta Q_{\text{enclosed}}]}{dt} \tag{6.67}$$

Similarly, as we shrink the height of the cuboid to enclose the surface only, the above equation reduces to

$$J_{1n} - J_{2n} = -\frac{d\rho_S}{dt} \tag{6.68}$$

and in the absence of ρ_S, or $\frac{d\rho_S}{dt}$ we can write $J_{1n} = J_{2n}$, or $\sigma_1 E_{1n} = \sigma_2 E_{2n}$, and again

$$\frac{\tan\theta_1}{\sigma_1} = \frac{\tan\theta_2}{\sigma_2} \tag{6.69}$$

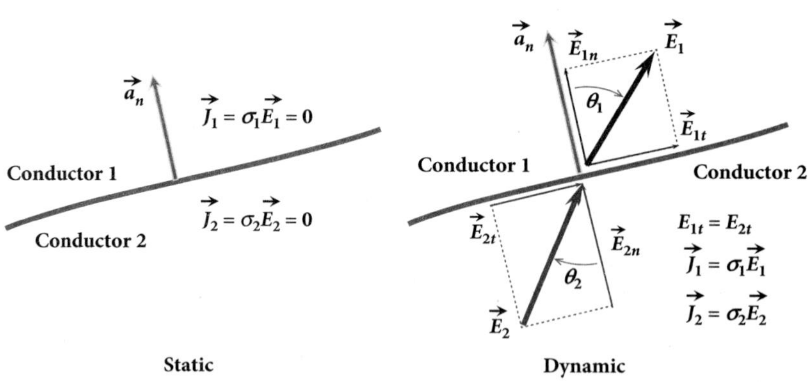

FIGURE 6.21 The boundary conditions at a conductor-conductor interface.

Example 6.5 At the interface between the two media shown in the figure, the current density in medium 1 is $10\sqrt{2}$ A/m^2. The conductivities in the two media are related by $\sigma_2 = 3\sigma_1$. Find the magnitude of the current density in medium 2, J_2, and the angle θ.

Solution

$$\frac{J_{1t}}{\sigma_1} = \frac{J_{2t}}{\sigma_2} \rightarrow \frac{J_1 \sin 45^\circ}{\sigma_1} = \frac{J_2 \sin\theta}{\sigma_2} \tag{1}$$

$$J_{1n} = J_{2n} \rightarrow J_1 \cos 45^\circ = J_2 \cos\theta \tag{2}$$

Dividing (1) by (2)

$$\frac{\tan 45^\circ}{\sigma_1} = \frac{\tan\theta}{\sigma_2} \rightarrow \tan\theta = \frac{\sigma_2}{\sigma_1}\tan 45^\circ = 3 \rightarrow \theta = \tan^{-1} 3 = 71.6^\circ$$

From (2) $10\sqrt{2}\cos 45^\circ = J_2 \cos 71.6^\circ \rightarrow J_2 = \dfrac{10\sqrt{2}\cos 45^\circ}{\cos 71.6^\circ} = 31.6 \text{ A/m}^2$

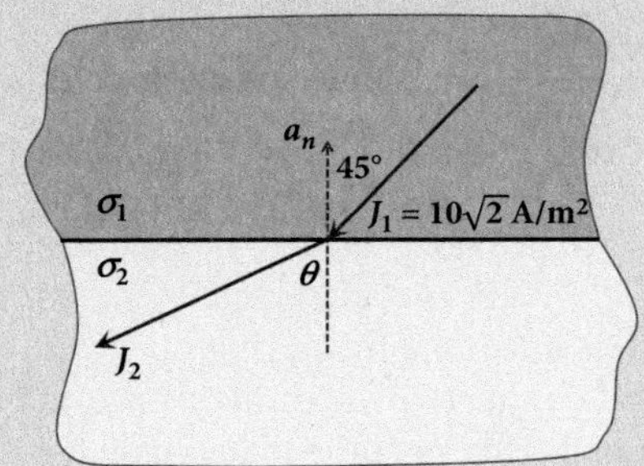

6.8.4 Conductor-Dielectric Interface

Again, we need to consider both static and dynamic cases since they would differ regarding possible flow of conduction current in the conductor under dynamic conditions (Figure 6.22).

Static Case

Under static conditions, no conduction currents in the conductor and hence E within the conductor region is zero. Again, with $E_{1t} = E_{2t}$, and since $E_{2t} = 0$, then $E_{1t} = 0$, and hence, $E_1 = E_{1n}$.

Following the flux analysis on the cuboid Gaussian surface, we can write

$$D_{1n} - D_{2n} = \rho_S \text{ leading to } \quad D_{1n} = \rho_S \text{ and } \quad \varepsilon_1 E_{1n} = \rho_S \tag{6.70}$$

Dynamic Case

Since no conduction current can flow in the insulating dielectric, then the current continuity analysis discussed earlier yields $J_{1n} = J_{2n} = 0$, and hence, only J_{2t} may exist. Consequently, $E_{2n} = 0$ and only E_{2t} may exist, that is, $E_2 = E_{2t}$. Meanwhile, the flux analysis can be used to show that $D_{1n} = \rho_S$ and $\varepsilon_1 E_{1n} = \rho_S$, and the conductor current density would be given by $\vec{J}_2 = \sigma_2 \vec{E}_{2t}$.

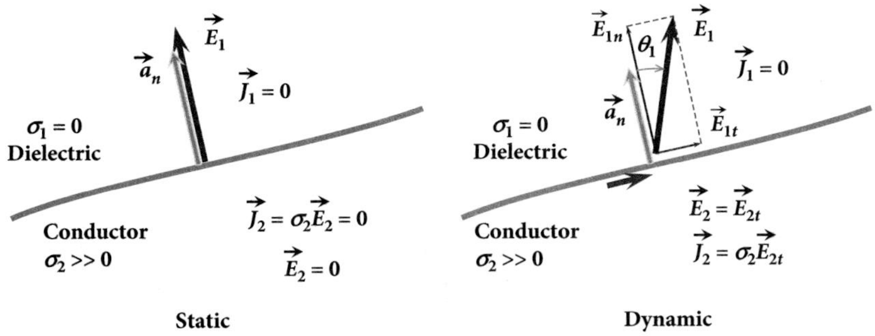

FIGURE 6.22 The boundary conditions at a conductor-dielectric interface.

Bound (Polarization) Charges at the Interface

Similar to our discussion earlier for the dielectric-dielectric interface, while replacing material 2 with a conductor, the resulting bound charge distribution would become

$$\rho_{Sbt} = \rho_{Sb1} = \vec{P}_1 \cdot [-\vec{a}_n] = -P_{1n} = -\frac{\varepsilon_{r1} - 1}{\varepsilon_{r1}} D_{1n} \tag{6.71}$$

And using Equation (6.70), we can write

$$\rho_{Sbt} = \rho_{Sb1} = -\frac{\varepsilon_{r1} - 1}{\varepsilon_{r1}} \rho_S \tag{6.72}$$

ADDENDUM 6A
RESISTANCE EVALUATION

6A.1 Resistance Evaluation

Although resistance is defined as the ratio between applied potential (voltage) and resulting current, the resistance of a "linear" material configuration depends only on its geometrical parameters as well as the material conductivity. In other words, the E field factors out in Equation (6.19). To demonstrate, let us consider a small element of the material of length $\Delta\ell$ (along the electric field) and cross section ΔS (normal to the field) (Figure 6A.1) the resistance of this element becomes

$$\Delta R = \frac{\Delta V}{\Delta I} = \frac{\vec{E} \cdot \overrightarrow{\Delta\ell}}{\sigma \vec{E} \cdot \overrightarrow{\Delta S}} = \frac{\Delta\ell}{\sigma \Delta S} \tag{6A.1}$$

Consequently, to evaluate the resistance of any material configuration, we have two choices. The first choice is based on the use of Equation (6.19), while the second is based on the use of Equation (6A.1).

6A.1.1 Using V and I for Resistance Evaluation [Equation (6.19)]

In this approach, we start by solving for the E field based on the applied potential difference V_{AB}. Once E is determined, we use Equation (6.3) to evaluate I_{AB}. Finally, R_{AB} is evaluated as the ratio V_{AB}/I_{AB}.

6A.1.2 Using ΔR for Resistance Evaluation [Equation (6A.1)]

Again, we start by solving for the E field based on the applied potential difference V_{AB}. The only thing needed out of this solution is the field map and not the field intensity. The field map is then used to divide the material configuration into incremental volumes, each of which is considered an incremental resistor $\Delta R_{\text{increment}} = \dfrac{d\ell}{\sigma dS}$.

Next, we workout the total resistance of the whole configuration by combining the incremental resistance in series/parallel using one of two possible scenarios:

1. *Series then parallel (Figure 6A.2):* In this scenario, we add the incremental resistance along the field line to evaluate the resistance on an incremental *tube* of the configuration: $\Delta R_{\text{tube}} = \int_A^B \dfrac{d\ell}{\sigma dS}$. Next, we combine the resistances of all the parallel tubes to obtain the overall resistance. $\dfrac{1}{R_{\text{total}}} = \iint_S \dfrac{1}{\int_A^B \frac{d\ell}{\sigma dS}}$

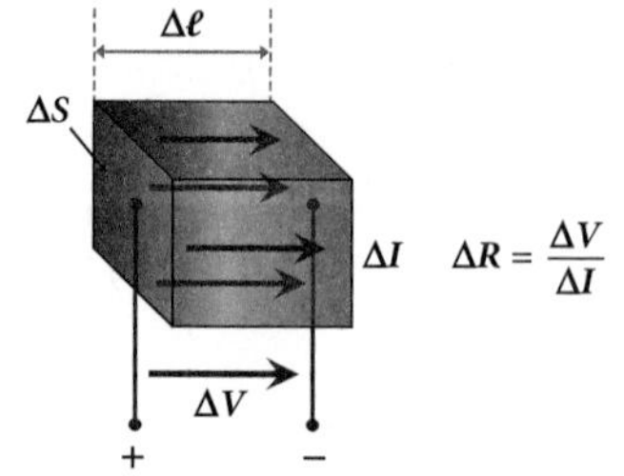

FIGURE 6A.1 Incremental volume to be used for resistance evaluation.

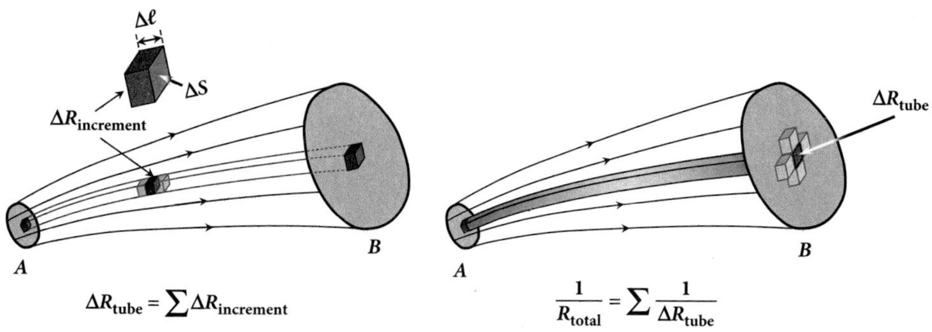

FIGURE 6A.2 Resistance evaluation starting with series summation (integration).

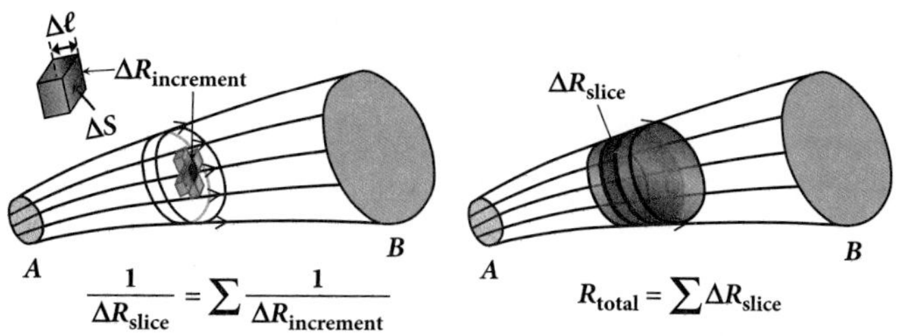

FIGURE 6A.3 Resistance evaluation starting with parallel summation (integration).

2. *Parallel then series (Figure 6A.3):* In this scenario, we start by the parallel combination of the incremental resistance in a general cross-sectional location (normal to the field lines, which is also an equipotential surface). The result is the resistance of an incremental *slice* of the configuration: $\frac{1}{\Delta R_{\text{slice}}} = \iint_S \frac{\sigma\, dS}{d\ell}$. Next, we combine the resistances of all the series slices to obtain the overall resistance $R_{\text{total}} = \int_A^B 1/\iint_S \frac{\sigma\, dS}{d\ell}$.

Now, we demonstrate the application of the above methods to the simple case of a uniform cross-sectional (cylindrical[1]) body of material with terminals "A" and "B" as shown in Figure 6A.4. We start by solving for the electric field due to an applied potential difference V_{AB}. For this simple geometrical configuration, it is obvious that the electric field lines are along the length of the cylinder and are uniform in intensity everywhere in the body of the cylinder.

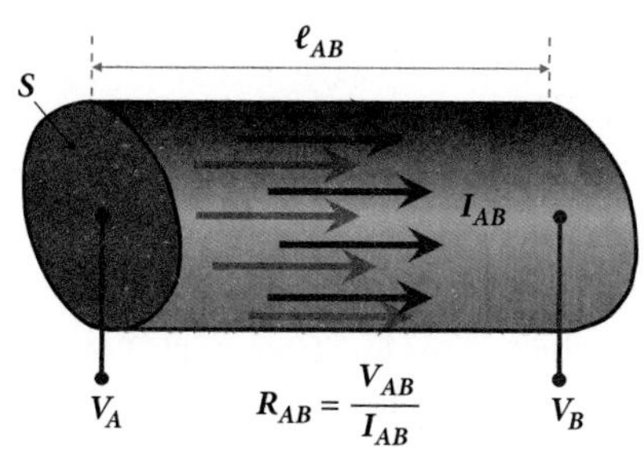

FIGURE 6A.4 Application of resistance evaluation schemes to a cylindrical body.

[1]A cylinder is a geometrical body that extends along a linear axis with a uniform cross section and does not have to have a circular cross section is to be called a cylinder.

Consequently,

$$V_{AB} = \int_A^B \vec{E} \cdot \vec{d\ell} = E\ell_{AB} \quad \text{and hence} \quad \vec{E} = \frac{V_{AB}}{\ell_{AB}} \vec{a}_\ell \tag{6A.2}$$

$$I_{AB} = \iint_S \sigma \vec{E} \cdot \vec{dS} = \sigma ES \tag{6A.3}$$

Thus, the resistance of this cylindrical body of material between the terminals "A" and "B" is given by

$$R_{AB} = \frac{V_{AB}}{I_{AB}} = \frac{E\ell_{AB}}{\sigma ES} = \frac{\ell_{AB}}{\sigma S} \tag{6A.4}$$

We now turn to the series parallel combination approach. We start by dividing the cylinder into small volumetric increments. Each increment has a length $d\ell$ along the length of the cylinder and a normal cross section dS. The resistance of each increment would be given by

$$\Delta R_{\text{increment}} = \frac{d\ell}{\sigma\, dS} \tag{6A.5}$$

Now, adding the resistors in series to form a tube of length ℓ_{AB} and a cross section of dS, the resistance of the incremental tube is

$$\Delta R_{\text{tube}} = \int_A^B \frac{d\ell}{\sigma\, dS} = \frac{\ell_{AB}}{\sigma\, dS} \tag{6A.6}$$

Next, we combine all the incremental tubes in parallel to get the total resistance:

$$\frac{1}{R_{AB}} = \iint_S \frac{1}{\dfrac{\ell_{AB}}{\sigma\, dS}} = \iint_S \frac{\sigma\, dS}{\ell_{AB}} = \frac{\sigma S}{\ell_{AB}}, \quad \text{hence} \quad R_{AB} = \frac{\ell_{AB}}{\sigma S} \tag{6A.7}$$

If we were to do the parallel combination first, we combine the incremental resistors to form a cylindrical slice of length $d\ell$ and a cross section S,

$$\frac{1}{\Delta R_{\text{slice}}} = \iint_S \frac{1}{\dfrac{d\ell}{\sigma\, dS}} = \iint_S \frac{\sigma\, dS}{d\ell} = \frac{\sigma S}{d\ell} \tag{6A.8}$$

$$\Delta R_{\text{slice}} = \frac{d\ell}{\sigma S} \tag{6A.9}$$

Next, we add the resistances of all the incremental slices in series to get the total resistance:

$$R_{AB} = \int_A^B \frac{d\ell}{\sigma S} = \frac{\ell_{AB}}{\sigma S} \tag{6A.10}$$

which is the same result as Equation (6A.4).

6A.2 Coaxial Cable Transmission Line R_z and G_z Parameters

Using the above results, we can evaluate the (DC or low frequency[2]) resistance of the coaxial cable transmission line (TL) conductors (Figure 6A.5). We have two conductors through which the TL current flows; an inner conductor and an outer conductor. The resistance of interest here is the resistance of both conductors along

[2]At DC and low frequencies, the electric field is uniform over the conductor cross section. However, as we will learn later, the field at high frequencies is not uniform over the conductors' cross section, and hence the TL resistance expression is different from that at low frequency.

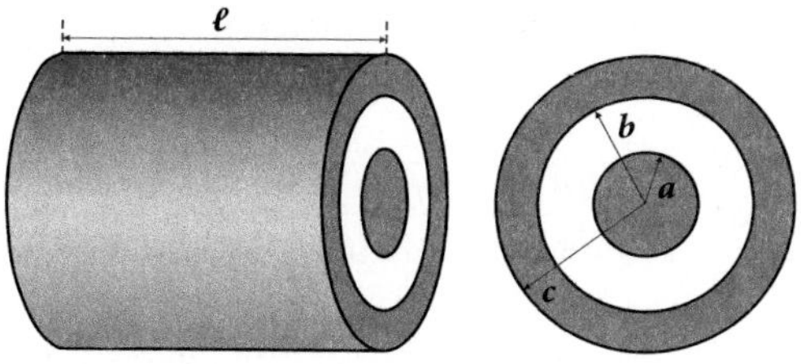

FIGURE 6A.5 Application of resistance evaluation schemes to the conductors of a coaxial cable.

their length; the z axis. Both are cylindrical in nature and hence the expression $R_\ell = \dfrac{\ell}{\sigma S}$ applies to both, consequently,

$$R_{\ell,\mathrm{TL}} = \frac{\ell}{\sigma_i S_i} + \frac{\ell}{\sigma_o S_o} \tag{6A.11}$$

where σ_i and σ_o are the conductivities of the materials of the inner and outer conductors, respectively, and S_i and S_o are the cross-sectional areas of the inner and outer conductors, respectively, and hence, the resistance per unit axial length parameter of the TL at DC, $R_{z,\mathrm{TL}}$

$$R_{z,\mathrm{TL,DC}} = \frac{1}{\sigma_i S_i} + \frac{1}{\sigma_o S_o} = \frac{1}{\sigma_i \pi a^2} + \frac{1}{\sigma_o \pi (c^2 - b^2)} \tag{6A.12}$$

6A.2.1 Conductance of TL Insulator, $G_{z,\mathrm{TL}}$

The conductance of interest here is that responsible for the leakage current between the inner and outer conductors through the insulating material. These leakage currents are caused by the potential difference between the inner and outer conductors. Due to the cylindrical symmetry, this potential difference causes a cylindrically radial field throughout the insulator (cylindrical shell). Hence, we can divide the insulator region into cylindrical volumetric increments as shown in Figure 6A.6, with the resistance of each increment given by

$$\Delta R_{\mathrm{increment}} = \frac{d\ell}{\sigma\, dS} = \frac{d\rho}{\sigma_d (\rho\, d\varphi)\cdot(dz)} \tag{6A.13}$$

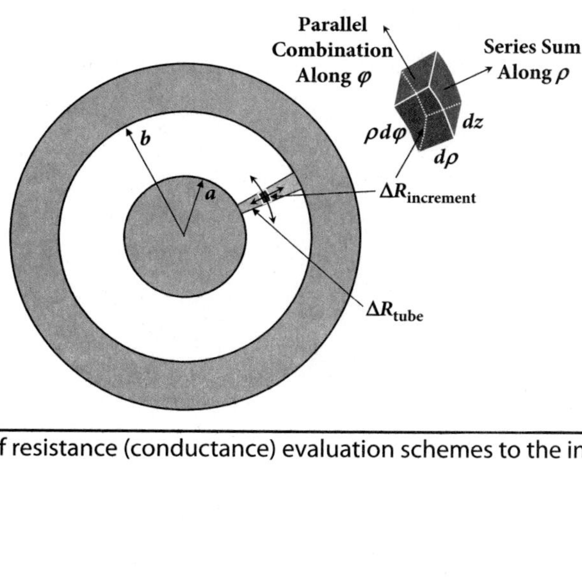

FIGURE 6A.6 Application of resistance (conductance) evaluation schemes to the insulating dielectric of a coaxial cable.

where σ_d is the conductivity of the insulating (dielectric) material

$$\Delta R_{z,\text{tube}} = \int_a^b \frac{d\rho}{\sigma_d(\rho\, d\varphi)\cdot(dz)} = \frac{\ln\left(\frac{b}{a}\right)}{\sigma_d(d\varphi)\cdot(dz)} \tag{6A.14}$$

$$G_{z,\text{TL}} = \frac{1}{R_{d \text{ per unit axial length}}} = \iint_S \frac{1}{\dfrac{\ln\left(\frac{b}{a}\right)}{\sigma_d(d\varphi)\cdot(dz)}} = \int_{z=0}^{1}\int_{\varphi=0}^{2\pi} \frac{\sigma_d(d\varphi)\cdot(dz)}{\ln\left(\frac{b}{a}\right)} = \frac{2\pi\sigma_d}{\ln\left(\frac{b}{a}\right)} \tag{6A.15}$$

ADDENDUM 6B CAPACITANCE EVALUATION

6B.1 Capacitance Evaluation

Again, the capacitance of a "linear" material configuration depends only on its geometrical parameters as well as the material permittivity. In other words, the E field factors out in Equation (6.44). To demonstrate, let us consider a small element of the material of length $\Delta\ell$ (along the electric field) and cross section ΔS (normal to the field) (Figure 6B.1), the capacitance of this element becomes

$$\Delta C = \frac{\Delta\psi}{\Delta V} = \frac{\varepsilon\vec{E}\cdot\overrightarrow{\Delta S}}{\vec{E}\cdot\overrightarrow{\Delta\ell}} = \frac{\varepsilon\,\Delta S}{\Delta\ell} \tag{6B.1}$$

Consequently, to evaluate the capacitance of any material configuration, we have two choices. The first choice is based on the use of Equation (6.44), while the second is based on the use of Equation (6B.1).

6B.1.1 Using V and ψ for Capacitance Evaluation

In this approach, we start by solving for the E field based on the applied potential difference V_{AB}. Once E is determined, we use Equations (4.7), (6.34) and (6.35) to evaluate ψ_{AB}. Finally, C_{AB} is evaluated as the ratio ψ_{AB}/V_{AB}.

6B.1.2 Using ΔC for Capacitance Evaluation

Again, we start by solving for the E field based on the applied potential difference V_{AB}. The only thing needed out of this solution is the field map and not the field intensity. The field map is then used to divide the material configuration into incremental volumes, each of which is considered an incremental capacitor $\Delta C_{\text{increment}} = \frac{\varepsilon\, dS}{d\ell}$. Next, we work out the total capacitance of the whole configuration by combining the incremental capacitance in series/parallel using one of two possible scenarios:

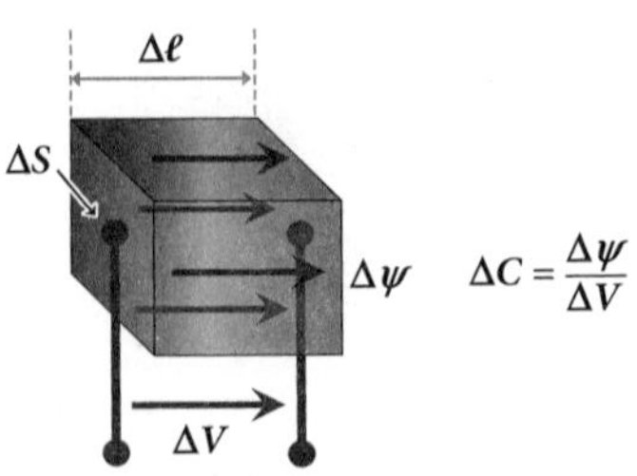

FIGURE 6B.1 Incremental volume to be used for capacitance evaluation.

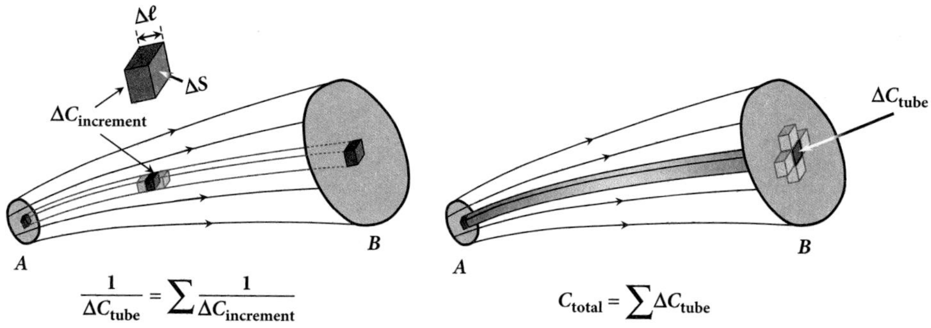

Figure 6B.2 Capacitance evaluation starting with series combinations.

1. *Series then parallel:* In this scenario, we combine the incremental capacitances along the field line to evaluate the capacitance on an incremental *tube* of the configuration: $\frac{1}{\Delta C_{\text{tube}}} = \int_A^B \frac{d\ell}{\varepsilon dS}$. Next, we combine the capacitances of all the parallel tubes to obtain the overall capacitance: $C_{\text{total}} = \iint_S \frac{1}{\int_A^B \frac{d\ell}{\varepsilon dS}}$ (Figure 6B.2).
2. *Parallel then series:* In this scenario, we start by the parallel combination of the incremental capacitances in a general cross-sectional location (normal to the field lines, which is also an equipotential surface). The result is the capacitance of an incremental *slice* of the configuration: $\Delta C_{\text{slice}} = \iint_S \frac{\varepsilon dS}{d\ell}$. Next, we combine the capacitances of all the series slices to obtain the overall capacitance: $\frac{1}{C_{\text{total}}} = \int_A^B \frac{1}{\iint_S \frac{\varepsilon dS}{d\ell}}$ (Figure 6B.3).

Now, we demonstrate the application of the above methods to the simple case of a uniform cross section (cylindrical) body of material with terminals "A" and "B" as shown in Figure 6B.4. We start by solving for the electric field due to an applied potential difference V_{AB}. For this simple geometrical configuration, it is obvious that the electric field lines are along the length of the cylinder and are uniform in intensity everywhere in the body of the cylinder.

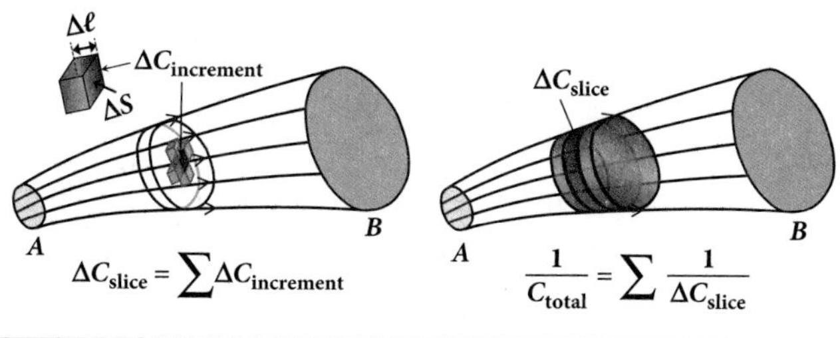

Figure 6B.3 Capacitance evaluation starting with parallel combinations.

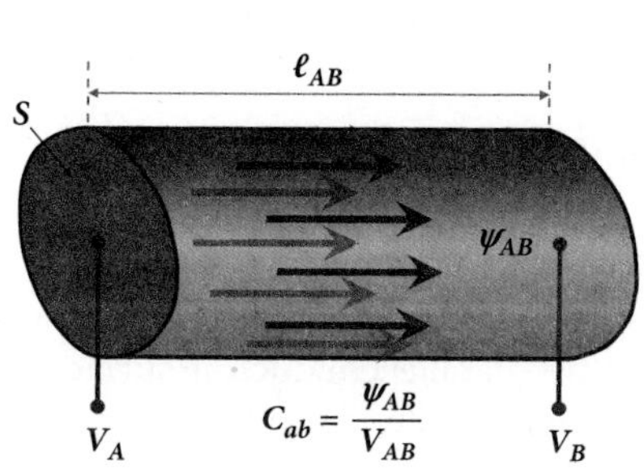

Figure 6B.4 Application of capacitance evaluation schemes to a cylindrical body.

Consequently,

$$V_{AB}=\int_A^B \vec{E}\cdot\vec{d\ell}=E\ell_{AB} \qquad \text{and hence} \qquad \vec{E}=\frac{V_{AB}}{\ell_{AB}}\vec{a}_\ell \tag{6B.2}$$

$$\psi_{AB}=\iint_S \varepsilon\vec{E}\cdot\vec{dS}=\varepsilon ES \tag{6B.3}$$

Thus, the capacitance of this cylindrical body of material between the terminals "A" and "B" is given by

$$C_{AB}=\frac{\psi_{AB}}{V_{AB}}=\frac{\varepsilon ES}{E\ell_{AB}}=\frac{\varepsilon S}{\ell_{AB}} \tag{6B.4}$$

We now turn to the series-parallel combination approach. We start by dividing the cylinder into small volumetric increments. Each increment has a length $d\ell$ along the length of the cylinder and a normal cross section dS. The capacitance of each increment would be given by

$$\Delta C_{\text{increment}}=\frac{\varepsilon dS}{d\ell} \tag{6B.5}$$

Now, adding the resistors in series to form a tube of length ℓ_{AB} and a cross section of dS, the capacitance of the incremental tube is

$$\frac{1}{\Delta C_{\text{tube}}}=\int_A^B \frac{1}{\Delta C_{\text{increment}}}=\int_A^B \frac{d\ell}{\varepsilon dS}=\frac{\ell_{AB}}{\varepsilon dS} \tag{6B.6}$$

Next, we combine all the incremental tubes in parallel to get the total capacitance:

$$C_{AB}=\iint_S \Delta C_{\text{tube}}=\iint_S \frac{1}{\frac{\ell_{AB}}{\varepsilon dS}}=\iint_S \frac{\varepsilon dS}{\ell_{AB}}=\frac{\varepsilon S}{\ell_{AB}} \tag{6B.7}$$

If we were to do the parallel combination first, we combine the incremental resistors to form a cylindrical slice of length $d\ell$ and a cross section S,

$$\Delta C_{\text{slice}}=\iint_S \Delta C_{\text{increment}}=\iint_S \frac{\varepsilon dS}{d\ell}=\frac{\varepsilon S}{d\ell} \tag{6B.8}$$

Next, we combine the capacitances of all the incremental slices in series to get the total capacitance:

$$\frac{1}{C_{AB}}=\int_A^B \frac{1}{\Delta C_{\text{slice}}}=\int_A^B \frac{d\ell}{\varepsilon S}=\frac{\ell_{AB}}{\varepsilon S} \qquad \text{or} \qquad C_{AB}=\frac{\varepsilon S}{\ell_{AB}} \tag{6B.9}$$

which is the same result as Equation (6B.4).

6B.2 Examples of "Controlled Geometry" Capacitances

6B.2.1 Parallel Plate Capacitance

One of the most basic and commonly discussed capacitance configurations is the parallel plate capacitance. Figure 6B.5 depicts this shape in which two parallel conducting surfaces "A" and "B" face each other in parallel with a separation distance d_{AB}. The common surface area of the two surfaces is denoted S_{AB}. Although the two conductors are shown to have thicknesses, the body of the two conductors and their configurations do not factor in the capacitance, which is solely determined by the flux bundle common to the two facing surfaces.

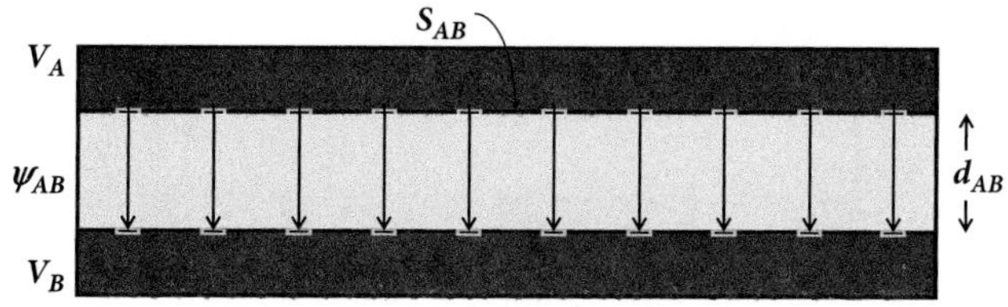

FIGURE 6B.5 Capacitance evaluation of a parallel plate capacitor.

The figure shows the flux lines departing the surface of conductor "A" and landing on the surface of conductor "B" in a normal fashion in compliance with the boundary conditions. However, the figure shows the flux lines to be uniform and parallel to each other all along the capacitance cross section. This would only be true if the two conductors have infinite common surfaces as was presented earlier in Figure 4A.32, Equations (4A.62). In practical cases, we can approximate such uniform flux density if the plate separation d_{AB} is much smaller the narrowest cross-sectional dimension.

In such case, the flux per unit cross-sectional area is given by Equation (4A.62)

$$\vec{D}(z) = \rho_S \vec{a}_z \tag{4A.62}$$

And using Equations (6.34) and (6.35), the electric field intensity vector becomes

$$\vec{E}(z) = \frac{1}{\varepsilon}\rho_S \vec{a}_z \tag{6B.10}$$

Consequently, the capacitance of the S_{AB} cross section separated by d_{AB} by a dielectric of permittivity ε is given by

$$C_{AB-SAB} = \frac{\Psi_{AB-SAB}}{V_{AB}} = \frac{\iint_{SAB} \vec{D}\cdot\vec{dS}}{\int_A^B \vec{E}\cdot\vec{d\ell}} = \frac{\iint_{SAB} (\rho_S \vec{a}_z)\cdot(dS\vec{a}_z)}{\int_A^B \left(\frac{1}{\varepsilon}\rho_S \vec{a}_z\right)\cdot(dz\vec{a}_z)} = \frac{\rho_S S_{AB}}{\frac{1}{\varepsilon}\rho_S d_{AB}} = \frac{\varepsilon S_{AB}}{d_{AB}} \tag{6B.11}$$

It is important to realize that the flux due to parallel plates with finite areas will not be uniform throughout. Figure 6B.6 demonstrates this case showing reasonable uniformity away from the edges of the plates with the presence of fringing flux extended beyond the common plates' area. In the absence of an exact analytical solution for this case, empirical formulae are typically used to improve the capacitance estimate through an added modification to Equation (6B.11).

6B.2.2 Coaxial Capacitor (Transmission Line C_z Parameter)

The coaxial capacitor is another common capacitance that exists in many applications. One particular application of interest in our daily lives is coaxial cables (transmission lines). The capacitance of interest here is the capacitance of the dielectric insulation between the inner and outer conductors (Figure 6B.7). Due to the cylindrical symmetry as was developed earlier at the end of Chapter 4, the electric flux density in the dielectric region due to a uniform charge density distribution ρ_ℓ is given by

$$\vec{D} = \begin{cases} 0, & \text{for } \rho < a \\ \dfrac{+\rho_\ell}{2\pi\rho}\vec{a}_\rho, & \text{for } b > \rho > a \\ 0, & \text{for } \rho > b \end{cases} \tag{4A.59}$$

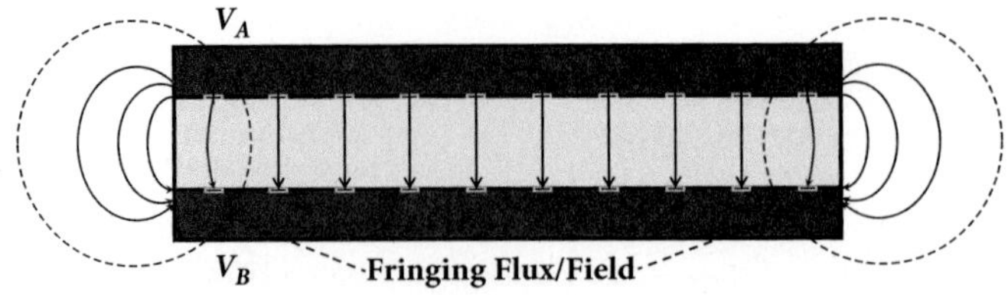

FIGURE 6B.6 Fringing fields (edge effects) for a parallel plate capacitor.

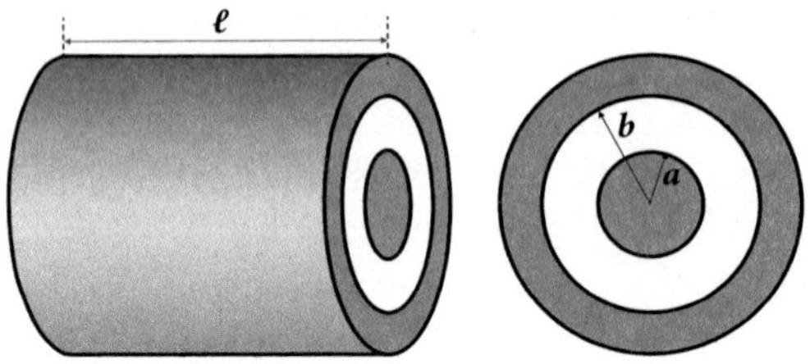

FIGURE 6B.7 Capacitance evaluation for a coaxial cable using the flux and potential difference scheme.

Similar to our discussion of fringing in the parallel plate capacitor case, an analytical expression for coaxial capacitor is available only for the special case of infinite axial extension. The approximate analytical expression for coaxial capacitances where the length ℓ_{AB} is much larger than the separation between the two conductors can obtained by referring to Figure 4A.31 and Equation (4A.59):

To evaluate the capacitance per unit length of the coaxial cable, we will need to derive an expression for the potential difference between the two TL conductors. So, starting with $D(\rho)$ inside the capacitor region, $a<\rho<b$:

$$\vec{D}(\rho)=\frac{\rho_\ell}{2\pi\rho}\vec{a}_\rho \tag{6B.12}$$

from which

$$\vec{E}(\rho)=\frac{\rho_\ell}{2\pi\varepsilon\rho}\vec{a}_\rho \tag{6B.13}$$

and the corresponding potential difference, V_{ab}, between the inner and outer conductors, is given by

$$V_{ab}=\int_a^b \vec{E}\cdot\vec{d\ell}=\int_a^b \frac{\rho_\ell}{2\pi\varepsilon\rho}\vec{a}_\rho\cdot\vec{d\rho}=\int_a^b \frac{\rho_\ell}{2\pi\varepsilon\rho}d\rho=\frac{\rho_\ell}{2\pi\varepsilon}\ln\left(\frac{b}{a}\right) \tag{6B.14}$$

and the corresponding flux per unit axial length is

$$\psi_{z,ab\text{ per unit axial length}}=\iint_{S\text{ per unit axial length}}\vec{D}\cdot\vec{dS}=\rho_\ell \tag{6B.15}$$

Hence, the expression for $C_{z,ab}$ for the TL is

$$C_{z,ab}=\frac{\psi_{z,ab\text{ per unit axial length}}}{V_{ab}}=\frac{2\pi\varepsilon}{\ln\left(b/a\right)}\text{ F/m} \tag{6B.16}$$

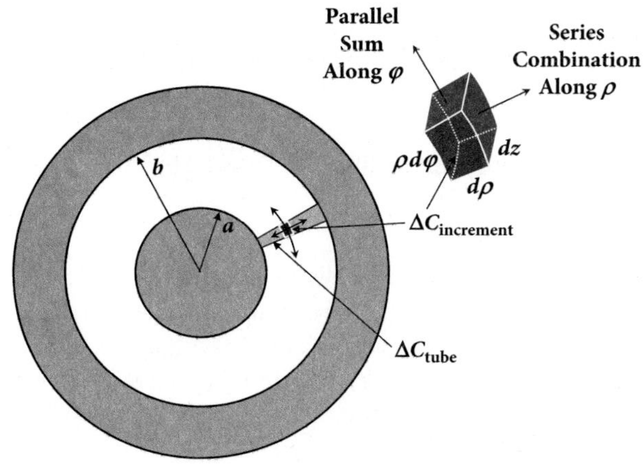

FIGURE 6B.8 Capacitance evaluation for a coaxial cable using the increment capacitors combination scheme.

Let us now practice the series-parallel and parallel-series approaches for the same purpose of evaluating the coaxial capacitance. This time, we are working in cylindrical coordinates as our increment is of the form of a pineapple chunk where the $\Delta C_{\text{increment}}$ is between the two curved surfaces Figure (6B.8):

$$\Delta C_{\text{increment}} = \frac{\varepsilon dS}{d\ell} = \frac{\varepsilon[\rho d\varphi][dz]}{d\rho} \tag{6B.17}$$

If we were to do the parallel combination first, we combine the incremental capacitors to form a cylindrical shell of thickness $d\rho$ and the cross section is the cylinder of radius ρ and a length of 1 m along the z axis,

$$\Delta C_{z,\text{slice}} = \iint_S \Delta C_{\text{increment}} = \iint_S \frac{\varepsilon dS}{d\ell} = \int_{z=z}^{z+1}\int_{\varphi=0}^{2\pi} \frac{\varepsilon[\rho d\varphi][dz]}{d\rho} = \frac{\varepsilon[2\pi\rho][1]}{d\rho} \tag{6B.18}$$

Next, we combine the capacitances of all the incremental slices in series to get the total capacitance:

$$\frac{1}{C_{z,\text{TL}}} = \int_a^b \frac{1}{\Delta C_{\text{slice}}} = \int_a^b \frac{d\rho}{\varepsilon[2\pi\rho]} = \frac{\ln\left(b/a\right)}{2\pi\varepsilon} \quad \text{or} \quad C_{z,\text{TL}} = \frac{2\pi\varepsilon}{\ln\left(b/a\right)} \tag{6B.19}$$

which is the same as Equation (6B.16)

Now, for the series combination first, we combine the incremental capacitors to form a cylindrical tube extending between the radii a and b with a cylindrical angle of $d\varphi$ and height dz,

$$\frac{1}{\Delta C_{\text{tube}}} = \int_a^b \frac{1}{\Delta C_{\text{increment}}} = \int_a^b \frac{d\rho}{\varepsilon[\rho d\varphi][dz]} = \frac{\ln\left(b/a\right)}{\varepsilon[d\varphi][dz]} \tag{6B.20}$$

$$C_{z,\text{TL}} = \iint_S \Delta C_{\text{tube}} = \int_{z=z}^{z+1}\int_{\varphi=0}^{2\pi} \frac{\varepsilon[d\varphi][dz]}{\ln\left(b/a\right)} = \frac{2\pi\varepsilon}{\ln\left(b/a\right)} \tag{6B.21}$$

which is again the same result as Equation (6B.16).

ADDENDUM 6C
RESISTORS AND CAPACITORS AS CIRCUIT ELEMENTS

6C.1 Resistance Circuit Relationships: Current, Voltage, Power, and Energy

6C.1.1 Current-Voltage Relationship

In circuit theory, the term electrical potential is typically referred to as the electrical voltage or simply voltage for short (in reference to the potential units of volt). Circuit analysis uses block models for the different circuit elements, which are represented by their corresponding current-voltage relationships (i-v in the time domain or I-V in the phasor domain) in the circuit's mathematical model (i.e., its equations).

The current flow in the resistance/conductance circuit is determined by the rate of flow of mobile (conduction) charges through the resistance/conductance body. Using Equation (6.17) and referring to Figure 6C.1, the Ohm's law expression for the current flow from "A" to "B" is of the form

$$i_{AB} = G_{AB} v_{AB} = \frac{1}{R_{AB}} v_{AB} \text{ (for instantaneous currents and voltages)} \tag{6C.1}$$

$$\bar{I}_{AB} = G_{AB} \bar{V}_{AB} = \frac{1}{R_{AB}} \bar{V}_{AB} \text{ (for phasor currents and voltages)} \tag{6C.2}$$

6C.1.2 Power in Resistance/Conductance

Starting with Equation (6.17) for the power dissipation in the volume v

$$P_{\text{total}} = \int_{\text{vol}} JE\, dv = \int_{\text{vol}} JE\, dS\, d\ell \tag{6C.3}$$

We can perform this integration for the resistance of Figure 6C.1 in two possible ways similar to what we have done earlier in the resistance evaluation process (Figures 6A.2 and 6A.3). Hence, we have a choice between integrating over the slice of surface S first

$$P_{\text{total}} = \int_\ell \left[\int_S J\, dS\right] E\, d\ell = \int_\ell [I_{AB}] E\, d\ell = [I_{AB}] \int_\ell E\, d\ell = [I_{AB}] \cdot [V_{AB}] \tag{6C.4}$$

or integrating along the tube length ℓ first

$$P_{\text{total}} = \int_S \left[\int_\ell E\, d\ell\right] J\, dS = \int_S [V_{AB}] J\, dS = [V_{AB}] \int_S J\, dS = [V_{AB}] \cdot [I_{AB}] \tag{6C.5}$$

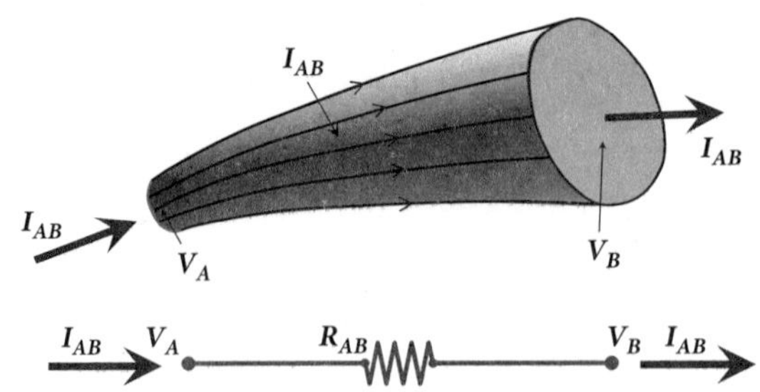

FIGURE 6C.1 Schematic representation of the resistance displaying the current through and the voltage difference across.

with both leading to the same exact answer, hence

$$P_{\text{total}} = [I_{AB}]\cdot[V_{AB}] = [I_{AB}]^2 \cdot [R_{AB}] = [V_{AB}]^2 \cdot [G_{AB}] \tag{6C.6}$$

The above formulae are good for steady currents; however, for instantaneous values of a time varying currents, we need to rewrite them as

$$P_{\text{total}}(\text{for any volume } v) = \text{average}\left\{\int_{\text{vol}} \sigma[E_{pk}\cos(\omega t + \varphi_E)]^2\, dv\right\} = \int_{\text{vol}} \sigma[E_{rms}]^2\, dv \tag{6C.7}$$

And for the resistance R_{AB},

$$P_{\text{total}} = \text{average}\left\{[I_{AB,pk}\cos(\omega t + \varphi_I)]\cdot[V_{AB,pk}\cos(\omega t + \varphi_V)]\right\} = \left(\frac{1}{2}\right) I_{AB,pk} V_{AB,pk} \cos(\varphi_V - \varphi_I) \tag{6C.8}$$

Since $\varphi_V = \varphi_I$ for the resistance R_{AB}, then the final form reduces to

$$P_{\text{total}} = \left(\frac{1}{2}\right) I_{AB,pk} V_{AB,pk} = I_{AB,rms} V_{AB,rms} = [I_{AB,rms}]^2 \cdot [R_{AB}] = [V_{AB,rms}]^2 \cdot [G_{AB}] \tag{6C.9}$$

where the subscripts *pk* and *rms* refer to the peak and root mean square values, respectively, for both the harmonic voltage and current.

6C.1.3 Resistances/Conductances in Series

Referring to Figure 6C.2, the cascade of current bundles (of equal current amounts) results in combining their individual resistance/conductance in series. In the figure, two current bundles $I_{AB} = I_{BC}$ cascade the two resistance R_{AB} and R_{BC} in series through the common terminal B. The total potential difference between the end terminal A and C equals the sum of the individual differences V_{AB} and V_{BC} and hence,

$$V_{AC} = V_{AB} + V_{BC} = R_{AB} I_{AB} + R_{BC} I_{AB} = I_{AB}[R_{AB} + R_{BC}] \tag{6C.10}$$

If we were to model the combination by a single resistance $R_{AC\text{-combined}}$ (or R_{AC-c} for short), then

$$V_{AC} = I_{AB}[R_{AB} + R_{BC}] = I_{AC-c} R_{AC-c} \tag{6C.11}$$

which implies that the combined resistance of the two series resistances is given by

$$R_{AC-c} = [R_{AB} + R_{BC}] \tag{6C.12}$$

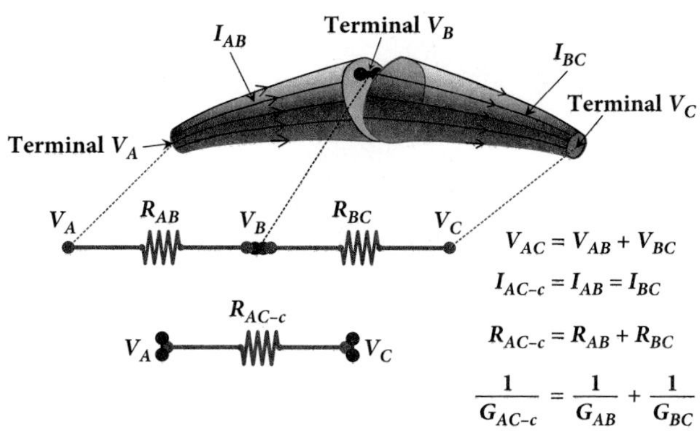

FIGURE 6C.2 Series combination of resistances.

and the corresponding conductance relationship is

$$\left[\frac{1}{G_{AC-c}}\right]=\left[\frac{1}{G_{AB}}+\frac{1}{G_{BC}}\right] \tag{6C.13}$$

which can be generalized for multiple series resistances as

$$R_{\text{series combination}}=\sum_{\text{all series}} R_{\text{individual}} \quad \text{and} \quad \frac{1}{G_{\text{series combination}}}=\sum_{\text{all series}}\frac{1}{G_{\text{individual}}} \tag{6C.14}$$

6C.1.4 Resistances/Conductances in Parallel

Referring to Figure 6C.3, the combination of currents flowing between the same two terminals results in combining their individual resistances/conductances in parallel. In the figure, two currents I_{AB1} and I_{AB2} flow between the same two terminals A and B. The two resistances R_{AB1} and R_{AB2} are then said to be combined in parallel.

The total current flowing between A and B is then equal to the sum of the individual currents and hence,

$$I_{AB}=I_{AB1}+I_{AB2}=V_{AB}/R_{AB1}+V_{AB}/R_{AB2}=\left[1/R_{AB1}+1/R_{AB2}\right]V_{AB} \tag{6C.15}$$

If we were to model the combination by a single resistance R_{AB-c}, then

$$I_{AB}=\left[1/R_{AB1}+1/R_{AB2}\right]V_{AB}=\left[1/R_{AB-c}\right]V_{AB} \tag{6C.16}$$

which implies that the combined resistance of the two parallel resistances is given by

$$\left[\frac{1}{R_{AB-c}}\right]=\left[\frac{1}{R_{AB1}}+\frac{1}{R_{AB2}}\right] \tag{6C.17}$$

and the corresponding conductance relationship is

$$\left[G_{AB-c}\right]=\left[G_{AB1}+G_{AB2}\right] \tag{6C.18}$$

which can be generalized for multiple parallel resistances as

$$\frac{1}{R_{\text{parallel combination}}}=\sum_{\text{all parallel}}\frac{1}{R_{\text{individual}}} \quad \text{or} \quad G_{\text{parallel combination}}=\sum_{\text{all parallel}} G_{\text{individual}} \tag{6C.19}$$

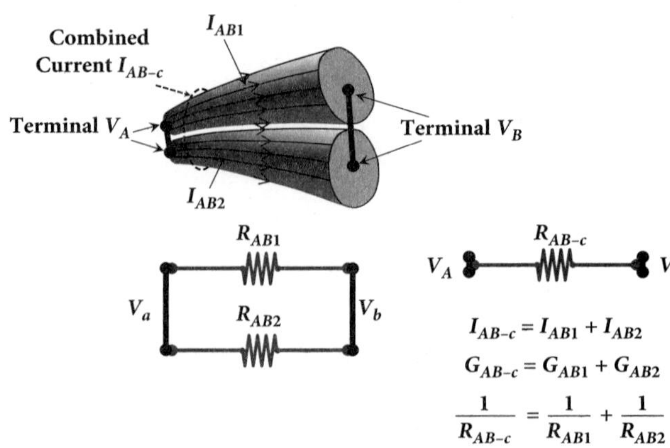

FIGURE 6C.3 Parallel combination of resistances.

6C.2 Capacitance Circuit Relationships: Current, Voltage, Power, and Energy

6C.2.1 Current-Voltage Relationship

Obviously, under static conditions, there will be no current flowing in the capacitance and the static voltage across its terminals will be governed by the amount of charges present on the terminals.

However, under dynamic conditions, the current flow in the *capacitance* circuit is determined by the rate of change of the conductor charge. Using Equation (6.4) and referring to Figure 6C.4, the current into conductor A can be expressed as the rate of increase in charge accumulation on that conductor. Hence using the lower-case notation for i and v in the time domain, and using Equation (6.45),

$$i_{AB} = \frac{dQ_{AB}}{dt} = \frac{d[C_{AB} v_{AB}]}{dt} = C_{AB}\frac{d[v_{AB}]}{dt} \text{ (for stationary fixed capacitances)} \tag{6C.20}$$

The corresponding phasor domain expression is

$$\bar{I}_{AB} = j\omega C_{AB} \bar{V}_{AB} \tag{6C.21}$$

It is to be noted that the capacitance current i_{AB} or $\bar{I}_{AB}$ is not a conduction current that physically "flows" in the capacitance dielectric region. A perfect dielectric should have zero conductivity and hence no conduction current can flow through it. This sounds like a mystery as we have just derived expressions for the currents i_{AB} and $\bar{I}_{AB}$. The explanation of this baffling current will be presented in adequate detail in a later chapter when we discuss the topic of displacement currents. However, a quick explanation lies in the dynamic change in polarization within the dielectric material as a result of the dynamic change in its internal field. A dynamic change in polarization implies dynamic "finite displacements" of the internal bound charges, and hence the "displacement" current.

This displacement current is equal to the conduction current flowing in the conductor leads to the terminals of the capacitance. Hence, from the circuit theory point of view, the current continuity implies the same current flows throughout the capacitance and its terminals' leads (Figure 6C.5).

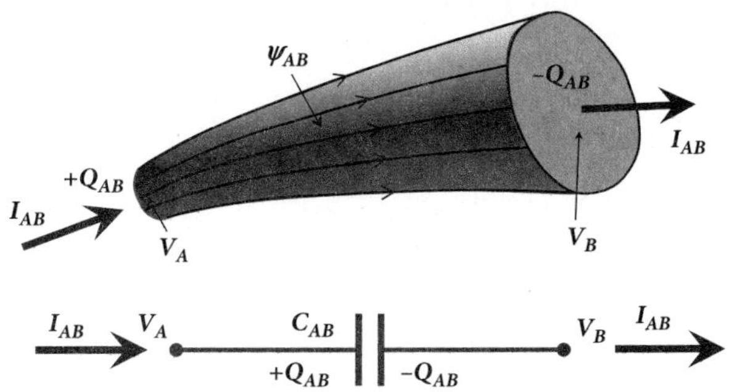

Figure 6C.4 Schematic representation of the capacitance displaying the current through and the voltage difference across.

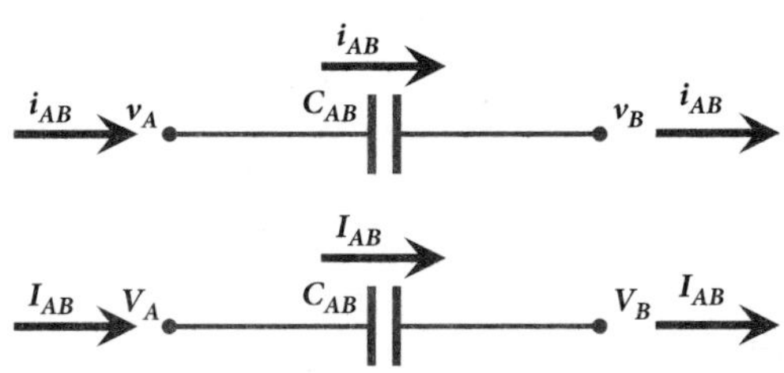

Figure 6C.5 Instantaneous and phasor currents and voltages for the capacitance.

6C.2.2 Power in Capacitance

Under static EM fields, the capacitance stores static energy and hence there is no associated power. Under dynamic situations, the energy stored fluctuates with time but results in no average power dissipation (or generation). This can be confirmed by using Equation (1A.16) in Addendum 1A

$$P_{AB,av} = Re\left\{\frac{1}{2}[(V_{AB})]\cdot[(I_{AB})]^*\right\} = Re\left\{\frac{1}{2}j\omega C_{AB}\left|\bar{V}_{AB}\right|^2\right\} = 0 \tag{6C.22}$$

6C.2.3 Energy Stored in Capacitance

According to Equation (5.82), the energy stored in a capacitance is given by

$$W_e = \int_{\text{vol}} \frac{1}{2}\{\vec{D}\cdot\vec{E}\}dv \tag{5.82}$$

Now, referring to Figure 6C.6, the integration's incremental volume dv can be chosen to have its length $\Delta\ell$ along the field (or flux) lines and its cross section normal to the flux lines. Consequently, we can write

$$W_e = \int_{\text{capacitor vol}} \frac{1}{2}\{\vec{D}\cdot\vec{E}\}dv = \int_{\text{cap vol}} \frac{1}{2}\{DE\}dS\,d\ell = \int_{\text{cap vol}} \frac{1}{2}[D\,dS][E\,d\ell] \tag{6C.23}$$

$$W_e = \frac{1}{2}\int_{\text{cap cross section}} [D\,dS]\int_{\text{cap length}} [E\,d\ell] = \frac{1}{2}\psi_{AB}V_{AB} = \frac{1}{2}Q_{AB}V_{AB} \tag{6C.24}$$

$$W_e = \frac{1}{2}Q_{AB}V_{AB} = \frac{1}{2}C_{AB}V_{AB}^2 = \frac{1}{2}\frac{Q_{AB}^2}{C_{AB}} \tag{6C.25}$$

This energy is stored in the charge polarization in the capacitance system, both the bound charge polarization within the dielectric material as well as the free charge polarization between the two conductor terminals.

6C.2.4 Capacitances in Series

Referring to Figure 6C.7, the cascade of flux bundles (of equal flux amounts) results in combining their individual capacitances in series. In the figure, two flux bundles $\psi_{AB} = \psi_{BC}$ cascade the two capacitances C_{AB} and C_{BC} in series through the common terminal B. The total potential difference between the end terminal A and C is the equals the sum of the individual differences V_{AB} and V_{BC} and hence,

$$V_{AC} = V_{AB} + V_{BC} = \frac{\psi_{AB}}{C_{AB}} + \frac{\psi_{AB}}{C_{BC}} = \psi_{AB}\left[\frac{1}{C_{AB}} + \frac{1}{C_{BC}}\right] = Q_{AB}\left[\frac{1}{C_{AB}} + \frac{1}{C_{BC}}\right] \tag{6C.26}$$

If we were to model the combination by a single capacitance $C_{AC\text{-combined}}$ (or $C_{AC\text{-}c}$ for short), then

$$V_{AC} = Q_{AB}\left[\frac{1}{C_{AB}} + \frac{1}{C_{BC}}\right] = Q_{AC-c}\left[\frac{1}{C_{AC-c}}\right] \tag{6C.27}$$

which implies that the combined capacitance of the two series capacitances is given by

$$\left[\frac{1}{C_{AC-c}}\right] = \left[\frac{1}{C_{AB}} + \frac{1}{C_{BC}}\right] \tag{6C.28}$$

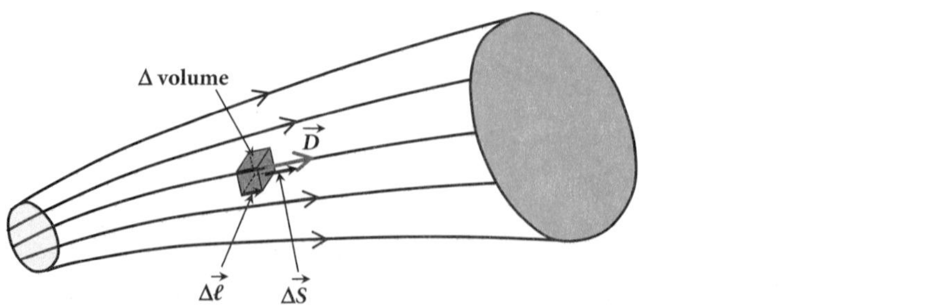

Figure 6C.6 Incremental volume used for the evaluation of the energy stored in the capacitance.

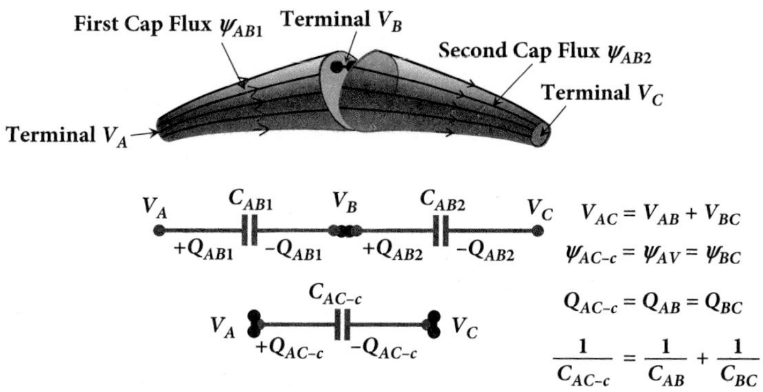

FIGURE 6C.7 Series combination of capacitances.

which can be generalized for multiple series capacitances as

$$\frac{1}{C_{\text{series combination}}} = \sum_{\text{all series}} \frac{1}{C_{\text{individual}}} \tag{6C.29}$$

6C.2.5 Capacitances in Parallel

Referring to Figure 6C.8, the combination of flux bundles flowing between the same two terminals results in combining their individual capacitances in parallel. In the figure, two flux bundles ψ_{AB1} and ψ_{AB2} flow between the same two terminals A and B. The two capacitances C_{AB1} and C_{AB2} are then said to be combined in parallel.

The total flux flowing between A and B is then equal to the sum of the individual fluxes and hence,

$$\psi_{AB} = \psi_{AB1} + \psi_{AB2} = C_{AB1}V_{AB} + C_{AB2}V_{AB} = [C_{AB1} + C_{AB2}]V_{AB} \tag{6C.30}$$

If we were to model the combination by a single capacitance C_{AB-c}, then

$$\psi_{AB} = Q_{AB} = [C_{AB1} + C_{AB2}]V_{AB} = [C_{AB-c}]V_{AB} \tag{6C.31}$$

which implies that the combined capacitance of the two parallel capacitances is given by

$$[C_{AB-c}] = [C_{AB1} + C_{AB2}] \tag{6C.32}$$

which can be generalized for multiple parallel capacitances as

$$C_{\text{parallel combination}} = \sum_{\text{all parallel}} C_{\text{individual}} \tag{6C.33}$$

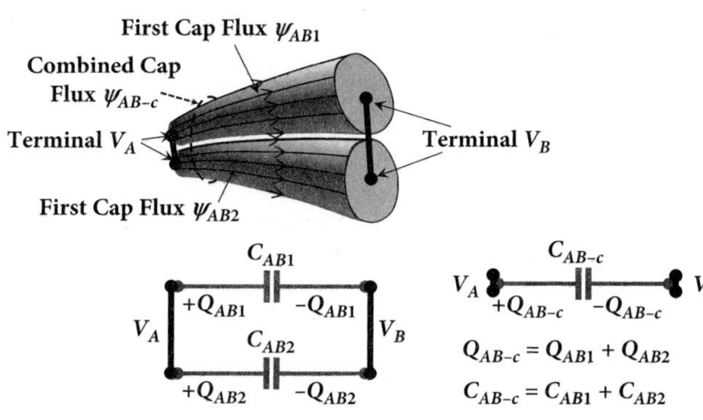

FIGURE 6C.8 Parallel combination of capacitances.

CHAPTER 6 PROBLEMS

- Problems 6.1 to 6.3: Problems dealing with conductor properties under static conditions
- Problems 6.4 to 6.11: Problems dealing with R and C evaluation
- Problems 6.12 to 6.18: Problems dealing with energy in capacitors
- Problems 6.19 to 6.21: Problems dealing with boundary conditions

6.1 Derive expressions for the electric field intensity vector in all regions of space ($0 < r < \infty$) for the concentric spherical setup shown, the radii are given by a for the inner conductor and "b and c," respectively, for the outer conductor's inner and outer surfaces. Starting with both conductors uncharged, a charge density Q was injected in the inner conductor while the outer conductor was grounded.

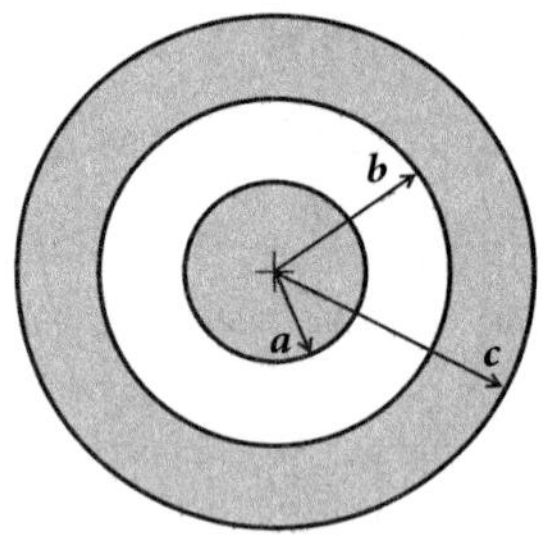

FIGURE 6P.1

6.2 Repeat if the outer conductor was not grounded during the charge injection process.

6.3 Derive expressions for the electric field intensity vector in all regions of space for the coaxial cylindrical cable with infinite extension, the cylindrical radii are given by a for the inner conductor and "b and c" for the outer conductor's inner and outer surfaces, respectively. Starting with both conductors uncharged, a uniform linear charge density ρ_ℓ was injected in the inner conductor.

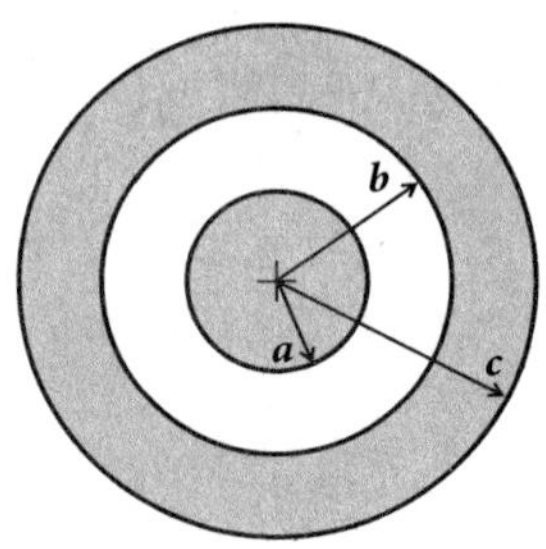

FIGURE 6P.3

6.4 Find an expression for the capacitance per unit length between two cylindrical wires of negligible radii located a distance d from center to center.

6.5 Two cylindrical conductors with infinite extension along the z axis (normal to the page), the conductors radii are given by a, b, and c as shown.

a. Find the capacitance C_{ab} per unit z axis length.

b. Discuss the capacitance $C_{c\infty}$.

c. Discuss the case if the outer conductor was grounded.

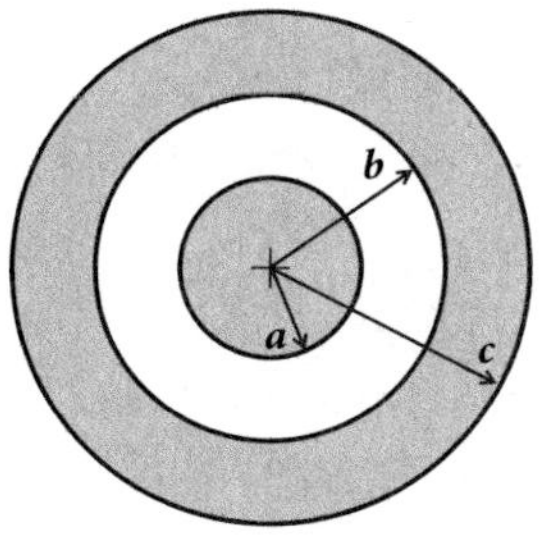

Figure 6P.5

6.6 Write down the expression for the capacitance between the two conductors of the dielectric-filled parallel plate capacitor with conductor surface area S, dielectric thickness d, and dielectric constant ε_r. Assume d is much too small when compared to the dimensions of S. Discuss the limitation of this expression.

6.7 Derive an expression for the capacitance between two concentric hemispherical conducting shapes as shown in the left part of the figure: $a<r<b$, $0<\theta<\frac{\pi}{2}$, and $0<\varphi<2\pi$, given that $r=a$ and $r=b$ are equipotential surfaces and the dielectric constant of the insulating material is ε_r. Discuss how your answer would change with the change in the conductor shapes as shown in the right part of the figure.

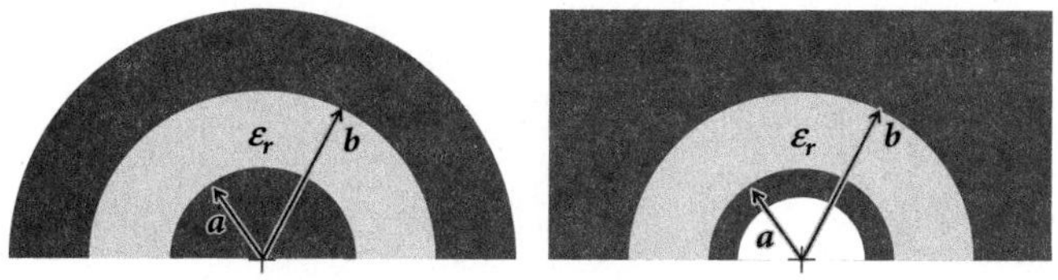

Figure 6P.7

6.8 Derive an expression for the resistance of the conducting hemispherical shell: $a<r<b$, $0<\theta<\frac{\pi}{2}$, and $0<\varphi<2\pi$, given that $r=a$ and $r=b$ are equipotential surfaces and the conductivity of the shell material is σ.

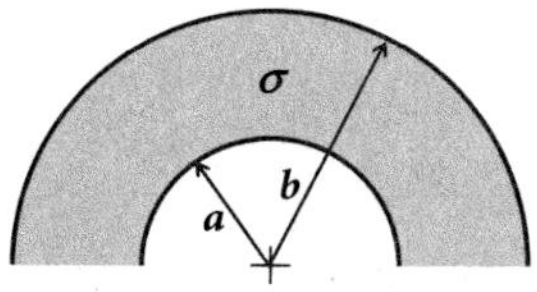

Figure 6P.8

6.9 Starting with a 3D increment and integrating capacitances in parallel and in series, evaluate the capacitance for the spherical dielectric shell defined by $r_1<r<r_2$, $\theta_1<\theta<\theta_2$, and $\varphi_1<\varphi<\varphi_2$ for each of the following conductor (equipotential) locations:

a. Conductors located on the r_1 and r_2 surfaces
b. Conductors located on the θ_1 and θ_2 surfaces
c. Conductors located on the φ_1 and φ_2 surfaces

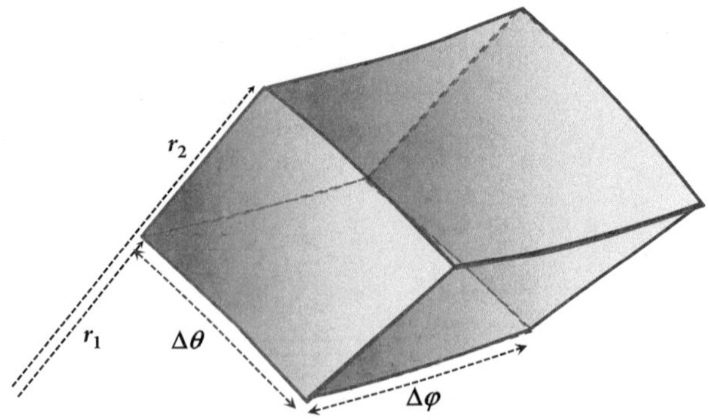

Figure 6P.9

6.10 Starting with a 3D increment and integrating capacitances in parallel and in series, evaluate the capacitance for the cylindrical dielectric shell defined by $\rho_1 < \rho < \rho_2$, $\varphi_1 < \varphi < \varphi_2$, and $z_1 < z < z_2$ for each of the following conductor (equipotential) locations:

a. Conductors located on the ρ_1 and ρ_2 surfaces

b. Conductors located on the φ_1 and φ_2 surfaces

c. Conductors located on the z_1 and z_2 surfaces

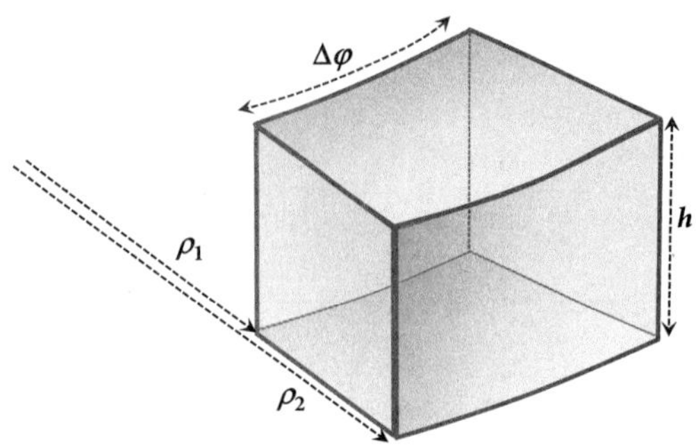

Figure 6P.10

6.11 Repeat Problems 6.8 and 6.9 with the dielectric shell replaced by a conducting shell with material conductivity σ.

6.12 A parallel plate capacitor is filled with a dielectric material of permittivity $2\varepsilon_o$ is shown in the left side of the figure. The dielectric strength of the dielectric is 10 × 106 V/m.

Find the capacitance, the energy stored for an applied voltage of 20 V, and the maximum voltage that can be applied across the plates.

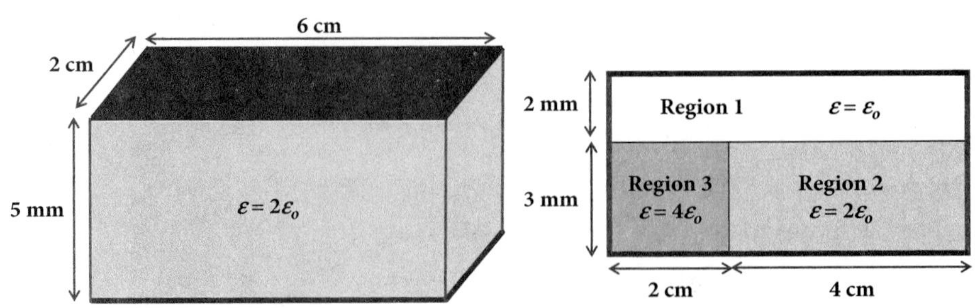

Figure 6P.12

If the dielectric in this capacitor is now replaced by a three-dielectric region configuration as shown the right side of the figure,

a. Find the equivalent capacitance.

b. Find the energy stored for an applied voltage of 20 V.

c. Find the overall breakdown voltage given that the dielectric strengths of three dielectrics are 3, 10, and 20 MV/m, respectively.

6.13 Given the following dielectric properties for the materials used in the three regions shown in the figure:

Region	Material	Dielectric Constant	Dielectric Strength
Region 1	Air	1	3 MV/m
Region 2	Teflon	2	20 MV/m
Region 3	Glass	4	10 MV/m

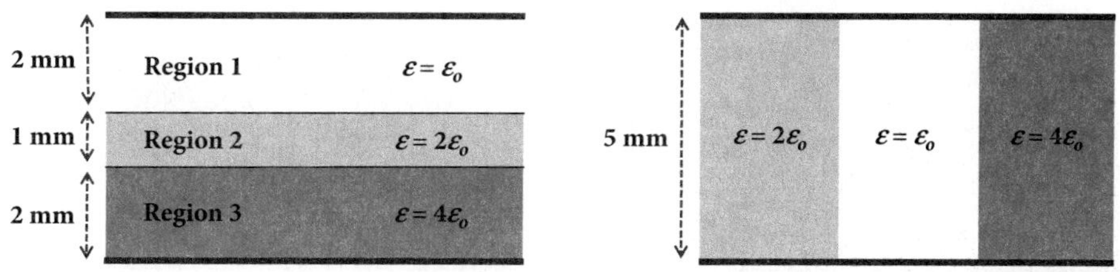

Figure 6P.13

a. Evaluate the capacitance of the parallel plate capacitors shown given the surface area of the plates is 120 cm^2. Also given that the three dielectrics in the right-side configuration have equal cross-sectional areas (40 cm^2 each).

b. Evaluate the energy stored in each dielectric region with an applied voltage of 50 V.

c. Evaluate the maximum voltage each of the two capacitors can withstand before breaking down.

6.14 For the two spherical capacitors shown, the two insulating dielectrics have the following properties:

Region	Material	Dielectric Constant	Dielectric Strength
Region 2	Teflon	2	20 MV/m
Region 3	Glass	4	10 MV/m

a. Evaluate the capacitance of the two capacitors given the radii r_1, r_2, and r_3 are 0.8, 1.0, 1.2 cm, respectively.

b. Evaluate the energy stored in each dielectric region with an applied voltage of 50 V.

c. Evaluate the maximum voltage each of the two capacitors can withstand before breaking down.

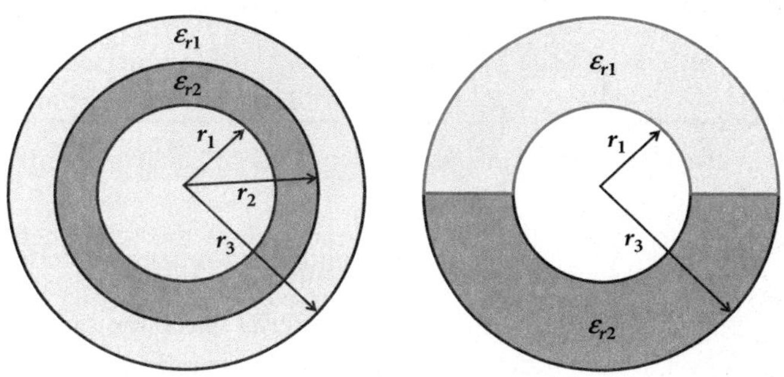

Figure 6P.14

6.15 For the two cylindrical capacitors shown, the two insulating dielectrics have the following properties:

Region	Material	Dielectric Constant	Dielectric Strength
Region 2	Teflon	2	20 MV/m
Region 3	Glass	4	10 MV/m

a. Evaluate the capacitance of the two capacitors given the radii ρ_1, ρ_2, and ρ_3 are 0.8, 1.0, 1.2 cm, respectively, and the capacitor is 5 cm in length.

b. Evaluate the energy stored in each dielectric region with an applied voltage of 50 V.

c. Evaluate the maximum voltage each of the two capacitors can withstand before breaking down.

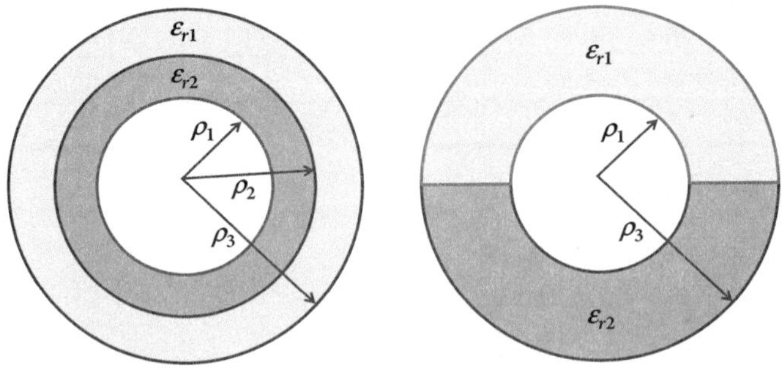

FIGURE 6P.15

6.16 Derive expressions for the electric field intensity vector in all regions of space ($0 < r < \infty$) for the concentric spherical setup shown, the radii are given by a for the inner conductor and "b and g," respectively, for the outer conductor's inner and outer surfaces. Starting with both conductors uncharged, a charge density Q was injected in the inner conductor while the outer conductor was grounded.

a. Find the total energy stored in the space between the inner and outer conductors.

b. Derive and expression for the capacitance between the two conductors as well as the self-capacitance of the outer conductor.

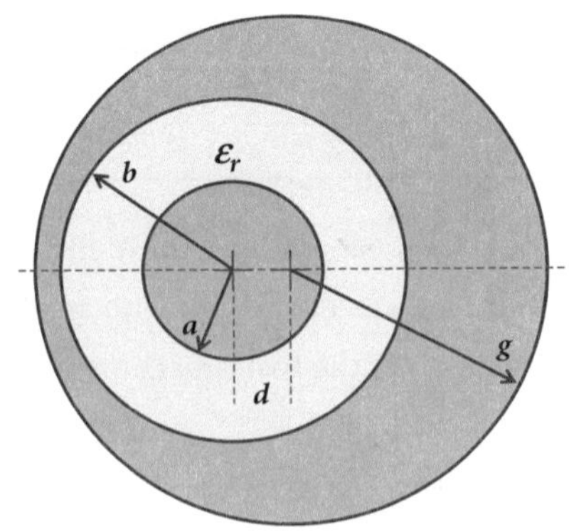

FIGURE 6P.16

6.17

a. Find the energy density in the dielectric region as well as the total energy stored in a parallel plate capacitor when connected to a battery of a constant voltage V.

b. Find the work done in removing the dielectric material while the capacitor is kept connected to the battery V. {Hint: Work done = difference in energy stored as ε is replaced by ε_o.}

c. Repeat part b above if the battery was disconnected first before removing the dielectric.

6.18 Two concentric spherical conductors with radii as shown. The space between the conductors is air. Using the energy stored approach, find the capacitance C_{ab} between the two conductors.

a. Find the work done in filling the space between the conductors with oil (dielectric constant = 7), if the voltages of the inner and outer conductors are kept constant at V_i and "V_o = zero," respectively.

b. Find the surface density of the free charges on the conductor surfaces in both the air and oil cases.

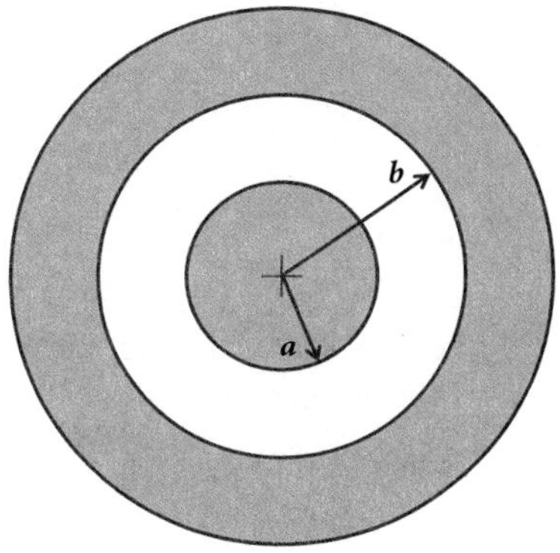

Figure 6P.18

6.19 Two dielectric media with permittivities $\varepsilon_o\varepsilon_{r1}$ and $\varepsilon_o\varepsilon_{r2}$ are separated by a charge-free boundary as shown in the figure. The electric field in medium 1 at the interface has a magnitude of $10\sqrt{3}$ V/m and angle 30° with the normal to the interface; $\varepsilon_{r1} = 1$ and $\varepsilon_{r2} = 3$. Apply the appropriate boundary conditions to find the magnitude and direction of the electric field in medium 2 at the interface, E_2 and θ.

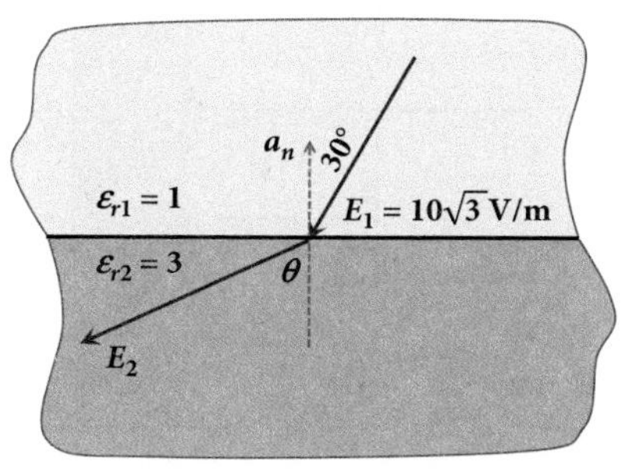

Figure 6P.19

6.20 An electric field is incident on a two dielectric slab arrangement as shown in the figure. The incidence angle is $\theta_o = 30°$. The two dielectric constants are ($\varepsilon_{r1} = 2$ and $\varepsilon_{r2} = 4$) and the corresponding thicknesses of the slabs are d_1, and d_2, respectively. Both slabs extend indefinitely in both x and z directions. Given the incident field intensity is 10 V/m,

a. Find electric field intensity as it refracts in the following regions.

b. Find the angles of refractions as labeled in the figure.

c. Repeat if the second slab was changed to a conducting slab with perfect conductivity.

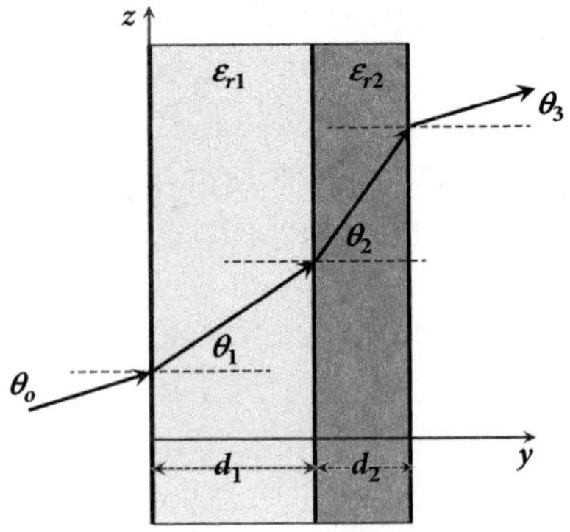

FIGURE 6P.20

6.21 Derive expressions for the electric field intensity vectors, the electric flux density vectors, as well as the volumetric and surface charge densities in all four regions of space $y < 0$, $0 < y < d_1$, $d_1 < y < d_1 + d_2$, $d_1 + d_2 < y$. Given the dielectric constants, are $\varepsilon_{r1} = 2$ and $\varepsilon_{r2} = 4$ and their corresponding thicknesses are $d_1 = 5$ mm and $d_2 = 2$ mm. Also, given that the applied field intensity vector is

a. $\vec{E} = 5\vec{a}_y$

b. $\vec{E} = 3\vec{a}_z + 5\vec{a}_y$

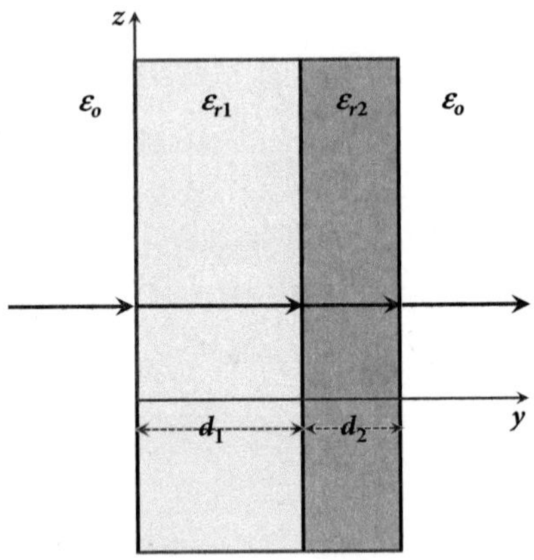

FIGURE 6P.21

CHAPTER 6 SUMMARY[3]

Conductors

- Conductors have mobile "free" charges — applied field forces mobile-charges to drift causing conduction current — average drift velocity = (charge carrier mobility) × (electric field)
- Conductors under static conditions:
  - The entire bulk of the conductor is equipotential.
  - Electric fields are nulled within the conductor bulk and can only exist external to the conductor.
  - Charges can only exist on the surface and never within the bulk of the conductor.
- Conductors under dynamic conditions:
  - Charges flow in the body of the conductor causing an electric current.
  - An internal electric field sustains the motion of the charge carriers to overcome the conductor's internal resistance due to the carriers' multiple collisions.
  - A potential difference appears across the terminals of the conductor as determined by the current flow/internal electric field.

Electric Current, Current Densities, and Resistance

Line, surface, and volume current (A): $I = dQ/dt$

Linear density of surface current (A/m): $\vec{K} = dI/d\ell_n \vec{a}_I$

Surface density of volume current (A/m^2): $\vec{J} = dI/dS_n \vec{a}_I$

$$I = \iint_S \vec{J} \cdot \overrightarrow{dS} \qquad \{I_{\text{closed surface } S}\} = -\frac{dQ_{\text{enclosed by } S}}{dt} \qquad \vec{\nabla} \cdot \vec{J} = -\frac{\partial \rho_v}{\partial t} \text{ [= 0 for statics]}$$

$$\vec{J} = \sigma \vec{E} \qquad \sigma = \left[\rho_v^+ \mu^+ - \rho_v^- \mu^-\right] = 1/\rho \qquad \rho \text{ units } \Omega \cdot \text{m}$$

$$R_{AB} = \frac{V_{AB}}{I_{AB}} = \frac{\int_A^B \vec{E} \cdot \overrightarrow{d\ell}}{\iint_S \vec{J} \cdot \overrightarrow{dS}} = \frac{\int_A^B \vec{E} \cdot \overrightarrow{d\ell}}{\iint_S \sigma \vec{E} \cdot \overrightarrow{dS}} = 1/\iint_S \frac{1}{\int_A^B \frac{d\ell}{\sigma\, dS}} = \int_A^B 1/\iint_S \frac{\sigma\, dS}{d\ell} = \frac{1}{G_{AB}} \qquad V_{AB} = R_{AB} I_{AB} \text{ [Ohm's law]}$$

$$\text{Coaxial TL (per unit axial length): } R_{z,\text{TL,DC}} = \frac{1}{\sigma_i \pi a^2} + \frac{1}{\sigma_o \pi (c^2 - b^2)} \qquad G_{z,\text{TL}} = \frac{2\pi\sigma_d}{\ln(b/a)}$$

Dielectrics

Dielectrics have bound charges (no mobile charges—no conduction currents).

Applied electric field causes bound charge polarization—dipole moments, $\vec{p}_d$

$$\text{Polarization vector, } \vec{P} = \lim_{\Delta v \to 0} \frac{\sum_{\Delta v} \vec{p}_d}{\Delta v} = \chi \varepsilon_o \vec{E} \qquad \vec{D} = \varepsilon_o \vec{E} + \vec{P} = \varepsilon_r \varepsilon_o \vec{E} = \varepsilon \vec{E}$$

Bound charge volumetric density: $\rho_{vb} = -\vec{\nabla} \cdot \vec{P}$ Surface density: $\rho_{Sb} = \vec{P} \cdot \vec{a}_S$

$$\text{Volumetric energy density: } W_e = \frac{1}{2}\{\vec{D} \cdot \vec{E}\} = \frac{1}{2}\varepsilon E^2 \quad \left[= \frac{1}{2}\varepsilon_o E^2 \text{ for free space}\right]$$

$$\text{Energy stored: } W_e = \int_{\text{vol}} \frac{1}{2}\{\vec{D} \cdot \vec{E}\}\, dv = \int_{\text{vol}} \frac{1}{2}\varepsilon E^2\, dv \quad \left[= \int_{\text{vol}} \frac{1}{2}\varepsilon_o E^2\, dv \text{ for free space}\right]$$

---

[3]***Important Warning to Students:*** Please do not use this summary without thorough study of the relevant material in this chapter. Using this summary and its formulas without adequate appreciation of the physics involved and conditions of validity can lead to wrong implementations and invalid conclusions.

Capacitance

$$C_{AB} = \frac{\Psi_{AB}}{V_{AB}} = \frac{\iint_S \varepsilon \vec{E} \cdot \vec{dS}}{\int_A^B \vec{E} \cdot \vec{d\ell}} = \iint_S \frac{1}{\int_A^B \frac{d\ell}{\varepsilon dS}} = 1/\int_A^B \frac{1}{\iint_S \frac{\varepsilon dS}{d\ell}} \qquad Q_{AB} = C_{AB} V_{AB} \qquad Q_i = \sum_{j=1}^{n} C_{ij} V_{ij}$$

Parallel plate capacitance: $C_{AB} = \dfrac{\varepsilon S_{AB}}{d_{AB}}$ Coaxial TL capacitance: $C_{z,\mathrm{TL}} = \dfrac{2\pi\varepsilon}{\ln\left(b/a\right)}$ (per unit axial length)

Boundary Conditions

Dielectric-dielectric BC:

$E_{1t} = E_{2t}$ $D_{1n} - D_{2n} = \rho_S$ If $\rho_S = 0, \tan\theta_1/\varepsilon_1 = \tan\theta_2/\varepsilon_2$ $\rho_{Sb1} = -\dfrac{\varepsilon_{r1} - 1}{\varepsilon_{r1}} D_{1n}$ $\rho_{Sb2} = \dfrac{\varepsilon_{r2} - 1}{\varepsilon_{r2}} D_{2n}$

$\rho_{Sb} = \rho_{Sb1} + \rho_{Sb2}$

Conductor-conductor:

Static: $E_1 = E_2 = 0$

Dynamic: $E_{1t} = E_{2t}$ $J_{1n} - J_{2n} = -\dfrac{d\rho_S}{dt}$ If $\dfrac{d\rho_S}{dt} = 0$, $\tan\theta_1/\sigma_1 = \tan\theta_2/\sigma_2$

Conductor-dielectric:

Static: $E_{dt} = E_{ct} = 0$ $E_{cn} = 0$ $D_{dn} = \rho_S$ $\rho_{Sbt} = \rho_{Sb1} = -\dfrac{\varepsilon_{r1} - 1}{\varepsilon_{r1}} \rho_S$

Dynamic: $E_{dt} = E_{ct}$ $E_{cn} = 0$ $D_{dn} = \rho_S$ $\vec{J}_c = \sigma_c \vec{E}_{ct}$

CHAPTER 7

Poisson's and Laplace's Equations and Solution Methods

Siméon-Denis Poisson (1781–1840)

Pierre-Simon marquis de Laplace (1749–1827)

Chapter Learning Objectives

- Derive Poisson's and Laplace's equations from the differential form of Gauss' law.
- Express the Laplacian operator using rectangular (Cartesian), cylindrical, and spherical coordinate systems.
- Solve Poisson's equation for symmetrical charge distributions.
- Solve Poisson's equation for nonsymmetrical charge distribution using three different approaches (superposition, Method of Images, and numerical methods).
- Use graphical methods to evaluate the capacitance and conductance of nonsymmetrical shapes.

7.1 Introduction

Earlier, in Chapter 4, we learned about the electric flux and its density vector as produced by different charge distributions. The powerful differential (point) form of Gauss' law culminated that development relating the flux density vector $\vec{D}$ to the volumetric charge density ρ_v

$$\vec{\nabla} \cdot \vec{D} = \rho_v \tag{4.24}$$

This equation can be used to find ρ_v for a given flux distribution by applying the divergence differential operator to the flux density distribution vector. Moreover, the differential equation (4.24) can also be solved for $\vec{D}$ given a volume charge density ρ_v. Solving such a differential equation analytically may be too complicated unless we deal with cases of special symmetries such as that of the example given earlier at the end of Chapter 4. Numerical solutions of this differential equation may be possible but again could be complicated because we have to deal with a vector unknown, which implies three unknown components in the general case.

In Chapter 5, we learned about the electric field intensity vector and its relationship to the electric flux density vector in free space; specifically,

$$\vec{E}(r) = \frac{1}{\varepsilon_o}\vec{D}(r) \quad \text{and} \quad \vec{D}(r) = \varepsilon_o \vec{E}(r) \tag{5.12}$$

This relationship was later generalized for all dielectric media to take the form

$$\vec{E}(r) = \frac{1}{\varepsilon}\vec{D}(r) \quad \text{and} \quad \vec{D}(r) = \varepsilon \vec{E}(r) \tag{6.39}$$

Both Equations (5.12) and (6.39) provide a tool for relating the vectors $\vec{E}$ and $\vec{D}$ to each other. Also, in Chapter 5, we introduced the electric scalar potential and developed its relationship to the electric field intensity vector

$$\vec{E} = -\vec{\nabla} V = -\text{grad } V \tag{5.64}$$

With this group of relationships, we tie the main electrostatic quantities to each other:

- The charge (and its density)
- The flux (and flux density)
- The field (and force)
- The potential (and energy)

Figure 7.1 summarizes how these quantities are related.

The arrows on this graph indicate the way to derive each of the shown EM quantities with knowledge of another. Some of these relationships are available in an integral form and others are given in a point form. As was discussed earlier, there is a preference for the point (differential) form as opposed to the integral form. Hence, we present the core elements of Figure 7.1 in Figure 7.2 where we remove all the integral form relationships.

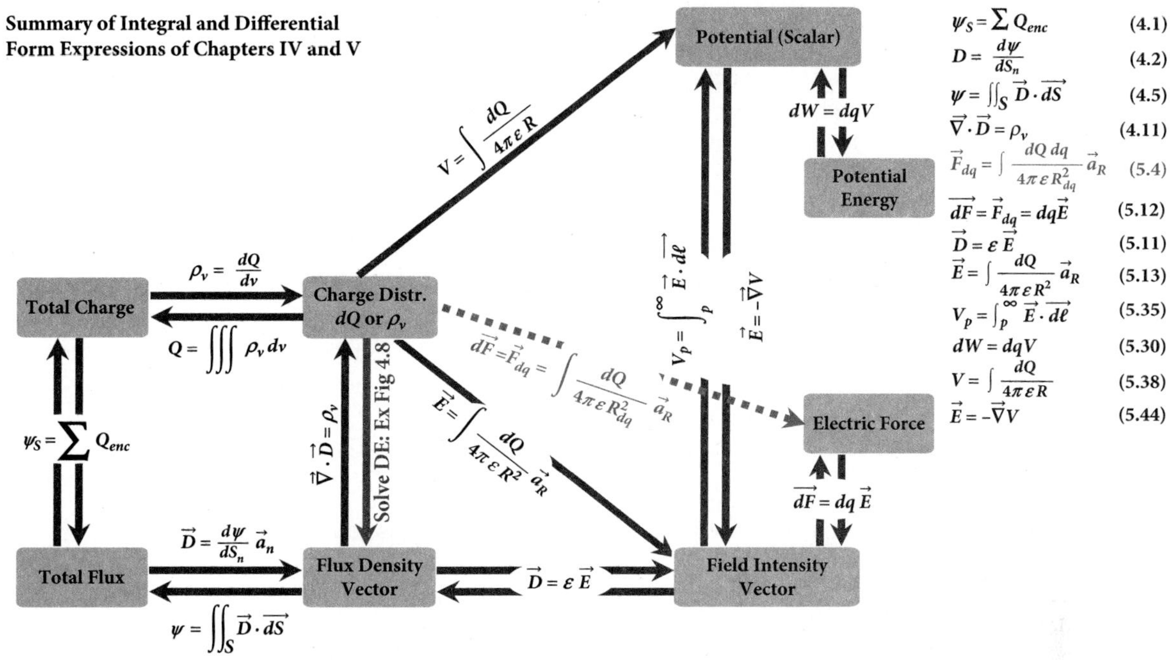

FIGURE 7.1 Summary of integral and differential form expressions for electrostatic quantities.

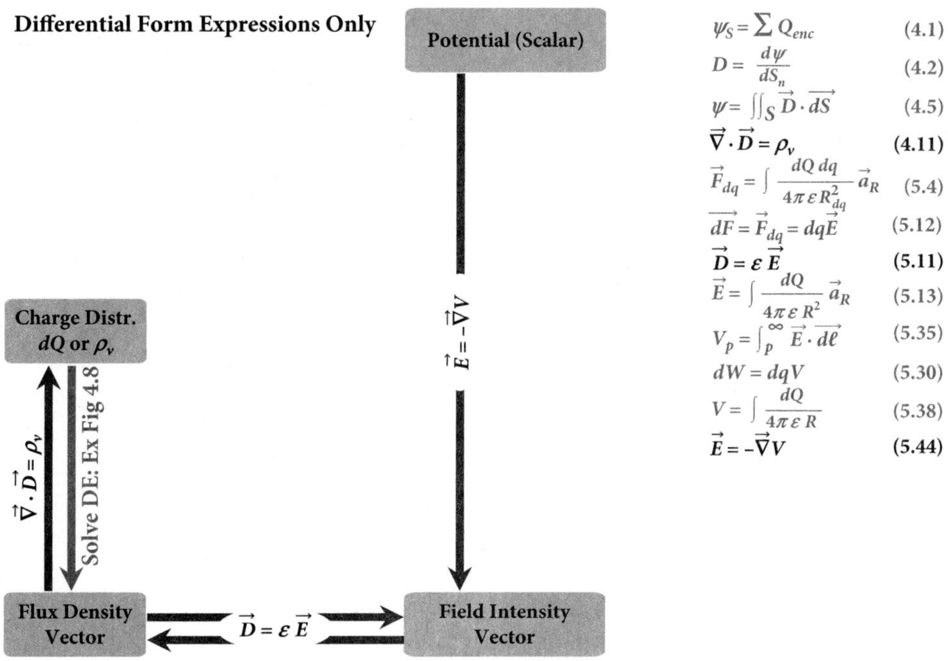

FIGURE 7.2 Some of the differential form expressions for electrostatic quantities.

7.2 Poisson's and Laplace's Equations

Figure 7.2 shows the need for a point form relationship between the potential V and the charge distribution (volumetric density ρ_v). This relationship is possible through combining the three-point form relationships (4.17), (5.12), and (5.64) as shown in the figure.

$$\vec{\nabla}\cdot\vec{D}=\rho_v=\vec{\nabla}\cdot\varepsilon\vec{E}=\vec{\nabla}\cdot\varepsilon(-\vec{\nabla}V) \tag{7.1}$$

Hence, and assuming a homogeneous isotropic medium (ε is not a function of position or direction), we obtain $\vec{\nabla}\cdot\varepsilon(-\vec{\nabla}V)=-\varepsilon\vec{\nabla}\cdot(\vec{\nabla}V)$, and the above manipulation takes the famous Poisson's equation form,

$$\vec{\nabla}\cdot\vec{\nabla}V=-\rho_v/\varepsilon, \quad \text{or} \quad \text{div}(\text{grad } V)=-\rho_v/\varepsilon \tag{7.2}$$

The latter can be rewritten using the Laplacian operator:

$$\vec{\nabla}\cdot\vec{\nabla}V=\text{div}(\text{grad})=\nabla^2 \tag{7.3}$$

$$\nabla^2V=-\rho_v/\varepsilon \tag{7.4}$$

For charge-free regions ($\rho_v=0$), Poisson's equation is called Laplace's equation and takes the form:

$$\nabla^2V=0 \tag{7.5}$$

Poisson's relationship in Equation (7.4) is a partial differential equation that can be solved for the potential as a function of position for a given charge distribution. On the other hand, a solution of the differential equation requires the specification of boundary conditions. If analytical solutions are not feasible, we can always use numerical methods, as we will demonstrate later in this chapter.

Now that we have added the point form of Poisson's equation to our collection of relationships, Figure 7.1 is augmented as shown in Figure 7.3.

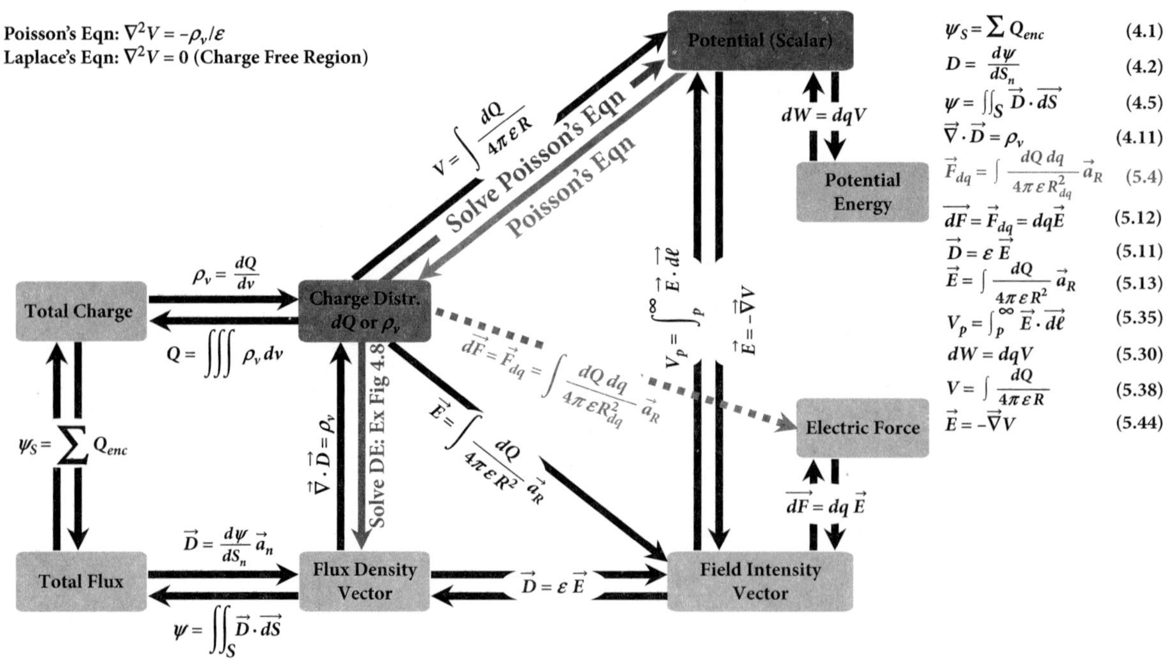

FIGURE 7.3 Summary of integral and differential form expressions for electrostatic quantities including Poisson's and Laplace's equations.

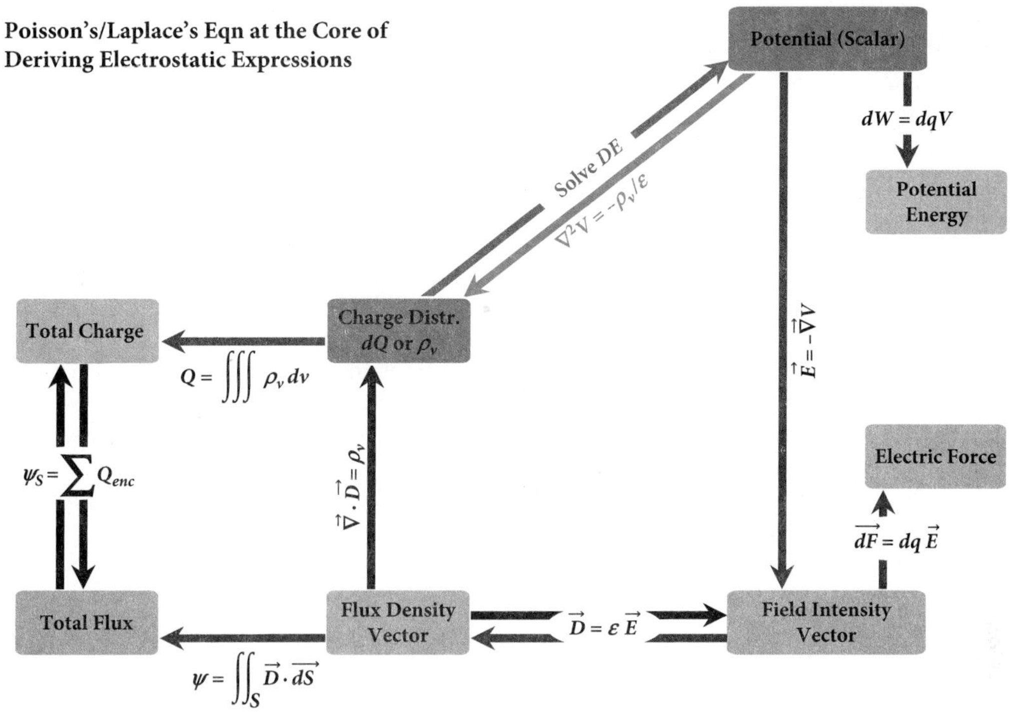

FIGURE 7.4 Poisson's and Laplace's equations at the core of deriving electrostatic expressions.

Next, in Figure 7.4 it is shown how Poisson's equation can be used as a starting point to deriving expressions for all other EM quantities.

This figure shows a completely different approach to solving electrostatic problems using the point form of Poisson's equation. In the following, we will demonstrate the main usage of Poisson's equation (both in direct and inverse manner). But before we do that we will need to derive expressions for the Laplacian operator.

7.3 The Laplacian Operator

The Laplacian operator as defined by Equation (7.3) is of the form

$$\nabla^2 V = \text{div}(\text{grad } V) = \vec{\nabla}\cdot\vec{\nabla}(V) \tag{7.3}$$

Using Equations (5.65) and (4.19), Equation (7.3) can be expressed in Cartesian coordinates as

$$\nabla^2 V(x,y,z) = \vec{\nabla}\cdot\left[\frac{\partial V}{\partial x}\vec{a}_x + \frac{\partial V}{\partial y}\vec{a}_y + \frac{\partial V}{\partial z}\vec{a}_z\right] = \frac{\partial}{\partial x}\left[\frac{\partial V}{\partial x}\right] + \frac{\partial}{\partial y}\left[\frac{\partial V}{\partial y}\right] + \frac{\partial}{\partial z}\left[\frac{\partial V}{\partial z}\right] = \frac{\partial^2 V}{\partial x^2} + \frac{\partial^2 V}{\partial y^2} + \frac{\partial^2 V}{\partial z^2} \tag{7.6}$$

The Laplacian of the electric potential in cylindrical and spherical coordinates is given by

$$\nabla^2 V(\rho,\varphi,z) = \frac{1}{\rho}\frac{\partial}{\partial \rho}\left(\rho\frac{\partial V}{\partial \rho}\right) + \frac{1}{\rho^2}\frac{\partial^2 V}{\partial \varphi^2} + \frac{\partial^2 V}{\partial z^2} \tag{7.7}$$

$$\nabla^2 V(r,\theta,\varphi) = \frac{1}{r^2}\frac{\partial}{\partial r}\left(r^2\frac{\partial V}{\partial r}\right) + \frac{1}{r^2 \sin\theta}\frac{\partial}{\partial \theta}\left(\sin\theta\frac{\partial V}{\partial \theta}\right) + \frac{1}{r^2\sin^2\theta}\frac{\partial^2 V}{\partial \varphi^2} \tag{7.8}$$

respectively.

7.4 Demonstration of Solving Poisson's Equation

The use of Poisson's equation to evaluate the charge distribution given the potential as a function of position should be *fairly* straightforward. In this regard, we apply the Laplacian differential operator to the potential function as given in Equations (7.6) to (7.8). The outcome in this case would be the volumetric charge density that can be integrated over the volume of interest to get the total charge. The inverse process, which is deriving the potential function given the charge distribution requires solving the corresponding Poisson's differential equation. In some cases, an analytical solution is feasible; more generally, however, a numerical method has to be used.

Next, we will demonstrate how we can solve Poisson's equation analytically for the electric potential V for a given charge distribution and use the result to obtain the electric field $\vec{E}$ and the electric flux density $\vec{D}$. Toward this goal, let us consider the "relatively simple" case shown in Figure 4.8, given here in Figure 7.5. This is the case of a spherical volume charge distribution with ρ_v allowed to vary with r only (no variations with θ or φ).

The spherical charge distribution is confined to the region $0 < r < b$. Beyond this region, $b < r < \infty$, there are no charge distributions, and hence we need to write two separate mathematical expressions for each of the two regions. Similar to what we did earlier in Chapter 4, we will refer to $0 < r < b$ as region I and $b < r < \infty$ as region II, and consequently denote the potential function V as V_I and V_{II} in the first and second region, respectively.

Applying Poisson's equation to the two regions, we write

$$\nabla^2 V_I = -\rho_v(r)/\varepsilon, \qquad 0 < r < b \tag{7.9}$$

where $\rho_v(r)$ describes the volumetric charge density distribution within the sphere (region I), and

$$\nabla^2 V_{II} = 0, \qquad b < r < \infty \tag{7.10}$$

outside. Next, we express the Laplacian of V in spherical coordinates using Equation (7.7)

$$\nabla^2 V = \frac{1}{r^2}\frac{\partial}{\partial r}\left(r^2\frac{\partial V}{\partial r}\right) + \frac{1}{r^2 \sin\theta}\frac{\partial}{\partial \theta}\left(\sin\theta\frac{\partial V}{\partial \theta}\right) + \frac{1}{r^2\sin^2\theta}\frac{\partial^2 V}{\partial \varphi^2} \tag{7.11}$$

Considering the spherical symmetry of the charge distribution (no variation with respect to θ and φ), we obtain

$$\nabla^2 V = \frac{1}{r^2}\frac{\partial}{\partial r}\left(r^2\frac{\partial V}{\partial r}\right) \tag{7.12}$$

and Equation (7.9) assumes the form

$$\nabla^2 V_I = \frac{1}{r^2}\frac{\partial}{\partial r}\left(r^2\frac{\partial V_I}{\partial r}\right) = -\rho_v(r)/\varepsilon \tag{7.13}$$

which can be rewritten as

$$\left[\frac{\partial}{\partial r}\left(r^2\frac{\partial V_I}{\partial r}\right)\right] = -r^2\rho_v(r)/\varepsilon \tag{7.14}$$

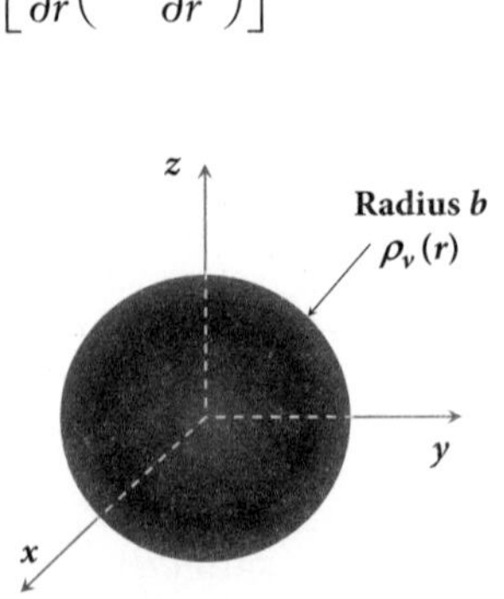

Figure 7.5 Demonstration of solving Poisson's equation for a spherical charge distribution.

Integration over r yields

$$\left(r^2\frac{\partial V_I}{\partial r}\right)=-\frac{1}{\varepsilon}\int r^2\rho_v(r)dr+C_1 \tag{7.15}$$

where C_1 is an integration constant. Division by r^2 results in the expression

$$\frac{\partial V_I}{\partial r}=\frac{-1}{\varepsilon r^2}\int r^2\rho_v(r)dr+\frac{C_1}{r^2} \tag{7.16}$$

An additional integration over r yields, finally, the potential

$$V_I=\int\left[\frac{-1}{\varepsilon r^2}\int r^2\rho_v(r)dr+\frac{C_1}{r^2}\right]dr+C_2=\int\left[\frac{-1}{\varepsilon r^2}\int r^2\rho_v(r)dr\right]dr-\frac{C_1}{r}+C_2 \tag{7.17}$$

Similarly, for region II where $\rho_v = 0$, the potential assumes the form:

$$V_{II}=\int\left[\frac{C_3}{r^2}\right]dr+C_4=-\frac{C_3}{r}+C_4 \tag{7.18}$$

Next, we use available boundary conditions to determine the four arbitrary constants C_1 to C_4.

BC#1: At r = 0 (in region I)

$$V_I=\int\left[\frac{-1}{\varepsilon r^2}\int r^2\rho_v(r)dr\right]dr-\frac{C_1}{r}+C_2 \tag{7.19}$$

At $r = 0$, the term $\frac{C_1}{r}$ blows up to infinity which is not physically possible in the absence of a singularity [for example, $\rho_v(0) = \infty$]. This implies that this term must be rejected as it does not represent a physically acceptable solution, and hence $C_1 = 0$.

$$V_I=\int\left[\frac{-1}{\varepsilon r^2}\int r^2\rho_v(r)dr\right]dr+C_2 \tag{7.20}$$

BC#2: At r = ∞ (in region II)

$$V_{II}=-\frac{C_3}{r}+C_4 \tag{7.21}$$

As $r \to \infty$, the expression for the potential reduces to C_4. Since the potential at infinity (far away from all charge distributions) should be zero, then C_4 must be zero as well.

$$V_{II}=-\frac{C_3}{r} \tag{7.22}$$

BC#3: At r = b (common to both regions I and II)

At $r = b$, we expect continuities of both the potential function as well as in the radial flux density between the two regions. This is the case assuming no singularities in the charge distribution at this interface (for example, surface or point charges that would cause $\rho_v(b) = \infty$). Hence, for the potential continuity:

$$\left[V_I=\int\left[-\frac{1}{\varepsilon r^2}\int r^2\rho_v(r)dr\right]dr+C_2\right]_{r=b}=\left[V_{II}=-\frac{C_3}{r}\right]_{r=b} \tag{7.23}$$

$$\left[\int\left[-\frac{1}{\varepsilon r^2}\int r^2\rho_v(r)dr\right]dr+C_2\right]_{r=b}=\left[-\frac{C_3}{b}\right] \tag{7.24}$$

And for the flux density expression, we first use Equations (5.12), (5.64), and (5.67) to express the radial flux density as

$$D_r = \varepsilon E_r = -\varepsilon \frac{\partial V}{\partial r} \tag{7.25}$$

$$\left[D_{rI} = -\varepsilon \frac{\partial V_I}{\partial r} = \frac{1}{r^2} \int r^2 \rho_v(r) dr \right]_{r=b} = \left[D_{rII} = -\varepsilon \frac{\partial V_{II}}{\partial r} = -\varepsilon \frac{C_3}{r^2} \right]_{r=b} \tag{7.26}$$

$$\left[\frac{1}{r^2} \int r^2 \rho_v(r) dr \right]_{r=b} = \left[-\frac{\varepsilon C_3}{b^2} \right] \tag{7.27}$$

Equations (7.24) and (7.27) can now be solved for the two remaining constants C_2 and C_3.

Now, let us apply the obtained solutions to the simple case of $\rho_v(r) = \rho_o$ (a constant). In this case, Equations (7.24) and (7.27) reduce to

$$\left[\int \left[\frac{-1}{\varepsilon r^2} \int r^2 \rho_o dr \right] dr + C_2 \right]_{r=b} = \left[\frac{-r^2 \rho_o}{6\varepsilon} + C_2 \right]_{r=b} = \left[\frac{-b^2 \rho_o}{6\varepsilon} + C_2 \right] = \left[\frac{-C_3}{b} \right] \tag{7.28}$$

$$\left[\frac{1}{r^2} \int r^2 \rho_o dr \right]_{r=b} = \left[\frac{r^3}{3r^2} \rho_o \right]_{r=b} = \left[\frac{b\rho_o}{3} \right] = \left[\frac{-\varepsilon C_3}{b^2} \right] \tag{7.29}$$

Equation (7.29) tells us that

$$C_3 = \left[-\frac{b^3 \rho_o}{3\varepsilon} \right] \tag{7.30}$$

while Equation (7.28) yields

$$C_2 = \left[\frac{b^2 \rho_o}{6\varepsilon} - \frac{C_3}{b} \right] = \left[\frac{b^2 \rho_o}{6\varepsilon} + \frac{b^2 \rho_o}{3\varepsilon} \right] = \left[\frac{b^2 \rho_o}{2\varepsilon} \right] \tag{7.31}$$

The final expression for the potential function takes the form

$$V = \begin{cases} \dfrac{b^2 \rho_o}{2\varepsilon} - \dfrac{r^2 \rho_o}{6\varepsilon}, & 0 < r < b \\ \dfrac{b^3 \rho_o}{3\varepsilon r}, & b < r < \infty \end{cases} \tag{7.32}$$

from which we can derive the expressions for the electric field intensity and electric flux density vectors as

$$\vec{E} = \begin{cases} \left(\dfrac{\rho_o r}{3\varepsilon} \right) \vec{a}_r, & 0 < r < b \\ \left(\dfrac{\rho_o b^3}{3\varepsilon r^2} \right) \vec{a}_r, & b < r < \infty \end{cases} \tag{7.33}$$

$$\vec{D} = \begin{cases} \left(\dfrac{\rho_o r}{3} \right) \vec{a}_r, & 0 < r < b \\ \left(\dfrac{\rho_o b^3}{3r^2} \right) \vec{a}_r, & b < r < \infty \end{cases} \tag{7.34}$$

These fields are in agreement with the previous results obtained earlier in Chapter 4.

7.5 Solving Poisson's Equation for Nonsymmetrical Charge Distributions

The spherical charge example we just concluded was chosen to demonstrate an analytical solution to Poisson's equation and how it can be used as the starting point for deriving expressions for the electrostatic quantities of interest (V, E, D, etc.). The symmetry in the charge distribution enabled us to carry out the analytical solution. The following question arises now: what if we don't have the symmetries that enable us to proceed analytically? In this regard, we present a few alternative approaches:

- Superposition
- The Method of Images
- Numerical methods

The method of superposition can be used in cases where the given charge distribution is in fact a collection of other charge distributions where each has its own symmetry. Breaking the problem to solve for each symmetrical distribution followed by superposition of the obtained solutions would allow us to obtain an analytical solution for the composite charge distribution.

Discussion of both the Method of Images and numerical methods follows in this chapter's addenda.

ADDENDUM 7A
THE METHOD OF IMAGES

7A.1 Uniqueness Theorem

Generally speaking, problems described by a system of mathematical equations can be classified into three categories:

1. System of equations with no solutions (solutions do not exist)
2. System of equations with many solutions either finite or infinite
3. System of equations with a single unique solution

Before attempting to solve, or even consider knowing how to solve, a system of equations, a preliminary step of verifying the *existence* of a solution has to be completed, that is, ensuring the equations belong to either category 2 or 3. For many problems, this can be done independently of the knowledge of how to obtain the actual solution.

If a problem is determined to be in category 1, further investigation of possible solution(s) to the problem becomes pointless. To pass the *existence* test, all that is required is to show that at least one solution exists.

To demonstrate the concept, let us consider the simple example for which a solution does not exist:

> Consider the system of two simultaneous equations in the two unknowns x and y: $x + y = 1$ and $x + y = 2$. It is obvious that we have two contradicting solutions for y, for any *assumed* possible solution for x (and vice versa). In other words, there are no solutions that exist for this system of equations.

Now, let us discuss problems under categories 2 and 3. After passing the existence test, another such investigational test is the *uniqueness* test, which ascertains whether there exists one *unique* solution or more solutions to the problem. If the system of equations was determined to have a unique solution, then whichever way we use to solve the problem, we will be destined to arrive at this same *unique* solution. Uniqueness applies to equations describing a physical reality whenever all parameters, initial and boundary conditions, and all constraints that influence this physical reality are fully defined (well-posed problem). Said another way, if there is only one reality, there can only be one *unique* solution.

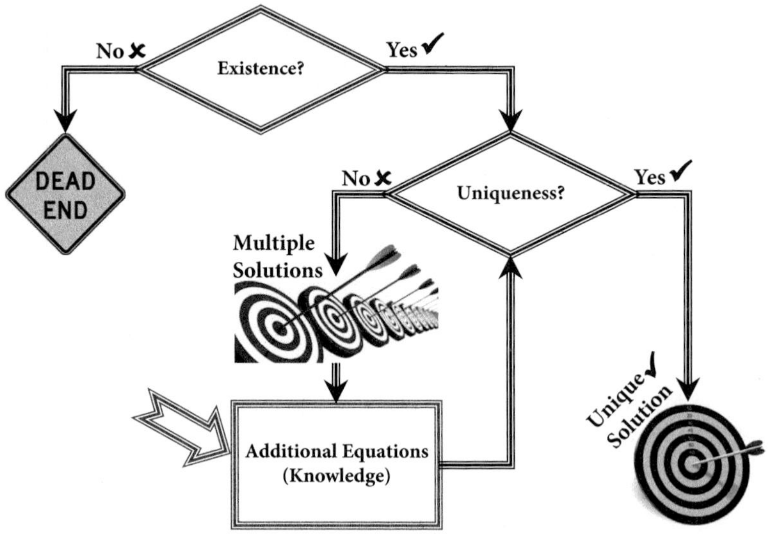

FIGURE 7A.1 A complete set of knowledge is required to achieve the system's unique solution.

A physical example that we can relate to is the determination of the time taken by an object to fall from a given height down to the ground. According to Newton's laws of motion and as witnessed by the repeatability in real-life observation, there is a definite unique solution given by

$$\text{Time} = \sqrt{2 \times \text{height/gravitational acceleration}}$$

Now let us consider an example for which multiple solutions exist, for example, $x + y = 3$, which is an equation of a straight line with infinite possible combinations of x and y. However, introducing a second information equation, for example, $x - y = 1$, uniquely defines one and only one solution, $x = 2$ and $y = 1$.

Hence, we can conclude that we can reduce the possible number of solutions of a system of equations in category 2 by adding more equations (additional information/knowledge). Adding a sufficient number of equations (information/knowledge) will result in a single unique solution (Figure 7A.1). However, adding more equations than necessary can lead to a system with no solutions; this case is referred to as the problem being *overdetermined*. In contrast, a system with a deficient number of equations leading to multiple solutions is referred to as being *underdetermined*.

7A.2 The Uniqueness Theorem for Poisson's Equation

A simple way to introduce the concept of the *Uniqueness theorem* as applied to Poisson's equation is to examine the example given in Figure 7A.2.

In this figure, we have two identical regions as indicated. They are internally identical in their geometrical configurations, in their content of material properties and distributions, in their internal charge distributions, as well as in their boundary potentials. The two outer regions could possibly be different in their geometries, material, or other boundary potentials. If we write Poisson's equation for either of the two identical regions, we will certain write the same exact equation:

$$\nabla^2 V = -\rho_v/\varepsilon \tag{7.4}$$

This is the case since ρ_v and ε as functions of position are identical in both regions. Solving this differential equation for either region will certainly yield the same identical *general* solution. The obtained general solution contains arbitrary constants (two for second-order differential equations) that can only be determined by the region's boundary conditions. Since the boundary conditions are identical for both regions, then the arbitrary constants of both regions will certainly be identical. Hence, the two identical regions will have one unique solution despite the fact that each is a part of a bigger region that could possibly be different from one case to another.

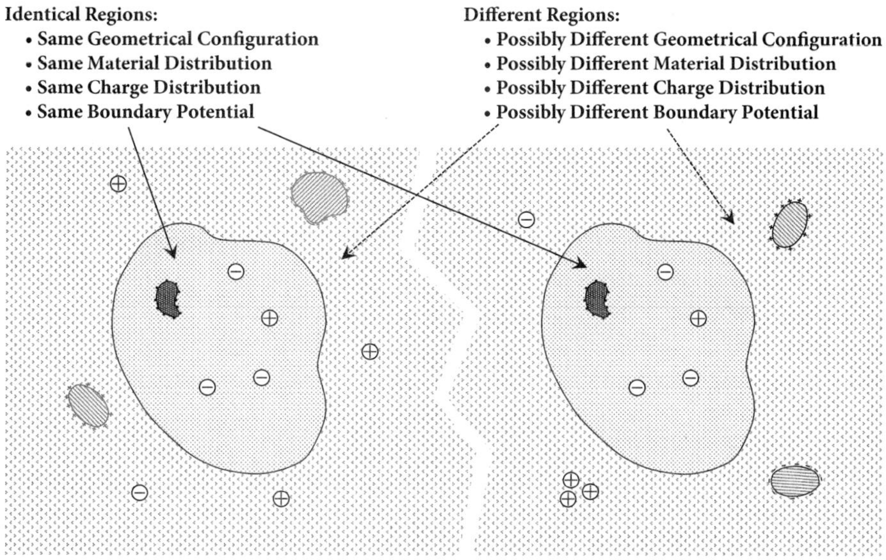

FIGURE 7A.2 Two identical regions with identical boundaries should have identical (unique) solutions.

Hence, the Uniqueness theorem tells us that a fully defined region with fully defined boundaries will have one unique solution irrespective of the properties of all other regions.

One powerful application of this theorem in electromagnetics is the use of the Method of Images in solving certain classes of problems, as we will demonstrate in this addendum.

7A.3 Example of the Use of the Method of Images

In this example, Figure 7A.3, we have a point charge (+Q) located at a height (h) above an infinitely extended grounded conducting plane. The question is can we derive expressions for the electric field intensity vector or the electric potential as a function of position in the upper half space? To answer that, we need to be mindful of the following.

We are not given any information about the lower half space. However, the infinite extension of the ground conductor totally isolates the two spaces from each other, and the fields (and potentials) of the lower half space do not influence their counterparts in the upper half space.

The presence of the charge in the upper half space will certainly induce surface charges on the upper surface of the ground conductor. Considering the neutrality of the total charge in the upper half space, the total surface charge induced on the upper conductor surface must be equal (and opposite) to the (+Q). This surface charge is not distributed uniformly and its distribution can only be determined through the knowledge of the field intensity as a function of position on the conductor surface, which is what we are trying to solve for.

This problem is an excellent example for demonstrating the use of the Method of Images combined with the Uniqueness theorem applied to Poisson's equation.

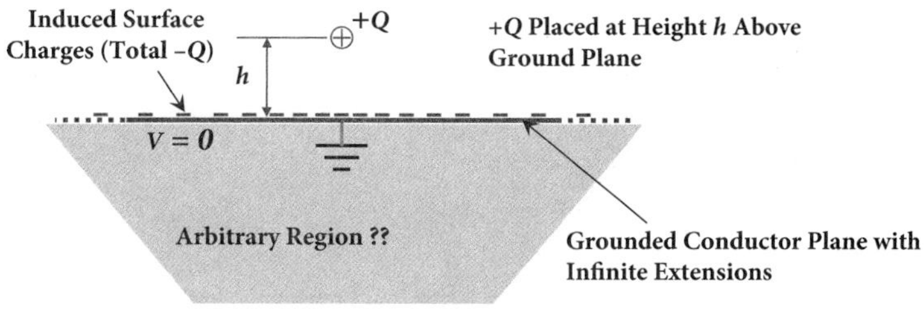

FIGURE 7A.3 The example of a point charge above a ground plane.

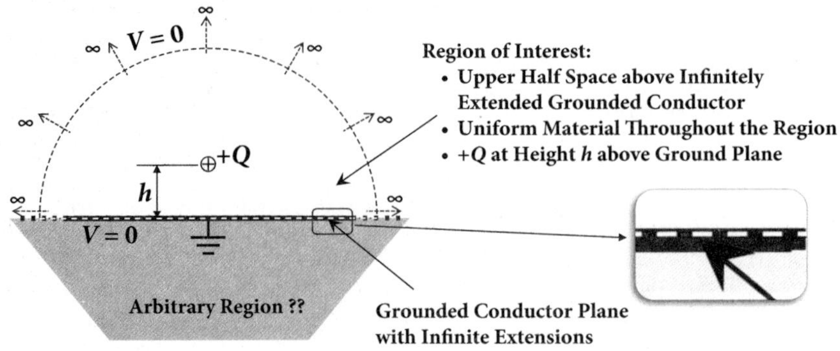

FIGURE 7A.4 Solving the example of a point charge above a ground plane using Poisson's equation.

Examining the upper half space in this light, we realize the following (Figure 7A.4):

- We have a well-defined region including the upper half space starting with an infinitely extending hypothetical plane just an increment beneath the upper surface of the conducting plane all the way to the upper half space infinite extension.
- We have complete information regarding material distribution throughout this region (free space).
- We have complete information regarding the free charge distribution (sources) throughout this region ($+Q$ at height h).
- We have complete information regarding the boundary conditions (potential is zero at all boundary points): The conductor plane is grounded and the infinitely extended spherical surface bounding the upper half of the space is located at infinity where the potential is zero.
- Hence, we can write Poisson's equation for this region for which we are certain there is one unique solution.

To solve this problem, we resort to the Method of Images where we identify another problem that would be easier to solve, while maintaining the same identical region of the original problem with identical Poisson's equation and identical boundary conditions.

Figure 7A.5 offers us the alternate problem that satisfies this constraint. In this figure, we compare the region of interest in the original problem (left) with its counter in the image problem (right).

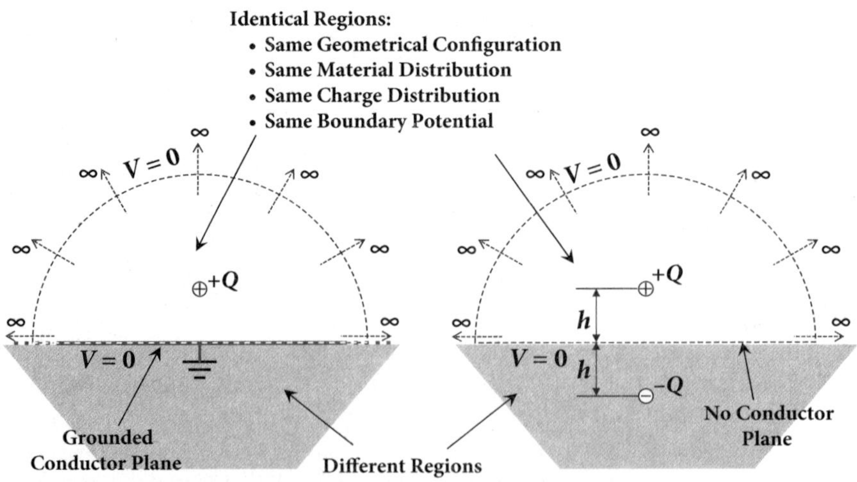

FIGURE 7A.5 Developing the image problem for the point charge above the ground plane.

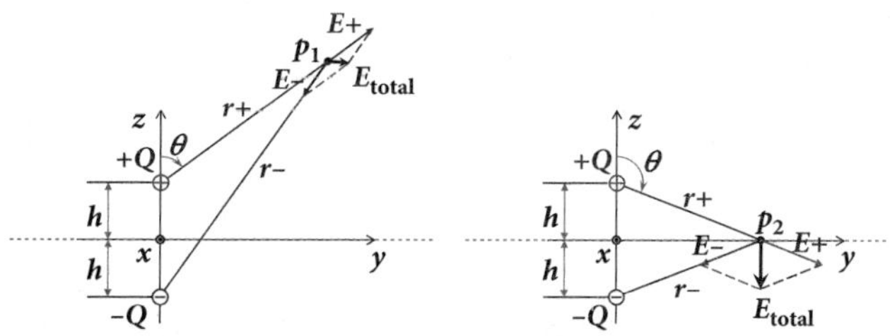

FIGURE 7A.6 Solving the example of a point charge above a ground plane using the Method of Images.

This region is bounded by the hypothetical surface just beneath the surface in the grounded conductor region and the infinitely extended half sphere enclosing the upper half space.

For this region, both the original and the alternate image problem are identical, as they both have the same geometrical configuration, the same material distribution, the same charge distribution, and most certainly same boundary potential. Hence both problems will have the same identical (unique) solution of the identical regions.

Obviously, solving the alternate (image) problem on the right is fairly straight forward as we can simply superimpose the field (and potential) expressions for the two point-charges $+Q$ and $-Q$ (Figure 7A.5).

In Figure 7A.6, we demonstrate the solution of the alternate problem by deriving expressions for E and V as functions of position. Only expressions within the uniqueness region (upper half space) are valid to claim for the original problem. Two sample figures demonstrate the results at a general location (p_1) and a point on the ground plane (p_2).

At (p_1), we can write

$$\vec{E}_{p1} = \frac{+Q}{4\pi\varepsilon_o r_+^2}\vec{a}_{r+} + \frac{-Q}{4\pi\varepsilon_o r_-^2}\vec{a}_{r-} \tag{7A.1}$$

$$V_{p1} = \frac{+Q}{4\pi\varepsilon_o r_+} + \frac{-Q}{4\pi\varepsilon_o r_-} \tag{7A.2}$$

While at (p_2), the E and V expressions reduce to

$$\vec{E}_{p2} = \frac{+Q}{4\pi\varepsilon_o r_+^2}\vec{a}_{r+} + \frac{-Q}{4\pi\varepsilon_o r_-^2}\vec{a}_{r-} = \left(\frac{Q\cos\theta}{2\pi\varepsilon_o r_+^2}\right)\vec{a}_z \tag{7A.3}$$

$$V_{p2} = \frac{+Q}{4\pi\varepsilon_o r_+} + \frac{-Q}{4\pi\varepsilon_o r_-} = 0 \tag{7A.4}$$

We can use Equation (7A.3) to derive an expression for the induced charge density on the surface of the ground conductor plane, using Equation (6.70):

$$\rho_S = D_n = \varepsilon_o E_n = \left(\frac{Q\cos\theta}{2\pi r_+^2}\right) \tag{7A.5}$$

7A.3.1 Example 1: Point Charge and Grounded Spherical Conductor

This example is depicted in the left diagram ("Original Problem") below. As shown here, we have a point charge "+Q" located outside a grounded conducting sphere. In this example, we will confine our work to solving for the potential function at all points of space. With the availability of the expression for the potential, it will be possible to derive expressions for the electric field vector and the surface charge distribution on the spherical conductor surface as needed.

Recognizing that the region within the conducting sphere is source free and that the sphere is grounded, we can safely state that the potential for all points in this region is zero. This leaves the region outside the sphere as the one we need to work on in order to derive the potential function expression. To solve for the potential function in this region, we resort to the Method of Images. In this regard, we identify an alternate configuration with the same identical geometry, charge and material distributions, and boundary conditions. The alternate configuration satisfying these requirements is shown in the right diagram ("Image Problem") below.

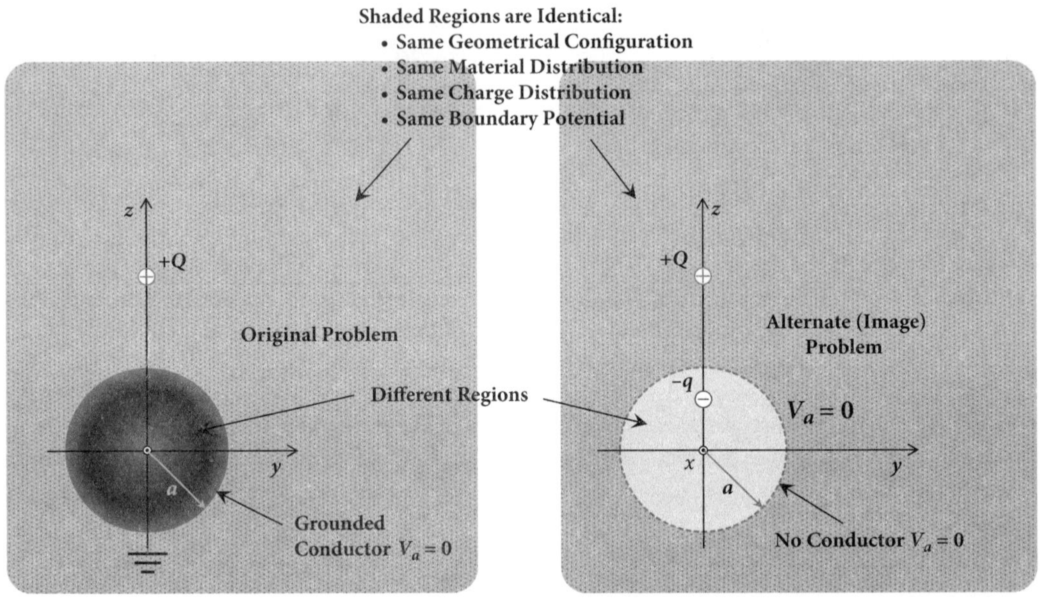

Original Problem

Image Problem

In these two diagrams, we compare the region of interest in the original problem (outside the sphere) with its counterpart in the image problem configuration.

In the image configuration, this region is bounded by a hypothetical spherical surface at the same exact location where the grounded conducting sphere used to be.

To achieve zero potential boundary condition at this hypothetical surface, a negative charge is needed to counter the potential contribution of the "+Q" charge on that surface.

Logically, this negative charge "$-q$" needs to be located outside the region of interest in the image problem, and hence inside the hypothetical spherical surface. Moreover, the problem symmetry implies that we locate the "$-q$" charge along the same radial line of the "$+Q$" charge.

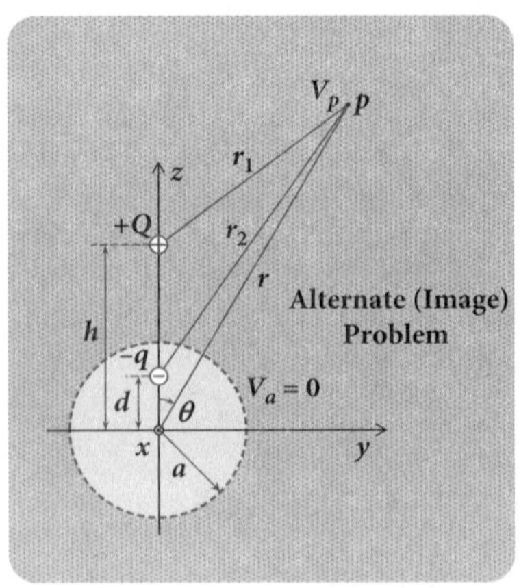

Image Problem Analysis

Now, before we can start our derivation of the expression for the potential, we need to determine the exact location for the "$-q$" charge as well as its value.

Both the location and the value of this charge should be determined so that they satisfy the zero potential condition on the hypothetical spherical surface in the image problem.

Toward this goal, we write the potential expression for the image problem and equate it to zero on the spherical surface as follows: Using spherical coordinates, and referring to the "Image Problem Analysis" diagram above, the potential function at a general point, $p(r, \theta, \varphi)$, is of the form:

$$V_p = \frac{Q}{4\pi\varepsilon_o r_1} - \frac{q}{4\pi\varepsilon_o r_2}$$

Now, using trigonometric identities, r_1 and r_2 can be expressed in terms of r, θ, z_1, z_2 to yield:

$$V_p = \frac{Q}{4\pi\varepsilon_o r_1} - \frac{q}{4\pi\varepsilon_o r_2} = \frac{1}{4\pi\varepsilon_o}\left\{\frac{Q}{\sqrt{r^2+h^2-2hr\cos\theta}} - \frac{q}{\sqrt{r^2+d^2-2dr\cos\theta}}\right\}$$

Now, at the $r = a$ surface, we write V_a as

$$V_a = 0 = \frac{1}{4\pi\varepsilon_o}\left\{\frac{Q}{\sqrt{a^2+h^2-2ha\cos\theta}} - \frac{q}{\sqrt{a^2+d^2-2da\cos\theta}}\right\}$$

which implies: $\dfrac{Q}{\sqrt{a^2+h^2-2ha\cos\theta}} = \dfrac{q}{\sqrt{a^2+d^2-2da\cos\theta}}$

$$q^2[a^2+h^2-2ha\cos\theta] = Q^2[a^2+d^2-2da\cos\theta]q^2(a^2+h^2) - Q^2(a^2+d^2) - 2a[q^2h - Q^2d]\cos\theta = 0$$

Now, for this equation to be valid for all θ values, we conclude

$$[q^2h - Q^2d] = 0$$

$$q^2(a^2+h^2) - Q^2(a^2+d^2) = 0$$

Equations (7A.12) and (7A.13) are to be solved simultaneously to determine the value "$-q$" and location d of the image charge. The results are

$$q = \frac{a}{h}Q \quad \text{and} \quad d = \frac{a^2}{h}$$

Finally, we can write the general solution for the potential function at all points in space outside the conducting sphere as

$$V_p = \frac{1}{4\pi\varepsilon_o}\left\{\frac{Q}{\sqrt{r^2+h^2-2hr\cos\theta}} - \frac{q}{\sqrt{r^2+d^2-2dr\cos\theta}}\right\}$$

$$= \frac{Q}{4\pi\varepsilon_o}\left\{\frac{1}{\sqrt{r^2+h^2-2hr\cos\theta}} - \frac{a}{h}\frac{1}{\sqrt{r^2+\left(\frac{a^2}{h}\right)^2 - 2\frac{a^2}{h}r\cos\theta}}\right\}$$

$$= \frac{Q}{4\pi\varepsilon_o}\left\{\frac{1}{\sqrt{r^2+h^2-2hr\cos\theta}} - \frac{a}{\sqrt{h^2r^2+a^4-2hra^2\cos\theta}}\right\}$$

ADDENDUM 7B
NUMERICAL METHODS

7B.1 Introduction

In most of our earlier presentations, we discussed cases where geometrical configurations and charge distributions could be expressed with "simple" analytical functions and expressions. Examples were geometries and distributions with spherical symmetries, infinite length cylinders with circular symmetries, and infinitely extended shapes with planar symmetries. This is not likely to be the case in practical applications. Planar, cylindrical, and spherical configurations and distributions are, at best, approximations to those of real problems. If the approximation is not perceived to be within acceptable tolerances, numerical methods are typically used.

In the past, when computational powers were limited, graphical methods were typically used for simple EM analysis problems. Today, we typically use graphical methods for mostly educational purposes as well as for visualization of the results obtained numerically.

The term "numerical methods" is typically used for solving for an unknown term in a mathematical expression while an explicit mathematical expression for this term is not feasible. Typical cases where numerical methods are used include but are not limited to

- Simultaneous linear equations and matrices
- Integration and differentiation
- Nonlinear equations and solving for roots
- Curve fitting regression and interpolation
- Solving differential equations

In some cases, analytical procedures may be possible to use in order to find symbolic expression solutions. In cases when such solutions are not possible or too complicated, numerical methods offer a preferred alternative.

In this addendum, we focus on the use of numerical methods for solving the point form Poisson's/Laplace's differential equations.

7B.2 Numerical Analysis of Electrostatic Problems

To apply numerical methods to an electrostatic problem, we need to segment the spatial region of interest into small increments, where each increment could be treated as a point in space. The increment should be small enough to ensure that field quantities would not have significant variations throughout the increment region. The segmentation process is typically referred to as meshing. The simplest form of meshing is to follow the preferred coordinate system for the problem and divide the problem region into equal segments. Figure 7B.1 demonstrates the concept in a two-dimensional region using uniform meshing in Cartesian coordinate.

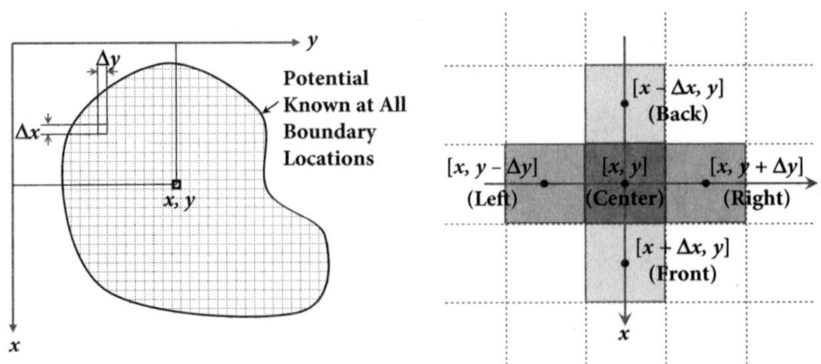

FIGURE 7B.1 Uniform 2D meshing of the problem region for Laplace's equation application.

Nonuniform meshing is typically used to allow better accuracy by increasing the density of increments in places where the variability of the field quantities is expected to be significant. In the following, we will only work with the simplest form of meshing in Cartesian coordinates in both 2D and 3D in order to demonstrate the concept.

7B.3 Demonstration of Numerical Solution of Laplace's Equation in 2D Problems

Let us demonstrate the use of numerical methods in a 2D Laplace's problem using uniform meshing in Cartesian coordinates.

What is meant by a 2D problem is the case of an *upright* cylindrical structure with an identical cross section all along its *height* implying an infinite height extension. Since the cross section is the same all along, the analysis of the cross-sectional shape is sufficient and is considered a 2D analysis. Cases where the analysis is carried over the cross section of an upright shape with large enough height far away from its ends can be approximated through a 2D analysis as well.

For the purpose of this 2D demonstration, we will assume an "upright" charge-free cylindrical structure with boundaries for which the potential function is well defined. The cross section of the region of interest is shown in Figure 7B.1. As the figure indicates, we divided the 2D region into equal Δx and Δy increments.

Consider the 2D (x-y plane) differential form of Laplace's equations (7.5) and (7.6)

$$\nabla^2 V = \frac{\partial^2 V}{\partial x^2} + \frac{\partial^2 V}{\partial y^2} = 0 \tag{7B.1}$$

Let us express the second-order derivatives by their corresponding difference forms. In this regard, let us start with the first-order derivative of one of the two terms, for example, $\frac{\partial V}{\partial x}$:

$$\frac{\partial V}{\partial x} = \lim_{\Delta x \to 0} \left(\frac{V(x+\Delta x, y) - V(x,y)}{\Delta x} \right) \tag{7B.2}$$

Consequently, we can write the second-order derivative as

$$\frac{\partial^2 V}{\partial x^2} = \frac{\partial}{\partial x}\left[\frac{\partial V}{\partial x}\right] = \lim_{\Delta x \to 0} \left[\frac{\lim\limits_{\Delta x \to 0} \left(\frac{V(x+\Delta x, y) - V(x,y)}{\Delta x} \right) - \lim\limits_{\Delta x \to 0} \left(\frac{V(x,y) - V(x-\Delta x, y)}{\Delta x} \right)}{\Delta x} \right] \tag{7B.3}$$

$$= \lim_{\Delta x \to 0} \left[\frac{V(x+\Delta x, y) + V(x-\Delta x, y) - 2V(x,y)}{(\Delta x)^2} \right]$$

Similarly,

$$\frac{\partial^2 V}{\partial y^2} = \lim_{\Delta y \to 0} \left[\frac{V(x, y+\Delta y) + V(x, y-\Delta y) - 2V(x,y)}{(\Delta y)^2} \right] \tag{7B.4}$$

Using both Equations (7B.3) and (7B.4), we can write

$$\frac{\partial^2 V}{\partial x^2} + \frac{\partial^2 V}{\partial y^2} = 0 = \lim_{\Delta x \to 0} \left[\frac{V(x+\Delta x, y) + V(x-\Delta x, y) - 2V(x,y)}{(\Delta x)^2} \right] + \lim_{\Delta y \to 0} \left[\frac{V(y+\Delta y, x) + V(y-\Delta y, x) - 2V(y,x)}{(\Delta y)^2} \right] \tag{7B.5}$$

Now, considering the uniform meshing, $\Delta x = \Delta y$, and using the approximation that both are small enough in size, the above expression reduces to

$$0 = \left[\frac{[V(x+\Delta x, y) + V(x-\Delta x, y) - 2V(x,y)] + [V(x, y+\Delta y) + V(x, y-\Delta y) - 2V(x,y)]}{(\Delta x)^2} \right]$$

or

$$\left[V(x+\Delta x,y)+V(x-\Delta x,y)+V(x,y+\Delta y)+V(x,y-\Delta y)-4V(x,y)\right]=0 \tag{7B.6}$$

and, finally,

$$V(x,y)=\frac{1}{4}\left[V(x+\Delta x,y)+V(x-\Delta x,y)+V(x,y+\Delta y)+V(x,y-\Delta y)\right] \tag{7B.7}$$

In other words, and referring to Figure 7B.2, the potential at the "center" cell is equal to the average of the four potential values of the adjacent cells, "front, back, right, and left." This is a powerful conclusion that we can use as the basis of achieving a numerical solution to the problem at hand.

The process starts by assigning initial arbitrary values to the potential of all the mesh cells while assigning the known potential values to the boundary cells. Next, we apply iterative methods to the mesh cells using Formula (7B.7). Iterations would bring the cell values to converge to the correct potential values. The iteration process can then be concluded when the cell potential values show little or no change.

It is worthy to state that for the case of a 3D problem the iteration formula is quite similar to that of Equation (7B.7), except that the averaging process now includes the third-dimension cells. In other words, the potential at the "center" cell is equal to the average of the six potential values of the adjacent cells, "front, back, right, left, up, and down (Figure 7B.2).

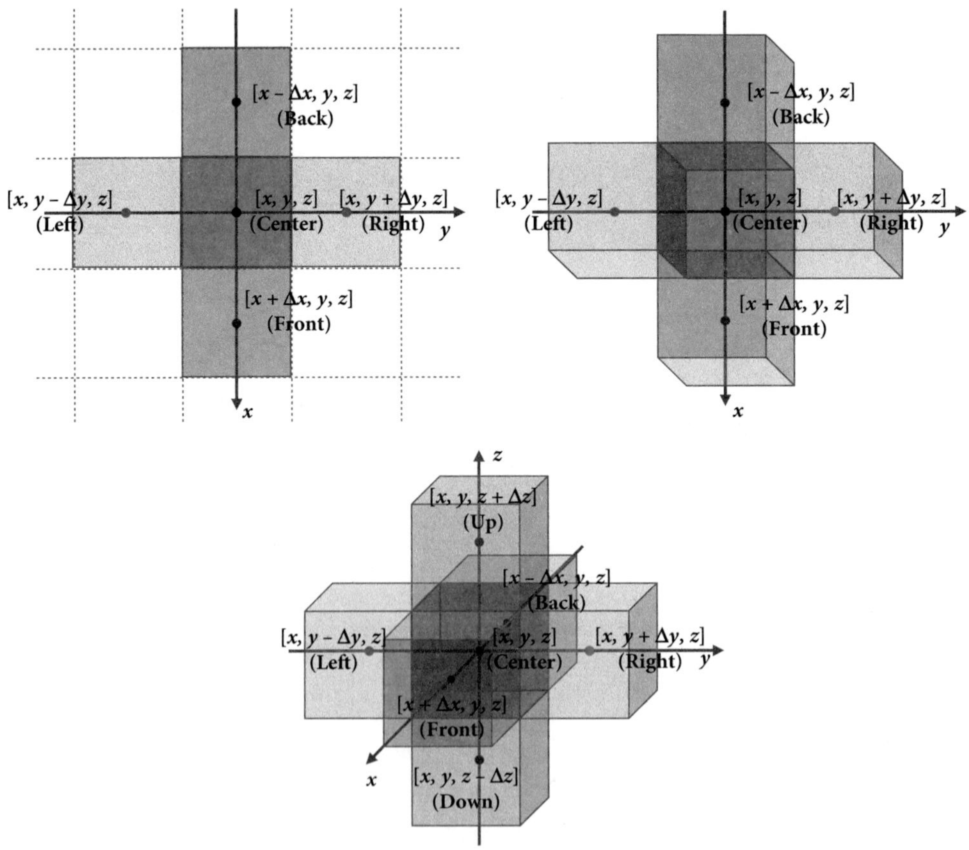

FIGURE 7B.2 Uniform 3D meshing for Laplace's equation application.

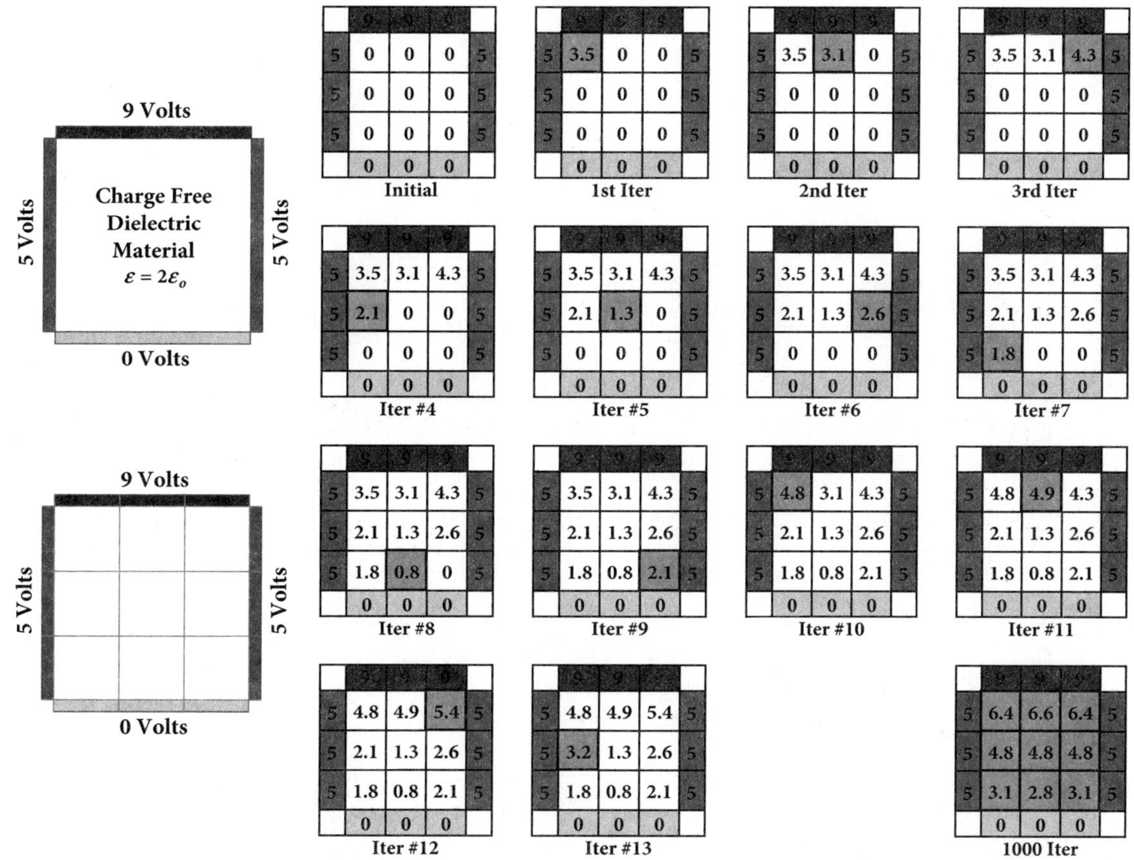

Figure 7B.3 Demonstration of uniform 2D Laplace's equation application to a square region.

7B.4 Demonstration of Iterative Solution of Laplace's Equation in 2D

First, let us consider a simple 2D configuration of a square charge-free dielectric region with four conductor sides each of which is set to a different equipotential (Figure 7B.3). The region is meshed into a 3 × 3 uniform mesh. Initially, all mesh cells are assigned zero values for their potential.

Next, we start the iterative process by applying Equation (7B.7) to the first cell (top-left). With the new value for this cell's potential in place, we move to the second iteration where the averaging Equation (7B.7) is now applied to the next cell (top-middle). The process continues until done with all the mesh cells after which other rounds of iteration are repeated cell by cell until potential values converge and show insignificant change in subsequent iterations. In the figure, we are demonstrating the first 13 iterations of the given example as well as showing the 1000th iteration which was concluded as the convergence answer in this case. The inaccurate solution is due to the sparsity of the mesh. The iterations in this demonstration were performed row-wise; however, other orders can be used as well.

Example 7B.1 Use the numerical method to solve Laplace's equation to evaluate the potential distribution for equation to the case of two coaxial conductors with a round inner conductor kept at 9 V and a square outer conductor grounded. The cross-sectional dimensions of the two conductors are 100 mm on a side for the outer conductor and 25 mm diameter for the inner conductor.

Solution Following the steps outlined in Section 7B.4, we created a spreadsheet with a 100 × 100 cells to match the outer conductor dimensions with 1 mm per cell. Each one of these cells is to represent the voltage value at that location in the matrix created (Figure 7B.4). Each one of these cells was programmed to be the

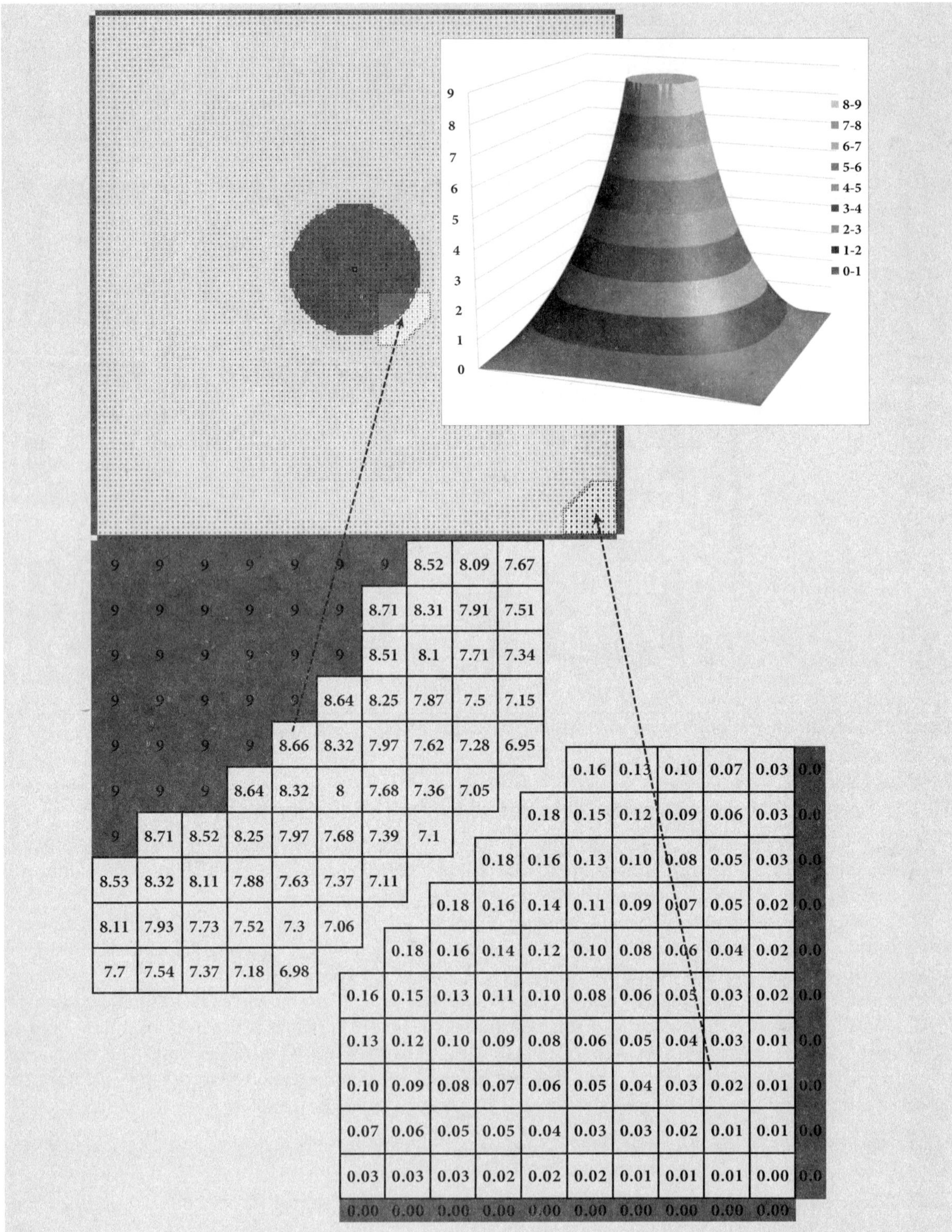

FIGURE 7B.4 Numerical application of Laplace's equation to the case of two coaxial conductors with the round inner conductor kept at 9 V and the square outer conductor grounded.

numerical average of the four adjacent cells according to Equation (7B.7), that is, the cell in the i row and j column is of the form

$$V_{i,j} = [V_{i-1,j} + V_{i+1,j} + V_{i,j-1} + V_{i,j+1}]/4$$

Zero-value border cells were added on the outside of the 100 × 100 matrix to enforce the zero potential of the outer conductor. The round inner conductor was approximated by a group of cells as shown in the figure and were assigned the value of the voltage of the inner conductor, that is, 9 V.

The spreadsheet was then allowed to do iterative calculations for the "formula" cells converging to realize that all the cells comply with the Laplace equation averaging relationship with the two conductor fixed voltages as boundary conditions.

With adequate number of iterations, the potential distribution is reached and is given in Figure 7B.4.

With the choice of 9 and 0 V for the boundaries while formatting the cells to display only one integer digit, the 2D 100 × 100 matrix will give a visual display of voltage contours as the figure demonstrate. In addition, a 3D chart with the voltage value on the vertical axis is also demonstrated in the figure.

7B.5 Graphical Methods

Graphical methods can be used either as stand-alone tools or as complementary to numerical (or even analytical) solutions. In the following, we will demonstrate the use of graphical methods in conjunction with the numerical results obtained from solving 2D Laplace's equation. The example shown in Figure 7B.5 involves two concentric square cross-section conductors separated by a dielectric region. The inner conductor is set to a 5-V potential while the other one is grounded at 0 V.

A numerical (iterative) solution was determined and the potential distribution is given in Figure 7B.6. The four contours, shown by different colors, are the cross sections of the equipotential surfaces with potentials 1, 2, 3, and 4 V, respectively.

Taking advantage of the configuration symmetry, we elect to proceed with the graphical solution applied to 1/8th of the shape, which can be mirrored to produce the full solution. This 1/8th section is shown in Figure 7B.7.

Starting with the equipotential lines in Figure 7B.7, we sketch in a set of *curvilinear* squares that match the equipotential contours. A curvilinear square is a squarish shape with all its four angles as 90° (Figure 7B.8). Since the electric field lines (and hence the corresponding flux lines) are supposed to be orthogonal to the equipotential surfaces, we need to fit each curvilinear square in the developed pattern so that two opposing

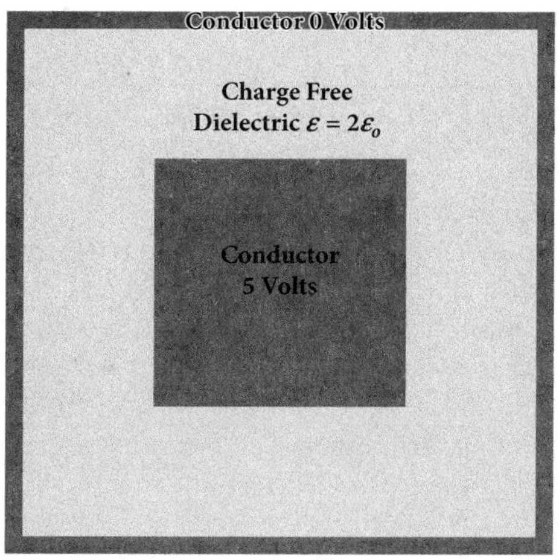

FIGURE 7B.5 Case of two coaxial square cross-section conductors used for the demonstration of the graphical method in Section 7B.5.

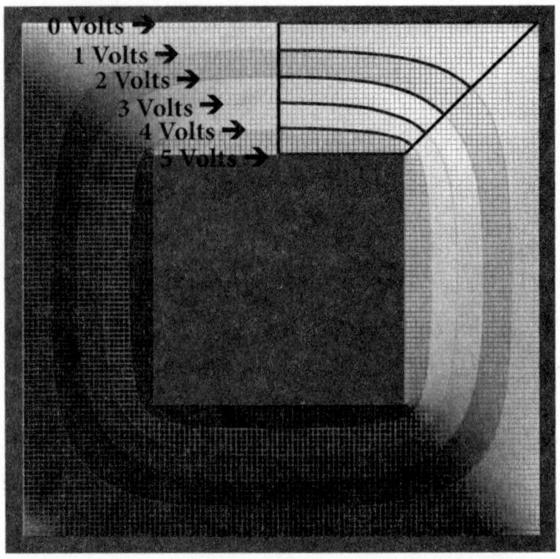

Figure 7B.6 Solution for the case of Figure 7B.5 obtained using the Laplace's equation numerical method.

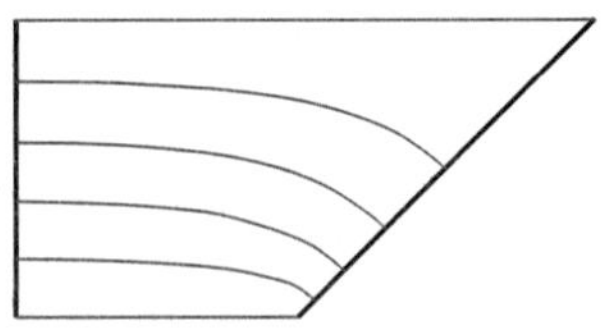

Figure 7B.7 A section of 1/8th of the full shape of the solution of Figure 7B.6.

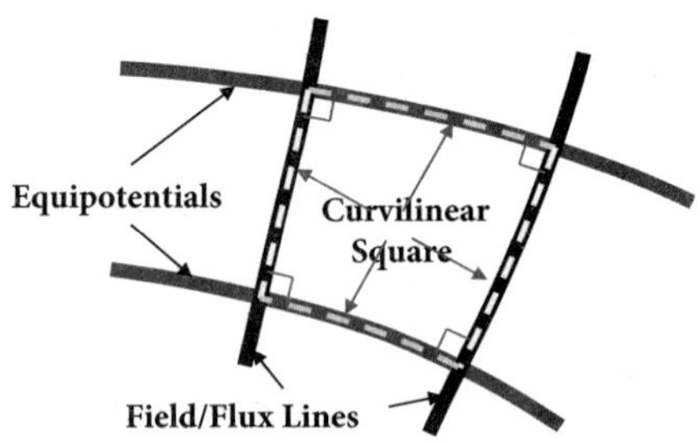

Figure 7B.8 Demonstration of the curvilinear squares formed with the equipotential surfaces and the field/flux lines.

sides coincide with two adjacent equipotential surfaces and the other two sides line up with the field lines in that region. This process is demonstrated in Figure 7B.9*a* through *d*.

In Figure 7B.9*a*, you notice that initial *proper* squares are fitted in between the equipotentials starting from the bottom left corner. We could have started from other corners as well. Fitting the squares to the curved lines results in different square dimensions. Next, in Figure 7B.9*b*, the squares are turned in place to better match their sides with the curved equipotentials. This is a trial and error process and requires some practice. Finally. In Figure 7B.9*c*, the field lines are approximated in between the squares while maintaining orthogonality with the equipotential lines. Depending on the "width" of the section, we may end up with a remainder region that would not fit whole squares. Instead, curvilinear rectangles would be used to fit in such region. In the case shown, after fitting six curvilinear squares across the equipotential lines, the remainder region was found to fit curvilinear rectangles with a 0.1 aspect ratio (width/length = 0.1).

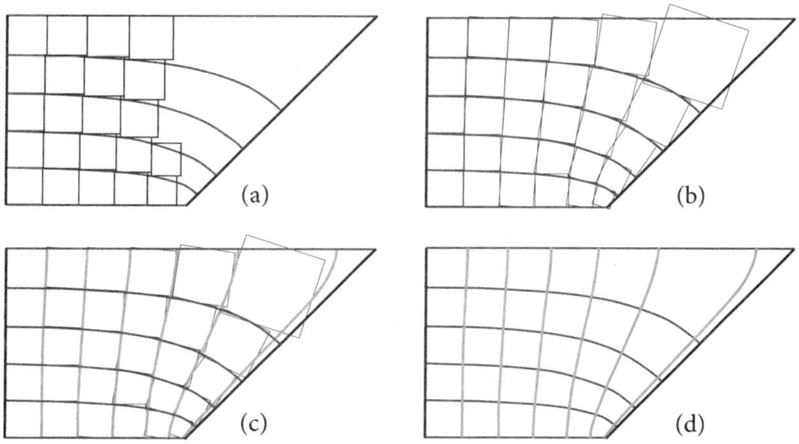

Figure 7B.9 Steps of graphical development of the curvilinear squares for the 1/8th section of the case of Figure 7B.6.

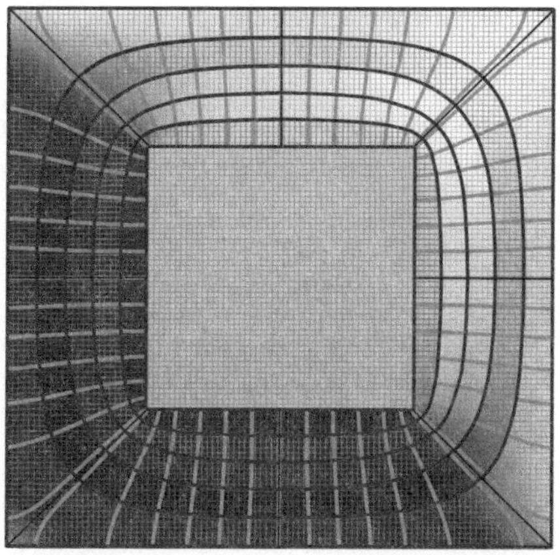

Figure 7B.10 The complete set of curvilinear squares obtained by mirroring the solution in Figure 7B.9*d* for the 1/8th section.

Figure 7B.9*d* shows the finished process where we have an arrangement of 6.1×5 curvilinear squares fitted to the section. While in Figure 7B.10, we show the solution for the whole dielectric region after mirroring the obtained solution of the 1/8th section of Figure 7B.9*d*.

7B.6 Field Intensity and Flux Density Evaluation

With this graphical solution on hand, we can now estimate the electric field intensity, the flux density, as well as the capacitance of the dielectric region.

The formulas for electric field intensity vector given the potential function was derived in Chapter 5

$$\vec{E} = -\vec{\nabla} V \tag{5.44}$$

Converting this differential form to a difference equation, we can write

$$E_\ell = -\frac{\partial V}{\partial \ell} = -\frac{\Delta V}{\Delta \ell} = \frac{V_a - V_b}{\ell_{ab}} \tag{7B.8}$$

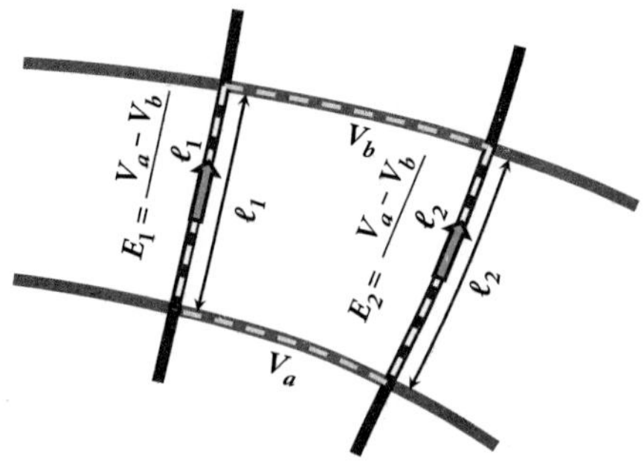

FIGURE 7B.11 Using the graphically developed curvilinear squares to evaluate the electric field intensity at the different locations of the graphical solution.

Also, using Equation (5.12), we can write

$$D_\ell = \varepsilon E_\ell = \varepsilon \frac{V_a - V_b}{\ell_{ab}} \tag{7B.9}$$

Using these equations, we can estimate the magnitudes of both E_ℓ and D_ℓ at the various cell locations (Figure 7B.11). The directions for these two vectors are along the side lines of the graphed curvilinear squares.

7B.7 Capacitance Evaluation

To evaluate the capacitance of the whole dielectric region, we start by evaluating the capacitances of its constituent cells. The total capacitance can then be evaluated by combing the cell capacitance in parallel and in series as will be demonstrated later.

Referring to Figure 7B.12, which shows the 2D shape of a typical cell, the capacitance of the cell is determined by the surface area of the end equipotential surfaces of the cell and the distance separating them. Since the cell dimensions vary from one end to the other as well as across, and since the cell is supposed to be

$$\Delta C_{cell} = \frac{\Psi_{ab}}{V_a - V_b} = \frac{D_{av}\Delta S_{av}}{V_a - V_b} = \frac{\varepsilon \Delta S_{av}}{\Delta \ell_{av}} = \frac{\varepsilon \Delta h \Delta w_{av}}{\Delta \ell_{av}}$$

Δh is The Height of an Increment of The Cylindrical Structure

For a Curvilinear Rectangle Cell ($\Delta \ell_{av} \neq \Delta w_{av}$)

$\Delta C_{cr} = \varepsilon \frac{\Delta h \Delta w_{av}}{\Delta \ell_{av}}$, **and**

$\Delta C_{cr}/\Delta h$ (Capacitance Per Unit Height)

$$= \varepsilon \frac{\Delta w_{av}}{\Delta \ell_{av}} \text{ (F/m)}$$

For a Curvilinear Square Cell ($\Delta \ell_{av} = \Delta w_{av}$)

$\Delta C_{cs} = \varepsilon \Delta h$, **and**

$\Delta C_{cr}/\Delta h$ (Capacitance Per Unit Height)

$$= \varepsilon \frac{\Delta w_{av}}{\Delta \ell_{av}} = \varepsilon \text{ (F/m)}$$

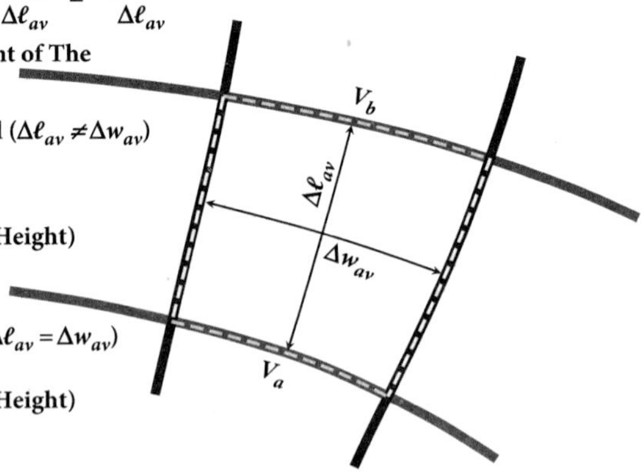

FIGURE 7B.12 Using the graphically developed curvilinear squares to evaluate the capacitance of each cell of the graphical solution.

"infinitesimally" small, we will use the cell's average dimensions in developing the capacitance estimates. Using the labels on the figure; Δw_{av} for the average cell width and $\Delta \ell_{av}$ for the average cell length, we can write

$$\Delta C_{\text{cell}} = \frac{\Psi_{ab}}{V_a - V_b} = \frac{D_{av}\Delta S_{av}}{V_a - V_b} = \frac{\varepsilon \Delta S_{av}}{\Delta \ell_{av}} = \frac{\varepsilon \Delta h \Delta w_{av}}{\Delta \ell_{av}} \tag{7B.10}$$

where $\Delta S_{av} = \Delta h \Delta w_{av}$ is the area of the cell capacitor cross section with a width of Δw_{av} and a height of Δh. The height is assumed to be the dimension off the page which is the third dimension of the 2D problem. This dimension is the dimension of uniformity of the problem structure and is assumed to be infinite in extension (or at least so far away from its ends or other transitions).

Thus, for a general-shaped curvilinear rectangle cell, the capacitance per unit height is given by

$$\frac{\Delta C_{cr}}{\Delta h} = \varepsilon \frac{\Delta w_{av}}{\Delta \ell_{av}} \text{ F/m} \tag{7B.11}$$

The same quantity for the case of a curvilinear square cell with an average length equal to the average width is given by

$$\frac{\Delta C_{cs}}{\Delta h} = \varepsilon \text{ F/m} \tag{7B.12}$$

With these two expressions, we now turn to the capacitance combination process as demonstrated in Figure 7B.13. We have a lattice of capacitances that forms the components of the overall capacitance. Combining the capacitors in any of the rows yields the total capacitance between any two consecutive equipotential surfaces.

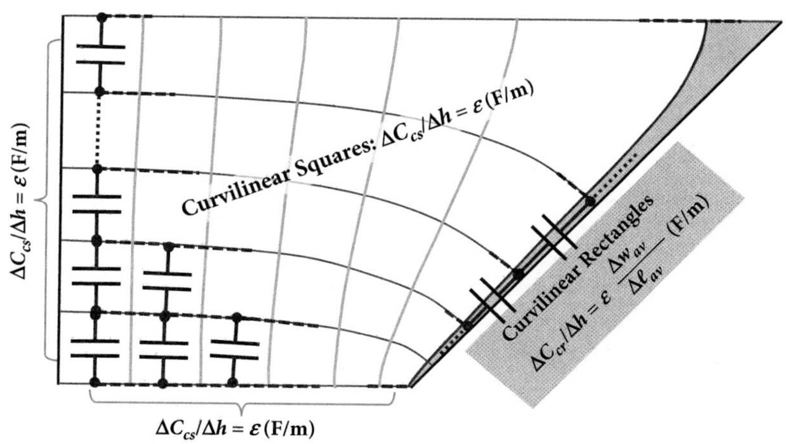

For Each Row, We Add $[6 \times \Delta C_{cs} + 1 \times \Delta C_{cr}]$ Capacitances in Parallel = 6.1 $\varepsilon \Delta h$ (F)
Next, We Combine 5 Such Rows of Capacitances in Series $\rightarrow C_{section} = (6.1\ \varepsilon \Delta h)/5$ (F)
Capacitance of Section Shown/$\Delta h = \varepsilon \times 6.1/5 = 1.22\ \varepsilon$ (F/m)
Capacitance of All 8 Sections/$\Delta h = 8 \times 1.22\ \varepsilon = 9.76\ \varepsilon = 19.52\ \varepsilon_o$ (F/m)

FIGURE 7B.13 The combination of cell capacitances is used to evaluate the total capacitance. The figure shows the concept applied to the 1/8th section.

Example 7B.2 Use the graphical method to evaluate the capacitance per unit length for two coaxial square cross-section conductors demonstrated in Figure 7B.5.

Solution Following the steps outlined in Section 7B.5, we arrive at the graphical map for the curvilinear squares as shown in Figure 7B.10.

To take advantage of the symmetry, let us evaluate the capacitance of 1/8th section of the capacitor shape as shown in Figure 7B.13.

As the figure shows, in this section, there are six complete curvilinear square cells in each row plus a fraction (~0.1) of a cell. While there are five of these rows between the inner and outer conductors.

The capacitance of each complete cell is

$$C_{\text{cell}} = \varepsilon \frac{S_{\text{cell}}}{d_{\text{cell}}} = \varepsilon \frac{(\text{width}_{\text{cell}})(\text{height}_{\text{cell}})}{(\text{depth}_{\text{cell}})}$$

$$C_{\text{cell, per unit heigth}} = \varepsilon \frac{(\text{width}_{\text{cell}})(1)}{(\text{depth}_{\text{cell}})}$$

Since the cells are formed to be curvilinear squares, i.e., $(\text{width}_{\text{cell}}) = (\text{depth}_{\text{cell}})$, then

$$C_{\text{cell, per unit heigth}} = \varepsilon$$

For the 6.1 cells in each row, the total capacitance is the parallel combination (addition of the capacitances of all the 6.1 cells), hence

$$C_{\text{row of cells, per unit heigth}} = 6.1\, C_{\text{cell, per unit heigth}} = 6.1\ \varepsilon$$

For the five rows in the 1/8th section, the total capacitance is the series combination of the five row capacitances, hence

$$C_{\text{1/8th section, per unit heigth}} = C_{5\ \text{row of cells, per unit heigth}} = \frac{1}{5} C_{\text{row, per unit heigth}} = \frac{6.1}{5}\ \varepsilon = 1.22\ \varepsilon$$

And finally, the total capacitance per unit length of the coaxial conductors is 8 times the capacitance of the 1/8th section (Figure 7B.14), i.e.,

$$C_{\text{total, per unit heigth}} = 8C_{\text{1/8th section, per unit heigth}} = (8)(1.22\varepsilon) = 19.52\varepsilon$$

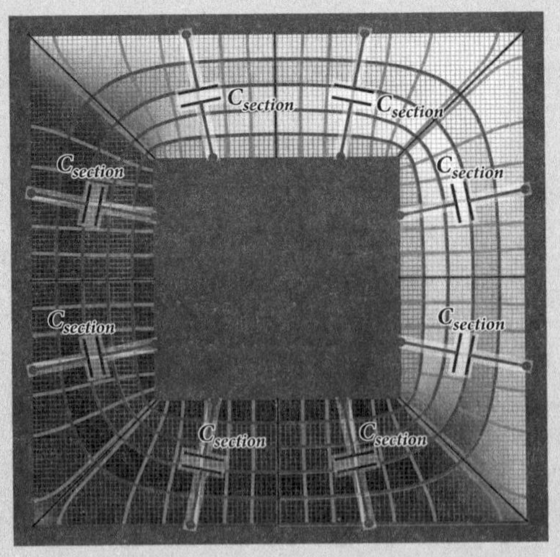

Figure 7B.14 The combination of section capacitances used to evaluate the total capacitance of the whole coaxial configuration.

CHAPTER 7 PROBLEMS

- Problems 7.1 to 7.5: Problems dealing with using Poisson's equation to find the charge distribution
- Problems 7.6 to 7.10: Problems dealing with solving Poisson's equation to find the potential function
- Problems 7.11 to 7.20: Problems dealing with using the Method of Images
- Problems 7.21 to 7.23: Problems dealing with the use of numerical and graphical techniques

7.1 Derive an expression for the total charge enclosed by the spherical volume $0 \le R \le R_o$, where R_o is the outer radius of the volume, $R_o \gg d$, given $\vec{E} = \dfrac{(qd)(2\cos\theta\,\vec{a}_r + \sin\theta\,\vec{a}_\theta)}{(4\pi\varepsilon_o R^3)}$.

7.2 Derive expressions for $\rho_v(r)$ given $V = V_o$, $r < a$; $V = \left(\dfrac{aV_o}{r}\right)$, $a < r < b$; and $V = 0, r > b$.

7.3 Find the electric flux on the closed surface of the sphere of unit radius centered at the origin given that the potential function covering this region is $V = x + 3y^2 + z^3$.

7.4 Find the electric flux on the outer surface of a hemisphere with radius of 1 cm centered at the origin and based in the $x-y$ plane given that the potential function covering this region is $V = 3x + y^2 + z$.

7.5 Derive an expression for the charge enclosed by the hemisphere with a radius of 1 m sitting in the upper space of the *x-y* plane associated with the electric field:

a. $\vec{E} = 5\vec{a}_x - 3\vec{a}_y + 2\vec{a}_z$

b. $\vec{E} = 5\vec{a}_x - 3\vec{a}_y + 2z\vec{a}_z$

Hint: $\int (\sin\theta)^2 \cdot \cos\theta\, d\theta = \dfrac{+(\sin\theta)^3}{3}$, $\int (\cos\theta)^2 \cdot \sin\theta\, d\theta = \dfrac{-(\cos\theta)^3}{3}$, $\int \cos\theta \cdot \sin\theta\, d\theta = \dfrac{(\sin\theta)^2}{2}$

7.6 Use Poisson's equation to derive expressions for the electric potential function and electric field intensity vector at all points $a < r < b$, given that the region is free of charges (source free) and that $V_{(r=a)} = V_a$, and $V_{(r=b)} = V_b$.

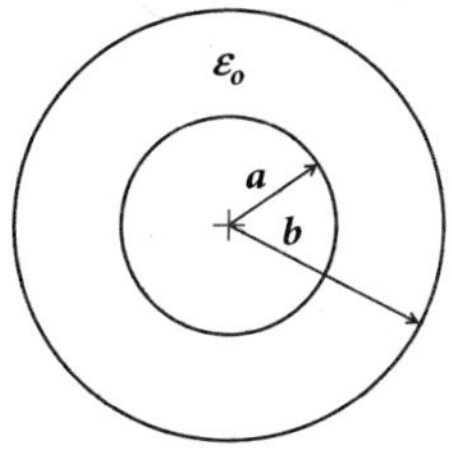

FIGURE 7P.6

7.7 Starting with Poisson's formulation, derive expressions for the electric potential function and the electric field intensity vector for the dielectric filling between two concentric conducting spheres. Assume the inner conductor has an outer radius a and is set at V_o while the outer conductor has an inner radius b and an outer radius c and is grounded ($V = 0$). Also, assume ε_r for the dielectric's relative permittivity.

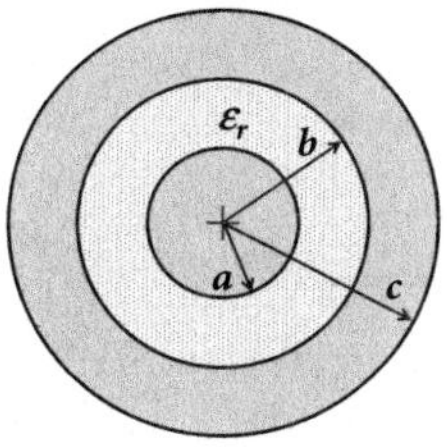

FIGURE 7P.7

7.8 Two cylindrical conductors with infinite extension along the z axis (normal to the page), the conductor radii are given by a, b, and g as shown. The voltages of the inner and outer conductors were set to V_i and "V_o = zero," respectively.

a. Use Poisson/Laplace's equation approach to derive expressions for the potential as a function of position for all locations.

b. Find the potentials at $P_1\left(\rho=\frac{a}{2}\right)$, $P_2\left(\rho=\frac{a+b}{2}\right)$, and $P_3\left(\rho=\frac{b+d+g}{2}\right)$.

c. Find the potential at $P_1\left(\rho=\frac{a}{2}\right)$, $P_2\left(\rho=\frac{a+b}{2}\right)$, and $P_3\left(\rho=\frac{b+g}{2}\right)$.

d. Find the surface charge densities on all three conductor surfaces.

e. Find the corresponding linear charge densities.

f. Find the capacitance C_{ab} per unit z axis length.

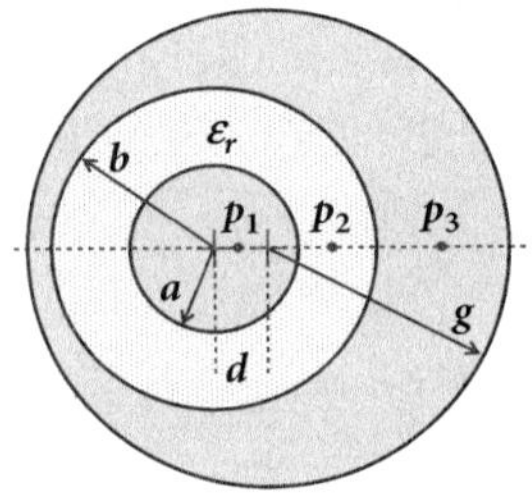

FIGURE 7P.8

7.9 Use Poisson's equation to derive expressions for the electric potential function and electric field intensity vector at all points $a < r < b$, given that the region is free of charges (source free) and that $V_{(r=a)} = V_a$, and $V_{(r=b)} = V_b$.

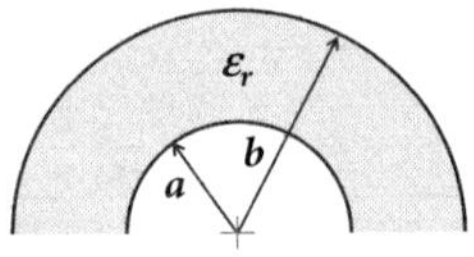

FIGURE 7P.9

7.10 Given the dielectric cylindrical shell: $a < r < b$, $0 < \varphi < 2\pi$, and $-\infty < z < +\infty$. Use Poisson's equation to derive expressions for the electric potential function and electric field intensity vector at all points $a < r < b$, given that the region is free of charges (source free) and that $V_{(\rho=a)} = V_1$, and $V_{(\rho=b)} = V_2$.

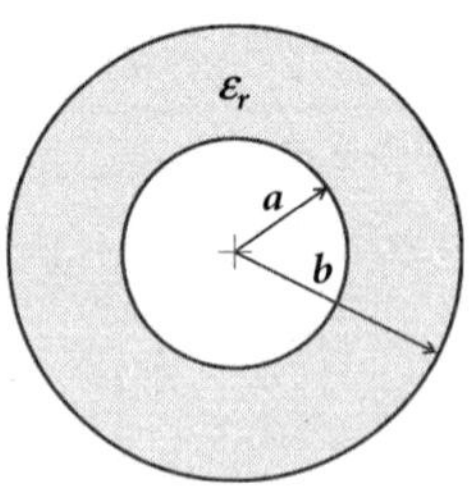

FIGURE 7P.10

7.11 Two point-charges, $+Q$ and $-Q$ C are placed a distance d above an infinite extension ground plane. The two charges are spaced a distance d apart. Use the image techniques based on the uniqueness of Poisson's equation solutions to solve for the electric field vector at all points on the surface of the ground plane in the shown configuration.

Derive an expression for the surface charge density on the upper and lower surfaces of the ground plane.

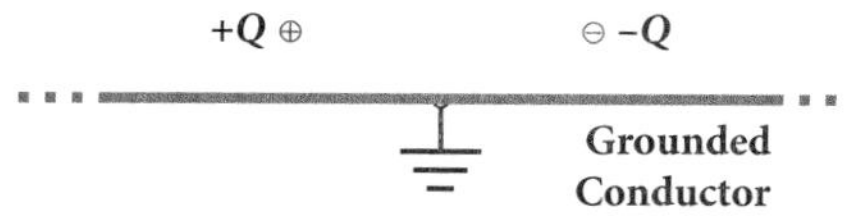

FIGURE 7P.11

7.12 An infinite line of charge, ρ_ℓ C/m is placed a distance d above an infinite extension ground plane. Use the image techniques based on the uniqueness of Poisson's equation solutions to solve for the electric field vector at all points on the surface of the ground plane in the shown configuration.

Derive an expression for the surface charge density on the upper and lower surfaces of the ground plane.

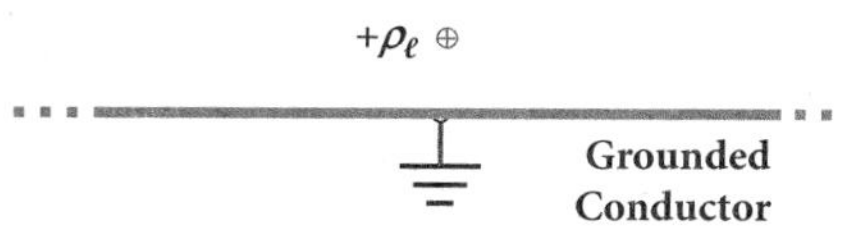

FIGURE 7P.12

7.13 A point charge, $+Q$ C is placed a distance d from the center of a perfectly conducting grounded sphere of radius a. Use the image techniques based on the uniqueness of Poisson's equation solutions to solve for the electric field vector at the top point of the sphere (closest to the charge).

Derive an expression for the surface charge density at the same point.

FIGURE 7P.13

7.14 An infinite line of charge, ρ_ℓ C/m is placed a distance d from the center of a perfectly conducting cylinder of infinite length and radius a. Use the image techniques based on the uniqueness of Poisson's equation solutions to solve for the electric field vector at the top point of the cylinder (closest to the charge).

Derive an expression for the surface charge density at the same point.

FIGURE 7P.14

7.15 A conductor has a spherical cavity of radius R. A dipole with dipole moment p is placed in the center of the cavity. Derive expressions for the potential function at all points in the cavity and for the induced charge distribution inside the cavity.

7.16 Electrical dipole of dipole moment p is placed at a distance r from the center of the conducting grounded sphere of radius R. The angle between direction of the dipole moment p and radius vector r is θ. Find the force acting on the dipole and its energy.

7.17 Consider a field produced by a charge q placed in the center of the capacitor formed by two parallel disks separated by distance d from each other. Assume that the radius R of the disks is large $R \gg d$ and ignore the edge effects. Find the energy that is required to bring this charge from the infinity.

7.18 Two charges, q_1 and q_2 are held at a fixed distance d apart at a fixed height from a conducting ground-plane. Derive expressions for the potential function at all points in space and for the charge distribution on the ground plane.

7.19 Two dipoles, p_1 and p_2 are held at a fixed distance d apart at a fixed height from a conducting ground-plane. Derive expressions for the potential function at all points in space and for the charge distribution on the ground plane.

7.20 Find the field generated by the charge q inside an insulating slab of thickness d with dielectric constant ε_r sandwiched between two ground conducting-planes. Find expressions for the charge distribution on the two ground planes.

7.21 Using numerical/graphical methods to solve Poisson's equation to estimate the capacitance per unit axial length between the inner and outer conductor of the square cross-section transmission line shown. Given that inner conductor 1×1 cm, outer conductor 2×2 cm, and the dielectric has $\varepsilon_r = 2$.

Comment on the usability of such transmission line in practical applications: (pros and cons).

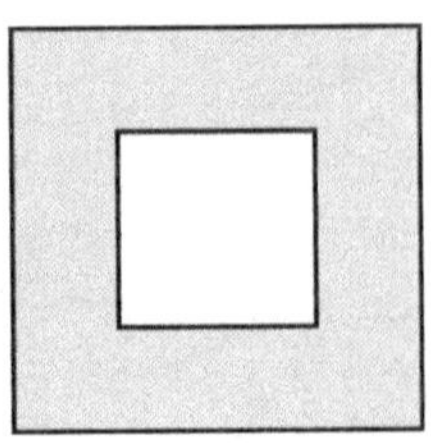

FIGURE 7P.21

7.22 Using numerical/graphical methods to solve Poisson's equation to estimate the capacitance per unit axial length for the stripline shown. Given that the strip width is 4 mm wide, has a negligible thickness, and is located midway within the dielectric sandwiched by the upper and lower conductors. The dielectric is 2 mm in thickness and has a dielectric constant, $\varepsilon_r = 4$.

FIGURE 7P.22

7.23 Using numerical/graphical methods to solve Poisson's equation to estimate the capacitance per unit axial length for off-center cylindrical transmission line shown. Given that the shaded area is that of the dielectric separating the two cylindrical conducting surfaces and has a dielectric constant, $\varepsilon_r = 6$.

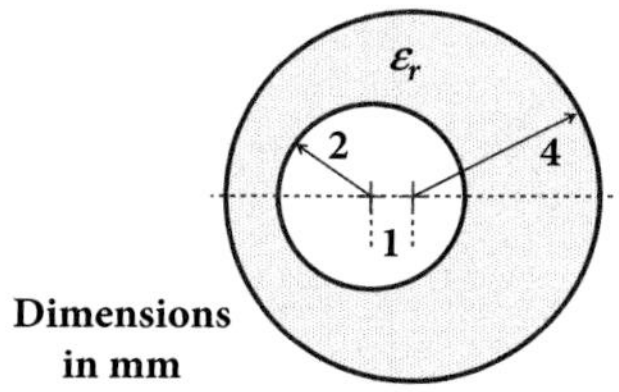

FIGURE 7P.23

CHAPTER 7 SUMMARY[1]

Summary of Electrostatic Relations and Poisson's and Laplace's Equations

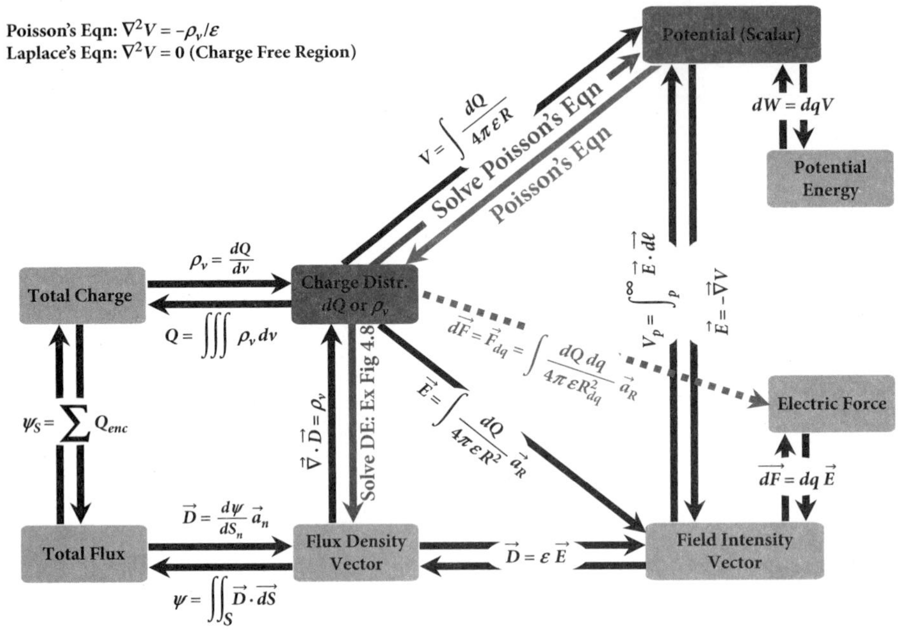

$\psi_S = \sum Q_{enc}$	(4.1)
$D = \frac{d\psi}{dS_n}$	(4.2)
$\psi = \iint_S \vec{D} \cdot \vec{dS}$	(4.5)
$\vec{\nabla} \cdot \vec{D} = \rho_v$	(4.11)
$\vec{F}_{dq} = \int \frac{dQ\, dq}{4\pi\varepsilon R_{dq}^2} \vec{a}_R$	(5.4)
$\vec{dF} = \vec{F}_{dq} = dq\vec{E}$	(5.12)
$\vec{D} = \varepsilon \vec{E}$	(5.11)
$\vec{E} = \int \frac{dQ}{4\pi\varepsilon R^2} \vec{a}_R$	(5.13)
$V_p = \int_p^\infty \vec{E} \cdot \vec{d\ell}$	(5.35)
$dW = dqV$	(5.30)
$V = \int \frac{dQ}{4\pi\varepsilon R}$	(5.38)
$\vec{E} = -\vec{\nabla}V$	(5.44)

Using Poisson's and Laplace's Equations to Solve for Electrostatic Quantities

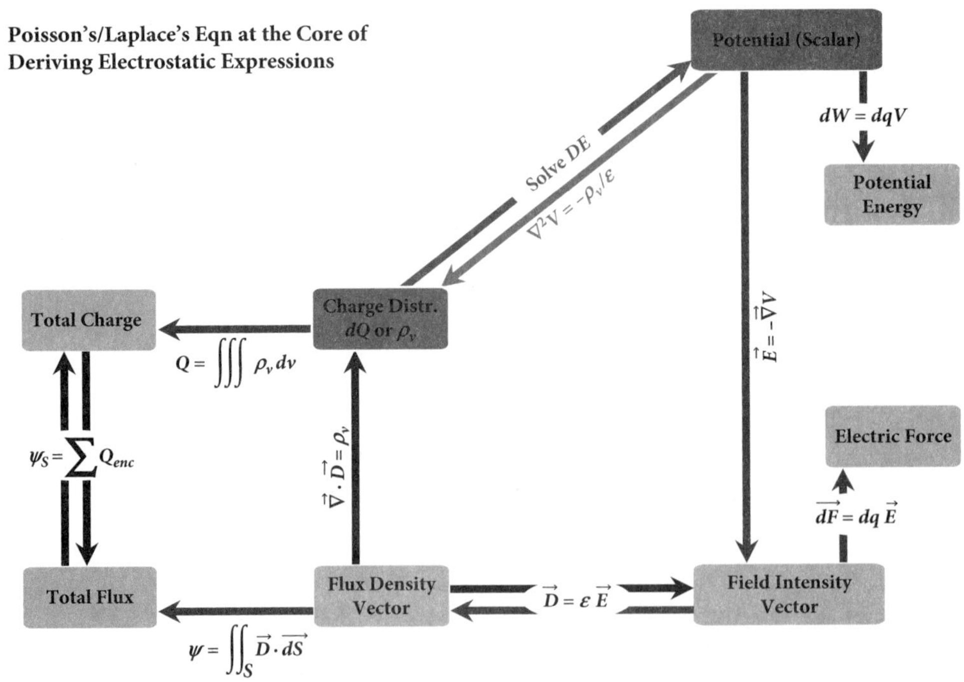

[1]***Important Warning to Students:*** Please do not use this summary without thorough study of the relevant material in this chapter. Using this summary and its formulas without adequate appreciation of the physics involved and conditions of validity can lead to wrong implementations and invalid conclusions.

The Laplacian Operator

$$\nabla^2 V(x,y,z) = \frac{\partial^2 V}{\partial x^2} + \frac{\partial^2 V}{\partial y^2} + \frac{\partial^2 V}{\partial z^2}$$

$$\nabla^2 V(\rho,\varphi,z) = \frac{1}{\rho}\frac{\partial}{\partial \rho}\left(\rho\frac{\partial V}{\partial \rho}\right) + \frac{1}{\rho^2}\frac{\partial^2 V}{\partial \varphi^2} + \frac{\partial^2 V}{\partial z^2}$$

$$\nabla^2 V(r,\theta,\varphi) = \frac{1}{r^2}\frac{\partial}{\partial r}\left(r^2\frac{\partial V}{\partial r}\right) + \frac{1}{r^2 \sin\theta}\frac{\partial}{\partial \theta}\left(\sin\theta\frac{\partial V}{\partial \theta}\right) + \frac{1}{r^2 \sin^2\theta}\frac{\partial^2 V}{\partial \varphi^2}$$

Uniqueness Theorem

A fully defined region (geometry, material and sources) with fully defined boundaries (geometry and potential) will have one unique solution irrespective of the properties of all other regions.

Method of Images

A region in the image problem having the same geometrical configuration, same material distribution, same charge distribution, and same boundary potential will have an identical solution to the identical region in the original problem.

CHAPTER 8
Magnetic Fields and Flux

Hans Christian Ørsted
(1777–1851)

André-Marie Ampère
(1775–1836)

Jean-Baptiste Biot
(1774–1862)

Félix Savart
(1791–1841)

Chapter Learning Objectives

- Calculate the magnetic force exerted by a magnetic field on different current distributions using Ampere's law for magnetic force.
- State Gauss' law for magnetism and explain how it is different from Gauss' law for electrostatics and why.
- Evaluate the magnetic field intensity due to different current distributions using the "incrementation" scheme. Five current distribution cases are considered (finite and infinite thin straight conductor, thin loop conductor, finite and infinite height circular solenoid).
- Evaluate the magnetic field intensity for symmetrical current distributions using Ampere's law. Six current distribution cases are considered under three main categories [cylindrical (axial/coaxial) symmetries, planar symmetries, and toroidal and solenoidal symmetries].
- Derive the magnetostatic point forms for Gauss' law in magnetism, and Ampere's circuital law.
- State static forms of Maxwell's equations in both integral and differential (point) forms.
- Apply Stokes' theorem to Maxwell's equations.
- Describe and derive the expressions for scalar and vector magnetic potentials.

8.1 Introduction

In Chapter 4, we identified the electric charge as the source of electrostatic fields. Specifically, the electric charge is responsible for producing the electric flux that emerges from positive charges and sinks into negative ones. An electric field was identified as the (per unit charge) normalized electric force between electric charges.

In this chapter, we introduce the magnetic flux and magnetic field. In a way, we see some "apparent" similarities to the work presented in electrostatics, but there are obviously differences. One essential difference is that the source of magnetism is an electric charge in motion. A static electric charge does not produce any magnetic phenomena. This essential difference in the nature of the source is responsible for the differences that we discuss.

The reader might think of some similarities based on forces of attraction and repulsion of *magnets* in a way similar to those of the Coulomb forces between charges. This view can only be supported through the assumption of the presence of magnetic poles (or magnetic charge). This assumption is not valid in the light of modern knowledge that magnetic poles/charges can only be assumed in the form of dipoles and cannot exist separately.

It was the work done by Orested in 1820 that revealed the true source of magnetism as the electric current. His experiments also showed that the resulting magnetic field forms closed circles surrounding the current-carrying wire. Figure 8.1 depicts an illustration of Orested's experiment with a current-carrying wire and how a compass changes its orientation as a result of the circular (rotational) magnetic field surrounding the wire. Further examination of these findings explains the physics of *magnetized* objects where internal microscopic currents due to orbiting and spinning electrons can add up to produce noticeable magnetic behavior. We will further explore this aspect of magnetism when we learn about magnetic materials in the follow-up chapter.

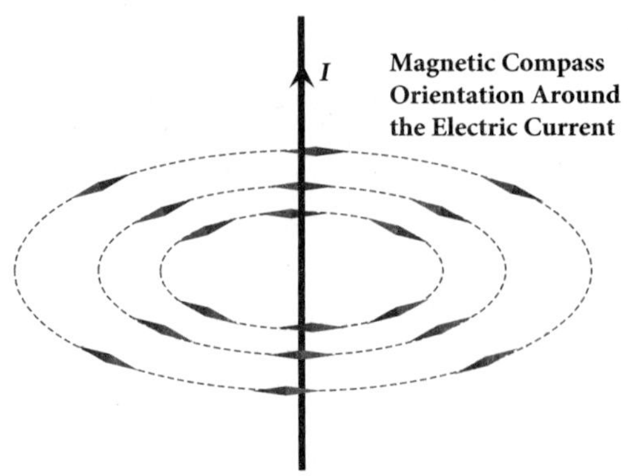

FIGURE 8.1 Magnetic campus orientation around an electric current.

In Chapters 4 and 5, similarly to what we have done in electrostatics, we will define the magnetic flux, the magnetic flux density vector, the magnetic force, and the magnetic field intensity vector. Again, depending on the magnetic properties of the media, we can write the relationship between the magnetic flux density and the magnetic field intensity. We will also define the magnetic energy as well as the magnetic potential. Finally, we will learn about magnetic materials and inductance.

Although the titles listed above are somehow similar to those of electrostatics, the underlying physics is quite different. The fact that the nature of the source is different in electrostatics and magnetostatics makes everything else different. The nature of the electrostatic fields and forces can be compared, to a limited extent, to those of gravitational fields and forces. However, the magnetic fields and forces are rotational in nature and can be compared, again to a limited extent, to those of fluid mechanics, for example, tornadoes and whirlpools.

At the conclusion of this chapter, we summarize the similarities and differences between the electrostatic and magnetostatic quantities, related physical phenomena and their mathematical expressions in Addendum 8A. The reader is encouraged to review this addendum more frequently to gain adequate insight in the underlying physics.

At this point, we are ready to introduce the analytical expression for the magnetic force, which was introduced by Ampere in 1820, and that for the magnetic field introduced by Biot and Savart.

8.2 Ampere's Law for Magnetic Force

André-Marie Ampère showed that currents running in the same directions in parallel wires attract each other while those running in opposite directions repel. The magnitude of the magnetic force between the two wires was found to be proportional to both the magnitude of the two currents and their wire lengths and inversely proportional to the distance between the two wires. He also pointed out the absence of forces if the wires are turned perpendicular to each other.

The mathematical form of the magnetic force has some resemblance to that of Coulomb's law that we introduced in Equations (5.1) to (5.5) earlier, except for the zero magnetic force between the perpendicular wires. If we were to write separate expressions for the parallel and perpendicular wires, we would formulate them as follows:

$$|F_m| = k_m \left| \frac{(Id\ell)_1 (Id\ell)_2}{R_{12}^2} \right| \text{ for the parallel wires} \tag{8.1}$$

$$|F_m| = 0 \text{ for the perpendicular wires} \tag{8.2}$$

Using vector format eloquently expresses both formulas into one relationship as

$$\overrightarrow{dF}_{m2} = k_m \frac{(\overrightarrow{Id\ell})_2 \times \left[(\overrightarrow{Id\ell})_1 \times \vec{a}_{12} \right]}{R_{12}^2} \tag{8.3}$$

This relationship is demonstrated in Figure 8.2 for the general case of current orientations as well as the two special cases of parallel and perpendicular currents.

Similar to what was done to the Coulomb constant in Equation (5.1), the constant k_m is chosen to take the form $\mu_o/(4\pi)$. The (4π) is introduced to simplify the mathematical expressions, especially when dealing with spherical and cylindrical configurations where π appears in surface area and circumference expressions.

$$\overrightarrow{dF}_{m2} = \mu_o \frac{(\overrightarrow{Id\ell})_2 \times [(\overrightarrow{Id\ell})_1 \times \vec{a}_{12}]}{4\pi R_{12}^2} \tag{8.4}$$

The constant (μ_o) is referred to as the free space permeability. This constant is to be replaced by the proper permeability constant (μ) for a material other than free space as will be discussed in a later chapter about magnetic materials.

In the SI system of units, the value of the constant μ_o is given by

$$\mu_o = 4\pi \times 10^{-7} \text{ N/A}^2 = \text{H/m (henries/meter)} \tag{8.5}$$

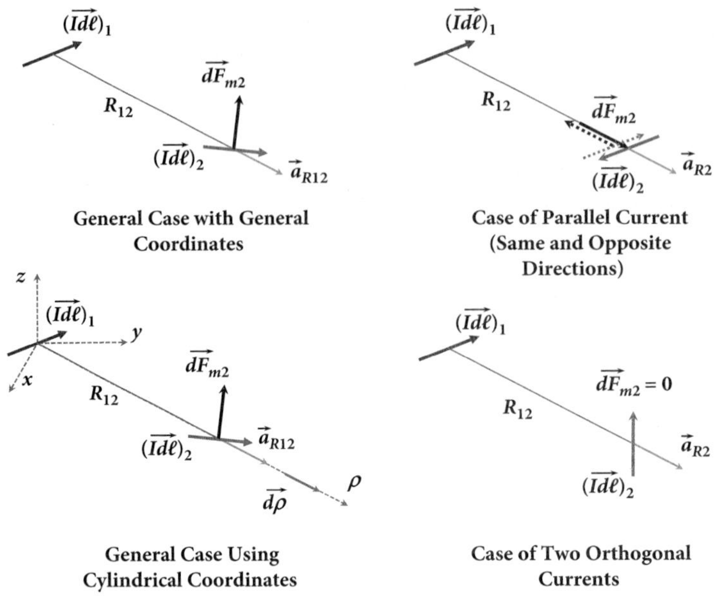

FIGURE 8.2 Ampere's law for magnetic force shown between the current elements.

8.3 Magnetic Field Intensity and Magnetic Flux Density

As a result of the quantitative experimental work of Jean-Baptiste Biot and Félix Savart, a mathematical expression was determined for the "magnetic field" at a point in space in terms of the electric current that produces this field. However, before we present this formula, we pose to discuss the concepts of field and flux in magnetism in the light of our earlier discussion in electrostatics.

In electrostatics, we introduced the electric flux and the electric field as if they were independent of each other, then later showed their direct relationship through the material permittivity, ε. In magnetics, we will follow a different approach, as we will define the magnetic flux and magnetic field with their dependency in mind. Analogous to the electrostatic relationship [Equation (5.12)],

$$\vec{D} = \varepsilon \vec{E} \tag{5.12}$$

we will have a similar relationship between the magnetic flux density vector, $\vec{B}$, and the magnetic field intensity vector, $\vec{H}$, as

$$\vec{B} = \mu \vec{H}, \text{ where } \mu = \mu_o \text{ for free space} \tag{8.6}$$

Comparing Equations (5.12) and (8.6), we can see that $\vec{B}$ is introduced in analogy to $\vec{D}$, and $\vec{H}$ to $\vec{E}$. Now, similar the electric flux and its density relationships given in Equations (4.3) to (4.7) of Chapter 4, we can write the corresponding magnetic flux, ψ_m, (over a given surface) as the surface integral of the flux density over the surface on interest

$$B = \frac{d\psi_m}{dS_{\text{normal}}} \tag{8.7}$$

$$d\psi_m = \vec{B} \cdot \overrightarrow{dS} \tag{8.8}$$

$$\psi_m = \iint_S \vec{B} \cdot \overrightarrow{dS} \tag{8.9}$$

Historically, the "magnetic field" was defined through the Biot-Savart law, as will be presented in the next section. Besides, if we were to start by defining the magnetic flux in a direct relationship (of equality) to the source similar to what we did in Chapter 4, we would not have the physical "logic" to support such concept. In electrostatics, we defined the flux put out by a source to be equal to that source and even have the same units. The flux was to emerge from (out of or into) the charge, and hence that definition made sense. In magnetostatics, the relationship of the magnetic flux to the source is a rotational one. As it will become clearer as we progress throughout this chapter, the magnetic flux forms closed lines linking the closed current path. As the result, the resulting magnetic flux is also rotational and in fact results in closed loops that link the current source loops. The simple equality between the magnetic flux and the source current is not fitting. We will learn about the proper relationship between the two when we study Ampere's circuital law later in this chapter.

8.4 Biot-Savart Law

In 1820, Jean-Baptiste Biot (1774-1862) and Félix Savart (1791-1841) carried on quantitative experiments to measure the force acting on a magnet placed near a wire carrying electric current. Based on their results, they found a mathematical expression for calculating what they referred to as the "magnetic field" at a point in space in terms of the electric current that produces this "field." Referring to Figure 8.3, the mathematical expression they arrived at, which is known as the Biot-Savart Law, is of the form

$$\overrightarrow{dB_p} = \frac{\mu_o}{4\pi} \frac{(\overrightarrow{Id\ell}) \times \vec{a}_R}{R^2_{d\ell-p}} \tag{8.10}$$

where μ_o is the permeability of free space and $R_{d\ell-p}$ is the distance between the differential current element $(Id\ell)$ acting as the magnetic source and the field observation point p.

You may have noticed that we used the Biot-Savart term of "magnetic field" with double quotes. The reason for that is $\vec{B}$ is what is used to denote the magnetic flux density while $\vec{H}$ is the magnetic field intensity. Hence, we should present the Biot-Savart law of Equation (8.10) as the law for the magnetic flux density vector, or alternatively, present the magnetic field intensity vector form of Biot-Savart as

$$\overrightarrow{dH_p} = \frac{1}{4\pi} \frac{(\overrightarrow{Id\ell}) \times \vec{a}_R}{R^2_{d\ell-p}} \tag{8.11}$$

Relating the Biot-Savart expression to that of Ampere, we notice that Biot-Savart relationship presents a normalized force expression in a manner similar to what we did in electrostatics in defining the electric field as the force per unit charge. However, there is a slight difference in the magnetic definition. It is the flux density vector $\vec{B}$ and not the field $\vec{H}$ that represents the analogous normalization.

Starting with Ampere's law of force, Equation (8.4), and recognizing the current element source as $(\overrightarrow{Id\ell})_1$, while the second current element $(\overrightarrow{Id\ell})_2$ senses the magnetic force influence created by $(\overrightarrow{Id\ell})_1$, we can define a normalized magnetic force $(nF)_m$, as the force per unit current element where $(Id\ell)_2 = 1$; hence,

$$\overrightarrow{d(nF)_m} = \mu_o \frac{\left[(\overrightarrow{Id\ell})_1 \times \vec{a}_{12}\right]}{4\pi R^2_{12}} \tag{8.12}$$

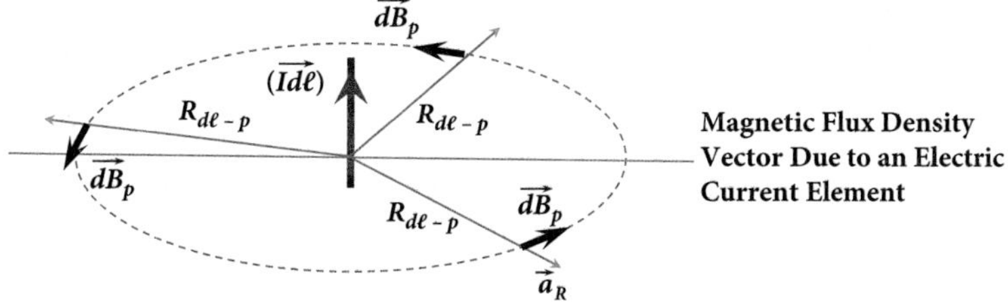

FIGURE 8.3 Biot-Savart law for magnetic flux density due to an electric current element.

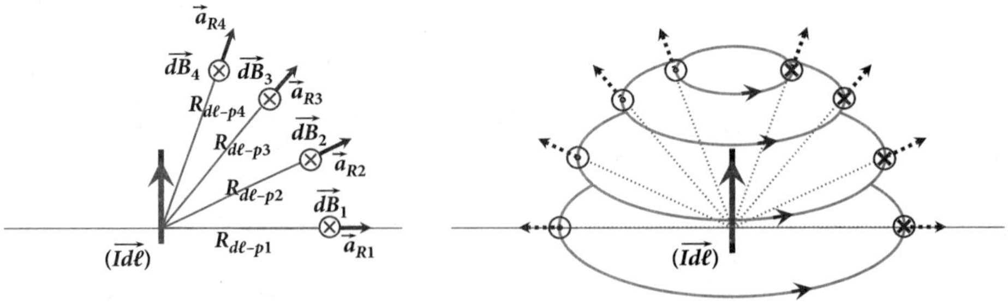

FIGURE 8.4 Magnetic flux lines form closed loops.

Comparing Equations (8.12) to (8.10) and (8.11), we realize that it is the magnetic flux density vector that represents the normalized force per unit current element. Alternatively, we can say that the magnetic field intensity represents a normalized force as well; however, the normalization is with respect to $\mu_o(Id\ell)$.

Consequently, and referring to Figure 8.4,

$$\overrightarrow{dF}_{m2} = (\overrightarrow{Id\ell})_2 \times \overrightarrow{dB} = \mu_o(\overrightarrow{Id\ell})_2 \times \overrightarrow{dH} \tag{8.13}$$

8.5 Magnetic Flux and Gauss' Law for Magnetism

Examining the Biot-Savart law for the magnetic flux density [Equation (8.10)] and the associated graph (Figure 8.3), we can argue that the magnetic flux due to the incremental current element forms closed lines. This argument is based on the observation of the closed circular paths of the magnetic flux lines. To explain, let us consider the demonstration shown in Figure 8.4. The figure to the left shows the magnetic flux vectors at different points due to the current element $Id\ell$. Only four points are shown in the demonstration for illustration purposes. The points are all chosen in the same plane as the current element and hence the resulting magnetic flux density vectors are all perpendicular to that plane. Now, if we were to rotate the plane of points around the current element axis, we would cover all points in space around the current element. This is shown in the figure on the right in which we can see the magnetic flux lines form closed circular loops.

To generalize, if we were to consider a general form of electrical current distribution, we could always use the incrementation approach and treat it as a superposition of incremental current elements each producing closed magnetic flux lines of its own.

Although there are more rigorous proofs available to show the nature of the magnetic flux, we will suffice with this demonstration to prove the point. The conclusion is that magnetic flux lines can only exist in the form of closed loops circling the current responsible for producing them. This is compatible with the facts that

- The magnetic field (and flux) is rotational in nature.
- The electric current is the source of magnetism.
- A single magnetic pole (charge) does not exist.

With this finding at hand, we are now ready to discuss "Gauss'" law for magnetic flux. Referring to Figure 8.5, Gauss' law for magnetic flux simply states that the magnetic flux over a closed surface is zero. Since

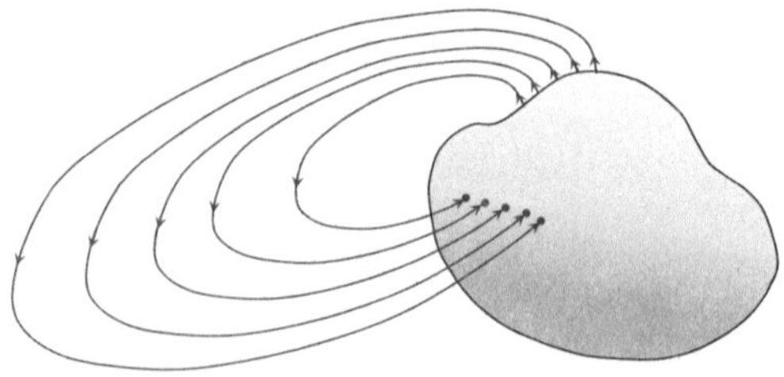

FIGURE 8.5 Gauss' law for magnetic flux simply states that the magnetic flux over a closed surface is zero.

all flux lines close on themselves, then a flux entering a closed surface is bound to exit at some other point on that surface, thus yielding a zero net-flux on that surface. In terms of the magnetic flux density vector, and using Equation (8.9), we write

$$\{\psi_{m,\text{ closed surface } S}\} = 0 = \oiint_S \vec{B} \cdot \overrightarrow{dS} \tag{8.14}$$

8.6 Ampere's Circuital Law

In 1826, Ampere discovered what we now call Ampere's circuital law that relates the circulation relationship between the electric current and the magnetic field. The proper derivation of this law is too complicated to present here, however, we will present a verification case that demonstrates the law and its usability. This will be the case of an electric current flowing in a straight wire of infinite length.

Ampere's circuital law simply states that the closed loop integration of the magnetic field times its path length is equal to the linkage current source. In vector notation, the integral form of Ampere's circuital law is

$$\oint \vec{H} \cdot \vec{d\ell} = \sum I_{\text{Linkage}} \tag{8.15}$$

To demonstrate, let us examine this law with the help of Figures 8.6 and 8.7. In Figure 8.6*a*, a closed loop path is shown linking three of the shown four current circuits (loops). A linkage is defined between two closed paths such that you cannot disengage the two loops without breaking either of them (Figure 8.7). Positive and negative linkage polarities are determined using the right-hand rule (Figure 8.6). The diagram in this figure shows that if we were to orient the thump of the right hand along the integration path, a current, I_p, flowing along the same direction as the rest of the fingers will have positive linkage. However, an opposite flowing current, I_n, is considered negative linkage current. Consequently, Ampere's law applied to the demonstration example of Figure 8.6 will have one positive linkage current, I_1, two negative linkage currents, I_2 and I_4, while I_3 does not link the integration path.

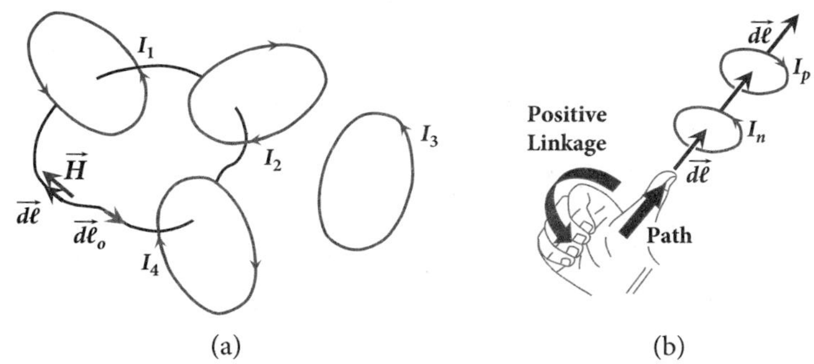

FIGURE 8.6 A demonstration of Ampere's circuital law and the right-hand-rule. (a) Ampere's circuital law demonstration. (b) Right-hand rule.

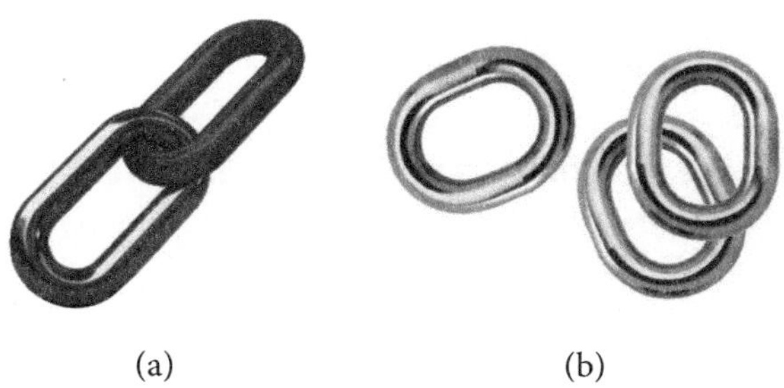

FIGURE 8.7 Demonstration of the concept of linkage. (a) Loops in linkage. (b) No linkage between the loops.

Hence, Ampere's circuital law applied to the example shown in Figure 8.6 yields

$$\oint \vec{H}\cdot\vec{d\ell} = \sum I_{\text{Linkage}} = I_1 - I_2 - I_4 \tag{8.16}$$

It is worth mentioning that if we were to integrate the same path in the opposite direction, $d\ell_o$, then we would use the opposite polarities for the current linkage.

$$\oint \vec{H}\cdot\vec{d\ell_o} = -I_1 + I_2 + I_4 = -\oint \vec{H}\cdot\vec{d\ell} \tag{8.17}$$

Physically speaking, the right-hand rule determines the polarity of linkage between the magnetic field and the current loops. However, the algebraic linkage polarity used in the mathematical expression is determined by the relationship between the integration path and the linkage current.

8.7 Magnetic Field Evaluation Schemes

At this point in our studies, we are ready to derive $\vec{H}$ and $\vec{B}$ expressions for different current distributions using the tools available to us

- The Biot-Savart law using the incrementation scheme
- Ampere's circuital law for current distributions with adequate symmetries

In the following, we present both schemes as well as provide demonstrative examples of their implementation. The chosen examples represent idealized configurations of practical current distributions that have use in real life applications.

8.8 Magnetic Field Evaluation Using the "Incrementation" Scheme

In this segment, we derive $\vec{H}$ and $\vec{B}$ expressions for different current distributions using the Biot-Savart law, Equations (8.10) and (8.11), together with the incrementation scheme. Similar to what we did in electrostatics, we will use the Biot-Savart law as applied to incremental elements of the source current and the total resulting field would be obtained by integrating the incremental result.

To demonstrate this approach, five *familiar* current distribution examples are analyzed in this chapter. We will cover the full detailed analysis of all five cases in Addendum 8B later on; however, we will also present a brief overview of the analysis and the final result for each case. The five current distribution cases considered here are

1. A finite length thin straight current-carrying conductor
2. An infinite length thin straight current-carrying conductor
3. A thin circular current-carrying (loop) conductor; also, the case of a segment of the loop
4. A finite height circular solenoid
5. An infinite height circular solenoid

8.8.1 Case 1: Magnetic Field due to a Finite Length Thin Straight Current-Carrying Conductor (Figure 8.8)

The magnetic field intensity due to a thin straight conductor carrying a constant current I, at the point $p(\rho,\varphi,z)$, can be expressed as

$$\vec{H}_p = \int \overrightarrow{dH_p} = \int_{z_1}^{z_2} \frac{1}{4\pi}\frac{\overrightarrow{Id\ell}\times\vec{a}_R}{R^2_{Id\ell-p}} \tag{8.18}$$

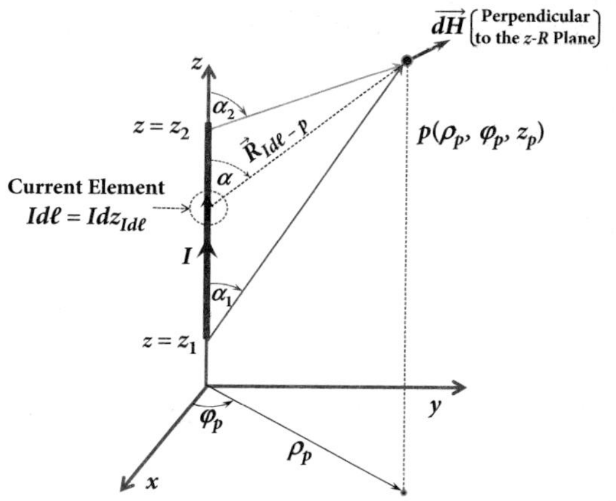

FIGURE 8.8 Using the incrementation scheme to derive an expression for the magnetic field due to a finite length thin straight current-carrying conductor.

$\vec{R}_{Id\ell-p}$ is the vector that points from the differential current element to the observation point. $R_{Id\ell-p} = \left|\vec{R}_{Id\ell-p}\right|$ and $\vec{a}_R$ is the unit vector along the direction of $\vec{R}_{Id\ell-p}$.

As shown in Example 8.1, and referring to Figure 8.8, the integration of Equation (8.18) yields the final expression for the magnetic field intensity vector in this case as

$$\vec{H}_p = \frac{I}{4\pi\rho_p}[\cos\alpha_1 - \cos\alpha_2]\vec{a}_\varphi \tag{8.19}$$

Example 8.1 Derive an expression for the magnetic field intensity vector at a general location (p) due to the finite length thin straight current-carrying conductor of Figure 8.8.

Solution Using the Biot Savart law

$$\vec{H} = \frac{1}{4\pi}\int_{y_l} \frac{\overrightarrow{Id\ell} \times \vec{a}_R}{\left|\vec{R}_{Id\ell-p}\right|^2} = \frac{1}{4\pi}\int_{y_l} \frac{\overrightarrow{Id\ell} \times \vec{R}_{Id\ell-p}}{\left|\vec{R}_{Id\ell-p}\right|^3}$$

$$\overrightarrow{Id\ell} = I(\vec{a}_z dz) \qquad \vec{R}_{Id\ell-p} = (z - z_p)\vec{a}_z + \rho_p\vec{a}_\rho \qquad R_{Id\ell-p} = \left|\vec{R}_{Id\ell-p}\right| = \sqrt{(z-z_p)^2 + \rho_p^2}$$

$$\overrightarrow{Id\ell} \times \vec{R}_{Id\ell-p} = I\left[(\vec{a}_z dz) \times ((z - z_p)\vec{a}_z + \rho_p\vec{a}_\rho)\right] = \vec{a}_\varphi I\rho_p dz$$

$$\vec{H} = \frac{1}{4\pi}\int_{z_1}^{z_2} \frac{\overrightarrow{Id\ell} \times \vec{a}_R}{\left|\vec{R}_{Id\ell-p}\right|^2} = \frac{1}{4\pi}\int_{z_1}^{z_2} \frac{\overrightarrow{Id\ell} \times \vec{R}_{Id\ell-p}}{\left|\vec{R}_{Id\ell-p}\right|^3} = \vec{a}_\varphi \frac{I\rho_p}{4\pi}\int_{z_1}^{z_2} \frac{dz}{\left|\vec{R}_{Id\ell-p}\right|^3} = \vec{a}_\varphi \frac{I\rho_p}{4\pi}\int_{z_1}^{z_1} \frac{dz}{\left((z-z_p)^2 + \rho_p^2\right)^{3/2}}$$

$$= \vec{H} = \vec{a}_\varphi \frac{I\rho_p}{4\pi}\left(\frac{1}{\rho_p^2}\frac{(z-z_p)}{\left((z-z_p)^2+\rho_p^2\right)^{1/2}}\right)_{z_1}^{z_2} = \vec{a}_\varphi \frac{I}{4\pi\rho_p}\left[\frac{z_2 - z_p}{\sqrt{(z_2-z_p)^2+\rho_p^2}} - \frac{z_1 - z_p}{\sqrt{(z_1-z_p)^2+\rho_p^2}}\right]$$

If we express $\frac{z_2 - z_p}{\sqrt{(z_2-z_p)^2+\rho_p^2}}$ as $-\cos(\alpha_2)$ and $\frac{z_1 - z_p}{\sqrt{(z_1-z_p)^2+\rho_p^2}}$ as $-\cos\alpha_1$, $\vec{H} = \vec{a}_\varphi \frac{I}{4\pi\rho_p}(\cos(\alpha_1) - \cos(\alpha_2))$

8.8.2 Case 2: Magnetic Field due to an Infinite Length Thin Straight Current-Carrying Conductor (Figure 8.8)

If we were to extend the current-carrying wire to an infinite length, the angles α_1 and α_2 converge to 0° and 180°, respectively. The expression for the magnetic field intensity vector reduces to

$$\vec{H}_p = \frac{I}{2\pi\rho_p}\vec{a}_\varphi \tag{8.20}$$

This result shows that the magnitude of the magnetic flux density is inversely proportional to the distance from the conductor. Figure 8.9 illustrates the change of the magnitude of $\vec{H}$ with the cylindrical radial distance from the wire current, ρ. This result is quite similar to those obtained earlier for the electric field due to lines of charges [Equations (5.34) and (5.35)].

Both solutions [Equations (8.19) and (8.20)] show that the field lines wrap around the current and form closed loops along the φ directions. These closed field lines link the current producing them. The corresponding magnetic flux lines do exactly the same thing as $\vec{B}$ is simply related to $\vec{H}$ through the permeability μ_o.

$$\vec{B}_p = \mu_o \vec{H}_p \tag{8.21}$$

One would ask where is the closed current loop to claim linkage with the magnetic field loop. The answer is that the case of a finite length wire is not physically realizable since the electric current can only flow in a closed electric circuit. Hence, the answer obtained in this case is for only a section of a straight wire that is part of a closed circuit, for example, check the example cases shown in Figure 8.10.

The case of an infinite wire is also a nonrealizable case. However, it can be thought of the case where the wire is long enough to apply the answer [Equation (8.20)] as an approximation. What is meant here is that $\alpha_1 \approx 0°$ and $\alpha_2 \approx 180°$, implying $z_2 - z_1 >> \rho$ (Figure 8.11), making it possible to close the *open* ends of the wire allowing a closed circuit of current.

Figure 8.9 also demonstrates the right-hand rule for determining the direction of the magnetic field direction due to the electric current source. It is interesting to observe that the rule works alternatively for both I and $\vec{H}$ (Figures 8.6 and 8.9). We demonstrate this again side by side in Figure 8.12.

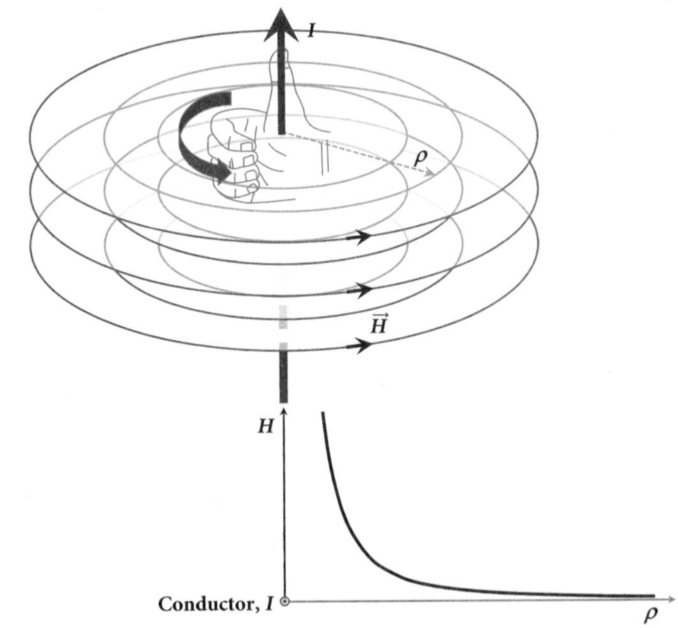

FIGURE 8.9 The magnetic field intensity versus distance due to an infinite length thin straight current-carrying conductor.

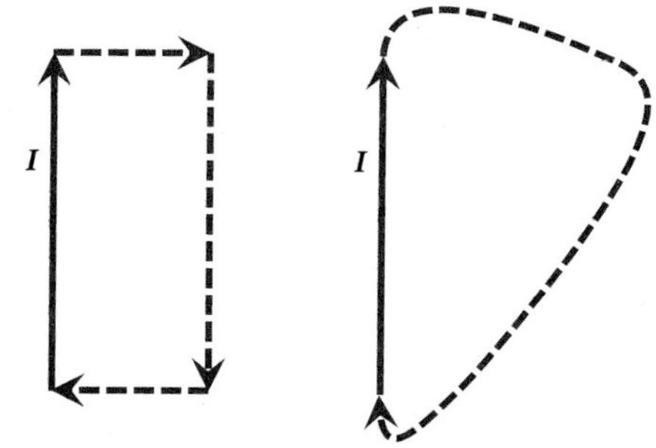

FIGURE 8.10 An electric current flow only in a closed circuit, and hence, the infinite length conductor of Figure 8.9 can be thought of as a section of infinite size circuit.

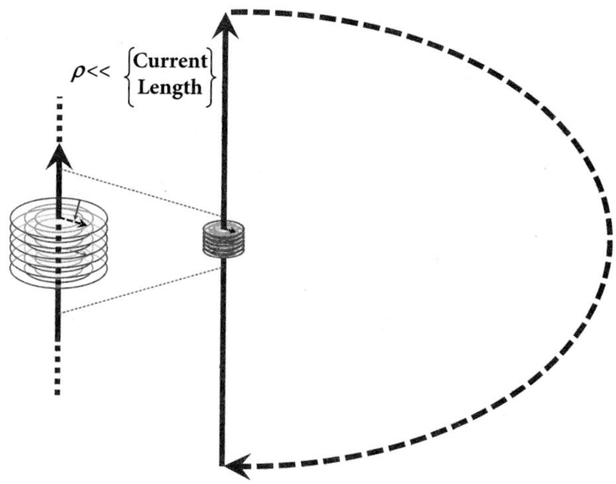

FIGURE 8.11 A demonstration of the linkage required for the use of Ampere's law between the magnetic field lines and the closed circuit of the infinite conductor.

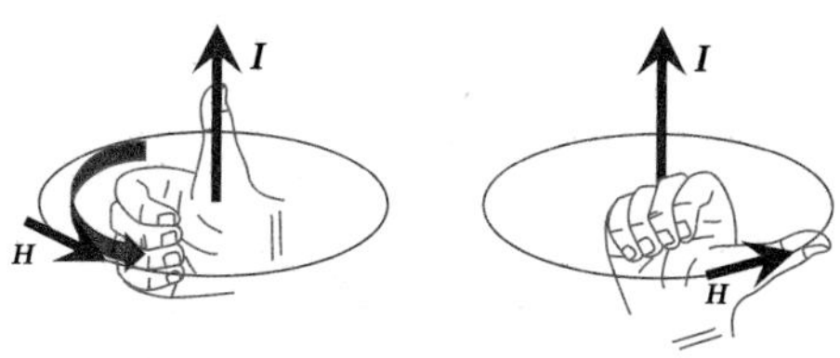

FIGURE 8.12 The right-hand rule applies alternatively for both I and $\vec{H}$.

Example 8.2 A very long thin wire carrying current has a right-angle bend, as shown. Using the Biot-Savart law, find the magnetic flux density, $\vec{B}$, at a point along the x axis.

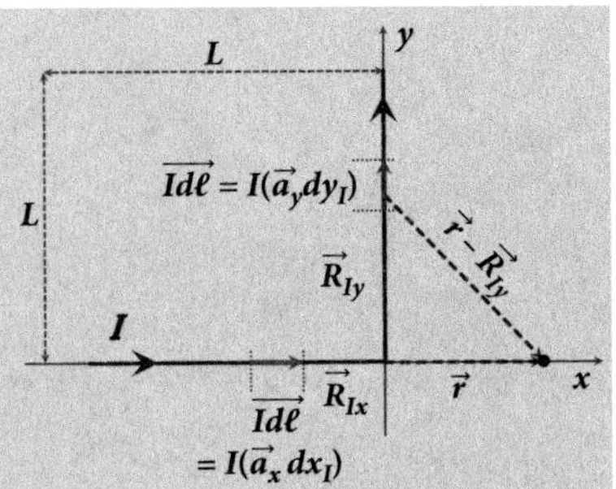

Solution For the section of the wire along the x axis:

$$\overrightarrow{Id\ell} = I(\vec{a}_x dx_I) \qquad \vec{r} = \vec{a}_x x \qquad \vec{R}_{Ix} = \vec{a}_x x_I$$

$$\vec{R}_{Id\ell-p} = \vec{r} - \vec{R}_{Ix} = \vec{a}_x x - \vec{a}_x x_I = \vec{a}_x (x - x_I)$$

$$\overrightarrow{Id\ell} \times \vec{R}_{Id\ell-p} = I[\vec{a}_x dx_I \times \vec{a}_x (x - x_I)] = 0$$

Therefore, this section of the wire doesn't contribute to the magnetic field at a point along the x axis. For the section of the wire along the y axis:

$$\overrightarrow{Id\ell} = I(\vec{a}_y dy_I) \qquad \vec{r} = \vec{a}_x x \qquad \vec{R}_{Iy} = \vec{a}_y y_I$$

$$\vec{R}_{Id\ell-p} = \vec{r} - \vec{R}_{Iy} = (\vec{a}_x x - \vec{a}_y y_I)$$

$$\overrightarrow{Id\ell} \times \vec{R}_{Id\ell-p} = I[\vec{a}_y dy_I \times (\vec{a}_x x - \vec{a}_y y_I)] = I[-\vec{a}_z x\, dy_I]$$

$$R_{Id\ell-p} = \left|\vec{R}_{Id\ell-p}\right| = \sqrt{x^2 + y_I^2}$$

$$\vec{B} = \frac{\mu_o}{4\pi}\int_{y_I} \frac{\overrightarrow{Id\ell} \times \vec{a}_R}{\left|\vec{R}_{Id\ell-p}\right|^2} = \frac{\mu_o}{4\pi}\int_{y_I} \frac{\overrightarrow{Id\ell} \times \vec{R}_{Id\ell-p}}{\left|\vec{R}_{Id\ell-p}\right|^3} = \frac{\mu_o I}{4\pi}\int_0^L \frac{-\vec{a}_z x\, dy_I}{\left|\vec{R}_{Id\ell-p}\right|^3} = -\vec{a}_z \frac{\mu_o I}{4\pi}\int_0^L \frac{x\, dy_I}{\left(x^2 + y_I^2\right)^{3/2}}$$

$$= -\vec{a}_z \frac{\mu_o I}{4\pi}\left(\frac{x y_I}{x^2\left(x^2 + y_I^2\right)^{1/2}}\right)\Bigg|_{y_I=0}^{y_I=L} = -\vec{a}_z \frac{\mu_o I}{4\pi} \frac{L}{x(x^2 + L^2)^{1/2}} \quad (T)$$

8.8.3 Case 3: Magnetic Field due to a Thin Circular Current-Carrying Conductor (Loop) (Figure 8.13)

As detailed in Addendum 8B, we conclude that the magnetic field vector at the point $p(0, 0, z_p)$ is given by

$$\vec{H}_P = \int d\vec{H}_P = \frac{1}{4\pi}\int \frac{\overrightarrow{Id\ell} \times \vec{a}_R}{\left|\vec{R}_{Id\ell-p}\right|^2} = \int \frac{\overrightarrow{Id\ell} \times \vec{R}_{Id\ell-p}}{4\pi R_{Id\ell-p}^3} = \frac{Ia^2}{2\left(a^2 + z_p^2\right)^{\frac{3}{2}}}\vec{a}_z \tag{8.22}$$

For the special point at the origin, $z_p = 0$, we find

$$\vec{H}_P = \frac{I}{2a}\vec{a}_z \tag{8.23}$$

Although we have not derived it, we are displaying a map of a cross section of the magnetic field (and flux) lines due to the current loop in Figure 8.14. The figure shows the field at the origin as well as on all points of the z axis to be in the $\vec{a}_z$ direction.

Now, for the case of a segment of the loop (Figure 8.15), we have

$$\vec{H}_P = \int_{\text{segment}} \frac{\overrightarrow{Id\ell} \times \vec{R}_{Id\ell-p}}{4\pi R_{Id\ell-p}^3} = \frac{Ia^2(\varphi_2 - \varphi_1)}{4\pi(a^2 + z^2)^{\frac{3}{2}}}\vec{a}_z + \frac{Iaz}{4\pi(a^2 + z^2)^{\frac{3}{2}}}\left[\vec{a}_x(\sin\varphi_2 - \sin\varphi_1) - \vec{a}_y(\cos\varphi_2 - \cos\varphi_1)\right] \tag{8.24}$$

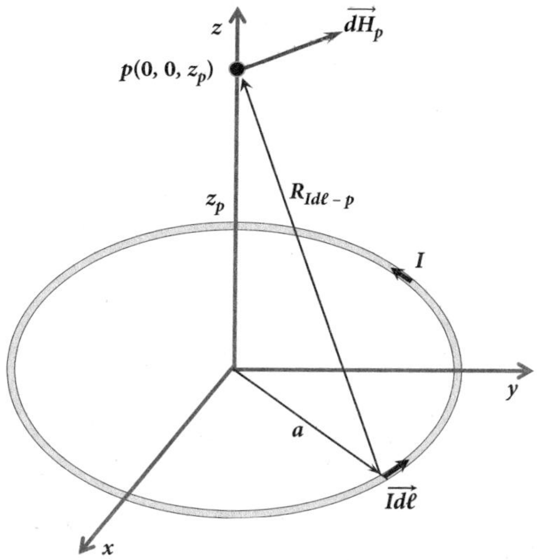

FIGURE 8.13 Using the incrementation scheme to derive an expression for the magnetic field due to a thin circular current-carrying conductor (loop).

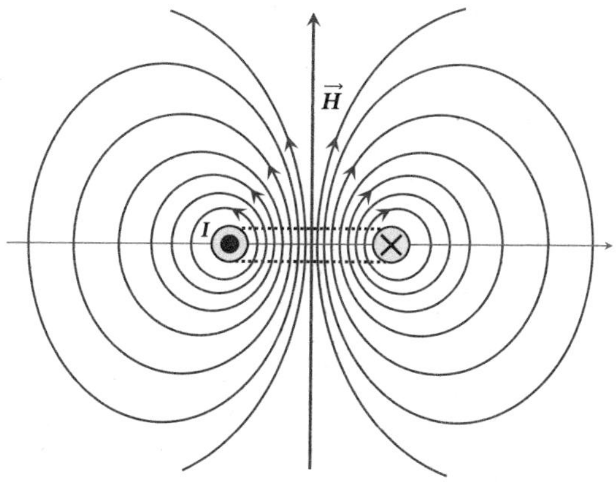

FIGURE 8.14 A map for magnetic field due to the circular current of Figure 8.13.

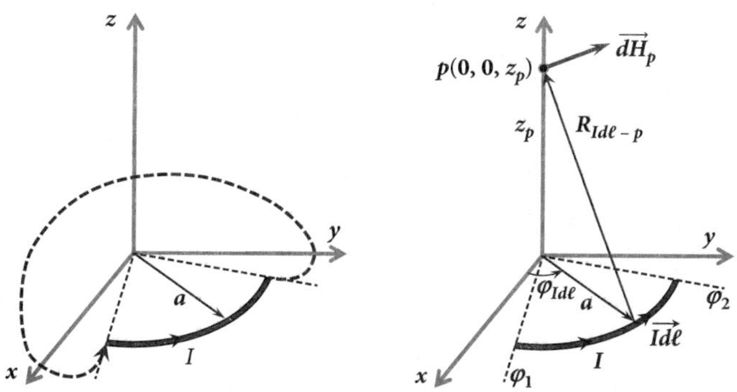

FIGURE 8.15 Using the incrementation scheme to derive an expression for the magnetic field due to a thin arc-shaped current-carrying conductor.

where φ_1 and φ_2 are the cylindrical coordinate φ angles for the beginning and end points of the segment, respectively. As for the field intensity at the origin where $z = 0$, we have

$$\bar{H}_p = \frac{Ia^2(\varphi_2 - \varphi_1)}{4\pi(a^2 + z^2)^{\frac{3}{2}}}\vec{a}_z \tag{8.25}$$

The answers in Equations (8.24) and (8.25) are consistent with those obtained earlier in Equations (8.22) and (8.23) when replacing φ_1 and φ_2 by zero and 2π, respectively.

8.8.4 Case 4: Magnetic Field due to a Finite Height Circular Solenoid (Figure 8.17)

The case of the solenoid is an important one, as it resembles coils that are typically used in many practical devices. Examples include wire-wound inductors, relays, actuators, and coil antennas (Figure 8.16).

The physical shape is a spiral wound wire along a straight axis as demonstrated in Figure 8.17. In this analysis, we will limit our discussion to the case of a uniform spiral pitch that is infinitesimally small that it can be neglected. As presented in Addendum 8B, such solenoid configuration can be thought of as a vertical stack of single-turn current-carrying loops. The incremental vector magnetic field due to the incremental current loop slice of height dz can be written as

$$\overrightarrow{dH}_p = \frac{dI_{\text{ring}}a^2}{2[a^2 + z_{\text{ring}-p}^2]^{\frac{3}{2}}}\vec{a}_z \tag{8.26}$$

where

$$dI_{\text{ring}} = (ndz)I \tag{8.27}$$

$$z_{\text{ring}-p} = z_p - z \tag{8.28}$$

where n is the number of turns per unit length.

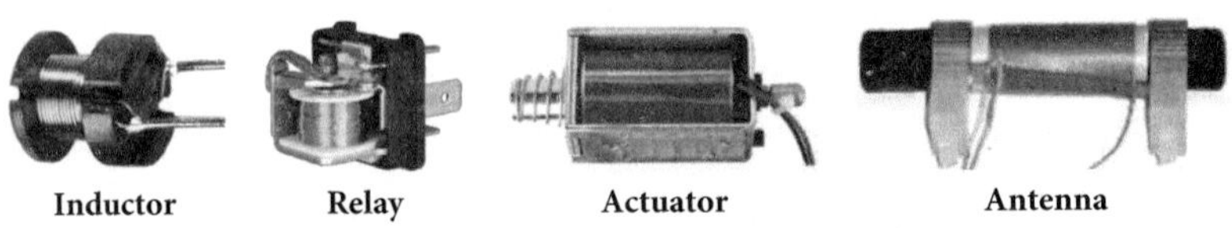

Figure 8.16 Examples of the application of solenoids in practical devices.

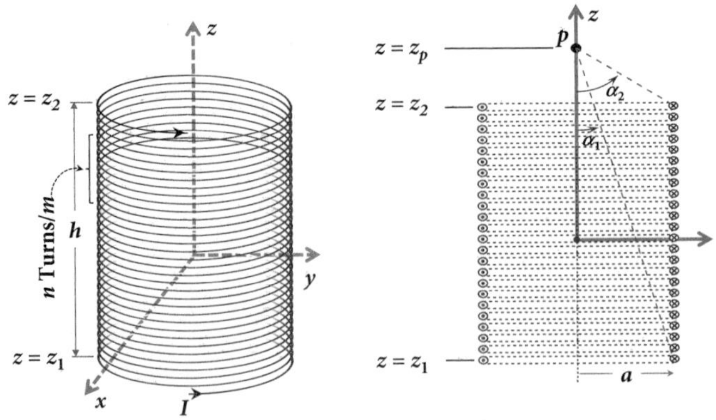

Figure 8.17 Using the magnetic field expression for a ring current to derive an expression for the magnetic field due to a finite height solenoid.

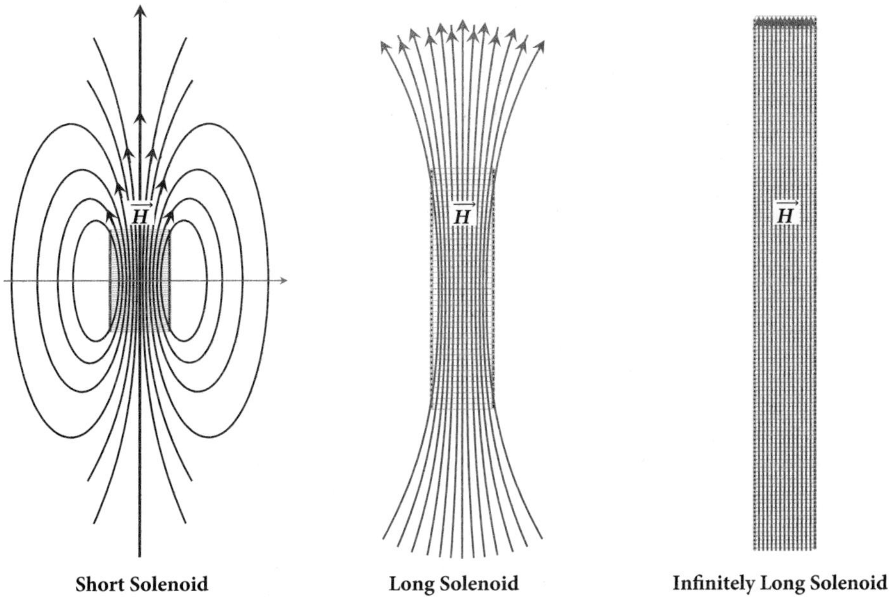

FIGURE 8.18 Magnetic field maps for solenoids of different heights.

The final answer for the magnetic field intensity vector at point p along the solenoid's axis is of the form

$$\vec{H}_P = \int \overrightarrow{dH}_P = \int_{z1}^{z2} \frac{(ndz)Ia^2}{2[a^2 + (z_p - z)^2]^{\frac{3}{2}}} \vec{a}_z = \frac{nI}{2}(\cos\alpha_1 - \cos\alpha_2)\vec{a}_z \tag{8.29}$$

Again, although we have derived the special case expression for points on the solenoid axis, we are presenting a field map for points off the axis in Figure 8.18. Three field maps are presented for short, long, and infinitely long solenoids.

8.8.5 Case 5: Magnetic Field due to an Infinite Height Circular Solenoid (Figure 8.17)

In this case, we consider the *infinitely* long solenoid. In practice, that would be the case of a long solenoid with its height much greater than its diameter. Using the result obtained for a finite solenoid and substituting the angles $\alpha_1 \to 0$ and $\alpha_2 \to 180°$, the result reduces to

$$\vec{H}_p = nI\,\vec{a}_z \tag{8.30}$$

It is interesting to observe that the answer for the infinite solenoid indicates a uniform magnetic field intensity all along the solenoid axis (Figure 8.18). As we will see later, this will not be only for values along the axis but also everywhere else within the solenoid. Furthermore, we will learn that the shape of the solenoid's cross section and size does not matter either. We will discuss both of these revelations in the following section.

8.9 Magnetic Field Evaluation Using Ampere's Circuital Law Scheme

The use of Ampere's circuital law for deriving expressions for $\vec{H}$ is in a way similar to the use of Gauss' law for the determination of the electric flux density that was discussed earlier in Chapter 4. The point is that the normal use of an equation of the form such as that of Equation (8.15) is to evaluate the left-hand side integration for a given $\vec{H}$ over a choice path to evaluate the linkage current in the right-hand side of the expression. At first sight, one cannot predict how such an equation can be used to do the inverse; that is to evaluate $\vec{H}$ given the current distribution.

Well, the use of Equation (8.15) to derive expressions for $\vec{H}$ is only possible for special cases of symmetry where we can predict the $\vec{H}$ field map and carry out the integrations in regions where $\vec{H}$ can be treated as a constant that can be factored outside the integration sign. The following examples are cases of special current distribution symmetries where we can use Ampere's circuital law to derive the corresponding expressions for the magnetic field intensity vector.

Again, to demonstrate the Ampere's circuital law approach, six cases of current distribution symmetries are analyzed in this chapter. We will cover the full detailed analysis of all six cases in Addendum 8C; however, we will also present a brief overview of the analysis procedure and the final result for each case in the following. The six current distribution cases considered here are

a. *Cylindrical (axial/coaxial) symmetries*
   - a1. An infinite length thin straight current-carrying conductor
   - a2. An infinite length thick straight current-carrying conductor
   - a3. An infinite length coaxial transmission line

b. *Planar symmetries*
   - b1. An infinite extension thin current sheet

c. *Toroidal and solenoidal symmetries*
   - c1. Toroid
   - c2. An infinite height solenoid

8.10 Category A: Magnetic Field due to Infinite Length Axial/Coaxial Current Distributions with Cylindrical Symmetries (Figure 8.19)

The first three cases in the list above are categorized under cylindrical (axial/coaxial) symmetries. The current distribution for this category is assumed to flow along an infinitely extending axis, for example, the z axis, with symmetries in both φ and z coordinates. In other words, the current distribution/density can only be a function of the cylindrical ρ coordinate (Figure 8.19).

Based on our studies and our physical intuition, we can predict the magnetic field map as circular loops concentric with the current spread. This is true for all locations in space both radially and vertically (all radii and all heights that are far away from the ends that are located at infinity). We can further predict the field directions using the right-hand rule as demonstrated in Figure 8.19. The intensities of those magnetic field lines are expected to vary only with the cylindrical radius ρ.

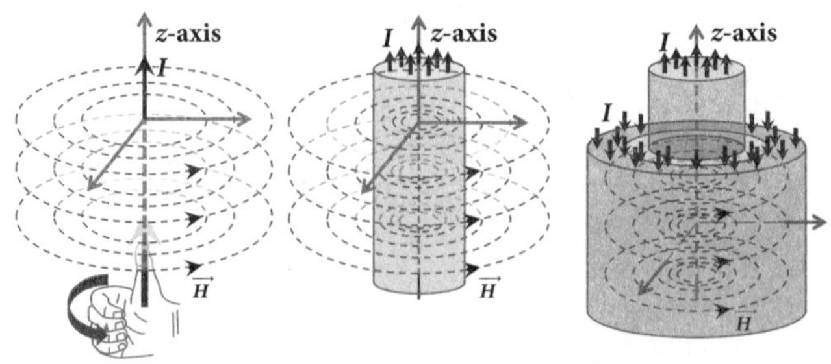

FIGURE 8.19 Using Ampere's law to derive expressions for the magnetic field due to infinite length axial/coaxial current distributions with cylindrical symmetries. (Left) Case 1: Thin current. (Right) Case 2: Thick current. (Center) Case 3: Coaxial currents.

For such cylindrical symmetries, as presented in Addendum 8B, we recognize that

$$\vec{H}(\rho,\varphi,z)=H_{\varphi}(\rho)\vec{a}_{\varphi} \quad (8.31)$$

$$\vec{d\ell}=d\ell\vec{a}_{\varphi} \quad (8.32)$$

$$\oint\vec{H}\cdot\vec{d\ell}=\oint(H_{\varphi}\vec{a}_{\varphi})\cdot(d\ell\vec{a}_{\varphi})=H_{\varphi}\oint d\ell=2\pi\rho H_{\varphi}=\sum I_{\text{Linkage}} \quad (8.33)$$

Hence,

$$\vec{H}(\rho)=\frac{\sum_{\rho}I_{\text{Linkage}}}{2\pi\rho}\vec{a}_{\varphi} \quad (8.34)$$

8.10.1 Case a1: Magnetic Field due to an Infinite Length Thin Straight Current-Carrying Conductor (Left of Figure 8.19)

As can be seen from the demonstration of this case in Figure 8.19, the linkage current is the same for all ρ radii, which is the current *I*. Hence, we can use the result [Equation (8.34)] to conclude that

$$\vec{H}(\rho)=\frac{I}{2\pi\rho}\vec{a}_{\varphi} \quad (8.35)$$

which is true for all points in space and is identical to the answer we obtained earlier, Equation (8.20), using the Biot-Savart incrementation scheme. The field intensity versus position for this case was given earlier in Figure 8.9.

8.10.2 Case a2: Magnetic Field due to an Infinite Length Thick Straight Current-Carrying Conductor (Center Figure 8.19)

In this case, the cylindrical symmetry current flows in a circular cross section conductor of radius a with current density distribution that can only vary with ρ.

Referring to the result [Equation (8.34)], we can express the field-intensity vector expressions for the two regions of space as follows:

$$\vec{H}(\rho)=\frac{\sum_{\rho}I_{\text{Linkage}}}{2\pi\rho}\vec{a}_{\varphi}=\frac{1}{2\pi\rho}\vec{a}_{\varphi}\begin{cases} I & \text{for } a<\rho<\infty \\ \int_{\varphi=0}^{2\pi}\int_{\rho=0}^{\rho}J(\rho)dS & \text{for } 0<\rho<a \end{cases} \quad (8.36)$$

where $J(\rho)$ is the surface current-density as a function of ρ.

Since the current is uniformly distributed over the cross section of the wire, the current density $J(\rho)$ can be expressed as $J(\rho)=\frac{I}{\pi a^2}\cdot(\pi\rho^2)=\frac{\rho^2}{a^2}$.

For the special case of a constant uniform current distribution over the cross section, $J(\rho)$ is constant

$$\vec{H}=\begin{cases} \frac{I}{2\pi\rho}\vec{a}_{\varphi} & \text{for } a<\rho<\infty \\ \frac{I\rho}{2\pi a^2}\vec{a}_{\varphi} & \text{for } 0<\rho<a \end{cases} \quad (8.37)$$

8.10.3 Case a3: Magnetic Field due to an Infinite Length Coaxial Transmission Line (Right of Figure 8.19)

This is the case of a coaxial transmission line composed of two *coaxial* conductors. Both conductors carry the same current in cylindrical symmetries but in opposite directions. In this case, we have four regions to investigate and use the current linkage expressions for each to determine the corresponding magnetic field intensity vectors according to Equation (8.34). The process is detailed in Addendum 8C, and final expressions are as follows:

$$\vec{H}(\rho)=\frac{\sum_{\rho} I_{\text{Linkage}}}{2\pi\rho}\vec{a}_{\varphi}=\frac{1}{2\pi\rho}\vec{a}_{\varphi}\begin{cases} +I & \text{for } a<\rho<b \\ +I-I=0 & \text{for } c<\rho<\infty \\ \int_{\varphi=0}^{2\pi}\int_{\rho=0}^{\rho} J_a(\rho)dS & \text{for } 0<\rho<a \\ +I-\int_{\varphi=0}^{2\pi}\int_{\rho=0}^{\rho} J_{bc}(\rho)dS & \text{for } b<\rho<c \end{cases} \tag{8.38}$$

Here, $J_a(\rho)$ and $J_{bc}(\rho)$ are the areal densities of the current as a function of ρ for the regions 0-a and b-c, respectively, as $J_a(\rho)=\frac{\rho^2}{a^2}$ and $J_{bc}(\rho)=\frac{I}{\pi(c^2-b^2)}\cdot(\pi(\rho^2-b^2))=\frac{\rho^2-b^2}{c^2-b^2}$.

For the special case of constant uniform current distributions over the two cross sections, we can write the final expressions as

$$\vec{H}=\begin{cases} \frac{I\rho}{2\pi a^2}\vec{a}_{\varphi} & \text{for } 0<\rho<a \\ \frac{I}{2\pi\rho}\vec{a}_{\varphi} & \text{for } a<\rho<b \\ \frac{I}{2\pi(c^2-b^2)}\left[\frac{c^2}{\rho}-\rho\right]\vec{a}_{\varphi} & \text{for } b<\rho<c \\ 0 & \text{for } c<\rho<\infty \end{cases} \tag{8.39}$$

The graph of Figure 8.20 gives a plot of the magnetic field intensity H_{φ} versus position ρ for all three cases above. We notice a linear rise for H_{φ} in the region between zero and a and inverse relationship $(1/\rho)$ for the region outside a. Obviously for the thin current case, the current radius is zero and hence the $(1/\rho)$ region starts at the origin. The linear relationship within "a" is due to the constant uniform current density in that region. For other symmetrical distribution with a ρ-varying current density, we expect other forms of relationships to exist.

As for the coaxial line, the field solution for the inner region $\rho< b$ is identical to that of same region of the thick conductor with the same inner conductor dimension. This is simply because the change of current linkage

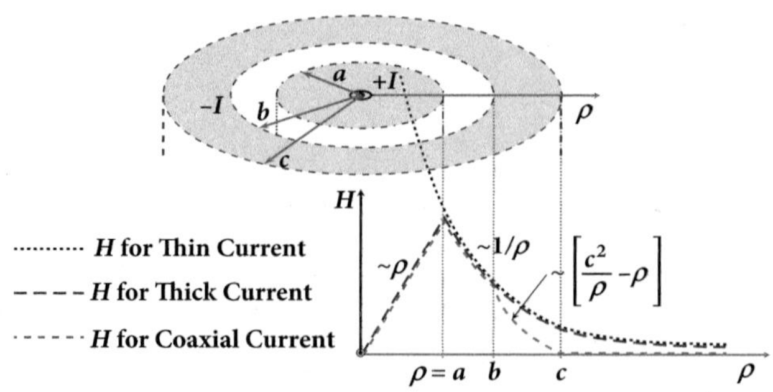

FIGURE 8.20 Magnetic field versus distance for infinite length axial/coaxial current distributions with cylindrical symmetries.

due to the outer current is only valid for the region $\rho > b$. This change is negative since the outer current flows in the opposite direction. As a result, the total current linkage decreases gradually and reaches a zero value at $\rho = c$. At this point, a full cancelation of the positive current linkage of the inner conductor occurs due to the equal but opposite current linkage of the outer conductor.

It is important to recognize the zero magnetic field for radii greater than the outer radius of the outer conductor, c. This is an indication of what is known as magnetic shielding. This magnetic shielding occurred as a result of the cancelation of the magnetic field contributions of the opposite currents flowing in the inner and outer conductors. In fact, cancelation occurs to the electric field of the coaxial transmission line in the same region as well. This was discussed earlier in Addendum 4A (Figure 4A.30). Hence, we claim a total electric and magnetic shielding in the region outside the outer conductor of the coaxial transmission line. In other words, electromagnetic activities occurring within the confines of a coaxial transmission line do not have any contributions outside its outer conductor. This finding has its great advantage in the field of communications. This is because in addition to providing the advantages of shielding, isolation, and security, it also preserves the signal energy from leaking and losing strength as it travels.

Example 8.3 Consider an infinitely long coaxial cable that carries the following current distribution:

$$j = \begin{cases} J_o e^{k\rho}\vec{a}_z & 0 < \rho \le a \\ 0 & a < \rho \le b \\ -J_o e^{-k\rho}\vec{a}_z & b < \rho \le c \\ 0 & \rho > c \end{cases}$$

Find the magnetic field everywhere.

Region 1 $\quad 0 < \rho \le a$

$$\oint \vec{H}\cdot\vec{dl} = I_{\text{enclosed}} = I_1 = \int \vec{J}\cdot\vec{dS} \qquad J = J_o e^{k\rho}\vec{a}_z \qquad \vec{dS} = \rho\, d\rho\, d\varphi\, \vec{a}_z \qquad \vec{J}\cdot\vec{dS} = J_o e^{k\rho}\rho\, d\rho\, d\varphi$$

$$H_\varphi 2\pi\rho = \int_0^{2\pi}\int_0^{\rho} J_o e^{k\rho}\rho\, d\rho\, d\varphi = 2\pi\int_0^{\rho} J_o \rho e^{k\rho} d\rho$$

$$= 2\pi J_o\left[\left(\frac{\rho e^{k\rho}}{k}\right)_0^{\rho} - \frac{1}{k}\int_0^{\rho} e^{k\rho}\, d\rho\right]$$

$$H_\varphi = \frac{2\pi J_o\left[\left(\frac{\rho e^{k\rho}}{k}\right) - \frac{(e^{k\rho}-1)}{k^2}\right]}{2\pi\rho} = \frac{J_o}{\rho}\left[\left(\frac{\rho e^{k\rho}}{k}\right) - \frac{(e^{k\rho}-1)}{k^2}\right]$$

Region 2 $\quad a < \rho \le b$

$$\oint \vec{H}\cdot\vec{dl} = I_{\text{enclosed}} = I_2 = I_1|_{\rho=a}$$

$$H_\varphi 2\pi\rho = I_1|_{\rho=a} = 2\pi J_o\left[\left(\frac{ae^{ka}}{k}\right) - \frac{(e^{ka}-1)}{k^2}\right]$$

$$H_\varphi = \frac{J_o}{\rho}\left[\left(\frac{ae^{ka}}{k}\right) - \frac{(e^{ka}-1)}{k^2}\right]$$

Region 3 $b<\rho\leq c$

$$\oint \vec{H}\cdot\vec{dl} = I_{\text{enclosed}} = I_3 = I_2 + 2\pi\int_b^{\rho} -J_o\rho e^{-k\rho}\,d\rho$$

$$I_3 = 2\pi J_o\left[\left(\frac{ae^{ka}}{k}\right)-\frac{(e^{ka}-1)}{k^2}\right]+2\pi J_o\left[\left(\frac{\rho e^{-k\rho}}{k}-\frac{be^{-kb}}{k}\right)-\frac{(e^{-kb}-e^{-k\rho})}{k^2}\right]$$

$$H_\varphi = \frac{I_3}{2\pi\rho} = \frac{J_o}{\rho}\left[\frac{ae^{ka}}{k}-\frac{(e^{ka}-1)}{k^2}+\left(\frac{\rho e^{-k\rho}}{k}-\frac{be^{-kb}}{k}\right)-\frac{(e^{-kb}-e^{-k\rho})}{k^2}\right]$$

Region 4 $\rho>c$

$$\oint \vec{H}\cdot\vec{dl} = I_{\text{enclosed}} = I_4 = I_3|_{\rho=c}$$

$$H_\varphi = \frac{I_4}{2\pi\rho} = \frac{J_o}{\rho}\left[\frac{ae^{ka}}{k}-\frac{(e^{ka}-1)}{k^2}+\left(\frac{ce^{-kc}}{k}-\frac{be^{-kb}}{k}\right)-\frac{(e^{-kb}-e^{-kc})}{k^2}\right]$$

8.11 Category B: Magnetic Field due to Planar Current Distributions with Planar Symmetries

Although we are citing only the example of a thin current sheet here, cases of thick and composite planar current distributions can be treated in a similar fashion as well. The examples of planar symmetries covered earlier in Chapter 4 can be used as a guide as to what other current distribution can be dealt with similarly.

8.11.1 Case b1: Magnetic Field due to an Infinite Extension Thin Current Sheet (Figure 8.21)

We now switch to a planar symmetry configuration. Two possible ways to consider in this case, a surface (laminar) flow of current in an infinitesimally thin conductor, or an array of infinitesimally thin wire conductors arranged in a flat plane. Both arrangements will lead to the same exact current distributions and hence the same final answers. The thinner the wires and the closer they are packed next to each other the closer we are to the thin current sheet case.

In this case, we will be constrained to the case of uniform linear density of the current, K ampere/meter, along the linear cross section of the current. The current is assumed to lie in the x-y plane with infinite extension in both directions (Figure 8.21).

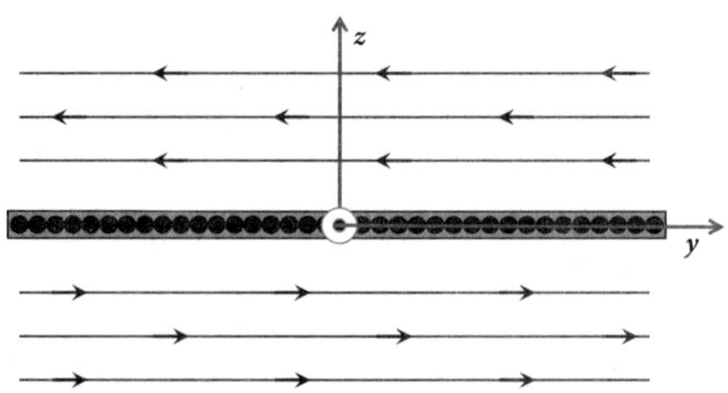

FIGURE 8.21 Predicted magnetic field map due to an infinite extension thin current sheet.

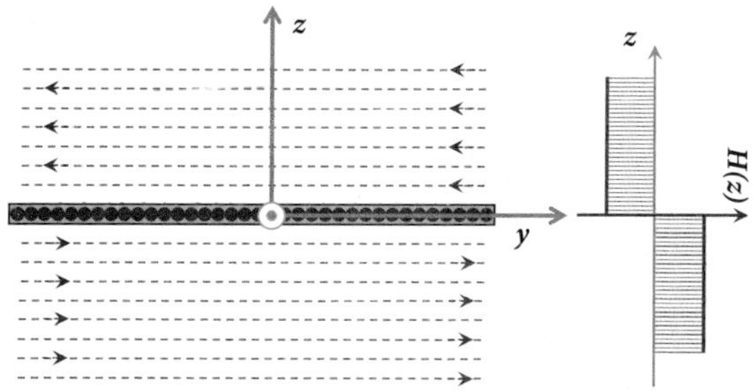

Figure 8.22 Magnetic field versus distance due to an infinite extension thin current sheet.

As detailed in Addendum 8C, the magnetic field intensity vector on either side of the current sheet is given by

$$\vec{H} = \begin{cases} -\dfrac{K}{2}\vec{a}_y & \text{for } z > 0 \\ +\dfrac{K}{2}\vec{a}_y & \text{for } 0 > z \end{cases} \tag{8.40}$$

This result is demonstrated graphically in Figure 8.22.

The reader may recall seeing a similar result when we discussed the infinite thins sheet of charges back in Chapter 4.

8.12 Category C: Magnetic Field due to Toroidal and Solenoidal Current Distributions with Uniform Linear Current Densities

The two cases presented in this category have one thing in common; basically, both are coils wound on a uniform cross section with the coil ends either close to each other or extending to infinity. In other words, the coil has the same exact cross section everywhere with no exposed or open ends. The coil is assumed to have the same linear current density all along its length. This implies that the coil turns are wound with the same number of turns per unit length all along the coil extension.

8.12.1 Case c1: Magnetic Field due to a Toroid (Figure 8.23)

A toroid coil can be described as a doughnut-shaped coil or as a solenoid wrapped around with both ends facing each other at a meeting point. The latter will be useful in our later discussion of the solenoid in Case 6 to follow. The toroid coil configuration is demonstrated in Figure 8.23. Although the example treated here assumes a circular toroidal shape, the approach followed in the analysis as well as the obtained results are valid

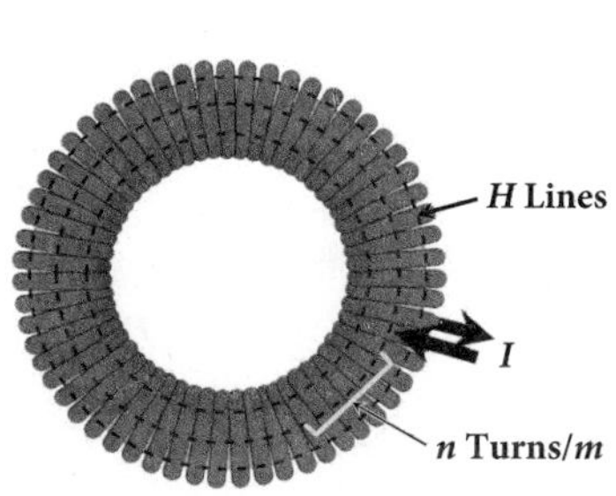

Figure 8.23 Using Ampere's law for deriving expressions for the magnetic field due to a toroidal current.

for other configurations. Furthermore, it is important to point out that the analysis developed here applies and produces the same answers for all cross-sectional shapes. In addition, the toroid's core material does not matter at this point of our study and hence, it will not be specified. In Chapter 9, we will learn about magnetic properties of materials and their impact on magnetic fields, flux, and other relevant quantities.

In our discussion of this case, we will assume the toroid's coil is tightly wound with closely packed thin wires. Such current-carrying windings would behave as a sheet of current flowing on the surface of the toroid's core. The earlier discussion for the sheet current case can be used to argue the smooth field lines tangential to the core and the sheet of current rather than curved lines contouring the coil turns.

Referring to Figure 8.23, the displayed toroid coil is wrapped around a circular donut-shaped core that is typically found in most practical applications. The critical condition here is to have a uniform cross section all around. The coil is assumed to have a uniform n turns per meter winding and carries a current I, thus making the equivalent linear current density $K = nI$.

The full analysis leading to the derivation of the magnetic field intensity vector is presented in Addendum 8C; however, a brief summary is given below.

For any of the chosen paths, we can write

$$\oint \vec{H} \cdot \vec{d\ell} = \oint H_\ell \, d\ell = H_\ell \ell = \sum I_{\text{Linkage}} = nI\ell_{av} \tag{8.41}$$

where ℓ is the total length of the $\vec{H}$ path and ℓ_{av} is the length over which n is defined.

Consequently,

$$\vec{H} = H_\ell \vec{a}_\ell = \begin{cases} \left[nI \dfrac{\ell_{av}}{\ell} \right] \vec{a}_\ell & \text{inside the toroid's core} \\ 0 & \text{outside the toroid's core} \end{cases} \tag{8.42}$$

For the special case of the circular toroid, the $\vec{H}$ lines will be circular as well and are expected to be concentric with the toroid axis, hence, we write

$$\vec{H} = \begin{cases} \left[nI \dfrac{\rho_{av}}{\rho} \right] \vec{a}_\varphi & \text{inside the toroid's core} \\ 0 & \text{outside the toroid's core} \end{cases} \tag{8.43}$$

where ρ_{av} and ℓ_{av} are the average radius and length of the toroid ring as demonstrated in Figure 8.24.

The conclusion is that field lines will only exist within the coil cross section of the toroid's coil and none outside it. This result is graphed in Figure 8.24. The field intensity as shown is inversely proportional to the length of the closed H line (or ρ for the cylindrical toroid). In cases where the coil's width cross-sectional dimensions are much smaller than the coil length, the magnetic field intensity variation would be small and an average value can be claimed.

$$H_{av} = nI \text{ inside the toroid's core} \tag{8.44}$$

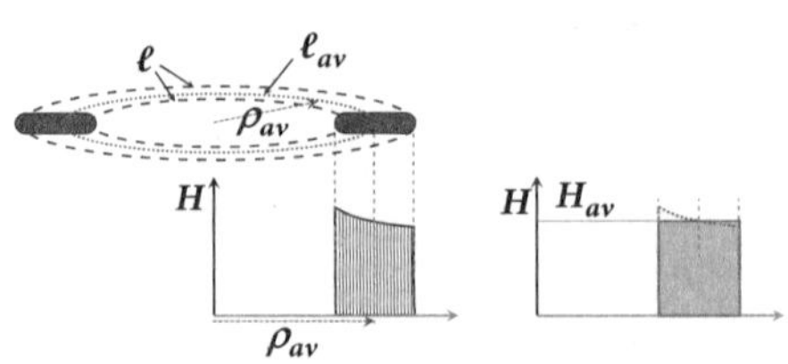

FIGURE 8.24 Magnetic field versus position for a toroidal current.

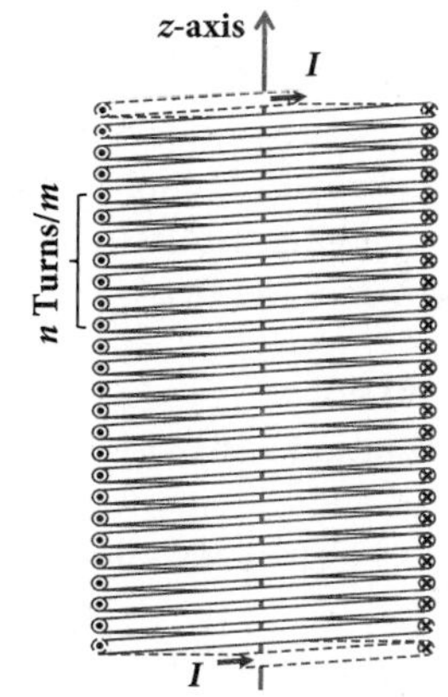

Figure 8.25 Using Ampere's law for deriving expressions for the magnetic field due to an infinite height solenoidal current.

8.12.2 Case c2: Magnetic Field due to an Infinite Height Solenoid (Discuss Shape of Cross Section) (Figure 8.25)

Finally, we discuss the case of an infinite length solenoid. This case was discussed earlier in this chapter using the Biot-Savart law. The solenoid configuration is given again in Figure 8.25 and is assumed to stretch along the z axis, have n turns per meter along its axis, and its coil carries a current I.

Again, the full analysis of both approaches is given in Addendum 8C yielding the following expression of the magnetic field intensity vector

$$\vec{H} = \begin{cases} [nI]\vec{a}_\ell = [nI]\vec{a}_z & \text{inside the solenoid's cross section} \\ 0 & \text{outside the solenoid's cross section} \end{cases} \tag{8.45}$$

This expression is valid for all solenoids irrespective of their cross-sectional configuration as demonstrated in Figure 8.26. It is important to recognize that this expression can also be used, in approximation, for the central region (away from the ends) of a fairly long solenoid with lengths much larger than the cross-sectional dimensions of the solenoid.

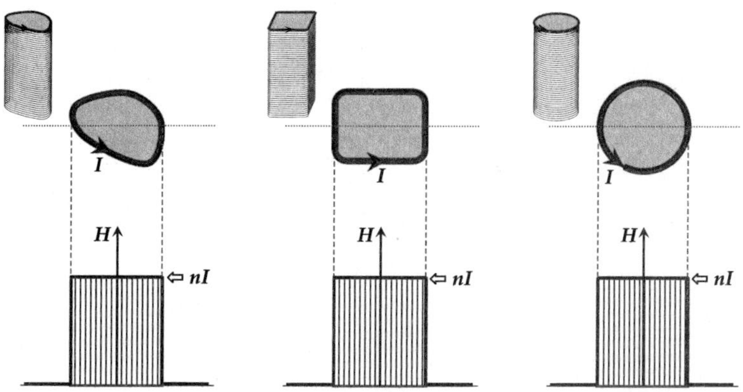

Figure 8.26 Magnetic field versus position for an infinite height solenoidal current.

8.13 Magnetostatic Differential (Point) Forms

Point forms provide powerful tools in electromagnetic analyses and can set the stage for the use of numerical techniques in cases where analytical solutions are difficult or impossible to achieve. Similar to what we did in electrostatics, we derive the point forms by applying the integral forms to an infinitesimally small increment and take the limit as the increment size diminishes to zero.

8.13.1 Point Form of Gauss' Law in Magnetism

To start, we consider Gauss' law for magnetostatics, with its integral form given in Equation (8.14)

$$\oiint_S \vec{B}\cdot\overrightarrow{dS}=0 \tag{8.14}$$

The derivation of the point form is identical to that carried earlier in Chapter 4 leading to the same form [Equation (4.24)] except to replace $\vec{D}$ by $\vec{B}$

$$\vec{\nabla}\cdot\vec{D}=\lim_{\Delta v\to 0}\left\{\frac{\oiint_S \vec{D}\cdot\overrightarrow{dS}}{\Delta v}\right\} \tag{4.24}$$

Hence, we write Gauss' law for magnetism in the differential point form as the divergenceless relationship

$$\lim_{\Delta v\to 0}\left\{\frac{\oiint_S \vec{B}\cdot\overrightarrow{dS}}{\Delta v}\right\}=\vec{\nabla}\cdot\vec{B}=0 \tag{8.46}$$

This is one of the four Maxwell's equations in the point form.

8.13.2 Point Form of Ampere's Circuital Law

We start with Ampere's circuital law in the integral form [Equation (8.15)] as applied to a small incremental current loop (Figure 8.27). Using Cartesian coordinates for convenience, we chose our increment at

$$\oint \vec{H}\cdot\overrightarrow{d\ell}=\int_a^b(\vec{H}_{ab})\cdot(\overrightarrow{d\ell}_{ab})+\int_b^c(\vec{H}_{bc})(\overrightarrow{d\ell}_{bc})+\int_c^d(\vec{H}_{cd})(\overrightarrow{d\ell}_{cd})+\int_d^a(\vec{H}_{da})(\overrightarrow{d\ell}_{da})=\sum I_{\text{Linkage}} \tag{8.47}$$

$$\int_a^b H_{x,ab}\,dx+\int_b^c H_{y,bc}\,dy-\int_c^d H_{x,cd}\,dx-\int_d^a H_{y,da}\,dy=J_z\Delta x\Delta y \tag{8.48}$$

Each of the four-side integrations can be approximated as the value for the field at mid-point of a given side multiplied by the length of that side. Meanwhile, the field value at mid-point can be expressed in terms of the differential increment of the value at the loop, for example,

$$\int_a^b H_{x,ab}\,dx=H_x\left(x,y-\frac{\Delta y}{2}\right)\Delta x=\left[H_x(x,y)-\frac{\partial H_x(x,y)}{\partial y}\left(\frac{\Delta y}{2}\right)\right]\Delta x \tag{8.49}$$

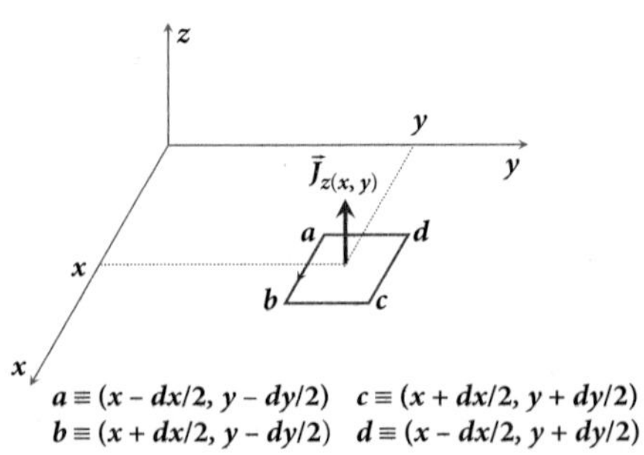

FIGURE 8.27 Incremental current loop used for the derivation of the point form of Ampere's circuital law.

Hence, Equation (8.48) reduces to

$$\left[H_x(x,y)-\frac{\partial H_x(x,y)}{\partial y}\left(\frac{\Delta y}{2}\right)\right]\Delta x+\left[H_y(x,y)+\frac{\partial H_y(x,y)}{\partial x}\left(\frac{\Delta x}{2}\right)\right]\Delta y$$
$$-\left[H_x(x,y)+\frac{\partial H_x(x,y)}{\partial y}\left(\frac{\Delta y}{2}\right)\right]\Delta x-\left[H_y(x,y)-\frac{\partial H_y(x,y)}{\partial x}\left(\frac{\Delta x}{2}\right)\right]\Delta y=J_z\Delta x\Delta y \tag{8.50}$$

With the cancelation of both the $H_x(x,y)$ and $H_y(x,y)$ terms, dividing both sides by $\Delta x\Delta y$, and taking the limit as both Δx and Δy approach zero, we can write

$$\lim_{\Delta x,\Delta y\to 0}\left\{\frac{\oint \vec{H}\cdot d\vec{\ell}}{\Delta x\Delta y}\right\}=-\left[\frac{\partial H_x(x,y)}{\partial y}\right]+\left[\frac{\partial H_y(x,y)}{\partial x}\right]=\left\{\frac{\partial H_y}{\partial x}-\frac{\partial H_x}{\partial y}\right\}=J_z \tag{8.51}$$

Examining this result, we observe that the left-hand side represents the circulation of the vector $\vec{H}$ (around an incremental loop, that is, at a "point") per unit area (that is normal to the current direction) is equal to the current density vector. This per unit normal area circulation is denoted as the curl. What we have in Equation (8.51) above is but the curl form in the special case of an x-y increment and the z current component.

To generalize Equation (8.51) in all the three Cartesian coordinates, we follow a similar procedure to include the other current density components J_x and J_y. The expected result in this case would be the general form for the curl expression

$$\text{curl } \vec{H}=\lim_{\Delta S_n\to 0}\left\{\frac{\oint \vec{H}\cdot d\vec{\ell}}{\Delta S_n}\right\}=\left\{\frac{\partial H_z}{\partial y}-\frac{\partial H_y}{\partial z}\right\}\vec{a}_x+\left\{\frac{\partial H_x}{\partial z}-\frac{\partial H_z}{\partial x}\right\}\vec{a}_y+\left\{\frac{\partial H_y}{\partial x}-\frac{\partial H_x}{\partial y}\right\}\vec{a}_z$$
$$=J_x\vec{a}_x+J_y\vec{a}_y+J_z\vec{a}_z=\vec{J} \tag{8.52}$$

which can be put in the form of the determinant

$$\text{curl } \vec{H}=\begin{vmatrix}\vec{a}_x & \vec{a}_y & \vec{a}_z\\ \dfrac{\partial}{\partial x} & \dfrac{\partial}{\partial y} & \dfrac{\partial}{\partial z}\\ H_x & H_y & H_z\end{vmatrix}=\vec{J} \tag{8.53}$$

In addition, it can take the form

$$\text{curl } \vec{H}=\left\{\vec{a}_x\frac{\partial}{\partial x}+\vec{a}_y\frac{\partial}{\partial y}+\vec{a}_z\frac{\partial}{\partial z}\right\}\times\{H_x\vec{a}_x+H_y\vec{a}_y+H_z\vec{a}_z\}=\vec{\nabla}\times\vec{H}=\vec{J} \tag{8.54}$$

where the differential vector operator $\vec{\nabla}$ in Cartesian coordinates is of the form

$$\vec{\nabla}=\left\{\vec{a}_x\frac{\partial}{\partial x}+\vec{a}_y\frac{\partial}{\partial y}+\vec{a}_z\frac{\partial}{\partial z}\right\} \tag{8.55}$$

The derivation of the curl operator in cylindrical and spherical coordinates is much more involved than that of Cartesian. In the following, we present the final expressions for both without presenting their derivations.

Cartesian:

$$\vec{\nabla}\times\vec{H}=\left\{\frac{\partial H_z}{\partial y}-\frac{\partial H_y}{\partial z}\right\}\vec{a}_x+\left\{\frac{\partial H_x}{\partial z}-\frac{\partial H_z}{\partial x}\right\}\vec{a}_y+\left\{\frac{\partial H_y}{\partial x}-\frac{\partial H_x}{\partial y}\right\}\vec{a}_z \tag{8.56}$$

Cylindrical:

$$\vec{\nabla}\times\vec{H}=\left\{\frac{1}{\rho}\frac{\partial H_z}{\partial\varphi}-\frac{\partial H_\varphi}{\partial z}\right\}\vec{a}_\rho+\left\{\frac{\partial H_\rho}{\partial z}-\frac{\partial H_z}{\partial\rho}\right\}\vec{a}_\varphi+\frac{1}{\rho}\left\{\frac{\partial(\rho H_\varphi)}{\partial\rho}-\frac{\partial H_\rho}{\partial\varphi}\right\}\vec{a}_z \tag{8.57}$$

Spherical:

$$\vec{\nabla}\times\vec{H}=\frac{1}{r\sin\theta}\left\{\frac{\partial(\sin\theta H_\varphi)}{\partial\theta}-\frac{\partial H_\theta}{\partial\varphi}\right\}\vec{a}_r+\frac{1}{r}\left\{\frac{1}{\sin\theta}\frac{\partial H_r}{\partial\varphi}-\frac{\partial(rH_\varphi)}{\partial r}\right\}\vec{a}_\theta+\frac{1}{r}\left\{\frac{\partial(rH_\theta)}{\partial r}-\frac{\partial H_r}{\partial\theta}\right\} \tag{8.58}$$

With this definition of the curl, we recall in Chapter 5, Equation (5.70) that the circulation of the (static) electric field intensity vector around a closed path was zero.

$$\oint\vec{E}\cdot\vec{d\ell}=0 \tag{5.70}$$

This is the case because of the conservation of energy in static electric fields, hence a charge moving around a closed path returns to its initial energy state with no net change in energy/potential. For this case, we claim that electrostatic fields are rotationless, that is, their curl is zero,

$$\vec{\nabla}\times\vec{E}=0 \tag{8.59}$$

8.14 Stokes' Theorem

Presenting Stokes' theorem at this point is a natural extension to the above discussion and the derivation of the point form of Ampere's circuital law. If we are to combine the two forms of Ampere's integral and point forms, we can write

$$\oint_C\vec{H}\cdot\vec{d\ell}=\sum I_{\text{Linkage}}=\iint_{S\text{ bound by }C}\vec{J}\cdot\vec{dS}=\iint_{S\text{ bound by }C}(\vec{\nabla}\times\vec{H})\cdot\vec{dS} \tag{8.60}$$

To exercise the derivation of this theorem in general (independent of Ampere's circuital law), we carry on the following analysis in which will even use a different notation for the vector field, for example, $\vec{G}$.

Referring to Figure 8.28, a general closed path C can be divided to incremental closed paths on any surface S bound by C. The closed loop integration of the left-hand side of Equation (8.60) can be expressed as the sum (or integration) of the incremental closed loop integrations around the incremental closed paths

$$\oint_C\vec{G}\cdot\vec{d\ell}=\sum_i\oint_{C_i}\vec{G}\cdot\vec{d\ell}=\sum_i\left[\frac{\oint_{C_i}\vec{G}\cdot\vec{d\ell}}{\Delta S_{\text{in}}}\right]\Delta S_{\text{in}}=\lim_{\Delta S_i\to 0}\sum_i[(\vec{\nabla}\times\vec{G})\cdot\vec{a}_{S_i}]\Delta S_{\text{in}}=\iint_S(\vec{\nabla}\times\vec{G})\cdot\vec{dS} \tag{8.61}$$

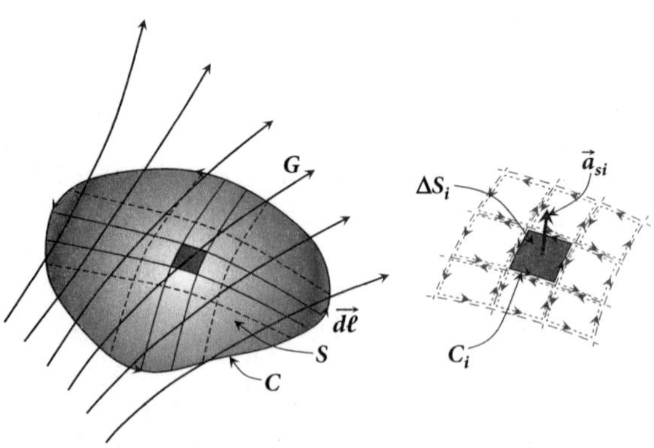

FIGURE 8.28 Supporting figure for the development of Stokes' theorem.

	Integral Form	Differential (Point) Form
Magnetic field circulation	$\oint \vec{H}\cdot\overrightarrow{d\ell}=\sum I_{\text{Linkage}}$	$\vec{\nabla}\times\vec{H}=\vec{J}$
Electric field circulation	$\oint \vec{E}\cdot\overrightarrow{d\ell}=0$	$\vec{\nabla}\times\vec{E}=0$
Electric flux divergence	$\oiint_S \vec{D}\cdot\overrightarrow{dS}=\sum Q_{\text{enclosed}}$	$\vec{\nabla}\cdot\vec{D}=\rho_v$
Magnetic flux divergence	$\oiint_S \vec{B}\cdot\overrightarrow{dS}=0$	$\vec{\nabla}\cdot\vec{B}=0$
Material permittivity	$\vec{D}=\varepsilon\vec{E}$	$\vec{D}=\varepsilon\vec{E}$
Material permeability	$\vec{B}=\mu\vec{H}$	$\vec{B}=\mu\vec{H}$
Material conductivity	$\vec{J}=\sigma\vec{E}$	$\vec{J}=\sigma\vec{E}$

TABLE 8.1 Summary of the Static Form of Maxwell's Equations

or

$$\oint_C \vec{G}\cdot\overrightarrow{d\ell}=\iint_S(\vec{\nabla}\times\vec{G})\cdot\overrightarrow{dS} \tag{8.62}$$

8.15 Static Form of Maxwell's Equations

At this point, it is important to highlight four important relationships that we encountered in static electric and magnetic fields in both their integral and differential (point) forms. These four equations are known as the static form of Maxwell's equations. A summary of these equations is given in Table 8.1.

8.16 Scalar and Vector Magnetic Potential

When we discussed the potential in the electrostatic section earlier, we did that in the context of energy. The physical meaning of potential was the normalized energy per unit charge. In magnetics, the definition will be made in mathematical analogy to that of electrostatics. As it turned out that such definition of potential will not be adequate for carrying advanced dynamic electromagnetic analysis and we will end up defining another form of magnetic potential. The one defined in mathematical analogy is the "scalar" potential while the one needed for "advanced" dynamic analysis is the vector potential.

8.16.1 Scalar Magnetic Potential

The scalar magnetic potential is defined in mathematical analogy to that of the electric potential defined earlier in Chapter 5 [Equation (5.59)]

$$V_{BA}=V_B-V_A=-\int_A^B \vec{E}\cdot\overrightarrow{d\ell} \tag{5.59}$$

which in view of Equation (5.70)

$$\int_A^A \vec{E}\cdot\overrightarrow{d\ell}=\oint \vec{E}\cdot\overrightarrow{d\ell}=0 \tag{5.70}$$

defines a conservative (single valued) function of position.

The corresponding relationships for the analogous scalar magnetic potential V_m for Equation (5.59) is

$$V_{m,BA} = V_{m,B} - V_{m,A} = -\int_A^B \vec{H} \cdot \vec{d\ell} \tag{8.63}$$

However, the closed loop integration of $\vec{H}$ is not necessarily zero,

$$\int_A^A \vec{H} \cdot \vec{d\ell} = \oint \vec{H} \cdot \vec{d\ell} = \sum I_{\text{Linkage}} \tag{8.64}$$

Consequently, the scalar magnetic potential is not generally conservative. It is only in the absence of linkage currents that we can define a single valued/conservative magnetic potential.

8.16.2 Vector Magnetic Potential

The vector magnetic potential definition is based on the fact that the magnetic field is divergenceless. Since, $\vec{\nabla} \cdot \vec{B} = 0$ and the divergence of a curl is zero, then $\vec{B}$ is of the form of a curl

$$\vec{B} = \vec{\nabla} \times \vec{A} \tag{8.65}$$

and since

$$\vec{B} = \mu_o \vec{H} \tag{8.6}$$

$$\vec{\nabla} \times \vec{B} = \vec{\nabla} \times \vec{\nabla} \times \vec{A} = \mu_o \vec{\nabla} \times \vec{H} = \mu_o \vec{J} \tag{8.66}$$

$$\vec{\nabla}(\vec{\nabla} \cdot \vec{A}) - \nabla^2 \vec{A} = \mu_o \vec{\nabla} \times \vec{H} = \mu_o \vec{J} \tag{8.67}$$

Recall the Poisson relationship for the electric potential in Chapter 7 [Equation (7.4)]

$$\nabla^2 V = -\rho_v / \varepsilon_o \tag{7.4}$$

This equation has elements of similarity with the expression in Equation (8.67), except for the presence of the term $\vec{\nabla}(\vec{\nabla} \cdot \vec{A})$ in the latter. However, our definition of $\vec{A}$ is not complete since Equation (8.65) only defines its curl (a derivative definition). Hence, we have one degree of freedom to add a constraint to achieve a complete definition. The logical choice is to make $(\vec{\nabla} \cdot \vec{A}) = 0$, and thus the complete definition of the vector magnetic potential $\vec{A}$ is

$$\vec{B} = \vec{\nabla} \times \vec{A} \quad \text{and} \quad \vec{\nabla} \cdot \vec{A} = 0 \tag{8.68}$$

with this definition, Equation (8.67) reduces to

$$\nabla^2 \vec{A} = -\mu_o \vec{J} \tag{8.69}$$

Also, since the expression

$$V_p = \int_Q \frac{dQ}{4\pi\varepsilon_o R_{dQ-p}} = \iiint_v \frac{\rho_v \, dv}{4\pi\varepsilon_o R_{dv-p}} \tag{5.69}$$

represent a solution to the Poisson's differential equation [Equation (7.4)], and then we can predict that the solution for the differential equation [Equation (8.69)] would be of the form

$$\vec{A}_p = \iiint_v \frac{\mu_o \vec{J} \, dv}{4\pi R_{dv-p}} = \oint_\ell \frac{\mu_o I \vec{d\ell}}{4\pi R_{d\ell-p}} \tag{8.70}$$

ADDENDUM 8A
ANALOGIES WITH ELECTROSTATIC QUANTITIES

As stated earlier, the physics of the electric and magnetic field and flux terms are different. While the electrostatic field and flux are divergent, the magnetic field and flux are rotational. However, some analogies can be used to add conveniences to our studies of the subject. We cite some of these analogies in Table 8A.1.

	Electrostatics	Magnetostatics	Comments
1	dQ	$\overrightarrow{Id\ell}$	Scalar/vector source element
2	$\vec{D}$	$\vec{B}$	Flux density vector
3	$\vec{E}$	$\vec{H}$	Field intensity vector
4	$\vec{D}=\varepsilon\vec{E}$	$\vec{B}=\mu\vec{H}$	Material constitutive relationship
5	$\varepsilon_o=8.8541878\times10^{-12}$ F/m	$\mu_o=4\pi\times10^{-7}$ H/m	Free space permittivity and permeability
6	$\oiint_S \vec{D}\cdot\overrightarrow{dS}=\sum Q_{\text{enclosed}}$	$\oiint_S \vec{B}\cdot\overrightarrow{dS}=0$	Gauss' law
7	$\oiint_S \vec{D}\cdot\overrightarrow{dS}=\sum Q_{\text{enclosed}}$	$\oint \vec{H}\cdot\overrightarrow{d\ell}=\sum I_{\text{Linkage}}$	Evaluation of *D*/*H* for symmetrical distributions
8	$\oint \vec{E}\cdot\overrightarrow{d\ell}=0$	$\oiint_S \vec{B}\cdot\overrightarrow{dS}=0$	Divergent *E* is rotationless while rotational *B* is divergentless
9	$\overrightarrow{dF}_{e2}=\dfrac{dq_2 dQ_1}{4\pi\varepsilon_o R_{12}^2}$	$\overrightarrow{dF}_{m2}=\mu_o\dfrac{(\overrightarrow{Id\ell})_2\times\left[(\overrightarrow{Id\ell})_1\times\vec{a}_{12}\right]}{4\pi R_{12}^2}$	Coulomb's and Ampere's laws for force
10	$\overrightarrow{dF}_{e2}=dq_2\overrightarrow{dE}$	$\overrightarrow{dF}_{m2}=(\overrightarrow{Id\ell})_2\times\overrightarrow{dB}$	Force-field relationships
11	$\overrightarrow{dE}_p=\dfrac{1}{4\pi\varepsilon_o}\dfrac{dQ}{R_p^2}$	$\overrightarrow{dH}_p=\dfrac{1}{4\pi}\dfrac{(\overrightarrow{Id\ell})\times\vec{a}_R}{R_{d\ell-p}^2}$	Evaluation of *E*/*H* using incrementation
12		$\overrightarrow{dB}_p=\dfrac{\mu_o}{4\pi}\dfrac{(\overrightarrow{Id\ell})\times\vec{a}_R}{R_{d\ell-p}^2}$	Biot-Savart law
13	$\vec{\nabla}\cdot\vec{D}=\rho_v$	$\vec{\nabla}\cdot\vec{B}=0$	Gauss' law divergence
14	$\vec{\nabla}\times\vec{E}=0$	$\vec{\nabla}\times\vec{H}=\vec{J}$	Circulations (curl equations)
15	$V_{BA}=-\int_A^B\vec{E}\cdot\overrightarrow{d\ell}$	$V_{m,BA}=-\int_A^B\vec{H}\cdot\overrightarrow{d\ell}$	Scalar potentials
16		$\vec{B}=\vec{\nabla}\times\vec{A}\;\&\;\vec{\nabla}\cdot\vec{A}=0$	Magnetic vector potential
17	$\nabla^2 V=-\rho_v/\varepsilon_o$	$\nabla^2\vec{A}=-\mu_o\vec{J}$	Potentials
18	$V_p=\iiint_v\dfrac{\rho_v\,dv}{4\pi\varepsilon_o R_{dv-p}}$	$\vec{A}_p=\oint_\ell\dfrac{\mu_o\overrightarrow{Id\ell}}{4\pi R_{d\ell-p}}$	Potentials

TABLE 8A.1 Electrostatic-Magnetostatic Analogies

ADDENDUM 8B
MAGNETIC FIELD EVALUATION USING THE "INCREMENTATION" SCHEME

In this section, the incrementation scheme using the Biot-Savart law will be demonstrated in deriving magnetic field expressions for the following six cases:

1. A finite length thin straight current-carrying conductor
2. An infinite length thin straight current-carrying conductor
3. A thin circular current-carrying (loop) conductor; also, the case of a segment of the loop
4. A segment of a thin circular current-carrying (loop) conductor
5. A finite height circular solenoid
6. An infinite height circular solenoid

8B.1 Case 1: Magnetic Field due to a Finite Length Thin Straight Current-Carrying Conductor

Consider a thin straight conductor carrying a constant current I and placed along the z axis as illustrated in Figure 8B.1. Our objective in this exercise is to demonstrate how to use the incrementation scheme to determine $\vec{H}_p$ at the point $p(\rho,\varphi,z)$ due to this current. Similar to what we did in a similar example in cylindrical coordinates in Chapter 5, Figures 5.7 and 5.8, we will add to our coordinate parameters the angle α (Figure 8B.1). Referring to Figure 8B.1, the current-carrying wire spans the range α_1 to α_2.

Now, we dice the current-carrying wire into infinitesimally small increments and use the Biot-Savart law to evaluate the incremental magnetic field vector components due to each incremental current element. The incremental field at p due to a current element $Id\ell$ located at a general location (0, 0, z) on the wire is given by

$$\overrightarrow{dH}_p = \frac{1}{4\pi}\frac{\overrightarrow{Id\ell}\times\vec{a}_R}{R_{Id\ell-p}^2} = \frac{1}{4\pi}\frac{\overrightarrow{Idz}\times\vec{a}_R}{R_{Id\ell-p}^2} = \frac{1}{4\pi}\frac{Idz\,\vec{a}_z\times[\cos\alpha\vec{a}_z+\sin\alpha\vec{a}_\rho]}{R_{Id\ell-p}^2} = \frac{1}{4\pi}\frac{Idz\sin\alpha}{R_{Id\ell-p}^2}\vec{a}_\varphi \tag{8B.1}$$

Hence, the total magnetic field intensity vector due to the entire current-carrying wire is obtained by integrating the incremental components of Equation (8B.1)

$$\vec{H}_p = \int \overrightarrow{dH}_p = \frac{I}{4\pi}\left[\int_{z_1}^{z_2}\frac{\sin\alpha}{R_{Id\ell-p}^2}dz\right]\vec{a}_\varphi \tag{8B.2}$$

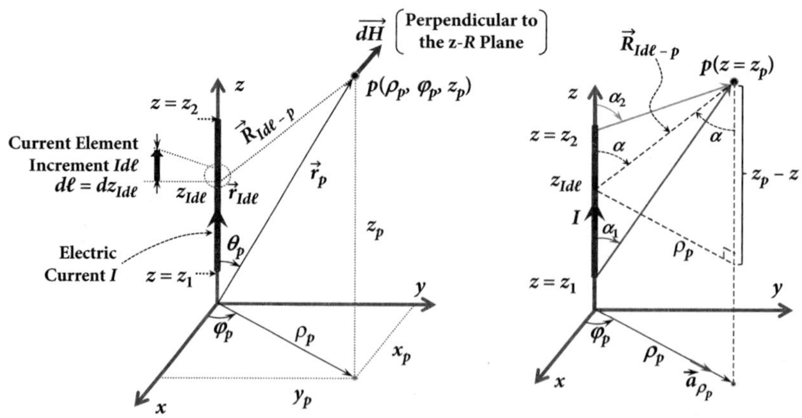

FIGURE 8B.1 Using the incrementation scheme to derive an expression for the magnetic field due to a finite length thin straight current-carrying conductor.

Examining the integration in Equation (8B.2), we make two main observations. The first one is that we factored the unit vector $\vec{a}_\varphi$ outside the integrand. Although $\vec{a}_\varphi$ is typically a variable vector in cylindrical coordinates, it happens to be a constant throughout our integration as we vary z from z_1 to z_2. The second observation is that our integral includes three (interrelated) variables, z, $R_{Id\ell-p}$, and α. Thus, we need to reduce its form to one variable only in order to be able to carry out this integration. From our past experience with the previously cited case in Chapter 5, we choose this one variable to be the angle α.

Referring to Figure 8B.1, we can write the interrelations between the three variables as $z_p - z = R_{Id\ell-p}\cos\alpha$ and $\rho_p = R_{Id\ell-p}\sin\alpha$; therefore,

$$z_p - z = \rho_p \frac{\cos\alpha}{\sin\alpha} = \rho_p \cot\alpha \tag{8B.3}$$

Hence, by differentiating the z-α relationship, we get

$$-dz = -\frac{\rho_p}{(\sin\alpha)^2} d\alpha \tag{8B.4}$$

Using Equations (8B.3) and (8B.4), the integration of Equation (8B.2) now takes the form

$$\vec{H}_p = \frac{I}{4\pi}\left[\int_{z_1}^{z_2} \frac{\sin\alpha}{R^2_{Id\ell-p}} dz\right]\vec{a}_\varphi = \frac{I}{4\pi}\left[\int_{\alpha_1}^{\alpha_2} \frac{\sin\alpha}{(\rho_p/\sin\alpha)^2}\left(\frac{\rho_p}{(\sin\alpha)^2}d\alpha\right)\right]\vec{a}_\varphi = \frac{I}{4\pi}\left[\int_{\alpha_1}^{\alpha_2} \frac{\sin\alpha}{\rho_p} d\alpha\right]\vec{a}_\varphi \tag{8B.5}$$

And finally,

$$\vec{H}_p = \frac{I}{4\pi\rho_p}[\cos\alpha_1 - \cos\alpha_2]\vec{a}_\varphi \tag{8B.6}$$

8B.2 Case 2: Magnetic Field due to an Infinite Length Thin Straight Current-Carrying Conductor

This case is identical to case 1 above except for the length of the current-carrying conductor. Figure 8B.2 shows the infinite length of the wire extending along the z axis from $z_1 = -\infty$ to $z_2 = +\infty$. As the figure shows, the angles α_1 and α_2 converge 0° and 180°, respectively. Using the result [Equation (8B.6)], we obtain the expression for the magnetic field intensity vector for the infinite wire as

$$\vec{H}_p = \frac{I}{4\pi\rho_p}[\cos\alpha_1 - \cos\alpha_2]\vec{a}_\varphi = \frac{I}{4\pi\rho_p}[1-(-1)]\vec{a}_\varphi = \frac{I}{2\pi\rho_p}\vec{a}_\varphi \tag{8B.7}$$

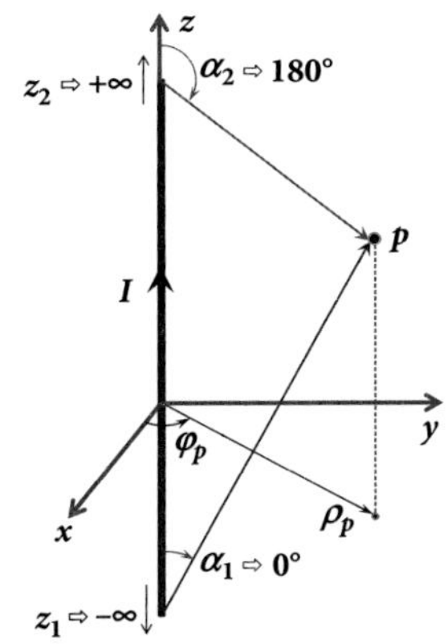

FIGURE 8B.2 Expressing the magnetic field due to a thin straight current-carrying conductor in terms of the α angles.

8B.3 Case 3: Magnetic Field due to a Thin Circular Current-Carrying Conductor (Loop)

In this case, we assume a circular current loop of radius a is located in the x-y plane and is centered at the origin. We will further assume that the current in the loop is constant and we will give it the notation I (amperes) (Figure 8B.3).

Our exercise here is to use the incrementation scheme with the Biot-Savart law to derive expressions for the magnetic field intensity vector at any point on the z axis. Considering general points in space (e.g., x, y, z) adds layers of complexity to the problem analysis that we prefer to avoid at this point and suffice with considering (0, 0, z) locations.

Therefore, we start by dicing up the current loop into small increments, $Id\ell$, each around the perimeter. In doing so, we recognize that the cylindrical coordinate system may add some conveniences to this analysis as the incremental length $d\ell$ corresponds to an increment $d\varphi$ of the cylindrical coordinate angle. We must keep in mind that working with cylindrical coordinates typically introduces variable unit vectors that could complicate the analysis. In this regard, we stay alerted to the possibility of switching back and forth between cylindrical and Cartesian coordinates as the need arises.

Next, we write the Biot-Savart expression for the incremental field vector at $p(0, 0, z)$ due to any of the current increments. The easiest and least confusing approach is to choose one of the increments in $(+x, +y)$ quadrant, $(0 < \varphi < \pi/2)$, Figure 8B.3. The choice increment starts at $(a, \varphi, 0)$ and ends at $(a, \varphi + d\varphi, 0)$ in cylindrical coordinates. In this case, the Biot-Savart expression takes the form

$$\overrightarrow{dH}_p = \frac{\overrightarrow{Id\ell} \times \vec{a}_R}{4\pi R^2_{Id\ell-p}} = \frac{\overrightarrow{Id\ell} \times \vec{R}_{Id\ell-p}}{4\pi R^3_{Id\ell-p}} \tag{8B.8}$$

This expression can be converted to all-cylindrical coordinates' terms by making the following substitutions

$$\overrightarrow{d\ell} = a\,d\varphi\,\vec{a}_\varphi \tag{8B.9}$$

$$\vec{R}_{Id\ell-p} = a(-\vec{a}_\rho) + z\vec{a}_z \tag{8B.10}$$

$$R^3_{Id\ell-p} = (a^2 + z^2)^{3/2} \tag{8B.11}$$

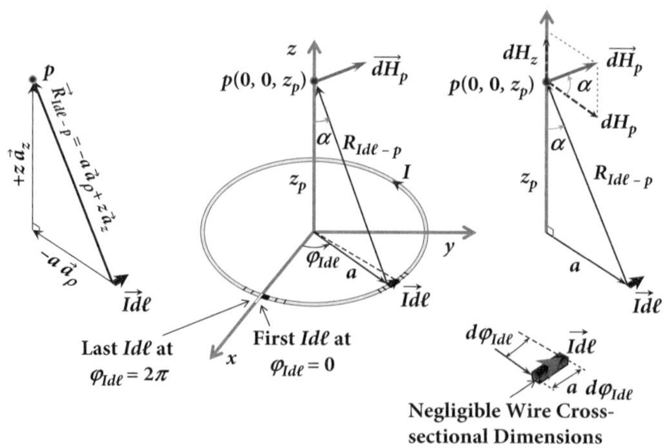

Figure 8B.3 Using the incrementation scheme to derive an expression for the magnetic field due to a thin circular current-carrying conductor (loop).

As a consequence, we obtain the following expression

$$\overrightarrow{dH}_p = \frac{\overrightarrow{Id\ell} \times \vec{R}_{Id\ell-p}}{4\pi R^3_{Id\ell-p}} = \frac{Iad\varphi \vec{a}_\varphi \times (-a\vec{a}_\rho + z\vec{a}_z)}{4\pi(a^2+z^2)^{\frac{3}{2}}} = \frac{Ia\,(+a\vec{a}_z + z\vec{a}_\rho)}{4\pi(a^2+z^2)^{\frac{3}{2}}} d\varphi \tag{8B.12}$$

$$\vec{H}_p = \int \overrightarrow{dH}_p = \int_0^{2\pi} \frac{Ia(a\vec{a}_z + z\vec{a}_\rho)}{4\pi(a^2+z^2)^{\frac{3}{2}}} d\varphi = \int_0^{2\pi} \frac{Ia(a\vec{a}_z)}{4\pi(a^2+z^2)^{\frac{3}{2}}} d\varphi + \int_0^{2\pi} \frac{Ia(z\vec{a}_\rho)}{4\pi(a^2+z^2)^{\frac{3}{2}}} d\varphi$$

$$= \frac{Ia(a\vec{a}_z)}{4\pi(a^2+z^2)^{\frac{3}{2}}} \int_0^{2\pi} d\varphi + \frac{Ia(z)}{4\pi(a^2+z^2)^{\frac{3}{2}}} \int_0^{2\pi} \vec{a}_\rho\, d\varphi = \frac{Ia^2(2\pi)}{4\pi(a^2+z^2)^{\frac{3}{2}}} \vec{a}_z + \frac{Ia(z)}{4\pi(a^2+z^2)^{\frac{3}{2}}}(0)$$

$$= \frac{Ia^2}{2(a^2+z^2)^{\frac{3}{2}}} \vec{a}_z \tag{8B.13}$$

Please note that the integration $\int_0^{2\pi} \vec{a}_\rho d\varphi = 0$ was discussed in Addendum 3D, Equation (3D.5) and Figure 3D.1.

For the special point at the origin, $z = 0$, we find

$$\vec{H}_p = \frac{I}{2a} \vec{a}_z \tag{8B.14}$$

An alternative approach to deriving the above expressions involves the use of trigonometric analysis of the field vectors along with physical and intuitive logic to reduce the analytical expressions. In this approach, we recognize the $\overrightarrow{dH}_p$ vector increment to be orthogonal to both $\overrightarrow{Id\ell}$ and $\vec{R}_{Id\ell-p}$, Figure 8B.3, hence, we break $\overrightarrow{dH}_p$ into its two cylindrical coordinate components dH_ρ and dH_z. First, we look at the radial components

$$\overrightarrow{dH}_\rho = dH_p \cos\alpha\, \vec{a}_\rho = \frac{Id\ell \cos\alpha}{4\pi R^2_{Id\ell-p}} \vec{a}_\rho = \frac{Ia\cos\alpha}{4\pi R^2_{Id\ell-p}} \vec{a}_\rho d\varphi \tag{8B.15}$$

$$\int \overrightarrow{dH}_\rho = \frac{Ia\cos\alpha}{4\pi R^2_{Id\ell-p}} \int_0^{2\pi} \vec{a}_\rho\, d\varphi = 0 \tag{8B.16}$$

which is obvious since all the radial components are of equal intensities distributed uniformly around $0 < \varphi < 2\pi$.

Next, we consider the z axis component

$$\overrightarrow{dH}_z = dH_p \sin\alpha\, \vec{a}_z = \frac{Id\ell \sin\alpha}{4\pi R^2_{Id\ell-p}} \vec{a}_z = \frac{Ia\sin\alpha}{4\pi R^2_{Id\ell-p}} \vec{a}_z d\varphi \tag{8B.17}$$

Consequently, the total vector field is given by

$$\vec{H}_p = \vec{H}_z = \int \overrightarrow{dH}_z = \frac{Ia\sin\alpha}{4\pi R^2_{Id\ell-p}} \vec{a}_z \int_0^{2\pi} d\varphi = \frac{Ia\sin\alpha}{4\pi R^2_{Id\ell-p}} \vec{a}_z(2\pi) = \frac{Ia^2}{2R^3_{Id\ell-p}} \vec{a}_z \tag{8B.18}$$

8B.4 Case 4: Magnetic Field due to a Segment of a Thin Circular Current-Carrying Conductor (Loop)

This is another case where we discuss the contribution of a "segment" of the current "circuit" (Figure 8B.4). For this case, the analysis takes the same form as that of Case 3 above except for limiting the angles of integration to that of the segment instead of 0-2π. Hence, we write

$$\vec{H}_p = \int_{\varphi 1}^{\varphi 2} \frac{Ia(a\vec{a}_z + z\vec{a}_\rho)}{4\pi(a^2+z^2)^{\frac{3}{2}}} d\varphi$$

$$= \frac{Ia(a\vec{a}_z)}{4\pi(a^2+z^2)^{\frac{3}{2}}} \int_{\varphi 1}^{\varphi 2} d\varphi + \frac{Ia(z)}{4\pi(a^2+z^2)^{\frac{3}{2}}} \int_{\varphi 1}^{\varphi 2} \vec{a}_\rho d\varphi$$

$$= \frac{Ia^2(\varphi_2-\varphi_1)}{4\pi(a^2+z^2)^{\frac{3}{2}}}\vec{a}_z + \frac{Ia(z)}{4\pi(a^2+z^2)^{\frac{3}{2}}}\left[\vec{a}_x \int_{\varphi 1}^{\varphi 2} (\cos\varphi) d\varphi + \vec{a}_y \int_{\varphi 1}^{\varphi 2} (\sin\varphi) d\varphi\right]$$

$$= \frac{Ia^2(\varphi_2-\varphi_1)}{4\pi(a^2+z^2)^{\frac{3}{2}}}\vec{a}_z + \frac{Iaz}{4\pi(a^2+z^2)^{\frac{3}{2}}}\left[\vec{a}_x(\sin\varphi_2 - \sin\varphi_1) - \vec{a}_y(\cos\varphi_2 - \cos\varphi_1)\right] \tag{8B.19}$$

As for the field intensity at the origin where $z = 0$, we have

$$\vec{H}_p = \frac{Ia^2(\varphi_2-\varphi_1)}{4\pi(a^2+z^2)^{\frac{3}{2}}}\vec{a}_z \tag{8B.20}$$

The answers in Equations (8B.19) and (8B.20) are consistent with those obtained earlier in Equations (8B.13) and (8B.14) when replacing φ_1 and φ_2 by zero and 2π, respectively.

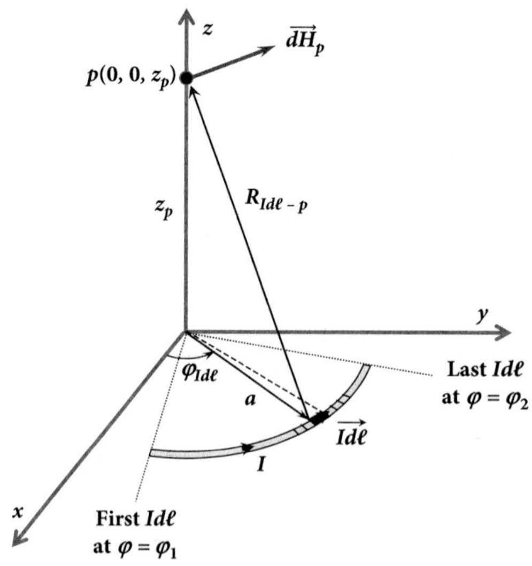

FIGURE 8B.4 Using the incrementation scheme to derive an expression for the magnetic field due to a thin arc-shaped current-carrying conductor.

8B.5 Case 5: Magnetic Field due to a Finite Height Circular Solenoid

The physical shape is a spiral wound wire along a straight axis as demonstrated in Figure 8B.5. In this analysis, we will limit our discussion to the case of a uniform spiral pitch. As the figure suggests, such solenoid configuration can be thought of as a vertical stack of single turn current-carrying loops.

Since we are practicing the incrementation scheme for the Biot-Savart law, we need to start by dicing this current distribution into infinitesimally small increments. The next step is to integrate the Biot-Savart expression for the incremental vector fields leading to the overall field expression.

Examining Figure 8B.6, we realize that integrating field contributions requires summing up the contributions of the incremental elements in two dimensions, the circular/ring direction and the vertical direction. As the figure suggests, this can be done into two possible orders

1. Integrating vertical elements first then the circular direction
2. Integrating the ring elements first then the vertical direction

Having gone through the single turn loop analysis in Case 3 above, we can claim that the integration around the ring direction first makes more sense in our case. In doing so, we can use the answers already obtained for the ring and proceed to the second dimension of integration by summing up the contributions of the rings vertically.

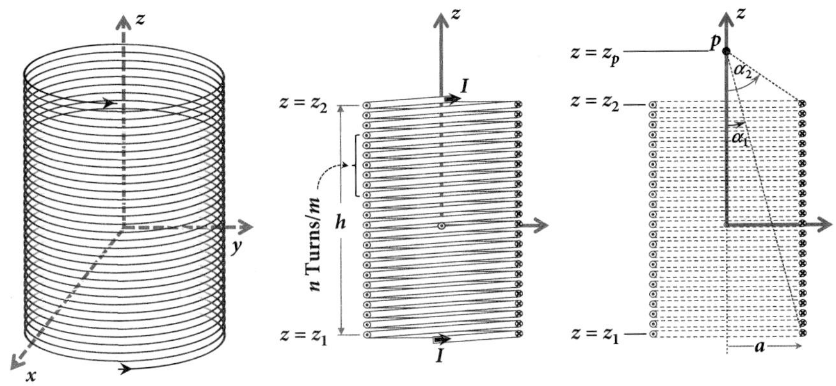

FIGURE 8B.5 Using the magnetic field expression for a ring current to derive an expression for the magnetic field due to a finite height solenoid.

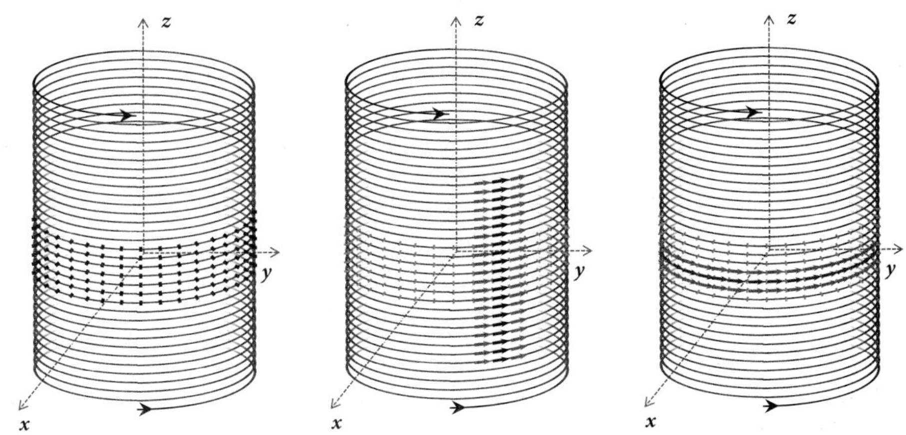

FIGURE 8B.6 Using the incrementation scheme to derive an expression for the magnetic field due to a finite height solenoid.

The answer we have for the magnetic field intensity vector at points on the z axis $p(0, 0, z)$ due to a ring current I of radius a lying in the $z = 0$ plane is

$$\vec{H}_p = \frac{I_{\text{ring}}\, a^2}{2\,(a^2 + z_{\text{ring}-p}^2)^{\frac{3}{2}}}\vec{a}_z \tag{8B.21}$$

Please note that we renamed the subscripts on both I and z. The reason is to allow this expression to adapt to general locations of the ring current other than the x-y plane. It is also to allow for other current values/ notations.

To start our analysis for the solenoid's magnetic field intensity, we consider the circular solenoid configuration carrying a current I flowing through its coil. We further specify the solenoid to have a radius a with uniformly spaced n turns/meter coil stretching over a length (height) h from $z = z_1$ to $z = z_2 (= z_1 + h)$ along the z axis as shown in Figure 8B.5. Again, and for the same reasons cited before, we will limit our interest to finding expressions for $\vec{H}$ (and $\vec{B}$) at points on the solenoid's axis, the z axis in this case.

First, we slice up the height of the solenoid into infinitesimally small increments along the z axis. Now, we choose a general location increment dz located at z to start our incrementation scheme (Figure 8B.7). Now, we adapt Equation (8B.13) to the chosen increment. To do so, we need to figure out the equivalent incremental "ring" current dI_{ring} of the chosen dz slice and distance between the slice and the point p (at which we are to evaluate $\vec{H}$).

$$dI_{\text{ring}} = (n\,dz)I \tag{8B.22}$$

$$z_{\text{ring}-p} = z_p - z \tag{8B.23}$$

Hence, we can write the incremental vector magnetic field due to the incremental slice dz as

$$\overrightarrow{dH}_p = \frac{dI_{\text{ring}}a^2}{2\left[a^2 + z_{\text{ring}-p}^2\right]^{\frac{3}{2}}}\vec{a}_z = \frac{(n\,dz)Ia^2}{2\left[a^2 + (z_p - z)^2\right]^{\frac{3}{2}}}\vec{a}_z \tag{8B.24}$$

and the total field obtainable through the integration becomes

$$\vec{H}_p = \int_{z_1}^{z_2} \frac{(n\,dz)Ia^2}{2\left[a^2 + (z_p - z)^2\right]^{\frac{3}{2}}}\vec{a}_z = \frac{nIa^2}{2}\vec{a}_z \int_{z_1}^{z_2} \frac{(dz)}{\left[a^2 + (z_p - z)^2\right]^{\frac{3}{2}}} \tag{8B.25}$$

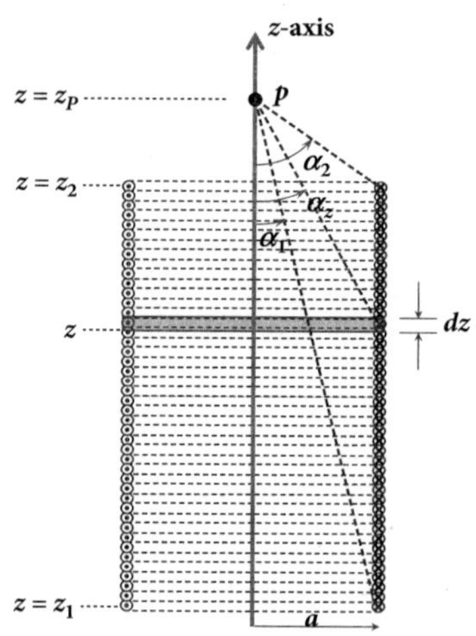

Figure 8B.7 Expressing the magnetic field due to a solenoidal current in terms of the α angles.

This integration can be simplified through the substitution

$$z_p - z = a\cot\alpha_z \tag{8B.26}$$

which, upon differentiation yields

$$-dz = -a(\csc\alpha_z)^2\, d\alpha_z \tag{8B.27}$$

With this substitution, Equation (8B.25) becomes

$$\begin{aligned}\vec{H}_p &= \frac{nIa^2}{2}\vec{a}_z\int_{z_1}^{z_2}\frac{dz}{\left[a^2+(z_p-z)^2\right]^{\frac{3}{2}}} = \frac{nIa^2}{2}\vec{a}_z\int_{\alpha_1}^{\alpha_2}\frac{a(\csc\alpha_z)^2\,d\alpha_z}{\left[a^2+(a\cot\alpha_z)^2\right]^{\frac{3}{2}}}\\ &= \frac{nIa^2}{2}\vec{a}_z\int_{\alpha_1}^{\alpha_2}\frac{a(\csc\alpha_z)^2\,d\alpha_z}{a^3\left[1+(\cot\alpha_z)^2\right]^{\frac{3}{2}}} = \frac{nI}{2}\vec{a}_z\int_{\alpha_1}^{\alpha_2}\frac{(\csc\alpha_z)^2\,d\alpha_z}{\left[(\csc\alpha_z)^2\right]^{\frac{3}{2}}}\\ &= \frac{nI}{2}\vec{a}_z\int_{\alpha_1}^{\alpha_2}\sin\alpha_z\,d\alpha_z = -\frac{nI}{2}(\cos\alpha_2-\cos\alpha_1)\vec{a}_z\end{aligned} \tag{8B.28}$$

Hence, the final answer for the magnetic field intensity vector at point p along the solenoid's axis is of the form

$$\vec{H}_p = \frac{nI}{2}(\cos\alpha_1 - \cos\alpha_2)\vec{a}_z \tag{8B.29}$$

8B.6 Case 6: Magnetic Field due to an Infinite Height Circular Solenoid

In this case, we consider the "infinitely" long solenoid. In practice, that would be the case of $h(=z_2-z_1) >> a$. This case is simply an extension of the finite solenoid case where we let $z_1 \to -\infty$ and $z_2 \to +\infty$ corresponding to $\alpha_1 \to 0$ and $\alpha_2 \to 180°$, respectively (Figure 8B.8). Hence, for the infinite length solenoid, Equation (8B.29) reduces to

$$\vec{H}_p = \frac{nI}{2}(\cos\alpha_1 - \cos\alpha_2)\vec{a}_z = \frac{nI}{2}(\cos 0 - \cos\pi)\vec{a}_z = nI\,\vec{a}_z \tag{8B.30}$$

It is interesting to observe that the answer for the infinite solenoid indicates a uniform magnetic field intensity all along the solenoid axis. As we will see later, this will not be only for values along the axis but also everywhere else within the solenoid. Furthermore, we will learn that the shape of the solenoid's cross section and size do not matter either. We will discuss both of these revelations in the following section.

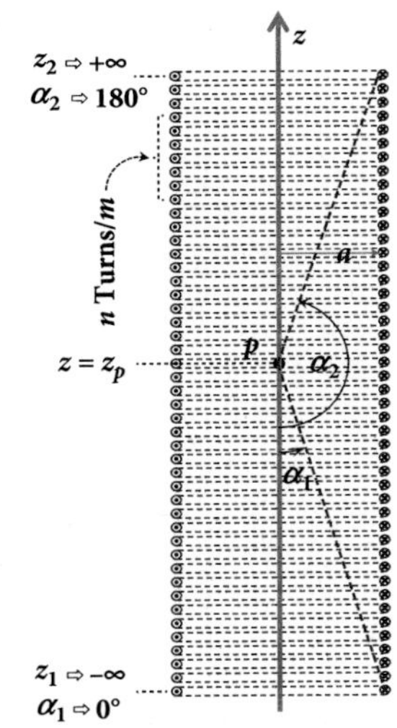

FIGURE 8B.8 For an infinite height solenoid, $\alpha_1 = 0°$ and $\alpha_2 = 180°$.

ADDENDUM 8C
MAGNETIC FIELD EVALUATION USING AMPERE'S CIRCUITAL LAW SCHEME

Ampere's circuital law was discussed earlier in this chapter and is of the form

$$\oint \vec{H} \cdot \vec{d\ell} = \sum I_{\text{Linkage}} \tag{8.15}$$

The use of Ampere's circuital law for deriving expressions for $\vec{H}$ is in a way similar to the use of Gauss' law for the determination of the electric flux density that was discussed earlier in Chapter 4. The point is that the normal use of an equation of the form such as that of Equation (8.15) is to evaluate the left-hand side integration for a given $\vec{H}$ over a choice path to evaluate the linkage current in the right-hand side of the expression. At first sight, one cannot predict how such an equation can be used to do the inverse; that is to evaluate $\vec{H}$ given the current distribution.

Well, the use of Equation (8.15) to derive expressions for $\vec{H}$ is only possible for special cases of symmetry where we can predict the $\vec{H}$ field map and carry out the integrations in regions where $\vec{H}$ can be treated as a constant that can be factored outside the integration sign. The following examples are cases of special current distribution symmetries where we can use Ampere's circuital law to derive the corresponding expressions for the magnetic field intensity vector

1. An infinite length thin straight current-carrying conductor
2. An infinite length thick straight current-carrying conductor
3. An infinite length coaxial transmission line
4. An infinite extension thin current sheet
5. Toroid
6. An infinite height solenoid

8C.1 Case 1: Magnetic Field due to an Infinite Length Thin Straight Current-Carrying Conductor

This case is demonstrated in Figure 8C.1. The current I flows in an infinitesimally thin line (e.g., conductor or charge beam) that extends indefinitely. Based on our studies and our physical intuition, we can predict the magnetic field map as circular loops concentric with the current-carrying wire. This is true for all locations in

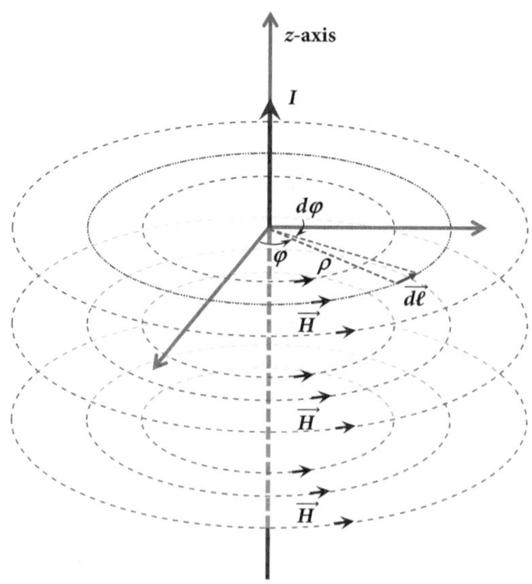

FIGURE 8C.1 Using Ampere's law to derive expressions for the magnetic field due to infinite length thin straight current-carrying conductor.

space both radially and vertically (all radii and all heights that are far away from the ends that are located at infinity). We can further predict the field directions using the right-hand rule. It is obvious that the use of cylindrical coordinates in this case would offer the most convenience in our mathematical expressions. By examining the field map of Figure 8C.1, we recognize the uniform field intensity along any of the circular lines in display. Choosing an arbitrary path along any of these field lines, we can carry out the closed line integration of Ampere's circuital law [Equation (8.15)]. Thus recognizing

$$\vec{H} = H_\varphi \vec{a}_\varphi \tag{8C.1}$$

and choosing

$$\vec{d\ell} = d\ell \vec{a}_\varphi \tag{8C.2}$$

We can write

$$\oint \vec{H} \cdot \vec{d\ell} = \int_0^{2\pi} (H_\varphi(\rho)\vec{a}_\varphi) \cdot (\rho\, d\varphi \vec{a}_\varphi) = \rho H_\varphi(\rho) \int_0^{2\pi} d\varphi = 2\pi\rho H_\varphi(\rho) = \sum I_{\text{Linkage}} = I \tag{8C.3}$$

Hence,

$$\vec{H} = H_\varphi \vec{a}_\varphi = \frac{I}{2\pi\rho} \vec{a}_\varphi \tag{8C.4}$$

which is true for all points in space and is identical to the answer we obtained earlier, Equation (8B.7), using the Biot-Savart incrementation scheme. The field intensity versus position for this case was given earlier in Figure 8.9.

8C.2 Case 2: Magnetic Field due to an Infinite Length Thick Straight Current-Carrying Conductor

This case is demonstrated in Figure 8C.2 as a current distribution flows in a circular cross section conductor of radius a. The constraint here is that the distribution of the current density J can only vary with ρ, that is, $J(\rho)$. If J varies with φ, we would not have the circular symmetry that is displayed in the figure and we would not be able to proceed with the use of Ampere's circuital law to derive expressions for $\vec{H}$ in such case. Again, using cylindrical coordinates and physical intuition based on the current distribution symmetries, the predicted field map is presented in the figure.

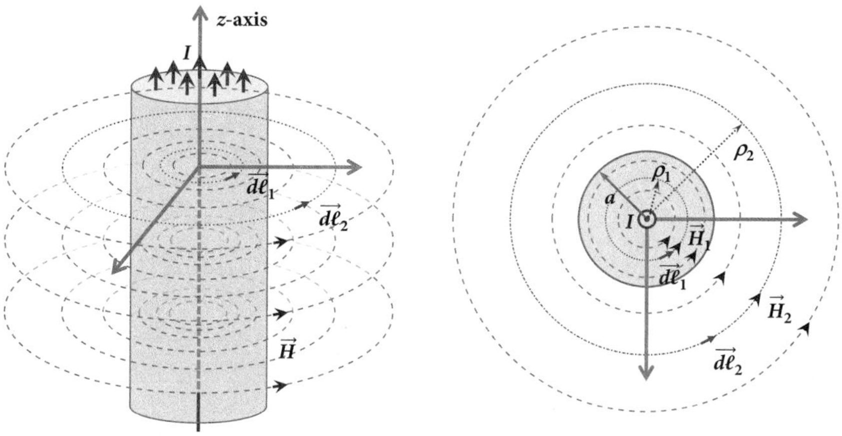

Figure 8C.2 Using Ampere's law to derive expressions for the magnetic field due to infinite length thick straight current-carrying conductor.

We can see the field map is quite similar to that of the thin wire except that we expect different expressions for the field within the radius of the current and that outside it. This is because there is a difference in the current linkage in the two regions.

Again, using the expressions of Equations (8C.1) and (8C.2) and applying Ampere's circuital law to two different paths for the two regions $0 < \rho < a$ and $a < \rho < \infty$, we can write

$$\oint \vec{H} \cdot \vec{d\ell} = \oint (H_\varphi \vec{a}_\varphi) \cdot (d\ell \vec{a}_\varphi) = H_\varphi \oint d\ell = 2\pi\rho H_\varphi = \sum I_{\text{Linkage}} \tag{8C.5}$$

Now, we recognize that for $a < \rho < \infty$, the closed integration path links all of the current I, while for $0 < \rho < a$ only the portion of I within the radius ρ is in linkage.

$$\sum I_{\text{Linkage}} = \begin{cases} I & \text{for } a < \rho < \infty \\ \int_{\varphi=0}^{2\pi} \int_{\rho=0}^{\rho} J(\rho) dS & \text{for } 0 < \rho < a \end{cases} \tag{8C.6}$$

where $J(\rho)$ is the surface current density as a function of ρ.

For the special case of a uniform current distribution over the cross section, $J(\rho)$ is constant given by

$$J(\rho) = I/(\pi a^2) \qquad \text{and} \qquad \int_{\varphi=0}^{2\pi} \int_{\rho=0}^{\rho} J(\rho) dS = I \frac{\pi\rho^2}{\pi a^2} \tag{8C.7}$$

yielding the final expressions

$$\vec{H} = \begin{cases} \dfrac{I}{2\pi\rho} \vec{a}_\varphi & \text{for } a < \rho < \infty \\ \dfrac{I\rho}{2\pi a^2} \vec{a}_\varphi & \text{for } 0 < \rho < a \end{cases} \tag{8C.8}$$

The graph of Figure 8C.3 gives a plot of the magnetic field intensity H_φ versus position ρ. We notice a linear rise for the region between zero and a and inverse relationship $(1/\rho)$ for the region outside a. The linear relationship within a is due to the constant uniform current density in that region. For other symmetrical distribution with a ρ-varying current density, we expect other forms of relationships to exist.

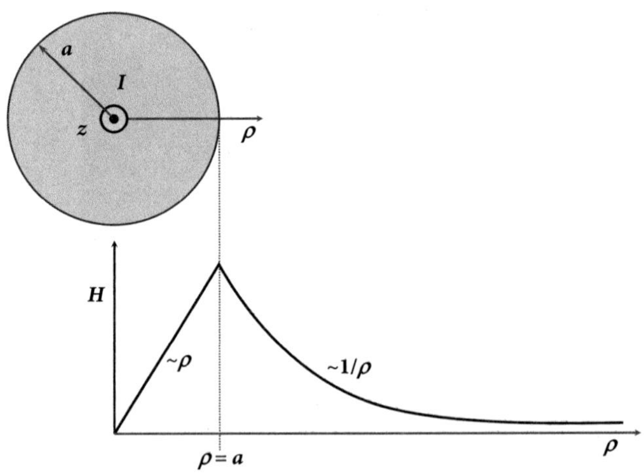

FIGURE 8C.3 The magnetic field intensity versus distance due to an infinite length thick straight current-carrying conductor.

8C.3 Case 3: Magnetic Field due to an Infinite Length Coaxial Transmission Line

In this case, we added an outer conductor carrying the opposite (return) current of the same magnitude to the thick conductor of case 2 above. The combination is what is known as a coaxial transmission line (Figure 8C.4). Our scheme of analysis will be exactly the same as in case 2 except we now have four regions to investigate. For all four regions, the process leading to and including the result of Equation (8C.5) is exactly the same.

$$\oint \vec{H}\cdot\vec{d\ell} = \oint (H_\varphi \vec{a}_\varphi)\cdot(d\ell\vec{a}_\varphi) = H_\varphi \oint d\ell = 2\pi\rho H_\varphi = \sum I_{\text{Linkage}} \tag{8C.9}$$

Now, we evaluate the linkage currents for all four regions

$$\sum I_{\text{Linkage}} = \begin{cases} +I & \text{for } a<\rho<b \\ +I-I=0 & \text{for } c<\rho<\infty \\ \int_{\varphi=0}^{2\pi}\int_{\rho=0}^{\rho} J_a(\rho)dS & \text{for } 0<\rho<a \\ +I-\int_{\varphi=0}^{2\pi}\int_{\rho=0}^{\varphi} J_{bc}(\rho)dS & \text{for } b<\rho<c \end{cases} \tag{8C.10}$$

where $J_a(\rho)$ and $J_{bc}(\rho)$ are the surface current densities as a function of ρ for the regions 0-a and b-c, respectively.

For the special case of uniform current distributions over the two cross sections, we can write

$$J_a(\rho) = I/(\pi a^2) \qquad \text{and} \qquad J_{bc}(\rho) = I/(\pi c^2 - \pi b^2) \tag{8C.11}$$

Hence,

$$\int_{\varphi=0}^{2\pi}\int_{\rho=0}^{\rho} J_a(\rho)dS = I\frac{\pi\rho^2}{\pi a^2} \qquad \text{and} \qquad \int_{\varphi=0}^{2\pi}\int_{\rho=0}^{\rho} J_{bc}(\rho)dS = I\frac{(\pi\rho^2-\pi b^2)}{(\pi c^2-\pi b^2)} \tag{8C.12}$$

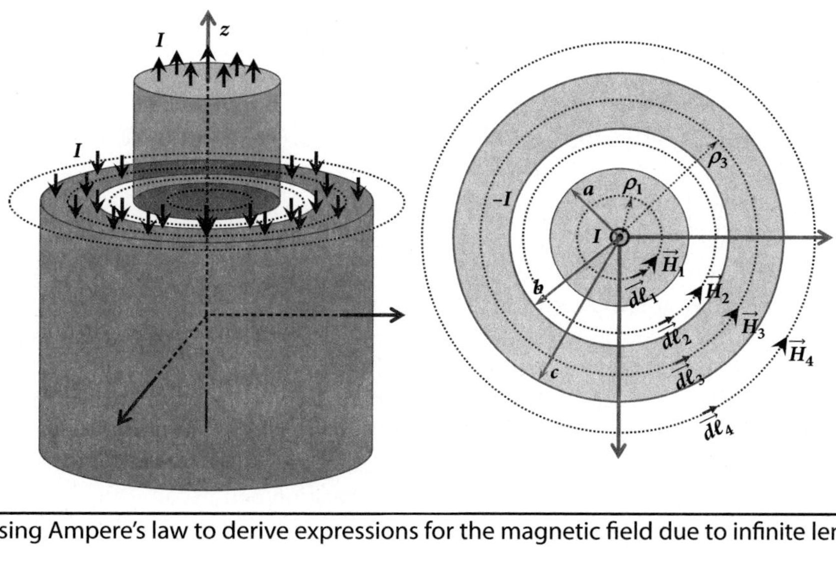

FIGURE 8C.4 Using Ampere's law to derive expressions for the magnetic field due to infinite length coaxial conductors.

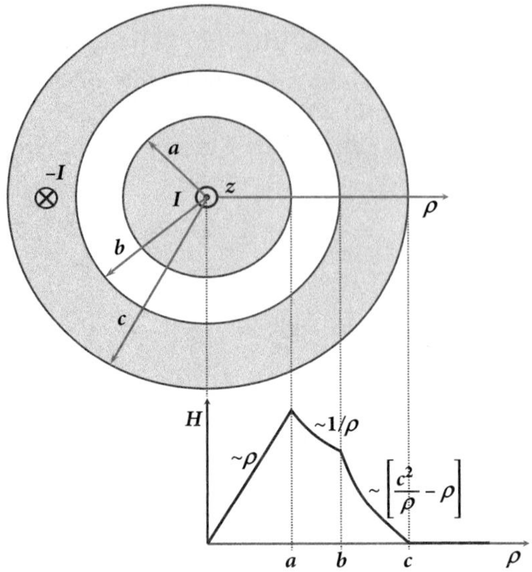

FIGURE 8C.5 The magnetic field intensity versus distance due to infinite length coaxial conductors.

yielding the final expressions

$$\vec{H} = \begin{cases} \dfrac{I\rho}{2\pi a^2}\vec{a}_\varphi & \text{for } 0<\rho<a \\ \dfrac{I}{2\pi\rho}\vec{a}_\varphi & \text{for } a<\rho<b \\ \dfrac{I}{2\pi(c^2-b^2)}\left[\dfrac{c^2}{\rho}-\rho\right]\vec{a}_\varphi & \text{for } b<\rho<c \\ 0 & \text{for } c<\rho<\infty \end{cases} \tag{8C.13}$$

The graph of Figure 8C.5 gives a plot of the magnetic field intensity H_φ versus position ρ due to constant current density distributions over the cross sections of the coaxial line. For other symmetrical distribution with ρ-varying current densities, we expect other forms of relationships to exist.

8C.4 Case 4: Magnetic Field due to an Infinite Extension Thin Current Sheet

We now switch to a planar symmetry configuration. Two possible ways to consider in this case, a surface (laminar) flow of current in an infinitesimally thin conductor, or an array of infinitesimally thin wire conductors arranged on a flat plane. Both arrangements will lead to the same exact current distributions and hence the same final answers. In this case, we will be constrained to the case of uniform linear density of the current, K ampere/meter, along the linear cross section of the current.

To argue for the array of wires case, we examine Figure 8C.6 in which we start with two closely spaced wires. If each wire by itself produces circular concentric field patterns, once close to the next wire, the patterns combine leading to field lines joining each other in the matter shown in the figure. The immediate lines to the conductors may maintain their closed loop nature around the individual wires but as we get further away from the wires, more and more joining of both wires take place. At some distance much far from the wires, we expect the field lines to lose the curvature around the wire location and become almost straight line-like in their paths. The thinner the wires and the closer they are packed next to each other the closer (they are to) the thin current sheet case.

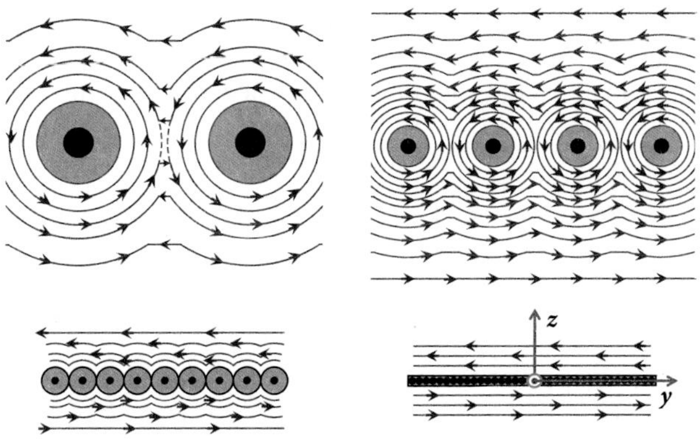

FIGURE 8C.6 Predicting the magnetic field map for a closely spaced array of current-carrying wires.

With infinite extension of this current sheet (or wire array) distribution in both right-left (y axis) and in-out of the page (z axis) we predict the magnetic field map as shown in Figures 8C.6 and 8C.7. We further predict based on the uniform current distribution and the configuration physical symmetry that this field will not vary with either x or y and can only vary with z. We can also state with confidence that the field pattern for $z > 0$ should be identical to that for $z < 0$ while maintaining polarity (right-hand rule) considerations. This can be easily affirmed by turning the figure(s) upside down and noticing the identical pattern distribution.

To proceed with the use of Ampere's circuital law to derive $\vec{H}$ everywhere, we need to follow a different tactic than that we followed in the above three cases. In the previous cases, we had circular symmetries and the integration around a closed circular loop was possible with one simple expression. In the present case, the field lines are planar and a closed loop along the field path requires its extension to infinity. To avoid potential complication, we will resort to the use of finite size rectangular loops. Segments of the loop will be chosen along the field lines where the field is expected to have constant intensity while the other loop closing segments will be chosen orthogonal to the field lines. Those orthogonal segments will be passing through potentially variable intensity field values; hence, $\vec{H}$ cannot be factored outside the integration. However, with the path orthogonal to the field for these segments, their dot product will be zero thus avoiding the potential variable $\vec{H}$ complexity.

Now that agreed that the plan is to have rectangular path integrations, the question is where we place them. Two possible locations are suggested in Figure 8C.7. One is off the x-y plane, a-b-c-d and does not link any current and the other is placed symmetrically with respect to the current plane, e-f-g-h linking a part of the current. Let us proceed with applying Ampere's circuital law to both path integrations

$$\oint \vec{H}\cdot\vec{d\ell} = \int_a^b \vec{H}_{ab}\cdot\vec{d\ell}_{ab} + \int_b^c \vec{H}_{bc}\cdot\vec{d\ell}_{bc} + \int_c^d \vec{H}_{cd}\cdot\vec{d\ell}_{cd} + \int_d^a \vec{H}_{da}\cdot\vec{d\ell}_{da} = \sum I_{\text{Linkage}} \tag{8C.14}$$

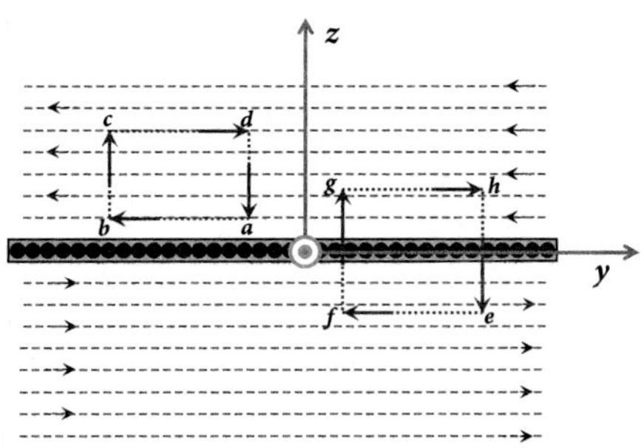

FIGURE 8C.7 Using Ampere's law to derive expressions for the magnetic field due to an infinite extension thin current sheet.

For this path the integrations *b-c* and *d-a* are both zeros due to the orthogonality between the field and paths. However, for the *a-b* path, the two vectors align and for *c-d* path, the two vectors are in opposing directions. Applying this information to Equation (8C.14) and including a zero for the linkage current, we get

$$\oint \vec{H} \cdot \overrightarrow{d\ell} = H_{ab}\ell_{ab} - H_{cd}\ell_{cd} = \sum I_{\text{Linkage}} = 0 \tag{8C.15}$$

This result is clearly stating that $H_{ab} = H_{cd}$ which means that field intensity is the same at both locations. Since the choice of the location of the *a-b-c-d* loop was made at an arbitrary height off the current sheet, then we can safely affirm the magnetic field intensity is the same for different *z* locations keeping in mind that its polarity on one side of the current sheet is opposite to its polarity on the other side. We may recall seeing a similar result when we discussed the infinite thin sheet of charges back in Chapter 4.

Repeating the same procedure for the *e-f-g-h* loop, and recognizing the magnitudes of H_{ef} and H_{gh} are the same, we can write

$$\begin{aligned}\oint \vec{H}\cdot\overrightarrow{d\ell} &= \int_e^f \vec{H}_{ef}\cdot\overrightarrow{d\ell}_{ef} + \int_f^g \vec{H}_{fg}\cdot\overrightarrow{d\ell}_{fg} + \int_g^h \vec{H}_{gh}\cdot\overrightarrow{d\ell}_{gh} + \int_h^e \vec{H}_{he}\cdot\overrightarrow{d\ell}_{he}\\ &= H_{ef}\ell_{ef} - H_{gh}\ell_{gh} = H_{ef}\ell_{ef} - (-H_{ef})\ell_{gh}\\ &= 2H_{ef}\ell_{ef} = \sum I_{\text{Linkage}} = K\ell_{ef}\end{aligned} \tag{8C.16}$$

where the linkage current was substituted as the linear current density *K* times the linked length of the current sheet ($\ell_{ef} = \ell_{gh}$). Hence, the final result for the magnetic field intensity vector due to a current sheet lying in the *x-y* plane with linear current density *K* along the *x* axis is given by

$$\vec{H} = \begin{cases} -\dfrac{K}{2}\vec{a}_y & \text{for } z > 0 \\ +\dfrac{K}{2}\vec{a}_y & \text{for } 0 > z \end{cases} \tag{8C.17}$$

Figure 8C.8 demonstrates this result graphically.

8C.5 Case 5: Magnetic Field due to a Toroid

Referring to Figure 8C.9, the displayed toroid coil is assumed to have *N* turns carrying a current *I* wrapped around a donut-shaped core with the inner and outer perimeter radii of curvature denoted by *a* and *b*, respectively. We will use the cylindrical coordinates to describe the configuration as well as in our analysis as shown in the figure. The shown shape of the core cross section is circular; however, as it turns out, the analysis developed here applies and produces the same answers for all other cross-sectional shapes. The toroid is assumed to lay horizontally in the *x-y* plane with its axis concentric with the *z* axis. Although we are displaying the toroid symmetrically located with respect to the *x-y* plane, its exact location on the *z* axis is immaterial. The critical condition here is to have a uniform cross section all around while maintaining cylindrical symmetry with respect to the *z* axis.

In our discussion of this case, we will assume a tightly wound coil where the *N* turns carrying the current *I* would behave as a sheet of current flowing on the surface of the donut core. The earlier discussion for the sheet

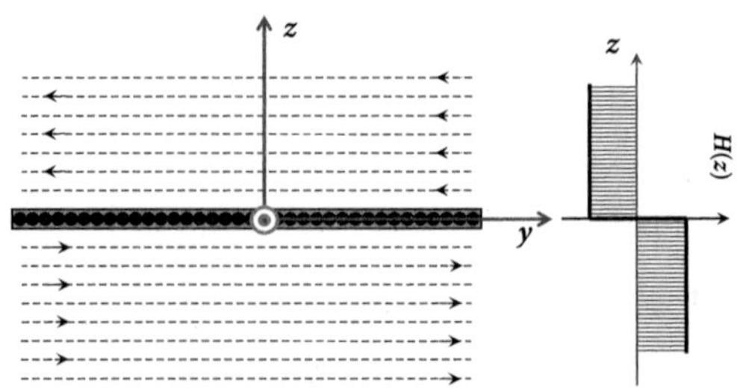

FIGURE 8C.8 Magnetic field versus position due to an infinite extension thin current sheet.

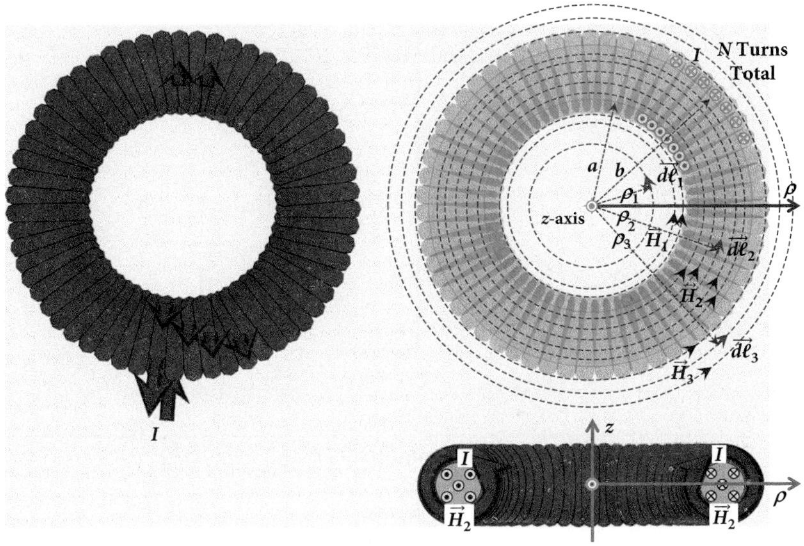

FIGURE 8C.9 Using Ampere's law for deriving expressions for the magnetic field due to a toroidal current.

current case can be used to argue the smooth field lines tangential to the sheet of current rather than curved lines contouring the coil turns.

By examining the symmetry of the toroidal coil assumed here, we can predict that the $\vec{H}$ field lines will form closed circular paths concentric with the z axis as demonstrated in Figure 8C.9. To apply Ampere's circuital law to this case, we choose similar circular paths concentric with the z axis as well. Three such paths are shown on the figure, ℓ_1, ℓ_2, and ℓ_3. One of these paths ℓ_1 links all the coil loops while the other two paths ℓ_2 and ℓ_3 do not link any.

For any of the chosen paths, we can write

$$\oint \vec{H}\cdot\vec{d\ell} = \oint (H_\varphi \vec{a}_\varphi)\cdot(d\ell\vec{a}_\varphi) = H_\varphi \oint d\ell = 2\pi\rho H_\varphi = \sum I_{\text{Linkage}} \tag{8C.18}$$

Hence, for paths 1 and 3

$$2\pi\rho H_\varphi = \sum I_{\text{Linkage}} = 0 \quad \text{or} \quad H_\varphi = H_1 = H_3 = 0 \tag{8C.19}$$

While

$$2\pi\rho_2 H_2 = \sum I_{\text{Linkage}} = NI \quad \text{or} \quad \vec{H}_i = [NI/(2\pi\rho_i)]\vec{a}_\varphi \tag{8C.20}$$

where the subscript "2" was changed to "i" to indicate inside the cross-section radii.

The conclusion is that field lines will only exist within the toroid coil as they experience full linkage with all of NI, while no field lines would exist otherwise. In other words, the magnetic field will exist only within the coil cross section and none outside it. This answer is independent of the z coordinate value as long as the location is inside the cross section of the current-carrying coil. This result is graphed in Figure 8C.10. The field intensity as shown is inversely proportional to the cylindrical coordinate radius ρ. In cases where the coil's width b-a is much smaller than a (and b), the magnetic field intensity variation would be small and an average value can be claimed.

$$\vec{H}_{av} = \frac{NI}{(2\pi\rho_{av})}\vec{a}_\varphi, \text{ where } \rho_{av} \cong (a+b)/2 \tag{8C.21}$$

This answer is demonstrated in Figure 8C.11.

It is important to recognize that the cross-sectional shape of the toroid's coil was not a factor in our analysis here. What mattered was the distinction between the two regions inside the cross section versus outside it.

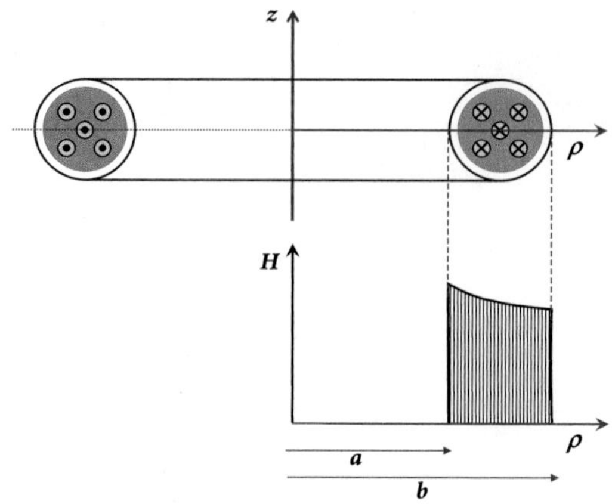

FIGURE 8C.10 Magnetic field versus position for a toroidal current.

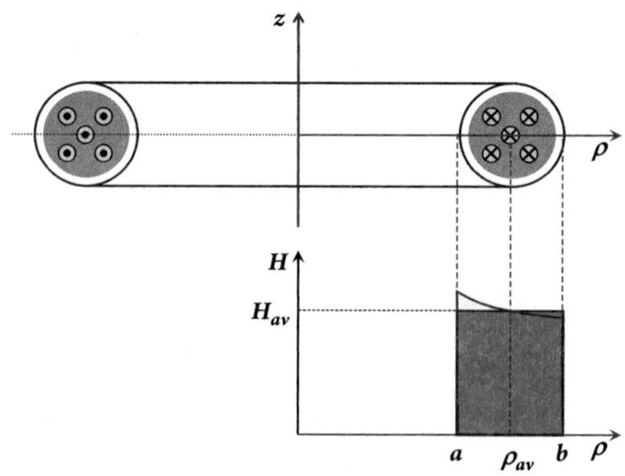

FIGURE 8C.11 Magnetic field average value for a toroidal current.

Hence, the conclusion is that the magnetic field intensity vector for a toroid irrespective of its cross-section geometry, for example, Figure 8C.12 is given by

$$\vec{H} = \begin{cases} \left[\dfrac{NI}{2\pi\rho_i}\right]\vec{a}_\varphi & \text{inside the solenoid's cross section} \\ 0 & \text{outside the solenoid's cross section} \end{cases} \tag{8C.22}$$

$$\vec{H}_{av} = \begin{cases} \left[\dfrac{NI}{2\pi\rho_{av}}\right]\vec{a}_\varphi & \text{inside the solenoid's cross section} \\ 0 & \text{outside the solenoid's cross section} \end{cases} \tag{8C.23}$$

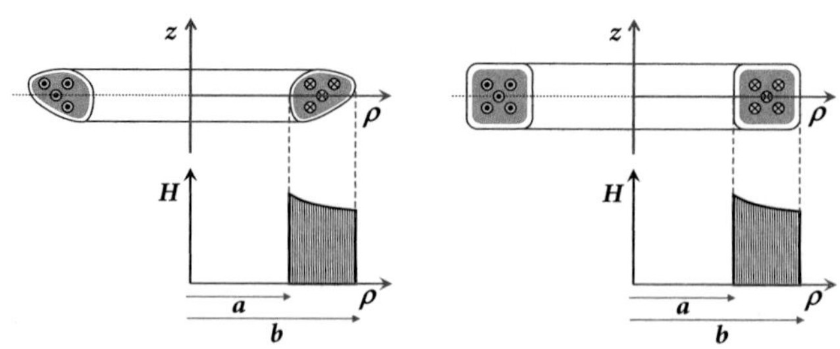

FIGURE 8C.12 The magnetic field intensity expressions [Equation (8C.22)] are valid for all φ-symmetry toroids irrespective of their cross-sectional geometry.

8C.6 Case 6: Magnetic Field due to an Infinite Height Solenoid

We are now ready to discuss the use of Ampere's circuital law in the case of an infinite length solenoid. This case was discussed earlier in this chapter using the Biot-Savart law. The solenoid configuration is given again in Figure 8C.13 and is assumed to stretch along the z axis, have n turns per meter along its axis, and its coil carries a current I. The cross-sectional shape of the solenoid as we will demonstrate is immaterial to our analysis and the obtained results.

We can approach the analysis in this case in two possible ways. Either we can use the results of the toroid as a starting point, or we can apply Ampere's circuital-law to chosen closed-paths similar to what we did before for the case of the thin current sheet.

Starting with the toroid result, Equation (8C.20), and assuming the radii of the toroid a and b to be extremely large (approach infinity) while maintaining a finite cross section size, we can claim that the toroid has taken a straight solenoid shape. For that case, we can write

$$\vec{H}_i = \left[\frac{NI}{(2\pi\rho_i)}\right]\vec{a}_\varphi = \left[\frac{N}{\ell_T}I\right]\vec{a}_\varphi = [nI]\vec{a}_\varphi \tag{8C.24}$$

where ℓ_T is the toroid coil length and $n = N/\ell_T$ is the number of coil-turns per unit length. To apply this result to the solenoid of Figure 8C.13, the toroid's "axial" unit vector along its circumference $\vec{a}_\varphi$ corresponds to the solenoid's "axial" unit vector along its axis $\vec{a}_z$, and hence the final result for the solenoid's internal magnetic field is

$$\vec{H}_i = [nI]\vec{a}_z \tag{8C.25}$$

Regarding the magnetic field outside the coil confines, it will be zero exactly as it was for the toroid case, and hence the final expression for the magnetic field intensity vector for the infinite solenoid is

$$\vec{H} = \begin{cases} [nI]\vec{a}_z & \text{inside the solenoid's cross section} \\ 0 & \text{outside the solenoid's cross section} \end{cases} \tag{8C.26}$$

Now, to discuss the closed path implementation of Ampere's circuital law, we are to start by predicting the magnetic field map of the solenoid as shown in Figure 8C.13. Recognizing the field lines to be straight lines parallel to the solenoid's axis and ignoring curvatures around the coil windings (assume dense coil windings), we can choose the closed loop integration paths to be rectangular similar to those we used before in the thin

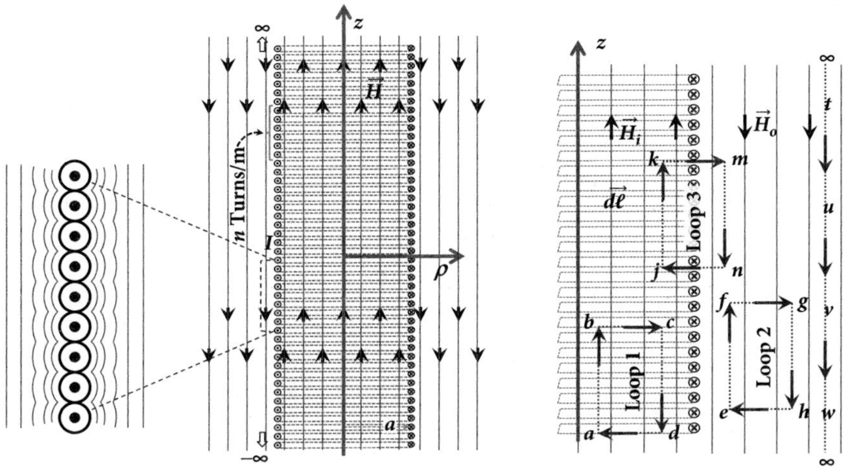

FIGURE 8C.13 Using Ampere's law for deriving expressions for the magnetic field due to an infinite height solenoidal current.

current sheet development. Again, we select several paths in the different regions we identify for the predicted field map.

The four paths shown in Figure 8C.13 are chosen to reveal relevant information about the field inside and outside the solenoid. The path *a-b-c-d* tells us of variations of the field intensity within the solenoid, while the path *e-f-g-h* does a similar thing for the outside region. For both these two paths, the current linkage is zero (both do not link the solenoid coil), and hence their integration would be identical to that of Equations (8C.14) and (8C.15) (Figure 8C.7). The results are that we will have no field intensity variability in either region. Hence, we conclude a uniform field intensity for the two regions as

$$\vec{H} = \begin{cases} H_i\,\vec{a}_z & \text{inside the solenoid's cross section} \\ -H_o\,\vec{a}_z & \text{outside the solenoid's cross section} \end{cases} \tag{8C.27}$$

Next, we apply Ampere's circuital law to the path (∞)-*t*-*u*-*v*-*w*-($-\infty$). This path is an infinitely extended path that follows the H_o field lines all along. We can claim this path to comply with the closed path requirement of Ampere's circuital law since its ends can be thought of to close on each other at infinity (both $+\infty$ and $-\infty$ can be argued to be the same far away location). The integration on this path yields the following:

$$\oint \vec{H}\cdot\vec{d\ell} = \int_{\infty}^{-\infty} \vec{H}_o \cdot \vec{d\ell}_{\infty tuvw\infty} = H_o \ell_{\infty tuvw\infty} = H_o(\infty) = \sum I_{\text{Linkage}} = 0 \tag{8C.28}$$

which confirms the same result obtained using the toroid approach, that is $H_o = 0$.

Finally, we consider the integration applied to the closed path *j-k-m-n*.

$$\begin{aligned} \oint \vec{H}\cdot\vec{d\ell} &= \int_j^k \vec{H}_{jk}\cdot\vec{d\ell}_{jk} + \int_k^m \vec{H}_{km}\cdot\vec{d\ell}_{km} + \int_m^n \vec{H}_{mn}\cdot\vec{d\ell}_{mn} + \int_n^j \vec{H}_{nj}\cdot\vec{d\ell}_{nj} \\ &= H_i \ell_{jk} - H_o \ell_{mn} = H_i \ell_{jk} = \sum I_{\text{Linkage}} = (n\ell_{jk})I \end{aligned} \tag{8C.29}$$

or

$$H_i = nI \tag{8C.30}$$

Hence, we can write the vector form expression for the magnetic field intensity for an infinite solenoid stretched along the *z* axis as

$$\vec{H} = \begin{cases} nI\,\vec{a}_z & \text{inside the solenoid's cross section} \\ 0 & \text{outside the solenoid's cross section} \end{cases} \tag{8C.31}$$

CHAPTER 8 PROBLEMS

- Problems 8.1 and 8.2: Problems dealing with magnetic flux
- Problems 8.3 to 8.8: Problems dealing with Biot-Savart law
- Problems 8.9 to 8.15: Problems dealing with Ampere's law integral form
- Problems 8.16 to 8.22: Problems dealing with Ampere's law point form
- Problem 8.23: Problems dealing with scalar magnetic potential

8.1 Given the magnetic field vector due to the infinite length filamentary current I flowing in free space along the z axis as $\vec{H}=\frac{I}{2\pi\rho}\vec{a}_{\varphi}$, find the expression for the magnetic flux density vector, $\vec{B}$. Using $\vec{B}$, derive an expression for the magnetic flux over the shaded rectangular area shown.

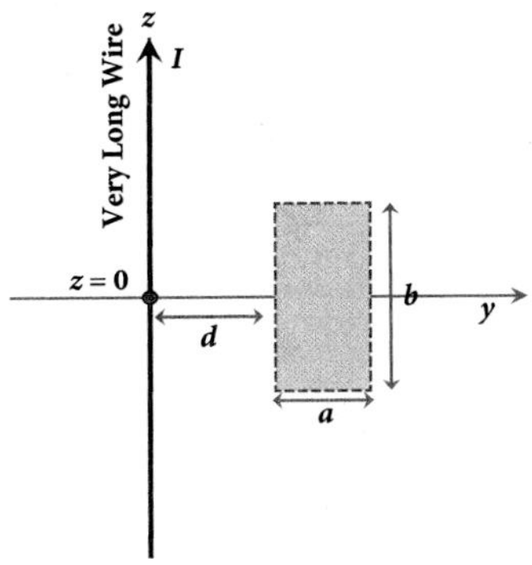

Figure 8P.1

8.2 Given the magnetic field intensity vector due to the infinite length coaxial cable carrying current I as

$$\vec{H}=\begin{cases}\dfrac{I\rho}{2\pi a^2}\vec{a}_{\varphi} & \text{for } 0<\rho<a\\[2mm] \dfrac{I}{2\pi\rho}\vec{a}_{\varphi} & \text{for } a<\rho<b\\[2mm] \dfrac{I}{2\pi(c^2-b^2)}\left[\dfrac{c^2}{\rho}-\rho\right]\vec{a}_{\varphi} & \text{for } b<\rho<c\\[2mm] 0 & \text{for } c<\rho<\infty\end{cases}$$

where a is the radius of the center conductor, while b and c are the inner and outer radii of the outer conductor.

Derive expressions for the magnetic flux per unit length over the four shaded rectangular areas shown.

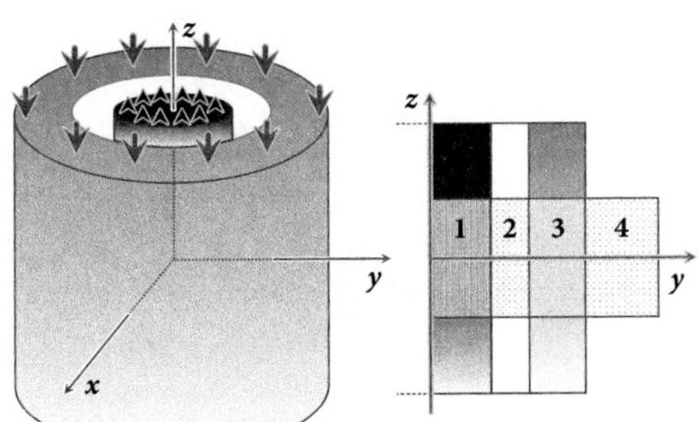

Figure 8P.2

8.3 Use Biot-Savart law to derive an expression for the magnetic field intensity vector at $p(0,0,z)$ due to the filamentary current arc $0<\varphi<\pi$ of radius $\rho = a$, current I, lying in the $z = 0$ plane, and centered at the origin.

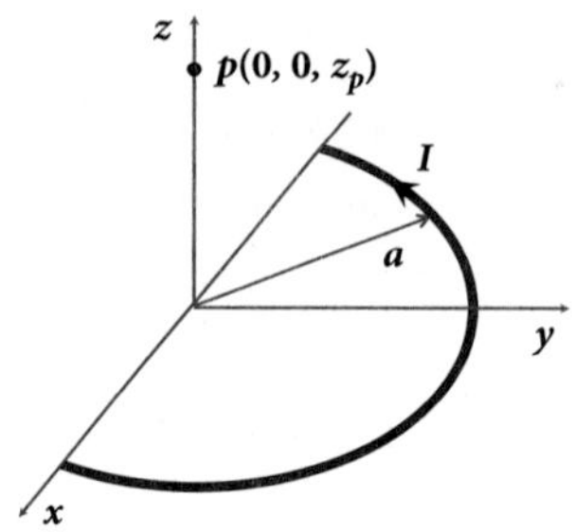

FIGURE 8P.3

8.4 Find the magnetic field intensity vector at $p(0,0,z)$ due to a circular filamentary current loop of radius $\rho = 10$ cm, current $I = 10$ A, lying in the $z = 0$ plane, and centered at the origin.

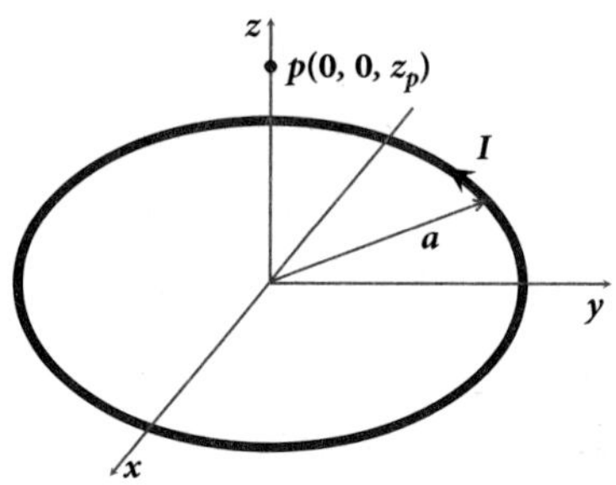

FIGURE 8P.4

8.5 Use Biot-Savart law to derive an expression for the magnetic field intensity vector at $P(x,y,z)$ due to a finite filamentary current $I = I_o$, flowing along the z axis from $z = -h$ to $z = +h$.

8.6 For the current loop shown in the figure, use the Biot-Savart law to derive an expression for the magnetic field vector at the origin.

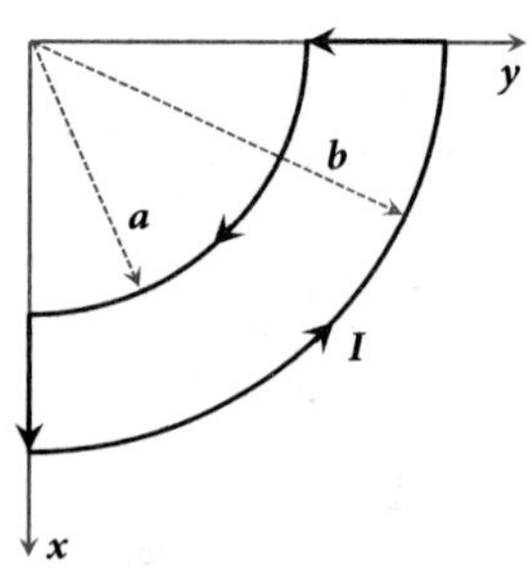

FIGURE 8P.6

8.7 Derive an expression for the magnetic moment of the uniformly charged solid sphere of radius a rotating around its axis with angular velocity ω.

8.8 A thin wire carrying current I has a right-angle bend, as shown in the figure. Using the Biot-Savart law, find the magnetic flux density, $\vec{B}$, at a point, $p(0, y, 0)$, along the y axis.

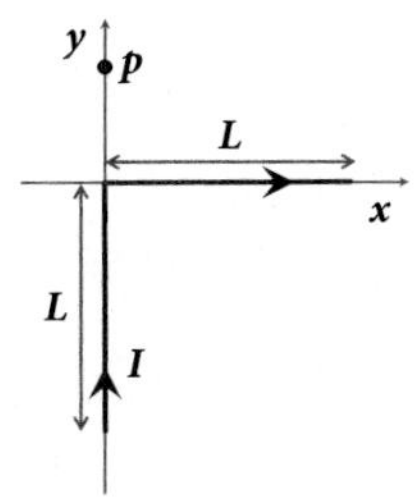

FIGURE 8P.8

8.9 A thin conducting plate lies in the x-z plane has large dimensions in both the x and y directions. A surface current with linear density K flows along z direction. Derive expressions for the magnetic field at points on the x axis.

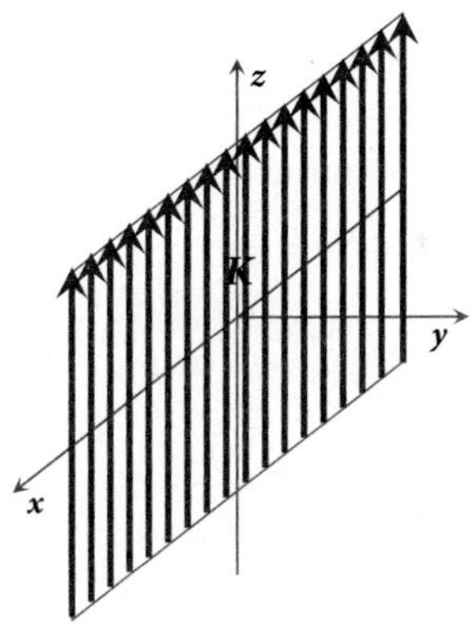

FIGURE 8P.9

8.10 For the "long" conductor pipe shown in the figure, with inner and outer radii of a and b, respectively, derive expressions for the magnetic field as a function of ρ at all points in the pipe's middle section (away from its edges) due to a uniform current density J flowing along the z axis in the body of the conductor. State your assumptions and the limitations of the derived expressions.

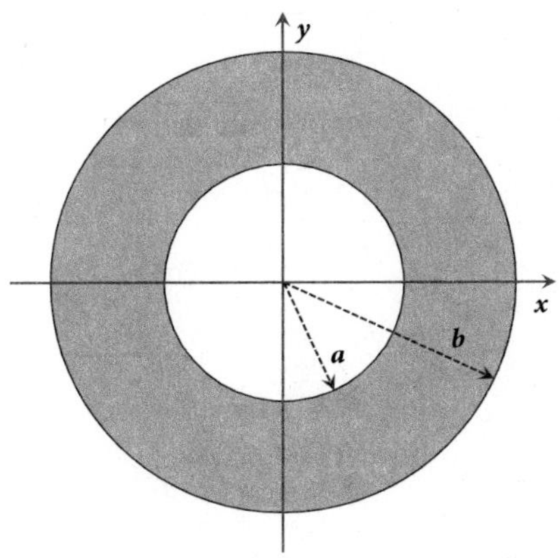

FIGURE 8P.10

8.11 Use Ampere's law to derive expressions for the magnetic field intensity vector $\vec{H}$ at all points of space due to the infinitely extended cylindrical current shell flowing along the z axis with uniform current density vector $J_o\vec{a}_z$.

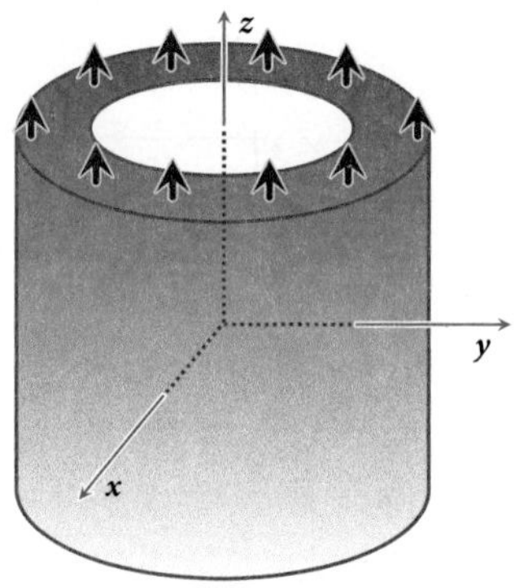

FIGURE 8P.11

8.12 Use Ampere's law to derive expressions for the magnetic field intensity vector $\vec{H}$ at all points of space due to the infinitely extended cylindrical coaxial cable shown (filamentary wire coaxial with conducting shell). The current in the center conductor is "I" flowing along the z axis while the same current flows uniformly over the cross section of the outer shell in the opposite direction.

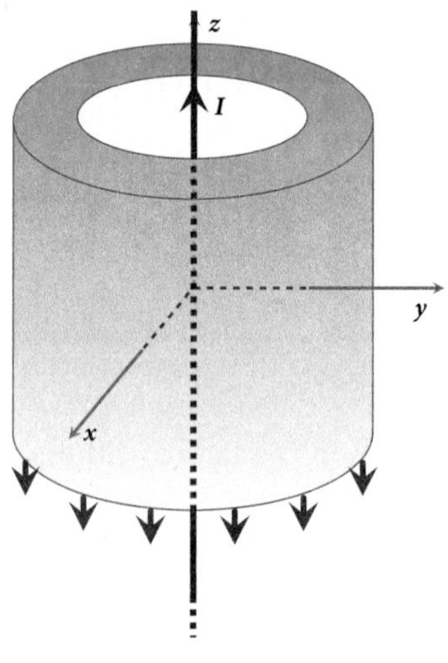

FIGURE 8P.12

8.13 A toroidal magnetic core with an average radius of 5 cm and a rectangular cross-sectional area of 1 cm width × 0.5 cm height are given. The relative permeability of the core is 900. A coil of 300 turns is wound around the core and carries a current of 6 A. Find $\vec{H}$ and $\vec{B}$ at all points in the core as well as integrate ψ_m for the core cross section.

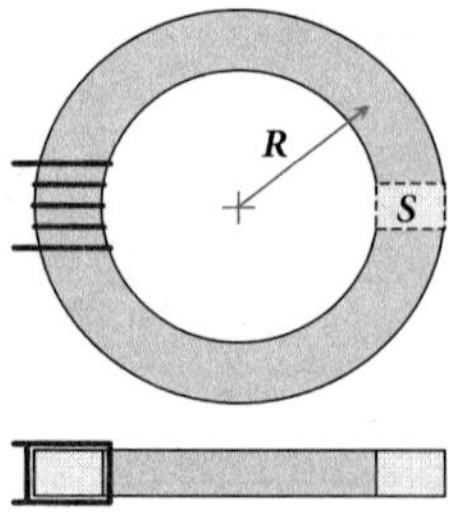

FIGURE 8P.13

8.14 Repeat for a circular cross-sectional area of 0.8 cm^2.

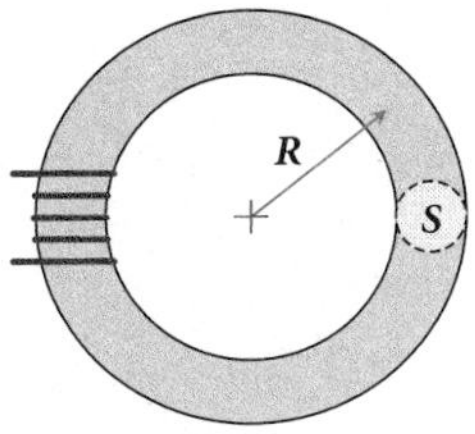

FIGURE 8P.14

8.15 A solenoid with a magnetic core with a length of 5 cm and a rectangular cross-sectional area of 1×0.5 cm. The relative permeability of the core is 900. The coil has 300 turns/m wound clockwise looking from the top of the solenoid and carries a current of 6 A. Find $\vec{H}$ and $\vec{B}$ at all points in the core as well as ψ_m for the core cross section.

FIGURE 8P.15

8.16 Use the point form of Ampere's law to derive expressions for the magnetic field intensity vector $\vec{H}$ at all points of space due to the infinitely extended current layer flowing along the x axis with uniform current density vector $J_o\vec{a}_x$. The layer extends from $-a$ to $+a$ along the z axis.

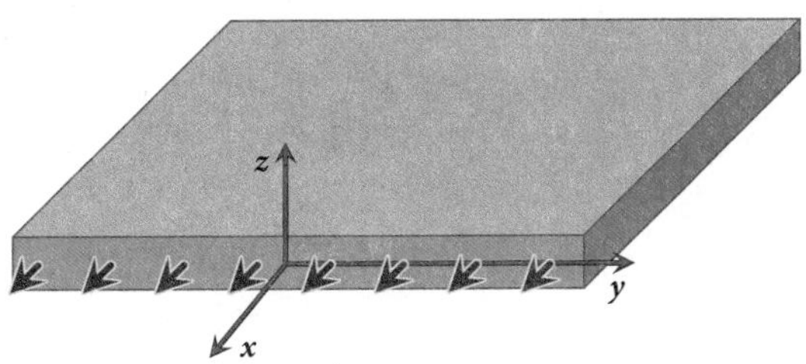

FIGURE 8P.16

8.17 Use the point form of Ampere's law to derive expressions for the magnetic field intensity vector $\vec{H}$ at all points of space due to the infinitely extended cylindrical current of radius a flowing along the z axis with uniform current density vector $J_o\vec{a}_z$.

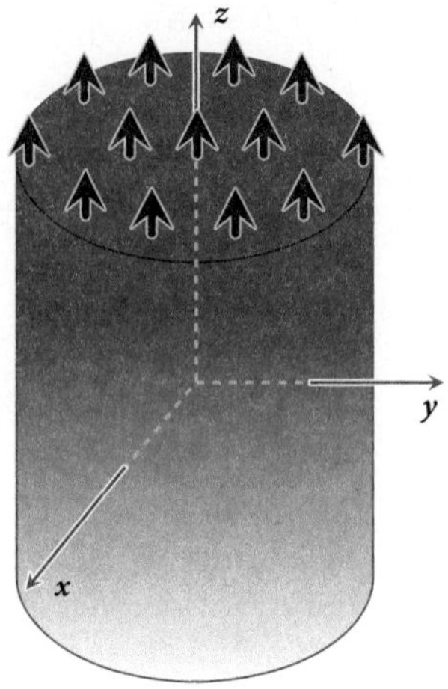

FIGURE 8P.17

8.18 Given the magnetic field intensity vector $\vec{H} = C_1\vec{a}_z$ for $0<\rho<a$ and $\vec{H}=0$ for $a<\rho<\infty$, use the point form of Ampere's law to derive an expression for the current distribution causing this field.

8.19 Given the magnetic field intensity vector $\vec{H} = C_1\rho\vec{a}_\varphi$ for $0<\rho<a$ and $\vec{H}=C_2/\rho\vec{a}_\varphi$ for $a<\rho<\infty$, use the point form of Ampere's law to derive an expression for the current distribution causing this field.

8.20 Given the magnetic field intensity vector $\vec{H} = C_1\rho\vec{a}_\varphi$ for $0<\rho<a$, $\vec{H}=\frac{C_2}{\rho}\vec{a}_\varphi$ for $a<\rho<b$, $\vec{H}=C_3\left(\frac{C_4}{\rho}-\rho\right)\vec{a}_\varphi$ for $b<\rho<c$, and $\vec{H}=0$ for $c<\rho<\infty$, use the point form of Ampere's law to derive an expression for the current distribution causing this field.

8.21 Given the magnetic field intensity vector $\vec{H} = C_1\vec{a}_z$ for $z>0$ and $\vec{H}=-C_1\vec{a}_z$ for $z<0$, use the point form of Ampere's law to derive an expression for the current distribution causing this field.

8.22 Use the point form of Ampere's law to derive expressions for the magnetic field intensity vector $\vec{H}$ at all points of space due to the infinitely extended cylindrical current of radius a flowing along the z axis with uniform current density vector $J_o\vec{a}_z$.

8.23 Derive an expression for the magnetic scalar potential difference $V_{m,BA} = -\int_A^B \vec{H}\cdot\overrightarrow{d\ell}$, for the two points (in cylindrical coordinates: $A(a, 0, 0)$ and $B(a, \pi, 0)$. Carry out the integration from A to B along three different paths:

1. The semicircle, radius a, and φ changes from 0 to π
2. The semicircle, radius a, and φ changes from 0 to $-\pi$
3. The straight line as x changes from $+a$ to $-a$

Explain the similarity/differences in the obtained results.

CHAPTER 8 SUMMARY[1]

Magnetostatics—Basic Laws

- Ampere's law for magnetic force: $\overrightarrow{dF}_{m2} = \mu_o \dfrac{(\overrightarrow{Id\ell})_2 \times [(\overrightarrow{Id\ell})_1 \times \vec{a}_{12}]}{4\pi R_{12}^2}$
- Free space permeability: $\mu_o = 4\pi \times 10^{-7}$ N/A^2 = H/m (henries/meter)
- Magnetic flux density: $\overrightarrow{dB}_P = \dfrac{\mu_o}{4\pi} \dfrac{(\overrightarrow{Id\ell}) \times \vec{a}_R}{R_P^2}$ = {magnetic normalized force with $(Id\ell)_2 = 1$}
- Magnetic field intensity: $\overrightarrow{dH}_P = \dfrac{1}{4\pi} \dfrac{(\overrightarrow{Id\ell}) \times \vec{a}_R}{R_P^2}$ – $\vec{B} = \mu \vec{H}$
- Magnetic flux: $\psi_m = \iint_S \vec{B} \cdot \overrightarrow{dS}$ – $B = \dfrac{d\psi_m}{dS_{\text{normal}}}$
- $\overrightarrow{dF}_m = (\overrightarrow{Id\ell}) \times \overrightarrow{dB} = \mu(\overrightarrow{Id\ell}) \times \overrightarrow{dH}$
- Gauss' law for magnetism: $\{\psi_{m,\,\text{closed surface } S}\} = 0 = \oiint_S \vec{B} \cdot \overrightarrow{dS}$
- Ampere's circuital law: $\oint \vec{H} \cdot \overrightarrow{d\ell} = \sum I_{\text{Linkage}}$

Magnetic Field Evaluation Using the "Incrementation" Scheme and Biot-Savart Law

- A finite length thin straight current-carrying conductor: $\vec{H}_p = \dfrac{I}{4\pi\rho_p}[\cos\alpha_1 - \cos\alpha_2]\vec{a}_\varphi$
- An infinite length thin straight current-carrying conductor: $\vec{H}_p = \dfrac{I}{2\pi\rho_p}\vec{a}_\varphi$
- A thin circular current-carrying (loop) conductor at points on the z axis: $\vec{H}_p = \dfrac{Ia^2}{2(a^2+z^2)^{\frac{3}{2}}}\vec{a}_z$
- Also, the case of a segment of a loop.
- A finite height circular solenoid at points on the z axis: $\vec{H}_p = \dfrac{nI}{2}(\cos\alpha_1 - \cos\alpha_2)\vec{a}_z$
- An infinite height circular solenoid inside the solenoid: $\vec{H}_p = nI\,\vec{a}_z$, and zero outside.

Magnetic Field Evaluation Using Ampere's Circuital Law Scheme

- ONLY for certain symmetrical current distributions and combinations thereof
- Cylindrical (axial/coaxial) symmetries

  An infinite length thin straight current-carrying conductor: $\vec{H} = [I/(2\pi\rho)]\vec{a}_\varphi$

  An infinite length thick straight current-carrying conductor: $\vec{H} = \begin{cases} [I/(2\pi\rho)]\vec{a}_\varphi, & a < \rho < \infty \\ [I\rho/(2\pi a^2)]\vec{a}_\varphi, & 0 < \rho < a \end{cases}$

---

[1]***Important Warning to Students:*** Please do not use this summary without thorough study of the relevant material in this chapter. Using this summary and its formulas without adequate appreciation of the physics involved and conditions of validity can lead to wrong implementations and invalid conclusions.

An infinite length coaxial transmission line: $\vec{H}=\begin{cases}\left[\dfrac{I\rho}{2\pi a^2}\right]\vec{a}_\varphi & \text{for } 0<\rho<a\\ \left[\dfrac{I}{2\pi\rho}\right]\vec{a}_\varphi & \text{for } a<\rho<b\\ \dfrac{I}{2\pi(c^2-b^2)}\left[\dfrac{c^2}{\rho}-\rho\right]\vec{a}_\varphi & \text{for } b<\rho<c\\ 0 & \text{for } c<\rho<\infty\end{cases}$

- Planar symmetries
  - An infinite extension thin current sheet: $\vec{H}=\begin{cases}-[K/2]\vec{a}_y & \text{for } z>0\\ +[K/2]\vec{a}_y & \text{for } 0>z\end{cases}$
- Toroidal and solenoidal symmetries

Toroid: $\vec{H}=\begin{cases}[(NI)/(2\pi\rho_i)]\vec{a}_\varphi & \text{inside the solenoid's cross section}\\ 0 & \text{outside the solenoid's cross section}\end{cases}$

An infinite height solenoid: $\vec{H}=\begin{cases}[nI]\vec{a}_z & \text{inside the solenoid's cross section}\\ 0 & \text{outside the solenoid's cross section}\end{cases}$

- Choose appropriate closed path along the physically predicted field lines. Segments of the closed path may be chosen perpendicular to the field lines.
- Perform the closed loop integration $\oint \vec{H}\cdot\vec{d\ell}=\sum I_{\text{Linkage}}$ with the integrand H factored outside the integration.
- Equate the integral to the current linkage to the closed loop.
- Evaluate $\vec{H}$.

Magnetostatic Differential (Point) Forms

- Point form of Gauss' law in magnetism: $\vec{\nabla}\cdot\vec{B}=0$
- Point form of Ampere's circuital law: $\vec{\nabla}\times\vec{H}=\vec{J}$
- The curl: $\vec{\nabla}=\left\{\vec{a}_x\dfrac{\partial}{\partial x}+\vec{a}_y\dfrac{\partial}{\partial y}+\vec{a}_z\dfrac{\partial}{\partial z}\right\}$

$$\vec{\nabla}\times\vec{H}=\left\{\frac{\partial H_z}{\partial y}-\frac{\partial H_y}{\partial z}\right\}\vec{a}_x+\left\{\frac{\partial H_x}{\partial z}-\frac{\partial H_z}{\partial x}\right\}\vec{a}_y+\left\{\frac{\partial H_y}{\partial x}-\frac{\partial H_x}{\partial y}\right\}\vec{a}_z$$

$$\vec{\nabla}\times\vec{H}=\left\{\frac{1}{\rho}\frac{\partial H_z}{\partial\varphi}-\frac{\partial H_\varphi}{\partial z}\right\}\vec{a}_\rho+\left\{\frac{\partial H_\rho}{\partial z}-\frac{\partial H_z}{\partial\rho}\right\}\vec{a}_\varphi+\frac{1}{\rho}\left\{\frac{\partial(\rho H_\varphi)}{\partial\rho}-\frac{\partial H_\rho}{\partial\varphi}\right\}\vec{a}_z$$

$$\vec{\nabla}\times\vec{H}=\frac{1}{r\sin\theta}\left\{\frac{\partial(\sin\theta H_\varphi)}{\partial\theta}-\frac{\partial H_\theta}{\partial\varphi}\right\}\vec{a}_r+\frac{1}{r}\left\{\frac{1}{\sin\theta}\frac{\partial H_r}{\partial\varphi}-\frac{\partial(rH_\varphi)}{\partial r}\right\}\vec{a}_\theta+\frac{1}{r}\left\{\frac{\partial(rH_\theta)}{\partial r}-\frac{\partial H_r}{\partial\theta}\right\}$$

Stokes' theorem: $\oint_C \vec{G}\cdot\vec{d\ell}=\iint_S(\vec{\nabla}\times\vec{G})\cdot\vec{dS}-\oint_C\vec{H}\cdot\vec{d\ell}=\iint_S(\vec{\nabla}\times\vec{H})\cdot\vec{dS}$

Static Form of Maxwell's Equations

	Integral Form	Differential (Point) Form
Magnetic field circulation	$\oint \vec{H} \cdot \overrightarrow{d\ell} = \sum I_{\text{Linkage}}$	$\vec{\nabla} \times \vec{H} = \vec{J}$
Electric field circulation	$\oint \vec{E} \cdot \overrightarrow{d\ell} = 0$	$\vec{\nabla} \times \vec{E} = 0$
Electric flux divergence	$\oiint_S \vec{D} \cdot \overrightarrow{dS} = \sum Q_{\text{enclosed}}$	$\vec{\nabla} \cdot \vec{D} = \rho_v$
Magnetic flux divergence	$\oiint_S \vec{B} \cdot \overrightarrow{dS} = 0$	$\vec{\nabla} \cdot \vec{B} = 0$
Constitutive relations	$\vec{D} = \varepsilon \vec{E}, \vec{B} = \mu \vec{H}, \vec{J} = \sigma \vec{E}$	$\vec{D} = \varepsilon \vec{E}, \vec{B} = \mu \vec{H}, \vec{J} = \sigma \vec{E}$

Scalar and Vector Magnetic Potential

- Scalar magnetic potential: $V_{m,BA} = V_{m,B} - V_{m,A} = -\int_A^B \vec{H} \cdot \overrightarrow{d\ell}$
- Vector magnetic potential: $\vec{B} = \vec{\nabla} \times \vec{A}$ and $\vec{\nabla} \cdot \vec{A} = 0$
- $\nabla^2 \vec{A} = -\mu_o \vec{J}$ and $\vec{A}_p = \iiint_v \frac{\mu_o \vec{J} dv}{4\pi R_{dv-p}} = \oint_\ell \frac{\mu_o I \overrightarrow{d\ell}}{4\pi R_{d\ell-p}}$

Electrostatic-Magnetostatic Analogies

Electrostatics	Magnetostatics	Comments
dQ	$(I\overrightarrow{d\ell})$	Scalar/vector source element
$\vec{D}$	$\vec{B}$	Flux density vector
$\vec{E}$	$\vec{H}$	Field intensity vector
$\vec{D} = \varepsilon \vec{E}$	$\vec{B} = \mu \vec{H}$	Material constitutive relationship
$\varepsilon_o = 8.8541878 \times 10^{-12}$ F/m	$\mu_o = 4\pi \times 10^{-7}$ H/m	Free space permittivity and permeability
$\oiint_S \vec{D} \cdot \overrightarrow{dS} = \sum Q_{\text{enclosed}}$	$\oiint_S \vec{B} \cdot \overrightarrow{dS} = 0$	Gauss' law
$\oiint_S \vec{D} \cdot \overrightarrow{dS} = \sum Q_{\text{enclosed}}$	$\oint \vec{H} . \overrightarrow{d\ell} = \sum I_{\text{Linkage}}$	Evaluation of *D*/*H* for symmetrical distributions
$\oint \vec{E} \cdot \overrightarrow{d\ell} = 0$	$\oiint_S \vec{B} \cdot \overrightarrow{dS} = 0$	Divergent *E* is rotationless while rotational *B* is divergentless
$\overrightarrow{dF}_{e2} = \frac{dq_2 dQ_1}{4\pi\varepsilon_o R_{12}^2}$	$\overrightarrow{dF}_{m2} = \mu_o \frac{(I\overrightarrow{d\ell})_2 \times \left[(I\overrightarrow{d\ell})_1 \times \vec{a}_{12}\right]}{4\pi R_{12}^2}$	Coulomb's and ampere's laws for force
$\overrightarrow{dF}_{e2} = dq_2 \overrightarrow{dE}$	$\overrightarrow{dF}_{m2} = (I\overrightarrow{d\ell})_2 \times \overrightarrow{dB}$	*E*/*B* as normalized forces per unit source element
$\overrightarrow{dE}_p = \frac{1}{4\pi\varepsilon_o} \frac{dQ}{R_p^2}$	$\overrightarrow{dH}_p = \frac{1}{4\pi} \frac{(I\overrightarrow{d\ell}) \times \vec{a}_R}{R_p^2}$	Evaluation of *E*/*H* using incrementation (Coulomb/Biot-Savart laws)

Electrostatic-Magnetostatic Analogies (*Continued*)

Electrostatics	Magnetostatics	Comments
$\vec{\nabla}\cdot\vec{D}=\rho_v$	$\vec{\nabla}\cdot\vec{B}=0$	Gauss' law divergence
$\vec{\nabla}\times\vec{E}=0$	$\vec{\nabla}\times\vec{H}=\vec{J}$	Circulations (curl equations)
$V_{BA}=-\int_A^B\vec{E}\cdot\overrightarrow{d\ell}$	$V_{m,BA}=-\int_A^B\vec{H}\cdot\overrightarrow{d\ell}$	Scalar potentials
	$\vec{B}=\vec{\nabla}\times\vec{A}\;\&\;\vec{\nabla}\cdot\vec{A}=0$	Magnetic vector potential
$\nabla^2V=-\rho_v/\varepsilon_o$	$\nabla^2\vec{A}=-\mu_o\vec{J}$	Potentials
$V_p=\iiint_v\frac{\rho_v dv}{4\pi\varepsilon_o R_{dv-p}}$	$\vec{A}_p=\oint_\ell\frac{\mu_o I\overrightarrow{d\ell}}{4\pi R_{d\ell-p}}$	Potentials

CHAPTER 9

Magnetic Materials, Magnetic Circuits, and Inductance

Joseph Henry
(1797–1878)

Wilhelm Eduard Weber
(1804–1891)

Chapter Learning Objectives

- Evaluate the magnetic force acting on a moving point charge.
- Evaluate the magnetic torque on current loops of different configurations.
- Discuss the energy stored in a magnetic field.
- Identify the different magnetic properties of materials. Identify examples for each type.
- Identify the boundary conditions at the interface between two materials of different magnetic properties for both the magnetic field intensities and magnetic flux densities.
- Analyze magnetic circuits using both Ampere's circuital law and the electrical circuit analogy approaches.
- Evaluate the magnetic pull in the air gap of a magnetic circuit.
- Define the inductance for a coil.
- Define flux linkage for the case of a multi-turn coil and a current flowing in thick conductors.
- Evaluate the self- and mutual inductances for different coil configurations using different approaches (flux-current relationship, energy-stored approach, inductance-reluctance relationship approach).

9.1 Introduction

In Chapter 8, we learned that the source of magnetic phenomena is the motion of an electric charge. We also studied the basic laws of magnetostatics due to Ampere, Biot, and Savart. We carried our discussion in some analogy to the electrostatics. We defined the magnetic field, magnetic flux, flux density, and magnetic potential. Furthermore, we developed both the integral and point forms of the magnetostatic form of Maxwell's equations.

In this chapter, we further our magnetostatic studies by learning about magnetic forces and torques on charges in motion and current loops. We will also learn about magnetic properties of materials; also, how an interface between two materials affects the magnetic field and flux in the region. In addition, we will discuss the energy stored in a magnetic field, magnetic circuits, and finally self- and mutual inductances.

9.2 Magnetic Force and Torque

In Chapter 8, we discussed the physics of the magnetic force. We also surveyed relevant expressions as presented by Ampere's law for force and the Biot-Savart law. In this section, we summarize the earlier material and add relevant details regarding the magnetic force and the corresponding magnetic torque.

9.2.1 Ampere's Law for Magnetic Force

As presented earlier, the Ampere's law for magnetic force between two current elements $(\overrightarrow{Id\ell})_1$ and $(\overrightarrow{Id\ell})_2$ separated by the spatial vector $\vec{R}_{12}$ is of the form:

$$\overrightarrow{dF}_{m2} = \mu_o \frac{(\overrightarrow{Id\ell})_2 \times \left[(\overrightarrow{Id\ell})_1 \times \vec{a}_{12}\right]}{4\pi R_{12}^2} \tag{8.4}$$

Using the Biot-Savart law, Equation (8.10), we were able to rewrite Equation (8.4) as

$$\overrightarrow{dF}_m = \overrightarrow{Id\ell} \times \vec{B} \tag{8.13}$$

This expression, Equation (8.13), gives the force on one current element immersed in a magnetic field with a magnetic flux density vector $\vec{B}$. This magnetic field could be the result of having one or more current sources causing the flux density $\vec{B}$ at the location of interest. This expression is demonstrated in Figure 9.1 showing the force vector $\overrightarrow{dF}_m$ to be orthogonal to both $\vec{B}$ and $(\overrightarrow{Id\ell})$.

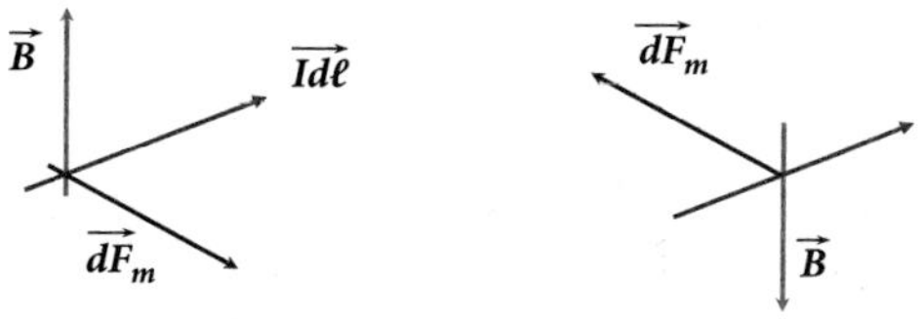

FIGURE 9.1 Magnetic force on a current element due to an applied magnetic field.

9.2.2 Magnetic Force on Moving Charge

In some practical applications, the current element sensing the force is to be expressed in terms of the causing charge in motion. In other words, instead of using I and $d\ell$, we would use the charge dq and its velocity v. In such case, and recognizing

$$\vec{Id\ell} = \frac{dq}{dt} d\ell \vec{a}_\ell \equiv dq \frac{d\ell}{dt} \vec{a}_\ell = dqv\vec{a}_v = dq\vec{v} \tag{9.1}$$

Then we can rewrite Equation (8.13) for the case of a charge dq moving at a velocity $\vec{v}$ in a magnetic field with a magnetic flux density $\vec{B}$ as

$$\vec{dF}_m = dq\vec{v} \times \vec{B} \tag{9.2}$$

Expression (9.2) is demonstrated in Figure 9.2. Again, the force vector $\vec{dF}_m$ is orthogonal to both $\vec{B}$ and $(dq\vec{v})$.

9.2.3 Magnetic Force and Torque on a Current Loop

As we are well aware of, electric currents flow in closed loops. Hence, we expect differing force directions on the various locations of the loop. To study the magnetic forces in such case, we examine Figure 9.3 where we consider two special cases for the direction of the magnetic flux density vector with respect to a planar current loop. In the case to the left in the figure, $\vec{B}$ is orthogonal to the loop, while in the case to the right, $\vec{B}$ lies on the same plane as the loop. In both cases, the direction of the force vectors as determined by the cross product of Equation (8.13) is displayed.

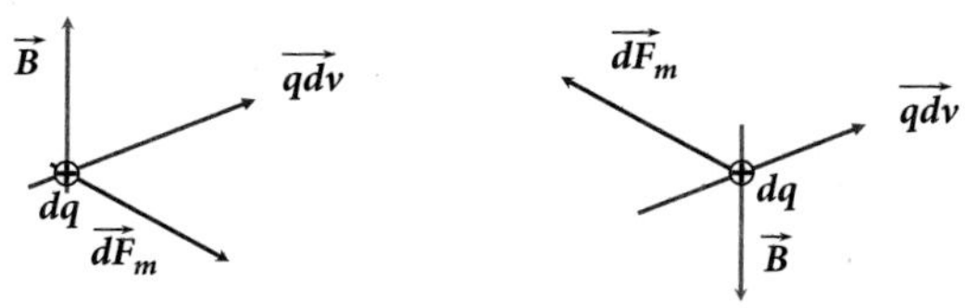

FIGURE 9.2 Magnetic force on a moving charge due to an applied magnetic field.

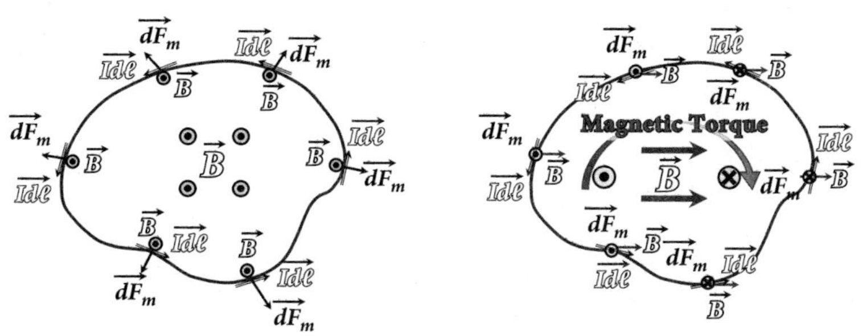

FIGURE 9.3 Magnetic force and torque on a current loop due to an applied magnetic field.

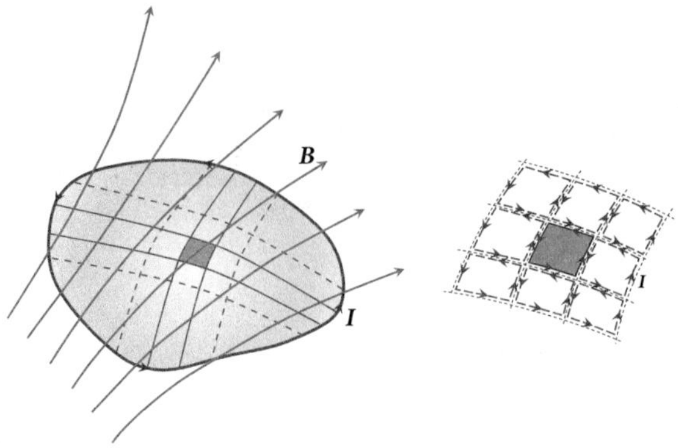

FIGURE 9.4 Incremental current loop used for the development of the magnetic torque expression for the bigger current loop.

Examining the two cases demonstrated in Figure 9.3, we recognize the presence of a pair of opposing forces operating on opposite locations of the loop. As the current circulates in the closed loop, there will always be two opposite places where the currents flow in opposite directions and hence the pair of opposite forces.

For the case to the left, all forces lie in the same plane of the current loop and hence the opposing force will be in directions as to force the loop's current elements apart. These forces try to expand the loop and stretch it outward. If the current-carrying wire of the loop is flexible, the uniform forces all around it would try to shape it into a uniform circle.

However, for the case to the right of the figure, all forces are normal to the plane of the loop. Some forces as shown are directed upward (out of the page) and the other forces operating on the loop's opposite locations are directed downward. The net result of these force couples is a torque operating on the current-carrying loop as illustrated in the figure.

To express the forces and torques on closed current loops, we consider the general case of a current loop located in a general-shaped magnetic flux distribution as shown in Figure 9.4. Next, we use the incrementation approach to determine the incremental forces and torques on a current loop *increment* as demonstrated in the figure. This incrementation is to be performed in a curvilinear square fashion such that each increment will have orthogonal sides. We do that for the convenience of the analyses that we will carry later in this section.

As we dice up the loop surface into small incremental surfaces, each increment can be considered as carrying the same current in the main loop. The logic is as you piece the adjacent increments together, opposite currents on the perimeters cancel leaving only the outside perimeter's current which would be that of the full closed loop.

By making the individual increments of Figure 9.4 infinitesimally small, we can treat each increment as a planar surface. The analysis of forces and torques on such an incremental current loop is demonstrated in Figure 9.5. In the figure, we chose a local Cartesian coordinate system with the z axis normal to the increment while the x and y axes are chosen to go along the sides of the increment. Now, we break the magnetic flux density vector into its three components: B_x, B_y, and B_z. As per our earlier discussion of the two cases in

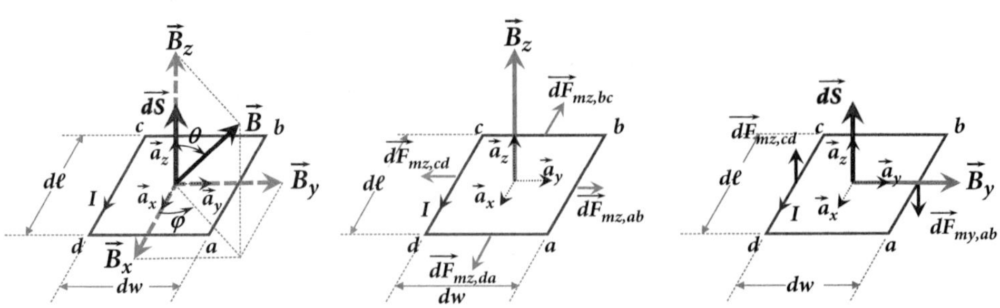

FIGURE 9.5 Magnetic forces and torque on the incremental current loop.

Figure 9.3, the normal component of $\vec{B}$, B_z, will produce forces lying on the same increment plane while the planar (tangential) magnetic flux density components, B_x and B_y, will be responsible for the normal forces resulting in a net torque on the incremental loop.

Referring to Figure 9.5, we can write the following expressions for the shown forces and the corresponding torque:

$$\vec{B}_z = B\cos\theta\,\vec{a}_z, \qquad \vec{B}_x = B\sin\theta\cos\varphi\,\vec{a}_x, \qquad \text{and} \qquad \vec{B}_y = B\sin\theta\sin\varphi\,\vec{a}_y \tag{9.3}$$

For the B_z component, we observe equal and opposite forces acting on the same lines, thus producing no torques, $\overrightarrow{dT}_{mz} = 0$:

$$\overrightarrow{dF}_{mz,ab} = -Id\ell\vec{a}_x \times B_z\vec{a}_z = Id\ell\, B_z\vec{a}_y \tag{9.4a}$$

$$\overrightarrow{dF}_{mz,bc} = -Idw\vec{a}_y \times B_z\vec{a}_z = -Idw\, B_z\vec{a}_x \tag{9.4b}$$

$$\overrightarrow{dF}_{mz,cd} = Id\ell\vec{a}_x \times B_z\vec{a}_z = -Id\ell\, B_z\vec{a}_y \tag{9.4c}$$

$$\overrightarrow{dF}_{mz,da} = Idw\vec{a}_y \times B_z\vec{a}_z = Idw\, B_z\vec{a}_x \tag{9.4d}$$

while for the B_y component,

$$\overrightarrow{dF}_{my,ab} = -Id\ell\vec{a}_x \times B_y\vec{a}_y = -Id\ell\, B_y\vec{a}_z \tag{9.5a}$$

$$\overrightarrow{dF}_{my,bc} = -Idw\vec{a}_y \times B_y\vec{a}_y = 0 \tag{9.5b}$$

$$\overrightarrow{dF}_{my,cd} = Id\ell\vec{a}_x \times B_y\vec{a}_y = Id\ell\, B_y\vec{a}_z \tag{9.5c}$$

$$\overrightarrow{dF}_{my,da} = Idw\vec{a}_y \times B_y\vec{a}_y = 0 \tag{9.5d}$$

which means a couple of $\vec{F}_z$ forces with a torque equals

$$\overrightarrow{dT}_{my} = \overrightarrow{dF}_{my,cd} \times dw\vec{a}_y = Id\ell\, B_y\vec{a}_z \times dw\vec{a}_y = I(d\ell dw)\vec{a}_z \times \vec{B}_y \tag{9.6}$$

Similarly (not shown in the figure),

$$\overrightarrow{dF}_{mx,ab} = -Id\ell\vec{a}_x \times B_x\vec{a}_x = 0 \tag{9.7a}$$

$$\overrightarrow{dF}_{mx,bc} = -Idw\vec{a}_y \times B_x\vec{a}_x = Idw\, B_x\vec{a}_z \tag{9.7b}$$

$$\overrightarrow{dF}_{mx,cd} = Id\ell\vec{a}_x \times B_x\vec{a}_x = 0 \tag{9.7c}$$

$$\overrightarrow{dF}_{mx,da} = Idw\vec{a}_y \times B_x\vec{a}_x = -Idw\, B_x\vec{a}_z \tag{9.7d}$$

which means a couple of F_z forces with a torque equals

$$\overrightarrow{dT}_{mx} = \overrightarrow{dF}_{my,bc} \times d\ell\vec{a}_x = Idw\, B_x\vec{a}_z \times d\ell\vec{a}_x = I(d\ell dw)\vec{a}_z \times \vec{B}_x \tag{9.8}$$

Thus, the total torque on the incremental loop takes the form:

$$\begin{aligned}\overrightarrow{dT}_{m,\text{total}} &= \overrightarrow{dT}_{mx} + \overrightarrow{dT}_{my} + \overrightarrow{dT}_{mz} = I(d\ell dw)\vec{a}_z \times \vec{B}_x + I(d\ell dw)\vec{a}_z \times \vec{B}_y + 0 \\ &= I(dS)\vec{a}_z \times (\vec{B}_x + \vec{B}_y) = I(dS)\vec{a}_z \times \vec{B}\end{aligned} \tag{9.9}$$

since the inclusion of B_z does not change the outcome of the expression since $\vec{a}_z \times \vec{B}_z = 0$. Thus, the final expression for the torque on the current loop increment is

$$\overrightarrow{dT}_{m,\text{total}} = \overrightarrow{IdS} \times \vec{B} \tag{9.10}$$

where the area vector, $\overrightarrow{dS}$, is taken along the same direction as the current, I, and hence, we will always group it with the current under the vector accent mark, $\overrightarrow{IdS}$.

Now, the total torque on the whole current loop would be the integration of the incremental contributions of the incremental loops, that is,

$$\vec{T}_{m,\text{loop}} = \int_{\text{loop}} d\vec{T}_{m,\text{increment}} = \iint_S \overrightarrow{IdS} \times \vec{B} \tag{9.11}$$

To summarize, electric current loops immersed in a magnetic field would experience either or both magnetic forces and magnetic torques. The net force on the loop can be obtained by integrating the incremental forces on the loop increments:

$$\vec{F}_{m,\text{loop}} = \int_{\text{loop}} d\vec{F}_{m,\text{increment}} = \oint_{\text{loop}} Id\vec{\ell} \times \vec{B} \tag{9.12}$$

while the net magnetic torque on the loop is as given by Equation (9.11).

Example 9.1 A dc current $I = 10$ A flows in a triangular loop in the XY plane as shown. Assuming a uniform magnetic flux density $\vec{B} = \vec{a}_y 0.5T$ in the region. Find the forces and torque on the loop. The dimensions are in cm.

Solution

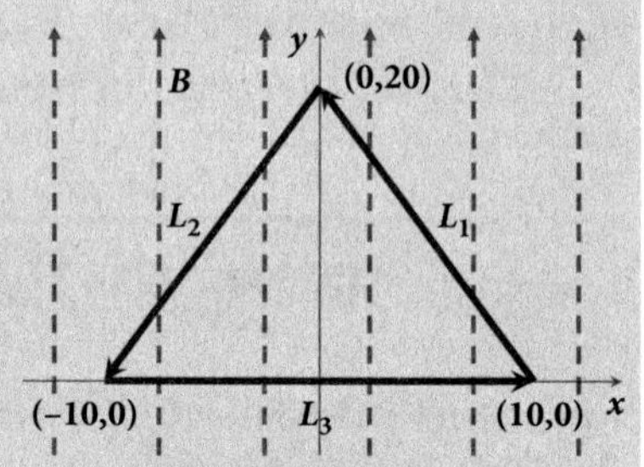

$$\vec{L}_1 = \vec{a}_x(0-0.1) + \vec{a}_y(0.2-0) = -\vec{a}_x 0.1 + \vec{a}_y 0.2$$

$$\vec{L}_2 = \vec{a}_x(-0.1-0) + \vec{a}_y(0-0.2) = -\vec{a}_x 0.1 - \vec{a}_y 0.2$$

$$\vec{L}_3 = \vec{a}_x(0.1+0.1) + \vec{a}_y(0-0) = \vec{a}_x 0.2$$

$$\vec{F}_1 = I\vec{L}_1 \times \vec{B} = 10(-\vec{a}_x 0.1 + \vec{a}_y 0.2) \times \vec{a}_y 0.5 = -\vec{a}_z 0.5N$$

$$\vec{F}_2 = I\vec{L}_2 \times \vec{B} = 10(-\vec{a}_x 0.1 - \vec{a}_y 0.2) \times \vec{a}_y 0.5 = -\vec{a}_z 0.5N$$

$$\vec{F}_3 = I\vec{L}_3 \times \vec{B} = 10(\vec{a}_x 0.2) \times \vec{a}_y 0.5 = \vec{a}_z 1N$$

notice $\vec{F}_1 + \vec{F}_2 + \vec{F}_3 = 0$

$$\vec{T} = \vec{m} \times \vec{B}$$

$$\vec{m} = \vec{a}_z IS = \vec{a}_z 10 \times \left(\frac{1}{2} \times 0.2 \times 0.2\right) = \vec{a}_z 0.2 \text{ Am}^2$$

$$\vec{T} = \vec{a}_z 0.2 \times \vec{a}_y 0.5 = -\vec{a}_x 0.1 \text{ Nm}$$

Example 9.2 Identify the forces and the torques (if any) on the 15 × 20 cm current loop shown. Calculate their numerical values. Given $I = 2.5$ A, $\vec{B} = 2\vec{a}_y T$. The current loop plane makes 45° with the x-z plane. The section of the loop parallel to the z axis is 15 cm.

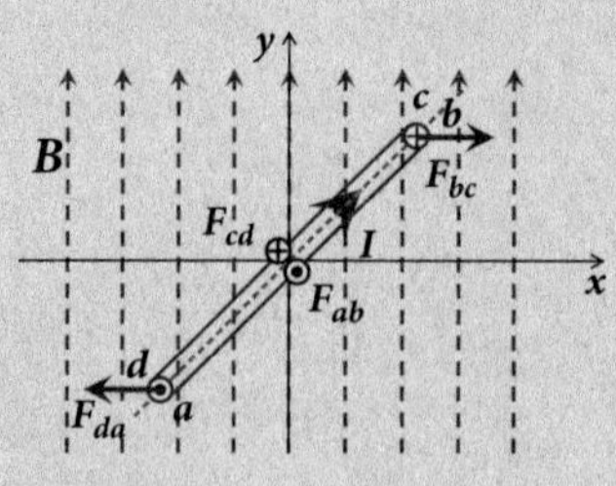

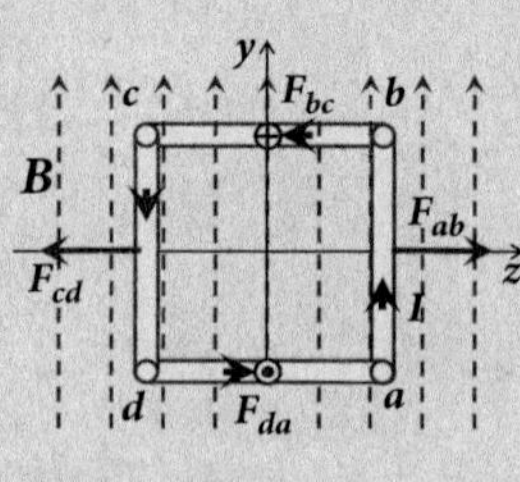

Solution Use $\vec{F} = I\vec{\ell} \times \vec{B}$ on the 4 sides and $\vec{T} = \overrightarrow{IS} \times \vec{B}$

Since the section of the loop $\vec{L}_{ab}$ makes 45° with the x-z plane, it can be expressed as

$$\vec{L}_{ab} = 0.2\left(\vec{a}_x \cos(45°) + \vec{a}_y \sin(45°)\right) = (0.2)\left(\frac{\vec{a}_x + \vec{a}_y}{\sqrt{2}}\right)$$

$$\vec{F}_{ab} = I\vec{L}_{ab} \times \vec{B} = (2.5)(0.2)\left(\frac{\vec{a}_x + \vec{a}_y}{\sqrt{2}}\right) \times (2)\vec{a}_y = 0.707\,\vec{a}_z N \text{ and } \vec{F}_{cd} = -0.707\vec{a}_z N$$

$$\vec{L}_{bc} = (0.15)(-\vec{a}_z) \rightarrow \vec{F}_{bc} = (2.5)(0.15)(-\vec{a}_z) \times (2)\vec{a}_y = 0.75\vec{a}_x N \text{ and } \vec{F}_{da} = -0.75\vec{a}_x N$$

The unit vector along the $\overrightarrow{IS}$ can be obtained from the cross product of the unit vector along $\vec{L}_{ab}$ and a unit vector along $\vec{L}_{bc}$

$$\vec{a}_{IS} = \left(\frac{\vec{a}_x + \vec{a}_y}{\sqrt{2}}\right) \times (-\vec{a}_z) = \left(\frac{-\vec{a}_x + \vec{a}_y}{\sqrt{2}}\right)$$

$$\vec{T} = (2.5)(0.2 \times 0.15)\left(\frac{-\vec{a}_x + \vec{a}_y}{\sqrt{2}}\right) \times (2)\vec{a}_y = -0.11\vec{a}_z \text{ Nm}$$

9.3 Energy Stored in Magnetic Field

From the discussion above, we realize the presence of forces and torques associated with the presence of magnetic fields. Consequently, it makes sense to expect the presence of magnetic energy storage. Unfortunately, the derivation of the magnetostatic energy-storage expressions is too complicated and is typically avoided in published introductory books and literature. Typical approaches for presenting this topic rely on familiar circuit theory relationships and Faraday's law that both fall in the domain of dynamics and not statics. In the following, we will use analogies to electrostatics to present the magnetostatic energy expressions. We must keep in mind that this presentation is not by any means a proof or a derivation of the expressions. The intent of this approach is to bring the magnetostatic energy expressions closer to our appreciation in a *simple* way that relies on our state of knowledge at this point in our studies.

To introduce the energy stored in a magnetic field, we will use the same definition we discussed in Chapter 5 for the energy in an electrostatic field. The definition was equating the energy in a system (of sources) as the work done to establish the system, or equally valid, the work released as the system disintegrates. The energy of the system of sources is in fact the energy stored in the fields generated by these sources and is housed within the media containing these sources.

Consequently, and likewise, the magnetic energy in a system of magnetic sources (electric currents) is the work done to construct the system or the work released as those sources separate infinite distances apart. Hence, similar to our derivation procedure in Chapter 5, we will sum up the work done by system's internal (magnetic) forces in releasing its current sources from to their positions to far away locations at infinity as a zero-energy reference location.

Recalling Figure 8.2, and starting with Ampere's law for the force on the current elements $(\overrightarrow{I\Delta\ell})_2$ due to the current element $(\overrightarrow{I\Delta\ell})_1$ with a separating distance R_{12}, we have

$$\overrightarrow{\Delta F}_{m2} = \mu_o \frac{(\overrightarrow{I\Delta\ell})_2 \times \left[(\overrightarrow{I\Delta\ell})_1 \times \vec{a}_{12}\right]}{4\pi R_{12}^2} \tag{9.13}$$

Now, let us choose to express this analysis in cylindrical coordinates where the vector $(\overrightarrow{I\Delta\ell})_1$ lines up with the z axis. Consequently, we express the current element $(\overrightarrow{I\Delta\ell})_2$ and the unit vector $\vec{a}_{12}$ in terms of their relative ρ, φ, and z components, hence

$$\begin{aligned}\overrightarrow{\Delta F}_{m2} &= \mu_o \frac{[(I\Delta\ell)_{2\rho}\vec{a}_\rho + (I\Delta\ell)_{2\varphi}\vec{a}_\varphi + (I\Delta\ell)_{2z}\vec{a}_z] \times [(I\Delta\ell)_1 \vec{a}_z \times (\sin\theta\,\vec{a}_\rho + \cos\theta\,\vec{a}_z)]}{4\pi R_{12}^2} \\ &= \mu_o \frac{[(I\Delta\ell)_{2\rho}\vec{a}_\rho + (I\Delta\ell)_{2\varphi}\vec{a}_\varphi + (I\Delta\ell)_{2z}\vec{a}_z] \times [(I\Delta\ell)_1 \sin\theta\,\vec{a}_\varphi]}{4\pi R_{12}^2} \\ &= \mu_o \frac{[(I\Delta\ell)_1 \sin\theta][(I\Delta\ell)_{2\rho}\vec{a}_z - (I\Delta\ell)_{2z}\vec{a}_\rho]}{4\pi R_{12}^2}\end{aligned} \tag{9.14}$$

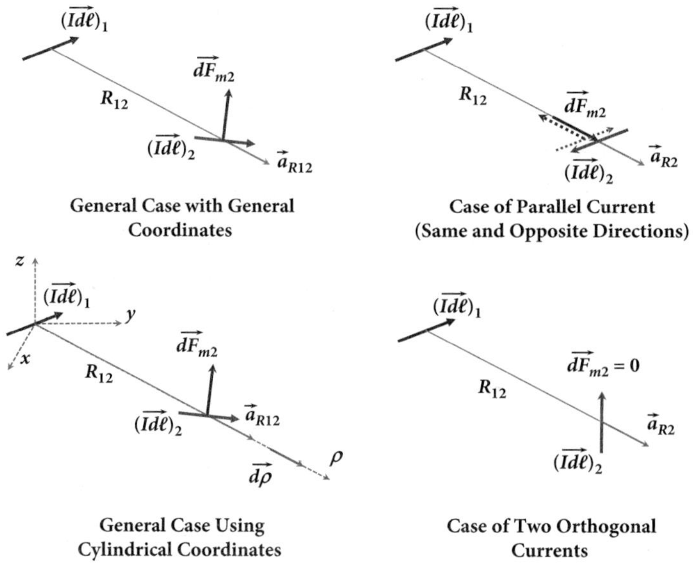

FIGURE 8.2 Ampere's law for magnetic force shown between two current elements. In the derivation of the energy storage expression, we used the cylindrical coordinate presentation as we evaluate the work done in moving $(Id\ell)_2$ from its location to infinity along the ρ coordinate.

Next, we write the expression for the work done as the current element $(\overline{I\Delta\ell})_2$ moves to position in order to evaluate the total magnetic energy in the system. Without loss of generality, and for the sake of convenience, we will choose the path of integration along the ρ coordinate, thus making the incremental distance moved $\overrightarrow{d\rho}$ (Figure 8.2).

$$\Delta W_{m,12} = \int d(\Delta W_{m12}) = \int_\rho^\infty \overrightarrow{\Delta F}_m \cdot \overrightarrow{d\rho} = \int_\rho^\infty \mu_o \frac{\{(I\Delta\ell)_1[\sin\theta]\}\left[(I\Delta\ell)_{2\rho}\vec{a}_z - (I\Delta\ell)_{2z}\vec{a}_\rho\right]}{4\pi R_{12}^2} \cdot \overrightarrow{d\rho}$$

$$= \int_\rho^\infty \mu_o \frac{\{(I\Delta\ell)_1[\rho/R_{12}]\}[-(I\Delta\ell)_{2z}d\rho]}{4\pi R_{12}^2} = -\frac{\mu_o(I\Delta\ell)_1(I\Delta\ell)_{2z}}{4\pi}\int_\rho^\infty \frac{\rho d\rho}{(\rho^2+z^2)^{\frac{3}{2}}}$$

$$= -\frac{\mu_o(I\Delta\ell)_1(I\Delta\ell)_{2z}}{4\pi}\left[\frac{-1}{\sqrt{\rho^2+z^2}}\right]_\rho^\infty = -\frac{\mu_o(I\Delta\ell)_1(I\Delta\ell)_{2z}}{4\pi R_{12}}$$

$$= -\frac{\mu_o\left[(\overline{I\Delta\ell})_2 \cdot (\overline{I\Delta\ell})_1\right]}{4\pi R_{12}} \tag{9.15}$$

Likewise, we can write the energy for three current elements as

$$\Delta W_{m,123} = -\frac{\mu_o\left[(\overline{I\Delta\ell})_1 \cdot (\overline{I\Delta\ell})_2\right]}{4\pi R_{12}} - \frac{\mu_o\left[(\overline{I\Delta\ell})_2 \cdot (\overline{I\Delta\ell})_3\right]}{4\pi R_{23}} - \frac{\mu_o\left[(\overline{I\Delta\ell})_3 \cdot (\overline{I\Delta\ell})_1\right]}{4\pi R_{31}} \tag{9.16}$$

and for multiple current elements

$$\Delta W_{m,1\ldots n} = -\sum_{j=i+1}^{n}\sum_{i=1}^{n}\frac{\mu_o\left[(\overline{I\Delta\ell})_j \cdot (\overline{I\Delta\ell})_i\right]}{4\pi R_{ij}} \quad \text{or} \tag{9.17}$$

$$\Delta W_{m,1\ldots n} = -\frac{1}{2}\sum_{j=1}^{n,\ j\neq i}\sum_{i=1}^{n}\frac{\mu_o\left[(\overline{I\Delta\ell})_j \cdot (\overline{I\Delta\ell})_i\right]}{4\pi R_{ij}} \tag{9.18}$$

where the ½ is included to compensate for the repetition of ij and ji terms.

Examining Equation (9.18) and comparing it to that of electrostatic energy storage, Equation (5.49)

$$W_{e,q_1\ldots q_n} = \frac{1}{2}\sum_{j=1}^{n,\,j\neq i}\sum_{i=1}^{n}\frac{q_j q_i}{4\pi\varepsilon_o R_{ij}} = \frac{1}{2}\sum_{j=1}^{n,\,j\neq i} q_j \sum_{i=1}^{n}\frac{q_i}{4\pi\varepsilon_o R_{ij}} \tag{5.49}$$

The two Equations (9.18) and (5.49) are quite similar to each other except for having a dot product of the vector sources in Equation (9.18) as well as a negative sign. This difference confirms that magnetic forces exist only between currents that are parallel to each other. Moreover, similar-direction currents attract and opposite-direction currents repel which is contrary to what happens with charges.

In fact, we can rewrite Equation (9.18) in terms of the only components of $(\overrightarrow{I\Delta\ell})_j$ that would survive the dot product with $(-\overrightarrow{I\Delta\ell})_i$ which would then become a scalar expression

$$\Delta W_{m,1\ldots n} = \frac{1}{2}\sum_{j=1}^{n,\,j\neq i}\sum_{i=1}^{n}\frac{\mu_o[(I\Delta\ell)_j]_{(-I\Delta\ell)_i}(I\Delta\ell)_i}{4\pi R_{ij}} = \frac{1}{2}\sum_{j=1}^{n,\,j\neq i}[(I\Delta\ell)_j]_{(-I\Delta\ell)_i}\sum_{i=1}^{n}\frac{\mu_o(I\Delta\ell)_i}{4\pi R_{ij}} \tag{9.19}$$

Now, we can see the full analogy between Equations (9.19) and (5.49), which we can use to "predict" the final expression of the magnetostatic energy storage as

$$W_m = \int_{\text{volume}}\frac{1}{2}\{\vec{B}\cdot\vec{H}\}dv = \int_{\text{volume}}\frac{1}{2}\mu_o H^2\,dv \tag{9.20}$$

similar to that obtained in Equation (5.77) in the electrostatic case

$$W_e = \int_{\text{volume}}\frac{1}{2}\{\vec{D}\cdot\vec{E}\}dv = \int_{\text{volume}}\frac{1}{2}\varepsilon_o E^2\,dv \tag{5.77}$$

As we stated earlier, proper proof and derivation of Equation (9.20) is not available using magnetostatic knowledge that we have covered at this point. However, we will visit this topic again in Chapter 10 to present an analytical justification for this expression using knowledge gained in that chapter.

Finally, based on Equation (9.20), we can write the volumetric density of the energy storage in magnetostatic field region as

$$w_m = \frac{1}{2}\{\vec{B}\cdot\vec{H}\} = \frac{1}{2}\mu_o H^2 = \frac{1}{2}\frac{B^2}{\mu_o} \tag{9.21}$$

Example 9.3 Find the magnetic energy stored in the coaxial cable shown per unit axial length. Given $a = 3$ mm, $b = 7$ mm, and the current $I = 1\times 10^{-3}\cos\omega t\,A$, with $\omega = 10$ kHz.

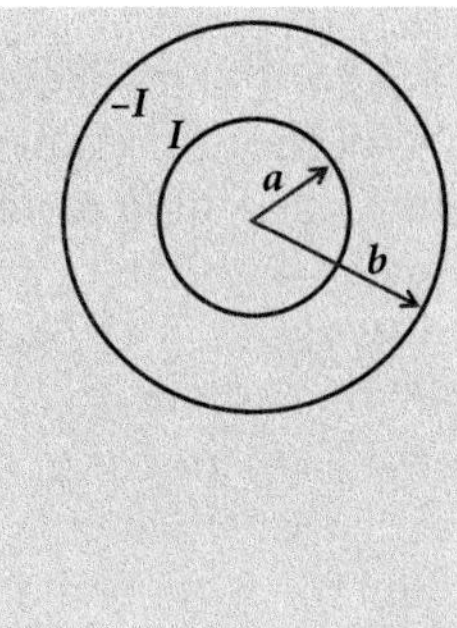

Solution Use

$$\vec{H} = \frac{I}{2\pi\rho}\vec{a}_\varphi \quad a\le\rho\le b,\ \vec{B} = \mu_o\vec{H},\ w = \frac{1}{2}\vec{H}\cdot\vec{B} = \frac{\mu_o}{2}\left(\frac{I}{2\pi\rho}\right)^2, \text{ and } W = \iiint w\,dv$$

$$W_m = \int_{z=0}^{z=1}\left\{\int_0^{2\pi}\left[\int_a^b w\rho\,d\rho\right]d\varphi\right\}dz = \int_{z=0}^{z=1}\left\{\int_0^{2\pi}\left[\int_a^b \frac{\mu_o}{2}\left(\frac{I}{2\pi\rho}\right)^2\rho\,d\rho\right]d\varphi\right\}dz$$

$$W_m = \frac{\mu_o I^2}{8\pi^2}\int_{z=0}^{z=1}\left\{\int_0^{2\pi}\left[\int_a^b \frac{1}{\rho}d\rho\right]d\varphi\right\}dz = \frac{\mu_o I^2}{8\pi^2}\ln\left(\frac{b}{a}\right)(2\pi)(1) = \frac{\mu_o I^2}{4\pi}\ln\left(\frac{b}{a}\right) = 8.47\times 10^{-14}\ \text{J/m}$$

9.4 Magnetic Properties of Materials

9.4.1 Introduction

By now, we know that the source of magnetic phenomena is the electric current, which is simply a charge in motion. We also know that matter contains charged particles some of which are in constant motion. Hence, we expect magnetic behavior of the different materials as a result of these microscopic/internal currents.

Since accurate presentation of this topic requires knowledge of quantum theory, we will resort to an oversimplified approach to explain the different magnetic behaviors of different materials. Although the following presentation contains some analytical components, it should be only viewed as a qualitative approach of explanation rather than an accurate analytical proof of the obtained results. Consequently, we will only limit our discussion in this chapter to the three main categories of magnetic behavior of materials, namely paramagnetism, diamagnetism, and ferromagnetism.

Both paramagnetic and diamagnetic responses are extremely weak and do not contribute to a serious alteration of the overall magnetization of the material. In fact, in most cases, we will treat both paramagnetic and diamagnetic materials as *nonmagnetic* materials, that is, free space. Moreover, both classes of materials do not have magnetic memory which means whenever an applied magnetic field is removed, the microscopic magnetic moments return to their original random orientations with zero net magnetization.

Ferromagnetism is a *form* of paramagnetism in which the material has a strong positive susceptibility to magnetization under the influence of an applied magnetic field. The magnetic susceptibility is so high that we can almost achieve full alignment of the material's dipole moments with strong applied fields.

9.4.2 Paramagnetism

From our physics background, we learned that there are two types of microscopic/internal electronic charge motions: one is the electronic orbital motion and the other is the electronic spin motion. Both contribute to microscopic magnetic dipole moments. Except for a material with permanent magnetic behavior, these internal dipole moments are in random distribution within the bulk of the material resulting in a null macroscopic result.

Under the influence of an external magnetic field, such a random distribution is altered giving rise to a resultant magnetization on the macroscopic level. In this regard, two response-phenomena are known to exist. The first one is when the magnetic torque caused by the applied magnetic field forces the magnetic dipole moments to rotate to some degree of alignment with the applied field. The second phenomenon is caused by the reduction of the orbiting speed of the electrons whose magnetic moments are affected by the applied field. These two phenomena result in two apposite magnetic responses. The net result would decide our categorization of the material as *paramagnetic* or *diamagnetic*. Paramagnetic materials have a positive net response that enhances the applied magnetic field while diamagnetic materials do the opposite.

9.4.3 Dipole Moments and Magnetization Vector

From the discussion above, we can model the magnetic behavior of paramagnetic materials by a corresponding distribution of magnetic dipole moments located in free space. Through this model, we can represent the material response to applied magnetic fields and the type of induced magnetization. Let us consider the illustrations given in Figures 9.6 and 9.7.

In Figure 9.6, we start with a representation depicting a random distribution of magnetic dipole moments in the absence of any applied magnetic fields. The next subfigure demonstrates the magnetic torques experienced by the dipole moments under the influence of an applied field. These torques would induce some alignment of the dipole moment vectors with the applied field depending on the strength of the field applied and the *susceptibility* of the material to this alignment. This is demonstrated in the third subfigure in the group.

Finally, the figure demonstrates the equivalent magnetization vector concept, which is defined as the resultant (vector sum) of the microscopic dipole moments in alignment with the applied field per unit volume.

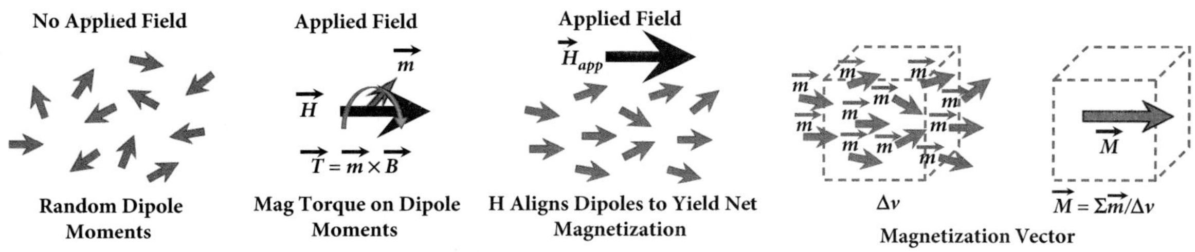

Figure 9.6 The magnetic dipole moments and the magnetization vector.

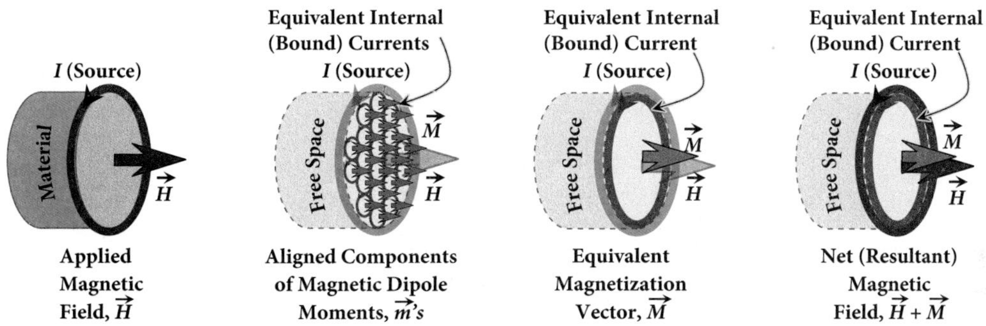

Figure 9.7 The magnetization vector is proportional to the applied magnetic field; see Equation (9.23).

On the macroscopic level, there will be a net induced magnetization vector density, which we define as

$$\vec{M} = \frac{\sum_{\Delta v} \vec{m}}{\Delta v} \tag{9.22}$$

Because of its relationship to the net dipole moment alignment, this magnetization vector density is also determined by the intensity of the applied magnetic field and the *susceptibility* of the material to *magnetization*. For linear materials, we expect a linear proportionality between the applied field and the magnetization vector. Hence, we write

$$\vec{M} = \chi_m \vec{H}_{\text{applied}} \tag{9.23}$$

where χ is a dimensionless constant parameter that we use to indicate the magnetic susceptibility of the material, that is, the material's positive response to magnetization.

Now, referring to Figure 9.7, we can say that $\vec{M}$ is the magnetic field contribution of the combined internal (bound) currents that are now reoriented as a result of the alignment of the magnetic dipole moments. As the figure suggests, we can attribute the magnetization field to an equivalent internal bound current that works side by side to the original current source that initiated the applied magnetic field in the first place. Hence, we can claim that the presence of the material media can be substituted by its equivalent magnetization field in free space. Hence, the total equivalent free space field would be given by

$$\vec{H}_{\text{free space equiv}} = \vec{H}_{\text{applied}} + \vec{M} \tag{9.24}$$

and the corresponding magnetic flux density expression would then become

$$\vec{B} = \mu_o \vec{H}_{\text{free space equiv}} = \mu_o\{\vec{H}_{\text{applied}} + \vec{M}\} = \mu_o\{\vec{H}_{\text{applied}} + \chi_m \vec{H}_{\text{applied}}\} = \mu_o\{1 + \chi_m\}\vec{H}_{\text{applied}} = \mu_o\{\mu_r\}\vec{H}_{\text{applied}} \tag{9.25}$$

which we can finally write as

$$\vec{B} = \mu\vec{H} = \mu_r\mu_o\vec{H} \tag{9.26}$$

This expression is analogous to the material permittivity expression that we developed for the electric field in Chapter 6. Basically, it replaces the free space μ_o by an unscripted μ for the permeability of the material. Moreover, the relative permeability term μ_r is defined as the ratio between the material permeability and that of free space.

$$\mu_r = \frac{\mu}{\mu_o} = 1 + \chi_m \tag{9.27}$$

Some of the known paramagnetic materials are aluminum, magnesium, tin, tungsten, and platinum. As we indicated earlier, the induced paramagnetic magnetization is extremely small, thus

$$\chi_{m,\text{paramagnetic}} > 0 \quad \text{and} \quad \left|\chi_{m,\text{paramagnetic}}\right| \ll 1 \tag{9.28}$$

and hence

$$\mu_{r,\text{paramagnetic}} = 1 + \chi_{m,\text{paramagnetic}} \approx 1 \quad \text{and} \quad \mu_{\text{paramagnetic}} \approx \mu_o \tag{9.29}$$

9.4.4 Diamagnetism

Now, we talk about diamagnetism. In the discussion above, we indicated that diamagnetism results when the net magnetic response of the orbiting electrons dominates over the net contribution of the electron spin. Under the influence of an applied magnetic field, orbiting speeds of the electrons are reduced resulting in an inverse net magnetization of the material. The *oversimplified* model that we used above for paramagnetism can still be used while recognizing that the magnetization vector $\vec{M}$ will be opposite to the applied field for diamagnetic materials. The expressions developed earlier, Equations (9.22) to (9.27) will still hold (algebraically). The only difference is that $\chi_{m,\text{diamagnetic}}$ is negative and $\mu_{r,\text{diamagnetic}}$ is slightly less than 1. Hence

$$\chi_{m,\text{diamagnetic}} < 0 \quad \text{and} \quad \left|\chi_{m,\text{diamagnetic}}\right| \ll 1 \tag{9.30}$$

$$\mu_{r,\text{diamagnetic}} = 1 + \chi_{m,\text{diamagnetic}} \approx 1, \quad \text{and} \quad \mu_{\text{diamagnetic}} \approx \mu_o \tag{9.31}$$

Some of the known diamagnetic materials are gold, silver, copper, lead, and carbon.

Among the strongest diamagnetic materials are bismuth and carbon graphite. They are about eight times stronger than silver. Other weaker diamagnetic materials include water, diamonds, and wood. The properties of materials table at the end of the book lists known values for magnetic properties of some materials. In addition, certain types of superconductors show perfect diamagnetism and they strongly repel a permanent magnet placed near them.

9.4.5 Ferromagnetism

As we stated earlier, ferromagnetism is a *form* of paramagnetism in which the material has a strong positive susceptibility to magnetization that almost full dipole alignment is achievable with strong applied fields. With close to full alignment, we observe a saturation behavior at which almost no further alignments are possible. Hence, the linear relationship [Equation (9.26)] is no longer valid; in other words, the material permeability is no longer a constant, and becomes a function of the applied magnetic field.

Another feature of a ferromagnetic material is its tendency (ability) to retain some of its induced magnetization after the removal of the applied field. This behavior of "residual" magnetization is known as *permanent* magnetism. This permanent (or more accurately long lasting) magnetism can be found in natural materials as well as can be artificially induced in ferromagnetic materials. Typically, two key phenomena lead to this strong magnetization response. One is the strong susceptibility of the spin magnetic dipole moments to the applied field. The other phenomenon is the presence of large numbers of atoms' dipole moments readily aligned in magnetic domains. These domains line up more readily under the influence of the applied magnetic field.

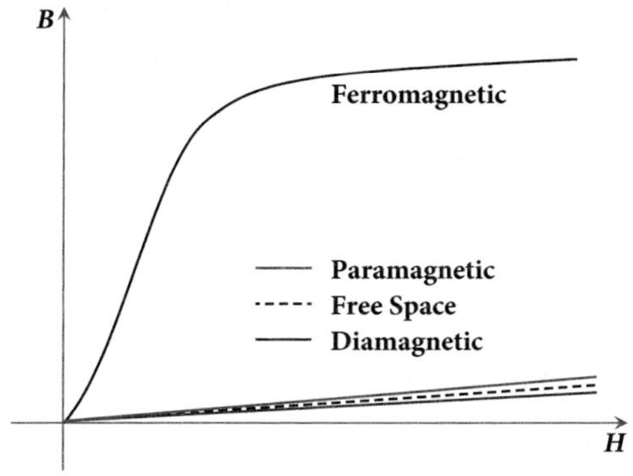

FIGURE 9.8 The *B-H* curve for the different magnetic classifications of materials.

There are only four ferromagnetic elements in the periodic table; they are iron, nickel, cobalt, and gadolinium. Other ferromagnetic materials are derived alloys containing one or more of these four elements.

Figure 9.8 shows typical magnetization performances of the three classes of material that we discussed earlier in the form of B versus H curves.

9.4.6 Residual Magnetism (Permanent Magnets)

Figure 9.9 demonstrates the residual magnetization behavior of ferromagnetic materials. The figure shows that as the material is magnetized through the application of an external magnetic field, the flux density B tends to saturate in line with the saturation of the magnetization density M. As we reduce the applied H field, some or most of the aligned dipole moments retain some or most of their alignment. With the complete removal of the applied field, a residual amount of magnetization remains in the material.

Upon the application of the magnetic field in the reverse direction, the internal dipole moments are now forced to reorient in the new field direction. With an increased field intensity, the dipole moments would tend to saturate in the opposite direction in alignment with the reversed field direction. At some point in the reversal process, there will be a zero net alignment as demonstrated in the figure. Cycling the field reversal results in what we call the "magnetic hysteresis" curve (Figure 9.10). Two forms of hysteresis curves are demonstrated in the figure, *hard* and *soft*, corresponding to larger and smaller residual magnetizations, respectively. Hard ferromagnetic materials are able to retain more of their long-lasting residual magnetization than the soft materials, and hence are more suitable for permanent magnet applications. An example of a hard ferromagnetic-material is steel while iron is considered a soft ferromagnetic-material.

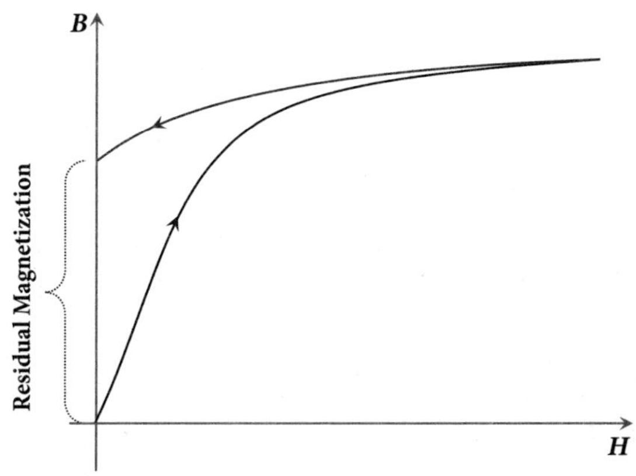

FIGURE 9.9 The *B-H* curve for a ferromagnetic material with residual magnetization.

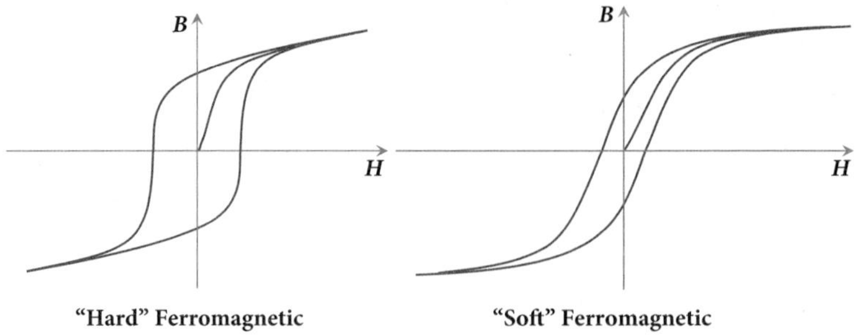

Figure 9.10 The hysteresis *B-H* curve for ferromagnetic materials with different degrees of ferromagnetic hardness.

Permanent magnets are typically used in many applications where a steady magnetic flux is needed. Examples are motors, generators, galvanometers, and magnetrons. However, for applications requiring variable magnetic flux, that can be controlled and switched on and off, we use extremely soft ferromagnetic materials. In this case, we use a current-carrying coil wrapped around the ferromagnetic core in a setup that we refer to as an electromagnet. Controlling the current in the coil results in a corresponding control of the induced magnetic flux.

In Addendum 9C, we discuss the use of ferromagnetic materials in magnetic pull and magnetic lift applications. Examples include circuits with both permanent magnets as well as electromagnets are given.

9.5 Magnetic Boundary Conditions

Similar to our discussion of boundary conditions in electrostatics, it is important to learn how the interface between two different materials affects the magnetic field intensity and magnetic flux density in the region. The procedure to follow is pretty much the same followed in Chapter 4. We will analyze the flux continuity/discontinuity through the application of Gauss' law to a small height closed cuboid shared between the two materials. On the other hand, we will analyze the field continuity/discontinuity through the application of Ampere's circuital law to a slim rectangular closed loop, again shared between the two materials.

Recognizing that the magnetic permeability is the only material parameter that will influence the continuities of the magnetic field and flux, we will categorize the interface materials into two main categories:

1. *Magnetic* materials where the relative permeability is greater (typically much greater) than 1. Examples include ferromagnetic and ferrimagnetic materials, for example, iron, nickel, and ferrites.
2. *Nonmagnetic* materials where the relative permeability is almost equal to 1. Examples include paramagnetic and diamagnetic material, for example, air, glass, aluminum, and copper.

With this "coarse" categorization, we can divide our study into three possible interfaces:

1. Interface between two magnetic materials
2. Interface between two nonmagnetic materials
3. Interface between a nonmagnetic and a magnetic material

In the following, we take these three cases one at a time.

9.5.1 Interface between Two Different Magnetic Materials

As we mentioned earlier, we will study the behavior at the interface through the two main laws of magnetics, Gauss' law and Ampere's circuital law. We start with the application of Gauss' law to a closed Gaussian surface as shown in Figure 9.11. The closed surface integration on this surface is to be equated to zero according to Equation (8.14).

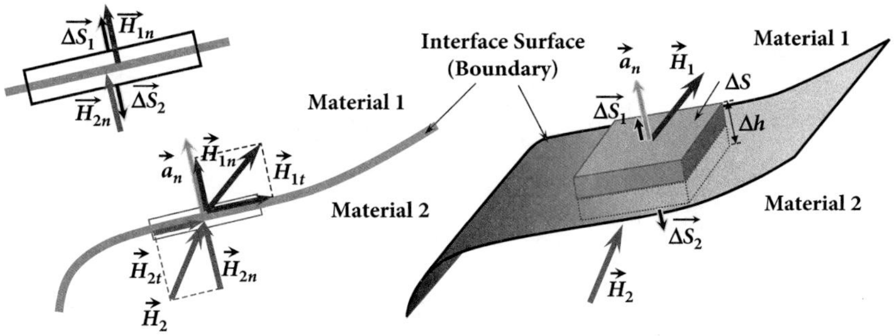

FIGURE 9.11 Deriving the magnetic flux density boundary conditions at the interface between two materials.

The chosen closed Gaussian surface is of the form of a cuboid (or any other upright cylindrical shape) of surface area ΔS and a height of Δh, part of which lies in material #1 and the rest lies in material #2. Obviously, the closed surface integration around this shape must be written in terms of three surface integrations: one on the top surface, another on the bottom surface, and a third one on the orthogonal sides. This latter integration (on the sides) will vanish as we take the limit of Δh to approach zero.

$$0=\lim_{\Delta h\to 0}\left\{\oiint_{\text{Cuboid}}\vec{B}\cdot\vec{dS}\right\}=\iint_{\text{Top surface}}\vec{B}_1\cdot\vec{dS}+\iint_{\text{Bottom surface}}\vec{B}_2\cdot\vec{dS}+\iint_{\text{Side surfaces}}\vec{B}\cdot\vec{dS}=B_{1n}\,\Delta S-B_{2n}\,\Delta S+0 \tag{9.32}$$

which leads to

$$B_{1n}=B_{2n} \tag{9.33}$$

What we have here in this result is that flux lines incident normal to the interface between two materials experience no change or discontinuity in their densities. We notice that we did not specify any information about the permeabilities of two materials at the interface while deriving this expression. The fact is that this result is always true for all material interfaces. It is logical to expect this result in the absence of what we might consider as a *single* magnetic charge or magnetic pole. In magnetism, all flux lines are always closed lines with no discontinuities or open ends.

As for the normal magnetic field intensity vectors, we can write

$$B_{1n}=\mu_1 H_{1n}=B_{2n}=\mu_2 H_{2n} \tag{9.34}$$

or

$$\mu_1 H_{1n}=\mu_2 H_{2n} \qquad \text{and hence} \qquad \frac{H_{2n}}{H_{1n}}=\frac{\mu_1}{\mu_2}=\frac{\mu_{r1}}{\mu_{r2}} \tag{9.35}$$

This result is telling us that that the normal magnetic field intensifies in the lower permeability material relative to its value in the higher permeability material. This finding is in line with the physical meaning of magnetic permeability. The higher μ region is more permeable to flowing the magnetic flux and hence requires relatively lower field intensity to do so.

Now, we turn our attention to the tangential components of the field and flux. For that, we will work with Ampere's circuital law and the closed loop integration demonstrated in Figure 9.12.

Similar to what we did in Chapter 6, the closed loop is assumed rectangular with length $\Delta\ell$ and height Δh. The loop height is shared between the two materials. As we apply the closed loop integration of Ampere's circuital law to the four sides of the rectangular shape, we get

$$\sum I_{\text{Linkage}}=\oint\vec{H}\cdot\vec{d\ell}=\int_a^b\vec{H}_1\cdot\vec{d\ell}+\int_b^c\vec{H}_1\cdot\vec{d\ell}+\int_c^d\vec{H}_1\cdot\vec{d\ell}+\int_d^e\vec{H}_2\cdot\vec{d\ell}+\int_e^f\vec{H}_2\cdot\vec{d\ell}+\int_f^a\vec{H}_2\cdot\vec{d\ell} \tag{9.36}$$

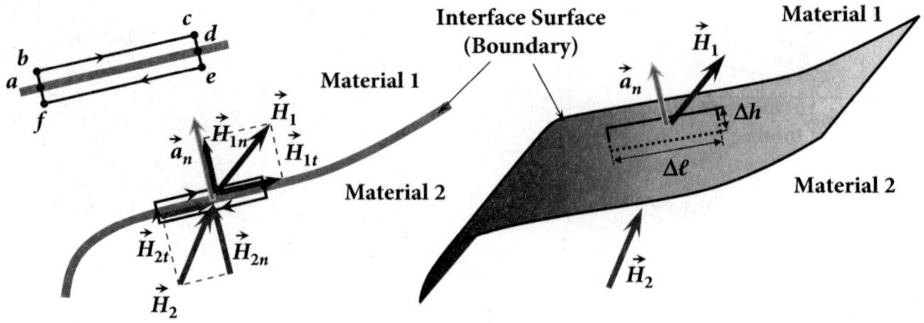

FIGURE 9.12 Deriving the magnetic field intensity boundary conditions at the interface between two materials.

However, in the limit as Δh goes to zero, the integrals *a-b*, *c-d*, *d-e*, and *f-a* will vanish, and the linkage current would be what possibly could be flowing on the interface surface leaving:

$$\sum I_{\text{Linkage}} = \Delta I_{\text{surface within}\,\Delta\ell} = \int_b^c \vec{H}_1 \cdot \vec{d\ell} + \int_e^f \vec{H}_2 \cdot \vec{d\ell} = [H_{1t} - H_{2t}]\Delta\ell \tag{9.37}$$

As we shrink $\Delta\ell$ to zero as well, we can write

$$[H_{1t} - H_{2t}] = \lim_{\Delta\ell \to 0} \frac{\Delta I}{\Delta\ell} = K \tag{9.38}$$

Hence, to summarize, at the interface between two magnetic materials:

$$[H_{1t} - H_{2t}] = K \quad \text{and} \quad B_{1n} = B_{2n} \tag{9.39}$$

And in the absence of any interface surface currents,

$$H_{1t} = H_{2t} \quad \text{and} \quad B_{1n} = B_{2n} \text{ which implies } \mu_1 H_{1n} = \mu_2 H_{2n} \tag{9.40}$$

Hence, in the case of no surface currents at the interface,

$$\frac{H_{1t}}{\mu_1 H_{1n}} = \frac{H_{2t}}{\mu_2 H_{2n}} \quad \text{or} \quad \frac{\tan\theta_1}{\mu_1} = \frac{\tan\theta_2}{\mu_2} \tag{9.41}$$

So, the angle of the *H* field vector (and hence *B*) changes as the field crosses the interface boundary into a different media (Figure 9.13). The angle of *refraction* depends on the relationship between the media permeabilities. In higher permeability media, the normal field component is smaller and the tangential field component dominates resulting into a larger angle of refraction.

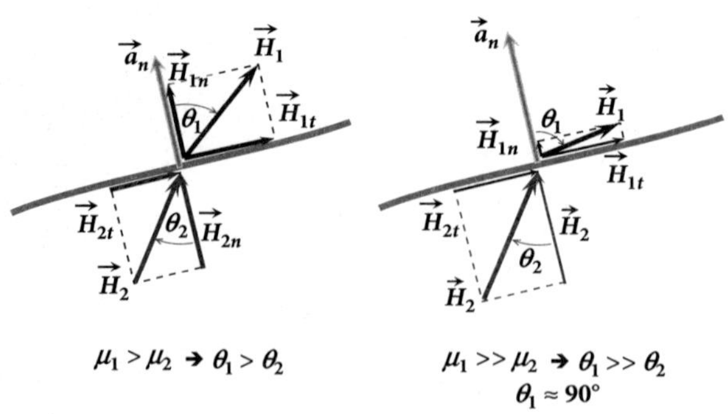

FIGURE 9.13 The magnetic boundary conditions at the interface between two materials in the absence of surface currents.

9.5.2 Interface between Two Nonmagnetic Materials (e.g., Paramagnetic/Diamagnetic with Paramagnetic/Diamagnetic)

To explore this interface, we simply replace both permeabilities with that of free space, as both materials would have relative permeabilities of almost 1. By doing do, we can reduce the results [Equations (9.39) to (9.41)] to

$$[H_{1t} - H_{2t}] = K \qquad \text{and} \qquad B_{1n} = B_{2n} \tag{9.39}$$

And in the absence of any interface surface currents,

$$H_{1t} = H_{2t} \text{ and } B_{1n} = B_{2n} \qquad \text{which implies} \qquad \mu_o H_{1n} = \mu_o H_{2n} \text{ or } H_{1n} = H_{2n} \tag{9.40}$$

Hence, in the case of no surface currents at the interface,

$$\frac{H_{1t}}{H_{1n}} = \frac{H_{2t}}{H_{2n}} \qquad \text{or} \qquad \theta_1 = \theta_2 \tag{9.42}$$

In other words, no change is expected across the boundary between two nonmagnetic materials except in the presence of surface currents. In the presence of surface currents, only the tangential components are expected to differ and this difference is owed to the presence of the current and not to any other differences in material properties.

9.5.3 Interface between Nonmagnetic and Magnetic Materials (e.g., Ferromagnetic with Paramagnetic/Diamagnetic)

For this interface, we assume material #2 to be the nonmagnetic material with relative permeability of almost 1, while material #1 is the magnetic material with relatively higher permeability. In other words, $\mu_{r1} >> \mu_{r2}$. Making the substitution in the results [Equation (9.41)], we conclude

$$\tan\theta_1 = \mu_{r1} \tan\theta_2 \tag{9.43}$$

9.5.4 Magnetic Flux Confinement in Magnetic Materials

With $\mu_{r1} >> 1$, we conclude that θ_1 would be fairly large approaching 90° except for the case of $\theta_2 = 0$. Based on that we can state the following:

- If the incidence from material #2 is normal to the interface between the two materials, that is, the incidence angle θ_2 is zero, the field continues in material #1 at the same angle of zero. The normal incident flux density does not change; however, the field intensity is significantly reduced by the ratio $H_{1n} = H_{2n}/\mu_{r1}$.
- If the incidence angle in material #2 is not zero, both normal and tangential components of the field and the flux are expected to exist. Again, the normal component of the field will be significantly reduced in material #1. On the other hand, the tangential component will stay the same resulting in a refraction angle close to 90°.
- The 90° refraction angle implies that the field and the flux continue within the confinement of the high-permeability material tangential to its boundaries. This feature is very useful in using high-permeability material in designing magnetic circuits. The high-permeability "core" material tends to confine (hug) the magnetic flux within its boundaries similar to what a hose does to the water flow and the conductors do to the electric current.
- To demonstrate this concept, we examine the presentation in Figure 9.14. The figure shows predicted flux maps for three different coil arrangements, a single turn coil, a short solenoid (multi-turn coil), and a relatively long solenoid. Two sets of flux maps are demonstrated, one set with the coils are located in air (or nonmagnetic material) media, and the other set where the coils are wrapped around a high-permeability core insert. As the figure demonstrates, the high-permeability core alters the flux map resulting in the confinement of most of the magnetic flux generated by the coil within the core boundaries.

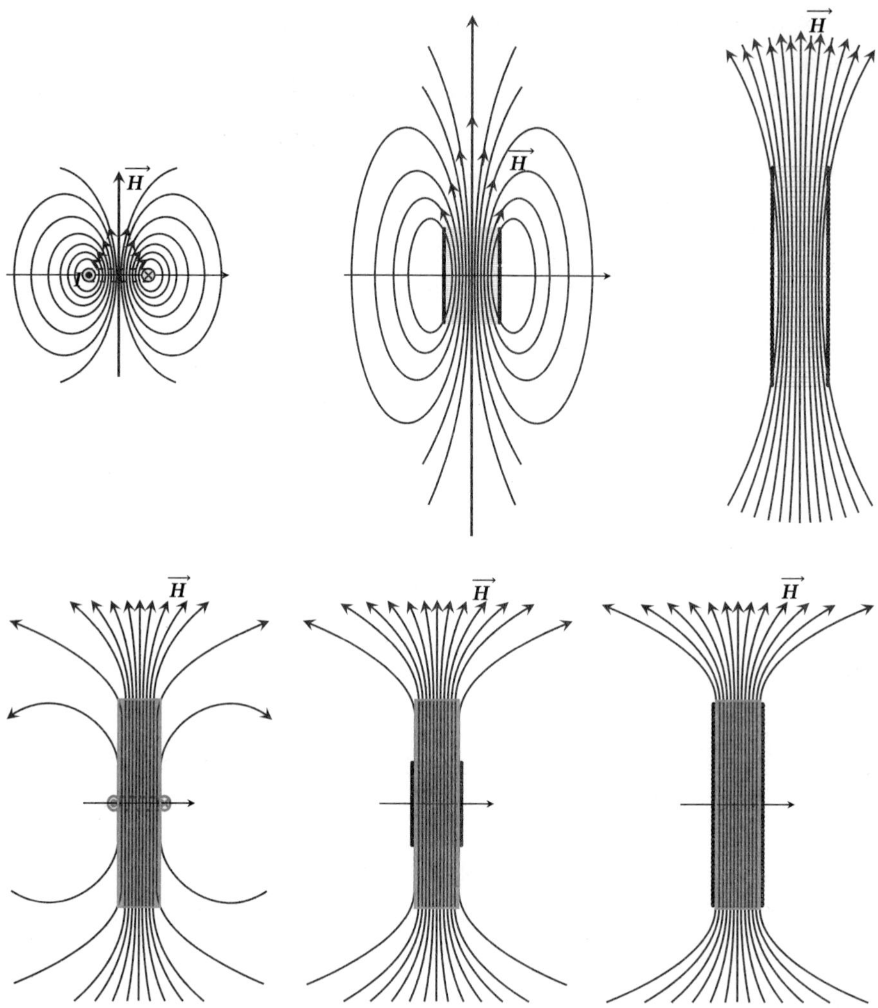

FIGURE 9.14 Magnetic flux confinement in ferromagnetic materials.

9.6 Magnetic Circuits

We use the term "magnetic circuits" to refer to devices that are designed to produce specific magnetic field distributions in specific regions of space. Such devices are common in many practical applications such as transformers, motors, generators, relays, and electromagnets. The two main ingredients of a magnetic circuit are a current-carrying coil (or coils) and a high-permeability "magnetic material" core (or cores). Typically, we use the high-permeability magnetic material(s) for part or all of the core material.

It is worth noting that we already covered two examples of magnetic circuits when we discussed the toroid and the solenoid. In both cases, we had a "one coil–one core" magnetic circuit. The only new information we can add in this section regarding both devices is the possibility of changing the core material to other than free space. In this case, and referring to the cases cited in Chapter VIII, the core field intensity and flux density take the form

$$H_c = nI \qquad \text{and} \qquad B_c = \mu_c H_c = \mu_c nI \tag{9.44}$$

where μ_c is the permeability of the core material.

In deriving the expressions for the core field intensity and flux density, we used the magnetostatic laws of Biot-Savart, Ampere's circuital, and Gauss. For most practical applications, the magnetic circuit geometries and current distributions may not lend themselves to the detailed analysis required by these tools. We need a much simpler and more practical analysis approach to deal with practical configurations such as a short nonuniform wound solenoid, a toroid with a noncircular shape, or a mixed material core coil. The answer is the use of the magnetic circuit analysis approach.

9.6.1 Magnetic Circuit Analysis Using the Electrical Circuit Analogy

The concept is to represent the magnetic circuit in a fashion analogous to that of the electric circuit. A simple DC circuit is composed of a DC voltage source, which acts as the electromotive force, emf, connected to one or more resistance(s) by means of high-conductivity conductors. The emf is responsible for pushing the current "flow" through the circuit conductors and resistances. In analogy, we represent the magnetic circuit with a "magnetomotive force, or *mmf*" which is a current-carrying coil as the source of the magnetic "flow or flux" into the magnetic circuit "reluctances" through high-permeability core material. The analogy is summarized as follows:

	Electric Circuit		Magnetic Circuit	
Source	Voltage source, emf	volts	*mmf*	amperes
Potential difference	Electric potential	volts	Scalar magnetic potential	amperes
Flow	Electric current, I	amperes	Magnetic flux, ψ	webers
Connectors	High conductivity conductors, internal resistance, R_c	ohms, or (Siemens)$^{-1}$	High-permeability core, internal reluctance, $\Re_C$	(henries)$^{-1}$
Load	Load resistance, R_L	ohms, or (Siemens)$^{-1}$	Load reluctance, $\Re_L$	(henries)$^{-1}$

Table 9.1 Analogy between Electric and Magnetic Circuits

9.6.2 Magnetic Reluctance

The magnetic "reluctance, *n*." is then a representation of the "reluctance, adj." of the magnetic circuit to the flow magnetic flux for a given *mmf*.

This concept of this analogy is depicted in Figure 9.15.

In this analogy, we will be able to use the likes of Ohm's law and both Kirchhoff's voltage and current loop and node equations to analyze magnetic circuit configurations. For example, as we are able to write Ohm's law for a simple DC circuit as

$$\{\text{Electric potential difference}\} = \{\text{Electric current}\} \times \{\text{Circuit resistance}\} \tag{6C.1}$$

our analogous magnetic circuit should then have a similar formula for its relationship

$$\{\text{Scalar magnetic potential difference}\} = \{\text{Magnetic flux}\} \times \{\text{Magnetic circuit reluctance}\} \tag{9.45}$$

To derive the elements of this analogy, we will follow an approach similar to what we used in our discussion of electric resistance in Addendum 6A of Chapter 6. Now, let us take the example of Figure 9.15 where an *mmf*

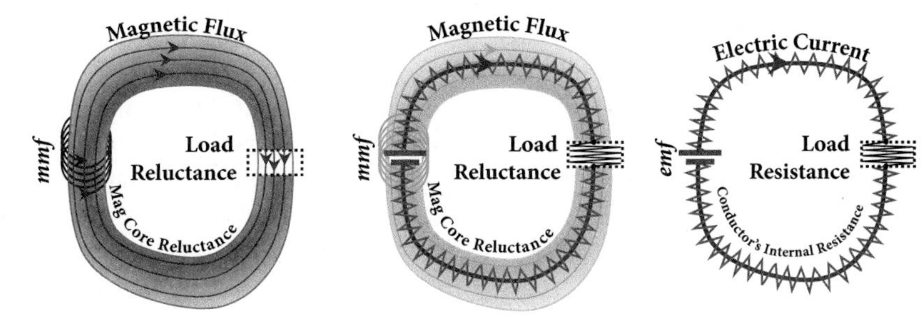

Figure 9.15 Using the analogy between magnetic and electric circuit to define the magnetic reluctance.

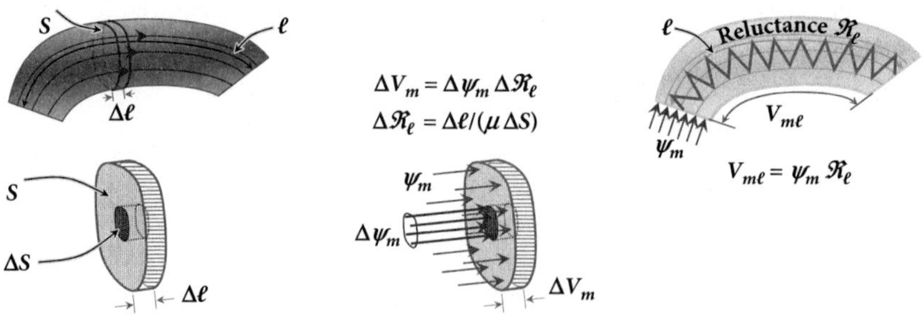

FIGURE 9.16 Using a segment of the magnetic circuit to demonstrate the analogous use of "Ohm's law" for magnetic circuit analysis.

(the current coil) drives a flux ψ_m into the magnetic circuit as shown. The "magnetic circuit" shown in the figure corresponds to a closed magnetic flux bundle that could very well be a portion of the bigger flux distribution produced by the *mmf* source. The boundaries of the magnetic circuit are defined by the confinement of the considered flux bundle.

Next, let us consider a segment of length ℓ of that magnetic circuit (Figure 9.16). This segment is chosen with both ends orthogonal to the magnetic flux flow; hence, they coincide with magnetic equipotential surfaces. The difference in scalar magnetic potential between the two ends "surfaces/terminals" of the segment is, according to Equation (8.64), given by

$$V_{m\ell} = \int_\ell dV_m = \int_\ell \vec{H} \cdot \vec{d\ell} \tag{9.46}$$

$$V_{m,\text{loop}} = \oint dV_m = \oint \vec{H} \cdot \vec{d\ell} = \sum\nolimits_{\text{loop}} I_{\text{Linkage}} = mmf \text{ of magnetic circuit}$$

The total magnetic potential for the complete magnetic circuit is

$$V_{m,\text{loop}} = \oint dV_m = \oint \vec{H} \cdot \vec{d\ell} = \sum\nolimits_{\text{loop}} I_{\text{Linkage}} = mmf \text{ of magnetic circuit} \tag{9.47}$$

Hence, $V_{m\ell}$ is a portion of the *mmf* responsible for driving the magnetic flux ψ_m into the segment of length ℓ. Thus, the relationship

$$V_{m\ell} = (mmf)_\ell = \psi_m R_\ell \tag{9.48}$$

where R_ℓ is the reluctance of the magnetic circuit segment.

Meanwhile, the flux over the cross section can be expressed in terms of the flux density distribution as follows:

$$\psi_m = \iint_S d\psi_m = \iint_S \vec{B} \cdot \vec{dS} \tag{9.49}$$

In general, the cross section S of the flux bundle varies from one location to another; however, the total flux is the same as per our definition of the chosen segment.

Now, let us consider an incremental volume within the chosen segment. The increment is to have a length $\Delta\ell$ along the flux lines and a cross-sectional area, ΔS, orthogonal to the flux lines. Hence, the two end surfaces of this increment coincide with magnetic equipotential surfaces with a difference in magnetic potential given by

$$\Delta V_m = \vec{H} \cdot \vec{\Delta\ell} = H\Delta\ell \tag{9.50}$$

The incremental flux flowing within the increment is

$$\Delta\psi_m = \vec{B} \cdot \vec{\Delta S} = B\Delta S = \mu H \Delta S \tag{9.51}$$

Combining the above two relationships [Equations (9.50) and (9.51)], we find

$$\Delta V_m = H\Delta\ell = \Delta\psi_m \frac{\Delta\ell}{\mu\Delta S} = \Delta\psi_m \Delta\Re \tag{9.52}$$

where

$$\Delta\Re = \frac{\Delta\ell}{\mu\Delta S} = \frac{\Delta V_m}{\Delta\psi_m} \text{ is the reluctance of the increment of the magnetic circuit} \tag{9.53}$$

This expression is quite similar to that of electrical resistance as was discussed in Chapter 6.

$$\Delta R = \frac{\Delta\ell}{\sigma\Delta S} = \frac{\Delta V}{\Delta I} \tag{6A.1}$$

Now, we seek the reluctance expression for the whole segment of length ℓ. So, we relate $V_{m\ell}$ to ψ_m. We can do that in two possible ways; write either $V_{m\ell}$ in terms of ψ_m or the opposite. Let us do the opposite first by relating ψ_m to ΔV_m of the magnetic circuit slice of length $\Delta\ell$.

$$\psi_m = \iint_S d\psi_m = \iint_S \mu H\, dS = \iint_S \mu \frac{\Delta V_m}{\Delta\ell} dS = \Delta V_m \iint_S \frac{\mu dS}{\Delta\ell} \tag{9.54}$$

Next, we integrate the incremental scalar magnetic potential for the whole length by adding ΔV_m for all the slices,

$$V_{m\ell} = \int_\ell dV_m = \int_\ell \frac{\psi_m}{\iint_S \frac{\mu dS}{d\ell}} = \psi_m \int_\ell \frac{1}{\iint_S \frac{\mu dS}{d\ell}} = \psi_m \Re_\ell \tag{9.55}$$

Now, we can see the relationship between the total scalar magnetic potential for the segment and the flux flow along this length. Hence, the total reluctance of the segment in consideration is

$$\Re_\ell = \frac{V_{m\ell}}{\psi} = \int_\ell \frac{1}{\iint_S \frac{\mu dS}{d\ell}} \tag{9.56}$$

An alternate approach is to start by integrating magnetic potential increments along ℓ, then integrate the flux over S. The resulting expression in this case is

$$V_{m\ell} = \int_\ell dV_m = \int_\ell \vec{H}\cdot\vec{d\ell} = \int_\ell H\, d\ell = \int_\ell \frac{B}{\mu} d\ell = \int_\ell \frac{B\Delta S}{\mu\Delta S} d\ell = \int_\ell \frac{\Delta\psi_m}{\mu\Delta S} d\ell = \Delta\psi_m \int_\ell \frac{d\ell}{\mu\Delta S} \tag{9.57}$$

$$\psi_m = \iint_S d\psi_m = \iint_S \frac{V_{m\ell}}{\int_\ell \frac{d\ell}{\mu dS}} = V_{m\ell} \iint_S \frac{1}{\int_\ell \frac{d\ell}{\mu dS}} = \frac{V_{m\ell}}{\Re_\ell} \tag{9.58}$$

$$\Re_\ell = \frac{V_{m\ell}}{\psi} = \frac{1}{\iint_S \frac{1}{\int_\ell \frac{d\ell}{\mu dS}}} \tag{9.59}$$

If we were to consider the complete closed flux bundle, then we would talk about $V_{m,\text{loop}}$ for the closed loop that would be given by

$$V_{m,\text{loop}} = \oint_\ell dV_m = \oint_\ell \vec{H}\cdot\vec{d\ell} = \sum_\ell I_{\text{Linkage}} = (mmf)_{\text{magnetic circuit}} \tag{9.60}$$

and, according to Equations (9.56) and (9.59), the total reluctance of the magnetic circuit in this case would be

$$\Re_{\ell,\text{loop}} = \frac{V_{m,\text{loop}}}{\psi} = \oint \frac{1}{\iint_S \frac{\mu dS}{d\ell}} \text{ or } = \frac{1}{\iint_S \frac{1}{\oint \frac{d\ell}{\mu dS}}} \tag{9.61}$$

In the following, we give two examples of magnetic circuit analysis using the electrical analogy approach. More examples of the same are given in Addendum 9A at the end of this chapter.

9.6.3 Examples of Magnetic Circuit Analysis Using the Electrical Circuit Analogy

We start with an example of a magnetic material core with an air gap. This typical configuration is used in applications where an air gap is needed to house a device needed to be subjected to a magnetic field in order to produce a specific effect. Examples include galvanometers, motors and generators, cathode ray tubes, and magnetrons.

The magnetic field is pushed into the core and air gap by a current-carrying coil (*mmf*). The high-permeability magnetic material core confines the magnetic flux within its boundaries, thus delivering almost all the coil's magnetic flux to flow through the air gap. The magnetic core simply acts like the conductor that delivers the electrical source potential to the load of the electric circuit.

Because of the relatively high permeability of the magnetic core (relative to its air/nonmagnetic surrounding), the magnetic field and flux within the core maintain its tangential flow within the core material. As the core curves and even bends, with high enough permeability, the field and flux stay confined to the core boundaries. Only when the terminal edges of the core meet the magnetic circuit air gap, that the magnetic field and flux become normal to the interface. The result is that the normal flux will continue through the air gap at approximately the same density. The reason we say approximately rather than exactly is the presence of fringing that takes place in the air gap. The larger the air gap the more pronounced the fringing would be. Because of fringing, the same flux leaving the cross section of the magnetic core is now distributed over a slightly larger *effective* cross-sectional area in the gap.

A simple configuration for this class of magnetic circuit is given in Figure 9.17. The figure also demonstrates the magnetic flux fringing in the air gap. The magnetic circuit "electrical-equivalent" model is given in Figure 9.18. In this figure, the current-carrying coil is replaced by its $mmf = NI$, the magnetic material section of the circuit is replaced by its reluctance

$$\Re_c = \int_\ell \frac{1}{\iint_S \frac{\mu dS}{d\ell}} = \int_\ell \frac{1}{\frac{\mu S}{d\ell}} = \int_\ell \frac{d\ell}{\mu S} = \frac{\ell_c}{\mu_c S_c} \tag{9.62}$$

The above integrations are simple ones because of the uniformity of the cross-sectional area for the entire magnetic core length.

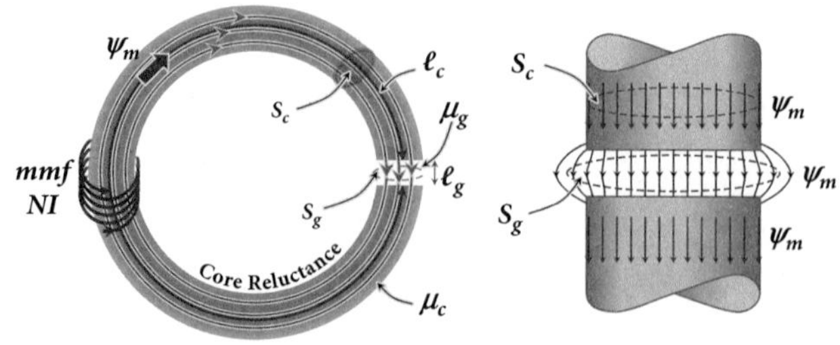

FIGURE 9.17 Typical use of magnetic circuits to produce a magnetic field in an air gap. The figure on the right demonstrates the magnetic flux fringing in the air gap.

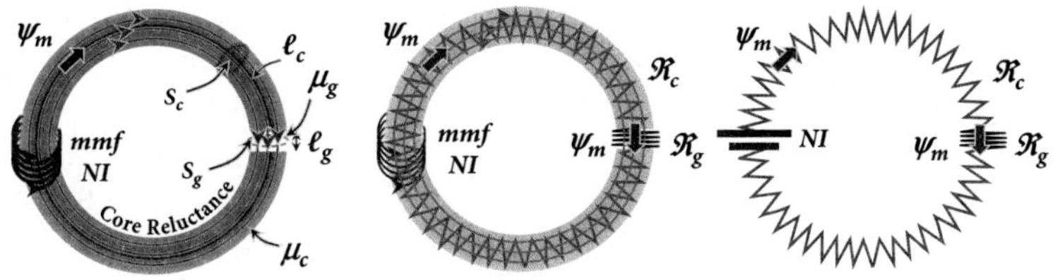

FIGURE 9.18 The magnetic circuit "electrical-equivalent" model.

Similarly, the reluctance of the air gap is

$$\mathfrak{R}_g = \frac{\ell_g}{\mu_g S_g} \tag{9.63}$$

Using the analog circuit model on the right side of Figure 9.18, we can write an expression for the total flux in the magnetic circuit as

$$\psi_m = \frac{NI}{\mathfrak{R}_c + \mathfrak{R}_g} \tag{9.64}$$

With ψ_m known, we can proceed to write "approximate" expressions for the flux densities and field intensities in the core and the gap

$$B_c = \frac{\psi_m}{S_c}, \quad B_g = \frac{\psi_m}{S_g}, \quad H_c = \frac{B_c}{\mu_c}, \quad \text{and} \quad H_g = \frac{B_g}{\mu_g} \tag{9.65}$$

It is important to recognize that these solutions are approximate and are sort of average values which is adequate for many practical applications. Accurate solutions were derived earlier for a few cases and revealed that you do not always get a uniform field and flux density solution over a cross section.

It is interesting to compare the reluctance of the core segment to that of the air gap given the typically huge difference in their lengths. With close cross-sectional areas, the reluctance ratio is given by

$$\frac{\mathfrak{R}_c}{\mathfrak{R}_g} = \frac{\ell_c}{\mu_c S_c} \cdot \frac{\mu_g S_g}{\ell_g} = \frac{\ell_c/\mu_{rc}}{\ell_g} \cdot \frac{S_g}{S_c} \approx \frac{\ell_c/\mu_{rc}}{\ell_g} \tag{9.66}$$

The relatively large permeability of the core segment counters its relatively long length and brings its reluctance down. In good designs, the core should serve as the "connector" that connects the source *mmf* to the load reluctance. Hence, the reluctance of the core section should be minimized compared to that of the air gap. Consequently, the core permeability needs to be chosen adequately high to reduce the core reluctance making the air gap's reluctance the dominant value in the magnetic circuit.

The second example in this section is typical of transformers where you would have two or more coils in one magnetic circuit. The example is shown in Figure 9.19 to have a magnetic circuit with three core branches each having its own linkage coil. We assume for simplicity that each branch has a constant cross-sectional area throughout its length.

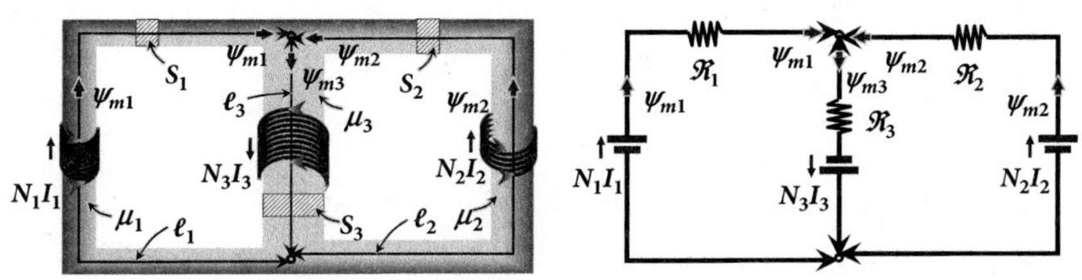

FIGURE 9.19 An example of a two-loop magnetic circuit and its "electrical-equivalent" model (schematic).

The analysis procedure in this example is pretty much the same as in the previous one. We divide the magnetic circuit into segments that can be simply modeled as reluctance elements in the analogous circuit model. We choose sections of the same cross-sectional areas and same permeability and model each as one reluctance.

Referring to the figure, we recognize the following relationships for the branch reluctances

$$\mathfrak{R}_1 = \frac{\ell_1}{\mu_1 S_1}, \quad \mathfrak{R}_2 = \frac{\ell_2}{\mu_2 S_2}, \quad \text{and} \quad \mathfrak{R}_3 = \frac{\ell_3}{\mu_3 S_3} \tag{9.67}$$

We also notice that each current-carrying coil is modeled with the appropriate mmf value in the appropriate polarity. Now, we can carry out the circuit analysis of the analogous electrical circuit using well-known circuit analysis approach based on KVL, KCL, and Ohm's law, for example,

$$\text{KCL: } \psi_{m3} = \psi_{m1} + \psi_{m2} \tag{9.68}$$

$$\text{KVL: } N_1 I_1 - \psi_{m1}\mathfrak{R}_1 = -N_3 I_3 + \psi_{m3}\mathfrak{R}_3 = N_2 I_2 - \psi_{m2}\mathfrak{R}_2 \tag{9.69}$$

These are three equations in three unknowns, ψ_{m1}, ψ_{m2}, and ψ_{m3}. With solutions for all the ψ terms on hand, we can proceed to find the average flux densities and field intensities in each branch of the magnetic circuit.

Example 9.4 Find B, H, and W in each of the segments of the magnetic circuit shown. Also, find the magnetic force in the air gap (between the two faces of the magnetic circuit core). Given: $N = 500$ T, $I = 1$ A, $\ell_1 = 12$ cm, $\ell_2 = \ell_5 = 8$ cm, $\ell_{gap} = 4$ mm, $\ell_3 = \ell_4 = 5.8$ cm, $S_1 = S_2 = S_3 = S_4 = S_5 = 1.5$ cm^2, and $\mu_{rc} = 750$. Consider a fringing factor of 120% (the ratio between the effective area in the gap and that of the core).

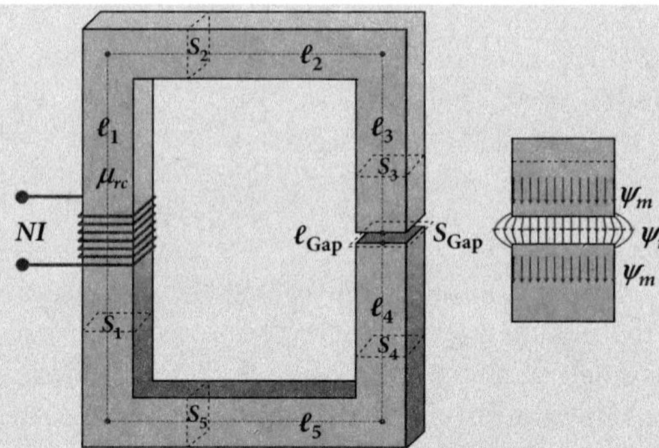

Solution 1 Form a total of 13 eqns. in 13 unknowns: {one loop eqn., 6 H eq ns., and 6 B eqns.}

$$\{\text{Ampere's circuital law: } \sum_{\text{loop}} H\ell = NI\} \;\&\; \{H_i|_{6\text{eqns}} = B_i/\mu_i\} \;\&\; \{B_i|_{6\text{eqns}} = \psi_m/S_i\}$$

$$\text{Ampere's circuital law: } H_1\ell_1 + H_2\ell_2 + H_3\ell_3 + H_4\ell_4 + H_5\ell_5 + H_g\ell_g = NI$$

Solve 13 simultaneous eqns. → Easy linear eqns. but require too much book keeping.

Solution 2 Calculate the reluctances for all segments: $\mathfrak{R}_i = \ell_i/m_i S_i$, $\mathfrak{R}_t = \sum_{\text{loop}} \mathfrak{R}_i$, $\psi_m = NI/\mathfrak{R}_t$

Use $\{H_i|_{6\text{eqns}} = B_i/\mu_i\}$ & $\{B_i|_{6\text{eqns}} = \psi_m/S_i\}$ to get all H's and B's

$$w_{mi} = 0.5 H_i B_i, \qquad W_{mi} = w_{mi}\text{vol}_i = w_{mi}\ell_i S_i, \qquad \text{and} \qquad F_g = w_{mg} S_g \tag{9.140}$$

	1	2 & 5	3 & 4	gap		1	2 & 5	3 & 4	gap
ℓ, m	0.12	0.08	0.058	0.004	B, Wb/m^2	162.7E–3	162.7E–3	162.7E–3	135.6E–3
S, m^2	1.5E–4	1.5E–4	1.5E–4	1.8E–4	H, A/m	172.7	172.7	172.7	107.9E+3
μ, H/m	942.5E–6	942.5E–6	942.5E–6	1.3E–6	w, J/m^3	14.0	14.0	14.0	7.3E+3
$\mathfrak{R}$, H^{-1}	848.8E+3	565.9E+3	410.3E+3	17.7E+6	W, J	252.8E–6	168.6E–6	122.2E–6	5.3E–3
$\mathfrak{R}_t$, H^{-1}	20.5E+6				F_g, N				1.3
ψ_m, Wb	24.4E–6								

9.7 Self- and Mutual Inductance

We defined resistance and capacitance before in Chapter 6. Resistance was inversely related to the ability of an electric potential to flow a current in a specific material of given configuration. Capacitance was an expression of the charge build up at the terminals of the given material configuration due to an applied electric potential. Now, it is time to define the third passive circuit element, the inductance. Simply put, inductance is the circuit element expressing the magnetic behavior of electric current flow. It is an indication of the magnetic "flux-linkage" resulting from flowing an electric current in a given conductor configuration. We will use the term "coil" to refer to the conductor configuration throughout this discussion since current only flows in a closed circuit (loop, or coil).

9.7.1 Flux-Linkage

In the statement above, we used the term "flux-linkage" instead of flux. This new term indicates the extent of interaction between the magnetic flux and the coil. A flux linking a multi-turn coil will have more impact on the coil circuit as compared to that of a single turn coil. We define the flux-linkage as the product of the flux and the number of times it links the coil and will use the uppercase Greek letter lambda, Λ, for its notation. Figure 9.20 demonstrates this concept.

In the example shown in the figure, two coils link the same flux ψ_m. Coil #1 has N_1 turns while coil #2 has N_2 turns. Each coil is assigned a direction of rotation based on the convention used for expressing the current flow direction in its circuit. This direction is shown in the figure with a black arrowhead. According to the rotation polarity of coil #1, it does have a positive linkage while that of coil #2 is negative. Hence,

$$\Lambda_1 = +N_1\psi_m \qquad \text{and} \qquad \Lambda_2 = -N_2\psi_m \tag{9.70}$$

It is relevant to state the linkage multiplier "N" is typically an integer greater than 1 as it represents the number of turns the coil wraps around the flux. However, at times the linkage multiplier could be fractional in cases where the flux lines only link a fraction of the current. We will experience a case of this nature later in this chapter when we deal with the internal inductance of a thick conductor.

9.7.2 Self- and Mutual Inductance

With the above definition of flux linkage, we define the (self-) inductance of a coil as

$$L_{\text{coil}} = \Lambda_{\text{coil due to own current}}/I_{\text{coil}} \tag{9.71}$$

we also define the (mutual) inductance between two coils as

$$L_{\text{mutual}} = \Lambda_{\text{coil due to other coil current}}/I_{\text{other coil}} \tag{9.72}$$

To explain, let us consider the two-coil arrangement shown in Figure 9.21. Let us further assume that a current I_1 flows in coil #1. The flux ψ_{11} produced by I_1 will all link coil #1, while a portion of that flux ψ_{21} links coil #2. Hence, the corresponding flux linkages Λ_{11} and Λ_{21} for both coils can be expressed as

$$\Lambda_{11} = N_1\psi_{m11} \qquad \text{and} \qquad \Lambda_{21} = N_2\psi_{m21} \tag{9.73}$$

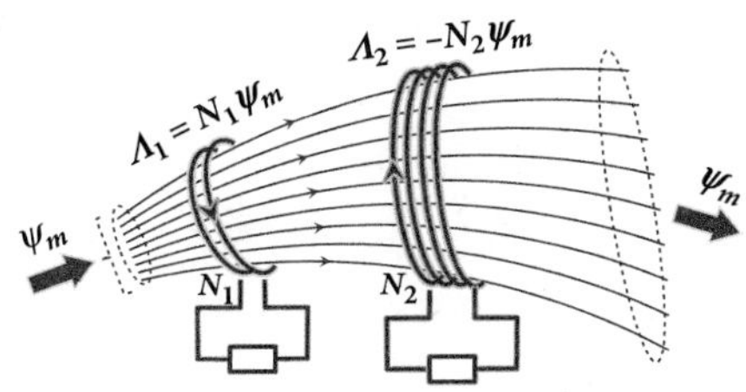

FIGURE 9.20 A demonstration to support the discussion of the flux-linkage concept.

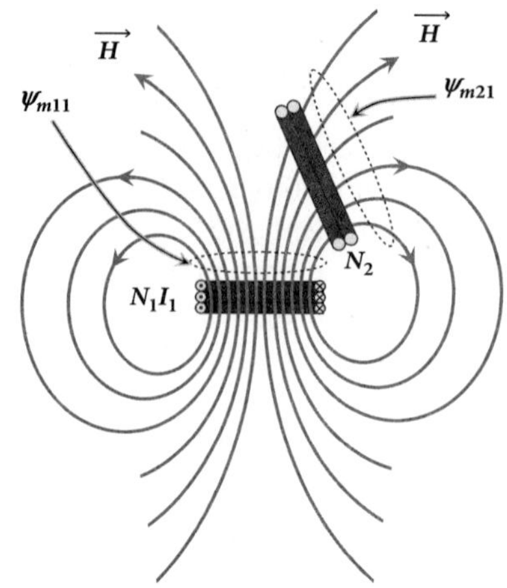

FIGURE 9.21 A two-coil demonstration to support the discussion of the concepts of self- and mutual inductances.

and hence, the expressions for self- and mutual inductances are given by

$$L_{11} = \frac{\Lambda_{11}}{I_1} = \frac{N_1 \psi_{m11}}{I_1} \quad \text{and} \quad L_{21} = \frac{\Lambda_{21}}{I_1} = \frac{N_2 \psi_{m21}}{I_1} \tag{9.74}$$

To generalize, let us consider the three-coil illustrative configuration of Figure 9.22 which we will use to extrapolate multiple coil relationships.

The figure shows three coils and their associated flux bundles. To simplify the figures, these flux bundles are shown in concept as shaded tube-like forms (the physical flux bundle do not have the sharp defined boundaries of the shown tubes). Taking advantage of the superposition concept for linear systems, each subfigure displays the flux bundles corresponding to one coil current at a time. The overall picture with all currents active would be the superposition of all flux bundles in the three subfigures. Two types of bundles are displayed in each subfigure, bundles that link the source coil alone and bundles that have linkage to each of the other two coils, for example, ψ_{m02} loops through coil #2 alone, while ψ_{m12} has common linkage to both coils #1 and #2. The reason for using the "0" subscript and not "1" for the flux bundle not linking other coils can be explained when we consider the total flux for the individual coils.

The total flux for each coil due to its own current is simply the algebraic sum of all flux bundles that flow through the specific coil's linkage area in its corresponding subfigure, for example,

$$\psi_{11} = \psi_{01} + \psi_{21} + \psi_{31} \tag{9.75}$$

or

$$\psi_{ii} = \psi_{0i} + \sum_{j=1,\ j\neq i}^{\text{all coils}} \psi_{ji} \tag{9.76}$$

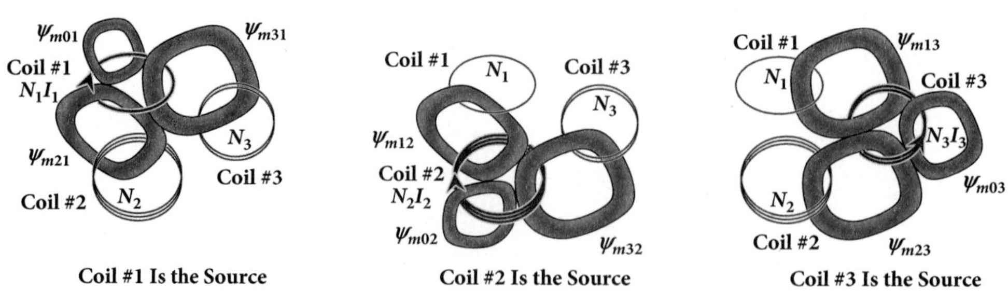

FIGURE 9.22 A three-coil demonstration to support the discussion of the concepts of self- and mutual inductances.

Hence, the subscript notations used are as follows:

- ψ_{ii} is the flux on coil #*i* due to its own current I_i (all the flux linking coil #*i*)
- ψ_{ji} is the flux on coil #*j* due to the current I_i (links both coils #*i* and #*j*)
- ψ_{0i} is the flux on coil #*i* due to its own current I_i (but does not link other coils)

Multiplying the coil flux by its number of turns, we can write the flux linkage expression as

$$\Lambda_{ii} = N_i \psi_{ii} = L_{ii} I_i \tag{9.77}$$

where L_{ii} is the self-inductance of coil #*i*.

Meanwhile, the flux-linkage with coil #2 due to the current in coil #1 is given by

$$\Lambda_{21} = N_2 \psi_{21} \tag{9.78}$$

or

$$\Lambda_{ij} = N_i \psi_{ij} = L_{ij} I_j \tag{9.79}$$

where L_{ji} is the mutual inductance between coils #*i* and #*j*.

In the general case, with all three coils active (carrying currents), the total flux-linkage expression would be the superposition given by

$$\Lambda_i = \sum_{j=1}^{\text{all coils}} \Lambda_{ij} = \sum_{j=1}^{\text{all coils}} L_{ij} I_j \tag{9.80}$$

So, for the three-coil arrangement, we can write

$$\Lambda_1 = L_{11} I_1 + L_{12} I_2 + L_{13} I_3 \tag{9.81a}$$

$$\Lambda_2 = L_{21} I_1 + L_{22} I_2 + L_{23} I_3 \tag{9.81b}$$

$$\Lambda_3 = L_{31} I_1 + L_{32} I_2 + L_{33} I_3 \tag{9.81c}$$

As we will show later in Addendum 9B, the mutual inductance between two coils is the same independent of which coil acts as the source, hence,

$$L_{ij} = L_{ji} \tag{9.82}$$

We often use the notation *M* instead of *L* to indicate mutual inductances in which case we use a single subscript on the *L* for self-inductance notation, for example,

$$\Lambda_1 = L_1 I_1 + M_{12} I_2 + M_{13} I_3 \tag{9.83a}$$

$$\Lambda_2 = M_{21} I_1 + L_2 I_2 + M_{23} I_3 \tag{9.83b}$$

$$\Lambda_3 = M_{31} I_1 + M_{32} I_2 + L_3 I_3 \tag{9.83c}$$

9.7.3 Inductance Relationship to Reluctance

These definitions for self- and mutual inductances can be related to the magnetic circuit reluctances as demonstrated in the two examples of Figure 9.23. Referring to the figure, we can write

$$\psi_{m11} = \frac{N_1 I_1}{\Re_{11}} \quad \text{and} \quad \psi_{m21} = \frac{N_1 I_1}{\Re_{21}} \tag{9.84}$$

where $\Re_{11}$ and $\Re_{21}$ are the reluctances experienced by the flux bundles ψ_{m11} and ψ_{m21}, respectively. Both flux bundles are shaded grey in the figure examples. Using Equations (9.77) and (9.79), we can conclude

$$L_{11} = \frac{\Lambda_{11}}{I_1} = \frac{N_1 \psi_{m11}}{I_1} = \frac{N_1^2}{\Re_{11}} \quad \text{and} \quad L_{21} = \frac{\Lambda_{21}}{I_1} = \frac{N_2 \psi_{m21}}{I_1} = \frac{N_2 N_1}{\Re_{21}} \tag{9.85}$$

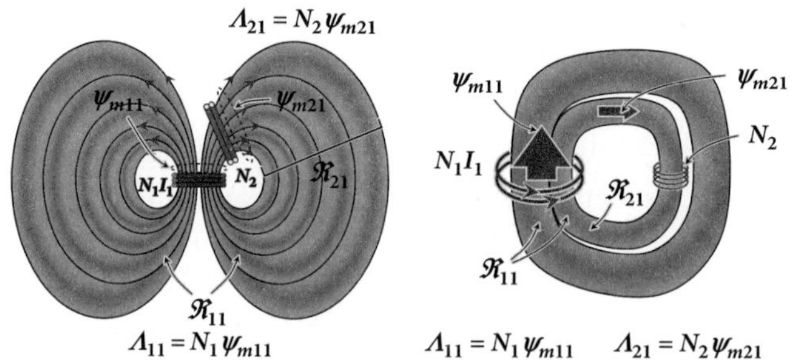

FIGURE 9.23 Two-coil demonstrations to support the discussion of the concepts of self- and mutual inductances.

9.7.4 Inductances as Circuit Elements

As we are aware, in direct current circuits where the current does not vary with time, no electrical potential develops across the terminals of an ideal inductor. The familiar $v = L\,di/dt$ expression will be the subject of our studies in the next chapter when we learn about time varying fields and Faraday's law. For the time being, we will defer the discussion of the time-varying current-voltage relationship until then.

For static magnetic fields, the inductor stores static energy and hence there is no associated power. As we will discuss later in detail that under dynamic situations the energy stored fluctuates with time but results in no average power dissipation (or generation).

9.7.5 Energy Stored in Inductance

According to Equation (9.20), the energy stored in an inductance is given by

$$W_m = \int_{\text{vol}} \frac{1}{2}\{\vec{B}\cdot\vec{H}\}dv \tag{9.20}$$

Now, referring to Figure 9.24, we can perform the integration in Equation (9.20) using an incremental volume dv chosen to have its length $\Delta\ell$ along the field (or flux) lines and its ΔS normal to the flux lines. Consequently, we can write

$$W_m = \int_{\text{vol}} \frac{1}{2}\{\vec{B}\cdot\vec{H}\}dv = \int_{\text{vol}} \frac{1}{2}\{BH\}dS\,d\ell = \frac{1}{2}\int_S [B\,dS]\oint[H\,d\ell] = \frac{1}{2}\psi_m NI = \frac{1}{2}\Lambda I \tag{9.86}$$

or

$$W_m = \frac{1}{2}\Lambda I = \frac{1}{2}LI^2 = \frac{1}{2}\Lambda^2/L \tag{9.87}$$

This energy is stored in the magnetic flux established in the inductance environment.

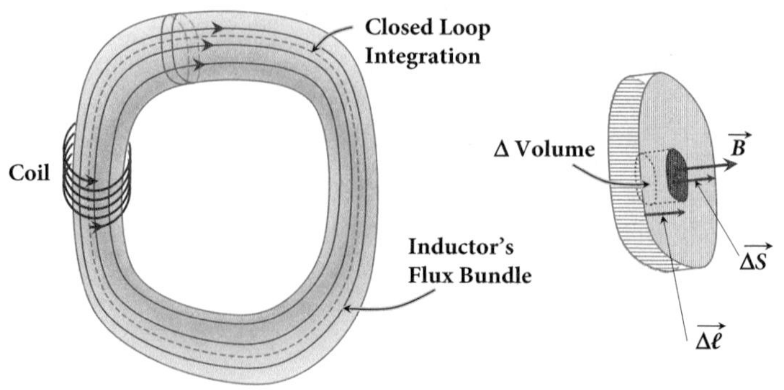

FIGURE 9.24 The energy stored in an inductance is the energy stored in its coil's magnetic flux linkage.

Example 9.5 Find the mutual inductance between two coplanar concentric circular-coils with N_1, radius R_1, and N_2 radius R_2, where $R_2 >> R_1$.

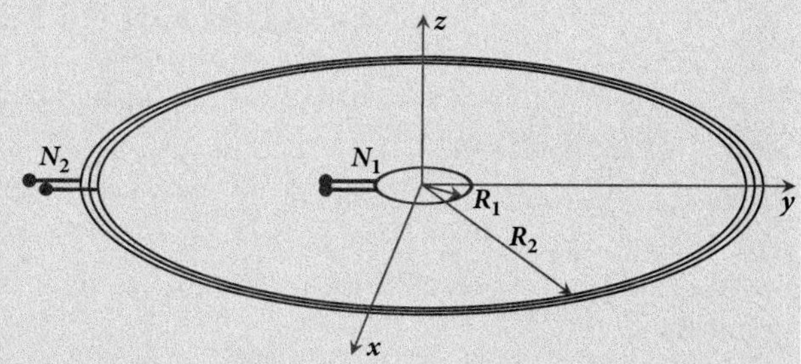

Solution Assume I_2 to flow in coil #2, using Equation (8.23), the magnetic field at the center point $\vec{H}_{2o} = \frac{N_2 I_2}{2R_2}\vec{a}_z$, since $R_1 << R_2$, we can assume H_2 to be almost uniform over the area of coil #1, hence the flux on coil 1 due to the current

$$I_2\psi_{m12} = (\mu_o H_{2o})(S_1) = \left(\mu_o \frac{N_2 I_2}{2R_2}\right)(\pi R_1^2)$$

And the linkage-flux $\Lambda_{12} = N_1\psi_{m12} = N_1\left(\mu_o \frac{N_2 I_2}{2R_2}\right)(\pi R_1^2) \rightarrow L_{12} = \frac{\Lambda_{12}}{I_2} = \left(\mu_o \frac{N_1 N_2}{2R_2}\right)(\pi R_1^2)$

Example 9.6

a. Find the self-inductance of the coil in Example 9.4.
b. In the same example, find the mutual inductance between the coil and a second coil (#2) wrapped around the same core with $N_2 = 200$.

Solution

a. Form the solution of Example 9.4, $L_{11} = \frac{N_1^2}{R_t} = \frac{500^2}{20.5E6} = 12.2$ mH

b. $L_{21} = \frac{\Lambda_{21}}{I_1} = \frac{N_2\psi_m}{I_1} = \frac{(200)(24.4E-6)}{1} = 4.88$ mH

Example 9.7 Determine the mutual inductance between a very long straight wire and a conducting equilateral triangle loop as shown in the figure.

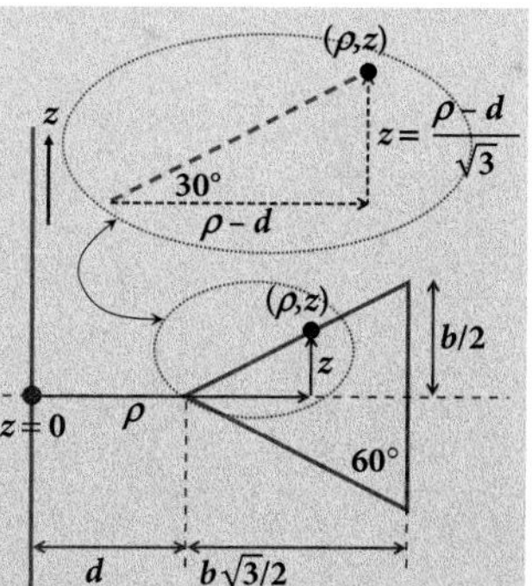

Solution The magnetic flux density due to the long wire

$$\oint \vec{B}\cdot d\vec{l} = \mu_o I_{\text{enclosed}}$$

$$B_\varphi 2\pi\rho = \mu_o I \rightarrow B_\varphi = \frac{\mu_o I}{2\pi\rho} = B_1$$

From symmetry, the magnetic flux linking the triangular loop can be calculated on the upper half of the triangle and then multiplied by 2

$$\psi_{m21} = \int \vec{B}_1\cdot d\vec{s}_2 = 2\int_d^{d+\frac{b}{2}\sqrt{3}}\int_0^{\frac{\rho-d}{\sqrt{3}}} \frac{\mu_o I}{2\pi\rho}dz\,d\rho = \frac{\mu_o I}{\pi}\int_d^{d+\frac{b}{2}\sqrt{3}} \frac{z\Big|_0^{\frac{\rho-d}{\sqrt{3}}}}{\rho}d\rho = \frac{\mu_o I}{\pi}\int_d^{d+\frac{b}{2}\sqrt{3}} \frac{\rho-d}{\sqrt{3}\rho}d\rho$$

$$= \frac{\mu_o I_o}{\pi\sqrt{3}}\left[\rho - d\ln\rho\right]_d^{d+\frac{b}{2}\sqrt{3}} = \frac{\mu_o I_o}{\pi\sqrt{3}}\left[\left(d+\frac{b}{2}\sqrt{3}-d\right) - d\ln\left(\frac{d+\frac{b}{2}\sqrt{3}}{d}\right)\right]$$

$$\psi_{m21} = \frac{\mu_o I}{\pi\sqrt{3}}\left[\frac{b}{2}\sqrt{3} - d\ln\left(1+\frac{b}{2d}\sqrt{3}\right)\right]$$

$$L_{21} = \frac{\psi_{m21}}{I} = \frac{\mu_o}{\pi\sqrt{3}}\left[\frac{b}{2}\sqrt{3} - d\ln\left(1+\frac{b}{2d}\sqrt{3}\right)\right](H)$$

ADDENDUM 9A
EVALUATION OF SELF-INDUCTANCE

9A.1 Introduction

We defined the inductance earlier in this chapter as the ratio between the magnetic flux linkage and the current causing the flux. For linear systems, this ratio is independent of both the current and flux values. Instead, the inductance of a coil configuration depends only on its geometrical parameters as well as the material permeability.

In this addendum, we present different approaches to a coil's self-inductance evaluation. Namely, we discuss the ratio between the flux-linkage and coil current approach, the use of magnetic reluctance approach, and the energy storage approach.

In the body of the chapter, we elaborated more on the flux-linkage to current approach and not on the other two approaches. In the following, we explain in more detail the other two approaches followed by examples of implementing all three approaches.

9A.2 Inductance Evaluation Using Magnetic Reluctance

This approach recognizes that the inductance of a coil configuration depends only on its geometrical parameters as well as the material permeability. To elaborate, let us take the simple configuration of Figure 9A.1. The figure presents a current-carrying coil along with the flux bundle produced by the current. As we already know, the flux produced by the current forms closed paths that link the current source. The flux distribution in the space around the coil will be determined by the material properties in the region. Without lack of generality, the produced flux is shown as confined to a finite region of space. The following development based on the finite region flux bundle applies just as well to an all space flux distribution.

Using the notation N for the number of coil-turns, I for the current in the coil, ψ_m for the total magnetic flux, S for the cross section of the bundle (variable), ℓ for the length of the flux closed path (variable), and $\mathfrak{R}$ as the reluctance seen by the flux bundle, we can write

$$mmf = NI = \psi_m \mathfrak{R} \qquad (9.47,8)$$

where

$$\mathfrak{R} = \int_\ell \frac{1}{\iint_S \frac{\mu dS}{d\ell}} \qquad (9.56)$$

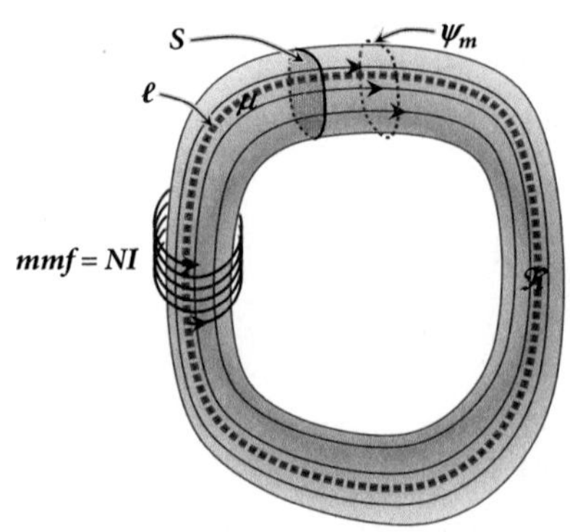

FIGURE 9A.1 Inductance evaluation using magnetic reluctance.

Now, with these relationships at hand, we can write the coil inductance as

$$L = \frac{\Lambda}{I} = \frac{N\psi_m}{I} = \frac{N\left(\frac{NI}{\mathfrak{R}}\right)}{I} = \frac{N^2}{\mathfrak{R}} \tag{9A.1}$$

Since $\mathfrak{R}$ for a linear media is independent of both I and Λ, it follows that the inductance is also independent of both.

The formula in Equation (9A.1) can be used to evaluate the inductance through the evaluation of the reluctance of the region of space housing the inductor's flux. However, there are two other approaches to inductance evaluation that we will present in this addendum. The three approaches for inductance evaluation presented in the following are

1. Inductance evaluation through reluctance evaluation
2. Inductance evaluation as the ratio between flux linkage and current
3. Inductance evaluation through energy storage

The procedures for reluctance evaluation and that for flux linkage were discussed earlier in this chapter; however, we did not present that for relating the energy stored to inductance. In the following, we learn about this relationship and how it can be used for inductance evaluation.

9A.3 Inductance Evaluation Using Magnetic Energy Storage

Consider the same simple representation of a current-carrying coil and its produced flux, Figure 9A.2. Again, without loss of generality we can write the energy stored in the flux bundle as

$$\begin{aligned} W_m &= \int_v \frac{1}{2}\{\vec{B}\cdot\vec{H}\}dv = \int_v \frac{1}{2}\{[BdS][H\,d\ell]\} = \frac{1}{2}\oint_\ell\left\{[H\,d\ell]\iint_S[B\,dS]\right\} = \frac{1}{2}\left\{[\psi_m]\oint[H\,d\ell]\right\} \\ &= \frac{1}{2}\{[\psi_m]NI\} = \frac{1}{2}\{[N\psi_m]I\} = \frac{1}{2}\Lambda I = \frac{1}{2}LI^2 \end{aligned} \tag{9A.2}$$

Hence, we can use the formula

$$W_m = \int_v \frac{1}{2}\{\vec{B}\cdot\vec{H}\}dv = \int_v \frac{1}{2}\mu H^2\,dv = \frac{1}{2}LI^2 \tag{9A.3}$$

to evaluate the inductance given the H distribution for the coil.

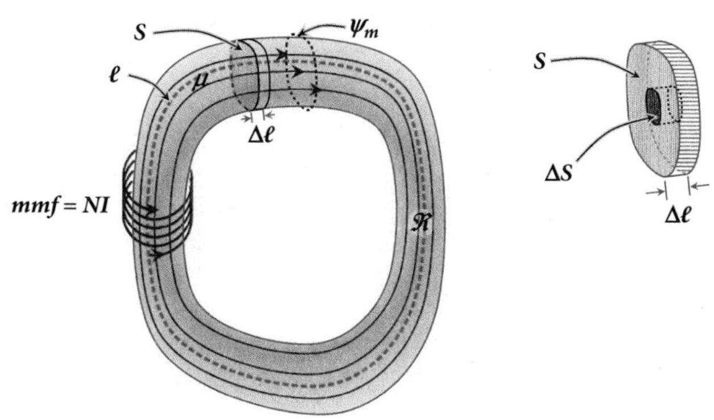

FIGURE 9A.2 Inductance evaluation using magnetic energy storage.

9A.4 Examples of Inductance Evaluation for Specific Coil Configurations

In the following, we will consider a few examples of coil configurations that we worked on earlier in Chapters 8 and 9 for the purpose of inductance evaluation. We will attempt some or all the three approaches cited above whenever feasible and meaningful. The examples considered are

1. Toroid
2. Infinite solenoid
3. Infinite wire
4. Infinite coaxial line
5. Magnetic circuit with and without air gap

9A.4.1 Example 1: The Toroid

The toroid configuration considered in this example is the same one we worked on before in Figure 8C.9. We will assume a rectangular shape cross section that is uniform throughout the toroid length for simplicity. The cross-sectional dimensions of the core are $(b\text{-}a)$ for width and (h) for height. We will attempt all three approaches for inductance evaluation in this example.

a. *The reluctance approach*: Start with the reluctance formula

$$\mathfrak{R} = \int_\ell \frac{1}{\iint_S \frac{\mu dS}{d\ell}} \tag{9.56}$$

The integration over the cross-sectional area is a simple one

$$\iint_S \frac{\mu dS}{d\ell} = \iint_S \frac{\mu dh d\rho}{\rho d\varphi} = \int_{\rho=a}^{b} \left[\int_0^h \frac{\mu h d\rho}{\rho d\varphi} \right] = \frac{\mu h \ln\left(\frac{b}{a}\right)}{d\varphi} \tag{9A.4}$$

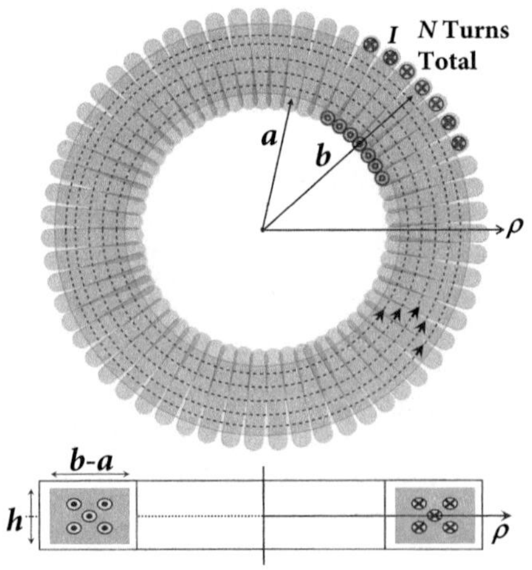

FIGURE 8C.9 Evaluation of the self-inductance of a toroid coil.

Now for the loop integration, we get

$$\Re = \int_{\varphi=0}^{2\pi} \frac{d\varphi}{\mu h \ln\left(\frac{b}{a}\right)} = \frac{2\pi}{\mu h \ln\left(\frac{b}{a}\right)} \tag{9A.5}$$

Hence, the inductance of the toroid is

$$L = \frac{N^2}{\Re} = \frac{N^2 \mu h \ln\left(\frac{b}{a}\right)}{2\pi} \tag{9A.6}$$

b. *The flux linkage approach*: Starting with the solution we had for H in the toroid, Equation (8C.22)

$$\vec{H} = \begin{cases} \left[\dfrac{NI}{2\pi\rho}\right]\vec{a}_\varphi & \text{inside the toroid's cross section} \\ 0 & \text{outside the toroid's cross section} \end{cases} \tag{8C.22}$$

and referring to Figure 9A.3, we can write

$$L = \frac{\Lambda}{I} = \frac{N\psi_m}{I} = \frac{N\iint_S B\,dS}{I} = \frac{N\iint_S \mu\left[\frac{NI}{2\pi\rho}\right][dh\,d\rho]}{I} = N\iint_S \mu\left[\frac{N}{2\pi\rho}\right]dh\,d\rho = \left[\frac{\mu N^2 h}{2\pi}\right]\ln\left(\frac{b}{a}\right) \tag{9A.7}$$

which is an identical answer to the one above.

c. *The energy storage approach*: Again, starting with Equation (8C.22), we write the energy stored in the toroid flux bundle, Equation (9A.3)

$$\frac{1}{2}LI^2 = W_m = \int_v \frac{1}{2}\{\vec{B}\cdot\vec{H}\}dv = \int_v \frac{1}{2}\mu H^2\,dv = \int_v \frac{1}{2}\mu\left[\frac{NI}{2\pi\rho}\right]^2 [dh\,d\rho\,\rho d\varphi]$$
$$= \int_v \frac{1}{2}\mu\left[\frac{NI}{2\pi}\right]^2 [h2\pi]\frac{d\rho}{\rho} = \frac{1}{2}I^2\left[\frac{\mu N^2 h}{2\pi}\right]\ln\left(\frac{b}{a}\right) \tag{9A.8}$$

Thus, getting the same answer for L

$$L = \left[\frac{\mu N^2 h}{2\pi}\right]\ln\left(\frac{b}{a}\right) \tag{9A.9}$$

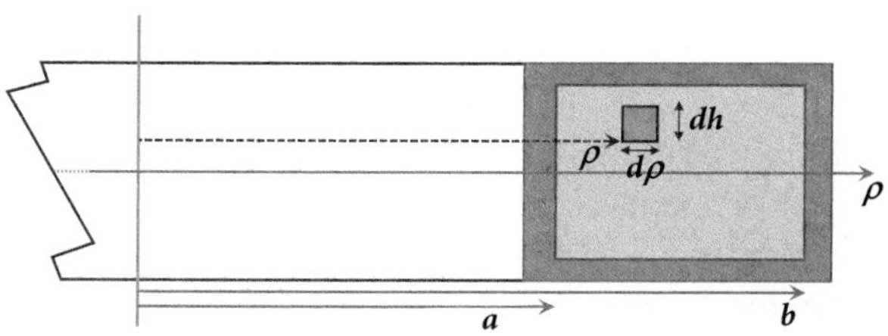

FIGURE 9A.3 Evaluation of the self-inductance of a toroid coil using the flux linkage approach.

9A.4.2 Example 2: The Infinite Solenoid

a. *The reluctance approach*: Start with the reluctance formula

$$\Re = \int_{\ell} \frac{1}{\iint_S \frac{\mu dS}{d\ell}} \tag{9.56}$$

The integration over the cross-sectional area is a simple one

$$\iint_S \frac{\mu dS}{d\ell} = \frac{\mu S}{dh} \tag{9A.10}$$

Now for the loop integration, we get

$$\Re = \int_h \frac{dh}{\mu S} = \frac{h}{\mu S} \tag{9A.11}$$

Hence, the inductance of the toroid is

$$L = \frac{N^2}{\Re} = \frac{N^2 \mu S}{h} = n^2 \mu Sh \tag{9A.12}$$

$$L_{\text{per meter}} = n^2 \mu S \tag{9A.13}$$

where n is the number of turns per unit length, n = N/h.

b. *The flux linkage approach*: Starting with the solution we had for H in the toroid, Equation (8C.22)

$$\vec{H} = \begin{cases} nI\vec{a}_z & \text{inside the solenoid's cross section} \\ 0 & \text{outside the solenoid's cross section} \end{cases} \tag{8.30}$$

we can write

$$L = \frac{\Lambda}{I} = \frac{N\psi_m}{I} = \frac{N\iint_S BdS}{I} = \frac{N\iint_S \mu[nI]dS}{I} = Nn\mu S = n^2 \mu Sh \tag{9A.14}$$

which is an identical answer to the one above.

c. *The energy storage approach*: Again, starting with Equation (8C.22), we write the energy stored in the toroid flux bundle, Equation (9A.3)

$$\frac{1}{2}LI^2 = W_m = \int_v \frac{1}{2}\{\vec{B}\cdot\vec{H}\}dv = \int_v \frac{1}{2}\mu H^2 dv = \int_v \frac{1}{2}\mu[nI]^2[dh\,dS] = \frac{1}{2}\mu[nI]^2[hS] \tag{9A.15}$$

Thus, getting the same answer for L

$$L = n^2 \mu Sh \tag{9A.16}$$

9A.4.3 Example 3: The Infinite Wire

In this example, we will consider a conductor of radius a infinitely extending along the z axis (Figure 9A.4). The flux distribution in this case forms circular loops concentric with the conductor axis. The flux distribution has different density formulas for the regions within a and outside a. Hence, we have two different flux bundles that form the whole flux produced by the conductor current. Another observation regarding the two different flux bundles is that they do not link similar currents. The flux lines outside a do link the entire current I.

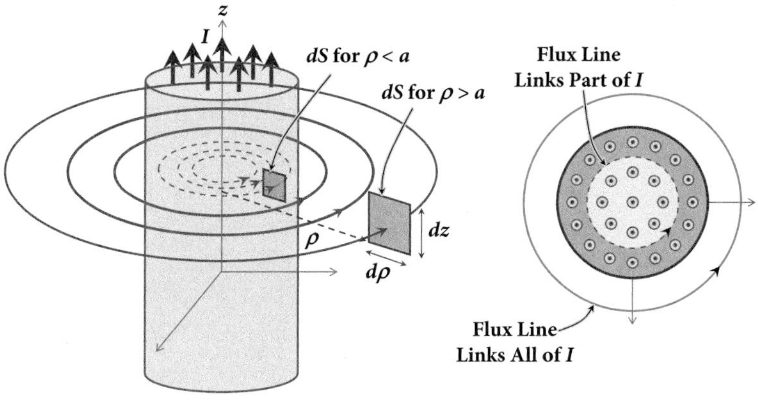

FIGURE 9A.4 Evaluation of the self-inductance of an infinite length thick straight current-carrying conductor using the flux linkage approach.

However, the flux lines within a link only part of I as demonstrated in the figure. For this reason, we will only use the linkage flux and energy storage approaches of inductance evaluation.

a. *The flux linkage approach*: We derived the expressions for the magnetic field intensity for this example earlier in Chapter 8. The obtained expressions were

$$\vec{H} = \begin{cases} \dfrac{I}{2\pi\rho}\vec{a}_\varphi & \text{for } a<\rho<\infty \\ \dfrac{I\rho}{2\pi a^2}\vec{a}_\varphi & \text{for } 0<\rho<a \end{cases} \tag{8C.8}$$

Referring to Figure 9A.4, we can evaluate the linkage flux expressions for the two regions ($\rho > a$ and $\rho < a$) by integrating the corresponding B over the flux distribution area. We start with the outside linkage flux

$$\Lambda_{\text{outside}} = N\psi_{m,\,\text{outside}} = \psi_{m,\,\text{outside}} \tag{9A.17}$$

This is the case since all flux lines in this region link all of the current I only once and hence $N = 1$ in this case. Consequently,

$$\Lambda_{\text{outside}} = \iint_{S\,\text{outside}} B\,dS = \int_{z=0}^{\infty}\int_{\rho=0}^{\infty}\mu\left[\frac{I}{2\pi\rho}\right]d\rho\,dz = \left[\frac{\mu I z}{2\pi}\right]_{z=0}^{\infty}\left[\ln(\rho)\right]_0^{\infty} = \infty \tag{9A.18}$$

For the expression of the flux per unit height length, we set the z integration limit to 1 m. The answer is still infinity

$$\psi_{m,\,\text{outside, per meter}} = \left[\frac{\mu I}{2\pi}\right]\left[\ln(\rho)\right]_0^{\infty} = \infty \tag{9A.19}$$

An infinite answer for the outside flux implies an infinite answer for the outside component of the self-inductance.

Now, let us deal with the internal flux. Flux lines in this region link a portion of the current I and not the whole I. Hence, the linkage coefficient N is a fraction. Typically, N is an integer representing how many times the flux links the current, for example, the number of current turns in the coil. However, in our case here, the flux lines link only a fraction of the current and hence N is a fraction.

If the current density is uniform over the current cross section, then this fraction is simply the ratio of the area of the linked current to that of the whole current. Hence

$$N=\frac{J\pi\rho^2}{J\pi a^2}=\frac{\rho^2}{a^2}, \text{ which varies with } \rho \tag{9A.20}$$

Consequently,

$$d\Lambda_{\text{internal}}=N\,d\psi_{m,\text{internal}}=\frac{\rho^2}{a^2}\,d\psi_{m,\text{internal}} \tag{9A.21}$$

$$\Lambda_{\text{internal}}=\iint_{S\text{ inside}}\frac{\rho^2}{a^2}B\,dS=\int_{z=0}^{\infty}\int_{\rho=0}^{a}\frac{\rho^2}{a^2}\mu\left[\frac{I\rho}{2\pi a^2}\right]d\rho\,dz=\frac{\mu I}{2\pi a^4}[z]_0^{\infty}\left[\frac{\rho^4}{4}\right]_0^a=\infty \tag{9A.22}$$

For the expression of the flux per unit height length, we set the z integration limit to 1 m. The answer is

$$\Lambda_{\text{internal per meter}}=\frac{\mu I}{8\pi} \tag{9A.23}$$

and the internal inductance per meter length of the wire is

$$L_{\text{internal per meter}}=\frac{\Lambda_{\text{internal per meter}}}{I}=\frac{\mu}{8\pi} \tag{9A.24}$$

for the case of a uniform current density over the conductor's cross section.

b. *The energy storage approach*: Again, we divide the problem to two regions, internal and external. Since we found the answer to be infinite for the external part, we will suffice to do the internal part only.

$$\begin{aligned}\frac{1}{2}L_{\text{internal per meter}}I^2&=W_{m,\text{ internal per meter}}=\int_v\frac{1}{2}\mu H^2\,dv=\int_v\frac{1}{2}\mu\left[\frac{I\rho}{2\pi a^2}\right]^2[dz\,d\rho\,\rho d\varphi]\\&=\frac{1}{2}\mu\left[\frac{I}{2\pi a^2}\right]^2[z]_0^1[\varphi]_0^{2\pi}\frac{\rho^4}{4}\Big|_0^a=\frac{1}{2}I^2\left[\frac{\mu}{8\pi}\right]\end{aligned} \tag{9A.25}$$

Thus, getting the same answer

$$L_{\text{internal per meter}}=\frac{\mu}{8\pi} \tag{9A.26}$$

9A.4.4 Example 4: The Infinite Coaxial Line

In this example, we will consider the coaxial arrangement of two conductors discussed in Chapter 8 and shown here in Figure 9A.5. The field expressions for this case were found to be

$$\vec{H}=\begin{cases}\dfrac{I\rho}{2\pi a^2}\vec{a}_\varphi & \text{for } 0<\rho<a\\[2ex] \dfrac{I}{2\pi\rho}\vec{a}_\varphi & \text{for } a<\rho<b\\[2ex] \dfrac{I}{2\pi(c^2-b^2)}\left[\dfrac{c^2}{\rho}-\rho\right]\vec{a}_\varphi & \text{for } b<\rho<c\\[2ex] 0 & \text{for } c<\rho<\infty\end{cases} \tag{8.39}$$

It is obvious that we have three inductances to evaluate. These are for the regions $\rho<a$, $a<\rho<b$, and $b<\rho<c$. The region $\rho>c$ has zero field and hence no magnetic flux or inductive behavior is expected. Since the line

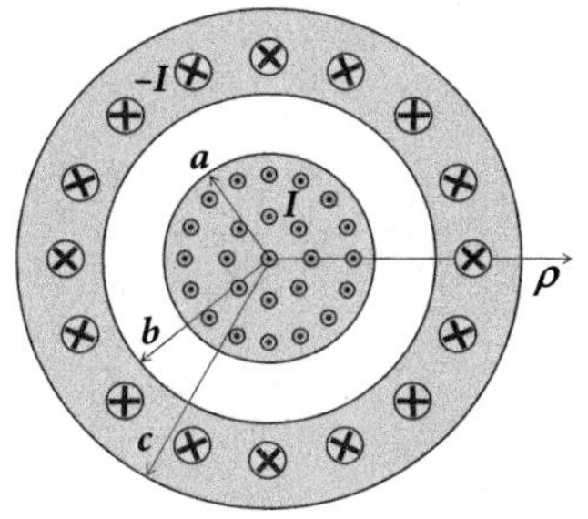

FIGURE 9A.5 Evaluation of the self-inductance of an infinite length coaxial line.

extends to infinity, all the remaining three inductances are expected to be infinite; hence, we evaluate the inductances per unit length of the line instead. The internal inductance for the region $\rho < a$ will have the same exact answer as that found in the case of the single conductor found earlier in the previous example. Therefore, in this section, we will derive the expressions for the inductances of the remaining two regions. For the sake of simplicity, we will do the region $a < \rho < b$ using the flux linkage approach and the region $b < \rho < c$ using energy storage approach.

a. Evaluation of $L_{a<\rho<b}$ using the flux linkage approach

In this region, the expression for the magnetic field is given by

$$\vec{H} = \frac{I}{2\pi\rho}\vec{a}_\varphi \tag{8.39}$$

Referring to Figure 9A.5, all flux lines in this region link the entire current I. Hence, we can write

$$\Lambda_{a<\rho<b \text{ per meter}} = \psi_{m,\, a<\rho<b \text{ per meter}} = \iint_{S\, a<\rho<b} B dS = \int_{z=0}^{1}\int_{\rho=a}^{b} \mu\left[\frac{I}{2\pi\rho}\right] d\rho\, dz$$
$$= \frac{\mu I}{2\pi}[z]_0^1\,[\ln(\rho)]_a^b = \frac{\mu I}{2\pi}\ln\left(\frac{b}{a}\right) \tag{9A.27}$$

Consequently, the inductance per unit axial length for this region is

$$L_{a<\rho<b \text{ per meter}} = \frac{\mu}{2\pi}\ln\left(\frac{b}{a}\right) \tag{9A.28}$$

Evaluation of $L_{b<\rho<c}$ using the energy storage approach

For this region, the magnetic field expression is given by

$$\vec{H} = \frac{I}{2\pi(c^2-b^2)}\left[\frac{c^2}{\rho} - \rho\right]\vec{a}_\varphi \tag{9A.29}$$

hence, we write

$$\frac{1}{2}L_{b<\rho<c \text{ per meter}} I^2 = W_{m,b<\rho<c \text{ per meter}} = \int_{v,\, b<\rho<c}\frac{1}{2}\mu H^2 dv = \int_v \frac{1}{2}\mu\left[\frac{I}{2\pi(c^2-b^2)}\left[\frac{c^2}{\rho}-\rho\right]\right]^2 [dz\, d\rho\, \rho\, d\varphi]$$
$$= \frac{1}{2}\mu\left[\frac{I}{2\pi(c^2-b^2)}\right]^2 [z]_0^1 [\varphi]_0^{2\pi}\left[c^4\ln\rho + \frac{\rho^4}{4} - c^2\rho^2\right]_b^c$$
$$= \frac{1}{2}\mu\left[\frac{I}{2\pi(c^2-b^2)}\right]^2 2\pi\left[c^4\ln\left(\frac{c}{b}\right) + \frac{c^4-b^4}{4} - c^2(c^2-b^2)\right] \tag{9A.30}$$

thus, yielding the following expression for the inductance per unit length for the internal region of the outer conductor

$$L_{b<\rho<c\,\text{per meter}} = \frac{\mu}{2\pi}\left[\frac{c^4}{(c^2-b^2)^2}\ln\left(\frac{c}{b}\right)+\frac{(c^2+b^2)}{4(c^2-b^2)}-\frac{c^2}{(c^2-b^2)}\right] \tag{9A.31}$$

9A.4.5 Example 5: The Magnetic Circuit with and without Air Gap

The simplest method to evaluate the inductance in the magnetic circuit case is the use of the reluctance approach. We derived the inductance expression earlier as

$$L = \frac{N^2}{\mathfrak{R}} \tag{9A.32}$$

Hence, all we need to do is evaluate the reluctance associated with the magnetic circuit flux bundle and use that in the expression above to get the inductance expression. In this regard, let us consider the three examples shown in Figure 9A.6.

For the first and second examples, the total reluctances are $\mathfrak{R}_t = \frac{\ell_c}{\mu_c S_c}$, and $\mathfrak{R}_t = \frac{\ell_c}{\mu_c S_c} + \frac{\ell_g}{\mu_g S_g}$, respectively. For the third example with the multiple-branch magnetic circuit, the three branch reluctances are

$$\mathfrak{R}_1 = \frac{\ell_1}{\mu_1 S_1}, \qquad \mathfrak{R}_2 = \frac{\ell_2}{\mu_2 S_2}, \qquad \text{and} \qquad \mathfrak{R}_3 = \frac{\ell_3}{\mu_3 S_3} \tag{9A.33}$$

Hence, the total (equivalent) reluctance as seen by the coil is

$$\mathfrak{R}_t = \mathfrak{R}_1 + \frac{\mathfrak{R}_2\mathfrak{R}_3}{\mathfrak{R}_2 + \mathfrak{R}_3} \tag{9A.34}$$

The coil inductances can then follow as

$$L_c = \frac{N^2}{\mathfrak{R}_t} \tag{9A.35}$$

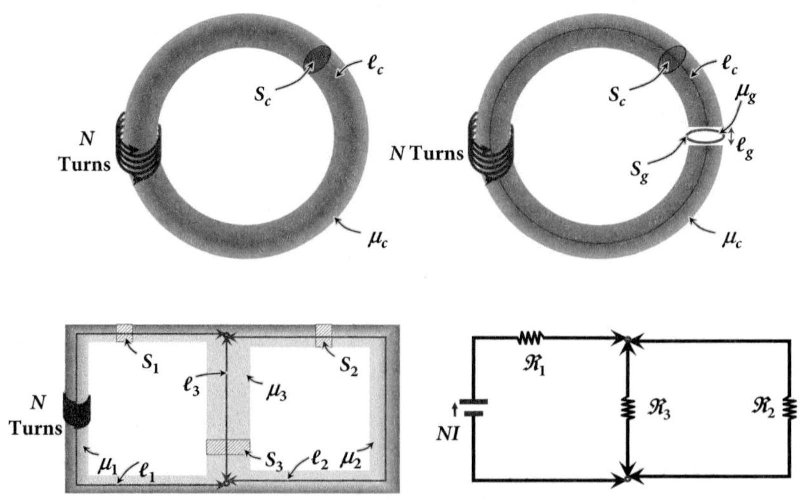

FIGURE 9A.6 Examples for the evaluation of the self-inductance of magnetic circuit coils using the reluctance approach.

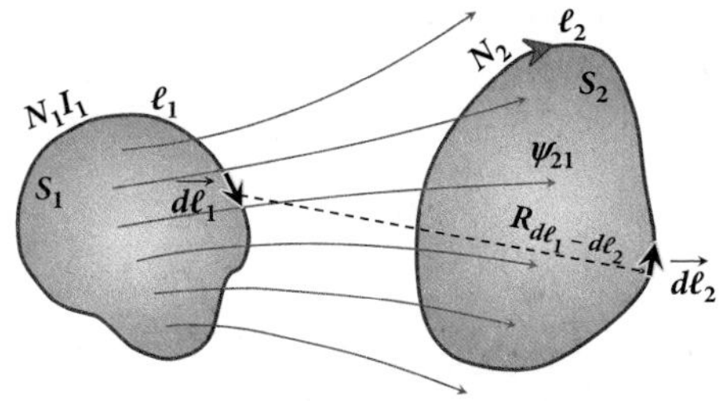

Figure 9B.1 An illustration used in the discussion of mutual inductance reciprocity.

ADDENDUM 9B EVALUATION OF MUTUAL INDUCTANCE

The mutual inductance between two coils, as we defined it earlier in this chapter, is the ratio between the magnetic flux linkage for one of a two-coil configuration and the other coil's current causing that flux. For linear systems, this ratio is independent of both the current and flux values. Instead, the mutual inductance of a two-coil configuration depends only on its geometrical parameters as well as the material permeabilities of the media encountered by the magnetic flux.

To prove the reciprocity feature for mutual inductance, let us consider the two-coil arrangement in Figure 9B.1. The figure shows two coils placed in proximity of each other. The two coils are assumed to have multiple turns of N_1 and N_2 in general. To derive an expression for the mutual inductance L_{21}, we assume a current I_1 to flow in the first coil and then evaluate the resulting flux linkage in the second coil. The ratio of the second coil linkage flux to the first coil's current yields our desired expression for L_{21}.

9B.1 Mutual Inductance Evaluation Using Flux-Linkage

We start by evaluating the total magnetic flux on coil #2 due to I1. We write the total flux as the integration of the flux density over the surface area of the second coil. In this regard, we note that there are infinite number of areas that we can choose for this integration, all are bound by the same coil *perimeter*. Using the vector magnetic potential definition [Equation (8.65)] and Stokes' theorem [Equation (8.62)], we can write

$$\psi_{21} = \iint_{S_2} \vec{B}_{21}\cdot d\vec{S}_2 = \iint_{S_2} (\vec{\nabla}\times\vec{A}_{21})\cdot d\vec{S}_2 = \oint_{\ell_2} \vec{A}_{21}\cdot d\vec{\ell}_2 \tag{9B.1}$$

Next, we use the magnetic vector potential solution expression as it relates to the electric current distribution, Equation (8.70)

$$\vec{A}_{21} = \oint_{\ell_1} \frac{\mu_o (I d\vec{\ell})_1}{4\pi R_{d\ell_1-d\ell_2}} \tag{8.70}$$

We develop Equation (9B.1) to take the form

$$\psi_{21} = \oint_{\ell_2} \vec{A}_{21}\cdot d\vec{\ell}_2 = \oint_{\ell_2}\oint_{\ell_1} \frac{\mu_o N_1 I_1 d\vec{\ell}_1}{4\pi R_{d\ell_1-d\ell_2}}\cdot d\vec{\ell}_2 = \frac{\mu_o N_1 I_1}{4\pi}\oint_{\ell_2}\oint_{\ell_1}\frac{d\vec{\ell}_1\cdot d\vec{\ell}_2}{R_{d\ell_1-d\ell_2}} \tag{9B.2}$$

and finally, we write the expression for the mutual inductance L_{21} as

$$L_{21} = \frac{\Lambda_{21}}{I_1} = \frac{N_2\psi_{21}}{I_1} = \frac{\mu_o N_2 N_1}{4\pi}\oint_{\ell_2}\oint_{\ell_1}\frac{d\vec{\ell}_1\cdot d\vec{\ell}_2}{R_{d\ell_1-d\ell_2}} \tag{9B.3}$$

Now, if we were to seek L_{12} instead, we would have started the process with flowing a current I_2 in coil #2 and evaluated the linkage flux over coil #1. It is clear from the development above that the final expression would be similar to that of Equation (9B.3) except for swapping the 1 and 2 subscripts, hence

$$L_{12} = \frac{\mu_o N_1 N_2}{4\pi} \oint_{\ell_1} \oint_{\ell_2} \frac{\overrightarrow{d\ell_2} \cdot \overrightarrow{d\ell_1}}{R_{d\ell_2 - d\ell_1}} \tag{9B.4}$$

which is in fact identical to Equation (9B.3), and hence,

$$L_{21} = L_{12} \tag{9B.5}$$

This finding adds important convenience to mutual inductance evaluation tasks. One convenience is that we would only need to evaluate either of the two terms. Another convenience is that we can choose which term is easier (sometimes possible) to derive and avoid the development of the harder/impossible to derive one.

9B.2 Examples of Mutual Inductance Evaluation

In this section, we will practice the development of mutual inductance expression for three typical configurations:

1. Two-coil toroid
2. Two-coil solenoid
3. Two-coil magnetic circuit

In this regard, we will only use the ratio between flux linkage and current approach in our analysis.

9B.2.1 Example 1: The Two-Coil Toroid

The toroid configuration considered in this example is the same one we worked on before in the self-inductance evaluation example Figure 9B.2. Coil #1 is formed of N_1 turns wound uniformly around the rectangular shape cross-section toroid. Meanwhile, coil #2 has N_2 turns wound around a section of coil #1. Coil #2 may or may not cover the entire length of the toroid.

To start the flux linkage approach to mutual inductance evaluation, we need to decide which coil to initiate a current in. It is clear that starting a current in coil #1 would be the way to go. We have expressions on hand for the flux density and flux resulting from a uniform wound coil current on toroidal core in Chapter 8. We used these expressions earlier in self-inductance evaluation as well. On the other hand, starting a current in

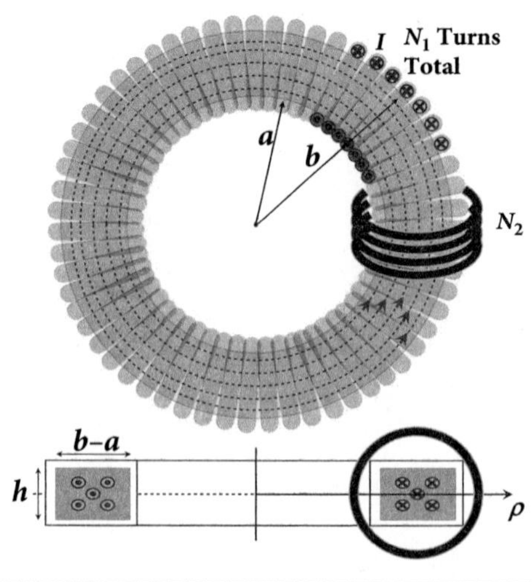

FIGURE 9B.2 An example of a two-coil toroid for mutual inductance evaluation.

coil #2 offers two challenges. First, is solving for a multi-turn "short" solenoid–like coil. The second challenge is that the solution is needed at all locations of coil #1.

Therefore, in a process similar to that of Equation (9.74), we start with I_1 and attempt the derivation of Λ_{21} as follows:

$$L_{21} = \frac{\Lambda_{21}}{I_1} = \frac{N_2 \Psi_{m21}}{I_1} = \frac{N_2 \iint_{S_2} \vec{B}_{21} \cdot \overrightarrow{dS}_2}{I_1} = \frac{N_2 \iint_{S_2} B_{21}\, dS_2}{I_1} \tag{9B.6}$$

Since we know the flux generated by I_1 to be all contained within the toroid cross-section S_1 and no other flux exist outside it. Then we conclude that integrating the flux of coil 2 over S_2 is the same as the flux linking coil 1 within the area S_1.

$$L_{21} = \frac{N_2 \iint_{S_2} B_{21}\, dS_2}{I_1} = \frac{N_2 \iint_{S_1} B_{21}\, dS_1}{I_1} = \frac{N_2 \iint_{S_1} \mu \left[\frac{N_1 I_1}{2\pi\rho}\right][dh\, d\rho]}{I_1} = \frac{\mu N_1 N_2}{2\pi} \iint_S \left[\frac{1}{\rho}\right] dh\, d\rho = \left[\frac{\mu N_1 N_2}{2\pi} h\right] \ln\left(\frac{b}{a}\right) \tag{9B.7}$$

9B.2.2 Example 2: Two-Coil "Long" Solenoid

This example is depicted in Figure 9B.3 in which a secondary "short" solenoid coil is wrapped around the mid-section of a long solenoid.

To decide which coil to flow the current in, we examine the predicted flux maps in both cases. The map to the left in the figure is for the current flowing in the short solenoid, while the map to the right if for the current flowing in the long solenoid. Although we did not derive exact expressions for either, it appears that the map to the right can be approximated by that of an infinite solenoid (Figure 8.18). Furthermore, we are only interested in its mid region where the second coil is located, which makes this approximation more accurate.

Consequently, the approach to follow is to assume the current to flow in the long solenoid, coil #1, and evaluate the linkage-flux over the short solenoid, coil #2. Hence, similar to our analysis of Equation (9A.14) for the self-inductance, we write

$$L_{21} = \frac{\Lambda_{21}}{I_1} = \frac{N_2 \Psi_{m21}}{I_1} = \frac{N_2 \iint_{S_2} B_{21}\, dS_2}{I_1} \tag{9B.8}$$

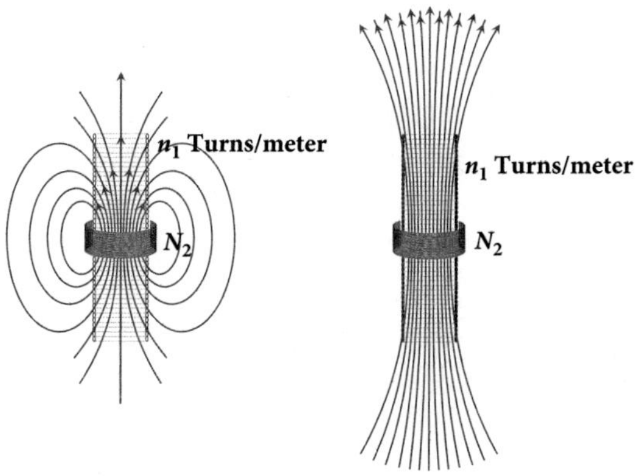

FIGURE 9B.3 An example of a two-coil solenoid for mutual inductance evaluation.

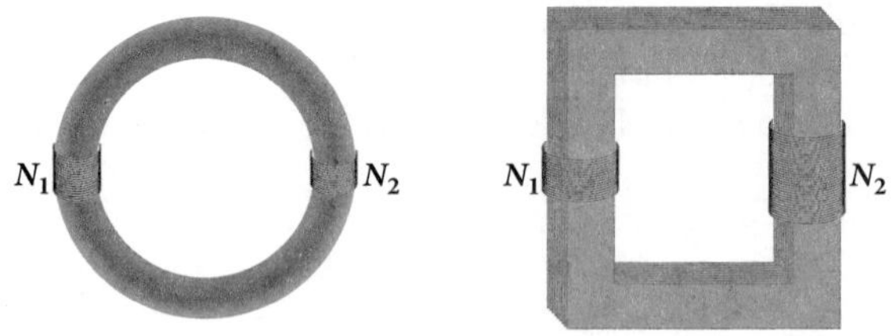

FIGURE 9B.4 An example of a two-coil magnetic circuit for mutual inductance evaluation.

Again, since we know the flux generated by I_1 to be all contained within the solenoid cross-section S_1 and no other flux exist outside it. Then we conclude that integrating the flux of coil 2 over S_2 is the same as the flux linking coil 1 within the area S_1, hence

$$L_{21} = \frac{N_2 \iint_{S_2} B_{21}\, dS_2}{I_1} = \frac{N_2 \iint_{S_1} B_{21}\, dS_1}{I_1} = \frac{N_2 B_{21} S_1}{I_1} = \frac{N_2 \mu [n_1 I_1] S_1}{I_1} = \mu n_1 N_2 S_1 \tag{9B.9}$$

9B.2.3 Example 3: Two-Coil Magnetic Circuit

In the previous two examples, the two coils were immersed in one uniform medium that extended everywhere, presumably air (free space). In this example, the two coils are linked to each other through a high-permeability core. Such is the typical configuration of what we call a *transformer*. The high-permeability core confines the magnetic flux within its boundaries ensuring its maximum linkage to both coils. Two such configurations are shown in Figure 9B.4. As the figure demonstrates, the core-confined magnetic flux is the same for both coils, hence we can write the flux equality for the two coils due to a current I_1 as

$$\psi_{m21} = \psi_{m11} \tag{9B.10}$$

Similarly, a current I_2 in coil #2 would produce equal fluxes again in both coils

$$\psi_{m12} = \psi_{m22} \tag{9B.11}$$

Using the magnetic circuit approach for analysis, we can write

$$\psi_{m21} = \frac{N_1 I_1}{\mathfrak{R}_{21}} \text{ where } \mathfrak{R}_{21} \text{ is the reluctance of the common core} \tag{9B.12}$$

and

$$L_{21} = \frac{\Lambda_{21}}{I_1} = \frac{N_2 \psi_{m21}}{I_1} = \frac{N_2 N_1 I_1}{I_1 \mathfrak{R}_{21}} = \frac{N_1 N_2}{\mathfrak{R}_{21}} \tag{9B.13}$$

ADDENDUM 9C MAGNETIC FORCES IN AIR GAPS: MAGNETIC PULL

9C.1 Introduction

One of the most typical applications of magnetism is the use of magnetic forces to pull or lift objects. These magnetic forces are generated in the air gap of a magnetic circuit containing both the magnetic source and the object. It is typical to use an electromagnet as the magnetic source in such applications in order to enable the control of the air gap force including turning it on and off. In this addendum, we derive the air-gap force expressions in two examples: magnetic pull and magnetic lift. In the lift example, we will discuss the use of both an electromagnet and a permanent magnet for lifting objects.

9C.2 Magnetic Forces in Air Gaps

Figure 9C.1 demonstrates a typical magnetic pull application. As the figure demonstrates, a magnetic circuit is formed of a ferromagnetic core part of which is a hinged arm. The magnetic source in this case is an electromagnet using a current-carrying coil wrapped around the fixed part of the core. An air gap between one end of the hinged arm and the fixed core is typically maintained by a mechanical spring. When a current is flown in the electromagnet coil, a magnetic force in the gap pulls the hinged arm causing the desired mechanical motion. Examples of use of such devices include mechanical motion of objects as well as opening and closing contacts in switches, relays, and circuit breakers.

To derive an expression for the pull force in the air gap, we will use the virtual work approach. Using Equation (9.86) for magnetic stored energy and Equation (9.64) for magnetic circuit analysis, we can write the total energy stored in the magnetic circuit as

$$W_m = \frac{1}{2}\psi_m NI = \frac{(NI)^2}{2\,(\mathfrak{R}_c + \mathfrak{R}_g)} \tag{9C.1}$$

where

$$\mathfrak{R}_c + \mathfrak{R}_g = \frac{\ell_c}{\mu_c S_c} + \frac{\ell_g}{\mu_g S_g} \tag{9C.2}$$

which correspond to the total reluctance of the magnetic circuit.

Next, we equate the virtual work in moving the hinged arm an incremental angle to the incremental change in energy storage. Since the incremental motion of the hinged arm correspond to an incremental change in the air gap length then we can write

$$\Delta W_m = \vec{F}_m \cdot \overrightarrow{\Delta \ell_g} = -F_m \Delta \ell_g \tag{9C.3}$$

where "$-\vec{F}_m$" is the magnetic pull force on the moving arm, and $\overrightarrow{\Delta \ell_g}$ is the incremental increase in the air gap length.

Now, differentiating Equation (9C.1), we get

$$\Delta W_m = \frac{\partial W_m}{\partial \ell_g}\Delta \ell_g = -\frac{(NI)^2}{2(\mathfrak{R}_c + \mathfrak{R}_g)^2}\frac{\partial(\mathfrak{R}_c + \mathfrak{R}_g)}{\partial \ell_g}\Delta \ell_g = -\frac{(NI)^2}{2(\mathfrak{R}_c + \mathfrak{R}_g)^2}\frac{1}{\mu_g S_g}\Delta \ell_g \tag{9C.4}$$

Equating Equation (9C.3) to (9C.4), we extract the expression for the magnetic pull force in the air gap as

$$F_m = \frac{(NI)^2}{2(\mathfrak{R}_c + \mathfrak{R}_g)^2}\frac{1}{\mu_g S_g} = \frac{(\psi_m)^2}{2\mu_g S_g} = \left\{\frac{1}{2}\frac{B_g^2}{\mu_g}\right\} S_g = w_{mg} S_g \tag{9C.5}$$

This last form for the force expresses the pull force per unit area as equal to the volumetric magnetic energy density in the air gap.

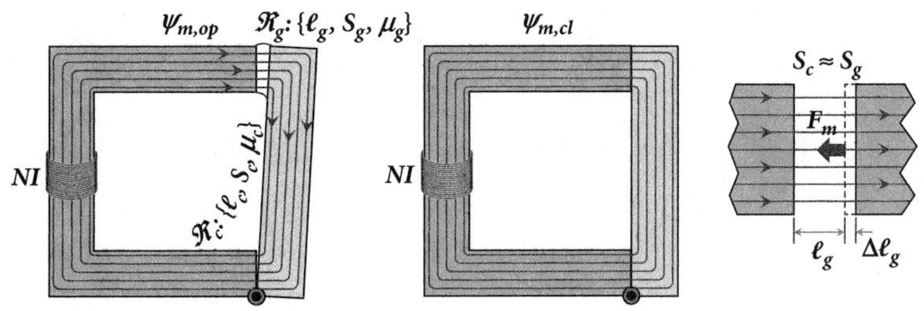

Figure 9C.1 An example of a magnetic circuit with an air gap for air gap magnetic force evaluation.

For design purposes, Equation (9C.5) can be used to determine the product NI required to produce the needed pull force in the air gap

$$NI = (\mathfrak{R}_c + \mathfrak{R}_g)\sqrt{2\mu_g S_g F_m} \tag{9C.6}$$

$$NI \cong \mathfrak{R}_g \sqrt{2\mu_g S_g F_m} \cong \ell_g \sqrt{\frac{2F_m}{\mu_g S_g}} \tag{9C.7}$$

9C.3 Magnetic Lift

Magnetic lifting of objects is a common practice in industrial automation as well as in recycling operations. The concept is the use of an electromagnet to generate the lifting (pull) force as the electromagnet approaches the object to be lifted. The object can then be moved to a desired destination where the magnetization is released by reducing (or turning off) the current in the electromagnet's coil. Figure 9C.2 demonstrates a schematic configuration of an electromagnet lift device.

The analysis is quite similar to the one above except for having two air gaps instead of one and the magnetic force is distributed between the two sides. Using Equation (9C.5) for the magnetic pull force, we can write

$$F_m = w_{mg} S_g \tag{9.140}$$

Meanwhile, the magnetic flux density in the gap can be evaluated using the *mmf* and reluctances of the magnetic circuit components

$$B_g S_g = \psi_m = \frac{NI}{(\mathfrak{R}_c + 2\mathfrak{R}_g + \mathfrak{R}_{wt})} \tag{9C.8}$$

Hence, the magnetic lifting force can be expressed as

$$F_{\text{lift}} = 2F_m = 2S_g w_{mg} = 2S_g \left\{ \frac{1}{2\mu_o} \left(\frac{\psi_m}{S_g} \right)^2 \right\} = \frac{1}{\mu_o S_g} \left\{ \frac{NI}{(\mathfrak{R}_c + 2\mathfrak{R}_g + \mathfrak{R}_{wt})} \right\}^2 \tag{9C.9}$$

Next, we analyze the same configuration with a permanent magnet instead of an electromagnet. Figure 9C.3 shows the magnetic circuit configuration for this case.

The analysis of this magnetic circuits using permanent magnets uses similar expressions to the one followed earlier in the chapter except for

- The absence of a current-carrying coil.
- The $\{B_{pm} - H_{pm}\}$ relationship is nonlinear. This relationship is typically available to us in graphical form as determined by the hysteresis curve for the magnetic material.

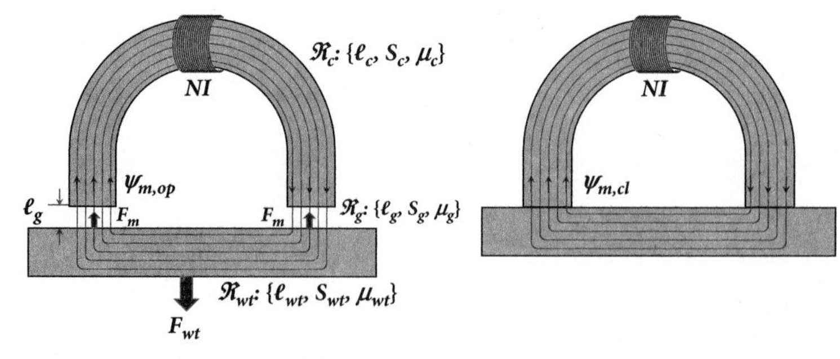

FIGURE 9C.2 An example of an electro-magnet magnetic circuit with an air gap for air gap magnetic force (pull) evaluation.

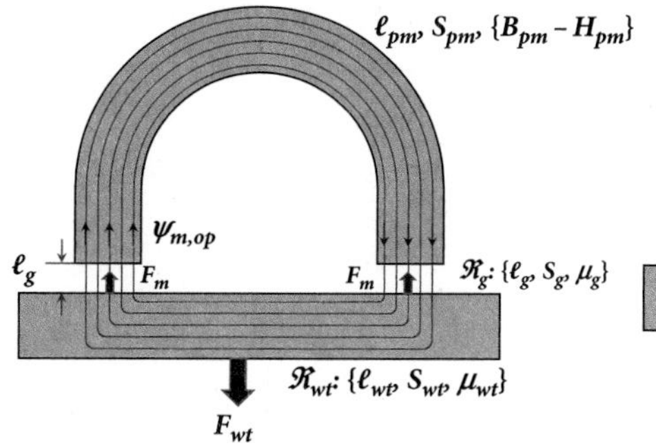

FIGURE 9C.3 An example of a permanent-magnet magnetic circuit with an air gap for air gap magnetic force (pull) evaluation.

Starting with Ampere's circuital law around the closed loop of the magnetic circuit

$$\oint \vec{H}\cdot\vec{d\ell} = H_{pm}\ell_{pm} + 2H_g\ell_g + H_{wt}\ell_{wt} = \sum NI = 0 \tag{9C.10}$$

It is obvious from this equation where the sum of three magnetic potential drops is equal to zero that at least one of the terms should be negative. A negative drop is in fact a source, and that is exactly the role the permanent magnet plays. By operating in the second quadrant of the magnetic hysteresis curve, the permanent magnet exhibits a negative magnetic field intensity and operates as the magnetic circuit source (Figure 9C.4).

Manipulating Equation (9C.10), we find

$$H_{pm}\ell_{pm} = -\{2H_g\ell_g + H_{wt}\ell_{wt}\} = -\left\{\frac{2B_g\ell_g}{\mu_g} + \frac{B_{wt}\ell_{wt}}{\mu_{wt}}\right\} = -(\psi_m)\left\{\frac{2\ell_g}{\mu_g S_g} + \frac{\ell_{wt}}{\mu_{wt} S_{wt}}\right\} = -(B_{pm}S_{pm})\{\mathfrak{R}_L\} \tag{9C.11}$$

$$\mathfrak{R}_L = \left\{\frac{2\ell_g}{\mu_g S_g} + \frac{\ell_{wt}}{\mu_{wt} S_{wt}}\right\} \tag{9C.12}$$

$$B_{pm} = -H_{pm}\frac{\ell_{pm}}{S_{pm}\mathfrak{R}_L} \tag{9C.13}$$

This is an equation of a straight line to be solved simultaneously with the B_{pm}-H_{pm} relationship given by the hysteresis curve of Figure 9C.4. The typical solution approach is to draw the "load" line of Equation (9C.13) on the same axes of the hysteresis curve as shown in Figure 9C.5. The intercept of the line and the B_{pm}-H_{pm} curve is the desired solution for the operating values of $B_{pm,\text{operating}}$ and $H_{pm,\text{operating}}$.

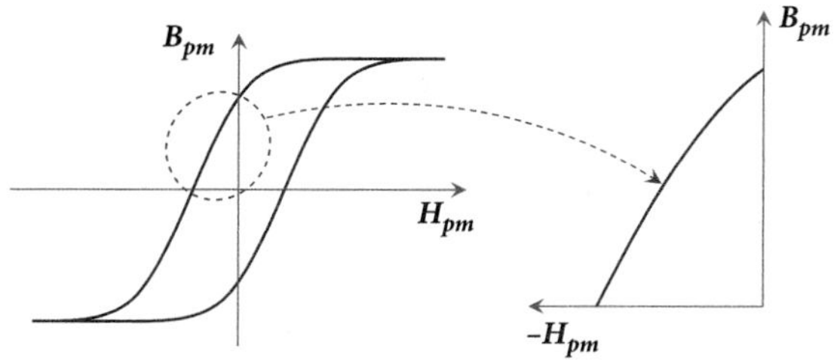

FIGURE 9C.4 The active region of the *B-H* hysteresis curve for magnetic circuits with permanent magnets.

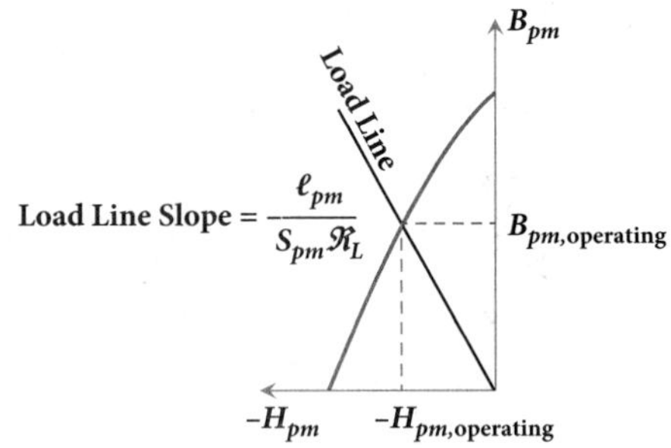

FIGURE 9C.5 An example of a graphical solution for the operation of a magnetic circuit with a permanent magnet.

CHAPTER 9 PROBLEMS

- Problems 9.1 to 9.7: Problems dealing with magnetic forces and torques
- Problems 9.8 to 9.13: Problems dealing with magnetic material and boundary conditions
- Problems 9.14 to 9.19: Problems dealing with magnetic circuits
- Problems 9.10, 9.11, 9.15, 9.16, and 9.19: Problems dealing with finding the magnetic energy storage
- Problems 9.14, 9.18, and 9.19: Problems dealing with magnetic pull
- Problems 9.20 and 9.21: Problems dealing with the magnetic flux-linkage
- Problems 9.22 to 9.32: Problems dealing with self- and mutual inductances

9.1 A particle with mass m carries charge q enters a uniform magnetic field confined to the region $0 < y < d$ with an initial velocity v_o along the y axis. Given the magnetic flux density B along the x axis, describe the motion of the particle for the following cases:

a. $d > \dfrac{mv}{qB}$

b. $d = \dfrac{mv}{qB}$

c. $d < \dfrac{mv}{qB}$

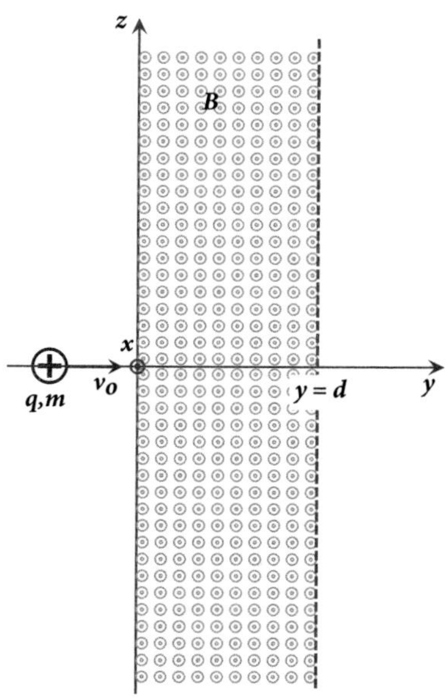

FIGURE 9P.1

9.2 Identify the forces and the torques (if any) on the current loop shown. Calculate their numerical values. Given $I = 1$ A, $B = 1T$, and the coil is 20×15 cm.

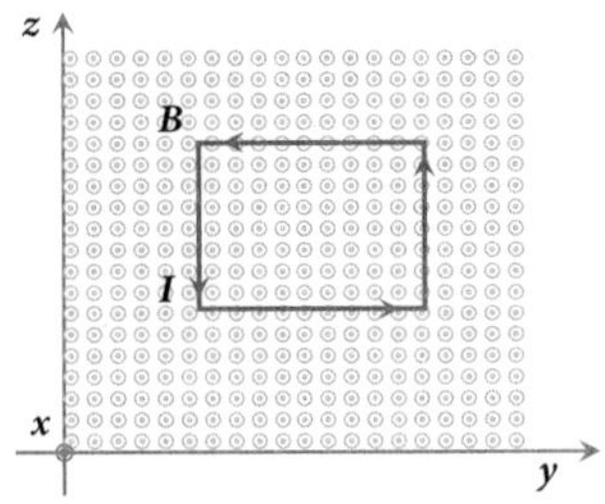

FIGURE 9P.2

9.3 Identify the forces and the torques (if any) on the current loop shown. Calculate their numerical values. Given $I = 1$ A, $B = 1T$.

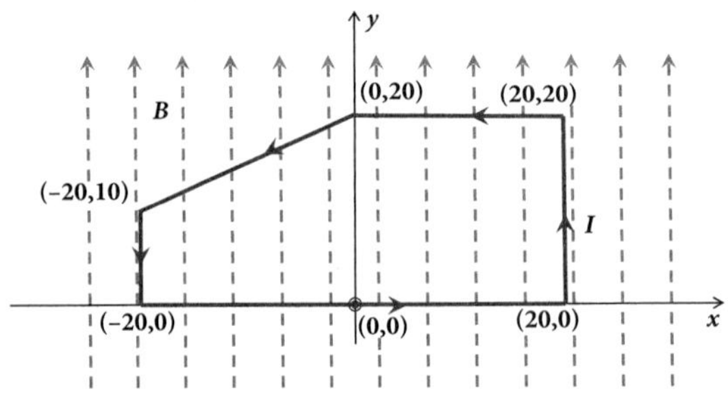

FIGURE 9P.3

9.4 For the same current loop in Problem 9.2 above, a DC current, $I = 10$ A flows in a loop in the x-y plane as shown. The dimensions are in centimeters. Assume a uniform magnetic flux density, $\vec{B}$, exists in the region where $\vec{B} = \vec{a}_y 1$ (*Tesla*).

a. Find the force on each wire section of the loop and the total force on the loop.

b. Determine the torque on the loop.

c. The torque will rotate the loop in the direction found in part b; in what plane will the loop be positioned after rotation and why?

d. The loop shown is initially placed in the x-y plane, in what plane should the loop be initially placed so that it experiences the maximum initial torque? What is the value of the maximum torque?

9.5 Identify the forces and the torques (if any) on the current loop shown. Calculate their numerical values. Given $I = 1$ A, $B = 1T$.

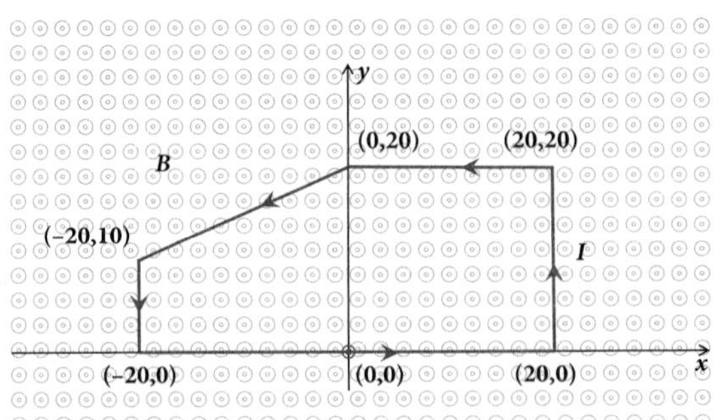

FIGURE 9P.5

9.6 Identify the forces and the torques (if any) on the current loop shown as a function of time. Calculate their numerical values. Given $I = 1$ A, $B = 1$ T, the coil is 20×15 cm and rotates at a constant angular velocity, $\theta(t) = \omega t$. The figure shows the coil at $t = 0$.

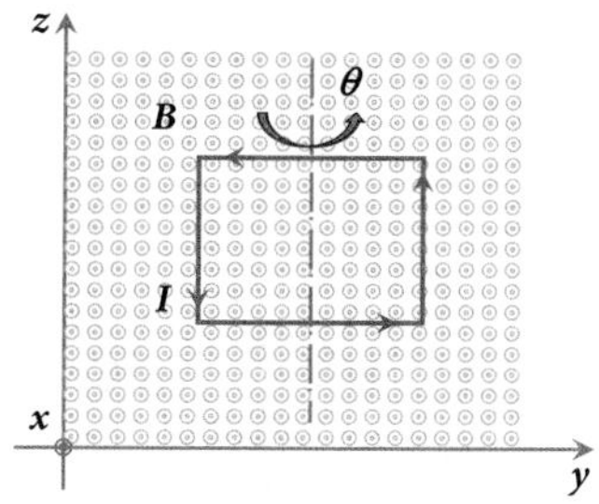

FIGURE 9P.6

9.7 Find the torque vector on the shown coil (lies in the y-z plane and carries current I in N turns) due to an applied uniform magnetic field intensity vector $H = \alpha \vec{a}_x + \beta \vec{a}_z \frac{A}{m}$. If the coil was free to move, describe its motion and final position (if any).

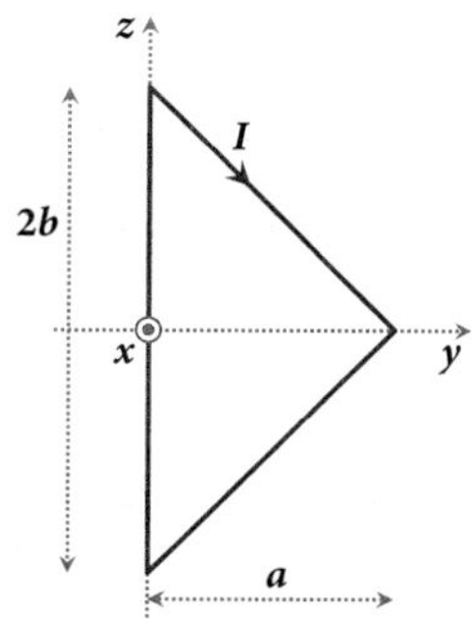

FIGURE 9P.7

9.8 Two concentric solenoids with radii $\rho = a$ and $\rho = b$. The core media for solenoid a has $\mu = \mu_1$ and the media filling the space between the two solenoids has $\mu = \mu_2$ and the outer region is air $\mu = \mu_o$. Derive expressions for H and B in all regions: $\rho < a$, $b > \rho > a$, and $\rho > b$ around the solenoids' central region away from the edges.

9.9 Find the magnetic field intensity and flux density inside an "infinite" extension slab of ferromagnetic material, $\mu_r >> 1$, immersed in a magnetic flux density vector placed at 45° as shown in the figure. Discuss the physics of the obtained results.

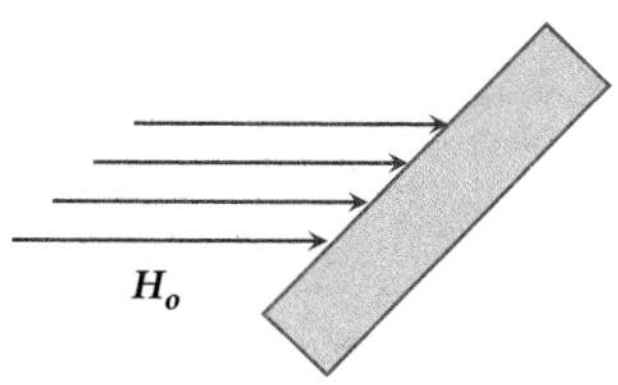

FIGURE 9P.9

9.10 A toroidal magnetic core with an average radius of 5 cm and a rectangular cross-sectional area of 1 cm width × 0.5 cm height. The relative permeability of the core is 900. A coil of 300 turns is wound around the core and carries a current of 6 A. Find H and B at all points in the core as well as integrate ψ_m for the core cross-section.

Also, find the energy stored in the core material.

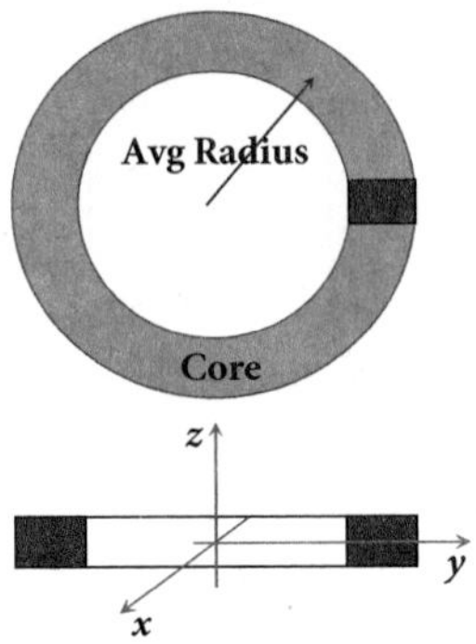

FIGURE 9P.10

9.11 A toroidal magnetic core with an average radius of 5 cm and a circular cross-sectional area of 0.8 cm^2. The relative permeability of the core is 900. A coil of 300 turns is wound around the core and carries a current of 6 A. Find the average values for H, B, ψ_m in the core. Also, find the energy stored in the core material.

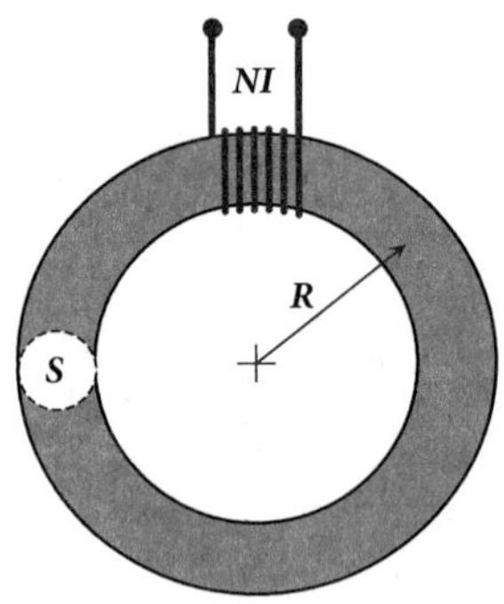

FIGURE 9P.11

9.12 Given the setup shown in the figure with a circular coil of radius b carrying a current I. A fairly long nonmagnetic metal core with circular cross section, radius a, is located concentric with the coil. Derive an expression for the magnetic field at the center point of the coil. Sketch the magnetic field map for the vertical projection (on the right side of the figure). Repeat the map sketching if the core is changed to a high-permeability magnetic material.

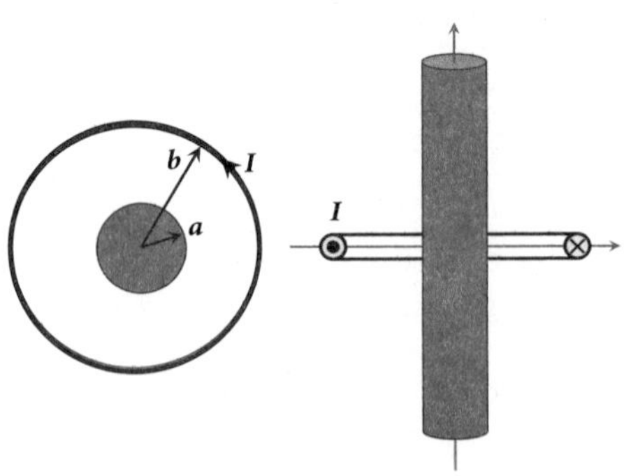

FIGURE 9P.12

9.13 Repeat Problem 9.11 for the case of a square cross-section core with dimensions $a \times a$.

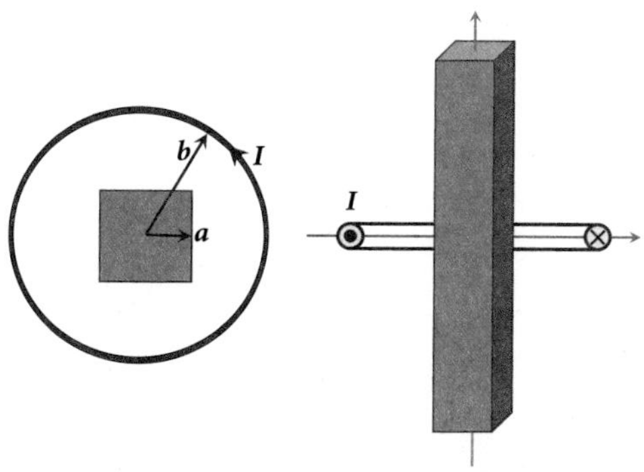

FIGURE 9P.13

9.14 The magnetic circuit shown is excited with an mmf of 2500 AT, find

a. B and H in the gap

b. B and H in the core material

c. Force on bottom piece to maintain gap

Given: $R = 5$ cm, $S_c = 1$ cm^2, $S_b = 2$ cm^2, $\ell_{gap} = 1$ mm, $\mu_{rc} = \mu_{rb} = 1000$

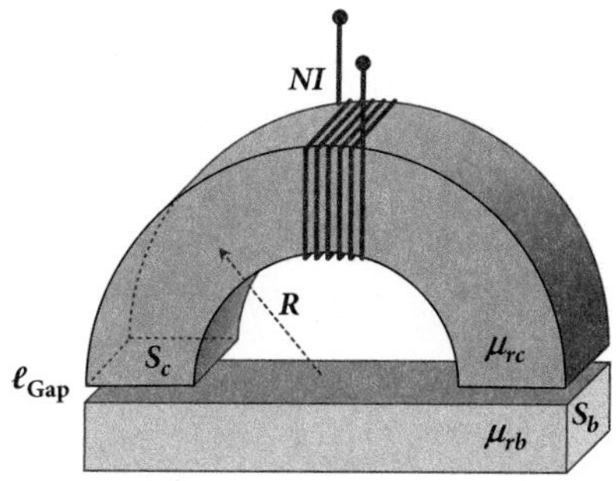

FIGURE 9P.14

9.15 Find the energy density in the magnetic core and in the air gap of the shown magnetic circuit. Evaluate the total magnetic energy in the magnetic circuit.

9.16 For the magnetic circuit shown, find

a. B and H in all the sections of the core material

b. Energy stored in each section

c. B and H in the gap

d. Energy stored in the air gap

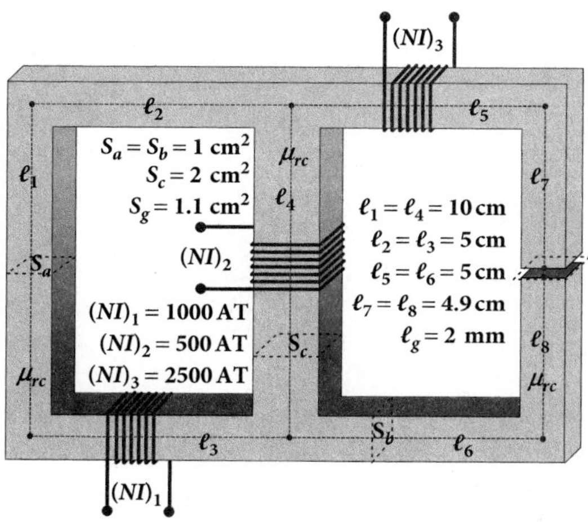

FIGURE 9P.16

9.17 Find the magnetic flux in the magnetic circuit shown in Problem 9.13 with $N = 500$ turns, $I = 2$ A, average radius $R = 10$ cm, cross-sectional area $S_c = 10\ \text{cm}^2$, and relative permeability of 1000. The air gap length ℓ_{gap} is 2 mm on either sides of the core, and has an effective cross-sectional area $S_g = 11\ \text{cm}^2$. The bottom section has cross-sectional area $S_b = 15\ \text{cm}^2$ and relative permeability $\mu_{rb} = 720$.

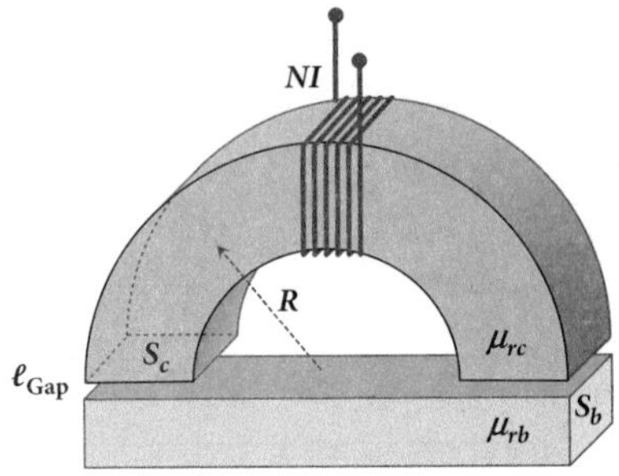

FIGURE 9P.17

9.18 The figure shows a *permanent* magnet in the form of a semicircular horse-shoe with an average radius of $R = 10$ cm and a cross-sectional area $S_c = 15\ \text{cm}^2$. The *B-H* curve shown represents the magnetic material of this material. The magnet is placed at 5 mm above an iron object with relative permeability $\mu_r = 720$.

a. Determine the maximum weight for the iron object that this magnet can lift.

b. Repeat for a 1-mm gap. Neglect fringing.

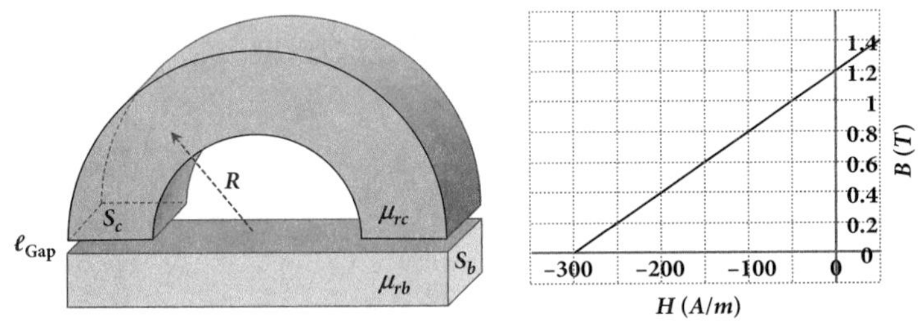

FIGURE 9P.18

9.19 For the magnetic circuit shown in Problem 9.13 with $N = 500$ turns and average radius of $R = 10$ cm. The bottom section has a cross-sectional area $S_b = 15\ \text{cm}^2$ and relative permeability $\mu_{rb} = 720$. Neglecting the reluctance of the core, find

a. The current needed to be able to lift a 5-kg mass with 1 mm air gap clearance

b. The magnetic energy stored in each (1 mm) gap as well as the magnetic energy stored in the bottom section

c. The self-inductance of the coil with and without the 1 mm gap

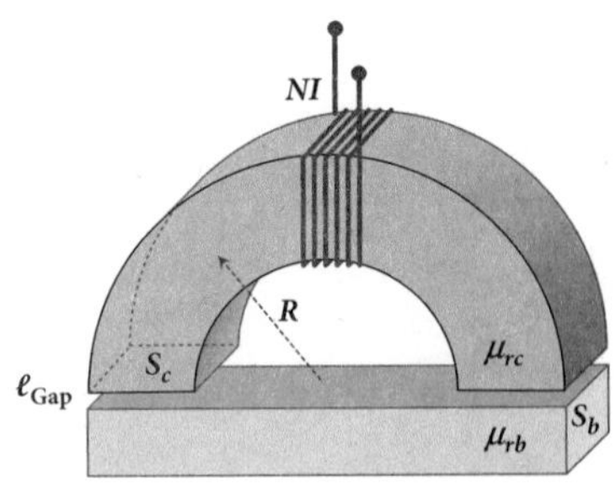

FIGURE 9P.19

9.20 Calculate the flux-linkage for the conducting loop shown due to the current I_1. Given $I_1 = 1$ A, $d = 15$ cm, $a = 10$ cm, and $b = 20$ cm.

Repeat if the loop rotates around the y axis with a constant angular velocity of 60 turns per second. The figure shows the loop at $t = 0$.

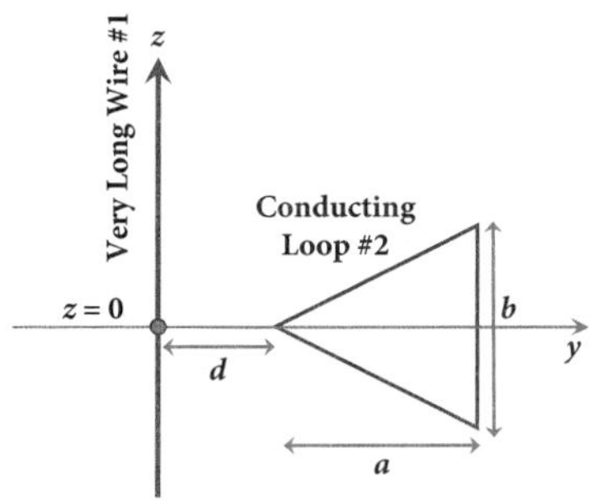

Figure 9P.20

9.21 Calculate the flux-linkage for the conducting loop shown due to the current I_1. Given $I_1 = 1$ A, $d = 15$ cm, $a = 20$ cm, and $b = 20$ cm.

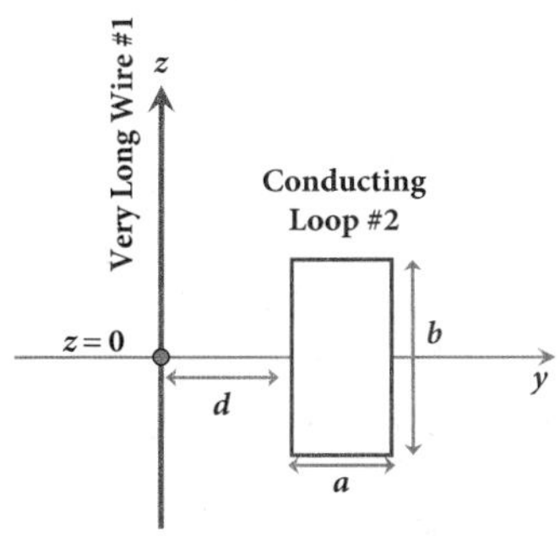

Figure 9P.21

9.22 Derive an expression for the inductance per unit axial length of the coaxial cable with thin (zero thickness) conductors with hollow inner conductor of radius a, and an outer conductor of radius b.

9.23 Find the self-inductance per unit length of a twin-conductor line consisting of two parallel circular cross-section wires (radius r each) located at a distance d (center to center) from each other. Wires are parts of an *infinitely* extended circuit, that is, carry opposite currents.

9.24 Two concentric solenoids of radii a and b each has n turns per meter along the z axis. Solenoid a carries a current I in the clockwise direction and the same current flows in solenoid b in the opposite direction.

Derive an expression for the inductance per unit axial length of the one coil formed by connecting the two solenoids as one circuit.

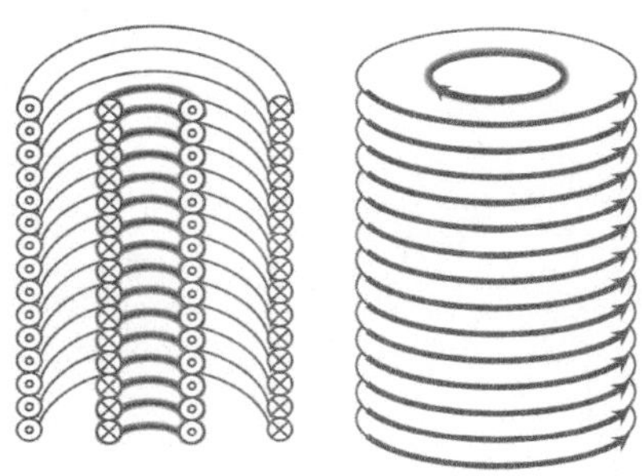

Figure 9P.24

9.25 Derive an expression for the mutual inductance L_{AB} between two concentric solenoids A and B, of radii A and B, respectively. Given the number of turns per unit axial length as n_A and n_B, respectively.

9.26 Derive an expression for the mutual inductance between two concentric solenoids: solenoid #1 has radius a meters, n_a turns/unit axial length, wound around a magnetic core with relative permeability of μ_r, and solenoid #2 has radius $(b > a)$ meters, n_b turns/unit axial length, wound around on a hollow plastic tube (nonmagnetic).

9.27 Derive an expression for the mutual inductance between:

- Coil #1: an infinite solenoid with a coil radius of a meters, an air core, and n turns per unit axial length, placed coaxially to the z axis
- Coil #2: a single turn coil with radius b meters $(b > 2a)$, placed at 45° angle with respect to the solenoid and centered at the origin

Repeat for the case $b < a$.

9.28 Find the mutual inductance between a toroidal coil (coil #2), and an infinite filamentary wire (coil #1), placed concentric with the toroid.

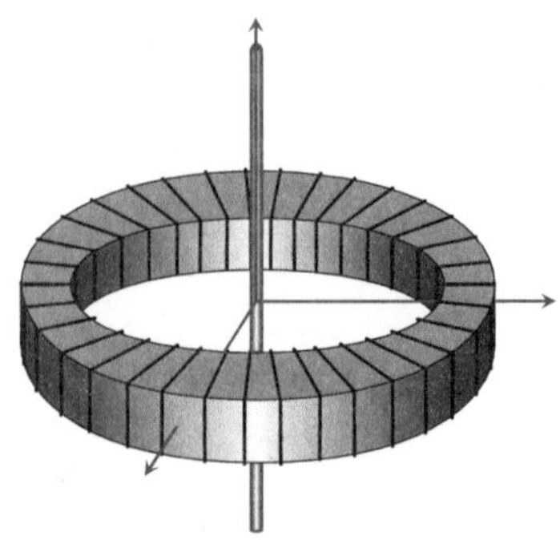

Figure 9P.28

9.29 Find the mutual inductance between the two coils of the transformer shown in the figure. The primary coil is 500 turns uniformly and tightly wound around the horse-shoe (semicircular)-shaped section of the magnetic core with an average radius of $R = 2$ cm. The secondary coil has 20 turns wound around the straight section of the core. The two sections are separated by a 2 mm gap. Both core sections have cross-sectional area $S = 1\ \text{cm}^2$ and relative permeabilities $\mu_r = 720$.

Repeat where the horse-shoe core of the primary coil is enclosed with a nonmagnetic 2 mm thick plastic casing. The figure to the right only shows the portion of the plastic casing that fills the separation gap.

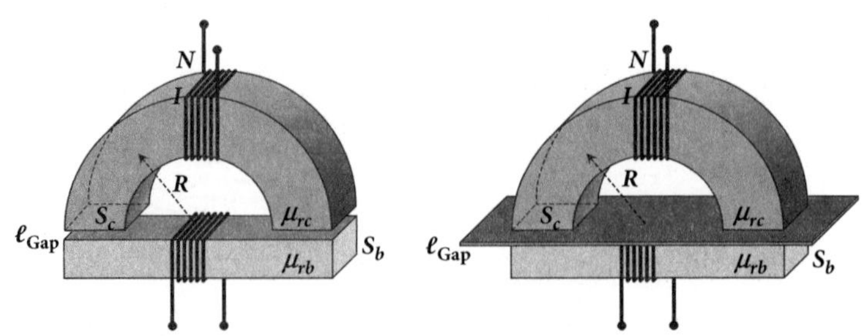

Figure 9P.29

9.30 For the figure of Problem 9.19, evaluate the mutual inductance between the long wire #1 and the conducting loop #2 shown. Given $d = 15$ cm, $a = 20$ cm, and $b = 20$ cm.

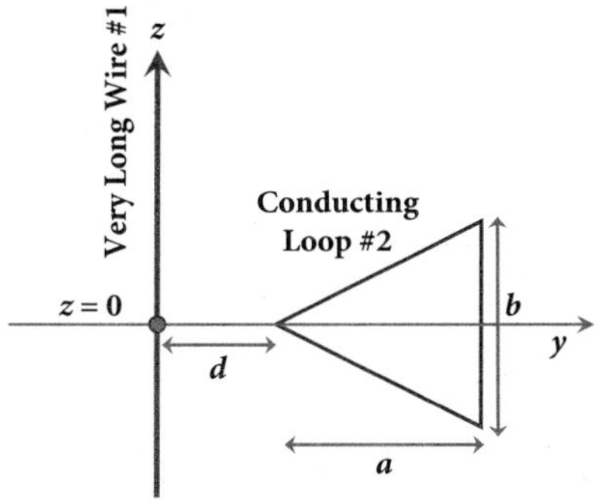

FIGURE 9P.30

9.31 For the figure of Problem 9.20, evaluate the mutual inductance between the long wire #1 and the conducting loop #2 shown. Given $d = 15$ cm, $a = 10$ cm, and $b = 20$ cm.

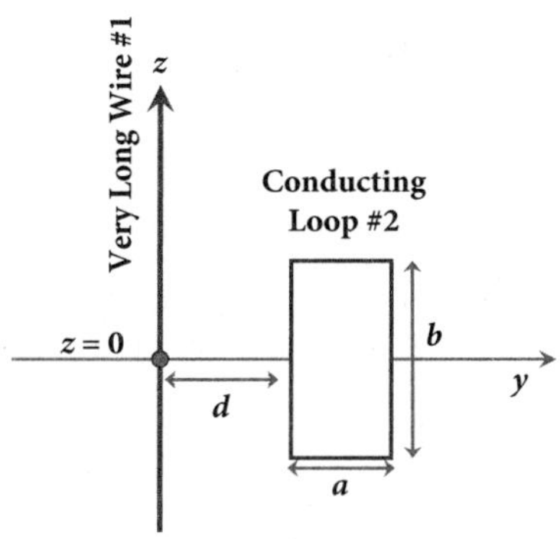

FIGURE 9P.31

9.32 Find expressions for the mutual inductance between the very long straight wire along the z axis and the conducting square loop shown in the three figures.

Considering that the three loops have equal area, discuss the difference between the three results in physical terms.

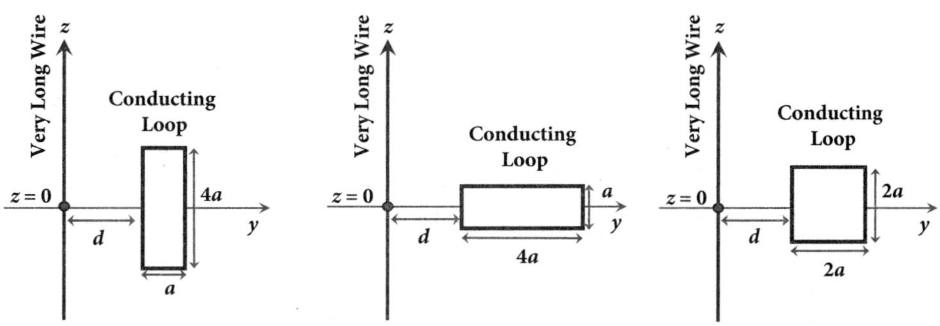

FIGURE 9P.32

CHAPTER 9 SUMMARY[1]

Magnetic Force and Torque

- Magnetic force on moving charge: $\overrightarrow{dF_m} = dq\vec{v} \times \vec{B}$
- Magnetic force and torque on current loop: $\overrightarrow{dT}_{m,\text{total}} = \overrightarrow{IdS} \times \vec{B}$
- $\vec{T}_{m,\text{loop}} = \iint_S \overrightarrow{IdS} \times \vec{B} \qquad \vec{F}_{m,\text{loop}} = \int_{\text{loop}} \overrightarrow{Id\ell} \times \vec{B}$
- Energy storage in magnetic fields: $W_m = \int_{\text{volume}} \frac{1}{2}\{\vec{B}\cdot\vec{H}\}dv = \int_{\text{volume}} \frac{1}{2}\mu_o H^2\, dv$
- Energy density in magnetic fields: $w_m = \frac{1}{2}\{\vec{B}\cdot\vec{H}\} = \frac{1}{2}\mu_o H^2 = \frac{1}{2}\frac{B^2}{\mu_o}$

Magnetic Properties of Materials

- Dipole moments and magnetization vector: $\vec{M} = \frac{\sum_{\Delta v} \vec{m}}{\Delta v} = \chi_m \vec{H}_{\text{applied}}$
- $\vec{B} = \mu_o\{\vec{H} + \vec{M}\} = \mu_o\{1 + \chi_m\}\vec{H} = \mu_o\{\mu_r\}\vec{H} = \mu\vec{H}$
- Paramagnetism (e.g., aluminum, tin, tungsten, and platinum), $\mu = \mu_o(1 + \text{tiny amount})$
- Diamagnetism (e.g., gold, silver, copper, lead, and carbon), $\mu = \mu_o(1 - \text{tiny amount}) \approx \mu_o$
- Ferromagnetism (four elements: iron, nickel, cobalt, and gadolinium), $\mu \gg \mu_o$
- Residual magnetism and permanent magnets: magnetized ferromagnetic material retains residual magnetic polarization after the removal of the magnetizing field, acts as "permanent" magnet.

Magnetic Boundary Conditions

- Interface between two materials: $\begin{cases} B_{1n} = B_{2n} & \text{(continuity of normal flux density)} \\ [H_{1t} - H_{2t}] = K & (K \text{ is linear density of surface current}) \end{cases}$

  If no surface currents: $\begin{cases} B_{1n} = B_{2n} \\ H_{1t} = H_{2t} \end{cases} \rightarrow \frac{\tan\theta_1}{\mu_1} = \frac{\tan\theta_2}{\mu_2}$

- Magnetic flux confinement in magnetic materials (due to interface with nonmagnetic material)
  If $\mu_1 >> \mu_2 \rightarrow \theta_1 \approx 90°$, flux is tangential to interface—zero normal flux density component

Magnetic Circuits

- Magnetic circuit analysis using the electrical circuit analogy: $V_{m\ell} = \psi_m \Re_\ell$
- [KVL]: $\sum_{\text{closed loop}} mmf = \sum_{\text{closed loop}} NI = \sum_{\text{closed loop}} \psi_m \Re_\ell$ [KCL]: $\sum_{\text{closed surface}} \psi_m = 0$
- Magnetic reluctance: $\Delta\Re = \frac{\Delta\ell}{\mu\Delta S} - \Re_\ell = \int_\ell \frac{1}{\iint_S \frac{\mu dS}{d\ell}} = \frac{1}{\iint_S \frac{1}{\int_\ell \frac{d\ell}{\mu dS}}}$

[1]***Important Warning to Students:*** Please do not use this summary without thorough study of the relevant material in this chapter. Using this summary and its formulas without adequate appreciation of the physics involved and conditions of validity can lead to wrong implementations and invalid conclusions.

Inductance

- Flux-linkage: Λ = Magnetic flux multiplied by the number of times it links a current (or current coil)
- The linkage number is a fraction if the flux links a portion of the current
- Self- and mutual inductances: $L_{\text{coil}} = \Lambda_{\text{coil due to own current}}/I_{\text{coil}}$, $L_{\text{mutual}} = \Lambda_{\text{coil due to other coil current}}/I_{\text{other coil}}$
- Inductance relationship to reluctance
- $L_{11} = \dfrac{N_1^2}{\mathfrak{R}_{11}}$ and $L_{21} = \dfrac{N_2 N_1}{\mathfrak{R}_{21}}$ for coils having N_1 and N_2 turns in linkage
- Energy stored in inductances: $W_m = \dfrac{1}{2}\Lambda I = \dfrac{1}{2}LI^2 = \dfrac{1}{2}\Lambda^2/L$

Addenda: Examples of Evaluation of Self- and Mutual Inductances and Magnetic Pull

- *A*: Self-inductance evaluation using flux-linkage, magnetic reluctance, and magnetic energy storage
- *B*: Mutual Inductance evaluation using flux-linkage, magnetic reluctance
- *C*: Magnetic forces in air gaps and magnetic lift

CHAPTER 10

Time-Varying Fields—Faraday's Law—Maxwell's Equations

Michael Faraday
(1791–1867)

Heinrich Friedrich Emil Lenz
(1804–1865)

James Clerk Maxwell
(1831–1879)

Oliver Heaviside
(1850–1925)

Chapter Learning Objectives

- Describe the trajectory of a charged particle injected into a magnetic field.
- Derive an expression for the Hall effect of a conducting slab placed in a magnetic field.
- Derive expressions for the induced *emf* in different coil configurations immersed in time-varying magnetic field.
- Derive expressions for the induced *emf* in different coil configurations moving in static magnetic field.
- Derive expressions for the induced *emf* in different coil configurations moving in time-varying magnetic field.
- Explain the principles of operation of the electric generator and the transformer.
- Express Faraday's law in both the integral and differential forms.
- Explain the physical concepts of the displacement current. Derive expressions for the displacement currents for a variety of cases.
- State Maxwell's equations including the time-varying modifications. Identify the physics involved in each equation.
- Derive expressions for the energy stored in the magnetic field produced in a coil and its relationship to coil inductance.

10.1 Introduction

In previous chapters, we discussed both electric and magnetic static fields. We learned that the mere presence of electric charges is responsible for producing electric forces and fields. We also learned that moving electric charges is the cause of magnetic forces and fields. Static electric fields are caused by constant charge distributions (static with no time variations). Meanwhile, static magnetic fields are caused by constant electric current distributions (DC, direct currents).

In this chapter, we begin exploring time-varying electromagnetic phenomena. In other words, we will study the general case of electromagnetic fields where both charges and current distributions vary with time. Our study will lead us to learn about Faraday's law and modify the static versions of Ampere's law and other static field and potential relationships, and at the end of the chapter, we will conclude with the famous Maxwell's equations.

Now, before we get into Faraday's law, let us arm ourselves with some useful relevant knowledge. We will start by examining the motion of electrical charges in magnetic fields. We will also study the physics of the Hall effect where an induced electric field is produced across conductors and semiconductors carrying an electrical current while being exposed to a transverse magnetic field.

10.2 Charge Trajectory in Magnetic Fields

This topic is of interest to a few applications where a moving charge carrier is launched in the domain of a magnetic field in an unconstrained space (no material to impede the motion of the charge carriers). Examples are the deflection of electron beams in cathode ray tubes and the circulation of electrons in traveling wave tubes and magnetrons.

To study this topic, let us assume that a single point-charge carrier is launched with some initial velocity on a plane that is orthogonal to a uniform magnetic field. This is demonstrated in Figure 10.1 where both positive and negative charges are launched on the plane of the page to which an out-of-the-page uniform magnetic field is applied. Let us further assume that we can ignore the effect of either applied or induced electric fields in the charge path and focus only on the forces due to the magnetic field as the dominant contributor to the changes in the charge trajectory. Using the notation v_+ and v_- for the velocities of the positive and negative charges, respectively, we can write expressions for the respective magnetic forces on both charges, F_{m+} and F_{m-}, as

$$\vec{F}_{m\pm} = q_\pm \vec{v}_\pm \times \vec{B} \tag{10.1}$$

where $\vec{B}$ denotes the magnetic flux density.

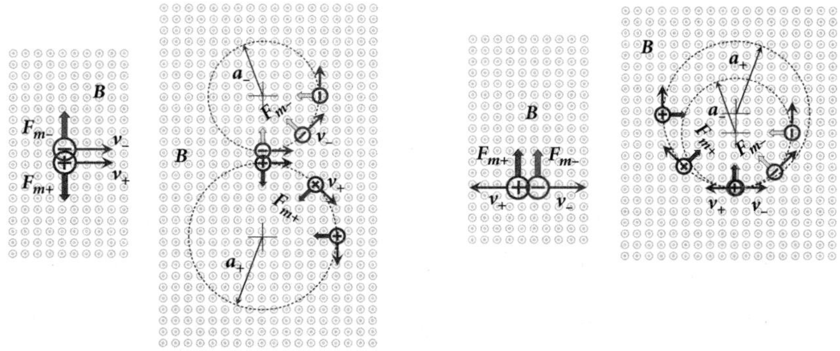

FIGURE 10.1 The trajectory of an electric charge injected in a magnetic field.

It is obvious from the cross product in Equation (10.1) that the magnetic forces on both charges are orthogonal to the charge velocities. These forces, having no components along the charge velocities will not contribute to any linear accelerations in the direction of motion. Hence, the charges will maintain their initial *linear* velocities to be the same value as those at the initial launch. Meanwhile, due to the orthogonal forces, angular/radial accelerations are developed causing continuous normal deflection to the charge trajectories. The result is charge trajectories in the form of circles with constant tangential velocities equal to the initial launch velocities and radii of $a_\pm$ given by

$$\left|\vec{F}_{m\pm}\right| = \left|q_\pm \vec{v}_\pm \times \vec{B}\right| = \left|q_\pm\right| v_\pm B = \frac{m_\pm (v_\pm)^2}{a_\pm}, \text{ hence } a_\pm = \frac{m_\pm v_\pm}{\left|q_\pm\right| B} \tag{10.2}$$

where $m_\pm$ are the masses of the charge particles.

Examining Figure 10.1, we notice that positive and negative charges launched in the same magnetic field and in the same direction will have opposite trajectories. This charge separation (polarization) is expected to induce an internal electric field between the positive and negative charges. Likewise, for charges moving in opposite directions, like the case of an electric current, we expect to observe positive and negative charges to have different trajectories. As per Equation (10.2), and due to the likely difference in the masses and/or the velocities of the two charge carriers, the two types of charges would have different trajectory radii. This, again, would result in an induced electric field.

In the following, we present the Hall effect in which the path of the charge flow is confined to a body of material. This will in turn limit the charge carrier trajectories by the material boundaries as well as its internal lattice structure. Nevertheless, we expect to observe charge polarization and hence an induced internal electric field.

10.3 Hall Effect

The Hall effect is observed in conductors and semiconductors carrying an electrical current while being exposed to a transverse magnetic field. The diagram of Figure 10.2 demonstrates the concept for a conductor or an n-type semiconductor where electrons represent the charge carriers responsible for the current flow. Under the influence of the transverse magnetic field, the electrons experience a magnetic force that deflects their path of motion and causes a negative charge concentration on one side of the conductor leaving a positive charge concentration of the other side. This charge *polarization* will be felt in the form of an induced internal electric field within the material and a corresponding electric potential difference across its two sides.

Obviously, this induced electric field will exert an electrical force on the charge carriers. This force will be in an opposite direction to that of the magnetic force and a force balance is reached when

$$q\vec{v} \times \vec{B} + q\vec{E}_i = 0 \tag{10.3}$$

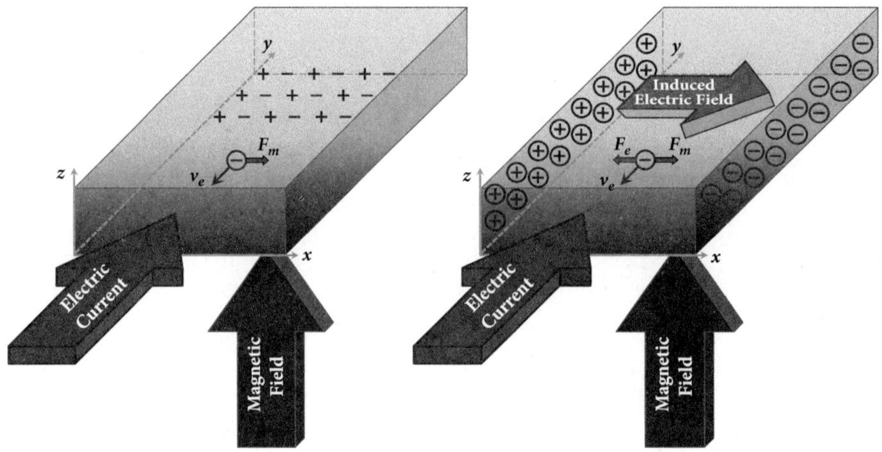

FIGURE 10.2 The Hall effect: a current flow orthogonal to a magnetic field results in charge polarization and hence an induced electric field.

hence,

$$\vec{E}_i = -\vec{v} \times \vec{B} \tag{10.4}$$

Now, let us add a twist to this discussion by considering a zero current conductor moving transversely to the magnetic field (Figure 10.3). In this case, no charges are moving within the conductor; however, the whole conductor including its internal charges is moving in the magnetic field. Consequently, we expect magnetic forces to develop on the charge carriers causing similar charge polarization as in the case of the current-carrying conductor. Again, there will be an induced electric field as a consequence of the charge polarization. The induced electric field is given by the same form [Equation (10.4)].

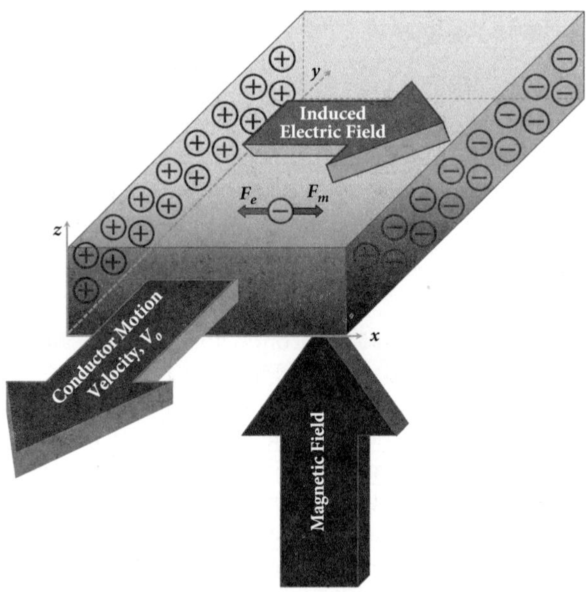

FIGURE 10.3 A conductor in motion across a magnetic field results in charge polarization and hence an induced electric field.

Example 10.1 One of the applications of the Hall effect is the measurement of the majority carrier concentration in a semiconductor material. In an N-type semiconductor material, the majority current carriers are electrons, while in a P-type semiconductor, the majority current carriers are positively charged holes.

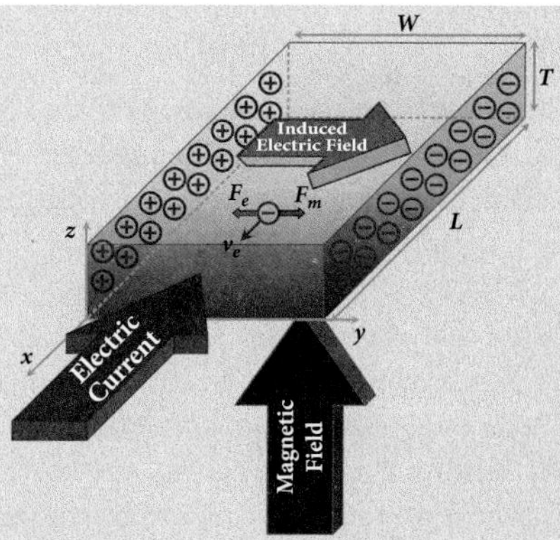

Solution Assume an N-type semiconductor bar as shown in the figure. The thickness of the semiconductor bar is T, the width is W, and the length is L. The current as shown in the figure is in the negative x direction, the magnetic field is in the positive z direction.

Starting from Equation (10.4)

$$\vec{E}_i = -\vec{v} \times \vec{B} \rightarrow -v_x\vec{a}_x \times B_z\vec{a}_z = v_xB_z\vec{a}_y = E_y\vec{a}_y \tag{1}$$

or

$$E_y = v_xB_z \tag{2}$$

where v_x is the drift velocity of the current carriers. For N-type semiconductors, v_x is the velocity of the electrons. For a current in the $-x$ direction, v_x is in the $+x$ direction. The current in the x direction is related to the drift velocity of the carrier as follows:

$$I_x = -qnv_xA \tag{3}$$

where A is the cross-sectional area, $A = WT$, q is the electronic charge, and n is the electron concentration in $\#/\text{cm}^3$. The negative sign accounts for the fact the direction of current flow is opposite to the direction of flow of the majority carriers, electrons in this case.

Substituting v_x from (3) into (2) and using $E_y = \frac{V_y}{W}$, where V_y is the Hall voltage induced

$$\frac{V_y}{W} = \frac{I_x}{-qnWT}B_z \rightarrow V_y = R_H\frac{I_xB_z}{T} \tag{4}$$

where R_H is referred to as the Hall coefficient $R_H = \frac{1}{-qn}$ for an N-type semiconductor. $R_H = \frac{1}{qp}$ for a P-type semiconductor. According to the equation above, measuring the Hall voltage can be used to find the majority carrier concentration n or p in a semiconductor material. Note that the Hall coefficient has opposite signs for an N-type and P-type semiconductor and therefore the induced Hall voltage will have opposite polarity.

Example 10.2 In an experiment to measure the doping concentration of an N-type semiconductor. The setup in the figure in Example 10.1 is used. A semiconductor bar of thickness 0.1 mm is placed in a magnetic field of $\vec{B} = 0.05\vec{a}_z$ tesla(T). The input current was –40 mA and the induced Hall voltage was measured to be 10 mV. Find the electron concentration.

$$V_y = R_H\frac{I_xB_z}{T} \quad \text{or} \quad R_H = \frac{V_yT}{I_xB_z} = \frac{0.01\times10^{-4}}{-.04\times0.05} = -0.0005$$

$$R_H = -\frac{1}{qn} \quad n = -\frac{1}{-0.0005\times1.6\times10^{-19}} = 1.25\times10^{22}\ \text{m}^{-3} = 1.25\times10^{16}\ \text{cm}^{-3}$$

10.4 Faraday's Law

The above discussion leads us to the presentation of Faraday's law. Motivated by Oersted's discovery that an electric current produces magnetism, Faraday worked hard to prove that the inverse is also possible. In 1831, Michael Faraday experimentally proved that an electric current could be generated in conductors immersed in changing magnetic fields.

To demonstrate the concept, we consider the diagrams in Figure 10.4. These diagrams represent the experiments that Faraday used to prove that a magnetic field could produce an electric current. In both diagrams, a magnet (permanent or electric) produces a magnetic flux that links the coil. The coil is a part of a closed circuit that includes a galvanometer to detect any current flow in the circuit. Whenever a change occurred to the magnetic flux linking the coil, a galvanometer reading was noted. The magnetic flux linkage changes were made through relative motion of the coil and the magnetic flux or by changing the current of the electromagnet.

Based on this experimentation, Faraday concluded that an electromotive force, *emf*, is induced in the coil as determined by the rate of change of magnetic flux linkage of the coil. Analytically, the *emf* is given by

$$emf = -\frac{\partial \Lambda}{\partial t}\left(= -\frac{\partial \psi_m}{\partial t} \text{ for single-turn coils}\right) \tag{10.5}$$

where Λ is the magnetic flux linkage of the coil, and the negative sign indicates the direction of the induced *emf*. The direction of this *emf* is the direction of the current that would flow in the coil to counter the change of the coil's magnetic flux linkage. In other words, if the flux linkage increases with time, the induced *emf* will be in the direction of the current that would produce an opposite magnetic flux in an attempt to counter the flux linkage increase.

To further this discussion, let us get back to the example of a conductor moving across a magnetic field, e.g., Figure 10.5. The figure demonstrates a conductor of length ℓ placed horizontally in the x-y plane with its length parallel to the x axis. The conductor is made to move at a constant velocity v_o along the y

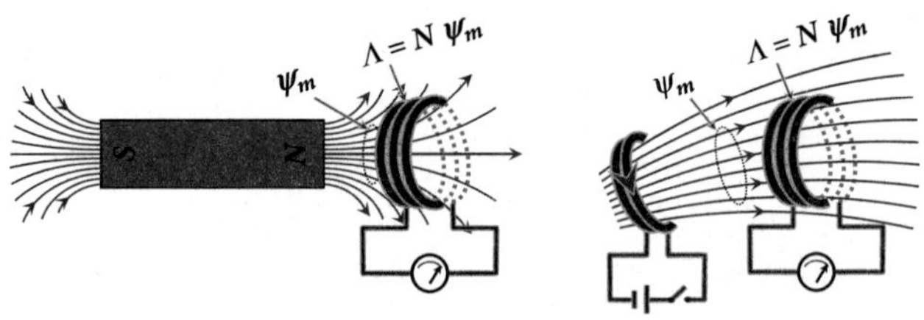

Figure 10.4 A demonstration of Faraday's discovery of an induced electric current in a coil immersed in a changing magnetic field. In the diagram to the left, a relative motion of a permanent magnet causes the change in the coil's flux linkage. In the right diagram, the change is caused by a time-varying (switching) current in a mutually coupled coil.

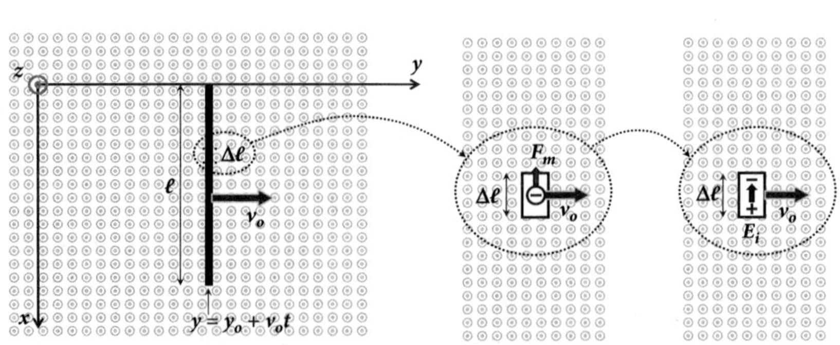

Figure 10.5 An example for the *emf* induced in a conductor moving across a magnetic field.

direction transversely across a uniform magnetic field with a flux density B pointing in the positive z direction. As the figure demonstrates, internal to the conductor, the charge carriers will undergo polarization under the influence of the magnetic force due to the conductor motion across the magnetic field. The induced electric field is then given by Equation (10.4). This induced electric field will be responsible for the development of an induced *emf* between the conductor terminals given by

$$emf = \int_{\ell} \vec{E}_i \cdot \vec{d\ell} = \int_{\ell} \{\vec{a}_x E_i\} \cdot \vec{a}_x \, d\ell = -\{\vec{a}_y v_o \times \vec{a}_z B\} \cdot \vec{a}_x \ell = -\{v_o B\}\ell \tag{10.6}$$

Now, we consider the moving conductor as part of a closed loop as shown in Figure 10.6. As we do that, we are now able to visualize the conductor motion in terms of a change in a magnetic flux linkage. The relationship between the conductor motion and the rate of change of the flux linkage for the moving conductor is given by

$$\frac{\Delta\psi_m}{\Delta t} = \frac{B\ell\Delta y}{\Delta t}, \text{ and hence } \frac{\partial\psi_m}{\partial t} = B\ell\frac{\partial y}{\partial t} = B\ell v_o \tag{10.7}$$

It is clear that both Equations (10.6) and (10.7) have the same right-hand side $B\ell v_o$; hence the two left-hand sides, *emf* and $\frac{\partial\psi_m}{\partial t}$, are supposed to be equal, which is what we expect based on Faraday's law except for missing the polarity. Well, as was discussed earlier, the polarity of the induced *emf* is expected in the direction of the current that would produce a magnetic flux in an attempt to counter the flux linkage change. In the figure, we notice the induced *emf*, which is positive on the lower end of the conductor, would push a clockwise current in the conductor loop. This current would then produce a negative z magnetic flux (into the page) which would counter the $\Delta\psi_m$ flux increase due to the conductor's motion. Hence, the result of our analysis in this case is in full agreement with Faraday's law in both magnitude and polarity.

Example 10.3 This is the same "example" depicted in Figure 10.6 repeated with more details on how to conclude the polarity of the induced *emf*, and also accounting for a general direction of the magnetic field.

Solution Using Equation (10.5) $emf = \oint \vec{E} \cdot \vec{d\ell} = -\frac{\partial\psi_m}{\partial t} = -\frac{\partial}{\partial t}\int \vec{B} \cdot \vec{dS}$

In the above expression, the direction of the circulation on the contour integral in $-\oint \vec{E} \cdot \vec{d\ell}$ and the direction of the differential element of area in $\int \vec{B} \cdot \vec{dS}$ are governed by the right-hand rule. Therefore, if the thumb is pointing in the direction of the vector $\vec{dS}$, the fingers of the right hand curl in the direction of the voltage drop in the *emf* source (from the positive terminal to the negative terminal), that is, in the direction of the assumed induced current in an external load connected across the terminals. If the induced *emf* was found to be negative, it means that the induced *emf* and current have polarities opposite to the assumption.

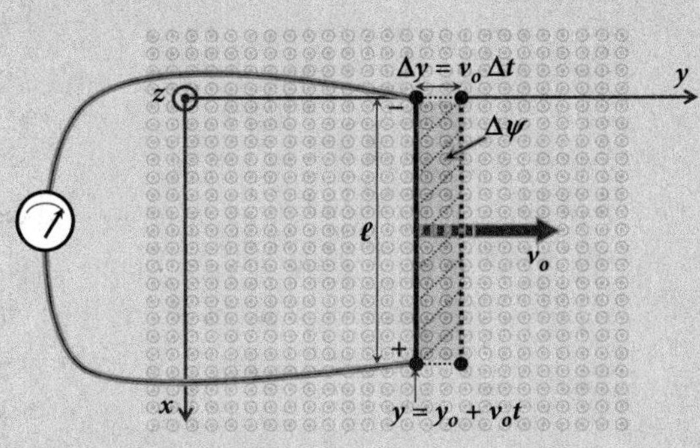

Figure 10.6 Demonstrating the relationship between the induced *emf* in the conductor moving across a magnetic field and the rate of change of the linkage flux.

For Figure 10.8, $\vec{B}\cdot\overline{dS} = B_z\vec{a}_z \cdot dx\,dy\,\vec{a}_z = B_z dx\, d_y = B_z dx(v_o dt)$

In the above expression, $\overline{dS}$ was assumed in the z direction, therefore the assumed direction of the induced current is counterclockwise. The polarity of the assumed *emf* is positive on the upper end of the moving conductor and negative on the lower end

$$emf = -\frac{\partial}{\partial t}\int \vec{B}\cdot\overline{dS} = -\frac{d}{dt}B_z v_o\, dt\int_o^{\ell} dx = -B_z v_o \ell$$

Since the *emf* obtained is negative, then it's opposite to our assumption. The correct polarity of the induced *emf* is negative on the upper end of the moving conductor and positive on the lower end. This will result in a current in the clockwise direction. In this case the magnetic flux increases as the conductor slides to the right and therefore the induced current generates a magnetic field in the opposite direction to the original field to oppose the increase in the magnetics flux linking the loop.

10.5 Faraday's Disc

In 1831, Michael Faraday, developed the first *homopolar* generator in the form of a rotating disc exposed to a magnetic field produced by a permanent magnet. Figure 10.7 demonstrates the Faraday disc setup with the metallic disc "D" rotating in the magnetic field produced by the horseshoe magnet "A." The electrical contacts for the induced "electricity" is available through the terminals "B" and "B′." The terminal B is connected to the center of the disc through its pivot while B′ is connected to the rim of the disc through a metallic "spring" contact "m."

To explain the concept, we start with the rotating conductor demonstration of Figure 10.8. The idea of an *emf* generated in a moving conductor was discussed above in Figure 10.6 and the associated Equations (10.6) and (10.7). Similarly, a rotating conductor is expected to develop an *emf* as well. Similar to Equation (10.7), the relationship between the conductor motion and the rate of change of the flux linkage for the moving conductor is

$$\frac{\Delta\psi_m}{\Delta t} = \frac{B\Delta S}{\Delta t} = \frac{B\frac{1}{2}\ell^2\Delta\varphi}{\Delta t}, \quad \text{and hence} \quad \frac{\partial\psi_m}{\partial t} = \frac{1}{2}B\ell^2\frac{\partial\varphi}{\partial t} = \frac{1}{2}B\ell^2\omega_o \tag{10.8}$$

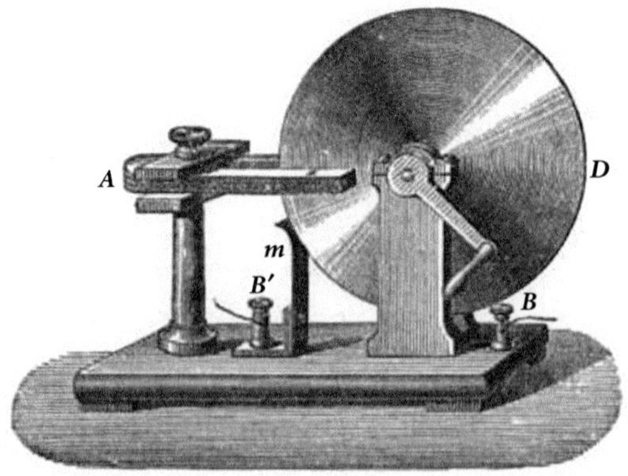

FIGURE 10.7 Faraday's "homopolar" generator in the form of a rotating disc exposed to a magnetic field produced by a permanent magnet. Also known as Faraday's disc.

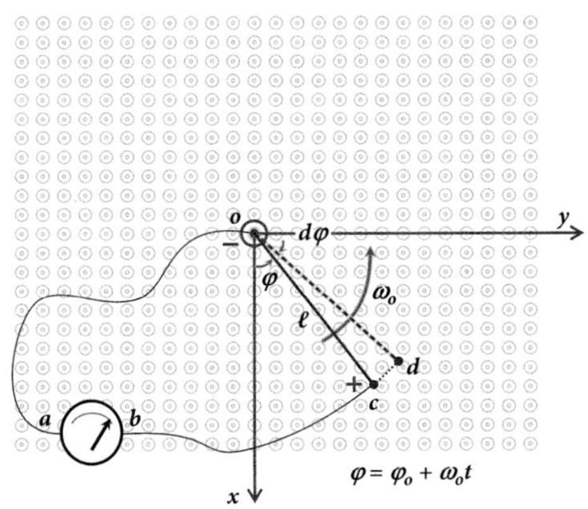

FIGURE 10.8 A simplified arrangement of a conductor rotating across a magnetic field to support the discussion of Faraday's disc-induced *emf*.

In this equation, we note that the incremental area ΔS covered by the incremental rotation $\Delta\varphi$ of the conductor ℓ can be approximated to the area of a triangle with a height ℓ and a base $\ell\Delta\varphi$, and hence $\Delta S = \frac{1}{2}(\ell)(\ell\Delta\varphi) = \frac{1}{2}\ell^2\Delta\varphi$. We also note that ω_o is the angular velocity of the rotating conductor, $\omega_o = \frac{\partial\varphi}{\partial t}$.

Now, based on Equation (10.8), the induced *emf* is given by

$$emf = -\frac{\partial\psi_m}{\partial t} = -\frac{1}{2}B\ell^2\omega_o \tag{10.9}$$

Now, considering that the rotating conductor of Figure 10.8 is simply one of the radii of the rotating disk as demonstrated in Figure 10.9, we can claim the same answer for the *emf* induced in both cases.

It is clear from the discussion above that an induced electric field and the associated *emf* occur as a result of two possible causes or a combination of both. One cause is a relative motion of the electrical circuit "coil/conductor" to a stationary "time-independent" magnetic field. The second cause is a time-varying magnetic field applied to a stationary coil. In a linear system, the total induced *emf* is simply the superposition of the contributions of the two causes.

$$emf = -\frac{\partial\psi_m}{\partial t} = -\frac{\partial\left(\iint_{S(t)} \vec{B}(t)\cdot\overrightarrow{dS}\right)}{\partial t} = emf_{\text{time-varying field}} + emf_{\text{motional}} \tag{10.10}$$

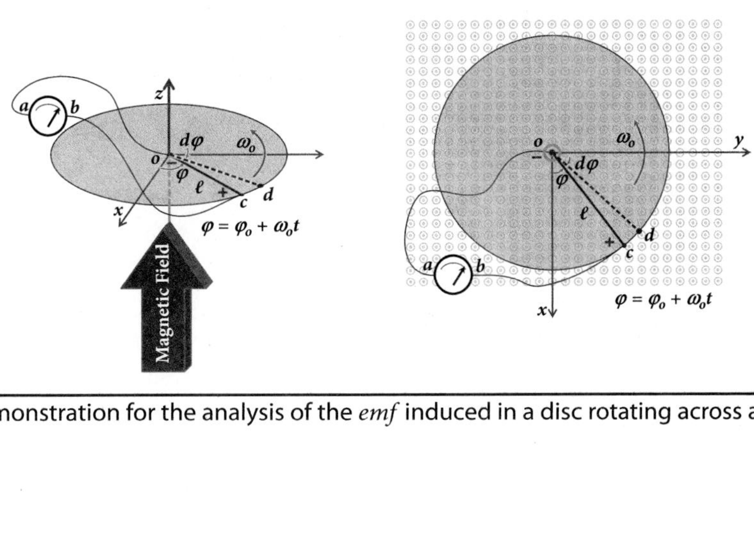

FIGURE 10.9 A demonstration for the analysis of the *emf* induced in a disc rotating across a magnetic field (Faraday's disc).

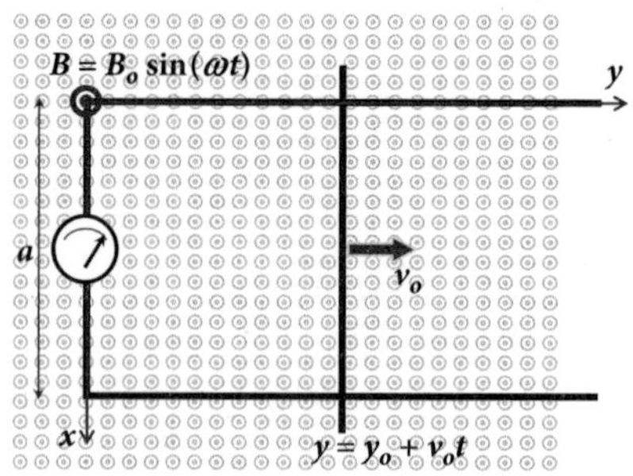

FIGURE 10.10 An example for the analysis of the *emf* induced in a conductor moving across a magnetic field.

10.6 An Example of a Moving Conductor in a Time-Varying Magnetic Field

To demonstrate a "combination" case, let us consider the setup of Figure 10.10. A sliding conductor is laid on a conducting loop as shown immersed in a time-varying magnetic field. The position of the sliding conductor which moves at a steady speed, v_o, along the y direction is given by

$$y = y_o + v_o t \tag{10.11}$$

while the time-varying magnetic field has a spatially uniform magnetic flux density given by

$$\vec{B} = B_o \sin(\omega t)\vec{a}_z \tag{10.12}$$

For this case, we can write

$$\begin{aligned} emf &= -\frac{\partial \psi_m}{\partial t} = -\frac{\partial [B(t)S(t)]}{\partial t} = -\frac{\partial [B_o \sin(\omega t) a(y_o + v_o t)]}{\partial t} \\ &= -aB_o[v_o \sin(\omega t) + (y_o + v_o t)\omega \cos(\omega t)] \end{aligned} \tag{10.13}$$

In the following, we will present examples of two cases in the form of commonly used devices, namely, the electric generator and the electric transformer.

Example 10.4 Consider an inductor formed by winding N turns of thin conducting wires in a square loop of side a is placed in a magnetic field with magnetic flux density $\vec{B} = B_o(4\vec{a}_x + 2\vec{a}_z)\sin\omega_o t$. The inductor is lying in the x-y plane and is connected to a load resistance $R = 10\ \text{k}\Omega$. Determine the following:

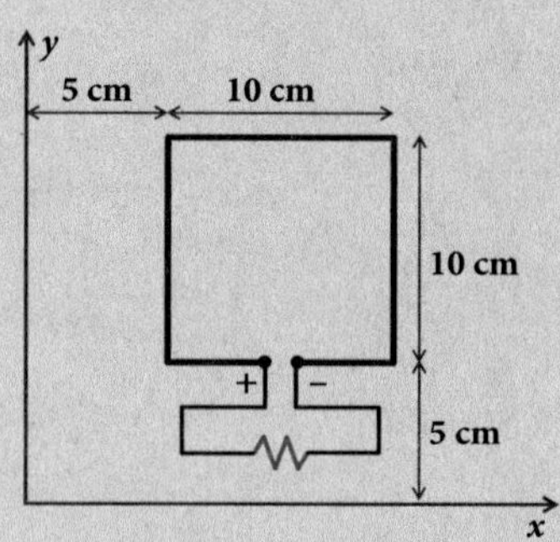

a. The magnetic flux linking a single loop of the inductor
b. The *emf* induced given $N = 10$ $\vec{B}_o = 0.1$ tesla $a = 10$ cm and $\omega_o = 10^4$ rad/s
c. The polarity of the induced *emf* at $t = 0$
d. The induced current in the circuit assuming that the wire resistance is much smaller than R and the resistance of the inductor can be ignored.

Solution

a. $\psi_m = \vec{B}\cdot\vec{dS}$ $\quad \Lambda = N\psi_m$ $\quad \vec{dS} = dx\,dy\,\vec{a}_z$

$\vec{dS}$ is assumed in the positive z direction and therefore the assumed direction of the induced current is counterclockwise and the assumed polarity of the induced *emf* is as shown in the figure.

$\psi_m = \vec{B}\cdot\vec{dS} = B_o(4\vec{a}_x + 2\vec{a}_z)\sin(\omega_o t)\cdot dx\,dy\,\hat{a}_z = 2B_o\sin(\omega_o t)dx\,dy$

b. $emf = -\frac{\partial \Lambda}{\partial t} = -N\frac{\partial}{\partial t}\int_S \vec{B}\cdot\vec{dS} = -N\frac{\partial}{\partial t}\int_S 2B_o \sin(\omega_o t)dx\,dy = -N\frac{d}{dt}(2B_o S\sin(\omega_o t))$

$= -N\frac{d}{dt}(2B_o a^2 \sin(\omega_o t)),$ where S is the area of the loop, $S = a^2$

$emf = -2NB_o a^2 \omega_o \cos(\omega_o t) = -2\times 10\times 0.1\times(0.1)^2\times 10^4 \cos(\omega_o t)$

$emf = -200 \cos(\omega_o t)$ V

c. For $0 \le t < \pi/2\omega_o$, the magnetic field is increasing, the induced *emf* is negative which means it's opposite to the assumed polarity, the induced current will be in the clock-wise direction which will generate a magnetic field in the negative z direction to oppose the increasing magnetic field.

At $t = 0$ $emf = -200$ V

d. $I = \frac{emf}{R} = -\frac{200\cos(\omega_o t)}{10k} = -20\cos(\omega_o t)$ mA

Example 10.5 The square wire loop shown has 10 turns and lies in the ρ-z plane. If the magnetic flux density in the region is given by: $\vec{B}(t) = \vec{a}_\varphi \frac{2\times 10^{-4}}{\rho}\sin(2\pi\times 10^4 t)$ Wb/m^2

a. Determine the induced *emf* across a small gap created in the loop as shown in the figure.

b. Determine the magnitude and direction of the current that would flow across a 100 Ω resistor connected across the gap at $t = 0$ and at $t = 50$ μs. Ignore the internal resistance of the loop.

Solution

$$\vec{B}(t) = \vec{a}_\varphi \frac{2\times 10^{-4}}{\rho}\sin(2\pi\times 10^4 t)\,\text{T}$$

a. The induced *emf* across a small gap created in the loop:

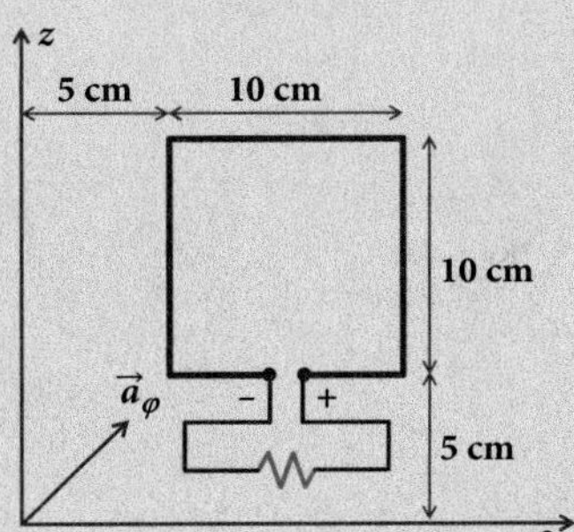

$emf = -N\frac{d\varphi}{dt}$

$\varphi = \int_S \vec{B}(t)\cdot\vec{dS} \qquad \vec{dS} = \vec{a}_\varphi\, d\rho\, dz$

$\varphi = \int_{z=0.06}^{z=0.16}\int_{\rho=0.05}^{\rho=0.15} \vec{a}_\varphi \frac{2\times 10^{-4}}{\rho}\sin(2\pi\times 10^4 t)\cdot\vec{a}_\varphi\, d\rho\, dz$

$\varphi = \int_{z=0.06}^{z=0.16}\int_{\rho=0.05}^{\rho=0.15} \frac{2\times 10^{-4}}{\rho}\sin(2\pi\times 10^4 t)d\rho\, dz$

$2\times 10^{-4}\sin(2\pi\times 10^4 t)\int_{\rho=0.05}^{\rho=0.15}\frac{1}{\rho}d\rho\int_{0.06}^{0.16}dz = 2\times 10^{-4}\sin(2\pi\times 10^4 t)[\ln(\rho)]_{0.05}^{0.15}[z]_{0.06}^{0.16}$

$\varphi = 2\times 10^{-4}\sin(2\pi\times 10^4 t)\ln\left(\frac{0.15}{0.05}\right)\times(0.16-0.06) = 2.2\times 10^{-5}\sin(2\pi\times 10^4 t)$ weber

$emf = -N\frac{d\varphi}{dt} = -10\times 2.2\times 10^{-5}\times 2\pi\times 10^4\times\cos(2\pi\times 10^4 t) = -13.82\cos(2\pi\times 10^4 t)$ V

Note that $\vec{dS}$ is assumed to be in the positive $\vec{a}_\varphi$ direction which is perpendicular to the page and going inward, this means that the contour bounding the area of the loop is traversed in the clock-wise direction. Using the right-hand rule and taking into account that the four fingers have to cross the gap in the direction of the voltage drop from the positive to the negative terminal, the polarity of the terminals and current direction will be as shown in figure.

At $t = 0$, $emf = -13.82\cos(0) = -13.82$ V. Because the *emf* is negative, then the polarity of the terminals is opposite to what is assumed in the figure and so is the current $I = \frac{emf}{R} = \frac{-13.82}{5} = -2.764$ A.

b. At $t = 50\ \mu s$, $emf = -13.82\cos(2\pi \times 10^4 \times 50 \times 10^{-6}) = -13.82\cos(\pi) = 13.82$ V. The *emf* is positive, therefore the polarity of the terminals will be the same as that assumed in the figure and so will the current $I = \frac{emf}{R} = \frac{13.82}{5} = 2.764$ A.

Note that at $t = 0$, the original magnetic field is increasing, and the induced *emf* will result in a magnetic field that is opposite to the original magnetic field. At $t = 50\ \mu s$, the original magnetic field is decreasing and therefore the induced currents will generate a magnetic in a direction that enhances the original magnetic field. The fact that the induced magnetic field opposes the change in the original magnetic field verifies Lenz's law.

10.7 The Electric Generator

The electric generator, in its simplest form, is a coil that moves relative to a time invariant magnetic field. The coil is typically wound on a high permeability armature that is immersed in the magnetic field of a permanent magnet. Figure 10.11 shows a cut away demonstration of a typical construction.

For analysis purposes, we present a one-turn coil demonstration in Figure 10.12. The figure depicts one turn of the coil rotating around its axis at an angular velocity ω. The permanent magnet is assumed to produce a uniform magnetic flux density, B, flooding the *effective* area of the coil. The effective area of the coil is the area in which the Faraday's *emf* is generated.

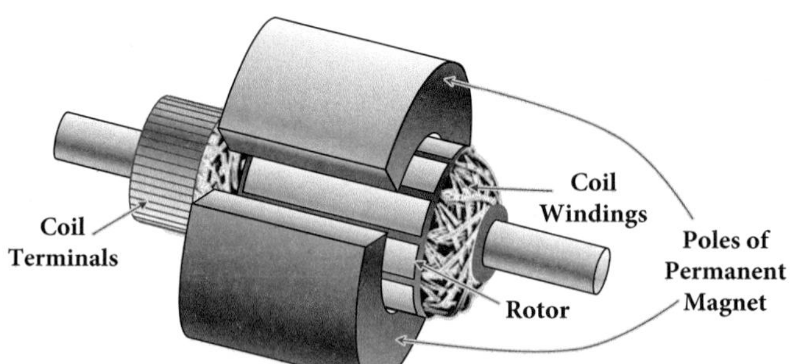

FIGURE 10.11 A cut away demonstration of the main parts of an electric generator.

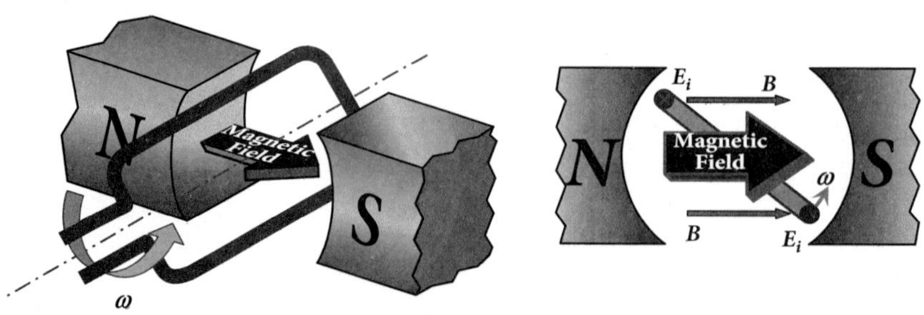

FIGURE 10.12 Demonstration of the electric generator concept through the analysis of a one-turn coil electric generator.

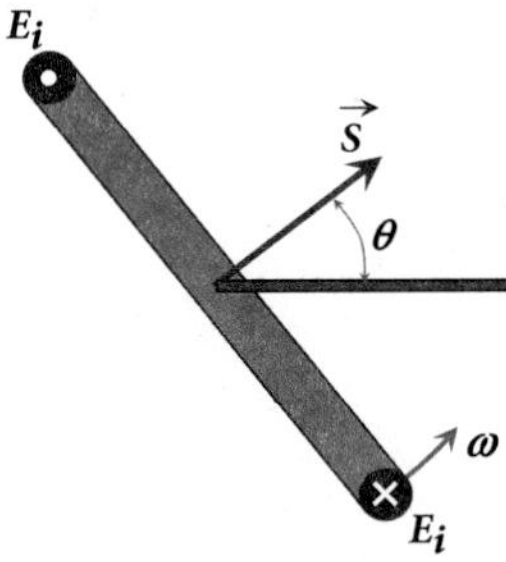

Figure 10.13 The analysis of the *emf* induced in a one-turn coil rotating across a magnetic field.

Referring to Figure 10.13, the effective area of the coil is represented with the vector $\vec{S}$ making an angle θ with the magnetic flux density vector, $\vec{B}$. Hence, the coil's linkage magnetic flux, ψ_m, over the effective area, S, is simply the dot product between the two vectors $\vec{B}$ and $\vec{S}$.

Now, as the coil rotates at the angular velocity ω starting from an initial angle, θ_o, the *emf* generated in this coil at the general position, $\theta = \omega t + \theta_o$, is given by

$$emf = -\frac{\partial \psi_m}{\partial t} = -\frac{\partial(\vec{B}\cdot\vec{S})}{\partial t} = -\frac{\partial(BS\cos\theta)}{\partial t} = BS\sin\theta\frac{\partial\theta}{\partial t} = \omega BS\sin\theta = \omega BS\sin(\omega t + \theta_o) \quad (10.14)$$

This result shows that the generated *emf* in the single-turn coil is harmonic (sinusoid) with the same frequency as that of the coil rotation. Practically, multi-turn coils with specific spatial winding distributions and specific connectivity of their end terminals are used in the design of both AC and DC generators.

10.8 The Transformer

Now, we turn our attention to the case of stationary coils placed in time-varying fields. The example we take in this case is that of the transformer. The transformer in its simplest form is a two-coil device with a mutual magnetic flux linkage. Three typical shapes of two-coil transformers are demonstrated in Figure 10.14.

As the figure shows, a magnetic core is typically used to confine the flux flow and provides maximum linkage between the two coils. This magnetic core is to be made of a high-permeability material to minimize its reluctance and hence maximize the flux linkage. The optimum transformer would have no flux leakage; all flux produced by the primary coil would link the secondary coil.

A typical physical shape for the first transformer arrangement on the left side of Figure 10.14 is shown in Figure 10.15. The two coils are wound on top of each other on the central arm of the core. One coil is typically called the primary while the other one is called the secondary.

As shown in Figure 10.16, the primary coil is the coil connected to a time-varying source while the secondary coil feeds the transformer output to the load. In some designs, more than one secondary coil is used to deliver signals/power to different loads.

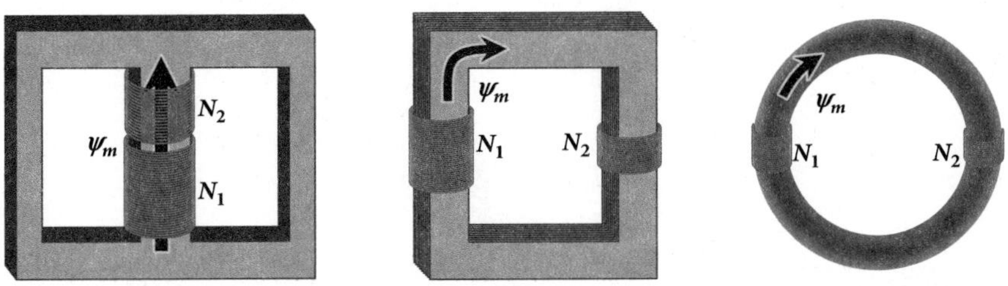

Figure 10.14 Simplified examples of two-coil transformers.

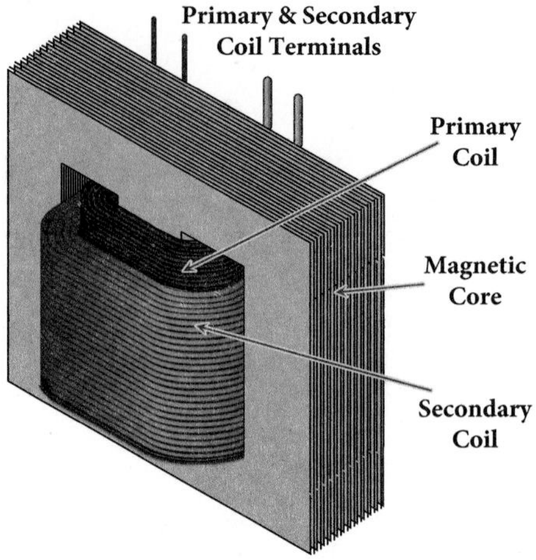

FIGURE 10.15 An example of a magnetic core transformer.

Referring to Figure 10.16, if we denote the common magnetic flux between the two coils $\psi_m(t)$, we can write the following expressions

$$emf_1(t) = -\frac{\partial \Lambda_1}{\partial t} = -\frac{\partial (N_1 \psi_m)}{\partial t} = -N_1 \frac{\partial \psi_m}{\partial t} \tag{10.15}$$

$$emf_2(t) = -\frac{\partial \Lambda_2}{\partial t} = -\frac{\partial (N_2 \psi_m)}{\partial t} = -N_2 \frac{\partial \psi_m}{\partial t} \tag{10.16}$$

The ratio between the two *emfs* is then the same as the ratio between the numbers of coil turns, typically called the "turns ratio"

$$\frac{emf_2(t)}{emf_1(t)} = \frac{N_2}{N_1} \quad \text{or} \quad emf_2(t) = \frac{N_2}{N_1} emf_1(t) \tag{10.17}$$

Now, replacing the *emf* notation by the primary and secondary time domain voltages, $v_1(t)$ and $v_2(t)$, respectively, we rewrite Equation (10.17) as

$$\frac{v_2(t)}{v_1(t)} = \frac{N_2}{N_1} \quad \text{or} \quad v_2(t) = \frac{N_2}{N_1} v_1(t) \tag{10.18}$$

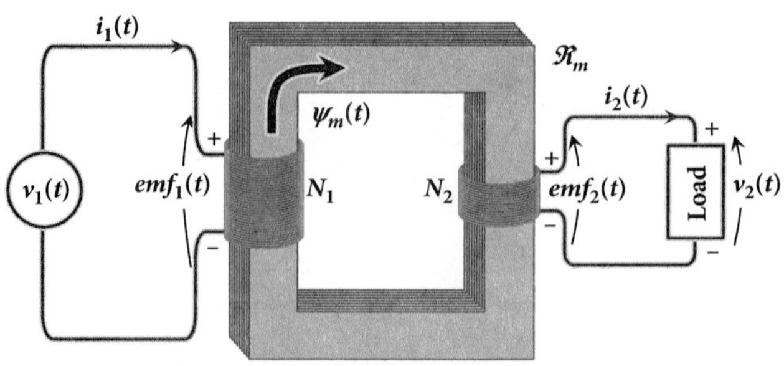

FIGURE 10.16 A simplified example of a two-coil transformer with the schematics of both the primary and secondary electrical circuits.

The time-domain expression (10.18) can be converted to the frequency domain, yielding the same turns ratio relationship between the frequency-domain expression primary and secondary voltage phasors, thus

$$\frac{V_2}{V_1}=\frac{N_2}{N_1} \quad \text{or} \quad V_2=\frac{N_2}{N_1}V_1 \tag{10.19}$$

Next, we write the magnetic circuit expression for the mutual flux and the common core reluctance as

$$N_1 i_1(t)-N_2 i_2(t)=\psi_m \mathfrak{R}_m \tag{10.20}$$

Notice the minus sign in this expression is because the induced *emf* in the secondary coil (and hence the secondary current) is to counter the change in the flux generated by the primary current.

For an ideal transformer with negligible core reluctance, Equation (10.20) reduces to

$$N_1 i_1(t)=N_2 i_2(t) \text{ and the phasor form is } N_1 I_1=N_2 I_2 \tag{10.21}$$

Combining Equations (10.19) and (10.21), we conclude that for ideal transformers

$$V_2 I_2=V_1 I_1 \quad \text{and} \quad Z_2=\left(\frac{V_2}{I_2}\right)=\left(\frac{N_2}{N_1}\right)^2\left(\frac{V_1}{I_1}\right)=\left(\frac{N_2}{N_1}\right)^2 Z_1 \tag{10.22}$$

The first part indicates a conservation of "power" while the second part shows an impedance level change.

Examining the above equations, we can predict the types of uses for transformer devices. Typically, they involve one or more of the following features

a. Changes to the voltage level (step-up/step-down)
b. Changes to the current level (step-up/step-down)
c. Changes to the impedance level (step-up/step-down)

10.9 Faraday's Maxwell's Equation

We are now ready to present Faraday's equation as one of the four famous Maxwell's equations. In this regard, we will be rewriting the earlier [Equations (10.5) and (10.10)] forms to explicitly relate the electric and magnetic fields to each other. We will present both the integral and differential (point) forms.

10.9.1 Integral Form

Consider the closed loop C of Figure 10.17 that has a flux linkage ψ_m.

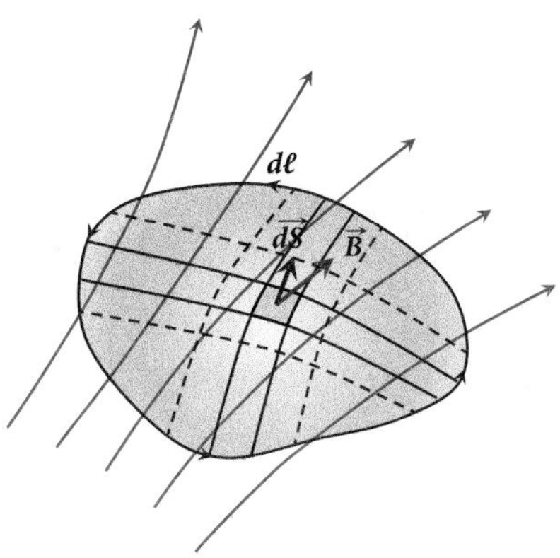

Figure 10.17 An illustration used in the development of the induced *emf* in a closed loop as it relates to time-varying flux linkage.

If we were to assume that this flux varies with time, then the corresponding induced *emf* around the loop would be given by

$$emf = -\frac{\partial \psi_m}{\partial t} \tag{10.5}$$

Meanwhile, we can express both the *emf* and ψ_m for the loop in terms of the electric field intensity and the magnetic flux density as

$$emf = \oint_C \vec{E}\cdot\vec{d\ell}, \qquad \text{and} \tag{10.23}$$

$$\psi_m = \iint_S \vec{B}\cdot\vec{dS} \tag{10.24}$$

Hence,

$$emf = \oint_C \vec{E}\cdot\vec{d\ell} = -\frac{\partial \psi_m}{\partial t} = -\frac{\partial\left(\iint_S \vec{B}\cdot\vec{dS}\right)}{\partial t} = -\iint_S \frac{\partial \vec{B}}{\partial t}\cdot dS \tag{10.25}$$

assuming that the loop is stationary. This leads us to the integral form of the Faraday-Maxwell equation as

$$\oint_C \vec{E}\cdot\vec{d\ell} = -\iint_S \frac{\partial \vec{B}}{\partial t}\cdot dS \tag{10.26}$$

10.9.2 Differential Form

Using Stokes' theorem, we can write

$$\oint_C \vec{E}\cdot\vec{d\ell} = \iint_S (\vec{\nabla}\times\vec{E})\cdot\vec{dS} \tag{10.27}$$

which, when combined with Equation (10.26), yields

$$\oint_C \vec{E}\cdot\vec{d\ell} = \iint_S (\vec{\nabla}\times\vec{E})\cdot d\vec{S} = -\iint_S \frac{\partial \vec{B}}{\partial t}\cdot dS \tag{10.28}$$

From this we conclude the differential form of the Faraday-Maxwell equation is

$$\vec{\nabla}\times\vec{E} = -\frac{\partial \vec{B}}{\partial t} \tag{10.29}$$

It is important to realize that time-varying magnetic fields are responsible for inducing an electric field as indicated by Equations (10.26) and (10.29). This is the case whether we have a conducting medium, a dielectric, or even an empty space. The discussion above used the conducting media to physically demonstrate the presence of the electric field and to develop the analytical expressions.

10.10 Revisiting Ampere's Law—The Displacement Current

Earlier in Chapter 5, we talked about the conservative electric field and potential. At that time, we were only discussing static field quantities.

$$\oint \vec{E}\cdot\vec{d\ell} = 0 \tag{5.70}$$

$$\vec{\nabla}\times\vec{E} = 0 \tag{8.59}$$

Now, in the current chapter, and in the light of our new knowledge of Faraday's law, we realize that time-varying electric fields are not conservative. The static field form [Equation (8.59)] is now modified in dynamics to take the form of Faraday's law

$$\vec{\nabla}\times\vec{E}=-\frac{\partial\vec{B}}{\partial t} \tag{10.30}$$

Hence, we need to go back and check on earlier statics relationships that need to be modified in order to become valid for dynamics as well. In this regard, we start with the static form of Ampere's law which we presented in Chapter 8. We need to explore what kind of modifications, if any, that it requires to stay valid for time-varying fields. Two previously presented formulae need to be recalled in this regard

$$\text{Ampere's law for statics: } \vec{\nabla}\times\vec{H}=\vec{J}, \text{ and} \tag{8.54}$$

$$\text{Current continuity relationship: } \vec{\nabla}\cdot\vec{J}=-\frac{\partial\rho_v}{\partial t} \tag{6.8}$$

By applying the divergence differential vector operator to Equation (8.54), and using Equation (6.8) for the result on the right-hand side, we find

$$\vec{\nabla}\cdot(\vec{\nabla}\times\vec{H})=\vec{\nabla}\cdot\vec{J}=-\frac{\partial\rho_v}{\partial t} \tag{10.31}$$

The left-hand side of Equation (10.31) is zero since the curl function is known to be divergenceless (differential vector identity). However, the right-hand side can only be zero under static conditions and may not be so in time-varying fields. We conclude, therefore, that Ampere's law in the form of Equation (8.54) is only good for static fields and needs an adjustment to accommodate dynamic conditions.

Recalling our earlier discussion in Chapter 6 when we talked about the capacitor current of Equation (6C.20). In that discussion, we noted that

> the capacitance current is not a conduction current that physically flows in the capacitance dielectric region. A perfect dielectric should have zero conductivity and hence no conduction current can flow through it. This sounds like a mystery as we have just derived expressions for the currents in the capacitance. The explanation of this baffling current will be presented in adequate detail in a later chapter when we discuss the topic of displacement currents. However, a quick explanation lies in the dynamic change in polarization within the dielectric material as a result of the dynamic change in its internal field. A dynamic change in polarization implies dynamic "finite displacements" of the internal bound charges, and hence the "displacement" current.

Well, this *other* current that we expected at that time needs to be included in Ampere's law [Equation (8.45)]. Hence, we propose to rewrite Equation (8.54) with an additional current term which we expect to accommodate the dynamic nature of time-varying fields as well as explains the "displacement" currents in dielectrics. The proposed rewrite of Equation (8.54) takes the form

$$\vec{\nabla}\times\vec{H}=\vec{J}+\vec{G} \tag{10.32}$$

where $\vec{G}$ represents the added current density vector. Now, let us apply the divergence operator to this new form

$$\vec{\nabla}\cdot(\vec{\nabla}\times\vec{H})=0=\vec{\nabla}\cdot\vec{J}+\vec{\nabla}\cdot\vec{G}=-\frac{\partial\rho_v}{\partial t}+\vec{\nabla}\cdot\vec{G} \tag{10.33}$$

which yields

$$\vec{\nabla}\cdot\vec{G}=\frac{\partial\rho_v}{\partial t} \tag{10.34}$$

But, using Gauss' law, $\vec{\nabla}\cdot\vec{D}=\rho_v$, we obtain

$$\vec{\nabla}\cdot\vec{G}=\frac{\partial(\vec{\nabla}\cdot\vec{D})}{\partial t}=\vec{\nabla}\cdot\frac{\partial\vec{D}}{\partial t} \tag{10.35}$$

hence,

$$\vec{G}=\frac{\partial\vec{D}}{\partial t} \tag{10.36}$$

and finally, we obtain the dynamic version of Ampere's law as

$$\vec{\nabla}\times\vec{H}=\vec{J}+\frac{\partial\vec{D}}{\partial t} \tag{10.37}$$

Similar to the case of Faraday's law point and integral forms in Equations (10.26) and (10.29), we can expect the integral form for the dynamic version of Ampere's to be

$$\oint_C \vec{H}\cdot\vec{d\ell}=\iint_{S\text{ contoured by }C}\left[\vec{J}+\frac{\partial\vec{D}}{\partial t}\right]\cdot dS \tag{10.38}$$

10.11 Revisiting Field and Potential Formulas—The Retarded Potentials

With the above modifications in mind, it behooves us to check into other relationships that were developed earlier for electrostatic and magnetostatic fields. Table 10.1 gives a list of the main relationships and their expected modifications required to make them valid for the general case of time-varying fields. Following the table, we present the justifications for the shown modifications.

Electrostatic/Magnetostatic Relationships		Time-Varying Modifications		
$\vec{\nabla}\cdot\vec{D}=\rho_v$	(4.17)		$\vec{\nabla}\cdot\vec{D}=\rho_v$	(4.17)
$\vec{\nabla}\cdot\vec{B}=0$	(8.46)		$\vec{\nabla}\cdot\vec{B}=0$	(8.46)
$\vec{\nabla}\times\vec{E}=0$	(8.59)		$\vec{\nabla}\times\vec{E}=-\frac{\partial\vec{B}}{\partial t}$	(10.29)
$\vec{\nabla}\times\vec{H}=\vec{J}$	(8.52)		$\vec{\nabla}\times\vec{H}=\vec{J}+\frac{\partial\vec{D}}{\partial t}$	(10.37)
$\vec{E}=-\vec{\nabla}V$	(5.64)	(*a*)	$\vec{E}=-\vec{\nabla}V-\frac{\partial\vec{A}}{\partial t}$	(10.41)
$\vec{B}=\vec{\nabla}\times\vec{A}$ & $\vec{\nabla}\cdot\vec{A}=0$	(8.65)	(*b*)	$\vec{B}=\vec{\nabla}\times\vec{A}$ & $\vec{\nabla}\cdot\vec{A}=-\mu\varepsilon\,\partial V/\partial t$	(10.46)
$\nabla^2 V=-\frac{\rho_v}{\varepsilon}$	(7.4)	(*c*)	$\nabla^2 V=-\frac{\rho_v}{\varepsilon}+\mu\varepsilon\frac{\partial^2 V}{\partial t^2}$	(10.50)
$\nabla^2\vec{A}=-\mu\vec{J}$	(8.69)	(*d*)	$\nabla^2\vec{A}=-\mu\vec{J}+\mu\varepsilon\frac{\partial^2\vec{A}}{\partial t^2}$	(10.47)

Table 10.1 Summary of Dynamic Electric and Magnetic Field and Potential Relationship (with the inclusion of time-varying modifications)

10.12 Derivation and Justifications for the Time-Varying Modifications of the Above Field and Potential Relationships

Let us now put the relationships (a)-(d) to the test. Starting with the electrostatic relationship (a),

$$\vec{E} = -\vec{\nabla} V \tag{5.64}$$

and applying the curl to both sides of Equation (5.64), we get

$$\vec{\nabla} \times \vec{E} = \vec{\nabla} \times (-\vec{\nabla} V) \tag{10.39}$$

The right-hand side is the curl of a gradient which is known to be zero (gradients are rotationless). This is only true when $-\frac{\partial \vec{B}}{\partial t} = 0$, that is, if B is not time varying. Thus, we conclude that the relationship in Equation (5.64) is only valid for statics and fails to represent dynamic cases.

To remedy this incompatibility, we need to modify the result of the curl operation to yield the right answer of $-\frac{\partial \vec{B}}{\partial t}$. We do that by adding the term $-\frac{\partial \vec{B}}{\partial t}$ to the right-hand side of Equation (10.39).

Hence, we rewrite Equation (10.39) as

$$\vec{\nabla} \times \vec{E} = \vec{\nabla} \times (-\vec{\nabla} V) - \frac{\partial \vec{B}}{\partial t} = \vec{\nabla} \times (-\vec{\nabla} V) - \frac{\partial (\vec{\nabla} \times \vec{A})}{\partial t} = \vec{\nabla} \times \left[-\vec{\nabla} V - \frac{\partial \vec{A}}{\partial t} \right] \tag{10.40}$$

which implies that the general form for the relationship [a: Equation (5.64)] in Table 10.1 should be

$$\vec{E} = \left[-\vec{\nabla} V - \frac{\partial \vec{A}}{\partial t} \right] \tag{10.41}$$

Next, we move on to the next relationship [b: Equation (8.65)]: $\vec{B} = \vec{\nabla} \times \vec{A}$ & $\vec{\nabla} \cdot \vec{A} = 0$.

To check into this "definition" of the magnetic vector potential and its compatibility with time-varying fields, we start with Faraday's law

$$\vec{\nabla} \times \vec{H} = \vec{J} + \frac{\partial \vec{D}}{\partial t} \tag{10.37}$$

from which we can write

$$\frac{1}{\mu} \vec{\nabla} \times \vec{B} = \vec{J} + \varepsilon \frac{\partial \vec{E}}{\partial t} \tag{10.42}$$

Now, using Equations (8.65) and (10.41), while multiplying both sides by μ, we obtain

$$\vec{\nabla} \times \vec{\nabla} \times \vec{A} = \mu \vec{J} + \mu\varepsilon \frac{\partial \left[-\vec{\nabla} V - \frac{\partial \vec{A}}{\partial t} \right]}{\partial t} \tag{10.43}$$

Next, using the curl-curl vector identity on the left-hand side of Equation (10.43), we get

$$\vec{\nabla} \times \vec{\nabla} \times \vec{A} = \vec{\nabla}(\vec{\nabla} \cdot \vec{A}) - \nabla^2 \vec{A} = \mu \vec{J} - \mu\varepsilon \left[\vec{\nabla} \frac{\partial V}{\partial t} + \frac{\partial^2 \vec{A}}{\partial t^2} \right] \tag{10.44}$$

Rearranging the terms, we can rewrite the Equation (10.44) relationship as

$$\nabla^2 \vec{A} = -\mu \vec{J} + \mu\varepsilon \frac{\partial^2 \vec{A}}{\partial t^2} + \vec{\nabla}(\vec{\nabla} \cdot \vec{A}) + \mu\varepsilon \vec{\nabla} \frac{\partial V}{\partial t} = -\mu \vec{J} + \mu\varepsilon \frac{\partial^2 \vec{A}}{\partial t^2} + \vec{\nabla} \left[\vec{\nabla} \cdot \vec{A} + \mu\varepsilon \frac{\partial V}{\partial t} \right] \tag{10.45}$$

Recall that the definition of $\vec{B}=\vec{\nabla}\times\vec{A}$ left for us the freedom to choose a value for $\vec{\nabla}\cdot\vec{A}$; this value was chosen to be zero at the time. That choice was reasonable considering that we were dealing with statics at the time. With time-varying fields, the above relationship suggests that choosing the quantity within the square brackets $\left[\vec{\nabla}\cdot\vec{A}+\mu\varepsilon\frac{\partial V}{\partial t}\right]$ as zero would be more appropriate. Thus, the new form for realtions (b) becomes

$$\vec{B}=\vec{\nabla}\times\vec{A} \qquad \text{and} \qquad \vec{\nabla}\cdot\vec{A}=-\mu\varepsilon\frac{\partial V}{\partial t} \tag{10.46}$$

which leaves us with the new form for the relationship (d)

$$\nabla^2\vec{A}=-\mu\vec{J}+\mu\varepsilon\frac{\partial^2\vec{A}}{\partial t^2} \tag{10.47}$$

Finally, to check the formula (c), we start with Gauss' law

$$\vec{\nabla}\cdot\vec{D}=\rho_v \tag{4.17}$$

Using Equation (10.41) while replacing $\vec{D}$ by $\varepsilon\vec{E}$, we get

$$\varepsilon\vec{\nabla}\cdot\vec{E}=\varepsilon\vec{\nabla}\cdot\left[-\vec{\nabla}V-\frac{\partial\vec{A}}{\partial t}\right]=-\varepsilon\left[\nabla^2V+\frac{\partial(\vec{\nabla}\cdot\vec{A})}{\partial t}\right]=\rho_v \tag{10.48}$$

Rearranging the terms, we can rewrite Equation (10.48) as

$$\nabla^2V=-\frac{\rho_v}{\varepsilon}-\frac{\partial(\vec{\nabla}\cdot\vec{A})}{\partial t}=-\frac{\rho_v}{\varepsilon}-\frac{\partial\left(-\mu\varepsilon\frac{\partial V}{\partial t}\right)}{\partial t}=-\frac{\rho_v}{\varepsilon}+\mu\varepsilon\frac{\partial^2V}{\partial t^2} \tag{10.49}$$

Hence the final "dynamic" form for (c) becomes

$$\nabla^2V=-\frac{\rho_v}{\varepsilon}+\mu\varepsilon\frac{\partial^2V}{\partial t^2} \tag{10.50}$$

10.13 Summary of Maxwell's Equations

Maxwell's Equations			
Equations	**Point Form**	**Integral Form**	**(10.51)**
Faraday's law	$\vec{\nabla}\times\vec{E}=-\frac{\partial\vec{B}}{\partial t}$	$\oint_C\vec{E}\cdot\vec{d\ell}=-\iint_{S\text{ contoured by C}}\frac{\partial\vec{B}}{\partial t}\cdot\vec{dS}$	(10.26), (10.29)
Ampere's law	$\vec{\nabla}\times\vec{H}=\vec{J}+\frac{\partial\vec{D}}{\partial t}$	$\oint_C\vec{H}\cdot\vec{d\ell}=\iint_{S\text{ contoured by C}}\left[\vec{J}+\frac{\partial\vec{D}}{\partial t}\right]\cdot\vec{dS}$	(10.37), (10.38)
Gauss' law for electric flux	$\vec{\nabla}\cdot\vec{D}=\rho_v$	$\oint\!\!\oint_S\vec{D}\cdot\vec{dS}=\iiint_{\text{volume enclosed by S}}\rho_v\,dv$	(4.17)
"Gauss' law" for magnetic flux	$\vec{\nabla}\cdot\vec{B}=0$	$\oint\!\!\oint_S\vec{B}\cdot\vec{dS}=0$	(8.46)
Constitutive Relationships			**(10.52)**
Material permittivity	$\vec{D}=\varepsilon\vec{E}$		(6.34)
Material permeability	$\vec{B}=\mu\vec{H}$		(8.6)
Material conductivity	$\vec{J}=\sigma\vec{E}$		(6.13)

Table 10.2 Summary of the "Dynamic" Maxwell's Equations and the Relevant Constitutive Relationships

ADDENDUM 10A
INDUCTANCE UNDER TIME-VARYING CURRENTS

10A.1 Introduction

In Chapter 9, we presented the concept of inductance as an electrical circuit element to express the magnetic behavior of a current-carrying coil. However, we were unable to explore the full picture of the inductor as an element of the electrical circuit because we were confined to "static" constant currents and static fields. With a constant current flowing in the coil, the resulting magnetic field and flux are static, and hence no *emf* is induced in the coil according to Faraday's law. We have learned in circuit theory that an inductor in DC circuits (constant currents) has no voltage drop across its terminals unless it has an internal resistance in which case, the drop will be due to the internal resistance and not the inductance.

In this addendum, we will study the behavior of the inductance under dynamic electrical currents. We will be able to explore the full performance in terms of voltage drop across the inductor terminals and the energy stored in the inductor. We will also be able to examine the inductor as a circuit element connected in series and in parallel with other inductors.

10A.2 Inductance under Dynamic Electrical Currents

In Chapter 9, we defined the self- and mutual inductances in Equation (9.74):

$$L_1 = \frac{\Lambda_{11}}{I_1} = \frac{N_1 \Psi_{m11}}{I_1} \quad \text{and} \quad M = L_{12} = L_{21} = \frac{\Lambda_{21}}{I_1} = \frac{N_2 \Psi_{m21}}{I_1} \tag{9.74}$$

To clarify the terms of this definition, we refer to Figure 10A.1. In this figure, the flux linkage of coil #1 due to its own current is Λ_{11}, which is equal to the product of the coil's number of turns and the magnetic flux through the coil turns. Likewise, the flux linkage of coil #2 due to i_1 is Λ_{21}, which is equal to the product of the coil's number of turns and the magnetic flux through the coil.

Now, for the two-coil arrangement of Figure 10A.1, we rewrite Equations (9.83) as follows:

$$\Lambda_1 = L_1 I_1 + M I_2 \tag{9.83a}$$

$$\Lambda_2 = M I_1 + L_2 I_2 \tag{9.83b}$$

10A.2.1 Current-Voltage Relationship

Next, we apply Faraday's law to the expression in Equation (9.83) assuming time-varying currents and, consequently, time-varying magnetic flux linkage. Specifically,

$$\left.\begin{aligned} emf_1(t) &= -\frac{d\Lambda_1(t)}{dt} = -L_1 \frac{di_1(t)}{dt} - M\frac{di_2(t)}{dt} \quad \text{and} \\ emf_2(t) &= -\frac{d\Lambda_2(t)}{dt} = -M\frac{di_1(t)}{dt} - L_2\frac{di_2(t)}{dt} \end{aligned}\right\} \tag{10A.1}$$

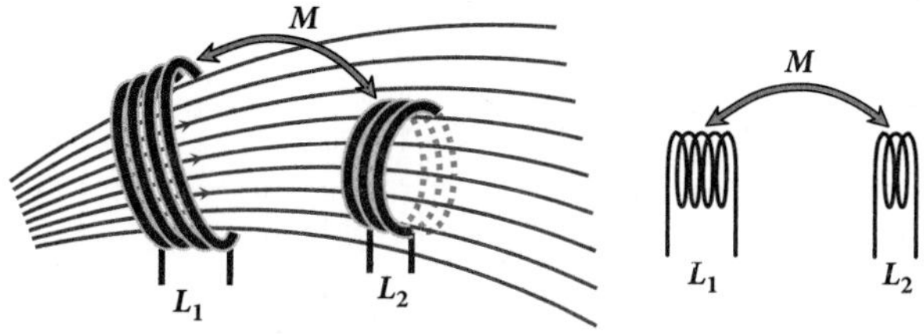

FIGURE 10A.1 A simple two-coil arrangement demonstrating the self and mutual flux linkages and inductances. To the right is the equivalent electrical circuit schematic.

These induced *emfs* work to counter the time-varying magnetic fluxes in the coils and hence the electrical currents producing them. The corresponding "voltage" drops occur across these coils in the same amounts as the induced *emfs* that will obviously be in the opposite polarities. Hence, the voltage drops for the above two equations assume the forms

$$\left.\begin{aligned} v_1(t) &= -emf_1(t) = L_1\frac{di_1(t)}{dt} + M\frac{di_2(t)}{dt} \quad \text{and} \\ v_2(t) &= -emf_2(t) = M\frac{di_1(t)}{dt} + L_2\frac{di_2(t)}{dt} \end{aligned}\right\} \tag{10A.2}$$

The corresponding phasor domain expressions are given as follows:

$$\left.\begin{aligned} \bar{V}_1 &= j\omega L_1\bar{I}_1 + j\omega M\bar{I}_2 \quad \text{and} \\ \bar{V}_2 &= j\omega M\bar{I}_1 + j\omega L_2\bar{I}_2 \end{aligned}\right\} \tag{10A.3}$$

10A.2.2 Inductances in Series and Parallel

Now, referring to Figure 10A.2, the current-voltage relationship for the single-coil inductor as a circuit element takes the form

$$v_L(t) = -emf_L(t) = L\frac{di_L(t)}{dt} \tag{10A.4}$$

Consequently, and referring to Figures 10A.3 and 10A.4, we can conclude that combinations of inductors connected in series and parallel will result in combined equivalencies as follows:

$$v_{c,\text{series}}(t) = v_1(t) + v_2(t) = L_1\frac{di_1(t)}{dt} + L_2\frac{di_2(t)}{dt} = (L_1 + L_2)\frac{di_c(t)}{dt} = L_{c,\text{series}}\frac{di_c(t)}{dt} \tag{10A.5}$$

Hence,

$$L_{c,\text{series}} = L_1 + L_2 \tag{10A.6}$$

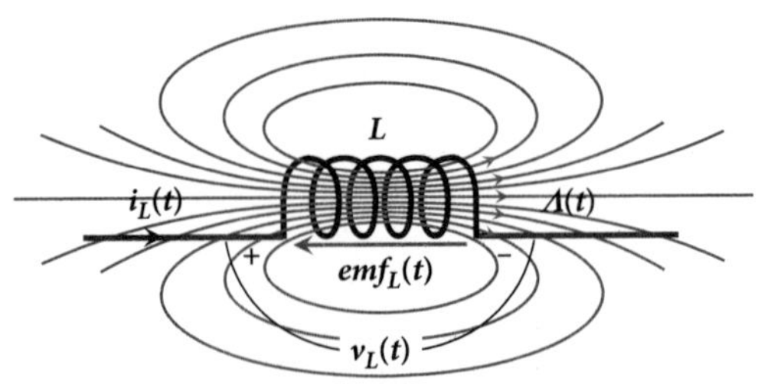

FIGURE 10A.2 An illustration of a coil (and its inductance) under dynamic conditions.

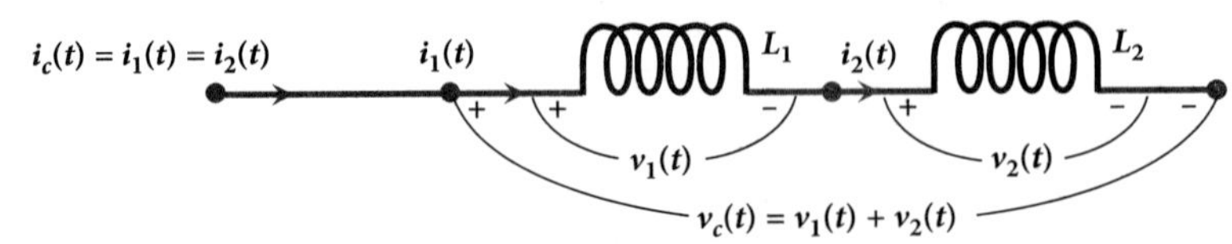

FIGURE 10A.3 A schematic of two inductors in series.

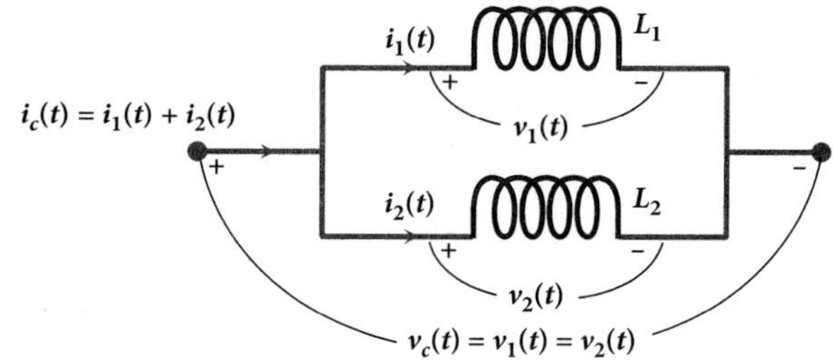

FIGURE 10A.4 A schematic of two inductors in parallel.

$$v_{c,\text{parallel}}(t) = v_1(t) = v_2(t) = L_1 \frac{di_1(t)}{dt} = L_2 \frac{di_2(t)}{dt} = L_{c,\text{parallel}} \frac{di_c(t)}{dt} \tag{10A.7}$$

Hence,

$$\frac{1}{L_{c,\text{parallel}}} = \frac{1}{L_1} + \frac{1}{L_2} \tag{10A.8}$$

10A.2.3 Energy Stored and Power in Inductances

For time-varying currents and fields, the inductor power and energy stored vary with time as well. The corresponding expressions are

$$\text{instantaneous power in an inductor} = p_L(t) = v_L(t) i_L(t), \quad \text{and} \tag{10A.9}$$

the instantaneous energy stored in inductor, $w_L(t)$, is given by

$$w_L(t) = \int_{-\infty}^{t} p_L(t)dt = \int_{-\infty}^{t} v_L(t) i_L(t) dt \tag{10A.10}$$

Using the voltage-current relationship [Equation (10A.4)], the above forms reduce to

$$p_L(t) = v_L(t) i_L(t) = L i_L(t) \frac{di_L(t)}{dt} = \frac{1}{2} L \frac{di_L^2(t)}{dt}, \quad \text{and} \tag{10A.11}$$

$$w_L(t) = \int_{-\infty}^{t} p_L(t)dt = \int_{-\infty}^{t} L i_L(t) \frac{di_L(t)}{dt} dt = \int_{0}^{i(t)} L i_L(t) di_L(t) = \frac{1}{2} L i_L^2(t) \tag{10A.12}$$

The corresponding phasor-domain expressions for both Equations (10A.11) and (10A.12) become

$$P_L = Re\left\{\frac{1}{2} j\omega L |\bar{I}_L|^2\right\} = 0, \quad \text{and} \tag{10A.13}$$

$$W_L = \frac{1}{2} L |\bar{I}_L|^2 \tag{10A.14}$$

which indicates that there is no average power dissipation (or generation) in ideal inductances or in magnetic fields in general.

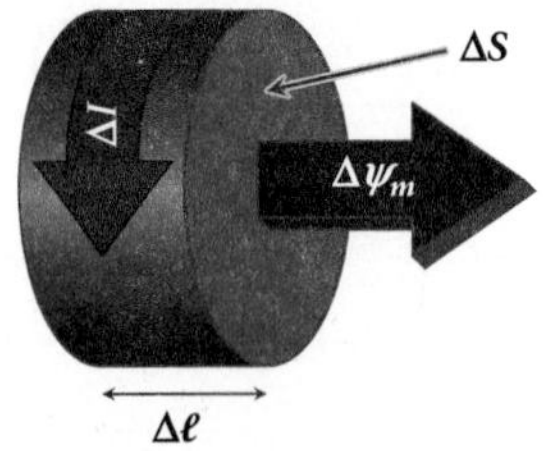

FIGURE 10A.5 Developing an expression for the energy stored in a magnetic field using previously developed expressions for the energy stored in (an incremental) solenoid.

10A.2.4 Energy Stored in Magnetic Fields

You recall that we discussed this point earlier in Chapter 9 for which we presented expressions (9.20) and (9.21) for the energy and energy density storage in magnetic fields. No solid proofs were given for the expressions that were reached at the time based on analogies between electric and magnetic cases.

In this section, we provide a derivation for these expressions based on the dynamic behavior and Faraday's law. Starting with Equation (10A.14), we can write

$$W_L = \frac{1}{2}L|\bar{I}_L|^2 = \frac{1}{2}|\Lambda_L||I_L| \tag{10A.15}$$

Next, we apply this expression to an incremental solenoid as shown in Figure 10A.5. Using Equation (8.45) for the magnetic field inside the solenoid while replacing nI by $\Delta I/\Delta \ell$ since both express the axial density of the solenoid's current, we write

$$H = \frac{\Delta I}{\Delta \ell} \quad \text{and hence,} \quad \Delta I = H\Delta \ell \tag{10A.16}$$

$$\Delta W_m = \frac{1}{2}|\Delta\Lambda||\Delta I| = \frac{1}{2}|B\Delta S||H\Delta\ell| = \frac{1}{2}BH\Delta v \tag{10A.17}$$

$$w_m = \frac{\Delta W_m}{\Delta v} = \frac{1}{2}BH = \frac{1}{2}\mu H^2 = \frac{1}{2}\frac{B^2}{\mu} \tag{10A.18}$$

$$W_m = \int_{\text{vol}} w_m \, dv \tag{10A.19}$$

This energy is stored in the magnetic field/flux linkage with the inductance.

CHAPTER 10 PROBLEMS

- Problems 10.1 and 10.2: Problems dealing with Hall effect
- Problems 10.3 to 10.11: Problems dealing with Faraday's law and the induced *emf*
- Problems 10.12 and 10.13: Problems dealing with the displacement current
- Problems 10.14 to 10.16: Problems dealing with Maxwell's equations and the magnetic vector potential
- Problems 10.17 and 10.18: Problems dealing with inductors under dynamic conditions

10.1 In an experiment to measure the doping concentration of a semiconductor bar, the setup in the figure shown is used. The semiconductor bar has a thickness 0.1 mm and is placed in a magnetic field of $\vec{B}=0.1\,\vec{a}_z$ tesla. The input current is −30 mA and the induced Hall voltage is measured to be −10 mV. Find the majority carrier concentration. Is the semiconductor bar *N*-type or *P*-type?

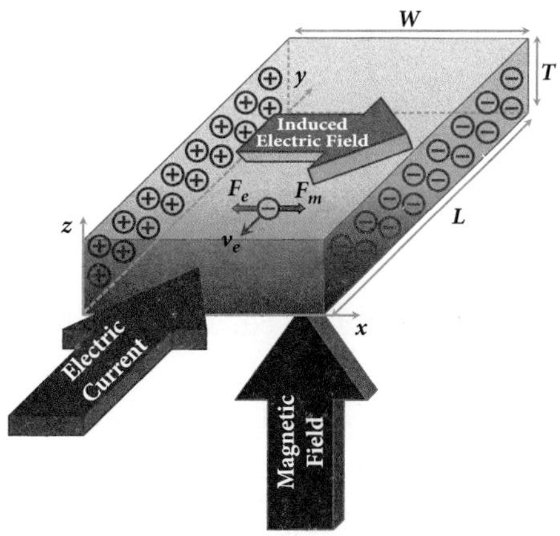

FIGURE 10P.1

10.2 In an experiment to measure the doping concentration of an *N*-type semiconductor bar, the measured Hall voltage had an added unknown offset. In order to eliminate the effect of the offset voltage, two measurements were taken at two different values of the input current. The measurement values are as shown in the table below. The semiconductor bar had a thickness of 0.1 mm and is placed in a magnetic field of $\vec{B}=0.1\vec{a}_z$ tesla. Find the electron concentration.

	Input Current, I_X	Measured Hall Voltage, V_H
a.	−40 mA	10 mV
b.	−20 mA	6 mA

10.3 A square wire loop of side $a=5$ cm in the *x-y* plane is a placed in magnetic field given by $\vec{B}=2te^{-0.2t}\,\vec{a}_z$ Wb/m^2. Find the magnitude of the induced *emf*.

10.4 A conductive rod of length $a=5$ cm along the *z* axis is placed in a magnetic field $\vec{B}=0.2\vec{a}_x$ Wb/m^2. One end of the rod is fixed at the origin while the other end is free to rotate in the *y-z* plane. If the rod rotates at a rate $\omega_o=100$ rad/s, find the induced *emf* between the two ends of the rod.

10.5 The square wire loop shown has 10 turns and lies in the ρ-z plane. If the magnetic flux density in the region is given by

$$\vec{B}(t) = (\vec{a}_\varphi + 0.1\vec{a}_z)\frac{2\times10^{-4}}{\rho}\cos(2\pi\times10^4 t)\,(\mathrm{Wb/m^2})$$

a. Determine the induced *emf* across a small gap created in the loop as shown in the figure.

b. Determine the magnitude and direction of the current that would flow across a 100-Ω resistor connected across the gap at $t = 0$ and at $t = 50\ \mu s$. Ignore the internal resistance of the loop.

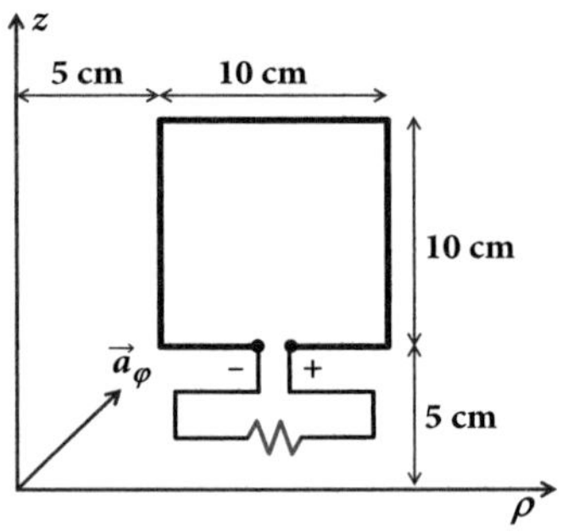

Figure 10P.5

10.6 Derive an expression for the *emf* induced in a circular wire loop, radius a, moving in a time-varying field given by $\vec{H}(t) = H_o \exp(-kt)\vec{a}_z$.

Assume wire was initially in the x-y plane and centered at the origin. Given that the coil motion is

a. Sliding along the x axis at a speed u

b. Rotating around the x axis at an angular speed ω

10.7 Derive an expression for the *emf* induced in a square coil with 50 turns and side $a = 10$ cm, moving in a time-varying field given by $\vec{B}(t) = (2\vec{a}_x + 3\vec{a}_z)\exp(-0.2t)\,\mathrm{Wb/m^2}$.

Assume the coil was initially in the x-y plane and centered at the origin. Given that the coil motion is

a. Sliding along the y axis at a speed u

b. Rotating around the z axis at an angular speed ω

10.8 Derive an expression for the *emf* induced in the single turn square coil ($a \times a$) rotating around the z axis in the positive φ direction at an angular speed ω in a time-varying field given by $\vec{H}_x(t) = H_o\cos(\omega t + \pi/3)$. Identify the *emf* polarity. Assume the coil was initially lying in the x-z plane ($\varphi = 0$) cornered at the origin.

10.9 Consider the rectangular loop shown in Figure 10.10 where one side of the loop is a conductive rod of length ($a = 4$ cm) moving in the y direction with a velocity $\vec{v} = 5\vec{a}_y \frac{\mathrm{m}}{\mathrm{s}}$. The magnetic field in the region is given by $\vec{B} = 0.1(2\vec{a}_x + 4\vec{a}_z)\sin(2\pi\times10^3 t)\ \mathrm{Wb/m^2}$. Find the *emf* induced in this loop. Assume that the initial position of the conductive rod was 5 cm from the x axis.

10.10 Derive an expression for the *emf* induced in the single-turn square coil of side $a = 10$ cm rotating around the z axis in the positive φ direction at an angular speed $\omega_o = 50\,\frac{\mathrm{rad}}{\mathrm{s}}$ in a time-varying field given by $\vec{B}(t) = 0.1\cos\left(2\pi\times10^3 t + \frac{\pi}{3}\right)\vec{a}_x\,\mathrm{Wb/m^3}$. Identify the *emf* polarity. Assume the coil was initially lying in the x-z plane ($\varphi = 0$) cornered at the origin.

10.11 The single-turn square loop shown in the figure of Example 10.4 is placed in the *x*-*y* plane. The magnetic flux density in the region is $\vec{B}=(\vec{a}_x+2\vec{a}_z)\times 10^{-5}\sin(2\pi\times 10^7 t)$ (tesla).

a. Determine the *emf* induced across a small gap in the loop as a function of time.

b. Determine the direction and magnitude of the current at $t=0$ and at $t=50$ ns for $R=25\ \Omega$. Show the polarity of the induced *emf* and the induced current direction.

c. Determine the direction and magnitude of the current at $t=12.5$ ns. Examine the direction of the induced current, and the induced magnetic field in the loop, check to see if Lenz's law is verified.

d. Qualitatively describe how the loop should be oriented relative to the applied magnetic field so that the induced *emf* is maximized.

10.12 In a material for which $\sigma=0$, $\varepsilon=4.5\ \varepsilon_o$, and $\mu=2\mu_o$, the displacement current density vector is $5\times 10^{-6}\cos(2\pi\times 10^8 t-2\pi z)\vec{a}_x$ A/m^2.

a. Use the definition of displacement current density vector to find $\vec{D}$ and $\vec{E}$.

b. Use the point form of Faraday's law and a time integration to find $\vec{H}$.

10.13 Two parallel conducting metal plates of cross-sectional area 100 cm^2 are separated by a distance of 5 mm using a dielectric material of $\varepsilon=10\varepsilon_o$. If a sinusoidal voltage source $v_s(t)=5\cos(2\pi\times 10^4 t)$V is applied between the plates, find the displacement current.

10.14 In a material for which $\sigma=500$, $\varepsilon=9\varepsilon_o$, and $\mu=\mu_o$, the electric current density vector is $50\cos(6\pi\times 10^8 t-2\pi z)\vec{a}_x$ A/m^2. Find the volumetric charge density distribution.

10.15 Given the magnetic vector potential as $\vec{A}(z,t)=\vec{a}_x\{A_o\cos(\omega t-\beta z)\}$ Wb/m. Find the corresponding $\vec{H}(z,t)$ and $\vec{E}(z,t)$ vectors.

10.16 Given the magnetic vector potential $\vec{A}=\vec{a}_y\left\{A_o\sin\left(\beta z-\frac{\pi}{3}\right)\right\}$ Wb/m. Find the corresponding $\vec{H}$ vector.

10.17 A harmonic voltage is applied to a coil with self-inductance *L*, given by $V=V_o\cos(5t)$. Write expressions for the following.

a. The magnetic flux linkage as a function of time

b. The current in the coil as a function of time

c. The magnetic energy stored in the coil as a function of time

Evaluate the quantities in a-c above if $V_o=1$ V, $f=10$ MHz, and $L=1$ mH.

10.18 An applied current of 12 sin(120πt) A is made to flow in coil #1 of a two-coil arrangement. Meanwhile, coil #2 is loaded with a 10 Ω resistor. If the mutual inductance between the two coils is 0.02 H with. Find:

a. The magnetic flux through coil 2

b. The *emf* induced in coil 2

c. The current in the resistor termination

CHAPTER 10 SUMMARY[1]

Introduction

- Charge trajectory in magnetic fields: $\vec{F}_{m\pm} = q_\pm \vec{v}_\pm \times \vec{B} = \frac{m_\pm (v_\pm)^2}{a_\pm}$
- Hall effect: $\vec{E}_i = -\vec{v} \times \vec{B}$

Faraday's Law

- Faraday's law: $emf = -\frac{\partial \Lambda}{\partial t} = emf_{\text{time-varying field}} + emf_{\text{motional}}$
- An example of a moving conductor in a time-varying magnetic field
- The electric generator: $emf = \omega BS \sin(\omega t + \theta_o)$
- The transformer: $\frac{emf_2(t)}{emf_1(t)} = \frac{v_2(t)}{v_1(t)} = \frac{N_2}{N_1}$

Faraday's Maxwell's Equation

- Integral form: $\oint_C \vec{E} \cdot \vec{d\ell} = -\iint_S \frac{\partial \vec{B}}{\partial t} \cdot dS$
- Differential form: $\vec{\nabla} \times \vec{E} = -\frac{\partial \vec{B}}{\partial t}$

Revisiting Field and Potential Formulas

- The displacement current: $\frac{\partial \vec{D}}{\partial t}$
- Revisiting Ampere's law: $\oint_C \vec{H} \cdot \vec{d\ell} = \iint_{S \text{ contoured by } C} \left[\vec{J} + \frac{\partial \vec{D}}{\partial t} \right] \cdot dS \cdot \vec{\nabla} \times \vec{H} = \vec{J} + \frac{\partial \vec{D}}{\partial t}$

Electrostatic/Magnetostatic Relationships	Time-Varying Modifications
$\vec{\nabla} \cdot \vec{D} = \rho_v$	$\vec{\nabla} \cdot \vec{D} = \rho_v$
$\vec{\nabla} \cdot \vec{B} = 0$	$\vec{\nabla} \cdot \vec{B} = 0$
$\vec{\nabla} \times \vec{E} = 0$	$\vec{\nabla} \times \vec{E} = -\frac{\partial \vec{B}}{\partial t}$
$\vec{\nabla} \times \vec{H} = \vec{J}$	$\vec{\nabla} \times \vec{H} = \vec{J} + \frac{\partial \vec{D}}{\partial t}$
$\vec{E} = -\vec{\nabla} V$	$\vec{E} = -\vec{\nabla} V - \frac{\partial \vec{A}}{\partial t}$

[1] ***Important Warning to Students:*** Please do not use this summary without thorough study of the relevant material in this chapter. Using this summary and its formulas without adequate appreciation of the physics involved and conditions of validity can lead to wrong implementations and invalid conclusions.

Electrostatic/Magnetostatic Relationships	Time-Varying Modifications
$\vec{B}=\vec{\nabla}\times\vec{A}$ & $\vec{\nabla}\cdot\vec{A}=0$	$\vec{B}=\vec{\nabla}\times\vec{A}$ & $\vec{\nabla}\cdot\vec{A}=-\mu\varepsilon\frac{\partial V}{\partial t}$
$\nabla^2 V=-\frac{\rho_v}{\varepsilon}$	$\nabla^2 V=-\frac{\rho_v}{\varepsilon}+\mu\varepsilon\frac{\partial^2 V}{\partial t^2}$
$\nabla^2\vec{A}=-\mu\vec{J}$	$\nabla^2\vec{A}=-\mu\vec{J}+\mu\varepsilon\frac{\partial^2\vec{A}}{\partial t^2}$

Summary of Maxwell's Equations

Equations	Point Form	Integral Form
Faraday's law	$\vec{\nabla}\times\vec{E}=-\frac{\partial\vec{B}}{\partial t}$	$\oint_C \vec{E}\cdot\vec{d\ell}=-\iint_{S\text{ contoured by }C}\frac{\partial\vec{B}}{\partial t}\cdot dS$
Ampere's law	$\vec{\nabla}\times\vec{H}=\vec{J}+\frac{\partial\vec{D}}{\partial t}$	$\oint_C \vec{H}\cdot\vec{d\ell}=\iint_{S\text{ contoured by }C}\left[\vec{J}+\frac{\partial\vec{D}}{\partial t}\right]\cdot dS$
Gauss' law for electric flux	$\vec{\nabla}\cdot\vec{D}=\rho_v$	$\oiint_S \vec{D}\cdot\vec{dS}=\iiint_{\text{volume enclosed by }S}\rho_v\, dv$
"Gauss' law" for magnetic flux	$\vec{\nabla}\cdot\vec{B}=0$	$\oiint_S \vec{B}\cdot\vec{dS}=0$
Constitutive relations	$\vec{D}=\varepsilon\vec{E},\ \vec{B}=\mu\vec{H},\ \vec{J}=\sigma\vec{E}$	

CHAPTER 11

Wave Propagation—Transmission Lines Revisited

Heinrich Rudolf Hertz
(1857–1894)

John Henry Poynting
(1852–1914)

Chapter Learning Objectives

- Describe and understand the physical insight of Maxwell's equations and be able to manipulate and apply them to electromagnetic (EM) problems, for example, derive the wave equation using Maxwell's equations.
- Use Maxwell equations to describe planar EM wave propagation in dielectrics. Describe wave propagation characteristics, for example, wave impedance, phase velocity, and wavelength.
- Describe and understand the physical insight of wave propagation characteristics in different materials.
- Describe and understand the physical insight of skin effect and shielding.
- Describe and understand the physical insight of loss tangent for energy-storage materials and devices.
- Investigate the power flow using Poynting vector, Poynting theorem.
- Describe and understand the physical insight of electromagnetic power flow in dielectrics and the role of conductors in providing guidance and confinement.
- Describe and understand the physical insight of plane waves and power flow in controlled-geometry transmission lines (for example, case of coaxial line).
- Study high-frequency coaxial line parameters.

11.1 Introduction

In previous chapters, we began our journey into time-varying EM fields. We learned about Faraday's law as well as the dynamic form of Ampere's law. We also developed the dynamic updates of other static field and potential relationships that were studied in earlier chapters. In the previous chapter, we presented the famous Maxwell's equations in the time domain both in the point (differential) form and the integral form.

In this chapter, we intend to proceed with the study of electromagnetic wave propagation in open (unguided) space/media. We will start with the simple case of one-dimensional propagation in a source-free free space. Next, we will tackle the cases of having lossless as well as lossy media replacing the free space. Hence, we will be able to discuss wave propagation characteristics in different types of materials ranging from lossless and low-loss dielectrics to high-conductivity conductors and perfect conductors.

This development will get us into discussions of the important topics of wave propagation, energy flow density, confinement and shielding of wave energy, skin effect, and the loss tangent of a dissipative medium.

11.2 Maxwell's Equations and the Wave Equation

As we have always tried throughout this book, we will attempt the simplest case for a new development before considering further details. The simplest case to start with for wave propagation is that of one-dimensional wave propagating in free space. We will constrain our study to that of a single-harmonic source. Moreover, we will assume the source to cause field excitations of uniform intensities over an infinitely extended plane in space, for example, the x-y plane in Cartesian coordinates. For analysis convenience, we will proceed with our development in the frequency domain using phasor expressions. Using Cartesian coordinates, we start our analysis by describing the above conditions/constraints in mathematical terms as identified in Table 11.1.

Source-free	$\vec{J}=0$ $\rho_v=0$	(11.1)
Free space	$\vec{D}=\varepsilon_o\vec{E}$ $\vec{B}=\mu_o\vec{H}$	(11.2)
One dimensional	Due to the assumed uniform field intensities over infinite extensions in both x and y directions, we expect the field intensities to be functions of z only $\vec{E}(x,y,z,t)\big\vert_{-\infty<x<\infty,\,-\infty<y<\infty}=\vec{E}(z,t)\big\vert_{-\infty<x<\infty,\,-\infty<y<\infty}$ $\vec{H}(x,y,z,t)\big\vert_{-\infty<x<\infty,\,-\infty<y<\infty}=\vec{H}(z,t)\big\vert_{-\infty<x<\infty,\,-\infty<y<\infty}$	(11.3)
Harmonic source	$E_x(z,t)=[E_x(z)\cdot\cos(\omega t+\varphi_{Ex})]\rightarrow\bar{E}_x(z)=E_x(z)\cdot e^{j\varphi_{Ex}}$ $H_x(z,t)=[H_x(z)\cdot\cos(\omega t+\varphi_{Hx})]\rightarrow\bar{H}_x(z)=H_x(z)\cdot e^{j\varphi_{Hx}}$ same for the y components	(11.4)

TABLE 11.1 Constraints for the Special Case of Harmonic Plane Wave Propagation in Free Space

Notation notes:

- The accent bars on top of the field notation symbols indicate phasor quantities while the top arrows indicate their vector nature.
- When we dealt with voltage and current waves on transmission lines in Chapter 2, we used lower case symbols for time-domain quantities and uppercase notation for phasors and amplitudes. However, for E and H field quantities, we will use uppercase notation in both the time and frequency domains. We do that to conform to the notation used by most references in the field. To demonstrate this point, we will use:

  $E_x(x,y,z)$ for the amplitude of E_x at the location (x,y,z).

  $\vec{E}(x,y,z,t)$ for the E field vector (in the time domain) as a function of position and time.

  $\bar{E}_x(x,y,z)$ for the E_x phasor (in the frequency domain) at the location (x,y,z).

  $\bar{\vec{E}}(x,y,z)$ for the E vector phasor (in the frequency domain) at the location (x,y,z).
- We will drop the (...) from our field expressions in the development in order to simplify the presentation.

Consequently, we proceed with the reduced forms of Maxwell's equations in source-free free space as presented in Table 11.2.

Maxwell's Equations in the Point Form (10.51)		
Faraday's law (10.29)	$\vec{\nabla}\times\bar{\vec{E}}=-j\omega\bar{\vec{B}}=-j\omega\mu_o\bar{\vec{H}}$	(11.5)
Ampere's law (10.38)	$\vec{\nabla}\times\bar{\vec{H}}=\vec{J}+j\omega\bar{\vec{D}}=j\omega\varepsilon_o\bar{\vec{E}}$	(11.6)
Gauss' law for electric flux (4.17)	$\vec{\nabla}\cdot\bar{\vec{D}}=\vec{\nabla}\cdot\varepsilon_o\bar{\vec{E}}=0$	(11.7)
Gauss' law for magnetic flux (8.46)	$\vec{\nabla}\cdot\bar{\vec{B}}=\vec{\nabla}\cdot\mu_o\bar{\vec{H}}=0$	(11.8)

TABLE 11.2 Maxwell's Equations Reduced for the Special Case of Harmonic Plane Wave Propagation in Source-Free Free Space

The set of equations in Table 11.2, in conjunction with the yet to be defined appropriate boundary conditions, fully characterize the spatial field distributions for the case in consideration. The differential equations of both Faraday's and Ampere's laws need to be simultaneously solved to obtain spatial expressions for both $\bar{\vec{E}}$ and $\bar{\vec{H}}$.

Now, let us take advantage of the one-dimensional simplification by expanding the two curl equations [Equations (11.5) and (11.6)] in Cartesian coordinates using the determinant approach:

$$\begin{vmatrix} \vec{a}_x & \vec{a}_y & \vec{a}_z \\ \dfrac{\partial}{\partial x} & \dfrac{\partial}{\partial y} & \dfrac{\partial}{\partial z} \\ \bar{E}_x & \bar{E}_y & \bar{E}_z \end{vmatrix} = -j\omega\mu_o \begin{bmatrix} \bar{H}_x\vec{a}_x \\ \bar{H}_y\vec{a}_y \\ \bar{H}_z\vec{a}_z \end{bmatrix} \tag{11.9}$$

and

$$\begin{vmatrix} \vec{a}_x & \vec{a}_y & \vec{a}_z \\ \dfrac{\partial}{\partial x} & \dfrac{\partial}{\partial y} & \dfrac{\partial}{\partial z} \\ \bar{H}_x & \bar{H}_y & \bar{H}_z \end{vmatrix} = j\omega\varepsilon_o \begin{bmatrix} \bar{E}_x\vec{a}_x \\ \bar{E}_y\vec{a}_y \\ \bar{E}_z\vec{a}_z \end{bmatrix} \tag{11.10}$$

Recognizing that the assumed uniform field intensities over the infinite extensions in both x and y directions imply that the field derivatives with respect to both x and y are zeros, that is, $\dfrac{\partial}{\partial x}=0$ and $\dfrac{\partial}{\partial y}=0$, we can write

$$\begin{vmatrix} \vec{a}_x & \vec{a}_y & \vec{a}_z \\ \dfrac{\partial}{\partial x} & \dfrac{\partial}{\partial y} & \dfrac{\partial}{\partial z} \\ \bar{E}_x & \bar{E}_y & \bar{E}_z \end{vmatrix} = \begin{vmatrix} \vec{a}_x & \vec{a}_y & \vec{a}_z \\ 0 & 0 & \dfrac{\partial}{\partial z} \\ \bar{E}_x & \bar{E}_y & \bar{E}_z \end{vmatrix} = \begin{bmatrix} -\dfrac{\partial\bar{E}_y}{\partial z}\vec{a}_x \\ \dfrac{\partial\bar{E}_x}{\partial z}\vec{a}_y \\ 0\,\vec{a}_z \end{bmatrix} = -j\omega\mu_o \begin{bmatrix} \bar{H}_x\vec{a}_x \\ \bar{H}_y\vec{a}_y \\ \bar{H}_z\vec{a}_z \end{bmatrix} \tag{11.11}$$

and

$$\begin{vmatrix} \vec{a}_x & \vec{a}_y & \vec{a}_z \\ \dfrac{\partial}{\partial x} & \dfrac{\partial}{\partial y} & \dfrac{\partial}{\partial z} \\ \bar{H}_x & \bar{H}_y & \bar{H}_z \end{vmatrix} = \begin{vmatrix} \vec{a}_x & \vec{a}_y & \vec{a}_z \\ 0 & 0 & \dfrac{\partial}{\partial z} \\ \bar{H}_x & \bar{H}_y & \bar{H}_z \end{vmatrix} = \begin{bmatrix} -\dfrac{\partial\bar{H}_y}{\partial z}\vec{a}_x \\ \dfrac{\partial\bar{H}_x}{\partial z}\vec{a}_y \\ 0\,\vec{a}_z \end{bmatrix} = j\omega\varepsilon_o \begin{bmatrix} \bar{E}_x\vec{a}_x \\ \bar{E}_y\vec{a}_y \\ \bar{E}_z\vec{a}_z \end{bmatrix} \tag{11.12}$$

from which we conclude

$$\left.\begin{aligned} -\frac{\partial\bar{E}_y}{\partial z} &= -j\omega\mu_o\bar{H}_x \\ \frac{\partial\bar{E}_x}{\partial z} &= -j\omega\mu_o\bar{H}_y \\ 0 &= \bar{H}_z \end{aligned}\right\} \tag{11.13}$$

$$\left.\begin{aligned} -\frac{\partial \bar{H}_y}{\partial z} &= j\omega\varepsilon_o \bar{E}_x \\ \frac{\partial \bar{H}_x}{\partial z} &= j\omega\varepsilon_o \bar{E}_y \\ 0 &= \bar{E}_z \end{aligned}\right\} \tag{11.14}$$

With both $\bar{E}_z = 0$ and $\bar{H}_z = 0$, both the $\vec{E}$ and $\vec{H}$ vector fields lie completely on the *x-y* plane, orthogonal to *z* coordinate. This characteristic prompts naming such wave propagation as *plane-wave propagation*.

If, without any lack of generality, we choose the *x* coordinate to line up with the $\vec{E}$ field vector, we set $\bar{E}_y = 0$ in the above equations, which is expected to yield much-simplified expressions to work with. Alternatively, we can get the same result if we were to consider rotating the EM source, and hence the resulting fields, in the *x-y* plane (around the *z* axis) such that the entire $\vec{E}$ field vector lines up with the *x* coordinate (leaving no components along the *y* coordinate).

Consequently, with this choice, we conclude from Equation (11.13) that

$$-\frac{\partial \bar{E}_y}{\partial z} = 0 = -j\omega\mu_o \bar{H}_x \rightarrow \bar{H}_x = 0 \tag{11.15}$$

and the only field components that do not vanish are $\bar{E}_x$ and $\bar{H}_y$, with the corresponding governing equations

$$\frac{\partial \bar{E}_x}{\partial z} = -j\omega\mu_o \bar{H}_y \tag{11.16}$$

and

$$\frac{\partial \bar{H}_y}{\partial z} = -j\omega\varepsilon_o \bar{E}_x \tag{11.17}$$

These are two coupled first-order differential equations in the two field variables. To resolve, we need to substitute one of the two variables from one of the equations into the other; let us choose to eliminate $\bar{H}_y$. To do that, we differentiate Equation (11.16) and substitute the result into Equation (11.17). Thus, we can write

$$\frac{\partial^2 \bar{E}_x}{\partial z^2} = \frac{\partial}{\partial z}[-j\omega\mu_o \bar{H}_y] = (-j\omega\mu_o)\left[\frac{\partial \bar{H}_y}{\partial z}\right] = (-j\omega\mu_o)[-j\omega\varepsilon_o \bar{E}_x] = -\omega^2\mu_o\varepsilon_o \bar{E}_x \tag{11.18}$$

We note that we were able to extract "μ_o" outside the spatial $\frac{\partial}{\partial z}$ derivative since the medium is homogeneous (free space with constant μ_o everywhere). Finally, we obtain the wave equation for the electric field intensity vector

$$\frac{\partial^2 \bar{E}_x}{\partial z^2} = -\omega^2\mu_o\varepsilon_o \bar{E}_x \tag{11.19}$$

11.3 Important Remarks

At this point, we pause to make a few important remarks:

1. It is important to realize that while solving the two coupled first-order differential equations, we ended up with one second-order differential equation in one variable, $\vec{E}$. In doing so, we had to differentiate one of the two relationships resulting in a "reduced-knowledge" relationship. To explain,

differentiation of the equation $f=5$ with respect to g yields $\frac{df}{dg}=0$ from which we can only conclude that f is a constant. The information available in the original equation $f=5$ is now lost and we have increased the number of unknowns in our system by 1.

2. To retrieve the *lost* knowledge required for a complete solution, we need to retain the original equation as part of our system of equations. Hence, our system of equations to be solved is now of the form:

$$\frac{\partial^2 \bar{E}_x}{\partial z^2} = -\omega^2 \mu_o \varepsilon_o \bar{E}_x \tag{11.20}$$

$$\frac{\partial \bar{E}_x}{\partial z} = -j\omega \mu_o \bar{H}_y \tag{11.21}$$

3. If we elected to reduce the above system of equations through the elimination of $\bar{E}_x$ instead of $\bar{H}_y$, we would get an identical relationship with $\bar{H}_y$ instead of $\bar{E}_x$; specifically,

$$\frac{\partial^2 \bar{H}_y}{\partial z^2} = -\omega^2 \mu_o \varepsilon_o \bar{H}_y \tag{11.22}$$

$$\frac{\partial \bar{H}_y}{\partial z} = -j\omega \varepsilon_o \bar{E}_x \tag{11.23}$$

4. The second-order differential equations above are referred to as wave equations. These equations are quite similar to those of the voltage and current phasors we dealt with in the transmission line analyses we carried earlier in Chapter 2. More on this to follow later in this chapter.

5. We need to be reminded that, in general, both $\bar{E}_x$ and $\bar{H}_y$ are functions of position, that is, functions of all of $x, y,$ and z. However, in our special case of working with one dimension with changes occurring only along the z coordinate, both $\bar{E}_x$ and $\bar{H}_y$ are functions of z only. Thus, we can write our system of Equations (11.18) and (11.16) in the form

$$\frac{\partial^2 \bar{E}_x(z)}{\partial z^2} = -\omega^2 \mu_o \varepsilon_o \bar{E}_x(z) \tag{11.16}$$

$$\frac{\partial \bar{E}_x(z)}{\partial z} = -j\omega \mu_o \bar{H}_y(z) \tag{11.14}$$

6. An alternate approach to the above derivations is given in Addendum 11A. In that approach, the 3D forms of Ampere's and Faraday's laws are used up front.

11.4 Solving the Wave Equation

To solve the wave equation, we will follow identical steps to those followed earlier in Chapter 2. Our wave equations for the field phasors are identical to those of voltage and current phasors developed in Chapter 2. The phasor $\bar{E}_x(z)$ is analogous to the phasor $V(z)$ while the phasor $\bar{H}_y(z)$ is analogous to the phasor $I(z)$. In the following, we will take advantage of these analogies in developing the field solutions for our wave equations.

In Table 11.3, we present the wave equation solution steps that we developed in Chapter 2 side by side along with the extrapolated analogous solution for the field problem at hand. We used the lossless transmission line (TL) solution since our field problem involves the lossless medium of free space. Hence, we are setting both R_z and G_z as zeros in the TL solutions presented in the table.

Lossless TL Equations		One-Dimensional Free Space Wave Equations	
Starter Equations (KVL and KCL)		**Starter Equations (Faraday and Ampere)**	
(2.8)	$\frac{dV(z)}{dz}=-\{\cancel{R_z}+j\omega L_z\}\cdot I(z)=-j\omega L_z\cdot I(z)$	$\frac{\partial \bar{E}_x(z)}{\partial z}=-j\omega\mu_o\bar{H}_y(z)$	(11.24)
(2.12)	$\frac{dI(z)}{dz}=-\{\cancel{G_z}+j\omega C_z\}\cdot V(z)=-j\omega C_z\cdot V(z)$	$\frac{\partial \bar{H}_y(z)}{\partial z}=-j\omega\varepsilon_o\bar{E}_x(z)$	(11.25)
Next, we follow up with the wave equation as the second-order differential equation resulting from combining the starter equations above. We also, present the characteristic parameters in both cases.			
Lossless TL Equations		One-Dimensional Free Space Wave Equations	
(2.16)	$\frac{d^2V(z)}{dz^2}=\{j\omega L_z\}\cdot\{j\omega C_z\}\cdot V(z)$	$\frac{\partial^2}{\partial z^2}\bar{E}_x(z)=-\omega^2\mu_o\varepsilon_o\bar{E}_x(z)$	(11.26)
Defining γ, the complex propagation constant		Defining γ, the complex propagation constant	
(2.19)	$\gamma^2=\{j\omega L_z\}\cdot\{j\omega C_z\}=-\omega^2L_zC_z$	$\gamma^2=-\omega^2\mu_o\varepsilon_o$	(11.27)
(2.20)	$\frac{d^2V(z)}{dz^2}=\gamma^2\cdot V(z)$	$\frac{\partial^2}{\partial z^2}\bar{E}_x(z)=\gamma^2\cdot\bar{E}_x(z)$	(11.28)
General Form Solution		General Form Solution	
(2.21)	$V(z)=V^+e^{-\gamma z}+V^-e^{+\gamma z}$	$\bar{E}_x(z)=\bar{E}_x^+e^{-\gamma z}+\bar{E}_x^-e^{+\gamma z}$	(11.29)
(2.23)	$I(z)=-\frac{1}{j\omega L_z}\cdot\frac{dV(z)}{dz}$	$\bar{H}_y(z)=-\frac{1}{j\omega\mu_o}\cdot\frac{\partial\bar{E}_x}{\partial z}$	(11.30)
(2.24)	$I(z)=\frac{\gamma}{j\omega L_z}\cdot\{V^+e^{-\gamma z}-V^-e^{+\gamma z}\}$	$\bar{H}_y(z)=\frac{\gamma}{j\omega\mu_o}\cdot\{\bar{E}_x^+e^{-\gamma z}-\bar{E}_x^-e^{+\gamma z}\}$	(11.31)
See Figure 11.1(a)		See Figure 11.1(b)	
Defining Z_o, the characteristic impedance of TL		Defining η_o, the intrinsic impedance of media	
(2.25)	$\frac{\gamma}{j\omega L_z}=\sqrt{\frac{C_z}{L_z}}=\frac{1}{Z_o},\ Z_o=\sqrt{\frac{L_z}{C_z}}$	$\frac{\gamma}{j\omega\mu_o}=\sqrt{\frac{\varepsilon_o}{\mu_o}}=\frac{1}{\eta_o},\ \eta_o=\sqrt{\frac{\mu_o}{\varepsilon_o}}$	(11.32)
(2.26)	$I(z)=\frac{1}{Z_o}\cdot\{V^+e^{-\gamma z}-V^-e^{+\gamma z}\}$	$\bar{H}_y(z)=\frac{1}{\eta_o}\cdot\{\bar{E}_x^+e^{-\gamma z}-\bar{E}_x^-e^{+\gamma z}\}$	(11.33)
(2.27)	$Z_o=\sqrt{L_z/C_z}$	$\eta_o=\sqrt{\mu_o/\varepsilon_o}\cong 377\ \Omega$	(11.34)
(2.28)	$\gamma=0+j\beta=j\omega\sqrt{L_z\cdot C_z}$	$\gamma=j\beta_o=\sqrt{-\omega^2\mu_o\varepsilon_o}=j\omega\sqrt{\mu_o\varepsilon_o}$	(11.35)
(2.29)	$V(z)=V^+e^{-\gamma z}+V^-e^{+\gamma z}$	$\bar{E}_x(z)=\bar{E}_x^+e^{-\gamma z}+\bar{E}_x^-e^{+\gamma z}$	(11.36)
(2.30)	$V(z)=V^+e^{-j\beta z}+V^-e^{+j\beta z}$	$\bar{E}_x(z)=\bar{E}_x^+e^{-j\beta_o z}+\bar{E}_x^-e^{+j\beta_o z}$	(11.37)
	$\beta=\omega\sqrt{L_z\cdot C_z}$	$\beta_o=\omega\sqrt{\mu_o\cdot\varepsilon_o}$	(11.38)
(2.31)	Transfer Back to Time Domain:	Transfer Back to Time Domain:	
(2.32)	$v(z,t)=\lvert V^+\rvert\cdot\cos(\omega t-\beta z+\varphi_v^+)+\lvert V^-\rvert\cdot\cos(\omega t+\beta z+\varphi_v^-)$	$E_x(z,t)=\lvert\bar{E}_x^+\rvert\cdot\cos(\omega t-\beta_o z+\varphi_E^+)+\lvert\bar{E}_x^-\rvert\cdot\cos(\omega t+\beta_o z+\varphi_E^-)$	(11.39)
See Figure 11.2(a)		See Figure 11.2(b)	
	$i(z,t)=\left\lvert\frac{V^+}{Z_o}\right\rvert\cdot\cos(\omega t-\beta z+\varphi_v^+)-\left\lvert\frac{V^-}{Z_o}\right\rvert\cdot\cos(\omega t+\beta z+\varphi_v^-)$	$H_y(z,t)=\left\lvert\frac{\bar{E}_x^+}{\eta_o}\right\rvert\cdot\cos(\omega t-\beta_o z+\varphi_H^+)-\left\lvert\frac{\bar{E}_x^-}{\eta_o}\right\rvert\cdot\cos(\omega t+\beta_o z+\varphi_H^-$	(11.40)
	$\beta=2\pi/\lambda$	$\beta=2\pi/\lambda$	(11.41)
	$c_{ph}=\frac{\omega}{\beta_o}=\lambda f$	$c_{ph}=\frac{\omega}{\beta_o}=\lambda f$	(11.42)
	$c_{ph}=1/\sqrt{L_z\cdot C_z}$	$c_{ph}=1/\sqrt{\mu_o\cdot\varepsilon_o}\cong 3\times10^8\ \text{m/s}$	(11.43)

Table 11.3 Solving the Wave Equation for the Special Case of Harmonic Plane Wave Propagation in Free Space with Analogy to the Case of Lossless TL

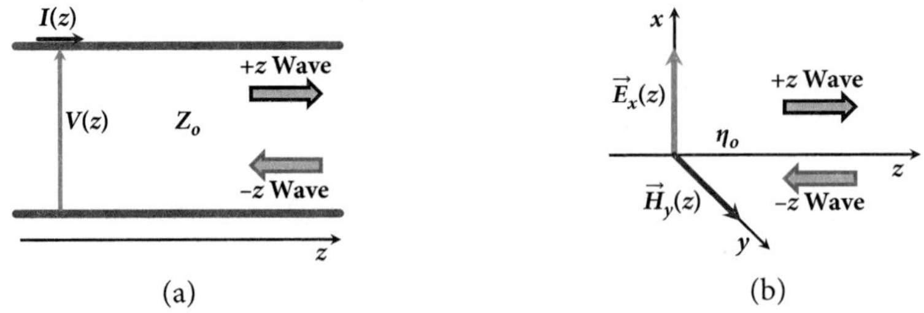

FIGURE 11.1 Forward and backward wave propagation. (a) Transmission line case. (b) Plane wave case.

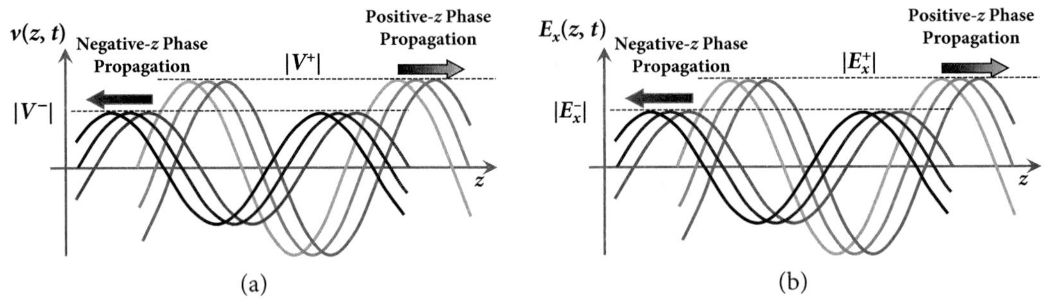

FIGURE 11.2 Instantaneous values of the harmonic wave versus position. (a) Transmission line case. (b) Plane wave case.

11.5 Physical Insight into the Obtained Wave Equation Solution

Examining the time-domain solution obtained above, Equations (11.39) and (11.40), we notice that the general solution for both the electric and magnetic fields is composed of two traveling waves that travel in both the positive and negative z directions (opposite to each other) (Figure 11.2). This is quite similar to what we found for voltage and current waves in TL analysis. To develop a physical insight in the obtained solution, it is best to examine each of the two sets of wave travel (the $+z$ and the $-z$ sets), separately.

Considering the $+z$ traveling wave as demonstrated in Figure 11.3, we can write

$$E_x^+(z,t)=\left|\bar{E}_x^+\right|\cdot\cos\left(\omega t-\beta_o z+\varphi_E^+\right) \quad \text{and} \quad H_y^+(z,t)=\left|\frac{\bar{E}_x^+}{\eta_o}\right|\cdot\cos\left(\omega t-\beta_o z+\varphi_H^+\right) \tag{11.44}$$

The phases $\varphi_{E,H}^+$ are equal in a lossless medium. We notice that

$$E_x^+(z,t)/H_y^+(z,t)=\eta_o \tag{11.45}$$

Analogously, for the $-z$ traveling set we obtain

$$E_x^-(z,t)=\left|\bar{E}_x^-\right|\cdot\cos(\omega t+\beta_o z+\varphi^-) \quad \text{and} \quad H_y^-(z,t)=-\left|\frac{\bar{E}_x^-}{\eta_o}\right|\cdot\cos(\omega t+\beta_o z+\varphi^-) \tag{11.46}$$

from which we notice that

$$E_x^-(z,t)/[-H_y^-(z,t)]=\eta_o \tag{11.47}$$

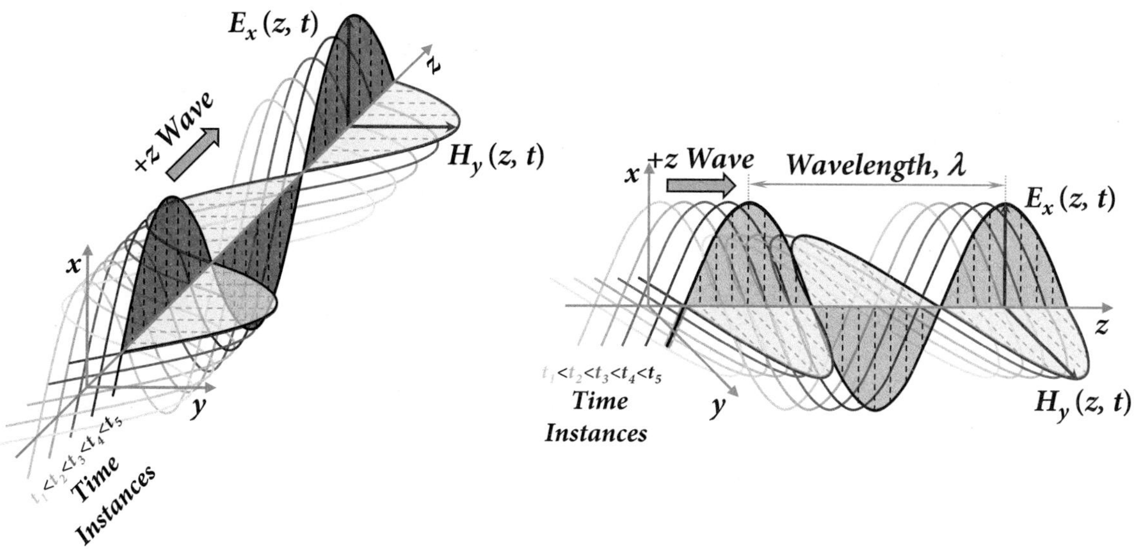

FIGURE 11.3 Electric and magnetic field for the $+z$ wave propagation shown versus distance at different time instances.

In this regard, we can make the following observations:

1. The two fields (electric and magnetic) are related to each other and occur in coexistence fashion (companionship). Hence, neither field can exist without the other. The ratio between the two fields traveling in the same direction is determined by the intrinsic impedance of the medium of travel (determined by the ratio of the permeability and permittivity of the medium) which is η_o for free space.
2. The two fields (electric and magnetic) are orthogonal to each other (E_x and H_y) and both lie on a plane transverse to the travel (z) axis. Hence, the composite traveling wave is referred to as a transverse electromagnetic wave (*TEM*).
3. The electric and magnetic field vectors traveling in both direction and their corresponding travel direction vector, form a right-hand orthogonal set. In other words, and referring to the displays in Figure 11.4, the vectors $\vec{E}_x^+, \vec{H}_y^+$, and $\vec{a}_z$ in this order form a right-hand set while $\vec{E}_x^-, \vec{H}_y^-$, and $-\vec{a}_z$ in this order also form a right-hand set. A right-hand set is characterized by having $\vec{a}_1 \times \vec{a}_2 = \vec{a}_3$ for their corresponding orthogonal unit vectors.
4. The *TEM* wave composed of the companion electric and magnetic field vectors travels along the z axis at a phase velocity equal to the speed of light in the medium, Equation (11.42). This phase velocity, c_{ph}, is determined by both the medium permeability and permittivity, Equation (11.43). The corresponding wavelength, λ, is determined by both the phase velocity and the wave frequency, Equation (11.42).

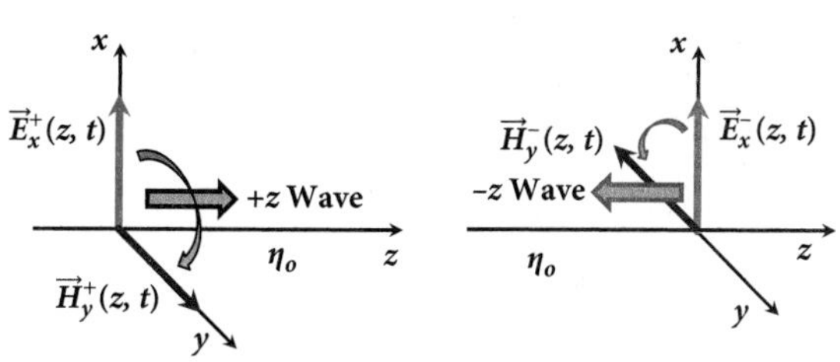

FIGURE 11.4 *TEM* wave propagation. Notice that E, H, and the propagation direction form right-hand sets for both the $+z$ and the $-z$ waves.

$\bar{E}_x(z)=\bar{E}_x^+e^{-j\beta z}+\bar{E}_x^-e^{+j\beta z}$	(11.48)
$\bar{H}_y(z)=\frac{1}{\eta}\cdot\{\bar{E}_x^+e^{-j\beta z}-\bar{E}_x^-e^{+j\beta z}\}$	(11.49)
$E_x(z,t)=\lvert\bar{E}_x^+\rvert\cdot\cos(\omega t-\beta z+\varphi_E^+)+\lvert\bar{E}_x^-\rvert\cdot\cos(\omega t+\beta z+\varphi_E^-)$	(11.50)
$H_y(z,t)=\frac{1}{\eta}\cdot\{\lvert\bar{E}_x^+\rvert\cdot\cos(\omega t-\beta z+\varphi_E^+)-\lvert\bar{E}_x^-\rvert\cdot\cos(\omega t+\beta z+\varphi_E^-)\}$	(11.51)
$\eta=\sqrt{\mu/\varepsilon}\ \rightarrow\ \eta=\eta_o\sqrt{\mu_r/\varepsilon_r}\ \rightarrow\ \eta$ real, $\lvert\eta\rvert=\eta$	(11.52)
$\beta=\omega\sqrt{\mu\cdot\varepsilon}\ \rightarrow\ \beta=\beta_o\sqrt{\mu_r\cdot\varepsilon_r}$	(11.53)
$\lambda=2\pi/\beta\ \rightarrow\ \lambda=\lambda_o/\sqrt{\mu_r\cdot\varepsilon_r}$	(11.54)
$c_{ph}=\frac{\omega}{\beta}=1/\sqrt{\mu\cdot\varepsilon}\ \rightarrow\ c_{ph}=c_o/\sqrt{\mu_r\cdot\varepsilon_r}$	(11.55)

TABLE 11.4 Summary of the *TEM* Wave Propagation Parameters for the Source-Free Lossless Medium (Perfect Dielectric)

Now, with *TEM* waves traveling in both the $+z$ and the $-z$ directions, we would have interference between the two waves similar to that discussed earlier in Chapter 2 for transmission lines.

Next, let us tackle the solution of the EM wave equations for more general media characteristics. In this regard, we will develop exact solutions for the two conductivity extremes: the perfect dielectric, $\sigma=0$, and the perfect conductor, $\sigma=\infty$. We will also develop approximate solutions for the two cases of near perfect (good/low loss) dielectric and near perfect (good) conductor. However, for the general case, we will present the equations governing the corresponding solutions.

11.5.1 Source-Free Case of a Lossless Medium (Perfect Dielectric)

The analysis of the case of source-free lossless medium is identical to the one we developed above for free-space except for using the proper expressions for the medium permittivity, ε, and permeability, μ. Hence, we can rewrite the wave solution expressions of Equations (11.29) to (11.43) as follows in Table 11.4.

Examining the above expressions, we notice that EM wave propagation in the medium occurs with shorter wavelengths and slower phase velocities. The reduction coefficient in both is equal to $1/\sqrt{\mu_r\cdot\varepsilon_r}$. Meanwhile, the phase constant is increased by the coefficient $\sqrt{\mu_r\cdot\varepsilon_r}$. However, the intrinsic impedance of the medium can go either way depending on the ratio $\sqrt{\mu_r/\varepsilon_r}$. For most material, with "nonmagnetic, $\mu_r=1$," the intrinsic impedance of the medium is reduced from that of free space by the coefficient $1/\sqrt{\varepsilon_r}$.

Example 11.1 The electric field of a wave propagating through a certain nonmagnetic material is given by $\vec{E}=\vec{a}_x 20\cos(10^8 t-0.5y)$ V/m. Find the following:

a. The direction of wave propagation
b. The frequency of the wave
c. The phase velocity
d. The wavelength in the material
e. The relative permittivity of the material
f. The magnetic field phasor

Solution

a. The direction of wave propagation: positive y direction

b. The frequency of the wave: $f=\frac{\omega}{2\pi}=\frac{10^8}{2\pi}=15.9\text{ MHz}$

c. The phase velocity: $c_{ph}=\frac{\omega}{\beta}=\frac{10^8}{0.5}=2\times10^8\text{ m/s}$

d. The wavelength in the material: $\lambda=\frac{2\pi}{\beta}=\frac{2\pi}{0.5}=4\pi=12.56\text{ m}$

The relative permittivity of the material can be obtained from its relationship to the phase velocity:

$$c_{ph}=2\times10^8\text{ m/s}=\frac{c}{\sqrt{\mu_r\epsilon_r}}=\frac{3\times10^8}{\sqrt{1.0\times\epsilon_r}}$$

e. The relative permittivity: $\sqrt{\epsilon_r}=\frac{3\times10^8}{2\times10^8}=1.5\rightarrow\epsilon_r=1.5^2=2.25$

f. The magnetic field phasor:

$$\vec{H}=\frac{1}{\eta}\vec{a}_k\times\vec{E}=\frac{1}{\eta}\vec{a}_y\times\vec{E}$$

$$\vec{E}=\vec{a}_x 20e^{-j0.5y}$$

$$\vec{H}=\frac{1}{\eta}\vec{a}_k\times\vec{E}=\frac{1}{\eta}\vec{a}_y\times\vec{E}=\frac{1}{\eta}\vec{a}_y\times\vec{a}_x 20e^{-j0.5y}=-\vec{a}_z\frac{20}{\eta}e^{-j0.5y}\text{ A/m}$$

$$\eta=120\pi\sqrt{\frac{\mu_r}{\epsilon_r}}=120\pi\sqrt{\frac{1.0}{2.25}}=80\pi=251.33\ \Omega$$

$$\vec{H}=-\vec{a}_z\frac{20}{\eta}e^{-j0.5y}=-\vec{a}_z\frac{20}{251.33}e^{-j0.5y}=-\vec{a}_z 0.0796e^{-j0.5y}\text{ A/m}$$

Example 11.2 The magnetic field phasor of a uniform plane wave is given by $\vec{H}=\vec{a}_y 10\,e^{j0.2z}$ mA/m. If the phase velocity of the wave is 10^8 m/s and the relative permeability of the medium μ_r is 2.4, find the following:

a. The direction of wave propagation
b. The wavelength in the material
c. The frequency of the wave
d. The relative permittivity of the material
e. The electric field phasor

Solution

a. The direction of wave propagation: negative z direction

b. The wavelength in the material: $\lambda=\frac{2\pi}{\beta}=\frac{2\pi}{0.2}=10\pi\text{ m}=31.4\text{ m}$

c. The frequency of the wave:

$$c_{ph}=\frac{\omega}{\beta}\rightarrow\omega=c_{ph}\times\beta=10^8\times0.2\rightarrow f=\frac{\omega}{2\pi}=\frac{10^8\times0.2}{2\pi}=3.183\text{ MHz}$$

d. The relative permittivity of the material:

$$c_{ph} = 10^8 \text{ m/s} = \frac{c_o}{\sqrt{\mu_r \epsilon_r}} = \frac{3\times 10^8}{\sqrt{2.4\times \epsilon_r}} \rightarrow \epsilon_r = \left(\frac{c_o}{c_{ph}\sqrt{\mu_r}}\right)^2 = \left(\frac{3\times 10^8}{10^8 \times \sqrt{2.4}}\right)^2 = 3.75$$

e. The electric field phasor:

The intrinsic impedance, $\eta = 120\pi\sqrt{\frac{\mu_r}{\epsilon_r}} = 120\pi\sqrt{\frac{2.4}{3.75}} = 96\pi = 301.6\ \Omega$

$$\vec{E} = -\eta\vec{a}_k \times \vec{H} = -\eta(-\vec{a}_z)\times \vec{H}$$

$$\vec{H} = \vec{a}_y 10\, e^{j0.2z} \text{ mA/m}$$

$$\vec{E} = -\eta(-\vec{a}_z)\times \vec{a}_y 10\, e^{j0.2z} = -\vec{a}_x 10\,\eta\, e^{j0.2z} = -\vec{a}_x 10\times 301.6\times e^{j0.2z} \text{ mV/m}$$

$$\vec{E} = -\vec{a}_x 3.016\, e^{j0.2z} \text{ V/m}$$

11.5.2 Case of Source-Free Lossy Media $\sigma \neq 0$ (General Material Properties)

Now, let us consider a more general case where the propagation medium is no longer lossless. Specifically, we will reintroduce the medium conductivity, σ, back to Maxwell's equations.

A clever way to put σ in the equations that does not require resolving them is to manipulate the equations with σ so that they take a form similar to that for the lossless case equations. We start with Faraday's equation for the general case in the form

$$\vec{\nabla}\times \vec{H} = \vec{J} + \frac{\partial \vec{D}}{\partial t} = \sigma\vec{E} + j\omega\varepsilon\vec{E} = (\sigma + j\omega\varepsilon)\vec{E} \tag{11.56}$$

then make the substitution $(\sigma + j\omega\varepsilon) = j\omega\varepsilon^*$, ending up with

$$\vec{\nabla}\times \vec{H} = (\sigma + j\omega\varepsilon)\vec{E} = j\omega\varepsilon^*\vec{E} \tag{11.57}$$

which is the same exact form as the lossless case while replacing ε with ε^*. It is to be noted that ε^*[1] is a complex quantity with components representing both the material's *real* dielectric permittivity ε and its conductivity, σ. We will refer to ε^* as the complex material permittivity;

$$\varepsilon^* = \frac{(\sigma + j\omega\varepsilon)}{j\omega} = \varepsilon - j\frac{\sigma}{\omega} = \varepsilon' - j\varepsilon'' \tag{11.58}$$

We notice that in Equation (11.58), we changed the notation for the *real* dielectric permittivity ε to ε' to eleminate the confusion with the expressions of the lossless case. In addition, we used ε'' to denote the imaginary component of the complex permittivity representing the loss element due to the conductivity σ.

In a similar fashion, replacing μ by μ^* can be used to accommodate magnetic losses in the media. The generalized expression for Ampere's law that accounts for magnetic losses in the media is of the form

$$\vec{\nabla}\times \vec{E} = -\frac{\partial \vec{B}}{\partial t} = -j\omega\vec{B} = -j\omega\mu^*\vec{H} \tag{11.59}$$

which is the same exact form as that of the lossless case while replacing μ by μ^*. Similarly, it is to be noted that μ^* is a complex quantity that we will refer to as the complex material permeability;

$$\mu^* = \mu' - j\mu'' \tag{11.60}$$

[1]We are using the (*) symbol to indicate a complex quantity. It is to be distinguished from the (∗) symbol which is typically used to indicated the conjugate of a complex quantity.

Again, we notice that in Equation (11.60) that we changed the notation for the *real* magnetic permeability μ to μ' to eleminate the confusion with the expressions of the lossless case. In addition, we used μ'' to denote the imaginary component of the complex permeability representing the material's magnetic losses.

Having done the substitutions of Equations (11.58) and (11.60), we can now write the solutions of the two Maxwell's equations [Equations (11.18) and (11.16)] based on the solutions obtained earlier for the lossless case. Table 11.5 demonstrates this development as well as shows the analogous TL solutions.

Lossless Material Expressions	**Lossy Material Expressions (General Material Properties)**	**Lossy Transmission Line Expressions**	
Media permittivity, ε	$\sigma + j\omega\varepsilon = j\omega\varepsilon^*,\ \varepsilon^* = \varepsilon - j\frac{\sigma}{\omega}$ Complex permittivity, $\varepsilon^* = \varepsilon' - j\varepsilon''$	Let $G_z + j\omega C_z = j\omega C_z^*$ Then $C_z^* = C_z - j\frac{G_z}{\omega}$	(11.61)
Media permeability, μ	Complex permeability, $\mu^* = \mu' - j\mu''$	Similarly, let $R_z + j\omega L_z = j\omega L_z^*$ Then $L_z^* = L_z - j\frac{R_z}{\omega}$	(11.62)
$\gamma = \sqrt{j\omega\mu \cdot j\omega\varepsilon} = j\omega\sqrt{\mu\varepsilon} = j\beta$	$\gamma = \alpha + j\beta = \sqrt{-\omega^2\mu^*\varepsilon^*}$	$\gamma = \alpha + j\beta = \sqrt{-\omega^2 L_z^* C_z^*}$	(11.63)
$\eta = \sqrt{\mu/\varepsilon}$	$\eta = \sqrt{\mu^*/\varepsilon^*} = \sqrt{(\mu' - j\mu'')/(\varepsilon' - j\varepsilon'')}$	$Z_o = \sqrt{L_z^*/C_z^*} = \sqrt{\{R_z + j\omega L_z\}/\{G_z + j\omega C_z\}}$	(11.64)
$\bar{E}_x(z) = \bar{E}_x^+ e^{-\gamma z} + \bar{E}_x^- e^{+\gamma z}$	$\bar{E}_x(z) = \bar{E}_x^+ e^{-\gamma z} + \bar{E}_x^- e^{+\gamma z}$	$V(z) = V^+ e^{-\gamma z} + V^- e^{+\gamma z}$	(11.65)
$\bar{H}_y = \frac{1}{\eta}\cdot\{\bar{E}_x^+ e^{-\gamma z} - \bar{E}_x^- e^{+\gamma z}\}$	$\bar{H}_y = \frac{1}{\eta}\cdot\{\bar{E}_x^+ e^{-\gamma z} - \bar{E}_x^- e^{+\gamma z}\}$	$I(z) = \frac{1}{Z_o}\cdot\{V^+ e^{-\gamma z} - V^- e^{+\gamma z}\}$	(11.66)
$\bar{E}_x(z) = \bar{E}_x^+ e^{-j\beta z} + \bar{E}_x^- e^{+j\beta z}$	$\bar{E}_x(z) = \bar{E}_x^+ e^{-\alpha z} e^{-j\beta z} + \bar{E}_x^- e^{+\alpha z} e^{+j\beta z}$	$V(z) = V^+ e^{-\alpha z} e^{-j\beta z} + V^- e^{+\alpha z} e^{+j\beta z}$	(11.67)
$\bar{H}_y(z) = \frac{\bar{E}_x^+}{\eta} e^{-j\beta z} - \frac{\bar{E}_x^-}{\eta} e^{+j\beta z}$	$\bar{H}_y(z) = \frac{\bar{E}_x^+}{\eta} e^{-\alpha z} e^{-j\beta z} - \frac{\bar{E}_x^-}{\eta} e^{+\alpha z} e^{+j\beta z}$	$I(z) = \frac{V^+}{Z_o} e^{-\alpha z} e^{-j\beta z} - \frac{V^-}{Z_o} e^{+\alpha z} e^{+j\beta z}$	(11.68)
Transfer Back to Time Domain:			
$E_x(z,t) = \lvert\bar{E}_x^+\rvert\cos(\omega t - \beta z + \varphi^+)$ $+ \lvert\bar{E}_x^-\rvert\cos(\omega t + \beta z + \varphi^-)$	$E_x(z,t) = \lvert\bar{E}_x^+\rvert e^{-\alpha z}\cos(\omega t - \beta z + \varphi_E^+)$ $+ \lvert\bar{E}_x^-\rvert e^{+\alpha z}\cos(\omega t + \beta z + \varphi_E^-)$	$v(z,t) = \lvert V^+\rvert e^{-\alpha z}\cos(\omega t - \beta z + \varphi_v^+)$ $+ \lvert V^-\rvert e^{+\alpha z}\cos(\omega t + \beta z + \varphi_v^-)$	(11.69)
$H_y(z,t) = \left\lvert\frac{\bar{E}_x^+}{\eta}\right\rvert\cos(\omega t - \beta z + \varphi^+)$ $- \left\lvert\frac{\bar{E}_x^-}{\eta}\right\rvert\cos(\omega t + \beta z + \varphi^-)$	$H_y(z,t) = \left\lvert\frac{\bar{E}_x^+}{\eta}\right\rvert e^{-\alpha z}\cos(\omega t - \beta z + \varphi_H^+)$ $- \left\lvert\frac{\bar{E}_x^-}{\eta}\right\rvert e^{+\alpha z}\cos(\omega t + \beta z + \varphi_H^-)$	$i(z,t) = \left\lvert\frac{V^+}{Z_o}\right\rvert e^{-\alpha z}\cos(\omega t - \beta z + \varphi_i^+)$ $- \left\lvert\frac{V^-}{Z_o}\right\rvert e^{+\alpha z}\cos(\omega t + \beta z + \varphi_i^-)$	(11.70)

Table 11.5 *TEM* Wave Solution Parameters for Lossy Medium in Analogy with the Case of a Lossy TL

From Table 11.5, we conclude that the solution to Maxwell's equations for the one-dimensional wave propagation in a source-free medium (general material properties) is summarized by the two expressions:

$$\bar{E}_x(z) = \bar{E}_x^+ e^{-\gamma z} + \bar{E}_x^- e^{+\gamma z} \quad \text{and} \quad \bar{H}_y = \frac{1}{\eta}\cdot\{\bar{E}_x^+ e^{-\gamma z} - \bar{E}_x^- e^{+\gamma z}\} \tag{11.71}$$

Here, $\bar{E}_x^+$ and $\bar{E}_x^-$ are the electric field phasors of the two wave components traveling forward and backward along the wave propagation axis, respectively. These two phasors are determined by the one-dimensional boundary conditions at the source and the load, for example, the field intensities of the sources and their respective internal impedances at the boundaries of the region of the solution.

On the other hand, γ and η are the complex propagation constant and complex wave impedance of the propagation medium as determined by its complex permittivity and complex permeability:

$$\gamma = \alpha + j\beta = \sqrt{-\omega^2 \mu^* \varepsilon^*} \qquad \text{and} \qquad \eta = \sqrt{\mu^* / \varepsilon^*} \tag{11.72}$$

11.5.3 Special Cases (Nonmagnetic)

The solution presented above is for the general material properties. In the following, we will consider special cases of most practical interest for the one-dimensional wave propagation in different media types. Three cases are of particular interest that we present them below:

1. The case of a low-loss dielectric, typically referred to as a good dielectric, for which $\sigma \ll \omega\varepsilon$, that is, $\varepsilon'' \ll \varepsilon'$
2. The case of a good conductor with $\sigma \gg \omega\varepsilon$, that is, $\varepsilon'' \gg \varepsilon'$ which can be thought of as a high-loss dielectric
3. The case of a perfect conductor with $\sigma = \infty$ implying $\varepsilon'' = \infty$.

For each of these cases, we will develop the corresponding (exact or approximate) expressions for the wave propagation parameters $\gamma = \alpha + j\beta$ and η. Table 11.6 presents this development along with the already developed expressions for the perfect (lossless) dielectric. Hence, this table presents the solutions for a wide range of material conductivities from zero to infinity.

Example 11.3 Consider a material with $\varepsilon = 10\varepsilon_o$, $\mu = \mu_o$, and $\sigma = 0.2$ S/m. The operating frequency is 500 MHz, find $\eta, \gamma, \alpha, \beta, \lambda$, and c_{ph}.

$$\eta = \sqrt{\frac{j\omega\mu}{\sigma + j\omega\varepsilon}} = \sqrt{\frac{j2\pi \times 500 \times 10^6 \times 4\pi \times 10^{-7}}{(0.2 + j2\pi \times 500 \times 10^6 \times 10 \times \frac{1}{36\pi} \times 10^{-9})}} = 107\angle 17.9^\circ\ \Omega$$

$$\gamma = \sqrt{j\omega\mu(\sigma + j\omega\varepsilon)}$$

$$\gamma = \sqrt{j2\pi \times 500 \times 10^6 \times 4\pi \times 10^{-7} \times \left(0.2 + j2\pi \times 500 \times 10^6 \times 10 \times \frac{1}{36\pi} \times 10^{-9}\right)}$$

$$\gamma = 11.3 + j35\ \text{m}^{-1}$$

$$\alpha = 11.3\ \text{Np/m} \qquad \beta = 35\ \text{rad/m}$$

$$\lambda = \frac{2\pi}{\beta} = \frac{2\pi}{35} = 0.18\ \text{m}$$

$$c_{ph} = \frac{\omega}{\beta} = \frac{2\pi \times 500 \times 10^6}{35} = 9 \times 10^7\ \text{m/s}$$

11.6 Physical Insight into the Obtained Solutions for Wave Propagation Parameters in Dielectrics and Conductors

Let us examine the results developed in Table 11.6 in order to develop a physical insight into the effects of the different material properties on plane wave propagation. To do so in an effective manner, let us consider the one-way propagation through a continuous medium with no discontinuities or reflections.

We start with the perfect dielectric in the left-most column. In the absence of material losses, both the forward and backward traveling waves propagate unattenuated with $\alpha = 0$ and with a propagation phase

Nonmagnetic Perfect Dielectric	Nonmagnetic Good Dielectric (Low Loss)	General Expressions	Nonmagnetic Good Conductor (High Loss)	Nonmagnetic Perfect Conductor
$\sigma=0$	$\sigma \ll \omega\varepsilon \to \varepsilon'' \ll \varepsilon'$	$\varepsilon'=\varepsilon,$ $\varepsilon''=\sigma/\omega$ (11.58)	$\sigma \gg \omega\varepsilon \to \varepsilon'' \gg \varepsilon'$	$\sigma=\infty \to \varepsilon''=\infty$
$\varepsilon^*=\varepsilon'=\varepsilon$ $\mu^*=\mu_o$	$\varepsilon^*=\varepsilon'-j\varepsilon''=\varepsilon'\left[1-j\frac{\varepsilon''}{\varepsilon'}\right]$ $\mu^*=\mu_o$	$\varepsilon^*=\varepsilon'-j\varepsilon''$ $\mu^*=\mu'-j\mu''$ (11.60)	$\varepsilon^*=\varepsilon'-j\varepsilon''=-j\varepsilon''\left[1+j\frac{\varepsilon'}{\varepsilon''}\right]$ $\mu^*=\mu_o$	$\varepsilon^*=\varepsilon_o-j\varepsilon''=\varepsilon_o-j\infty$ $\mu^*=\mu_o$
$\gamma^2=-\omega^2\mu_o\varepsilon$	$\gamma^2=-\omega^2\mu_o\varepsilon'\left[1-j\frac{\varepsilon''}{\varepsilon'}\right]$	$\gamma^2=-\omega^2\mu^*\varepsilon^*$ (11.63)	$\gamma^2=-\omega^2\mu_o\left\{-j\varepsilon''\left[1+j\frac{\varepsilon'}{\varepsilon''}\right]\right\}$ $=j\omega^2\mu_o\varepsilon''\left[1+j\frac{\varepsilon'}{\varepsilon''}\right]$	$\gamma^2=j\omega^2\mu_o\varepsilon''$
$\gamma=\alpha+j\beta=\sqrt{-\omega^2\mu_o\varepsilon}$ $\alpha=0$ $\beta=\omega\sqrt{\mu_o\varepsilon}$ $\beta=\omega\sqrt{\mu_o\varepsilon_o\varepsilon_r}$ $\beta=\omega\sqrt{\mu_o\varepsilon}$ $=\beta_o\sqrt{\varepsilon_r}$ $\lambda=\frac{2\pi}{\beta}=\lambda_o/\sqrt{\varepsilon_r}$ $c_{ph}=\frac{\omega}{\beta}=c_o/\sqrt{\varepsilon_r}$	$\gamma=\alpha+j\beta=\sqrt{-\omega^2\mu_o(\varepsilon'-j\varepsilon'')}$ $\gamma=\sqrt{[-\omega^2\mu_o\varepsilon']}\cdot[1-j(\varepsilon''/\varepsilon')]^{1/2}$ $[1-j(\varepsilon''/\varepsilon')]^{1/2}\cong\left[1-j\left(\frac{\varepsilon''}{2\varepsilon'}\right)+\cdots\right]$ $\alpha+j\beta\cong j\omega\sqrt{\mu_o\varepsilon'}\cdot\left[1-j\left(\frac{\varepsilon''}{2\varepsilon'}\right)+\cdots\right]$ $\alpha\cong\omega\sqrt{\mu_o\varepsilon'}\cdot\left(\frac{\varepsilon''}{2\varepsilon'}\right)=\left(\frac{\omega\varepsilon''}{2}\right)\cdot\sqrt{\mu_o/\varepsilon'}$ $\alpha\cong\left(\frac{\omega\varepsilon''\eta}{2}\right)=\left(\frac{\sigma\eta_o}{2\sqrt{\varepsilon_r}}\right)$ $\beta\cong\omega\sqrt{\mu_o\varepsilon'}=\beta_o\sqrt{\varepsilon_r}$ $\lambda=\frac{2\pi}{\beta}=\lambda_o/\sqrt{\varepsilon_r}$ $c_{ph}=\frac{\omega}{\beta}=c_o/\sqrt{\varepsilon_r}$	$\gamma=\alpha+j\beta$ $=\sqrt{-\omega^2\mu^*\varepsilon^*}$ (11.73)	$\gamma=\alpha+j\beta=\sqrt{-\omega^2\mu_o(\varepsilon'-j\varepsilon'')}$ $\alpha+j\beta=\sqrt{j\omega^2\mu_o\varepsilon''(1+j\varepsilon'/\varepsilon'')}$ $\alpha+j\beta=\sqrt{j}\cdot\omega\cdot\sqrt{\mu_o\varepsilon''}\cdot[1+j(\varepsilon'/\varepsilon'')]^{1/2}$ $\sqrt{j}=(1+j)/\sqrt{2}$ $[1+j(\varepsilon'/\varepsilon'')]^{1/2}\cong[1+j(\varepsilon'/2\varepsilon'')+\cdots]$ $\alpha+j\beta\cong\frac{1+j}{\sqrt{2}}\cdot\omega\cdot\sqrt{\mu_o\varepsilon''}\cdot[1+j(\varepsilon'/2\varepsilon'')+\cdots]$ $\alpha\cong\beta\cong\omega\cdot\sqrt{\frac{\mu_o\varepsilon''}{2}}=\sqrt{\frac{\omega\mu_o(\omega\varepsilon'')}{2}}=\sqrt{\pi f\mu_o\sigma}$ $\lambda=\frac{2\pi}{\beta}=\lambda_o\sqrt{2\omega\varepsilon_o/\sigma}$ $c_{ph}=\frac{\omega}{\beta}=c_o\sqrt{2\omega\varepsilon_o/\sigma}$	$\gamma=\alpha+j\beta$ $=\sqrt{-\omega^2\mu_o(\varepsilon'-j\varepsilon'')}$ $=\sqrt{[-\omega^2\mu_o\varepsilon_o]+j\left[\omega^2\left(\mu_o\frac{\sigma}{\omega}\right)\right]}$ $\alpha=\beta=\infty$ $\lambda=\frac{2\pi}{\beta}=0$ $c_{ph}=\frac{\omega}{\beta}=0$
$\eta=\sqrt{\mu_o/\varepsilon}$ $\eta=\sqrt{\mu_o/\varepsilon_o}/\sqrt{\varepsilon_r}$ $\eta=\sqrt{\mu_o/\varepsilon}$ $=\eta_o/\sqrt{\varepsilon_r}$	$\eta=\sqrt{\mu_o/(\varepsilon'-j\varepsilon'')}$ $=\sqrt{\mu_o/\varepsilon'}\cdot[1-j(\varepsilon''/\varepsilon')]^{-1/2}$ $[1-j(\varepsilon''/\varepsilon')]^{-1/2}\cong[1+j(\varepsilon''/2\varepsilon')+\cdots]$ $\eta\cong\sqrt{\mu_o/\varepsilon'}\cdot[1+j(\varepsilon''/2\varepsilon')+\cdots]$ $\eta\cong\sqrt{\frac{\mu_o}{\varepsilon'}}=\eta_o/\sqrt{\varepsilon_r}$	$\eta=\sqrt{\mu^*/\varepsilon^*}$ (11.74)	$\eta=\sqrt{\mu_o/\varepsilon^*}=\sqrt{\mu_o/(\varepsilon'-j\varepsilon'')}$ $=\sqrt{\mu_o/[-j\varepsilon''(1+j\varepsilon'/\varepsilon'')]}$ $\eta=\sqrt{j\mu_o/\varepsilon''}/\sqrt{(1+j\varepsilon'/\varepsilon'')}$ $\eta\cong\frac{1+j}{\sqrt{2}}\sqrt{\mu_o/\varepsilon''}\cdot[1+j(\varepsilon'/2\varepsilon'')+\cdots]$ $\eta\cong(1+j)\sqrt{\mu_o/2\varepsilon''}=\eta_o(1+j)\sqrt{\omega\varepsilon_o/(2\sigma)}$	$\eta=\sqrt{\mu_o/\varepsilon^*}$ $=\sqrt{\mu_o/(\varepsilon'-j\varepsilon'')}$ $=\sqrt{\mu_o/[-j\infty]}$ $\eta=0$

TABLE 11.6 *TEM* Wave Solution Parameters for Different Materials Covering the Cases of Perfect Dielectric, Good Dielectric, Perfect Conductor, Good Conductor, and a General Material

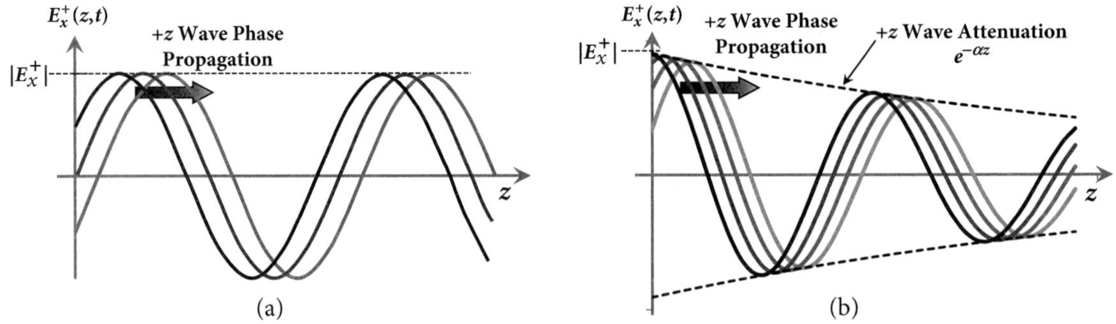

FIGURE 11.5 Lossless wave propagation in a perfect dielectric versus decaying wave propagation in a low-loss (good) dielectric. (a) Perfect dielectric. (b) Good dielectric.

constant of $\beta = \omega\sqrt{\mu_o \varepsilon} = \beta_o \sqrt{\varepsilon_r}$. The corresponding wavelength and phase velocity are reduced from those of free space by the coefficient $1/\sqrt{\varepsilon_r}$. Figure 11.5*a* demonstrates lossless propagation of the $+z$ wave in a perfect dielectric.

11.6.1 Plane Waves in a Good Dielectric

Next, we take on the low-loss good dielectric case in the second column from the left. The presence of material losses will cause both forward and backward traveling waves to attenuate as they propagate in the medium. For $\sigma \ll \omega\varepsilon$, the attenuation coefficient α can be approximated to $\alpha \cong \left(\dfrac{\omega\varepsilon''\eta}{2}\right)$. This low loss does not cause any serious change to the propagation phase constant, the corresponding wavelength, the phase velocity, or the wave impedance from their corresponding ones in the lossless case. Figure 11.5*b* demonstrates the slow decay of the $+z$ wave propagation in a good dielectric.

Let us now jump to the case of a perfect (ideal) conductor where the conductivity $\sigma = \infty$. For this case, we find that the attenuation coefficient α is infinite, $\alpha = \infty$. This indicates a full attenuation of the waves to a zero level at all location within the material, $e^{-\alpha z} = e^{-\infty} = 0$ (Figure 11.6*a*). In other words, the wave cannot propagate in perfect conductors, that is, an incident wave fails to penetrate a perfect conducting material. In confirmation to this, we notice that the perfect conductor wave impedance is zero, which is basically a short circuit to an incoming wave. We also notice a zero value for both the wavelength and the phase velocity. Figure 11.6*b* demonstrates the infinite attenuation of wave propagation in perfect conductors.

11.6.2 Uniform Plane Waves in a Good Conductor

Finally, we discuss the case of a good (typical/practical) conductor as presented in the second column from the right. In this case the material conductivity is fairly high, $\sigma \gg \omega\varepsilon$. As per our development in Table 11.6, we find that the material attenuation is very high resulting in a strong attenuation of the wave in the conductor medium; $\alpha \cong \beta \cong \sqrt{\pi f \mu_o \sigma}$, $\lambda = \lambda_o\sqrt{2\omega\varepsilon_o/\sigma}$, $c_{ph} = c_o\sqrt{2\omega\varepsilon_o/\sigma}$, and $\eta \cong \eta_o(1+j)\sqrt{\omega\varepsilon_o/(2\sigma)}$. Figure 11.6 demonstrates the strong attenuation of wave propagation in a good conductor as well as the infinite attenuation (complete shut-off) in a perfect conductor.

Figure 11.7 summarizes the above cases of attenuations ranging from zero to infinity in one graph.

To further examine the case of wave propagation in conductors, we examine the exponential decay function, $e^{-\alpha z}$, for large values of α as demonstrated in Figure 11.8. As we can see from the figure, the field intensity decays fairly fast to reach about 36.8% for e^{-1} and less than 1% (0.674%) for e^{-5}. Beyond the e^{-5} point we can claim the field intensity to be almost zero and that no further wave propagation can take place. The e^{-1} point corresponds to $\alpha z = 1$ or $z = \delta = 1/\alpha$ and the e^{-5} point correspond to $z = 5\delta$.

$$\delta = 1/\alpha \cong 1/\sqrt{\pi f \mu_o \sigma} \tag{11.75}$$

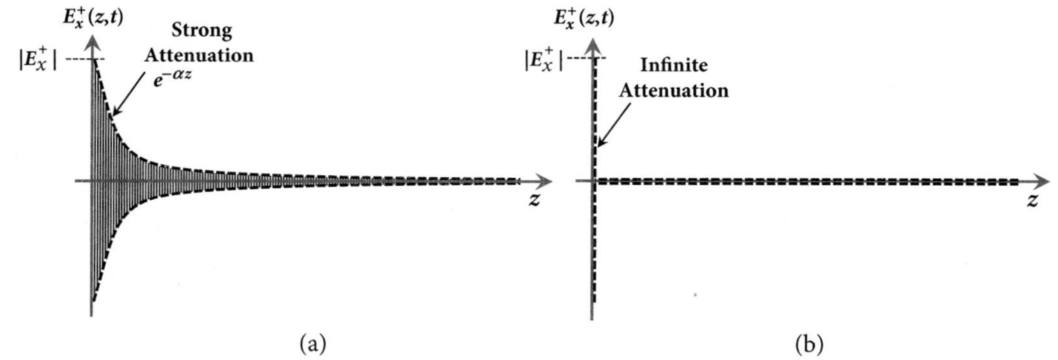

FIGURE 11.6 Waves are strongly attenuated in a good conductor and are completely shut off in a perfect conductor. (a) Good conductor. (b) Perfect conductor.

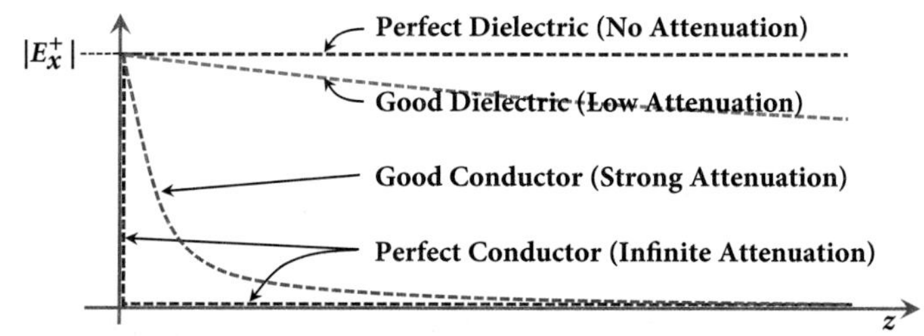

FIGURE 11.7 Demonstration of the different attenuation coefficients covering the cases of perfect dielectric, good dielectric, good conductor, and perfect conductor.

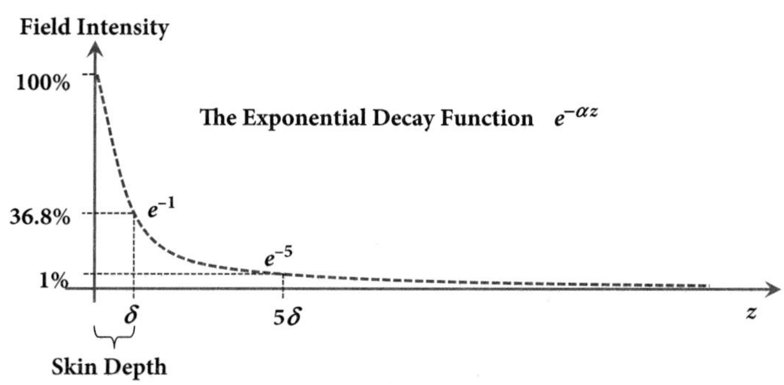

FIGURE 11.8 The skin depth, δ, is defined as the distance of wave penetration in the material before its field intensity drops below $1/e$ (36.8%) of its value at the surface. Notice that the field intensity drops below 1% after a distance of 5δ.

Since the wave energy is related to the square of the field intensity, we find that the wave energy decays at a much faster rate as the wave "travels" through the conducting medium. The wave energy at the $z = \delta$ point is now about 13.5% [=(36.8%)^2] of its original intensity. Hence, for all practical purposes, the convention is to choose this point to mark the extent of the wave penetration into the conducting medium. The thickness $z = \delta$ is then refered to as the *skin depth*. As we demonstrate in Figure 11.9, the skin depth is the equivalent penetration region of the electromagnetic wave into the conducting medium.

In the above discussions, we used the forward traveling wave to demonstrate the effect of the material's attenuation coefficient on the wave propagation in the medium. As we have seen in Chapter 2, as well as in this chapter's developed solutions, both forward and backward wave propagations are possible depending on the conditions at the source and load ends (boundary conditions). Hence, in general, both forward and backward waves may coexist resulting in an interference causing a possible combination of both traveling and standing waves. Similar to our earlier discussion in Chapter 2, Figures 2.9 through 2.15, the expected forms of traveling and standing waves for both perfect and good dielectrics are displayed in Figures 11.10 and 11.11, respectively.

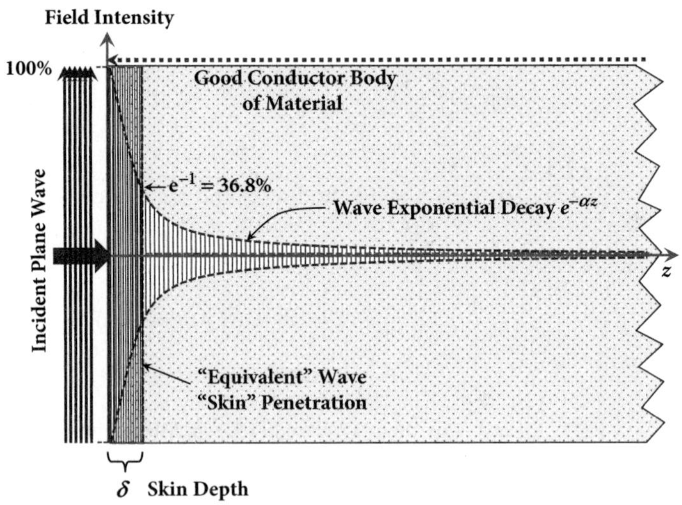

FIGURE 11.9 The skin depth, δ, is the equivalent penetration region of the electromagnetic wave into the conducting medium.

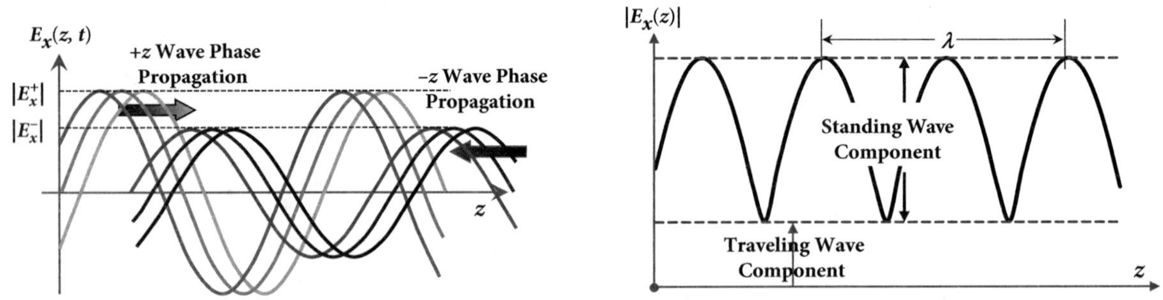

FIGURE 11.10 In a perfect dielectric, both forward and backward waves experience no attenuation.

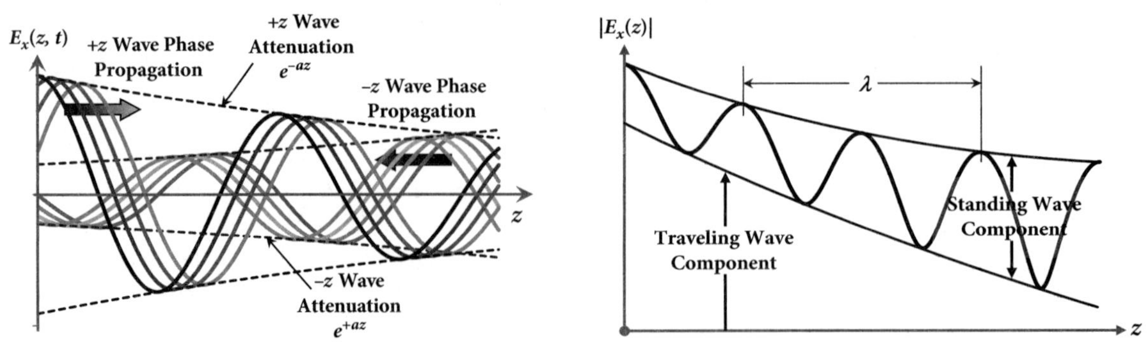

FIGURE 11.11 In a good dielectric, both forward and backward waves experience attenuation.

Example 11.4 Consider the case of a y-polarized wave traveling in free space along the z direction, where the electric field phasor is expressed as $\vec{E}(z)=\bar{E}_o^+e^{-\gamma z}+\bar{E}_o^-e^{+\gamma z}\,\vec{a}_y$. Write an expression for the instantaneous electric field vector, the corresponding magnetic field phasor and the instantaneous magnetic field vector. Assume $\bar{E}_o^+=5$ V/m and $\bar{E}_o^-=0.2$ mV/m. The operating frequency is 1 GHz.

Since the medium is lossless (free space) $\rightarrow \alpha=0$

$$\beta=\omega\sqrt{\mu_o\varepsilon_o}=\frac{\omega}{c}=\frac{2\pi\times 10^9}{3\times 10^8}=20.9 \text{ rad/m}$$

$$\vec{E}(z)=5e^{-20.9z}+0.2e^{+20.9z}\,\vec{a}_y \text{ V/m}$$

The instantaneous expression of the electric field vector is given by

$$\vec{E}(z,t)=5\cos(2\pi\times 10^9 t-20.9z)+0.2\cos(2\pi\times 10^9 t+20.9z)\vec{a}_y \text{ V/m}$$

To find the corresponding magnetic field phasor:

For the forward going component

$$\vec{H}^+(z)=\frac{1}{\eta}\vec{a}_z\times\vec{E}^+(z)=\frac{1}{\eta}\vec{a}_z\times 5e^{-20.9z}\,\vec{a}_y=\frac{5}{\eta}e^{-20.9z}(-\vec{a}_x)$$

$\vec{a}_z$ represents a unit vector in the direction of propagation of the foreword wave

$$\eta=\eta_0=\sqrt{\frac{\mu_o}{\varepsilon_o}}=\sqrt{\frac{4\pi\times 10^7}{\frac{1}{36\pi}\times 10^{-9}}}=120\pi\ \Omega$$

$$\vec{H}^+(z)=\frac{5}{120\pi}e^{-20.9z}(-\vec{a}_x)=-13.3e^{-20.9z}\,\vec{a}_x \text{ mA/m}$$

For the backward going component

$$\vec{H}^-(z)=\frac{1}{\eta}(-\vec{a}_z)\times\vec{E}^-(z)=\frac{1}{\eta}(-\vec{a}_z)\times 0.2e^{+20.9z}\,\vec{a}_y=\frac{0.2}{120\pi}e^{+20.9z}\vec{a}_x=0.53e^{+20.9z}\vec{a}_x \text{ mA/m}$$

$$\vec{H}(z)=\vec{H}^+(z)+\vec{H}^-(z)=(-13.3e^{-20.9z}+0.53e^{+20.9z})\vec{a}_x \text{ mA/m}$$

$$\vec{H}(z,t)=-13.3\cos(2\pi\times 10^9 t-20.9z)+0.53\cos(2\pi\times 10^9 t+20.9z)\vec{a}_y \text{ V/m}$$

Example 11.5 In Example 11.1, we found the intrinsic impedance and the propagation constant for a medium with constitutive parameters $\varepsilon=10\varepsilon_o$, $\mu=\mu_o$, and $\sigma=0.2$ S/m at an operating frequency of 500 MHz. Write an expression for an electric field polarized along the z direction and propagating in the positive x direction. The peak amplitude of the electric field at $x=0$ is 0.1 V/m. Write an expression for the corresponding magnetic field both as a phasor and as time domain expression.

From the solution of the Example 11.1, we have

$$\alpha=11.3 \text{ Np/m} \qquad \beta=35 \text{ rad/m} \qquad \eta=107\angle 17.9^\circ\ \Omega$$

The phasor form of the electric field

$$\vec{E}(x)=0.1e^{-\alpha x}e^{-j\beta x}\,\vec{a}_z=0.1e^{-11.3x}e^{-j35x}\,\vec{a}_z \text{ V/m}$$

The instantaneous electric field expression is given by

$$\vec{E}(x,t)=0.1e^{-11.3x}\cos(2\pi\times5\times10^{8}t-35x)\vec{a}_z \text{ V/m}$$

The phasor form of the magnetic field can be obtained as

$$\vec{H}(x)=\frac{1}{\eta}\vec{a}_x\times\vec{E}(x)=\frac{1}{107\angle17.9^\circ}\,\vec{a}_x\times0.1e^{-11.3x}e^{-j35x}\vec{a}_z \text{ V/m}$$

$$\vec{H}(x)=-0.93e^{-11.3x}e^{-j35x}e^{-j17.9^\circ}\vec{a}_y \text{ mA/m}$$

The instantaneous magnetic field expression will be given by

$$\vec{H}(x,t)=-0.93e^{-11.3x}\cos(2\pi\times5\times10^{8}t-35x-17.9^\circ)\vec{a}_y$$

Or

$$\vec{H}(x,t)=0.93e^{-11.3x}\cos(2\pi\times5\times10^{8}t-35x-17.9^\circ+180^\circ)\vec{a}_y$$

$$=0.93e^{-11.3x}\cos(2\pi\times5\times10^{8}t-35x+160.1^\circ)\vec{a}_y$$

11.7 Poynting Vector–Poynting Theorem

In this section, we investigate the power flow associated with plane wave propagation. In this regard, we need to develop what is known as the Poynting theorem named after the English physicist John Henry Poynting which he derived in 1884. Poynting's theorem states that the power flow density vector, $\vec{\mathcal{P}}$, is defined as

$$\vec{\mathcal{P}}=\vec{E}\times\vec{H} \tag{11.76}$$

Before we go through the proof/derivation of the Poynting theorem, let us examine the field solution we developed above for plane wave propagation. We notice that both the electric and magnetic field vectors are orthogonal to each other and both lie in a transverse plane to the direction of propagation. Hence, we can see that the cross product $\vec{E}\times\vec{H}$ results in a vector directed along the direction of wave propagation. For the special case above, we have $\vec{E}$ along the x axis and $\vec{H}$ along the y axis with the wave propagation along the z axis, $\vec{a}_x\times\vec{a}_y=\vec{a}_z$.

If we examine the units of the product $\vec{E}\times\vec{H}$, we find (volts/meter) multiplied by (amperes/meter), yielding (watts/squared meter) for the product, which is certainly a surface power density. Hence, we can get an appreciation to the Poynting vector in this case, $\vec{\mathcal{P}}_z=\vec{E}_x\times\vec{H}_y$, as the plane wave power flow density along the wave propagation direction.

Now, for the development/proof of the Poynting theorem, we start with the vector identity:

$$\vec{\nabla}\cdot[\vec{\mathcal{P}}]=\vec{\nabla}\cdot[\vec{E}\times\vec{H}]=-\vec{E}\cdot[\vec{\nabla}\times\vec{H}]+\vec{H}\cdot[\vec{\nabla}\times\vec{E}] \tag{11.77}$$

Next, we substitute in the time domain form of Maxwell equations for the two curl terms yielding

$$\vec{\nabla}\cdot[\vec{\mathcal{P}}]=-\vec{E}\cdot\left[\vec{J}+\frac{\partial\vec{D}}{\partial t}\right]+\vec{H}\cdot\left[-\frac{\partial\vec{B}}{\partial t}\right] \tag{11.78}$$

$$\vec{\nabla}\cdot[\vec{\mathcal{P}}]=-\vec{E}\cdot\vec{J}-\vec{E}\cdot\left[\frac{\partial\vec{D}}{\partial t}\right]-\vec{H}\cdot\left[\frac{\partial\vec{B}}{\partial t}\right] \tag{11.79}$$

$$\vec{\nabla}\cdot[\vec{\mathcal{P}}] = -\vec{E}\cdot\vec{J} - \left[\vec{E}\cdot\frac{\partial \varepsilon \vec{E}}{\partial t}\right] - \left[\vec{H}\cdot\frac{\partial \mu \vec{H}}{\partial t}\right] \tag{11.80}$$

Now, recognizing that

$$\vec{E}\cdot\frac{\partial \varepsilon \vec{E}}{\partial t} = E\vec{a}_E\cdot\frac{\partial \varepsilon E}{\partial t}\vec{a}_E = E\frac{\partial \varepsilon E}{\partial t} = \frac{\partial}{\partial t}\left[\frac{\varepsilon E^2}{2}\right] \tag{11.81}$$

and likewise, for the H term, $\vec{H}\cdot\frac{\partial \mu \vec{H}}{\partial t} = \frac{\partial}{\partial t}\left[\frac{\mu H^2}{2}\right]$, we obtain

$$\vec{\nabla}\cdot[\vec{\mathcal{P}}] = -\vec{E}\cdot\vec{J} - \frac{\partial}{\partial t}\left[\frac{\varepsilon E^2}{2}\right] - \frac{\partial}{\partial t}\left[\frac{\mu H^2}{2}\right] \tag{11.82}$$

Now, applying the divergence theorem [Equation (4.27)]

$$\oiint_S \vec{\mathcal{P}}\cdot\overrightarrow{dS} = \int_{\text{volume enclosed by } S}[\vec{\nabla}\cdot\vec{\mathcal{P}}]dv \tag{11.83}$$

to Equation (11.82), we get

$$\oiint_S \vec{\mathcal{P}}\cdot\overrightarrow{dS} = \int_{\text{vol}}\left\{-\vec{E}\cdot\vec{J} - \frac{\partial}{\partial t}\left[\frac{\varepsilon E^2}{2}\right] - \frac{\partial}{\partial t}\left[\frac{\mu H^2}{2}\right]\right\}dv \tag{11.84}$$

The left-hand side of this equation is the total outflow of the term $\vec{\mathcal{P}}\cdot\overrightarrow{dS}$ integrated over the entire closed surface of a given volume. While, the right-hand side is the volume integration of the term $\left\{-\vec{E}\cdot\vec{J} - \frac{\partial}{\partial t}\left[\frac{\varepsilon E^2}{2}\right] - \frac{\partial}{\partial t}\left[\frac{\mu H^2}{2}\right]\right\}$ over the enclosed volume.

$$\oiint_S \vec{\mathcal{P}}\cdot\overrightarrow{dS} = -\int_{\text{vol}}\{\vec{E}\cdot\sigma\vec{E}\}dv - \int_{\text{vol}}\left\{\frac{\partial}{\partial t}\left[\frac{\varepsilon E^2}{2}\right]\right\}dv - \int_{\text{vol}}\left\{\frac{\partial}{\partial t}\left[\frac{\mu H^2}{2}\right]\right\}dv \tag{11.85}$$

Next, exchanging the volume integration operation with that of the time derivative for a stationary system, we write

$$\oiint_S \vec{\mathcal{P}}\cdot\overrightarrow{dS} = -\left\{\int_{\text{vol}}\sigma E^2\,dv\right\} - \frac{\partial}{\partial t}\left\{\int_{\text{vol}}\frac{\varepsilon E^2}{2}dv\right\} - \frac{\partial}{\partial t}\left\{\int_{\text{vol}}\frac{\mu H^2}{2}dv\right\} \tag{11.86}$$

Now, using Equations (6.17) and (9.20), we recognize

- $p_e = \sigma E^2$ W/m^3 (11.87)

  is the dissipated power per unit volume within the enclosure due to the material conductivity (ohmic losses).

- $-P_e = -\int_{\text{vol}} p_e\,dv$ W (11.88)

  is the reduction in power dissipated within the volume enclosure indicating a power outflow from within the volume across its surface boundaries.

- $$w_e = \frac{1}{2}\varepsilon E^2 \text{ J/m}^3 \tag{11.89}$$

  is the electric energy per unit volume stored within the enclosure due to the material permittivity (capacitive energy).

- $$-\frac{\partial}{\partial t}\left\{\int_{\text{vol}} \frac{\varepsilon E^2}{2} dv\right\} = -\frac{\partial}{\partial t}\left\{\int_{\text{vol}} w_e\, dv\right\} = -\frac{\partial}{\partial t}\{W_e\} \text{ W} \tag{11.90}$$

  is the rate of decrease of electric energy stored within the volume enclosure indicating a power outflow from within the volume across its surface boundaries.

- $$w_m = \frac{1}{2}\mu H^2 \text{ J/m}^3 \tag{11.91}$$

  is the magnetic energy per unit volume stored within the enclosure due to the material permeability (inductive energy).

- $$-\frac{\partial}{\partial t}\left\{\int_{\text{vol}} \frac{\mu H^2}{2} dv\right\} = -\frac{\partial}{\partial t}\{W_m\} \text{ W} \tag{11.92}$$

  is the rate of decrease of magnetic energy stored within the volume enclosure indicating a power outflow from within the volume across its surface boundaries.

Finally, we have the final form

$$\oiint_S \vec{\mathcal{P}} \cdot \overrightarrow{dS} = -\{P_e\} - \frac{\partial}{\partial t}\{W_e\} - \frac{\partial}{\partial t}\{W_m\} \tag{11.93}$$

Consequently, we realize that the three negative terms of the right-hand side of Equation (11.93) add up to the total outflow of power across the surface of the volume enclosure.

The left-hand side of this equation term is actually the total outflow of power across the surface of the volume enclosure. This makes the Poynting vector $\vec{\mathcal{P}}$ the outflow power (surface) density at any point in the electromagnetic field.

To iterate, the Poynting theorem states that $\vec{\mathcal{P}} = \vec{E} \times \vec{H}$ is the electromagnetic vector power flow (surface) density at any point in space.

11.8 The Complex Poynting Theorem

The above development for the Poynting theorem defines the time domain form of the Poynting vector. In the following, we will present the corresponding definition in the phasor/frequency domain form. Such development will be useful for treating harmonic electromagnetic field radiation problems.

Starting with the time domain (real) Poynting vector definition as defined in terms of the real electric and magnetic field intensities:

$$\vec{\mathcal{P}}(\vec{r}, t) = \vec{E}(\vec{r}, t) \times \vec{H}(\vec{r}, t) \tag{11.94}$$

We now consider the special case of monochromatic (single harmonic) fields, the real Poynting vector can be written as

$$\vec{\mathcal{P}}(\vec{r}, t) = Re\{\vec{\bar{E}}(\vec{r})e^{j\omega t}\} \times Re\{\vec{\bar{H}}(\vec{r})e^{j\omega t}\} \tag{11.95}$$

in terms of the complex electric and magnetic field phasors. The last expression above can be rewritten as follows:

$$\vec{\mathcal{P}}(\vec{r},t)=\left(\frac{\bar{\vec{E}}(\vec{r})e^{j\omega t}+\bar{\vec{E}}^*(\vec{r})e^{-j\omega t}}{2}\right)\times\left(\frac{\bar{\vec{H}}(\vec{r})e^{j\omega t}+\bar{\vec{H}}^*(\vec{r})e^{-j\omega t}}{2}\right)$$

$$=\frac{1}{4}(\bar{\vec{E}}(\vec{r})\times\bar{\vec{H}}(\vec{r})e^{2j\omega t}+\bar{\vec{E}}^*(\vec{r})\times\bar{\vec{H}}^*(\vec{r})e^{-2j\omega t})+\frac{1}{2}\left(\frac{\bar{\vec{E}}(\vec{r})\times\bar{\vec{H}}^*(\vec{r})+\bar{\vec{E}}^*(\vec{r})\times\bar{\vec{H}}(\vec{r})}{2}\right)$$

$$=\frac{1}{4}(\bar{\vec{E}}(\vec{r})\times\bar{\vec{H}}(\vec{r})e^{2j\omega t}+\bar{\vec{E}}^*(\vec{r})\times\bar{\vec{H}}^*(\vec{r})e^{-2j\omega t})+\frac{1}{2}Re\{\bar{\vec{E}}(\vec{r})\times\bar{\vec{H}}^*(\vec{r})\} \tag{11.96}$$

where the "*" superscript indicates the complex conjugate of the field phasor.

Upon integration of this expression over a period $T=2\pi/\omega$, both the terms containing $e^{2j\omega t}$ and $e^{-2j\omega t}$ average to zeros leaving the time-averaged Poynting vector as given by

$$\vec{\mathcal{P}}_{\text{avg}}(\vec{r})=\frac{1}{2}Re\{\bar{\vec{E}}(\vec{r})\times\bar{\vec{H}}^*(\vec{r})\} \tag{11.97}$$

which provides the time-averaged surface power density for monochromatic fields.

11.9 The Complex Poynting Vector

In view of Equation (11.97), let us define the complex Poynting vector by the expression

$$\bar{\vec{\mathcal{P}}}(\vec{r})=\frac{1}{2}\bar{\vec{E}}(\vec{r})\times\bar{\vec{H}}^*(\vec{r}) \tag{11.98}$$

This form is analogous to that of the power expression $\bar{P}=\frac{1}{2}\bar{V}\cdot\bar{I}^*$ which we are acquainted with in electrical circuits' analyses.

Now, applying the divergence to Equation (11.98), we get

$$\vec{\nabla}\cdot\bar{\vec{\mathcal{P}}}(\vec{r})=\frac{1}{2}\bar{\vec{H}}^*(\vec{r})\cdot\vec{\nabla}\times\bar{\vec{E}}(\vec{r})-\frac{1}{2}\bar{\vec{E}}(\vec{r})\cdot\vec{\nabla}\times\bar{\vec{H}}^*(\vec{r}) \tag{11.99}$$

Next, we replace the curl terms in Equation (11.99) by their equivalent terms from the frequency domain forms of Maxwell equations [Equations (11.57) and (11.59)] yielding

$$\vec{\nabla}\cdot\bar{\vec{\mathcal{P}}}(\vec{r})=\frac{1}{2}[\bar{\vec{H}}]^*(\vec{r})\cdot[-j\omega\mu^*\bar{\vec{H}}(\vec{r})]-\frac{1}{2}\bar{\vec{E}}(\vec{r})\cdot[j\omega\varepsilon^*\bar{\vec{E}}(\vec{r})]^*$$

$$=-\frac{1}{2}j\omega\mu^*\left|\bar{\vec{H}}(\vec{r})\right|^2+\frac{1}{2}j\omega[\varepsilon^*]^*\left|\bar{\vec{E}}(\vec{r})\right|^2 \tag{11.100}$$

For materials characterized by both electric and magnetic losses, we have the complex permittivity $\varepsilon^*=\varepsilon'-j\varepsilon''$ and the complex permeability $\mu^*=\mu'-j\mu''$, Equations (11.58) and (11.60), respectively. We substitute, next, these expressions in Equation (11.100):

$$\vec{\nabla}\cdot\bar{\vec{\mathcal{P}}}(\vec{r})=-\frac{1}{2}j\omega(\mu'-j\mu'')\left|\bar{\vec{H}}(\vec{r})\right|^2+\frac{1}{2}j\omega(\varepsilon'+j\varepsilon'')\left|\bar{\vec{E}}(\vec{r})\right|^2$$

$$=\frac{1}{2}j\omega\left[\varepsilon'\left|\bar{\vec{E}}(\vec{r})\right|^2-\mu'\left|\bar{\vec{H}}(\vec{r})\right|^2\right]-\frac{1}{2}\omega\left[\varepsilon''\left|\bar{\vec{E}}(\vec{r})\right|^2+\mu''\left|\bar{\vec{H}}(\vec{r})\right|^2\right] \tag{11.101}$$

Equating the real and imaginary parts on both sides, we obtain

$$\vec{\nabla}\cdot\vec{\bar{\mathcal{P}}}_{re}(\vec{r})=\vec{\nabla}\cdot Re\left\{\frac{1}{2}\vec{\bar{E}}(\vec{r})\times\vec{\bar{H}}^{*}(\vec{r})\right\}=\vec{\nabla}\cdot\vec{\mathcal{P}}_{\text{avg}}(\vec{r})=-\frac{1}{2}\omega\left[\varepsilon''\left|\vec{\bar{E}}(\vec{r})\right|^{2}+\mu''\left|\vec{\bar{H}}(\vec{r})\right|^{2}\right] \tag{11.102}$$

$$\vec{\nabla}\cdot\vec{\bar{\mathcal{P}}}_{im}(\vec{r})=\vec{\nabla}\cdot Im\left\{\frac{1}{2}\vec{\bar{E}}(\vec{r})\times\vec{\bar{H}}^{*}(\vec{r})\right\}=\frac{1}{2}\omega\left[\varepsilon'\left|\vec{\bar{E}}(\vec{r})\right|^{2}-\mu'\left|\vec{\bar{H}}(\vec{r})\right|^{2}\right] \tag{11.103}$$

Equation (11.103) states that the real power flowing into a unit volume, $-\vec{\nabla}\cdot\vec{\bar{\mathcal{P}}}_{re}(\vec{r})$, is equal to the power loss per unit volume $\frac{1}{2}\omega\left[\varepsilon''\left|\vec{\bar{E}}(\vec{r})\right|^{2}+\mu''\left|\vec{\bar{H}}(\vec{r})\right|^{2}\right]$. On the other hand, $-\vec{\nabla}\cdot\vec{\bar{\mathcal{P}}}_{im}(\vec{r})$ represents the reactive power per unit volume due to the power flow and the (time-averaged) stored electric and magnetic energy densities. The latter are given by $w_e(\vec{r})=\frac{1}{4}\varepsilon'\left|\vec{\bar{E}}(\vec{r})\right|^{2}$ and $w_m(\vec{r})=\frac{1}{4}\mu'\left|\vec{\bar{H}}(\vec{r})\right|^{2}$, respectively.

To summarize, we conclude the following four components of the complex power density as a result of the above development of the complex Poynting vector:

- $$p_e=\frac{1}{2}\omega\varepsilon''\left|\vec{\bar{E}}(\vec{r})\right|^{2}=\frac{1}{2}\sigma\left|\vec{\bar{E}}(\vec{r})\right|^{2}\ \text{W/m}^3 \tag{11.104}$$

  is the active power volumetric density dissipated due to the conductivity (electric losses) of the medium

- $$p_m=\frac{1}{2}\omega\mu''\left|\vec{\bar{H}}(\vec{r})\right|^{2}\ \text{W/m}^3 \tag{11.105}$$

  is the active power volumetric density dissipated due to the magnetic losses of the medium

- $$w_e(\vec{r})=\frac{1}{4}\varepsilon'\left|\vec{\bar{E}}(\vec{r})\right|^{2}\ \text{J/m}^3 \tag{11.106}$$

  is the volumetric density of the electric energy storage (reactive power) associated with the permittivity of the medium.

- $$w_m(\vec{r})=\frac{1}{4}\mu'\left|\vec{\bar{H}}(\vec{r})\right|^{2}\ \text{J/m}^3 \tag{11.107}$$

  is the volumetric density of the magnetic energy storage (reactive power) associated with the permeability of the medium.

Comparing the results obtained in the above expressions [Equations (11.104) to (11.107)] to those obtained earlier [Equations (11.87) to (11.92)] we notice the following:

- Equations (11.87) to (11.92) are for the time-domain field quantities while those of Equations (11.104) to (11.107) are for the phasor field quantities. Hence, we notice the ½ coefficient in Equations (11.104) to (11.107) for the root mean square (effective power/energy) of the harmonic fields.
- Equations (11.87) and (11.104) refer to the same volumetric density of the ohmic losses due to the medium's conductivity.
- Equations (11.89) and (11.106) refer to the same volumetric density of electric energy storage due to the medium's permittivity (capacitive behavior).
- Equations (11.91) and (11.107) refer to the same volumetric density of magnetic energy storage due to the medium's permeability (inductive behavior).
- Equation (11.105) refers to the volumetric density of magnetic losses that were assumed absent in the analysis leading to Equations (11.87) to (11.92).

11.10 Plane Waves Power Flow

Using the Poynting expression [Equation (11.98)] for the power density, and substituting the field expressions from Table 11.5, we can derive expressions for the $\bar{\bar{\mathcal{P}}}^{+}$ and $\bar{\bar{\mathcal{P}}}^{-}$ power density vector phasors corresponding to the forward and backward wave propagations, respectively, as follows:

$$\begin{aligned}\bar{\bar{\mathcal{P}}} &= \frac{1}{2}\bar{\bar{E}}\times\bar{\bar{H}}^{*} = \frac{1}{2}\left[\bar{E}_x(z)\right]\vec{a}_x\times\left[\bar{H}_y(z)\right]^{*}\vec{a}_y = \frac{1}{2}\vec{a}_z\left[\bar{E}_x(z)\right]\left[\bar{H}_y(z)\right]^{*}\\ &= \vec{a}_z\frac{1}{2\eta^{*}}\left[\left\{\bar{E}_x^{+}e^{-\gamma z}\right\}+\left\{\bar{E}_x^{-}e^{+\gamma z}\right\}\right]\left[\left\{\bar{E}_x^{+}e^{-\gamma z}\right\}^{*}-\left\{\bar{E}_x^{-}e^{+\gamma z}\right\}^{*}\right]\\ &= \vec{a}_z\frac{1}{2\eta^{*}}\left[\left|\bar{E}_x^{+}e^{-\alpha z}\right|^{2}-\left|\bar{E}_x^{-}e^{+\alpha z}\right|^{2}+\left\{\bar{E}_x^{-}\bar{E}_x^{+*}e^{+2j\beta z}\right\}-\left\{\bar{E}_x^{+}\bar{E}_x^{-*}e^{-2j\beta z}\right\}\right]\end{aligned} \tag{11.108}$$

Noting that the last two terms in the above equation, $\left\{\bar{E}_x^{+}\bar{E}_x^{-*}e^{-2j\beta z}\right\}$ and $\left\{\bar{E}_x^{-}\bar{E}_x^{+*}e^{+2j\beta z}\right\}$ are the complex conjugate of each other, the difference between the two terms is then equal to twice the imaginary part of the positive term in the subtraction. Hence, we can write

$$\begin{aligned}\bar{\bar{\mathcal{P}}} &= \vec{a}_z\frac{1}{2\eta^{*}}\left[\left|\bar{E}_x^{+}\right|^{2}e^{-2\alpha z}-\left|\bar{E}_x^{-}\right|^{2}e^{+2\alpha z}+\left\{\bar{E}_x^{-}\bar{E}_x^{+*}e^{+2j\beta z}\right\}-\left\{\bar{E}_x^{-}\bar{E}_x^{+*}e^{+2j\beta z}\right\}^{*}\right]\\ &= \vec{a}_z\frac{1}{2\eta^{*}}\left[\left|\bar{E}_x^{+}\right|^{2}e^{-2\alpha z}-\left|\bar{E}_x^{-}\right|^{2}e^{+2\alpha z}+2j\,Im\left\{\bar{E}_x^{-}\bar{E}_x^{+*}e^{+2j\beta z}\right\}\right]\end{aligned} \tag{11.109}$$

Finally, we can express the average power-flow-density vector as the real part of the phasor expression in Equation (11.109)

$$\vec{\mathcal{P}}_{\text{avg}} = \vec{a}_z\,Re\left\{\frac{1}{2\eta^{*}}\left[\left|\bar{E}_x^{+}\right|^{2}e^{-2\alpha z}-\left|\bar{E}_x^{-}\right|^{2}e^{+2\alpha z}+2j\,Im\left\{\bar{E}_x^{-}\bar{E}_x^{+*}e^{+2j\beta z}\right\}\right]\right\} \tag{11.110}$$

For low-loss dielectric materials where the intrinsic impedance η is real, Equation (11.110) reduces to

$$\vec{\mathcal{P}}_{\text{avg}} = \vec{a}_z\frac{1}{2\eta}\left[\left|\bar{E}_x^{+}\right|^{2}e^{-2\alpha z}-\left|\bar{E}_x^{-}\right|^{2}e^{+2\alpha z}\right] \tag{11.111}$$

Hence, we deduce that forward and backward power flow densities for low loss (and lossless) media can be separated as

$$\mathcal{P}_{\text{avg}}^{+} = \frac{1}{2\eta}\left|\bar{E}_x^{+}\right|^{2}e^{-2\alpha z}\ \text{W/m}^2\text{, flows in the } +z \text{ direction} \tag{11.112}$$

$$\mathcal{P}_{\text{avg}}^{-} = \frac{1}{2\eta}\left|\bar{E}_x^{-}\right|^{2}e^{+2\alpha z}\ \text{W/m}^2\text{, flows in the } -z \text{ direction} \tag{11.113}$$

These expressions demonstrate the forward and backward power flows corresponding to the wave propagation directions (Figure 11.12).

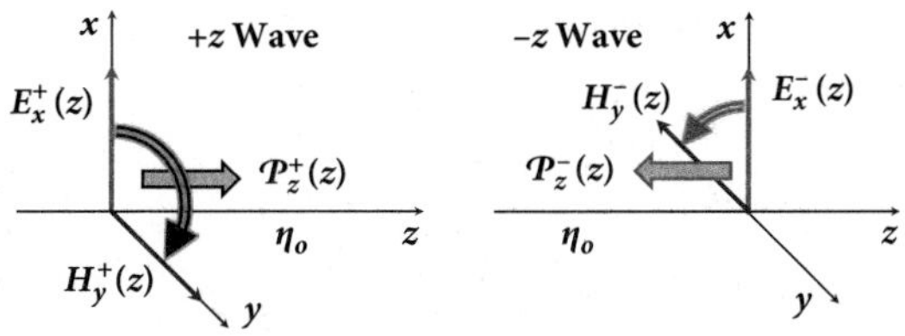

FIGURE 11.12 The Poynting vector for both forward and backward waves, where E, H, and $\mathcal{P}$ form a right-hand set.

Example 11.6 Consider a uniform plane wave traveling downward in the negative z direction into seawater. The electric field expression in the seawater is given by $\vec{E}(z,t)=0.1e^{\alpha z}\cos(2\pi\times10^4 t+\beta z)\vec{a}_y$ for $z<0$. Given that $z=0$ is the surface of the water. The constitutive parameters of the seawater are $\mu_r=1, \varepsilon_r=80$, and $\sigma=5$ S/m

a. Write an expression for the magnetic field intensity vector $\vec{H}(z,t)$
b. Find the average power density
c. Determine the depth where the average power density is 0.1% of its value at $z=0$

Solution

Solution Part a

$$\frac{\sigma}{\omega\varepsilon}=\frac{\sigma}{\omega\varepsilon_r\varepsilon_o}=\frac{5}{2\pi\times10^4\times80\times8.854\times10^{-12}}=11.25\times10^4\gg1$$

Since $\sigma\gg\omega\varepsilon$ at the given frequency, the medium acts as a good conductor and we can use the good conductor approximations in Table 11.6

$$\eta\approx\frac{1+j}{\sqrt{2}}\sqrt{\frac{\mu_o}{\varepsilon''}}=\Omega$$

$$\varepsilon''=\frac{\sigma}{\omega}\rightarrow\eta\approx\frac{1+j}{\sqrt{2}}\sqrt{\frac{\mu_o\omega}{\sigma}}=\frac{1+j}{\sqrt{2}}\sqrt{\frac{4\pi\times10^{-7}\times2\pi\times10^4}{5}}=\frac{0.1257}{\sqrt{2}}(1+j)=0.126<45^\circ\ \Omega$$

Note that the angle of the intrinsic impedance in a good conductor is 45° which indicates that the electric field vector will be leading the magnetic field vector by 45°.

For a good conductor we also have

$$\alpha\cong\beta\cong\sqrt{\frac{\omega\mu_o(\omega\varepsilon'')}{2}}=\sqrt{\frac{\omega\mu_o\sigma}{2}}=\sqrt{\frac{2\pi\times10^4\times4\pi\times10^{-7}\times5}{2}}=0.44$$

$$\alpha=0.44\text{ Np/m}\qquad\beta=0.44\text{ rad/m}$$

$$\vec{E}(z,t)=0.1e^{\alpha z}\cos(2\pi\times10^4 t+\beta z)\vec{a}_y\quad z<0$$

The phasor form of the electric field can be expressed as

$$\vec{E}(z)=0.1e^{\alpha z}e^{j\beta z}\vec{a}_y=0.1e^{0.44z}e^{j0.44z}\vec{a}_y\text{ V/m}$$

$$\vec{H}(z)=\frac{1}{\eta}(-\vec{a}_z)\times\vec{E}(z)=\frac{1}{0.126\angle45^\circ}(-\vec{a}_z)\times\left(0.1e^{0.44z}e^{j0.44z}\vec{a}_y\right)\text{ V/m}$$

$$\vec{H}(z)=0.79e^{0.44z}e^{j0.44z}e^{-j45^\circ}\vec{a}_x\text{ A/m}$$

$$\vec{H}(z,t)=0.79e^{0.44z}\cos(2\pi\times10^4 t+0.44z-45^\circ)\vec{a}_x\quad z<0$$

Solution Part b

$$\vec{P}_{\text{avg}}(z)=\frac{1}{2}Re\{\vec{E}(z)\times\vec{H}^*(z)\}=\frac{1}{2}Re\left\{0.1e^{0.44z}e^{j0.44z}\vec{a}_y\times0.79e^{0.44z}e^{-j0.44z}e^{j45^\circ}\vec{a}_x\right\}$$

$$\vec{P}_{\text{avg}}(z)=\frac{1}{2}Re\left\{0.079e^{0.88z}e^{j45^\circ}(-\vec{a}_z)\right\}=-0.04e^{0.88z}\cos(45^\circ)\vec{a}_z=-0.028e^{0.88z}\vec{a}_z\text{ W/m}^2$$

Solution Part c

$$\vec{P}_{\text{avg}}(z)\Big|_{z<0}=-0.028e^{0.88z}\vec{a}_z\text{ W/m}^2\ \left|\vec{P}_{\text{avg}}(z=0)\right|=0.028\text{ W/m}^2$$

$$\left|\vec{P}_{\text{avg}}(z_1)\right|=0.028e^{0.88z_1}=\frac{0.1}{100}\times0.028\rightarrow z_1=\frac{1}{0.88}\ln(0.001)=-7.85\text{ m}$$

Example 11.7 A 3 GHz, y polarized uniform plane wave propagates in the $+x$ direction in a nonmagnetic material, $\mu_r = 1$, having a relative dielectric constant ε_r of 2.5.

a. Determine the intrinsic impedance, the wavelength, the phase velocity of the wave in the medium.
b. Assuming at $x = 0$, $\vec{E} = \vec{a}_y 50 \sin\left(6\pi \times 10^9 t + \frac{\pi}{3}\right)$ V/m, write the instantaneous expressions for $\vec{E}$ and $\vec{H}$ for all t and x.
c. Determine the average power density in the propagating wave.

Solution

Solution Part a

$$\eta = 120\pi\sqrt{\frac{\mu_r}{\epsilon_r}} = 120\pi\sqrt{\frac{1}{2.5}} = 238.4\,\Omega$$

$$\lambda = \frac{c}{f\sqrt{\mu_r\epsilon_r}} = \frac{3\times10^8}{3\times10^9\sqrt{1\times2.5}} = 6.32\text{ cm} \qquad c_{ph} = \frac{\omega}{\beta} = \frac{c}{\sqrt{\mu_r\epsilon_r}} = 1.9\times10^8\text{ m/s}$$

Solution Part b At $x = 0$, $\vec{E}(x=0,t) = \vec{a}_y 50 \sin\left(6\pi\times10^9 t + \frac{\pi}{3}\right)$ V/m and with the wave propagating in the positive x direction, the general electric field expression can be written as

$$\vec{E} = \vec{a}_y 50 \sin\left(6\pi\times10^9 t - \beta x + \frac{\pi}{3}\right) = \vec{a}_y 50 \cos\left(6\pi\times10^9 t - \beta x + \frac{\pi}{3} - \frac{\pi}{2}\right)$$
$$= \vec{a}_y 50 \cos\left(6\pi\times10^9 t - \beta x - \frac{\pi}{6}\right)$$

$$\vec{E} = \vec{a}_y 50 e^{-j\beta x} e^{-\frac{j\pi}{6}} \text{ where } \beta = \frac{2\pi}{\lambda} = \frac{2\pi}{0.0632} = 99.35\text{ rad/m} \rightarrow \vec{E} = \vec{a}_y 50 e^{-j99.35x} e^{-\frac{j\pi}{6}}$$

$$\vec{E}(x,t) = \vec{a}_y 50 \cos\left(6\pi\times10^9 t - 99.35x - \frac{\pi}{6}\right)\text{ V/m}$$

$$\vec{H} = \frac{1}{\eta}\,\vec{a}_k \times \vec{E} = \frac{1}{\eta}(\vec{a}_x)\times\vec{a}_y 50 e^{-j99.35x} e^{-\frac{j\pi}{6}} = \vec{a}_z \frac{50}{238.4} e^{-j99.35x} e^{-\frac{j\pi}{6}} = \vec{a}_z 0.21 e^{-j99.35x} e^{-\frac{j\pi}{6}}$$

$$\vec{H}(x,t) = Re\left\{\vec{a}_z 0.21 e^{-j99.35x} e^{-\frac{j\pi}{6}} e^{j\omega t}\right\} = \vec{a}_z 0.21 \cos\left(6\pi\times10^9 t - 99.35x - \frac{\pi}{6}\right)\text{ A/m}$$

Solution Part c The average power density in the propagating wave.

$$P = \frac{\vec{a}_k}{2\eta}\left|\vec{E}\right|^2 = \vec{a}_x \frac{50^2}{2\times238.4} = \vec{a}_x\, 5.25\text{ W/m}^2$$

or

$$P = \frac{1}{2}Re\{\vec{E}\times\vec{H}^*\} = \frac{1}{2}Re\left\{\left(\vec{a}_y 50 e^{-j99.35x} e^{-\frac{j\pi}{6}}\right)\times\left(\vec{a}_z 0.21 e^{+j99.35x} e^{\frac{j\pi}{6}}\right)\right\}$$
$$= \vec{a}_x 5.25\text{ W/m}^2$$

11.11 Plane Waves in Controlled-Geometry Transmission Lines

In Chapter 2, we discussed wave propagation for controlled-geometry transmission lines using the physically based RLGC circuit model. However, as we indicated back at that time, the circuit theory approach in such analysis can only yield a constrained solution with limitations. The proper way for such an analysis is the use of a dynamic electromagnetic fields approach which requires the use of Maxwell's equations. That is what we stated back in Chapter 2, and here is an excerpt of what we stated back then

> The accurate and comprehensive way to analyze the performance of transmission lines is to do the analysis as an electromagnetic fields problem. This requires the use of Maxwell's equations to develop a full EM field model and study the field waves as they propagate along the TL. …

Now, that we have been through Maxwell's equations and their usage in analyzing plane wave propagation, let us take on the challenge of using them in solving wave propagation along controlled geometry transmission lines. In this regard, we will confine our study to the simplest of cases; that of two conductors with uniform cross section (material and geometry do not vary along the TL length).

The line will be assumed straight with no curves or bends. We will further limit ourselves to the case where the TL material, that is, both conductors and the insulating dielectric(s) are linear (properties do not change with field intensity), and frequency independent (properties do not change with frequency).

Let us further limit the analysis (at least initially) to the case of the coaxial transmission line. We do that because we already have developed its static field solutions back in Chapters 4 and 8.

11.11.1 Case of Coaxial Transmission Line

We start by bringing back the shape of the coaxial configuration presented earlier in Figures 4A.30 and 8.40. This time, we present it in Figure 11.13 with the line being coaxial with the z axis. Assuming the two TL conductors are perfect conductors and the dielectric insulator is a perfect dielectric, we recall the static solution derived earlier in Chapters 4 and 8 as follows:

$$\vec{D} = \varepsilon\vec{E} = \frac{\rho_\ell}{2\pi\rho}\vec{a}_\rho \qquad \text{for } a < \rho < b \tag{4A.59}$$

$$\vec{H} = \frac{I}{2\pi\rho}\vec{a}_\varphi \qquad \text{for } a < \rho < b \tag{8C.13}$$

and

$$\vec{D} = \vec{H} = 0 \qquad \text{for } \rho < a \text{ and } \rho > b \tag{11.114}$$

Although these expressions do not represent the full time-varying field solutions, they are simply valid solutions that meet the boundary conditions imposed by the coaxial conductors and the dielectric spacing in between. To generalize the two expressions, we replace them with the following ones in which we maintain the spatial dependence while setting arbitrary levels through the use of the multipliers C_1 and C_2.

$$\vec{E} = \frac{C_1}{\rho}\vec{a}_\rho \qquad \text{for } a < \rho < b \tag{11.115}$$

$$\vec{H} = \frac{C_2}{\rho}\vec{a}_\varphi \qquad \text{for } a < \rho < b \tag{11.116}$$

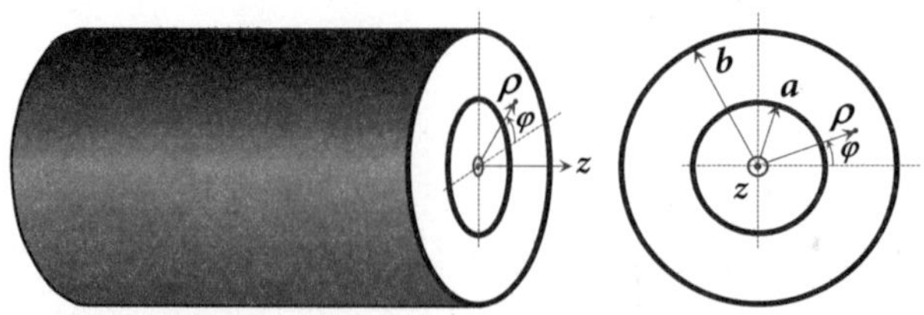

FIGURE 11.13 A simplified configuration of a coaxial transmission line.

These expressions represent what we refer to as quasi-static solutions and do in fact represent a plane wave (*TEM*) solution as we examine the corresponding Poynting vector as

$$\vec{\mathcal{P}} = \vec{E} \times \vec{H} = \left[\frac{C_1}{\rho}\,\vec{a}_\rho\right] \times \left[\frac{C_2}{\rho}\,\vec{a}_\varphi\right] = \left(\frac{C_1 C_2}{\rho^2}\right)\vec{a}_z \tag{11.117}$$

which indicates power flow along the transmission line (z coordinate).

A full field analysis of the coaxial transmission line would result in more than one possible field solution (field pattern). Each one of these solutions represents a possible mode of wave propagation along the line. The general solution is then set as a superposition of all possible modes of propagation with the line's boundary conditions (including that of the source and load) determining the operating modes and their intensities.

Multimode propagation of electromagnetic waves is not recommended in communication systems as it results in an undesired signal dispersion. Fortunately, for coaxial lines (and other forms of transmission lines as well), the *TEM* plane wave mode is the lowest order mode that operates in the base-band (starting from DC). Higher order modes can only exist above certain frequencies which we would consider as the upper cut off frequencies for the line's operation in single mode. In our discussion here, we will limit ourselves to coaxial line operation in the *TEM* mode, thus allowing only plane wave propagation.

For the *TEM* mode, we recognize orthogonal E_ρ and H_φ field vectors that are responsible for producing wave propagation along the z axis. The quasi-static expressions [Equations (11.115) and (11.116)] reveal the ρ dependence of both vectors with no φ dependence. The dynamic solution for the field propagation is expected to have a z dependence in both field expressions. Hence, setting the stage for the use of the separation of variables scheme for deriving the ρ and z dependence field expressions, we can write $E_\rho(\rho,z)$ and $H_\varphi(\rho,z)$ in the product forms:

$$E_\rho = f_\rho(\rho) f_z(z) \tag{11.118}$$

$$H_\varphi = g_\rho(\rho) g_z(z) \tag{11.119}$$

with the functional dependencies on ρ in fact already set by Equations (11.115) and (11.116), that is,

$$E_\rho = f_\rho(\rho) f_z(z) = \frac{C_1}{\rho} f_z(z) \tag{11.120}$$

$$H_\varphi = g_\rho(\rho) g_z(z) = \frac{C_2}{\rho} g_z(z) \tag{11.121}$$

To develop the z dependencies of these field expressions, we refer to Maxwell's equations for the coaxial configuration in cylindrical coordinates [Equation (8.57)]

$$\begin{aligned}\vec{\nabla} \times \vec{E} &= \left\{\frac{1}{\rho}\frac{\partial E_z}{\partial \varphi} - \frac{\partial E_\varphi}{\partial z}\right\}\vec{a}_\rho + \left\{\frac{\partial E_\rho}{\partial z} - \frac{\partial E_z}{\partial \rho}\right\}\vec{a}_\varphi + \frac{1}{\rho}\left\{\frac{\partial(\rho E_\varphi)}{\partial \rho} - \frac{\partial E_\rho}{\partial \varphi}\right\}\vec{a}_z = -j\omega\mu^*\,\vec{H}\\ &= -j\omega\mu^*\{H_\rho \vec{a}_\rho + H_\varphi \vec{a}_\varphi + H_z \vec{a}_z\}\end{aligned} \tag{11.122}$$

$$\begin{aligned}\vec{\nabla} \times \vec{H} &= \left\{\frac{1}{\rho}\frac{\partial H_z}{\partial \varphi} - \frac{\partial H_\varphi}{\partial z}\right\}\vec{a}_\rho + \left\{\frac{\partial H_\rho}{\partial z} - \frac{\partial H_z}{\partial \rho}\right\}\vec{a}_\varphi + \frac{1}{\rho}\left\{\frac{\partial(\rho H_\varphi)}{\partial \rho} - \frac{\partial H_\rho}{\partial \varphi}\right\}\vec{a}_z = j\omega\varepsilon^*\,\vec{E}\\ &= j\omega\varepsilon^*\{E_\rho \vec{a}_\rho + E_\varphi \vec{a}_\varphi + E_z \vec{a}_z\}\end{aligned} \tag{11.123}$$

By removing all field components except for the E_ρ and H_φ terms, the two Maxwell's equations reduce to

$$\left\{\frac{\partial E_\rho}{\partial z}\right\}\vec{a}_\varphi = \{-j\omega\mu^*\,H_\varphi \vec{a}_\varphi\} \tag{11.124}$$

$$\left\{-\frac{\partial H_\varphi}{\partial z}\right\}\vec{a}_\rho = \left\{j\omega\varepsilon^* E_\rho \vec{a}_\rho\right\} \tag{11.125}$$

Now, we differentiate the first equation with respect to z and make the substitution in the second, we get the second-order differential equation:

$$\left\{\frac{\partial^2 E_\rho}{\partial z^2}\right\} = -j\omega\mu^* \left\{\frac{\partial H_\varphi}{\partial z}\right\} = -j\omega\mu^* \left\{-j\omega\varepsilon^* E_\rho\right\} = -\omega^2\mu^*\varepsilon^*\{E_\rho\} = \gamma^2\{E_\rho\} \tag{11.126}$$

which takes on the familiar form of the wave equation.

Substituting Equation (11.120) in (11.126), we can write

$$\frac{\partial^2}{\partial z^2}\left\{\frac{C_1}{\rho} f_z(z)\right\} = \frac{C_1}{\rho}\frac{\partial^2}{\partial z^2}\{f_z(z)\} = \gamma^2\left\{\frac{C_1}{\rho} f_z(z)\right\} = \frac{C_1}{\rho}\gamma^2\{f_z(z)\} \tag{11.127}$$

Dividing both sides by $\frac{C_1}{\rho}$, we conclude

$$\frac{\partial^2}{\partial z^2}\{f_z(z)\} = \gamma^2\{f_z(z)\} \tag{11.128}$$

yielding the familiar solution

$$f_z(z) = C_3 e^{-\gamma z} + C_4 e^{+\gamma z} \tag{11.129}$$

and hence,

$$\begin{aligned} E_\rho(\rho, z) &= \left\{\frac{C_1}{\rho}C_3\right\}e^{-\gamma z} + \left\{\frac{C_1}{\rho}C_4\right\}e^{+\gamma z} = \left\{\frac{C^+}{\rho}\right\}e^{-\gamma z} + \left\{\frac{C^-}{\rho}\right\}e^{+\gamma z} \\ &= \left\{E_\rho^+(\rho)\right\}e^{-\gamma z} + \left\{E_\rho^-(\rho)\right\}e^{+\gamma z} \end{aligned} \tag{11.130}$$

Using Equation (11.124), we can derive the H_φ expression as

$$\begin{aligned} \{H_\varphi\} &= \frac{1}{-j\omega\mu^*}\left\{\frac{\partial E_\rho}{\partial z}\right\} = \frac{1}{-j\omega\mu^*}\left\{-\gamma\left\{E_\rho^+(\rho)\right\}e^{-\gamma z} + \gamma\left\{E_\rho^-(\rho)\right\}e^{+\gamma z}\right\} \\ &= \frac{\gamma}{j\omega\mu^*}\left\{\left\{E_\rho^+(\rho)\right\}e^{-\gamma z} - \left\{E_\rho^-(\rho)\right\}e^{+\gamma z}\right\} \end{aligned} \tag{11.131}$$

Substituting

$$\frac{\gamma}{j\omega\mu^*} = \frac{\sqrt{(j\omega\mu^*)(j\omega\varepsilon^*)}}{j\omega\mu^*} = \sqrt{\frac{j\omega\varepsilon^*}{j\omega\mu^*}} = \sqrt{\frac{\varepsilon^*}{\mu^*}} = \frac{1}{\eta} \tag{11.132}$$

$$\{H_\varphi\} = \left\{\frac{E_\rho^+(\rho)}{\eta}\right\}e^{-\gamma z} - \left\{\frac{E_\rho^-(\rho)}{\eta}\right\}e^{+\gamma z} = \left\{H_\varphi^+(\rho)\right\}e^{-\gamma z} - \left\{H_\varphi^-(\rho)\right\}e^{+\gamma z} \tag{11.133}$$

To summarize, the *TEM* wave solution in the coaxial line has the following field phasor expressions

$$E_\rho(\rho, z) = \left\{E_\rho^+(\rho)\right\}e^{-\gamma z} + \left\{E_\rho^-(\rho)\right\}e^{+\gamma z} \tag{11.134}$$

$$H_\varphi(\rho, z) = \left\{\frac{E_\rho^+(\rho)}{\eta}\right\}e^{-\gamma z} - \left\{\frac{E_\rho^-(\rho)}{\eta}\right\}e^{+\gamma z} \tag{11.135}$$

while

$$\gamma = \sqrt{-\omega^2\mu^*\varepsilon^*} = \alpha + j\beta \tag{11.136}$$

$$\eta = \sqrt{\frac{\mu^*}{\varepsilon^*}} \tag{11.137}$$

Although the above discussion was related to controlled-geometry coaxial lines, an almost-identical development can be made for *TEM* (plane wave) modes for all controlled-geometry transmission lines. The final results would be similar to those of Equations (11.134) and (11.135) except for replacing $E_\rho^+(\rho)$ and $E_\rho^-(\rho)$ by the proper ρ and φ dependent functions if using cylindrical coordinates (or x and y if working with Cartesian ones).

We notice that the solution [Equations (11.134) and (11.135)] resembles the one we derived in Chapter 2 using a circuit theory approach as listed in Table 11.3

$$V(z) = V^+ e^{-\gamma z} + V^- e^{+\gamma z} \tag{11.138}$$

$$I(z) = \frac{1}{Z_o} \cdot \{V^+ e^{-\gamma z} - V^- e^{+\gamma z}\} \tag{11.139}$$

$$\gamma = \sqrt{-\omega^2 L_z^* C_z^*} = \alpha + j\beta \tag{11.140}$$

$$Z_o = \sqrt{L_z^* / C_z^*} \tag{11.141}$$

To show the relationship between the two solutions, let us, without loss of generality, take on the $+z$ propagation terms on a lossless coaxial line for the sake of simplicity. The electric field term $E_\rho^+(\rho)$ simply corresponds to the V^+ through the quasi-static formula [Equation (5.56)].

$$V_{BA} = -\int_A^B \vec{E} \cdot \overrightarrow{d\ell} \tag{5.56}$$

Hence,

$$V^+ = V_{ab} = -\int_b^a E_\rho^+(\rho)\vec{a}_\rho \cdot \vec{d\rho} = \int_a^b \frac{C^+}{\rho} d\rho = \int_a^b \frac{C^+}{\rho} d\rho = C^+ \ln\frac{b}{a} \tag{11.142}$$

thus yielding

$$C^+ = V^+ / \ln\frac{b}{a} \tag{11.143}$$

making the field solutions of the form:

$$E_\rho^+(\rho, z) = \left\{ \frac{V^+}{\ln\frac{b}{a}} \cdot \frac{1}{\rho} \right\} e^{-\gamma z} \tag{11.144}$$

$$H_\varphi^+(\rho, z) = \left\{ \frac{1}{\eta} \cdot \frac{V^+}{\ln\frac{b}{a}} \cdot \frac{1}{\rho} \right\} e^{-\gamma z} \tag{11.145}$$

Furthermore, for the lossless coaxial line,

$$Z_o = \sqrt{L_z / C_z} \tag{11.146}$$

By using Equations (9A.28) and (6B.16) for the inductance and capacitance terms per unit length of the coaxial line, we get

$$L_{a<\rho<b \text{ per meter}} = \frac{\mu}{2\pi}\ln\left(\frac{b}{a}\right) \text{ H/m} \tag{9A.28}$$

$$C_{z,ab} = \frac{\psi_{z,ab \text{ per unit axial length}}}{V_{ab}} = \frac{2\pi\varepsilon}{\ln\left(b/a\right)} \text{ F/m} \tag{6B.16}$$

$$Z_o = \sqrt{\frac{\frac{\mu}{2\pi}\ln\left(\frac{b}{a}\right)}{\frac{2\pi\varepsilon}{\ln\left(b/a\right)}}} = \frac{\ln\left(b/a\right)}{2\pi}\sqrt{\frac{\mu}{\varepsilon}} = \eta\frac{\ln\left(b/a\right)}{2\pi} \text{ ohms} \tag{11.147}$$

11.11.2 Power Flow in Coaxial Lines

At this point, it is worthy to pursue the power flow in the coaxial line using the above derived field expressions [Equations (11.144) and (11.145)]. The power flow density as expressed by the Poynting vector is given by

$$\mathcal{P}_z^+ = E_\rho^+ \cdot H_\varphi^+ = \frac{1}{\eta}\cdot\left\{\frac{V^+}{\ln\frac{b}{a}}\cdot\frac{1}{\rho}\right\}^2 e^{-2\gamma z} = \frac{1}{\eta}\cdot\left\{\frac{V^+}{\ln\frac{b}{a}}\right\}^2\cdot\frac{1}{\rho^2}e^{-2\gamma z} \tag{11.148}$$

This expression reveals to us that the power flow is not uniform over the coaxial line cross section; it varies as the inverse square of the cylindrical radial distance from the axis, $\left(\frac{1}{\rho^2}\right)$ (Figure 11.14). This means that the power density is much higher closer to the inner conductor than what travels closer to the outer conductor. This finding explains why commercial coaxial cables are designed with higher conductivity inner conductors compared to the outer conductors.

Now, integrating the power flow density Equation (11.148) over the coaxial line cross section, $a<\rho<b$, we get the power flow as a function of the z position along the line

$$\begin{aligned} P_z^+ &= \iint_{\text{cross section}} [\mathcal{P}_z^+][dS] = \int_{\rho=a}^{\rho=b}\left[\frac{1}{\eta}\cdot\left\{\frac{V^+}{\ln\frac{b}{a}}\right\}^2\frac{1}{\rho^2}e^{-2\gamma z}\,[2\pi\rho\, d\rho]\right] \\ &= \frac{2\pi}{\eta}\left\{\frac{V^+}{\ln\frac{b}{a}}\right\}^2 e^{-2\gamma z}\int_{\rho=a}^{\rho=b}\left[\frac{1}{\eta}d\rho\right] = \frac{2\pi\cdot\{V^+\}^2}{\eta\cdot\ln\frac{b}{a}}e^{-2\gamma z} \end{aligned} \tag{11.149}$$

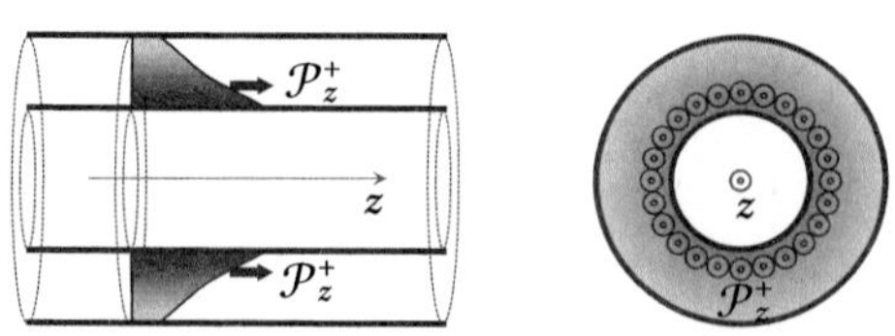

Figure 11.14 The distribution of power flow density over the TL cross section. Notice that the power flow density is much higher closer to the inner conductor than what travels closer to the outer conductor.

Finally, using Equation (11C.3), we conclude

$$P_z^+ = \frac{2\pi \cdot \{V^+\}^2}{\eta \cdot \ln\frac{b}{a}} e^{-2\gamma z} = \frac{\{V^+\}^2}{Z_o} e^{-2\gamma z} \tag{11.150}$$

which is the same expression as that derived using the circuit theory approach in Chapter 2.

Table 11.7 summarizes the $+z$ propagation terms on a lossless coaxial line.

	Circuit Solution	Field Solution	
Electric field and potential	$V^+(z) = V^+e^{-\gamma z}$	$E_\rho^+(\rho,z) = \left\{\frac{V^+}{\ln\left(\frac{b}{a}\right)} \cdot \frac{1}{\rho}\right\} e^{-\gamma z}$	(11.144)
Magnetic field and current	$I^+(z) = \frac{1}{Z_o}\{V^+e^{-\gamma z}\}$	$H_\varphi^+(\rho,z) = \left\{\frac{1}{\eta} \cdot \frac{V^+}{\ln\left(\frac{b}{a}\right)} \cdot \frac{1}{\rho}\right\} e^{-\gamma z}$	(11.145)
Capacitance and permittivity	$C_z = \frac{2\pi\varepsilon}{\ln(b/a)}$	ε	
Inductance and permeability	$L_z = \frac{\mu}{2\pi}\ln\left(\frac{b}{a}\right)$	μ	
Propagation constant	$\gamma = \sqrt{-\omega^2 L_z C_z} = \sqrt{-\omega^2\mu\varepsilon} = j\beta$	$\gamma = \sqrt{-\omega^2\mu\varepsilon} = j\beta$	
Characteristic and intrinsic impedance	$Z_o = \frac{\ln(b/a)}{2\pi}\sqrt{\frac{\mu}{\varepsilon}} = \eta\frac{\ln(b/a)}{2\pi}$	$\eta = \sqrt{\frac{\mu}{\varepsilon}}$	
Power flow and power flow density	$P_z^+ = \frac{\{V^+\}^2}{Z_o}e^{-2\gamma z}$	$\mathcal{P}_z^+ = \frac{1}{\eta} \cdot \left\{\frac{V^+}{\ln\frac{b}{a}} \cdot \frac{1}{\rho}\right\}^2 e^{-2\gamma z}$	(11.148)

TABLE 11.7 Summary of the Forward Propagation Terms on a Lossless Coaxial Line

11.12 Important Observation

11.12.1 Electromagnetic Power Flows in Dielectric Media

At this point in our study of electromagnetic wave propagation, it is important to highlight the finding that associated power flow occurs in dielectrics. It is clear from the Poynting vector Equation (11.76) that the power density occurs where the electric and magnetic fields coexist (in a nonparallel form) in a dielectric region. The role of the conductors in this process is to guide (and in some cases guide and confine) the electromagnetic wave propagation.

11.12.2 Conductors Provide Guidance (and Confinement)

This is contrary to misconceptions typically associated with the use of circuit theory models. Using circuit theory models, with its limited physical insight, we get the impression that the conductors carry the power since they are responsible for the presence of the voltage and current terms. The fact of the matter is that electromagnetic waves occur within nonperfect conductors and result in power dissipation and signal attenuation. The bulk of power flow is then what travels within the dielectric region.

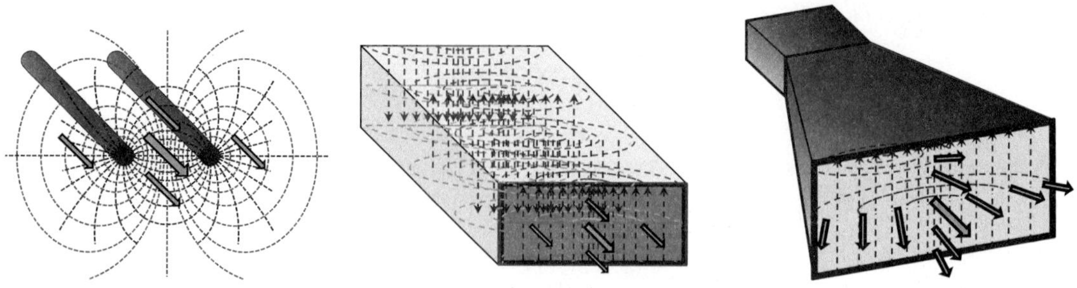

Parallel Wire Transmission Line Power Flow in Dielectric Conductors Provide Guidance

Rectangular Waveguide Power Flow in Dielectric Filling – Conductors Provide Guidance and Confinement

Rectangular Horn Antenna Power Flow in Dielectric Conductors Provide Directivity (Initial Guidance)

FIGURE 11.15 Examples to demonstrate that electromagnetic power flows only in dielectric media (not in the conductor) and that conductors are only used to provide guidance (and confinement).

The above point is graphically demonstrated in Figures 11.12 and 11.14. In Figure 11.12, the power flow in the dielectric media with no conductors present in its domain. While in Figure 11.14, we have a coaxial conductor that serves to guide and confine the electromagnetic wave propagation along the transmission line within the enclosed dielectric region.

Figure 11.15 demonstrates some forms of electromagnetic wave propagation in dielectrics, including some examples of conductors used in guiding /guiding and confining the power flow.

ADDENDUM 11A
DERIVATION OF THE LAPLACIAN FORM OF THE WAVE EQUATION

11A.1 Introduction

In the body of this chapter, we used the one-dimensional form of Maxwell's equations to develop the wave equation. In the following, we will start with the 3D curl forms of Maxwell's equation for the same development. The purpose is to derive the more general (Laplacian) form of the wave equation.

In this regard, we start with Maxwell's equations for the general form using both the complex permittivity and permeability:

$$\vec{\nabla} \times \vec{H} = j\omega\varepsilon^* \vec{E} \tag{11A.1}$$

$$\vec{\nabla} \times \vec{E} = -j\omega\mu^* \vec{H} \tag{11A.2}$$

$$\vec{\nabla} \cdot \vec{D} = 0 = \vec{\nabla} \cdot \vec{E} \tag{11A.3}$$

$$\vec{\nabla} \cdot \vec{B} = 0 = \vec{\nabla} \cdot \vec{H} \tag{11A.4}$$

The above set of equations, in conjunction with the yet to be defined appropriate boundary conditions, fully characterize the spatial field distribution for the case in consideration. We start by manipulating the two differential equations [Equations (11A.1) and (11A.2)] in order to eliminate one of the two fields, $\vec{E}$ or $\vec{H}$. In the following, we chose to eliminate the magnetic field $\vec{H}$. To do that, we apply the curl differential operator to Equation (11A.2) and use the curl-curl vector identity to obtain

$$\vec{\nabla} \times \vec{\nabla} \times \vec{E} = \vec{\nabla} \times [-j\omega\mu^* \vec{H}] \tag{11A.5}$$

$$\vec{\nabla}[\vec{\nabla} \cdot \vec{E}] - \nabla^2[\vec{E}] = -j\omega\mu^* [\vec{\nabla} \times \vec{H}] \tag{11A.6}$$

and, using Equation (11A.3), we can write

$$-\nabla^2\vec{E} = -j\omega\mu^*[\vec{\nabla}\times\vec{H}] \tag{11A.7}$$

It is relevant to remember that we are assuming that the medium under consideration is homogeneous (with constant ε^* and μ^* everywhere), hence, we can extract these constant material parameters outside the spatial differential operations (grad, div, or curl).

Finally, substituting the $\vec{\nabla}\times\vec{H}$ of Ampere's law [Equation (11A.1)], we arrive at the second-order spatial differential equation for the vector electric field phasor, $\vec{E}$

$$-\nabla^2\vec{E} = -j\omega\mu^*[\vec{\nabla}\times\vec{H}] = -j\omega\mu^*[j\omega\varepsilon^*\vec{E}] = \omega^2\varepsilon^*\mu^*\vec{E} \tag{11A.8}$$

$$\nabla^2\vec{E} = -\omega^2\varepsilon^*\mu^*\vec{E} \tag{11A.9}$$

Using the same definition of the complex propagation constant $\gamma = \sqrt{-\omega^2\mu^*\varepsilon^*}$ as was defined earlier in Table 11.5, we can rewrite the above wave equation in its *final* form

$$\nabla^2\vec{E} = \gamma^2\vec{E} \tag{11A.10}$$

which is the Laplacian form of the wave equation for the electric field phasor vector $\vec{E}$. The one-dimensional version derived earlier [Equation (11.19)] is a special case of Equation (11A.9).

In a similar fashion, if we were to eliminate $\vec{E}$ instead of $\vec{H}$, we would have obtained a similar form for the wave equation for the magnetic field phasor vector $\vec{H}$; specifically,

$$\nabla^2\vec{H} = \gamma^2\vec{H} \tag{11A.11}$$

11A.2 One-Dimensional Form of Wave Equation

In this section, we take on the special case of plane wave propagation in homogeneous media that led to the forms developed earlier in Table 11.5.

Since we are working with a 1D case, we can follow similar steps to those used in our earlier discussion in the development of Equations (11.9) through (11.19).

To do so, we express Equation (11A.10) in Cartesian coordinate as follows:

$$\nabla^2\begin{bmatrix}\bar{E}_x(x,y,z)\vec{a}_x\\ \bar{E}_y(x,y,z)\vec{a}_y\\ \bar{E}_z(x,y,z)\vec{a}_z\end{bmatrix} = \nabla^2\begin{bmatrix}\bar{E}_x(z)\vec{a}_x\\ \bar{E}_y(z)\vec{a}_y\\ \bar{E}_z(z)\vec{a}_z\end{bmatrix} = \gamma^2\begin{bmatrix}\bar{E}_x(z)\vec{a}_x\\ \bar{E}_y(z)\vec{a}_y\\ \bar{E}_z(z)\vec{a}_z\end{bmatrix} \tag{11A.12}$$

We notice that we removed the functional dependence on x and y in the above expressions due to the infinite extension of the source and media in both transverse coordinates.

Next, we remove the unit vectors from the expressions as they factor out of both sides of the equation. We will also express the Laplacian operator in its Cartesian form while eliminating the field derivatives with respect to both x and y since field changes are assumed to occur only along the z coordinate.

$$\nabla^2\begin{bmatrix}\bar{E}_x(z)\\ \bar{E}_y(z)\\ \bar{E}_z(z)\end{bmatrix} = \left(\frac{\partial^2}{\partial x^2}+\frac{\partial^2}{\partial y^2}+\frac{\partial^2}{\partial z^2}\right)\begin{bmatrix}\bar{E}_x(z)\\ \bar{E}_y(z)\\ \bar{E}_z(z)\end{bmatrix} = \left(\frac{\partial^2}{\partial z^2}\right)\begin{bmatrix}\bar{E}_x(z)\\ \bar{E}_y(z)\\ \bar{E}_z(z)\end{bmatrix} = \gamma^2\begin{bmatrix}\bar{E}_x(z)\\ \bar{E}_y(z)\\ \bar{E}_z(z)\end{bmatrix} \tag{11A.13}$$

So far, this Equation (11A.13) is identical to that developed earlier in Equation (11.20). The remaining steps to conclude the final governing expressions for all electric and magnetic fields are identical to those developed earlier:

- Using Equations (11.9) through (11.14) to concluded that both $\bar{E}_z$ and $\bar{H}_z$ are zeros.
- Choosing the x coordinate to line up with the $\vec{E}$ field vector, we set $\bar{E}_y = 0$.
- According to Equation (11.15), $\bar{H}_x$ also becomes zero.

Hence, we can finally conclude that Equation (11A.13) reduces to

$$\left(\frac{\partial^2}{\partial z^2}\right)\begin{bmatrix} \bar{E}_x(z) \\ 0 \\ 0 \end{bmatrix} = \gamma^2 \begin{bmatrix} \bar{E}_x(z) \\ 0 \\ 0 \end{bmatrix} \tag{11A.14}$$

and our system of equations is now reduced to the two equations

$$\frac{\partial^2}{\partial z^2}\bar{E}_x(z) = \gamma^2 \cdot \bar{E}_x(z) \tag{11A.15}$$

$$\bar{H}_y(z) = \frac{j}{\omega\mu^*}\frac{\partial \bar{E}_x(z)}{\partial z} \tag{11A.16}$$

which are identical to the earlier developed forms [Equations (11.20) and (11.21)].

ADDENDUM 11B
SKIN EFFECT AND SHIELDING

11B.1 Skin Depth

By now, we should have recognized that electromagnetic waves have limited abilities in penetrating (and propagating) in highly conductive media. The reason is that the waves attenuate severely as they are launched into conductive media. In fact, they cannot penetrate (or exist in) perfect conductors. A measure of wave penetration into the media is given by the "skin depth, δ," which decreases with an increase of either the material conductivity, σ, or the operating frequency, f, (or both!).

$$\delta \cong 1/\sqrt{\pi f \mu_o \sigma} \tag{11.62}$$

11B.2 Shielding

This finding has a very important application in shielding electromagnetic waves from reaching an enclosure of space. An equally important application is the confinement of electromagnetic waves in an enclosure of space. An example of the first one is an RF screen room, in which a room enclosure is kept free from outside electromagnetic activities. Such room is used for performing RF experiments and measurements free from interference of other EM waves and noise as demonstrated in Figure 11B.1.

Conversely, the screen room will also protect the external surroundings from EM activities within the room. This concept is demonstrated in the microwave oven example presented next.

The microwave oven, demonstrated in Figure 11B.2, is an example of an enclosure that confines electromagnetic waves within. This way, we maximize the oven cooking efficiency by not allowing EM energy to leak out, in the meantime keeping the environment outside free of hazardous EM radiation that is harmful to humans.

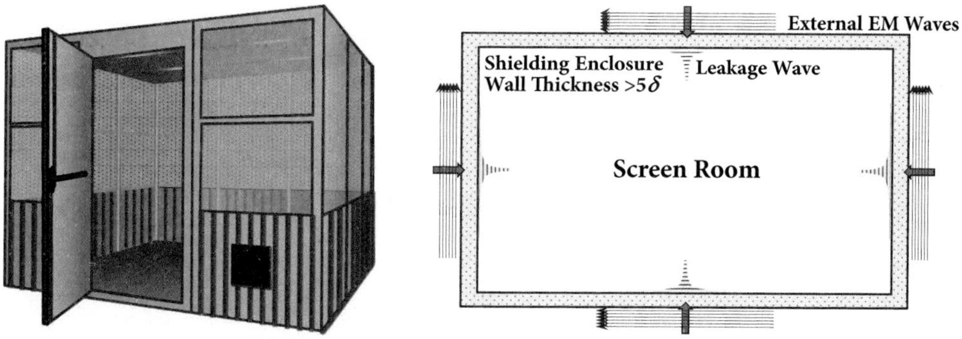

FIGURE 11B.1 A demonstration of an RF screen room where the enclosure is kept free from outside electromagnetic activities.

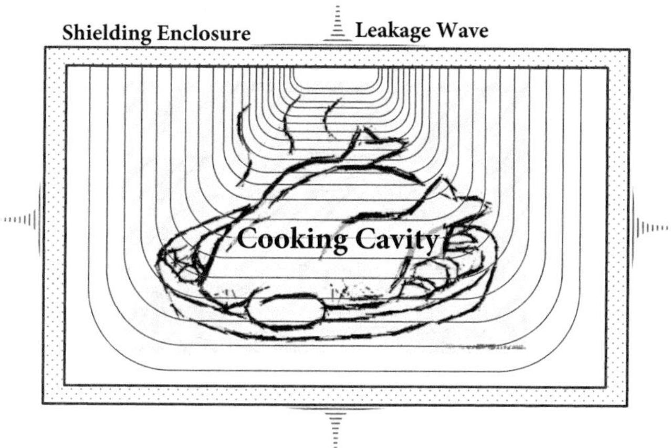

FIGURE 11B.2 The microwave oven is an example of an enclosure that confines electromagnetic waves within.

Another relevant application of shielding is that of coaxial transmission lines which are typically used in communication systems, for example, cable TV transmission. As demonstrated in Figure 11B.3, EM waves propagating within the space confined by the coaxial conductors are isolated from external EM waves. In such arrangement, we minimize energy leakage and cross talk between the inside and outside zones of the coaxial line.

For effective shielding, the conductivity and thickness of the shielding conductors must be chosen properly. Figure 11B.4 demonstrates the effect of the thickness on the shielding performance. Typically, it is best to design the conductor thickness to be greater than 5δ, five times the skin depth parameter.

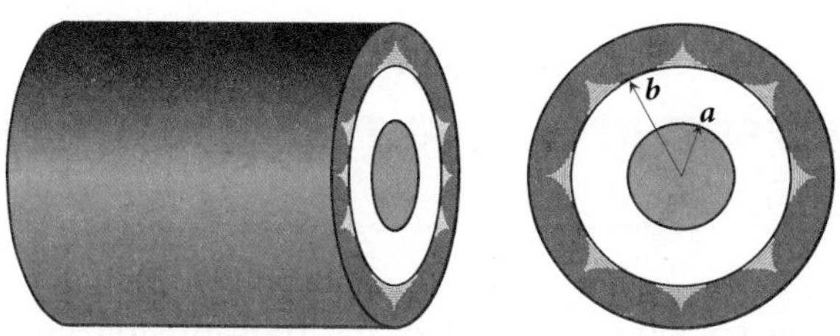

FIGURE 11B.3 The outer conductor of a coaxial line needs to be much thicker than the skin depth to ensure minimum leakage of its waves to its outer vicinity.

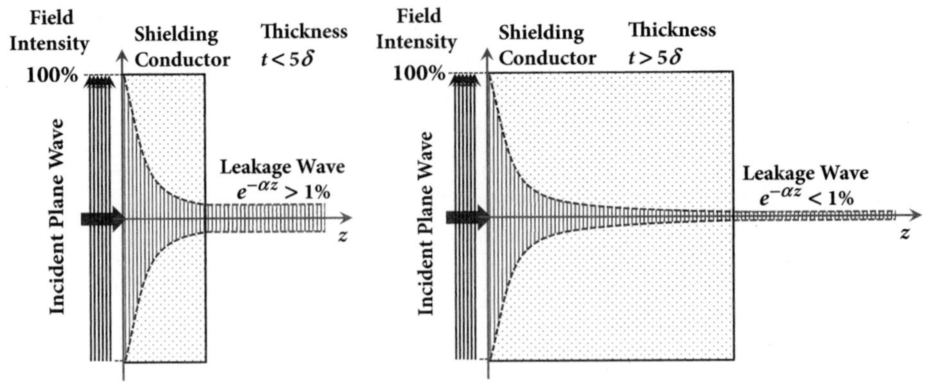

FIGURE 11B.4 The conductor thickness needs to be much thicker than 5δ to ensure that leakage waves drop below 1%.

Conductor Material		Silver	Copper	Gold	Aluminum	Units
Conductivity, $\sigma (\times 10^7)$		6.3	5.8	4.4	3.6	S/m
	Frequency					
Skin depth, δ	1 MHz	0.063	0.066	0.076	0.084	mm
	10 MHz	0.020	0.021	0.024	0.027	mm
	100 MHz	6.34	6.61	7.59	8.39	μm
	1 GHz	2.01	2.09	2.40	2.65	μm
	10 GHz	0.63	0.66	0.76	0.84	μm
	100 GHz	0.20	0.21	0.24	0.27	μm
Minimum conductor thickness, 5δ	1 MHz	0.317	0.330	0.379	0.419	mm
	10 MHz	0.100	0.104	0.120	0.133	mm
	100 MHz	31.70	33.04	37.94	41.94	μm
	1 GHz	10.03	10.45	12.00	13.26	μm
	10 GHz	3.17	3.30	3.79	4.19	μm
	100 GHz	1.00	1.04	1.20	1.33	μm

TABLE 11B.1 A Tabulation of the Skin Depth 5δ and the 5δ Values for Commonly Used (Good) Conductors for Different Frequencies

Table 11B.1 demonstrates the approximate minimum thicknesses for shielding for some of the good conductors. The table presents the conductor thicknesses for operating frequencies ranging from 1 MHz to 100 GHz.

ADDENDUM 11C
SKIN EFFECT IN COAXIAL TRANSMISSION LINES

11C.1 Introduction

In this addendum, we discuss the skin effect in coaxial transmission lines and its effect on the line performance. Specifically, we will use the approximate plane wave skin effect model to derive modified expressions for the line's RLGC parameters. These modified expressions are expected to be frequency

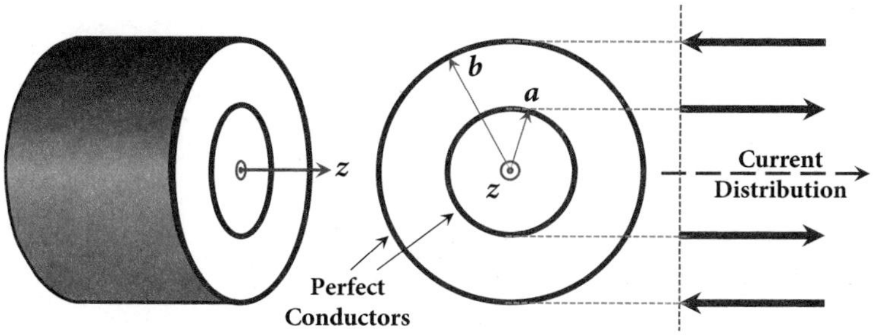

FIGURE 11C.1 Current distribution in a perfect conductor coaxial line.

dependent and so will be the corresponding complex propagation and characteristic impedance parameters.

To start, we remind ourselves that the wave solution presented in this chapter assumed the line conductors to be perfect. Hence, the coaxial line depicted earlier in Figure 11.13 was shown with "practically" no thickness as its infinite conductivity would block out EM wave penetration in any conductor regions where $\rho < a$ and $\rho > b$. With "effectively zero" conductor thicknesses, all electric currents will flow on the surfaces $\rho = a$ and $\rho = b$ and no currents will exist within the rest of conductor regions $\rho < a$ and $\rho > b$. This concept is demonstrated in Figure 11C.1.

Practically speaking, the coaxial line will use *good* conductors (rather than perfect conductors) with nonzero thicknesses (Figure 11C.2). These conductors will possess *finite* conductivities and hence the EM waves will have *some* presence within conductor regions (beyond the surfaces) of both the inner and outer conductors. At low operating frequencies, with DC as the extreme, the EM waves will have significant to full penetration in the conducting regions. Electric currents would exist throughout the conductors' entire (or significant parts of the) cross sections. At DC, the current distributions would have uniform densities over the conductors' cross sections. However, at relatively high operating frequencies, the wave penetration would have limited penetration within the conducting bodies (skin effect). Hence, the electric currents are expected to have decaying densities over limited thicknesses of the conductors next to the $\rho = a$ and $\rho = b$ surfaces. Figure 11C.2 demonstrates current distribution models at both DC (and very low frequencies) as well as at relatively high frequencies.

Using the skin depth model discussed earlier in Figure 11.9, we can model the current distribution at relatively high frequencies as uniform over the "equivalent wave skin penetration" thickness, δ. This is demonstrated in Figure 11C.3.

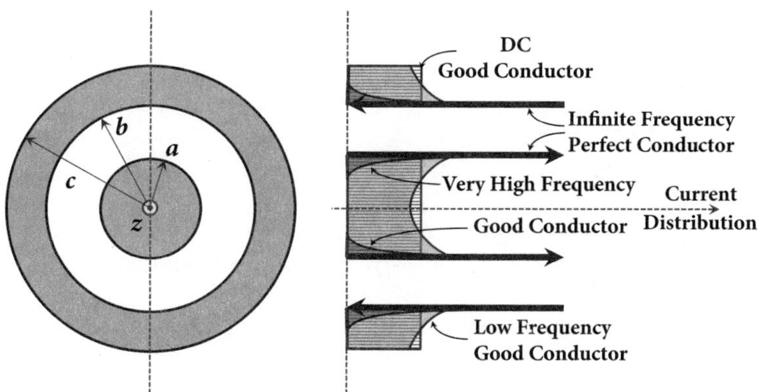

FIGURE 11C.2 Current distribution in a good conductor coaxial line.

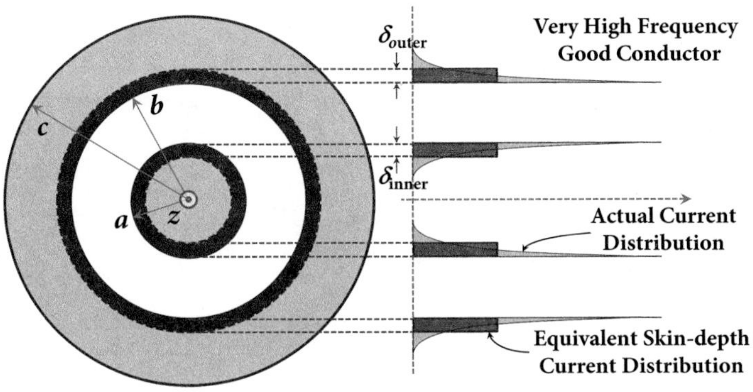

FIGURE 11C.3 Equivalent RF current distribution in a good conductor coaxial line is modeled as uniform over the skin depth.

11C.2 High-Frequency Coaxial Line Parameters

Using the skin-depth current distribution "equivalency" as demonstrated above, and referring to Figure 11C.4, we can approximate the high-frequency coaxial line parameters as follows:

R_z: is the sum of both $R_{z,\text{ inner}}$ and $R_{z,\text{ outer}}$ corresponding to the per unit length resistance of the inner and outer conductors, respectively. Now, let us recall the low-frequency (and DC) expression for R_z that was given in Equation (6A.12)

$$R_{z,\text{DC}} = \frac{1}{\sigma_{\text{inner}} S_{\text{inner}}} + \frac{1}{\sigma_{\text{outer}} S_{\text{outer}}} = \frac{1}{\sigma_{\text{inner}} \pi a^2} + \frac{1}{\sigma_{\text{outer}} \pi (c^2 - b^2)} \tag{11C.1}$$

We now have different expressions for S_{inner} and S_{outer} since the two resistances $R_{z,\text{ inner}}$ and $R_{z,\text{ outer}}$ correspond to the dark-shaded cross-sectional areas around the conductor perimeters with thicknesses equal to the relative conductor's skin depth Figure 11C.4.

$$R_{z,\text{ inner}} = \frac{1}{\sigma_{\text{inner}} \cdot S_{\text{inner}}} \cong \frac{1}{\sigma_{\text{inner}} \cdot (2\pi a \cdot \delta_{\text{inner}})} \tag{11C.2}$$

$$R_{z,\text{ outer}} = \frac{1}{\sigma_{\text{outer}} \cdot S_{\text{outer}}} \cong \frac{1}{\sigma_{\text{outer}} \cdot (2\pi b \cdot \delta_{\text{outer}})} \tag{11C.3}$$

$$R_z = R_{z,\text{ inner}} + R_{z,\text{ outer}} \cong \frac{1}{\sigma_{\text{inner}} \cdot (2\pi a \cdot \delta_{\text{inner}})} + \frac{1}{\sigma_{\text{outer}} \cdot (2\pi b \cdot \delta_{\text{outer}})} \ \Omega/\text{m} \tag{11C.4}$$

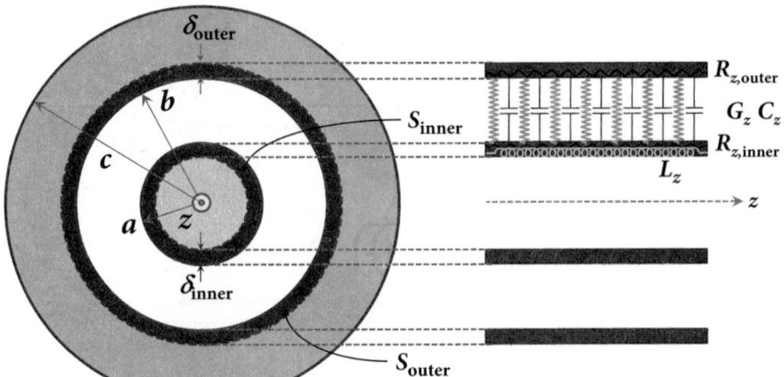

FIGURE 11C.4 Evaluation of the resistance of the TL conductors per unit axial length.

	DC and Low Frequencies	High Frequencies	Units	
R_z	$\frac{1}{\sigma_i(\pi a^2)}+\frac{1}{\sigma_o(\pi[c^2-b^2])}$	$\cong\frac{1}{\sigma_i(2\pi a\delta_i)}+\frac{1}{\sigma_o(2\pi b\delta_o)}$	Ω/m	(11C.4)
L_z	$\frac{\mu}{2\pi}\ln\left(\frac{b}{a}\right)$	$\cong\frac{\mu}{2\pi}\ln\left(\frac{b}{a}\right)$	H/m	(9A.28)
G_z	$\frac{2\pi\sigma_d}{\ln\left(\frac{b}{a}\right)}$	$\frac{2\pi\sigma_d}{\ln\left(\frac{b}{a}\right)}$	S/m	(6A.15)
C_z	$\frac{2\pi\varepsilon}{\ln(b/a)}$	$\frac{2\pi\varepsilon}{\ln(b/a)}$	F/m	(6B.16)

Table 11C.1 Summary of the RLCG Parameters of a Good Conductor TL at Both Low and High Frequencies

L_z: is the per unit length inductance of the coaxial arrangement. In principle, it is the sum of three inductances representing the magnetic energy presence in the three regions $a<\rho<b$, $a-\delta_{\text{inner}}<\rho<a$, and $b<\rho<b+\delta_{\text{outer}}$. With the *effective* conductor thicknesses δ_{inner} and δ_{outer} fairly small, the internal inductances of these regions can be neglected. Thus, L_z can be approximated to its "$a<\rho<b$" expression derived earlier in Equation (9A.28).

$$L_z \cong \frac{\mu}{2\pi}\ln\left(\frac{b}{a}\right)\ \text{H/m} \tag{9A.28}$$

Finally, expressions for G_z and C_z are not expected to be altered from their earlier forms (DC, low frequency, or quasi-static) since they represent the performance of the dielectric region, $a<\rho<b$. Hence, we borrow their expressions from

$$G_z = \frac{2\pi\sigma_d}{\ln\left(\frac{b}{a}\right)}\ \text{S/m} \tag{6A.15}$$

$$C_z = \frac{2\pi\varepsilon}{\ln(b/a)}\ \text{F/m} \tag{6B.16}$$

To summarize, Table 11C.1 gives all the RLGC model expressions at both operating frequency extremes.

ADDENDUM 11D
LOSS TANGENT FOR ENERGY-STORAGE MEDIA (MATERIALS) AND DEVICES

11D.1 Introduction

In Table 11.5, we introduced the complex permittivity and complex permeability of materials to facilitate the analysis of wave propagation in lossy materials. The expressions we used for these complex material properties were the complex permittivity ($\varepsilon^* = \varepsilon' - j\varepsilon''$) and the complex permeability ($\mu^* = \mu' - j\mu''$).

In the same table, Table 11.5, we showed analogous expressions for complex expressions for both the capacitance and inductance terms of the RLGC transmission line model: $C_z^* = C_z - j\frac{G_z}{\omega}$ and $L_z^* = L_z - j\frac{R_z}{\omega}$

that include their respective losses along with the real capacitance and inductance values, respectively. In principle, we can generalize the concept of the complex capacitance and complex inductance for all forms of practical (lossy) capacitors and inductors and not just those associated with the transmission line model. Hence, we write the analogous expressions

$$C^* = C - j\frac{G}{\omega} \quad \text{and} \quad L^* = L - j\frac{R}{\omega} \tag{11D.1}$$

and their corresponding admittance and impedance expressions are

$$Y_{C^*} = j\omega C^* = G + j\omega C \quad \text{and} \quad Z_{L^*} = j\omega L^* = R + j\omega L \tag{11D.2}$$

To further our understanding of this analogy, we need to realize that it is not just a mathematical analogy; it is truly a physical one as well. To explain, let us consider the case of a parallel plate capacitor as an example. Figure 11D.1 depicts an idealized parallel plate capacitor with plate area S and a spacing d. The capacitor's dielectric medium has a complex dielectric permittivity of $\varepsilon^* = \varepsilon' - j\varepsilon''$.

There are two field related phenomena associated with the two components ε' and ε''. First, the ε' component is associated with the enegy storage in the capacitor structure, and hence the capacitive circuit model expression as per our earlier derivation [Equation (6B.1)]

$$C = \varepsilon'\left(\frac{S}{d}\right) \tag{11D.3}$$

Meanwhile, the material's ε'' (due to its conductivity) is responsible for power dissipation in the capacitor device that we model as a parallel conductance, G, given by

$$G = \sigma\left(\frac{S}{d}\right) = \omega\varepsilon''\left(\frac{S}{d}\right) \tag{11D.4}$$

Hence, we can write the combined (complex) admittance of the lossy capacitor device as

$$Y = G + j\omega C = \omega\varepsilon''(S/d) + j\omega\varepsilon'(S/d) = j\omega[\varepsilon' - j\varepsilon''](S/d) = j\omega\varepsilon^*(S/d) \tag{11D.5}$$

This suggests the use of the following expression of the complex capacitance, C^*,

$$Y = G + j\omega C = j\omega[\varepsilon' - j\varepsilon''](S/d) = j\omega\varepsilon^*(S/d) = j\omega C^* \tag{11D.6}$$

Dividing the above equation by $j\omega$, we can wirte

$$C^* = \varepsilon^*(S/d) = [\varepsilon' - j\varepsilon''](S/d) = C - j\frac{G}{\omega} = C' - jC'' \tag{11D.7}$$

In which we expressed C^* as the complex expression

$$C^* = C' - jC'' \tag{11D.8}$$

Now, we notice the imaginary term

$$C'' = \frac{G}{\omega} = \frac{\sigma}{\omega}(S/d) = \varepsilon''(S/d) \tag{11D.9}$$

If we were to consider other capacitor configurations, the only change in the above equation would be the term S/d which would be replaced by the proper configuration (geometrical) term. The analogy remains valid as demonstrated in Table 11D.1.

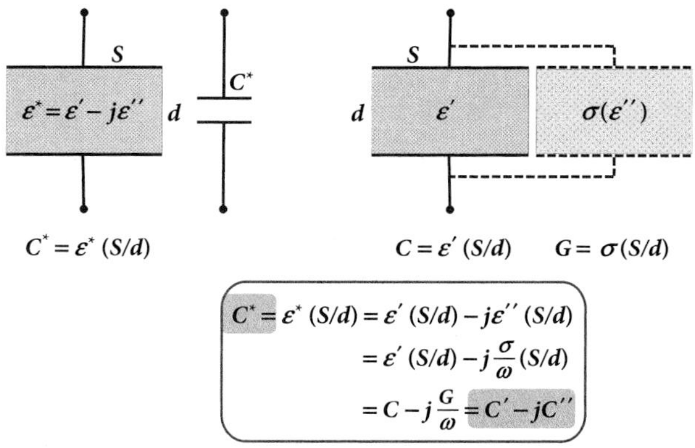

FIGURE 11D.1 The complex circuit model of a capacitor using good dielectric showing both the capacitance and the internal conductance.

	Material Property	Device Property	
Energy storage (permittivity/capacitance)	ε'	C'	
Power dissipation (conductivity/conductance)	$\varepsilon'' = \dfrac{\sigma}{\omega}$	$C'' = \dfrac{G}{\omega}$	
Complex permittivity complex capacitance	$\varepsilon^* = \varepsilon' - j\varepsilon''$	$C^* = C' - jC''$	(11D.7)

TABLE 11D.1 Analogy between Material Bulk Properties and Those of Capacitive Devices

11D.2 The Loss Tangent

The loss tangent is an expression defined to express the quality of an energy-storage material (or device) in terms of its deviation from the perfect lossless state. Figures 11D.2 and 11D.3 demonstrate the loss tangent definitions for both electrical and magnetic energy-storage materials/devices, respectively. In both figures, the energy-storage and the power-dissipation components are represented by two orthogonal phasors with the resultant phasor making a "preferably small" angle delta with the energy-storage component. For a perfect lossless energy-storage material/device, the angle delta should be zero.

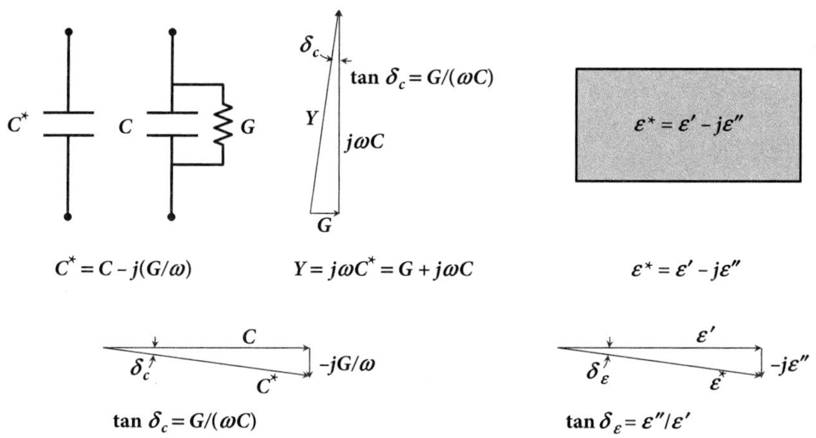

FIGURE 11D.2 The loss tangent (tan δ) of a capacitor is the same as that of its dielectric.

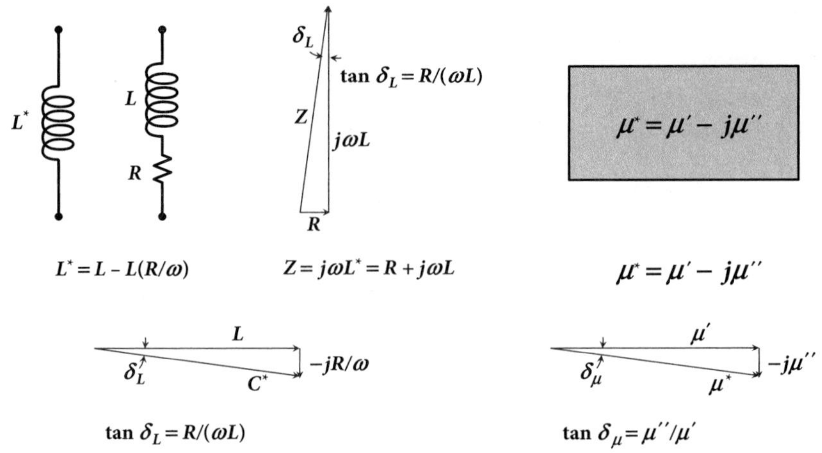

FIGURE 11D.3 The loss tangent (tan δ) of an inductor is the same as that of its core material.

The tangent of the angle δ representing the ratio between the power-dissipation and the energy-storage components in the phasor diagrams is what we use to define the loss tangent (aka the loss factor or the loss coefficient). High-quality (near perfect) energy-storage materials/devices should have extremely small loss tangent.

The case of electrical energy-storage dielectric material/capacitive device is demonstrated in the Figure 11D.2. In this figure, we can see that the complex phasor diagrams of ε^* and C^* demonstrates the loss tangent definition as

$$\tan\delta_\varepsilon = \varepsilon''/\varepsilon' \quad \text{and} \quad \tan\delta_C = G/(\omega C) \tag{11D.10}$$

For the case of magnetic energy-storage magnetic material/inductive device as shown in Figure 11D.3, the complex phasor diagrams of μ^* and L^* yields the loss tangent definition as

$$\tan\delta_\mu = \frac{\mu''}{\mu'} \quad \text{and} \quad \tan\delta_L = R/(\omega L) \tag{11D.11}$$

CHAPTER 11 PROBLEMS

- Problems 11.1 to 11.7: Problems dealing with wave propagation parameters
- Problems 11.8 to 11.10: Problems dealing with skin depth and shielding
- Problems 11.11 to 11.18: Problems dealing with Poynting vector and power flow
- Problems 11.19 to 11.24: Problems dealing with high frequency TL

11.1 Given the electric field vector of a uniform plane wave propagating in a dielectric medium $(\varepsilon_r^* \varepsilon_o, \mu_o)$ as: $\vec{E}(z,t) = \vec{a}_x Re\{\exp(j2\pi \times 10^9 t)[6\exp(-j30z) - 4\exp(+j30z)]\}$ V/m. Find

a. The operating frequency
b. The propagation phase coefficient
c. The phase velocity
d. The attenuation coefficient
e. The wavelength
f. The complex relative permittivity
g. The magnetic field vector phasor

11.2 Consider a uniform electromagnetic plane wave with in a homogeneous dielectric medium with permittivity $\varepsilon = \varepsilon_r \varepsilon_o$ and $\mu = \mu_o$ with $\vec{E}(x,y,z,t) = \cos(6\pi \times 10^8 t - 10\pi y)\vec{a}_x + \sin(6\pi \times 10^8 t - 10\pi y)\vec{a}_z$.

a. Determine the associated magnetic field in phasor form.
b. Determine the wavelength of this wave.
c. Determine the relative dielectric constant ε_r of the medium.
d. Determine the direction of propagation of the electromagnetic wave in this medium.

11.3 Given the magnetic field vector of a uniform plane wave propagating in a dielectric medium $(\varepsilon_r^* \varepsilon_o, \mu_o)$ as

$$\vec{H}(z,t) = \vec{a}_x \left\{ 0.5\exp(-0.01z)\cos\left(4\pi \cdot 10^9 t - 60z + \frac{\pi}{2}\right) - \exp(0.01z)\cos\left(4\pi \cdot 10^9 t + 60z + \frac{\pi}{3}\right) \right\} \text{ A/m.}$$

Find

a. The operating frequency
b. The propagation phase coefficient
c. The phase velocity
d. The attenuation coefficient
e. The wavelength
f. The complex relative permittivity
g. The electric field vector phasor

11.4 Consider a uniform plane wave propagating in a lossless nonmagnetic material with $\varepsilon = 25\varepsilon_o$. The magnetic field vector is given by $\vec{H}(x,t) = \vec{a}_y \left\{ 0.2\cos\left(\omega t - \pi x + \frac{\pi}{2}\right) \right\}$. Find

a. The direction of propagation
b. The phase velocity
c. The operating frequency
d. The wavelength
e. The instantaneous electric field vector

11.5 The electric field of a 1-MHz plane wave traveling in the positive z direction in air is polarized in the x direction. The peak value of the electric field is 1.2π V/m at $t = 0$ and $z = 50$ m. Determine the following:

a. The propagation phase coefficient

b. The wavelength

c. Expressions for the instantaneous and phasor electric field vector

d. Expressions for the instantaneous and phasor magnetic field vector

11.6 For each of the following combinations of parameters, determine if the material is a low-loss dielectric, a quasi-conductor, or a good conductor and then calculate α, β, c_{ph} and η

Hint: A good conductor has $\left(\frac{\sigma}{\omega\varepsilon} = \frac{\varepsilon''}{\varepsilon'}\right) > 100$, a low-loss dielectric has $\left(\frac{\sigma}{\omega\varepsilon} = \frac{\varepsilon''}{\varepsilon'}\right) < 0.01$ and a quasi-conductor has $0.01 \le \left(\frac{\sigma}{\omega\varepsilon} = \frac{\varepsilon''}{\varepsilon'}\right) \le 100$.

a. Glass with $\varepsilon_r = 5$, $\mu_r = 1$, and $\sigma = 10^{-12}$ S/m at 10 GHz

b. Wood with $\varepsilon_r = 3$, $\mu_r = 1$, and $\sigma = 10^{-4}$ S/m at 1 kHz

c. Dry concrete, $\varepsilon_r' = 7$, $\tan\delta = 0.4$, $\mu = \mu_o$

d. Wet concrete, $\varepsilon_r' = 100$, $\tan\delta = 1$, $\mu = \mu_o$

e. Distilled water (use the material properties given in Appendix C)

f. Sea water (use the material properties given in Appendix C)

11.7 In a nonmagnetic, lossy dielectric medium, a 300-MHz plane wave is characterized by the electric field phasor: $\bar{E}(y) = (\hat{a}_x - j\hat{a}_z)e^{-2y}e^{-j9y}$ V/m.

Find

a. ε', ε'' and the complex intrinsic impedance of the material, η.

b. Obtain a time-domain, instantaneous, expressions for the electric and magnetic field vectors.

c. At what distance does the magnitude of the electric field drop to 10% of its value at $y = 0$?

11.8 A uniform plane wave is incident normal to a concrete wall. Determine the depth of penetration at 0.1, 1, and 10 GHz, given that $\varepsilon_r' = 100$, $\tan\delta = 1$, and $\mu = \mu_o$. Repeat for the following cases:

a. Dry concrete, $\varepsilon_r' = 7$, $\tan\delta = 0.4$, $\mu = \mu_o$

b. Wet concrete, $\varepsilon_r' = 100$, $\tan\delta = 1$, $\mu = \mu_o$

11.9 A uniform plane wave is incident normal to a dry concrete wall, 10 cm thick. Determine the power level of the transmission (in dB relative to the incident wave) at the other side of the wall at 1 GHz. Repeat for wet concrete. (Ignore reflections at the discontinuities.)

a. Dry concrete, $\varepsilon_r' = 7$, $\tan\delta = 0.4$, $\mu = \mu_o$

c. Wet concrete, $\varepsilon_r' = 100$, $\tan\delta = 1$, $\mu = \mu_o$

Hint: You can assume that the wave has an electric field polarized in the x direction and is propagating in the positive z direction. The amplitude of the incident electric field is 1 V/m at the surface of the concrete wall. Find $\vec{E}(z,t)$ and $\vec{P}_{avg}(z)$. Assume $z = 0$ is the surface of the concrete wall. $\vec{P}_{avg}(z = 0)$ is the average power on the surface and $\vec{P}_{avg}(z = 10 \text{ cm})$ is the average power on the other side of the wall.

11.10 Calculate the depth of penetration for a 1-GHz plane wave incident on the following list of materials. Also, calculate the minimum thickness needed to achieve good shielding.

a. An aluminum sheet

b. An iron sheet

c. A glass sheet

d. Surface of the earth

e. Surface of the ocean

Use the material properties given in Appendix C.

11.11 A uniform plane wave is traveling downward in the +z direction in fresh water. $z=0$ is just below the surface. $\varepsilon_r=80$, $\mu_r=1$, and $=2\times10^{-4}$ S/m. If the magnetic field vector at $z=0$ is given by $\vec{H}(0,t)=\vec{a}_y 100\cos(10\pi\times10^7 t)$ mA/m. Find

a. The instantaneous magnetic field vector, $\vec{H}(z,t)$

b. The instantaneous electric field vector, $\vec{E}(z,t)$

c. The average power density in the propagating wave, $\vec{P}_{avg}(z)$

11.12 Consider two 3.0 GHz uniform electromagnetic plane waves propagating simultaneously in the +z direction in free space. The phasor of the electric field of plane wave #1 is measured at $z=-2.5$ cm to be 1.0 V/m along the x axis and the phasor of the magnetic field of plane wave #2 is measured at $z=0$ cm to be $1/(120\ \pi)$ along the x axis. Find

a. The phasor forms of the electric and magnetic field of wave #1

b. The phasor forms of the electric and magnetic field of wave #2

c. The wavelength of these waves

d. The Poynting's vector of the total field (the sum of these two individual plane waves) at $z=0.0$ cm

11.13 The electric field of a uniform electromagnetic plane wave propagating in free space is given in phasor form $\vec{E}(x,y,z)=\left(2e^{j0.75\pi}\vec{a}_y-e^{j0.25\pi}\vec{a}_z\right)e^{j\pi x}$ V/m. Determine the following quantities:

a. The direction of propagation of the plane wave

b. The wavelength of the plane wave

c. The frequency of the plane wave

d. The polarization of the plane wave

e. The associated magnetic field phasor

f. The average Poynting's vector of the plane wave

11.14 Find the Poynting vector phasors for Problems 11.1 and 11.2, and 11.3.

11.15 A coaxial transmission line is formed of two perfectly conducting cylinders with radii of 3 mm and 7 mm separated by an air dielectric. The line is excited by a *TEM* wave at 1 GHz. The resulting $\vec{E}$ field is given by $[(1.18/\rho)\cos(\omega t-\beta z)]\vec{a}_\rho$ V/m. Find the corresponding $\vec{H}(\rho,z,t)$, $\vec{\mathcal{P}}(\rho,z,t)$, and the average power density transmitted.

11.16 A coaxial transmission line is formed of two perfectly conducting cylinders with radii a and b separated by an ideal dielectric with relative permittivity of ε_r. The line is excited by a *TEM* wave at 1 GHz. The resulting $\vec{H}$ field is given by $[(H^+/\rho)\cos(\omega t-\beta z)\vec{a}_\varphi]$ A/m. Find the corresponding $\vec{E}(\rho,z,t)$, $\vec{\mathcal{P}}(\rho,z,t)$, and the average power density transmitted.

11.17 Describe briefly what is meant by polarization of a wave. What are the conditions for a wave to be circularly polarized?

The electric field vector of a uniform plane wave traveling in free space is given by $\vec{E} = (\vec{a}_x(2jE_o) - \vec{a}_y E_o)e^{j\beta z}$ V/m. Determine the following:

a. The direction of propagation

b. The type of polarization of the wave

c. The corresponding magnetic field phasor

d. The instantaneous electric field vector

e. The average power density in the propagating wave

11.18 The electric field of a uniform plane wave propagating through a certain nonmagnetic material is given by: $\vec{E} = \vec{a}_x 20\cos\left(\pi \times 10^7 t - \frac{\pi}{12} y - \frac{\pi}{4}\right)$ V/m

Determine the following:

a. The type of polarization of the wave and the direction of wave propagation

b. The phase velocity

c. The wavelength in the material

d. The relative permittivity of the material

e. The intrinsic impedance of the material

f. The electric and magnetic field phasors

g. The average power density of the wave

11.19 Calculate the low-frequency values for *R*, *L*, *C*, and *G* of a coaxial transmission line with copper conductors and Teflon dielectric, given that

- Outer diameter of center conductor is 1 mm
- Inner diameter of outer conductor is 3 mm
- Outer diameter of outer conductor is 4 mm

Repeat for $f = 10$ GHz.

11.20 For the coaxial transmission line, derive an expression for the inductance per unit axial length at low frequencies. The center conductor is made of a solid conductor, radius *a* that would carry a uniform current density over its cross section at low frequencies. The outer conductor is extremely thin that would allow us to ignore its thickness and it has a radius *b*. Both the conductors and the dielectric filling are non-magnetic ($\mu = \mu_o$). Discuss how the inductance changes at radio frequencies.

11.21 Derive an expression for the capacitance per unit axial length of the same cable of Problem 11.18. Any change with frequency? Why?

11.22 Derive expressions for the (high frequency) characteristic impedance of the coaxial transmission line by first deriving the corresponding *L* and *C* per unit axial length (starting with their EM fields' definition). The cylindrical radii are given by *a* for the inner conductor and *b* and *c*, respectively, for the outer conductor's inner and outer surfaces.

11.23 In the above problem, if the space between the conductors was changed (from air) to a material with a dielectric constant ε_r and a relative magnetic permeability μ_r, find

a. The change in the total energy stored in the cable per unit axial length

b. The new expression for the characteristic impedance

11.24 Design a 50-Ω copper coaxial line with Teflon filling (dielectric constant is 2) with the outer diameter of the inner conductor is 1 mm. Estimate the line's attenuation coefficient at 10 GHz.

CHAPTER 11 SUMMARY[2]

Maxwell's Equations and the Wave Equation

Maxwell's Equations in the Phasor Point Form			
Faraday's law	$\vec{\nabla}\times\bar{\vec{E}}=-j\omega\bar{\vec{B}}$	Gauss' law for electric flux	$\vec{\nabla}\cdot\bar{\vec{D}}=\bar{\rho}_v$
Ampere's law	$\vec{\nabla}\times\bar{\vec{H}}=\bar{\vec{J}}+j\omega\bar{\vec{D}}$	Gauss' law for magnetic flux	$\vec{\nabla}\cdot\bar{\vec{B}}=0$

- Forming the wave equation for source free space, $\bar{\vec{J}}=0$ and $\bar{\rho}_v=0$

$$\nabla^2\begin{pmatrix}\vec{E}\\ \vec{H}\end{pmatrix}=-\omega^2\varepsilon\mu\begin{pmatrix}\vec{E}\\ \vec{H}\end{pmatrix}=\gamma^2\begin{pmatrix}\vec{E}\\ \vec{H}\end{pmatrix}$$

- The Laplacian operator:

$$\nabla^2 f(x,y,z)=\frac{\partial^2 f}{\partial x^2}+\frac{\partial^2 f}{\partial y^2}+\frac{\partial^2 f}{\partial z^2}$$

$$\nabla^2 f(\rho,\varphi,z)=\frac{1}{\rho}\frac{\partial}{\partial\rho}\left(\rho\frac{\partial f}{\partial\rho}\right)+\frac{1}{\rho^2}\frac{\partial^2 f}{\partial\varphi^2}+\frac{\partial^2 f}{\partial z^2}$$

Solving Wave Equation

- Solving the wave equation in one-dimension for source free space

$$\bar{E}_x(z)=\bar{E}_x^+e^{-\gamma z}+\bar{E}_x^-e^{+\gamma z}\quad\text{and}\quad\bar{H}_y(z)=\frac{1}{\eta_o}\cdot\left\{\bar{E}_x^+e^{-\gamma z}-\bar{E}_x^-e^{+\gamma z}\right\}$$

$$\gamma=j\beta=j\omega\sqrt{\mu_o\varepsilon_o}\quad\text{and}\quad\eta_o=\sqrt{\mu_o/\varepsilon_o}\cong 377\text{ ohms}$$

$$\beta=2\pi/\lambda\quad\text{and}\quad c_{ph}=\frac{\omega}{\beta}=\lambda f=1/\sqrt{\mu_o\cdot\varepsilon_o}\cong 3\times10^8\text{ m/s}$$

- Source-free case of lossless media (perfect dielectric)

Same as free space solution except for use of ε and μ instead of ε_o and μ_o

- Case of source-free lossy media $\sigma\neq 0$ (general material properties)

Same expressions but we now use ε^* and μ^*

$$\varepsilon^*=\frac{(\sigma+j\omega\varepsilon)}{j\omega}=\varepsilon-j\frac{\sigma}{\omega}=\varepsilon'-j\varepsilon''\quad\text{and}\quad\mu^*=\mu'-j\mu''$$

[2]*Important Warning to Students:* Please do not use this summary without thorough study of the relevant material in this chapter. Using this summary and its formulas without adequate appreciation of the physics involved and conditions of validity can lead to wrong implementations and invalid conclusions.

Wave Equation Parameters for Different Materials

- General form for wave equation solution in one-dimension for source free material

$$\bar{E}_x(z)=\bar{E}_x^+e^{-\gamma z}+\bar{E}_x^-e^{+\gamma z} \quad \text{and} \quad \bar{H}_y(z)=\frac{1}{\eta_o}\cdot\{\bar{E}_x^+e^{-\gamma z}-\bar{E}_x^-e^{+\gamma z}\}$$

	Non-Mag. Perfect Diel.	Non-Mag. Good Diel. (Low Loss)	General Material	Non-Mag. Good Cond. (High Loss)	Non-Mag. Perfect Cond.
σ	$\sigma=0$	$\sigma\ll\omega\varepsilon\to\varepsilon''\ll\varepsilon'$	$\varepsilon'=\varepsilon,\varepsilon''=\sigma/\omega$	$\sigma\gg\omega\varepsilon\to\varepsilon''\gg\varepsilon'$	$\sigma=\infty\to\varepsilon''=\infty$
ε^*	$=\varepsilon'=\varepsilon$	$\varepsilon'\left[1-j\frac{\varepsilon''}{\varepsilon'}\right]$	$\varepsilon'-j\varepsilon''$	$-j\varepsilon''\left[1-j\frac{\varepsilon'}{\varepsilon''}\right]$	$\varepsilon_o-j\infty$
μ^*	μ_o	μ_o	$\mu'-j\mu''$	μ_o	μ_o
γ^2	$-\omega^2\mu_o\varepsilon$	$-\omega^2\mu_o\varepsilon'\left[1-j\frac{\varepsilon''}{\varepsilon'}\right]$	$-\omega^2\mu^*\varepsilon^*$	$j\omega^2\mu_o\varepsilon''\left[1-j\frac{\varepsilon'}{\varepsilon''}\right]$	∞
α	0	$\cong\left(\frac{\omega\varepsilon''\eta}{2}\right)\cong\left(\frac{\sigma\eta_o}{2\sqrt{\varepsilon_r}}\right)$	$Re\left\{\sqrt{-\omega^2\mu^*\varepsilon^*}\right\}$	$\cong\sqrt{\frac{\omega\mu_o(\omega\varepsilon'')}{2}}\cong\sqrt{\pi f\mu_o\sigma}$	∞
β	$=\omega\sqrt{\mu_o\varepsilon}=\beta_o\sqrt{\varepsilon_r}$	$\cong\omega\sqrt{\mu_o\varepsilon'}\cong\beta_o\sqrt{\varepsilon_r}$	$Im\left\{\sqrt{-\omega^2\mu^*\varepsilon^*}\right\}$	$\cong\sqrt{\frac{\omega\mu_o(\omega\varepsilon'')}{2}}\cong\sqrt{\pi f\mu_o\sigma}$	∞
λ	$\lambda_o/\sqrt{\varepsilon_r}$	$\lambda_o/\sqrt{\varepsilon_r}$	$\frac{2\pi}{\beta}$	$\lambda_o\sqrt{2\omega\varepsilon_o/\sigma}$	0
c_{ph}	$c_o/\sqrt{\varepsilon_r}$	$c_o/\sqrt{\varepsilon_r}$		$c_o\sqrt{2\omega\varepsilon_o/\sigma}$	0
η	$=\sqrt{\frac{\mu_o}{\varepsilon}}=\eta_o/\sqrt{\varepsilon_r}$	$\cong\sqrt{\frac{\mu_o}{\varepsilon'}}\cong\eta_o/\sqrt{\varepsilon_r}$	$\sqrt{\mu^*/\varepsilon^*}$	$\cong(1+j)\sqrt{\mu_o/2\varepsilon''}$ $\cong\eta_o(1+j)\sqrt{\omega\varepsilon_o/(2\sigma)}$	0

Physical Insight in Wave Propagation in Dielectrics and Conductors

- Plane waves in good dielectric: practically same β, λ, c_{ph}, η, with small attenuation $\alpha\cong\left(\frac{\omega\varepsilon''\eta}{2}\right)$

Loss tangents: $\tan\delta_\varepsilon=\varepsilon''/\varepsilon'$, $\tan\delta_\mu=\frac{\mu''}{\mu'}$

- Uniform plane waves in good conductor:

Strong decay (attenuation) $\alpha\cong\beta\cong\sqrt{\pi f\mu_o\sigma}$. Limit wave penetration to conductor "skin"
Skin depth: Depth from surface for penetrating field to reach ($1/e$ = 36.8%) of its original intensity

$$\delta=1/\alpha\cong1/\sqrt{\pi f\mu_o\sigma}$$

- Shielding: Use of high conductivity conductor to block wave propagation. Conductor shield thickness needs to be >> skin depth (typically > 5δ, leakage < 1%)

Poynting Vector–Poynting Theorem

- Power flow density vector: $\vec{\mathcal{P}}=\vec{E}\times\vec{H}$
- The complex Poynting vector: $\bar{\vec{\mathcal{P}}}(\vec{r})=\frac{1}{2}\bar{\vec{E}}(\vec{r})\times\bar{\vec{H}}^*(\vec{r})$
- Time-averaged Poynting vector: $\vec{\mathcal{P}}_{\text{avg}}(\vec{r})=\frac{1}{2}Re\{\bar{\vec{E}}(\vec{r})\times\bar{\vec{H}}^*(\vec{r})\}$

Plane Wave Power Flow Density

- Average power flow density vector: $\vec{\mathcal{P}}_{avg} = \vec{a}_z Re\left\{\frac{1}{2\eta^*}\left[\left|\bar{E}_x^+\right|^2 e^{-2\alpha z} - \left|\bar{E}_x^-\right|^2 e^{+2\alpha z} + 2j\,Im\left\{\bar{E}_x^- \bar{E}_x^{+*} e^{+2j\beta z}\right\}\right]\right\}$
- For low-loss dielectric: $\vec{\mathcal{P}}_{avg} = \mathcal{P}^+_{avg} - \mathcal{P}^-_{avg} = \vec{a}_z\left[\frac{1}{2\eta}\left|\bar{E}_x^+\right|^2 e^{-2\alpha z} - \frac{1}{2\eta}\left|\bar{E}_x^-\right|^2 e^{+2\alpha z}\right]$

Plane Waves in Controlled-Geometry Transmission Lines

- *TEM* lossless wave solution: $E_\rho^+(\rho,z) = \left\{\frac{V^+}{\ln\frac{b}{a}}\cdot\frac{1}{\rho}\right\}e^{-j\beta z},\ H_\varphi^+(\rho,z) = \left\{\frac{1}{\eta}\cdot\frac{V^+}{\ln\frac{b}{a}}\cdot\frac{1}{\rho}\right\}e^{-j\beta z}$

$$L_{a<\rho<b \text{ per meter}} = \frac{\mu}{2\pi}\ln\left(\frac{b}{a}\right),\ C_{z,ab} = \frac{2\pi\varepsilon}{\ln\left(b/a\right)},\ Z_o = \sqrt{L_z/C_z} = \eta\frac{\ln\left(b/a\right)}{2\pi}$$

- Power flow in coaxial lines: $\mathcal{P}_z^+ = \frac{1}{\eta}\cdot\left\{\frac{V^+}{\ln\frac{b}{a}}\right\}^2\cdot\frac{1}{\rho^2}e^{-2\gamma z},\ P_z^+ = \frac{\{V^+\}^2}{Z_o}e^{-2\gamma z}$
- High-frequency coaxial line parameters (skin effect)

$$R_z \cong \frac{1}{\sigma_{\text{inner}}\cdot(2\pi a\cdot\delta_{\text{inner}})} + \frac{1}{\sigma_{\text{outer}}\cdot(2\pi b\cdot\delta_{\text{outer}})},\ L_z \cong \frac{\mu}{2\pi}\ln\left(\frac{b}{a}\right),\ G_z = \frac{2\pi\sigma_d}{\ln\left(\frac{b}{a}\right)},\ C_z = \frac{2\pi\varepsilon}{\ln\left(b/a\right)}$$

Important Observation

- In circuit theory analysis, we are often misled to think that power flow occurs in the interconnecting circuit conductors. The truth is that electromagnetic power flows in the surrounding dielectric media and not in the conductors. The conductors provide guidance (and confinement) to the propagating waves.

CHAPTER 12

Wave Polarization and Propagation in Multiple Layers

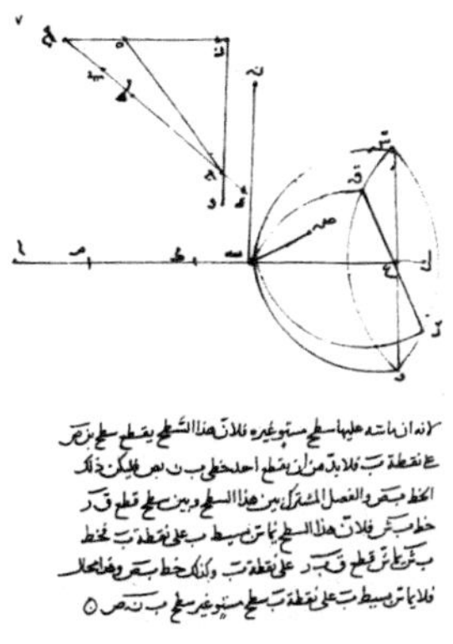

Abū Sa'd al-'Alā' ibn Sahl (Ibn Sahl) (940–1000)

Willebrord Snellius (Snell) (1580–1626)

Sir David Brewster (1781–1868)

Chapter Learning Objectives

- Discuss the different types of polarizations of electromagnetic waves (linear polarization, circular polarization, and elliptical polarization) and their physical insights.
- Describe the transmission and reflection of uniform plane waves for normal and oblique incidence for the case of two dielectrics with a planar interface. Explain related mathematical expressions.
- Describe the transmission and reflection of uniform plane waves for normal and oblique incidence for the case of a dielectric conductor interface. Explain related mathematical expressions.
- Explain the physical insight of Snell's law. Demonstrate how to derive its mathematical expression.
- Describe the concept of total reflection and hence the critical angle. Explain related mathematical expressions.
- Describe the transmission and reflection of both parallel and perpendicular polarization waves in the case of oblique incidence.
- Discuss the concept of Brewster angle for parallel and perpendicular polarizations. Explain related mathematical expressions.

12.1 Introduction

In the previous chapter, we learned about plane wave propagation in open (unguided) space/media. We confined our discussion to the simple case of one-dimensional propagation in a source-free free space. Next, we extended our analysis to include different types of materials ranging from lossless and low-loss dielectrics to high-conductivity conductors and perfect conductors.

Later in the chapter, we introduced the Poynting theorem, which enabled us to examine the power flow-density vector associated with the propagating wave. We concluded with a return to the analysis of plane wave propagation in controlled-geometry transmission lines. Earlier in Chapter 2, we carried out the analysis of transmission lines using circuit theory concepts; however, in Chapter 11, we had a chance to revisit the same subject using an EM field theory approach.

In this chapter, we discuss types of polarizations of electromagnetic waves, as well as the propagation of transverse electromagnetic (*TEM*) waves in multiple dielectric layers. Both topics are important and have significant practical implications. In this regard, we will give examples of practical applications in demonstration of the relevant concepts.

12.2 Wave Polarization

Wave polarization is an important attribute describing how the fields change orientation as a wave propagates. *TEM* waves described in Chapter 11 earlier are linearly polarized, which means that both the electric and magnetic field vectors maintain the same orientation all the time and at all locations. Figure 12.1 depicts a linearly polarized wave where the electric field is always oriented along the x axis while the magnetic field is always oriented along the y axis. In this figure, we opted to show the case of a lossless medium (with no attenuation) in order to simplify the graphical presentation.

12.2.1 Linear Polarization

The mathematical expression of the electric field for a linearly polarized wave propagating along the z axis in its general form can be written as

$$\vec{E}(z,t)=[E_1(0)\vec{a}_x+E_2(0)\vec{a}_y]e^{-\alpha z}\cos(\omega t-\beta z) \tag{12.1}$$

$$\bar{\vec{E}}(z)=[E_1(0)\vec{a}_x+E_2(0)\vec{a}_y]e^{-\gamma z} \tag{12.2}$$

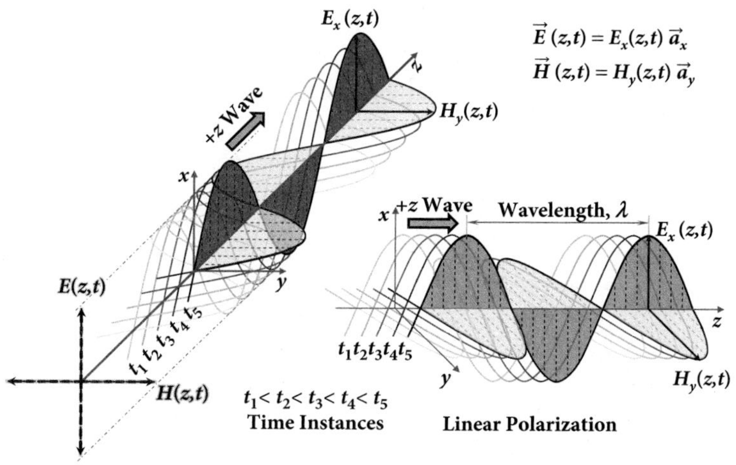

FIGURE 12.1 An illustration of fields of a linearly polarized wave propagating along the z axis.

In Figure 12.1, we chose to set our axes such that the electric field vector is in full alignment with the x axis, and hence

$$\vec{E}(z,t) = [E_x(0)\vec{a}_x]e^{-\alpha z}\cos(\omega t - \beta z) \tag{12.3}$$

$$\vec{\bar{E}}(z) = [E_x(0)\vec{a}_x]e^{-\gamma z} \tag{12.4}$$

The corresponding magnetic field vector has the form

$$\vec{H}(z,t) = [H_y(0)\vec{a}_y]e^{-\alpha z}\cos(\omega t - \beta z) \tag{12.5}$$

$$\vec{\bar{H}}(z) = [H_y(0)\vec{a}_y]e^{-\gamma z} \tag{12.6}$$

Two other commonly discussed types of polarization are circular and elliptical; the former is shown in Figure 12.2 and both are depicted in Figure 12.3. Again, in both figures, we opted to show the case of a lossless medium (with no attenuation) in order to simplify the graphical presentation.

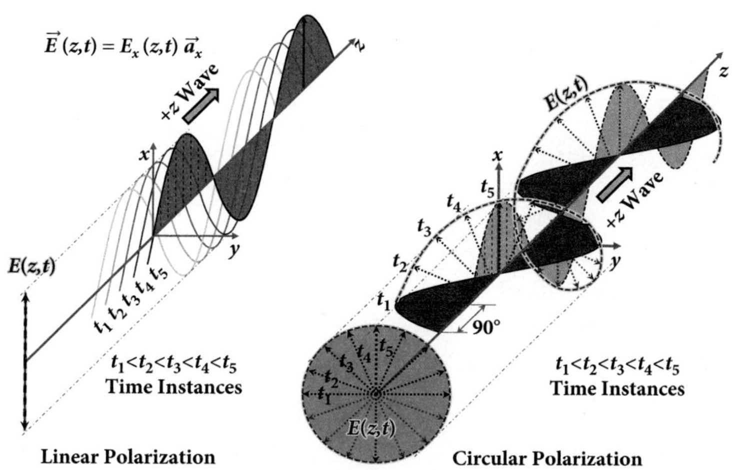

FIGURE 12.2 An illustration of the electric field of a circularly polarized wave propagating along the z axis.

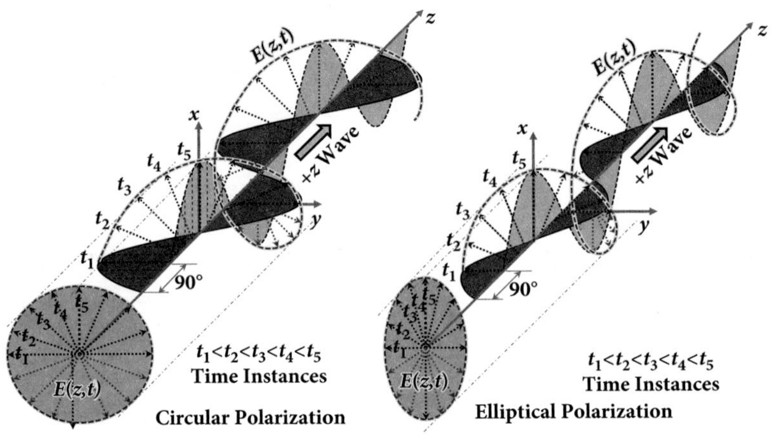

FIGURE 12.3 An illustration of the electric field showing the difference between circular and elliptical polarizations.

Example 12.1 The electric field of a traveling electromagnetic wave is given by

$$\vec{E} = \vec{a}_z \cos\left(\pi \times 10^7 t + \frac{\pi}{15} y + \frac{\pi}{6}\right) \text{ V/m}$$

Determine the following:

a. The direction of wave propagation
b. The type of polarization of the wave
c. The wave frequency
d. The wavelength
e. The phase velocity
f. The corresponding magnetic field

The average power density in the propagating wave

Solution

a. The direction of wave propagation: negative y direction
b. The type of polarization of the wave: linearly polarized in the z direction
c. The wave frequency $\omega = 2\pi f = \pi \times 10^7 \rightarrow \quad f = \dfrac{\pi \times 10^7}{2\pi} = 5 \text{ MHz}$
d. The wavelength $\lambda = \dfrac{2\pi}{\beta} = \dfrac{2\pi}{\pi/15} = 30 \text{ m}$
e. The phase velocity $c_{ph} = \dfrac{\omega}{\beta} = \dfrac{\pi \times 10^7}{\pi/15} = 1.5 \times 10^8 \text{ m/s}$
f. The corresponding magnetic field

$$\vec{E} = \vec{a}_z e^{\frac{j\pi}{15}y} e^{\frac{j\pi}{6}} \quad \vec{H} = \frac{1}{\eta}\,\vec{a}_k \times \vec{E} = \frac{1}{\eta}(-\vec{a}_y) \times \vec{a}_z e^{\frac{j\pi}{15}y} e^{\frac{j\pi}{6}} = -\frac{1}{\eta}\vec{a}_x e^{\frac{j\pi}{15}y} e^{\frac{j\pi}{6}}$$

g. The average power density in the propagating wave

$$P = \frac{\vec{a}_k}{2\eta}\left|\vec{E}\right|^2 = \frac{-\vec{a}_y}{2\eta}(1^2) = \frac{-1}{2\eta}\vec{a}_y \text{ W/m}^2$$

or

$$P=\frac{1}{2}Re\{\vec{E}\times\vec{H}^*\}=\frac{1}{2}Re\left\{\vec{a}_z e^{\frac{j\pi}{15}y}e^{\frac{j\pi}{6}}\times\left(-\frac{1}{\eta}\vec{a}_x e^{\frac{-j\pi}{15}y}e^{\frac{-j\pi}{6}}\right)\right\}=-\frac{1}{2\eta}\vec{a}_y \text{ W/m}^2$$

12.2.2 Circular Polarization

In circular polarization, the vector fields rotate uniformly as the wave advances along its propagation axis. Both electric and magnetic field vectors describe a spiral path whose axis is aligned with the wave propagation axis. In the absence of loss, both vector fields maintain constant magnitudes throughout their propagation, and thus the projection of each spiral in the case of circular polarization is a circle (Figure 12.2). In this figure, we are only showing the electric field vector in order to simplify the graphical presentation.

As shown in the figure, the circularly polarized wave can be seen as the superposition of two linearly polarized waves that are 90° apart in phase, that is,

$$\vec{E}(z,t)=E(0)[\cos(\omega t-\beta z)\vec{a}_x+\sin(\omega t-\beta z)\vec{a}_y]e^{-\alpha z} \tag{12.7}$$

$$\vec{\bar{E}}(z)=E(0)[\vec{a}_x-j\vec{a}_y]e^{-\gamma z} \tag{12.8}$$

12.2.3 Elliptical Polarization

The elliptical polarization is demonstrated in Figure 12.3. As the figure shows, in contrast to the circular polarization, the spiral path has an elliptic projection. The magnitude of both electric and magnetic field vectors change *elliptically* as the wave propagates. Again, we are only showing the electric field vector for presentation simplicity; however, the magnetic field is expected to behave in a similar fashion.

The expressions for an elliptically polarized wave are identical to those of the circularly polarized wave except for the inequality of the x and y field components, and hence

$$\vec{E}(z,t)=[E_x(0)\cos(\omega t-\beta z)\vec{a}_x+E_y(0)\sin(\omega t-\beta z)\vec{a}_y]e^{-\alpha z} \tag{12.9}$$

$$\vec{\bar{E}}(z)=[E_x(0)\vec{a}_x-jE_y(0)\vec{a}_y]e^{-\gamma z} \tag{12.10}$$

It is worthy to mention that both the circular and elliptical polarization cases presented above are what we refer to as right-handed polarizations from the point of view of the source. This is so since looking down the propagation path from the source side, we observe that the polarized fields are in clockwise/right-handed rotation. For left-handed polarization, the fields would be rotating in a counter-clockwise fashion. The equations describing left-handed polarizations are similar to those above Equations (12.7) to (12.10) with the reversal of the polarity associated with the E_y terms.

Example 12.2 Consider a uniform plane wave propagating in the positive z direction has an electric field phasor given by $\vec{\bar{E}}(z)=\left[E_x(0)\vec{a}_x+E_y(0)e^{j\theta}\vec{a}_y\right]e^{-j\beta z}$, where θ represents the phase shift between the x and y components of the electric field. Determine the polarization of the wave in the following cases

a. $E_x(0)=E_y(0)=E_o$ and $\theta=\frac{\pi}{2}$

b. $E_x(0)=E_o$, $E_y(0)=2E_o$, and $\theta=-\frac{\pi}{2}$

c. $E_x(0)=E_o$, $E_y(0)=2E_o$, and $\theta=-\pi$

Answer: Part a

$$E_x(0) = E_y(0) = E_o \text{ and } \theta = \frac{\pi}{2}$$

$$\vec{\bar{E}}(z) = \left[E_o\vec{a}_x + E_o e^{j\frac{\pi}{2}}\vec{a}_y\right]e^{-j\beta z}$$

$$\vec{E}(z,t) = \left[E_o\cos(\omega t - \beta z)\vec{a}_x + E_o\cos\left(\omega t - \beta z + \frac{\pi}{2}\right)\vec{a}_y\right]$$

$$\vec{E}(z,t) = [E_o\cos(\omega t - \beta z)\vec{a}_x - E_o\sin(\omega t - \beta z)\vec{a}_y]$$

The x component of the electric field vector is given by $E_x = E_o\cos(\omega t - \beta z)$ and the y component $E_y = -E_o \sin(\omega t - \beta z)$.

The magnitude of the electric field vector is given by $|\vec{E}(z,t)| = \sqrt{E_x^2 + E_y^2} = \sqrt{E_o^2\cos^2(\omega t - \beta z) + E_o^2\sin^2(\omega t - \beta z)} = \sqrt{2}E_o$.

We can also write $\left(\frac{E_x}{E_o}\right)^2 + \left(\frac{E_y}{E_o}\right)^2 = \cos^2(\omega t - \beta z) + \sin^2(\omega t - \beta z) = 1.$

The equation above represents the locus traced by the tip of the electric field over time. In this case, the locus is circle with radius E_o. Therefore, the wave is circularly polarized. To determine the direction of rotation, we need to find how the angle that the electric field vector makes with the x axis changes over time.

The angle that the tip of the electric field vector makes with the x axis is given by

$$\varphi = \tan^{-1}\frac{E_y}{E_x} = \tan^{-1}\frac{-E_o\sin(\omega t - \beta z)}{E_o\cos(\omega t - \beta z)} = \tan^{-1}(-\tan(\omega t - \beta z)) = -(\omega t - \beta z)$$

To find the sense of rotation, it's easier to consider a certain point in space, say $z = 0$.

In this case $\varphi = -\omega t$. The expression shows that the angle with the x axis becomes more negative over time which implies that the tip of the electric field rotates in the clockwise direction. If the fingers of the right-hand are curled in the direction of rotation of the electric field vector, the thumb will point in the negative z direction which is *opposite* to the given direction of propagation and the wave is said to be left-hand polarized. Notice that the direction of rotation of the electric field vector will in general be from the leading electric field component toward the lagging field component. In this case, the y component is leading the x component by an angle of $\pi/2$ and so the rotation will be from the y direction to the x direction (clockwise).

Answer: Part b

$$E_x(0) = E_o,\ E_y(0) = 2E_o \text{ and } \theta = -\frac{\pi}{2}$$

$$\vec{\bar{E}}(z) = \left[E_o\vec{a}_x + 2E_o e^{-j\frac{\pi}{2}}\vec{a}_y\right]e^{-j\beta z}$$

$$\vec{E}(z,t) = \left[E_o\cos(\omega t - \beta z)\vec{a}_x + 2E_o\cos\left(\omega t - \beta z - \frac{\pi}{2}\right)\vec{a}_y\right]$$

$$\vec{E}(z,t) = \left[E_o\cos(\omega t - \beta z)\vec{a}_x + 2E_o\sin(\omega t - \beta z)\vec{a}_y\right]$$

$$E_x = E_o\cos(\omega t - \beta z) E_y = 2E_o\sin(\omega t - \beta z)$$

$$\left(\frac{E_x}{E_o}\right)^2 + \left(\frac{E_y}{2E_o}\right)^2 = \cos^2(\omega t - \beta z) + \sin^2(\omega t - \beta z) = 1$$

The above equation represents the equation of an ellipse with major axes of lengths E_o and $2E_o$ along the x and y directions, respectively. The tip of the electric field vector traces an ellipse over time and the wave is said to be elliptically polarized. The sense of rotation can be obtained as in part a, in this case the x component of the electric field is leading by an angle of $\pi/2$. Curling the fingers of the right-hand from the x toward the y direction, the thumb point in the positive z direction which matches the direction of propagation and therefore the wave is right-hand elliptical.

Note that in this case the ellipse has the major axes along the x and y axes, if the phase shift between the electric field components is different from $\pm\pi/2$, the trace of the tip of the electric field vector will be a tilted ellipse.

Answer: Part c

$$E_x(0)=E_o,\ E_y(0)=2E_o \text{ and } \theta=-\pi$$

$$\vec{E}(z)=\left[E_o\vec{a}_x+2E_oe^{-j\pi}\vec{a}_y\right]e^{-j\beta z}$$

$$\vec{E}(z,t)=\left[E_o\cos(\omega t-\beta z)\vec{a}_x+2E_o\cos(\omega t-\beta z+\pi)\vec{a}_y\right]$$

$$\vec{E}(z,t)=\left[E_o\cos(\omega t-\beta z)\vec{a}_x-2E_o\cos(\omega t-\beta z)\vec{a}_y\right]$$

$$E_x=E_o\cos(\omega t-\beta z)\qquad E_y=-2E_o\cos(\omega t-\beta z)$$

In this case, $E_y=-2E_x$

$$\left|\vec{E}(z,t)\right|=\sqrt{E_x^2+E_y^2}=\sqrt{5}E_o\cos(\omega t-\beta z)$$

and $\varphi=\tan^{-1}\dfrac{E_y}{Ex}=\tan^{-1}(-2)=-63.3°$

The relationship between the y and x components of the electric field is linear, and the wave is said to be linearly polarized along a direction, which is $-63.3°$ from the x axis.

We can conclude from the above analysis that the elliptical polarization is the most general type. In the special case when the phase shift between E_x and E_y is either $\theta=0$ or $\theta=\pi$, the wave becomes linearly polarized. Circular polarization is also a special case where the amplitudes of E_x and E_y are the same and the phase shift is $\pm\pi/2$.

Example 12.3 Elliptically polarized electromagnetic waves are used in radar and some communication systems. Suppose that you need to detect a circularly polarized wave but that your detectors only detect elliptically polarized waves. Show that any arbitrary circularly polarized wave may be expressed as the sum of two oppositely rotating elliptically polarized components.

Solution A right-hand elliptical wave can be represented as $\vec{E}_1=\left(\vec{a}_x0.5E_o-\vec{a}_yj2E_o\right)e^{-jkz}$

A left-hand elliptical wave can be represented as $\vec{E}_2=\left(\vec{a}_x0.5E_o+\vec{a}_yjE_o\right)e^{-jkz}$

$\vec{E}_1+\vec{E}_2=\left(\vec{a}_xE_o-\vec{a}_yjE_o\right)e^{-jkz}=\left(\vec{a}_xE_o+\vec{a}_ye^{-j\frac{\pi}{2}}E_o\right)e^{-jkz}$ which is right-hand circular.

Therefore, the sum of a right-hand elliptical wave and a left-hand elliptical wave resulted in a right-hand circular wave in this case. However

A right-hand elliptical wave represented as $\vec{E}_1=\left(\vec{a}_x0.5E_o-\vec{a}_yjE_o\right)e^{-jkz}$

and a left-hand elliptical wave represented as $\vec{E}_2=\left(\vec{a}_x0.5E_o+\vec{a}_yj2E_o\right)e^{-jkz}$

$\vec{E}_1+\vec{E}_2=\left(\vec{a}_xE_o+j\vec{a}_yE_o\right)e^{-jkz}$ which is left-hand circular.

Example 12.4 A uniform plane wave traveling in free space given by $\vec{E}=(-\vec{a}_x E_{ox}+\vec{a}_y jE_{oy})e^{-jkz}$ V/m, determine the following:

a. The type of polarization of the wave
b. The corresponding magnetic field vector phasor
c. Expressions for the instantaneous electric and magnetic field intensities
d. The average power density in the propagating wave

Solution

a. The type of polarization of the wave $\vec{E}=-(\vec{a}_x E_{ox}-\vec{a}_y jE_{oy})e^{-jkz}$. The x component leads the y component; therefore, the tip of the electric field will rotate from the x to the y direction or counterclockwise. Using the right-hand rule, the thumb points in the positive z direction, which is the same as direction of propagation → right-hand. Because the magnitudes of the x component and the y component are different, the wave is right-hand elliptical.

b. The corresponding magnetic field vector phasor.

$$\vec{H}=\frac{1}{\eta}\vec{a}_k\times\vec{E}=\frac{1}{\eta}\vec{a}_z\times[-(\vec{a}_x E_{ox}-\vec{a}_y jE_{oy})]e^{-jkz}$$

$$\vec{H}=\frac{-1}{\eta}(\vec{a}_y E_{ox}+\vec{a}_x jE_{oy})e^{-jkz}=\frac{-1}{\eta}(\vec{a}_x jE_{oy}+\vec{a}_y E_{ox})e^{-jkz}$$

c. Expressions for the instantaneous electric and magnetic field intensities

$$\vec{E}(z,t)=Re\left\{(-\vec{a}_x E_{ox}+\vec{a}_y jE_{ox})e^{-jkz}e^{j\omega t}\right\}\text{ V/m}$$

$$\vec{E}(z,t)=Re\left\{\left(-\vec{a}_x E_{ox}+\vec{a}_y e^{j\frac{\pi}{2}}E_{oy}\right)e^{-jkz}e^{j\omega t}\right\}\text{ V/m}$$

$$\vec{E}(z,t)=\left(-\vec{a}_x E_{ox}\cos(\omega t-kz)+\vec{a}_y E_{oy}\cos\left(\omega t-kz+\frac{\pi}{2}\right)\right)\text{ V/m}$$

$$\vec{H}(z,t)=Re\left\{\frac{-1}{\eta}(\vec{a}_x jE_{oy}+\vec{a}_y E_{ox})e^{-jkz}e^{j\omega t}\right\}=Re\left\{\frac{-1}{\eta}\left(\vec{a}_x e^{j\frac{\pi}{2}}E_{oy}+\vec{a}_y E_{ox}\right)e^{-jkz}e^{j\omega t}\right\}$$

$$\vec{H}(z,t)=\frac{-1}{\eta}\left(\vec{a}_x E_{oy}\cos\left(\omega t-kz+\frac{\pi}{2}\right)+\vec{a}_y E_{ox}\cos(\omega t-kz)\right)\text{ A/m}$$

d. The average power density in the propagating wave

$$P=\frac{1}{2}Re\{\vec{E}\times\vec{H}^*\}=\frac{1}{2}Re\left\{\left[-(\vec{a}_x E_{ox}-\vec{a}_y jE_{oy})\right]e^{-jkz}\times\frac{-1}{\eta}(\vec{a}_x(-j)E_{oy}+\vec{a}_y E_{ox})e^{+jkz}\right\}$$

$$=\frac{\vec{a}_z}{2\eta}\left(E_{ox}^2+E_{oy}^2\right)\text{ W/m}^2$$

or

$$P=\frac{\vec{a}_k}{2\eta}|\vec{E}|^2=\frac{\vec{a}_z}{2\eta}\left(E_{ox}^2+E_{oy}^2\right)\text{ W/m}^2$$

12.2.4 Physical Insight

To appreciate wave polarization and its implications, let us consider the simple example of light received from the sun. Typically, this light is in the form of a mixture of electromagnetic wave polarizations that we would consider unpolarized. When passing through a polarizing filter, we would only receive the one polarization that the filter allows to pass through.

One simple experiment that the reader can perform using two polarizing filters such as the lenses of polarizing sunglasses is demonstrated in Figure 12.4. By using the first polarizing lens, one wave polarization passes through. Now, subjecting the resulting polarized wave to the second lens will have an interesting result. If the second lens polarization axis is in alignment with that of the first lens, the wave will pass through unaffected as well (part *a* of the figure).

By rotating the second lens polarizing axis 90° relative to that of the first lens, the polarized light will not be able to pass through, (part *c* of the figure). Now, a partial rotation of the second lens would result in only passing the component of the polarized wave that is in alignment with the second lens polarization axis (part *b* of the figure).

Another interesting observation is related to the alignment of the antenna of a radio receiver (or other communication devices) relative to that of a linear polarization transmitter antenna. We demonstrate the discussion here in Figure 12.5. In part *a* of the figure, we are showing a receiver antenna aligned with that of the

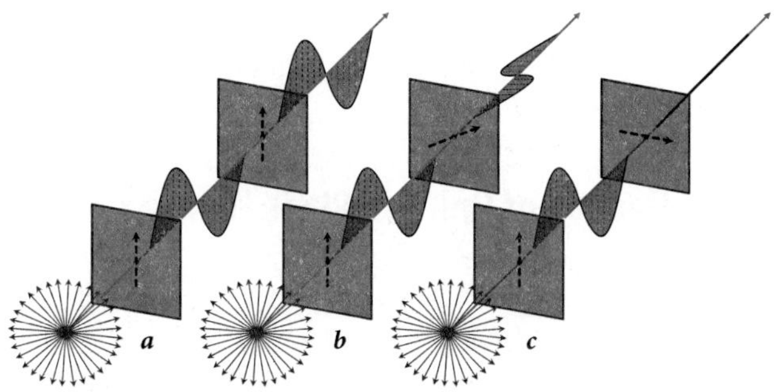

FIGURE 12.4 In this demonstration, the first polarization filter passed through a linearly polarized wave. The second polarizing filter controls the through-wave intensity depending on the its alignment relative to that of the polarized wave.

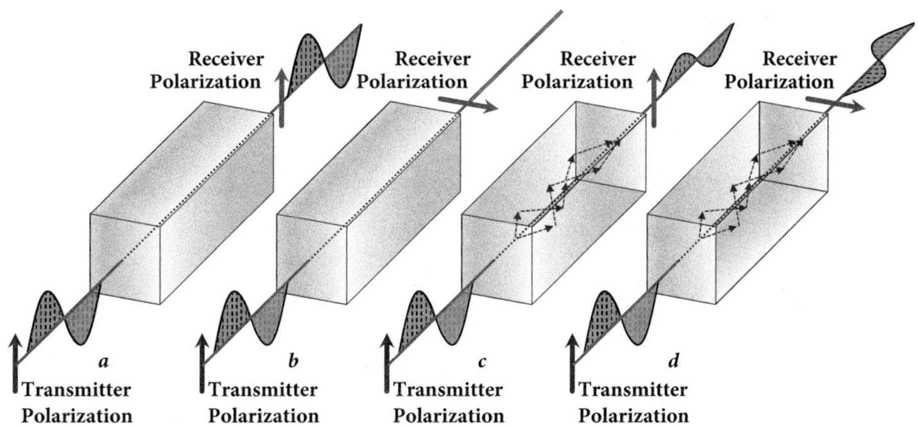

FIGURE 12.5 The polarizing characteristics of a communication channel will affect the transmitted signal and hence the quality of the reception at the receiver end.

transmitter while the communication channel in between is a line-of-site one with no scattering objects within. In such a case, the wave is expected to travel through with no change in polarization, resulting in full alignment with the receiver antenna, and hence achieving the best signal reception conditions.

In part *b* of the figure, the receiver antenna is set orthogonal to that of the transmitter with the same communication channel of part *a*. This results in a zero component of the transmitted polarized wave to match the receiver antenna, and hence achieving no signal reception in this receiver antenna position.

In parts *c* and *d* of the figure, we are assuming the communication channel to have multiple reflection and scattering objects that would alter the linear polarization of the transmitted wave. In both cases, there is likely that a component of the wave at the receiver end would be in alignment with the receiver antenna. In this case, we would expect a weaker signal reception depending on the orientation of the receiving antenna.

12.3 Transmission and Reflection of Uniform Plane Waves in Multilayer Media

In the last chapter, we studied the propagation of one-dimensional electromagnetic waves in a homogenous unguided medium with infinite extension in all spatial dimensions. In this chapter, we will consider cases of wave propagation in multiple materials with planar interfaces. We will start with the simple case of two materials with a planar interface, with an EM wave incident normally at the interface. Next, we will consider the case of oblique incidence, and finally, we will talk briefly about cases involving multilayer (a multi-material stack) media. In all these cases, we will confine our discussion to *TEM* (uniform plane) waves; the type which was the subject of our studies in the last chapter. In this regard, we will be assuming infinite extension of the media orthogonal to the wave propagation direction. For waves propagating along the z direction, all media will have infinite extensions in both the x and y directions.

12.4 Transmission and Reflection of Uniform Plane Waves: Normal Incidence

The case of normal incidence is depicted in Figure 12.6.

As seen in this figure, the entire space is shared by two media, labeled dielectric #1 and dielectric #2. The two media have infinite extensions in both the x and y directions. Along the z direction, they have a planar interface at $z = 0$ while extending indefinitely along the z coordinate. It is obvious that such an arrangement and choice of coordinates is to simplify the mathematical analysis.

We assume a planar *TEM* wave incident normally to the interface, that is, along the positive z direction.

On the bottom of Figure 12.6, we show the transmission line model analogous to the two dielectric media. As we will demonstrate in the following, the analysis of the normal incidence case greatly resembles that of the two-transmission line model shown.

We now start our analysis of the plane wave propagation as it goes through the two- dielectric arrangement. To examine what happens to the field components of the *TEM* wave, we consider the illustration in Figure 12.7.

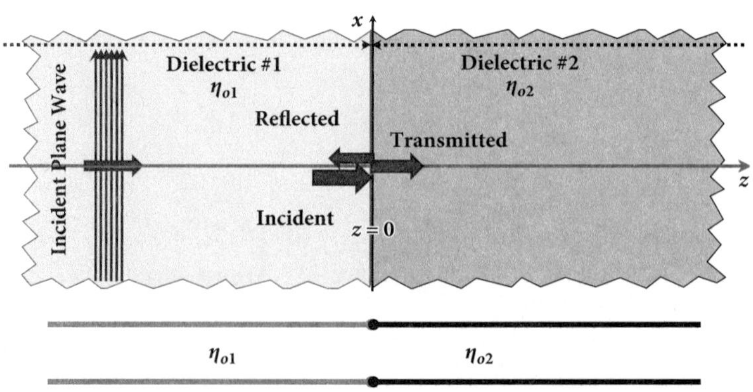

Figure 12.6 Transmission and reflection of uniform plane-waves incident normal to the interface between two dielectrics. Also shown: the *TL* equivalent model.

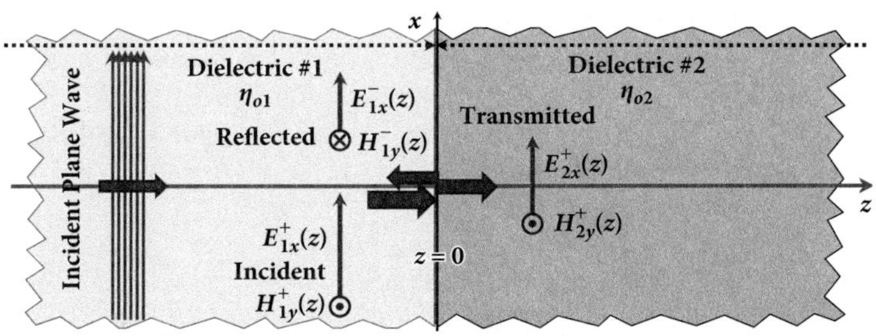

FIGURE 12.7 Electric and magnetic fields of incident, reflected, and transmitted waves incident normal to the interface between two dielectrics.

In this demonstration, we chose our coordinates such that the incident electric field component is in alignment with the x axis and the magnetic field component is aligned with the y axis. Logically, we expect two resulting *TEM* waves to result as a consequence to the incident wave; a reflected wave and a transmitted wave. The field components of both waves are expected to maintain the same alignments as those of the incident wave field components as we assume both dielectric materials to be isotropic.

Using $z = 0$ as the reference plane, we can express the field components of all three waves as follows:

Incident wave (*travels in the positive z direction*):

$$\bar{E}_{1x}^+(z) = \bar{E}_{1x}^+(0) \cdot e^{-\gamma_1 z} \tag{12.11}$$

$$\bar{H}_{1y}^+(z) = \bar{H}_{1y}^+(0) \cdot e^{-\gamma_1 z} \tag{12.12}$$

Reflected wave (*travels in the negative z direction*):

$$\bar{E}_{1x}^-(z) = \bar{E}_{1x}^-(0) \cdot e^{+\gamma_1 z} \tag{12.13}$$

$$\bar{H}_{1y}^-(z) = \bar{H}_{1y}^-(0) \cdot e^{+\gamma_1 z} \tag{12.14}$$

Transmitted wave (*travels in the positive z direction*):

$$\bar{E}_{2x}^+(z) = \bar{E}_{2x}^+(0) \cdot e^{-\gamma_2 z} \tag{12.15}$$

$$\bar{H}_{2y}^+(z) = \bar{H}_{2y}^+(0) \cdot e^{-\gamma_2 z} \tag{12.16}$$

Using the boundary conditions at the planar interface, $z = 0$, we can relate these field components to each other. As we have learned in earlier chapters, the tangential electric and magnetic fields experience no discontinuities in the absence of any surface charges or surface currents at the interface between two materials. Hence, we state that the total (sum) of tangential field components on either side of the interface $z = 0$ plane are equal, and hence we can write

$$\bar{E}_{1x}^+(0) + \bar{E}_{1x}^-(0) = \bar{E}_{2x}^+(0) \tag{12.17}$$

$$\bar{H}_{1y}^+(0) + \bar{H}_{1y}^-(0) = \bar{H}_{2y}^+(0) \tag{12.18}$$

In keeping with the tradition of expressing most of the analysis in terms of the electric field components, we use Equations (11.43) and (11.45) to rewrite Equation (12.18) in the form:

$$\frac{\bar{E}_{1x}^+(0)}{\bar{\eta}_{o1}} - \frac{\bar{E}_{1x}^-(0)}{\bar{\eta}_{o1}} = \frac{\bar{E}_{2x}^+(0)}{\bar{\eta}_{o2}} \tag{12.19}$$

where $\bar{\eta}_{o1}$ and $\bar{\eta}_{o2}$ are the intrinsic (phasor) impedances of dielectrics #1 and #2, respectively.

Now, if we concern ourselves for the time being with what happens at $z = 0$ upon the incidence of the *TEM* wave, we can state that the incident field component, $\bar{E}_{1x}^{+}(0)$, has now resulted in causing two new field components: a reflected component, $\bar{E}_{1x}^{-}(0)$, and a transmitted component, $\bar{E}_{2x}^{+}(0)$. Similarly to what we did in the case of transmission line discontinuities, we now define a phasor reflection coefficient, $\bar{\Gamma}(0)$, and phasor transmission coefficient, $\bar{\tau}(0)$, at the interface as

$$\bar{\Gamma}(0)=\frac{\bar{E}_{1x}^{-}(0)}{\bar{E}_{1x}^{+}(0)} \tag{12.20}$$

$$\bar{\tau}(0)=\frac{\bar{E}_{2x}^{+}(0)}{\bar{E}_{1x}^{+}(0)} \tag{12.21}$$

Dividing both Equations (12.17) and (12.19) by $\bar{E}_{1x}^{+}(0)$ and applying the above definitions, we can write

$$1+\frac{\bar{E}_{1x}^{-}(0)}{\bar{E}_{1x}^{+}(0)}=\frac{\bar{E}_{2x}^{+}(0)}{\bar{E}_{1x}^{+}(0)} \tag{12.22}$$

$$1+\bar{\Gamma}(0)=\bar{\tau}(0) \tag{12.23}$$

$$\frac{1}{\bar{\eta}_{o1}}-\frac{\bar{\Gamma}(0)}{\bar{\eta}_{o1}}=\frac{\bar{\tau}(0)}{\bar{\eta}_{o2}} \tag{12.24}$$

$$1-\bar{\Gamma}(0)=\frac{\bar{\eta}_{o1}}{\bar{\eta}_{o2}}\bar{\tau}(0) \tag{12.25}$$

Solving Equations (12.23) and (12.25) simultaneously for $\bar{\Gamma}(0)$ and $\bar{\tau}(0)$, we get

$$\bar{\Gamma}(0)=\frac{\bar{\eta}_{o2}-\bar{\eta}_{o1}}{\bar{\eta}_{o2}+\bar{\eta}_{o1}} \tag{12.26}$$

$$\bar{\tau}(0)=1+\bar{\Gamma}(0)=\frac{2\bar{\eta}_{o2}}{\bar{\eta}_{o2}+\bar{\eta}_{o1}} \tag{12.27}$$

It is worth noting here that Equation (12.26) for the reflection coefficient $\bar{\Gamma}(0)$ is similar to that derived earlier for transmission line discontinuities, Equation (2.62).

Now, going back to the reflected and transmitted wave [Equations (12.13) to (12.16)], we can rewrite them all in terms of $\bar{E}_{1x}^{+}(0)$ and $\bar{\Gamma}(0)$ as follows:

$$\bar{E}_{1x}^{-}(z)=\bar{E}_{1x}^{-}(0)\cdot e^{+\gamma_1 z}=\bar{\Gamma}(0)\cdot\bar{E}_{1x}^{+}(0)\cdot e^{+\gamma_1 z} \tag{12.28}$$

$$\bar{H}_{1y}^{-}(z)=\bar{H}_{1y}^{-}(0)\cdot e^{+\gamma_1 z}=-\frac{1}{\bar{\eta}_{o1}}\cdot\bar{E}_{1x}^{-}(z)=-\frac{1}{\bar{\eta}_{o1}}\cdot\bar{\Gamma}(0)\cdot\bar{E}_{1x}^{+}(0)\cdot e^{+\gamma_1 z} \tag{12.29}$$

$$\bar{E}_{2x}^{+}(z)=\bar{E}_{2x}^{+}(0)\cdot e^{-\gamma_2 z}=[1+\bar{\Gamma}(0)]\cdot\bar{E}_{1x}^{+}(0)\cdot e^{-\gamma_2 z} \tag{12.30}$$

$$\bar{H}_{2y}^{+}(z)=\bar{H}_{2y}^{+}(0)\cdot e^{-\gamma_2 z}=\frac{1}{\bar{\eta}_{o2}}\cdot\bar{E}_{2x}^{+}(z)=\frac{1}{\bar{\eta}_{o2}}\cdot[1+\bar{\Gamma}(0)]\cdot\bar{E}_{1x}^{+}(0)\cdot e^{-\gamma_2 z} \tag{12.31}$$

The total fields at all $z-$ and $z+$ locations would be the superposition of all existing field components, that is,

$$\begin{aligned}\bar{E}_{1x}(z)\big|_{z\le 0}&=\bar{E}_{1x}^{+}(z)+\bar{E}_{1x}^{-}(z)=\bar{E}_{1x}^{+}(0)\cdot e^{-\gamma_1 z}+\bar{\Gamma}(0)\cdot\bar{E}_{1x}^{+}(0)\cdot e^{+\gamma_1 z}\\&=\bar{E}_{1x}^{+}(0)\cdot\{e^{-\gamma_1 z}+\bar{\Gamma}(0)e^{+\gamma_1 z}\}=\bar{E}_{1x}^{+}(z)\cdot\{1+\bar{\Gamma}(0)\cdot e^{+2\gamma_1 z}\}\end{aligned} \tag{12.32}$$

$$\bar{H}_{1y}(z)\Big|_{z\le 0} = \bar{H}_{1y}^{+}(z)+\bar{H}_{1y}^{-}(z)=\frac{1}{\bar{\eta}_{o1}}\cdot\left\{\bar{E}_{1x}^{+}(0)\cdot e^{-\gamma_1 z}-\bar{\Gamma}(0)\cdot\bar{E}_{1x}^{+}(0)\cdot e^{+\gamma_1 z}\right\}$$

$$=\frac{1}{\bar{\eta}_{o1}}\cdot\bar{E}_{1x}^{+}(0)\cdot\{e^{-\gamma_1 z}-\bar{\Gamma}(0)e^{+\gamma_1 z}\}=\frac{1}{\bar{\eta}_{o1}}\cdot\bar{E}_{1x}^{+}(z)\cdot\{1-\bar{\Gamma}(0)\cdot e^{+2\gamma_1 z}\} \tag{12.33}$$

$$\bar{E}_{2x}(z)\Big|_{z\ge 0} = \bar{E}_{2x}^{+}(z)=\bar{E}_{2x}^{+}(0)\cdot e^{-\gamma_2 z}=[1+\bar{\Gamma}(0)]\cdot\bar{E}_{1x}^{+}(0)\cdot e^{-\gamma_2 z} \tag{12.34}$$

$$\bar{H}_{2y}(z)\Big|_{z\ge 0} = \bar{H}_{2y}^{+}(z)=\bar{H}_{2y}^{+}(0)\cdot e^{-\gamma_2 z}=\frac{1}{\bar{\eta}_{o2}}\cdot[1+\bar{\Gamma}(0)]\cdot\bar{E}_{1x}^{+}(0)\cdot e^{-\gamma_2 z} \tag{12.35}$$

Using Equations (12.11) and (12.28), we can define the general expression for the reflection coefficient at a general location z, as

$$\bar{\Gamma}(z)=\frac{\bar{E}_{1x}^{-}(z)}{\bar{E}_{1x}^{+}(z)}=\frac{\bar{\Gamma}(0)\cdot\bar{E}_{1x}^{+}(0)\cdot e^{+\gamma_1 z}}{\bar{E}_{1x}^{+}(0)\cdot e^{-\gamma_1 z}}=\bar{\Gamma}(0)\cdot e^{+2\gamma_1 z} \tag{12.36}$$

Using this definition, we can rewrite Equations (12.32) and (12.33) as follows:

$$\bar{E}_{1x}(z)\Big|_{z\le 0} = \bar{E}_{1x}^{+}(z)+\bar{E}_{1x}^{-}(z)=\bar{E}_{1x}^{+}(z)+\bar{\Gamma}(z)\cdot\bar{E}_{1x}^{+}(z)=\bar{E}_{1x}^{+}(z)\cdot\{1+\bar{\Gamma}(z)\} \tag{12.37}$$

likewise,

$$\bar{H}_{1y}(z)\Big|_{z\le 0} = \frac{1}{\bar{\eta}_{o1}}\cdot\bar{E}_{1x}^{+}(z)\cdot\{1-\bar{\Gamma}(z)\} \tag{12.38}$$

12.5 Physical Insight—Special Cases

Now that we have derived the full mathematical solution for the fields for a normally incident *TEM* wave at the discontinuity of Figure 12.7, let us consider some special cases in order to gain some physical insight. In this regard, we will consider three cases: (1) the case of two perfect dielectrics, (2) the case of two lossy dielectrics, and finally, (3) the case of a dielectric-conductor interface.

12.5.1 The Case of Two Perfect Dielectrics (Figure 12.8)

Based on our findings back in Chapter 2 for transmission lines, we expect wave propagation in perfect dielectrics to be similar to that in lossless transmission lines. The coexistence of waves traveling in opposite directions throughout dielectric #1 is expected to result in the creation of standing waves. Depending on the relative amplitudes of the two waves, a mixture of traveling and standing waves will be the case as demonstrated at the bottom of Figure 12.8. Meanwhile, in dielectric #2, with only one wave traveling in the positive z direction, a pure traveling wave is expected as demonstrated in the same figure.

We now, carry on the analytical development for this special case. For perfect dielectrics, both γ's are expected to be pure imaginary quantities, $\gamma = j\beta$ while the $\bar{\eta}_o$'s are expected to be real, η_o. Consequently, and according to Equation (12.26), $\bar{\Gamma}(0)$ should be a real quantity. So, it follows that Equation (12.36) can be written in the form:

$$\bar{\Gamma}(z)=\Gamma(0)\cdot e^{+2j\beta_1 z}=\frac{\eta_{o2}-\eta_{o1}}{\eta_{o2}+\eta_{o1}}\cdot e^{+2j\beta_1 z} \tag{12.39}$$

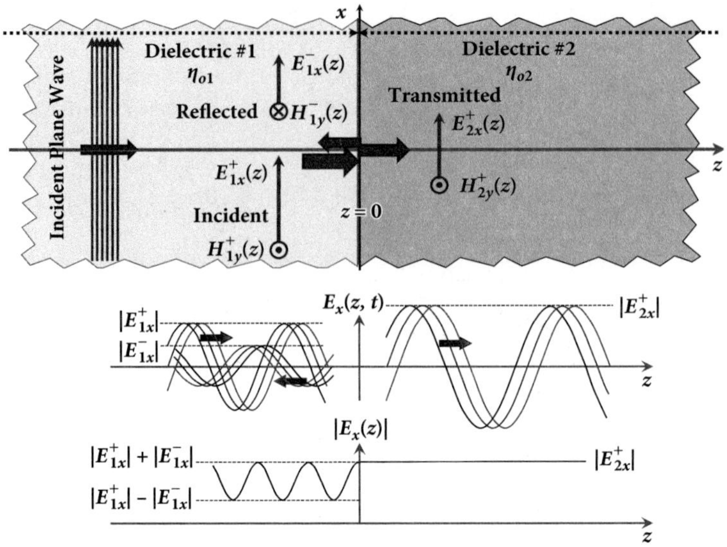

FIGURE 12.8 The reflected wave in the incidence dielectric will result in the formation of standing waves while the transmitted wave will be a travelling one.

Moreover, we can now develop Equation (12.32) as follows:

$$\begin{aligned}\bar{E}_{1x}(z)\Big|_{z\le 0} &= \bar{E}_{1x}^+(z)\cdot\{1+\bar{\Gamma}(z)\} = \bar{E}_{1x}^+(0)\cdot\{e^{-j\beta_1 z}+\Gamma(0)\cdot e^{+j\beta_1 z}\}\\ &= \bar{E}_{1x}^+(0)\{e^{-j\beta_1 z}+\Gamma(0)e^{+j\beta_1 z}+[\Gamma(0)e^{-j\beta_1 z}-\Gamma(0)e^{-j\beta_1 z}]\}\\ &= \bar{E}_{1x}^+(0)\{[1-\Gamma(0)]e^{-j\beta_1 z}+\Gamma(0)[e^{+j\beta_1 z}+e^{-j\beta_1 z}]\}\\ &= \bar{E}_{1x}^+(0)\{[1-\Gamma(0)]e^{-j\beta_1 z}+2\Gamma(0)[\cos(\beta_1 z)]\}\\ &= \left\{[1-\Gamma(0)]\bar{E}_{1x}^+(0)\right\}e^{-j\beta_1 z}+\left\{2\Gamma(0)\bar{E}_{1x}^+(0)\right\}\cos(\beta_1 z)\end{aligned} \tag{12.40}$$

$$\bar{E}_{1x}(z)\Big|_{z\le 0} = \left\{[1-\Gamma(0)]\bar{E}_{1x}^+(0)\right\}e^{-j\beta_1 z}+\left\{2\Gamma(0)\bar{E}_{1x}^+(0)\right\}\cos(\beta_1 z) \tag{12.41}$$

And, the corresponding time-domain form of Equation (12.41) is

$$\begin{aligned}E_{1x}(z,t)\Big|_{z\le 0} &= \left\{[1-\Gamma(0)]\left|\bar{E}_{1x}^+(0)\right|\right\}\cos(\omega t-\beta_1 z+\varphi_{E1x})\\ &\quad+\left\{2\Gamma(0)\left|\bar{E}_{1x}^+(0)\right|\right\}\cos(\omega t+\varphi_{E1x})\cos(\beta_1 z)\end{aligned} \tag{12.42}$$

which is similar to Equation (2.75) developed earlier for the case of a lossless transmission line discontinuity (with the removal of the arbitrary phases φ_v^+, φ_1, and φ_2):

$$v(z,t) = \left\{\left[\left|V^+\right|-\left|V^-\right|\right]\cos(\omega t-\beta z)\right\}+\left\{2\left|V^-\right|[\cos(\omega t)\cos(\beta z)]\right\} \tag{2.75}$$

Similar to our earlier discussion in Chapter 2 of the solution (2.75), the field solution given in Equations (12.41) and (12.42) can be viewed as the sum of two *TEM* waves; a traveling wave and a standing wave:

$$\bar{E}_{1x}(z)\Big|_{z\le 0,\,\text{Traveling}} = \left\{[1-\Gamma(0)]\bar{E}_{1x}^+(0)\right\}e^{-j\beta_1 z} \tag{12.43}$$

$$E_{1x}(z,t)\Big|_{z\le 0,\,\text{Traveling}} = \left\{[1-\Gamma(0)]\left|\bar{E}_{1x}^+(0)\right|\right\}\cos(\omega t-\beta_1 z+\varphi_{E1x}) \tag{12.44}$$

$$\bar{E}_{1x}(z)\Big|_{z\le 0,\ \text{Standing}} = \left\{2\Gamma(0)\ \bar{E}_{1x}^{+}(0)\right\}\cos(\beta_1 z) \tag{12.45}$$

$$E_{1x}(z,t)\Big|_{z\le 0,\ \text{Standing}} = \left\{2\Gamma(0)\left|\bar{E}_{1x}^{+}(0)\right|\right\}\cos(\omega t+\varphi_{E1x})\cos(\beta_1 z) \tag{12.46}$$

and the magnitude of the total electric field intensity in dielectric #1 is given by

$$\left|\bar{E}_{1x}(z)\right|\Big|_{z\le 0} = \left|\bar{E}_{1x}^{+}(0)\right|\left|[1-\Gamma(0)]e^{-j\beta_1 z}+2\Gamma(0)\cos(\beta_1 z)\right| \tag{12.47}$$

If we normalize the field expression with respect to the intensity of the incident wave at the interface, $\bar{E}_{1x}^{+}(0)$, we obtain

$$\left.\frac{\left|\bar{E}_{1x}(z)\right|}{\left|\bar{E}_{1x}^{+}(0)\right|}\right|_{z\le 0} = \left|[1-\Gamma(0)]e^{-j\beta_1 z}+2\Gamma(0)\cos(\beta_1 z)\right| \tag{12.48}$$

The transmitted wave in dielectric #2 is totally a traveling one, according to Equation (12.34)

$$\bar{E}_{2x}(z)\Big|_{z\ge 0} = [1+\Gamma(0)]\cdot\bar{E}_{1x}^{+}(0)\cdot e^{-\gamma_2 z} = [1+\Gamma(0)]\bar{E}_{1x}^{+}(0)e^{-j\beta_2 z} \tag{12.49}$$

$$E_{2x}(z,t)\Big|_{z\ge 0} = [1+\Gamma(0)]\left|\bar{E}_{1x}^{+}(0)\right|\cos(\omega t-\beta_2 z+\varphi_{E1x}) \tag{12.50}$$

$$\left|\bar{E}_{2x}(z)\right|\Big|_{z\ge 0} = [1+\Gamma(0)]\left|\bar{E}_{1x}^{+}(0)\right| \tag{12.51}$$

with the normalized expression for the field in dielectric #2 given by

$$\left.\frac{\left|\bar{E}_{2x}(z)\right|}{\left|\bar{E}_{1x}^{+}(0)\right|}\right|_{z\ge 0} = [1+\Gamma(0)] \tag{12.52}$$

It is worth noting that both Equations (12.48) and (12.52) have the same exact value at $z = 0$, which is implied by the continuity of the tangential electric field at the interface,

$$\left.\frac{\left|\bar{E}_{1x}(z)\right|}{\left|\bar{E}_{1x}^{+}(0)\right|}\right|_{z\le 0} = \{[1-\Gamma(0)]e^{-j0}+2\Gamma(0)\cos(0)\} = [1+\Gamma(0)] = \left.\frac{\left|\bar{E}_{2x}(z)\right|}{\left|\bar{E}_{1x}^{+}(0)\right|}\right|_{z\ge 0} \tag{12.53}$$

Similarly, we can develop the expressions for the magnetic field intensity phasor and its magnitude, the results for which are given as follows:

$$\bar{H}_{1y}(z)\Big|_{z\le 0,\ \text{Traveling}} = \bar{H}_{1y}^{+}(0)\cdot\{[1-\Gamma(0)]\}e^{-j\beta_1 z} \tag{12.54}$$

$$\bar{H}_{1y}(z)\Big|_{z\le 0,\ \text{Standing}} = \bar{H}_{1y}^{+}(0)\cdot\{-2j\Gamma(0)\}\sin(\beta_1 z) \tag{12.55}$$

$$\left|\bar{H}_{1y}(z)\right|\Big|_{z\le 0} = \bar{H}_{1y}^{+}(0)\cdot\left\{[1-\Gamma(0)]e^{-j\beta_1 z}-2j\Gamma(0)\sin(\beta_1 z)\right\} \tag{12.56}$$

with the normalized expression for the field in dielectric #1 given by

$$\left.\frac{\left|\bar{H}_{1y}(z)\right|}{\left|\bar{H}_{1y}^{+}(0)\right|}\right|_{z\le 0} = \left|[1-\Gamma(0)]e^{-j\beta_1 z}-2j\Gamma(0)\sin(\beta_1 z)\right| \tag{12.57}$$

and for dielectric #2, we can derive

$$\bar{H}_{2y}(z)\Big|_{z\geq 0} = [1-\Gamma(0)]\cdot \bar{H}^+_{1y}(0)\cdot e^{-\gamma_2 z} = [1-\Gamma(0)]\bar{H}^+_{1y}(0)\, e^{-j\beta_2 z} \tag{12.58}$$

$$\left|\bar{H}_{2y}(z)\right|\Big|_{z\geq 0} = [1-\Gamma(0)]\left|\bar{H}^+_{1y}(0)\right| \tag{12.59}$$

with the normalized expression for the field in dielectric #2 given by

$$\left.\frac{\left|\bar{H}_{2y}(z)\right|}{\left|\bar{H}^+_{1y}(0)\right|}\right|_{z\geq 0} = [1-\Gamma(0)] \tag{12.60}$$

A plot of the normalized magnitudes of both electric and magnetic field intensities as functions of position is shown in Figures 12.9 and 12.10 for the cases of $\eta_{o1} < \eta_{o2}$ and $\eta_{o1} > \eta_{o2}$, respectively. For the case of $\eta_{o1} < \eta_{o2}$, Figure 12.9, $\Gamma(0)$ has a positive real value which makes $1 + \Gamma(0)$ greater than 1 and hence $\left|\bar{E}_{2x}(0)/\bar{E}^+_{1x}(0)\right| > 1$. The opposite is true for the case of $\eta_{o1} > \eta_{o2}$, Figure 12.10, where $\Gamma(0)$ has a negative real value with $1 + \Gamma(0)$ less than 1 and $\left|\bar{E}_{2x}(0)/\bar{E}^+_{1x}(0)\right| < 1$.

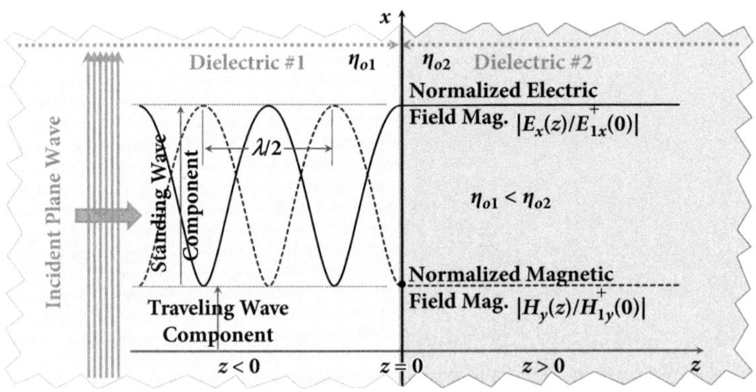

FIGURE 12.9 The normalized electric and magnetic fields magnitudes on both sides of the interface for the case of lossless dielectrics with $\eta_{o1} < \eta_{o2}$.

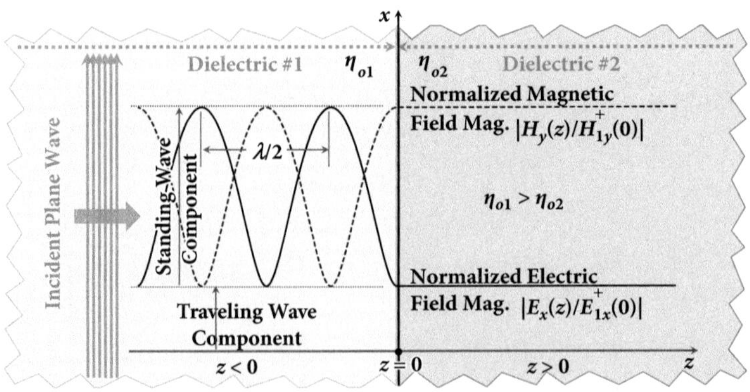

FIGURE 12.10 The normalized electric and magnetic fields magnitudes on both sides of the interface for the case of lossless dielectrics with $\eta_{o1} < \eta_{o2}$.

12.5.2 The Case of Two Lossy Dielectrics (Figure 12.11)

In this case, we will assume that both materials are low loss dielctrics, both γ's are expected to be complex quantities, $\gamma = \alpha + j\beta$, with relatively small α values, while the $\bar{\eta}_o$'s are expected to be almost pure real, η_o. Consequently, and according to Equation (12.26), $\bar{\Gamma}(0)$ can be treated as a real quantity, $\Gamma(0)$. So, it follows that Equation (12.36) can be written in the form:

$$\bar{\Gamma}(z) = \Gamma(0)e^{+2\gamma_1 z} = \frac{\eta_{o2} - \eta_{o1}}{\eta_{o2} + \eta_{o1}} e^{+2\gamma_1 z} \tag{12.61}$$

while Equation (12.32) can now be developed as follows:

$$\begin{aligned}\bar{E}_{1x}(z)\Big|_{z\le 0} &= \bar{E}_{1x}^{+}(z)\cdot\{1+\bar{\Gamma}(z)\} = \bar{E}_{1x}^{+}(0)\cdot\{e^{-\gamma_1 z} + \Gamma(0)e^{+\gamma_1 z}\}\\ &= \bar{E}_{1x}^{+}(0)\{e^{-\gamma_1 z} + \Gamma(0)e^{+\gamma_1 z} + [\Gamma(0)e^{-\gamma_1 z} - \Gamma(0)e^{-\gamma_1 z}]\}\\ &= \bar{E}_{1x}^{+}(0)\{[1-\Gamma(0)]e^{-\gamma_1 z} + \Gamma(0)[e^{+\gamma_1 z} + e^{-\gamma_1 z}]\}\\ &= \bar{E}_{1x}^{+}(0)\{[1-\Gamma(0)]e^{-\gamma_1 z} + 2\Gamma(0)[\cosh(\gamma_1 z)]\}\end{aligned} \tag{12.62}$$

$$\bar{E}_{1x}(z)\Big|_{z\le 0} = \left\{[1-\Gamma(0)]\bar{E}_{1x}^{+}(0)\right\}e^{-\gamma_1 z} + \left\{2\Gamma(0)\bar{E}_{1x}^{+}(0)\right\}\cosh(\gamma_1 z) \tag{12.63}$$

Similar to our earlier discussion in the case of perfect dielectrics, the field solution given in Equation (12.63) can be viewed as the sum of two *TEM* waves; a traveling wave and a standing wave:

$$\bar{E}_{1x}(z)\Big|_{z\le 0,\ \text{Traveling}} = \left\{[1-\Gamma(0)]\bar{E}_{1x}^{+}(0)\right\}e^{-\gamma_1 z} \tag{12.64}$$

$$\bar{E}_{1x}(z)\Big|_{z\le 0,\ \text{Standing}} = \left\{2\Gamma(0)\bar{E}_{1x}^{+}(0)\right\}\cosh(\gamma_1 z) \tag{12.65}$$

and the normalized magnitude of the total electric field intensity in dielectric #1 is given by

$$\left.\frac{\left|\bar{E}_{1x}(z)\right|}{\left|\bar{E}_{1x}^{+}(0)\right|}\right|_{z\le 0} = \left|[1-\Gamma(0)]e^{-\gamma_1 z} + 2\Gamma(0)[\cosh(\gamma_1 z)]\right| \tag{12.66}$$

The transmitted wave in dielectric #2 is totally a traveling one, according to Equation (12.34)

$$\bar{E}_{2x}(z)\Big|_{z\ge 0} = [1+\bar{\Gamma}(0)]\cdot \bar{E}_{1x}^{+}(0)\cdot e^{-\gamma_2 z} \tag{12.67}$$

with the normalized expression for the field in dielectric #2 given by

$$\left.\frac{\left|\bar{E}_{2x}(z)\right|}{\left|\bar{E}_{1x}^{+}(0)\right|}\right|_{z\ge 0} = [1+\Gamma(0)]\cdot e^{-\alpha_2 z} \tag{12.68}$$

Similarly, we can develop the expressions for the magnetic field intensity phasor and its magnitude. The results for the normalized magnetic field expressions are of the form:

$$\left.\frac{\left|\bar{H}_{1y}(z)\right|}{\left|\bar{H}_{1y}^{+}(0)\right|}\right|_{z\le 0} = \left|[1-\Gamma(0)]e^{-\gamma_1 z} - 2\Gamma(0)[\sinh(\gamma_1 z)]\right| \tag{12.69}$$

$$\left.\frac{\left|\bar{H}_{2y}(z)\right|}{\left|\bar{H}_{1y}^{+}(0)\right|}\right|_{z\ge 0} = [1-\Gamma(0)]\cdot e^{-\alpha_2 z} \tag{12.70}$$

A plot of the normalized magnitudes of both electric and magnetic field intensity as functions of position for the case of low-loss dielectrics is shown in Figure 12.11 for the case of $\eta_{o1} > \eta_{o2}$.

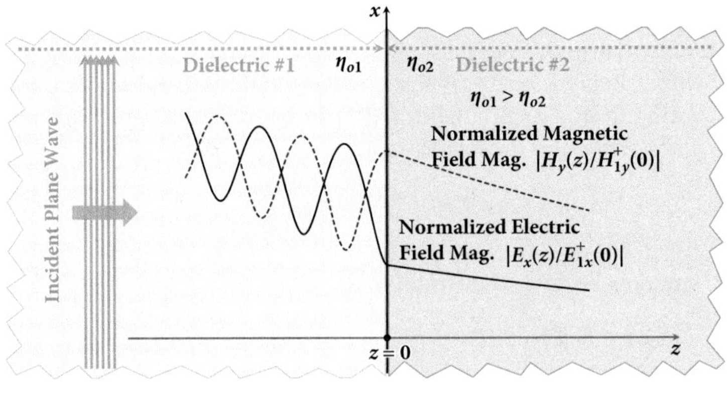

FIGURE 12.11 The normalized electric and magnetic fields magnitudes on both sides of the interface for the case of lossy dielectrics with $\eta_{o1} < \eta_{o2}$.

12.5.3 The Case of a Dielectric-Conductor Interface (Figure 12.12)

Now, we study the important case of a dielectric-conductor interface. The study of this case is useful in order to appreciate what happens to wave propagation in the presence of highly conductive discontinuities. Examples of such are vehicles, metallic faced buildings, or even steel reinforced concrete structures. For the sake of mathematical simplicity, we consider the extreme case of a *TEM* wave and a dielectric-perfect conductor interface as demonstrated in Figure 12.12. Again, the transmission-line equivalent model is displayed where we expect to see pure standing waves resulting due to the 100% reflection at the discontinuity.

Now, we turn to the mathematical analysis of the case. All we need to do in order to derive the corresponding field expressions is to set $\bar{\eta}_o$ for the conductor material to 0 (Table 11.6). Consequently, and according to Equation (12.26), $\bar{\Gamma}(0) = -1$, from which the following expressions are obtained:

$$\bar{\Gamma}(z) = \Gamma(0)e^{+2\gamma_1 z} = -e^{+2\gamma_1 z} \tag{12.71}$$

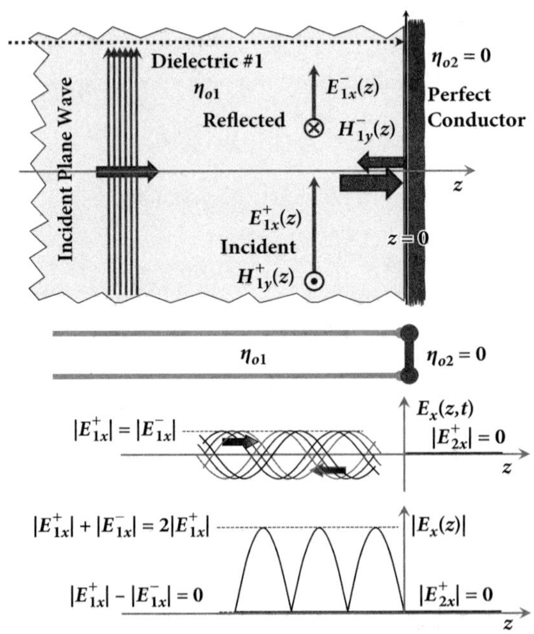

FIGURE 12.12 Fields of incident and reflected waves for normal incidence to the interface between a lossless dielectric and a perfect conductor.

$$\bar{E}_{1x}(z)\Big|_{z\le 0} = \bar{E}^+_{1x}(z)\cdot\{1+\bar{\Gamma}(z)\} = -\bar{E}^+_{1x}(0)\{2\,\sinh(\gamma_1 z)\} \tag{12.72}$$

$$\bar{E}_{2x}(z)\Big|_{z\ge 0} = 0 \tag{12.73}$$

and the normalized magnitude of the total electric and magnetic field intensities in dielectric #1 are given by

$$\left.\frac{|\bar{E}_{1x}(z)|}{|\bar{E}^+_{1x}(0)|}\right|_{z\le 0,\ \text{lossy dielectric}} = |2\,\sinh(\gamma_1 z)| \tag{12.74}$$

$$\left.\frac{|\bar{E}_{1x}(z)|}{|\bar{E}^+_{1x}(0)|}\right|_{z\le 0,\ \text{lossless dielectric}} = |2\,\sin(\beta_1 z)| \tag{12.75}$$

$$\left.\frac{|\bar{E}_{2x}(z)|}{|\bar{E}^+_{1x}(0)|}\right|_{z\ge 0} = 0 \tag{12.76}$$

$$\left.\frac{|\bar{H}_{1y}(z)|}{|\bar{H}^+_{1y}(0)|}\right|_{z\le 0,\ \text{lossy dielectric}} = |2\,\cosh(\gamma_1 z)| \tag{12.77}$$

$$\left.\frac{|\bar{H}_{1y}(z)|}{|\bar{H}^+_{1y}(0)|}\right|_{z\le 0,\ \text{lossless dielectric}} = |2\,\cos(\beta_1 z)| \tag{12.78}$$

$$\left.\frac{|\bar{H}_{2y}(z)|}{|\bar{H}^+_{1y}(0)|}\right|_{z\ge 0} = 0 \tag{12.79}$$

A plot of the normalized magnitudes of both electric and magnetic field intensity as functions of position for the cases of lossless as well as low loss dielectrics are shown in Figure 12.13.

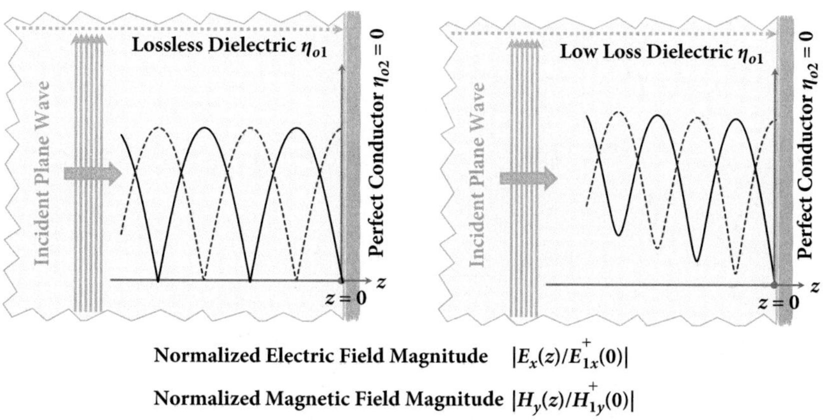

FIGURE 12.13 Standing waves formed in the case of normal incidence to the interface between a dielectric and a perfect conductor. To the left is the case of a lossless dielectric and to the right is the case of a low-loss dielectric.

Example 12.5 A plane wave traveling in medium 1 with $\varepsilon_{r1} = 2.25$ is normally incident upon medium 2 with $\varepsilon_{r2} = 4$. Both media are nonmagnetic and nonconducting. If the electric field of the incident wave is given by $\vec{E}_1^+(x,t) = \vec{a}_y[8\cos(6\pi \times 10^9 t - \beta_1 x)]$ V/m.

a. Obtain time domain expressions for the electric and magnetic field vectors in each of the two media.
b. Find the standing wave ratio and the location of the first electric field maximum and minimum from the interface at $x = 0$.
c. Determine the average power densities of the incident, reflected and transmitted waves.
d. Determine the fraction of the incident power that is transmitted to medium 2.

Solution Part a

$$\beta_1 = \frac{\omega}{c_{ph_1}} = \frac{\omega}{c} \times \sqrt{u_{r1}\varepsilon_{r1}} = \frac{6\pi \times 10^9}{3 \times 10^8} \times \sqrt{2.25} = 30\pi \text{ rad/m}$$

$$\vec{E}_1^+(x,t) = \vec{a}_y[8\cos(6\pi \times 10^9 t - 30\pi x)]$$

$$\bar{\vec{E}}_1^+(x) = \vec{a}_y 8e^{-j30\pi x} \qquad \bar{\vec{E}}_1^-(x) = \vec{a}_y \bar{\Gamma}(0).8e^{+j30\pi x} \qquad \bar{\Gamma}(0) = \frac{\bar{\eta}_{o2} - \bar{\eta}_{o1}}{\bar{\eta}_{o2} + \bar{\eta}_{o1}}$$

$$\bar{\eta}_{o2} = \sqrt{\frac{\mu_2}{\varepsilon_2}} = 120\pi\sqrt{\frac{\mu_{r2}}{\varepsilon_{r2}}} = \frac{120\pi}{2} = 60\pi\ \Omega$$

$$\bar{\eta}_{o1} = \sqrt{\frac{\mu_1}{\varepsilon_1}} = 120\pi\sqrt{\frac{\mu_{r1}}{\varepsilon_{r1}}} = \frac{120\pi}{1.5} = 80\pi\ \Omega$$

$$\bar{\Gamma}(0) = \frac{60\pi - 80\pi}{60\pi + 80\pi} = -\frac{1}{7}$$

$$\bar{\vec{E}}_1^-(x) = \vec{a}_y\left(-\frac{8}{7}\right)e^{j30\pi x}$$

$$\bar{\vec{E}}_1(x) = \bar{\vec{E}}_1^+(x) + \bar{\vec{E}}_1^-(x) = \vec{a}_y 8\left[e^{-j30\pi x} - \frac{1}{7}e^{j30\pi x}\right]$$

$$\vec{E}_1(x,t) = \vec{a}_y 8\left[\cos(6\pi \times 10^9 t - 30\pi x) - \frac{1}{7}\cos(6\pi \times 10^9 t + 30\pi x)\right] \text{ V/m}$$

$$\bar{\vec{E}}_2(x) = (1 + \bar{\Gamma}(0))\bar{\vec{E}}_1^+(0)e^{-j\beta_2 x}$$

$$\beta_2 = \frac{\omega}{c_{ph_2}} = \frac{\omega}{c} \times \sqrt{u_{r2}\varepsilon_{r2}} = \frac{6\pi \times 10^9}{3 \times 10^8} \times \sqrt{4} = 40\pi \text{ rad/m}$$

$$\bar{\vec{E}}_2(x) = \left(1 + \left(-\frac{1}{7}\right)\right)(\vec{a}_y 8)e^{-j40\pi x} = \vec{a}_y \frac{48}{7}e^{-j40\pi x} \text{ V/m}$$

$$\vec{E}_2(x,t) = \vec{a}_y \frac{48}{7}[\cos(6\pi \times 10^9 t - 40\pi x)] \text{ V/m}$$

Expressions for the magnetic field phasors can be obtained from the corresponding electric field phasors using

$$\bar{\vec{H}} = \frac{1}{\bar{\eta}}\vec{a}_\beta \times \bar{\vec{E}}$$

where $\vec{a}_\beta$ represents a unit vector in the direction of propagation. In this case, $\vec{a}_\beta = \vec{a}_x$ for the forward going waves in both medium 1 and medium 2, and $\vec{a}_\beta = -\vec{a}_x$ for the reflected wave in medium 1.

$$\bar{\bar{H}}_1^+(x) = \frac{1}{\eta_{o1}}\vec{a}_x \times \bar{\bar{E}}_1^+(x) = \frac{1}{\eta_{o1}}\vec{a}_x \times \vec{a}_y 8e^{-j30\pi x} = \vec{a}_z \frac{8}{80\pi}e^{-j30\pi x} = \vec{a}_z \frac{1}{10\pi}e^{-j30\pi x} \text{ A/m}$$

$$\bar{\bar{H}}_1^-(x) = \frac{1}{\eta_{o1}}(-\vec{a}_x) \times \bar{\bar{E}}_1^-(x) = \frac{1}{\eta_{o1}}(-\vec{a}_x) \times \vec{a}_y\left(-\frac{8}{7}\right)e^{j30\pi x} = \vec{a}_z \frac{8/7}{80\pi}e^{j30\pi x} = \vec{a}_z \frac{1}{70\pi}e^{j30\pi x}$$

$$\bar{\bar{H}}_1(x) = \bar{\bar{H}}_1^+(x) + \bar{\bar{H}}_1^-(x) = \vec{a}_z \frac{1}{10\pi}\left[e^{-j30\pi x} + \frac{1}{7}e^{j30\pi x}\right] \text{ A/m}$$

$$\vec{H}_1(x,t) = \vec{a}_z \frac{1}{10\pi}\left[\cos(6\pi \times 10^9 t - 30\pi x) + \frac{1}{7}\cos(6\pi \times 10^9 t + 30\pi x)\right] \text{ A/m}$$

$$\bar{\bar{H}}_2(x) = \frac{1}{\eta_{o2}}\vec{a}_x \times \bar{\bar{E}}_2(x) = \vec{a}_z \frac{48/7}{60\pi}e^{-j40\pi x} = \vec{a}_z \frac{4}{35\pi}e^{-j40\pi x} \text{ A/m}$$

$$\vec{H}_2(x,t) = \vec{a}_z \frac{4}{35\pi}[\cos(6\pi \times 10^9 t - 40\pi x)] \text{ A/m}$$

Solution Part b The standing wave ratio and the location of the electric field maximum and minimum can be obtained using Equations (2.76) and (2.81) derived in Chapter 2 for a standing wave pattern

$$SWR = \frac{1+|\bar{\Gamma}|}{1-|\bar{\Gamma}|} = \frac{1+1/7}{1-1/7} = 1.33$$

The location of the maxima can be obtained using $d_{\max} = \frac{\lambda_1}{4\pi}\varphi_o \pm \frac{n\lambda_1}{2}$, where φ_o is the angle of the reflection coefficient at the interface, and λ_1 is the wavelength in medium 1, and $n = 0, 1, 2, \ldots$

$$\bar{\Gamma}(0) = -\frac{1}{7} = \frac{1}{7}\angle\pi \qquad \lambda_1 = \frac{c_{ph1}}{f} = \frac{3\times10^8/\sqrt{2.25}}{3\times10^9} = 6.67 \text{ cm}$$

$$d_{\max 1} = \frac{6.67}{4\pi}\pi \pm n\frac{6.67}{2} = 1.67 \text{ cm} \qquad \text{and} \qquad d_{\min 1} = d_{\max 1} \pm \frac{\lambda_1}{4} = 0 \text{ and } 3.33 \text{ cm}$$

Note that since the reflection coefficient is real and negative ($\varphi_o = \pi$) the first minimum will occur at the location of the interface, and the first maximum will be $\lambda_1/4$ away from the interface.

Solution Part c The average power densities of the incident, reflected and transmitted waves can be obtained as follows:

$$\bar{\bar{P}}_{\text{avg1}}^+(x) = \frac{1}{2}Re\left(\bar{\bar{E}}_1^+(x) \times \bar{\bar{H}}_1^{+*}(x)\right) = \vec{a}_x \frac{1}{2}\left|\bar{\bar{E}}_1^+(x)\right|^2 Re\left\{\frac{1}{\eta_{o1}^*}\right\} = \vec{a}_x \frac{1}{2}8^2\left(\frac{1}{80\pi}\right) = \vec{a}_x \frac{2}{5\pi} \text{ W/m}^2$$

In the above expression, we used the fact that the direction of power flow is the same as the direction of propagation and that the electric and magnetic fields are related by

$$\bar{H}_z = \frac{\bar{E}_y}{\eta}$$

$$\bar{\bar{P}}_{\text{avg1}}^-(x) = \frac{1}{2}Re\left(\bar{\bar{E}}_1^-(x) \times \bar{\bar{H}}_1^{-*}(x)\right) = \vec{a}_x \frac{1}{2}\left|\bar{\bar{E}}_1^-(x)\right|^2 Re\left\{\frac{1}{\eta_{o1}^*}\right\} = \vec{a}_x \frac{1}{2}\left(\frac{8}{7}\right)^2\left(\frac{1}{80\pi}\right) = \vec{a}_x \frac{2}{245\pi} \text{ W/m}^2$$

$$\bar{\bar{P}}^{-}_{avg2}(x) = \frac{1}{2}Re\left(\bar{\bar{E}}_2(x) \times \bar{\bar{H}}_1^*(x)\right) = \bar{a}_x \frac{1}{2}\left|\bar{\bar{E}}_2(x)\right|^2 Re\left\{\frac{1}{\bar{\eta}^*_{o2}}\right\} = \bar{a}_x \frac{1}{2}\left(\frac{48}{7}\right)^2 \left(\frac{1}{60\pi}\right) = \bar{a}_x \frac{96}{245\pi} \text{ W/m}^2$$

We can confirm that the difference between the incident average power-density and the reflected average power-density is equal to the transmitted average power-density

$$\bar{\bar{P}}^{+}_{avg1}(x) - \bar{\bar{P}}^{-}_{avg1}(x) = \bar{\bar{P}}_{avg2}(x)$$

Solution Part d The fraction of the incident power that is transmitted to medium 2

$$\frac{|\bar{\bar{P}}_{avg2}(x)|}{|\bar{\bar{P}}^{+}_{avg1}(x)|} = \frac{\frac{96}{245\pi}}{\frac{2}{5\pi}} = \frac{96}{98} = 98\%$$

A general expression of the fraction of the incident power transmitted to medium 2 can be obtained in terms of the transmission coefficient

$$\frac{\left|\bar{\bar{P}}_{avg2}(x)\right|}{\left|\bar{\bar{P}}^{+}_{avg1}(x)\right|} = \frac{\left|\bar{\bar{E}}_2(x)\right|^2 Re\{1/\bar{\eta}^*_{o2}\}}{\left|\bar{\bar{E}}^{+}_1(x)\right|^2 Re\{1/\bar{\eta}^*_{o1}\}}$$

For the given example where the two media are lossless

$$\frac{\left|\bar{\bar{P}}_{avg2}(x)\right|}{\left|\bar{\bar{P}}^{+}_{avg1}(x)\right|} = |\tau(0)|^2 \frac{\eta_{o1}}{\eta_{o2}} = \left(1 - \frac{1}{7}\right)^2 \frac{80\pi}{60\pi} = 98\%$$

12.6 Reflection of Uniform Plane Waves: Normal Incidence on Multiple Layers

In many practical cases, wave propagation encounters the presence of multiple dielectric layers (Figure 12.14). As a result, we would expect multiple reflection-transmission to occur causing standing and traveling waves in the different layers. The study of a general case is expected to be demanding and challenging. For this reason, we will limit our discussion here to demonstrating the concept in an oversimplified case; specifically, the case of one-layer dielectric sandwiched between two unbounded media as shown on the right side of Figure 12.14. Carrying out the analysis of this case will enable us to develop an insight in how to deal with the more complex structures.

To proceed with the analysis of this case, we present three possible approaches; (1) the field analysis approach, (2) the impedance (circuit model) approach, and (3) using the bounce diagram approach. In a way, this is similar to circuit analysis where one has more than one possible approach to deal with. In the following, we will do the field analysis approach and include the other two approaches in Addendum 12A.

12.6.1 The Field Analysis Approach for Normal Incidence on Multiple Layers (Figure 12.15)

We start by writing the expected expressions for the field components in all three regions of space. Next, we apply the boundary conditions at the interfaces. The result is the following set of equations:

Region 1, incident wave:

$$\bar{E}^{+}_{1x}(z) = \bar{E}^{+}_{1x}(0) \cdot e^{-\gamma_1 z} \tag{12.80}$$

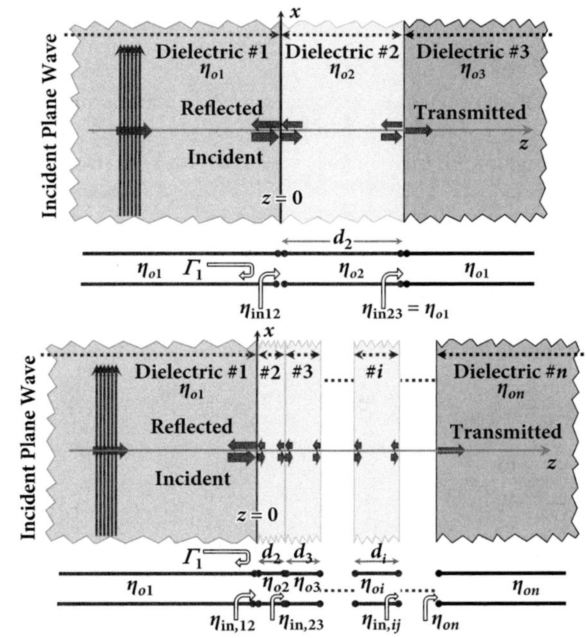

FIGURE 12.14 The general case of normal incidence on multiple layers can be treated using its equivalent *TL* model.

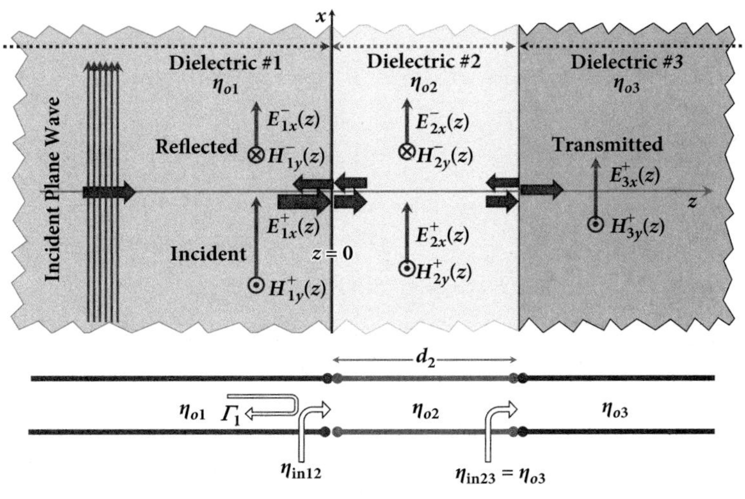

FIGURE 12.15 The analysis of normal incidence in the case of three-layers of dielectrics.

$$\bar{H}_{1y}^{+}(z) = \bar{H}_{1y}^{+}(0) \cdot e^{-\gamma_1 z} \tag{12.81}$$

Region 1, reflected wave (travels in the negative z direction):

$$\bar{E}_{1x}^{-}(z) = \bar{E}_{1x}^{-}(0) \cdot e^{+\gamma_1 z} \tag{12.82}$$

$$\bar{H}_{1y}^{-}(z) = \bar{H}_{1y}^{-}(0) \cdot e^{+\gamma_1 z} \tag{12.83}$$

Region 2, transmitted wave (travels in the positive z direction):

$$\overline{E}_{2x}^{+}(z)=\overline{E}_{2x}^{+}(0)\cdot e^{-\gamma_2 z} \tag{12.84}$$

$$\overline{H}_{2y}^{+}(z)=\overline{H}_{2y}^{+}(0)\cdot e^{-\gamma_2 z} \tag{12.85}$$

Region 2, reflected wave (travels in the negative z direction):

$$\overline{E}_{2x}^{-}(z)=\overline{E}_{2x}^{-}(0)\cdot e^{+\gamma_2 z} \tag{12.86}$$

$$\overline{H}_{2y}^{-}(z)=\overline{H}_{2y}^{-}(0)\cdot e^{+\gamma_2 z} \tag{12.87}$$

Region 3, transmitted wave (travels in the positive z direction):

$$\overline{E}_{3x}^{+}(z)=\overline{E}_{3x}^{+}(d)\cdot e^{-\gamma_3 (z-d)} \tag{12.88}$$

$$\overline{H}_{3y}^{+}(z)=\overline{H}_{3y}^{+}(d)\cdot e^{-\gamma_3 (z-d)} \tag{12.89}$$

We notice that region 3 does not have a reflected wave since it has infinite extension in the z direction. We also notice that we referred the z propagation expressions in this region to the beginning of the region at $z = d$.

We now write the boundary conditions at both interfaces.

Interface at $z = 0$:

$$\overline{E}_{1x}^{+}(0)+\overline{E}_{1x}^{-}(0)=\overline{E}_{2x}^{+}(0)+\overline{E}_{2x}^{-}(0) \tag{12.90}$$

$$\overline{H}_{1y}^{+}(0)+\overline{H}_{1y}^{-}(0)=\overline{H}_{2y}^{+}(0)+\overline{H}_{2y}^{-}(0) \tag{12.91}$$

with Equation (12.91) rewritten as

$$\frac{\overline{E}_{1x}^{+}(0)}{\overline{\eta}_{o1}}-\frac{\overline{E}_{1x}^{-}(0)}{\overline{\eta}_{o1}}=\frac{\overline{E}_{2x}^{+}(0)}{\overline{\eta}_{o2}}-\frac{\overline{E}_{2x}^{-}(0)}{\overline{\eta}_{o2}} \tag{12.92}$$

Interface at $z = d$:

$$\overline{E}_{2x}^{+}(d)+\overline{E}_{2x}^{-}(d)=\overline{E}_{3x}^{+}(d) \tag{12.93}$$

$$\overline{H}_{2y}^{+}(d)+\overline{H}_{2y}^{-}(d)=\overline{H}_{3y}^{+}(d) \tag{12.94}$$

$$\frac{\overline{E}_{2x}^{+}(d)}{\overline{\eta}_{o2}}-\frac{\overline{E}_{2x}^{-}(d)}{\overline{\eta}_{o2}}=\frac{\overline{E}_{3x}^{+}(d)}{\overline{\eta}_{o3}} \tag{12.95}$$

Equations (12.93) and (12.95) are rewritten as

$$\overline{E}_{2x}^{+}(0)\cdot e^{-\gamma_2 d}+\overline{E}_{2x}^{-}(0)\cdot e^{+\gamma_2 d}=\overline{E}_{3x}^{+}(d) \tag{12.96}$$

$$\frac{\overline{E}_{2x}^{+}(0)\cdot e^{-\gamma_2 d}}{\overline{\eta}_{o2}}-\frac{\overline{E}_{2x}^{-}(0)\cdot e^{+\gamma_2 d}}{\overline{\eta}_{o2}}=\frac{\overline{E}_{3x}^{+}(d)}{\overline{\eta}_{o3}} \tag{12.97}$$

Now, we normalize all field expressions with respect to $\overline{E}_{1x}^{+}(0)$; hence, we rewrite Equations (12.90), (12.92), (12.96), and (12.97), respectively, as

$$1+\frac{\overline{E}_{1x}^{-}(0)}{\overline{E}_{1x}^{+}(0)}=\frac{\overline{E}_{2x}^{+}(0)}{\overline{E}_{1x}^{+}(0)}+\frac{\overline{E}_{2x}^{-}(0)}{\overline{E}_{1x}^{+}(0)} \tag{12.98}$$

$$\frac{1}{\overline{\eta}_{o1}}\left[1-\frac{\overline{E}_{1x}^{-}(0)}{\overline{E}_{1x}^{+}(0)}\right]=\frac{1}{\overline{\eta}_{o2}}\left[\frac{\overline{E}_{2x}^{+}(0)}{\overline{E}_{1x}^{+}(0)}-\frac{\overline{E}_{2x}^{-}(0)}{\overline{E}_{1x}^{+}(0)}\right] \tag{12.99}$$

$$\frac{\overline{E}_{2x}^{+}(0)}{\overline{E}_{1x}^{+}(0)}\cdot e^{-\gamma_2 d}+\frac{\overline{E}_{2x}^{-}(0)}{\overline{E}_{1x}^{+}(0)}\cdot e^{+\gamma_2 d}=\frac{\overline{E}_{3x}^{+}(d)}{\overline{E}_{1x}^{+}(0)} \tag{12.100}$$

$$\frac{1}{\overline{\eta}_{o2}}\left[\frac{\overline{E}_{2x}^{+}(0)}{\overline{E}_{1x}^{+}(0)}e^{-\gamma_2 d}-\frac{\overline{E}_{2x}^{-}(0)}{\overline{E}_{1x}^{+}(0)}e^{+\gamma_2 d}\right]=\frac{1}{\overline{\eta}_{o3}}\left[\frac{\overline{E}_{3x}^{+}(d)}{\overline{E}_{1x}^{+}(0)}\right] \tag{12.101}$$

In the above four expressions, we recognize the following physical descriptions:

$$\Gamma(0)=\frac{\overline{E}_{1x}^{-}(0)}{\overline{E}_{1x}^{+}(0)}=\text{normalized reflected field signal in material 1 at } z=0 \tag{12.102}$$

$$\frac{\overline{E}_{2x}^{+}(0)}{\overline{E}_{1x}^{+}(0)}=\text{normalized forward field signal in material 2 at } z=0 \tag{12.103}$$

$$\frac{\overline{E}_{2x}^{-}(0)}{\overline{E}_{1x}^{+}(0)}=\text{normalized backward field signal in material 2 at } z=0 \tag{12.104}$$

$$\frac{\overline{E}_{3x}^{+}(d)}{\overline{E}_{1x}^{+}(0)}=\text{normalized traveling field signal in material 3 at } z=d \tag{12.105}$$

Finally, we write the solution of the four equations (12.98) to (12.101) in the four *unknown* quantities, $\Gamma(0)=\frac{\overline{E}_{1x}^{-}(0)}{\overline{E}_{1x}^{+}(0)}, \frac{\overline{E}_{2x}^{+}(0)}{\overline{E}_{1x}^{+}(0)}, \frac{\overline{E}_{2x}^{-}(0)}{\overline{E}_{1x}^{+}(0)}$, and $\frac{\overline{E}_{3x}^{+}(d)}{\overline{E}_{1x}^{+}(0)}$ defined in Equations (12.102) to (12.105) as follows:

$$\Gamma(0)=\frac{\overline{E}_{1x}^{-}(0)}{\overline{E}_{1x}^{+}(0)}=\frac{P_{12}+P_{23}\cdot e^{-2\gamma_2 d}}{1+P_{12}\cdot P_{23}\cdot e^{-2\gamma_2 d}} \tag{12.106}$$

$$\frac{\overline{E}_{2x}^{+}(0)}{\overline{E}_{1x}^{+}(0)}=\frac{1+P_{12}}{1+P_{12}\cdot P_{23}\cdot e^{-2\gamma_2 d}} \tag{12.107}$$

$$\frac{\overline{E}_{2x}^{-}(0)}{\overline{E}_{1x}^{+}(0)}=\frac{(1+P_{12})\cdot P_{23}\cdot e^{-2\gamma_2 d}}{1+P_{12}\cdot P_{23}\cdot e^{-2\gamma_2 d}} \tag{12.108}$$

$$\frac{\overline{E}_{3x}^{+}(d)}{\overline{E}_{1x}^{+}(0)}=\frac{(1+P_{12})\cdot(1+P_{23})\cdot e^{-\gamma_2 d}}{1+P_{12}\cdot P_{23}\cdot e^{-2\gamma_2 d}} \tag{12.109}$$

where P_{12} and P_{23} (with P the upper case Greek letter rho) are the "local/instantaneous" reflection coefficients of the harmonic waves at the two discontinuities 12 and 23, respectively. The physical insight related to the definitions of these reflection coefficients is further demonstrated in Addendum 12A, Equations (12A.11) and (12A.12)

$$P_{12}=\frac{\overline{\eta}_{o2}-\overline{\eta}_{o1}}{\overline{\eta}_{o2}+\overline{\eta}_{o1}} \tag{12A.11}$$

$$P_{23}=\frac{\overline{\eta}_{o3}-\overline{\eta}_{o2}}{\overline{\eta}_{o3}+\overline{\eta}_{o2}} \tag{12A.12}$$

As stated earlier, two other approaches for solving this case are presented in Addendum 12A with developed expressions for the four identified physical quantities, Equations (12.102) to (12.105).

Example 12.6 A dielectric layer of thickness d and intrinsic impedance η_2 is placed between two lossless dielectric media 1 and 3 with intrinsic impedances η_1 and η_3, respectively. Determine d and η_2 such that no reflection occurs when a uniform plane wave in medium 1 impinges normally on the interface with medium 2.

Solution: Using Equation (12.106)

$$\Gamma(0)=\frac{\bar{E}_{1x}^{-}(0)}{\bar{E}_{1x}^{+}(0)}=\frac{P_{12}+P_{23}\cdot e^{-2\gamma_2 d}}{1+P_{12}\cdot P_{23}\cdot e^{-2\gamma_2 d}}=0$$

$$P_{12}+P_{23}\cdot e^{-2\gamma_2 d}=0\rightarrow P_{12}+P_{23}\cdot e^{-j2\beta_2 d}=0$$

The above equation can be expressed as

$$P_{12}\cdot e^{j\beta_2 d}=-P_{23}\cdot e^{-j\beta_2 d}$$

where $P_{12}=\dfrac{\eta_2-\eta_1}{\eta_2+\eta_1}$ and $P_{23}=\dfrac{\eta_3-\eta_2}{\eta_3+\eta_2}$

Substituting for P_{12} and P_{23}

$$\frac{\eta_2-\eta_1}{\eta_2+\eta_1}\cdot e^{j\beta_2 d}=-\frac{\eta_3-\eta_2}{\eta_3+\eta_2}\cdot e^{-j\beta_2 d}$$

$$\frac{\eta_2-\eta_1}{\eta_2+\eta_1}\cdot[\cos(\beta_2 d)+j\sin(\beta_2 d)]=-\frac{\eta_3-\eta_2}{\eta_3+\eta_2}\cdot[\cos(\beta_2 d)-j\sin(\beta_2 d)]$$

Equating the real part on both sides of the equation

$$\frac{\eta_2-\eta_1}{\eta_2+\eta_1}\cdot[\cos(\beta_2 d)]=\frac{\eta_2-\eta_3}{\eta_2+\eta_3}\cdot[\cos(\beta_2 d)] \quad (1)$$

Equating the imaginary parts on both sides of the equation

$$\frac{\eta_2-\eta_1}{\eta_2+\eta_1}\cdot[\sin(\beta_2 d)]=\frac{\eta_3-\eta_2}{\eta_3+\eta_2}\cdot[\sin(\beta_2 d)] \quad (2)$$

For equations (1) and (2) to be satisfied there are two possible scenarios, I and II as described below:

I. Medium 1 and medium 3 are the same $\rightarrow \eta_1=\eta_3$
In this case equation (1) will automatically be satisfied and equation (2) will be in the form

$$\frac{\eta_2-\eta_1}{\eta_2+\eta_1}\cdot[\sin(\beta_2 d)]=\frac{\eta_1-\eta_2}{\eta_1+\eta_2}\cdot[\sin(\beta_2 d)] \quad \text{or} \quad \sin(\beta_2 d)=-\sin(\beta_2 d)$$

The above equation will be satisfied only if

$$\sin(\beta_2 d)=0 \quad \text{or} \quad \frac{2\pi}{\lambda_2}d=n\pi\rightarrow d=n\frac{\lambda_2}{2} \quad n=1,2,3,\ldots \text{ where } \lambda_2=\frac{c_{ph_2}}{f}=\frac{c}{f\sqrt{\mu_{r2}\varepsilon_{r2}}}$$

Therefore, if the thickness of the dielectric layer inserted between medium 1 and medium 2 is an integral multiple of $\lambda_2/2$, there will be no reflection. The concept is applied in the design of a Radome, an enclosure used to protect a radar antenna. If the zero reflection coefficient is satisfied by adjusting the thickness of the enclosure, the enclosure will be effectively transparent to the radio wave.

II. Medium 1 and medium 3 are different $\rightarrow \eta_1\neq\eta_3$
In this case, equation (1) can only be satisfied if the term $\cos(\beta_2 d)$ is zero,

$$\cos(\beta_2 d)=0 \quad \text{or} \quad \frac{2\pi}{\lambda_2}d=(2n+1)\frac{\pi}{2}\rightarrow d=(2n+1)\frac{\lambda_2}{4} \quad n=0,1,2,3,\ldots$$

Since $\cos(\beta_2 d) = 0$, then $\sin(\beta_2 d) \neq 0$ and for equation (2) to be satisfied,

$$\frac{\eta_2 - \eta_1}{\eta_2 + \eta_1} = \frac{\eta_3 - \eta_2}{\eta_3 + \eta_2} \rightarrow \eta_2^2 = \eta_1 \eta_3$$

The condition of zero reflection is satisfied in this case, if the intrinsic impedance of the middle layer is equal to the geometric mean of the intrinsic impedances of the outside layers and the thickness of the middle layer is an odd multiple of $\frac{\lambda_2}{4}$.

The above result can also be explained using the analysis in Addendum 12A where a cascaded three-transmission line sections are used to model the three-media arrangement. The characteristic impedances of the lines are the intrinsic impedances of their corresponding material. In this case, the medium in the middle is equivalent to a matching quarter wavelength transmission line section.

12.7 Reflection of Uniform Plane Waves: Oblique Incidence

We now deviate from the idealized case of normal incidence to consider oblique incidence of *TEM* waves on the interface between two infinitely extended materials. The oblique incidence in consideration is depicted in Figure 12.16.

In this figure, we assume an incident *TEM* wave on the planar interface between two dielectrics at $z = 0$. The direction of the incident wave (which is the same as its Poynting vector) makes an angle θ_i with the z axis. Consequently, we expect both a reflected wave and a transmitted wave to result. The reflected wave is assumed to reflect with an angle of θ_r and the angle of the transmitted wave is named θ_t as shown in the figure.

Guided by our knowledge of Snell's law in ray optics, we expect the angle of reflection to equal that of incidence and that the angle of transmission depends on the indices of refraction of the two materials. In Addendum 12B of this chapter, we present the derivation of Snell's law using two different approaches; one based on the EM fields and the other using a ray theory approach.

Referring to Figure 12.17 and following the derivations in the addendum, the relations between the three angles are given by

$$\theta_i = \theta_r = \theta_1 \tag{12.110}$$

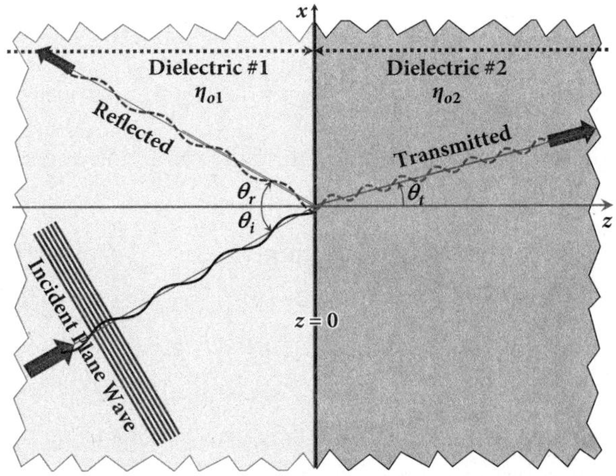

Figure 12.16 Oblique incidence of *TEM* waves on the interface between two infinitely extended materials.

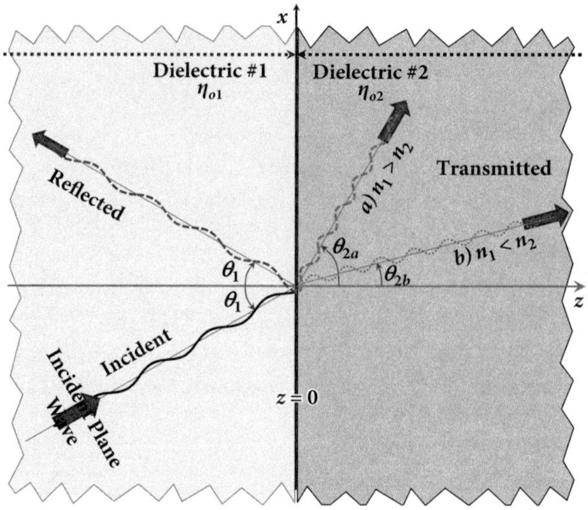

FIGURE 12.17 Snell's law for the angle of the transmitted wave.

$$\frac{\sin\theta_2}{\sin\theta_1} = \frac{\sin\theta_t}{\sin\theta_i} = \frac{\beta_1}{\beta_2} = \frac{\sqrt{\mu_1\varepsilon_1}}{\sqrt{\mu_2\varepsilon_2}} = \frac{\sqrt{\mu_{r1}\,\varepsilon_{r1}}}{\sqrt{\mu_{r2}\varepsilon_{r2}}} = \frac{n_1}{n_2} \tag{12.111}$$

where n is defined as $\sqrt{\mu_r\varepsilon_r}$ and is called the index of refraction for the dielectric material.

The figure demonstrates two cases for the transmitted (refraction) angle based on the relationship between the indices of refraction of the two materials; case a: $n_1 > n_2$ for which $\theta_2 > \theta_1$, and case b: $n_1 < n_2$ for which $\theta_2 < \theta_1$.

12.8 Total Reflection—Critical Angle

Based on the above observation, as well as the fact that the refraction angle cannot physically exceed 90°, we can conclude that for the case of $n_1 > n_2$ there will be a critical (cutoff) angle of incidence at which the refraction angle reaches 90°. For this case, incidence angles greater than this critical angle will result in nonphysical values for the refraction angle (greater than 90°), which physically implies that no refraction can exist due to such incidence and hence total reflection will occur. We demonstrate this finding in Figure 12.18.

Starting with Snell's law written as

$$\sin\theta_2 = \frac{n_1}{n_2}\sin\theta_1 \tag{12.112}$$

we derive the expression for this critical angle, θ_{1c}, by setting the refraction angle, θ_{2c}, to its maximum value of 90° for which $\sin\theta_{2c} = 1$, hence

$$\sin\theta_{1c} = \frac{n_2}{n_1}\sin\theta_{2c} = \frac{n_2}{n_1} \tag{12.113}$$

$$\theta_{1c} = \sin^{-1}\left(\frac{n_2}{n_1}\right) \tag{12.114}$$

As shown in the figure, the critical incidence angle is labeled θ_{1c} for which the incident wave will have both a reflection at an angle of θ_{1c} and a transmission at $\theta_{2c} = 90°$. Incidence, reflection and transmission for this angle are labeled, I_C, R_C and T_C, respectively. Incidence at angles less than θ_{1c} is demonstrated by the

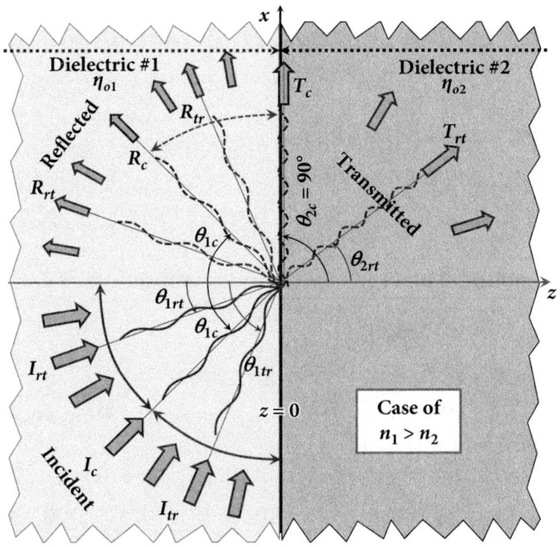

FIGURE 12.18 Complete reflection of the incident wave (with zero transmission) occurs when the angle on incidence exceeds a critical value.

subscript "*rt*" indicating the presence of both *reflection* and *transmission*. While incidence with angles greater than θ_{1c} is subscripted by "*tr*" indicating *total reflection*.

To appreciate relevant orders of magnitudes, we consider two practical cases of glass to air interface and water to air interface. Typical value for the index of refraction of glass is 1.52 while that of water is 1.33. With air having an index of refraction of 1.0, this makes the incidence critical angle for the glass-air interface $\theta_{1c} = 48.75°$, and for the water-air interface, $\theta_{1c} = 41.14°$.

12.9 Physical Insight

The concept of total reflection has many practical applications and we present a few here in this chapter. Several of these applications are related to optics from which two are presented in Addendum 12C. However, an important one related to radio signal propagation is presented below.

This application has to do with the transmission of short wave radio signals to remote locations that are not in the line-of-sight of the transmitting antenna. Taking advantage of the changes of the index of refraction of the upper atmospheric layers, radio waves are reflected off at angles greater than the critical angle, thus achieving total reflection. The totally reflected signal would have adequate strength to reach those remote locations that are not in the line of sight with the transmitting antenna. Other reflections occurring at angles smaller than the critical one are expected to have weaker intensities as the bulk of the incident power would transmit through the atmospheric layers. The gradual/small changes in the values of the intrinsic impedances (and refraction indices) of these layers is expected to result in a relatively small reflection coefficients and a large transmission coefficient. A simplified demonstration of this application is shown in Figure 12.19.

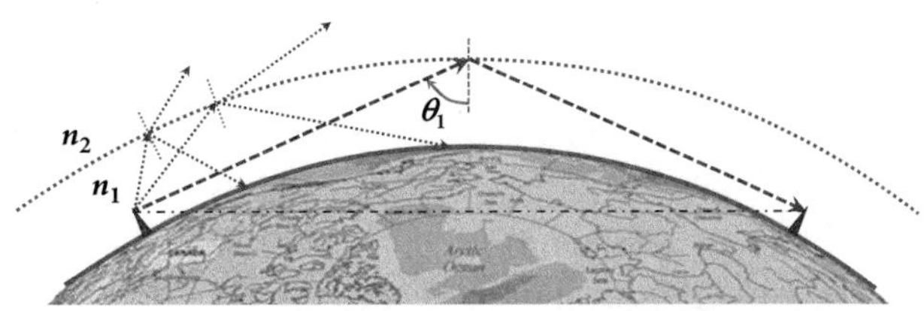

FIGURE 12.19 The concept of total reflection is used in radio transmission to locations beyond the line of sight of the transmitting antenna.

12.10 Analysis of Wave Reflection and Refraction for Oblique Incidence

In the analysis of oblique incidence, we recognize that the polarization of the incident wave may not necessarily coincide with the chosen/presented coordinate system. However, we can always think of the general case of polarization as the superposition of two waves with two different polarization components: a parallel polarization component and a perpendicular polarization component. The designations parallel and perpendicular refer to the direction of the electric field vector with respect to the x-z plane. We demonstrate these two polarizations in Figure 12.20.

In the following, we will carry out the field analysis and propagation of each polarization wave separately. We start with the case of parallel polarization.

12.10.1 Case of Oblique Incidence with Parallel Polarization

This case is depicted in Figure 12.21. As seen in the figure, the electric field vector lies parallel to the x-z plane while the magnetic field vector is normal to it along the y axis.

The incident wave approaches the $z = 0$ interface and makes an angle θ_1 with the z-axis.

Recalling the normal incidence expressions of Figure 12.7 developed earlier in Equations (12.11) to (12.16), we can write the parallel polarization oblique incidence ones for Figure 12.21 as follows:

$$\vec{\bar{E}}_1^+(x,z) = \bar{E}_1^+(0)(\vec{a}_x \cos\theta_1 - \vec{a}_z \sin\theta_1)e^{-\gamma_1(x\sin\theta_1 + z\cos\theta_1)} \tag{12.115}$$

$$\vec{\bar{E}}_1^-(x,z) = \bar{E}_1^-(0)(\vec{a}_x \cos\theta_1 + \vec{a}_z \sin\theta_1)e^{-\gamma_1(x\sin\theta_1 - z\cos\theta_1)} \tag{12.116}$$

$$\vec{\bar{E}}_1^+(x,z) = \bar{E}_2^+(0)(\vec{a}_x \cos\theta_2 - \vec{a}_z \sin\theta_2)e^{-\gamma_2(x\sin\theta_2 + z\cos\theta_2)} \tag{12.117}$$

We notice in these expressions that we accounted for the oblique distance traveled by the wave as the sum of both its x and z components, see the insert in Figure 12.21. Furthermore, we also expressed the electric field vectors in terms of their corresponding x and z components $(\vec{a}_x \cos\theta_{1,2} \mp \vec{a}_z \sin\theta_{1,2})$.

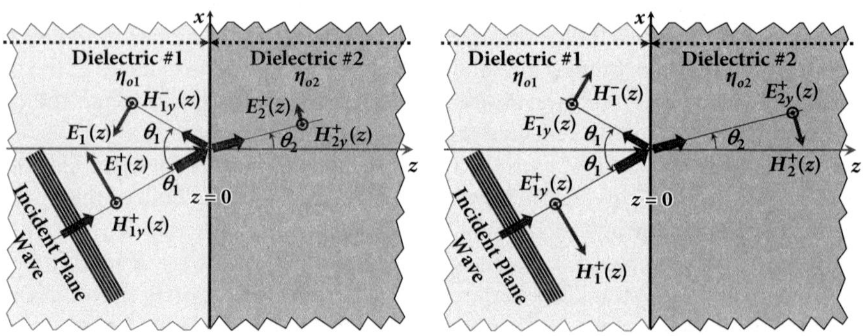

Parallel Polarization
E vector is parallel to the page (plane formed by the incident, reflected and transmitted Poynting vectors), i.e. the x-z plane

Perpendicular Polarization
E vector is perpendicular to the page (plane formed by the incident, reflected and transmitted Poynting vectors), i.e. the x-z plane

Figure 12.20 For analysis convenience for oblique incidence of *TEM* waves, we break a general case of wave polarization into the superposition of two orthogonal cases: the case of parallel polarization and the case perpendicular polarization.

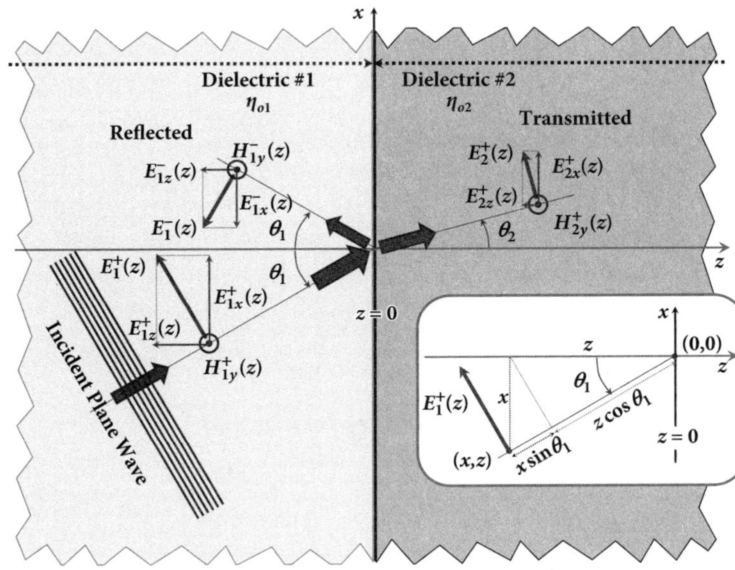

FIGURE 12.21 Case of oblique incidence with parallel polarization. For parallel polarization, the *E* field vector is parallel to the page (plane formed by the incident, reflected and transmitted Poynting vectors), i.e., the *x*-*z* plane.

Meanwhile, the magnetic field vectors are all along y coordinate and can be expressed as

$$\bar{\vec{H}}_1^+(x,z)=\vec{a}_y\frac{\bar{E}_1^+(0)}{\bar{\eta}_{o1}}e^{-\gamma_1(x\sin\theta_1+z\cos\theta_1)} \tag{12.118}$$

$$\bar{\vec{H}}_1^-(x,z)=-\vec{a}_y\frac{\bar{E}_1^-(0)}{\bar{\eta}_{o1}}e^{-\gamma_1(x\sin\theta_1-z\cos\theta_1)} \tag{12.119}$$

$$\bar{\vec{H}}_1^+(x,z)=\vec{a}_y\frac{\bar{E}_2^+(0)}{\eta_{o2}}e^{-\gamma_2(x\sin\theta_2+z\cos\theta_2)} \tag{12.120}$$

Applying the boundary conditions at $z = 0$ regarding the continuity of the tangential (x component) electric fields, we write

$$\bar{E}_1^+(0)\cos\theta_1+\bar{E}_1^-(0)\cos\theta_1=\bar{E}_2^+(0)\cos\theta_2 \tag{12.121}$$

similarly, the continuity of the tangential (y component) magnetic fields yields

$$\bar{H}_{1y}^+(0)+\bar{H}_{1y}^-(0)=\bar{H}_{2y}^+(0) \tag{12.122}$$

using Equations (12.118) to (12.120), Equation (12.122) takes the form

$$\frac{\bar{E}_1^+(0)}{\bar{\eta}_{o1}}-\frac{\bar{E}_1^-(0)}{\bar{\eta}_{o1}}=\frac{\bar{E}_2^+(0)}{\bar{\eta}_{o2}} \tag{12.123}$$

Normalizing the field expressions of Equations (12.121) and (12.124) by dividing all terms by $\bar{E}_1^+(0)$, then solving the resulting equations to derive the expressions of the reflection $\bar{\Gamma}(0)\big|_{\parallel}=\frac{\bar{E}_1^-(0)}{\bar{E}_1^+(0)}$ and transmission $\bar{\tau}(0)\big|_{\parallel}=\frac{\bar{E}_2^+(0)}{\bar{E}_1^+(0)}$ coefficients at the interface between the two materials, we write

$$\left[1+\frac{\bar{E}_1^-(0)}{\bar{E}_1^+(0)}\right]\cos\theta_1=\left[\frac{\bar{E}_2^+(0)}{\bar{E}_1^+(0)}\right]\cos\theta_2 \tag{12.124}$$

$$\frac{1}{\bar{\eta}_{o1}}\left[1-\frac{\bar{E}_1^-(0)}{\bar{E}_1^+(0)}\right]=\frac{1}{\bar{\eta}_{o2}}\left[\frac{\bar{E}_2^+(0)}{\bar{E}_1^+(0)}\right] \tag{12.125}$$

$$\bar{\Gamma}(0)\big|_{\|}=\frac{\bar{E}_1^-(0)}{\bar{E}_1^+(0)}=\frac{\bar{\eta}_{o2}\cos\theta_2-\bar{\eta}_{o1}\cos\theta_1}{\bar{\eta}_{o2}\cos\theta_2+\bar{\eta}_{o1}\cos\theta_1} \tag{12.126}$$

$$\bar{\tau}(0)\big|_{\|}=\frac{\bar{E}_2^+(0)}{\bar{E}_1^+(0)}=\frac{\cos\theta_1}{\cos\theta_2}\left\{1+\bar{\Gamma}(0)\big|_{\|}\right\}=\frac{2\,\bar{\eta}_{o2}\cos\theta_1}{\bar{\eta}_{o2}\cos\theta_2+\bar{\eta}_{o1}\cos\theta_1} \tag{12.127}$$

12.10.2 Case of Oblique Incidence with Perpendicular Polarization

Similar to the procedure followed for the case of parallel polarization, we demonstrate the case of perpendicular polarization in Figure 12.22. In the figure, the electric field vector is now perpendicular to the x-z plane along the y axis while the magnetic field vector is the one that lies in the x-z plane.

Hence, referring to Figure 12.22, we can write the vector electric and magnetic field components as follows:

$$\vec{\bar{E}}_1^+(x,z)=\vec{a}_y\bar{E}_1^+(0)e^{-\gamma_1(x\sin\theta_1+z\cos\theta_1)} \tag{12.128}$$

$$\vec{\bar{E}}_1^-(x,z)=\vec{a}_y\bar{E}_1^-(0)e^{-\gamma_1(x\sin\theta_1-z\cos\theta_1)} \tag{12.129}$$

$$\vec{\bar{E}}_1^+(x,z)=\vec{a}_y\bar{E}_2^+(0)e^{-\gamma_2(x\sin\theta_2+z\cos\theta_2)} \tag{12.130}$$

$$\vec{\bar{H}}_1^+(x,z)=\frac{\bar{E}_1^+(0)}{\bar{\eta}_{o1}}(-\vec{a}_x\cos\theta_1+\vec{a}_z\sin\theta_1)e^{-\gamma_1(x\sin\theta_1+z\cos\theta_1)} \tag{12.131}$$

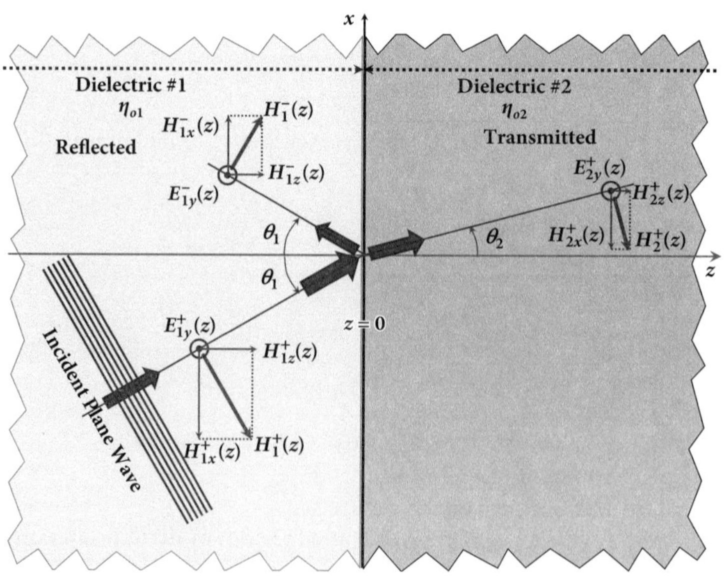

FIGURE 12.22 Case of oblique incidence with perpendicular polarization. For perpendicular polarization, the E field vector is perpendicular to the page (plane formed by the incident, reflected and transmitted Poynting vectors), i.e., the x-z plane.

$$\bar{\bar{H}}_1^-(x,z) = \frac{\bar{E}_1^-(0)}{\bar{\eta}_{o1}}(\vec{a}_x\cos\theta_1 + \vec{a}_z\sin\theta_1)e^{-\gamma_1(x\sin\theta_1 - z\cos\theta_1)} \quad (12.132)$$

$$\bar{\bar{H}}_1^+(x,z) = \frac{\bar{E}_2^+(0)}{\bar{\eta}_{o2}}(-\vec{a}_x\cos\theta_2 + \vec{a}_z\sin\theta_2)e^{-\gamma_2(x\sin\theta_2 + z\cos\theta_2)} \quad (12.133)$$

Again, applying the boundary conditions at $z = 0$ regarding the continuity of the tangential (y component) electric fields, and the continuity of the tangential (x component) magnetic fields we write

$$\bar{E}_1^+(0) + \bar{E}_1^-(0) = \bar{E}_2^+(0) \quad (12.134)$$

and

$$-\frac{\bar{E}_1^+(0)}{\bar{\eta}_{o1}}\cos\theta_1 + \frac{\bar{E}_1^-(0)}{\bar{\eta}_{o1}}\cos\theta_1 = -\frac{\bar{E}_2^+(0)}{\bar{\eta}_{o2}}\cos\theta_2 \quad (12.135)$$

Finally, normalizing the field expressions of Equations (12.134) and (12.135) by dividing all terms by $\bar{E}_1^+(0)$ and solving the resulting equations to derive the expressions of the reflection and transmission coefficients at the interface, we conclude,

$$\bar{\Gamma}(0)\Big|_\perp = \frac{\bar{E}_1^-(0)}{\bar{E}_1^+(0)} = \frac{\bar{\eta}_{o2}/\cos\theta_2 - \bar{\eta}_{o1}/\cos\theta_1}{\bar{\eta}_{o2}/\cos\theta_2 + \bar{\eta}_{o1}/\cos\theta_1} \quad (12.136)$$

$$\bar{\tau}(0)\Big|_\perp = \frac{\bar{E}_2^+(0)}{\bar{E}_1^+(0)} = 1 + \bar{\Gamma}(0)\Big|_\perp = \frac{2\,\bar{\eta}_{o2}/\cos\theta_2}{\bar{\eta}_{o2}/\cos\theta_2 + \bar{\eta}_{o1}/\cos\theta_1} \quad (12.137)$$

Example 12.7 Consider a uniform plane wave at 150 MHz is obliquely incident from air into a semi-infinite dielectric layer with planar interface. The dielectric layer is lossless and nonmagnetic with $\varepsilon_r = 4$. The amplitude of the incident magnetic field is 5 mA/m. The angle of incidence is 30°. If the wave has parallel polarization with the electric field in the x-z plane as shown in Figure 12.21, write expressions for the incident, reflected and transmitted electric and magnetic fields.

Solution Referring to Figure 12.21, for oblique incidence, the propagation factor can be expressed as $e^{-j\beta(\vec{a}_\beta \cdot \vec{r})}$, where $\vec{a}_\beta$ is a unit vector in the direction of propagation and $\vec{r}$ is the position vector given by

$$\vec{r} = \vec{a}_x x + \vec{a}_y y + \vec{a}_z z$$

For the incident wave, the unit vector in the direction of propagation is given by

$$\vec{a}_\beta^+ = \vec{a}_x\sin\theta_1 + \vec{a}_z\cos\theta_1 \qquad \text{and} \qquad \vec{a}_\beta^+ \cdot \vec{r} = x\sin\theta_1 + z\cos\theta_1$$

Since the magnetic field has only one component in the y direction, we will start by writing expressions for the incident, reflected and transmitted magnetic fields and then obtain the corresponding electric field phasors. The incident magnetic field phasor can be expressed as

$$\bar{\bar{H}}_1^+(x,z) = \vec{a}_y\bar{H}_1^+(0)e^{-j\beta_1(x\sin\theta_1 + z\cos\theta_1)}$$

$$\beta_1 = \frac{\omega}{c_{ph_1}} = \frac{\omega}{c} = \frac{2\pi \times 150 \times 10^6}{3 \times 10^8} = \pi \text{ rad/m}$$

$$\bar{\bar{H}}_1^+(x,z) = \vec{a}_y 5e^{-j\beta_1(x\sin\theta_1 + z\cos\theta_1)} \text{ mA/m}$$

$$\theta_1 = 30^\circ \qquad \cos(\theta_1) = \frac{\sqrt{3}}{2} \qquad \sin(\theta_1) = \frac{1}{2}$$

$$\bar{\vec{H}}_1^+(x,z) = \vec{a}_y 5e^{-j\pi(0.5x+0.866z)} \text{ mA/m}$$

Using Equations (12.118) and (12.119), we can conclude $\dfrac{\bar{H}_1^-(0)}{\bar{H}_1^+(0)} = -\dfrac{\bar{E}_1^-(0)}{\bar{E}_1^+(0)} = -\bar{\Gamma}(0)\big|_{\parallel}$.

We can then write an expression for the reflected magnetic field

For the reflected wave $\vec{a}_\beta^- = \vec{a}_x \sin\theta_1 - \vec{a}_z \cos\theta_1$ and $\vec{a}_\beta^- \cdot \vec{r} = x\sin\theta_1 - z\cos\theta_1$

$$\bar{\vec{H}}_1^-(x,z) = -\vec{a}_y \bar{\Gamma}(0)\big|_{\parallel} \bar{H}_1^+(0)e^{-j\pi(0.5x-0.866z)} \text{ mA/m}$$

$$\bar{\Gamma}(0)\big|_{\parallel} = \frac{\bar{\eta}_{o2}\cos\theta_2 - \bar{\eta}_{o1}\cos\theta_1}{\bar{\eta}_{o2}\cos\theta_2 + \bar{\eta}_{o1}\cos\theta_1}$$

$$\bar{\eta}_{o2} = \sqrt{\frac{\mu_2}{\varepsilon_2}} = 120\pi\sqrt{\frac{\mu_{r2}}{\varepsilon_{r2}}} = \frac{120\pi}{2} = 60\pi\,\Omega$$

$$\bar{\eta}_{o1} = \sqrt{\frac{\mu_1}{\varepsilon_1}} = 120\pi\ \Omega$$

Applying Snell's law $n_1\sin(\theta_1) = n_2\sin(\theta_2) \rightarrow \frac{1}{2} = \sqrt{4}\sin(\theta_2) \qquad \theta_2 = \sin^{-1}\frac{1}{4} = 14.48^\circ$

$$\bar{\Gamma}(0)\big|_{\parallel} = \frac{60\pi\cdot(\cos 14.48^\circ) - 120\pi\left(\frac{\sqrt{3}}{2}\right)}{60\pi\cdot(\cos 14.48^\circ) + 120\pi\left(\frac{\sqrt{3}}{2}\right)} = -0.28$$

$$\bar{\vec{H}}_1^-(x,z) = -\vec{a}_y \bar{\Gamma}(0)\big|_{\parallel} \bar{H}_1^+(0)e^{-j\pi(0.5x-0.866z)} \text{ mA/m}$$

$$= -\vec{a}_y(-0.28)5e^{-j\pi(0.5x-0.866z)} \text{ mA/m}$$

$$\bar{\vec{H}}_1^-(x,z) = \vec{a}_y 1.4e^{-j\pi(0.5x-0.866z)} \text{ mA/m}$$

Using the continuity of the tangential magnetic field condition, we can write an expression for the magnetic field in medium 2 at the interface

$$\bar{\vec{H}}_2(x=0, z=0) = \bar{\vec{H}}_1^+(0) + \bar{\vec{H}}_1^-(0) = \vec{a}_y(5+1.4) = \vec{a}_y 6.4 \text{ mA/m}$$

$$\bar{\vec{H}}_2(z) = \vec{a}_y \bar{\vec{H}}_2(0)e^{-j\beta_2(x\sin\theta_2 + z\cos\theta_2)}$$

$$\beta_2 = \frac{\omega}{c_{ph_2}} = \frac{\omega}{c}\sqrt{\mu_{r2}\varepsilon_{r2}} = \frac{2\pi\times 150\times 10^6}{3\times 10^8}\sqrt{4} = 2\pi \text{ rad/m}$$

$$\bar{\vec{H}}_2(z) = \vec{a}_y 6.4e^{-j2\pi(0.25x+0.97z)}$$

The expressions for the electric field can be obtained from the corresponding magnetic field expressions using the following:

$$\bar{\vec{E}}(z) = -\eta\vec{a}_\beta \times \bar{\vec{H}}(z)$$

where $\vec{a}_\beta$ is a unit vector in the direction of propagation and η is the intrinsic impedance of the medium.

For the incident wave, $\vec{a}_\beta^+ = \vec{a}_x \sin\theta_1 + \vec{a}_z \cos\theta_1$

$$\bar{\bar{E}}_1^+(z) = -\eta_{o1}(\vec{a}_x \sin\theta_1 + \vec{a}_z \cos\theta_1) \times \bar{\bar{H}}_1^+(z)$$

$$\bar{\bar{E}}_1^+(z) = -\eta_{o1}(\vec{a}_x \sin\theta_1 + \vec{a}_z \cos\theta_1) \times \vec{a}_y \bar{H}_1^+(0) e^{-j\pi(0.5x+0.866z)}$$

$$\bar{\bar{E}}_1^+(z) = \eta_{o1}(\vec{a}_x \cos\theta_1 - \vec{a}_z \sin\theta_1) \bar{H}_1^+(0) e^{-j\pi(0.5x+0.866z)}$$

$$\bar{\bar{E}}_1^+(z) = 0.6\pi(\vec{a}_x 0.866 - \vec{a}_z 0.5) e^{-j\pi(0.5x+0.866z)} \text{ V/m}$$

For the reflected wave $\vec{a}_\beta^- = \vec{a}_x \sin\theta_1 - \vec{a}_z \cos\theta_1$

$$\bar{\bar{E}}_1^-(z) = -\eta_{o1}(\vec{a}_x \sin\theta_1 - \vec{a}_z \cos\theta_1) \times \bar{\bar{H}}_1^-(z)$$

$$\bar{\bar{E}}_1^-(z) = -\eta_{o1}(\vec{a}_x \sin\theta_1 - \vec{a}_z \cos\theta_1) \times \vec{a}_y 1.4 e^{-j\pi(0.5x-0.866z)}$$

$$\bar{\bar{E}}_1^-(z) = -0.168\pi(\vec{a}_x \cos\theta_1 + \vec{a}_z \sin\theta_1) e^{-j\pi(0.5x-0.866z)}$$

$$\bar{\bar{E}}_1^-(z) = -0.168\pi(\vec{a}_x 0.866 + \vec{a}_z 0.5) e^{-j\pi(0.5x-0.866z)} \text{ V/m}$$

The electric field expression in medium 2 can also be found as

$$\bar{\bar{E}}_2(z) = -\eta_{o2}\vec{a}_\beta \times \bar{\bar{H}}_2(z) = -\eta_{o2}(\vec{a}_x \sin\theta_2 + \vec{a}_z \cos\theta_2) \times \vec{a}_y \bar{\bar{H}}_2(0) e^{-j\beta_2(x \sin\theta_2 + z\cos\theta_2)}$$

$$\bar{\bar{E}}_2(z) = 60\pi(\vec{a}_x \cos\theta_2 - \vec{a}_z \sin\theta_2) 6.4 e^{-j\beta_2(x \sin\theta_2 + z\cos\theta_2)}$$

$$\bar{\bar{E}}_2(z) = 1.206(\vec{a}_x 0.97 - \vec{a}_z 0.25) e^{-j2\pi(0.25x+0.97z)} \text{ V/m}$$

12.11 The Brewster Angle

An interesting phenomenon that has useful applications is what we refer to as the Brewster angle. This has to do with the fact that certain oblique incident waves can go through the interface without experiencing any reflection; in other words, the waves will go through the interface with total transmission. The angle at which a wave is undergoing a total transmission and no reflection is called the Brewster angle in reference to the Scottish physicist Sir David Brewster (1781–1868). This concept is demonstrated in Figure 12.23.

At the Brewster angle, the reflection coefficient $\bar{\Gamma}(0)\big|_B$ should be zero,

$$\bar{\Gamma}(0)\big|_B = 0 \tag{12.138}$$

Now, let us explore this condition for both parallel and perpendicular oblique incidences. For parallel polarization, we write

$$\bar{\Gamma}(0)\big|_{\parallel} = \frac{\bar{\eta}_{o2}\cos\theta_2 - \bar{\eta}_{o1}\cos\theta_1}{\bar{\eta}_{o2}\cos\theta_2 + \bar{\eta}_{o1}\cos\theta_1} = 0 \tag{12.139}$$

while for the perpendicular polarization, we write

$$\bar{\Gamma}(0)\big|_{\perp} = \frac{\bar{\eta}_{o2}/\cos\theta_2 - \bar{\eta}_{o1}/\cos\theta_1}{\bar{\eta}_{o2}/\cos\theta_2 + \bar{\eta}_{o1}/\cos\theta_1} = 0 \tag{12.140}$$

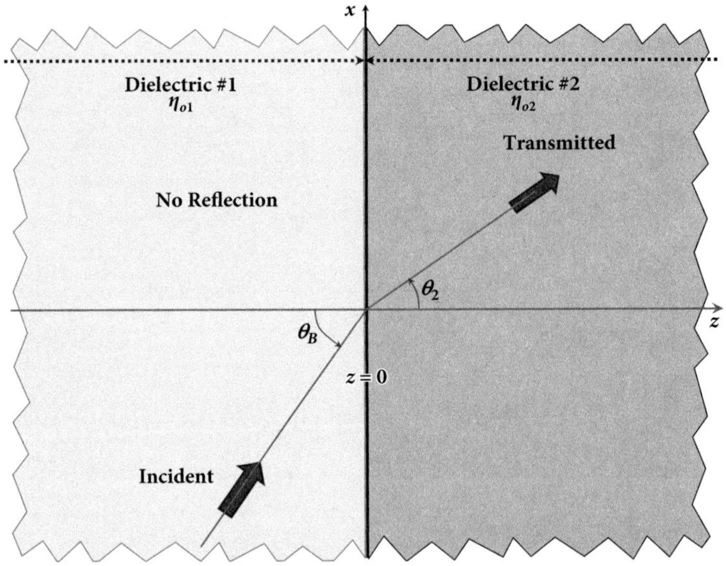

FIGURE 12.23 The Brewster angle is the angle at which an incident wave experiences complete transmission with no reflections.

Starting with Equation (12.139) for parallel polarization waves, we can write

$$\bar{\eta}_{o2}\cos\theta_2 = \bar{\eta}_{o1}\cos\theta_1 \tag{12.141}$$

from which, together with Snell's law, we conclude

$$\sin\theta_{1B\|} = \sqrt{\frac{1-\dfrac{\mu_2\varepsilon_1}{\mu_1\varepsilon_2}}{1-\left(\dfrac{\varepsilon_1}{\varepsilon_2}\right)^2}} \quad \text{(for the case of lossless dielectrics)} \tag{12.142}$$

We further simplify this expression for most practical cases of nonmagnetic materials where $\mu_1 = \mu_2$ to yield

$$\sin\theta_{1B\|}\Big|_{\mu_1=\mu_2} = \frac{1}{\sqrt{1+\left(\dfrac{\varepsilon_1}{\varepsilon_2}\right)}} \tag{12.143}$$

Similarly, for the perpendicular polarization waves, the condition of zero reflection as seen from Equation (12.140) is

$$\bar{\eta}_{o2}\cos\theta_1 = \bar{\eta}_{o1}\cos\theta_2 \tag{12.144}$$

which, together with Snell's law, yields

$$\sin\theta_{1B\perp} = \sqrt{\frac{1-\dfrac{\mu_1\varepsilon_2}{\mu_2\varepsilon_1}}{1-\left(\dfrac{\mu_1}{\mu_2}\right)^2}} \quad \text{(for the case of lossless dielectrics)} \tag{12.145}$$

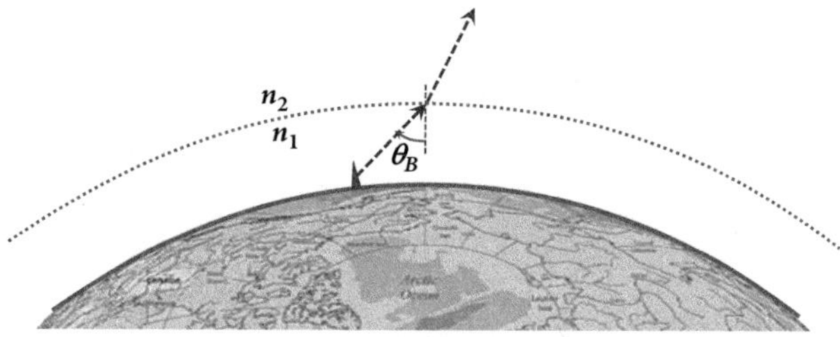

FIGURE 12.24 The concept of Brewster angle and the corresponding total transmission used to transmit waves through the ionospheric layer to outer space targets.

This result implies that a Brewster angle for nonmagnetic materials with $\mu_1 = \mu_2$ is not realizable.

$$\sin\theta_{1B\perp}\Big|_{\mu_1=\mu_2} \text{ does not exist} \tag{12.146}$$

The detailed development of the above equations is presented in Addendum 12D.

12.12 Physical Insight

One important application for the Brewster angle concept is demonstrated in Figure 12.24. In this application, radio wave signals can be sent to outer layers of space with practically total transmission when aimed at the proper Brewster angle through these layers.

ADDENDUM 12A
DERIVATION OF REFLECTION AND TRANSMISSION COEFFICIENTS FOR NORMAL INCIDENCE ON MULTIPLE LAYERS

As discussed in the body of this chapter, we present in this addendum two additional approaches to the derivation of reflection and transmission coefficients for normal incidence on multiple layers.

12A.1 The Impedance Approach

In this approach, we use the equivalent-circuit concept to reduce the *circuit* to a single impedance at the *source/input* side. Once this reduction is completed, we can proceed with signal analysis of the simplified circuit. Finally, we proceed with signal analysis of the intermediate circuits followed in the reduction steps, in reverse order, until we reach the complete signals of the original configuration.

The example given in Figure 12A.1 is to demonstrate this concept. In this example, we start with a three-transmission line concatenation to model the three-media arrangement. The characteristic impedances of the lines are the intrinsic impedances of their corresponding materials as demonstrated in the figure.

Next, we replace the third transmission line by its equivalent driving point (input) impedance. Since the line is infinite in length, its input impedance is equal to its characteristic impedance,

$$\eta_{\text{in}23} = \eta_{o3} \tag{12A.1}$$

Now, we have a reduced circuit with two transmission lines and a termination. Let us refer to this circuit as the *intermediate* equivalent circuit. We can then proceed to the next reduction step by replacing the second transmission line by its driving-point input impedance. Since this line is terminated in η_{o3}, according to Equation (2A.3) we can write its driving-point input impedance as

$$\eta_{\text{in}12} = \eta_{o2}\frac{[\eta_{o3} + \eta_{o2}\tanh(\gamma_2 d_2)]}{[\eta_{o2} + \eta_{o3}\tanh(\gamma_2 d_2)]} \tag{12A.2}$$

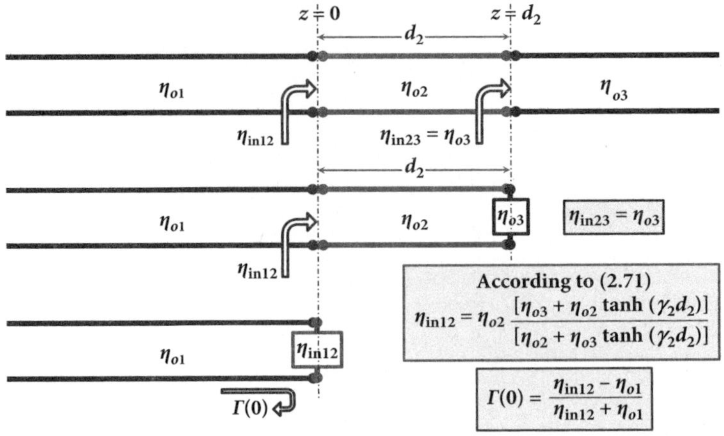

FIGURE 12A.1 Using the equivalent transmission line modeling for the analysis of wave incidence in the three-layer dielectrics case.

So, finally, we have a single transmission line reduced circuit. This circuit is composed of the first transmission line terminated in η_{in12}.

We now start with the signal analysis of the final reduced circuit. In this regard, we recognize the reflection coefficient $\Gamma(0)$ at the termination as given by

$$\Gamma(0)=\frac{\eta_{in12}-\eta_{o1}}{\eta_{in12}+\eta_{o1}} \tag{12A.3}$$

This reflection coefficient is in fact the ratio of the reflected signal to the input signal at the interface between material 1 and material 2:

$$\Gamma(0)=\frac{\bar{E}_{1x}^{-}(0)}{\bar{E}_{1x}^{+}(0)} \tag{12A.4}$$

Next as demonstrated in Figure 12A.2, we consider the signal at the interface $z = 0$. At this location, two sets of signals coexist; the first set is composed of the incident and reflected signals in region 1, and the second set consists of the forward and backward signals in region 2 at $z = 0$. The sum of the signals in each set represents the total tangential field in its corresponding region. Both tangential fields are supposed to be equal according to the boundary conditions at that interface. Hence, in view of Equations (12.90), (12.98), and (12.102), we can write

$$1+\Gamma(0)=\frac{\bar{E}_{1x}^{+}(0)+\bar{E}_{1x}^{-}(0)}{\bar{E}_{1x}^{+}(0)}=\frac{\bar{E}_{2x}^{+}(0)+\bar{E}_{2x}^{-}(0)}{\bar{E}_{1x}^{+}(0)}=\frac{\bar{E}_{2x}^{+}(0)\left[1+\dfrac{\bar{E}_{2x}^{-}(0)}{\bar{E}_{2x}^{+}(0)}\right]}{\bar{E}_{1x}^{+}(0)} \tag{12A.5}$$

also, using Equations (12.84) and (12.86), we can write

$$\frac{\bar{E}_{2x}^{-}(0)}{\bar{E}_{2x}^{+}(0)}=\frac{\bar{E}_{2x}^{-}(d_2)e^{-\gamma_2 d_2}}{\bar{E}_{2x}^{+}(d_2)e^{+\gamma_2 d_2}}=\Gamma_{23}e^{-2\gamma_2 d_2} \tag{12A.6}$$

Consequently, we can rewrite Equation (12A.5) as

$$1+\Gamma(0)=\frac{\bar{E}_{2x}^{+}(0)}{\bar{E}_{1x}^{+}(0)}\left[1+\Gamma_{23}e^{-2\gamma_2 d_2}\right] \tag{12A.7}$$

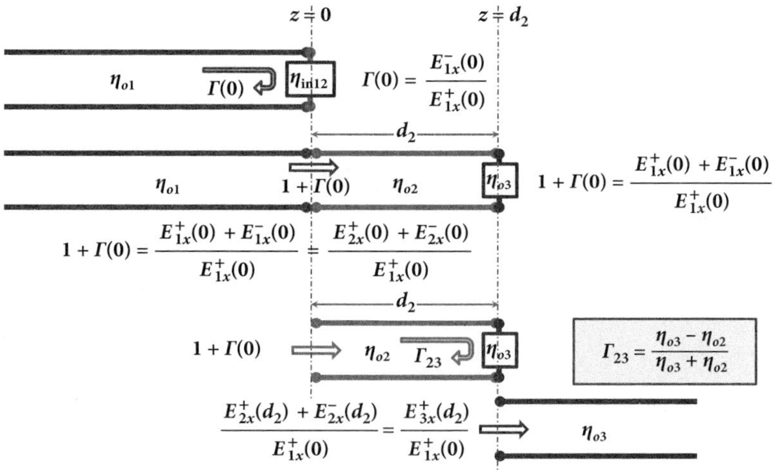

FIGURE 12A.2 The sequence of analysis for the three layer dielectric case of Figure 12A.1.

From which we conclude that the forward signal in region 2 at the interface $z = 0$ is given by

$$\frac{\bar{E}_{2x}^{+}(0)}{\bar{E}_{1x}^{+}(0)} = \frac{1+\Gamma(0)}{\left[1+\Gamma_{23}e^{-2\gamma_2 d_2}\right]} \tag{12A.8}$$

from which and using Equation (12A.6), we conclude the backward signal in region 2 at the same $z = 0$ interface to be of the form

$$\frac{\bar{E}_{2x}^{-}(0)}{\bar{E}_{1x}^{+}(0)} = \frac{\bar{E}_{2x}^{-}(0)}{\bar{E}_{2x}^{+}(0)}\frac{\bar{E}_{2x}^{+}(0)}{\bar{E}_{1x}^{+}(0)} = \Gamma_{23}e^{-2\gamma_2 d_2}\frac{1+\Gamma(0)}{\left[1+\Gamma_{23}e^{-2\gamma_2 d_2}\right]} \tag{12A.9}$$

Finally, we can write the expression for the traveling wave in region 3 at the $z = d_2$ interface as

$$\begin{aligned}\frac{\bar{E}_{3x}^{+}(d_2)}{\bar{E}_{1x}^{+}(0)} &= \frac{\bar{E}_{2x}^{+}(d_2)+\bar{E}_{2x}^{-}(d_2)}{\bar{E}_{1x}^{+}(0)} = \frac{1+\Gamma(0)}{\left[1+\Gamma_{23}e^{-2\gamma_2 d_2}\right]} + \frac{\Gamma_{23}e^{-2\gamma_2 d_2}[1+\Gamma(0)]}{\left[1+\Gamma_{23}e^{-2\gamma_2 d_2}\right]} \\ &= \frac{\left[1+\Gamma_{23}e^{-2\gamma_2 d_2}\right][1+\Gamma(0)]}{\left[1+\Gamma_{23}e^{-2\gamma_2 d_2}\right]}\end{aligned} \tag{12A.10}$$

12A.2 The Bounce Diagram Approach

We now proceed to the third approach of using the bounce diagram. In this regard, it would be useful for the reader to refer to Addendum 2C of Chapter 2 where we first discussed this subject in detail.

In this approach, we follow the same procedure described earlier in Chapter 2 to construct the bounce diagram. Again, the three materials are represented by their transmission line equivalencies as shown in Figure 12A.3.

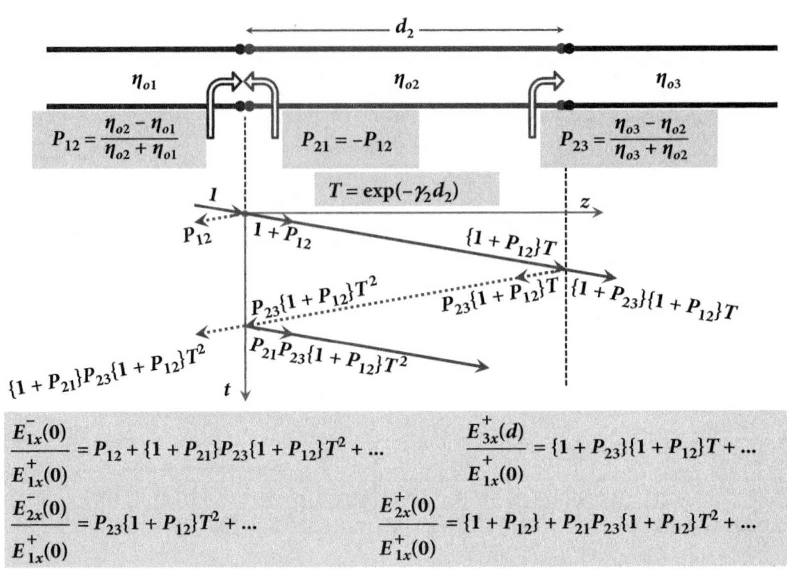

FIGURE 12A.3 Using the frequency-domain bounce-diagram for the analysis of the transmission line models of Figure 12A.1.

In this figure, we are using the *local* reflection coefficients ($P \equiv$ upper case rho) as the *instantaneous* reflection coefficient of the harmonic wave at the time it arrives at a discontinuity. The expressions for the different coefficients at the two interfaces are given by

$$P_{12} = \frac{\eta_{o2} - \eta_{o1}}{\eta_{o2} + \eta_{o1}} = -P_{21} \tag{12A.11}$$

$$P_{23} = \frac{\eta_{o3} - \eta_{o2}}{\eta_{o3} + \eta_{o2}} \tag{12A.12}$$

We also use the transmission propagation coefficient in region 2 as

$$T = \exp(-\gamma_2 d_2) \tag{12A.13}$$

Hence, as derived from the bounce diagram, we can write the normalized field expressions at the two interfaces as

$$\begin{aligned}\frac{\bar{E}_{1x}^-(0)}{\bar{E}_{1x}^+(0)} &= P_{12} + \{1+P_{21}\}P_{23}\{1+P_{12}\}T^2 + \{1+P_{21}\}P_{23}(P_{21}P_{23})\{1+P_{12}\}T^4 + \cdots \\ &= \frac{P_{12} + \{1+P_{12}+P_{21}\}P_{23}T^2}{1-P_{21}P_{23}T^2} = \frac{P_{12}+P_{23}T^2}{1-P_{21}P_{23}T^2}\end{aligned} \tag{12A.14}$$

$$\frac{\bar{E}_{2x}^+(0)}{\bar{E}_{1x}^+(0)} = \{1+P_{12}\} + P_{21}P_{23}\{1+P_{12}\}T^2 + (P_{21}P_{23})^2\{1+P_{12}\}T^4 + \cdots = \frac{\{1+P_{12}\}}{1-P_{21}P_{23}T^2} \tag{12A.15}$$

$$\frac{\bar{E}_{2x}^-(0)}{\bar{E}_{1x}^+(0)} = P_{23}\{1+P_{12}\}T^2 + P_{23}[P_{21}P_{23}\{1+P_{12}\}]T^4 + \cdots = \frac{P_{23}\{1+P_{12}\}T^2}{1-P_{21}P_{23}T^2} \tag{12A.16}$$

$$\frac{\bar{E}_{3x}^+(d)}{\bar{E}_{1x}^+(0)} = \{1+P_{23}\}\{1+P_{12}\}T + \{1+P_{23}\}P_{21}P_{23}\{1+P_{12}\}T^3 + \cdots = \frac{\{1+P_{23}\}\{1+P_{12}\}T}{1-P_{21}P_{23}T^2} \tag{12A.17}$$

Using Equations (12A.11) to (12A.13), it can be shown that the above four expressions [Equations (12A.14) to (12A.17)] are identical to those developed earlier in Equations (12.106) to (12.109).

ADDENDUM 12B
DERIVATION OF SNELL'S LAW

In the following, we will present two approaches to derive Snell's law. One using wave propagation expressions and the other is a graphical approach using a ray theoretical approach.

12B.1 Wave Propagation Derivation of Snell's Law

We choose to carry on this derivation for the perpendicular polarization. Similar development for the case of parallel polarization should lead to the same exact forms of Snell's Law as present here in this addendum.

Referring to Figure 12B.1, let us express the incident wave propagation along a general direction $\bar{a}_r$, where

$$\bar{a}_r = \sin\theta_i\,\bar{a}_x + \cos\theta_i\,\bar{a}_z \tag{12B.1}$$

$$r_i = x\sin\theta_i + z\cos\theta_i \tag{12B.2}$$

The electric field vector describing the plane wave propagation can be expressed as

$$\bar{E}_{1y}^{+}(x,z) = \bar{E}_{1y}^{+}(0)\cdot e^{-\gamma_1 r_i} = \bar{E}_{1y}^{+}(0)\cdot e^{-\gamma_1(x\sin\theta_i + z\cos\theta_i)} \tag{12B.3}$$

$$\bar{E}_{1y}^{-}(x,z) = \bar{E}_{1y}^{-}(0)\cdot e^{-\gamma_1 r_r} = \bar{E}_{1y}^{-}(0)\cdot e^{-\gamma_1(x\sin\theta_r - z\cos\theta_r)} \tag{12B.4}$$

$$\bar{E}_{2y}^{+}(x,z) = \bar{E}_{2y}^{+}(0)\cdot e^{-\gamma_2 r_t} = \bar{E}_{2y}^{+}(0)\cdot e^{-\gamma_2(x\sin\theta_t + z\cos\theta_t)} \tag{12B.5}$$

Now, applying the boundary condition of the continuity of the tangential electric field at $z = 0$, we can write

$$\bar{E}_{1y}^{+}(x,0) + \bar{E}_{1y}^{-}(x,0) = \bar{E}_{2y}^{+}(x,0) \tag{12B.6}$$

$$E_{1y}^{+}(0)\cdot e^{-\gamma_1(x\sin\theta_i)} + E_{1y}^{-}(0)\cdot e^{-\gamma_1(x\sin\theta_r)} = E_{2y}^{+}(0)\cdot e^{-\gamma_2(x\sin\theta_t)} \tag{12B.7}$$

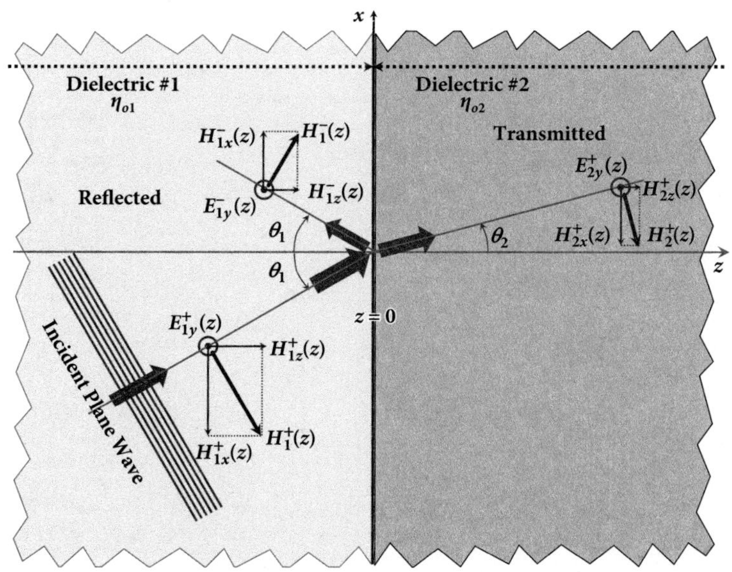

FIGURE 12B.1 Plane wave oblique incidence diagram.

For this equality to realize for all values of x, the x dependent terms must all factor out. This implies that the coefficients of x in all three exponent terms should be all the same leading to

$$-\gamma_1 \sin\theta_i = -\gamma_1 \sin\theta_r = -\gamma_2 \sin\theta_t \tag{12B.8}$$

$$\theta_i = \theta_r \tag{12B.9}$$

$$\gamma_1 \sin\theta_i = \gamma_2 \sin\theta_t \tag{12B.10}$$

Since γ_1 and γ_2 are generally complex quantities and θ_i is real, it is expected to encounter cases where Equation (12B.10) would lead to a complex value for $\sin\theta_t$. We will set such cases aside for now and will deal with them later in Addendum 12E.

For now, let us consider the case of lossless dielectrics where both θ_i and θ_t are real.

For lossless dielectrics, Equation (12B.10) takes the form

$$j\beta_1 \sin\theta_i = j\beta_2 \sin\theta_t \tag{12B.11}$$

$$\frac{\sin\theta_t}{\sin\theta_i} = \frac{\beta_1}{\beta_2} = \frac{\sqrt{\mu_1\varepsilon_1}}{\sqrt{\mu_2\varepsilon_2}} = \frac{\sqrt{\mu_{r1}\,\varepsilon_{r1}}}{\sqrt{\mu_{r2}\varepsilon_{r2}}} = \frac{n_1}{n_2} \tag{12B.12}$$

where n is defined as $\sqrt{\mu_r\varepsilon_r}$ and is called the index of refraction.

Equation (12B.12) presents the familiar form for Snell's law for lossless dielectrics, while Equation (12B.10) presents its general *complex* form which we plan to further explore in Addendum 12E.

12B.2 Graphical Derivation of Snell's Law

The simplest way to derive the relationships between the three angles is the graphical procedure demonstrated in Figures 12B.2 to 12B.5. In Figure 12B.2, we draw a set of parallel lines (rays) representing the incident wave propagation and we also draw a set of planes orthogonal to these lines. These planes are shown as dashed lines in the figure and can be thought of as wave fronts of the incident wave. The incident plane wave rays travel in phase and will have the same exact phase at all points on any of the orthogonal wave fronts. In the figure, we chose to show the planes at the peaks of the field intensity at a given instant of time; consequently, these planes are equally spaced at one wavelength, λ_1, apart along the direction of wave propagation.

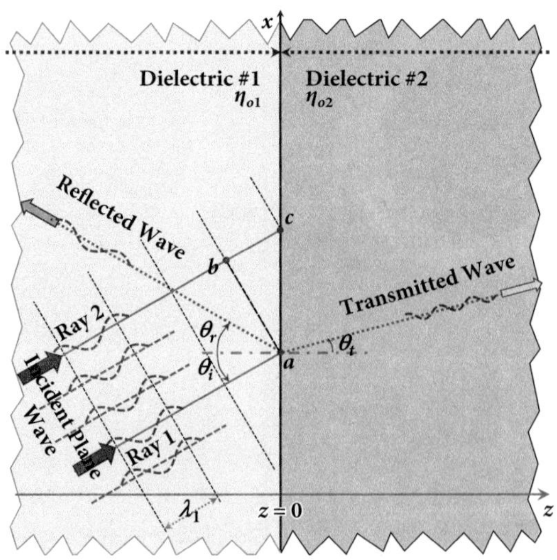

FIGURE 12B.2 Graphical presentation of the incident wave and its "wave-front-planes" in the case of oblique incidence.

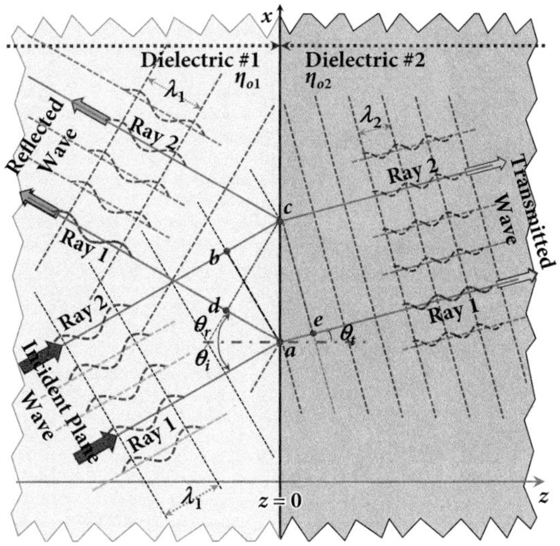

FIGURE 12B.3 Graphical presentation for the oblique incidence case to relate the angle of incidence to the angles of reflection and transmission.

Among the incident wave lines, we chose to highlight two of them, Ray 1 and Ray 2 that would arrive at the interface plane λ_1 apart. The chosen rays are those arriving at points *a* and *c* as marked in the figure. These two rays travel in phase and have the same exact phase (peak amplitude) at their corresponding points *a* and *b*.

When Ray 1 arrives at point *a*, it initiates its corresponding reflected and transmitted rays, while Ray 2 continues to travel an additional distance λ_1 before it reaches the interface plane. By the time Ray 2 reaches point *c*, the reflected Ray 1 would have reached point *d* and the transmitted Ray 1 would reach point *e* as demonstrated in Figure 12B.3.

By examining the wave reflection of both Rays, we recognize that the reflected wave fronts are again λ_1 apart which is the same wavelength of the incident wave since they both occur in the same material with the same phase velocity. We also recognize that Rays 1 and 2 will have the same exact phase at points *d* and *c*, respectively, since they lie on the same wave-front plane.

Meanwhile, for the wave transmission of both Rays, wave fronts in the second material are λ_2 apart. We also recognize that the transmitted Rays 1 and 2 will have the same exact phase at points *e* and *c*, respectively, since they lie on the same wave-front plane.

Based on the above discussion, we conclude that points *a* and *b* have the same identical phase while points *d*, *e*, and *c* are also identical in phase. Hence, the phase changes *ad*, *ae*, and *bc* are all equal.

Now, we move on to Figure 12B.4 where we highlight three triangles (*abc*), (*cda*), and (*aec*). These three triangles are shown separately in Figure 12B.5 for the sake of clarity. By examining the middle part of this figure, we can easily recognize that the two triangles (*abc*) and (*cda*) are congruent, and hence both the incidence and reflection angles θ_i and θ_r are equal.

$$\theta_r = \theta_i \tag{12B.13}$$

also, by examining the right most part of the figure, we can conclude that

$$\frac{\sin\theta_t}{\sin\theta_i} = \frac{\lambda_2}{\lambda_1} = \frac{\beta_1}{\beta_2} = \frac{\sqrt{\mu_1\varepsilon_1}}{\sqrt{\mu_2\varepsilon_2}} = \frac{\sqrt{\mu_{r1}\varepsilon_{r1}}}{\sqrt{\mu_{r2}\varepsilon_{r2}}} = \frac{n_1}{n_2} \tag{12B.14}$$

Finally, by expressing the material wavelengths in terms of its dielectric constants, we can rewrite Equation (12B.14) as

$$\frac{\sin\theta_t}{\sin\theta_i} = \frac{\lambda_2}{\lambda_1} = \frac{c_2}{c_1} = \frac{\sqrt{\mu_1\varepsilon_1}}{\sqrt{\mu_2\varepsilon_2}} = \frac{n_1}{n_2} \tag{12B.15}$$

where n_1 and n_2 are the corresponding indices of refraction of dielectrics 1 and 2, respectively.

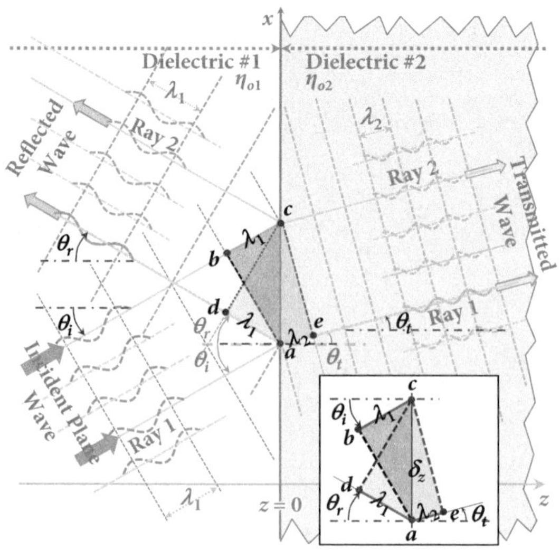

FIGURE 12B.4 Graphical presentation for the oblique incidence case used in the derivation of Snell's law.

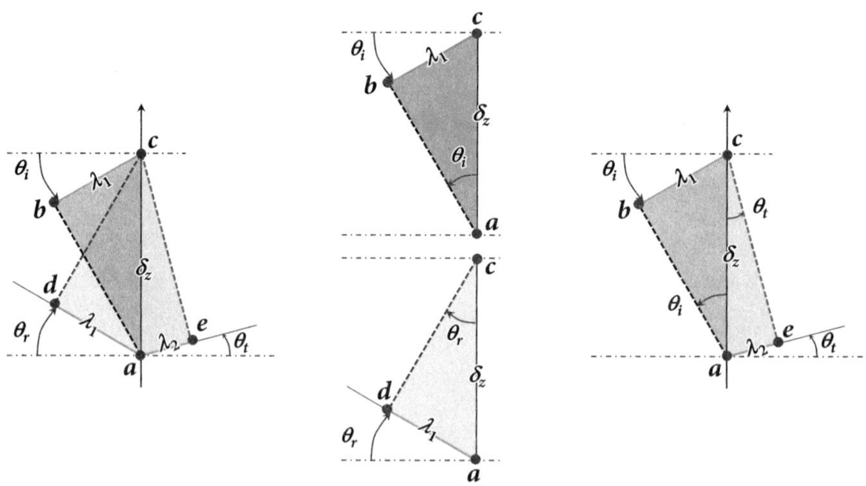

FIGURE 12B.5 Graphical presentation focused on the shaded triangles used in the derivation of Snell's law.

ADDENDUM 12C
TOTAL REFLECTION: PHYSICAL APPLICATIONS

Two physical applications of total reflection are presented in this addendum for the sake of bringing some physical insight to the concept and develop an appreciation to its implications.

The first application is that of using dielectric guides in the transmission of electromagnetic waves; typically for millimeter wave frequencies as well as optical signals. In the following, we will present the optical fiber as it is more commonly used.

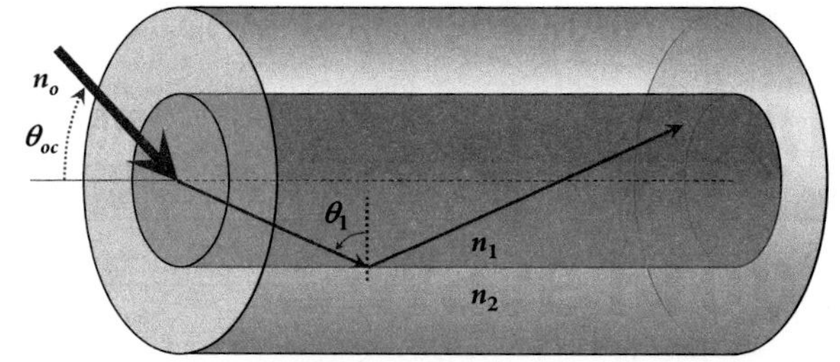

FIGURE 12C.1 The application of the concept of total reflection in "step-index" optical fibers.

We start with Figure 12C.1 where we show the launch of an optical signal at the end of an optical fiber. The fiber is composed of a cylindrical dielectric core with an index of refraction n_1 surrounded by another dielectric shell (cladding) with a smaller index of refraction n_2. We further assume the launch medium to have an index of refraction, n_o, typically air with $n_o = 1$.

The idea is to keep the propagating wave confined to the fiber core and not lose its energy through refraction outside it. This implies that the angle of incidence of the wave at the core-cladding interface should be larger than the corresponding critical angle. In doing so, we would maintain total reflection of the EM wave within the fiber core throughout its transmission along the fiber axis.

Consequently, the maximum value for the angle of incidence, θ_{1c}, is given by Equation (12.114),

$$\theta_{1c} = \sin^{-1}\left(\frac{n_2}{n_1}\right) \tag{12.114}$$

Working things backward to the launch end of the fiber, and applying Snell's law for the interface between the two media n_o and n_1, with the angle of refraction in the fiber core as the complementary angle to that of θ_{1c}, we can write

$$n_o \sin\theta_{oc} = n_1 \sin\left(\frac{\pi}{2} - \theta_{1c}\right) \tag{12C.1}$$

Now, using Equation (12.114), we can rewrite Equation (12C.1) to yield

$$\sin\theta_{oc} = \left(\frac{1}{n_o}\sqrt{(n_1)^2 - (n_2)^2}\right) \tag{12C.2}$$

The term $\sin\theta_{oc}$ is typically referred to as the numerical aperture (NA) of the fiber; hence, for $n_o = 1$ in air, we write

$$\text{NA} = \sin\theta_{oc} = \sqrt{(n_1)^2 - (n_2)^2} \tag{12C.3}$$

To summarize, all optical waves incident at the launch end of the fiber within its numerical aperture would result in total reflection at the fiber cladding interface. This is demonstrated in Figure 12C.2 showing full confinement of the EM wave energy within the fiber core ensuring full transmission of the signal along the optical fiber.

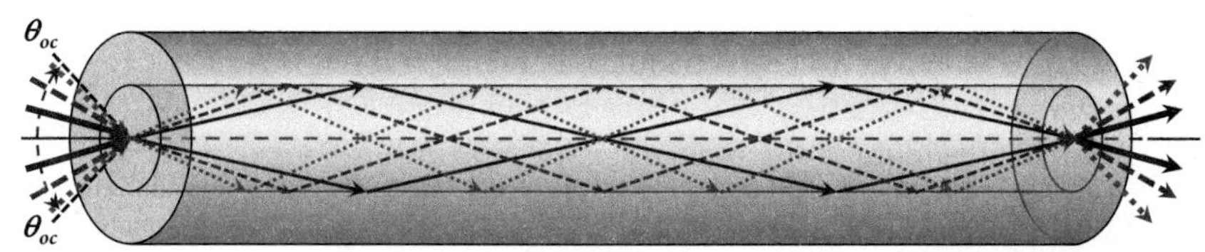

FIGURE 12C.2 A demonstration of the numerical aperture for a "step-index" optical fiber.

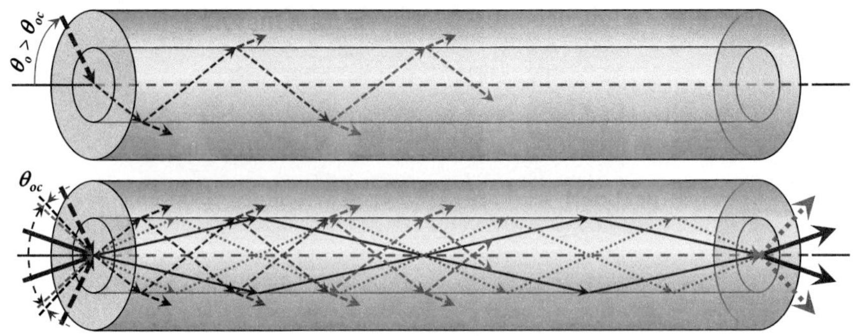

FIGURE 12C.3 A demonstration of leakage waves incident at angles outside the numerical aperture for a "step-index" optical fiber.

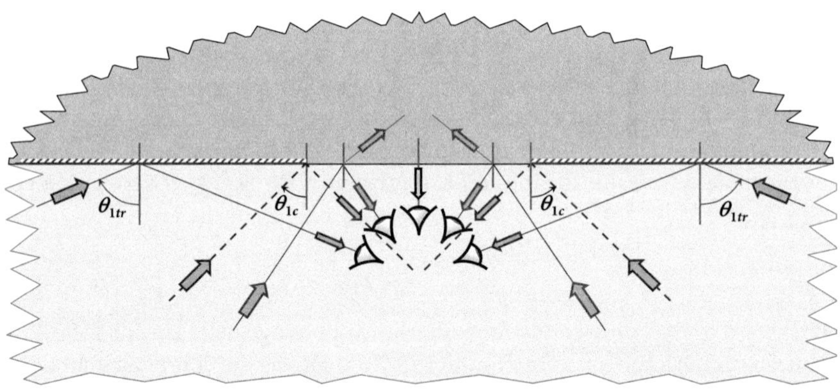

FIGURE 12C.4 A demonstration of the limited scope of visibility for an underwater observer to see above water objects.

However, waves that are incident at angles greater than the numerical aperture will certainly meet the fiber cladding interface at angles less than the corresponding critical angle. This will result in partial reflection as some of the signal energy will refract in the cladding causing continuous leakage of signal energy and hence a decaying transmission. Figure 12C.3 demonstrates this energy leakage of signals incident outside the fiber's numerical aperture.

Another physical observation is related to total reflection as observed by a person swimming under water. This is demonstrated in Figure 12C.4 where we show the person's eyes under water looking around in different directions. The dashed lines show the critical angle for total reflection. All views at any angle exceeding this critical angle will correspond to total reflection making the water surface for this region act as a perfect mirror. This region is marked by hashed lines in the figure. Views at other angles will experience both partial transmission and partial reflection as demonstrated.

ADDENDUM 12D
DERIVATION OF BREWSTER ANGLE EXPRESSIONS

As defined earlier in the body of the chapter, the Brewster angle is the angle at which certain oblique incidence waves can go through the interface experiencing total transmission without any reflection. Mathematically, we are talking about the angle at which $\overline{\Gamma}(0) = 0$.

Now, we will use the $\overline{\Gamma}(0)$ expression for both parallel and perpendicular oblique incidences to derive the expressions for the corresponding Brewster angles.

12D.1 Brewster Angle for the Parallel Polarization Case

Starting with Equation (12.139) for the parallel case, we write

$$\bar{\Gamma}(0)\Big|_{\|} = \frac{\bar{\eta}_{o2}\cos\theta_2 - \bar{\eta}_{o1}\cos\theta_1}{\bar{\eta}_{o2}\cos\theta_2 + \bar{\eta}_{o1}\cos\theta_1} = 0 \tag{12.139}$$

which implies that

$$\eta_{o1}\cos\theta_{1B\|} = \eta_{o2}\cos\theta_{2B\|} \tag{12D.1}$$

where the added subscript "$B\|$" is to indicate the Brewster condition for the parallel polarization case. Now, expressing the intrinsic impedance of the material in terms of its permittivity and permeability, we can write

$$\frac{\mu_1}{\varepsilon_1}(\cos\theta_{1B\|})^2 = \frac{\mu_2}{\varepsilon_2}(\cos\theta_{2B\|})^2 \tag{12D.2}$$

and using Snell's law

$$n_1 \sin\theta_{1B\|} = n_2 \sin\theta_{2B\|} \tag{12D.3}$$

and again replacing the indices of refraction in terms of the material permittivity and permeability, we get

$$(\mu_1\varepsilon_1)(\sin\theta_{1B\|})^2 = (\mu_2\varepsilon_2)(\sin\theta_{2B\|})^2 \tag{12D.4}$$

Next, we solve Equations (12D.2) and (12D.4) simultaneously to obtain an expression for $\theta_{1B\|}$

$$\frac{\mu_1}{\varepsilon_1}\left[1-(\sin\theta_{1B\|})^2\right] = \frac{\mu_2}{\varepsilon_2}\left[1-(\sin\theta_{2B\|})^2\right] = \frac{\mu_2}{\varepsilon_2}\left[1-\left(\frac{\mu_1\varepsilon_1}{\mu_2\varepsilon_2}\right)(\sin\theta_{1B\|})^2\right] \tag{12D.5}$$

$$\left[\frac{\mu_1}{\varepsilon_1} - \frac{\mu_1}{\varepsilon_1}(\sin\theta_{1B\|})^2\right] = \left[\frac{\mu_2}{\varepsilon_2} - \frac{\mu_2}{\varepsilon_2}\left(\frac{\mu_1\varepsilon_1}{\mu_2\varepsilon_2}\right)(\sin\theta_{1B\|})^2\right] \tag{12D.6}$$

$$\left[\frac{\mu_2}{\varepsilon_2}\left(\frac{\mu_1\varepsilon_1}{\mu_2\varepsilon_2}\right) - \frac{\mu_1}{\varepsilon_1}\right](\sin\theta_{1B\|})^2 = \frac{\mu_2}{\varepsilon_2} - \frac{\mu_1}{\varepsilon_1} \tag{12D.7}$$

to conclude that

$$\sin\theta_{1B\|} = \sqrt{\frac{\dfrac{\mu_2}{\varepsilon_2} - \dfrac{\mu_1}{\varepsilon_1}}{\left[\dfrac{\mu_2}{\varepsilon_2}\left(\dfrac{\mu_1\varepsilon_1}{\mu_2\varepsilon_2}\right) - \dfrac{\mu_1}{\varepsilon_1}\right]}} \tag{12D.8}$$

or, finally,

$$\sin\theta_{1B\|} = \sqrt{\frac{1-\dfrac{\mu_2\varepsilon_1}{\mu_1\varepsilon_2}}{1-\left(\dfrac{\varepsilon_1}{\varepsilon_2}\right)^2}} \tag{12D.9}$$

For the case of nonmagnetic media, we can write

$$\sin\theta_{1B\|}\Big|_{\mu_1=\mu_2} = \frac{1}{\sqrt{1+\left(\dfrac{\varepsilon_1}{\varepsilon_2}\right)}} \tag{12D.10}$$

12D.2 Brewster Angle for the Perpendicular Polarization Case

We now turn to the perpendicular polarization case. For zero reflection Equation (12.140) yields the condition

$$\Gamma(0)\big|_{\perp} = \frac{\bar{\eta}_{o2}/\cos\theta_2 - \bar{\eta}_{o1}/\cos\theta_1}{\bar{\eta}_{o2}/\cos\theta_2 + \bar{\eta}_{o1}/\cos\theta_1} = 0 \tag{12.140}$$

from which we extract that the numerator must equal to zero,

$$\bar{\eta}_{o2}/\cos\theta_{2B\perp} - \bar{\eta}_{o1}/\cos\theta_{1B\perp} \tag{12D.11}$$

The added subscript "$B\perp$" is to indicate the Brewster condition for the perpendicular polarization case. Next, we express the intrinsic impedances of the materials in terms of their corresponding permittivities and permeabilities. As a result, we get

$$\frac{\mu_1}{\varepsilon_1}(\cos\theta_{2B\perp})^2 = \frac{\mu_2}{\varepsilon_2}(\cos\theta_{1B\perp})^2 \tag{12D.12}$$

Now, we are set to solve Equation (12D.12) simultaneously with that of Snell's law earlier expressed in Equation (12D.4). We go through the following development

$$\frac{\mu_1}{\varepsilon_1}\left[1-(\sin\theta_{2B\perp})^2\right] = \frac{\mu_2}{\varepsilon_2}(\cos\theta_{1B\perp})^2 = \frac{\mu_1}{\varepsilon_1}\left[1-\left(\frac{\mu_1\varepsilon_1}{\mu_2\varepsilon_2}\right)(\sin\theta_{1B\perp})^2\right] \tag{12D.13}$$

$$\frac{\mu_2}{\varepsilon_2}\left[1-(\sin\theta_{1B\perp})^2\right] = \frac{\mu_1}{\varepsilon_1}\left[1-\left(\frac{\mu_1\varepsilon_1}{\mu_2\varepsilon_2}\right)(\sin\theta_{1B\perp})^2\right] \tag{12D.14}$$

$$\left[\frac{\mu_2}{\varepsilon_2} - \frac{\mu_2}{\varepsilon_2}(\sin\theta_{1B\perp})^2\right] = \left[\frac{\mu_1}{\varepsilon_1} - \frac{\mu_1}{\varepsilon_1}\left(\frac{\mu_1\varepsilon_1}{\mu_2\varepsilon_2}\right)(\sin\theta_{1B\perp})^2\right] \tag{12D.15}$$

$$\left[\frac{\mu_1}{\varepsilon_1}\left(\frac{\mu_1\varepsilon_1}{\mu_2\varepsilon_2}\right) - \frac{\mu_2}{\varepsilon_2}\right](\sin\theta_{1B\perp})^2 = \frac{\mu_1}{\varepsilon_1} - \frac{\mu_2}{\varepsilon_2} \tag{12D.16}$$

$$\sin\theta_{1B\perp} = \sqrt{\frac{\dfrac{\mu_1}{\varepsilon_1} - \dfrac{\mu_2}{\varepsilon_2}}{\left[\dfrac{\mu_1}{\varepsilon_1}\left(\dfrac{\mu_1\varepsilon_1}{\mu_2\varepsilon_2}\right) - \dfrac{\mu_2}{\varepsilon_2}\right]}} \tag{12D.17}$$

which can be brought to the form

$$\sin\theta_{1B\perp} = \sqrt{\frac{1-\dfrac{\mu_1\varepsilon_2}{\mu_2\varepsilon_1}}{1-\left(\dfrac{\mu_1}{\mu_2}\right)^2}} \tag{12D.18}$$

Finally, it can be seen from Equation (12D.18) that in most practical cases involving nonmagnetic media, where $\mu_1 = \mu_2$, the Brewster condition for total transmission is not realizable.

$$\sin\theta_{1B\perp}\Big|_{\mu_1=\mu_2} \text{ does not exist} \tag{12D.19}$$

ADDENDUM 12E
THE "COMPLEX" SNELL'S LAW AND THE MYSTERY OF THE "COMPLEX" ANGLE OF REFRACTION

12E.1 Oblique Incidence: The Case of a Perfect Dielectric–Lossy Dielectric Interface

Going back to the development of Snell's law that we covered in Addendum 12B, we recall the general form of Snell's law in Equation (12B.10). At the time, we confined our presentation to the interface between two perfect dielectrics, and hence Equation (12B.11).

In this addendum, we will take on the case of an interface between a perfect dielectric and a lossy one. The case of two lossy dielectrics can be treated in a similar fashion but will not be presented.

The reason for postponing the discussion of this case is that Equation (12B.10) in the presence of loss would lead to a complex value for $\sin\theta_t$ and, hence, to unphysical values for the refraction angle, θ_t. The logical conclusion is that the mathematical model we used for the wave transmission in a lossy medium is not appropriate. In the following, we will demonstrate this point and present a physically realizable alternative model for the propagation waves.

For our case here, the incident wave is launched in a perfect dielectric, #1, and is transmitted through a lossy dielectric, #2, as depicted in Figure 12E.1. The figure shows the case of oblique incidence with perpendicular polarization. Again, a similar development for the case of parallel polarization should lead to the same exact conclusions.

We start with Equations (12B.3) to (12B.5) describing the vector electric fields of the propagating waves with the replacement $\gamma_1 \to j\beta_1$

$$\overline{E}_{1y}^{+}(x,z) = \overline{E}_{1y}^{+}(0)\cdot e^{-j\beta_1(x\sin\theta_i + z\cos\theta_i)} \tag{12E.1}$$

$$\overline{E}_{1y}^{-}(x,z) = \overline{E}_{1y}^{-}(0)\cdot e^{-j\beta_1(x\sin\theta_r - z\cos\theta_r)} \tag{12E.2}$$

$$\overline{E}_{2y}^{+}(x,z) = \overline{E}_{2y}^{+}(0)\cdot e^{-\gamma_2(x\sin\theta_t + z\cos\theta_t)} \tag{12E.3}$$

Similar to our development that we carried in Addendum 12B, we can write the *complex* form of Snell's law for the present case by the modification $\gamma_1 \to j\beta_1$ in Equation (12B.10), and hence

$$j\beta_1\sin\theta_i = \gamma_2\sin\theta_t = (\alpha_2 + j\beta_2)\sin\theta_t \tag{12E.4}$$

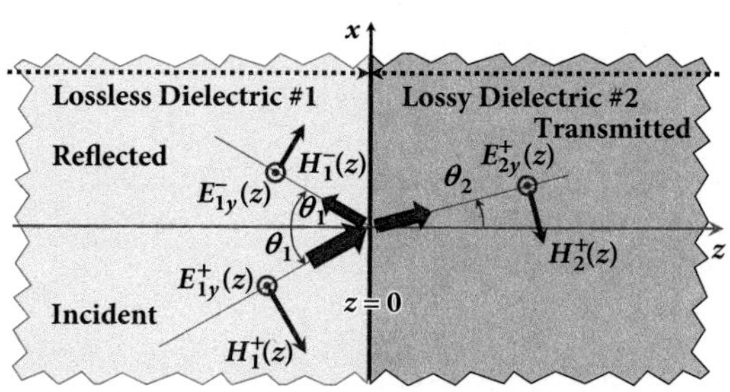

FIGURE 12E.1 A graphical illustration for the application of Snell's law for wave refraction in a lossy dielectric. The mathematical analysis concludes that the angle of refraction is complex. Hence the need for a modified "complex" form for Snell's law.

This result implies a complex value for $\sin\theta_t$, which in turn suggests that the model proposed for the transmitted wave in the lossy region, Equation (12E.3), is not *physically realizable* and the suggestion that θ_t is the angle of refraction is not proper. In the following, we will continue to use the expression in Equation (12E.3) as a mathematically proper model in order to derive an alternative expression that is physically realizable.

Let us manipulate Equation (12E.3), while using the information of Equation (12E.4), as follows:

$$\begin{aligned} \bar{E}_{2y}^{+}(x,z) &= \bar{E}_{2y}^{+}(0)e^{-(\gamma_2 \sin\theta_t)x-(\gamma_2\cos\theta_t)z} = \bar{E}_{2y}^{+}(0)e^{-(j\beta_1\sin\theta_i)x-\left(\gamma_2\sqrt{1-(\sin\theta_t)^2}\right)z} \\ &= \bar{E}_{2y}^{+}(0)e^{-(j\beta_1\sin\theta_i)x-(u+jv)z} = \bar{E}_{2y}^{+}(0)\cdot e^{-uz}e^{-j(\beta_1\sin\theta_i x+vz)} \end{aligned} \tag{12E.5}$$

where both u and v are both real quantities and

$$u+jv = \gamma_2\sqrt{1-(\sin\theta_t)^2} \tag{12E.6}$$

Further development to Equation (12E.6) leads us to

$$\begin{aligned} u+jv &= \gamma_2\sqrt{1-(\sin\theta_t)^2} = \sqrt{(\gamma_2)^2-(\gamma_2\sin\theta_t)^2} \\ &= \sqrt{(\alpha_2+j\beta_2)^2-(j\beta_1\sin\theta_i)^2} = \sqrt{\alpha_2^2-\beta_2^2+(\beta_1\sin\theta_i)^2+j2\alpha_2\beta_2} \end{aligned} \tag{12E.7}$$

$$u = \text{Real part of}\left[\sqrt{\alpha_2^2-\beta_2^2+(\beta_1\sin\theta_i)^2+j2\alpha_2\beta_2}\right] \tag{12E.8}$$

$$v = \text{Imaginary part of}\left[\sqrt{\alpha_2^2-\beta_2^2+(\beta_1\sin\theta_i)^2+j2\alpha_2\beta_2}\right] \tag{12E.9}$$

Now, examining Equation (12E.5), we recognize the following about the refracted wave into the lossy region:

a. The exponent term e^{-uz} indicates an attenuation to the propagating wave with an attenuation coefficient, u. This attenuation is only in the z direction and is not along the wave propagation direction, which is oblique.

   This is not the typical homogeneous plane wave nature we have been discussing all along. What we have here, as described by Equation (12E.5), is what is referred to as an *inhomogeneous plane wave*.

b. The exponent term $e^{-j(\beta_1\sin\theta_i x+vz)}$ in Equation (12E.5) indicates the phase propagation term of the refracted wave. This term can be expressed in the typical form

$$e^{-j(\beta_1\sin\theta_i x+vz)} = e^{-j\beta_{2,ph}(x\sin\theta_{t,ph}+z\cos\theta_{t,ph})} \tag{12E.10}$$

where $\beta_{2,ph}$, represent the effective phase propagation constant along the wave propagation direction and $\theta_{t,ph}$ is in fact the true physical angle of refraction in region 2.

Equating the exponents in Equation (12E.10), we can write

$$(\beta_1\sin\theta_i x+vz) = \beta_{2,ph}(x\sin\theta_{t,ph}+z\cos\theta_{t,ph}) \tag{12E.11}$$

$$\beta_{2,ph}\sin\theta_{t,ph} = \beta_1\sin\theta_i \quad \text{and} \quad \beta_{2,ph}\cos\theta_{t,ph} = v \tag{12E.12}$$

Hence, we arrive at the following expressions for $\beta_{2,ph}$ and $\theta_{t,ph}$:

$$\beta_{2,ph} = \sqrt{(\beta_1\sin\theta_i)^2+v^2} \tag{12E.13}$$

$$\sin\theta_{t,ph} = \frac{\beta_1\sin\theta_i}{\beta_{2,ph}} = \frac{\beta_1\sin\theta_i}{\sqrt{(\beta_1\sin\theta_i)^2+v^2}} \tag{12E.14}$$

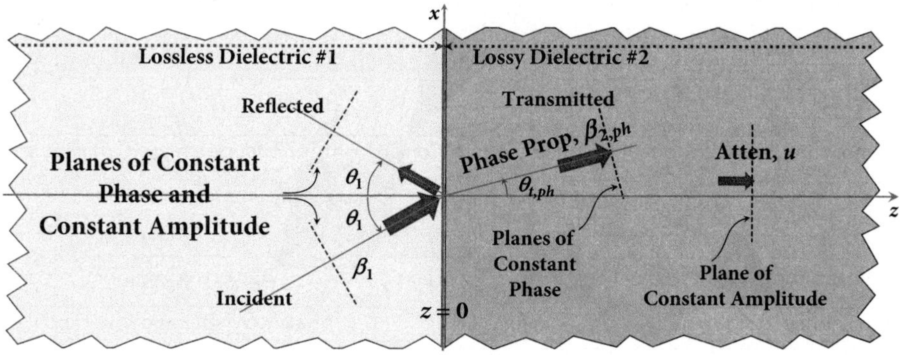

FIGURE 12E.2 A graphical illustration for the development of the *complex* Snell's law.

Based on the two observations above, *a* and *b*, we can depict the nature of the refracted *inhomogeneous plane wave* as demonstrated in Figure 12E.2.

The figure shows a wave refracted into region 2 with a refraction angle, $\theta_{t,ph}$, and a propagation phase constant, $\beta_{2,ph}$, along the angle $\theta_{t,ph}$. The corresponding planes of constant phase are shown to be orthogonal to the direction of propagation. Meanwhile, the refracted wave attenuation takes place along the *z* direction only according to the e^{-uz} exponent. The attenuation coefficient is *u*, and the planes of constant amplitudes are orthogonal to the *z* direction and hence parallel to the *x*-*y* plane.

12E.2 Low-Loss Approximations

In the following, we will derive low-loss approximations for the above expressions as well as provide numerical demonstrations.

Starting with Equation (12E.7) and assuming low loss, $\alpha_2 \ll \beta_2$, for the lossy dielectric region #2, we can write

$$
\begin{aligned}
u + jv &= \sqrt{\alpha_2^2 - \beta_2^2 + (\beta_1 \sin\theta_i)^2 + j2\alpha_2\beta_2} \\
&= \sqrt{-\left(\beta_2^2 - (\beta_1 \sin\theta_i)^2 - \alpha_2^2\right)}\sqrt{1 - j\frac{2\alpha_2\beta_2}{\left(\beta_2^2 - (\beta_1 \sin\theta_i)^2 - \alpha_2^2\right)}} \cong j\sqrt{\left(\beta_2^2 - (\beta_1 \sin\theta_i)^2 - \alpha_2^2\right)} \\
&\quad \left\{ \begin{aligned} &1 + \frac{1}{2}\left[-j\frac{2\alpha_2\beta_2}{\left(\beta_2^2 - (\beta_1 \sin\theta_i)^2 - \alpha_2^2\right)}\right] - \frac{1}{8}\left[-j\frac{2\alpha_2\beta_2}{\left(\beta_2^2 - (\beta_1 \sin\theta_i)^2 - \alpha_2^2\right)}\right]^2 \\ &+ \frac{1}{16}\left[-j\frac{2\alpha_2\beta_2}{\left(\beta_2^2 - (\beta_1 \sin\theta_i)^2 - \alpha_2^2\right)}\right]^3 + \cdots \end{aligned} \right\}
\end{aligned}
\tag{12E.15}
$$

Now, separating the real and imaginary terms in Equation (12E.15), we get

$$
u \cong \frac{\alpha_2\beta_2}{\sqrt{\left(\beta_2^2 - (\beta_1 \sin\theta_i)^2 - \alpha_2^2\right)}}\left\{1 - \frac{1}{2}\left[\frac{\alpha_2\beta_2}{\left(\beta_2^2 - (\beta_1 \sin\theta_i)^2 - \alpha_2^2\right)}\right]^2\right\}
\tag{12E.16}
$$

$$
v \cong \sqrt{\left(\beta_2^2 - (\beta_1 \sin\theta_i)^2 - \alpha_2^2\right)}\left\{1 + \frac{1}{2}\left[\frac{\alpha_2\beta_2}{\left(\beta_2^2 - (\beta_1 \sin\theta_i)^2 - \alpha_2^2\right)}\right]^2\right\}
\tag{12E.17}
$$

Quantity	Lossless	Extremely Low Loss	Very Low Loss	Low Loss
$\gamma_1 = j\beta_1$	$j314.159$	$j314.159$	$j314.159$	$j314.159$
γ_2	$j628.319$	$1+j628.319$	$5+j628.319$	$25+j628.319$
$\sin\theta_t$	0.353553	0.35355 $+j0.000563$	0.35353 $+j0.002813$	0.352995 $+j0.01404$
$\cos\theta_t$	0.935414	0.93542 $-j0.0002127$	0.93543 $-j0.001063$	0.93575 $-j0.005298$
$\gamma_2\cos\theta_t$	$j587.738$	1.06904 $+j587.738$	5.34520 $+j587.741$	26.7227 $+j587.814$
u	0	1.06904	5.34520	26.7227
v	587.738	587.738	587.741	587.814
$\sin\theta_{t,ph}$	0.353553	0.353553	0.353552	0.353513
$\theta_{t,ph}$ (rad)	0.361367	0.379268	0.379265	0.379213
$\theta_{t,ph}$ (deg)	20.7048°	21.7304°	21.7303°	21.7273°
$\beta_{2,ph}$	628.319	628.319	628.321	628.389
$\alpha_{2,z} = u$	0	1.06904	5.34520	26.7227

TABLE 12E.1 A Numerical Example of the Application of the Complex Snell's Law and the Corresponding Complex Angle of Refraction

12E.3 Numerical Demonstration

The following numerical demonstration is for a 45° oblique angle of incidence on the interface between a perfect dielectric and a lossy one. The following parameters are assumed:

$$\beta_1 = 100\pi \text{ rad/m}$$
$$\beta_2 = 200\pi \text{ rad/m}$$
$$\alpha_2 = 0, 1, 5, \text{ and } 25 \text{ Nep/m}$$
$$\theta_i = \pi/4 = 45°$$

Using Equations (12E.4) to (12E.14), Table 12E.1 is constructed to yield the physical values for the attenuation, $\alpha_{2,z} = u$, the phase coefficient, $\beta_{2,ph}$, as well as the physical angle of refraction, $\theta_{t,ph}$ of the refracted inhomogeneous plane wave.

CHAPTER 12 PROBLEMS

- Problems 12.1 to 12.3: Problems dealing with wave polarization
- Problems 12.4 to 12.7: Problems dealing with normal incidence dielectric-dielectric interface
- Problems 12.8 to 12.11: Problems dealing with normal incidence dielectric-conductor interface
- Problems 12.12 to 12.16: Problems dealing with normal incidence multilayer interface
- Problems 12.17 to 12.19: Problems dealing with oblique incidence
- Problems 12.20 to 12.23: Problems dealing with critical angle and Brewster angle

12.1 A linearly polarized wave passes through two polarizing plates. The characteristic direction of the second plate is orthogonal to that of the wave. What should the characteristic direction of the first plate be to allow 50% of the incident wave intensity to pass through the two plates?

12.2 An unpolarized wave passes through three stacked polarizing plates. The first and third are crossed: the one between has its axis at 45° to the axes of the other two. What fraction of the intensity of the incident wave is transmitted through?

12.3 Determine the polarization and direction of propagation of each of the following plane waves:

a. $\vec{E} = \vec{a}_z 10e^{-j3x}$ V/m

b. $\vec{E}(y,t) = \vec{a}_x 20\cos\left(\omega t + \frac{\pi}{3}y\right)$ V/m

c. $\vec{E} = \left(\vec{a}_x E_{ox} + \vec{a}_y 2E_{oy}\right)e^{+j\beta z}$

d. $\vec{E}(z,t) = \vec{a}_x 10\cos\left(\pi\times 10^7 t - \frac{\pi}{4}z\right) - \vec{a}_y 10\sin\left(\pi\times 10^7 t - \frac{\pi}{4}z\right)$

12.4 A 1-GHz plane wave in free space is incident normally on a large plane dielectric region (assumed of infinite thickness) with $\varepsilon_r = 4$. Given that the intensity of the incident electric field is 15 V/m, find

a. the electric field intensity of the reflected wave.

b. the electric field intensity of the transmitted wave.

12.5 For the same problem above, find

a. the magnetic field intensities of the incident, reflected, and transmitted waves.

b. the Poynting vector associated with all three waves.

12.6 Repeat the above two problems for a lossy dielectric with $\varepsilon_r' = 9$ and $\varepsilon_r'' = 0.1$.

12.7 A plane wave in free space is incident normally on a large plane dielectric region ($z \geq 0$) with complex relative permittivity of $3 - j0.01$. Given that the electric field of the incident wave is $E = 2.5\cos[2\pi(3\cdot 10^9 t - 10z)]$ V/m, find

a. the operating frequency, wavelength, propagation constant of the wave in both regions.

b. the corresponding mathematical expressions for E, H, and $\mathcal{P}$ at all points in space (in both regions).

12.8 For normal incidence,

a. prove that the reflected magnetic field from a conducting surface is in phase with the incident magnetic field at the interface.

b. prove that the reflected electric field from a conducting surface is in opposite phase with the incident electric field at the interface.

c. what is the total magnetic field intensity at the interface?

d. what is the total electric field intensity at the interface?

e. what is the surface current density (if any) on the conductor surface?

12.9 A plane wave in free space is incident along the z axis normally on a large conducting slab located in the x-y plane at $z = 0$. If the incident wave is $E = 15\sin(2\pi \cdot 10^9 t - \beta z + \varphi)$ V/m. find the total electric and magnetic fields at all locations in free space.

12.10 Calculate the complex reflection coefficient for a 1-GHz plane wave in air which is incident normally on

a. a large aluminum sheet

b. a large iron sheet

c. a large glass sheet (assume infinite thickness)

d. surface of the earth

e. surface of the ocean

Use the material properties given in Appendix C.

12.11 A plane wave with Poynting vector $\mathcal{P}_i$ is normally incident on the surface of the sea ($\sigma = 4$, $\varepsilon_r = 81$, and $\mu_r = 1$). Calculate the reflected and transmitted Poynting vectors $\mathcal{P}_r$ and $\mathcal{P}_t$, respectively if the frequency of the incident wave is (a) $f = 1$ MHz and (b) $f = 10$ GHz.

12.12 A uniform plane wave is incident normal to a dry concrete wall, 10 cm thick. Determine the power level of the transmission (in dB relative to the incident wave) at the other side of the wall at 1 GHz. Repeat for wet concrete. (Ignore reflections at discontinuities.)

a. Dry concrete, $\varepsilon_r' = 9$, $\tan\delta = 0.1$, $\mu = \mu_o$.

b. Wet concrete, $\varepsilon_r' = 100$, $\tan\delta = 1$, $\mu = \mu_o$.

c. Estimate the error resulting from ignoring the reflections in the case of dry concrete.

12.13 A 1 GHz 1 kW microwave beam is normally incident upon a 1 m^2 sheet of copper foil of 10 μm thickness. Calculate the power

a. Reflected from the foil

b. Transmitted in to the foil

c. Reemerges on the other side of the foil

12.14 Repeat the above problem for a 5 cm-thick granite block.

12.15 Calculate the percentage of incident energy transmitted through a dielectric panel ($\varepsilon_r = 2.5$) at 10 GHz. The thickness of the panel is 1 cm.

12.16 Find the thickness and relative permittivity of a quarter-wave matching dielectric layer to be used to eliminate reflections at an air-dielectric region interface at 1 GHz for the following cases:

a. The dielectric region is Teflon with $\varepsilon_r = 2.1$

b. The dielectric region is glass with $\varepsilon_r = 4 - j0.002$

12.17 A perpendicularly polarized plane wave is obliquely incident on an air-mica ($\varepsilon_r = 6$) interface with incidence angle of 60°. Calculate both reflection and transmission coefficients at the interface. Repeat for parallel polarization.

12.18 A parallel polarized plane wave propagates in dielectric #1 and is obliquely incident on dielectric #2 as shown in Figure 12.21, and the incidence angle is 45°. Calculate both reflection and transmission coefficients at the interface for the following cases:

a. Dielectric #1: $\varepsilon_r = 6$, and dielectric #2: $\varepsilon_r = 9$

b. Dielectric #1: $\varepsilon_r = 6 - j0.06$, and dielectric #2: $\varepsilon_r = 9 - j0.09$

c. Discuss the effect of the loss components on the obtained results

12.19 The magnetic field of a plane wave which is incident on a large, plane, perfect conductor at angle ($\theta_i = 45°$ is $\vec{H} = H_o \vec{a}_y \cos\left(10^6 \pi t - \vec{\beta}_1 \cdot \vec{r}\right)$. Calculate the density of average power flow in the x direction.

12.20 A uniform plane wave from an underwater source is incident on water-air interface at θ_i. Given:

- The plane of wave propagation is the y-z plane of Cartesian coordinates
- $\vec{H}_{\text{incident}}(z,t) = H_o(r,t)\vec{a}_x$ A/m
- $\varepsilon_{r\,\text{water}} = 81, \sigma_{\text{water}} = 0$, and $\mu_{r\,\text{water}} = 1$

a. Identify the type of polarization and the type of incidence of this wave.

b. Find the critical angle, also state the definition of the critical angle.

c. Find the Brewster angle, also state the definition of the Brewster angle.

d. For $\theta_i = 20°$, find θ_r and θ_t.

e. Find the reflection and transmission coefficients for this incidence.

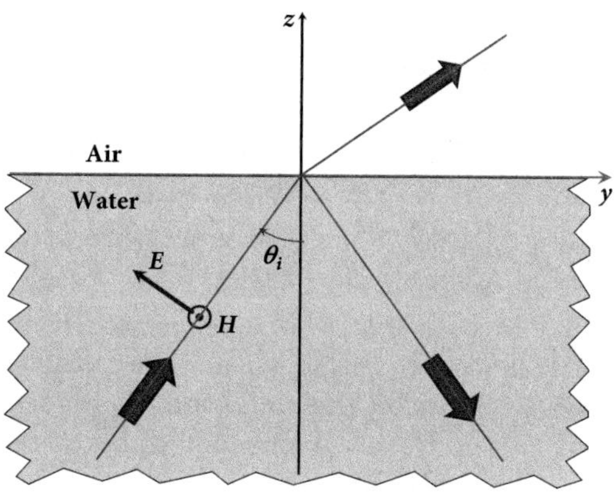

12.21 A uniform plane wave with $\vec{E}_{\text{incident}}(z,t) = \vec{a}_x E_{io} \exp\left\{j\left(2\pi \times 10^9 t - \beta_1 z\right)\right\}$ in the lossless dielectric medium 1, (ε_o, μ_o), is incident normally onto a lossless dielectric medium 2, ((9 – j0.09)ε_o, μ_o). Find

a. Γ at $z = 0$

b. τ at $z = 0$

c. Electric field vector phasor in medium 2

d. Critical angle

e. Brewster angle

Repeat while reversing the incidence direction (from 2 to 1).

12.22 A uniform plane wave from is incident on glass-air interface at θ_i. Given: the plane of wave propagation is the *y*-*z* plane of Cartesian coordinates, and $\vec{E}_{\text{incident}}(r,t) = E_o(r,t)\vec{a}_x$ A/m, and $\varepsilon_{r\,\text{glass}} = 4$, $\sigma_{\text{glass}} = 0$,

a. Identify the type of polarization and the type of incidence of this wave.

b. Define and find both the critical and Brewster angles.

c. For $\theta_i = 20°$, find θ_r and θ_t.

d. Find the reflection and transmission coefficients for this incidence.

Repeat part d above for the case of normal incidence.

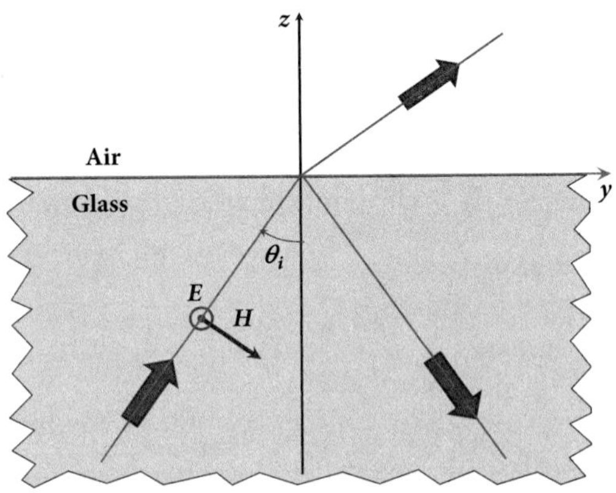

12.23 A uniform plane wave is incident on the air-fiber interface at θ_i as shown in the figure. Given $\varepsilon_{r1} = 5$, $\varepsilon_{r2} = 2.5$, and $\sigma_1 = \sigma_2 = 0$. Identify the type of polarization and the type of incidence of this wave:

a. Determine the critical angle for the $n_1 - n_2$ interface. Determine the corresponding θ_i.

b. Determine the Brewster angle for the $n_o - n_1$ interface. Determine the corresponding θ_1, θ_2, and θ_3.

c. For $\theta_i = 65.9°$, find θ_1, θ_2, and θ_3. Find the reflection and transmission coefficients for this incidence.

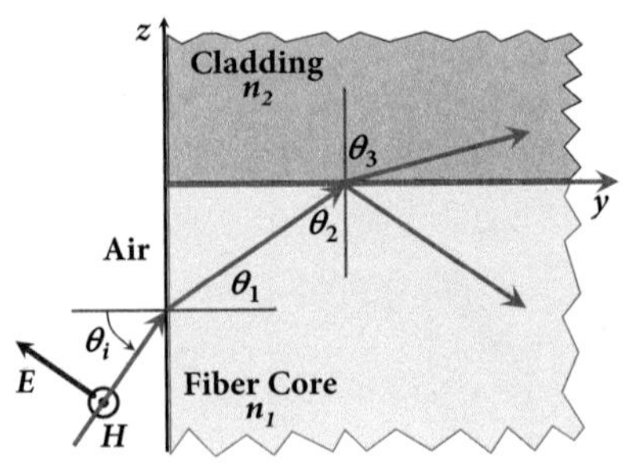

CHAPTER 12 SUMMARY[1]

Wave Polarization

- Linear polarization: $\vec{E}(z,t) = [E_x(0)\vec{a}_x]e^{-\alpha z}\cos(\omega t - \beta z)$, $\bar{\vec{E}}(z) = [E_x(0)\vec{a}_x]e^{-\gamma z}$
- Circular polarization:
  - $\vec{E}(z,t) = E(0)[\cos(\omega t - \beta z)\vec{a}_x + \sin(\omega t - \beta z)\vec{a}_y]e^{-\alpha z}$, $\bar{\vec{E}}(z) = E(0)[\vec{a}_x + j\,\vec{a}_y]e^{-\gamma z}$
- Elliptical polarization:
  - $\vec{E}(z,t) = [E_x(0)\cos(\omega t - \beta z)\vec{a}_x + E_y(0)\sin(\omega t - \beta z)\,\vec{a}_y]e^{-\alpha z}$, $\bar{\vec{E}}(z) = [E_x(0)\vec{a}_x + jE_y(0)\vec{a}_y]e^{-\gamma z}$

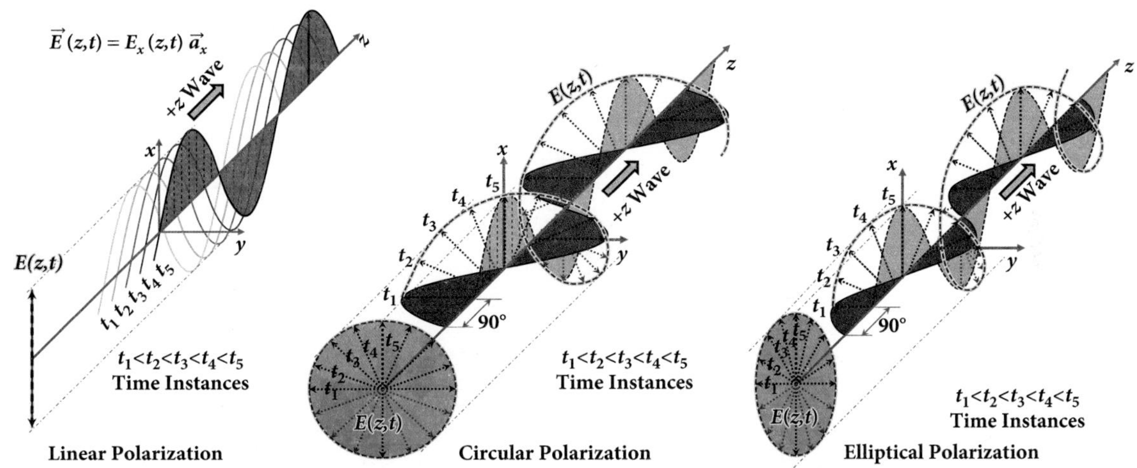

[1] ***Important Warning to Students:*** Please do not use this summary without thorough study of the relevant material in this chapter. Using this summary and its formulas without adequate appreciation of the physics involved and conditions of validity can lead to wrong implementations and invalid conclusions.

Transmission and Reflection of Uniform Plane Waves—Normal Incidence

- $\overline{E}_{1x}^{+}(z)=\overline{E}_{1x}^{+}(0)\cdot e^{-\gamma_1 z},\ \overline{E}_{1x}^{-}(z)=\overline{E}_{1x}^{-}(0)\cdot e^{+\gamma_1 z},\ \overline{E}_{2x}^{+}(z)=\overline{E}_{2x}^{+}(0)\cdot e^{-\gamma_2 z}$
- $\overline{\Gamma}(0)=\dfrac{\overline{E}_{1x}^{-}(0)}{\overline{E}_{1x}^{+}(0)}=\dfrac{\overline{\eta}_{o2}-\overline{\eta}_{o1}}{\overline{\eta}_{o2}+\overline{\eta}_{o1}},\ \overline{\tau}(0)=\dfrac{\overline{E}_{2x}^{+}(0)}{\overline{E}_{1x}^{+}(0)}=1+\overline{\Gamma}(0)=\dfrac{2\,\overline{\eta}_{o2}}{\overline{\eta}_{o2}+\overline{\eta}_{o1}}$

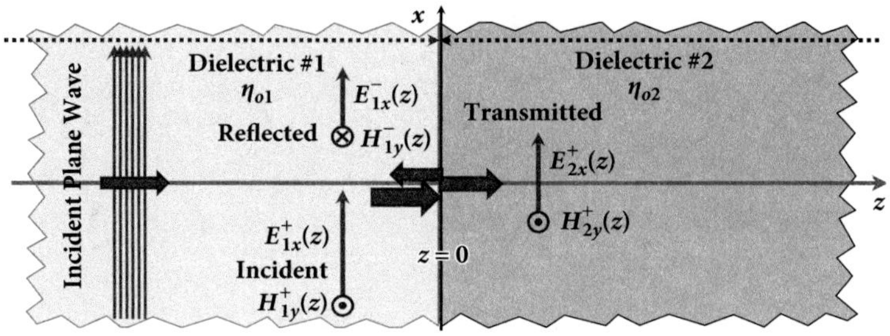

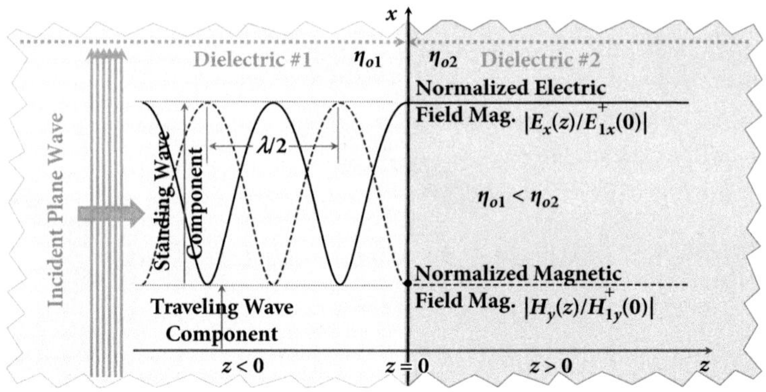

Two Lossless Dielectrics

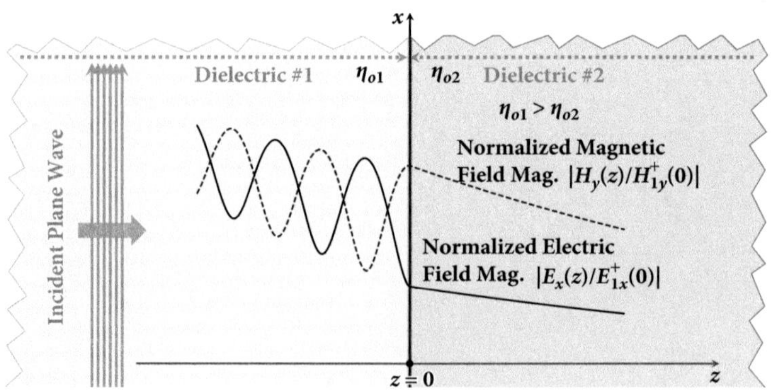

Two Lossy Dielectrics

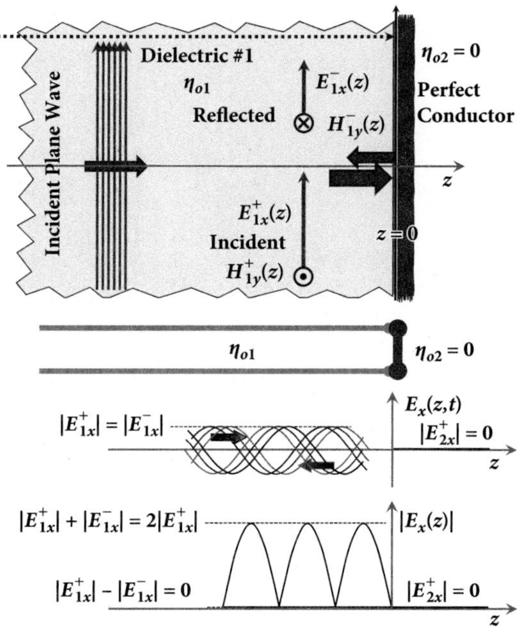

Lossless Diel. – Cond. Interface

Transmission and Reflection of Uniform Plane Waves—Multiple Layers

- $\bar{E}_{1x}^+(z) = \bar{E}_{1x}^+(0)\cdot e^{-\gamma_1 z}$, $\bar{E}_{1x}^-(z) = \bar{E}_{1x}^-(0)\cdot e^{+\gamma_1 z}$, $\bar{E}_{2x}^+(z) = \bar{E}_{2x}^+(0)\cdot e^{-\gamma_2 z}$, $\bar{E}_{2x}^-(z) = \bar{E}_{2x}^-(0)\cdot e^{+\gamma_2 z}$
- $\bar{E}_{3x}^+(z) = \bar{E}_{3x}^+(d)\cdot e^{-\gamma_3(z-d)}$
  - $\dfrac{\bar{E}_{1x}^-(0)}{\bar{E}_{1x}^+(0)} = \dfrac{P_{12} + P_{23}\cdot e^{-2\gamma_2 d}}{1 + P_{12}\cdot P_{23}\cdot e^{-2\gamma_2 d}}$
  - $\dfrac{\bar{E}_{2x}^+(0)}{\bar{E}_{1x}^+(0)} = \dfrac{1 + P_{12}}{1 + P_{12}\cdot P_{23}\cdot e^{-2\gamma_2 d}}$
  - $\dfrac{\bar{E}_{2x}^-(0)}{\bar{E}_{1x}^+(0)} = \dfrac{(1 + P_{12})\cdot P_{23}\cdot e^{-2\gamma_2 d}}{1 + P_{12}\cdot P_{23}\cdot e^{-2\gamma_2 d}}$
  - $\dfrac{\bar{E}_{3x}^+(d)}{\bar{E}_{1x}^+(0)} = \dfrac{(1 + P_{12})\cdot(1 + P_{23})\cdot e^{-\gamma_2 d}}{1 + P_{12}\cdot P_{23}\cdot e^{-2\gamma_2 d}}$
  - $P_{12} = \dfrac{\bar{\eta}_{o2} - \bar{\eta}_{o1}}{\bar{\eta}_{o2} + \bar{\eta}_{o1}}$ and $P_{23} = \dfrac{\bar{\eta}_{o3} - \bar{\eta}_{o2}}{\bar{\eta}_{o3} + \bar{\eta}_{o2}}$

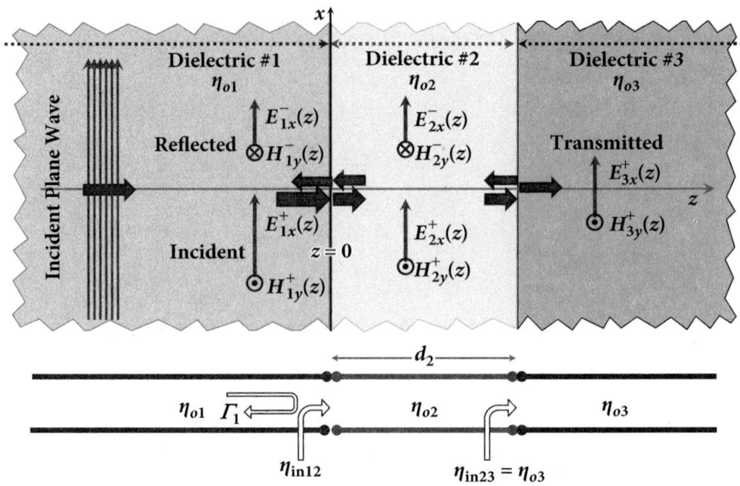

Transmission and Reflection of Uniform Plane Waves—Oblique Incidence

- $\bar{E}_{1x}^+(z) = \bar{E}_{1x}^+(0)\cdot e^{-\gamma_1 z}$, $\bar{E}_{1x}^-(z) = \bar{E}_{1x}^-(0)\cdot e^{+\gamma_1 z}$, $\bar{E}_{2x}^+(z) = \bar{E}_{2x}^+(0)\cdot e^{-\gamma_2 z}$
- $\bar{\Gamma}(0) = \dfrac{\bar{E}_{1x}^-(0)}{\bar{E}_{1x}^+(0)} = \dfrac{\bar{\eta}_{o2} - \bar{\eta}_{o1}}{\bar{\eta}_{o2} + \bar{\eta}_{o1}}$, $\bar{\tau}(0) = \dfrac{\bar{E}_{2x}^+(0)}{\bar{E}_{1x}^+(0)} = 1 + \bar{\Gamma}(0) = \dfrac{2\,\bar{\eta}_{o2}}{\bar{\eta}_{o2} + \bar{\eta}_{o1}}$

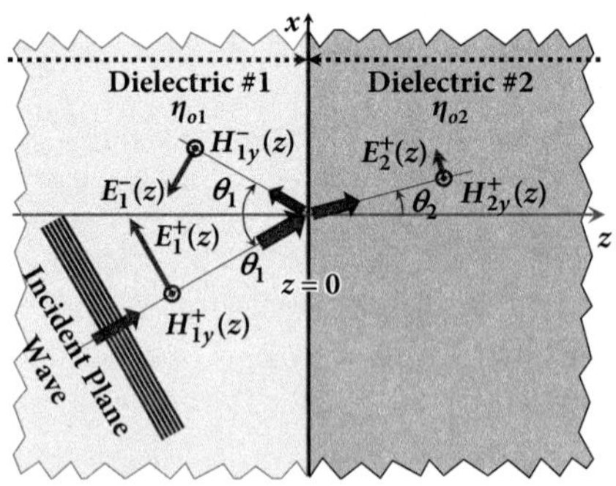

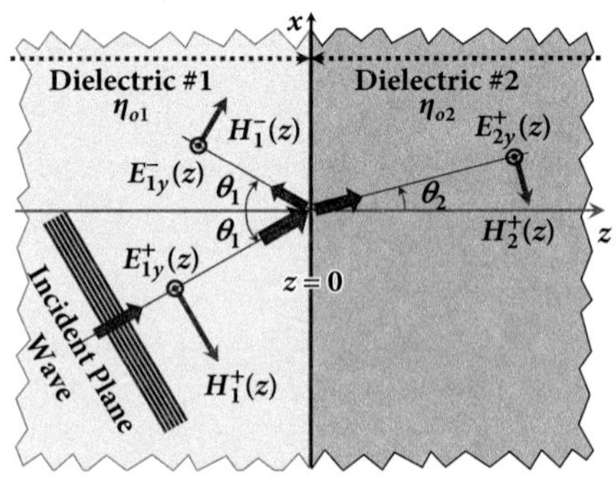

Parallel Polarization

$$\bar{\Gamma}(0)\big|_{\parallel} = \frac{\bar{\eta}_{o2}\cos\theta_2 - \bar{\eta}_{o1}\cos\theta_1}{\bar{\eta}_{o2}\cos\theta_2 + \bar{\eta}_{o1}\cos\theta_1}$$

$$\bar{\tau}(0)\big|_{\parallel} = \frac{2\,\bar{\eta}_{o2}\cos\theta_1}{\bar{\eta}_{o2}\cos\theta_2 + \bar{\eta}_{o1}\cos\theta_1}$$

Perpendicular Polarization

$$\bar{\Gamma}(0)\big|_{\perp} = \frac{\bar{\eta}_{o2}/\cos\theta_2 - \bar{\eta}_{o1}/\cos\theta_1}{\bar{\eta}_{o2}/\cos\theta_2 + \bar{\eta}_{o1}/\cos\theta_1}$$

$$\bar{\tau}(0)\big|_{\perp} = \frac{2\,\bar{\eta}_{o2}/\cos\theta_2}{\bar{\eta}_{o2}/\cos\theta_2 + \bar{\eta}_{o1}/\cos\theta_1}$$

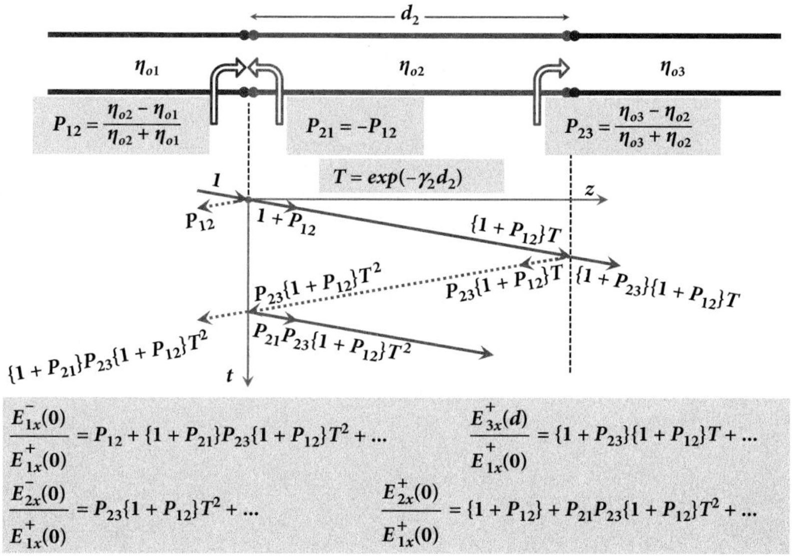

Snell's Law and Total Reflection

- Snell's law: $\dfrac{\sin\theta_t}{\sin\theta_i} = \dfrac{\beta_1}{\beta_2} = \dfrac{\sqrt{\mu_1\varepsilon_1}}{\sqrt{\mu_2\varepsilon_2}} = \dfrac{\sqrt{\mu_{r1}\varepsilon_{r1}}}{\sqrt{\mu_{r2}\varepsilon_{r2}}} = \dfrac{n_1}{n_2}$
  - The index of refraction: $n = \sqrt{\mu_r\ \varepsilon_r}$
- Total reflection occurs when $n_1 > n_2$ and the incident angle exceeds a critical value, $\theta_1 \geq \theta_{1c}$
  - The critical angle: $\theta_{1c} = \sin^{-1}\left(\dfrac{n_2}{n_1}\right)$

The Brewster Angle

- At the Brewster angle, θ_B, there is no reflection for an incident wave (total transmission): $\overline{\Gamma}(0) = 0$
  - For parallel polarization:

$$\sin\theta_{1B\|} = \sqrt{\frac{1 - \dfrac{\mu_2\varepsilon_1}{\mu_1\varepsilon_2}}{1 - \left(\dfrac{\varepsilon_1}{\varepsilon_2}\right)}} \rightarrow \text{for nonmagnetic media: } \sin\theta_{1B\|}\big|_{\mu_1=\mu_2} = \frac{1}{\sqrt{1 + \left(\dfrac{\varepsilon_1}{\varepsilon_2}\right)}}$$

  - For perpendicular polarization:

$$\sin\theta_{1B\perp} = \sqrt{\frac{1 - \dfrac{\mu_1\varepsilon_2}{\mu_2\varepsilon_1}}{1 - \left(\dfrac{\mu_1}{\mu_2}\right)}} \rightarrow \text{for nonmagnetic media: } \theta_{1B\perp} \text{ is not realizable.}$$

CHAPTER 13

Waveguides

Joseph John Thomson
(1856–1940)

Sir Oliver Joseph Lodge
(1851–1940)

John William Strutt,
Lord Rayleigh (1842–1919)

Chapter Learning Objectives

- Demonstrate the different types of waveguides.
- Develop "generic" field equations (in the point form) for a typical dielectric filled waveguide section.
- Apply the derived equations to metallic rectangular waveguides. Study the modes, cutoff frequencies.
- Determine the active (operating) modes for a given waveguide at a given frequency.
- Analyze rectangular waveguides propagating transverse electric (*TE*), transverse magnetic (*TM*), or transverse electromagnetic (*TEM*) modes.
- Describe propagation modes, cutoff frequency, impedance, and wave propagation in rectangular waveguide.
- Distinguish between *TEM*, *TE*, and *TM* modes.
- Derive expression for the power flow density for the dominant mode.
- Describe and understand the physical insight of modal dispersion in multimode propagation in waveguides.
- Describe and understand the physical insight of the characteristics of rectangular cavity resonators.
- Derive expressions for quality factor and bandwidth for the dominant mode of a rectangular cavity resonator.

13.1 Introduction

In previous chapters, we discussed plane wave propagation in open space. We would refer to that as unguided wave propagation. Such propagation lacks shielding and wave confinement required for reduced interference, communication security, and maintaining signal strength. Guided wave propagation using transmission lines was also covered in earlier chapters. With transmission lines, we can enhance shielding and confinement; however, there are limitations of high-frequency losses and power handling capabilities.

In this chapter, we discuss waveguides in general and provide preliminary analysis of one typical configuration as an example. We chose the metallic rectangular waveguide for its *reasonable* analysis complexity while introducing us to most of the concepts and features we need to learn about waveguides.

In this chapter, we use of Maxwell's equations for the formation and solution of the wave equations governing wave propagation in waveguides. Solving the wave equations and applying the boundary conditions will reveal to us the possible existence of more than one mode of operation. Each mode corresponds to its own specific field pattern and power flow distribution in the guide. We will learn that only modes operating at frequencies above a specific cutoff frequency can be sustained in the guide while those at lower frequencies will be in cutoff.

Later in the chapter, we discuss the propagation characteristics of a waveguide, including waveguide wavelength, phase velocity, impedance, and power flow. In addition, we discuss the logic of having phase velocities greater than the speed of light. Finally, we present the concept of modal dispersion in waveguides and the need for single-mode operation.

13.2 Unguided Propagation versus Guided Transmission

As stated above, unguided wave propagation lacks the shielding and confinement that can support communication systems and guard against interference and insecure communication. To deliver a communication signal to target receivers, it is necessary to use guided transmission. At low frequencies, transmission lines are typically used. At high frequencies, however, particularly with high-power requirement, different forms of waveguides are used.

In fact, transmission lines are simply a form of waveguides characterized by the presence of more than one conductor. One serious misconception regarding transmission lines is that the conductors carry the signal from one end to the other. The fact is that time-varying signals generate radiating electromagnetic waves that propagate through the dielectric medium of the transmission line. The conductors of the transmission line provide the boundaries guiding the wave propagation. Only in DC, the conductors can be said to carry the source power to the load end. Furthermore, EM waves cannot penetrate ideal conductors and are confined to the surrounding dielectric media.

What we typically refer to as a waveguide is a dielectric channel that carries the wave propagation using either a confining conductor or dielectric boundaries. This is analogous to a river channel, a garden hose, or a water pipe that are examples of *water guides* facilitating the transmission of water from the water source (e.g., lake, water tank, or reservoir) to the receiving end (e.g., ocean, house use, or agriculture field). By shaping the combination of conductors and dielectrics, we can control/guide wave propagation (transmission) characteristics to achieve the desired shielding and confinement.

Typically, metallic waveguides are used for microwave frequency transmission up to mm waves. Conductor losses and manufacturing complexity and cost makes their use prohibitive at the millimeter wave frequencies. For millimeter wave and optical frequencies, dielectric guides are the ones to use. An example of such a waveguide is an optical fiber that is commonly used for many communication applications in this era.

13.3 Why Waveguides?

We typically use transmission lines for the transmission of broadband signals up to radio frequencies. At the higher end of this band, signal radiation becomes an issue due to increased interference, resulting in loss of signal strength and integrity, as well as lack of security. In such case, we typically use the coaxial form of transmission lines to ensure shielding and signal confinement.

To use coaxial lines at microwave and millimeter wave frequencies, their cross-sectional size must be small enough to ensure single-mode operation. Operating with higher-order modes causes dispersion, which limits the useful bandwidth for the communication channel. Moreover, the smaller the coaxial dimensions are the higher the conductor losses (conductor resistance), especially those due to the small cross sections of the center conductors.

The elimination of the center conductor can help offset this disadvantage and allows us to use such a line at higher frequencies. Such a line is what we refer to as a metallic waveguide. The elimination of the outer conductor should further reduce the transmission losses, as we will be left with the guide's dielectric. It is an established fact that we can find extremely low-loss quality microwave, millimeter, and optical wave dielectrics to use as transmission channels.

13.3.1 Typical Waveguide Configurations

Figure 13.1 demonstrates four typical shapes for waveguides: two metallic (conductor) ones and two dielectric configurations. As the figure suggests, waveguides are typically made in both rectangular and circular shapes. Typically, the electromagnetic performance of rectangular guides is far better than those of the circular one. However, the circular guides are much easier and less costly to manufacture.

13.4 Field Analysis of Guide Filling/Core Region

We start with developing *generic* field equations (in the point form) for a typical dielectric filling/core of a waveguide section. This is typically a source-free low-loss dielectric (no charges or currents). Once these equations are developed, they can be used together with relevant boundary conditions for the dielectric region to solve for the electromagnetic fields within the guide region, and hence the guide's propagation characteristics. The boundary conditions reflect the geometrical configuration of the guide as well as the material properties of the guide wall/cladding.

Similarly, to our earlier discussion for the plane wave propagation in dielectric media, we start with the point form of Maxwell's equations in the frequency domain

$$\vec{\nabla} \times \vec{\bar{E}} = -j\omega \vec{\bar{B}} \tag{13.1}$$

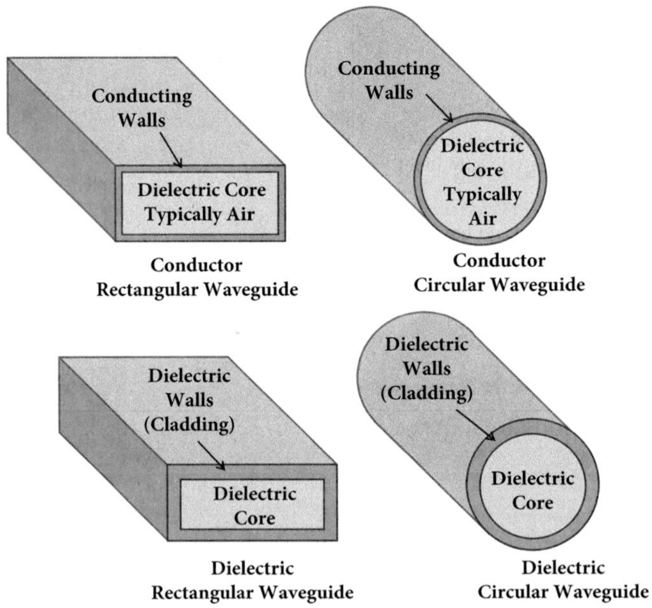

FIGURE 13.1 Typical waveguide configurations.

$$\vec{\nabla} \times \vec{H} = \vec{J} + j\omega \vec{D} \tag{13.2}$$

$$\vec{\nabla} \cdot \vec{D} = \rho_v \tag{13.3}$$

$$\vec{\nabla} \cdot \vec{B} = 0 \tag{13.4}$$

together with the material constitutive relationships

$$\vec{B} = \mu \vec{H} \tag{13.5}$$

$$\vec{D} = \varepsilon \vec{E} \tag{13.6}$$

$$\vec{J} = \sigma \vec{E} \tag{13.7}$$

Using the material constitutive relationships [Equations (13.5) to (13.7)], Equations (13.1) and (13.2) take the forms:

$$\vec{\nabla} \times \vec{E} = -j\omega\mu \vec{H} \tag{13.8}$$

$$\vec{\nabla} \times \vec{H} = \vec{J} + j\omega\varepsilon \vec{E} \tag{13.9}$$

Next, we set $\rho_v = 0$ and $\vec{J} = 0$ for the assumed source-free dielectric core (no free charges or currents), hence

$$\vec{\nabla} \cdot \vec{D} = \rho_v = 0 \tag{13.10}$$

$$\vec{\nabla} \times \vec{H} = j\omega\varepsilon \vec{E} \tag{13.11}$$

Also, because of the dielectric homogeneity, we can factor both ε and μ outside the differential operator, as they are both constants for all locations within the region described. As a consequence, Equations (13.3) and (13.4) become

$$\vec{\nabla} \cdot \vec{D} = 0 = \vec{\nabla} \cdot (\varepsilon \vec{E}) = \varepsilon \vec{\nabla} \cdot \vec{E} = 0 \quad \text{and} \quad \vec{\nabla} \cdot \vec{B} = \vec{\nabla} \cdot (\mu \vec{H}) = \mu \vec{\nabla} \cdot \vec{H} = 0 \tag{13.12}$$

which simplify to

$$\vec{\nabla}\cdot\vec{\bar{E}}=0 \qquad \text{and} \qquad \vec{\nabla}\cdot\vec{\bar{H}}=0 \tag{13.13}$$

Next, we apply the curl vector operator to both sides of Equation (13.8)

$$\vec{\nabla}\times\vec{\nabla}\times\vec{\bar{E}}=-j\omega(\vec{\nabla}\times\mu\vec{\bar{H}})=-j\omega\mu(\vec{\nabla}\times\vec{\bar{H}}) \tag{13.14}$$

Substituting $\vec{\nabla}\times\vec{\bar{H}}$ from Equation (13.11), we get

$$\vec{\nabla}\times\vec{\nabla}\times\vec{\bar{E}}=-j\omega\mu(\vec{\nabla}\times\vec{\bar{H}})=-j\omega\mu(j\omega\varepsilon\vec{\bar{E}})=\omega^2\mu\varepsilon\vec{\bar{E}} \tag{13.15}$$

Next, using the curl-curl vector identity of the form

$$\vec{\nabla}\times\vec{\nabla}\times\vec{\bar{E}}=\vec{\nabla}(\vec{\nabla}\cdot\vec{\bar{E}})-\nabla^2\vec{\bar{E}}=-\nabla^2\vec{\bar{E}} \qquad (\text{since } \vec{\nabla}\cdot\vec{\bar{E}}=0) \tag{13.16}$$

Hence,

$$\vec{\nabla}\times\vec{\nabla}\times\vec{\bar{E}}=\omega^2\mu\varepsilon\vec{\bar{E}}=-\nabla^2\vec{\bar{E}} \tag{13.17}$$

or

$$\nabla^2\vec{\bar{E}}=-\omega^2\mu\varepsilon\vec{\bar{E}} \tag{13.18}$$

This second-order differential equation is known as the Helmholtz equation, which we have seen before in Chapter 2, Equations (2.16) and (2.20). Likewise, if we were to carry on a similar procedure for the magnetic field starting with applying the curl to Equation (13.11), we would arrive at a similar equation for the magnetic field; specifically,

$$\nabla^2\vec{\bar{H}}=-\omega^2\mu\varepsilon\vec{\bar{H}} \tag{13.19}$$

Unlike the coupled Equations (13.8) and (13.11), Equations (13.18) and (13.19) are uncoupled and can be solved *independently* from each other. However, each solution will contain unknown *integration* constants that are in fact interrelated. Equations (13.8) and (13.11) are still needed to identify those interrelationships.

Equations (13.18) and (13.19) are second-order differential equations obtained from the original Maxwell's equations. The process for this derivation involved the application of the curl vector operation to the original equations, which implies a differentiation operation. This differentiation resulted in the elimination of *constant* terms that were part of the original set of knowledge. This elimination means that we lost information that is no longer available as part of the obtained second-order differential equations. The only way to recover that missing information is by using the original Maxwell's equations as part of the set of equations to be solved.

In the following, we will demonstrate how all of the four Equations (13.8), (13.11), (13.18), and (13.19) will be used to solve for all electrical and magnetic vector fields inside (metallic) rectangular waveguides.

13.5 (Metallic) Rectangular Waveguides

We confine our studies in this section to a uniform rectangular cross-section metallic waveguide. The guide we consider is assumed straight along the direction of propagation with perfectly conducting walls and dielectric filling (ε and μ). Figure 13.2 shows a sketch of a rectangular waveguide section lying on its broad side on the x-z plane of the Cartesian system. The guide's cross-sectional dimensions are "a" along the x axis and "b" along the y axis. The guide extends along the z axis between a source connected at one end and a load at the other.

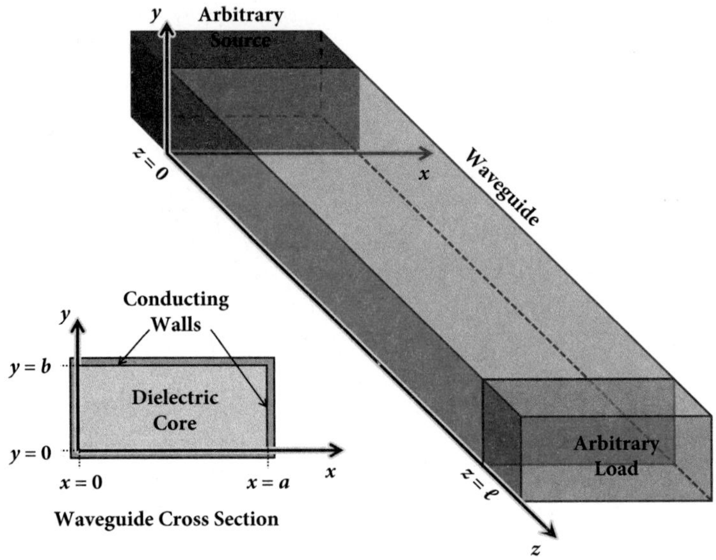

FIGURE 13.2 An illustration of a metallic rectangular waveguide section with arbitrary source and load terminations.

We start our analysis by solving for the fields inside the waveguide section not including the source and load sections. In this regard, all relevant governing equations are listed in Equations (13.8), (13.11), (13.18), and (13.19). These equations are to be solved together with the boundary conditions of the dielectric/conductor interface. We will introduce these boundary conditions in due time.

In Addendum 13A, we present the detailed procedure toward solving for the fields in the above-described guide. The analyses are done using Cartesian coordinates, which is the logical choice, given the rectangular guide configuration. The addendum concludes the analysis with the following findings regarding the z component of the electric field phasor, $\overline{E}_z$,

$$\overline{E}_{z\ m,n}(x,y,z)=\left[\sin\left(\frac{m\pi}{a}x\right)\sin\left(\frac{n\pi}{b}y\right)\right]\left[\overline{E}^{+}_{o\ m,n}e^{-jk_{z\ m,n}z}+\overline{E}^{-}_{o\ m,n}e^{+jk_{z\ m,n}z}\right] \tag{13.20}$$

Here, "m" and "n" are arbitrary integers and $\overline{E}^{+}_{o\ m,n}$ and $\overline{E}^{-}_{o\ m,n}$ are arbitrary constant representing field intensity phasors. $k_{z\ m,n}$ is the z component of the wave propagation constant (wave number).

We notice that the solution in Equation (13.20) is composed to two main components: a cross-sectional distribution as revealed by the product $\left[\sin\left(\frac{m\pi}{a}x\right)\sin\left(\frac{n\pi}{b}y\right)\right]$ and the z axis variation term $\left[\overline{E}^{+}_{o\ m,n}e^{-jk_{z\ m,n}z}+\overline{E}^{-}_{o\ m,n}e^{+jk_{z\ m,n}z}\right]$.

The z axis part has two elements corresponding to both the positive and negative z directions. The field intensity phasors of these two elements can only be determined with knowledge of the source and load connected at the guide $z = 0$ and $z = \ell$ ends.

We also note that we are using the subscripts "m,n" for the field solution and all its affiliated parameters. This is because that there are likely more than one m and n combinations that can produce a solution that meets the cross-sectional boundary conditions for the guide. Each combination of m and n corresponds to a possible solution for the field distribution inside the guide. It is typical to refer to these different m and n combinations, and hence, field solutions as *modes*. In other words, we will refer to each m and n combination as the "m, n" mode (or field solution/distribution).

According to our analysis in Addendum 13A, $k_{z\ m,n}$ is the z component of the wave propagation constant (wave number), the square of which is given by

$$k^2_{z\ m,n}=\omega^2\mu\varepsilon-\left[\left(\frac{m\pi}{a}\right)^2+\left(\frac{n\pi}{b}\right)^2\right]=\omega^2\mu\varepsilon-k^2_{c\ m,n} \tag{13.21}$$

where $k_{c\,m,n}$ is the cross-sectional component of the wave propagation constant. Later on, we will learn that we can use the subscript c to imply a "critical"/"cutoff" value besides its use for "cross sectional." In addition, $k_{c\,m,n}$ can be expressed in terms of a corresponding $\omega_{c\,m,n}$ frequency as

$$k_{c\,m,n} = \sqrt{\left(\frac{m\pi}{a}\right)^2 + \left(\frac{n\pi}{b}\right)^2} = \sqrt{\mu\varepsilon}\,\omega_{c\,m,n} \tag{13.22}$$

defined as

$$\omega_{c\,m,n} = 2\pi f_{c\,m,n} = \frac{1}{\sqrt{\mu\varepsilon}}\sqrt{\left(\frac{m\pi}{a}\right)^2 + \left(\frac{n\pi}{b}\right)^2} \tag{13.23}$$

which makes $k_{z\,m,n}$

$$k_{z\,m,n} = \sqrt{\mu\varepsilon(\omega^2 - \omega_{c\,m,n}^2)} = \omega\sqrt{\mu\varepsilon}\sqrt{1 - \left(\frac{f_{c\,m,n}}{f}\right)^2}\ ;\ f = \omega/(2\pi) \tag{13.24}$$

Now, let us examine the cross-sectional term, $\left[\sin\left(\frac{m\pi}{a}x\right)\sin\left(\frac{n\pi}{b}y\right)\right]$, of Equation (13.20). In this term, we recognize a periodic stationary pattern in both x and y (Figure 13.3).

For further illustration, we show in Figure 13.4 the field distribution related to this term of the solution over the guide's cross section for the two combinations (modes): ($m=1, n=1$) and ($m=3, n=2$), respectively. The dots and crosses in the figure are an indication of the $\bar{E}_{z\,m,n}(x,y)$ field distribution over the cross section.

These field patterns are in fact two of *many* possible solutions (modes) for the field distribution over the guide's cross section. For each one of these modes, the arbitrary constants $\bar{E}^+_{o\,m,n}$ and $\bar{E}^-_{o\,m,n}$ determining the intensity (magnitude) of each of these solution waves are in general different. We can argue that the general solution is then by a linear sum (superposition) of all possible "m,n" mode solutions, and hence

$$\bar{E}_z(x,y,z) = \sum_m \sum_n \bar{E}_{z\,m,n}(x,y,z) = \sum_m \sum_n \left[\sin\left(\frac{m\pi}{a}x\right)\sin\left(\frac{n\pi}{b}y\right)\right]\left[\bar{E}^+_{o\,m,n}e^{-jk_{z\,m,n}z} + \bar{E}^-_{o\,m,n}e^{+jk_{z\,m,n}z}\right] \tag{13.25}$$

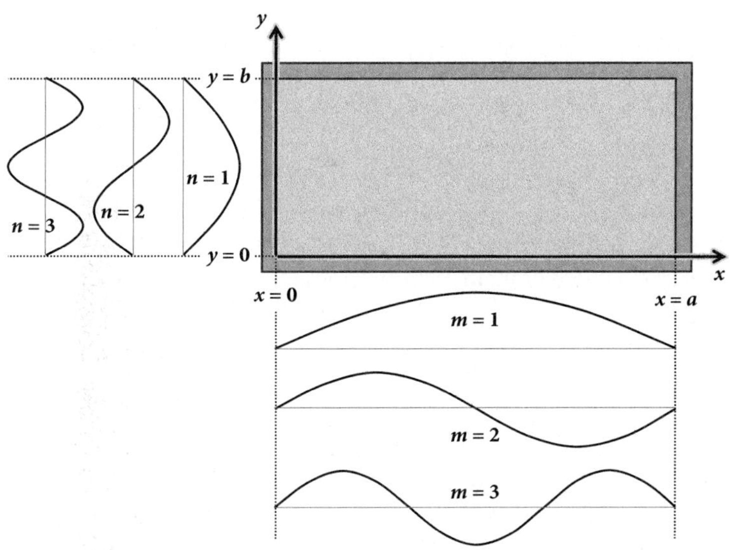

FIGURE 13.3 Possible electric field distributions over the waveguide's cross section.

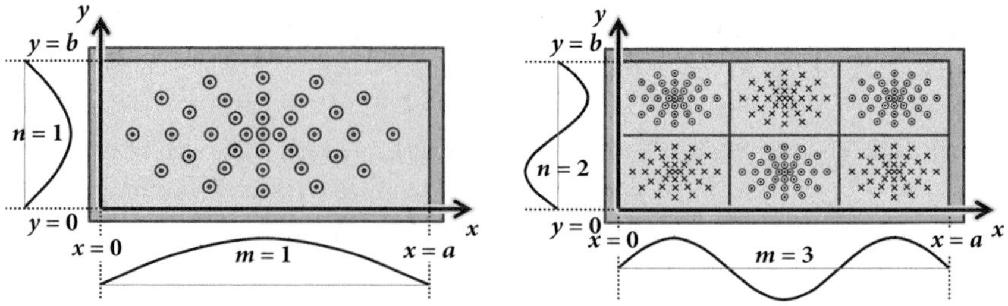

FIGURE 13.4 Sample E_z field maps for two different modes (11 and 32).

13.6 Modes and Cutoff Frequencies

The question now arises as what will determine the values of m and n and which modes out of the infinite combinations will exist as part of the final solution and which modes do not. For that, we refer back to Equation (13.21) which reveals to us how m and n determine the nature of $k_{z\ m,n}$ and hence the impact of the cross-sectional parameters m and n on the z-axis variation term.

$$k_{z\ m,n}^2 = \omega^2\mu\varepsilon - \left[\left(\frac{m\pi}{a}\right)^2 + \left(\frac{n\pi}{b}\right)^2\right] \tag{13.21}$$

This equation shows that for certain low orders of m and n, where $\omega^2\mu\varepsilon > \left[\left(\frac{m\pi}{a}\right)^2 + \left(\frac{n\pi}{b}\right)^2\right]$, (or $\omega > \omega_{c\ m,n}$), $k_{z\ m,n}^2$ will be positive, resulting in a real value for $k_{z\ m,n}$. On the other hand, for certain high orders of m and n, where $\omega^2\mu\varepsilon < \left[\left(\frac{m\pi}{a}\right)^2 + \left(\frac{n\pi}{b}\right)^2\right]$, (or $\omega < \omega_{c\ m,n}$), $k_{z\ m,n}^2$ will be negative, yielding an imaginary value for $k_{z\ m,n}$.

Next, we ask what does it mean to have real or imaginary values for $k_{z\ m,n}$. The answer lies in examining the terms of the z axis variation term $\left[\overline{E}_{o\ m,n}^+ e^{-jk_{z\ m,n}z} + \overline{E}_{o\ m,n}^- e^{+jk_{z\ m,n}z}\right]$.

13.7 Propagation Modes: Case of $\omega^2\mu\varepsilon > \left[\left(\frac{m\pi}{a}\right)^2 + \left(\frac{n\pi}{b}\right)^2\right]$, (or $\omega > \omega_{c\ m,n}, f > f_{c\ m,n}$)

With a real value for $k_{z\ m,n}$, the two terms $e^{-jk_{z\ m,n}z}$ and $e^{+jk_{z\ m,n}z}$ describe constant (nondecaying) magnitude *traveling* waves in the $+z$ and $-z$ directions, respectively. This is why we used the "+" and "−" superscript notations for their respective phasor intensities, $\overline{E}_{o\ m,n}^+$ and $\overline{E}_{o\ m,n}^-$. To demonstrate the physical nature of these traveling waves, we resort to the time domain forms for these two terms. To do so, we take the real part of the product of the phasor expression and $e^{j\omega t}$. The resulting time-domain expression for the z-axis variation term is of the form:

$$\begin{aligned} E_{z\ m,n}(x,y,z,t) &= Re\left\{\left[\overline{E}_{o\ m,n}^+ e^{-jk_{z\ m,n}z} + \overline{E}_{o\ m,n}^- e^{+jk_{z\ m,n}z}\right]e^{j\omega t}\right\} \\ &= \left|\overline{E}_{o\ m,n}^+\right|\cos(\omega t - k_{z\ m,n}z + \varphi_{E^+}) + \left|\overline{E}_{o\ m,n}^-\right|\cos(\omega t + k_{z\ m,n}z + \varphi_{E^-}) \end{aligned} \tag{13.26}$$

Figure 13.5 demonstrates the time-domain variation of the propagating waves in the same way that we demonstrated in Chapter 2, Figure 2.10. In this case, this real $k_{z\ m,n}$ term refers to the phase constant of the

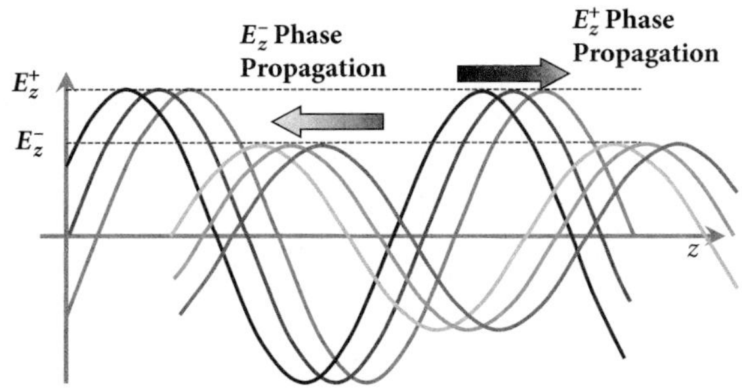

FIGURE 13.5 Forward and backward traveling waves along the waveguide's z axis. The combination is known to produce a standing wave component.

propagating waves expressing the rate of phase change with propagation distance (along the z axis). In this regard, we will replace $k_{z\ m,n}$ in our future expressions by $\beta_{g\ m,n}$, which is the typical notation used in waveguide propagation analyses. Hence, we rewrite Equations (13.21) and (13.24) with the new notation as

$$\beta_{g\ m,n} = \sqrt{\omega^2\mu\varepsilon - \left[\left(\frac{m\pi}{a}\right)^2 + \left(\frac{n\pi}{b}\right)^2\right]} = \omega\sqrt{\mu\varepsilon}\sqrt{1-\left(\frac{\omega_{c\ m,n}}{\omega}\right)^2} = \omega\sqrt{\mu\varepsilon}\sqrt{1-\left(\frac{f_{c\ m,n}}{f}\right)^2} \quad (13.27)$$

In addition, we rewrite Equation (13.25) and its time domain form for the case of $\omega^2\mu\varepsilon > \left[\left(\frac{m\pi}{a}\right)^2 + \left(\frac{n\pi}{b}\right)^2\right]$, (or $\omega > \omega_{c\ m,n}$) as

$$\bar{E}_z(x,y,z) = \sum_m \sum_n \left[\sin\left(\frac{m\pi}{a}x\right)\sin\left(\frac{n\pi}{b}y\right)\right]\left[\bar{E}^+_{o\ m,n}e^{-j\beta_{g\ m,n}z} + \bar{E}^-_{o\ m,n}e^{+j\beta_{g\ m,n}z}\right] \quad (13.28)$$

$$E_z(x,y,z,t) = \sum_m \sum_n \left[\sin\left(\frac{m\pi}{a}x\right)\sin\left(\frac{n\pi}{b}y\right)\right]\left\{\left|\bar{E}^+_{o\ m,n}\right|\cos(\omega t - \beta_{g\ m,n}z + \varphi_{E^+}) + \left|\bar{E}^-_{o\ m,n}\right|\cos(\omega t + \beta_{g\ m,n}z + \varphi_{E^-})\right\} \quad (13.29)$$

13.8 Cutoff Modes: Case of $\omega^2\mu\varepsilon < \left[\left(\frac{m\pi}{a}\right)^2 + \left(\frac{n\pi}{b}\right)^2\right]$, (or $\omega < \omega_{c\ m,n}$, $f < f_{c\ m,n}$)

Now, we discuss the case of $\omega^2\mu\varepsilon < \left[\left(\frac{m\pi}{a}\right)^2 + \left(\frac{n\pi}{b}\right)^2\right]$, (or $f < f_{c\ m,n}$), which yields an imaginary value for $k_{z\ m,n}$. In this case, the exponents of the two terms $e^{-jk_{z\ m,n}z}$ and $e^{+jk_{z\ m,n}z}$ are real and, hence, describe exponentially varying (physically decaying) magnitude fields along the z coordinate, as demonstrated by the time domain expression for the z variation term:

$$Re\left\{\left[\bar{E}^+_{o\ m,n}e^{-|k_{z\ m,n}|z} + \bar{E}^-_{o\ m,n}e^{+|k_{z\ m,n}|z}\right]e^{j\omega t}\right\} = \left|\bar{E}^+_{o\ m,n}\right|e^{-|k_{z\ m,n}|z}\cos(\omega t + \varphi_{E^+}) + \left|\bar{E}^-_{o\ m,n}\right|e^{+|k_{z\ m,n}|z}\cos(\omega t + \varphi_{E^-}) \quad (13.30)$$

Such modes are highly attenuated in the waveguide and do not represent traveling waves. The waveguide is said to be in *cutoff* in this case; it cannot support wave propagation for combinations of m and n that result in imaginary values for $k_{z\ m,n}$. In this case, it is typical to use the notation $\alpha_{g\ m,n}$, the attenuation constant to replace the "$jk_{z\ m,n} = |k_{z\ m,n}|$" term which is a measure of the rate of attenuation/decay of the field intensity along the z axis. Hence, we rewrite Equation (13.25) and its time domain form for the case $\omega^2\mu\varepsilon < \left[\left(\frac{m\pi}{a}\right)^2 + \left(\frac{n\pi}{b}\right)^2\right]$ as

$$\bar{E}_z(x,y,z) = \sum_m \sum_n \left[\sin\left(\frac{m\pi}{a}x\right)\sin\left(\frac{n\pi}{b}y\right)\right]\left[\bar{E}^+_{o\,m,n}e^{-\alpha_{g\,m,n}z} + \bar{E}^-_{o\,m,n}e^{+\alpha_{g\,m,n}z}\right] \tag{13.31}$$

$$E_z(x,y,z,t) = \sum_m \sum_n \left[\sin\left(\frac{m\pi}{a}x\right)\sin\left(\frac{n\pi}{b}y\right)\right]\left\{\left|\bar{E}^+_{o\,m,n}\right|e^{-\alpha_{g\,m,n}z}\cos(\omega t + \varphi_{E^+}) + \left|\bar{E}^-_{o\,m,n}\right|e^{+\alpha_{g\,m,n}z}\cos(\omega t + \varphi_{E^-})\right\} \tag{13.32}$$

Consequently, we can conclude that a waveguide can support traveling waves for operating frequencies above a certain "critical/cutoff" frequency below which the guide will be in a cutoff state with no wave propagation. This critical/cutoff frequency is determined by the condition, $f = f_{c\,m,n}$, and, hence, we can now refer to the subscript c as an abbreviation to *cutoff* in addition to the one stated earlier as *cross sectional.*

Using Equation (13.24), we can now write the expression for the cutoff frequency, $f_{c\,m,n}$, for the mode "m,n" as

$$f_{c\,m,n} = \frac{\omega_{c\,m,n}}{2\pi} = \frac{1}{\sqrt{\mu\varepsilon}}\sqrt{\left[\left(\frac{m}{2a}\right)^2 + \left(\frac{n}{2b}\right)^2\right]} = \frac{c_o}{\sqrt{\mu_r\varepsilon_r}}\sqrt{\left[\left(\frac{m}{2a}\right)^2 + \left(\frac{n}{2b}\right)^2\right]} \tag{13.33}$$

where c_o is the velocity (speed) of light in free space, $c_o = \frac{1}{\sqrt{\mu_o\varepsilon_o}}$.

We demonstrate this concept in Figure 13.6 for an arbitrary combination of m and n. The figure shows the critical values of $\omega_{c\,m,n}$ and the corresponding $f_{c\,m,n}$ below which the guide is cutoff and above which the particular "m,n" mode can propagate.

Now, recognizing the reality that there are infinite possible combinations of m and n in the field solution, the guide's operation frequency will be the determining factor as to which combination will result in wave propagation and which will be in cutoff. Basically, all modes with $f_{c\,m,n}$ below the guide's operating frequency will be sustained yielding propagating waves while all modes with $f_{c\,m,n}$ above the operating frequency will be in cutoff. We demonstrate this concept in Figure 13.7.

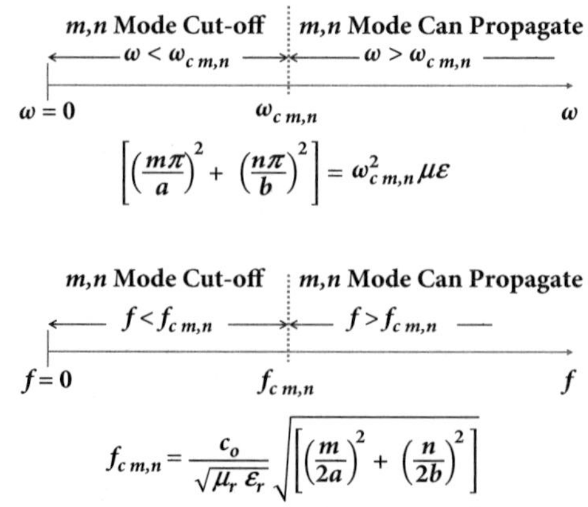

FIGURE 13.6 Waves operating at frequencies above the cutoff of the "mn" mode can propagate this mode while waves operating at lower frequencies will be cutoff.

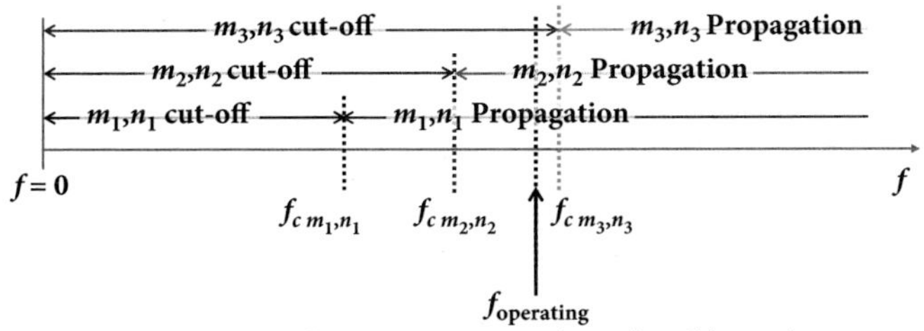

FIGURE 13.7 An example of multimode propagation. A wave operating at the shown frequency will have two possible modes for propagation (m_1,n_1 and m_2,n_2). Meanwhile, the (m_3,n_3) mode and all higher-order modes will be in cutoff.

13.9 Physical Insight: The Guided Wavelength and Phase Velocity

Examining Equations (13.27) to (13.29) as expressions of two traveling waves (traveling in opposite directions), we recognize the wavelength, $\lambda_{g\ m,n}$, and the phase velocity, $c_{ph\ g\ m,n}$, for each traveling wave in the guide as

$$\beta_{g\ m,n} = \frac{2\pi}{\lambda_{g\ m,n}} \quad \text{and hence,} \quad \lambda_{g\ m,n} = \frac{2\pi}{\beta_{g\ m,n}} = \frac{2\pi}{\omega\sqrt{\mu\varepsilon}\sqrt{1-\left(\frac{f_{c\ m,n}}{f}\right)^2}} \tag{13.34}$$

and

$$c_{ph\ g\ m,n} = \frac{\omega}{\beta_{g\ m,n}} = \frac{\omega}{\omega\sqrt{\mu\varepsilon}\sqrt{1-\left(\frac{\omega_{c\ m,n}}{\omega}\right)^2}} = \frac{1}{\sqrt{\mu\varepsilon}\sqrt{1-\left(\frac{f_{c\ m,n}}{f}\right)^2}} \tag{13.35}$$

In these expressions, we are using the subscript g for all values inside the guide in contrast to the subscript ug which we used for the *unguided* (or plane wave) propagation in the same medium (same μ and ε). In this regard, we recall from Chapter 11, that Equations (11.54) and (13.55) give us the expressions for the (unguided) wavelength and phase velocity of plane wave propagation as

$$\lambda = \frac{2\pi}{\beta_{ug}} = \frac{2\pi}{\omega\sqrt{\mu\cdot\varepsilon}} = \lambda_{ug} \tag{11.54}$$

$$c_{ph} = \frac{\omega}{\beta_{ug}} = \frac{1}{\sqrt{\mu\cdot\varepsilon}} = c_{ph\ ug} = \lambda_{ug} f \tag{11.55}$$

Dividing Equation (13.34) by Equation (11.54) and Equation (13.35) by Equation (11.55), we get

$$\frac{\lambda_{g\ m,n}}{\lambda_{ug}} = \frac{1}{\sqrt{1-\left(\frac{f_{c\ m,n}}{f}\right)^2}} \tag{13.36}$$

$$\frac{c_{ph\ g\ m,n}}{c_{ph\ ug}} = \frac{1}{\sqrt{1-\left(\frac{f_{c\ m,n}}{f}\right)^2}} \tag{13.37}$$

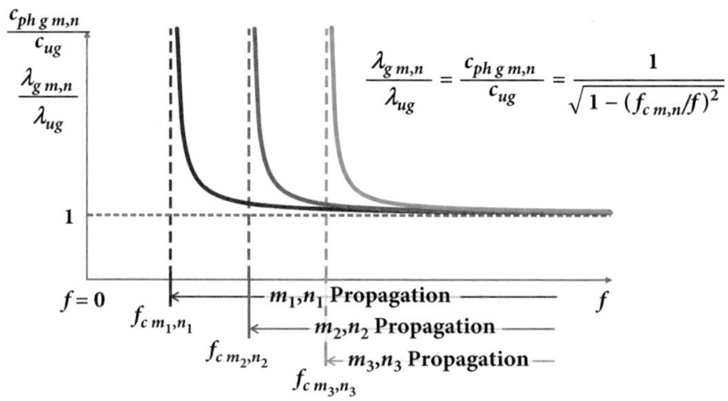

FIGURE 13.8 The functional dependence of the guide's phase velocity and wavelength versus frequency for the different possible modes.

from which we notice that both ratios yield the same value. Also, we can write

$$\frac{\beta_{g\,m,n}}{\beta_{ug}} = \frac{\lambda_{ug}}{\lambda_{g\,m,n}} = \sqrt{1-\left(\frac{f_{c\,m,n}}{f}\right)^2} \quad \text{or} \quad \beta_{g\,m,n} = \beta_{ug}\sqrt{1-\left(\frac{f_{c\,m,n}}{f}\right)^2} = \frac{\omega}{c_{ph\,ug}}\sqrt{1-\left(\frac{f_{c\,m,n}}{f}\right)^2} \tag{13.38}$$

Figure 13.8 demonstrates the functional variations of both guided phase velocity and guided wavelength with respect to the operating frequency as related to us by Equations (13.36) and (13.37).

Examining Equation (13.37) and its complementary graph in Figure 13.8, we discover, surprisingly, that the phase velocity inside the guide is larger than its unguided counterpart. Since the unguided phase velocity is that of the velocity of light in the medium, then our finding here implies that the guided phase velocity is faster than the velocity of light in the same medium. This surprising finding deserves a closer look as it raises the question whether it is possible to have velocities greater than the speed of light.

This point is discussed in some detail in Addendum 13B. The conclusion is that it is physically possible for phase velocities to be greater than the velocity of light in the same medium. The phase velocity being larger than the velocity of light is not in conflict with theory of relativity as there is no information carried with this speed. Signal information in a waveguide is carried along with another velocity, known as *group velocity*, which is smaller than the speed of light in the dielectric region. Addendum 13B introduces the concept of the group velocity as well as presents the derivation for its analytical expression.

13.10 Continuation of the Field Analysis

We now go back to continue our field solution for wave propagation in the rectangular waveguide. We present the details of this work in Addendum 13C.

In a similar way to solving for $\bar{E}_z$, we proceed for the solution for the z component of the magnetic field, $\bar{H}_z$, by solving the wave Equation (13.19) and applying the appropriate cross-sectional boundary conditions, we conclude that

$$\bar{H}_{z\,m,n} = \left[\cos\left(\frac{m\pi}{a}x\right)\cos\left(\frac{n\pi}{b}y\right)\right]\left[\bar{H}^{+}_{o\,m,n}e^{-jk_{z\,m,n}z} + \bar{H}^{-}_{o\,m,n}e^{+jk_{z\,m,n}z}\right] \tag{13.39}$$

Having expressions for both $\bar{E}_z$ and $\bar{H}_z$, we are now ready to derive the expressions for the remainder four field components. In this regard, we will only consider the solution for the active (propagation) modes. Furthermore, to simplify our math development working only with the positive z-propagation term. We can always reconstruct the full (two-way propagation) solution by adding the negative z-propagation term later. We first write the modified expressions for $\bar{E}_z$ and $\bar{H}_z$ as

$$\bar{E}^+_{z\ m,n} = \left[\sin\left(\frac{m\pi}{a}x\right)\sin\left(\frac{n\pi}{b}y\right)\right]\left[\bar{E}^+_{o\ m,n}e^{-j\beta_{g\ m,n}z}\right] \tag{13.40}$$

$$\bar{H}^+_{z\ m,n} = \left[\cos\left(\frac{m\pi}{a}x\right)\cos\left(\frac{n\pi}{b}y\right)\right]\left[\bar{H}^+_{o\ m,n}e^{-j\beta_{g\ m,n}z}\right] \tag{13.41}$$

Next, using Equations (13C.21) to (13C.24) in Addendum 13C, we can express the remaining four field components as follows:

$$\bar{E}^+_{x\ m,n}(x,y,z) = \frac{-j\pi}{k^2_{c\ m,n}}\left\{\frac{m\beta_{g\ m,n}}{a}\bar{E}^+_{o\ m,n} - \frac{n\omega\mu}{b}\bar{H}^+_{o\ m,n}\right\}\left[\cos\left(\frac{m\pi}{a}x\right)\sin\left(\frac{n\pi}{b}y\right)e^{-j\beta_{g\ m,n}z}\right] \tag{13.42}$$

$$\bar{E}^+_{y\ m,n}(x,y,z) = \frac{-j\pi}{k^2_{c\ m,n}}\left\{\frac{n\beta_{g\ m,n}}{b}\bar{E}^+_{o\ m,n} + \frac{m\omega\mu}{a}\bar{H}^+_{o\ m,n}\right\}\left[\sin\left(\frac{m\pi}{a}x\right)\cos\left(\frac{n\pi}{b}y\right)e^{-j\beta_{g\ m,n}z}\right] \tag{13.43}$$

$$\bar{H}^+_{x\ m,n}(x,y,z) = \frac{j\pi}{k^2_{c\ m,n}}\left\{\frac{n\omega\varepsilon}{b}\bar{E}^+_{o\ m,n} + \frac{m\beta_{g\ m,n}}{a}\bar{H}^+_{o\ m,n}\right\}\left[\sin\left(\frac{m\pi}{a}x\right)\cos\left(\frac{n\pi}{b}y\right)e^{-j\beta_{g\ m,n}z}\right] \tag{13.44}$$

$$\bar{H}^+_{y\ m,n}(x,y,z) = \frac{-j\pi}{k^2_{c\ m,n}}\left\{\frac{m\omega\varepsilon}{a}\bar{E}^+_{o\ m,n} - \frac{n\beta_{g\ m,n}}{b}\bar{H}^+_{o\ m,n}\right\}\left[\cos\left(\frac{m\pi}{a}x\right)\sin\left(\frac{n\pi}{b}y\right)e^{-j\beta_{g\ m,n}z}\right] \tag{13.45}$$

We find these expressions too cumbersome and not practical to use. Instead, the traditional approach is to break the final solutions into two sets of solutions that we would combine as needed in a more convenient fashion. These two sets of solutions may or may not coexist. They correspond to having either $\bar{E}_z$ or $\bar{H}_z$ is equal to zero. In doing so, the analysis and manipulation of the solutions for either of the two sets becomes much more manageable.

The solutions corresponding to $\bar{E}_z = 0$ will be referred to as the transverse electric (*TE*), solutions (modes), while the $\bar{H}_z = 0$ solutions are called the transverse magnetic (*TM*), solutions (modes). The combined solution, where neither is zero, will be referred to as hybrid mode solution.

Most of the literature starts this breakdown of the two sets of solutions at a much earlier stage in the analysis. If we were to do that, the reader would have wondered why and would have been uneasy about the reason for the breakdown. Now, that we presented the full solution, the reader can appreciate the simplification that we are about to do by breaking the two sets apart.

13.11 Waveguides and *TEM* Modes

At this point, it is important to recognize that the case of both $\bar{E}_z = 0$ and $\bar{H}_z = 0$, which is that of the transverse electromagnetic (*TEM*), waves cannot exist in our waveguide. According to Equations (13.42) to (13.45), setting both $\bar{E}_z$ and $\bar{H}_z$ to zero results in zero values for all other four field components, and hence, no fields can exist in the guide in a *TEM* mode.

In the following, we will treat the two sets, *TE* and *TM* mode solutions one at a time.

13.12 Transverse Electric (*TE*) Modes ($\bar{E}_z = 0$)

In this set of solutions (modes), the longitudinal field component, $\bar{E}_z$, is zero, and hence, the electric field exists only in the transverse coordinates, x and y. That is why we call this set of solutions the *TE* modes. For these modes, Equations (13.40) to (13.45) reduce to

$$\bar{E}^+_{z\ TE\ m,n} = 0 \tag{13.46}$$

$$\bar{H}^+_{z\ TE\ m,n} = \left[\cos\left(\frac{m\pi}{a}x\right)\cos\left(\frac{n\pi}{b}y\right)\right]\left[\bar{H}^+_{o\ m,n}e^{-j\beta_{g\ m,n}z}\right] \tag{13.47}$$

$$\bar{E}^+_{x\ TE\ m,n} = \frac{j\omega\mu}{k^2_{c\ m,n}}\left(\frac{n\pi}{b}\right)\left[\cos\left(\frac{m\pi}{a}x\right)\sin\left(\frac{n\pi}{b}y\right)\right]\left[\bar{H}^+_{o\ m,n}e^{-j\beta_{g\ m,n}z}\right] \tag{13.48}$$

$$\bar{E}^+_{y\ TE\ m,n} = \frac{-j\omega\mu}{k^2_{c\ m,n}}\left(\frac{m\pi}{a}\right)\left[\sin\left(\frac{m\pi}{a}x\right)\cos\left(\frac{n\pi}{b}y\right)\right]\left[\bar{H}^+_{o\ m,n}e^{-j\beta_{g\ m,n}z}\right] \tag{13.49}$$

$$\bar{H}^+_{x\ TE\ m,n} = \frac{j\beta_{g\ m,n}}{k^2_{c\ m,n}}\left(\frac{m\pi}{a}\right)\left[\sin\left(\frac{m\pi}{a}x\right)\cos\left(\frac{n\pi}{b}y\right)\right]\left[\bar{H}^+_{o\ m,n}e^{-j\beta_{g\ m,n}z}\right] = -\frac{\beta_{g\ m,n}}{\omega\mu}\bar{E}^+_{y\ TE\ m,n} \tag{13.50}$$

$$\bar{H}^+_{y\ TE\ m,n} = \frac{j\beta_{g\ m,n}}{k^2_{c\ m,n}}\left(\frac{n\pi}{b}\right)\left[\cos\left(\frac{m\pi}{a}x\right)\sin\left(\frac{n\pi}{b}y\right)\right]\left[\bar{H}^+_{o\ m,n}e^{-j\beta_{g\ m,n}z}\right] = \frac{\beta_{g\ m,n}}{\omega\mu}\bar{E}^+_{x\ TE\ m,n} \tag{13.51}$$

It is important to remind ourselves here that this solution is only for the active (propagation) modes, that is, modes operating at frequencies higher than the guide's cutoff frequency.

13.12.1 Example of *TE* Modes: TE_{10} Mode ($m = 1$ and $n = 0$)

In the following, we will demonstrate the *TE* mode field solution by considering the TE_{10} mode where $m=1$ and $n=0$. For simplicity, we will consider the guide to be air filled, that is, we will use free space material parameters. In Addendum 13D, we present more details regarding the field distributions, field maps, and power flow.

To start, we determine the cutoff frequency for this mode, as well as the related solution parameters. Using Equations (13.33), (13.22), and (13.38), we can write

$$f_{c\ 1,0} = \frac{c_o}{\sqrt{\mu_r\varepsilon_r}}\sqrt{\left[\left(\frac{m}{2a}\right)^2 + \left(\frac{n}{2b}\right)^2\right]} = \frac{c_o}{2a} \tag{13.52}$$

$$k_{c\ 1,0} = \sqrt{\left(\frac{m\pi}{a}\right)^2 + \left(\frac{n\pi}{b}\right)^2} = \frac{\pi}{a} \tag{13.53}$$

$$\beta_{g\ 1,0} = \beta_{ug}\sqrt{1 - \left(\frac{f_{c\ m,n}}{f}\right)^2} = \frac{\omega}{c_o}\sqrt{1 - \left(\frac{f_{c\ 1,0}}{f}\right)^2} \tag{13.54}$$

$$\bar{E}^+_{z\ TE\ 1,0} = 0 \tag{13.55}$$

$$\bar{H}^+_{z\ TE\ 1,0} = \cos\left(\frac{\pi}{a}x\right)\left[\bar{H}^+_{o\ 1,0}e^{-j\beta_{g\ 1,0}z}\right] \tag{13.56}$$

$$\bar{E}^+_{x\ TE\ 1,0} = 0 \tag{13.57}$$

$$\bar{E}^+_{y\ TE\ 1,0} = \frac{-j\omega\mu}{k^2_{c\ 1,0}}\left(\frac{\pi}{a}\right)\sin\left(\frac{\pi}{a}x\right)\left[\bar{H}^+_{o\ 1,0}e^{-j\beta_{g\ 1,0}z}\right] = -j\omega\mu\left(\frac{a}{\pi}\right)\sin\left(\frac{\pi}{a}x\right)\left[\bar{H}^+_{o\ 1,0}e^{-j\beta_{g\ 1,0}z}\right] \tag{13.58}$$

$$\bar{H}^+_{x\ TE\ 1,0} = \frac{j\beta_{g\ 1,0}}{k^2_{c\ 1,0}}\left(\frac{\pi}{a}\right)\sin\left(\frac{\pi}{a}x\right)\left[\bar{H}^+_{o\ 1,0}e^{-j\beta_{g\ 1,0}z}\right] = j\beta_{g\ 1,0}\left(\frac{a}{\pi}\right)\sin\left(\frac{\pi}{a}x\right)\left[\bar{H}^+_{o\ 1,0}e^{-j\beta_{g\ 1,0}z}\right] \tag{13.59}$$

$$\bar{H}^+_{y\ TE\ 1,0} = 0 \tag{13.60}$$

In Figure 13.9, we graph the field distributions over the cross section of the guide according to Equations (13.55) to (13.60). The functional dependence on both the x and y coordinates is also shown.

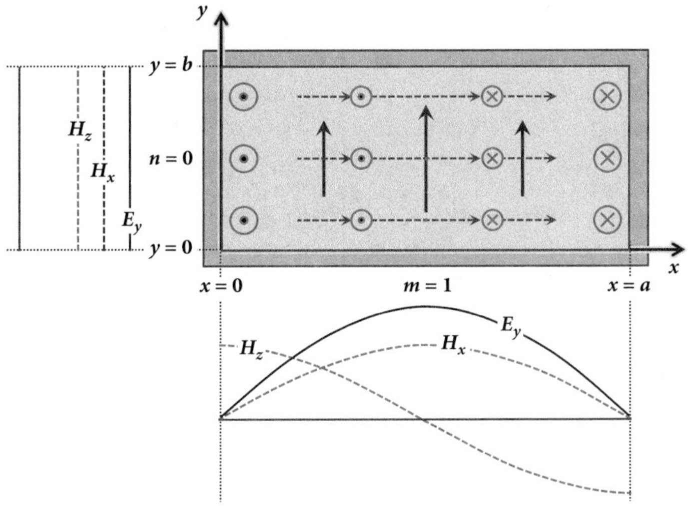

Figure 13.9 Fields map for the TE_{10} mode.

13.13 Transverse Magnetic (TM) Modes ($\bar{H}_z = 0$)

Using Equations (13.40) to (13.45), we can write the EM field solutions for TM modes as follows:

$$\bar{H}^+_{z\ TM\ m,n} = 0 \tag{13.61}$$

$$\bar{E}^+_{z\ TM\ m,n} = \sin\left(\frac{m\pi}{a}x\right)\sin\left(\frac{n\pi}{b}y\right)\bar{E}^+_{o\ m,n}e^{-j\beta_{g\ m,n}z} \tag{13.62}$$

$$\bar{E}^+_{x\ TM\ m,n} = \frac{-j\beta_{g\ m,n}}{k^2_{c\ m,n}}\left(\frac{m\pi}{a}\right)\cos\left(\frac{m\pi}{a}x\right)\sin\left(\frac{n\pi}{b}y\right)\bar{E}^+_{o\ m,n}e^{-j\beta_{g\ m,n}z} \tag{13.63}$$

$$\bar{E}^+_{y\ TM\ m,n} = \frac{-j\beta_{g\ m,n}}{k^2_{c\ m,n}}\left(\frac{n\pi}{b}\right)\sin\left(\frac{m\pi}{a}x\right)\cos\left(\frac{n\pi}{b}y\right)\bar{E}^+_{o\ m,n}e^{-j\beta_{g\ m,n}z} \tag{13.64}$$

$$\bar{H}^+_{x\ TM\ m,n} = \frac{j\omega\varepsilon}{k^2_{c\ m,n}}\left(\frac{n\pi}{b}\right)\sin\left(\frac{m\pi}{a}x\right)\cos\left(\frac{n\pi}{b}y\right)\bar{E}^+_{o\ m,n}e^{-j\beta_{g\ m,n}z} = -\frac{\omega\varepsilon}{\beta_{g\ m,n}}\bar{E}^+_{y\ TM\ m,n} \tag{13.65}$$

$$\bar{H}^+_{y\ TM\ m,n} = \frac{-j\omega\varepsilon}{k^2_{c\ m,n}}\left(\frac{m\pi}{a}\right)\cos\left(\frac{m\pi}{a}x\right)\sin\left(\frac{n\pi}{b}y\right)\bar{E}^+_{o\ m,n}e^{-j\beta_{g\ m,n}z} = \frac{\omega\varepsilon}{\beta_{g\ m,n}}\bar{E}^+_{x\ TM\ m,n} \tag{13.66}$$

We notice here that neither m nor n can have zero values; otherwise, all field components will be zero and hence no field solution exists for such modes.

13.13.1 Example of TM Modes: TM_{11} Mode ($m = 1$ and $n = 1$)

The following is a discussion of the TM_{11} mode. This demonstrates the mode field solution in the simple case of an air-filled guide. More details of this case are provided in Addendum 13D.

Using Equations (13.22), (13.33), and (13.38), we can determine the cutoff frequency for this mode, as well as the related solution parameters as follows:

$$f_{c\ 1,1} = \frac{c_o}{\sqrt{\mu_r\varepsilon_r}}\sqrt{\left[\left(\frac{m}{2a}\right)^2 + \left(\frac{n}{2b}\right)^2\right]} = \frac{c_o}{2}\sqrt{\left(\frac{1}{a}\right)^2 + \left(\frac{1}{b}\right)^2} \tag{13.67}$$

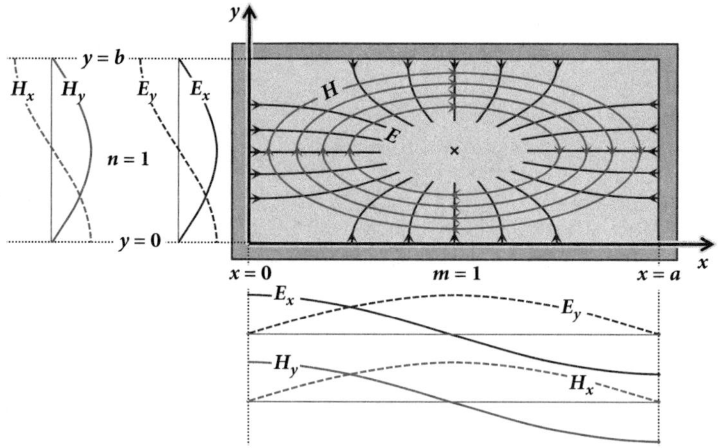

FIGURE 13.10 Fields map for the TM_{11} mode.

$$k_{c\ 1,1} = \sqrt{\left(\frac{\pi}{a}\right)^2 + \left(\frac{\pi}{b}\right)^2} \tag{13.68}$$

$$\beta_{g\ 1,1} = \beta_{ug}\sqrt{1 - \left(\frac{f_{c\ m,n}}{f}\right)^2} = \frac{\omega}{c_o}\sqrt{1 - \left(\frac{f_{c\ 1,1}}{f}\right)^2} \tag{13.69}$$

$$\bar{H}^+_{z\ TM\ 1,1} = 0 \tag{13.70}$$

$$\bar{E}^+_{z\ TM\ 1,1} = \sin\left(\frac{\pi}{a}x\right)\sin\left(\frac{\pi}{b}y\right)\bar{E}^+_{o\ 1,1}e^{-j\beta_{g\ 1,1}z} \tag{13.71}$$

$$\bar{E}^+_{x\ TM\ 1,1} = \frac{-j\beta_{g\ 1,1}}{k^2_{c\ 1,1}}\left(\frac{\pi}{a}\right)\cos\left(\frac{\pi}{a}x\right)\sin\left(\frac{\pi}{b}y\right)\bar{E}^+_{o\ 1,1}e^{-j\beta_{g\ 1,1}z} \tag{13.72}$$

$$\bar{E}^+_{y\ TM\ 1,1} = \frac{-j\beta_{g\ 1,1}}{k^2_{c\ 1,1}}\left(\frac{\pi}{b}\right)\sin\left(\frac{\pi}{a}x\right)\cos\left(\frac{\pi}{b}y\right)\bar{E}^+_{o\ 1,1}e^{-j\beta_{g\ 1,1}z} \tag{13.73}$$

$$\bar{H}^+_{x\ TM\ 1,1} = \frac{j\omega\varepsilon}{k^2_{c\ 1,1}}\left(\frac{\pi}{b}\right)\sin\left(\frac{\pi}{a}x\right)\cos\left(\frac{\pi}{b}y\right)\bar{E}^+_{o\ 1,1}e^{-j\beta_{g\ 1,1}z} = -\frac{\omega\varepsilon}{\beta_{g\ 1,1}}\bar{E}^+_{y\ TM\ 1,1} \tag{13.74}$$

$$\bar{H}^+_{y\ TM\ 1,1} = \frac{-j\omega\varepsilon}{k^2_{c\ 1,1}}\left(\frac{\pi}{a}\right)\cos\left(\frac{\pi}{a}x\right)\sin\left(\frac{\pi}{b}y\right)\bar{E}^+_{o\ 1,1}e^{-j\beta_{g\ 1,1}z} = \frac{\omega\varepsilon}{\beta_{g\ 1,1}}\bar{E}^+_{x\ TM\ 1,1} \tag{13.75}$$

In Figure 13.10, we give the fields distributions over the cross section of the guide according to Equations (13.70) to (13.75).

13.14 Waveguide Impedance

Examining both the *TE* and *TM* solutions, we notice that there are two sets of orthogonal electric and magnetic fields in each of the two solutions; $\{\bar{E}^+_{x\ m,n}, \bar{H}^+_{y\ m,n}\}$ and $\{\bar{E}^+_{y\ m,n}, \bar{H}^+_{x\ m,n}\}$. In each of these sets, the transverse $\bar{E}$ and $\bar{H}$ field phasors have the same cross-sectional functional field distribution and are related to each other by a *constant*. This is evident from the *TE* Equations (13.50) and (13.51) and the *TM* Equations (13.74) and (13.75).

Starting with the TE modes, Equations (13.50) and (13.51), we extract the relationships

$$\left[\bar{E}^{+}_{x\ TE\ m,n}(x,y,z)\right]=\frac{\omega\mu}{\beta_{g\ m,n}}\left[\bar{H}^{+}_{y\ TE\ m,n}(x,y,z)\right] \tag{13.76}$$

$$\left[\bar{E}^{+}_{y\ TE\ m,n}(x,y,z)\right]=\frac{\omega\mu}{\beta_{g\ m,n}}\left[-\bar{H}^{+}_{x\ TE\ m,n}(x,y,z)\right] \tag{13.77}$$

which prompts us to write the ratio

$$\frac{\left[\bar{E}^{+}_{x\ TE\ m,n}(x,y,z)\right]}{\left[\bar{H}^{+}_{y\ TE\ m,n}(x,y,z)\right]}=\frac{\left[\bar{E}^{+}_{y\ TE\ m,n}(x,y,z)\right]}{\left[-\bar{H}^{+}_{x\ TE\ m,n}(x,y,z)\right]}=\frac{\omega\mu}{\beta_{g\ m,n}}=\bar{\eta}_{g\ TE\ m,n} \tag{13.78}$$

In fact, this ratio is the wave impedance in the guide for the TE mode m,n. It is the ratio between the transverse E and H components, which correspond to this mode of propagation. We notice that $\{\bar{E}^{+}_{x\ m,n}, \bar{H}^{+}_{y\ m,n}\}$ form a right-handed system with the z direction of propagation, and hence their ratio yields the proper polarity for the impedance (positive). Meanwhile, the negative polarity of $-\bar{H}^{+}_{x\ m,n}$ in the pair $\{\bar{E}^{+}_{y\ m,n}, -\bar{H}^{+}_{x\ m,n}\}$ makes it form a right-handed system with the positive z wave travel.

If we compare this wave impedance to the medium (unguided) intrinsic impedance, $\bar{\eta}_{ug}$, ($\bar{\eta}_{ug}=377\ \Omega$ for free space), we get

$$\frac{\bar{\eta}_{g\ TE\ m,n}}{\bar{\eta}_{ug}}=\frac{\omega\mu}{\beta_{g\ m,n}}\div\frac{\omega\mu}{\beta_{ug}}=\frac{\beta_{ug}}{\beta_{g\ m,n}}=\frac{1}{\sqrt{1-\left[\frac{f_{c\ m,n}}{f}\right]^{2}}} \tag{13.79}$$

$$\bar{\eta}_{g\ TE\ m,n}=\frac{\bar{\eta}_{ug}}{\sqrt{1-\left[\frac{f_{c\ m,n}}{f}\right]^{2}}} \tag{13.80}$$

Similarly, for the TM modes, using Equations (13.65) and (13.66), we can write

$$\left[\bar{E}^{+}_{x\ TM\ m,n}\right]=\frac{\beta_{g\ m,n}}{\omega\varepsilon}\left[\bar{H}^{+}_{y\ TM\ m,n}\right] \tag{13.81}$$

$$\left[\bar{E}^{+}_{y\ TM\ m,n}\right]=\frac{\beta_{g\ m,n}}{\omega\varepsilon}\left[-\bar{H}^{+}_{x\ TM\ m,n}\right] \tag{13.82}$$

Hence,

$$\frac{\left[\bar{E}^{+}_{x\ TM\ m,n}(x,y,z)\right]}{\left[\bar{H}^{+}_{y\ TM\ m,n}(x,y,z)\right]}=\frac{\left[\bar{E}^{+}_{y\ TM\ m,n}(x,y,z)\right]}{\left[-\bar{H}^{+}_{x\ TM\ m,n}(x,y,z)\right]}=\frac{\beta_{g\ m,n}}{\omega\varepsilon}=\bar{\eta}_{g\ TM\ m,n} \tag{13.83}$$

and

$$\frac{\bar{\eta}_{g\ TM\ m,n}}{\bar{\eta}_{ug}}=\frac{\beta_{g\ m,n}}{\omega\varepsilon}\div\frac{\omega\mu}{\beta_{ug}}=\frac{\beta_{g\ m,n}}{\beta_{ug}}=\sqrt{1-\left[\frac{f_{c\ m,n}}{f}\right]^{2}} \tag{13.84}$$

$$\bar{\eta}_{g\ TM\ m,n}=\bar{\eta}_{ug}\sqrt{1-\left[\frac{f_{c\ m,n}}{f}\right]^{2}} \tag{13.85}$$

Figure 13.11 demonstrates the functional dependence of both waveguide impedances with frequency for different "m and n" modes.

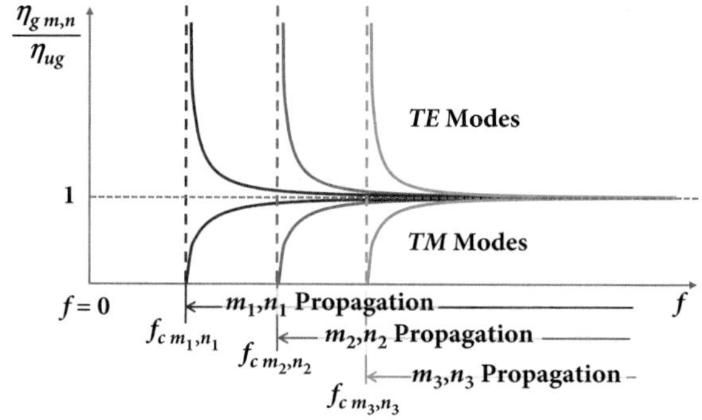

FIGURE 13.11 The functional dependence of the guide's wave impedance versus frequency for the different possible *TE* and *TM* modes.

13.15 Active and Dominant Mode Identification

In Addendum 13E, we discuss how to determine the active (operating) modes for a given waveguide at a given frequency. The process starts with figuring out the cutoff frequencies of the guide as determined by the guide dimensions and dielectric filling. Once the cutoff frequencies of the different m and n are determined, then the active modes are those with cutoff frequencies lower than the guide's operating frequency.

In the addendum, we present two approaches for active and dominant mode identification. In one approach, we use calculations of the cutoff frequencies for all values of m, n of interest. Then we tabulate the data and sort them out in an ascending order. In the second approach, we use a graphical scheme to identify the order of the cutoff frequencies for the different values of m and n.

As the addendum concludes, we can see that the TE_{10} mode has the lowest cutoff frequency of all possible modes. We can also state undoubtedly that the TM_{11} mode will have the lowest cutoff frequency of all *TM* modes.

The lowest cutoff frequency modes are called the dominant modes and, hence, we refer to the TE_{10} mode as the dominant mode of both *TE* and *TM* modes, for $a > b$. We also refer to the TM_{11} mode as the dominant *TM* mode.

A typical sorted order of cutoff frequencies for the case $a \geq 2b$ is demonstrated in Figure 13.12.

Example 13.1 Find all possible modes that can propagate in a WGR112 ($a = 1.122$ in, $b = 0.497$ in) rectangular waveguide operating at 14 GHz. For each mode, find the corresponding parameters:

a. Cutoff frequency	b. Phase constant	c. Wavelength
d. WG impedance	e. Phase velocity	f. Group velocity

Solution

a. Use Equation (13E.1) for the cutoff frequency.

b. Calculate $f_{c\ m,n}$ the different m and n combinations.

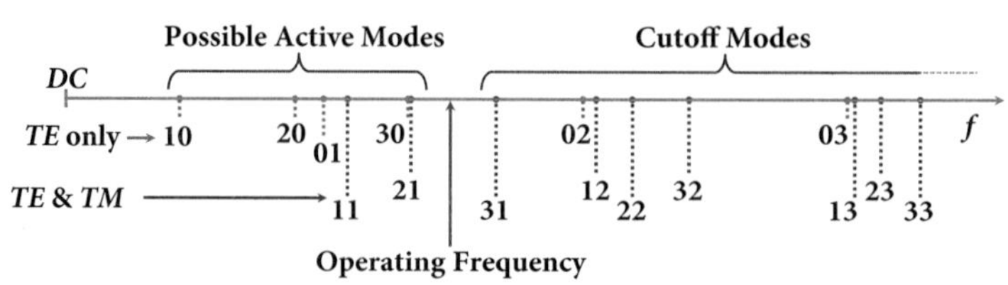

FIGURE 13.12 A Typical sorted order of cutoff frequencies for the case $a \geq 2b$.

c. Sort the results on the frequency scale.

d. Possible modes are the ones with cutoff frequencies below the 14 GHz operating frequency.

e. Use Equations (13.33), (13.27), (13.34), (13.37), (13B.2), (13.80), and (13.85) to get the required values

mn	$f_{c\,m,n}$ GHz	$TE-TM$	$\beta_{g\,m,n}$ rad/m	$\lambda_{g\,m,n}$ m	$C_{ph\,g\,m,n}$ ($\times10^8$)m/s	$C_{gr\,g\,m,n}$ ($\times10^8$)m/s	$\eta_{g\,m,n}$ Ω
10	5.263	TE	271.89	0.0231	3.24	2.78	406.6
20	10.526	TE	193.45	0.0325	4.55	1.98	571.4
01	11.886	TE	155.04	0.0405	5.68	1.59	714.0
11	12.999	TE & TM	108.95	0.0577	8.08	1.11	1014.6 & 139.9
30	15.789	TE					
21	15.877	TE & TM					

13.16 Wave Propagation: Power Flow

To study the wave propagation in a waveguide, we look at the power flow density vector (Poynting vector).

$$\vec{\mathcal{P}}_{\text{avg}}(x,y,z)=\frac{1}{2}Re\left\{\bar{\bar{E}}(x,y,z)\times\bar{\bar{H}}^*(x,y,z)\right\} \tag{13.86}$$

For illustration purposes, let us consider the forward wave ($+z$) power flow density in the TE_{mn} modes. A similar procedure can be followed to derive the backward wave ($-z$) expression. The net power flow density expressions would be the superposition of the two components.

In this regard, we recall that the ($+z$) field components [Equations (13.46) to (13.51)] for these modes are

$$\bar{E}^+_{z\;TE\;m,n}=0 \tag{13.46}$$

$$\bar{H}^+_{z\;TE\;m,n}=\left[\cos\left(\frac{m\pi}{a}x\right)\cos\left(\frac{n\pi}{b}y\right)\right]\left[\bar{H}^+_{o\;m,n}e^{-j\beta_{g\;m,n}z}\right] \tag{13.47}$$

$$\bar{E}^+_{x\;TE\;m,n}=\left[\frac{j\omega\mu}{k^2_{c\;m,n}}\left(\frac{n\pi}{b}\right)\cos\left(\frac{m\pi}{a}x\right)\sin\left(\frac{n\pi}{b}y\right)\right]\left[\bar{H}^+_{o\;m,n}e^{-j\beta_{g\;m,n}z}\right] \tag{13.48}$$

$$\bar{E}^+_{y\;TE\;m,n}=-\left[\frac{j\omega\mu}{k^2_{c\;m,n}}\left(\frac{m\pi}{a}\right)\sin\left(\frac{m\pi}{a}x\right)\cos\left(\frac{n\pi}{b}y\right)\right]\left[\bar{H}^+_{o\;m,n}e^{-j\beta_{g\;m,n}z}\right] \tag{13.49}$$

$$\bar{H}^+_{x\;TE\;m,n}=\left[\frac{j\beta_{g\;m,n}}{k^2_{c\;m,n}}\left(\frac{m\pi}{a}\right)\sin\left(\frac{m\pi}{a}x\right)\cos\left(\frac{n\pi}{b}y\right)\right]\left[\bar{H}^+_{o\;m,n}e^{-j\beta_{g\;m,n}z}\right] \tag{13.50}$$

$$\bar{H}^+_{y\;TE\;m,n}=\left[\frac{j\beta_{g\;m,n}}{k^2_{c\;m,n}}\left(\frac{n\pi}{b}\right)\cos\left(\frac{m\pi}{a}x\right)\sin\left(\frac{n\pi}{b}y\right)\right]\left[\bar{H}^+_{o\;m,n}e^{-j\beta_{g\;m,n}z}\right] \tag{13.51}$$

Now, substituting Equations (13.46) to (13.51) into Equation (13.86), we get

$$\vec{\mathcal{P}}^+_{\text{avg}\;TE\;m,n}(x,y,z)=\frac{1}{2}Re\left\{\bar{\bar{E}}^+_{TE\;m,n}(x,y,z)\times\bar{\bar{H}}^{+*}_{TE\;m,n}(x,y,z)\right\}$$

$$=\frac{1}{2}Re\left\{\left[\bar{H}^+_{o\;m,n}e^{-j\beta_{g\;m,n}z}\right]\left\{\left[\frac{j\omega\mu}{k^2_{c\;m,n}}\left(\frac{n\pi}{b}\right)\cos\left(\frac{m\pi}{a}x\right)\sin\left(\frac{n\pi}{b}y\right)\right]\vec{a}_x\right.\right.$$

$$-\left[\frac{j\omega\mu}{k_{c\,m,n}^2}\left(\frac{m\pi}{a}\right)\sin\left(\frac{m\pi}{a}x\right)\cos\left(\frac{n\pi}{b}y\right)\right]\vec{a}_y\Bigg\}$$

$$\times\left[\bar{H}_{o\,m,n}^{+}e^{-j\beta_{g\,m,n}z}\right]^*\left\{\left[\frac{j\beta_{g\,m,n}}{k_{c\,m,n}^2}\left(\frac{m\pi}{a}\right)\sin\left(\frac{m\pi}{a}x\right)\cos\left(\frac{n\pi}{b}y\right)\right]^*\vec{a}_x\right.$$

$$\left.+\left[\frac{j\beta_{g\,m,n}}{k_{c\,m,n}^2}\left(\frac{n\pi}{b}\right)\cos\left(\frac{m\pi}{a}x\right)\sin\left(\frac{n\pi}{b}y\right)\right]^*\vec{a}_y+\left[\cos\left(\frac{m\pi}{a}x\right)\cos\left(\frac{n\pi}{b}y\right)\right]\vec{a}_z\right\}\Bigg\}$$

$$=\mathcal{P}_{x,\,\text{avg}\,TE\,m,n}^{+}(x,y,z)\vec{a}_x+\mathcal{P}_{y,\,\text{avg}\,TE\,m,n}^{+}(x,y,z)\vec{a}_y+\mathcal{P}_{z,\,\text{avg}\,TE\,m,n}^{+}(x,y,z)\vec{a}_z \quad (13.87)$$

which breaks down to

$$\mathcal{P}_{x,\,\text{avg}\,TE\,m,n}^{+}(x,y,z)$$

$$=\frac{1}{2}\left|\bar{H}_{o\,m,n}^{+}\right|^2 Re\left\{\left[\frac{\omega\mu}{k_{c\,m,n}^2}\left(\frac{m\pi}{a}\right)\sin\left(\frac{m\pi}{a}x\right)\cos\left(\frac{n\pi}{b}y\right)\right]\cdot\left[-j\cos\left(\frac{m\pi}{a}x\right)\cos\left(\frac{n\pi}{b}y\right)\vec{a}_x\right]\right\}=0 \quad (13.88)$$

$$\mathcal{P}_{y,\,\text{avg}\,TE\,m,n}^{+}(x,y,z)$$

$$=\frac{1}{2}\left|\bar{H}_{o\,m,n}^{+}\right|^2 Re\left\{\left[\frac{\omega\mu}{k_{c\,m,n}^2}\left(\frac{n\pi}{b}\right)\cos\left(\frac{m\pi}{a}x\right)\sin\left(\frac{n\pi}{b}y\right)\right]\cdot\left[-j\cos\left(\frac{m\pi}{a}x\right)\cos\left(\frac{n\pi}{b}y\right)\vec{a}_y\right]\right\}=0 \quad (13.89)$$

$$\mathcal{P}_{z,\,\text{avg}\,TE\,m,n}^{+}(x,y,z)$$

$$=\frac{1}{2}\left|\bar{H}_{o\,m,n}^{+}\right|^2 Re\left\{\left[\frac{\omega\mu}{k_{c\,m,n}^2}\left(\frac{n\pi}{b}\right)\cos\left(\frac{m\pi}{a}x\right)\sin\left(\frac{n\pi}{b}y\right)\right]\cdot\left[\frac{\beta_{g\,m,n}}{k_{c\,m,n}^2}\left(\frac{n\pi}{b}\right)\cos\left(\frac{m\pi}{a}x\right)\sin\left(\frac{n\pi}{b}y\right)\right]\right.$$

$$\left.+\left[\frac{\omega\mu}{k_{c\,m,n}^2}\left(\frac{m\pi}{a}\right)\sin\left(\frac{m\pi}{a}x\right)\cos\left(\frac{n\pi}{b}y\right)\right]\cdot\left[\frac{\beta_{g\,m,n}}{k_{c\,m,n}^2}\left(\frac{m\pi}{a}\right)\sin\left(\frac{m\pi}{a}x\right)\cos\left(\frac{n\pi}{b}y\right)\right]\right\} \quad (13.90)$$

thus, yielding the final result

$$\vec{\mathcal{P}}_{\text{avg}\,TE\,m,n}^{+}(x,\,y,\,z)=$$

$$\frac{1}{2}\frac{\omega\mu\beta_{g\,m,n}}{k_{c\,m,n}^4}\left|\bar{H}_{o\,m,n}^{+}\right|^2\left\{\left(\frac{n\pi}{b}\right)^2\left[\cos\left(\frac{m\pi}{a}x\right)\sin\left(\frac{n\pi}{b}y\right)\right]^2+\left(\frac{m\pi}{a}\right)^2\left[\sin\left(\frac{m\pi}{a}x\right)\cos\left(\frac{n\pi}{b}y\right)\right]^2\right\}\vec{a}_z \quad (13.91)$$

This result explicitly demonstrates that the power flow within the guide is along its axis; the z axis in our case. Examining the interim expressions [Equations (13.88) and (13.89)], we notice reactive power components in the transverse directions ($\vec{a}_x$ and $\vec{a}_y$). These reactive powers are simply standing waves bouncing back and forth between the guide's walls.

A similar derivation for the TM_{mn} modes leads to the following results for the power flow-density expression:

$$\vec{\mathcal{P}}_{\text{avg}\,TM\,m,n}^{+}(x,y,z)=$$

$$\frac{1}{2}\frac{\omega\varepsilon\beta_{g\,m,n}}{k_{c\,m,n}^4}\left|\bar{E}_{o\,m,n}^{+}\right|^2\left\{\left(\frac{m\pi}{a}\right)^2\cos^2\left(\frac{m\pi}{a}x\right)\sin^2\left(\frac{n\pi}{b}y\right)+\left(\frac{n\pi}{b}\right)^2\sin^2\left(\frac{m\pi}{a}x\right)\cos^2\left(\frac{n\pi}{b}y\right)\right\}\vec{a} \quad (13.92)$$

13.16.1 Power Flow for the TE_{10} Dominant Mode

For the special case of the TE_{10} mode, the power flow-density Equation (13.91) reduces to

$$\vec{\mathcal{P}}^{+}_{\text{avg } TE\ 1,0}(x,y,z) = \frac{1}{2}\frac{\omega\mu\beta_{g\ 1,0}}{k^4_{c\ 1,0}}\left|\bar{H}^{+}_{o\ 1,0}\right|^2 \left\{ +\left(\frac{\pi}{a}\right)^2 \left[\sin\left(\frac{\pi}{a}x\right)\right]^2 \right\}\vec{a}_z = \frac{1}{2}\omega\mu\beta_{g\ 1,0}\left(\frac{a}{\pi}\right)^2 \sin^2\left(\frac{\pi}{a}x\right)\left|\bar{H}^{+}_{o\ 1,0}\right|^2 \vec{a}_z \tag{13.93}$$

Integration of Equation (13.93) over the guide's cross-sectional area yields the total power flow through the guide for the TE_{10} mode. The resulting expression is given by

$$\vec{P}^{+}_{\text{avg } TE\ 1,0} = \iint_{0,0}^{a,b} \vec{\mathcal{P}}^{+}_{\text{avg } TE\ 1,0}(x,y,z)\,dx\,dy$$

$$= \iint_{a,b}^{a,b} \frac{1}{2}\omega\mu\beta_{g\ 1,0}\left(\frac{a}{\pi}\right)^2 \sin^2\left(\frac{\pi}{a}x\right)\left|\bar{H}^{+}_{o\ 1,0}\right|^2 \vec{a}_z\,dx\,dy = \frac{ab}{4}\omega\mu\beta_{g\ 1,0}\left(\frac{a}{\pi}\right)^2 \left|\bar{H}^{+}_{o\ 1,0}\right|^2 \vec{a}_z \tag{13.94}$$

The field maps given in Addendum 13D for the TE_{10} mode, Figures 13D.3 and 13D.4, demonstrate the above findings. The figures show a resultant power flow along the guide's transmission axis (the z axis) with the presence of standing waves between the guide's side walls (the x axis). Also, Figure 13D.4 shows power flow density distribution over the guide's cross section that varies in the x coordinate $\left[\sin^2\left(\frac{\pi}{a}x\right)\right]$ with no variation along the y coordinate as per the findings of Equation (13.89).

13.16.2 Time-Domain Derivation of the Power Flow Density for the TE_{10} Mode

In the following, we will go through the exercise of deriving the power flow-density Equation (13.93) using the time-domain approach. In this regard, we start with the time-domain field expressions for the TE_{10} mode. Guided by the corresponding phasor Equations (13.56) to (13.59), the corresponding time-domain expressions are

$$H^{+}_{z\ TE\ 1,0}(x,y,z,t) = H^{+}_{o\ 1,0}(x,y,z,t)\left[\cos\left(\frac{\pi}{a}x\right)\cos(\omega t - \beta_{g\ 1,0}z)\right] \tag{13.95}$$

$$E^{+}_{y\ TE\ 1,0}(x,y,z,t) = -\omega\mu\left(\frac{a}{\pi}\right)H^{+}_{o\ 1,0}(x,y,z,t)\left[\sin\left(\frac{\pi}{a}x\right)\sin(\omega t - \beta_{g\ 1,0}z)\right] \tag{13.96}$$

$$H^{+}_{x\ TE\ 1,0}(x,y,z,t) = \beta_{g\ 1,0}\left(\frac{a}{\pi}\right)H^{+}_{o\ 1,0}(x,y,z,t)\left[\sin\left(\frac{\pi}{a}x\right)\sin(\omega t - \beta_{g\ 1,0}z)\right] \tag{13.97}$$

Using these expressions, we write the time-domain form for the Poynting vector as

$$\begin{aligned}
\vec{\mathcal{P}}^{+}_{TE\ 1,0}(x,y,z,t) &= \vec{E}^{+}_{TE\ 1,0}(x,y,z,t) \times \vec{H}^{+}_{TE\ 1,0}(x,y,z,t) \\
&= \left\{-\omega\mu\left(\frac{a}{\pi}\right)H^{+}_{o\ 1,0}(x,y,z,t)\left[\sin\left(\frac{\pi}{a}x\right)\sin(\omega t - \beta_{g\ 1,0}z)\right]\vec{a}_y\right\} \\
&\quad \times\left\{\beta_{g\ 1,0}\left(\frac{a}{\pi}\right)H^{+}_{o\ 1,0}(x,y,z,t)\left[\sin\left(\frac{\pi}{a}x\right)\sin(\omega t - \beta_{g\ 1,0}z)\right]\vec{a}_x\right. \\
&\quad \left. + H^{+}_{o\ 1,0}(x,y,z,t)\left[\cos\left(\frac{\pi}{a}x\right)\cos(\omega t - \beta_{g\ 1,0}z)\right]\vec{a}_z\right\} \\
&= \mathcal{P}^{+}_{x,\ TE\ 1,0}(x,y,z,t)\vec{a}_x + \mathcal{P}^{+}_{z,\ TE\ 1,0}(x,y,z,t)\vec{a}_z
\end{aligned} \tag{13.98}$$

which breaks down to

$$\mathcal{P}^+_{x,\,TE\;1,0}(x,y,z,t)=\left\{-\omega\mu\left(\frac{a}{\pi}\right)H^+_{o\;1,0}(x,y,z,t)\left[\sin\left(\frac{\pi}{a}x\right)\sin(\omega t-\beta_{g\;1,0}z)\right]\right\}$$

$$\times\left\{+H^+_{o\;1,0}(x,y,z,t)\left[\cos\left(\frac{\pi}{a}x\right)\cos(\omega t-\beta_{g\;1,0}z)\right]\right\}$$

$$=-\left\{\omega\mu\left(\frac{a}{\pi}\right)\left[H^+_{o\;1,0}(x,y,z,t)\right]^2 s\left[\text{in}\left(\frac{\pi}{a}x\right)\cos\left(\frac{\pi}{a}x\right)\right]\right\}\left\{\left(\frac{1}{2}\right)\sin(2\omega t-2\beta_{g\;1,0}z)\right\} \tag{13.99}$$

$$\mathcal{P}^+_{z,\,TE\;1,0}(x,y,z,t)=\left\{\omega\mu\left(\frac{a}{\pi}\right)H^+_{o\;1,0}(x,y,z,t)\left[\sin\left(\frac{\pi}{a}x\right)\sin(\omega t-\beta_{g\;1,0}z)\right]\right\}$$

$$\times\left\{\beta_{g\;1,0}\left(\frac{a}{\pi}\right)H^+_{o\;1,0}(x,y,z,t)\left[\sin\left(\frac{\pi}{a}x\right)\sin(\omega t-\beta_{g\;1,0}z)\right]\right\}$$

$$=\left\{\omega\mu\beta_{g\;1,0}\left(\frac{a}{\pi}\right)^2\left[H^+_{o\;1,0}(x,y,z,t)\right]^2\sin^2\left(\frac{\pi}{a}x\right)\right\}\left\{\left(\frac{1}{2}\right)\left[1-\cos(2\omega t-2\beta_{g\;1,0}z)\right]\right\} \tag{13.100}$$

Using Equations (13.99) and (13.100), we extract the real power flow density vector as the time average of the time-dependent expression, which yields

$$\vec{\mathcal{P}}^+_{\text{avg}\;TE\;1,0}(x,y,z)=0\vec{a}_x+\left(\frac{1}{2}\right)\left\{\omega\mu\beta_{g\;1,0}\left(\frac{a}{\pi}\right)^2\left[H^+_{o\;1,0}(x,y,z,t)\right]^2\sin^2\left(\frac{\pi}{a}x\right)\right\}\vec{a}_z \tag{13.101}$$

The resulting Equation (13.101) is identical to that of Equation (13.93) that we derived earlier using the frequency domain approach.

13.16.3 A Physical View of Wave Propagation and Power Flow in Waveguides

In Addendum 13F, we present an illustration to help us visualize the concepts of wave propagation and power flow in waveguides. The illustration applies to a parallel plate guide and not the rectangular waveguide. However, the concepts demonstrated here give adequate insight that should parallel those of the rectangular waveguide.

Example 13.2 For the WGR112 rectangular waveguide operating in the TE_{10} mode at 8 GHz, find the power flow density vector as well as the total power flow in the guide. Given the peak of the excitation electric field is 100 V/m.

Solution

a. Similar to Example (13.1), find operating modes → TE_{10} only (f_{c10} = 5.263 GHz)
b. Use Equation (13.58) to relate $H^+_{o\;1,0}$ to $E^+_{yo\;1,0}$ (given as 100 V/m)

$$\left|\bar{E}^+_{yo\;1,0}\right|=\left|\bar{E}^+_{y\;TE\;1,0}\right|_{x=\frac{a}{2},z=0}=\left|-j\omega\mu\left(\frac{a}{\pi}\right)\sin\left(\frac{\pi}{a}x\right)\left[\bar{H}^+_{o\;1,0}e^{-j\beta_{g\;1,0}z}\right]\right|_{x=\frac{a}{2},z=0}=\omega\mu\left(\frac{a}{\pi}\right)\left|\bar{H}^+_{o\;1,0}\right|$$

$$\left|\bar{H}^+_{o\;1,0}\right|=\frac{\left|\bar{E}^+_{yo\;1,0}\right|}{\omega\mu\left(\frac{a}{\pi}\right)}=100/573.03=0.175\text{ A/m}$$

c. Use Equations (13.93) and (13.94) to get the values for the Poynting vector and power flow

$$\vec{\mathcal{P}}^+_{\text{avg}\;TE\;1,0}(x,y,z)=\frac{1}{2}\omega\mu\beta_{g\;1,0}\left(\frac{a}{\pi}\right)^2\sin^2\left(\frac{\pi}{a}x\right)\left|\bar{H}^+_{o\;1,0}\right|^2\vec{a}_z \tag{13.93}$$

$$\vec{P}^{+}_{\text{avg } TE\ 1,0} = \frac{ab}{4}\omega\mu\beta_{g\ 1,0}\left(\frac{a}{\pi}\right)^2 \left|\bar{H}^{+}_{o\ 1,0}\right|^2 \vec{a}_z \tag{13.94}$$

$$\beta_{g\ 1,0} = \beta_{ug}\sqrt{1-\left(\frac{f_{c\ 1,0}}{f}\right)^2} = \frac{2\pi f}{c_o}\sqrt{1-\left(\frac{f_{c\ 1,0}}{f}\right)^2} = 126.18 \text{ rad/m}$$

$$\vec{\mathcal{P}}^{+}_{\text{avg } TE\ 1,0}(x,y,z) = \frac{1}{2}\omega\mu\beta_{g\ 1,0}\left(\frac{a}{\pi}\right)^2 \sin^2\left(\frac{\pi}{a}x\right)\left|\bar{H}^{+}_{o\ 1,0}\right|^2 \vec{a}_z = 9.995\sin^2\left(\frac{\pi}{a}x\right)\vec{a}_z \text{ W/m}^2$$

$$\vec{P}^{+}_{\text{avg } TE\ 1,0} = \frac{ab}{4}\omega\mu\beta_{g\ 1,0}\left(\frac{a}{\pi}\right)^2 \left|\bar{H}^{+}_{o\ 1,0}\right|^2 \vec{a}_z = 1.797\vec{a}_z \text{ mW}$$

13.17 Modal Dispersion and Waveguide Bandwidth

Operating the waveguide in multimodes causes modal dispersion to the transmitted signals. In the following, we present a simplified explanation for this behavior.

Let us assume an incident communication signal in the form of a single pulse is launched to propagate through a multimode guide as demonstrated in Figure 13.13. For illustrative purposes, let us further consider the propagation through two modes: namely, mode 1 and mode 2. The launched pulse is then to excite both modes in different proportions depending on the launch boundary conditions and the guide characteristics. As the figure demonstrates, each mode will propagate its own portion of the pulse as a result.

Since each one of the two modes has its own guide propagation characteristics (its own wavelength, its own cutoff frequency, etc.), both modes will have different propagation delays. Consequently, at the receiving end of the guide the two modes will deliver their own transmitted pulses at different times. The result is that the received signal will be a distorted version of the launched pulse. This type of distortion is what we refer to as dispersion (pulse spreading). Since this dispersion is due to multimode propagation, we refer to it as modal dispersion to distinguish it from other forms of dispersion, for example, material dispersion due to both the dielectric filling and the conductor walls.

The main issue with dispersion is that it limits the speed at which the guide can be used for digital communication, and hence, its useful bandwidth in this regard. Consequently, we prefer to operate waveguides in a single-mode state. For our rectangular guide, the dominant mode is the TE_{10} mode, which has the lowest cutoff frequency. Referring to Figures 13.12 and 13.14, the next mode up is either the TE_{20} mode or the TE_{01} mode, depending on the relative values of a and b.

$$f_{c\ 1,0} = \frac{c_{ug}}{2a}, \quad f_{c\ 2,0} = \frac{c_{ug}}{a}, \quad \text{and} \quad f_{c\ 0,1} = \frac{c_{ug}}{2b} \tag{13.102}$$

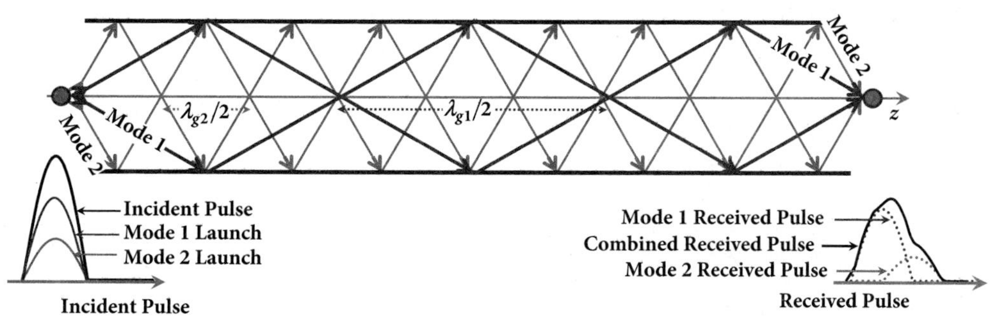

Figure 13.13 A demonstration of modal dispersion in a waveguide operating with two modes.

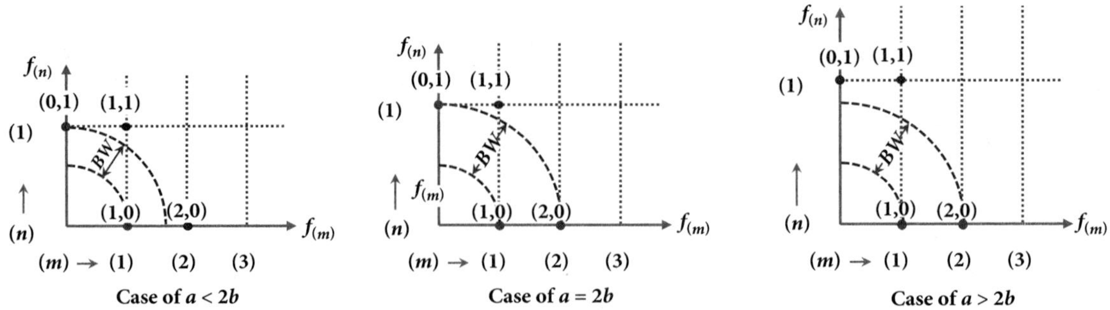

FIGURE 13.14 Waveguide bandwidth is defined as the frequency band for single-mode operation.

Hence, the guide single-mode bandwidth will be the smaller of the two values

$$BW_a = f_{c\ 2,0} - f_{c\ 1,0} = c_{ug}\frac{1}{2a}, \text{ and} \tag{13.103}$$

$$BW_b = f_{c\ 0,1} - f_{c\ 1,0} = c_{ug}\left(\frac{1}{2b} - \frac{1}{2a}\right) = c_{ug}\frac{1}{2a}\left(\frac{a}{b} - 1\right) \tag{13.104}$$

Comparing the two possible bandwidths, we realize that the second bandwidth, BW_b, will be smaller than BW_a, unless $a \geq 2b$. Hence, we have to design the guide with $a \geq 2b$ for maximum bandwidth in order to operate in the single-mode domain of the TE_{10} dominant mode.

Example 13.3 Graph a sorted order of the cutoff modes for the following guide dimensions: $a = 2.25b$, $a = 2b$, and $a = 1.75b$. Discuss the bandwidth of the guide for single-mode operation.

Solution

a. Use Equation 13E.1 for the normalized cutoff frequency.

b. Let us use $b = 1$ for convenience.

c. Calculate $f_{cn\ m,n}$ the different m and n combinations.

d. Sort the results on the frequency scale.

$$f_{cn\ m,n} = \frac{f_{c\ m,n}}{\left[\dfrac{c_o}{\sqrt{\mu_r \varepsilon_r}}\right]} = \frac{f_{c\ m,n}}{c_{ug}} = \sqrt{\left[\left(\frac{m}{2a}\right)^2 + \left(\frac{n}{2b}\right)^2\right]} \tag{13E.1}$$

Before Sorting				
m	*n*	**2.25**	**2.0**	**1.75**
1	0	0.222	0.250	0.286
2	0	0.444	0.500	0.571
3	0	0.667	0.750	0.857
4	0	0.889	1.000	1.143
5	0	1.111	1.250	1.429
0	1	0.500	0.500	0.500
1	1	0.547	0.559	0.576
2	1	0.669	0.707	0.759
3	1	0.833	0.901	0.992
...	...	...	...	...

After Sorting					
mn	**2.25**	*mn*	**2.0**	*mn*	**1.75**
10	0.222	10	0.250	10	0.286
20	0.444	20	0.500	01	0.500
01	0.500	01	0.500	20	0.571
11	0.547	11	0.559	11	0.576
30	0.667	21	0.707	21	0.759
21	0.669	30	0.750	30	0.857
31	0.833	31	0.901	31	0.992
40	0.889	40	1.000	02	1.000
02	1.000	02	1.000	12	1.040
...	...	...	...	...	...

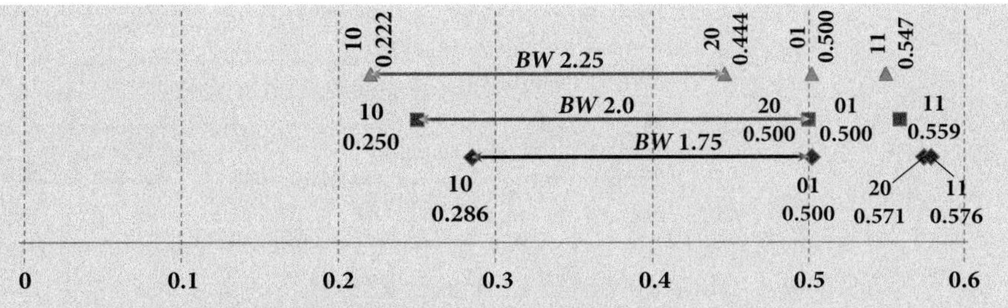

Case	a & b	Dom Mode	$f_{cn,\text{Dom}}$	Next mode	$f_{cn,\text{Nxt}}$	$f_{c,N}/f_{c,D}$
a	a = 2.25b	10	0.222	20	0.444	2.0
b	a = 2.0b	10	0.25	20 & 01	0.5	2.0
c	a = 1.75b	10	0.286	01	0.5	1.75

We conclude that $a \geq 2b$ yields the maximum $f_{c,N}/f_{c,D} = 2.0$ while maintaining single-mode operation and avoiding modal dispersion in the guide.

ADDENDUM 13A WAVE EQUATION SOLUTION FOR METALLIC RECTANGULAR WAVEGUIDES: THE LONGITUDINAL COMPONENT OF THE ELECTRIC FIELD PHASOR

13A.1 Introduction

In the following, we will take the challenge of solving for the fields in rectangular metallic waveguides. For that, Cartesian coordinates are the logical choice. For convenience, the following is a recap of all relevant equations expressed in Cartesian coordinates.

Wave equations:

$$\nabla^2 \vec{\bar{E}} = \nabla^2\{\bar{E}_x\vec{a}_x + \bar{E}_y\vec{a}_y + \bar{E}_z\vec{a}_z\} = -\omega^2\mu\varepsilon\{\bar{E}_x\vec{a}_x + \bar{E}_y\vec{a}_y + \bar{E}_z\vec{a}_z\}, \text{ or} \tag{13A.1}$$

$$\nabla^2 \bar{E}_x = -\omega^2\mu\varepsilon\bar{E}_x \tag{13A.2}$$

$$\nabla^2 \bar{E}_y = -\omega^2\mu\varepsilon\bar{E}_y \tag{13A.3}$$

$$\nabla^2 \bar{E}_z = -\omega^2\mu\varepsilon\bar{E}_z \tag{13A.4}$$

$$\nabla^2 \vec{\bar{H}} = \nabla^2\{\bar{H}_x\vec{a}_x + \bar{H}_y\vec{a}_y + \bar{H}_z\vec{a}_z\} = -\omega^2\mu\varepsilon\{\bar{H}_x\vec{a}_x + \bar{H}_y\vec{a}_y + \bar{H}_z\vec{a}_z\}, \text{ or} \tag{13A.5}$$

$$\nabla^2 \bar{H}_x = -\omega^2\mu\varepsilon\bar{H}_x \tag{13A.6}$$

$$\nabla^2 \bar{H}_y = -\omega^2\mu\varepsilon\bar{H}_y \tag{13A.7}$$

$$\nabla^2 \bar{H}_z = -\omega^2\mu\varepsilon\bar{H}_z \tag{13A.8}$$

Maxwell's curl equations:

$$\vec{\nabla} \times \{\bar{E}_x\vec{a}_x + \bar{E}_y\vec{a}_y + \bar{E}_z\vec{a}_z\} = -j\omega\mu\{\bar{H}_x\vec{a}_x + \bar{H}_y\vec{a}_y + \bar{H}_z\vec{a}_z\}, \text{ or} \tag{13A.9}$$

$$\frac{\partial \bar{E}_z}{\partial y} - \frac{\partial \bar{E}_y}{\partial z} = -j\omega\mu\bar{H}_x \tag{13A.10}$$

$$\frac{\partial \bar{E}_x}{\partial z} - \frac{\partial \bar{E}_z}{\partial x} = -j\omega\mu \bar{H}_y \tag{13A.11}$$

$$\frac{\partial \bar{E}_y}{\partial x} - \frac{\partial \bar{E}_x}{\partial y} = -j\omega\mu \bar{H}_z \tag{13A.12}$$

$$\vec{\nabla} \times \{\bar{H}_x \vec{a}_x + \bar{H}_y \vec{a}_y + \bar{H}_z \vec{a}_z\} = -j\omega\mu \{\bar{E}_x \vec{a}_x + \bar{E}_y \vec{a}_y + \bar{E}_z \vec{a}_z\}, \text{ or} \tag{13A.13}$$

$$\frac{\partial \bar{H}_z}{\partial y} - \frac{\partial \bar{H}_y}{\partial z} = j\omega\varepsilon \bar{E}_x \tag{13A.14}$$

$$\frac{\partial \bar{H}_x}{\partial z} - \frac{\partial \bar{H}_z}{\partial x} = j\omega\varepsilon \bar{E}_y \tag{13A.15}$$

$$\frac{\partial \bar{H}_y}{\partial x} - \frac{\partial \bar{H}_x}{\partial y} = j\omega\varepsilon \bar{E}_z \tag{13A.16}$$

13A.2 (Metallic) Rectangular Waveguides

We will confine our studies in this section to a uniform rectangular cross-section metallic waveguide. The guide we consider is assumed straight along the direction of propagation, with perfectly conducting walls and dielectric filling. Figure 13.2 shows a sketch of a rectangular waveguide section lying on its broad side on the x-z plane of the Cartesian system. The guide extends along the z axis between a source connected at one end and a load at the other.

We start our analysis by solving for the fields inside the waveguide section not including the source and load sections. In this regard, all relevant governing equations are listed above in Equations (13A.1) to (13A.16). These equations are to be solved together with the boundary conditions of the dielectric/conductor interfaces. We will introduce these boundary conditions in due time.

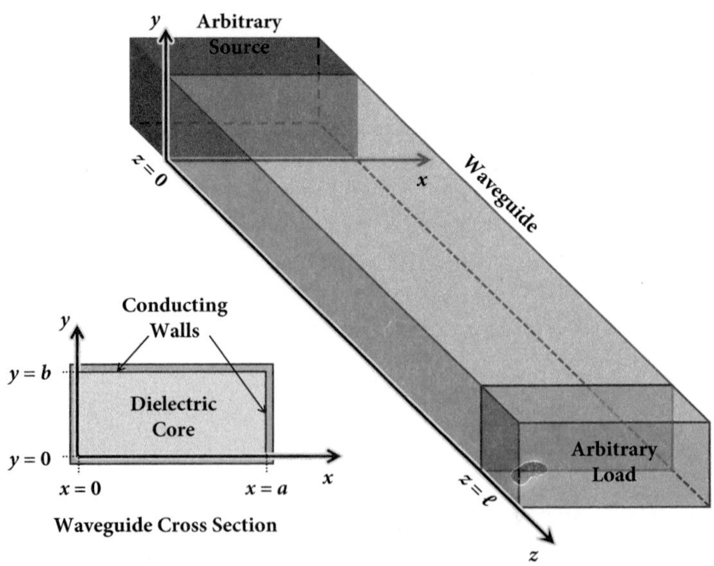

FIGURE 13.2 An illustration of a metallic rectangular waveguide section with arbitrary source and load terminations.

As we stated earlier, we will start our solution process by working with the wave equation(s). Since we have six, "basically identical," wave equations [Equations (13A.2) to (13A.4) and (13A.6) to (13A.8)], we can use the solution of any of them to predict the solution for all the others. Hence, we will initially use a generic field phasor $\overline{F}$ to represent any of the six field components in order to derive a generic solution that we can apply later to any of the six field components. So, let us start by solving the generic wave equation,

$$\nabla^2 \overline{F} = -\omega^2 \mu \varepsilon \overline{F} \tag{13A.17}$$

13A.3 Solution of the Generic Wave Equation $\nabla^2 \overline{F} = -\omega^2 \mu \varepsilon \overline{F}$

We start by defining the wave number k_{ug} for plane wave propagation in the same medium as that of the guide filling,

$$k_{ug}^2 = \omega^2 \mu \varepsilon \qquad \text{and} \qquad k_{ug} = \omega \sqrt{\mu \varepsilon} \tag{13A.18}$$

In this definition, we added the subscript "*ug*" to remind ourselves that this is the wave number of the *unguided* plane wave as we expect the guide solution to involve different values for its wave number(s). Hence, we rewrite Equation (13A.17) as

$$\nabla^2 \overline{F} = -k_{ug}^2 \overline{F} \tag{13A.19}$$

We recognize that for each of the six field components the phasor $\overline{F}$ is in general a function of position, that is a function of all three coordinates x, y, and z. Now, using the separation of variables technique for solving differential equations, we write $\overline{F}$ as the product of three independent functions, each of which depends only on one of the coordinate variables, x, y, and z, respectively. Hence,

$$\overline{F} = \overline{F}(x, y, z) = \overline{f}_x(x) \overline{f}_y(y) \overline{f}_z(z) \tag{13.20}$$

where $\overline{f}_x(x)$ is a function of x alone, $\overline{f}_y(y)$ is a function of y alone, and $\overline{f}_z(z)$ is a function of z alone. Applying Equation (13A.20) to Equation (13A.19), we write

$$\begin{aligned} \nabla^2 \overline{F} &= \frac{\partial^2 \overline{F}}{\partial x^2} + \frac{\partial^2 \overline{F}}{\partial y^2} + \frac{\partial^2 \overline{F}}{\partial z^2} = \frac{\partial^2 \overline{f}_x(x)}{\partial x^2} \overline{f}_y(y) \overline{f}_z(z) + \overline{f}_x(x) \frac{\partial^2 \overline{f}_y(y)}{\partial y^2} \overline{f}_z(z) + \overline{f}_x(x) \overline{f}_y(y) \frac{\partial^2 \overline{f}_z(z)}{\partial z^2} \\ &= -k_{ug}^2 \overline{f}_x(x) \overline{f}_y(y) \overline{f}_z(z) \end{aligned} \tag{13A.21}$$

For short, we write

$$\overline{f}_x'' \overline{f}_y \overline{f}_z + \overline{f}_x \overline{f}_y'' \overline{f}_z + \overline{f}_x \overline{f}_y \overline{f}_z'' = -k_{ug}^2 \overline{f}_x \overline{f}_y \overline{f}_z \tag{13A.22}$$

where the double prime superscript (″) is to indicate the function's second derivative with respect to its coordinate variable. Dividing both sides of the equation by $\overline{f}_x \overline{f}_y \overline{f}_z$, we arrive at

$$\frac{\overline{f}_x''}{\overline{f}_x} + \frac{\overline{f}_y''}{\overline{f}_y} + \frac{\overline{f}_z''}{\overline{f}_z} = -k_{ug}^2 \tag{13A.23}$$

The right side of Equation (13A.23) is a constant at all locations within the guide. In other words, it is not a function of either x, y, or z. On the other hand, the left side of the equation is the sum of three independent functions: $\frac{\overline{f}_x''}{\overline{f}_x}$ a function of x only, $\frac{\overline{f}_y''}{\overline{f}_y}$ a function of y only, and $\frac{\overline{f}_z''}{\overline{f}_z}$ a function of z only. Using simple logic, the sum of three independent functions can never be a constant unless each of the three independent functions is a constant in its own. Hence, we write

$$\frac{\overline{f}_x''}{\overline{f}_x} = -k_x^2 \tag{13A.24}$$

$$\frac{\bar{f}_y''}{\bar{f}_y} = -k_y^2 \tag{13A.25}$$

$$\frac{\bar{f}_z''}{\bar{f}_z} = -k_z^2 \tag{13A.26}$$

where $-k_x^2$, $-k_y^2$, and $-k_z^2$ are arbitrarily chosen constants, such as their sum equals $-k_{ug}^2$, that is,

$$k_x^2 + k_y^2 + k_z^2 = k_{ug}^2 \tag{13A.27}$$

Now, we can write the generic solution for second-order differential equations of the forms [Equations (13A.24) to (13.26)] as

$$\bar{f}_x = \bar{C}_{xa}e^{-jk_xx} + \bar{C}_{xb}e^{+jk_xx}, \quad \text{or alternatively,} \quad \bar{f}_x = \bar{C}_{xc}\cos(k_xx) + \bar{C}_{xd}\sin(k_xx) \tag{13A.28}$$

$$\bar{f}_y = \bar{C}_{ya}e^{-jk_yy} + \bar{C}_{yb}e^{+jk_yy}, \quad \text{or alternatively,} \quad \bar{f}_y = \bar{C}_{yc}\cos(k_yy) + \bar{C}_{yd}\sin(k_yy) \tag{13A.29}$$

$$\bar{f}_z = \bar{C}_{za}e^{-jk_zz} + \bar{C}_{zb}e^{+jk_zz}, \quad \text{or alternatively,} \quad \bar{f}_z = \bar{C}_{zc}\cos(k_zz) + \bar{C}_{zd}\sin(k_zz) \tag{13A.30}$$

The upper case $\bar{C}$'s are arbitrary constants expected to be determined using the waveguide boundary conditions.

Now, the expression for $\bar{F}$ according to Equation (13A.20) can be written as the product of the three solutions [Equations (13A.28) to (13A.30)], for example,

$$\bar{F} = [\bar{C}_{xc}\cos(k_xx) + \bar{C}_{xd}\sin(k_xx)][\bar{C}_{yc}\cos(k_yy) + \bar{C}_{yd}\sin(k_yy)]\left[\bar{C}_{za}e^{-jk_zz} + \bar{C}_{zb}e^{+jk_zz}\right] \tag{13A.31}$$

In this form, we chose a mix of sines/cosines and exponentials for a convenience that will be apparent later in the development. If we choose any other mix, the final result will certainly be the same but may take a few extra math manipulations to arrive at.

The derived (generic) solution is applicable to all six field components governed by the wave equations [Equations (13A.2) to (13A.4) and (13A.6) to (13A.8)].

We notice that the solution given in Equation (13A.31) is composed of two main components: a cross-sectional functional dependence and a z axis functional dependence. Each term includes arbitrary constants that we expect to resolve from the boundary conditions. The cross-sectional boundary conditions will have to do with the interface between the dielectric filling and the rectangular conducting sides (at $x = 0, a$ and $y = 0, b$). On the other hand, the z axis boundary conditions can only be determined with knowledge of the source and load connected at the guide $z = 0$ and ℓ ends.

It is logical to work with the cross-sectional boundary conditions first since the cross-sectional information is well defined at this point while the source and load are expected to vary depending on the application. In this regard, we choose to start with the z–electric field component, $\bar{E}_z$, since it has the simplest cross-sectional boundary conditions of all. As we recall from Chapter 6, Figure 6.32, all tangential electric field components must vanish on the interface with conductor surfaces.

We consider the solution for the wave equation for the electric field component, $\bar{E}_z$, Equation (13A.4), and based on the generic solution [Equation (13A.31)], we can write the solution for $\bar{E}_z$ as

$$\bar{E}_z = [\bar{C}_{x1}\cos(k_xx) + \bar{C}_{x2}\sin(k_xx)][\bar{C}_{y1}\cos(k_yy) + \bar{C}_{y2}\sin(k_yy)]\left[\bar{C}_{z1}e^{-jk_zz} + \bar{C}_{z2}e^{+jk_zz}\right] \tag{13A.32}$$

Now, we introduce the conditions at the boundaries of the solution region (the dielectric filling of the waveguide) as determined by the dielectric-conductor (guide walls) interface. Assuming perfect conductors for the walls, we can affirm that the tangential electric field at the four walls is zero (Figure 13A.1) and hence, we can write

$$\bar{E}_z\Big|_{x=0} = 0 \tag{13A.33}$$

$$\bar{E}_z\Big|_{x=a} = 0 \tag{13A.34}$$

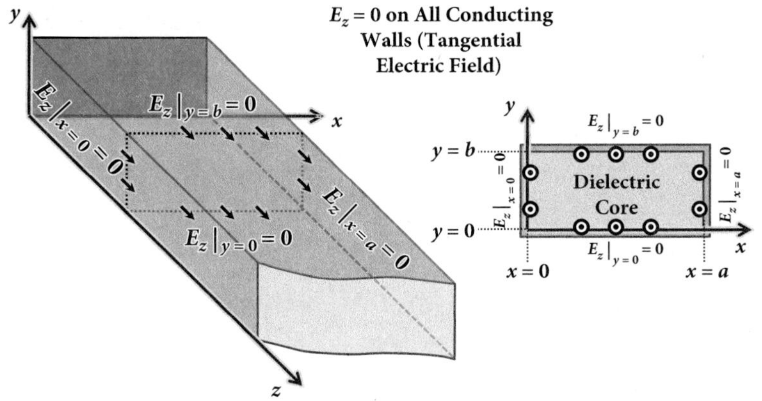

Figure 13A.1 Boundary conditions for the field solution in the guide's dielectric filling region: the electric field is zero at the dielectric-conductor interfaces.

$$\bar{E}_z\Big|_{y=0} = 0 \tag{13A.35}$$

$$\bar{E}_z\Big|_{y=b} = 0 \tag{13A.36}$$

By using Equations (13A.33) and (13A.35), we can conclude that the two constants $\bar{C}_{x1}$ and $\bar{C}_{y1}$ must vanish as the cosine solutions do not vanish for zero arguments. Equation (13A.32) now reduces to

$$\bar{E}_z = [\bar{C}_{x2}\sin(k_x x)][\bar{C}_{y2}\sin(k_y y)]\left[\bar{C}_{z1}e^{-jk_z z} + \bar{C}_{z2}e^{+jk_z z}\right] = [\sin(k_x x)\sin(k_y y)][\bar{C}_{x2}\bar{C}_{y2}]\left[\bar{C}_{z1}e^{-jk_z z} + \bar{C}_{z2}e^{+jk_z z}\right] \tag{13A.37}$$

or

$$\bar{E}_z = [\sin(k_x x)\sin(k_y y)]\left[\bar{E}_o^{+}e^{-jk_z z} + \bar{E}_o^{-}e^{+jk_z z}\right] \tag{13A.38}$$

where $\bar{E}_o^+ = \bar{C}_{x2}\bar{C}_{y2}\bar{C}_{z1}$ and $\bar{E}_o^- = \bar{C}_{x2}\bar{C}_{y2}\bar{C}_{z2}$ are both constants and have the units of electric fields. Since they are the coefficients of the functional variations in the z coordinate, they require knowledge of the z-parameter boundary conditions. In other words, it will be the boundary conditions at the guide terminals (source end and load end) that will determine these two constants. The notations of the superscripts, "+" and "–", are chosen for the same reason that was described earlier in the transmission line analysis in Chapter 2 and we explain that again later in this chapter.

Next, we use the remaining two boundary conditions [Equations (13A.33) and (13A.36)]. Applying these two equations to Equation (13A.38), we conclude that

$$[\sin(k_x x)\sin(k_y y)]_{x=a} = 0 \quad \text{and} \quad [\sin(k_x x)\sin(k_y y)]_{y=b} = 0 \tag{13A.39}$$

or

$$\sin(k_x a) = 0 \quad \text{and} \quad \sin(k_y b) = 0 \tag{13A.40}$$

Hence,

$$k_x = \frac{m\pi}{a} \quad \text{and} \quad k_y = \frac{n\pi}{b} \tag{13A.41}$$

where m and n are arbitrary integers. Equation (13A.38), then, takes the form:

$$\bar{E}_{z\,m,n}(x,y,z) = \left[\sin\left(\frac{m\pi}{a}x\right)\sin\left(\frac{n\pi}{b}y\right)\right]\left[\bar{E}^{+}_{o\,m,n}e^{-jk_{z\,m,n}z} + \bar{E}^{-}_{o\,m,n}e^{+jk_{z\,m,n}z}\right] \tag{13A.42}$$

We have added the subscript "m,n" to the field solution because there are likely more than one m and n combinations that can produce a solution meeting the boundary conditions used so far. We can justify adding the same subscripts also to $\overline{E}_o^+$ and $\overline{E}_o^-$ as they are indicators of the intensities of the particular solutions for the different m and n combinations. We have added the same subscripts to k_z as well since, according to Equation (13A.27), k_z depends on k_x and k_y and hence m and n,

$$k_{z\,m,n}^2 = k_{ug}^2 - \left[k_x^2 + k_y^2\right] = \omega^2\mu\varepsilon - \left[\left(\frac{m\pi}{a}\right)^2 + \left(\frac{n\pi}{b}\right)^2\right] = \omega^2\mu\varepsilon - k_{c\,m,n}^2 \tag{13A.43}$$

In this expression, we used $k_{c\,m,n}$ to denote the cross-sectional wave number for the combination "m,n," that is,

$$k_{c\,m,n}^2 = \left[\left(\frac{m\pi}{a}\right)^2 + \left(\frac{n\pi}{b}\right)^2\right] \tag{13A.44}$$

Later on, we will have an important physical implication of the use of the subscript c in addition to denoting "cross-sectional."

In the following, we present different forms for Equation (13A.43) that we can use depending on which one is of interest in our analysis.

$$k_{c\,m,n} = \sqrt{\left(\frac{m\pi}{a}\right)^2 + \left(\frac{n\pi}{b}\right)^2} \tag{13A.45}$$

$$k_{z\,m,n}^2 = \omega^2\mu\varepsilon - k_{c\,m,n}^2 = \mu\varepsilon\left(\omega^2 - \frac{k_{c\,m,n}^2}{\mu\varepsilon}\right) = \mu\varepsilon(\omega^2 - \omega_{c\,m,n}^2) \tag{13A.46}$$

where we defined $\omega_{c\,m,n}^2$ to replace $\frac{k_{c\,m,n}^2}{\mu\varepsilon}$; hence

$$k_{c\,m,n}^2 = \omega^2\mu\varepsilon - k_{z\,m,n}^2 = \mu\varepsilon\omega_{c\,m,n}^2 \tag{13A.47}$$

and

$$\omega_{c\,m,n} = \frac{k_{c\,m,n}}{\sqrt{\mu\varepsilon}} = \frac{1}{\sqrt{\mu\varepsilon}}\sqrt{\left(\frac{m\pi}{a}\right)^2 + \left(\frac{n\pi}{b}\right)^2} \tag{13A.48}$$

thus, allowing us to express $k_{z\,m,n}$ as

$$k_{z\,m,n} = \sqrt{\mu\varepsilon\left(\omega^2 - \omega_{c\,m,n}^2\right)} = \omega\sqrt{\mu\varepsilon}\sqrt{1 - \left(\frac{\omega_{c\,m,n}}{\omega}\right)^2} \tag{13A.49}$$

ADDENDUM 13B
PHASE AND GROUP VELOCITIES

13B.1 Phase Velocity

As presented earlier, the expression for the phase velocity was given in Equation (13.37),

$$c_{ph\,g\,m,n} = c_{ph\,ug} \Big/ \sqrt{1 - \left(\frac{f_{c\,m,n}}{f}\right)^2} \tag{13.37}$$

This expression indicates that the phase velocity inside the guide is larger than its unguided counterpart, implying that the guided phase velocity is faster than the velocity of light in the same medium. The phase velocity being larger than the velocity of light is not in conflict with theory of relativity as there is no information carried with this speed. Signal information in a waveguide is carried along with another velocity, known as *group veocity*, which is smaller than the speed of light in the dielectric region. In the following, we will introduce the concept of the group velocity and derive its analytical expression in the waveguide. Later on, we will present an analogy in an attempt to bring some physical meaning to the phase velocity and how it can possibly exceed the velocity of light.

13B.2 Group Velocity

Unlike the phase velocity, the group velocity is that of information travel, that is, the velocity of signal communication through the transmission medium. The name group comes from the fact that an information signal is composed of a group of harmonics. While the phase velocity is that of the phase of a single harmonic, the group velocity relates to having more than one harmonic.

The simplest case to present in this regard is that of an amplitude modulated signal of a carrier frequency ω_o modulated by an information signal with an angular frequency $\Delta\omega$ (Figure 13B.1).

The result of this amplitude modulation is the creation of two harmonic "sidebands" of angular frequencies $\omega_o - \Delta\omega$ and $\omega_o + \Delta\omega$. Let us denote the corresponding phase constants of these two harmonics by $\beta_o - \Delta\beta$ and $\beta_o + \Delta\beta$.

As a result, the electric field wave expressions should be the superposition of the two sideband fields, hence the form:

$$\begin{aligned} E &= E_o \cos[(\omega_o - \Delta\omega)t - (\beta_o - \Delta\beta)z] + E_o \cos[(\omega_o + \Delta\omega)t - (\beta_o + \Delta\beta)z] \\ &= 2E_o \cos[\Delta\omega t - \Delta\beta z]\cos[\omega_o t - \beta_o z] \end{aligned} \tag{13B.1}$$

We see in this expression a product of two space-time dependent functions. One of them is the carrier harmonic, $\cos[\omega_o t - \beta_o z]$, bounded by the amplitude-varying envelope $2E_o\cos[\Delta\omega t - \Delta\beta z]$. It is clear that the transmitted information is contained in the envelope. This product is depicted in Figure 13B.2.

Examining the time-distance dependence of the envelope (information signal), we can write the envelope (group/signal) propagation velocity as

$$c_{gr} = \frac{\Delta\omega}{\Delta\beta} \text{ which would become } c_{gr} = \frac{d\omega}{d\beta} \text{ in the limit as } \Delta\omega \Rightarrow 0.$$

Now, using Equation (13.38) for the mode m,n in a rectangular waveguide, we can carry out the derivative $\frac{d\omega}{d\beta}$ to reach the following expression for the group velocity in the guide:

$$c_{gr\,g\,m,n} = \frac{d\omega}{d\beta_{g\,m,n}} = \frac{\beta_{g\,m,n}}{\omega} c_{ph\,ug}^2 = \frac{1}{c_{ph\,g\,m,n}} c_{ph\,ug}^2 = c_{ph\,ug}\sqrt{1 - \left(\frac{f_{c\,m,n}}{f}\right)^2} \tag{13B.2}$$

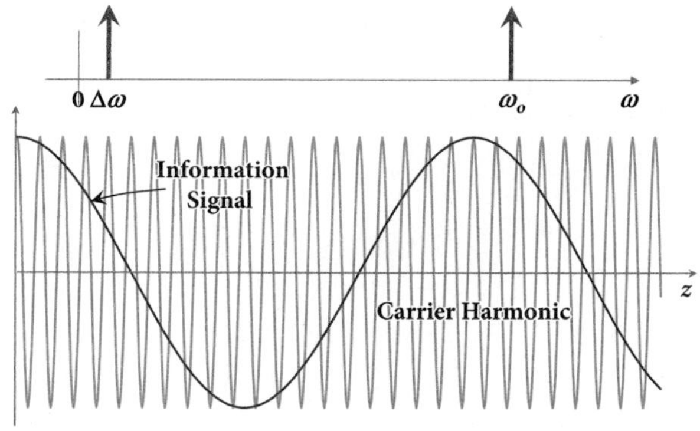

Figure 13B.1 Information signal and carrier harmonic for an amplitude-modulated signal used in the analysis of the group velocity for the waveguide.

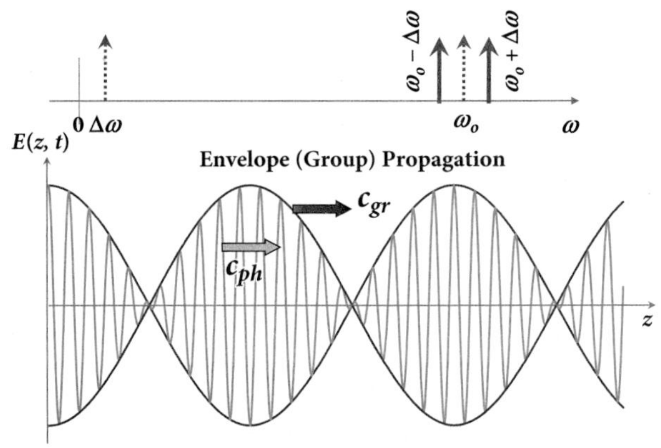

FIGURE 13B.2 A demonstration of phase and group velocities using an amplitude-modulated signal traveling in the waveguide.

We can conclude two observations from this final expression [Equation (13B.2)]:

1. The group velocity in a waveguide (information propagation velocity) smaller than the velocity of light in the same medium.
2. The product $(c_{gr\,g\,m,n}) \cdot (c_{ph\,g\,m,n}) = c_{ph\,ug}^2 = c_o^2$
3. For plane wave propagation (no cutoff), $c_{gr\,ug} = c_{ph\,ug} = c_o$

13B.2.1 Alternate Definition and Derivation of the Group Velocity

The group velocity (in the case of the waveguide the signal or energy transport velocity) is defined as follows:

$$c_{gr} = \frac{\text{Time-averaged total power}}{\text{Time-averaged total energy per unit length}} = \frac{\langle P_a \rangle_{cs}}{\langle W_a \rangle_v} \tag{13B.3}$$

This definition expresses the velocity corresponding to the crossing/flow of the energy contained in a unit length section of the waveguide through its edge cross-sectional surface (Figure 13B.3).

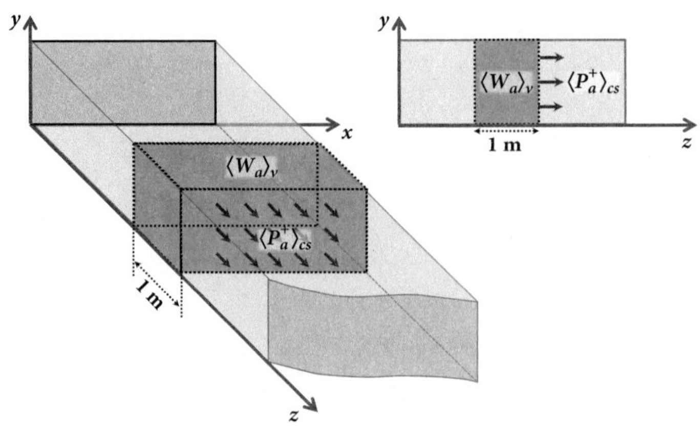

FIGURE 13B.3 An alternate definition of the group velocity in a waveguide as the rate of flow of the energy contained in a section of the waveguide per unit length.

Referring to Equation (11.112), the complex Poynting vector for plane waves in a lossless medium is given by

$$\mathcal{P}^{+}_{\text{avg}} = \frac{1}{2\eta}\left|\bar{E}^{+}_{x}\right|^{2} \tag{11.112}$$

Guided by Equation (11.112), we can write the following expression for the time-averaged total power for the forward traveling *TE* modes in the waveguide as

$$\left\langle P^{+}_{a}\right\rangle_{cs} = \iint_{cs} \mathcal{P}^{+}_{\text{avg}} dx\, dy = \frac{1}{2}\frac{1}{\eta_{TE\, g\, mn}} \iint_{cs} \left|\bar{E}^{+}_{x}\right|^{2} dx\, dy \tag{13B.4}$$

where the surface integration is performed over the waveguide's cross section (*cs*).

We can also write the time-averaged total energy for the volume of a waveguide section of unit length as

$$\left\langle W^{+}_{a}\right\rangle_{v} = \iiint_{v} \frac{1}{2}\varepsilon\left|\bar{E}^{+}_{x}\right|^{2} dx\, dy\, dz = \frac{1}{2}\varepsilon\int_{z}^{(z+1)} dz \iint_{cs}\left|\bar{E}^{+}_{x}\right|^{2} dx\, dy = \frac{1}{2}\varepsilon\iint_{cs}\left|\bar{E}^{+}_{x}\right|^{2} dx\, dy \tag{13B.5}$$

Therefore, the ratio between the two Equations (13B.4) and (13B.5) yields:

$$c_{gr\, g\, mn} = \frac{\left\langle P^{+}_{a}\right\rangle_{cs}}{\left\langle W^{+}_{a}\right\rangle_{v}} = \frac{1}{\varepsilon\eta_{TE\, g\, mn}} = \frac{1}{\varepsilon}\frac{1}{\dfrac{\omega\mu}{\beta_{g,mn}}} = \frac{1}{\varepsilon\mu}\frac{\beta_{g,mn}}{\omega} = c^{2}_{ph\, ug}\frac{1}{c_{ph\, g\, mn}} \tag{13B.6}$$

13B.2.2 Physical Insight: Can the Phase Velocity Exceed the Velocity of Light?

To appreciate the concept of the phase velocity and how it can possibly exceed the velocity of light, let us consider the following analogy. Suppose we have two particles A and B that are allowed to move up and down (vertically), while maintaining a fixed horizontal spacing d (Figure 13B.4). Let us further assume that their vertical motion is as described in the right-hand side diagram in the figure, that is, either particle moves from its top most height "$+h$" to its lowest "$-h$" in a time interval "τ" in a constant velocity motion. The particles' vertical velocities are then given by $|c_v| = \dfrac{2h}{\tau}$; this velocity cannot be faster than the velocity of light.

Meanwhile, we can argue that the peak of the *particles* moved from the location a to the location b in time interval τ. Hence, we can describe a "peak motion" velocity of $c_{pm} = \dfrac{d}{\tau}$. This velocity can become greater than the velocity of light, c_o, since d could very well be any value including values greater than $c_o\tau$.

This peak-motion velocity is in fact a phase velocity as it describes the velocity at which the peak's phase moved from location a to location b.

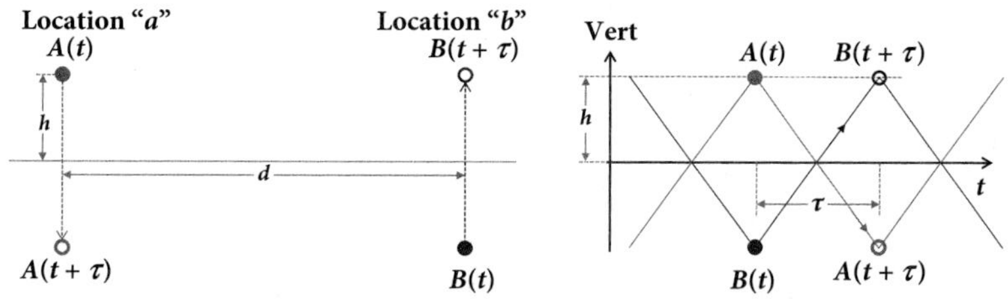

FIGURE 13B.4 An illustration associated with the discussion of Section 13B.2.2.

ADDENDUM 13C
WAVE EQUATION SOLUTION FOR METALLIC RECTANGULAR WAVEGUIDES: CONTINUATION FOR ALL FIELD COMPONENTS

13C.1 Continuation of the Field Analysis

We now go back to continue our field solution for wave propagation in a rectangular waveguide. In this regard, we need to remind ourselves of the wave Equations (13A.2) to (13A.4) and (13A.6) to (13A.8), where we choose to start the solution by working with the generic field phasor of Equation (13A.17). All six equations have identical forms, and hence, the general form solution that we derived for $\bar{F}$ in Equation (13A.31) would apply to all six field components. By applying this solution to $\bar{E}_z$, we were able to identify the different guide cutoff and propagation modes.

Now, let us extend this solution to the $\bar{H}_z$ field component and apply the relevant boundary conditions. Using Equation (13A.31), we can write the general solution for $\bar{H}_z$ as

$$\bar{H}_z = [\bar{C}_{x3}\cos(k_x x)+\bar{C}_{x4}\sin(k_x x)][\bar{C}_{y3}\cos(k_y y)+\bar{C}_{y4}\sin(k_y y)]\left[\bar{C}_{z3}e^{-jk_z z}+\bar{C}_{z4}e^{+jk_z z}\right] \tag{13C.1}$$

The corresponding boundary conditions for this magnetic field are based on the fact that the perfect conducting guide walls can only support (tangential) surface currents. These currents can only result from tangential magnetic fields and hence no normal magnetic fields can exist on the conductor surface. In addition, these conducting walls do not sustain any tangential electric fields. Consequently, we can write for the $x = 0$ wall,

$$\bar{H}_x\Big|_{x=0} = 0 \quad \text{and} \quad \bar{E}_y\Big|_{x=0} = 0 \tag{13C.2}$$

and using Equation (13A.15), we can conclude that

$$\frac{\partial \bar{H}_x}{\partial z} - \frac{\partial \bar{H}_z}{\partial x} = 0 - \frac{\partial \bar{H}_z}{\partial x} = j\omega\varepsilon \bar{E}_y = 0 \tag{13C.3}$$

thus, arriving at the following boundary condition regarding $\bar{H}_z$ at the $y = 0$ wall:

$$\frac{\partial \bar{H}_z}{\partial x}\Big|_{y=0} = 0 \tag{13C.4}$$

Similar expressions are obtained for the $y = b$ and $y = 0$ walls:

$$\frac{\partial \bar{H}_z}{\partial x}\Big|_{x=a} = 0, \qquad \frac{\partial \bar{H}_z}{\partial y}\Big|_{y=0} = 0, \qquad \text{and} \qquad \frac{\partial \bar{H}_z}{\partial y}\Big|_{y=b} = 0 \tag{13C.5}$$

Applying these conditions to the general expression for $\bar{H}_z$, Equation (13C.1), we arrive at expressions similar to those obtained earlier for $\bar{E}_z$, Equations (13A.42), except for having cosine distributions instead of the sine ones,

$$\bar{H}_{z\,m,n} = \left[\cos\left(\frac{m\pi}{a}x\right)\cos\left(\frac{n\pi}{b}y\right)\right]\left[\bar{H}^{+}_{o\,m,n}e^{-jk_{z\,m,n}z} + \bar{H}^{-}_{o\,m,n}e^{+jk_{z\,m,n}z}\right] \tag{13C.6}$$

We notice here that the boundary conditions for the magnetic field component, $\bar{H}_{z\,m,n}$, yielded the same k_x, k_y, and k_z constants (wave numbers) as those derived earlier for $\bar{E}_{z\,m,n}$. As we will see later, the remaining four field components will have the same "wave number" constants as well.

Having expressions for both $\bar{E}_z$ and $\bar{H}_z$, we are now ready to derive the expressions for the remainder four field components. In this regard, we will only consider the solution for the active (propagation) modes.

Furthermore, to simplify our math development by working only with the positive z-propagation term. We can always reconstruct the full (two-way propagation) solution by adding the negative z-propagation term later. First, we write the modified expressions for $\bar{E}_z$ and $\bar{H}_z$ as

$$\bar{E}^+_{z\ m,n} = \left[\sin\left(\frac{m\pi}{a}x\right)\sin\left(\frac{n\pi}{b}y\right)\right]\left[\bar{E}^+_{o\ m,n}e^{-j\beta_{g\ m,n}z}\right] \tag{13C.7}$$

$$\bar{H}^+_{z\ m,n} = \left[\cos\left(\frac{m\pi}{a}x\right)\cos\left(\frac{n\pi}{b}y\right)\right]\left[\bar{H}^+_{o\ m,n}e^{-j\beta_{g\ m,n}z}\right] \tag{13C.8}$$

and we can predict the other four field components to be of the forms

$$\bar{E}^+_{x\ m,n} = \left[\text{Mix of sines and cosines of }\left(\frac{m\pi}{a}x\right)\text{ and }\left(\frac{n\pi}{b}y\right)\right]\left[\bar{E}^+_{xo\ m,n}e^{-j\beta_{g\ m,n}z}\right] \tag{13C.9}$$

$$\bar{E}^+_{y\ m,n} = \left[\text{Mix of sines and cosines of }\left(\frac{m\pi}{a}x\right)\text{ and }\left(\frac{n\pi}{b}y\right)\right]\left[\bar{E}^+_{yo\ m,n}e^{-j\beta_{g\ m,n}z}\right] \tag{13C.10}$$

$$\bar{H}^+_{x\ m,n} = \left[\text{Mix of sines and cosines of }\left(\frac{m\pi}{a}x\right)\text{ and }\left(\frac{n\pi}{b}y\right)\right]\left[\bar{H}^+_{xo\ m,n}e^{-j\beta_{g\ m,n}z}\right] \tag{13C.11}$$

$$\bar{H}^+_{y\ m,n} = \left[\text{Mix of sines and cosines of }\left(\frac{m\pi}{a}x\right)\text{ and }\left(\frac{n\pi}{b}y\right)\right]\left[\bar{H}^+_{yo\ m,n}e^{-j\beta_{g\ m,n}z}\right] \tag{13C.12}$$

We have two choices on how to proceed next. One approach is to apply the boundary conditions for each on the four field components, and the other is to use Maxwell's equations to relate them all to the already defined $\bar{E}^+_{z\ m,n}$ and $\bar{H}^+_{z\ m,n}$. The fact is that the boundary condition approach is time consuming and will leave us with unrelated arbitrary parameters for each obtained solution. Therefore, we will proceed with the Maxwell's equations approach. In this regard, we will need to manipulate Equations (13A.10) to (13A.12) and (13A.14) to (13A.16) to serve our purpose by recognizing that all the $\partial/\partial z$ derivatives can be replaced by $-j\beta_{g\ m,n}$ as it is clear from the Equations (13C.7) to (13A.12). The resulting expressions are given as follows:

$$\frac{\partial \bar{E}_{z\ m,n}}{\partial y} + j\beta_{g\ m,n}\bar{E}_{y\ m,n} = -j\omega\mu\bar{H}_{x\ m,n} \tag{13C.13}$$

$$-j\beta_{g\ m,n}\bar{E}_{x\ m,n} - \frac{\partial \bar{E}_{z\ m,n}}{\partial x} = -j\omega\mu\bar{H}_{y\ m,n} \tag{13C.14}$$

$$\frac{\partial \bar{E}_{y\ m,n}}{\partial x} - \frac{\partial \bar{E}_{x\ m,n}}{\partial y} = -j\omega\mu\bar{H}_{z\ m,n} \tag{13C.15}$$

$$\frac{\partial \bar{H}_{z\ m,n}}{\partial y} + j\beta_{g\ m,n}\bar{H}_{y\ m,n} = j\omega\varepsilon\bar{E}_{x\ m,n} \tag{13C.16}$$

$$-j\beta_{g\ m,n}\bar{H}_{x\ m,n} - \frac{\partial \bar{H}_{z\ m,n}}{\partial x} = j\omega\varepsilon\bar{E}_{y\ m,n} \tag{13C.17}$$

$$\frac{\partial \bar{H}_{y\ m,n}}{\partial x} - \frac{\partial \bar{H}_{x\ m,n}}{\partial y} = j\omega\varepsilon\bar{E}_{z\ m,n} \tag{13C.18}$$

If we multiply Equation (13C.16) by "$j\omega\mu$" and Equation (13C.14) by "$-j\beta_{g\,m,n}$" and add the two products, we obtain

$$j\omega\mu\frac{\partial\bar{H}_{z\,m,n}}{\partial y}-\omega\mu\beta_{g\,m,n}\bar{H}_{y\,m,n}-\beta_{g\,m,n}^{2}\bar{E}_{x\,m,n}+j\beta_{g\,m,n}\frac{\partial\bar{E}_{z\,m,n}}{\partial x}=-\omega^{2}\mu\varepsilon\bar{E}_{x\,m,n}-\omega\mu\beta_{g\,m,n}\bar{H}_{y\,m,n} \tag{13C.19}$$

$$j\omega\mu\frac{\partial\bar{H}_{z\,m,n}}{\partial y}+j\beta_{g\,m,n}\frac{\partial\bar{E}_{z\,m,n}}{\partial x}=-\left(\omega^{2}\mu\varepsilon-\beta_{g\,m,n}^{2}\right)\bar{E}_{x\,m,n}=-k_{c\,m,n}^{2}\bar{E}_{x\,m,n} \tag{13C.20}$$

$$\bar{E}_{x\,m,n}=\frac{-j}{k_{c\,m,n}^{2}}\left(\beta_{g\,m,n}\frac{\partial\bar{E}_{z\,m,n}}{\partial x}+\omega\mu\frac{\partial\bar{H}_{z\,m,n}}{\partial y}\right) \tag{13C.21}$$

Similarly, we can derive the other three field components:

$$\bar{E}_{y\,m,n}=\frac{j}{k_{c\,m,n}^{2}}\left(-\beta_{g\,m,n}\frac{\partial\bar{E}_{z\,m,n}}{\partial y}+\omega\mu\frac{\partial\bar{H}_{z\,m,n}}{\partial x}\right) \tag{13C.22}$$

$$\bar{H}_{x\,m,n}=\frac{j}{k_{c\,m,n}^{2}}\left(\omega\varepsilon\frac{\partial\bar{E}_{z\,m,n}}{\partial y}-\beta_{g\,m,n}\frac{\partial\bar{H}_{z\,m,n}}{\partial x}\right) \tag{13C.23}$$

$$\bar{H}_{y\,m,n}=\frac{-j}{k_{c\,m,n}^{2}}\left(\omega\varepsilon\frac{\partial\bar{E}_{z\,m,n}}{\partial x}+\beta_{g\,m,n}\frac{\partial\bar{H}_{z\,m,n}}{\partial y}\right) \tag{13C.24}$$

Substitution of Equations (13C.7) and (13C.8) in Equations (13C.19) to (13C.22) enables us to write complete expressions for the positive z wave of the four field components. The results are as follows:

$$\bar{E}_{x\,m,n}^{+}(x,y,z)=\frac{-j\pi}{k_{c\,m,n}^{2}}\left\{\frac{m\beta_{g\,m,n}}{a}\bar{E}_{o\,m,n}^{+}-\frac{n\omega\mu}{b}\bar{H}_{o\,m,n}^{+}\right\}\left[\cos\left(\frac{m\pi}{a}x\right)\sin\left(\frac{n\pi}{b}y\right)e^{-j\beta_{g\,m,n}z}\right] \tag{13C.25}$$

$$\bar{E}_{y\,m,n}^{+}(x,y,z)=\frac{-j\pi}{k_{c\,m,n}^{2}}\left\{\frac{n\beta_{g\,m,n}}{b}\bar{E}_{o\,m,n}^{+}+\frac{m\omega\mu}{a}\bar{H}_{o\,m,n}^{+}\right\}\left[\sin\left(\frac{m\pi}{a}x\right)\cos\left(\frac{n\pi}{b}y\right)e^{-j\beta_{g\,m,n}z}\right] \tag{13C.26}$$

$$\bar{H}_{x\,m,n}^{+}(x,y,z)=\frac{j\pi}{k_{c\,m,n}^{2}}\left\{\frac{n\omega\varepsilon}{b}\bar{E}_{o\,m,n}^{+}+\frac{m\beta_{g\,m,n}}{a}\bar{H}_{o\,m,n}^{+}\right\}\left[\sin\left(\frac{m\pi}{a}x\right)\cos\left(\frac{n\pi}{b}y\right)e^{-j\beta_{g\,m,n}z}\right] \tag{13C.27}$$

$$\bar{H}_{y\,m,n}^{+}(x,y,z)=\frac{-j\pi}{k_{c\,m,n}^{2}}\left\{\frac{m\omega\varepsilon}{a}\bar{E}_{o\,m,n}^{+}-\frac{n\beta_{g\,m,n}}{b}\bar{H}_{o\,m,n}^{+}\right\}\left[\cos\left(\frac{m\pi}{a}x\right)\sin\left(\frac{n\pi}{b}y\right)e^{-j\beta_{g\,m,n}z}\right] \tag{13C.28}$$

ADDENDUM 13D FIELD MAPS FOR THE TE_{10} AND TM_{11} MODES

13D.1 Field Maps for the TE_{10} Mode

The following diagrams are self-explanatory. In each, we graph arrows representing the different field components. We graph the arrow directions as dictated by the field distribution according to the solutions obtained earlier. The arrow density/size is to reflect the field intensity at the different locations. Three views are given in each graph, the cross-sectional view, the side view, and top view of the guide (Figure 13D.1).

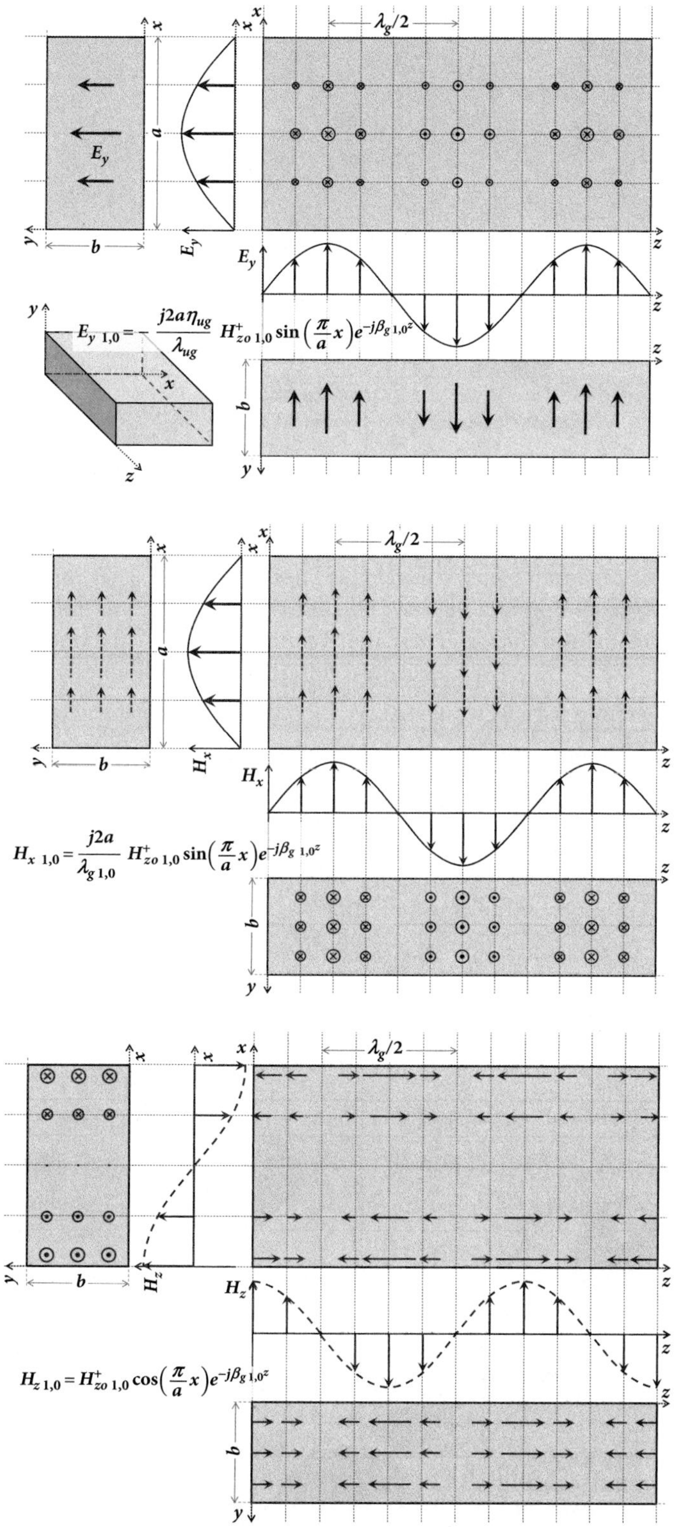

FIGURE 13D.1 Steps for the construction of the fields map for the TE_{10} mode.

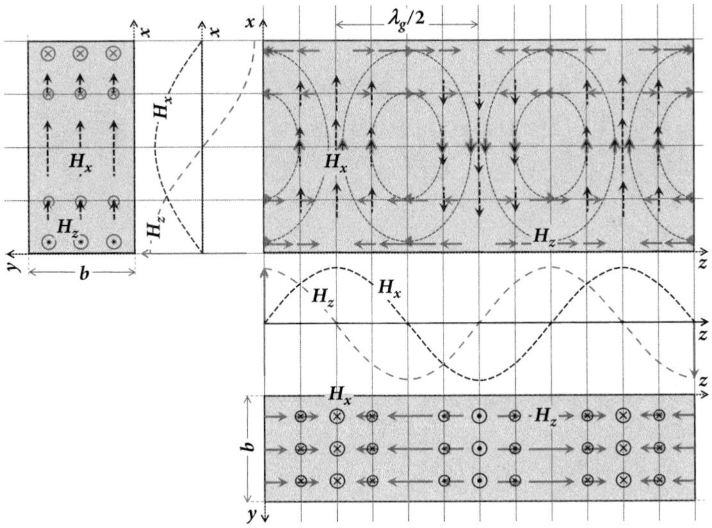

FIGURE 13D.2 The combined E and H fields map for the TE_{10} mode.

In Figure 13D.2, we combine the two components of the magnetic field recognizing that magnetic field lines form closed loops.

In the next two figures, we give a map of the power flow density vector (the Poynting vector). This vector is determined by the cross product of the electric field vector with that of the magnetic field,

$$\vec{\vec{\mathcal{P}}} = \vec{\vec{E}} \times \vec{\vec{H}}$$

Hence, in Figure 13D.3, at the different location, a short bold arrow is placed corresponding to the cross product at that location. It is important to recognize that we need to cross the total electric field vector by the total magnetic field vector at each location.

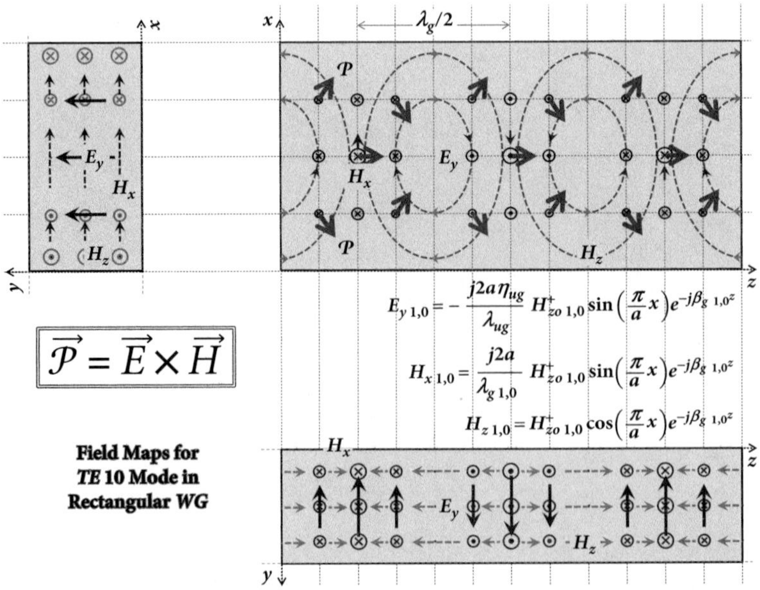

FIGURE 13D.3 The map of the Poynting vector overlaid on the combined E and H fields map for the TE_{10} mode.

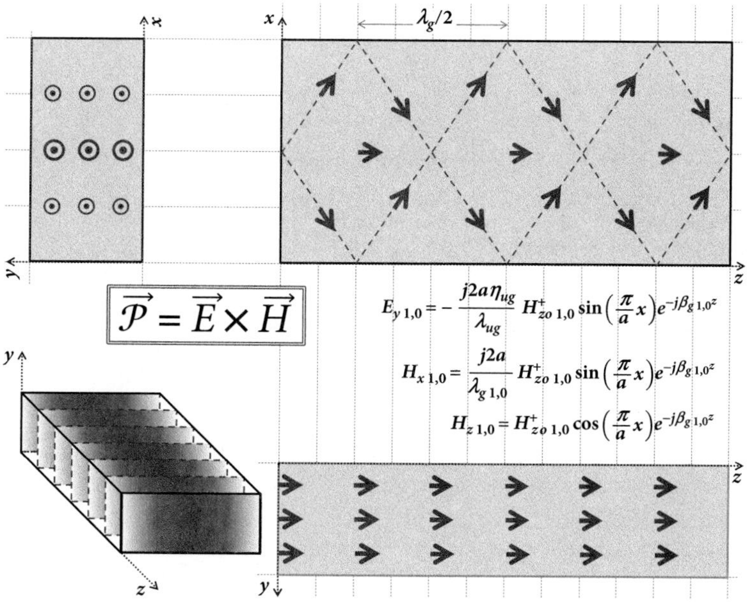

FIGURE 13D.4 The map of the Poynting vector for the TE_{10} mode showing a net power flow along the z axis of the guide.

In Figure 13D.4, we connect the short bold power flow-density arrows to trace the power flow through the guide. We can see, as a result, the net power flow along the guide axis. We can also see the bouncing waves off the guide sides corresponding to the standing waves along the z axis.

13D.2 Field Maps for the TM_{11} Mode

Similarly, in the following diagrams we graph arrows representing the different field components for the TM_{11} mode. Again, we graph the arrow directions as dictated by the field distribution according to the solutions obtained earlier. The arrow density/size is to reflect the field intensity at the different locations. Three views are given in each graph, the cross-sectional view, the side view and top view of the guide (Figures 13D.5 and 13D.6).

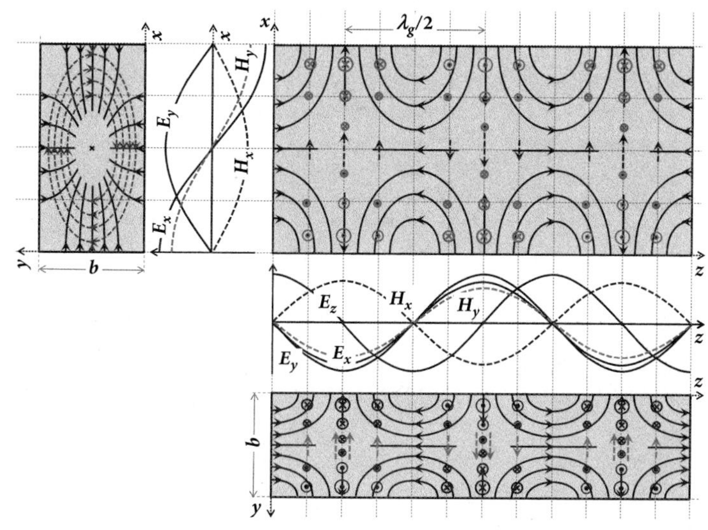

FIGURE 13D.5 The combined E and H fields map for the TM_{11} mode.

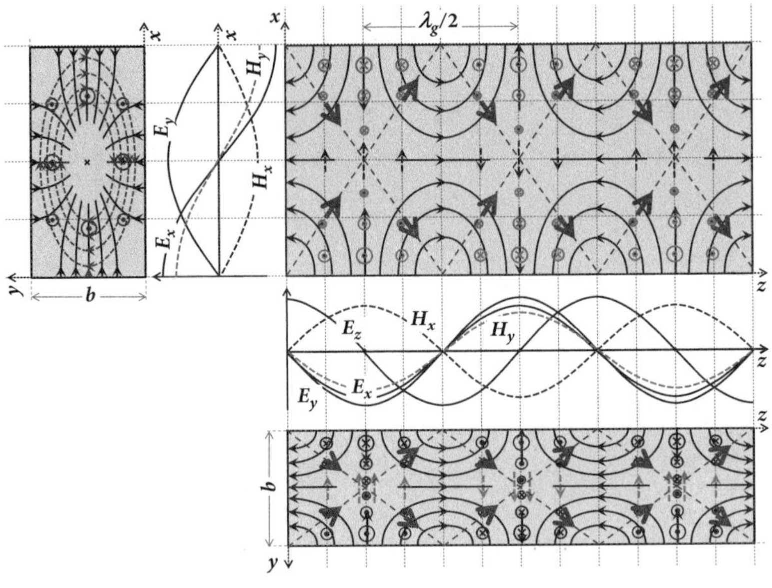

FIGURE 13D.6 The map of the Poynting vector overlaid on the combined E and H fields map for the TM_{11} mode.

ADDENDUM 13E
ACTIVE AND DOMINANT MODE IDENTIFICATION

In the following, we will examine how to determine the active (operating) modes for a given waveguide at a given frequency. The process starts with figuring out the cutoff frequencies of the guide as determined by the guide dimensions and dielectric filling. Once the cutoff frequencies of the different m and n are determined, then the active modes are those with cutoff frequencies lower than the guide's operating frequency.

To make sure that we do this process in an orderly manner, we present two approaches in the following to facilitate the calculations; a tabulation approach and a graphical one.

13E.1 The Tabulation Approach for Mode Identification

In this approach, we generate a table for the different m and n combinations and, for each, we evaluate the corresponding cutoff frequency as well as examine both the TE and TM field components. In the following, we show an example of such table, Table 13E.1, where we use m and n combinations from zero to two and list the cutoff frequencies as normalized to C_{ug}. Using Equation (13.33), we write

$$f_{cn\ m,n} = \frac{f_{c\ m,n}}{\left[\dfrac{c_o}{\sqrt{\mu_r \varepsilon_r}}\right]} = \frac{f_{c\ m,n}}{c_{ug}} = \sqrt{\left[\left(\frac{m}{2a}\right)^2 + \left(\frac{n}{2b}\right)^2\right]} \tag{13E.1}$$

In this table, we notice that the guide cannot support TM modes for zero values of either m or n. This was discussed earlier in the light of Equations (13A.42) to (13A.45) where all field components reduce to zero if both H_z and H_z are zeros.

The next step is to sort the table rows in an ascending order with respect to the cutoff frequencies. The resulting sorted order is likely to be different from the order used for m and n. We can see from the normalized cutoff frequency column that the sorted order depends on the guide's a and b dimensions.

m	n	$f_{cn} = f_c / c_{ug}$	$TE\ H_z$	$TE\ E_x, H_y$	$TE\ E_y, H_x$	$TM\ E_z$	$TM\ E_x, H_y$	$TM\ E_y, H_x$	Conclusion
0	0	0		0	0		0	0	*TEM* not possible
1	0	$(1/2a)$	(cos)(cos)	0	(sin)(cos)	0	0	0	Only *TE* possible
2	0	$(1/a)$	(cos)(cos)	0	(sin)(cos)	0	0	0	Only *TE* possible
0	1	$(1/2b)$	(cos)(cos)	(cos)(sin)	0	0	0	0	Only *TE* possible
1	1	$\sqrt{\left[\left(\frac{1}{2a}\right)^2+\left(\frac{1}{2b}\right)^2\right]}$	(cos)(cos)	(cos)(sin)	(sin)(cos)	(sin)(sin)	(cos)(sin)	(sin)(cos)	*TE* and *TM* possible
2	1	$\sqrt{\left[\left(\frac{1}{a}\right)^2+\left(\frac{1}{2b}\right)^2\right]}$	(cos)(cos)	(cos)(sin)	(sin)(cos)	(sin)(sin)	(cos)(sin)	(sin)(cos)	*TE* and *TM* possible
0	2	$(1/b)$	(cos)(cos)	(cos)(sin)	0	0	0	0	Only *TE* possible
1	2	$\sqrt{\left[\left(\frac{1}{2a}\right)^2+\left(\frac{1}{b}\right)^2\right]}$	(cos)(cos)	(cos)(sin)	(sin)(cos)	(sin)(sin)	(cos)(sin)	(sin)(cos)	*TE* and *TM* possible
2	2	$\sqrt{\left[\left(\frac{1}{a}\right)^2+\left(\frac{1}{b}\right)^2\right]}$	(cos)(cos)	(cos)(sin)	(sin)(cos)	(sin)(sin)	(cos)(sin)	(sin)(cos)	*TE* and *TM* possible

TABLE 13E.1 Tabulation of the Different m and n Combinations and the Corresponding Waveguide Modes

It is typical to use a for the broader side of the guide (as we used in all our figures), and, hence, we can examine the table terms with the condition of $a > b$. Accordingly, we can see that the TE_{10} mode will have the lowest cutoff frequency of all possible modes. We can also state undoubtedly that the TM_{11} mode will have the lowest cutoff frequency of all *TM* modes.

The lowest cutoff frequency modes are called the dominant modes and, hence, we refer to the TE_{10} mode as the dominant mode of both *TE* and *TM* modes, for $a > b$. We also refer to the TM_{11} mode as the dominant *TM* mode.

It is also typical to design rectangular waveguides with of $a \geq 2b$. The reason for this design criterion is to maximize the guide's frequency bandwidth of operation.

A typical sorted order of cutoff frequencies for the case $a \geq 2b$ is demonstrated in Figure 13.12.

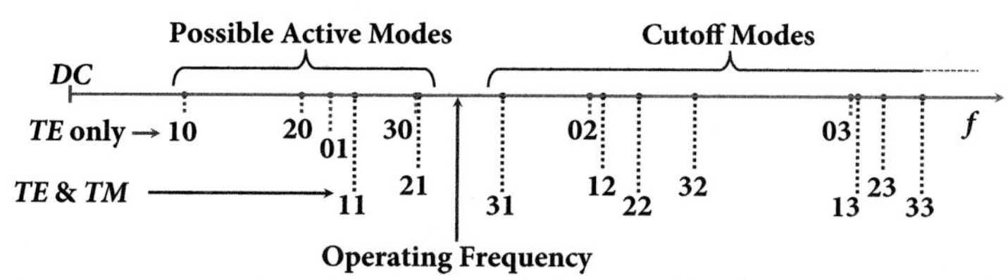

FIGURE 13.12 A typical sorted order of cutoff frequencies for the case $a \geq 2b$.

13E.2 The Graphical Approach for Mode Identification

In this approach, we create a graph as shown in Figures 13E.1 and 13E.2. On the horizontal axis of the graph we plot to a reasonable scale multiple of $c_{ug}/2a$. These multiples correspond to $c_{ug}(m/2a)$ for the different values of m. On the vertical axis, and using the same scale, we mark multiples of $c_{ug}/2b$ corresponding to $c_{ug}(n/2b)$ for the different values of n. Next, we establish the grid of lines corresponding to integer values of m and n. The points of intercepts on this grid correspond to the different combinations of m and n. The distance from the origin to any of these intercept points should equal to the cutoff frequency of the m and n mode.

$$f_{c\,m,n} = \sqrt{\left[\left(m\frac{c_{ug}}{2a}\right)^2 + \left(n\frac{c_{ug}}{2b}\right)^2\right]}$$

Next, we plot an arc centered at the origin with a radius equal to the operating frequency (using the same graph scale of course). All modes that appear inside the operating frequency arc are possible active modes since their $f_{c\,m,n}$ are less than the operating frequency. Meanwhile, all other modes outside the arc are certainly in cutoff.

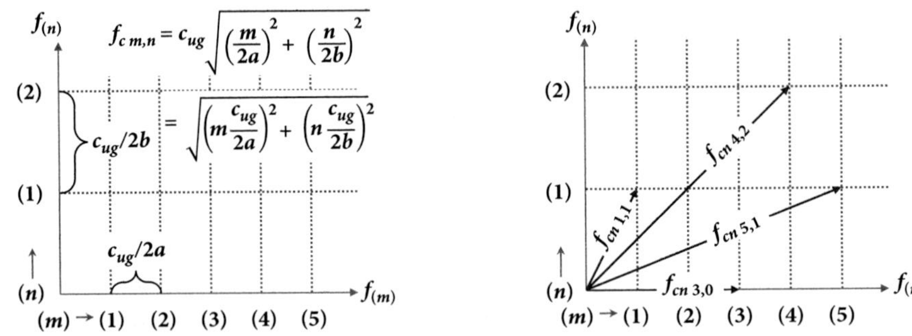

FIGURE 13E.1 The graphical approach for mode identification.

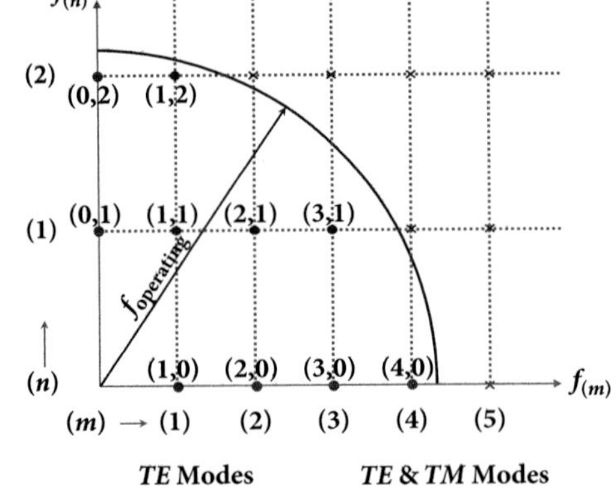

FIGURE 13E.2 The graphical approach for determining which mode are possibly active and which modes are definitely in cutoff.

ADDENDUM 13F
PHYSICAL INSIGHT IN GUIDED WAVE PROPAGATION

13F.1 A Physical View of Wave Propagation and Power Flow in Waveguides

We present the following illustration to help us visualize the concepts of wave propagation and power flow in waveguides. The illustration applies to a parallel plate guide and not the rectangular waveguide. However, the concepts demonstrated here give adequate insight that should parallel those of the rectangular waveguide.

We start the illustration by assuming a uniform plane wave that is launched at an angle θ with the z axis as shown in Figure 13F.1. In the figure, we mark the axis of wave propagation along its Poynting vector (similar to optical ray tracing). We also mark the uniform wave fronts of the propagating wave at the specific locations where the wave's electric field is at its maximum and minimum values. As we show in the figure, we label the peak wave fronts with positive signs and the valley ones with negative signs.

Now, let us assume that a second wave is introduced making the exact opposite angle with the z axis. We demonstrate this in Figure 13F.2 where we show similar markings and labeling.

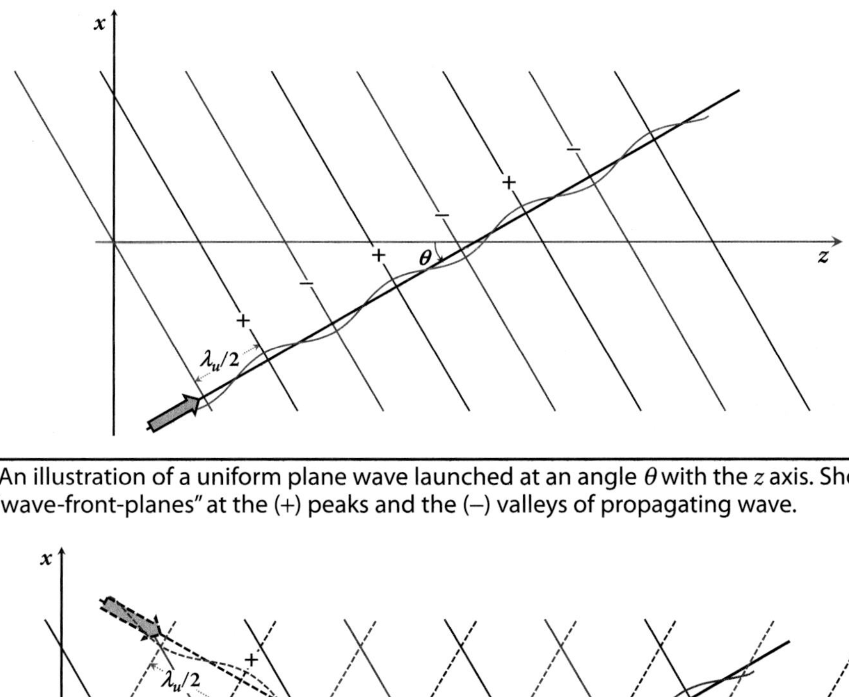

Figure 13F.1 An illustration of a uniform plane wave launched at an angle θ with the z axis. Shown on the graph are the "wave-front-planes" at the (+) peaks and the (–) valleys of propagating wave.

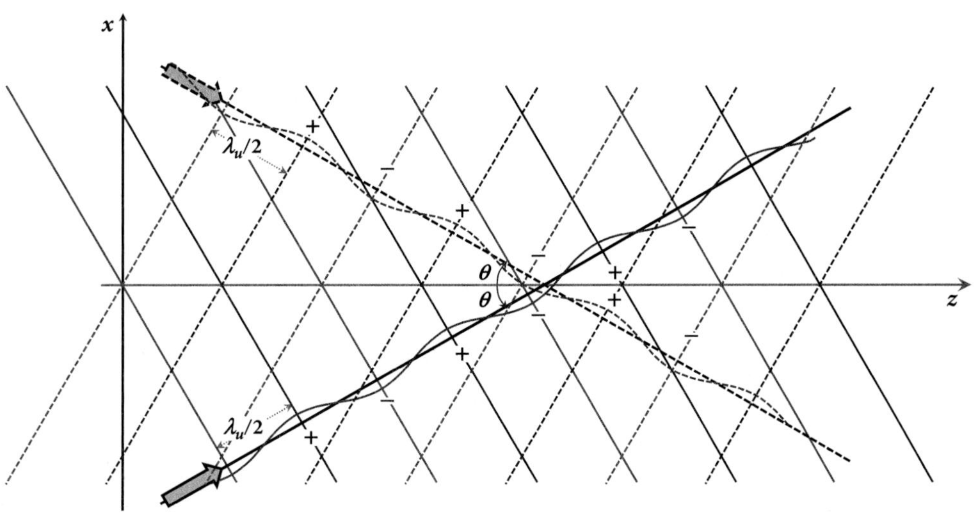

Figure 13F.2 An identical wave is now added at an opposite (mirror) angle to the first wave of Figure 13F.1.

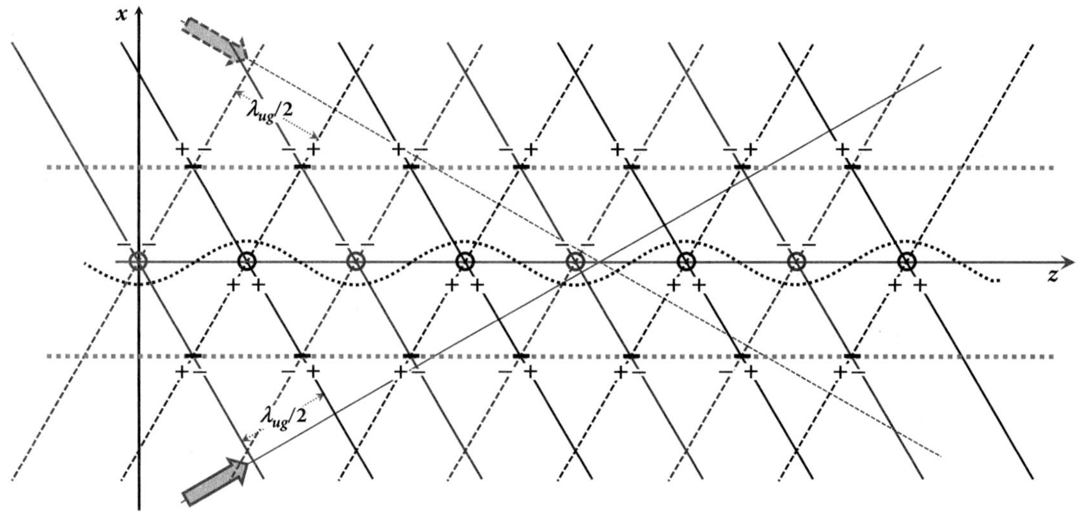

FIGURE 13F.3 The superposition of the two waves of Figure 13F.2 results in a z-propagating combination wave at the place $x = 0$. The dotted planes shown at other x values will be where the two waves cancel each other resulting in zero potentials.

In Figure 13F.3, we demonstrate the interference results of having both waves propagating simultaneously in the open space. The places where a positive wave front meets a negative one will have a zero net. Meanwhile, two positive wave fronts will combine into peak values and likewise, the locations where we have two negative wave fronts; we will have valley values.

As the figure demonstrates, we can see a *net* wave propagation along the z axis. In the meantime, the places with zero nets form symmetrically placed planes above and below the horizontal plane. Next, we consider Figure 13F.4 where we highlighted the region of the net wave propagation confined between two zero net planes. The figure as displayed has a great deal of similarity to a guided wave traveling between two perfect conducting planes (Figure 13F.5).

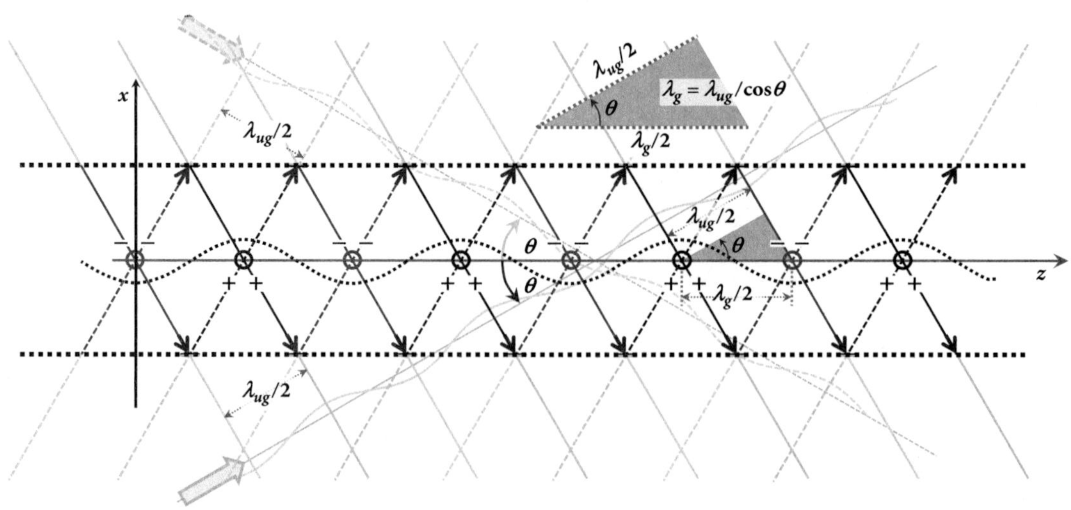

FIGURE 13F.4 The wavelength of the resulting combination wave, λ_g, is longer than the wavelength of the original waves, λ_{ug}, $\lambda_g = \lambda_{ug}/\cos\theta$.

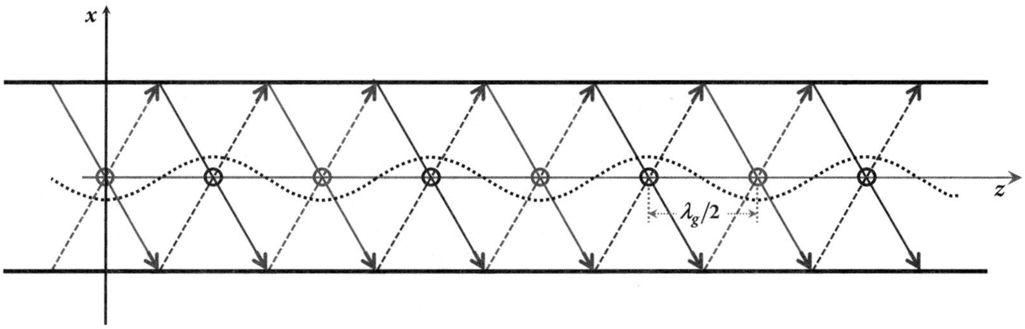

FIGURE 13F.5 Inserting conducting planes at the planes of zero potential does not alter the solution of the fields in the region in between. The two conducting planes form a parallel plate waveguide. The wavelength of the propagating wave in this region is the guide wavelength, λ_g and the wavelength of the original launched waves in free space is the unguided wavelength, λ_{ug}.

In fact, we can argue that using the uniqueness theorem that the region of the net wave propagation confined between two zero net planes and that of the guided wave between two parallel conductors have identical solutions. The justification is that the two regions have the same exact characteristics; same material properties and the same sources. In addition, the two regions have identical (null) boundary conditions at the parallel planes confining the two regions.

Going back to Figure 13F.4, we have labeled the distance between a peak and a valley along net wave propagation axis as $\lambda_g/2$, which is half the wavelength of the guided wave. Compared to the corresponding plane wave (unguided) half wavelength (also marked on the figure), it is clearly larger. As the figure shows, we can write the relationship between the two wavelengths as

$$\lambda_g = \lambda_{ug}/\cos\theta > \lambda_{ug} \tag{13F.1}$$

which carries some resemblance to the rectangular waveguide wavelength relationship, Equation (13.36).

$$\lambda_{g\ m,n} = \frac{\lambda_{ug}}{\sqrt{1-\left(\frac{f_{c\ m,n}}{f}\right)^2}} > \lambda_{ug} \tag{13.36}$$

ADDENDUM 13G METALLIC RECTANGULAR CAVITY RESONATORS

13G.1 Introduction

Cavity resonators in waveguide systems play the role that a tuned circuit does in wired/transmission line circuits. A cavity is tuned to a specific frequency and acts as a filter to enhance the system performance at that given frequency and reject signals of all other frequencies. One difference, as will be shown later in this addendum, is that cavity resonators, although designed to operate at a specific frequency will have more than one resonance frequency.

In the following, we will confine our study to the features and characteristics of metallic rectangular cavity resonators. To start, we choose the simplest form of a rectangular resonator as a finite length section of a rectangular waveguide with metallic walls placed at both ends (Figure 13G.1).

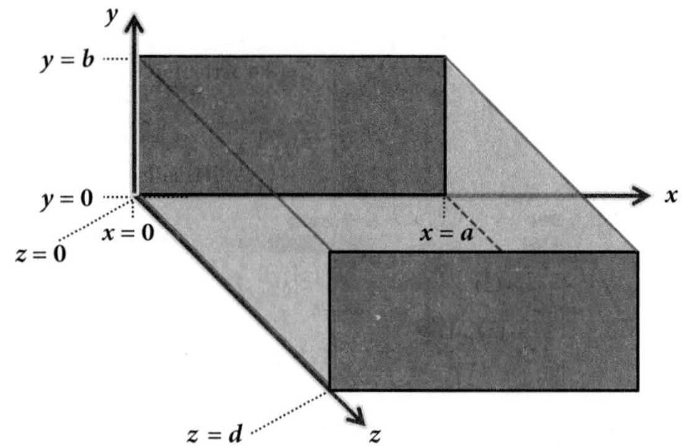

FIGURE 13G.1 An illustration of a metallic rectangular-waveguide cavity-resonator.

Referring to Figure 13G.1, and in harmony with the earlier development in this chapter, we will maintain that the cavity is laid down with its broader "a" side on the x-z plane and its narrower "b" side along the y axis. We will further assume that the z length of the cavity "d" is the largest of all three cavity dimensions, hence

$$d > a > b \tag{13G.1}$$

Like our findings in the waveguide study earlier in this chapter, a full wave analysis is lengthy with many details. Previously, we ended up breaking the full solution into two main components, the *TE* field solution and the *TM* field solution; each contained multiple possible modes of existence. In our cavity study, and to keep a limited size presentation of the topic at hand, we will only discuss the *TE* mode solution within the cavity. Referring to Figure 13G.1, by the *TE* modes in the cavity, we mean those with zero electric field component along the z axis.

13G.2 Cavity Field Analysis

To study the characteristics of this cavity, we recall the field solutions we derived earlier for the rectangular guide, specifically Equations (13.20), (13.39), and (13.21).

$$\bar{E}_{z\ m,n}(x,y,z) = \left[\sin\left(\frac{m\pi}{a}x\right)\sin\left(\frac{n\pi}{b}y\right)\right]\left[\bar{E}^{+}_{o\ m,n}e^{-jk_{z\ m,n}z} + \bar{E}^{-}_{o\ m,n}e^{+jk_{z\ m,n}z}\right] \tag{13G.2}$$

$$\bar{H}_{z\ m,n}(x,y,z) = \left[\cos\left(\frac{m\pi}{a}x\right)\cos\left(\frac{n\pi}{b}y\right)\right]\left[\bar{H}^{+}_{o\ m,n}e^{-jk_{z\ m,n}z} + \bar{H}^{-}_{o\ m,n}e^{+jk_{z\ m,n}z}\right] \tag{13G.3}$$

$$k^2_{z\ m,n} = \omega^2\mu\varepsilon - \left[\left(\frac{m\pi}{a}\right)^2 + \left(\frac{n\pi}{b}\right)^2\right] \tag{13G.4}$$

While developing the waveguide solution earlier, we emphasized the waveguide's propagation characteristics. This is why we chose to express the z axis component of the solution in the form $e^{\pm jk_{z\ m,n}z}$. Furthermore, when we wrote down the x and y components for both the electric and magnetic fields, we only considered the $+z$-axis propagation ($e^{-j\beta_{g\,m,n}z}$, with $k_{z\ m,n} = \beta_{g\ m,n}$), as seen in Equations (13.40) to (13.45).

In the cavity case before us here, the z axis is bounded by conducting walls at both ends; hence, it is not physically different from the x and y axes except in length. Consequently, and for convenience of the mathematical development, we need to use both exponential terms in both the $+z$ and $-z$ directions not just the $+z$ term as we did in Equations (13.40) to (13.45) for a one-way wave propagation. Furthermore, we need to change the exponential expressions along the z axis by appropriate sine and cosine terms in view of the bounding conducting walls similar to their use in the x- and y-axis expressions. For example, we need to rewrite the term $[\bar{H}^{+}_{o\ m,n}e^{-jk_{z\ m,n}z} + \bar{H}^{-}_{o\ m,n}e^{+jk_{z\ m,n}z}]$ in Equation (13.39) in terms of the sines and cosines as

$$\begin{aligned}\left[\bar{H}^{+}_{o\ m,n}e^{-jk_{z\ m,n}z} + \bar{H}^{-}_{o\ m,n}e^{+jk_{z\ m,n}z}\right] &= \left[\bar{H}^{+}_{o\ m,n}(\cos k_{z\ m,n}z - j\sin k_{z\ m,n}z) + \bar{H}^{-}_{o\ m,n}(\cos k_{z\ m,n}z + j\sin k_{z\ m,n}z)\right]\\ &= \left[(\bar{H}^{+}_{o\ m,n} + \bar{H}^{-}_{o\ m,n})\cos k_{z\ m,n}z + j(\bar{H}^{-}_{o\ m,n} - \bar{H}^{+}_{o\ m,n})\sin k_{z\ m,n}z\right]\\ &= \left[\bar{H}_{c\ m,n}\cos k_{z\ m,n}z + j\bar{H}_{s\ m,n}\sin k_{z\ m,n}z\right]\end{aligned} \tag{13G.2}$$

where $\bar{H}_{s\ m,n}$ and $\bar{H}_{c\ m,n}$ are the coefficients for the sine and cosine terms replacing the coefficients of the exponential terms. This manipulation enables us to rewrite Equation (13.39) in the form:

$$\bar{H}_{z\ m,n}(x,y,z) = \left[\cos\left(\frac{m\pi}{a}x\right)\cos\left(\frac{n\pi}{b}y\right)\right]\left[\bar{H}_{zc\ m,n}\cos k_{z\ m,n}z + j\bar{H}_{zs\ m,n}\sin k_{z\ m,n}z\right] \tag{13G.3}$$

Similarly, and guided by the Equations (13.20) and (13.42) to (13.45), we can rewrite the remaining components of the electric and magnetic fields solutions as

$$\bar{H}_{x\ m,n}(x,y,z) = \left[\sin\left(\frac{m\pi}{a}x\right)\cos\left(\frac{n\pi}{b}y\right)\right]\left[\bar{H}_{xc\ m,n}\cos k_{z\ m,n}z + j\bar{H}_{xs\ m,n}\sin k_{z\ m,n}z\right] \tag{13G.4}$$

$$\bar{H}_{y\ m,n}(x,y,z) = \left[\cos\left(\frac{m\pi}{a}x\right)\sin\left(\frac{n\pi}{b}y\right)\right]\left[\bar{H}_{yc\ m,n}\cos k_{z\ m,n}z + j\bar{H}_{ys\ m,n}\sin k_{z\ m,n}z\right] \tag{13G.5}$$

$$\bar{E}_{x\ m,n}(x,y,z) = \left[\cos\left(\frac{m\pi}{a}x\right)\sin\left(\frac{n\pi}{b}y\right)\right]\left[\bar{E}_{xc\ m,n}\cos k_{z\ m,n}z + j\bar{E}_{xs\ m,n}\sin k_{z\ m,n}z\right] \tag{13G.6}$$

$$\bar{E}_{y\ m,n}(x,y,z) = \left[\sin\left(\frac{m\pi}{a}x\right)\cos\left(\frac{n\pi}{b}y\right)\right]\left[\bar{E}_{yc\ m,n}\cos k_{z\ m,n}z + j\bar{E}_{ys\ m,n}\sin k_{z\ m,n}z\right] \tag{13G.7}$$

$$\bar{E}_{z\ m,n}(x,y,z) = \left[\sin\left(\frac{m\pi}{a}x\right)\sin\left(\frac{n\pi}{b}y\right)\right]\left[\bar{E}_{zc\ m,n}\cos k_{z\ m,n}z + j\bar{E}_{zs\ m,n}\sin k_{z\ m,n}z\right] \tag{13G.8}$$

where all the subscripted $\bar{E}$ and $\bar{H}$ coefficients within the rightmost square brackets are the alternate coefficients for the sine and cosine terms subject to determination by the cavity boundary conditions as well as the inter-relationships as determined by Maxwell's equations [Equations (13A.9) to (13A.16)].

13G.3 Applying Cavity Boundary Conditions and Cavity Resonance Frequency

Now, let us use the additional cavity boundary conditions of having perfect conducting walls at $z = 0$ and $z = d$ as we write

At $z = 0$

$$\bar{E}_{x\ m,n}(x,y,0) = 0 = \left[\cos\left(\frac{m\pi}{a}x\right)\sin\left(\frac{n\pi}{b}y\right)\right]\left[\bar{E}_{xc\ m,n} + j(0)\right] \tag{13G.9}$$

$$\bar{E}_{y\ m,n}(x,y,0) = 0 = \left[\sin\left(\frac{m\pi}{a}x\right)\cos\left(\frac{n\pi}{b}y\right)\right]\left[\bar{E}_{yc\ m,n} + j(0)\right] \tag{13G.10}$$

from which we conclude

$$\bar{E}_{xc\ m,n} = 0 \tag{13G.11}$$

$$\bar{E}_{yc\ m,n} = 0 \tag{13G.12}$$

and at $z = d$

$$\bar{E}_{x\ m,n}(x,y,d)=0=\left[\cos\left(\frac{m\pi}{a}x\right)\sin\left(\frac{n\pi}{b}y\right)\right]\left[0+j\bar{E}_{xs\ m,n}\sin k_{z\ m,n}d\right] \tag{13G.13}$$

$$\bar{E}_{y\ m,n}(x,y,d)=0=\left[\sin\left(\frac{m\pi}{a}x\right)\cos\left(\frac{n\pi}{b}y\right)\right]\left[0+j\bar{E}_{ys\ m,n}\sin k_{z\ m,n}d\right] \tag{13G.14}$$

from which we conclude that

$$\sin k_{z\ m,n}d=0 \tag{13G.15}$$

which makes

$$k_{z\ m,n}=\frac{\ell\pi}{d} \tag{13G.16}$$

where ℓ is an integer.

Now, this makes Equation (13.21)

$$k_{z\ m,n}^2=\left(\frac{\ell\pi}{d}\right)^2=\omega^2\mu\varepsilon-\left[\left(\frac{m\pi}{a}\right)^2+\left(\frac{n\pi}{b}\right)^2\right] \tag{13G.17}$$

which cannot be true for all values of ω. Only specific ω's corresponding to integer values for m, n, and ℓ can satisfy this condition, which leads us to write

$$\omega_{o\ m,n,\ell}^2\mu\varepsilon=\left[\left(\frac{m\pi}{a}\right)^2+\left(\frac{n\pi}{b}\right)^2+\left(\frac{\ell\pi}{d}\right)^2\right] \tag{13G.18}$$

or

$$f_{o\ m,n,\ell}=\frac{\omega_{o\ m,n,\ell}}{2\pi}=\frac{1}{\sqrt{\mu\varepsilon}}\sqrt{\left[\left(\frac{m}{a}\right)^2+\left(\frac{n}{b}\right)^2+\left(\frac{\ell}{d}\right)^2\right]} \tag{13G.19}$$

In the above frequency expressions, we added the subscripts "o" and "m,n,ℓ" to emphasize two important findings:

a. Field solutions can only exist at these specific frequencies and not at any other frequencies in between. This is exactly the behavior of resonance circuits/devices, and hence the use of the "o" subscript to denote a resonance frequency.
b. There are more than one resonance frequency corresponding to all possible integer combinations of "m, n, and ℓ."

Similarly, we will add the "ℓ" subscript to the m and n subscripts for all field components in the cavity from now on. However, for the sake of brevity of the mathematical expressions, we will drop writing all three subscripts "m,n,ℓ" in most of the intermediate steps of the development and recover them in relevantly important and final expressions. In addition, we will do the same for the "(x,y,z)" dependences of the field expressions.

13G.4 Cavity *TE* Modes

Now, we will get back to continuing our discussion of the field expressions inside the cavity. In the following, as we stated earlier, we will confine our discussions to the transverse electric solutions in the cavity. This means assuming no longitudinal z component to the electric field, hence

$$\bar{E}_{z\ m,n,\ell}(x,y,z)=0 \tag{13G.20}$$

Also, based on our findings [Equations (13G.11) and (13G.12)], we write the transverse electric field components as

$$\bar{E}_{x\ m,n,\ell}(x,y,z)=\left[\cos\left(\frac{m\pi}{a}x\right)\sin\left(\frac{n\pi}{b}y\right)\right]\left[j\bar{E}_{xs\ m,n,\ell}\sin\left(\frac{\ell\pi}{d}z\right)\right] \tag{13G.21}$$

$$\bar{E}_{y\ m,n,\ell}(x,y,z)=\left[\sin\left(\frac{m\pi}{a}x\right)\cos\left(\frac{n\pi}{b}y\right)\right]\left[j\bar{E}_{ys\ m,n,\ell}\sin\left(\frac{\ell\pi}{d}z\right)\right] \tag{13G.22}$$

Next, using Maxwell's curl [Equations (13A.10) to (13A.12)], while setting $\bar{E}_z=0$, we write

$$-\frac{\partial\bar{E}_y}{\partial z}=-j\omega\mu\bar{H}_x \tag{13G.23}$$

$$\frac{\partial\bar{E}_x}{\partial z}-=-j\omega\mu\bar{H}_y \tag{13G.24}$$

$$\frac{\partial\bar{E}_y}{\partial x}-\frac{\partial\bar{E}_x}{\partial y}=-j\omega\mu\bar{H}_z \tag{13G.25}$$

Now, substituting Equations (13G.21) and (13G.22) into Equations (13G.23) to (13G.25), we can write

$$-\frac{\partial\bar{E}_y}{\partial z}=-j\omega\mu\bar{H}_x=-\left[\sin\left(\frac{m\pi}{a}x\right)\cos\left(\frac{n\pi}{b}y\right)\right]\left[j\left(\frac{\ell\pi}{d}\right)\bar{E}_{ys}\cos\left(\frac{\ell\pi}{d}z\right)\right] \tag{13G.26}$$

or,

$$\bar{H}_x=\left[\frac{1}{\omega\mu}\left(\frac{\ell\pi}{d}\right)\bar{E}_{ys}\right]\left[\sin\left(\frac{m\pi}{a}x\right)\cos\left(\frac{n\pi}{b}y\right)\cos\left(\frac{\ell\pi}{d}z\right)\right] \tag{13G.27}$$

also,

$$\frac{\partial\bar{\bar{E}}_x}{\partial z}=-j\omega\mu\bar{H}_y=\left[\cos\left(\frac{m\pi}{a}x\right)\sin\left(\frac{n\pi}{b}y\right)\right]\left[j\left(\frac{\ell\pi}{d}\right)\bar{E}_{xs}\cos\left(\frac{\ell\pi}{d}z\right)\right] \tag{13G.28}$$

or,

$$\bar{H}_y=\left[\frac{-1}{\omega\mu}\left(\frac{\ell\pi}{d}\right)\bar{E}_{xs}\right]\left[\cos\left(\frac{m\pi}{a}x\right)\sin\left(\frac{n\pi}{b}y\right)\cos\left(\frac{\ell\pi}{d}z\right)\right] \tag{13G.29}$$

and finally,

$$\begin{aligned}\frac{\partial\bar{E}_y}{\partial x}-\frac{\partial\bar{E}_x}{\partial y}=-j\omega\mu\bar{H}_z&=\frac{m\pi}{a}\left[\cos\left(\frac{m\pi}{a}x\right)\cos\left(\frac{n\pi}{b}y\right)\right]\left[j\bar{E}_{ys}\sin\left(\frac{\ell\pi}{d}z\right)\right]\\&-\frac{n\pi}{b}\left[\cos\left(\frac{m\pi}{a}x\right)\cos\left(\frac{n\pi}{b}y\right)\right]\left[j\bar{E}_{xs}\sin\left(\frac{\ell\pi}{d}z\right)\right]\\&=\left[j\frac{m\pi}{a}\bar{E}_{ys}-j\frac{n\pi}{b}\bar{E}_{xs}\right]\left[\cos\left(\frac{m\pi}{a}x\right)\cos\left(\frac{n\pi}{b}y\right)\sin\left(\frac{\ell\pi}{d}z\right)\right]\end{aligned} \tag{13G.30}$$

or,

$$\bar{H}_z=\frac{1}{\omega\mu}\left[\frac{n\pi}{b}\bar{E}_{xs}-\frac{m\pi}{a}\bar{E}_{ys}\right]\left[\cos\left(\frac{m\pi}{a}x\right)\cos\left(\frac{n\pi}{b}y\right)\sin\left(\frac{\ell\pi}{d}z\right)\right] \tag{13G.31}$$

To summarize, the field solutions for the *TE* modes in the cavity are given by

$$\bar{E}_{x\ m,n,\ell}(x,y,z)=\left\{\left[\cos\left(\frac{m\pi}{a}x\right)\sin\left(\frac{n\pi}{b}y\right)\sin\left(\frac{\ell\pi}{d}z\right)\right]\left[j\bar{E}_{xs\ m,n,\ell}\right]\right\} \tag{13G.221}$$

$$\bar{E}_{y\ m,n,\ell}(x,y,z)=\left\{\left[\sin\left(\frac{m\pi}{a}x\right)\cos\left(\frac{n\pi}{b}y\right)\sin\left(\frac{\ell\pi}{d}z\right)\right]\left[j\bar{E}_{ys\ m,n,\ell}\right]\right\} \tag{13G.222}$$

$$\bar{H}_{x\ m,n,\ell}(x,y,z)=\left\{\left[\sin\left(\frac{m\pi}{a}x\right)\cos\left(\frac{n\pi}{b}y\right)\cos\left(\frac{\ell\pi}{d}z\right)\right]\left[\frac{1}{\omega\mu}\left(\frac{\ell\pi}{d}\right)\bar{E}_{ys\ m,n,\ell}\right]\right\} \tag{13G.227}$$

$$\bar{H}_{y\ m,n,\ell}(x,y,z)=\left\{\left[\cos\left(\frac{m\pi}{a}x\right)\sin\left(\frac{n\pi}{b}y\right)\cos\left(\frac{\ell\pi}{d}z\right)\right]\left[\frac{-1}{\omega\mu}\left(\frac{\ell\pi}{d}\right)\bar{E}_{xs\ m,n,\ell}\right]\right\} \tag{13G.229}$$

$$\bar{H}_{z\ m,n,\ell}(x,y,z)=\left\{\left[\cos\left(\frac{m\pi}{a}x\right)\cos\left(\frac{n\pi}{b}y\right)\sin\left(\frac{\ell\pi}{d}z\right)\right]\frac{1}{\omega\mu}\left[\frac{n\pi}{b}\bar{E}_{xs\ m,n,\ell}-\frac{m\pi}{a}\bar{E}_{ys\ m,n,\ell}\right]\right\} \tag{13G.231}$$

Similar to our earlier discussion leading to Equation (13.25) for waveguides, the general solution for the *TE* modes inside the cavity resonator can be written as a summation of all possible combinations of "m,n,ℓ" values. Hence, the summation operators, $\Sigma_m \Sigma_n \Sigma_\ell\{\ \}$, should be introduced to all the above expressions to yield the complete form of the cavity *TE* fields expressions.

13G.5 Cavity Dominant Mode

Next, we need to identify the cavity's lowest order (dominant) mode, that is, the lowest resonant frequency. In this regard, we consider the smallest possible values for all the three integers, m,n,ℓ. These values would be 0 and 1, and Table 13G.1 shows the corresponding resonance frequencies and field expressions for these combinations.

Comparing resonance frequencies in Table 13G.1, we can conclude that for the cavity dimensions $d>a>b$, $f_{o\ 1,0,1}<f_{o\ 0,1,1}<f_{o\ 1,1,1}$, and hence the dominant mode is the $TE_{1,0,1}$ mode.

To summarize, the solution for the dominant *TE* mode is given as

$$f_{o\ 101}=\frac{1}{\sqrt{\mu\varepsilon}}\sqrt{\left[\left(\frac{1}{a}\right)^2+\left(\frac{1}{d}\right)^2\right]} \quad \text{>>> Lowest resonance frequency: the } TE_{101} \text{ mode} \tag{13G.32}$$

$$\bar{E}_{x\ 101}(x,y,z)=0 \tag{13G.33}$$

$$\bar{H}_{y\ 101}(x,y,z)=0 \tag{13G.34}$$

$$\bar{E}_{y\ 101}(x,y,z)=\left[\sin\left(\frac{\pi}{a}x\right)\sin\left(\frac{\pi}{d}z\right)\right]\left[\bar{E}_{yo}\right] \tag{13G.35}$$

$$\bar{H}_{x\ 101}(x,y,z)=\left[\sin\left(\frac{\pi}{a}x\right)\cos\left(\frac{\pi}{d}z\right)\right]\left[\frac{-j\pi}{\omega\mu d}\bar{E}_{yo}\right] \tag{13G.36}$$

$$\bar{H}_{z\ 101}(x,y,z)=\left[\cos\left(\frac{\pi}{a}x\right)\sin\left(\frac{\pi}{d}z\right)\right]\left[\frac{j\pi}{\omega\mu a}\bar{E}_{yo}\right] \tag{13G.37}$$

where the arbitrary constant $\bar{E}_{yo}$ replace $j\bar{E}_{ys\ 1,0,1}$ for notation convenience.

Figure 13G.2 details the field map for the $TE_{1,0,1}$ mode.

m	n	ℓ	$f_{o\,m,n,\ell}=\frac{1}{\sqrt{\mu\omega}}\sqrt{\left[\left(\frac{m}{a}\right)^2+\left(\frac{n}{b}\right)^2+\left(\frac{\ell}{d}\right)^2\right]}$	**Field components**
0	0	0	0	$\ell=0$ $\sin\left(\frac{\ell\pi}{d}z\right)=0$, hence $\bar{E}_x=\bar{E}_y=\bar{H}_z=0$ Also $\left(\frac{\ell\pi}{d}\right)=0$, hence $\bar{H}_x=\bar{H}_y=0$ All field components = 0 TE modes are not supported for $\ell=0$
1	0	0	$f_{o\,1,0,0}=\frac{1}{\sqrt{\mu\varepsilon}}\left(\frac{1}{a}\right)$	
0	1	0	$f_{o\,0,1,0}=\frac{1}{\sqrt{\mu\varepsilon}}\left(\frac{1}{a}\right)$	
1	1	0	$f_{o\,1,1,0}=\frac{1}{\sqrt{\mu\varepsilon}}\sqrt{\left[\left(\frac{1}{a}\right)^2+\left(\frac{1}{b}\right)^2\right]}$	
0	0	1	$f_{o\,0,0,1}=\frac{1}{\sqrt{\mu\varepsilon}}\left(\frac{1}{a}\right)$	$m=n=0\rightarrow$ All field components = 0 TE modes are not supported for $m=n=0$
1	0	1	$f_{o\,1,0,1}=\frac{1}{\sqrt{\mu\varepsilon}}\sqrt{\left[\left(\frac{1}{a}\right)^2+\left(\frac{1}{d}\right)^2\right]}$	$m=1, n=0\rightarrow\bar{E}_x=\bar{H}_y=0$ $TE_{1,0,1}$ mode is supported ($\bar{E}_y, \bar{H}_x, \bar{H}_z$ can exist)
0	1	1	$f_{o\,0,1,1}=\frac{1}{\sqrt{\mu\varepsilon}}\sqrt{\left[\left(\frac{1}{b}\right)^2+\left(\frac{1}{d}\right)^2\right]}$	$m=0, n=1 >> \bar{E}_y=\bar{H}_x=0$ $TE_{0,1,1}$ mode is supported ($\bar{E}_x, \bar{H}_y, \bar{H}_z$ can exist)
1	1	1	$f_{o\,1,1,1}=\frac{1}{\sqrt{\mu\varepsilon}}\sqrt{\left[\left(\frac{1}{a}\right)^2+\left(\frac{1}{b}\right)^2+\left(\frac{1}{d}\right)^2\right]}$	$m=n=\ell=1 >>$ All field components = 0 $TE_{1,1,1}$ mode is supported ($\bar{E}_x, \bar{H}_y, \bar{E}_y, \bar{H}_x, \bar{H}_z$ can exist)

TABLE 13G.1 Tabulation of the Different m, n, and ℓ Combinations and the Corresponding Cavity Resonator Modes

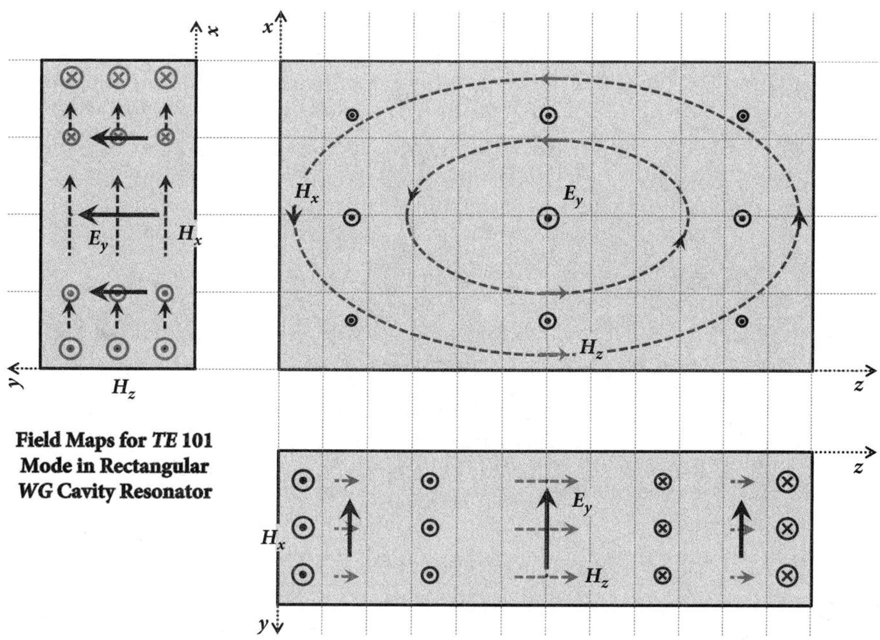

FIGURE 13G.2 The combined E and H fields map for the TE_{101} mode.

13G.6 Cavity Resonator Quality Factor

The main utility of a cavity resonator is to act as a frequency selective device that is tuned to a specific frequency while rejecting all other frequencies. The cavity selectivity is measured by what is referred to as the quality factor, *Q*. Two forms of definitions are typical for the quality factor of a resonator. The first is defined as the ratio, $Q = f_o/BW$, where f_o is the resonance frequency of the resonator and *BW* is the 3 dB bandwidth of the circuit around its resonance (Figure 13G.3).

However, the definition that we plan to use here is related to the cavity's energy storage and power dissipation

$$Q = 2\omega_o(W_{\text{stored}})/(P_{\text{dissipated}}) \tag{13G.38}$$

So, let us start with the derivation of the stored energy. In this regard, we recognize that there are two types of stored energy, an electric stored energy in the electric field and a magnetic stored energy in the magnetic field. We also recall that at resonance the energy storage alternates between the two forms; the electric and the magnetic, with both having the same exact value. Hence, it suffices to develop the expression for either form of energy, for example, the stored electric energy in the cavity at resonance. In the following, we will confine our work to the derivation of the stored electric energy for the $TE_{1,0,1}$ dominant resonance mode while briefly presenting the magnetic energy expression.

13G.6.1 Energy Stored in Cavity

Recall the expressions for the volumetric density of the electric and magnetic energy storage derived earlier in Chapter 11, Equations (11.106) and (11.107)

$$w_e(\vec{r}) = \frac{1}{4}\varepsilon'\left|\bar{E}(\vec{r})\right|^2 \text{ J/m}^3 \tag{11.106}$$

$$w_m(\vec{r}) = \frac{1}{4}\mu'\left|\bar{H}(\vec{r})\right|^2 \text{ J/m}^3 \tag{11.107}$$

Now, to derive the expression for the total electric energy storage, $W_{e\ 101}$, for the $TE_{1,0,1}$ mode, we integrate the density expression Equation (11.106) over the cavity volume while using Equation (13G.35) for the electric field phasor.

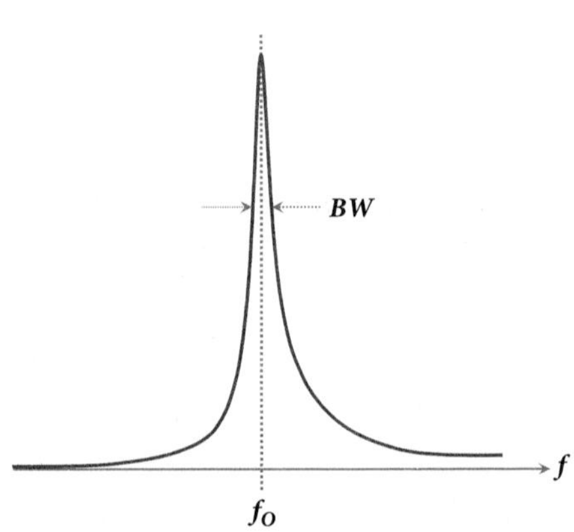

FIGURE 13G.3 A typical frequency response of a cavity resonator showing the peaking at the resonance frequency and the bandwidth of the cavity, *BW*, around the resonance.

$$
\begin{aligned}
W_{e\,101} &= \iiint_{\text{volume}} w_{e\,101}(x,y,z)dv = \iiint_{\text{volume}} \frac{1}{4}\varepsilon'\left|\bar{E}_{y\,101}(x,y,z)\right|^2 dx\,dy\,dz \\
&= \frac{1}{4}\varepsilon'\int_0^d\int_0^b\int_0^a \left|\left[\sin\left(\frac{\pi}{a}x\right)\sin\left(\frac{\pi}{d}z\right)\right][\bar{E}_{yo}]\right|^2 dx\,dy\,dz \\
&= \frac{1}{4}\varepsilon'\left|\bar{E}_{yo}\right|^2\int_0^d\int_0^b\int_0^a \left[\sin^2\left(\frac{\pi}{a}x\right)\sin^2\left(\frac{\pi}{d}z\right)\right]dx\,dy\,dz \\
&= \frac{1}{4}\varepsilon'\left|\bar{E}_{yo}\right|^2\int_0^d\int_0^b\int_0^a \left[\frac{1}{2}-\cos\left(\frac{2\pi}{a}x\right)\right]\left[\frac{1}{2}-\cos\left(\frac{2\pi}{d}z\right)\right]dx\,dy\,dz \\
&= \frac{1}{16}\varepsilon'\left|\bar{E}_{yo}\right|^2[abd]
\end{aligned}
\tag{13G.39}
$$

Similarly, we can derive the expression for total magnetic energy storage, $W_{m\,101}$, and show that it yields a result identical to that of the total electric energy in the cavity.

$$
\begin{aligned}
W_{m\,101} &= \iiint_{\text{volume}} w_{m\,101}(x,y,z)dv = \iiint_{\text{volume}} \frac{1}{4}\mu'\left\{\left|\bar{H}_{101}(x,y,z)\right|^2\right\}dx\,dy\,dz \\
&= \iiint_{\text{volume}} \frac{1}{4}\mu'\left\{\left|\bar{H}_{x\,101}(x,y,z)\right|^2+\left|\bar{H}_{z\,101}(x,y,z)\right|^2\right\}dx\,dy\,dz \rightarrow \frac{1}{16}\varepsilon'\left|\bar{E}_{yo}\right|^2[abd]
\end{aligned}
\tag{13G.40}
$$

13G.6.2 Energy Dissipated in Cavity

Next, we need to derive expressions for the power dissipated in the cavity. Such power dissipation may only take place if either or both the conducting walls and the dielectric filling of the cavity are not ideal. For perfect conducting walls and perfect dielectric filling, the power dissipation in the cavity is zero and, hence, we have a perfect cavity resonator with perfect (infinite) quality.

Now, to evaluate the power dissipation (losses) in a cavity with nonideal conductor and dielectrics, we need to have the proper expressions for the cavity fields for such cases. The solutions obtained thus far are only valid for a perfect cavity.

In the absence of proper solutions for the cavity fields with nonideal materials, we can resort to a perturbation scheme to approximate and estimate the losses. In the perturbation scheme, we use the available perfect material solution while introducing the material losses. The concept here is that a small/limited change/perturbation to the material properties should not significantly alter the solution from that available for the perfect material case. This scheme yields reasonably good estimates to the losses for the cases of near-ideal material but loses its accuracy as the losses are higher.

We will work with the perfect material solution given in Equations (13G.33) to (13G.37) in order to estimate the cavity power dissipation assuming good conductor walls as well as low-loss dielectric filling. We will evaluate each power dissipation component separately; $P_{c\,101}$ for the conductor power dissipation, and $P_{d\,101}$ for the dielectric power dissipation. The total power dissipation in the cavity $P_{t\,101}$ is then the sum of the two components.

13G.6.3 Cavity Conductor Losses

We start with the conductor losses. As was discussed earlier in Chapter 11, for high frequencies at which the cavity resonators operate, the electromagnetic waves will only penetrate the finite conductivity walls up to a finite/small skin depth, thus making the resistance of the conducting walls causing power dissipation to be that of the sheet resistance of the skin.

Referring to Figure 13G.4, guided by the development of Equation (11C.2) and using Equation (11.75) for the skin depth, δ, the incremental sheet resistance, $\Delta R_{\Delta w \Delta \ell \delta}$, of an element $(\Delta w \times \Delta \ell \times \delta)$ is given by

$$
\Delta R_{\Delta w\Delta\ell\delta} = \frac{\Delta\ell}{\sigma\,\Delta w\,\delta} = \frac{\Delta\ell\sqrt{\pi f \mu_o \sigma}}{\sigma\,\Delta w}
\tag{13G.41}
$$

where σ is the conductivity.

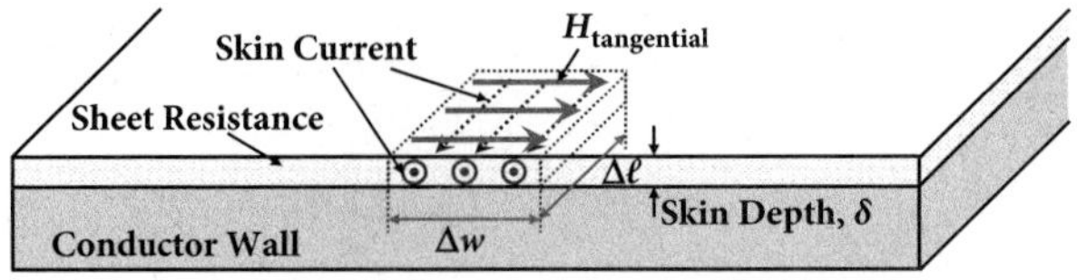

FIGURE 13G.4 A demonstration to support the analysis of evaluating the power dissipation in the skin depth of the conducting walls.

The skin current flow in this increment, $\Delta I_{\Delta w}$, is given by the product of the surface current linear density, K, as a result of the tangential magnetic field, $H_{\text{tangential}}$, and the width of the increment

$$\Delta I_{\Delta w} = K\Delta w = H_{\text{tangential}}\Delta w \tag{13G.42}$$

Hence, the incremental power dissipation, $\Delta P_{\Delta w\Delta\ell\delta}$, in the element $(\Delta w \times \Delta\ell \times \delta)$ is given by

$$\Delta P_{\Delta w\Delta\ell\delta} = (\Delta I_{\Delta w})^2 \Delta R_{\Delta w\Delta\ell\delta} = [H_{\text{tangential}}\Delta w]^2\left[\frac{\Delta\ell\sqrt{\pi f\mu_o\sigma}}{\sigma\Delta w}\right] = [H_{\text{tangential}}]^2\left[\sqrt{\frac{\pi f\mu_o}{\sigma}}\Delta\ell\Delta w\right] \tag{13G.43}$$

Now, we are ready to evaluate the power dissipation, $P_{c\ 101}$, in the cavity conducting walls for the TE_{101} mode by integrating Equation (13G.43) for all six conductor walls of the cavity (Figure 13G.5). In principle, all these six walls are expected to contribute to the conductor losses. Since every two opposite walls should have equal power dissipation we will only evaluate the losses in the three faces shown in the left diagram of the figure and multiply the obtained answer by two to obtain the total conductor power loss in all six faces. Using Equations (13G.34), (13G.36), and (13G.37) for the tangential H field components for the TE_{101} mode, we can write

$$P_{c\ 101} = 2\{P_{c\ xy\text{-face}} + P_{c\ yz\text{-face}} + P_{c\ xz\text{-face}}\} = 2\left\{\int_0^b\int_0^a dP_{c\ xy\text{-face}} + \int_0^d\int_0^b dP_{c\ yz\text{-face}} + \int_0^d\int_0^a dP_{c\ xz\text{-face}}\right\}$$

$$= 2\left\{\int_0^b\int_0^a |\bar{H}_{x\ 101}|^2\left[\sqrt{\frac{\pi f\mu_o}{\sigma}}dx\,dy\right] + \int_0^d\int_0^b |\bar{H}_{z\ 101}|^2\left[\sqrt{\frac{\pi f\mu_o}{\sigma}}dy\,dz\right] + \int_0^d\int_0^a\left[|\bar{H}_{x\ 101}|^2 + |\bar{H}_{z\ 101}|^2\right]\left[\sqrt{\frac{\pi f\mu_o}{\sigma}}dx\,dz\right]\right\}$$

$$= 2\sqrt{\frac{\pi f\mu_o}{\sigma}}\left[\frac{\pi}{\omega\mu_o}\right]^2 |\bar{E}_{yo}|^2\left\{\int_0^b\int_0^a \frac{1}{d^2}\left[\sin\left(\frac{\pi}{a}x\right)\cos\left(\frac{\pi}{d}z\right)\right]^2 dx\,dy + \int_0^d\int_0^b \frac{1}{a^2}\left[\cos\left(\frac{\pi}{a}x\right)\sin\left(\frac{\pi}{d}z\right)\right]^2 dy\,dz\right.$$

$$\left. + \int_0^d\int_0^a\left[\frac{1}{d^2}\left[\sin\left(\frac{\pi}{a}x\right)\cos\left(\frac{\pi}{d}z\right)\right]^2 + \frac{1}{a^2}\left[\cos\left(\frac{\pi}{a}x\right)\sin\left(\frac{\pi}{d}z\right)\right]^2\right]dx\,dz\right\}$$

$$= \frac{1}{2}\sqrt{\frac{\pi}{\sigma(f\mu_o)^3}}|\bar{E}_{yo}|^2\left\{\frac{ab}{4d^2} + \frac{bd}{4a^2} + \left[\frac{ad}{8d^2} + \frac{ad}{8a^2}\right]\right\} \tag{13G.44}$$

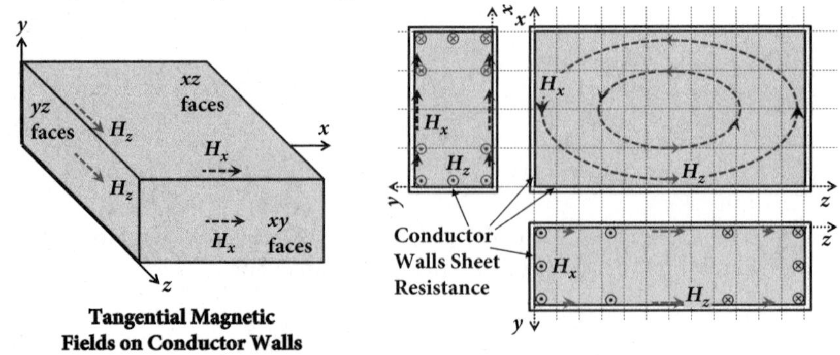

FIGURE 13G.5 Identifying the surface currents in the conducting walls as a result of the tangential magnetic fields at these surfaces.

13G.6.4 Cavity Dielectric Losses

Now, for the dielectric losses, $P_{d\,101}$, we need to integrate the power dissipation density as given by Equation (11.104), $p_{d\,101}$, over the cavity filling volume

$$P_{d\,101} = \iiint_{\text{cavity volume}} p_{d\,101}\,dv = \iiint_{\text{volume}} \frac{1}{2}\omega\varepsilon''\left|\overline{\overline{E}}_{101}(\vec{r})\right|^2 dv = \int_0^d\int_0^b\int_0^a \frac{1}{2}\omega\varepsilon''\left|\overline{E}_{y\,101}(x,y,z)\right|^2 dx\,dy\,dz$$

$$= \int_0^d\int_0^b\int_0^a \frac{1}{2}\omega\varepsilon''\left|\overline{E}_{yo}\right|^2\left|\sin\left(\frac{\pi}{a}x\right)\sin\left(\frac{\pi}{d}z\right)\right|^2 dx\,dy\,dz = \omega\varepsilon''\left|\overline{E}_{yo}\right|^2\frac{abd}{8} \tag{13G.45}$$

Finally, we can write the total power dissipation in the cavity as

$$P_{t\,101} = P_{c\,101} + P_{d\,101} = \frac{1}{8}\sqrt{\frac{\pi}{\sigma(f\mu_o)^3}}\left|\overline{E}_{yo}\right|^2\left\{\frac{ab}{d^2}+\frac{bd}{a^2}+\left[\frac{ad}{2d^2}+\frac{ad}{2a^2}\right]\right\}+\omega\varepsilon''\left|\overline{E}_{yo}\right|^2\frac{abd}{8} \tag{13G.46}$$

yielding the quality factor, Q_{101}, for the cavity operating in the TE_{101} mode as

$$Q_{101} = \frac{2\omega_o W_{101}}{P_{t\,101}} = \frac{\frac{1}{8}\omega_o\varepsilon'\left|\overline{E}_{yo}\right|^2[abd]}{\frac{1}{8}\sqrt{\frac{\pi}{\sigma(f\mu_o)^3}}\left|\overline{E}_{yo}\right|^2\left\{\frac{ab}{d^2}+\frac{bd}{a^2}+\left[\frac{ad}{2d^2}+\frac{ad}{2a^2}\right]\right\}+\omega\varepsilon''\left|\overline{E}_{yo}\right|^2\frac{abd}{8}}$$

$$= \frac{\omega_o\varepsilon'}{\sqrt{\frac{\pi}{\sigma(f\mu_o)^3}}\left\{\frac{1}{d^3}+\frac{1}{a^3}+\left[\frac{1}{2bd^2}+\frac{1}{2ba^2}\right]\right\}+\omega\varepsilon''} \tag{13G.47}$$

Example 13.4 Given a dielectric filled rectangular waveguide cavity with dimensions of 3.0 × 1.4 × 4.6 cm made out of copper with conductivity of 5.96E7 and the dielectric filling has $\varepsilon' = 2.23$ and $\varepsilon'' = 0.0005$. Find the dominant mode of operation, the corresponding resonance frequency, the cavity's quality factor and bandwidth.

Solution The dominant mode of a rectangular WG resonator is the TE_{101} for which

$$f_{o\,101} = \frac{1}{\sqrt{\mu\varepsilon}}\sqrt{\left[\left(\frac{1}{a}\right)^2+\left(\frac{1}{d}\right)^2\right]},\ Q_{101} = \frac{\omega_o\varepsilon'}{\sqrt{\frac{\pi}{\sigma(f\mu_o)^3}}\left\{\frac{1}{d^3}+\frac{1}{a^3}+\left[\frac{1}{2bd^2}+\frac{1}{2ba^2}\right]\right\}+\omega\varepsilon''},\ BW_{101} = f_{o\,101}/Q_{t\,101}$$

Dominant resonance frequency	$f_{o\,101}$	7.995	GHz
Skin depth	δ	7.328×10^{-7}	m
Conductor Q (assuming $\varepsilon'' = 0$)	$Q_{c\,101}$	12490	
Dielectric Q (assuming $\sigma = \infty$)	$Q_{d\,101}$	4460	
Cavity Q	$Q_{t\,101}$	3286	
Cavity bandwidth	$BW_{101} = f_{o\,101}/Q_{t\,101}$	2.433	MHz

CHAPTER 13 PROBLEMS

- Problems 13.1 to 13.14: Problems dealing with guide parameters and modes
- Problems 13.15 to 13.19: Problems dealing with power flow
- Problems 13.20 to 13.22: Problems dealing with modal dispersion
- Problem 13.23: Problem dealing with attenuation coefficient
- Problems 13.24 and 13.25: Problems dealing with impedance matching
- Problems 13.26 and 13.27: Problems dealing with cavity resonators

13.1 What is meant by the dominant mode for waveguides? Explain what happens if a waveguide is operated in that mode and what happens if operated otherwise.

13.2 What defines a waveguide's bandwidth? Explain what happens to signals excited outside this bandwidth as compared to one within it. How is it different in the case of plane wave propagation or a transmission line? Use a few short bullets (sentences) to explain.

13.3 What is the dominant mode for rectangular waveguides ($TE_{(mn)}/TM_{(mn)}$)? What does m and n represent in either case? Give physical insight.

13.4 Write the field equations and sketch the field maps for the modes TE_{12} and TM_{12}

13.5 An air-filled rectangular waveguide has $b = 0.5a$ and a dominant-mode cutoff frequency of 0.9 GHz. If the measured guide wavelength is 40 cm, find the operating frequency, the guide dimensions, and the cutoff frequency for the dominant mode.

Also, find the lowest frequency at which a TE_{21} mode would propagate.

13.6 Evaluate the dimensions of a rectangular waveguide, which operates at the mid-point of its single-mode frequency band at 10 GHz. Estimate the bandwidth of this guide.

13.7 An air-filled rectangular guide with $a/b = 2$ is used in the 12.4 to 18.8 GHz range for dominant-mode operation with a cutoff frequency of 9.49 GHz. What are the inside dimensions?

13.8 Find all possible modes that can propagate in a WGR90 rectangular waveguide [$a = 0.9''$ and $b = 0.4''$] operating at 18 GHz. For each mode, find the corresponding parameters:

a. Cutoff frequency

b. WG impedance

c. Phase constant

d. Wavelength

e. Phase velocity

f. Group velocity

13.9 A TE wave propagating in a dielectric filled waveguide of unknown permittivity has dimensions $a = 5$ cm and $b = 3$ cm. If the x component of its electric field is given by

$$E_x = -36\cos(40\pi x)\sin(100\pi y)\sin(2.4\pi\cdot 10^{10}t - 52.9\pi z)\ \text{V/m}.$$

Determine

a. The mode number

b. The relative permittivity of the material in the guide

c. The cutoff frequency for this mode

d. The expression for H_y

13.10 A waveguide filled with a material whose $\varepsilon_r = 2.25$ has dimensions $a = 2$ cm, $b = 1.4$ cm. If the guide is to transmit a 10.5-GHz signal, what possible modes can be used for transmission.

13.11 An air-filled C-band waveguide for use between 3.95 and 5.85 GHz measures 1.872" × 0.872" operates at 4.5 GHz. Calculate the dominant mode cutoff frequency and the guide wavelength, phase velocity, and wave impedance.

Redo if the guide is filled with a lossless dielectric of $\varepsilon_r = 2.1$.

13.12 Find the operating frequency range for the following guides to operate in single mode.
Guide #1: $a = 4$ cm, $b = 2$ cm, air filled
Guide #2: $a = 4$ cm, $b = 2$ cm, dielectric filled, dielectric constant = 3

13.13 a. A V-band waveguide for use between 26.5 and 40 GHz has inside dimensions 0.28" × 0.14". Calculate the dominant-mode cutoff frequency for air dielectric.

b. If the measured guide wavelength is 1.39 cm, what is the operating frequency?

13.14 A waveguide, with dimensions $a = 1$ cm and $b = 0.7$, is to be used at 20 GHz. Determine the wave impedance of the dominant mode when

a. The guide is empty.

b. The guide is filled with polyethylene (whose $\varepsilon_r = 2.25$).

13.15 A lossless air-dielectric waveguide for an S-band radar system operating at 3 GHz has the dimensions $a = 2.84$" and $b = 1.34$". Find the average power transmitted if the E field is 10 kV/m.

13.16 If a lossless dielectric having $\varepsilon_r = 2.1$ is inserted in the waveguide of above problem. Find the average power transmitted for the same E field of 10 kV/m.

13.17 A dielectric filled waveguide with $\varepsilon_r = 2$ and (3 cm × 1.5 cm) cross-section operating at 85 GHz. Which mode(s) is (are) possible. For the dominant mode, find λ_g, Z_{og}, c_{ph}, and c_{gr}.

If the peak of the electric field inside the guide was measured as 5 kV/m. Find the peak magnetic field inside the guide as well as the peak power flow density.

13.18 In a lossless air-dielectric waveguide for an S-band radar system operating at 3 GHz has the dimensions $a = 2.84$" and $b = 1.34$" the transmitted power in the dominant mode is 100 W. Find the level of excitation for the electric field.

13.19 If the waveguide of Problem 13.18 is now filled with a perfect dielectric with $\varepsilon_r = 2.1$. Find the level of excitation for the electric field for the same power transmitted of 100 W.

13.20 Calculate the lowest and highest useable frequencies for a rectangular waveguide with dimensions of 0.9" and 0.4", to avoid modal dispersion.

13.21 Explain what happens to a pulse $[A\delta(t)]$ when propagated in a circular waveguide. How is it different in the case of plane wave propagation or transmission line propagation? Use a few short bullets (sentences) to explain.

13.22 Radars use RF (modulated) pulses and not base band pulses. Explain why.

13.23 The inside dimensions of an X-band waveguide are $a = 0.9$" and $b = 0.4$". The guide operates at 10 GHz. Find the attenuation coefficient, given that the guide walls are made of brass and is filled with a low-loss dielectric with ($\varepsilon_r = 2.1$, and $\tan\delta = 0.01$).

13.24 Find the operating frequency range for the following guides to operate in single mode.
Guide 1 : $a = 4$ cm, $b = 2$ cm, air filled
Guide 2 : $a = 4$ cm, $b = 2$ cm, dielectric filled, dielectric constant = 3

If the operating frequency was set at the mid-point of that range, design a quarter wavelength WG matching section to match between the two guides.

13.25 Design a quarter wavelength WG matching section to match between the two guides at 6 GHz.
Guide 1: $a = 3.75$ cm, $b = 2$ cm, air filled
Guide 2: $a = 3$ cm, $b = 2$ cm, air filled

13.26 Given is an air-filled rectangular metallic cavity with dimension $2 \times 3 \times 5$ cm. Identify the three lowest order resonant modes and their corresponding resonance frequencies. Repeat if it now filled with a dielectric material with $\varepsilon_r = 2$.

13.27 Given is a dielectric-filled rectangular metallic cavity with dimension $3 \times 1.4 \times 4.6$ cm. Identify the three lowest order resonant modes and their corresponding resonance frequencies. The dielectric filling has $\varepsilon_r' = 2.23$ and $\varepsilon_r'' = 0.0005$ and the cavity walls is made out of copper. Find the cavity quality factor and bandwidth when operating at the dominant mode frequency. Repeat if the cavity is to be silver coated for an improved quality factor. Discuss the improvement of the quality factor caused by the improved conductivity of the cavity walls.

CHAPTER 13 SUMMARY[1]

Solution of EM Fields in Metallic Rectangular Waveguides

- General form of field solution:

$$\begin{pmatrix} \bar{E} \\ \bar{H} \end{pmatrix} = \sum_m \sum_n \begin{pmatrix} \bar{E}_{m,n} \\ \bar{H}_{m,n} \end{pmatrix}$$

$$= \sum_m \sum_n \left[\begin{pmatrix} \sin \\ \cos \end{pmatrix} \left(\frac{m\pi}{a} x \right) \begin{pmatrix} \sin \\ \cos \end{pmatrix} \left(\frac{n\pi}{b} y \right) \right] \left[\begin{pmatrix} \bar{E}^{+}_{o\,m,n} \\ \bar{H}^{+}_{o\,m,n} \end{pmatrix} e^{-jk_{z\,m,n} z} + \begin{pmatrix} \bar{E}^{-}_{o\,m,n} \\ \bar{H}^{-}_{o\,m,n} \end{pmatrix} e^{+jk_{z\,m,n} z} \right]$$

$$k^2_{z\,m,n} = \omega^2 \mu\varepsilon - \left[\left(\frac{m\pi}{a} \right)^2 + \left(\frac{n\pi}{b} \right)^2 \right] = \omega^2 \mu\varepsilon - k^2_{c\,m,n} = 4\pi^2 \mu\varepsilon \left\{ f^2 - f^2_{c\,m,n} \right\}$$

Wave propagation: $f > f_{c\,m,n} \rightarrow k_{z\,m,n}$ is real $\rightarrow \beta_{g\,m,n} = k_{z\,m,n}$

Wave cutoff: $f < f_{c\,m,n} \rightarrow k_{z\,m,n}$ is imaginary $\rightarrow \alpha_{g\,m,n} = jk_{z\,m,n}$

Rectangular Waveguide Propagation and Cutoff Modes

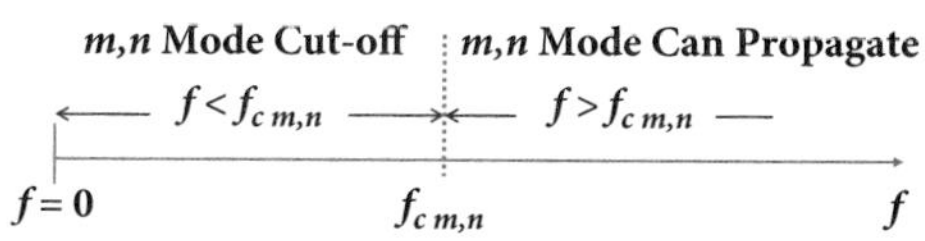

$$f_{c\,m,n} = \frac{c_o}{\sqrt{\mu_r\, \varepsilon_r}} \sqrt{\left[\left(\frac{m}{2a} \right)^2 + \left(\frac{n}{2b} \right)^2 \right]}$$

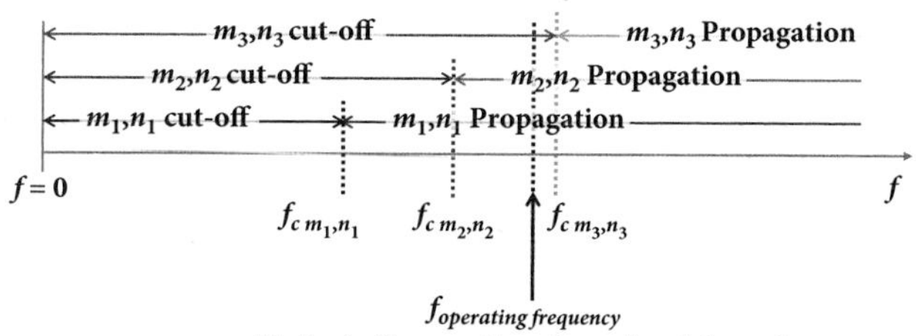

Modes in Propagation: (m_1,n_1) and (m_2,n_2)
Modes in Cut-off: (m_3,n_3) and Higher

Physical Insight: Guide Propagation Parameters

- $\dfrac{\beta_{g\,m,n}}{\beta_{ug}} = \sqrt{1 - \left(\dfrac{f_{c\,m,n}}{f} \right)^2}, \ \beta_{ug} = \omega\sqrt{\mu \cdot \varepsilon}$
- $\dfrac{\lambda_{g\,m,n}}{\lambda_{ug}} = \dfrac{c_{ph\,g\,m,n}}{c_{ph\,ug}} = \dfrac{1}{\sqrt{1 - \left(\dfrac{f_{c\,m,n}}{f} \right)^2}}$
- $\lambda_{ug} = \dfrac{2\pi}{\omega\sqrt{\mu \cdot \varepsilon}}, \ c_{ph\,ug} = \dfrac{1}{\sqrt{\mu \cdot \varepsilon}}$
- $\eta_{g\,TE\,m,n} = \dfrac{\omega\mu}{\beta_{g\,m,n}} = \dfrac{\eta_{ug}}{\sqrt{1 - \left[\dfrac{f_{c\,m,n}}{f} \right]^2}}$
- $\eta_{g\,TM\,m,n} = \dfrac{\beta_{g\,m,n}}{\omega\varepsilon} = \eta_{ug}\sqrt{1 - \left[\dfrac{f_{c\,m,n}}{f} \right]^2}$
- $\eta_{ug} = \sqrt{\dfrac{\mu}{\varepsilon}}$

---

[1] ***Important Warning to Students:*** Please do not use this summary without thorough study of the relevant material in this chapter. Using this summary and its formulas without adequate appreciation of the physics involved and conditions of validity can lead to wrong implementations and invalid conclusions.

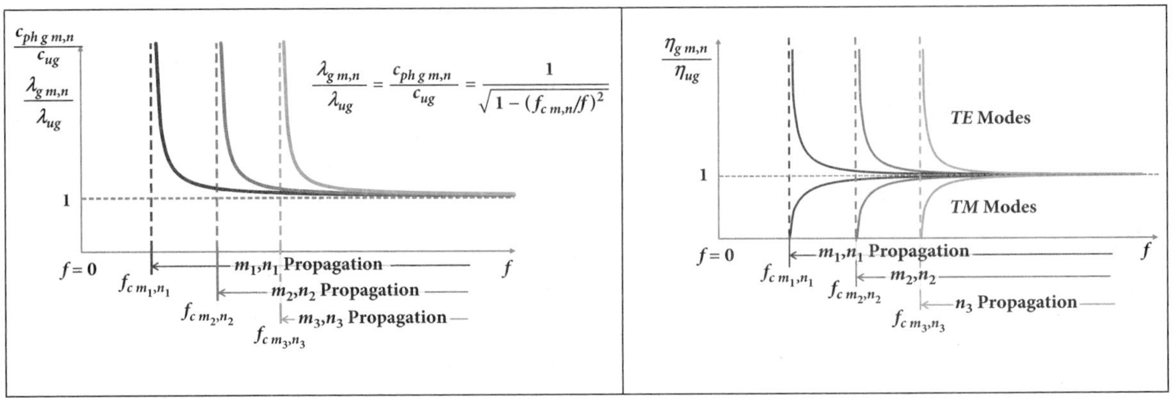

Active and Dominant Mode Identification

Dominant Mode Has the Lowest Cutoff Frequency

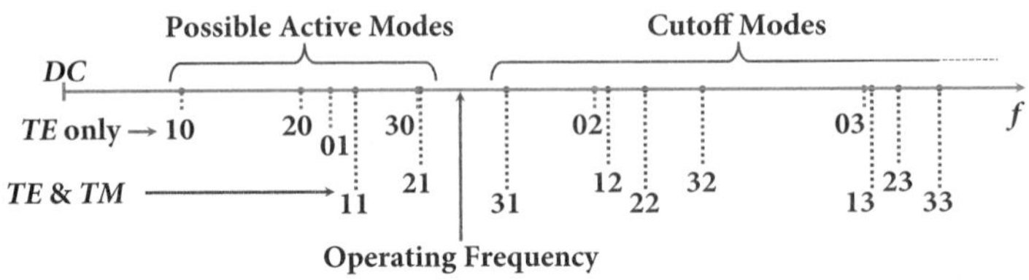

Wave Propagation: Power Flow

- $\vec{\mathcal{P}}^{+}_{\text{avg } TE\ m,n}$

$$= \frac{1}{2}\frac{\omega\mu\beta_{g\ m,n}}{k^4_{c\ m,n}}\left|\bar{H}^{+}_{o\ m,n}\right|^2 \left\{\left(\frac{n\pi}{b}\right)^2\left[\cos\left(\frac{m\pi}{a}x\right)\sin\left(\frac{n\pi}{b}y\right)\right]^2 + \left(\frac{m\pi}{a}\right)^2\left[\sin\left(\frac{m\pi}{a}x\right)\cos\left(\frac{n\pi}{b}y\right)\right]^2\right\}\vec{a}_z$$

- $\vec{\mathcal{P}}^{+}_{\text{avg } TM\ m,n}$

$$= \frac{1}{2}\frac{\omega\varepsilon\beta_{g\ m,n}}{k^4_{c\ m,n}}\left|\bar{E}^{+}_{o\ m,n}\right|^2 \left\{\left(\frac{m\pi}{a}\right)^2 \cos^2\left(\frac{m\pi}{a}x\right)\sin^2\left(\frac{n\pi}{b}y\right) + \left(\frac{n\pi}{b}\right)^2 \sin^2\left(\frac{m\pi}{a}x\right)\cos^2\left(\frac{n\pi}{b}y\right)\right\}\vec{a}_z$$

- Power flow for dominant mode TE_{10}

$$\vec{\mathcal{P}}^{+}_{\text{avg } TE\ 1,0} = \frac{1}{2}\frac{\omega\mu\beta_{g\ 1,0}}{k^4_{c\ 1,0}}\left|\bar{H}^{+}_{o\ 1,0}\right|^2 \left\{+\left(\frac{\pi}{a}\right)^2\left[\sin\left(\frac{\pi}{a}x\right)\right]^2\right\}\vec{a}_z = \frac{1}{2}\omega\mu\beta_{g\ 1,0}\left(\frac{a}{\pi}\right)^2 \sin^2\left(\frac{\pi}{a}x\right)\left|\bar{H}^{+}_{o\ 1,0}\right|^2 \vec{a}_z$$

$$\vec{P}^{+}_{\text{avg } TE\ 1,0} = \iint_{0,0}^{a,b} \vec{\mathcal{P}}^{+}_{\text{avg } TE\ 1,0}(x,y,z)\,dx\,dy = \frac{ab}{4}\omega\mu\beta_{g\ 1,0}\left(\frac{a}{\pi}\right)^2\left|\bar{H}^{+}_{o\ 1,0}\right|^2 \vec{a}_z$$

TE and *TM* Modes

- ***TE Modes***

$$\rightarrow \bar{E}^{+}_{z\ TE\ m,n}=0 \qquad k_{c\ m,n}=\sqrt{\left(\frac{m\pi}{a}\right)^2+\left(\frac{n\pi}{b}\right)^2}$$

$$\rightarrow \bar{H}^{+}_{z\ TE\ m,n}=\left[\cos\left(\frac{m\pi}{a}x\right)\cos\left(\frac{n\pi}{b}y\right)\right]\left[\bar{H}^{+}_{o\ m,n}e^{-j\beta_{g\ m,n}z}\right]$$

$$\rightarrow \bar{E}^{+}_{x\ TE\ m,n}=\frac{j\omega\mu}{k^2_{c\ m,n}}\left(\frac{n\pi}{b}\right)\left[\cos\left(\frac{m\pi}{a}x\right)\sin\left(\frac{n\pi}{b}y\right)\right]\left[\bar{H}^{+}_{o\ m,n}e^{-j\beta_{g\ m,n}z}\right]$$

$$\rightarrow \bar{E}^{+}_{y\ TE\ m,n}=\frac{-j\omega\mu}{k^2_{c\ m,n}}\left(\frac{m\pi}{a}\right)\left[\sin\left(\frac{m\pi}{a}x\right)\cos\left(\frac{n\pi}{b}y\right)\right]\left[\bar{H}^{+}_{o\ m,n}e^{-j\beta_{g\ m,n}z}\right]$$

$$\rightarrow \bar{H}^{+}_{x\ TE\ m,n}=\frac{j\beta_{g\ m,n}}{k^2_{c\ m,n}}\left(\frac{m\pi}{a}\right)\left[\sin\left(\frac{m\pi}{a}x\right)\cos\left(\frac{n\pi}{b}y\right)\right]\left[\bar{H}^{+}_{o\ m,n}e^{-j\beta_{g\ m,n}z}\right]=-\frac{1}{\eta_{g\ TE\ m,n}}\bar{E}^{+}_{y\ TE\ m,n}$$

$$\rightarrow \bar{H}^{+}_{y\ TE\ m,n}=\frac{j\beta_{g\ m,n}}{k^2_{c\ m,n}}\left(\frac{n\pi}{b}\right)\left[\cos\left(\frac{m\pi}{a}x\right)\sin\left(\frac{n\pi}{b}y\right)\right]\left[\bar{H}^{+}_{o\ m,n}e^{-j\beta_{g\ m,n}z}\right]=\frac{1}{\eta_{g\ TE\ m,n}}\bar{E}^{+}_{x\ TE\ m,n}$$

- ***TE$_{10}$ Mode***

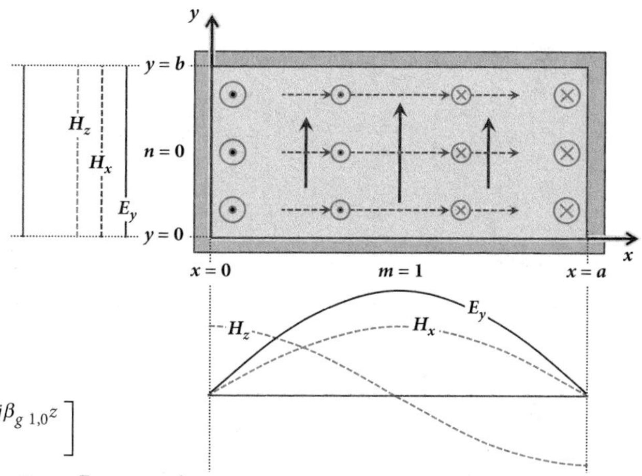

$$\rightarrow f_{c\ 1,0}=\frac{c_o}{2a}$$

$$\rightarrow \beta_{g\ 1,0}=\frac{\omega}{c_o}\sqrt{1-\left(\frac{c_o}{2af}\right)^2}$$

$$\rightarrow \bar{E}^{+}_{z\ TE\ 1,0}=\bar{E}^{+}_{x\ TE\ 1,0}=\bar{H}^{+}_{y\ TE\ 1,0}=0$$

$$\rightarrow \bar{H}^{+}_{z\ TE\ 1,0}=\cos\left(\frac{\pi}{a}x\right)\left[\bar{H}^{+}_{o\ 1,0}e^{-j\beta_{g\ 1,0}z}\right]$$

$$\rightarrow \bar{E}^{+}_{y\ TE\ 1,0}=-j\omega\mu\left(\frac{a}{\pi}\right)\sin\left(\frac{\pi}{a}x\right)\left[\bar{H}^{+}_{o\ 1,0}e^{-j\beta_{g\ 1,0}z}\right]$$

$$\rightarrow \bar{H}^{+}_{x\ TE\ 1,0}=j\beta_{g\ 1,0}\left(\frac{a}{\pi}\right)\sin\left(\frac{\pi}{a}x\right)\left[\bar{H}^{+}_{o\ 1,0}e^{-j\beta_{g\ 1,0}z}\right]=-\frac{1}{\eta_{g\ TE\ 1,0}}\bar{E}^{+}_{y\ TE\ 1,0}$$

- ***TM Modes***

$$\rightarrow \bar{H}^{+}_{z\ TM\ m,n}=0$$

$$\rightarrow \bar{E}^{+}_{z\ TM\ m,n}=\sin\left(\frac{m\pi}{a}x\right)\sin\left(\frac{n\pi}{b}y\right)\bar{E}^{+}_{o\ m,n}e^{-j\beta_{g\ m,n}z}$$

$$\rightarrow \bar{E}^{+}_{x\ TM\ m,n}=\frac{-j\beta_{g\ m,n}}{k^2_{c\ m,n}}\left(\frac{m\pi}{a}\right)\cos\left(\frac{m\pi}{a}x\right)\sin\left(\frac{n\pi}{b}y\right)\bar{E}^{+}_{o\ m,n}e^{-j\beta_{g\ m,n}z}$$

$$\rightarrow \bar{E}^{+}_{y\ TM\ m,n}=\frac{-j\beta_{g\ m,n}}{k^2_{c\ m,n}}\left(\frac{n\pi}{b}\right)\sin\left(\frac{m\pi}{a}x\right)\cos\left(\frac{n\pi}{b}y\right)\bar{E}^{+}_{o\ m,n}e^{-j\beta_{g\ m,n}z}$$

$$\rightarrow \bar{H}^+_{x\ TM\ m,n} = \frac{j\omega\varepsilon}{k^2_{c\ m,n}}\left(\frac{n\pi}{b}\right)\sin\left(\frac{m\pi}{a}x\right)\cos\left(\frac{n\pi}{b}y\right)\bar{E}^+_{o\ m,n}e^{-j\beta_{g\ m,n}z} = -\frac{1}{\eta_{g\ TM\ m,n}}\bar{E}^+_{y\ TM\ m,n}$$

$$\rightarrow \bar{H}^+_{y\ TM\ m,n} = \frac{-j\omega\varepsilon}{k^2_{c\ m,n}}\left(\frac{m\pi}{a}\right)\cos\left(\frac{m\pi}{a}x\right)\sin\left(\frac{n\pi}{b}y\right)\bar{E}^+_{o\ m,n}e^{-j\beta_{g\ m,n}z} = \frac{1}{\eta_{g\ TM\ m,n}}\bar{E}^+_{x\ TM\ m,n}$$

- ***TM$_{11}$ Mode***

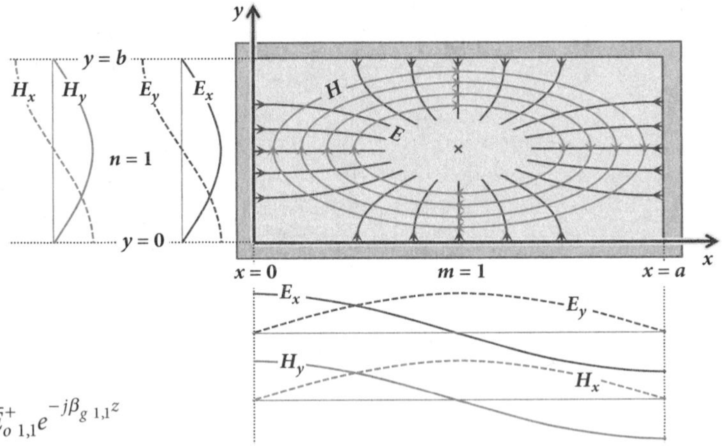

$$\rightarrow f_{c\ 1,1} = \frac{c_o}{2}\sqrt{\left(\frac{1}{a}\right)^2 + \left(\frac{1}{b}\right)^2}$$

$$\rightarrow k_{c\ 1,1} = \pi\sqrt{\left(\frac{1}{a}\right)^2 + \left(\frac{1}{b}\right)^2}$$

$$\rightarrow \beta_{g\ 1,1} = \frac{\omega}{c_o}\sqrt{1-\left(\frac{f_{c\ 1,1}}{f}\right)^2}$$

$$\rightarrow \bar{H}^+_{z\ TM\ 1,1} = 0$$

$$\rightarrow \bar{E}^+_{z\ TM\ 1,1} = \sin\left(\frac{\pi}{a}x\right)\sin\left(\frac{\pi}{b}y\right)\bar{E}^+_{o\ 1,1}e^{-j\beta_{g\ 1,1}z}$$

$$\rightarrow \bar{E}^+_{x\ TM\ 1,1} = \frac{-j\beta_{g\ 1,1}}{k^2_{c\ 1,1}}\left(\frac{\pi}{a}\right)\cos\left(\frac{\pi}{a}x\right)\sin\left(\frac{\pi}{b}y\right)\bar{E}^+_{o\ 1,1}e^{-j\beta_{g\ 1,1}z}$$

$$\rightarrow \bar{E}^+_{y\ TM\ 1,1} = \frac{-j\beta_{g\ 1,1}}{k^2_{c\ 1,1}}\left(\frac{\pi}{b}\right)\sin\left(\frac{\pi}{a}x\right)\cos\left(\frac{\pi}{b}y\right)\bar{E}^+_{o\ 1,1}e^{-j\beta_{g\ 1,1}z}$$

$$\rightarrow \bar{H}^+_{x\ TM\ 1,1} = \frac{j\omega\varepsilon}{k^2_{c\ 1,1}}\left(\frac{\pi}{b}\right)\sin\left(\frac{\pi}{a}x\right)\cos\left(\frac{\pi}{b}y\right)\bar{E}^+_{o\ 1,1}e^{-j\beta_{g\ 1,1}z} = -\frac{1}{\eta_{g\ TM\ 1,1}}\bar{E}^+_{y\ TM\ 1,1}$$

$$\rightarrow \bar{H}^+_{y\ TM\ 1,1} = \frac{-j\omega\varepsilon}{k^2_{c\ 1,1}}\left(\frac{\pi}{a}\right)\cos\left(\frac{\pi}{a}x\right)\sin\left(\frac{\pi}{b}y\right)\bar{E}^+_{o\ 1,1}e^{-j\beta_{g\ 1,1}z} = \frac{1}{\eta_{g\ TM\ 1,1}}\bar{E}^+_{x\ TM\ 1,1}$$

Modal Dispersion and Waveguide Bandwidth

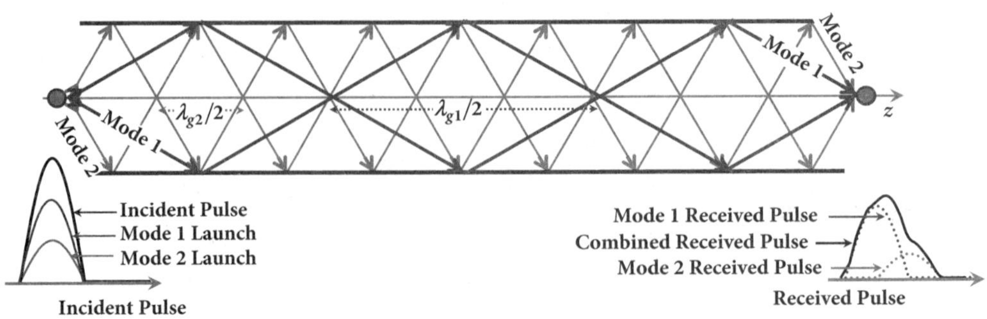

Field Maps for the TE_{10} and TM_{11} Modes

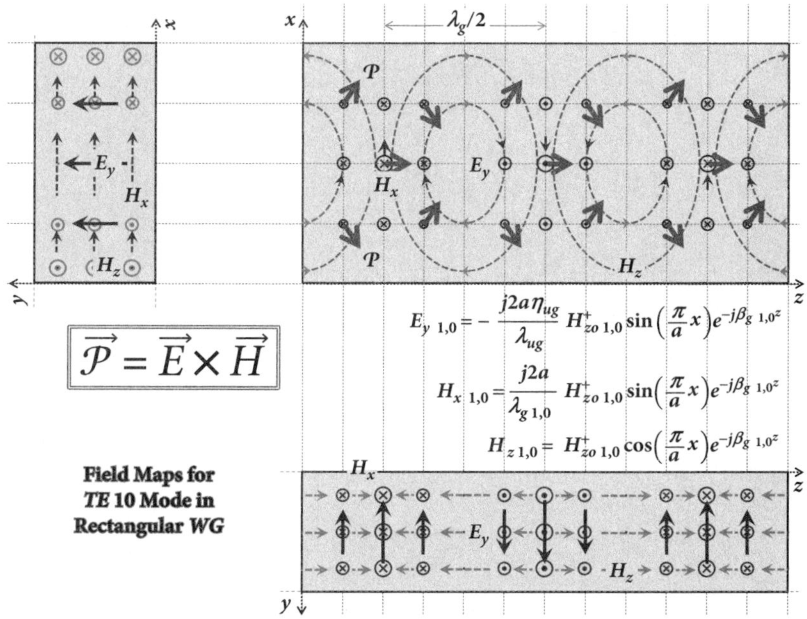

Field Maps for *TE* 10 Mode in Rectangular *WG*

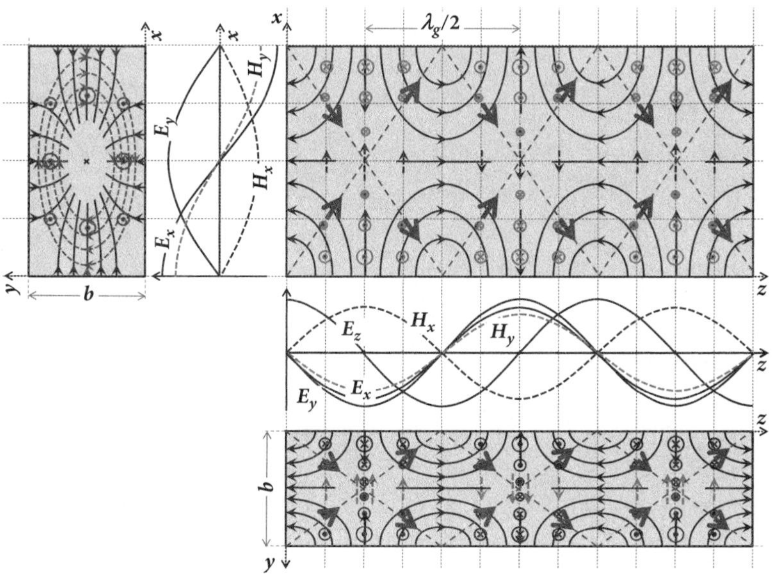

Phase and Group Velocities

- Phase velocity: $c_{ph\ g\ m,n} = \dfrac{\omega}{\beta_{g\ m,n}} = \dfrac{c_{ph\ ug}}{\sqrt{1-\left(\dfrac{f_{c\ m,n}}{f}\right)^2}}$

- Group velocity: $c_{gr\ g\ m,n} = \dfrac{d\omega}{d\beta_{g\ m,n}} = c_{ph\ ug}\sqrt{1-\left(\dfrac{f_{c\ m,n}}{f}\right)^2}$ (information propagation velocity)

  → The product $(c_{gr\ g\ m,n})\cdot(c_{ph\ g\ m,n}) = c_{ph\ ug}^2 = c_o^2$

  → For plane wave propagation (no cutoff), $c_{gr\ ug} = c_{ph\ ug} = c_o$

Waveguide Cavity Resonators

- Cavity resonant frequencies: $f_{o\ m,n,\ell} = \dfrac{\omega_{o\ m,n,\ell}}{2\pi} = \dfrac{1}{\sqrt{\mu\varepsilon}}\sqrt{\left[\left(\dfrac{m}{a}\right)^2+\left(\dfrac{n}{b}\right)^2+\left(\dfrac{\ell}{d}\right)^2\right]}$

- Cavity *TE* modes

  → $\bar{E}_{x\ m,n,\ell}(x,y,z) = \left\{\left[\cos\left(\dfrac{m\pi}{a}x\right)\sin\left(\dfrac{n\pi}{b}y\right)\sin\left(\dfrac{\ell\pi}{d}z\right)\right]\left[j\bar{E}_{xs\ m,n,\ell}\right]\right\}$

  → $\bar{E}_{y\ m,n,\ell}(x,y,z) = \left\{\left[\sin\left(\dfrac{m\pi}{a}x\right)\cos\left(\dfrac{n\pi}{b}y\right)\sin\left(\dfrac{\ell\pi}{d}z\right)\right]\left[j\bar{E}_{ys\ m,n,\ell}\right]\right\}$

  → $\bar{H}_{x\ m,n,\ell}(x,y,z) = \left\{\left[\sin\left(\dfrac{m\pi}{a}x\right)\cos\left(\dfrac{n\pi}{b}y\right)\cos\left(\dfrac{\ell\pi}{d}z\right)\right]\left[\dfrac{1}{\omega\mu}\left(\dfrac{\ell\pi}{d}\right)\bar{E}_{ys\ m,n,\ell}\right]\right\}$

  → $\bar{H}_{y\ m,n,\ell}(x,y,z) = \left\{\left[\cos\left(\dfrac{m\pi}{a}x\right)\sin\left(\dfrac{n\pi}{b}y\right)\cos\left(\dfrac{\ell\pi}{d}z\right)\right]\left[\dfrac{-1}{\omega\mu}\left(\dfrac{\ell\pi}{d}\right)\bar{E}_{xs\ m,n,\ell}\right]\right\}$

  → $\bar{H}_{z\ m,n,\ell}(x,y,z) = \left\{\left[\cos\left(\dfrac{m\pi}{a}x\right)\cos\left(\dfrac{n\pi}{b}y\right)\sin\left(\dfrac{\ell\pi}{d}z\right)\right]\dfrac{1}{\omega\mu}\left[\dfrac{n\pi}{b}\bar{E}_{xs\ m,n,\ell} - \dfrac{m\pi}{a}\bar{E}_{ys\ m,n,\ell}\right]\right\}$

- Cavity dominant (lowest resonance frequency): the TE_{101} mode

- $f_{o\ 101} = \dfrac{1}{\sqrt{\mu\varepsilon}}\sqrt{\left[\left(\dfrac{1}{a}\right)^2+\left(\dfrac{1}{d}\right)^2\right]}$

  → $\bar{E}_{x\ 101}(x,y,z) = \bar{H}_{y\ 101}(x,y,z) = 0$

  → $\bar{E}_{y\ 101}(x,y,z) = \left[\sin\left(\dfrac{\pi}{a}x\right)\sin\left(\dfrac{\pi}{d}z\right)\right]\left[\bar{E}_{yo}\right]$

  → $\bar{H}_{x\ 101}(x,y,z) = \left[\sin\left(\dfrac{\pi}{a}x\right)\cos\left(\dfrac{\pi}{d}z\right)\right]\left[\dfrac{-j\pi}{\omega\mu d}\bar{E}_{yo}\right]$

  → $\bar{H}_{z\ 101}(x,y,z) = \left[\cos\left(\dfrac{\pi}{a}x\right)\sin\left(\dfrac{\pi}{d}z\right)\right]\left[\dfrac{j\pi}{\omega\mu a}\bar{E}_{yo}\right]$

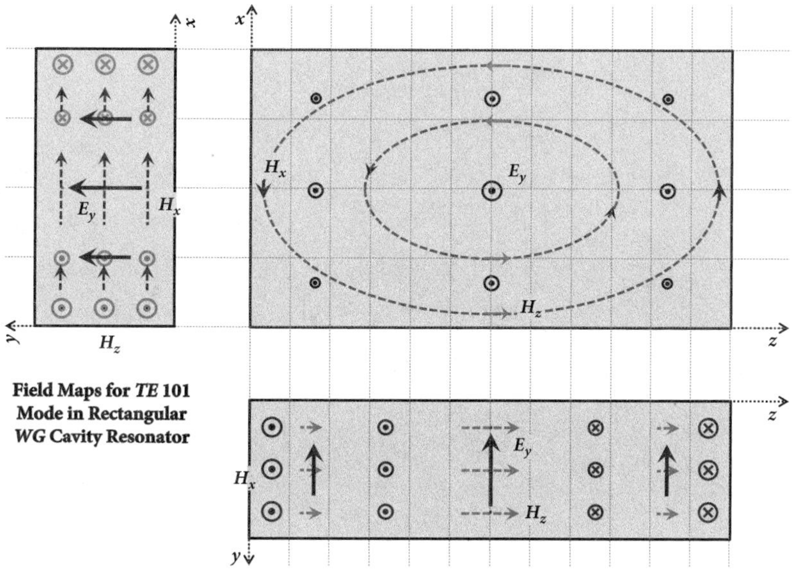

Field Maps for *TE* 101 Mode in Rectangular *WG* Cavity Resonator

- Cavity quality factor for the dominant mode

$$\rightarrow W_{e\ 101} = W_{m\ 101} = \frac{1}{16}\varepsilon'\left|\bar{E}_{yo}\right|^2[abd]$$

$$\rightarrow P_{c\ 101} = \frac{1}{2}\sqrt{\frac{\pi}{\sigma(f\mu_o)^3}}\left|\bar{E}_{yo}\right|^2\left\{\frac{ab}{4d^2}+\frac{bd}{4a^2}+\left[\frac{ad}{8d^2}+\frac{ad}{8a^2}\right]\right\}$$

$$\rightarrow P_{d\ 101} = \omega\varepsilon''\left|\bar{E}_{yo}\right|^2\frac{abd}{8}$$

$$\rightarrow Q_{101} = \frac{2\omega_o W_{101}}{P_{t\ 101}} = \frac{\omega_o\varepsilon'}{\sqrt{\frac{\pi}{\sigma(f\mu_o)^3}}\left\{\frac{1}{d^3}+\frac{1}{a^3}+\left[\frac{1}{2bd^2}+\frac{1}{2ba^2}\right]\right\}+\omega\varepsilon''}$$

APPENDIX A
Symbols and Units

Symbol	Quantity	SI Unit	Comments
Uppercase Symbols			
A	Magnetic vector potential	Wb/m	weber/meter
B	Magnetic flux density	T (or Wb/m^2)	tesla (or weber/meter^2)
B	Susceptance	S	siemens
C	Capacitance	F	farad
C_z	Capacitance per unit length	F/m	farad/meter
D	Electric flux density	C/m^2	coulomb/meter^2
D	Electric displacement	C/m^2	coulomb/meter^2
E	Electric field intensity	V/m	volt/meter
F	Force	N	newton
G	Conductance	S	siemens
G_z	Conductance per unit length	S/m	siemens/m
H	Magnetic field intensity	A/m	ampere/meter
I	Current	A	ampere
J	Cross-sectional density of volume current	A/m^2	ampere/meter^2
K	Linear density of surface current	A/m	ampere/meter
L	Inductance (self and mutual)	H	henry
L	Length	m	meter
L_z	Inductance per unit length	H/m	henry/meter
M	Magnetization intensity	A/m	ampere/meter
M	Mutual inductance	H	henry
N	Number of turns	—	dimensionless
P	Electric polarization density	C/m^2	coulomb/meter^2
P	Power	W	watt
$\mathcal{P}$	Poynting vector	W/m^2	watt/meter^2
Q	Electric charge	C	coulomb
R	Radial distance	m	meter
R	Resistance	Ω	ohm
R_z	Resistance per unit length	Ω/m	ohm/meter

Table A.1 Symbols and Units

Symbol	Quantity	SI Unit	Comments
$\mathfrak{R}$	Reluctance	H^{-1}	$henry^{-1}$
S	Area	m^2	$meter^2$
SWR	Standing wave ratio	—	dimensionless ratio
T	One-way propagation coefficient	—	dimensionless ratio
T	Time period	s	second
T	Torque	N·m	newton·meter
U	Energy stored	J	joule
V	Electric potential	V	volt
V	Voltage	V	volt
V_m	Magnetic scalar potential	A	ampere
W	Work, energy	J	joule
X	Reactance	Ω	ohm
Y	Admittance	S	siemens
Y_n	Normalized admittance	—	dimensionless ratio
Z	Impedance	Ω	ohm
Z_n	Normalized impedance	—	dimensionless ratio
Z_o	Characteristic impedance	Ω	ohm
	Lowercase Symbols		
a	Unit vector	—	dimensionless
b	Normalized susceptance	—	dimensionless ratio
c	Velocity	m/s	meter/second
c_g	Group velocity	m/s	meter/second
c_o	Velocity of light in free space	m/s	meter/second
c_{ph}	Phase velocity	m/s	meter/second
d	Length, distance	m	meter
e	Electronic charge	C	coulomb
emf	Electromotive force (emf)	V	volt
f	Frequency	Hz	hertz
f_c	Cutoff frequency	Hz	hertz
g	Normalized conductance	—	dimensionless ratio
i	Current, instantaneous	A	Ampere
k	Wave number	rad/m	radian/meter
k_c	Cutoff wave number	rad/s	radian/second
ℓ	Length	m	meter
m	Magnetic dipole moment	$A \cdot m^2$	$ampere \cdot meter^2$
mmf	Magnetomotive force	A	ampere
n	Index of refraction	—	dimensionless ratio
n	Number of turns per unit length	m^{-1}	$meter^{-1}$
p	Electric dipole moment	C·m	coulomb·meter
q	Electric charge (point)	C	Coulomb
r	Normalized resistance	—	dimensionless ratio
r	Radial distance	m	meter

Table A.1 Symbols and Units (*Continued*)

Symbol	Quantity	SI Unit	Comments
r	Spherical coordinate distance	m	meter
t	Time	s	second
$\tan(\delta)$	Loss tangent	—	dimensionless
u	Parametric variable	varies	varies
v	Velocity	m/s	meter/second
v	Voltage, instantaneous	V	volt
v	Volume	m^3	$meter^3$
w	Energy density	J/m^3	$joule/meter^3$
x	Cartesian coordinate distance	m	meter
x	Normalized reactance	—	dimensionless ratio
y	Cartesian coordinate distance	m	meter
z	Cartesian coordinate distance	m	meter
	Greek Symbols		
α	Attenuation coefficient	Np/m	neper/meter
β	Phase constant (wave number)	rad/m	radian/meter
γ	Propagation constant	m^{-1}	$meter^{-1}$
Γ	Reflection coefficient	—	dimensionless ratio
δ	Skin depth	m	meter
ε	Permittivity	F/m	farad/meter
ε^*	Complex permittivity	F/m	farad/meter
ε_o	Permittivity of free space	F/m	farad/meter
ε_r	Dielectric constant	—	dimensionless ratio
ε_r	Relative permittivity	—	dimensionless ratio
η	Intrinsic impedance	Ω	ohm
θ	Angle of incidence	rad	radian
θ	Spherical coordinate	rad	radian
λ	Wavelength	m	meter
λ_g	Wavelength in waveguides	m	meter
Λ	Magnetic flux linkage	Wb	weber
μ	Permeability	H/m	henry/meter
μ^*	Complex permeability	H/m	henry/meter
μ_o	Permeability of free space	H/m	henry/meter
μ_r	Relative permeability	—	dimensionless ratio
μ^{+-}	Mobility (electron, hole)	$m^2/(V \cdot s)$	$meter^2/(volt \cdot second)$
ρ	Cylindrical coordinate radius	m	meter
ρ	Resistivity	$\Omega \cdot m$	ohm · meter
ρ_L	Charge density (linear)	C/m	coulomb/meter
ρ_S	Charge density (surface)	C/m^2	$coulomb/meter^2$
ρ_v	Charge density (volume)	C/m^3	$coulomb/meter^3$
ρ	Reflection coefficient, local (time domain)	—	dimensionless ratio
P	Reflection coefficient phasor, local	—	dimensionless ratio

Table A.1 Symbols and Units (*Continued*)

Symbol	Quantity	SI Unit	Comments
σ	Conductivity	S/m	siemens/meter
τ	Time delay	s	second
τ	Transmission coefficient	—	dimensionless ratio
φ	Cylindrical and spherical coordinate	rad	radian
φ	Phase angle	rad	radian
χ_e	Electric susceptibility	—	dimensionless ratio
χ_m	Magnetic susceptibility	—	dimensionless ratio
ψ_e	Electric flux	C	coulomb
ψ_m	Magnetic flux	Wb	weber
ω	Angular velocity	rad/s	radian/second
ω	Radian frequency	rad/s	radian/second

TABLE A.1 Symbols and Units (*Continued*)

APPENDIX B

Constants and SI Units

Symbol	Quantity	Value	SI Unit
Electron charge	e	$1.60217733 \times 10^{-19}$	C
Electron mass	m	$9.1093897 \times 10^{-31}$	kg
Permittivity of free space	ε_o	$8.854187817 \times 10^{-12}$	F/m
Permeability of free space	μ_o	$4\pi \times 10^{-7}$	H/m
Velocity of light in free space	c_o	2.99792458×10^{8}	m/s
Intrinsic impedance of free space	η_o	$376.7 \sim 120\pi$	ohm

TABLE B.1 Constants

Physical Quantity	Unit	Symbol
Mass	kilogram	kg
Length	meter	m
Time	second	s
Temperature	kelvin	K
Amount of substance	mole	mol
Electric current	ampere	A
Luminous intensity	candela	cd

TABLE B.2 Fundamental SI (Rationalized MKS) Units

Multiplication Factor	Power of 10	Prefix	Symbol
$1{,}000{,}000{,}000{,}000{,}000{,}000 = 10^{+18}$	+18	exa	E
$1{,}000{,}000{,}000{,}000{,}000 = 10^{+15}$	+15	peta	P
$1{,}000{,}000{,}000{,}000 = 10^{+12}$	+12	tera	T
$1{,}000{,}000{,}000 = 10^{+9}$	+9	giga	G
$1{,}000{,}000 = 10^{+6}$	+6	mega	M
$1{,}000 = 10^{+3}$	+3	kilo	k
$100 = 10^{+2}$	+2	hecto	h
$10 = 10^{+1}$	+1	deka	da
$0.1 = 10^{-1}$	−1	deci	d
$0.01 = 10^{-2}$	−2	centi	c
$0.001 = 10^{-3}$	−3	milli	m
$0.000{,}001 = 10^{-6}$	−6	micro	μ
$0.000{,}000{,}001 = 10^{-9}$	−9	nano	n
$0.000{,}000{,}000{,}001 = 10^{-12}$	−12	pico	p
$0.000{,}000{,}000{,}000{,}001 = 10^{-15}$	−15	femto	f
$0.000{,}000{,}000{,}000{,}000{,}001 = 10^{-18}$	−18	atto	a

Table B.3 Power of 10 Prefixes

APPENDIX C
Material Properties

Note that the constitutive parameters of some of the materials are frequency and temperature dependent. The listed values are average low-frequency values at room temperature.

Sorted Alphabetically			Sorted Descending σ		
Material	**Scientific Format**	**Engineering Format**	**Scientific Format**	**Engineering Format**	**Material**
Aluminum	3.54E+07	35.40E+06	6.17E+07	6.17E+07	Silver
Bakelite	1.00E–09	1.00E–09	5.80E+07	5.80E+07	Copper
Brass	1.57E+07	15.70E+06	4.10E+07	4.10E+07	Gold
Bronze	1.00E+07	10.00E+06	3.54E+07	3.54E+07	Aluminum
Carbon	3.00E+04	30.00E+03	1.82E+07	1.82E+07	Tungsten
Clay	5.00E–03	5.00E–03	1.67E+07	1.67E+07	Zinc
Constantan	2.26E+06	2.26E+06	1.57E+07	1.57E+07	Brass
Copper	5.80E+07	58.00E+06	1.45E+07	1.45E+07	Nickel
Diamond	2.00E–13	200.00E–15	1.03E+07	1.03E+07	Iron
Ferrite (typical)	1.00E+02	100.00E+00	1.00E+07	1.00E+07	Bronze
Germanium	2.20E+06	2.20E+06	1.00E+07	1.00E+07	Phosphor bronze
Germanium (pure)	2.20E+00	2.20E+00	9.00E+06	9.00E+06	Tin
Glass	1.00E–12	1.00E–12	7.00E+06	7.00E+06	Solder
Gold	4.10E+07	41.00E+06	6.00E+06	6.00E+06	Steel (carbon)
Granite	1.00E–06	1.00E–06	5.00E+06	5.00E+06	Lead
Graphite	7.00E+04	70.00E+03	2.27E+06	2.27E+06	Manganin
Iron	1.03E+07	10.30E+06	2.26E+06	2.26E+06	Constantan
Lead	5.00E+06	5.00E+06	2.20E+06	2.20E+06	Germanium
Limestone	1.00E–02	10.00E–03	1.10E+06	1.10E+06	Steel (stainless)
Manganin	2.27E+06	2.27E+06	1.00E+06	1.00E+06	Mercury
Marble	1.00E–08	10.00E–09	1.00E+06	1.00E+06	Nichrome
Mercury	1.00E+06	1.00E+06	7.00E+04	7.00E+04	Graphite
Mica	1.00E–15	1.00E–15	3.00E+04	3.00E+04	Carbon
Nichrome	1.00E+06	1.00E+06	2.30E+03	2.30E+03	Silicon

TABLE C.1 Material Conductivities (σ), S/m

Sorted Alphabetically		
Material	**Scientific Format**	**Engineering Format**
Nickel	1.45E+07	14.50E+06
Oil (transformer)	1.00E−11	10.00E−12
Paraffin	1.00E−15	1.00E−15
Phosphor bronze	1.00E+07	10.00E+06
Polystyrene	1.00E−16	100.00E−18
Porcelain	2.00E−13	200.00E−15
Quartz	1.00E−17	10.00E−18
Rubber	1.00E−15	1.00E−15
Silicon	2.30E+03	2.30E+03
Silicon (pure)	4.40E−04	440.00E−06
Silver	6.17E+07	61.70E+06
Soil (dry) ~	1.00E−04	100.00E−06
Soil (sandy) ~	1.00E−05	10.00E−06
Soil (wet) ~	1.00E−02	10.00E−03
Solder	7.00E+06	7.00E+06
Steel (carbon)	6.00E+06	6.00E+06
Steel (stainless)	1.10E+06	1.10E+06
Tin	9.00E+06	9.00E+06
Tungsten	1.82E+07	18.20E+06
Water (distilled) ~	1.00E−04	100.00E−06
Water (fresh) ~	1.00E−03	1.00E−03
Water (sea)	4.00E+00	4.00E+00
Wax	1.00E−17	10.00E−18

Sorted Descending σ		
Scientific Format	**Engineering Format**	**Material**
1.00E+02	1.00E+02	Ferrite (typical)
4.00E+00	4.00E+00	Water (sea)
2.20E+00	2.20E+00	Germanium (pure)
1.00E−02	1.00E−02	Limestone
1.00E−02	1.00E−02	Soil (wet) ~
5.00E−03	5.00E−03	Clay
1.00E−03	1.00E−03	Water (fresh) ~
4.40E−04	4.40E−04	Silicon (pure)
1.00E−04	1.00E−04	Soil (dry) ~
1.00E−04	1.00E−04	Water (distilled) ~
1.00E−05	1.00E−05	Soil (sandy) ~
1.00E−06	1.00E−06	Granite
1.00E−08	1.00E−08	Marble
1.00E−09	1.00E−09	Bakelite
1.00E−11	1.00E−11	Oil (transformer)
1.00E−12	1.00E−12	Glass
2.00E−13	2.00E−13	Diamond
2.00E−13	2.00E−13	Porcelain
1.00E−15	1.00E−15	Mica
1.00E−15	1.00E−15	Paraffin
1.00E−15	1.00E−15	Rubber
1.00E−16	1.00E−16	Polystyrene
1.00E−17	1.00E−17	Quartz

Table C.1 Material Conductivities (σ), S/m (*Continued*)

Sorted Alphabetically		
Material	ε_r'	$\varepsilon_r'/\varepsilon_r''$
Air	1.0	0.00000
Alcohol (ethyl)	25.0	0.10000
Aluminum oxide	8.8	0.00060
Amber	2.7	0.00200
Ammonia	22.0	
Bakelite	4.7–5.0	0.02200
Barium titanate	1200.0	0.01300
Carbon dioxide	1.0	

Sorted Ascending ε_r'		
ε_r'	$\varepsilon_r'/\varepsilon_r''$	**Material**
1.0	0.00000	Vacuum
1.0	0.00000	Water (dehydrated)
1.0	0.00000	Air
1.0		Carbon dioxide
1.0	0.00010	Styrofoam
1.5–4.0	0.01000	Wood (dry)
2.0–4.0	0.00800	Paper
2.1		Petroleum oil

$\varepsilon = \varepsilon_r \cdot \varepsilon_o = (\varepsilon_r' - j\varepsilon_r'') \cdot \varepsilon_o$, $\varepsilon_o = 8.854 \times 10^{-12}$ F/m.

Table C.2 Material Permittivities, F/m

Sorted Alphabetically		
Material	ε_r'	$\varepsilon_r'/\varepsilon_r''$
Dry soil	2.5–3.5	
Dry soil	2.8–4.0	0.05000
Ferrite (NiZn)	12.4	0.00025
Formica	6.0	
Germanium	16.0	
Glass	4.0–7.0	0.00200
Ice	4.2	0.05000
Mica	5.4–6.0	0.00060
Neoprene	6.6	0.01100
Nylon	3.5	0.02000
Oil	2.3	
Paper	2.0–4.0	0.00800
Paraffin wax	2.2	
Petroleum oil	2.1	
Plexiglas	3.5	0.03000
Polyethylene	2.3	0.00020
Polypropylene	2.3	0.00030
Polystyrene	2.6	0.00005
Porcelain	5.7	0.01400
Pyranol	4.4	0.00050
Pyrex glass	4.0	0.00060
Quartz	3.8–5.0	0.00075
Rubber	2.3–4.0	0.00200
Seawater	72.0–80.0	4.00000
Silica	3.8	0.00075
Silicon	11.8	
Snow	3.3	0.50000
Sodium chloride	5.9	0.00010
Steatite	5.8	0.00300
Styrofoam	1.0	0.00010
Teflon	2.1	0.00030
Titanium dioxide	100.0	0.00150
Vacuum	1.0	0.00000
Water (dehydrated)	1.0	0.00000
Water (distilled)	81	0.04000
Wood (dry)	1.5–4.0	0.01000

Sorted Ascending ε_r'		
ε_r'	$\varepsilon_r'/\varepsilon_r''$	**Material**
2.1	0.00030	Teflon
2.2		Paraffin wax
2.3	0.00030	Polypropylene
2.3	0.00020	Polyethylene
2.3		Oil
2.3–4.0	0.00200	Rubber
2.5–3.5		Dry soil
2.6	0.00005	Polystyrene
2.7	0.00200	Amber
2.8–4.0	0.05000	Dry soil
3.3	0.50000	Snow
3.5	0.03000	Plexiglas
3.5	0.02000	Nylon
3.8–5.0	0.00075	Quartz
3.8	0.00075	Silica
4.0–7.0	0.00200	Glass
4.0	0.00060	Pyrex glass
4.2	0.05000	Ice
4.4	0.00050	Pyranol
4.7–5.0	0.02200	Bakelite
5.4–6.0	0.00060	Mica
5.7	0.01400	Porcelain
5.8	0.00300	Steatite
5.9	0.00010	Sodium chloride
6.0		Formica
6.6	0.01100	Neoprene
8.8	0.00060	Aluminum oxide
11.8		Silicon
12.4	0.00025	Ferrite (NiZn)
16.0		Germanium
22.0		Ammonia
25.0	0.10000	Alcohol, ethyl
72.0–80.0	4.00000	Seawater
81	0.04000	Water (distilled)
100.0	0.00150	Titanium dioxide
1200.0	0.01300	Barium titanate

Table C.2 Material Permittivities, F/m (*Continued*)

Sorted Alphabetically	
Material	μ_r
Diamagnetic ≈ 1.0	
Bismuth	0.99983400
Copper	0.99999400
Gold	0.99996000
Mercury	0.99997000
Paraffin	0.99999942
Sapphire	0.99999976
Silver	0.99998000
Water	0.99999200
Paramagnetic ≈ 1.0	
Air	1.00000037
Aluminum	1.00002200
Beryllium	1.00000079
Magnesium	1.00001200
Manganese sulfate	1.00000000
Nickel chloride	1.00004000
Palladium	1.00082000
Platinum	1.00026500
Titanium	1.00018000
Tungsten	1.00008000
Wood	1.00000043
Ferromagnetic (nonlinear)	
Carbon steel	100.0
Cast iron	60.0
Cobalt	100.0
Ferrite (typical)	1,000.0
Iron (pure)	5,000.0
Machine steel	300.0
Mild steel	2,000.0
Mumetal	20,000.0
Nickel	70.0
Permalloy 45	2,500.0
Powdered iron	100.0
Purified iron	200,000.0
Sendust	30,000.0
Silicon iron	7,000.0
Supermalloy	100,000.0
Transformer iron	3,000.0

Sorted Ascending μ_r	
μ_r	**Material**
Diamagnetic ≈ 1.0	
0.99983400	Bismuth
0.99996000	Gold
0.99997000	Mercury
0.99998000	Silver
0.99999200	Water
0.99999400	Copper
0.99999942	Paraffin
0.99999976	Sapphire
Paramagnetic ≈ 1.0	
1.00000000	Manganese sulfate
1.00000037	Air
1.00000043	Wood
1.00000079	Beryllium
1.00001200	Magnesium
1.00002200	Aluminum
1.00004000	Nickel chloride
1.00008000	Tungsten
1.00018000	Titanium
1.00026500	Platinum
1.00082000	Palladium
Ferromagnetic (nonlinear)	
60.0	Cast iron
70.0	Nickel
100.0	Cobalt
100.0	Carbon steel
100.0	Powdered iron
300.0	Machine steel
1,000.0	Ferrite (typical)
2,000.0	Mild steel
2,500.0	Permalloy 45
3,000.0	Transformer iron
5,000.0	Iron (pure)
7,000.0	Silicon iron
20,000.0	Mumetal
30,000.0	Sendust
100,000.0	Supermalloy
200,000.0	Purified iron

Ferromagnetic materials are highly nonlinear. Listed values for low magnetic flux densities. Maximum permeabilities may be much higher.

$$\mu = \mu_r \cdot \mu_o,\ \mu_o = 4\pi \times 10^{-7}\ \text{H/m}$$

TABLE C.3 Material Permeabilities, H/m

APPENDIX D

Vector Identities

The Gradient		
Cartesian	$\vec{\nabla} = \vec{a}_x \frac{\partial}{\partial x} + \vec{a}_y \frac{\partial}{\partial y} + \vec{a}_z \frac{\partial}{\partial z}$	(5.65)
Cylindrical	$\vec{\nabla} = \vec{a}_\rho \frac{\partial}{\partial \rho} + \vec{a}_\varphi \frac{1}{\rho}\frac{\partial}{\partial \varphi} + \vec{a}_z \frac{\partial}{\partial z}$	(5.66)
Spherical	$\vec{\nabla} = \vec{a}_r \frac{\partial}{\partial r} + \vec{a}_\theta \frac{1}{r}\frac{\partial}{\partial \theta} + \vec{a}_\varphi \frac{1}{r\sin\theta}\frac{\partial}{\partial \varphi}$	(5.67)
The Divergence		
Cartesian	$\vec{\nabla}\cdot\vec{D} = \left[\frac{\partial}{\partial x}D_x + \frac{\partial}{\partial y}D_y + \frac{\partial}{\partial z}D_z\right]$	(4.19)
Cylindrical	$\vec{\nabla}\cdot\vec{D} = \left[\frac{1}{\rho}\frac{\partial}{\partial \rho}(\rho D_\rho) + \frac{1}{\rho}\frac{\partial}{\partial \varphi}D_\varphi + \frac{\partial}{\partial z}D_z\right]$	(4.20)
Spherical	$\vec{\nabla}\cdot\vec{D} = \left[\frac{1}{r^2}\frac{\partial}{\partial r}(r^2 D_r) + \frac{1}{r\sin\theta}\frac{\partial}{\partial \theta}(\sin\theta D_\theta) + \frac{1}{r\sin\theta}\frac{\partial}{\partial \varphi}D_\varphi\right]$	(4.21)
The Curl		
Cartesian*	$\vec{\nabla}\times\vec{H} = \vec{a}_x\left\{\frac{\partial H_z}{\partial y} - \frac{\partial H_y}{\partial z}\right\} + \vec{a}_y\left\{\frac{\partial H_x}{\partial z} - \frac{\partial H_z}{\partial x}\right\} + \vec{a}_z\left\{\frac{\partial H_y}{\partial x} - \frac{\partial H_x}{\partial y}\right\}$	(8.56)
Cylindrical	$\vec{\nabla}\times\vec{H} = \vec{a}_\rho\left\{\frac{1}{\rho}\frac{\partial H_z}{\partial \varphi} - \frac{\partial H_\varphi}{\partial z}\right\} + \vec{a}_\varphi\left\{\frac{\partial H_\rho}{\partial z} - \frac{\partial H_z}{\partial \rho}\right\} + \frac{1}{\rho}\vec{a}_z\left\{\frac{\partial(\rho H_\varphi)}{\partial \rho} - \frac{\partial H_\rho}{\partial \varphi}\right\}$	(8.57)
Spherical	$\vec{\nabla}\times\vec{H} = \vec{a}_r\frac{1}{r\sin\theta}\left\{\frac{\partial(\sin\theta\, H_\varphi)}{\partial \theta} - \frac{\partial H_\theta}{\partial \varphi}\right\} + \vec{a}_\theta\frac{1}{r}\left\{\frac{1}{\sin\theta}\frac{\partial H_r}{\partial \varphi} - \frac{\partial(r\, H_\varphi)}{\partial r}\right\} + \vec{a}_\theta\frac{1}{r}\left\{\frac{\partial(r\, H_\theta)}{\partial r} - \frac{\partial H_r}{\partial \theta}\right\}$	(8.58)
The Laplacian Operator		
Cartesian	$\nabla^2 V(x,y,z) = \frac{\partial^2 V}{\partial x^2} + \frac{\partial^2 V}{\partial y^2} + \frac{\partial^2 V}{\partial z^2}$	(7.6)
Cylindrical	$\nabla^2 V(\rho,\varphi,z) = \frac{1}{\rho}\frac{\partial}{\partial \rho}\left(\rho\frac{\partial V}{\partial \rho}\right) + \frac{1}{\rho^2}\frac{\partial^2 V}{\partial \varphi^2} + \frac{\partial^2 V}{\partial z^2}$	(7.7)
Spherical	$\nabla^2 V(r,\theta,\varphi) = \frac{1}{r^2}\frac{\partial}{\partial r}\left(r^2\frac{\partial V}{\partial r}\right) + \frac{1}{r^2\sin\theta}\frac{\partial}{\partial \theta}\left(\sin\theta\frac{\partial V}{\partial \theta}\right) + \frac{1}{r^2\sin^2\theta}\frac{\partial^2 V}{\partial \varphi^2}$	(7.8)

*Also, the Cartesian curl can be expressed as

$$\text{curl } \vec{H} = \begin{vmatrix} \vec{a}_x & \vec{a}_y & \vec{a}_z \\ \frac{\partial}{\partial x} & \frac{\partial}{\partial y} & \frac{\partial}{\partial z} \\ H_x & H_y & H_z \end{vmatrix} \tag{8.53}$$

Table D.1 Differential Vector Operators

$\vec{\nabla}(V+W)=\vec{\nabla}V+\vec{\nabla}W$
$\vec{\nabla}\cdot(\vec{A}+\vec{B})=\vec{\nabla}\cdot\vec{A}+\vec{\nabla}\cdot\vec{B}$
$\vec{\nabla}\times(\vec{A}+\vec{B})=\vec{\nabla}\times\vec{A}+\vec{\nabla}\times\vec{B}$
$\vec{\nabla}(VW)=V\vec{\nabla}W+W\vec{\nabla}V$
$\vec{\nabla}\cdot(V\vec{A})=\vec{A}\cdot\vec{\nabla}V+V\vec{\nabla}\cdot\vec{A}$
$\vec{\nabla}\times(V\vec{A})=\vec{\nabla}V\times\vec{A}+V\vec{\nabla}\times\vec{A}$
$(\vec{A}\times\vec{B})\cdot\vec{C}=(\vec{B}\times\vec{C})\cdot\vec{A}=(\vec{C}\times\vec{A})\cdot\vec{B}$
$\vec{A}\times(\vec{B}\times\vec{C})=(\vec{A}\cdot\vec{C})\vec{B}-(\vec{A}\cdot\vec{B})\vec{C}$
$\vec{\nabla}\cdot(\vec{A}\times\vec{B})=\vec{B}\cdot(\vec{\nabla}\times\vec{A})-\vec{A}\cdot(\vec{\nabla}\times\vec{B})$
$\vec{\nabla}(\vec{A}\cdot\vec{B})=(\vec{A}\cdot\vec{\nabla})\vec{B}+(\vec{B}\cdot\vec{\nabla})\vec{A}+\vec{A}\times(\vec{\nabla}\times\vec{B})+\vec{B}\times(\vec{\nabla}\times\vec{A})$
$\vec{\nabla}\times(\vec{A}\times\vec{B})=\vec{A}(\vec{\nabla}\cdot\vec{B})-\vec{B}(\vec{\nabla}\cdot\vec{A})+(\vec{B}\cdot\vec{\nabla})\vec{A}-(\vec{A}\cdot\vec{\nabla})\vec{B}$
$(\vec{\nabla}\cdot\vec{\nabla})V=\nabla^2V$
$\vec{\nabla}\cdot\vec{\nabla}\times\vec{A}=0$
$\vec{\nabla}\times\vec{\nabla}V=0$
$\vec{\nabla}\times\vec{\nabla}\times\vec{A}=\vec{\nabla}(\vec{\nabla}\cdot\vec{A})-\nabla^2\vec{A}$

TABLE D.2 Vector Identities

APPENDIX E

Summary of EM Relationships

Maxwell's Equations

Maxwell's Equations	In the Integral Form	In the Point Form
Magnetic field circulation [Eqs. (8.15) & (8.52)]	$\oint \vec{H} \cdot \overrightarrow{d\ell} = \sum I_{\text{linkage}}$	$\vec{\nabla} \times \vec{H} = \vec{J}$
Electric field circulation [Eqs. (5.70) & (8.59)]	$\oint \vec{E} \cdot \overrightarrow{d\ell} = 0$	$\vec{\nabla} \times \vec{E} = 0$
Electric flux divergence [Eqs. (4.8) & (4.17)]	$\oiint_S \vec{D} \cdot \overrightarrow{dS} = \sum Q_{\text{enclosed}}$	$\vec{\nabla} \cdot \vec{D} = \rho_v$
Magnetic flux divergence [Eqs. (8.14) & (8.46)]	$\oiint_S \vec{B} \cdot \overrightarrow{dS} = 0$	$\vec{\nabla} \cdot \vec{B} = 0$

Table E.1 Static Forms of Maxwell's Equations

Maxwell's Equations	In the Integral Form	In the Point Form
Faraday's law [Eqs. (10.26) & (10.29)]	$\oint_C \vec{E} \cdot \overrightarrow{d\ell} = -\iint_{S \text{ contoured by } C} \frac{\partial \vec{B}}{\partial t} \cdot dS$	$\vec{\nabla} \times \vec{E} = -\frac{\partial \vec{B}}{\partial t}$
Ampere's law [Eqs. (10.38) & (10.37)]	$\oint_C \vec{H} \cdot \overrightarrow{d\ell} = \iint_{S \text{ contoured by } C} \left[\vec{J} + \frac{\partial \vec{D}}{\partial t} \right] \cdot dS$	$\vec{\nabla} \times \vec{H} = \vec{J} + \frac{\partial \vec{D}}{\partial t}$
Gauss' law for electric flux [Eqs. (4.10) & (4.17)]	$\oiint_S \vec{D} \cdot \overrightarrow{dS} = \iiint_{\text{volume enclosed by } S} \rho_v \, dv$	$\vec{\nabla} \cdot \vec{D} = \rho_v$
Gauss' law for magnetic flux [Eqs. (8.14) & (8.46)]	$\oiint_S \vec{B} \cdot \overrightarrow{dS} = 0$	$\vec{\nabla} \cdot \vec{B} = 0$

Table E.2 Dynamic Forms of Maxwell's Equations

Material permittivity [Eqs. (5.12) & (6.34)]	$\vec{D} = \varepsilon \vec{E}$
Material permeability [Eq. (8.6)]	$\vec{B} = \mu \vec{H}$
Material conductivity [Eq. (6.13)]	$\vec{J} = \sigma \vec{E}$

Table E.3 Constitutive Relationships

Electrostatic/Magnetostatic Relationships		Time-Varying Modifications	
$\vec{E}=-\vec{\nabla}V$	Eq. (5.64)	$\vec{E}=-\vec{\nabla}V-\frac{\partial\vec{A}}{\partial t}$	Eq. (10.41)
$\vec{B}=\vec{\nabla}\times\vec{A}$ & $\vec{\nabla}\cdot\vec{A}=0$	Eq. (8.68)	$\vec{B}=\vec{\nabla}\times\vec{A}$ & $\vec{\nabla}\cdot\vec{A}=-\mu\varepsilon\frac{\partial V}{\partial t}$	Eq. (10.46)
$\nabla^2 V=-\frac{\rho_v}{\varepsilon}$	Eq. (7.4)	$\nabla^2 V=-\frac{\rho_v}{\varepsilon}+\mu\varepsilon\frac{\partial^2 V}{\partial t^2}$	Eq. (10.50)
$\nabla^2\vec{A}=-\mu\vec{J}$	Eq. (8.69)	$\nabla^2\vec{A}=-\mu\vec{J}+\mu\varepsilon\frac{\partial^2\vec{A}}{\partial t^2}$	Eq. (10.47)

TABLE E.4 Potential Relationships

The Divergence Theorem

$$\oint\!\!\!\oint_S \vec{D}\cdot\vec{dS}=\iiint_{\text{volume enclosed by }S}\vec{\nabla}\cdot\vec{D}\,dv \tag{4.26}$$

Stokes' Theorem

$$\oint_C \vec{H}\cdot\vec{d\ell}=\iint_{S\text{ bound by }C}(\vec{\nabla}\times\vec{H})\cdot\vec{dS} \tag{8.60}$$

Continuity Equation

$$\vec{\nabla}\cdot\vec{J}=-\frac{\partial\rho_v}{\partial t} \tag{6.8}$$

Scalar and Vector Potentials

$$V_p=\int_Q\frac{dQ}{4\pi\varepsilon_o R_{dQ-p}}=\iiint_v\frac{\rho_v\,dv}{4\pi\varepsilon_o R_{dv-p}} \tag{5.69}$$

$$\vec{A}_p=\iiint_v\frac{\mu_o\vec{J}\,dv}{4\pi R_{dv-p}}=\oint_\ell\frac{\mu_o I\vec{d\ell}}{4\pi R_{d\ell-p}} \tag{8.70}$$

Complex Material Properties

$$\varepsilon^*=\varepsilon'-j\frac{\sigma}{\omega}=\varepsilon'-j\varepsilon'' \tag{11.58}$$

$$\mu^*=\mu'-j\mu'' \tag{11.60}$$

$$\tan\delta_\varepsilon=\varepsilon''/\varepsilon' \tag{11D.10}$$

$$\tan\delta_\mu=\mu''/\mu' \tag{11D.11}$$

Complex Wave Parameters

$$\gamma=\alpha+j\beta=\sqrt{-\omega^2\mu^*\varepsilon^*}\quad\text{and}\quad\eta=\sqrt{\mu^*/\varepsilon^*} \tag{11.72}$$

Skin Depth

$$\delta=1/\alpha\cong 1/\sqrt{\pi f\mu_o\sigma} \tag{11.75}$$

Poynting Vector

$$\vec{\mathcal{P}} = \vec{E} \times \vec{H} \tag{11.76}$$

Coaxial Transmission Line Parameters

	DC and Low Frequencies	High Frequencies	Units
R_z [Eq. (11C.4)]	$\frac{1}{\sigma_i(\pi a^2)} + \frac{1}{\sigma_o(\pi[c^2 - b^2])}$	$\cong \frac{1}{\sigma_i(2\pi a\delta_i)} + \frac{1}{\sigma_o(2\pi b\delta_o)}$	Ω/m
L_z [Eq. (9A.28)]	$\frac{\mu}{2\pi}\ln(b/a)$	$\cong \frac{\mu}{2\pi}\ln(b/a)$	H/m
G_z [Eq. (6A.15)]	$\frac{2\pi\sigma_d}{\ln(b/a)}$	$\frac{2\pi\sigma_d}{\ln(b/a)}$	S/m
C_z [Eq. (6B.16)]	$\frac{2\pi\varepsilon}{\ln(b/a)}$	$\frac{2\pi\varepsilon}{\ln(b/a)}$	F/m
Z_o [Eq. (11.147)]	$\frac{\ln(b/a)}{2\pi}\sqrt{\frac{\mu}{\varepsilon}} = \eta\frac{\ln(b/a)}{2\pi}$		Ω

Table E.5 Coaxial Transmission Line Parameters

Normal Incidence Relationships

Reflection and Transmission Coefficients

$$\bar{\Gamma}(0) = \frac{\bar{\eta}_{o2} - \bar{\eta}_{o1}}{\bar{\eta}_{o2} + \bar{\eta}_{o1}} \tag{12.26}$$

$$\bar{\tau}(0) = 1 + \bar{\Gamma}(0) = \frac{2\bar{\eta}_{o2}}{\bar{\eta}_{o2} + \bar{\eta}_{o1}} \tag{12.27}$$

Oblique Incidence Relationships

Snell's Law

$$n_1 \sin\theta_i = n_2 \sin\theta_t, \; n = \sqrt{\mu_r \varepsilon_r} \tag{12B.12}$$

Reflection and Transmission Coefficients

$$\bar{\Gamma}(0)\Big|_{\parallel} = \frac{\bar{\eta}_{o2}\cos\theta_2 - \bar{\eta}_{o1}\cos\theta_1}{\bar{\eta}_{o2}\cos\theta_2 + \bar{\eta}_{o1}\cos\theta_1} \tag{12.126}$$

$$\bar{\tau}(0)\Big|_{\parallel} = \frac{\cos\theta_1}{\cos\theta_2}\left\{1 + \bar{\Gamma}(0)\Big|_{\parallel}\right\} = \frac{2\bar{\eta}_{o2}\cos\theta_1}{\bar{\eta}_{o2}\cos\theta_2 + \bar{\eta}_{o1}\cos\theta_1} \tag{12.127}$$

$$\bar{\Gamma}(0)\Big|_{\perp} = \frac{\bar{\eta}_{o2}/\cos\theta_2 - \bar{\eta}_{o1}/\cos\theta_1}{\bar{\eta}_{o2}/\cos\theta_2 + \bar{\eta}_{o1}/\cos\theta_1} \tag{12.136}$$

$$\bar{\tau}(0)\Big|_{\perp} = 1 + \bar{\Gamma}(0)\Big|_{\perp} = \frac{2\bar{\eta}_{o2}/\cos\theta_2}{\bar{\eta}_{o2}/\cos\theta_2 + \bar{\eta}_{o1}/\cos\theta_1} \tag{12.137}$$

The Brewster Angle

$$\sin\theta_{1B\|} = \sqrt{\frac{1-(\mu_2\varepsilon_1/\mu_1\varepsilon_2)}{1-(\varepsilon_1/\varepsilon_2)^2}} \tag{12.142}$$

$$\sin\theta_{1B\perp} = \sqrt{\frac{1-(\mu_1\varepsilon_2/\mu_2\varepsilon_1)}{1-(\mu_1/\mu_2)^2}} \tag{12.145}$$

Rectangular Waveguide Relationships

$$f_{c\,m,n} = \frac{c_o}{\sqrt{\mu_r\varepsilon_r}}\sqrt{\left[\left(\frac{m}{2a}\right)^2+\left(\frac{n}{2b}\right)^2\right]} \tag{13.33}$$

$$\beta_{g\,m,n} = \beta_{ug}\sqrt{1-(f_{c\,m,n}/f)^2} \tag{13.38}$$

$$\lambda_{g\,m,n} = \lambda_{ug}/\sqrt{1-(f_{c\,m,n}/f)^2} \tag{13.36}$$

$$c_{ph\,g\,m,n} = c_{ph\,ug}/\sqrt{1-(f_{c\,m,n}/f)^2} \tag{13.37}$$

$$\bar{\eta}_{g\,TE\,m,n} = \bar{\eta}_{ug}/\sqrt{1-(f_{c\,m,n}/f)^2} \tag{13.80}$$

$$\bar{\eta}_{g\,TM\,m,n} = \bar{\eta}_{ug}\sqrt{1-\left[\frac{f_{c\,m,n}}{f}\right]^2} \tag{13.85}$$

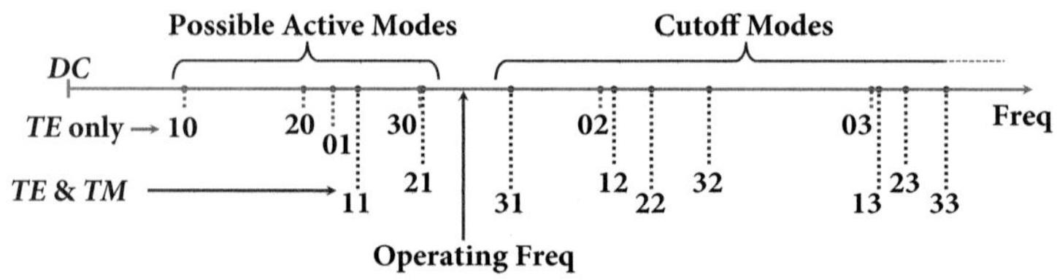

FIGURE 13.12 A typical sorted order of cutoff frequencies for the case $a \geq 2b$.

APPENDIX F

Historical Review of EM Scientists

In the following, we present brief historical information on a selected list of famous scientists who contributed to the advancements in the field of electromagnetics over the past few centuries. The reader is encouraged to learn more about these (and other) scientists and their contributions in the published literature. See the footnote below for some references containing such knowledge.[1]

The list is provided in a "timeline" and highlights the most notable discoveries in the history of electromagnetics that have influenced our understanding of electromagnetic theory today. In addition, the list includes selected scientists and inventors in related physics and mathematical fields whose work had significant impact on the progress and advancement of electromagnetic theory and applications.

Thales of Miletus (624–546 BC), a Greek philosopher, who is known for his work on static electricity. Around 585 BC, he was the first to discover the attraction of dry light material by rubbed amber.

Thales of Miletus

[1]https://www.britannica.com/science/electromagnetism/Historical-survey
https://en.wikipedia.org/wiki/Timeline_of_electromagnetism_and_classical_optics#cite_ref-moller_1-0
https://en.wikipedia.org/wiki/History_of_electromagnetic_theory

Muḥammad ibn Mūsā al-Khwārizmī (محمد بن موسى خوارزمي) (780–850), a mathematician and astronomer of the Islamic Golden Age. He is known as the founder (or father) of algebra. He developed the basic concepts of algebra as well as introduced the Hindu-Arabic numerals.

al-Khwārizmī

Ibn Sahl [Abū Sa'd al-'Alā' ibn Sahl (أبو سعد العلاء ابن سهل); 940–1000], a mathematician and physicist of the Islamic Golden Age. He is well known for his work in optics, and mathematics. He was credited as the first to discover the law of refraction known today as Snell's law. He used the law of refraction to derive lens shapes that can focus light with no geometric aberrations.

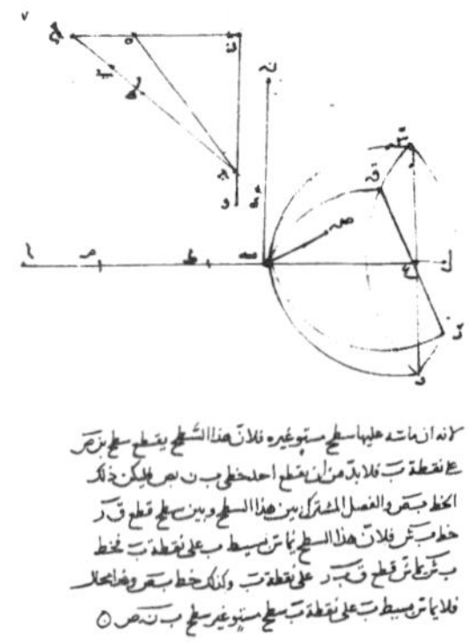

Ibn Sahl

Willebrord Snellius (Snell; 1580–1626), a Dutch astronomer and mathematician. He mathematically formulated the refraction of light rays at material interfaces. The formulation is known as Snell's law. It is acknowledged that this law was already known to Ibn Sahl in 984.

Willebrord Snellius

Around 1600, **William Gilbert (1554–1603)**, an English physician, who is credited by many as the "father of electricity and magnetism," proposed that the earth is a magnetic planet with North and South Poles. He became a pioneer researcher on magnetism. He was also the first to use the term electric attraction.

William Gilbert

In 1733, **Charles-François du Fay (1698–1739)**, a French chemist, made a notable discovery attempting to explain electrostatic attraction and repulsion when he first introduced what is known today as positive and negative charges. He also noted the difference between conductors and insulators.

Charles-François du Fay

In 1752, **Benjamin Franklin (1706–1790)**, an American scientist, discovered the link between lightening and electricity. He also discovered the principle of conservation of electric charge.

Benjamin Franklin

In 1785, **Charles-Augustin de Coulomb (1736–1806)**, a French physicist, formulated Coulomb's law, the inverse square law of attraction and repulsion of unlike and like electrical charges. Coulomb law states that the force between two charges is proportional to the product of the charges and inversely proportional to the square of the distance between them.

Charles-Augustin de Coulomb

In 1791, **Luigi Galvani (1737–1798)**, an Italian physician and physicist and a pioneer in studying bioelectricity, discovered electricity in animal tissues. His research led to the invention of the voltaic cell.

Luigi Galvani

In 1799, **Alessandro Volta (1745–1827)**, an Italian physicist, followed Galvani's discoveries and created a voltaic cell producing a continuous electric current using a chemical action. His invention proved that electricity can be generated chemically which eventually led to the development of the field of electrochemistry.

Alessandro Volta

Jean-Baptiste Joseph Fourier (1768–1830), a French mathematician and physicist. He is best known for his pioneer work in representing functions using trigonometric series. His theory was used in the solution of boundary value problems in partial differential equations.

Jean-Baptiste Joseph Fourier

Pierre-Simon marquis de Laplace (1749–1827), a French scholar who is known to be one of the most influential scientists. His work has influenced the development of engineering, mathematics, statistics, physics, and astronomy. He formulated Laplace's equation, and pioneered the Laplace transform which has several applications in many fields.

Pierre-Simon marquis de Laplace

In 1813, **Siméon-Denis Poisson (1781–1840)**, a French mathematician, published the Poisson equation which describes the relationship between the electric potential and a given charge distribution.

Siméon-Denis Poisson

Georg Simon Ohm (1789–1854), a German physicist and mathematician. He discovered Ohm's law, which states that the current in a conductor is directly proportional to the potential difference and inversely proportional to the resistance.

Georg Simon Ohm

In 1820, **Hans Christian Ørsted (1777–1851)**, a Danish physicist, discovered that electric current in a wire generates a magnetic field making the first connection between electricity and magnetism that later inspired the development of the electromagnetic theory.

Hans Christian Ørsted

Shortly after the discoveries of Ørsted, **André-Marie Ampère (1775–1836)**, a French physicist, founded the theory of electromagnetism. He was one of the first to develop techniques for electrical measurements. Named after him is the SI unit of measurement of electric current, the ampere.

André-Marie Ampère

Around the same time, **Jean-Baptiste Biot (1774–1862)**, a French physicist, discovered that the intensity of the magnetic field generated by a current in a wire is inversely proportional to the distance from the wire. He formulated with **Félix Savart** the Biot-Savart law that describes the magnetic field induced by a wire segment in terms of the current distribution.

Jean-Baptiste Biot

Félix Savart (1791–1841), a French physicist and mathematician who is primarily known for the Biot–Savart law of electromagnetism, which he discovered together with his colleague Jean-Baptiste Biot. The Biot–Savart law relates magnetic fields to electric currents.

Félix Savart

Sir David Brewster (1781–1868), a British scientist, inventor, and author. In science, his work was concentrated in physical optics. He made significant discoveries about the polarization of light including the discovery of the Brewster angle also known as the polarization angle.

Sir David Brewster

In 1831, **Michael Faraday (1791–1867)**, an English physicist and chemist, discovered electromagnetic induction. His experiments significantly contributed to the fundamental understanding of electromagnetism. He invented the first electric motor and electric generator.

Michael Faraday

In 1827, **Joseph Henry (1797–1878)**, an American scientist, invented the electromagnetic relay which was the basis of the electric telegraph. He also discovered the electromagnetic phenomenon of self-inductance. The SI unit of inductance, the henry, is named after him.

Joseph Henry

Gustav Robert Kirchhoff (1824–1887), a German physicist who is known for his contributions to the fundamentals of the theory of electrical circuits and the understanding of thermal emission. He also made significant contributions to spectroscopy. In circuit theory, Kirchhoff formulated two fundamental circuit laws. Kirchhoff's first law states that the algebraic sum of currents in a network of conductors meeting at node is zero. The second law states that in a closed circuit, the sum of the voltage drops is zero.

Gustav Robert Kirchhoff

In 1833, **Wilhelm Eduard Weber (1804–1891)**, a German physicist who worked closely with Carl Gauss and together built one of the first electromagnetics telegraph. The SI unit for measuring magnetic flux, the weber, is named after him. He developed and enhanced several devices for detecting and measuring magnetic fields and electric currents.

Wilhelm Eduard Weber

In 1834, **Heinrich Friedrich Emil Lenz (1804–1865)**, a Russian physicist, deduced Lenz's law, which indicated that the direction of flow of an induced current is such that it opposes the change that induced it.

Heinrich Friedrich Emil Lenz

In 1835, **Carl Friedrich Gauss (1777–1855)**, a German mathematician, astronomer, and physicist, formulated Gauss' law, which describes the electric flux flowing through a closed surface in terms of the electric charge enclosed. He invented an early type of magnetometer used for measuring the direction and strength of a magnetic field. With Wilhelm Weber, he built one of the first electromagnetic telegraphs.

Carl Friedrich Gauss

Ernst Werner Siemens (1816–1892), known to be a pioneer in applying scientific discoveries to industrial processes. The SI unit of electrical conductance, the siemens (S), was named after him.

Ernst Werner Siemens

In 1873, **James Clerk Maxwell (1831–1879)**, a Scottish scientist, made a significant contribution by his most notable achievement, the formulation of classical theory of electromagnetic radiation, combining the work of Faraday, Ampere, and Ørsted in one unified theory. He is known for making the connection between light waves and electromagnetic waves.

James Clerk Maxwell

In 1884, **Oliver Heaviside (1850–1925)**, a British physicist, reformulated the original Maxwell equations, 20 equations in 20 unknowns, to the modern form known today, four vector equations. He predicted the existence of the ionospheric layer and described its role in reflecting radio waves.

Oliver Heaviside

Around the same time, **John Henry Poynting (1852–1914)**, a British physicist, introduced the Poynting theorem which describes the rate of flow of energy associated with an electromagnetic wave. His research contributed to the understanding of radiation-pressure effects in the solar system.

John Henry Poynting

In 1887, **Heinrich Rudolf Hertz (1857–1894)**, a German physicist, confirmed Maxwell's theory of electromagnetism by building a system that was capable of generating and detecting electromagnetic waves at radio frequencies. He also showed that light and heat are electromagnetic radiations.

Heinrich Rudolf Hertz

In 1888, **Nikola Tesla (1856–1943)**, a Serbian-American inventor and engineer, invented the alternating current electric motor and was famous for his contribution to three phase systems. Tesla is described as one of the most important scientists and inventors of the modern age. The SI unit for measuring the magnetic flux density, the tesla, is named after him.

Nikola Tesla

In 1893, **Joseph John Thomson (1856–1940)**, an English physicist, proposed the first structure for guiding waves. Later, in 1897, he revolutionized the understanding of atomic structure by discovering the electron. He demonstrated a device that changed the path of electrons using electric and magnetic fields. He was also able to measure the ratio of the electron's charge to its mass.

Joseph John Thomson

In 1894, **Sir Oliver Joseph Lodge (1851–1940)**, a British physicist, was the first to experimentally test a waveguide. He was also involved in the development and application of electromagnetic radiation and wave propagation. Lodge demonstrated an early radio wave detector he named the "coherer."

Sir Oliver Joseph Lodge

In 1897, **Lord Rayleigh (John William Strutt, 3rd Baron Rayleigh, 1842–1919)**, a British scientist, performed the first mathematical analysis of electromagnetic waves in a metal cylinder. He made numerous fundamental discoveries related to the theory of wave propagation and scattering. He made extensive contributions to both theoretical and experimental physics. He received the Nobel Prize for Physics in 1904.

Lord Rayleigh

Phillip Hagar Smith (1905–1987), an American electrical engineer, who became famous for his invention of the Smith chart, a graphical tool for designing, analyzing, and visualizing the performance of transmission line networks.

Phillip Hagar Smith

Index

Note: Page numbers followed by *f* denote figures; page numbers followed by *t* denote tables; italicized page numbers denote images.

D

E

N

O

T

U

V

Z